DNA Synthesis
Present and Future

NATO ADVANCED STUDY INSTITUTES SERIES

A series of edited volumes comprising multifaceted studies of contemporary scientific issues by some of the best scientific minds in the world, assembled in cooperation with NATO Scientific Affairs Division.

Series A: Life Sciences

Recent Volumes in this Series

The series is published by an international board of publishers in conjunction with NATO Scientific Affairs Division

A	Life Sciences	Plenum Publishing Corporation
B	Physics	New York and London
C	Mathematical and Physical Sciences	D. Reidel Publishing Company Dordrecht and Boston
D	Behavioral and Social Sciences	Sijthoff International Publishing Company Leiden
E	Applied Sciences	Noordhoff International Publishing Leiden

DNA Synthesis
Present and Future

Edited by

Ian Molineux
Imperial Cancer Research Fund
London, England

and

Masamichi Kohiyama
University of Paris
Paris, France

PLENUM PRESS • NEW YORK AND LONDON
Published in cooperation with NATO Scientific Affairs Division

Library of Congress Cataloging in Publication Data

Nato Advanced Study Institute, Santa Flavia, Sicily, 1977.
  DNA synthesis.
  (NATO advanced study institutes series: Series A, Life sciences; v. 17)
  "Proceedings of the NATO Advanced Study Institute held in Santa Flavia, Sicily,
June 20–29, 1977."
  Includes index.
  1. Deoxyribunocleic acid synthesis–Congresses. I. Molineux, Ian. II. Kohiyama,
Masamichi. III. Title. IV. Series.
QP624.N37 1977   574.8'732   78-2832
ISBN-13: 978-1-4684-0846-1   e-ISBN-13: 978-1-4684-0844-7
DOI: 1007/978-1-4684-0844-7

Proceedings of the NATO Advanced Study Institute held in Santa Flavia,
Sicily, June 20–29, 1977

© 1978 Plenum Press, New York
Softcover reprint of the hardcover 1st edition 1978

A Division of Plenum Publishing Corporation
227 West 17th Street, New York, N.Y. 10011

PREFACE

This book represents the proceedings of the NATO Advanced
Study Institute held in Santa Flavia, Sicily from the 20 - 29th
June, 1977.

In addition to the review talks given by the Lecturers at
the Institute it proved feasible for other topics to be
splendidly reviewed. This has led to a much wider subject
coverage than would otherwise have been possible. The discussion
sessions which followed these review talks were extremely
valuable and almost all the participants played an active role.
Essentially all of the verbal contributions presented at this
ASI were subsequently put into written format, which is why these
proceedings are so extensive. They do, however, provide an
up-to-date summary of DNA synthesis in a wide variety of subjects
with many of the remaining problems clearly expressed.

The editing of these contributions has been essentially
confined to alterations in style and presentation. We have taken
some liberties in the re-organization of the papers into related
sections.

We express our thanks to those who helped organize the ASI
and to the session conveners who attempted to confine and
contain those who became too verbose. We are indebted to NATO,
Scientific Affairs Division for the financial support that made
this ASI possible. Finally, we express our gratitude to
Miss Brenda Marriott. She typed all seventy five papers in this
book, which was originally estimated to be less than half its
present length and which just grew and grew. She deserves our
special thanks.

CONTENTS

PART II

THE BIOSYNTHESIS OF CHROMOSOMAL DNA

PART III

DNA STRUCTURE, DNA BINDING AND UNWINDING PROTEINS

PART VI

REPAIR PATHWAYS

CONTROL OF DNA REPLICATION IN BACTERIA

R.H. Pritchard

Department of Genetics
University of Leicester
Leicester LE1 7RH
England

INTRODUCTION

The rate of DNA synthesis by a bacterial culture under steady
state growth conditions is determined by the rate of initiation of
rounds of chromosome replication. It is independent of the rate of
DNA chain elongation. (For an extensive elaboration of the point
see Pritchard, 1974). The process of initiation of a round of
replication can be considered from at least two points of view.
One can ask about the biochemistry of the process; what elements
are involved and what they do. Alternatively it is possible to ask
about the regulation of initiation; how the rate of initiation is
matched to the rate of accumulation of mass by the culture. This
article is concerned primarily with the latter question.

There is no reason at present to believe that it will not be
possible to account for control of initiation in terms of the
models that have been used so successfully to explain enzyme
induction, repression, inhibition and activation in bacteria. In the
case of chromosome replication however, **an additional factor must be**
considered. Within the domain of a single cell the number of copies
of at least one component of the system - the chromosome origin -
is small, sometimes as small as one, and the 'rate' of reaction is
such that only one or very few initiations take place within one
cell generation. Consequently, stochastic considerations may
become very important in understanding the response of the chromosome origin to a given concentration of effectors of initiation, and
also the way in which the concentration of these effectors is
influenced in turn by the concentration of chromosome origins.
Similar considerations may also be important in understanding the
timing of plasmid replication. The differences between so-called

'relaxed' and 'stringent' plasmids, which are discussed later, may be a consequence simply of differences in the number of plasmid copies per cell[1] rather than a reflection of different mechanisms of regulation.

The first attempt at a formal model for control of initiation was the replicon hypothesis of Jacob et al. (1963). These authors drew attention to two features of replicons which they sought to account for within a unifying framework. One was the stable inheritance of plasmids such as F despite their low copy number. The second was incompatibility between genetically distinguishable variants of the same plasmid. Both properties were ascribed to attachment of replicons to membranes at specific sites. The membrane grew in relation to these sites in such a way as to provide an analogue of the spindle in eukaryotes and ensured an ordered segregation of sister-plasmids. Incompatibility arose because membrane attachment was necessary for replication and an incoming plasmid was excluded from replication because it found all available sites occupied. The frequency of replication and the timing of its occurrence within a cell cycle were presumably determined by the frequency and timing of the generation of attachment sites but the model made no predictions about this. In addition the replicon hypothesis did not adequately account for a puzzling and significant feature of the interaction of the plasmid F and the bacterial chromosome which I want to refer to as switch-off. This was the apparent failure of F to initiate rounds of replication when integrated into the chromosome in Hfr strains. Switch-off could be understood in terms of the replicon hypothesis only by assuming that integration of F into the chromosome caused detachment from the membrane, but since this would leave an attachment site unoccupied such an explanation would predict that Hfr strains should lose incompatibility towards an incoming F particle. They do not.

---

[1] *Copy number has been used in the literature to describe the average number of copies of a plasmid either per genome equivalent or per cell. In this article it will be used to denote the number in unit volume of cytoplasm. Defined in this way it is, in effect, the plasmid concentration. The same term will be used to define the concentration of chromosome origins. The copy number of a plasmid and a plasmid origin will be virtually identical. The copy number of a chromosome and a chromosome origin may differ substantially because every replicating chromosome has at least two origins but consists of less than two genome equivalents of chromosomal DNA (cf. Pritchard, 1974). The initiation concentration of a replicon is the average copy number at the time of initiation of a round of replication. The reciprocal of the initiation concentration is the initiation volume (or initiation mass) as used in earlier papers.*

In an attempt to understand switch-off and incompatibility, and to draw attention to the need for any model for the control of initiation of replication to make *quantitative* predictions about the frequency and timing of initiation events, I suggested an alternative way of thinking about this problem (Pritchard, 1968, 1969a; Pritchard et al., 1969). It was based on changes in the concentration of a repressor-like DNA binding protein. Subsequently it has become apparent that it fits many of the facts as we understand them at present about chromosome and plasmid replication rather well. It therefore seems appropriate to summarise our ideas and update them in the light of more recent data and reflection.

It was proposed that initiation of replication, by whatever enzymes are involved in this process, could be inhibited by the binding of a repressor-like protein to the DNA near to the origin of replication. Synthesis of this protein was coupled to initiation such that each initiation event led to the production of a burst of inhibitor molecules. Initiation would consequently raise inhibitor concentration and reduce the probability of further initiations. Growth would reduce inhibitor concentration by dilution and so increase the probability of initiation. Under steady state conditions a simple regulatory loop of this sort would maintain a constant average number of copies of the chromosome origin per unit volume of cytoplasm in the culture as a whole.

The usefulness of the model seemed to be that it appeared to explain without further assumptions the occurrence of switch-off. It defined the conditions under which this would occur and it predicted which of the components of a composite replicon would be the one to suffer switch-off. It also predicted the occurrence of switch-on, one manifestation of which is called integrative suppression (Nishimura et al., 1971). It provided a straight-forward interpretation of incompatibility and it clarified to some extent the differences in behaviour of 'stringent' and 'relaxed' plasmids (cf. Clowes, 1972). I shall discuss each of these features of replication in more detail. Before doing so it is essential to note that they are not accounted for only by the inhibitor dilution model. A variety of mechanisms, as we later realised, which influence the frequency of initiation via the concentration of a cytoplasmic effector so as to maintain a constant copy number of replicon origins make similar predictions. Indeed, we drew attention to an alternative mechanism involving continuous synthesis of an unstable inhibitor (Pritchard et al., 1969) and other volume or mass titration mechanisms with similar properties have subsequently been described (Fantes et al., 1975). Nevertheless, where it seems clearer to talk in terms of a simple specific volume titration mechanism, I shall use the inhibitor-dilution model.

It is also appropriate at this point to consider one feature

of the inhibitor-dilution model (and other volume titration
mechanisms) which has an important influence on the timing of
initiation events in individual cells. This is the relationship
between inhibitor concentration and the probability of initiation.
At one extreme there could be a sharp transition between a
probability of zero and one for initiation over a very narrow range
of inhibitor concentrations. In cells containing only one copy of
a replicon before replication, initiation would be sharply phased
to occur over a narrow range of cell volumes and inhibitor
concentration would oscillate over a twofold range during the cell
cycle (Pritchard et al., 1969). At the other extreme, if the
probability of initiation increased linearly as inhibitor
concentration decreased, then initiation events would be poorly
phased with respect to cell volume and inhibitor concentration
would oscillate over a greater than twofold range. In our original
discussion of this model we assumed that initiation was sharply
phased in *E.coli* because in experiments with age fractionated
cultures (Helmstetter et al., 1968) it had been shown that
initiation of rounds of chromosome replication occurred during a
relatively small segment of the cell cycle (and hence over a small
range of cell volumes). Some doubt has more recently been cast on
the interpretation of data from age fractionated cultures and
synchronous cultures (see Koch, 1977; and the section in this
article on Relaxed Replication). However, our original assumption
that initiation was well phased in the cell cycle seems justified
by recent microscopic observations on the distribution of initiation
events among different cell size classes from exponential cultures
of *E.coli* with doubling times of about 100 minutes (L. Koppes;
personal communication from K. Woldringh). Thus it appears that the
regulatory mechanism which influences the timing of initiation is
sensitive to much less than a twofold difference in the copy number
of chromosome origins. For the inhibitor-dilution model to
generate such sensitivity it would be necessary to assume a
cooperative interaction between several protomers and the chromosome
origin to give a sharp transition, or switch, between a high and a
low probability of initiation.

Switch-Off

The essence of our proposal was that a replicon will have a
high probability of initiating a round of replication only when its
copy number falls to a level specific to that replicon. Suppose
that the plasmid F has a lower initiation concentration than that
of the bacterial chromosome. This means that it is diluted out by
growth to a greater extent than the chromosome origin before
initiation of a new round of replication takes place. Now consider
the consequence of integration of F into the chromosome. Since
initiation of chromosome replication will occur before that of F
the latter may be passively replicated by replication forks

originating at the chromosome origin before its copy number has
fallen to that which triggers its own initiation. In fact its
copy number may never fall to the initiation concentration and
its autonomous replication will be inhibited.

In general we predict that if two replicons forming a
composite have identical copy numbers both will initiate rounds of
replication with equal probability. If they have different copy
numbers the one with the smaller copy number will suffer switch-
off. The severity of switch-off will depend on how different the
two copy numbers are. It will also depend on how sensitive is
the control system of the replicon suffering switch-off to its own
copy number. In the case of replicons of very different size, such
as a composite of the plasmid F and the bacterial chromosome, there
is an additional complexity which is illustrated diagrammatically
in Fig. 1. It is arbitrarily assumed for illustrative purposes

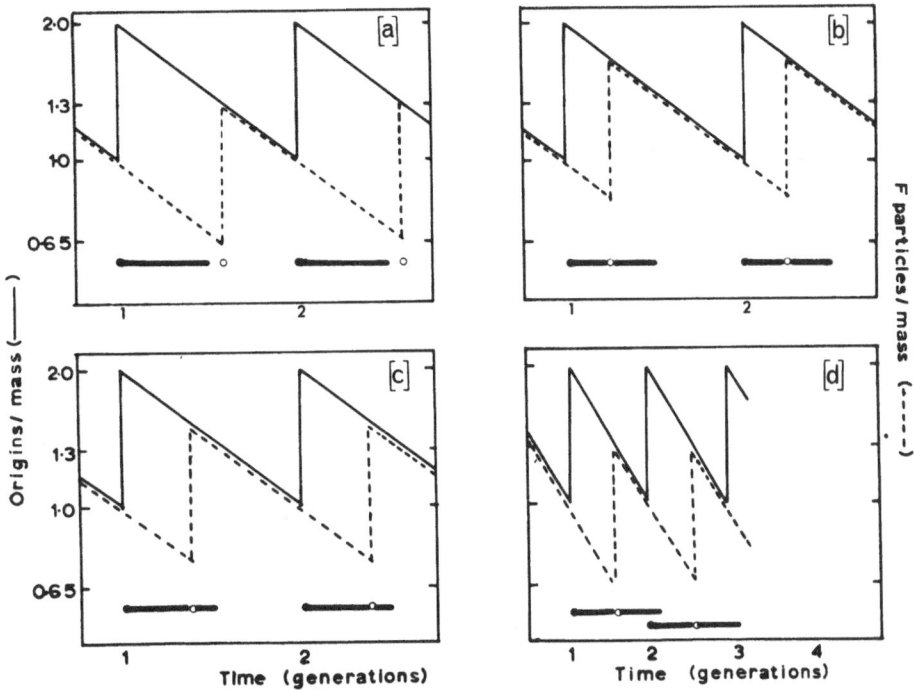

*Figure 1. Effect of growth rate and the chromosome location of
insertion on switch-off of an integrated F plasmid. The solid bar
represents a half chromosome with length equal to the replication
time (C = 40 minutes). The open circle represents F. The solid curve
shows the concentration of chromosome origins and the dashed curve
the concentration of F particles. The initiation concentration for
the chromosome origin is 1.0 (in arbitrary units) and that for F
is 0.67 (in the same units).*

that the initiation concentration of F is two-thirds that of the
chromosome. Comparison of panels (b) and (c) shows that the further
downstream from the chromosome origin that F is inserted, the less
effective switch-off will be. This is because passive replication
of F will occur at a cell volume which progressively approaches its
autonomous initiation volume. Panels (b) and (d) show, in addition,
that at any given site of insertion the effectiveness of switch-off
would decrease as the growth rate is increased for the same reason.
It can readily be calculated that to ensure switch-off at the
fastest growth rates in an Hfr in which F was inserted near the
chromosome terminus, the initiation concentration of F would need
to be less than one third of that of the chromosome origin. This
would lead to an additional potential complication. Cells of
$E.coli$ decrease in size as the growth rate decreases (Helmstetter
et al., 1968). At a doubling time of 60 minutes initiation of
chromosome replication probably occurs at about the time of
division (Cooper and Helmstetter, 1968). Cells at birth therefore
have a volume corresponding to the initiation volume for the
chromosome and just before division have twice this volume.
Consequently, if the initiation volume of F were more than twice
that of the chromosome, F would have failed to replicate by the
time cell division occurred and it would be lost from half of the
daughter cells. This does not occur - F maintenance is very
stable at all growth rates. It follows either that our hypothesis
is wrong, or that the relationship between cell size and growth
rate is not the same in F^+ and F^- cells, or that the initiation
volume of F is not constant but decreases with decreasing growth
rate.

     In order to test these rather specific predictions one of
which must be fulfilled for the behaviour of F to be consistent
with our interpretation of switch-off we determined the ratio of
F particles to chromosome origins in F-lac^+ cultures of $E.coli$
under conditions of steady-state exponential growth (Collins and
Pritchard, 1973; Pritchard et al., 1975). The data, which are
summarised in Fig. 2, show that the copy number of F is about
half that of the chromosome origin at fast growth rates but is
equal to it at slow growth rates. So far so good, but the
relationship shown in Fig. 2 leads to the further prediction that
in slow growing cultures switch-off of F should not be very
effective. A re-examination of the data of Nagata(1963) in the
light of our current knowledge of the bidirectionality of
chromosome replication in $E.coli$ suggests strongly that this
additional prediction will be upheld. Thus it seems that the
replication behaviour of F in Hfr strains and its copy number in
the autonomous state are consistent with each other in terms of my
interpretation of switch-off.

     The trend in F:origin ratios shown in Fig. 2 is that expected
to give both stable maintenance of F and switch-off in Hfr strains.
The actual ratios found on the other hand, suggest that switch-off

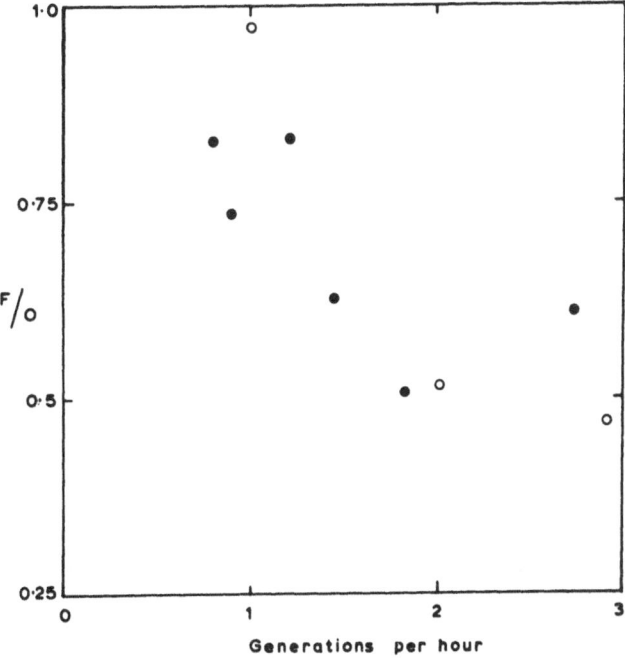

Figure 2. *The ratio of F particles to chromosome origins at different growth rates.* ○ *Data of Collins and Pritchard (1973). The quantity measured by hybridisation was the fraction of plasmid DNA. To calculate the ratio the following were assumed: m.wt. of Flac+ plasmid, 95 x 10^6; m.wt. of chromosome, 2.5 x 10^9; C = 40 mins.* ● *Data of Pritchard et al. (1975). The measured quantity was the basal level of β-galactosidase in an F-lac+ Δlac and an F− lac+ culture. Additional assumptions required to calculate the ratio were location of chromosome origin, 83 minutes; location of lac gene, 8 minutes; total map length, 100 minutes.*

will not be very effective even at fast growth rates in Hfr strains with F inserted in the distal half of the chromosome. Unpublished experiments in this laboratory with *E.coli* B/r gave a steeper gradient of F:origin ratios versus growth rate from 1.2 at one doubling per hour to 0.3 at 2.7 doublings per hour. A more direct determination of the F:origin ratio at different growth rates and a careful analysis of the frequency of initiation by the F replicon inserted at various locations in the bacterial chromosome would provide a rigorous quantitative test of the model and a useful measure of the sensitivity of the copy number control mechanism. A careful comparison of the copy number of F compared with that of other conjugative plasmids which do not normally form stable Hfr strains would also be useful, since it has been suggested (Pritchard, 1969b) that such plasmids fail to form stable Hfr strains because they have a higher copy number than F. If this were

so they could switch off the chromosome origin and elevate the
concentration of chromosomal DNA to a level which might disturb
growth.

An analogous example of switch-off in a composite replicon has
been described by Cabello et al. (1976). The plasmids ColEl and
pSC101 have about 18 and 5 copies per cell respectively. Both are
cleaved at a single site by the restriction enzyme EcoRl and a
composite plasmid can therefore be constructed. The composite is
present at about 18 copies per cell and in 36 replicating plasmid
molecules examined, the replication origin being used was that
belonging to ColEl.

                               Switch-On

If switch-off is due to passive replication of one replicon by
another with a higher copy number, it follows that should
initiation by the higher copy number component of a composite
replicon be blocked the low copy number component will switch-on
again after it has been diluted out to an appropriate concentration
by growth.

The phenomenon described as integrative suppression (Nishimura
et al., 1971), which should perhaps more accurately be called
suppressive integration and which was first reported by Nandadasa
(1970), is one example of switch-on. The initial observation was
that in a high proportion of thermoresistant revertants of F^+
derivatives of thermosensitive *dnaA* mutants of *E.coli* the plasmid
had become integrated to form an Hfr. The implication was that
chromosome replication could be driven by the F replicon if
initiation at the normal chromosome origin was blocked as it is in
dnaA mutants. In the case of suppressive integration by the plasmid
R100.1, it has been shown that replication is initiated at the site
of plasmid insertion at the non-permissive temperature (Bird et al.,
1976).

It has been argued (Nishimura et al., 1971) that additional
mutations in either the plasmid or the host genome are necessary
for suppressive integration to be expressed. At first sight this is
at variance with the notion that it is only necessary for
chromosome replication to be blocked in order for F replication to
be unblocked. It must therefore be emphasised in the first place
that the switch-on hypothesis makes specific predictions only
about DNA replication. It is not possible to predict whether
chromosome replication from the site of insertion of a plasmid,
and under the control of the initiation system of that plasmid,
will permit sufficiently normal cell growth and division for the
formation of a visible colony. In fact our own studies
(Tresguerres et al., 1975), using *dnaA* Hfr strains isolated at 30°C.

without selection, suggest that the primary reason why additional mutations may be found to be necessary for colony formation at non-permissive temperatures is that the plasmid itself is thermo-sensitive for replication in *dnaA* mutants. We presented evidence that F replication was blocked in a *dnaA* strain at 42°C but not at 40°C. Chromosome replication on the other hand was blocked at both temperatures. Hfr derivatives of this strain could form colonies at 40°C but not at 42°C *without selection for additional mutations*. In other words, suppression of the *dnaA* phenotype could be observed in the absence of additional mutations but only within a narrow thermal window.

Another example of switch-on is provided by the behaviour of the ColE1-pSC101 composite plasmid described above. Its replication is normally under the control of the ColE1 component but in a *polA* strain, in which ColE1 replication is blocked, the plasmid continues to replicate and uses the pSC101 origin to do so. The copy number also falls to that expected if the composite were replicating under pSC101 control.

Yet another significant example of switch-on has been demonstrated by Chandler et al. (1976). Chromosome replication in a *thy*⁻ Hfr strain was terminated by amino-acid starvation. The culture was then allowed to grow again by restoring the amino-acids required for growth but it was deprived of thymine to block replication. DNA replication was only permitted (by restoring thymine) when the culture mass (= volume) had doubled. About 30% of initiations which occurred when DNA replication was released from this inhibition occurred from the site of insertion of F. This is the response predicted by our interpretation of switch-on in the light of what we know about the initiation volumes of the chromosome origin and F (Fig. 2). It is pertinent to point out the implications of this observation for the interpretation of studies (e.g., Perlman and Rownd, 1976) on the replication pattern of composite plasmids. Pre-incubation in very low thymine concentrations is used to enrich for replicating molecules, the location of whose origin of replication is then determined microscopically. This treatment may show that replication of the plasmid can occur from more than one location but such studies do not provide quantitative or even qualitative information about origin usage under normal growth conditions. There is an interesting analogy here between thymine starvation in bacteria and FUDR treatment of mammalian cells, both of which are used to enrich for replicating DNA molecules. It has been shown that prolonging FUDR treatment progressively reduces the inter-origin distance on DNA fibres suggesting that more and more potential 'origins' are being brought into activity by cell growth in the absence of DNA synthesis (see Taylor - this Symposium).

Incompatibility

If initiation of replication is controlled by the concentration of cytoplasmic effectors it follows that all copies of a replicon in one cell will be on an equal footing. If autonomous replication of F, for example, is blocked in an Hfr strain because its copy number is maintained at too high a level to permit frequent initiation at the F origin then an incoming F particle will likewise find itself in an environment which will reduce its replication frequency. Consequently, it will not be maintained. This kind of incompatibility is a straightforward consequence of cytoplasmic control of copy number and requires no additional assumptions, or functions, to account for it.

Consider now infection of a cell, containing a plasmid, with a genetically distinguishable copy of the same plasmid. In the simplest possible case where there is an average copy number of one at birth and two at division the introduction of an additional plasmid cannot lead to the production of a stable mixed clone. Control of copy number inevitably means that progeny of the infected cell will ultimately be born with an average of one plasmid copy. How soon this happens will depend on the kinetics with which perturbations in copy number are corrected and on the mechanism of plasmid segregation.

In the case of multicopy plasmids the situation may be more complex but the end result will be the same - control of copy number is a sufficient precondition for incompatibility. Consider a situation in which a cell is born with two copies of a plasmid and receives a third distinguishable copy by infection. If segregation were random, statistical considerations would ensure that pure clones were generated from the mixed parent. If segregation were not random the extreme situation would be one in which sister plasmids always segregated to different cells at division. Clearly in this case insertion of an additional plasmid copy (or, indeed, mutation of an existing plasmid) would lead to the production of permanently heterozygous clones but such a system of segregation can be discounted precisely because it would be inconsistent with the occurrence of incompatibility. Any less rigid segregation system which ensures only that cells are not born with zero copies of a plasmid is bound to lead to the production of pure clones from a mixed parent sooner or later.

There is much evidence consistent with the hypothesis that incompatibility is solely a consequence of cytoplasmic control of copy number. (1) All copy number mutants in a large sample isolated by Uhlin and Nordström (1975) in the plasmid R1 had altered incompatibility properties. (2) In the case of the composite plasmid ColE1-pSC1O1 the severity of incompatibility between it and ColE1 is similar to that between two ColE1 particles but the

incompatibility between it and pSC101 is much more extreme than that
between two pSC101 particles (Fig. 3). The situation facing the incoming pSC101 plasmid differs in the cell containing the composite
as compared to a cell containing pSC101 because the plasmid concentration is higher in the former case. The situation is similar in
some respects to that facing an F plasmid in an Hfr strain. (3) De
Vries et al. (1975) isolated 72 mutants from matings between an

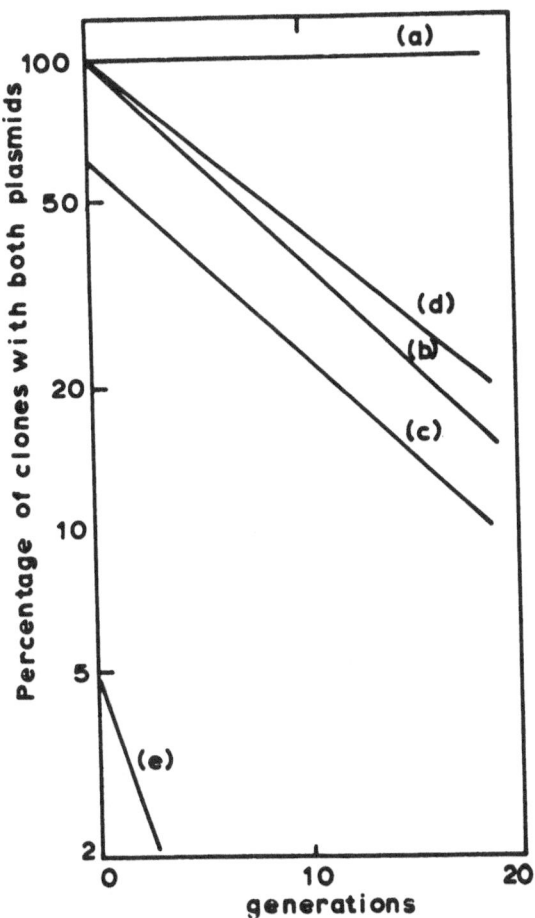

Figure 3. *Intensity of incompatibility between CoIE, pSC101 and a composite between them. Data taken from Cabello et al. (1976). (a) ColE and pSC101. (b) pSC101 and pSC101. (c) ColE and ColE. (d) ColE-pSC101 and COlE. (e) ColE-pSC101 and pSC101. The plasmids carried antibiotic resistance markers and cells containing incompatible plasmids were maintained by appropriate antibiotic selection. The rate of segregation was followed after removal of this selection pressure.*

F-prime donor and an Hfr recipient which could maintain the superinfecting plasmid. In all 72 strains the incompatibility (*inc⁻*)
mutation was in the recipient genome, probably in the integrated
F particle, and in 71 of these the integrated plasmid had become
transfer defective and appeared to have suffered extensive deletion
of genetic information. The inhibitor dilution model predicts that
single plasmid mutations which abolish incompatibility should be of
two types only. They should reduce or abolish either inhibitor
activity or sensitivity of the DNA binding site to inhibitor. In
either case there would be an increase in copy number, switch-on
of F if it were integrated, and a selectively disadvantageous,
possibly lethal, overproduction of plasmid DNA. It would therefore
be a strong prediction of this model - and, I suspect, of any model
in which incompatibility is a reflection of copy number control-
that *the great majority (vide infra) of inc⁻* mutants with total loss
of incompatibility should have aberrant replication. Consequently,
in the type of experiment described by De Vries et al., the mutations should be confined to the integrated plasmid, which can be
maintained passively as part of the host chromosome, and the mutants should be replication defective, which they appear to be.
In a similar search for *inc⁻* mutants in *Staphylococcus aureus* by
Wyman and Novick (1974) in which two genetically marked plasmids
could be maintained in the same cell without selection only one
was found. It is striking that this solitary *inc⁻* plasmid had
simultaneously become integrated into the host chromosome and replication defective.

     On the other hand, one of the mutants isolated by De Vries
et al. retained its capacity for autonomous replication and also its
incompatibility properties when in the autonomous state. This
phenotype is not understandable in terms of the hypothesis discussed
in this section but without a fuller understanding of its genetic
basis it seems reasonable to place more weight on the 71 mutants
with properties that are predicted than on the one whose properties
are not. The selective pressures available to the experimenter in
a genetic system like that of *E.coli* make it hazardous to draw
conclusions from the isolation of rare strains with particular
properties that have been specifically selected without a full
analysis of the underlying genetic change. It is in this light
that the isolated instances of success in selecting for compatible
F-primes (Palchaudhuri and Iyer, 1971; De Vries et al., 1974; San
Blas et al., 1974) should be viewed at present. The finding of one
copy mutant of R1 (out of a total of 100 such mutants) which
appeared to have simultaneously lost the expression of incompatibility (Uhlin and Nordström, 1975) is more disconcerting (because
loss of incompatibility was not selected for). In all these cases
however, the possibility that the plasmids concerned are composites
with cryptic origins suggests itself as an obvious one, complicating
the interpretation of the mutants isolated.

Relaxed Replication

Plasmids have been divided into two types on the basis of their replication behaviour. Based on his studies of replication of the plasmid NRl in *Proteus mirabilis*, Rownd (1969) **introduced the term** relaxed replication. It has been used widely (cf. Clowes, 1972; Arai and Clowes, 1975) to describe the behaviour of plasmids which exhibit one or more of the following features: (a) many copies per cell; (b) failure to stop replicating abruptly either when their hosts enter the stationary phase of growth or when protein synthesis is blocked; (c) random selection for replication so that individual plasmids may undergo two replications while others undergo none and (d) random replication in relation to the cell age of the host.

Whether it is useful or valid to classify plasmids into distinct relaxed and stringent classes on the basis of these criteria is questionable since the low copy number plasmid F is known to exhibit property (b) (Bazaral and Helinski, 1970), and both F and Rl have been claimed to exhibit properties (c) and (d) (Gustafsson and Nordström, 1975; Gustafsson, 1977). Nevertheless, the existence of multicopy plasmids has prompted a consideration of the extent to which their replication properties are consistent with control of copy number by volume titration.

In the case of a cell which is born with a single copy of a plasmid which replicates once during the cell cycle, the inhibitor dilution hypothesis predicts ideally that the concentration of inhibitor will oscillate over a twofold range. Now consider a multicopy situation (which for simplicity we might imagine is the result of a mutation which reduces the activity of inhibitor tenfold, or is the result of transfer to a new host in which transcription of the inhibitor gene is less efficient, as might be the case when NRl is transferred from *E.coli* to *P.mirabilis* (cf. Rownd, 1969). If plasmid replication could be synchronized perfectly the pulse of inhibitor would be as large as before (Fig. 4a). Such synchrony could not persist in the face of statistical fluctuations in the timing of individual initiation events. Moreover any spread in the timing of individual replications would be self amplifying because an 'early' replication would raise the inhibitor concentration and thus delay subsequent replications. Thus at equilibrium, in the case of a high copy number, plasmid replications would be distributed approximately at random with respect to cell age and probably with respect to parentage as well. Equally, inhibitor concentration would tend to become constant (Fig. 4c).

Thus we see that an important aspect of models of control of this sort is that in unicopy situations the timing of replication may be determined primarily by inhibitor concentration and may occur over a **narrow range of volume:replicon** ratios whereas in multicopy situations inhibitor concentration will tend to become

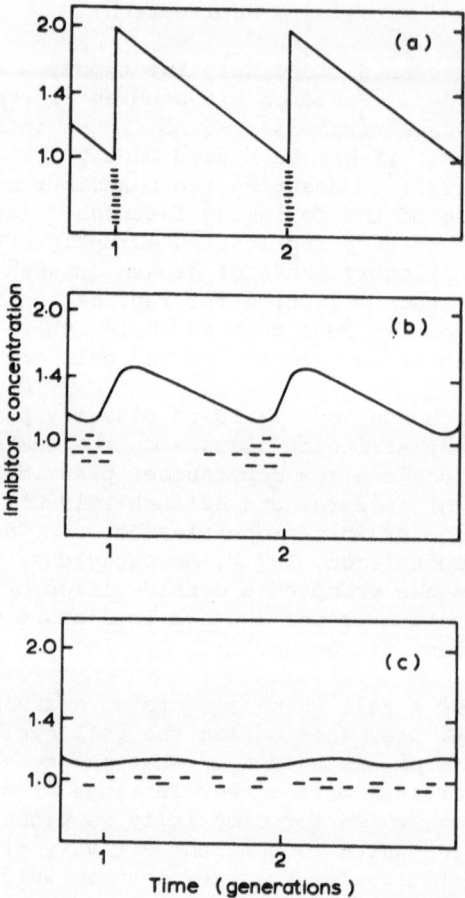

Figure 4. Effect of decay in synchrony on variation in inhibitor concentration in a multicopy situation. The diagrams are schematic only.

constant, and the timing of initiation will approach randomness. It is not truly random of course, because every initiation event reduces the probability of a subsequent one. In the light of the use of the term 'relaxed' in connection with this type of replication, it should be noted that copy number is still 'controlled' in the sense that average copy number will remain constant under equilibrium conditions. If in one time interval there is a greater than average number of replications, then the concentration of inhibitor will rise above its average level. There will consequently be fewer replications in the following time interval.

If the concentration of inhibitor becomes constant in multi-copy situations and the frequency of initiation therefore becomes constant with time, there is no reason in principle why replication should not continue indefinitely following inhibition of protein synthesis or entry into stationary phase. In practice, of course, the extent of continued replication will depend on whether inhibitor synthesis ceases, and if so, how soon other cell components involved in DNA replication become limiting. Nevertheless, the general point can be made that the higher the copy number the lower the effective inhibitor concentration and the greater the probability of continued replication in the absence of growth.

The conclusion I wish to come to at this point is that the properties of 'relaxed' plasmids as described earlier make sense in terms of the inhibitor dilution hypothesis (and it should be noted again, in terms of other models). The same mechanism could lead to stringent or relaxed behaviour depending on copy number.

The new question posed by this discussion is how can synchronous initiation by more than one copy of the same replicon in a single cell be assured since, as we have seen, cytoplasmic control of copy number predicts what should perhaps be described as antisynchronous replication. Indeed it has recently been questioned whether there are any examples of synchronous initiation in *E.coli* (Gustafsson, 1977). The evidence for synchronous initiation in multicopy situations comes from studies with synchronous cultures or from populations fractionated by age or size. In cultures with doubling times less than 60 minutes, cells have on average two chromosome origins at birth. With doubling times less than 30 minutes they have four origins at birth (Cooper and Helmstetter, 1968). Likewise F^+ cultures at all doubling times less than 60 minutes have two plasmid copies at birth (Collins and Pritchard, 1973). These are the only multicopy replicons so far as I know where initiation has been adequately studied in synchronous cultures. The fact that the timing of initiation of chromosome replication (Cooper and Helmstetter, 1968) and of F replication (Zeuthen and Pato, 1971; Davis and Helmstetter, 1973) can be identified by these techniques as occurring other than at random with respect to cell age is evidence for synchronous initiation.

In the case of F the conclusion that replication is synchronous with respect to cell age and, by implication, with respect to different plasmid copies in one cell has been challenged by Gustafsson (1977). Gustafsson and Nordström (1975) had previously shown for the stringent plasmid R1 that in density shift experiments with exponential cultures a substantial proportion of plasmid particles had not replicated and an equal proportion had replicated twice after one doulbing of total DNA. In other words these

stringent plasmids appear to be chosen at random for replication
as found by Rownd (1969) for a relaxed plasmid. In a further
interesting experiment, Gustafsson (1977) has presented data which
seem to show that replication of Rl and F is not only random
with respect to plasmid parentage but also with respect to time and
to cell volume. This conclusion is in direct conflict with those
derived from experiments with synchronous cultures and it is not
clear to me how the two sets of data are to be resolved. One
possibility stems from the choice of culture medium, giving a
generation time of about 60 minutes, by Gustafsson. According to
our studies (Collins and Pritchard, 1973; Pritchard, Chandler and
Collins, 1975), F replication will tend to coincide with cell
division at this growth rate. If F replication occurs before cell
division in a significant fraction of the population and
segregation is asymmetric in some cells, giving duaghters with
three and one plasmid copies respectively, then copy control via
volume titration could lead to results of the type found by
Gustafsson even though initiation is synchronous with respect to
cell volume in the bulk of the population. In other words,
Gustafsson's observations might be a function of infidelity in the
segregation of F.

    Since the pattern of F replication may nevertheless prove to
be asynchronous, despite the evidence from age fractionated
cultures, doubt is inevitably cast on the conclusion that initiation
of chromosome replication is synchronous in cells with more than
one copy of the chromosome origin, since this conclusion is based
on the same kinds of experimental observation. I believe it to be
unlikely that initiation at different chromosome origins is
grossly asynchronous and involves random choice of chromosome
origins since the consequence of this would be an extreme asymmetry
in the distribution of DNA to daughter cells and much greater
variation in cell size than is found in cultures of $E.coli$. However,
data can be found (Eberle, 1968) which could lead to the opposite
conclusion.

    In conclusion then, a rather important question which seems to
be emerging is whether it has been demonstrated convincingly for
any multicopy replicon (chromosome or plasmid) that initiation of
replication is synchronous. If it is synchronous, how is this to be
explained in terms of cytoplasmic control of copy number ? Presumably
there would need to be a significant delay between an initiation
event and the associated change in effector concentration, and a
significant dead-time between successive initiations by one
replicon, to ensure synchronous replication. In the case of a
large replicon like the bacterial chromosome, the delay in the
response of effector concentration to initiation could be achieved
by placing the control gene downstream from the origin.

    If initiation is $asynchronous$ then how is a low copy-number

plasmid like F maintained with such high stability ? It seems
necessary to invoke a mechanism which blocks cell division until
plasmid replication has taken place as was suggested earlier
(Pritchard, 1969a). Also if F replication is asynchronous (i.e.,
if the F copy number control system is not very sensitive) what is
the explanation for switch-off ? Perhaps it has become necessary
to ask how convincing *is* the evidence for switch-off of F in Hfr
strains.

Contrary Observations

There are a number of instances reported in which it is
observed that plasmid copy number seems to be determined by the
total amount of plasmid DNA rather than by the number of plasmid
particles (i.e., plasmid origins). A well studied example is that
described by Goebel and his colleagues (Goebel and Bonewald, 1975;
Boidol et al., 1977; Luibrand et al., 1977; Mayer et al., 1977). They
started with a copy mutant of the plasmid Rl isolated by Nordström
et al. (1972) with a fourfold increase in copy number. This mutant
plasmid (Rl-B2) generates small segregant plasmids with a very high
copy number. The total amount of plasmid DNA (about 15% of total
DNA) is about the same for the mini-segregants as for Rl-B2.
Furthermore, when a composite between a mini-segregant plasmid (ca.
100 copies per cell) and the mini-ColEl plasmid (Hershfield et al.,
1976) with *ca.* 120 copies per cell was constructed, the composite
again contributed about 15% to total DNA indicating that the copy
number of both component replicons was less than that of their
respective parents. Other composites involving mini-segregants
from the Rl copy mutant gave similar results. Thus not only does
it seem that total plasmid DNA content is what is determined but
the rule proposed earlier - that replication and copy number would
be under the control of the component with the higher copy number -
seems to be broken. In considering these observations, the first
point to note is that they are not the norm. The min-F and min-
R6-5 plasmids, for example, constructed by Timmis et al., (1975)
with less than 20% of the molecular length of their parent plasmids
F and R6-5 have unaltered copy numbers. It is therefore pertinent
to ask under what circumstances total plasmid DNA becomes important
in determining copy number. A possible interpretation of the
behaviour of the mini-Rl segregants is as follows.

In the original Rl copy mutant control of the frequency of
initiation is lost or greatly reduced. Copy number rises until
another cellular component involved in DNA synthesis becomes rate-
limiting for plasmid replication. The total amount of plasmid DNA
is thereby limited to 15%. If this new rate limiting component is
involved in the replication of all DNA species in the cell there
will be a disturbance of host cell growth as is found (Engberg et
al., 1975). Indeed Nordström et al. (1974) have presented evidence

suggesting that the presence of R1-B2 leads to competition for DNA
polymerase III. The small segregant plasmids will express more
fully the loss of control of initiation frequency but the plasmid
DNA content will again stabilise at 15 per cent of total DNA.
Similarly, if the mini-segregants are enlarged by insertion of
additional DNA the total amount of plasmid DNA will not be altered
but copy number will fall. Thus although it is copy number which is
normally controlled, conditions can be created in copy mutants in
which some other component involved in DNA synthesis becomes rate-
limiting as Mayer et al. (1977) also suggest.

The conclusion I wish to come to is that false deductions may
stem from the use of copy mutants to investigate the nature of the
quantitative controls which determine the copy number of non-mutant
replicons under steady state growth conditions.

Control Models

The simplest conceivable system which could lead to a constant
copy number would be one in which initiation was the consequence of
an interaction between a host-specified protein and a replicon. The
initiation frequency would follow the rules for a bimolecular
reaction with the particular feature that the reaction product would
be an additional copy of the substrate.

$$\text{Protein (P)} + \text{Replicon (R)} \rightleftharpoons PR \xrightarrow{\text{precursors}} 2R$$

The frequency of initiation would presumably reach equilibrium
when the host protein was saturated. Copy number would be
proportional to the concentration of the host protein and inversely
proportional to growth rate (because the number of initiations would
be constant with time not generations). Replication would be
relaxed in terms of the criteria described earlier.

Such a system might be described as *passive* with respect to
control of copy number. An *active* control of copy number would be
one in which deviations from the average in both directions were
'sensed' and the frequency of initiation adjusted accordingly.

Conceivable candidates for passive control of replication are
simple plasmids like ColEl in which initiation of DNA replication
depends on a host-specified protein and which exhibit extensive and
accelerated replication in the absence of protein synthesis
(Bazaral and Helinski, 1970; Kingsbury and Helinski, 1970). In the
case of a number of plasmids with similar properties found in
Enterobacter cloacae, Zandvliet and Jansz (1976) have shown that
after a prolonged period of plasmid replication in the absence of

protein synthesis, there is an immediate inhibition of replication
if the inhibition of protein synthesis is relieved. This suggests
that there is an *active* response to the abnormally high copy number
created by the replication in the absence of protein synthesis.
Moreover the initial acceleration in the rate of ColEl replication
induced by aminoacid starvation (Bazaral and Helinski, 1970) and the
continuous replication under these conditions tempts the obvious
conclusion that control is *negative* in the sense that the frequency-
limiting element in the system is an inhibitor of initiation.

     Neither of these conclusions is unambiguous however. If control
were passive, the rate limiting host protein might also be involved
in chromosome replication. Inhibition of chromosome replication by
aminoacid starvation or chloramphenicol treatment might increase its
availability to the plasmid and thus increase its rate of
replication. Reinitiation of chromosome replication could have the
opposite effect. Thus it is not inconceivable that the observed
pattern of replication during and after inhibition of protein
synthesis is the outcome of a *passive* determination of copy number
by a mechanism which is essentially *positive*.

     The situation in relation to the plasmid pSClOl seems clearer
at first sight. Switch-off of replication at the pSClOl origin in a
composite between it and ColEl as described earlier seems to
provide conclusive evidence for active control of copy number.
Cabello et al. (1976) have suggested further that control is
negative on the basis of this evidence. Again, it is possible to
argue that a control mechanism which was passive and positive could
give the same result. If initiation required the interaction of
several molecules of a host specified protein with the plasmid DNA
then raising the copy number might partition available initiator
between a greater number of plasmids thereby reducing the
probability of any one of them initiating a round of replication.

     Thus I conclude from these two examples that although it seems
likely that even in the case of simple plasmids replication is
actively and negatively controlled, the use of physiological studies
to try to demonstrate this unequivocally may be an unprofitable
exercise. It seems that one needs to understand more about the
underlying biochemical events involved in initiation before this
kind of approach can throw light on the basis of temporal control
of these events.

     For only one replicon is anything significant known at the
biochemical level concerning the temporal control of its frequency
of initiation. This is the laboratory plasmid λdv (Berg, 1974;
Matsubara, 1976). This plasmid in its simplest form seems to
consist of a single operon coding for three proteins (Fig. 5). One
(product of the gene *tof*)) is an auto-repressor of transcription

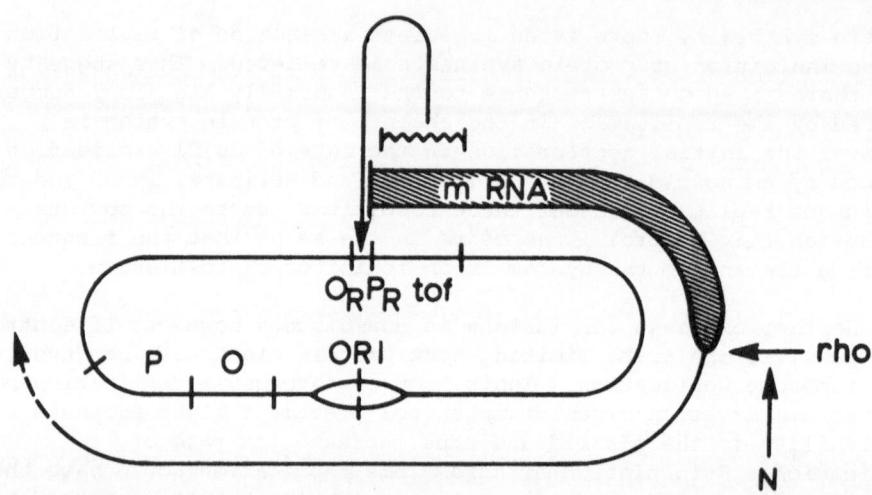

Figure 5. Diagram of λdv and the tof operon. O_R and P_R are operator and promoter sites. tof, O and P are three genes described in text. ORI is the plasmid replication origin. rho is host transcription termination protein and N is λ specified anti-termination protein.

of the operon. The other two (the products of genes O and P) are necessary for plasmid replication but they appear not to determine the frequency of replication (i.e., they are not present in rate limiting concentrations). It is suggested by both Berg and Matsubara that initiation of replication is the consequence of "transcriptional activation" of the plasmid origin. This suggestion is based on earlier observations (see Dove et al., 1971) suggesting that initiation of replication is triggered by a structural change in the DNA at the chromosome origin as a result of transcription in its vicinity. Thus it is supposed that *tof* is part of an auto-repression system which will maintain a constant concentration of *tof* protein and hence a transcription frequency proportional to growth rate. If initiation is coupled to this transcription the frequency of initiation of replication will likewise be proportional to growth rate. Thus it is suggested that maintaining a constant concentration of *tof* protein will lead to a constant concentration of plasmids. There seems to be good evidence that the plasmid copy number is related to the transcription frequency of the operon since mutants with altered *tof* protein or an altered operator show the expected changes in copy number. In addition evidence has been presented (Berg and Kellenberger, 1974) that the presence of the N gene product of λ increases copy number by antagonizing termination of transcription of the operon by the host protein *rho* at a site between the *tof* gene and the plasmid origin.

This is such an interesting system that it is to be hoped that
a quantitative simulation of it will be made to determine whether it
gives a stable control of copy number when realistic properties are
assumed for the elements involved. However, it does not seem to
provide a realistic model for plasmids like F and pSC1Ol since it
does not appear to predict switch-off. Thus inhibition of λdv
replication depends on the production of *tof* protein and production
of *tof* protein requires transcription which leads to initiation
of replication.

An interesting feature of the λdv system is that it indicates
how mutations in the host genome (e.g., affecting *rho* protein) might
influence the copy number of a plasmid.

Copy Number and Growth Rate

At fast growth rates the copy number (copies in unit volume) of
four plasmids, ColEl, F, Rl and prophage Pl, is less than it is at
slow growth rates (Clewell and Helinski, 1972; Collins and Pritchard,
1973; Engberg and Nordström, 1975; Prentki et al., 1977). On the
basis of more indirect evidence it seems probable that the copy
number of the chromosome origin is constant (*cf*. Donachie, 1968;
Pritchard et al., 1969). Consequently the plasmid:chromosome origin
ratio will be lower at fast growth rates than at slow growth rates
(Fig. 2) and this has been confirmed more directly in the case of
Pl (Prentki et al., 1977).

In the case of ColEl, it has been shown that addition of cyclic
AMP to glucose-grown cultures increases the rates of plasmid
synthesis and raises the copy number (Katz and Helinski, 1974). In
this case, therefore, it seems possible that the growth rate effect
is due to catabolite repression of the copy number control system.
It is not known whether cyclic AMP has a similar effect on the other
plasmids which exhibit an effect of growth rate on copy number.
Nevertheless the growth rate effect on F prompted me to consider
again the predictions of the inhibitor-dilution model (and other
models of this type). It was proposed initially that the output of
inhibitor per initiation was constant at all growth rates in order
to ensure that the initiation volume would be constant, since this
seemed to be the implication of the observations of Cooper and
Helmstetter (1968) and the earlier observations of Schaechter et
al. (1968) for the chromosome. If the relationship between
inhibitor concentration and initiation probability was a step
function in which the probability was either zero or one, the
assumption that a constant output of inhibitor would yield a constant
initiation volume would be correct. If a more realistic
relationship is assumed, however, the assumption would no longer be
valid. It has already been noted how, in the case of passive control
of a multicopy plasmid in which the specific activity of an

initiator is constant, copy number will increase as the growth rate
decreases because initiation frequency is constant with time. The
same will be true in the case of a multicopy plasmid under negative
control of initiation. The frequency of initiation will likewise
depend not only on inhibitor concentration but also on time. In
this case, however, the relationship is not linear because the
increased copy number will cause an increase in inhibitor
concentration. The shape of the curve relating copy number and
growth rate will therefore depend on the relationship between
inhibitor concentration and the probability of initiation. The
relationship between copy number and growth rate will presumably
be some kind of power function.

     The conclusion I wish to come to is that an increase in copy
number with a decrease in growth rate will be predicted by any
initiation control mechanism in which the frequency of initiation
is determined by the concentration of cytoplasmic effectors. I
do not wish to imply that this *is* the explanation for the growth
rate effect; only that such a response will be expected in addition
to any physiological effect of growth rate on the regulatory system
determining copy number. If the growth rate effect were primarily
due to such a direct physiological effect of growth rate on the
concentration of regulatory effectors then continued replication of
a plasmid after cessation of growth could be a reflection of this -
the cell might 'sense' that growth rate had become infinitely
slow and attempt to adjust copy number accordingly.

     Whatever the explanation for the growth rate effect on plasmid
copy number, an interesting question that arises is how the cell
maintains a constant concentration of chromosome origins. One
possibility which should be borne in mind is that the control genes
for initiation of chromosome replication might be located downstream
from the chromosome origin. It has been shown previously (Chandler
and Pritchard, 1975) that the concentration of any gene distal to
the chromosome origin decreases as the growth rate increases, the
effect increasing in magnitude the further downstream the gene is.
Consequently, the control gene could be so located that the increase
in initiation volume that would otherwise occur with increasing
growth rate is compensated for by decreasing the concentration of
the inhibitor gene. Such a strategy would also ensure synchronous
initiation as discussed earlier.

                              CONCLUSION

     This article has two themes. One has been to restate and
amplify the hypothesis we proposed some years ago, now that it is
possible to be more confident that the general idea, that
initiation frequency is determined by the concentration of cyto-

plasmic effectors, is valid. The relationships that have been suggested between control of copy number, switch-off and switch-on, relaxed and stringent replication, and incompatibility are doubtless at best an oversimplification. But I hope that they are helpful if only in provoking the execution of experiments that can refute them.

The second more depressing theme is that we still know very little about the temporal control of initiation of replication. It is encouraging to note that the "replicon hypothesis" no longer has much appeal. Progress in elucidating the biochemical basis of the temporal control of initiation should now be more rapid, particularly if the new techniques for constructing composite replicons are exploited.

An important question which has emerged from this discussion is the sensitivity of replication of plasmids like F to copy number. With an answer to this question, it should be possible to make quantitative predictions about switch-off, synchrony, and the influence of plasmid replication on cell division, thus testing more rigorously the ideas that have been put forward here.

Finally, I want to emphasise that I have discussed the various aspects of the behaviour of chromosomes and plasmids and composite replicons in terms of a particular simple model of cytoplasmic control of copy number in order to simplify argument. Preliminary experiments by M.C. Chandler and myself have suggested that the copy number of F may be continuously monitored. The inhibitor dilution mechanism could not do this. The unstable inhibitor model which we suggested as an alternative (Pritchard et al., 1969), could.

ACKNOWLEDGEMENTS

I am grateful to many friends for reading a draft of this article and making very valuable comments.

REFERENCES

Arai, T. and Clowes, R. (1975) In: Microbiology - 1974. 'ed. David Schlessinger) Soc. Microbiol., Washington, pp141-155

Bazaral, M. and Helinski, D.R. (1970) Biochemistry 9:399-406

Berg, D.E. (1974) Virology 62:224-233

Berg, D.E. and Kellenberger-Gujer, G. (1974) Virology 62:234-241

Bird, R.E., Chandler, M. and Caro, L. (1976) J. Bact. 126:1215-
    1223

Boidol, W., Siewert, G., Lindenmaier, W., Luibrand, G. and Goebel,
    W. (1977) Molec. Gen. Genet. 152:231-237

Cabello, F., Timmis, K. and Cohen, S.N. (1976) Nature, 259:285-290

Chandler, M. and Pritchard, R.H. (1975) Molec. Gen. Genet. 138:
    127-141

Chandler, M., Silver, L., Roth, Y. and Caro, L. (1976) J. Mol. Biol.
    104:517-523

Clewell, D.B. and Helinski, D.R. (1972) J. Bact. 110:1135-1146

Clowes, R.C. (1972) Bacteriol. Rev. 36:361-405

Collins, J. and Pritchard, R.H. (1973) J. Mol. Biol. 78:143-155

Cooper, S. and Helmstetter, C.E. (1968) J. Mol. Biol. 31:519-540

Davis, D.B. and Helmstetter, C.E. (1973) J. Bact. 114:294-299

De Vries, J.K., Pfister, A., Haenni, C., Palchaudhuri, S. and Maas,
    W. (1975) In: Microbiology - 1974. (ed. David Schlessinger).
    American Soc. Microbiol., Washington, pp166-170

Donachie, W.D. (1968) Nature, Lond. 219:1077-1079

Dove, W., Inokuchi, H. and Stevens, W. (1971) In: The Bacteriophage
    Lambda. (ed. A.D. Hershey) Cold Spring Harbor Press, Cold
    Spring Harbor, N.Y., pp747-771

Eberle, H. (1968) J. Mol. Biol. 31:149-152

Engberg, B., Hjalmarsson, K. and Nordström, K. (1975) J. Bact.
    124:633-640

Engberg, B. and Nordström, K. (1975) J. Bact. 123:179-186

Fantes, P.A., Grant, W.D., Pritchard, R.H., Sudbery, P.E. and Wheals,
    A.E. (1975) J. Theor. Biol. 50:213-244

Goebel, W. and Bonewald, R. (1975) J. Bact. 123:658-665

Gustafsson, P. (1977) Ph.D. thesis, University of Umea

Gustafsson, P. and Nordström, K. (1975) J. Bact. 123:443-448

Helmstetter, C.E., Cooper, S., Pierucci, O. and Revelas, E. (1968)
    Cold Spring Harbor Symp. Quant. Biol. 33:809-822

Herschfield, V., Boyer, H.W., Chow, L. and Helinski, D.R. (1976)
    J. Bact. 126:447-453

Katz, L. and Helinski, D.R. (1974) J. Bact. 119:450-460

Kingsbury, D.R. and Helinski, D.R. (1970) Biochem. Biophys. Res.
    Commun. 41:1538-1544

Koch, A.L. (1977) IN: Advances in Microbial Physiology. (eds.
    A.H. Rose and D.W. Tempest) Academic Press, Lond., in press

Jacob, F., Brenner, S. and Cuzin, F. (1963) Cold Spring Harbor
    Symp. Quant. Biol. 28:329-348

Luibrand, G., Blohm, D., Mayer, H. and Goebel, W. (1977) Molec.
    Gen. G enet. 152:43-51

Matsubara, K. (1976) J. Mol. Biol. 102:427-439

Mayer, H., Luibrand, G. and Goebel, G. (1977) Molec. Gen. Genet.
    152:145-152

Nagata, T. (1963) Proc. Nat. Acad. Sci. USA 49:551-559

Nandadasa, H.G. (1970) J. Gen. Microbiol. 63:viii

Nishimura, Y., Caro, L., Berg, C.M. and Hirota, Y. (1971) J. Mol.
    Biol. 55:441-456

Nordström, K., Ingram, L.C. and Lundbäck, A. (1972) J. Bact.
    110:526-569

Nordström, U.M., Engberg, B. and Nordström, K. (1974) Molec. Gen.
    Genet. 135:185-190

Palchaudhuri, S.R. and Iyer, V.N. (1971) J. Mol. Biol. 57:319-333

Perlman, D. and Rownd, R.H. (1976) Nature 259:281-284

Prentki, P., Chandler, M. and Caro, L. (1977) Molec. Gen. Genet.
    152:71-76

Pritchard, R.H. (1968) Heredity, Lond. 23:472-473

Pritchard, R.H. (1969a) Ciba Foundation Symp. on Bacterial
    Episomes and Plasmids. (eds. G.E.W. Wolstenholme and Maeve
    O'Connor) J. and A. Churchill, Ltd. London, pp65-74

Pritchard, R.H. (1969b) Ciba Foundation Symp. op.cit. pp251-2

Pritchard, R.H. (1974) Philosoph. Trans. Roy. Soc., B. 267:303-336

Pritchard, R.H., Barth, P.T. and Collins, J. (1969) XIX Microbial Growth. Symp. Soc. Gen. Microbiol. pp263-297

Pritchard, R.H., Chandler, M.G. and Collins, J. (1975) Molec. Gen. Genet. 138:143-155

Rownd, R. (1969) J. Mol. Biol. 44:387-402

San Blas, F., Thompson, R. and Broda, P. (1974) Molec. Gen. Genet. 130:153-163

Schaecter, M., Maaløe, O. and Kjeldgaard, N.O. (1958) J. Gen. Microbiol. 19:592-606

Timmis, K., Cabello, F. and Cohen, S.N. (1975) Proc. Nat. Acad. Sci. USA 72:2242-2246

Tresguerres, E.F., Nandadasa, H.G. and Pritchard, R.H. (1975) J. Bact. 121:554-561

Uhlin, B.E. and Nordström, K. (1975) J. Bact. 124:641-649

Wyman, L. and Novick, R.P. (1974) Molec. Gen. Genet. 135:149-161

Zandvliet, G.M. and Jansz, H.S. (1976) Eur. J. Biochem. 62:439-449

Zeuthen, J. and Pato, M.L. (1971) Molec. Gen. Genet. 111:242-255

MAP POSITION OF THE REPLICATION ORIGIN OF THE <u>E.COLI</u> K12

CHROMOSOME

Olivier Fayet[*] and Jean-Michel Louarn

Centre de Biochimie et de Génétique Cellulaires
du   CNRS
Toulouse, France

A biochemical approach of the initiation step of chromosomal replication in <u>E.coli</u> will very probably require the isolation of DNA sequences enriched in the unique site where initiation occurs, <u>ori</u>.

In order to obtain information that could help in the molecular cloning of <u>ori</u>, we have reinvestigated its location on the genetic map. To date, the best published estimation of the location of <u>ori</u> is ± 1 min. around the <u>ilv</u> operon (1,2), which is in a region with many closely-linked genes and which in particular contains the two initiation genes <u>dnaA</u> and <u>dnaP</u> (3,4).

The Mapping System

The following strategy was used:

(a) We have introduced Mu-1 prophage into the <u>ilv</u> region, the recipient strain also being a <u>dnaA</u> temperature sensitive mutant.

(b) From a given Mu-1 lysogen, we have derived integratively-suppressed Hfrs, chosen such as the <u>ori</u> region is replicated clock-wise (Hfr G) or counterclockwise (Hfr D) from the plasmid origin, as indicated in Figure 1 (5,6).

(c) The direction of replication of the Mu-1 prophage was characterized in the parental strain and the two derived Hfrs, by determining to which prophage strand the 5' growing end belongs when a replication fork passes the prophage. To this aim, the

*A.S.I. Participant

27

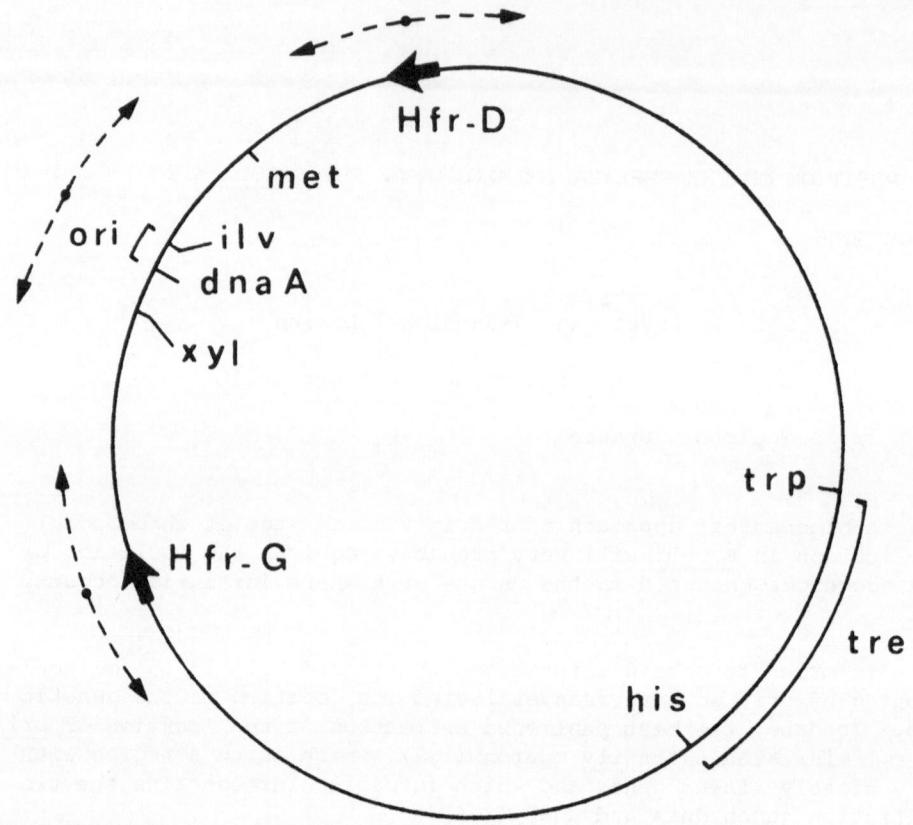

Figure 1. *Approximate map location on the E.coli K12 chromosome of the pAR 132 plasmid in the two types of integratively suppressed Hfrs used. As indicated by the two large arrows, the type D Hfrs transfer clockwise and the type G Hfrs counterclockwise. In the parental ts strains replication starts at* ori, *whereas in the Hfrs, growing at non permissive temperature, it starts within the inserted plasmid. In all cases, once initiated, replication goes on bi-directionally up to a fixed termination site (*tre*) located between the* trp *and* his *genes, as demonstrated by Louarn (6).*

strand preference of Mu-1 specific Okazaki fragments was established for the three strains (6).

The strand preference should be the same when the prophage is replicated from ori or from the plasmid located on the same side of the prophage as ori. It should be reversed when the replication is initiated on the plasmid located on the other side of the prophage.

Isolation of the Mu-1 Lysogens

In a population of bacteria harboring the dnaA 46 mutation and rendered lysogenic for Mu-1, we have selected mutants in the following genes: ilv operon, pho S, bgl B or C, rbs K or P. The unique association of a Mu-1 prophage with the mutated site was checked, as well as the order of the different genes, which was found to fit with the map of Bachmann et al. (7) (Fig. 2).

A resistance transfer factor, pAR 132, derivative of R100-1 (5,6), was used to promote integrative suppression. The location of the integrated plasmid was subsequently checked by conjugational assay.

Strand Preference Among Mu-1 Specific Okazaki Pieces

For a given Mu-1 insertion, exponentially growing cultures of the parental temperature sensitive strain and of its Hfr D and G derivatives were pulse labelled with tritiated thymidine. The Okazaki pieces smaller than 10S were subsequently purified and hybridized to isolated Mu-1 strands. The data obtained were used to calculate the amount of Mu-1 strand r present among Mu-1 specific Okazaki pieces. The results are presented in Table 1.

The strand preferences observed clearly indicate that the ilv and rbs loci are replicated from ori in the same direction as from the plasmid in Hfr G, i.e., clockwise, whereas the pho S and bgl loci are replicated from ori as from the plasmid in Hfr D, i.e., in the counterclockwise direction. The site ori is therefore located between the bgl/pho S and rbs genes.

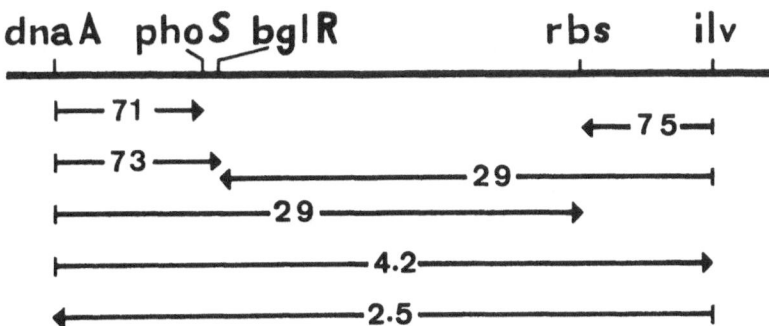

Figure 2. *Gene order in the dnaA - ilv region. The five indicated loci have been ordered by two factors crosses using P1 transduction. Cotransfer values are given. The arrowheads indicate the unselected markers. Note: No Mu-1 induced mutation was used for this mapping.*

TABLE 1

Site of integration of Mu-1	Ratio $\dfrac{\text{Mu-r strand}}{\text{Mu-r strand + Mu-}l\text{ strand}}$ among Okazaki pieces extracted from		
	parental ts strains	integratively suppressed Hfrs	
		type D	type G
Pho S	0,78	0,76	0,31
bgl B or C	0,32	0,32	0,68
rbs K or P	0,38	0,79	0,43
ilv	0,33	0,79	0,37

Strand preference of Mu-1 specific Okazaki pieces in different Mu-1 lysogens and their Hfrs derivatives

Exponential cultures of the indicated Mu-1 lysogens and their Hfrs D and G derivatives were pulse labelled with tritiated thymidine (20 sec at 30°C for the ts strains and 15 sec at 40°C for the Hfrs) and lysed as previously described (6). The lysates were submitted to sedimentation in alkaline sucrose gradients, and the labelled material shorter than 10S was pooled, purified by isopycnic centrifugation and incubated with filters loaded with isolated Mu-r and Mu-l strands. The ratio Mu-r DNA/total Mu-1 DNA was then calculated.

DISCUSSION

The present location of <u>ori</u> is entirely consistent with that proposed by Louarn et al. (2), using different means. The site <u>ori</u> thus appears to be separated from the two initiation genes <u>dnaA</u> or <u>dnaP</u> by several genes and operons. The two genes limiting the

segment bearing ori are separated by approximately 1/2 min. on the
genetic map, that is by 20 kb. We have attempted to refine this
map, using uncA or B Mu-1 induced mutants. Unfortunately, this
locus, placed between bgl B-C and rbs K-P on Bachmann's map,
appeared in our hand to be loosely linked to ilv, but outside the
dnaA - ilv segment.

Two attempts have been recently made in order to isolate ori.
Masters (8) postulated that a diploid for ori should lower cell
viability and isolated F' factors producing such an effect. The
chromosomal sequence involved appeared to be the pyr E   - bgl B-C
segment. Hiraga (9) supposed that a plasmid containing ori could
overcome incompatibility conditions and isolated F' factors able to
maintain themselves, although poorly, in Hfrs strains. The
chromosomal sequence involved appeared to be the dnaA - bglA
segment. Our mapping of the site ori renders questionable the
interpretations proposed by Masters and by Hiraga for the effects
that they have observed.

REFERENCES

(1) Hohlfeld,R., Vielmetter, W. (1973) Nature (Lond.) New Biol.
    242:130-132

(2) Louarn, J., Funderburgh, M. and Bird, R.E. (1974) J. Bacteriol.
    120:1-5

(3) Hirota, Y., Mordoh, J. and Jacob, F. (1970) J. Mol. Biol.
    53:369-387

(4) Wada, C. and Yura, T. (1974) Genetics 77:199-220

(5) Bird, R.E., Chandler, M. and Caro, L. (1976) J. Bact. 126:
    1215-1223

(6) Louarn, J., Patte, J. and Louarn, J.M. (1977) J. Mol. Biol.
    in press.

(7) Bachmann, B.J., Brooks Low, K. and Taylor, A.L. (1976) Bact.
    Rev. 40:116-167

(8) Masters, M. (1975) Molec. gen. Genet. 143:105-111

(9) Hiraga, S. (1976) Proc. Nat. Acad. Sci., U.S.A. 73:198-202

AN APPROACH TO THE ISOLATION OF THE REPLICATION ORIGIN OF

BACILLUS SUBTILIS[*]

Scott Winston[**] and Noboru Sueoka

Department of Molecular, Cellular and Developmental
Biology
University of Colorado
Boulder, Colorado, 80309, U.S.A.

INTRODUCTION

The initiation of DNA replication in bacteria may be the major regulatory mechanism controlling DNA synthesis, yet little is actually known about the process of initiation itself (1). To further understand the initiation process we have attempted to isolate the region of DNA where initiation occurs, the chromosomal origin. The organism we have chosen to study is B.subtilis, because of its synchronized initiation system using germinating spores (2) and its transformation system which allows a genetic characterization of any region of the chromosome where genetic markers are available (3).

RESULTS

Our origin isolation has been a two step process. Firstly, we have isolated origin-region DNA from B. subtilis using germinating spores. The timing of initiation can be controlled by starving for a required base, thymine, for 90 minutes post germination, during which time the spores accumulate initiation potential.

Initiation is begun by adding bromouracil and thymine, and replication is allowed to proceed for a short period of time. The density labeled origin DNA is separated from the rest of the chromosomal DNA by CsCl equilibrium centrifugation. As shown in

*This work was supported by the National Institues of Health, grant GM-20352
**A.S.I. Participant

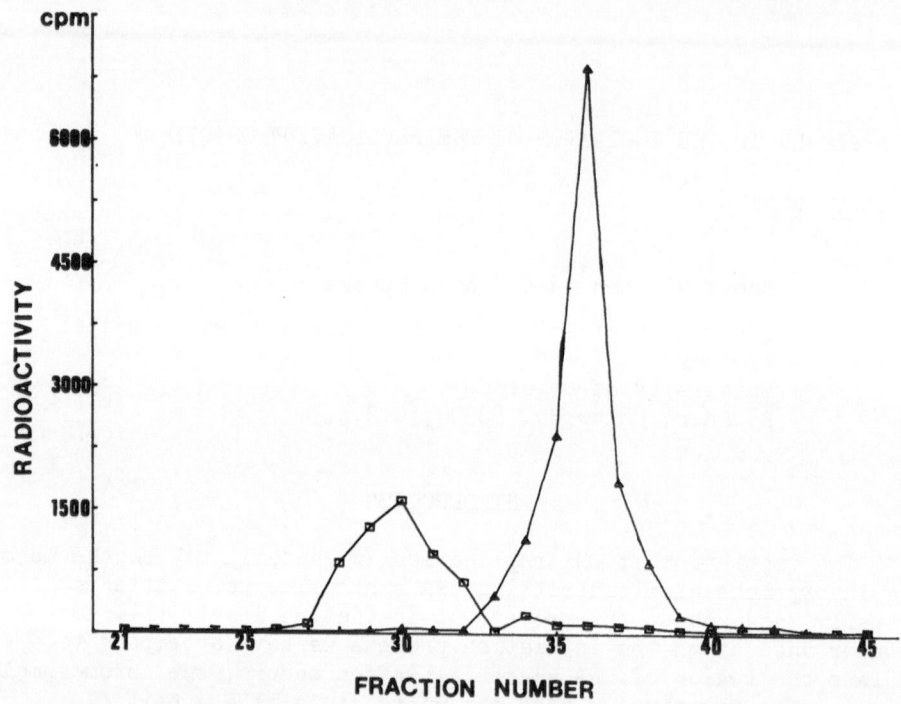

<u>Figure 1</u>.   *Isolation of Origin-Region DNA.*

An aliquot (0.05 ml) of spores (4 x 10[9] spores/ml) of <u>B. subtilis</u> strain 168 (<u>thy</u>,<u>trp</u>), uniformly labeled with ^{3}H-thymine were germinated in 10 ml of Spizizen's medium C+- GM-11 with <u>trp</u>, but without <u>thy</u> for 90 min at 37°C (11,12). Bromouracil and thymine were then added to final concentrations of 9 µg/ml and 1 µg/ml respectively, along with ^{14}C-thymine (final conc. 2 µc/ml; spec. activity 65 mc/mM). After 5 min replication was stopped by the addition of cold KCN to .01M on ice. The cells were then harvested and resuspended in 0.15 ml of Tris (.04M), EDTA (0.1M), sucrose (0.5M), pH 8.0 and 0.05 ml of 5 mg/ml lysozyme. The cells were lysed for 45 min at 37°C and then osmotically burst by the addition of 0.5 ml Tris (.05M), EDTA (0.1M), sucrose (0.15M), pH 8.0 at 20°C. Brij-58 was then added (0.1 ml of 1% Brij). The mixture was incubated for 10 min on ice, and brought up to a final volume of 3.74 ml with Tris (.01M), EDTA (.001M), pH 8.5. CsCl gradient centrifugation, fractionation, and radioactive counting were done as described previously (11).

△ - [3]H   □ - [14]C; *profile is left to right; bottom to top of gradient.*

Figure 1, approximately 1% of the uniformly labeled chromosomal DNA
has been transferred to hybrid regions after 5 minutes of
replication. The genetic markers ts8132 and purA, which lie
approximately 1 to 2 percent from the origin (4,13,14), show 13
and 8% transfer, respectively, to the hybrid region, while markers
from the middle or terminal regions of the chromosome show no
detectable transfer. The hybrid DNA molecules ranged in size from
$10-20 \times 10^6$ daltons on the basis of neutral sucrose sedimentation.
Thus, the density labeled DNA consisted of a heterogeneous set of
origin-region DNA fragments.

The second part of the origin isolation process has centered
on identifying and amplifying a specific origin-containing fragment
from the heterogeneous set of origin fragments generated during the
density transfer experiments. A restriction enzyme analysis of
the origin region yielded only nanogram quantities of DNA fragments,
and more importantly, did not allow identification of the fragment
which contains the replication origin.

To select the restriction fragment which contains the origin,
we have used a procedure which was successful in selecting and
cloning the origin of the F factor and R6K (5) (Figure 2). The
rationale behind this approach is to attach a non-replicative
restriction fragment which codes for some identifiable phenotype
to a set of restriction fragments, one of which contains an origin
of replication. To express its phenotype, the non-replicative
fragment must replicate autonomously from the chromosome as a
plasmid, so only those fragments attached to an origin-containing
fragment can do so.

Specifically, we have purified the EcoRI-generated, non-
replicative DNA fragment from pSC122, which codes for ampicillin
resistance (Ap^r) (5). The plasmid pSC122 is a composite of
pSC101 and an EcoRI restriction fragment from Staphylococcus
aureus which codes for Ap^r. The Ap^r fragment can be purified
either on the basis of its buoyant density in CsCl gradients
(pSC101 density = 1.710 g/cm^3; Ap^r fragment density = 1.692 g/cm^3)
or by electrophoretic mobility in agarose gels (pSC101 m.w. =
5.8×10^6 daltons; Ap^r fragment m.w. = 4.2×10^6 daltons) (5).
We chose Ap^r because of its availability on an EcoRI restriction
fragment, and because the B. subtilis genome does not contain the
genetic information for a β-lactamase. The latter reason
eliminates the possibility of spontaneous mutations which might
interfere with out genetic selection.

The selection of a specific B. subtilis origin fragment was
performed in the following manner. Origin-region DNA was cleaved
by EcoRI restriction enzyme and added to the purified EcoRI
fragment from pSC122 coding for Ap^r. The mixture was heated to
70°C to inactivate the restriction enzyme and slow-cooled to allow

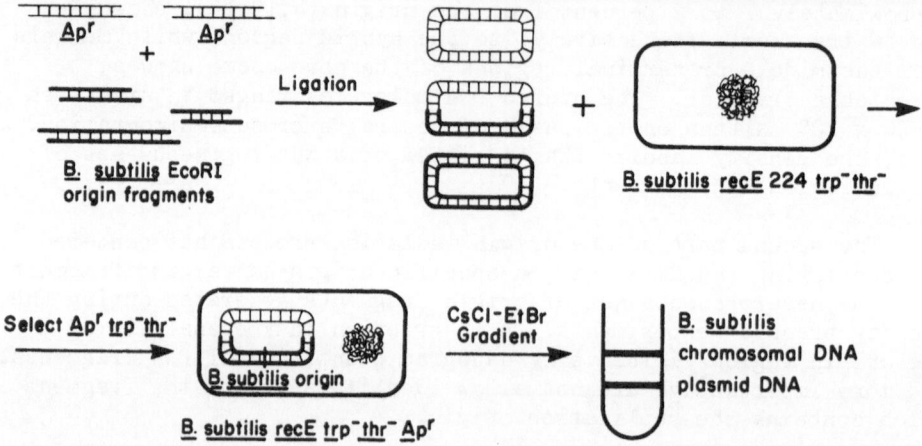

Figure 2.   *Experimental Protocol for Origin Fragment Identification*
              and Isolation.

Equal amounts (1 µg each) of the EcoRI generated Ap^r fragment and
EcoRI treated B. subtilis origin region DNA were incubated for 15
min at 70°C, 37°C, 20°C and 0°C in 0.05 ml of a solution
containing Tris (33 mM) MgCl$_2$(6.6 mM), ATP (.006 mM), DTT (6.6 mM).
Three units of T$_4$ ligase (1 unit/1 µ 1) were then added and the
reaction mixture was incubated 24 hr at 0°C.   The entire reaction
mixture was then added to 1 ml of competent B. subtilis strain
recE224 (trp thr) and shaken for 30 min at 37°C (11). The
transformed cells were mixed with 50 ml of seed layer soft agar
solution, 5 ml of which was layered on each base agar plate (10).
After 4 hours at 37°C, 10 ml of overlay containing 5 µg/ml
ampicillin was added.   Ap^r colonies were subsequently examined
for trp and thr phenotypes on appropriate selection plates. A
250 ml culture of one Ap^r thy thr clone was grown in Spizizen's
medium C+ containing 5 µg/ml Ap till stationary phase.   The cells
were harvested, lysed and prepared for CsCl-EtBr gradient
centrifugation using the cleared lysate method (8).   The two bands
observed by UV light after centrifugation were removed by syringe.
The solutions containing the bands were layered on a Dowex column
(5 mM x 50 mM) equilibrated with 0.1 X SSC to remove the EtBr. The
DNA was eluted with 0.1 x SSC and dialyzed for 48 hr against 6
liters of 0.1 M ammonium carbonate.   The dialyzed solution was
lyophilized to dryness, resuspended in EcoRI reaction mixture and
used for further analysis.

the EcoRI generated "sticky ends" to anneal. T_4 ligase was added and incubated 24 hrs at 0°C. The DNA mixture was then used to transform a competent culture of B. subtilis recE224 trp⁻ thr⁻ (6)

Ap^r colonies were selected by the overlay seed technique at 5 µg/ml Ap after allowing 4 hours expression time at 37°C (7). Twelve Ap^r colonies were found which were both trp⁻ and thr⁻ and positive for gram stain.

The B. subtilis recE224 trp⁻ thr⁻ strain was used as the host in these experiments because it is totally deficient in recombination. This prevents any of the donor DNA from integrating into the host chromosome. Control experiments were performed in which wild-type B. subtilis DNA and the Ap^r fragment were separately used to transform the recE strain to either trp and thr prototrophy or ampicillin resistance. Less than 1 in 10^9 cells were found to be trp$^+$ or thr$^+$, and no Ap^r colonies were detected. Thus, for Ap^r to be expressed in the recE host it must replicate autonomously from the host chromosome using the B. subtilis origin fragment for replication.

One Ap^r colony was examined for the presence of plasmid DNA. Plasmid DNA was isolated using a cleared lysate method similar to that for E.coli (8). The supernatant was run in a CsCl-ethidium-bromide gradient and examined with long-wave UV light. Two bands were observed, and both were removed for further study. The DNA was purified by removing the EtBr on a Dowex-SSC column, dialyzed, concentrated and resuspended in EcoRI reaction buffer. The upper band could transform a competent B. subtilis strain purA16, leu8, metB5 to prototrophy. This along with previous results indicates the Ap^r clone is the original B. subtilis trp⁻ thr⁻ strain, although the presence of the original recE phenotype has not been shown.

Preliminary results have shown that the lower band from the CsCl-EtBr gradients consisted of a circular covalently-closed DNA which could be cleaved by EcoRI. Further characterization of the Ap^r B. subtilis clones was stopped in order to conform to the NIH Guidelines on Recombinant Molecules (9). Since our present physical containment facilities prevent us from continuing these experiments under the NIH Guidelines, we have designed future experiments along the same lines, but have replaced our non-replicative Ap^r vector with a safer DNA fragment from B. subtilis.

At present we are purifying EcoRI fragments of B. subtilis which express some identifiable phenotype. Specific restriction fragments of B. subtilis can be selectively enriched by preparative gel electrophoresis (10). We are presently purifying EcoRI restriction fragments from B. subtilis which can either restore prototrophy to an auxotrophic recE strain or which can confer an antibiotic resistance to the same strain. This approach constitutes no biohazard, as we are constructing plasmid from B. subtilis DNA

which can grow in B. subtilis using the B. subtilis chromosomal
origin.

DISCUSSION

Our approach to selecting and amplifying a restriction fragment
from B. subtilis DNA enriched in origin sequences relies on that
fragment functioning in vivo as an origin and replicating
autonomously from the host chromosome. Although our preliminary
results are promising, we have no evidence that the plasmid-like
DNA isolated from the Ap^r B. subtilis clones contain the origin of
replication from B. subtilis. It is possible the Ap^r fragment
recombined with a preexisting plasmid in the host cell, although
no plasmid could be detected in these cells before transformation,
and addition of purified Ap^r fragment alone produced no Ap^r
colonies. Another possible explanation is that under our
experimental conditions, another sequence of DNA can function as an
origin of replication. A detailed analysis of the plasmid DNA
would help to eliminate this possibility, but under existing NIH
Guidelines (9) these types of experiments cannot be continued in
our laboratory. However, our results do suggest that by combining
B. subtilis origin DNA with the proper non-replicative DNA fragment
containing an assayable phenotype, we can construct a plasmid which
can replicate in B. subtilis and hopefully contain the host origin
of replication.

REFERENCES

(1) GEFTER, M. (1975) Ann. Rev. Biochem. 44,45

(2) OISHI, M., YOSHIKAWA, H. and SUEOKA, N. (1964) Nature, 204,
     1069

(3) SUEOKA, N. (1975) Stadler Symp. 7,71

(4) HARFORD, N., LEPESANT-KEJZLAROVA, J.L., LEPESANT, J.A., HAMERS,
     R. and DEDONDER, R. (1976) Microbiology (Schlessinger, D. ed.)
     pps 28-34

(5) TIMMIS, K., CABELLO, F. and COHEN, S.N. (1975) Proc. Nat. Acad.
     72,2242

(6) DUBNAU, D. and CIRIGLIANO, C. (1974) J. Bacteriol. 117,488

(7) HARFORD, N. and SUEOKA, N. (1970) J. Mol. Biol. 51,262

(8) TIMMIS, K., CABELLO, F. and COHEN, S.N. (1974) Proc. Nat. Acad.
     Sci. 71,4556

(9)   NIH Guidelines on Recombinant DNA Research (1976) Federal
      Register, 41, No 131, 27902

(10) HARRIS WARWICK, R.M., ELKANA, Y., EHRLICH, S.D. and LEDERBERG,
      J. (1975) Proc. Nat. Acad. Sci. 72,2207

(11) WINSTON, S. and MATSUSHITA, T. (1975) J. Bacteriol. 123,921

(12) HARFORD, N. (1975) J. Bacteriol. 121,835

(13) HARA, H. and YOSHIKAWA, H. (1973) Nature New Biol. 44,200

(14) O'SULLIVAN, A. and SUEOKA, N. (1967) J. Mol. Biol. 27,349

STABILITY OF ORIGIN-RNA AND ITS IMPLICATIONS ON THE STRUCTURE OF THE ORIGIN OF REPLICATION IN *E.COLI*.

John E. Womack[*] and Walter Messer

Max-Planck-Institut für molekulare Genetik
Abt. Trautner
Ihnestrasse 63/73
1000 Berlin 33, W. Germany

INTRODUCTION

The initiation of chromosomal replication is the initial event in the bacterial cell cycle and all subsequent steps occur at a specific time after initiation. Thus the frequency of initiation determines the overall rate of DNA synthesis per cell, the amount of DNA per cell, and the rate of cell division (1).

As one would expect from such an important cellular event, the steps involved in initiation have proven to be extremely complicated. Mutant studies have yielded five initiation specific genes: *dnaA* (2), *dnaC* (3), *dnaI* (4), *dnaP* (5), and possibly *dnaH* (6,7). Recently a hitherto unmapped initiation mutant has been shown to map in the *dnaB* locus (8). Other types of studies have shown that RNA polymerase also participates in the initiation reaction (9,10). Thus there are a minimum of six proteins involved in initiation. Since this number is based on mutant studies it is probably an underestimate.

The biochemical information about initiation has not kept pace with the genetics, since very little is known about the mechanisms involved or the function of any of the above gene products with the exception of RNA polymerase, *dnaA* and *dnaC*. The only things that seem to be well defined are that RNA polymerase initiates the initiation process (11,12,13,14) and that the *dnaA* product regulates the initiation event (11,14, G.Zahn, personal communication). For the in vitro initiation of some small phages with single stranded DNA a *dnaC* - *dnaB* complex is required (15,16).

[*] *A.S.I. Participant*

41

In order to get some insight into the initiation process we analyzed the RNA polymerase-mediated reaction - the synthesis of origin RNA (o-RNA) (11,12). This report presents experiments to demonstrate that o-RNA is a stable fixture in both strands of the *E.coli* genome.

RESULTS AND DISCUSSION

o-RNA isolation. The analysis of any RNA covalently bound to DNA has proven to be very tedious since these RNAs represent a very small fraction of the cell's total RNA and there is often a problem of trapping nonspecific RNA with the DNA (17).   For that reason we avoided the separation of nucleic acids by density gradient centrifugation and used instead Sepharose 2B column chromatography to separate the DNA-RNA copolymer from RNA.

The cells are first grown to the desired titer and then labeled with 6-^{3}H-Uridine (^{3}H-UR). Then the cells are lysed in a boiling lysis mix and phenolized. At this stage the deproteinized nucleic acids are heated to 100° to melt all hydrogen bonds, quickly cooled and applied to the Sepharose 2B columns (11, with modifications to be published).

The elution profile of such an experiment is shown in Fig. 1. Here a *pyrG* strain, JM463 (12) was continuously labelled with ^{14}C-thymine (spec.act. 0.15 μCi/μg) and then labelled for 90 min with 50 μCi/ml ^{3}H-uridine (spec.act. 40 Ci/mM). At this point the cell titer was 8.7 x 10^7 and the culture was harvested by centri- fugation, the nucleic acids were prepared as described and then applied to a 2.5 cm x 50 cm Sepharose 2B column. The profile in Fig. 1 shows three distinct tritium peaks and one ^{14}C peak. The tritium peak corresponding to the ^{14}C peak is the radioactivity incorporated into o-RNA which is covalently bound to DNA (no ^{3}H- uridine is converted to DNA precursors in this strain (12)). The middle tritium peak is composed of all the cellular RNA and the third peak contains radioactive nucleotides. Thus the columns give a very good separation between the DNA, containing o-RNA, and the rest of the cell's RNA. The RNA found with the DNA after this treatment and column separation has been shown to be covalently bound o-RNA (11,12).

In experiments using strains without the marker *pyrG* there are many tritium counts in DNA which come from the dCTP pools (12). Thus our standard procedure for the analysis of o-RNA requires an alkaline hydrolysis of this material, an alkaline phosphatase treatment, and analysis of the resulting ribonucleosides by thin layer chromatography. The radioactivity in the ribonucleoside spots is then determined.

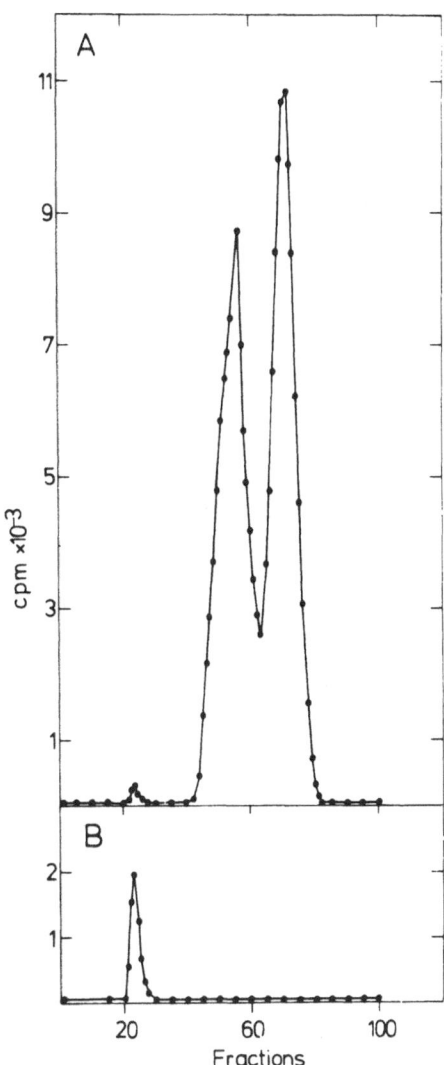

Figure 1. *The elution profile of nucleic acids from JM463 labelled with* ^{14}C-*thymine and* ^{3}H-*uridine. (A) cpm* ^{3}H-*uridine; (B) cpm* ^{14}C- *thymine.*

o-RNA synthesis and initiation mutants. The foregoing information told us that this piece of RNA is covalently bound to DNA; however, it gave no information on whether or not it was initiation specific. In order to test this we measured the synthesis of o-RNA at the nonpermissive temperature in DNA initiation defective mutants, i.e., whether the mutants were blocked before or after the synthesis of o-RNA. Using an isogenic background we found that mutants *dnaA46* and *dnaA5* made very little o-RNA at the non-permissive temperature, whereas a strain carrying the *dnaC201* mutation made a full complement of o-RNA at 42°C. Thus the *dnaA* mutation affects initiation before o-RNA synthesis and the *dnaC* mutation affects the elongation of o-RNA with DNA (11).

Stability. Several experiments indicated that the o-RNA pieces are not removed immediately after the start of DNA elongation. Therefore we determined the stability of o-RNA in a pulse-chase experiment. WM301 (4) was grown to a cell density of 5.5 x 10^7 cells/ml and pulse labelled with 50 μCi/ml ^{3}H-UR for 4 min. o-RNA was determined after a chase with 100 μg/ml uridine (plus 20 μg/ml thymidine (18)) of 1,15,35, and 55 min. Table 1 shows that there is very little loss in o-RNA during the course of the experiment.

Thus the o-RNA pieces that were made during the 4 min pulse persisted for about one and one half generations after the end of the labelling period. This experiment shows that o-RNA is probably a permanent fixture of the origin.

o-RNA in the origin region. These experiments showed that there is o-RNA covalently bound to DNA at the onset of initiation and is not immediately removed. They do not tell whether or not the o-RNA is attached to one new strand or to both strands. In order to test these two alternatives, we determined the RNase A sensitivity of the origin region.

The first step in this analysis was the labelling of DNA in the origin region with tritium. This was done by shifting WM430, which carries *dnaC201* (4) to 42° C for 90 min. After all chromosomes had finished elongation they were switched from unlabelled thymine to ^{3}H-thymidine (18), and then shifted back to 30° C to allow reinitiation. Incorporation was terminated when the increase in the amount of DNA reached approximately 1%. Thus, ^{3}H-thymidine was incorporated only into DNA at the origin. Double stranded DNA was prepared by a lysozyme, EDTA lysis (19). Protein was removed with phenol, and the DNA (sheared to approximately 10-20 x 10^6 daltons) was tested for RNase sensitivity. 100 μl were incubated for 60 min in buffer (100 mM Tris · HCl, pH 7.5, 10 mM $MgCl_2$) with 50 μg/ml RNase A (treated at 100° C to inactivate contaminating DNase) in the presence and absence of 0.28 M NaCl. In high salt buffers RNase A will not cleave RNA that is hydrogen bonded to either RNA or DNA; whereas in low salt such an RNA is hydrolyzed (20). The RNase treated samples were subjected to electrophoresis on a 200 mm, 0.6% agarose gel. After electrophoresis the gel was cut into 1 mm slices and counted.

The results of such an experiment are shown in Fig. 2. The molecular weight of origin DNA, treated with RNase A in the presence of 0.28 M NaCl was identical to the untreated control The DNA

Table 1

Stability of o-RNA

Min. after start of chase	Cell No. x 10^7	^{3}H-UMP remaining in o-RNA (pmoles/10^8 cells)
1	5.5	8.1×10^{-3}
15	7.1	8.0×10^{-3}
35	9.9	7.2×10^{-3}
55	14.0	6.9×10^{-3}

$E.coli$ WM301 was pulsed for 4 min with ^{3}H-UR and
then o-RNA measured after various times of chase
with µg/ml uridine and 20 µg/ml thymidine.

samples treated with RNase A at low salt, however, migrated to a
position in the gel which corresponds to a much lower molecular
weight.

These results show that o-RNA must be hydrogen bonded. The
most probable candidate being the complementary strand (Fig. 3).
o-RNA must be present in both strands since the removal of o-RNA
from only one strand by RNase would not split the origin in half.
There is a possibility that removal of RNA from one strand would
expose a shear-sensitive region and thereby yield a reduction of
the molecular weight. However, we feel this is highly unlikely
since no shear forces were applied after the RNase treatment.

This gives three possible structures (Fig. 3) for the origin:
The o-RNA pieces can be either directly self-complementary (Fig. 3A),
partially self-complementary (Fig. 3B), or not complementary to
each other (Fig. 3C). These data do not allow us to decide which
is the correct case; however, a segment of DNA-DNA-hybridization
(X in Fig. 3C) longer than 14 base pairs would stabilize the linear
structure (21). Thus if Fig. 3B is the correct structure then the
length of the DNA-DNA base paired region is less than 14 base pairs.

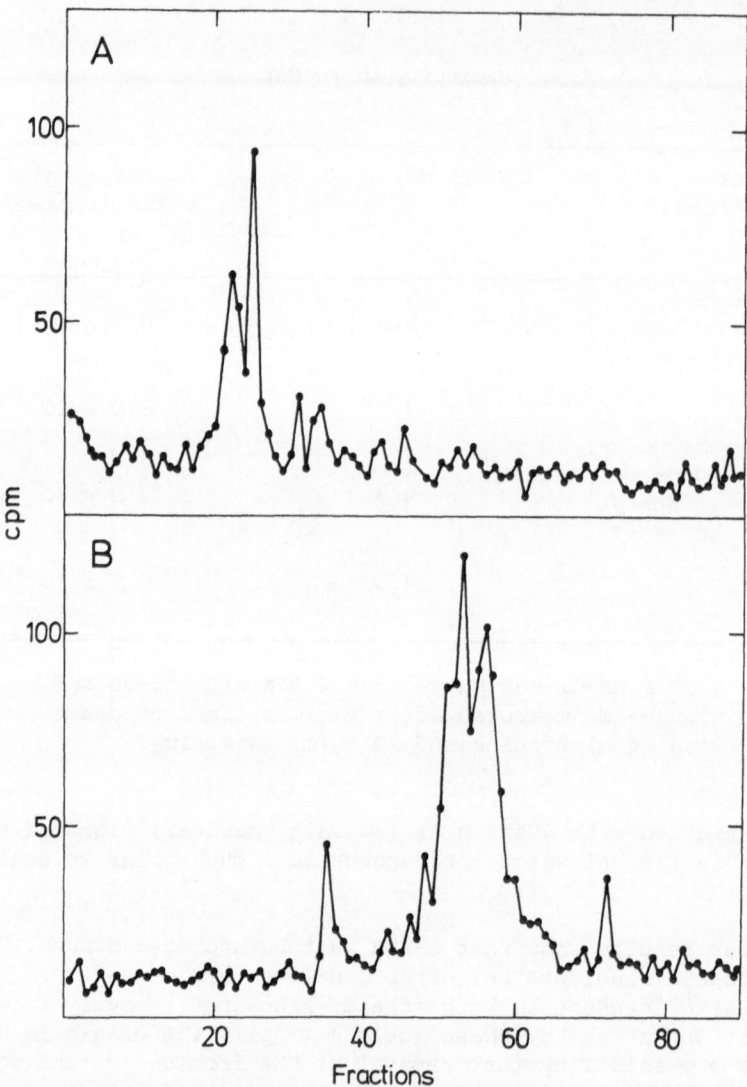

Figure 2. RNase sensitivity of radioactive DNA from the origin of replication in the presence (A) and absence (B) of 0.28 M NaCl.

CONCLUSIONS

We have described experiments which strongly indicate that o-RNA is present on both strands and once synthesized is a permanent fixture of the DNA. The observation that we could lower the molecular weight of all molecules with RNase A is in itself strong evidence for the stability of o-RNA. If, for example, the o-RNA were replaced during the first subsequent initiation, then none of

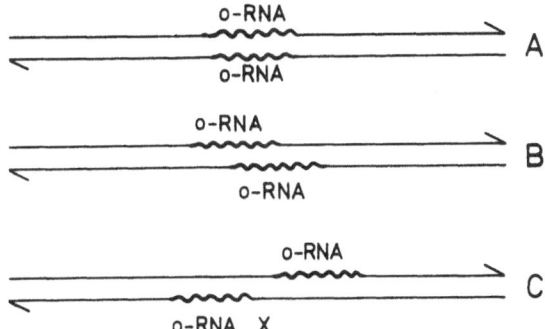

<u>Figure 3.</u> *Three possible structures for o-RNA on both strands at the origin.*

these molecules would have been RNase sensitive. If o-RNA were replaced by DNA during the second generation after the labelling then one half on this DNA should be insensitive to RNase A. Using this logic and the data in Fig. 2, we can conclude that o-RNA is stable for at least three generations.

The initial event of replication is the synthesis of a copy of o-RNA for each of the two replication directions of bi-directional replication, resulting in a newly synthesized o-RNA molecule for each new strand of double-stranded DNA. This synthesis is under negative control and requires the action of the *dnaA* product (11). For the following elongation of o-RNA, by DNA, the *dnaC* proeuct and probably all or most of the replication proteins are required.

ACKNOWLEDGEMENT

We thank Michael Hearne, Kayron Dube and Anna Köhne for their excellent technical assistance. We also thank T.A. Trautner and the Wednesday night discussion group for their interesting and valuable suggestions.

REFERENCES

1.   Helmstetter, C.E., Cooper, S., Pierucci, D. and Revelas, E. (1968) Cold Spring Harbor Symp. Quant. Biol. 33:809

2.   Carl, P.C. (1970) Molec. Gen. Genet. 109:107

3.   Wechsler, J.A. and Gross, J.D. (1971) Molec. Gen. Genet. 113:273

4.   Beyersmann, D., Messerm W. and Schlicht, M. (1974) J.
     Bacteriol. 118:783

5.   Wada, C. and Yura, T. (1974) Genetics 77:199

6.   Sakai, H., Hashimoto, S. and Komano, T. (1974) J. Bacteriol.
     119:811

7.   Derstine, P.L. and Dumas, L.B. (1976) J. Bacteriol. 128:801

8.   Zyskind, J.W. and Smith, D.W. (1977) J. Bacteriol. 129:1476

9.   Messer, W. (1972) ˙J. Bacteriol. 112:7

10. Lark, K.G. (1972) J. Mol. Biol. 64:47

11. Messer, W., Dankwarth, L., Schindler, R.T.-, Womack, J.E. and
     Zahn, G. (1975) In: DNA Synthesis and Its Regulation. (eds.
     M. Goulian and P. Hanawalt) W.A. Benjamin, Inc. p602

12. Womack, J.E. and Messer, W. (1975) In: DNA Synthesis and Its
     Regulation (eds. M. Goulian and P. Hanawalt) W.A. Benjamin, Inc.
     p618

13. Hanna, M.H. and Carl, P.L. (1975) J. Bacteriol. 121:219

14. Zyskind, J.W., Deen, L.T. and S-ith, D.W. (1977) J. Bacteriol.
     129:1466

15. Schekman, R., Weiner, A., and Kornberg, A. (1974) Science
     186:987

16. Wickner, S. and Hurwitz, J. (1975) Proc. Nat. Acad. Sci. USA
     72:921

17. Okazaki, R., Okazaki, T., Hirose, S., Sugino, A., Ogawa, T.,
     Kurosawa, Y., Shinozaki, K., Tamonoi, F., Seki, T., Machida,
     Y., Fujiyama, A. and Kohara, A. (1975) In: DNA Synthesis and
     Its Regulation (eds. M. Goulian and P. Hanawalt) W.A. Benjamin
     Inc., p832

18. Womack, J.E. Molec. Gen. Genet. (submitted)

19. Bonhoeffer, F. and Gierer, A. (1963) J. Mol. Biol. 7:534

20. Fdy, V.G., Szekely, M., Loviny, T. and Dreyer, C. (1976) Eur.
     J. Biochem. 61:563

21. Mertz, J.E. and Davis, R.W. (1972) Proc. Nat. Acad. Sci. USA
     69:3370

THE GENETICS OF <u>E.COLI</u> DNA REPLICATION

James A. Wechsler

Department of Biological Sciences
Columbia University
New York, New York 10027, U.S.A.

INTRODUCTION

The elucidation of the structure of DNA was elegant, perhaps primarily, for the simplicity with which the double helix appeared to explain replication (1). Indeed, the molecular architecture of the genetic material would seem to be magnificently, if not expressly, designed for the replication of genes. Despite this apparent simplicity of molecular design, determining the intricacies of the replication process has proved a fertile, though hard, ground for both geneticists and biochemists. Although this review is from a genetics perspective, biochemistry and genetics are but two horses pulling the same cart. The biochemical horse has been driven across this ground often and well of late (2-5) whereas the genetic steed has been comparatively neglected.

I shall attempt to summarize the genetics of E.coli DNA replication with a constant eye on the biochemistry. An excellent, earlier review from a similar perspective contains some information which has changed little in the intervening years and which will not be repeated here (6). I apologize to other organisms, their friends and patrons for limiting this review to E.coli.

The chromosome of E.coli is a circular DNA double helix that is folded into a series of supercoiled loops. Since a single nick does not result in relaxation of the entire molecule, but rather relieves supercoiling in only one loop, there must be some mechanical linkage separating the loops and restricting free rotation of the helix (7).

Replication of the chromosome proceeds bidirectionally from an

origin in the 81-83 minute region of the revised genetic map (8,9).
Strong evidence that the origin is a fixed point comes from the
isolation of a series of F' factors that can be maintained in Hfr
cells. The rationale for the isolation of these F's was that an F'
carrying the chromosomal origin (poh) would be able to replicate from
that origin, and, thus, bypass F incompatibility which otherwise
excludes the replication of an F' in an Hfr (10). The technique
for isolating such F's is straightforward and has proven to be
relatively easy (10,11; M. Zdzienicka, personal communication). The
inclusion or deletion of poh from the chromosomal material on F's
carrying the region between 81 and 84-minutes on the chromosome is
similar to that of other genes and is consistent with its being a
specific locus. There is some selection against the maintenance of
poh (10), probably because cells diploid for this region grow slowly
and are larger than cells carrying a single copy of the origin (12).
As might be expected, these F's are resistant to curing by acridine
orange; presumably replication begins at poh when F-directed
replication is inhibited by the dye (11).

Initiation of chromosome replication comprises an event, or
series of events, at poh and is required for beginning each round of
chromosome replication. Though several gene products are known to
be required for initiation, the mechanism of this process is unclear
and will probably remain so until an in vitro initiation assay is
devised (13-17).

From antibiotic treatment of synchronized cells, three events
during the cell cycles which are apparently necessary for subsequent
initiation have been defined. Two of these are protein synthesis
requirements which are differentiated by their susceptibility to
different drug concentrations (18,19). Theoretical questions
regarding the validity of concluding that there are two rather than
one protein synthesis requirement based on such experiments have been
raised (20). At times just prior to initiation when chloramphenicol
addition is not inhibitory, rifampicin will block initiation. Thus,
there is a requirement for RNA synthesis just prior to, or coincident
with, initiation (21).

The propagation of the deoxynucleotide chains from the origin
to the terminus of replication is better understood. The problem
of how the anti-parallel strands could be replicated in unison when
DNA polymerase I, and all subsequently identified polymerases
polymerize exclusively in the 5' $\rightarrow$ 3' direction was resolved by the
discovery of discontinuous DNA synthesis (5,23). In principle the
polymerization of the strand which has its 5' end at the replication
origin and its 3' end at the replication terminus can be continuous
while polymerization in the opposite direction must be discontinuous.
There are many results indicating that the synthesis of both strands
is discontinuous and opposing results in favor of discontinuities in
only one strand.

The genetic map of E.coli showing the genes involved in DNA replication is shown in Figure 1.

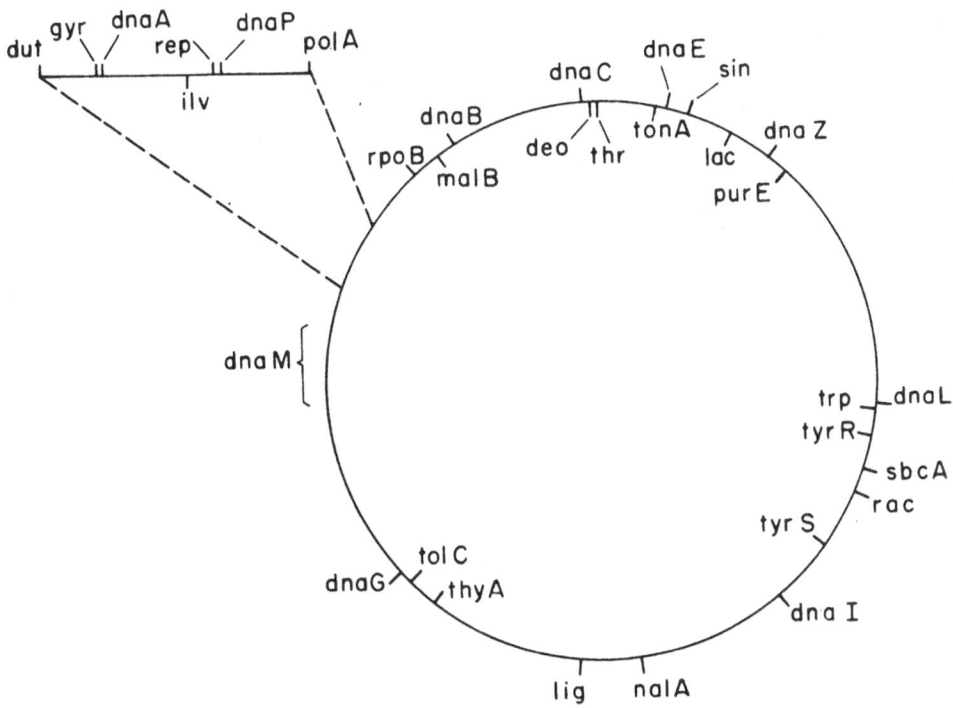

Figure 1. *The genetic map of E.coli showing the position of genes involved in DNA replication. Several loci not involved in replication are shown for orientation. New loci not shown on the revised genetic map (9) are described in the text.*

Chain Elongation

In vitro assays for chain elongation have generated a wealth

of information over the last several years and for this reason mutants
defective in chain elongation will be examined before those deficient
in initiation.

dnaE

The primary enzymatic requirement in DNA replication is for a
DNA polymerase. It is, thus, fitting that a DNA polymerase activity
was the first to be correlated with a particular dna gene. The
identification of DNA polymerase III as the dnaE gene product (23,24)
and the demonstration that some dnaE mutants instantly and completely
cease DNA synthesis at restrictive temperature (15) show that DNA
polymerase III is the major replication polymerase.

Two dnaE alleles, dnaE486 and dnaE511, show a moderate mutator
activity. The alterations of pol III in these strains increase
several transversions, T/A ➡ A/T, C/G ➡ G/C, and G/C ➡ T/A, but
probably do not cause transitions nor do they reverse frameshift
mutations (25). As in the case of the T4 polymerase this result
implies that pol III is involved in base selection (26,27), but there
is no evidence that the resultant mutator activity is due to a
defective editing function as is the case for T4 gene 43 mutators
(28,29).

From in vitro studies utilizing single-strand bacteriophage DNA-
directed systems, many components of the DNA replication machinery
have been purified. It was from these studies that the functional
pol III was termed by one laboratory to be a complex of what was
called pol III* and co-pol III* (30). It now seems likely that
co-pol III* of Kornberg and colleagues is identical to DNA
elongation factor I of Wickner and Hurwitz (31) and that pol III*
is equivalent to pol III plus elongation factor II. DNA elongation
factor II turns out to have at least two components one of which is
the product of the dnaZ gene and the other has been designated
elongation factor III (31).

dnaZ

The dnaZ2016 mutant is defective in chain elongation. This
mutation was initially designated dnaH (32).

In vitro elongation of primed single-stranded DNA template
requires dnaZ protein. Elongation factor I is bound to the DNA
template in a reaction which requires the dnaZ product, elongation
factor III and ATP. Pol III binds to this elongation factor I-primed
template complex (33). If this is an accurate reflection of the in
vivo function of dnaZ protein, the recent observation that the dnaZ
mutant is temperature sensitive for the continued synthesis both of

Okazaki pieces and of high molecular weight DNA (J.R. Walker, personal communication) implies either that synthesis along both strands is discontinuous or that synthesis of the two strands is tightly coupled.

All stages (SS → RF, RF → RF and RF → SS) of M13 and ΦX174 *in vivo* replication require *dnaZ* product in strains which are wild-type for all of the known polymerases (J.R. Walker, personal communication).

The dna*Z* mutant displays some breakdown of DNA following the shift to restrictive temperature.

dnaG

Mutants in *dnaG* cease replication immediately after the shift to restrictive temperature (15). The *dnaG* locus has been shown *in vitro* to be involved specifically in the initiation of Okazaki piece synthesis (34). Though it has been suggested that the *dnaG* product is a rifampicin-resistant RNA polymerase (35), recent data demonstrates that the *dnaG* protein is involved in the synthesis of a polynucleotide primer that can utilize deoxyribonucleoside or ribonucleoside triphosphates with equal efficiency (36).

Analysis of the two, smaller than parental-size DNAs produced in the cellophane disc *in vitro* system has shown that the 38S DNA and the 10S Okazaki piece DNA are synthesized from opposite strands of the helix (37). Though *dnaG* mutants are defective only in 10S synthesis *in vitro* (34) and probably primarily affect Okazaki piece synthesis *in vivo* (38), *dnaG* mutants quickly terminate all DNA replication at high temperature. These results imply, again, that synthesis of both strands in the region of the replication fork is tightly coupled.

Lig and *polA*

The small DNA intermediates synthesized during discontinuous replication must be joined to form large, parental size DNA. Polynucleotide ligase, the product of the *lig* gene, catalyzes the formation of a phosphodiester bond between abutting polynucleotide chains (39,40).

The *lig-7* allele is a temperature-sensitive lethal mutation though cells incubated at restrictive temperature retain some (∿ 1%) ligase activity (41,42,43). When the DNA of this mutant is pulse-labeled at restrictive temperature nearly all of the DNA is isolated as short pieces in alkaline sucrose gradients. This small DNA does chase into parental DNA but far more slowly than is the case in wild-type cells. DNA synthesis, however, continues for long

times at restrictive temperatures though it eventually ceases;
presumably, when a replication fork reaches a region where one strand
of the template is discontinuous, replication cannot continue. If
replication requires that both strands be replicated in the region
of the replication fork the displacement of short pieces would
eliminate one template strand and replication would halt.
Alternatively, if supercoiling is necessary for replication, discontinuities in one strand would block synthesis by relaxing the
supercoiled structure.

The lig-4 mutation is also a temperature-sensitive allele in the
lig gene but this mutation does not cause lethality. Combination of
lig-4 with polA1 is, however, lethal and results in a defect similar
to, but more severe than, lig-7 alone when nascent DNA is analyzed
by alkaline sucrose centrifugation (42). The polA1 mutation lowers
the polymerase activity of DNA polI to 1% or less of wild-type levels
but has little or no effect on the associated nuclease activities
(44). If the single defect in the polA1 gene product affects only
the polymerizing activity, the results with lig-4 polA1 strains
suggest that polI is a normal part of the replication process. This
conclusion is in agreement with the observation that lig$^+$ polA1
strains are also somewhat slow to join Okazaki pieces (45). It has
been noted before and is restated here that the inability to find a
lethal mutation which affects polI polymerase activity implies only
that there is not an absolute requirement for this activity in DNA
replication (3,5) but the data definitely points to its normally
having a role in the replication process. Since polII and polIII
can substitute for polI during repair (45a,45b), they may also be
able to do so during replication. The implication of this information
is that Okazaki pieces are separated by gaps, rather than nicks,
and that these gaps are filled by polI prior to ligation.

When synthesis is opposite to the direction of replication fork
travel, the piece being synthesized will be polymerized toward the
5'-end of the piece just completed. Based on the simplest
biochemical considerations, polymerization of the two neighboring
pieces would leave them separated only by a nick. A barrier which
prevents synthesis of the nascent Okazaki piece from proceeding
all the way to the 5'-deoxynucleotide end of the previously
synthesized piece would generate a gap. Since the known DNA
polymerases will not synthesize DNA de novo but require a primer,
it has been suggested that the barrier is an RNA primer at the 5'
end of all Okazaki pieces (5,35). An elegant model has been
constructed on the basis of this hypothesis. According to the model,
the polI 5' ➤ 3' exonuclease activity would remove the RNA primer
while the polymerase activity fills in the gap.

This mechanism predicts that mutants lacking the 5' ➤ 3'
exonuclease activity would be lethal. Several conditionally lethal
mutants deficient in this activity have been isolated (46,47). These

mutants retain a low level of exonuclease activity and their
conditional lethality is not complete; a polA480(ex) (formerly
polA exl) mutant plates at 42C with an efficiency of 100% on
minimal media and 0.08% on nutrient, and an isogenic polA4113 mutant
has an e.o.p. of .03% on minimal media and .01% on nutrient media
(W. Kelley, personal communication). Lethality is a gross
measurement of the inability to form colonies; though the lethality
clearly shows that a loss of the 5' ➡ 3' exonuclease is a severe
detriment to cell survival, it cannot be concluded that the observed
lethality is a direct consequence of a defect in DNA replication.

Consistent with this hypothesized requirement for the polI
exonuclease is the evidence that two of the other known activities
from E.coli which are capable of removing RNA primer (RNase H and
exonuclease III) cannot remove the last 5' ribonucleotide covalently
attached to the DNA chain (48,49). On the other hand, a strain lacking
the polI 5' ➡ 3' exonuclease is deficient in M13 synthesis, but the
known and conjectural RNA priming steps during M13 replication occur
at stages different from those inhibited in the mutant (50).

There is substantial evidence for RNA primers in viral DNA
synthesis and in the replication of some plasmids. A primer RNA in
polyoma replication has been sized (about 10 nucleotides long) and
its base composition determined (the base composition generally
reflects the viral base composition) (51).

Evidence for RNA primers during chromosome synthesis is based
on a series of elegant experiments which have not been reproduced
in other laboratories (52). However, by using a mutant deficient in
the 5' ➡ 3' polI exonuclease, evidence has been obtained for the
attachment of RNA to a portion of the small DNA intermediates
(D. Denhardt, personal communication).

dut

A complication in the analysis of results based on the isolation
of DNA intermediates has recently been raised by studies of mutants
lacking dUTPase activity (dut, sof, or dnaS mutants - all of these
genetic notations are equivalent) (53,54). Alkaline sucrose gradient
analysis of pulse-labeled DNA from these mutants shows that most of
the DNA is found as very small pieces (approximately 4S). The
conclusion is that dUTP is erroneously inserted in the growing DNA
chain and subsequently repaired by excision and synthesis (54). In
dut mutants the dUTP pool is presumed to be unusually large and,
therefore, the frequency of dUTP insertion into the DNA should be
increased. The implication, however, is that insertion of dUTP
during wild-type synthesis and the subsequent excision of the
incorrect base will generate small DNA fragments which are breakdown
products of the nascent DNA and not true intermediates in DNA

replication (54). Such a mechanism can be invoked in support of one
strand being synthesized continuously in spite of the finding that all
pulse-labeled DNA in lig-4, polA1 strains can be isolated as small
pieces. This complication jeopardizes conclusions drawn from the slow
joining of small nascent DNA as the excision of the deoxyuridine could
result in a replenishing of the small fragment material from larger
material. Based on the current metabolic maps of pyrimidine
nucleotide synthesis it does not appear possible to engineer a strain
that will have no dUTP pool (55, 56). Approaching this problem from
the opposite end, two nucleases are known which will excise deoxyuridine from DNA and no mutant is yet available for one of these (57,58)

dnaB

In mass isolations of dna mutants the most common group are
dnaB mutants (15,59). Of the dnaB mutants available several years
ago, all of seventeen tested were shown to be allelic (60) and it
seems likely that their preponderance among isolated mutants is a
consequence of the aggregation of dnaB products into a multimer. The
dnaB product aggregates in vitro, giving rise to several active
molecular weight species (61), and there is strong in vivo evidence
consistent with the conclusion that the dnaB protein functions as
a multimer (62).

The replication of λ bacteriophage also requires the dnaB
product. In this case there is good evidence that the products of
the host dnaB gene and the λ P gene interact. A series of mutants
(termed GroP⁻) of E.coli which restrict λ growth and the efficiency
of plating of the virus but which do not restrict λπ mutants have
been identified as dnaB mutants. The π mutations were also shown
to lie within the P gene of λ (63). This GroP⁻ restriction of λ
is at permissive temperatures for the host; at temperatures non-
permissive for the dnaB host, no λ growth is possible. In a
subsequent screen of dnaB mutants it was found that some dnaB mutants
restrict λπ more than λP+ (Wechsler, unpublished). These results
strongly imply that the two gene products function together and that
when both are altered they can mutually compensate for each other
63).

P1 bacteriophage contains a gene which produces a dnaB product
analogue (64,65). This gene, ban, (dnaB analogue) is under control
of a cis-acting locus, bac (dnaB analogue control). P1 bac ban⁺
lysogens of dnaB strains are able to grow at temperatures restrictive
for the parental dnaB mutant. By exploiting the fact that dnaB266
is an amber mutation it was shown that the ban gene does produce a
dnaB-like gene product. The GroP⁻ phenotype is also suppressed by
these P1 mutants though suppression of this character is not
functional at all temperatures. There was an indication in these
experiments that the host dnaB protein interacts with the P1 ban

protein (65).

The most severe dnaB mutants cease replication almost
instantaneously at high temperature. In some cases breakdown of
labeled DNA is observed during incubations at 42C. This breakdown
is not a definitive characteristic of particular mutants as it may
or may not occur in repeat experiments (15). When it does occur,
breakdown proceeds from the replication fork towards the origin and
is effected by the recBC nuclease, exonuclease V (66,67).

In some dnaB mutants a small DNA intermediate that has been
nominally designated 4S is observed. These small pieces are easily
detected in vitro and can be isolated from cells pulse-labeled in
vivo (62). In a minority of experiments a 4S peak has been detected
in wild-type cells. In vitro pulse-chase experiments led to the
conclusion that these 4S pieces are intermediate in the synthesis of
the 38S fragments characteristic of the cellophane disc in vitro
system and not precursors of 10S Okazaki fragments. The 38S and 10S
fragments appear to be copied from opposite template strands (37).
Though hybridization of 4S versus 10S, and versus 38S fragments to
prove their genesis from separate strands has not yet been done, it
seems likely that the dnaB function is involved mainly in the
synthesis of the opposite strand from that dependent on the dnaG
product (34). If this is so, the fast-stop kinetics characteristic
of dnaB mutants implies, once again, that synthesis of the two
strands at the replication fork are interdependent.

These in vitro and in vivo experiments with dnaB mutants also
concluded that many 4s pieces were joined together to form the 38S
fragments and that inactivation of DNA ligase by mutation or by
addition of NMN in vitro did not inhibit this conversion. The
simplest conclusion, from these data is that there is a second
joining mechanism in E.coli (62). In vivo experiments currently in
progress in our laboratory suggest that the presence of certain dnaB
alleles antagonizes ligase function (R. Sclafani, unpublished).

A single mutant mapping at or near dnaB has been shown to
affect initiation (68) Though this mutation has not been shown to
be in dnaB by complementation, we will assume that it is a dnaB
mutation and will consider the meaning of this finding later.

dnaC, elongation defective

Mutations at the dnaC locus are defective in either chromosome
initiation or chain elongation (14,69,70). When the defect is in
chain elongation, it is not possible to ascertain whether the protein
also fails to fulfill its role in initiation. Of the dnaC mutants
isolated, two are clearly defective in chain elongation (70). As
is the case for some dnaB mutants, cessation of DNA synthesis in

these dnaC mutants is followed by degradation. It has not been
possible to show that this degradation proceeds from the replication
fork toward the origin as is true of dnaB mutants which exhibit
breakdown. There is, however, genetic and biochemical evidence that
the dnaC and dnaB products are required for the same process during
chain elongation. One of the dnaC elongation mutants can be
suppressed by Pl bac mutants which suppress dnaB mutations (65) and
the dnaC and dnaB products can be isolated as a complex (71). Recent
in vivo experiments with a dnaB, dnaC double mutant also support the
idea that the products of these two genes interact during replication
(72).

DNA gyrase

    The role of chromosome structure in DNA synthesis has not been
widely investigated because of the lack of a suitable approach. The
recent discovery of a new activity, DNA gyrase, which supercoils DNA
presents a possible entry to such studies (73,74). Several gene
products are apparently required for DNA gyrase activity
(N.Cozzarelli, personal communication).

    One of the original mutants studied by us and tentatively
classed as dnaA at that time behaved unusually for a dnaA mutant
and has now been tentatively classified as having a defect in DNA
gyrase (75). The mutant appears to be leaky and it is not clear that
it is defective in chain elongation (Wechsler, unpublished), however
because of hypotheses concerning the role of supercoiling in
replication, it is placed in this category (3).

Replication Intermediates and Chromosome Structure

    Since evidence has now been offered that chromosome structure
may play an important role in DNA replication, several interesting
correlations can reasonably be mentioned. In the cellophane-disc
and the folded chromosome directed in vitro DNA replication systems
two intermediate DNA synthesis peaks are observed in alkaline sucrose
(76,77). One of these is at the usual Okazaki piece position and the
other is a broad peak centered at approximately 38S. These peaks
have been isolated from the cellophane-disc system, and it has been
shown that DNA in the 10S peak hybridizes with DNA from the 38S peak
but that the DNA from a single peak does not self-hybridize (37).
Thus, the 10S and 38S DNA are apparently from opposite strands. The
measurement of the number of supercoiled loops in the E.coli
chromosome is imprecise and varies from 12 to 80 (7). Recent
estimates from electron micrographs give an estimate of approximately
100 loops (78). If 100 is close to the correct number, the 38S DNA
correlates very well with the average loop size.

The loops are supercoiled and the average distance between supercoil twists, taking into account the revised unwinding angle due to ethidium bromide intercalation, is 200 bases (7,79). As noted earlier, the 4S DNA observed with dnaB extracts chases into the 38S, and not the 10S, peak (62). Recent measurement of the small DNA in dnaB mutants places the sedimentation coefficient very near 4.5S which is equivalent to 203 bases (80).

If these correlations are other than accidental, the hypotheses to explain them would be that the progression of the replication fork in the in vitro systems is retarded between supercoiled loops and that, therefore, 38s DNA is observed. Going a step further, these numbers raise the possibility that inactivation of the dnaB protein specifically retards replication fork travel through a supercoil twist.

Additional Chain Elongation Mutants

The GroPC phenotype (formerly GroPAB) is similar to the other GroP phenotypes (termed GroPA and GroPB) exhibited by many dnaB mutants (63,81,82). It does not permit growth of λP+ but does allow various λπ mutants to grow and plate efficiently. The GroPC host strains are themselves temperature-sensitive for growth and DNA synthesis appears to be more affected than RNA synthesis at restrictive temperatures (81,82). The tentative assumption is that the gene which is mutated in GroPC cells codes for a gene product which interacts with the λgene P product. By extension, it is reasonable to suppose that this host gene product may also interact with the dnaB gene product. This supposition is supported by the observation that P1 bac mutants suppress the original GroPAB phenotype (65). One peculiarity of these mutants, if they are in fact dna mutants, is that protein synthesis terminates sooner at restrictive temperatures than it does in any other class of dna mutant. Only the probable gyrase mutant among the other known dna mutants appears to affect protein synthesis other than indirectly. The GroPC mutations map between thr and leu on the chromosome and are, therefore, distinct from any known dna locus. It is not clear whether the several GroPC mutations are in one or more cistrons. Recently, one allele which displays the GroPC phenotype has been designated dnaK (100).

A vast series of mutants has been isolated by a computerized automatic screening procedure (83). The two new loci obtained from this work are shown in Figure 1 as dnaL and dnaM. Both types stop synthesizing DNA rapidly and are, therefore, defective in chain elongation. (In the original manuscript dnaL was denoted dnaK but this has now been changed to prevent confusion with the dnaK locus of the GroPC phenotype).

Initiation Mutants

Initiation is defined as the beginning of a round of chromosome replication and occurs at the origin. If it is postulated that termination of a round of replication is necessary for subsequent initiations under steady-state conditions, there is no useful experimental method for differentiating initiation-defective from termination-defective phenotypes. For the purpose of this review, all mutants that are deficient in the ability to begin new rounds of DNA replication are classified as initiation defective. Termination will be dealt with briefly in the final section.

Several methods have been employed to differentiate initiation mutants from leaky chain elongation mutants among the group of dna mutants that synthesize substantial amounts of DNA following a shift to high temperature. These methods included (1) density transfer, (2) incubation at restrictive temperatures in medium that does not permit DNA synthesis and subsequent readdition of necessary nutrients and (3) marker frequency analysis. The last of these is tedious in E.coli where P1 transduction or DNA-DNA hybridization must be used but is clearly the method of choice in Bacillus subtilis and other Gram positive bacteria where efficient transformation systems exist.

dnaA

The first set of initiation mutations to be defined were the dnaA mutants and this group has been studied extensively (13).

The origin of chromosome replication and dnaA map close together on the bacterial chromosome, however, they are clearly not identical. Using criteria for the presence of the chromosomal origin on an F' it was found that the origin and dnaA were separable (10). Also, the demonstration that the dnaA508 allele is recessive to dnaA+ shows that the dnaA gene produces a product (11). Conjecture that this product may act as a multimer or in a complex with other proteins is based on the partial dominance of some dnaA alleles (84).

The dnaA mutations can be suppressed by integration of any of a series of replicons into the chromosome (85,86,15). The plasmids and bacteriophage shown to effect integrative suppression include F, F'-merogenotes, R-factors, P1 and P2 (85-87). With the exception of phage P2 all of these replicons are plasmids that are normally maintained at a low copy number, 1 to 2, per chromosome in the non-integrated state. (P2 forms stable lysogens by integration). Integrative suppression by conjugative plasmids results in formation of an Hfr. In the case of integrative suppression by R100-1, the origin of replication is located within the plasmid and not at the normal host origin (88). Similarly, the origin in a P2 integrative suppressed strain is within the prophage (P. Kuempel, personal communication).

Since integrative suppression by F or F'-factors occurs at a
lower frequency than Hfr formation and since it is possible to
transduce dnaA into an Hfr and obtain a normal frequency of
temperature-sensitive transductants (85,15), integration of the
plasmid is not sufficient in itself to effect suppression. The
nature of the additional requirement for integrative suppression
is obscure (86).

P2 prophage that act to integratively suppress dnaA mutations
contain insertions within the P2 genome (87). P2 replication
requires the cis-acting protein encoded by gene A of the phage and
integrative suppression by P2 also requires that this gene be active
(82). In addition, the host rep gene does not seem to be required
for bacterial chromosome replication but is required for P2
replication and for integrative suppression by P2 (Lindahl, personal
communication). In a rep mutant, however, the rate of replication
fork travel appears to be decreased during chromosome replication
(89). As evidence has recently been presented that the rep gene
product is involved in the melting of the DNA at the replication fork
(90) this result suggests that the structure of the replication fork
as well as the mechanism of initiation may be altered during
chromosome replication under P2 control (G. Lindahl, personal
communication).

Since dnaA function can be bypassed by inserting other
replicons into the bacterial genome and replication then appears to
be under control of the integrated replicon, it can be concluded that
dnaA defines or partially defines the bacterial replicon.

Extragenic Suppression of dnaA (sin, das, rpoB)

Several classes of external suppressors of dnaA have been
isolated. A set of suppressor mutations called sin which map
between proA and metD have been isolated and characterized (G.Lindahl,
personal communciation). Suppression by sin is locus specific but not
allele specific (i.e., all dnaA alleles that have been tested are
suppressed but dnaC mutations are not). The observation that the
expression of sin is sensitive to the medium suggests that these
mutations may affect the membrane or a hypothetical membrane-
associated complex (G. Lindahl, personal communication).

Another series of dnaA suppressors that appear to affect the
cell membrane have also been reported (91). These suppressors,
termed das, resulted from a mutagenesis treatment designed to produce
lesions near the origin. Several of these das mutations are external
to dnaA and the available evidence suggests that there are several
das genes (Zdzienicka and Wechsler, unpublished). Unlike the sin

mutations the various das suppressors that have been adequately
tested appear to be allele specific.

It has been suggested that both sin and das mutations alter
the cell membrane in some unspecified way (91; G. Lindahl, personal
communication). The phenotypic perturbation of dnaA and dnaC
mutants by inhibition of unsaturated fatty acid synthesis also
implies that the initiation function is dependent on the condition
of the membrane (92). Though none of these results discriminates
between direct and indirect effects, the allele-specificity of some
das mutations implies that the interaction of dnaA and membrane are
more than casual.

The best insight into the function of the dnaA product also
comes from suppression studies. It has been clearly demonstrated
that some mutations in the rpoB gene, the cistron for the RNA
polymerase β subunit, suppress dnaA mutations (93). Tests to
determine if this suppression is dnaA allele-specific have not been
completed (93). This evidence suggests that the dnaA product and
RNA polymerase interact and further suggests that the dnaA product
is involved in the transcriptional event known to be required for
initiation (21). This dramatic result correlates with the suggestive
evidence that dnaA functions before or during the obligatory RNA step
in initiation (94) and before or in the synthesis of "origin-RNA"
(95).

dnaC, Initiation Defective

Several mutants at the dnaC locus are initiation defective
(14,69). As these mutations are allelic with the elongation-
defective dnaC mutations, the product of this gene functions in
both processes (70). Presumably, the requirements for the role in
initiation are more exacting and some alterations of the dnaC
protein destroy only this function.

The reversibility of dnaC initiation mutants is not affected by
rifampicin and, thus, the step in initiation appears to be subsequent
to that requiring dnaA function or transcription (94). Evidence that
inactivation of dnaC does not affect the production of "origin-RNA"
is consistent with this sequence (95).

The single dnaB allele that affects initiation, dnaB252 (68),
can be understood if it is assumed that the dnaC and dnaB products
interact during initiation as they do in elongation. The altered
dnaB protein could then affect initiation indirectly by distorting
a normal dnaC product. This explanation is supported by the
demonstration that even when the dnaB mutation in a dnaB, dnaC
strain (the dnaC allele is apparently of the initiation defective
class) is suppressed by P1 bac, the strain continues to display an

elongation-defective phenotype (72). In the absence of the dnaC
mutation, Pl bac efficiently suppresses the dnaB allele (72).

dnaI

One mutant defective in initiation of DNA synthesis has been
isolated in an E.coli B/r derivative and defines the dnaI locus. The
dnaI gene is present on F'148 and therefore is in the 37.25' - 40'
or 41.5' - 44' region of the chromosome map (Projan and Wechsler,
unpublished). Cotransductional placement of dnaI has so far been
unsuccessful.

This is one of two mutant classes (the other is dnaP) not
recovered in the automatic mutant screening procedure (83). Also,
no dnaC mutants have been shown to be integratively suppressible,
in spite of considerable effort, except in this E.coli B/r background (16). In addition, the pattern of integrative suppression
of dnaA mutants in this background is at variance with the work
of others (15,16,85,86). This catalogue of peculiarities plus the
difficulty in mapping may be unimportant, but a simple explanation
for them is that the E.coli B/r parent used contains an altered
initiation protein due to a non-lethal mutation at an unspecified
locus. If this hypothetical protein were to interact with the other
gene products involved in initiation all of the above peculiarities
could be easily rationalized.

dnaP

The single dnaP mutant was isolated among a series of mutants
selected for resistance to phenethyl alcohol at 30°C (17). From
reversion studies it appears that the phenethyl alcohol resistance
and the DNA temperature-sensitivity are the consequence of a single
mutation which is recessive to dnaP+. Presumably the mutation
affects the membrane as the cell envelope appears to be the target
for phenethyl alcohol, but no specific membrane defect could be
attributed directly to the dnaP mutation (17). The varying expression
of this mutation when it is transferred to other genetic backgrounds
is consistent with its being a membrane mutation as laboratory
strains are likely to have acquired a variety of innocuous membrane
alterations over the years (T. Yura, personal communication). The
temperature sensitivity of the dnaP mutation is reversible and the
reversibility is not inhibited by chloramphenicol or rifampicin.
At restrictive temperatures the dnaP mutant forms non-septated
filaments and the nucleoids appear dispersed (17).

dnaH

It appears that the dnaH initiation defective mutation (96) is actually two mutations, one resulting in an unusually high thymine requirement and the other being a dnaA allele (97). Several mutations isolated by the automated screening procedure have been tentatively classed as dnaH but no information on the genetics of these strains is yet available (83). This dnaH should not be confused with the dnaH mutation of Filip et al (32) which is now dnaZ2016

Termination

In the simplest use of the word, termination means only that the entire chromosome has been replicated. It is usually assumed that termination results in covalent closure of the circular chromosome; and since the entire chromosome can be transferred linearly by Hfr cells, the genome must be covalently closed. Since it is known that multiple initiations can occur within the span of one replication time, it is clear that termination of one round is not necessary for initiation of the next (98). Completion of replication is obviously required for division if the daughter cells are to receive complete chromosomes. There is no convincing evidence for a role of any of the dna gene products in termination. On the other hand, the region surrounding the apparent terminus of bidirectional replication has several interesting characteristics (8,9).

This region from tyrR to tyrS is coincident with the largest gap on the genetic map, and this gap is three times the size of any other. The estimate of the size of this gap is based on conjugation transfer time of this region. If the DNA at the terminus had an unusual structure, the replication process required for conjugation could be slowed and the estimate of the gap would be larger than the physical length of the DNA in this region. New information showing that replication in a P2 integratively suppressed strain is retarded in this region is consistent with such a notion (99).

A second feature of this region is that two markers, sbcA and rac, which lie in this region appear to be absent from some genetic lines. It is not clear that this reflects any special feature of this region as spontaneous deletions are not uncommon and the result may reflect, instead, the small number of strains analyzed. Alternatively, this region may be particularly prone to deletion and addition of genetic material because of a hypothetical unusual structure such as a nick or gap which might lead to a localized "hyper-rec" condition (46).

ACKNOWLEGEMENTS

I am grateful to N. Cozzarelli, D. Glaser, P. Kuempel, G. Lindahl, W. Kelley, J. Walker, and S. Wickner for communication of results prior to publication.

REFERENCES

(1) WATSON, J.D. and CRICK, F.H.C.(1953) Nature 171,737

(2) KLEIN, A. and BONHOEFFER, F.(1972) Ann. Rev. Biochem. 41,301

(3) GEFTER, M.L. (1975) Ann. Rev. Biochem. 44,45

(4) GEIDER, K. (1976) Curr. Top. Microbiol, Immunol. 74,55

(5) KORNBERG, A. (1974) DNA Synthesis, W.H. Freeman and Co., San Francisco.

(6) GROSS, J.D. (1972) Curr. Top. Microbiol. Immunol. 57,39

(7) WORCEL, A. and BURGI, E. (1972) J. Mol. Biol. 71,127

(8) BIRD, R.E., LOUARN, J., MARTUSCELLI, J. and CARO, L. (1972) J. Mol. Biol. 70,549

(9) BACHMANN, B.J., LOW, K.B. and TAYLOR, A.L. (1976) Bacteriol.Revs. 40,116

(10) HIRAGA, S. (1976) Proc. Nat. Acad. Sci., U.S.A. 73,198

(11) GOTFRIED, F. and WECHSLER, J.A. (1977) J. Bacteriol. 130,963

(12) MASTERS, M. (1975) Molec. Gen. Genetics. 143,105

(13) HIROTA, Y., RYTER, A. and JACOB, F. (1968) Cold Spring Harb. Symp. Quant. Biol. 33,677

(14) CARL, P. (1970) Molec. Gen. Genetics 109, 107

(15) WECHSLER, J.A. and GROSS, J.D. (1971) Molec. Gen. Genetics 113,273

(16) BEYERSMANN, D., MESSER, W. and SCHILCHT, M. (1974) J. Bacteriol. 118,783

(17) WADA, C. and YURA, T. (1974) Genetics 77,199

(18) LARK, K.G. and RENGER, H. (1969) J. Mol. Biol. 42,221

(19) WARD, C.B. and GLASER, D.A. (1969) Proc. Nat. Acad. Sci. U.S.A. 64, 905

(20) COOPER, S. (1974) J. Theor. Biol. 46,117

(21) LARK, K.G. (1972) J. Mol. Biol. 64,47

(22) OKAZAKI, R., OKAZAKI, T., SAKABE,K., SUGINO, K. and SUGINO, A. (1968) Proc. Nat, Acad. Sci., U.S.A. 59,598

(23) GEFTER, M.L., HIROTA, Y., KORNBERG, T., WECHSLER, J.A. and BARNOUX, C. (1971) Proc. Nat. Acad. Sci. U.S.A. 68,3150

(24) NUSSLEIN, V., OTTO, B., BONHOEFFER, F. and SCHALLER, H. (1971) Nature New Biol. 234,285

(25) HALL, R.M. and BRAMMER, W.J. (1973) Molec. Gen. Genetics. 121, 271

(26) SPEYER, J. (1965) Biochem. Biophys. Res. Commun. 21,6

(27) DRAKE, J.W. and ALLEN, E.F. (1968) Cold Spring Harb. Symp. Quant. Biol. 33,339

(28) HERSHFIELD, M.S. and NOSSAL, N.G. (1972) J. Biol. Chem. 247,3393

(29) MUZYCZKA, N., POLAND, R.L. and BESSMAN, M.J.(1972) J. Biol. Chem. 247,7116

(30) WICKNER, W., SCHEKMAN, R., GEIDER, K. and KORNBERG, A. (1973) Proc. Nat. Acad. Sci., U.S.A. 70,1764

(31) WICKNER, S. and HURWITZ, J. (1976) Proc. Nat, Acad. Sci., U.S.A. 73,1053

(32) FILIP, C.C., ALLEN, J.S., GUSTAFSON, R.A.,·ALLEN, R.G. and WALKER, J.R. (1974) J. Bacteriol. 119,443

(33) WICKNER, S. (1976) Proc. Nat. Acad. Sci., U.S.A. 73,3511

(34) LARK, K.G. (1972) Nature New Biol. 240,237

(35) BOUCHE, J-P., ZECHEL, K. and KORNBERG, A. (1975) J, Biol. Chem. 250,5995

(36) WICKNER, S. Proc. Nat. Acad. Sci., U.S.A. in press

(37) HERRMANN, R., HUF, J. and BONHOEFFER, F. (1972) Nature New Biol. 240,235

(38) LOUARN, J. (1974) Molec. Gen. Genetics 133,193

(39) GELLERT, M., LITTLE, J.W., OSHINSKY, C.K. and ZIMMERMAN, S.B. (1968) Cold Spring Harb. Symp. Quant. Biol. 33,21

(40) OLIVERA, B.M., HALL, Z.W., ANRAKU, Y., CHIEN, J.R. and LEHMAN, I.R. (1968) Cold Spring Harb. Symp. Quant. Biol. 33,27

(41) PAULING, C. and HAMM, L. (1968) Proc. Nat. Acad, Sci., U.S.A. 64,1195

(42) GOTTESMAN, M.M., HICKS, M.L. and GELLERT, M. (1973) J. Mol. Biol. 77,531

(43) KONRAD, E.B., MODRICH, P. and LEHMAN, I.R. (1973) J. Mol. Biol. 77,519

(44) LEHMAN, I.R. and CHIEN, J.R. (1973) In DNA Synthesis In Vitro (Wells, R.D. and Inman, R.B., eds.) pp 3-11 University Park Press, Baltimore, Md.

(45) KUEMPEL, P.L. and VEOMETT, G.E. (1970) Biochem. Biophys. Res. Commun. 41,973

(45a) YOUNGS, D.A. and SMITH, K.C. (1973) Nature New Biol. 244,240

(45b) TAIT, R.C., HARRIS, A.L. and SMITH, D.W. (1974) P.N.A.S. 71,675

(46) KONRAD, E.B. and LEHMAN, I.R. (1974) Proc. Nat. Acad. Sci., U.S.A. 71,2048

(47) OLIVERA, B.M. and BONHOEFFER, F. (1974) Nature 250,513

(48) BERKOWER, I, LEIS, J. and HURWITZ, J. (1973) J. Biol. Chem. 248,5914

(49) KELLER, W. and CROUCH, R. (1972) Proc. Nat. Acad. Sci., U.S.A. 69,3360

(50) CHEN, T.-C. and RAY, D.S. (1976) J. Mol. Biol. 106,589

(51) REICHARD, P., ELIASSON, R. and SÖDERMAN, G. (1974) Proc. Nat. Acad. Sci. U.S.A. 71,4901

(52) OGAWA, T., HIROSE, S., OKAZAKI, T. and OKAZAKI, R. (1977) J. Mol. Biol. 112,121

(53) KONRAD, E.B. and LEHMAN, I.R. (1975) Proc. Nat. Acad. Sci., U.S.A. 72,2150

(54) TYE, B-K, NYMAN, P.-O., LEHMAN, I.R., HOCHHAUSER, S. and WEISS, B. (1977) Proc. Nat. Acad. Sci., U.S.A. 74,154

(55) BECK, C.F., INGRAHAM, J.L., NEUHARD, J. and THOMASSEN, E. (1972) J.Bacteriol. 110,219

(56) BECK, C.F., NEUHARD, J. and THOMASSEN, E. (1977) J. Bacteriol. 129,305

(57) LINDAHL, T. (1974) Proc. Nat. Acad. Sci. U.S.A. 71,3649

(58) GATES, F.T. III and LINN, S. (1977) J. Biol.Chem. 252,1647

(59) WECHSLER, J.A., NUSSLEIN, V., OTTO, B., KLEIN, A., BONHOEFFER, F., HERRMANN, R., GLOGER, L. and SCHALLER, H. (1973) J.Bacteriol. 113,1381

(60) WECHSLER, J.A. (1973) In: DNA Synthesis In Vitro (Wells, R.D. and Inman, R.B., eds) pp375-382 University Park Press, Baltimore, Md.

(61) WICKNER, S., WRIGHT, M. and HURWITZ, J. (1974) Proc. Nat. Acad. Sci. U.S.A. 71,783

(62) LARK, K.G. and WECHSLER, J.A. (1975) J. Mol. Biol. 92,145

(63) GEORGOPOULOS, C.P. and HERSKOWITZ, I. (1971) In: The Bacteriophage Lambda (Hersey, A., ed.) pp 553-564. Cold Spring Harbor Laboratory. Cold Spring Harbor, New York.

(64) OGAWA, T. (1975) J. Mol. Biol. 94,327

(65) D'ARI, R., JAFFÉ-BRACHET, A., TOUATI-SCHWARTZ, D. and YARMOLINSKY, M. (1975) J. Mol. Biol. 94,341

(66) MIKOLAJCZK, M. and SCHUSTER, H. (1971) Molec. Gen. Genetics 110,272

(67) BUTTIN, G. and WRIGHT, M. (1968) Cold Spring Harb. Symp. Quant. Biol. 33,259

(68) ZYSKIND, J.W. and SMITH, D.W. (1977) J. Bacteriol. 129,1476

(69) SCHUBACH, W.H ., WHITMER, J.D. and DAVERN, C.I. (1973) J. Mol. Biol. 74,205

(70) WECHSLER, J.A. (1975) J. Bacteriol. 121,594

(71) WICKNER, S. and HURWITZ, J. (1975) Proc. Nat. Acad. Sci., U.S.A. 72,921

(72) SCHUSTER, M., SCHLICHT, M., LANKA, E., MIKOLAJCZYK, M. and EDELBLUTH, C. (1977) Molec. Gen. Genetics 151,11

(73) GELLERT, M., MIZUUCHI, K., O'DEA, M.H. and NASH, H.A. (1976) Proc. Nat. Acad. Sci., U.S.A. 73,3872

(74) GELLERT, M., O'DEA, M.H., ITOH, T. and TOMIZAWA, J.-I. (1976) Proc. Nat. Acad. Sci., U.S.A. 73,4474

(75) PROJAN, S.J. and WECHSLER, J.A. (1977) this symposium

(76) SCHALLER, H., OTTO, B., NUSSLEIN, V., HUF, J., HERRMANN, R. and BONHOEFFER, F. (1972) J. Mol. Biol. 63,183

(77) KORNBERG, T., LOCKWOOD, A. and WORCEL, A. (1974) Proc. Nat. Acad. Sci. U.S.A. 71, 3189

(78) KAVENOFF, R. and BOWEN, B.C. (1977) Chromosoma 59,89

(79) WANG, J.C. (1974) J. Mol. Biol. 89,783

(80) STUDIER, F.W. (1965) J. Mol. Biol. 11,373

(81) GEORGOPOULOS, C.P. (1977) Molec. Gen. Genetics 151,35

(82) SUNSHINE, M., FEISS, M., STUART, J. and YOCHEM, J. (1977) Molec. Gen. Genetics 151,27

(83) SEVASTOPOULOS, C.G., WEHR, C.T. and GLASER, D.A. Proc. Nat. Acad. Sci., U.S.A. in press

(84) ZAHN, G., TIPPE-SCHINDLER, R. and MESSER, W. in press

(85) NISHIMURA, Y., CARO, L., BERG, C.M. and HIROTA, Y. (1971) J. Mol. Biol. 55,441

(86) TRESGUERRES, E.F., NANDADASA, H.G. and PRITCHARD, R.H. (1975) J. Bacteriol. 121,554

(87) LINDAHL, G., HIROTA, Y. and JACOB, F. (1971) Proc. Nat. Acad. Sci., U.S.A. 68,2407

(88) BIRD, R.E., CHANDLER, M. and CARO, L. (1976) J. Bacteriol. 126,1215

(89) LANE, H.E.D., DENHARDT, D.T. (1975) J. Mol. Biol. 97,99

(90) SCOTT, J.F., EISENBERG, S., BERTSCH, L.L. and KORNBERG, A. (1977) Proc. Nat. Acad. Sci., U.S.A. 74,193

(91) WECHSLER, J.A. and ZDZIENICKA (1975) In: DNA Synthesis and its
     Regulation (Goulian, M., Hanawalt, P.C. and Fox, C.F., eds.)
     pp 624-639 W.A. Benjamin, Inc., Menlo Park, California.

(92) FRALICK, J.A. and LARK, K.G. (1973) J. Mol. Biol. 80,459

(93) BAGDASARIAN, M.M., IZAKOWSKA, M. and BAGDASARIAN, M. (1977) J.
     Bacteriol. 130,577

(94) ZYSKIND, J.W., DEEN, L.T. and SMITH, D.W. (1977) J. Bacteriol.
     129,1466

(95) MESSER, W., DANKWARTH, L., TIPPE-SCHINDLER, R., WOMACK, J.E.
     and ZAHN, G. (1975) In: DNA Synthesis and its Regulation
     (Goulian, M., Hanawalt, P.C. and Fox, C.F., eds.) pp 602-617
     W.A. Benjamin Inc., Menlo Park, California.

(96) SAKAI, H., HASHIMOTO, S. and KOMANO, T. (1974) J. Bacteriol.
     119,811

(97) DERSTINE, P.L. and DUMAS, L.B. (1976) J. Bacteriol. 128,801

(98) OISHI, M., YOSHIKAWA, H. and SUEOKA, N. (1964) Nature, 204,1069

(99) KUEMPEL, P.L., DUERR, S.A. and SEELEY, N.R. (1977) Proc. Nat.
     Acad. Sci., U.S.A. in press.

(100) SAITO, H. and UCHIDA, H. (1977) J. Mol. Biol. 113,1

CHARACTERIZATION OF CHROMOSOMAL REPLICATION IN A *dna*A MUTANT

Joe A. Fralick

Department of Microbiology
Texas Tech University School of Medicine
Lubbock, Texas 79409, U.S.A.

INTRODUCTION

It has been well documented that chromosomal replication is tightly coupled to cell division (Maaløe and Kjeldgaard, 1966; Helmstetter et al., 1968; Yoshikawa and Haas, 1968; Donachie, 1968; Bleecken, 1969; Pritchard et al., 1969; Pritchard, 1974; Somapayrac and Maaløe, 1974) in the bacterium *Escherichia coli* and plays an important role in the cell cycle of this organism. Unfortunately the means by which the cell regulates this important process has remained obscure.

The *in vivo* replication of the bacterial chromosome is complex requiring a minimum of some 12 distinct gene products (see Geider, 1976 for a review) and an active RNA polymerizing system (Lark, 1972; Messer, 1972). The rate limiting step in this process appears to be the frequency at which new rounds of chromosomal replication are initiated (Maaløe and Kjeldgaard, 1966; Cooper and Helmstetter, 1968; Bleecken, 1969; Pritchard et al., 1969) and for this reason most models attempting to explain how DNA replication is regulated have focused on the initiation event.

Physiological and genetic studies have demonstrated a requirement for *de novo* protein (Maaløe and Hanawalt, 1961; Lark et al., 1963; Lark and Renger, 1969; Ward and Glaser, 1969) and RNA (Lark, 1972; Messer, 1972) synthesis for the initiation of each round of chromosomal duplication. These studies suggest a working model in which the replication of the bacterial chromosome is initiated at a specific site, the replicative origin (Maaløe and Hanawalt, 1961; Lark et al., 1963; Abe and Tomizawa, 1967; Louarn et al., 1974), by the assemblage of a group of proteins (and

71

perhaps RNA) into a functional replication complex which may be
associated with the bacterial envelope (Jacob et al., 1963; Bleeken,
1971; Fralick and Lark, 1973; Helmstetter, 1974) and which is
somehow activated at the beginning and inactivated at the end of
each round of chromosomal replication (Kogoma and Lark, 1975).
Specific models for initiation have emphasized either positive
control of this activation process by means of an initiator protein
(Jacob et al., 1963; Helmstetter et al., 1968) or negative control
by a repressor (Rosenberg et al., 1963; Pritchard et al., 1969;
Blau and Mordoh, 1972; Sompayrac and Maaløe, 1973; Messer et al.,
1975). The complexity of this activation process is illustrated by
the isolation of 5 genetically distinct, temperature-sensitive,
conditionally lethal, *dna* initiation mutants (*dnaA*, C, H, I and P)
(Hirota et al., 1974; Beyerson et al., 1974; Wada and Yura, 1974;
Geider, 1976). Although little is known about the function of the
gene products of these mutants, it seems likely that they must
somehow be involved in controlling the process of DNA replication.
This paper describes some of the characteristics of chromosomal
replication in a *dnaA*, initiation mutant growing at permissive
temperatures.

RESULTS AND DISCUSSION

The rate of DNA synthesis, the amount of DNA per cell and the
frequency of cell division are determined by the rate at which new
rounds of chromosomal replication are initiated (Helmstetter et al.,
1968; Pritchard, 1974). We have examined these parameters in a
well characterized conditionally lethal, temperature sensitive,
dnaA mutant, N167 (Abe and Tomizawa, 1971); Hiraga and Saitoh, 1974,
1975; Saitoh and Hiraga, 1974), at growth temperatures ranging
from 30 to 38°C. At these temperatures this mutant grows with
balanced growth (Campbell, 1957; Zaritsky and Pritchard, 1974) as
determined by the rate of increase in cell number (particle
number), cell mass (OD_{550}) and DNA content (steady-state ^{14}C
thymine incorporation) of an exponentially growing culture (Fig. 1).
The results of this study are summarized in Fig. 2 and can be
described as follows:

(1) The rate of DNA synthesis on a per cell basis, as
    measured by ^3H thymidine incorporation kinetics,
    decreases sharply over the temperature range of
    30 to 35°C.

(2) The amount of DNA per cell, based on ^{14}C thymine
    steady-state labeling, decreases by approximately
    two fold, over this same temperature spectrum.

(3) The rate of cell division increases with increasing
    temperature between 30 and 35°C. Only at higher

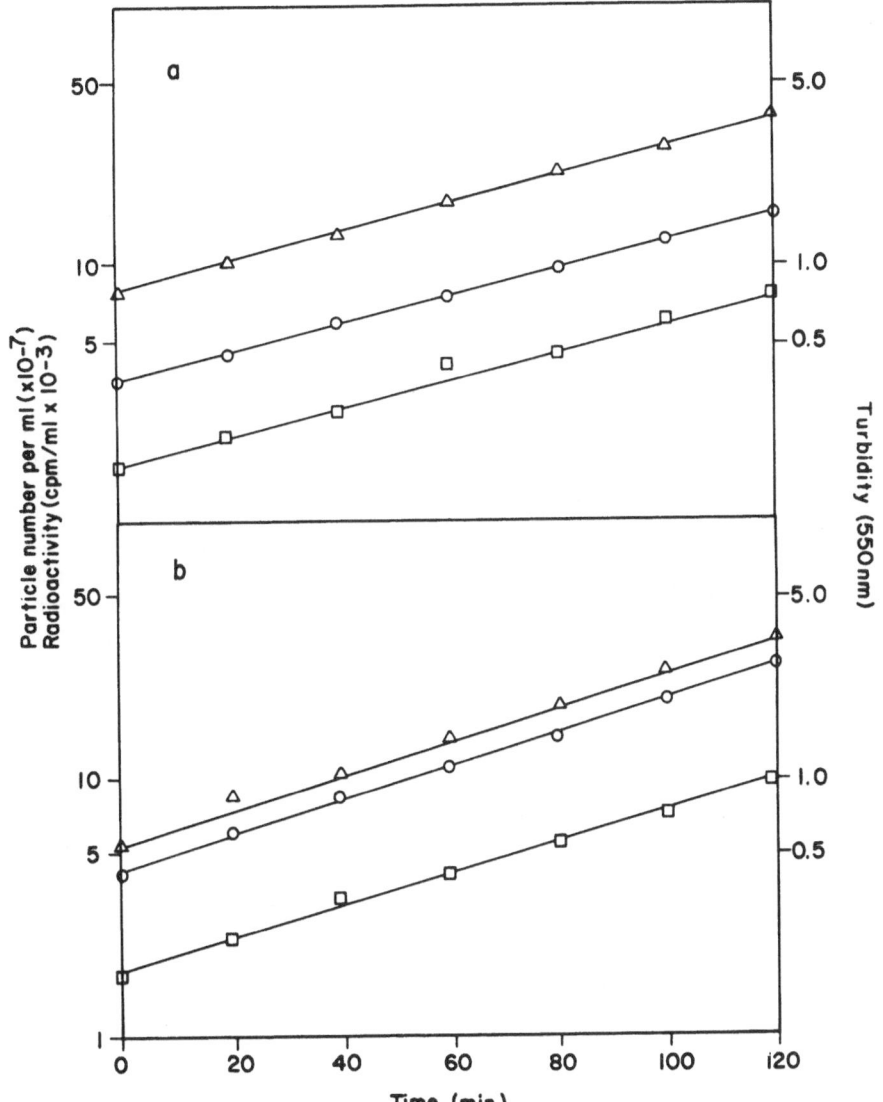

Figure 1. Growth kinetics in N167 (Hfr H thy⁻ met⁻ dna167) cells.
N167 cells were grown at 30 (a) and 38°C (b) in minimal media A
medium (Miller, 1972) supplemented with 0.4% glucose, 0.1%
casamino acids, thiamine (2μg/ml) and ¹⁴C thymine (0.5μCi/5μg/ml)
for 4 doublings before starting the experiment. Samples were
removed at the indicated intervals and the rate of increase in
absorbance at 550 nm (□), particle number (0) as determined by
a Coulter counter; and ¹⁴C thymine incorporation (△) was
determined.

Figure 2.   *Comparison of the effect of growth temperature on DNA*
synthesis, DNA content and cell division rate in N167
cells.

The normalized incorporation rate (□-- □) was determined by adding
^3H *thymidine (20μCi/2μg/ml) to cultures of N167 cells which had been*
growing for more than 5 doublings at the indicated temperature and
sampling at 2 minute intervals for 10 minutes to determine the
initial incorporation rate (k) of ^3H *thymidine into acid insoluble*
material. The exponential growth rate constant (μ) was determined
with a coulter counter. The normalized incorporation rate was
calculated by dividing the incorporation rate constant by the
exponential growth rate constant (k/μ).

The effect of growth temperature on the DNA content of N167 cells
was determined by measuring the radioactivity (O--O) of these cells
which had grown for more than 5 doublings in the presence of ^{14}C
thymine (0.5μCi/5μg/ml) at the indicated growth temperature. Cell
number was determined with a Coulter counter. Each data point is
the average of 4 or more independent determinations.

The exponential growth rate (Δ--Δ) for N167 cells growing at the
indicated temperature was determined with a coulter counter. The
N167 cells had grown for more than 5 doublings at the indicated
temperature before each experiment was started. Greater than 5
different sampling times were used to calculate the growth rate of
each culture.

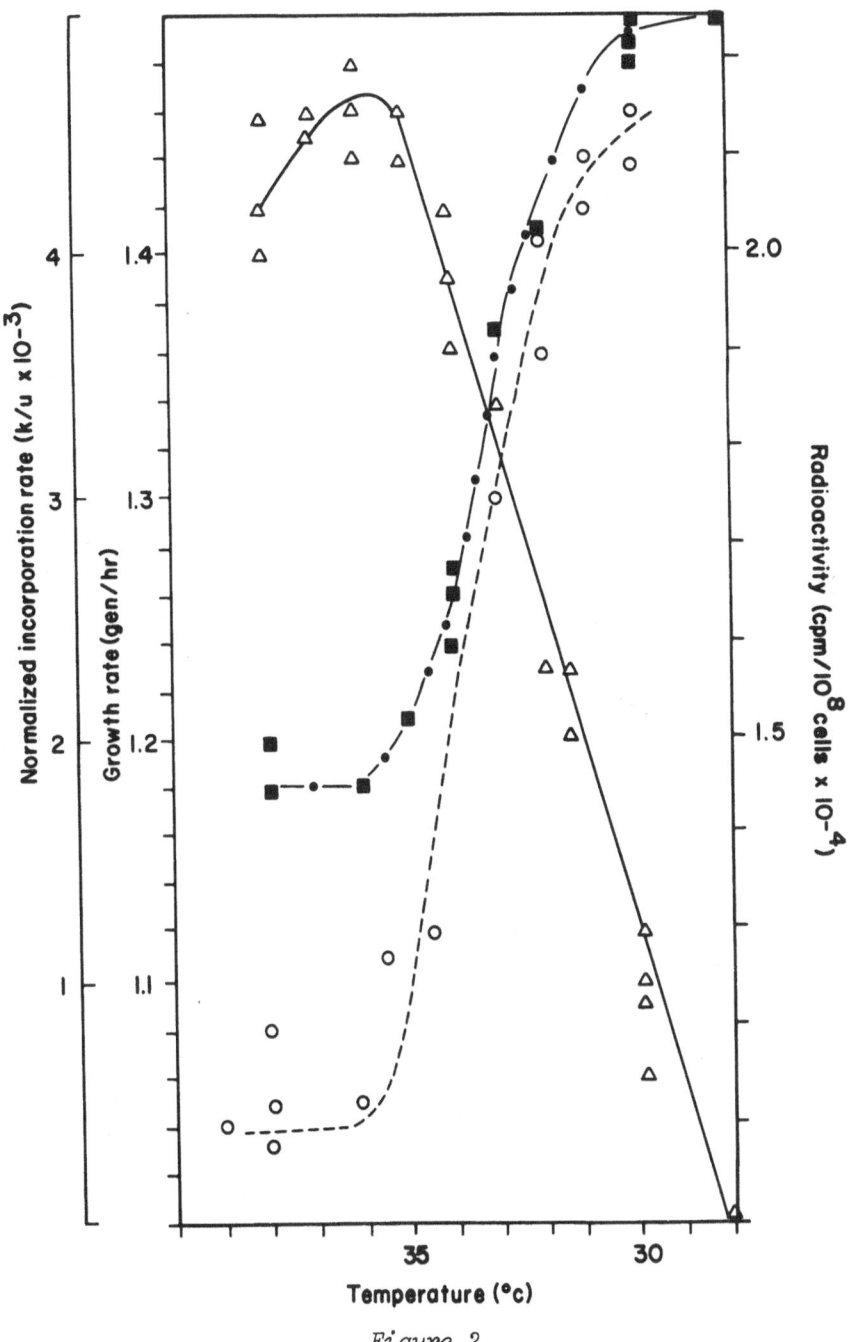

Figure 2

temperatures does one begin to see a decrease
in growth rate.

From these results it appears that the *dnaA* gene product
becomes limiting with respect to the rate of DNA synthesis and the
amount of DNA per cell over a temperature range at which the rate
of cell division is increasing. Since under balanced growth
conditions the rate of cell division is directly proportional to
the frequency of initiation of chromosomal replication these results
seriously question the likelihood that the *dnaA* gene product is a
regulatory molecule involved in triggering the initiation of
chromosomal replication.

An alternative explanation would be that the *dnaA* gene
product somehow affects the velocity of polynucleotide chain
elongation, causing it to slow down at higher temperatures. If
this were the case, however, there is no reason to expect N167 cells
growing at 30°C to have a greater amount of DNA per cell than those
growing at 38°C. In fact, one might expect the opposite since the
amount of DNA per cell and the DNA to mass ratio under balanced
growth conditions are inversely proportional to the replication
velocity (Pritchard and Zaritsky, 1970). However, Table 1
illustrates that the amount of DNA per cell and the DNA to
protein ratio of N167 cells growing at 38°C is approximately one
half the amount it is in these same cells growing at 30°C.

Perhaps the most plausible interpretation of this data assumes
that the *dnaA* product determines the number of functional
replication complexes (i.e., the number of replicating chromosomes)
within a cell. Our results could then be interpreted in the
following way. At 30°C, all of the potential replicating sites
(complexes) are functional and can support chromosomal replication.
As the growth temperature increases, however, more and more of the
dnaA product becomes inactive until at the non-permissive
temperature all of the *dnaA* gene product is non-functional and,
therefore, all of the replicative complexes are inoperative.
However, since this gene product would only determine the number of
functional replication complexes and not their frequency of
activation, the effect of increasing growth temperature
(inactivation of the *dnaA* gene product) would be to decrease the
number of replicative forks per chromosome or replicating
chromosomes per cell. This, in turn, would alter the rate of DNA
synthesis and the amount of DNA per cell but not necessarily affect
the frequency of cell division until some critical temperature is
reached, at which time the number of functional replication
complexes within a cell becomes limiting. For the *dnaA*167 allele
the critical temperature appears to be 35°C. This explanation is
further supported by the quantitative data of Table 1 which
demonstrates that N167 cells growing at 30°C contain enough DNA for
two replicating chromosomes while these same cells growing at 38°C

Table 1

Strain	Growth temperature ($^\circ$C)	DNA/cell (μg $\times$ 10^{-9})	Specific activity (1) of DNA (cpm/μg)	Chromosome (2) equivalents per cell	Replicating chromosome equivalents per cell
N167	38°C	7.2	2990	1.7	1.18
	30°C	12.5	3032	2.96	2.05

DNA content and DNA to mass ratio of N167 cells growing at 30 and 38°C

N167 cells growing at 30 and 38°C were harvested by centrifugation at a population density of approximately 2 $\times$ 10^8 cells per ml. and analyzed for deoxyribose using the modified diphenylamine method of Giles and Myers (1965).

(1) Chromosome equivalents were calculated by assuming the mass of a unit chromosome to be 4.22 $\times$ 10^{-15} grams which corresponds to a molecular weight of 2.5 $\times$ 10^9 daltons for the _E. coli_ chromosome.

(2) For comparative purposes the value of 1.44 unit chromosomes per replicating chromosome was used in this calculation. This value was obtained from the equation of chromosomal replication (Sueoka and Yoshikawa, 1965) assuming the average number of replicative forks (pairs) per chromosome to be one.

contain only enough for one such molecule. That this decrease in
DNA per cell at 38°C is not due to a decrease in the number of
replicative forks per chromosome can be seen from the data of
Table 2. The amount of residual DNA synthesized at the non-
permissive temperature is a measure of the relative number of
replicative forks on the chromosome and theoretically should
equal 39% if only one pair of replicative forks are traversing
the chromosome at any given time (Sueoka et al., 1965).

In summary the *dnaA* gene product appears to regulate the
number of chromosomes the cell can replicate. The means by which
this is accomplished is unknown. However, it is possible to explain
our data using a model previously offered to describe the function
of the *dnaA* product (Messer et al., 1975). In this scheme the *dnaA*
gene was involved in controlling an origin-RNA (O-RNA) operon
required for the initiation of chromosomal replication. This operon
is envisioned to be negatively controlled by a DNA bound repressor.
It was suggested that the *dnaA* gene product might be either
required for the release of the repressor (i.e., act as an anti-
repressor or inducer) or might be the repressor molecule itself,
providing a simple but effective system for the temporal control of
initiation by the *dnaA* gene product. This model could account for
our data as well as that of Messer et al. (1975) if one stipulated
that the *dnaA* product was a constitutively synthesized repressor
of the O-RNA operon and that the temperature sensitive lesion

Table   2

Growth temperature before shift to 43°C	% increase in residual DNA synthesized at 43°C
30°C	40.8
38°C	39.2

Increase of the amount of DNA synthesized at 43°C

N167 cells growing at the indicated temperature for more than 4
doublings in media containing ^{14}C thymine (0.5μCi/5μg/ml) were
placed at 43°C for 120 minutes and the residual increase in acid
insoluble ^{14}C thymine determined.

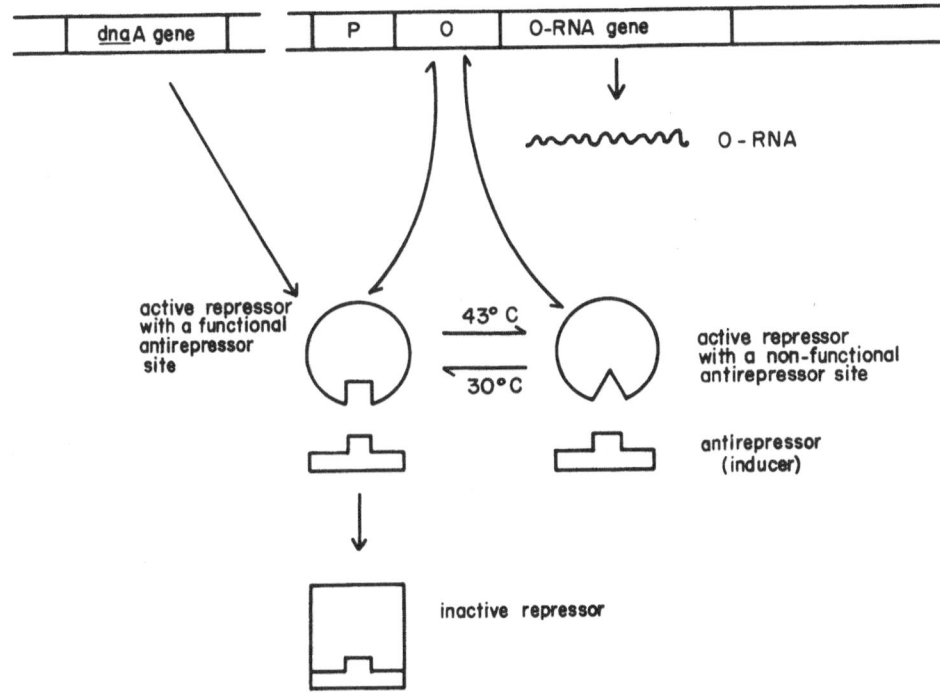

Figure 3. A modified version of the O-RNA operon (Messer et al. 1975).

The dnaA gene synthesizes a repressor protein which binds to the operator site and prevents the transcription of the O-RNA. The active form of the repressor is represented by a circle, with an insertion site to which inducer binds (anti-repressor site). When inducer is added, it binds to the repressor protein converting it to an inactive form, represented by a square. This depresses the operon and allows for the synthesis of Origin-RNA which in turn primes the replication complex for the initiation of chromosomal replication. When the repressor contains a thermo-sensitive anti-repressor binding site (dnaA^{ts} protein) the O-RNA operon will only be de-repressed at permissive temperatures. The degree to which de-repression can occur, therefore, will be temperature dependent.

occurred at the anti repressor recognition site. One could then
explain our observations in the context of this model as follows.
At the restrictive temperature, none of the repressor molecules
would be inactivated and, therefore, the operon and hence the
initiation of chromosomal replication would be repressed. At
intermediate temperatures, however, a portion of these repressor
molecules would be capable of binding inducer and therefore a
fraction of these operons would be inducible, the number of which
would depend upon the growth temperature. Finally at 30°C or at
the permissive temperatures, all of the repressors will be sensitive
to the inducer and hence all of the O-RNA operons will be
potentially inducible (derepressible). Note that the frequency
of induction of this operon and therefore of initiation of
chromosomal replication will not be dependent upon the functional
state of the anti-repressor site but will depend upon the
concentration of the inducer molecules within the cell. The
potential number of inducible operons and therefore replicatable
chromosomes, however, will be determined by the relative fraction
of the repressor molecules which are sensitive to inducers. This
modified scheme is shown in Fig. 3. It is interesting to note
that such a model would predict the temperature sensitive mutation
in the anti-repressor recognition site to be dominant over wild
type (Wilson et al., 1963). This has been observed in two different
studies (Beyersmann et al., 1974; Zahn et al., 1977) although
this conclusion is not unanimous (Wechsler and Gross, 1971;
Gotfried and Wechsler, 1977). Also a recent report (Bagdasarian
et al., 1977) has shown suppression of the *dnaA* phenotype by a
mutation of the *rpoB* cistron suggesting an interaction between the
dnaA product and RNA polymerase. This is what one might expect
if the *dnaA* product was involved in controlling the transcription
of the O.RNA operon. Hopefully, further characterization of the
genetics of the *dnaA* mutants and their suppressors will clarify the
role of this gene product in chromosomal replication.

ACKNOWLEDGEMENTS

The author is grateful to Dr. T.M. Joys for critically reading
this manuscript and for valuable discussions, and to Mr. A.M. Davis
for his excellent technical assistance. I also thank Mrs L.J.
Ragland, Mrs K.L. Beebe and Miss P. Lee for their assistance in
preparing the manuscript. This research was supported in part by
a grant from the Public Health Service (GM 22653).

REFERENCES

Abe, M. and Tomizawa, J. (1967) Proc. Nat. Acad. Sci. U.S.A. 58:
1911-1918

Abe, M. and Tomizawa, J. (1971) Genetics 69:1-15

Bachmann, B.J., Low, K.B. and Taylor, A.L. (1976) Bacteriol. Rev. 40:116-167

Bagdasarian, M.M., Izakowska, M. and Bagdasarian, M. (1977) J. Bacteriol. 130:577-582

Beyersmann, D., Messer, W. and Schlicht, M. (1974) J. Bact. 118: 783-789

Blau, S., Mordoh, J. (1972) Proc. Nat. Acad. Sci (Wash.) 69:2895-2898

Bleecken, S. (1969) J. Theor. Biol. 25:137-158

Bleecken, S. (1971) J. Theor. Biol. 32:81-92

Campbell, A. (1957) Bact. Rev. 21:263-272

Caro, L.G. (1970) J. Mol. Biol. 48:320-338

Cooper, S. and Helmstetter, C.E. (1968) J. Mol. Biol. 31:519-540

Donachie, W.D. (1968) Nature (London) 219:1077-1079

Fralick, J.A. and Lark, K.G. (1973) J. Mol. Biol. 80:459-475

Geider, K. (1976) Curr. Topics in Microbiol. and Immunol. 74: 55-112

Giles, K. and Myers, A. (1965) Nature (London) 206:93

Gotfried, F. and Wechsler, J.A. (1977) J. Bacteriol. 130:963-964

Helmstetter, C., Cooper, S., Pierucci, O. and Revelas, E. (1968) Cold Spring Harbor Symp. Quant. Biol. 33:809-822

Helmstetter, C.E. and Cooper, S. (1968) J. Mol. Biol. 31:507-518

Helmstetter, C.E. (1974) J. Mol. Biol. 84:21-36

Hiraga, S. and Saitoh, T. (1974) Mol. gen. Genet. 132:49-62

Hiraga, S. and Saitoh, T. (1975) Mol. gen. Genet. 137:239-248

Hirota, Y., Mordoh, J. and Jacob, F. (1970) J. Mol. Biol. 53:369-388

Jacob, F., Brenner, S. and Cuzin, F. (1963) Cold Spring Harbor
     Symp. Quant. Biol. 28:329-348

Kogoma, T. and Lark, K.G. (1975) J. Mol. Biol. 94:243-256

Lark, K.G., Repko, T., Hoffman, E.J. (1963) Biochim. Biophys. Acta
     76:9-24

Lark, K.G. and Lark, C. (1966) J. Mol. Biol. 20:9-19

Lark, C. (1968) Methods in Enzymology XIIb, 616-625

Lark, K.G. and Renger, H. (1969) J. Mol. Biol. 42:221-235

Lark, K.G. (1972) J. Mol. Biol. 44:217-231

Louarn, J., Furdeaburgh, M. and Bird, R. (1974) J. Bacteriol. 120:
     1-13

Lowry, O.H., Rosebrough, A., Farr, A.L. and Randall, R.J. (1951)
     J. Biol. Chem. 193:265-275

Maaløe, O. and Hanawalt, P.C. (1961) J. Mol. Biol. 3:144-155

Maaløe, O. and Kjeldgaard, N.O. (1966) Macromolecular Synthesis.
     W.A. Benjamin Co., Inc., New York

Messer, W. (1972) J. Bact. 112:7-12

Messer, W., Dankworth, L., Tippe-Schinder, R., Womack, J.E. and
     Zahn, G. (1975) In: DNA Synthesis and its Regulation (Goulian,
     M. and Hanawalt, P., eds) pp. 602-617. W.A. Benjamin, Inc.
     California.

Miller, J.H. (1972) Experiments in Molecular Genetics. Cold Spring
     Harbor Laboratory, Cold Spring Harbor, New York.

Pritchard, R.H., Barth, P.T. and Collins, J. (1969) XIX Microbiol.
     growth. Symp. Soc. Gen. Microbiol. pp.263-297

Pritchard, R.H. and Zaritsky, A. (1970) Nature (London) 226:126-131

Pritchard, R.H. (1974) Philosophical Trans. Royal Soc. Lond. 267:
     303-336

Rosenberg, B., Cavalieri, L. and Ungers, G. (1969) Proc. Nat. Acad.
     Sci. U.S.A. 63:1410-1417

Saitoh, T. and Hiraga, S. (1975) Molec. gen. Genet. 137:249-261

Sompayrac, L. and Maaløe, O. (1973) Nature (London) New Biol. 241: 133-135

Sueoka, N. and Yoshikawa, H. (1965) Genetics 52:747-748

Wada, C. and Yura, T. (1974) Genetics 77:199-220

Ward, C.B. and Glaser, D.A. (1969) Proc. Nat. Acad. Sci. U.S.A. 64:905-912

Wechsler, J.A. and Zdzienicka, (1975) In: DNA Synthesis and its Regulation (Goulian, M. and Hanawalt, P. eds) pp.602-617, W.A. Benjamin Inc., California.

Wechsler, J.A. and Gross, J.D. (1971) Mol. gen. Genet. 113:273-284

Willson, C., Perrin, D., Cohn, M., Jacob, F. and Monod, J. (1964) J. Mol. Biol. 8:582-592

Yoshikawa, H. and Haas, M. (1968) Cold Spring Harbor Symp. Quant. Biol. 33:843-855

Zahn, G., Tippe-Schindler, R. and Messer, W. (1977) Mol. gen. Genet. 153:45-49

Zaritsky, A. and Pritchard, R.H. (1974) J. Bacteriol. 114:824-837

GENETIC AND PHYSIOLOGICAL PROPERTIES OF AN *ESCHERICHIA COLI*

STRAIN CARRYING THE *dnaA* MUTATION T46

E. Orr[*], P.A. Meacock and R.H. Pritchard[*]

Department of Genetics
University of Leicester
Leicester LE1 7RH, England

INTRODUCTION

Initiation of DNA replication is a crucial event in the control of chromosome duplication in bacteria (for review see Pritchard, 1974) and possibly of other events in the cell cycle like cell division (Helmstetter et al., 1968; Donachie, 1968; Pritchard et al., 1969); Pritchard, 1974).

One approach which has been used to study this process is the isolation of temperature conditional mutants which have lost the ability to initiate chromosome replication when shifted to the restrictive temperature (Hirota et al., 1968; Carl, 1970; Wechsler and Gross, 1971). Several mutants of this type in *Escherichia coli* have been examined extensively during the last decade and one group of mutations designated as belonging to the *dnaA* locus has been mapped close to the isoleucine-valine region on the bacterial chromosome (Bachmann et al., 1976). However, very little is yet known about the biochemical function of the products made by these genes.

It appears that *dnaA* mutants differ from each other in several properties. Some, but not all, mutants produce products which are inactivated at the restrictive temperature but function again when shifted back to the permissive temperature even in the presence of chloramphenicol (Abe and Tomizawa, 1971; Blay and Mordoh, 1972; Hanna and Carl, 1975; Messer et al., 1975; Saitoh and Hiraga, 1975) or rifampicin (Hanna and Carl, 1975; Messer et al., 1975). Several *dnaA* mutations have been found to be dominant (Beyersmann et al., 1974; Zahn et al., 1977) while others are reported to be recessive

[*]*A.S.I. Participants*

(Wechsler and Gross, 1971; Wehr et al., 1975; Derstine and Dumas, 1976; Gotfried and Wechsler, 1977).

It has been shown by several groups that replication of some F-like plasmids is affected by changes in the activity of the *dnaA* product *in vivo* although not immediately (Goebel, 1973; Zeuten and Pato, 1971; Hiraga and Saitoh, 1975; Tresguerres et al., 1975; Brunt et al., 1977). The dependence of ColEl replication on this product needs to be clarified in view of the controversial reports which have been published (Goebel, 1973; Collins et al., 1975; Mayer et al., 1977). *dnaA* mutants can be suppressed by F-like plasmids by a mechanism which has been discussed before (Nishimura et al., 1971; Tresguerres et al., 1975; Bird et al., 1977) and by Pritchard in this Symposium.

In this paper we would like to present some observations on the physiological behaviour and the genetic properties of a *dnaA* mutant which carries the mutation T46 (Kohiyama et al., 1966).

RESULTS

Growth of A3 at Various Temperatures

The *dnaA* strain used in this work (A3) is derived from CRT46 (Kohiyama et al., 1966). Strain CRT46 grows very poorly on synthetic media, carries defects that render it recombination deficient and is sensitive to methylmethane sulfonate (Tresguerres et al., 1975). A3 is a recombinant from a cross between CRT46 and an Hfr Cavalli strain (Hayes, 1968) selected for rapid growth on synthetic medium. It is recombination proficient and not sensitive to methylmethane sulfonate. To study their growth behaviour, strains A3 and $A3^+$ (Table 1) were grown at various temperatures. Fig. 1 shows the increase in mass, particle number and viable particles of A3 at 37°C. The parallel increase in these parameters and in total DNA (data not shown) indicates that at this temperature steady-state exponential growth can be maintained. This was also found at 34°C, 31°C and for $A3^+$ at all temperatures checked. The relationship between incubation temperature, DNA concentration and cell size of A3 and $A3^+$ is shown in Fig. 2. The decrease found in the DNA concentration in A3 could reflect an alteration in the replication time (C) or a change in the initiation volume. To distinguish between these possibilities we compared the time C of exponentially growing cultures of A3 and $A3^+$ by measuring the increment (ΔG) of DNA (Pritchard and Zaritsky, 1970) in run out experiments in the presence of rifampicin. The run out times and the increment of DNA (Table 2) suggest that there are no substantial differences in the C times between A3 and $A3^+$. This conclusion was reinforced by comparing the differential rate of synthesis of two inducible

Table 1

Strains used.   (All are *E.coli* K12)

Strain	Genotype	Source
A3	F⁻ *thr leu thi lacY supE dnaA* T46	See text
A3⁺	A3 *dnaA*⁺	P1 transduction
LE309	A3 *bgl*⁺	Spontaneous
LE234	F⁻ *met arg ilv tna supE*(?)	P.A. Meacock
LE236	F⁻ *met arg ilv his tna bgl*⁺	P.A. Meacock
KL16	Hfr *thi relA*	P.A. Meacock
LE310	A3 *relA*	A3 x KL16
CO101	HfrC *thi dgoD relA*	R.A. Cooper
CO503	*mtl ptsF fpk pro met his cysI glp*⁺ *dgoD*	R.A. Cooper

enzymes, tryptophanase and β-galactosidase at various temperatures. It has been shown (Chandler and Pritchard, 1975) that in steady state exponential cultures, the differential rate of synthesis of these enzymes is proportional to the average number of copies of the corresponding structural genes per unit mass. The output of the two genes (*tna* and *lac*) which are located near to the chromosome origin and some distance from it, respectively (Masters and Broda, 1971; Bird et al., 1972) was reduced by approximately the same amount with increasing temperature, as would be expected if there were no change in the relative concentration of the two genes and consequently no change in C. It is therefore likely that the decrease in the DNA concentration at intermediate temperatures is caused by a defect in initiation rather than by a change in the C time. These experiments will be published elsewhere in detail.

Cold Sensitivity of A3

When a culture of A3 was grown on glucose minimal medium

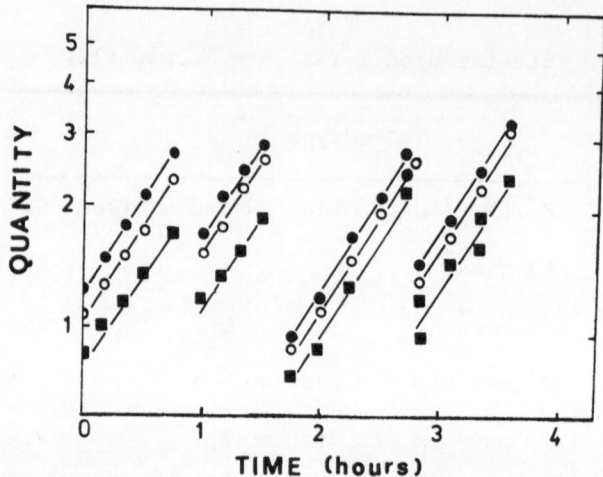

Figure 1. Growth of A3 at 37°C. Strain A3 was grown in M9 glucose minimal medium supplemented with 50 μg/ml of the required amino acids, 10 μg/ml thiamine and 0.5% casamino acids. Growth was followed during several dilutions.

●————● 10^1 x absorbance at 450 nm.

○————○ 10^7 particles/ml estimated by Coulter counter

■————■ 10^7 viable particles/ml assayed on M9 glucose minimal plates with appropriate requirements at 34°C.

Table   2

Increment of DNA (ΔG) in presence of rifampicin

Temperature	τ (min)		G (%)		C_1 (min)		C_2 (min)	
	A3	A3$^+$	A3	A3$^+$	A3	A3$^+$	A3	A3$^+$
31°C	50	43	83	85.5	70	55	95	81
34°C	35	30	53	84	45	50	42	60
37°C	42	25	43	96	40	40	47	55

C_2 was calculated from the values of ΔG and C_1 was estimated from the run out times.

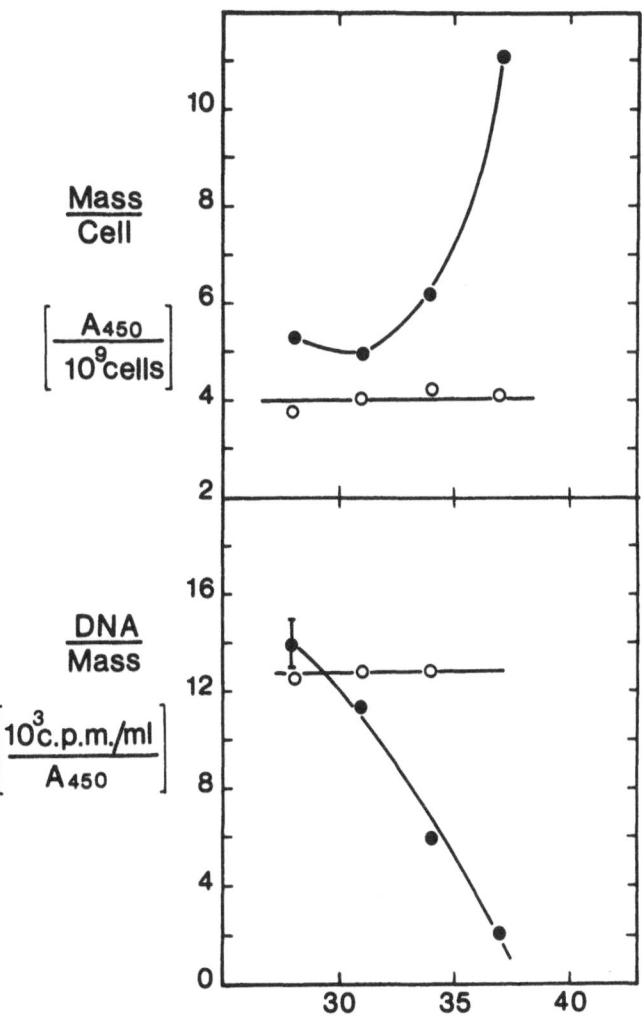

Figure 2. Relation between the incubation temperature, DNA concentration and mass per cell. Data for this figure were obtained from experiments performed at various temperatures as described in the legend of Fig. 1.

●————● A3;   ○————○ A3⁺.

supplemented with 0.5% casamino acids at 28°C and plated at 34°C,
a remarkable decrease in viability was observed (Fig. 3), although
mass, particle number (Fig. 3) and DNA (Fig. 4) still continued to
increase. The A3⁺ strains showed steady-state growth under the
same conditions. If cultures were grown at 34°C and plated at 26°C
(Table 3) there was likewise a loss of viability. These experiments
show that the loss of viability during incubation in liquid at 28°C
is irreversible. In addition, the data in Table 3 demonstrate
that cryosensitivity is not expressed on glucose minimal plates. It
was also found that cryosensitivity was not evident during
incubation in glucose minimal medium (data not shown).

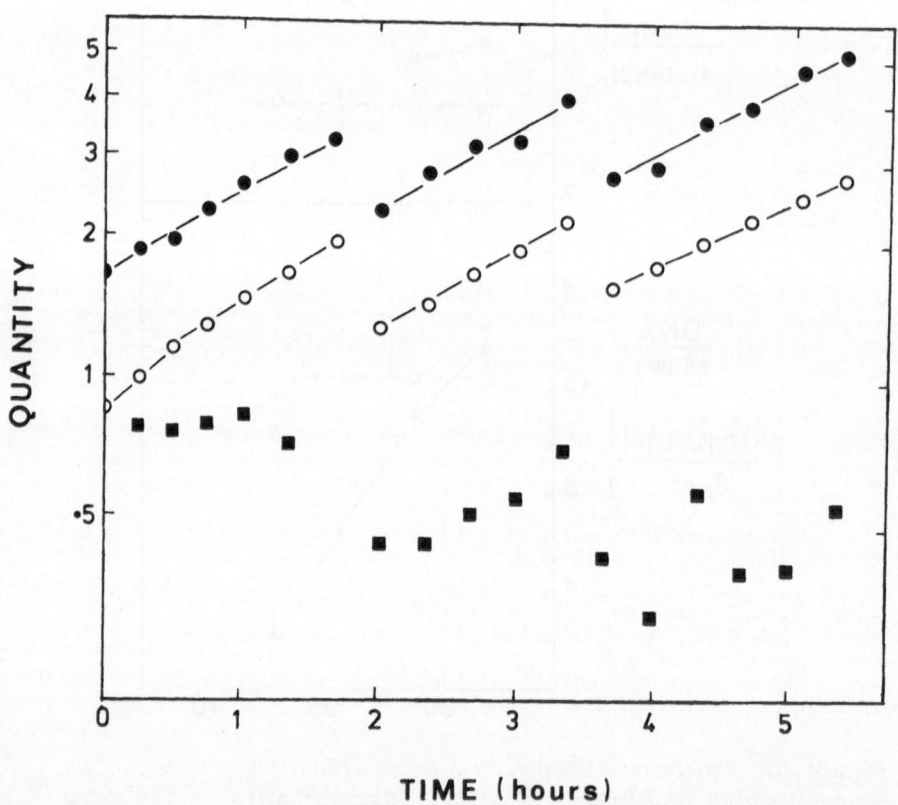

TIME (hours)

*Figure 3. Growth of A3 at 28°C. The experimental conditions were
the same as described for Fig. 1.*

■————■ 10^7 *viable particles/ml,*

●————● 10^7 *particles/ml,*

○————○ 10^1 *x absorbance (450 nm)*

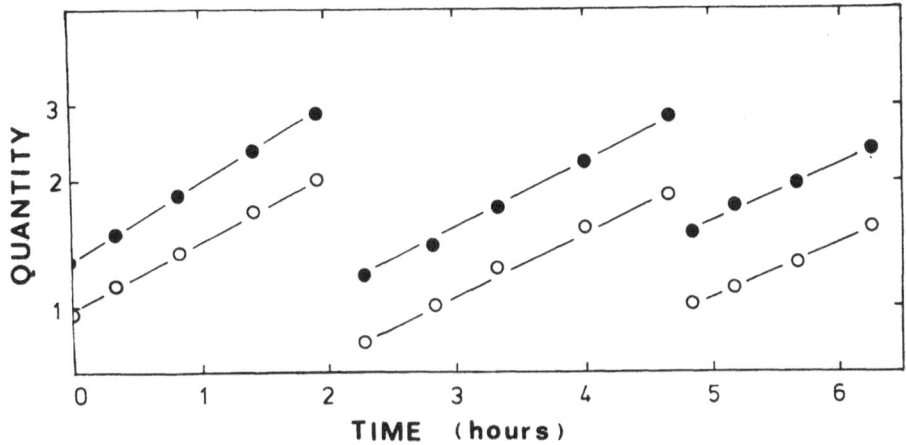

Figure 4.   *DNA synthesis in A3 at 28°C.   Strain A3 was grown under the same conditions as described in Fig. 1, except that the growth medium contained 1.5 mM uridine and 0.08 µCi/ml ¹⁴C thymidine (0.04 µCi/µg).   The culture was grown at least eight generations in the radioactive medium before the beginning of the experiment. Radioactivity was estimated as described previously (Tresguerres et al. 1975).*

O————————O 10^1 x absorbance (450 nm)

●————————● 10^3 c.p.m./ml.

Table 3

Viability[(a)] of A3 at Different Temperatures

Temperature (°C)	Glucose Minimal Medium	Nutrient Broth
26	56	3×10^{-4}
30	100	39
34	100	100
37	100	100
42	1.5×10^{-7}	10^{-6}

(a)   Percentage of viable counts was calculated from the maximum values obtained at 37°C and 34°C. The viability of A3[+] at all temperatures on both media was 100%.

Genetic Analysis of Cold Sensitivity

Reversion tests were first used to check whether the cryolethality and the temperature sensitivity are pleiotropic effects of the same mutation. Unfortunately, *dnaA* is suppressed easily at 42°C giving a relatively high frequency of revertant colonies (10^{-4} - 10^{-6}) and we were unable to find revertants which grew normally at 26°C. On the contrary, most of these revertants were even more sensitive to the low temperature than A3. However, cold resistant revertants isolated at 26°C grew normally at 42°C.

An attempt was made to separate cryolethality from thermo-sensitivity by transducing out the *dnaA* allele using an *ilv⁻* recipient and selecting for *ilv⁺* colonies. As shown in Table 4, the cotransduction of *dnaA* with *ilv* was less than 2%, but no separation of the cold sensitivity phenotype from temperature sensitivity could be found. In a complementary test, the *dnaA⁺* allele was transduced into A3. Again no separation between the two phenotypes could be detected (Table 4). Recently we have used a mutant isolated by Dr. R.A. Cooper as defective in its ability to

Table 4

Linkage Between Thermosensitivity and Cryolethality in A3

Donor	Recipient	Selected Marker	*dnaA*,Cry	*dnaA⁺*,Cry	*dnaA⁺*	*tna⁺*	*bgl⁺*	*ilv*
LE209	LE234	*ilv⁺*	4	0	285	13	154	-
A3	CO503	*dgoD⁺*	214	0	46	-	-	-
A3	CO101	*dgoD⁺*	187	0	21	-	-	-
LE236	A3	*dnaA⁺*	-	0	453	33	378	8

Cryolethality (Cry) was checked by the ability to form colonies at 26°C and nutrient broth plates. Tryptophanase activity was measured as described by Chandler and Pritchard (1975). The *bgl⁺* alleles were isolated independently and are possibly located in different genes (c.f. Bachmann et al., 1976).

ferment D-galactonic acid (*dgoD*). The mutation involved maps near
ilv (Cooper, personal communication) and was found to be
cotransducible with *dnaA* (Table 4). Transducing the *dnaA* allele
into *dgoD* strains again resulted in no separation between cryolethality and temperature sensitivity. Thus these results strongly
suggest that the two phenotypes are pleiotropic effects of the
same mutation.

Effect of Rifampicin

It has been found in our laboratory and recently reported
elsewhere (Bagdasarian et al., 1977) that some of the suppressors
of *dnaA* CRT46 are located in the gene specifying for the β subunit
of RNA polymerase. To examine the involvement of the β subunit in
the activity of the *dnaA* product we checked the sensitivity of A3
and A3$^+$ to rifampicin. Fig. 5a shows that at 37°C the ability of
A3 to form colonies in the presence of low concentrations of
rifampicin is much reduced in comparison to A3$^+$. On the other hand,
the same concentrations restore the ability of A3 to form colonies
at the low temperatures (Fig. 5b). The low concentrations of
rifampicin (0.5 - 4.0 µg/ml) used at 37°C do not affect mass
increase of A3 in liquid medium at least for several generations
but strongly decrease the rate of DNA synthesis shortly after the
addition of the drug (Orr, Unpublished results).

To eliminate the possibility that changes in the cell envelope
cause the sensitivity of A3 to rifampicin, we have tested the effect
of other drugs and detergents on the growth of A3 and A3$^+$. The
inhibitory effect of actinomycin D, nalidixic acid, chloramphenicol,
tetracycline, sodium dodecylsulphate, deoxycholic acid and sarkosyl
was no greater on A3 than on A3$^+$ at any temperature.

Effect of Chloramphenicol (CAM)

Some reports have suggested that CAM can stimulate the rate of
synthesis of DNA in various *dnaA* strains at permissive temperatures
(Blau and Mordoh, 1972; Messer et al., 1975). We have confirmed
this observation and also found that the increment in DNA is larger
with CAM at permissive temperatures than when incubation is at
42°C without CAM. We conclude from this, and the previous evidence
that the C time is normal in this strain, that the effect of CAM is
on initiation. We were unable to get the same stimulation during
amino acid starvation (Fig. 6a). To investigate the basis of this
difference we constructed a strain which carries the *relA* mutation
as well as *dnaA*. In this strain amino acid starvation and CAM both
stimulated DNA synthesis (Fig. 6b). Moreover, when CAM was added
to the stringent A3 in the absence of the required amino acids, the

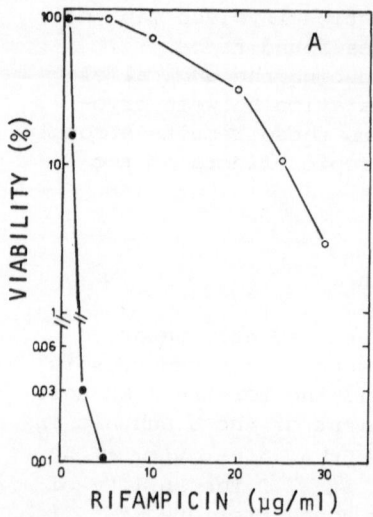 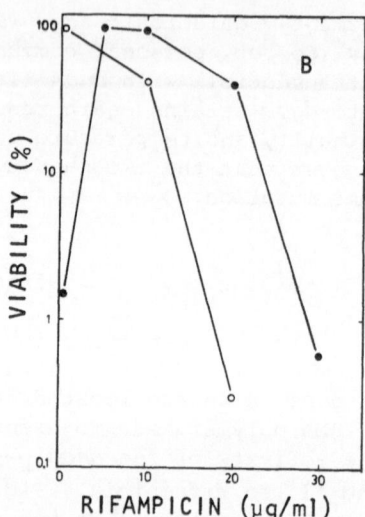

Figure 5. _Effect of rifampicin on the viability of A3 and A3$^+$ at_
37°C and 30°C. _Strains A3 and A3$^+$ were grown in M9 glucose minimal_
medium at 34°C. _Colony forming ability was checked on nutrient_
broth plates containing various concentrations of rifampicin at (a)
37°C and (b) 30°C. Percentage of viable counts for A3 at 37°C and
for A3$^+$ at both temperatures was calculated from the maximum values
obtained without rifampicin. _The values for A3 at 30°C were_
calculated from the number of colonies grown in the presence of
5 μg/ml rifampicin.   ●———● _A3;_   ○———○ _A3$^+$._

same stimulation as caused by CAM alone could be detected (data not
shown). This stimulation of synthesis could not be found in A3$^+$
and was sensitive to rifampicin. We conclude from these experiments
that the stimulation of DNA synthesis is related to the stimulation
of RNA synthesis, which also occurs under similar conditions (see
Discussion).

DISCUSSION

Data presented in this paper suggest that the frequency of
initiation of chromosome replication in the mutant strain A3 is
influenced by the rate of synthesis of an RNA species. This RNA
would be characterized by an accelerated rate of synthesis either
in the presence of CAM or during amino acid starvation of a _relA_
dnaA strain. It has previously been shown that CAM stimulates the
synthesis of several stable RNA species such as rRNA and tRNA
(Kurland and Maaløe, 1962; Shen and Bremer, 1977). The presence of
a mutation in the _relA_ gene likewise permits continued synthesis
of the same RNA species in the absence of required amino acids

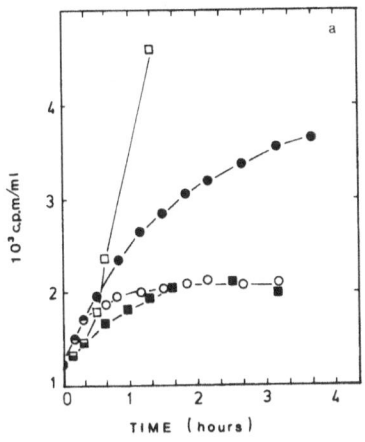
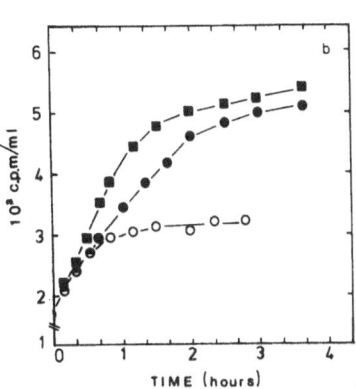

Figure 6. Effect of chloramphenicol and amino acid starvation on DNA synthesis in A3 and A3 relA. Strains A3 and LE310 were grown in glucose minimal medium (containing the appropriate requirements) with 0.5% casamino acids, 1.5 mM uridine, and 0.075 µCi/ml ³H methyl thymidine (0.05 µCi/µg. The cultures were grown at least for eight generations in the radioactive medium before the beginning of the experiment. At time 0 the cultures were filtered washed and resuspended in buffer. Aliquots were added to equal volumes of pre-warmed medium and treated as indicated: (a) A3 pre-grown at 35°C; ◯——◯, shifted to 42°C; □——□, maintained at the pre-growth temperature; ●——●, maintained at the pre-growth temperature in the presence of 150 µg/ml chloramphenicol; ■——■, maintained at the pre-growth temperature without the required amino acids and without casamino acids; (b) LE310, pre-grown at 33°C. Symbols are the same as in (a) except that the culture without treatment was omitted.

(Borek et al., 1956; Cashel, 1975). It would be interesting to know whether the I-RNA described by Messer et al. (1975) gives a similar response in *dnaA* strains to CAM and in *relA dnaA* strains to amino acid starvation.

What is the biochemical basis for the stimulation of initiation of DNA replication in the *dnaA* strain ? There are two observations which suggest that the *dnaA* product interacts directly with RNA polymerase. One is the discovery that some suppressors of *dnaA* T46 map in the β subunit of the enzyme (Bagdasarian et al. 1977; Orr, unpublished results). The other is our observation that strain A3 is sensitive to concentrations of rifampicin which at 37°C greatly reduce the rate of DNA synthesis before there is any detectable effect on RNA synthesis. The same concentrations of rifampicin have no detectable effect on DNA synthesis in A3⁺. It

is therefore attractive to consider the possibility that a *dnaA*
product - RNA polymerase complex influences the rate of synthesis
of an RNA species specifically involved in initiation. If this
complex has low activity in *dnaA* strains the effect of CAM might
be to divert RNA synthesizing capacity to this RNA, thus
stimulating DNA synthesis. CAM would not stimulate initiation
in wild type cells on this hypothesis, because the complex is not
present at a rate-limiting concentration.

Is the *dnaA* gene the "control" gene for initiation of
chromosome replication ? For example, is its product a repressor
of initiation as Messer et al. (1975) have suggested ?
Alternatively, is the *dnaA* gene product involved at a later stage
in initiation ? The fact that *dnaA* mutants accumulate capacity to
initiate rounds of replication at permissive temperatures which
is released by CAM argues for the latter alternative. It suggests
that even at permissive temperatures initiation of replication has
already been triggered in many cells but cannot be expressed. The
alternative explanation for this effect is that the *dnaA* gene
product is an unstable repressor of initiation the concentration of
which decays when its synthesis is blocked by CAM. This seems to
us unlikely for several reasons. Firstly, amino acid starvation
does not release initiation capacity in A3. Secondly, CAM does
not permit initiation in wild-type strains; consequently one
would have to assume that only the mutant repressor is unstable
and would then be driven to ask why such mutants are defective
in initiation. Thirdly, recessive initiation defective *dnaA* alleles
should not be found, but such mutants have been reported (Wechsler
and Gross, 1971; Gotfried and Wechsler, 1977) although it should
perhaps be emphasized that it has not yet been shown that all *dnaA*
mutations are located within a single gene. Finally, the delayed
inhibitory effect of *dnaA* mutations at the restrictive temperature
on the replication of other replicons (see Introduction) also
argues against the hypothesis that the *dnaA* product is a repressor.

An interpretation of the stimulation of DNA synthesis by CAM
in *dnaA* strains in terms of the hypothesis that the *dnaA* gene
product is a positive control element, also seems unlikely. This is
because such a product must be present at a rate limiting
concentration if it is to have control properties. Consequently, a
stimulation of DNA synthesis is expected in wild type cells by this
hypothesis.

Finally, the fact that plasmids (see Introduction), which do
not replicate in synchrony with chromosome initiation, appear to
require *dnaA* product for their replication also argues against the
hypothesis that *dnaA* is the gene which controls the timing of
initiation.

ACKNOWLEDGEMENTS

This work was supported by a grant to R.H.P. from the Science Research Council and a Fellowship to E.O. from the European Molecular Biology Organization.

REFERENCES

Abe, M. and Tomizawa, J. (1971) Genetics 69:1-15

Bachmann, B.J., Low, K.B. and Taylor, A.L. (1976) Bacteriol. Rev. 40:116-167

Bagdasarian, M.M., Izakowska, M. and Bagdasarian, M. (1977) J. Bacteriol. 130:577-582

Beyersmann, D., Messer, W., and Schlicht, M. (1974) J. Bacteriol. 118:783-789

Bird, R.E., Louarn, J.M., Martuscelli, J. and Caro, L. (1972) J. Mol. Biol. 70:549-566

Bird, R.E., Chandler, M. and Caro, L. (1976) J. Bacteriol. 126: 1215-1223

Blau, S., and Mordoh, J. (1972) Proc. Nat. Acad. Sci. U.S.A. 69: 2896-2898

Borek, W., Rockenbach, J. and Ryan, A. (1956) J. Bacteriol. 71: 318-323

Brunt, J.V., Waggoner, B.T. and Pato, M.L. (1977) Molec. gen. Genet. 150:285-292

Carl, P.L. (1970) Molec. gen. Genet. 109:107-122

Cashel, M. (1975) Ann. Rev. Microbiol. 29:301-318

Chandler, M.G. and Pritchard, R.H. (1975) Molec. gen. Genet. 138:127-141

Collins, J., Williams, P. and Helsinki, D.R. (1975) Molec. gen. Genet. 136:273-289

Derstine, P.L. and Dumas, L.B. (1976) J. Bacteriol. 128:801-809

Goebel, W. (1973) Biochem. Biophys. Res. Commun. 51:1000-1007

Gotfried,F. and Wechsler, J.A. (1977) J. Bacteriol. 130:963-964

Hanna, M.H. and Carl, P.L. (1975) J. Bacteriol. 121:219-226

Hayes, W. (1968) The Genetics of Bacteria and their Viruses. 2nd
    Edn. Blackwell Scientific Publication, Oxford & Edinburgh

Helmstetter, C., Cooper, S., Pierucci, O. and Revelas, E. (1968)
    Cold Spring Harbor Symp. Quant. Biol. 33:809-822

Hiraga, S. and Saitoh, T. (1975) Molec. gen. Genet. 137:239:248

Hirota, Y., Ryter, A. and Jacob, F. (1968) Cold Spring Harbor
    Symp. Quant. Biol. 33:677-693

Kohiyama, M., Cousin, D., Ryter, A. and Jacob, F. (1966) Ann.
    Inst. Pasteur, 110:465-486

Kurland, C.G. and Maaløe, O.(1962) J. Mol. Biol. 4:193-210

Masters, M. and Broda, P. (1971) Nature (London) New Biol.
    232:137-140

Mayer, H., Luibrand, G. and Goebel, W. (1977) Molec. gen. Genet.
    152:145-152

Messer, W., Dankworth, L., Tippe-Schnidler, R., Womack, J.E. and
    Zahn, G. (1975)   In: M. Goulian and P. Hanawalt
    (eds.) DNA SYnthesis and Its Regulation. W.A. Benjamin,
    Menlow Park, Calif. pp602-617

Nishimura, Y.,   Caro, L., Berg, C. and Hirota, Y. (1971) J. Mol.
    Biol. 55:441-456

Pritchard, R.H., Barth, P.T. and Collins, J. (1969) Symp. Soc.
    Gen. Microbiol. 19:263-297

Pritchard, R.H. and Zaritsky, A. (1970) Nature 226:126-131

Pritchard, R.H. (1974) Phil. Trans. Roy. Soc. B, 267:303-336

Saitoh, T. and Hiraga, S. (1975) Molec. gen. Genet. 137:249-261

Shen, V. and Bremer, H. (1977) J. Bacteriol. 130:1098-1108

Tresguerres, E.F., Nandadasa, H.G. and Pritchard, R.H. (1975)
    J. Bacteriol. 121:554-561

Wechsler, J.A. and Gross, J.D. (1971) Molec. gen. Genet. 113:
    273-284

Wechsler, J.A. and Zdzienicka, M. (1975)   In M. Goulian and
    P. Hanawalt (eds.) DNA Synthesis and Its Regulation. W.A.
    Benjamin, Menlo Park, Calif. pp 624-639

Wehr, C.T., Waskell, L. and Glaser, D.A. (1975) J. Bacteriol.
    121:99-107

Zahn, G., Tippe-Schindler, R. and Messer, W. (1977) Molec. gen.
    Genet. 153:45-49

Zeuthen, J. and Pato, M.L. (1971) Molec. gen. Genet. 3:242-255

THE FUNCTION OF RNA POLYMERASE AND *dnaA* IN THE INTITIATION OF CHROMOSOME REPLICATION IN *Escherichia coli* AND *Salmonella typhimurium*

Miroslawa M. Bagdasarian, Maryla Izakowska, Renata Natorff and Michael Bagdasarian

Department of Biochemistry
Warsaw Medical School and Institute of Biochemistry and Biophysics
Polish Academy of Sciences, Warsaw, Poland

The replication of bacterial chromosome is regulated at the step of initiation (Maaløe and Hanawalt, 1961; Maaløe and Kjeldgaard, 1966; Helmstetter, Cooper, Pierucci and Revelas, 1968). Initiation occurs at a unique site on the chromosome termed the "replicator" in the original replicon hypothesis of Jacob, Brenner and Cuzin (1963). At present it is most frequently referred to as the "chromosomal origin" or the "origin of replication" (Louarn, Funderburgh and Bird, 1974; Bird, Chandler and Caro, 1976; Fujisawa and Eisenstark, 1973; Gyurasits and Wake, 1973).

Considerable evidence has accumulated so far suggesting that replication is initiated as the synthesis of a short polyribonucleotide fragment which serves as primer for the initiation and elongation of the DNA chain (Lark, 1972; Messer, 1972; Messer et al., 1975).

Conditional lethal mutations in several loci affect the initiation of replication in *Escherichia coli*. Temperature-sensitive mutations in *dnaA* and *dnaC* (Hirota, Mordoh and Jacob, 1970), *dnaI* (Beyersman, Messer and Schlicht, 1974), *dnaP* (Wada and Yura, 1974) and *dnaB252* (Zyskind and Smith, 1977) block the initiation of a new round of chromosome replication at elevated temperatures. Mutations in the *dnaA* locus have also been studied in *Salmonella typhimurium* (Bagdasarian et al., 1975).

The multiplicity of genetic loci which provide functions essential for initiation suggests that, during the normal initiation process, the interaction of several factors must take place at the chromosomal origin.

101

It has been suggested that the product of *dnaA* gene was involved in the early stages of initiation (Messer et al., 1975; Zyskind et al., 1977). It is possible therefore that the function of this product might be directly involved in the regulation of the synthesis of the origin-RNA (o-RNA).

Functional Relation Between the RNA Polymerase and the *dnaA* Gene Product

The sensitivity of the initiation of chromosome replication to rifampicin suggests that RNA polymerase is responsible for the synthesis of o-RNA in bacteria (Lark, 1972; Messer, 1972; Laurent, 1973; Saitoh and Hiraga, 1975), bacterial viruses (Wickner et al., 1972; Hayes and Szybalski, 1975) and plasmids (Helinski et al., 1975; Sakakibara and Tomizawa, 1974).

Examination of the effect of mutations in the *rpoB* cistron - the structural gene for the beta subunit of RNA polymerase - on the phenotypic expression of *dnaA* mutations resulted in the recognition of a specific class of *rpoB* mutations. These mutations were able to suppress the thermosensitivity of chromosome initiation in *dnaA* mutants (Bagdasarian et al., 1977). As a result of this suppression the double mutants: *dnaA rpoB* could form colonies at 40-42° as opposed to non-suppressed *dnaA* mutants, which do not grow on agar plates at elevated temperatures. The results presented in Table 1 show that mutations *rpoB3003, 3004* and *3006* suppress the thermosensitivity of three *dnaA* mutants while mutations *rpoB3007* and *3008* have no effect on thermosensitivity of the *dnaA* mutants tested. These data also indicate that the suppressors of *dnaA* present in the strains *rpoB3003, 3004* and *3006* map in the *rpoB* locus or very close to it. The finding that the suppressive mutants *rpoB* isolated in *Salmonella typhimurium* contain RNA polymerase resistant to rifampicin *in vitro* confirmed the above suggestion that the phenomenon of suppression was indeed due to the structural alteration of RNA polymerase (Bagdasarian et al., 1977). Further support to this notion was obtained by analyzing the *rpoB-3012* mutation in *E.coli*. This mutation exhibited the property of suppressing the *dnaA46* mutation, but only in the presence of rifampicin (25 µg per ml., Table 2).

The fact that an alteration of RNA polymerase can suppress the effect of mutations in *dnaA* was interpreted as an indication that *dnaA* product might be directly involved in the RNA polymerase - catalyzed reaction of the synthesis of o-RNA. In this case a direct recognition of the wild type *dnaA* gene product by the wild type RNA polymerase was assumed to take place during normal initiation of a round of replication. The defect of initiation in *dnaA* mutants could then be explained as the result of inability to

Table 1

The effect of *rpoB* mutations on the
thermosensitivity of *dnaA* strains

Pertinent Genotype			Number of Recombinants				
Donor[*]	Recipient		$argE^+$	*rpoB* tr	$rpoB^+$ ts	*rpoB* ts	$rpoB^+$ tr
rpoB3003	*argE*	*dnaA46*	100	17	79	2	0
	argE · *dnaA177*		100	40	60	0	0
	argE	*dnaA47*	100	39	61	0	0
rpoB3004	*argE*	*dnaA46*	60	22	38	0	0
	argE	*dnaA177*	100	48	52	0	0
rpoB3006	*argE*	*dnaA46*	100	32	68	0	0
	argE	*dnaA177*	100	36	61	3	0
	argE	*dnaA47*	100	41	59	0	0
rpoB3007	*argE*	*dnaA46*	50	0	27	23	0
rpoB3008	*argE*	*dnaA46*	50	0	34	16	0

[*]Pl vir[s] grown on the *rpoB* mutant indicated
was used as donor.

Isogenic strains carrying different *dnaA*
alleles indicated and the same *argE* allele
were used as recipients.

Table 2

The effect of rifampicin on the suppression
of *dnaA46* by *rpoB3012*

Recipient [*] Pertinent genotype	Conditions of testing	Number of Recombinants		
		$argE^+$	*rpoB* tr	*rpoB* ts
argE dnaA46	Nutrient Broth	50	0	21
	+ rifampicin/ 25 µg/ml	50	21	0

[*] Pl vir^s grown on the strain rpoB3012 was used as donor.

recognize the defective *dnaA* product by the normal RNA polymerase.
The suppressive *rpoB* mutants would be those mutants in which the
mutational alteration of RNA polymerase makes the recognition of
an altered *dnaA* product possible. This model predicts the
possibility of finding yet another class of *rpoB* mutations - a class
which, contrary to the positive complementation described above,
would exhibit negative complementation with the defective *dnaA*
product and would therefore have a lethal effect when introduced
into *dnaA* mutants. This type of *rpoB* mutations was sought out and
found among several rifampicin-resistant *E.coli* mutants isolated in
a *dnaA*$^+$ strain. This mutation - *rpoB30013* - does not visibly affect
growth of a *dnaA*$^+$ strain and gives the expected cotransduction
frequency with the *argE* locus but gives virtually no *rpoB* transductants with an isogenic *dnaA argE* strain used as recipient
(Table 3). These results can only be explained by assuming that
the mutation *rpoB3013* cannot coexist with either *dnaA46* or *dnaA177*
even at low temperatures.

The Specificity of Suppression of Different *dnaA*
alleles by *rpoB* mutations

If the suppression of *dnaA* phenotype by *rpoB* mutations,
described in the preceding section, were the result of a positive
complementation in a system of protein-protein interaction of the
dnaA gene product with RNA polymerase similar to the system
described for the interaction of bacteriophage T4 gene products and
some host proteins (Takahashi et al., 1975; Coppo et al., 1975a, b),

Table 3

Cotransduction frequency of the *rpoB3013* mutation
with *argE* in a *dnaA* and *dnaA*[+] background

| Recipient[*] | Number of Recombinants | | |
Pertinent Genotype	$argE^+$	$rpoB^-$ tr	$rpoB^-$ ts
argE dnaA46	400	O	O
argE dnaA177	100	O	1
argE dnaA[+]	80	53	O

[*] Pl vir[s] grown on the strain *rpoB3013* was used as
donor.

it might be expected that a strict if not absolute specificity
should exist between the different alleles of the suppressor
(*rpoB*) and of the suppressed mutation (*dnaA*). However, as may be
observed from the result of Table 1, each of the three different
independently isolated *rpoB* mutations, which suppressed *dnaA46*
suppressed also two other *dnaA* mutations – *dnaA47* and *177*.

It may be argued that most of the independently isolated *dnaA*
mutations, and in particular those studied in the present work, are
similar in the sense that they result in the similar alteration
of the *dnaA* gene product. Therefore an alteration of RNA polymerase
which complements one of the mutated *dnaA* alleles would also
complement the others. There is however no indication at present
about the type of defect that results in the dnaA phenotype. Moreover the high frequency with which the suppressive *rpoB* mutants
appear as spontaneous mutants resistant to rifampicin even in *dnaA*[+]
strains – about 10% in *E.coli* (Bagdasarian et al., 1977) – suggests
that the alteration of RNA polymerase might suppress the initiation
defect in *dnaA* by mechanisms other than a direct recognition of an
altered *dnaA* gene product.

The effect of *rpoB* mutations on polarity

One of the classes of *rpoB* mutants isolated by Hong, Smith and Ames (1971) in *S.typhimurium* exhibited an unusually low frequency of lysogenization. Wild type P22 phage formed clear plaques on these mutants compared to turbid plaques formed on wild type *S.typhimurium*. Similar phenomenon has been observed in *E.coli* strains carrying mutations in the transcription termination factor Rho. These mutants were also affected in their ability to form stable lysogens of temperate phages λ and P1 (Das, Court and Adhya, 1976). It is possible that *rpoB* mutations described by Hong et al. (1971) affected the RNA polymerase in such a way as to make it relatively insensitive to the action of the termination factor Rho.

Mutants *rpoB* which suppressed the *dnaA* phenotype in *S.typhimurium* also gave clear plaques when used as hosts for the wild type P22 phage. Their effect on polarity caused by an amber mutation in the histidine operon was therefore determined. Mutations *rpoB705* and *706*, which suppressed the thermosensitivity of the *dnaA1* mutation were introduced by transduction into a strain carrying a histidine regulatory mutation *hisT1504* and an amber mutation *hisC3216*. The polar effect of this mutation lowers the expression of all genes distal to *hisC* including *hisB* - the gene coding for the enzyme histidinol phosphate phosphohydrolase - 7 to 10 times, reducing its specific activity to about the level found in wild type strains. As shown in Table 4 the presence of the *rpoB705* mutation in the *hisC* strain increased the expression of the *hisB* gene by a factor of 2 and the *rpoB706* by about 1.6. In contrast to this three other independent *rpoB* mutations, *rpoB711*, *712* and *713* isolated in the strain *hisT1504 hisC2316* showed no increase in the activity of histidinol phosphate phosphohydrolase, i.e., no change in the polarity caused by the *hisC* amber mutation. It was found that mutations *rpoB711*, *712* and *713* could not be introduced into the strain *dnaA1 thi-502* by cotransduction with the *thi* locus. It seems therefore that these mutations are of the same type as the mutation *rpoB3013* isolated in *E.coli* and described in Table 3, i.e, they cannot coexist with the *dnaA1* mutation. These results indicate that at least two distinct classes of *rpoB* mutations exist - mutations which affect polarity exerted by amber mutations and mutations which do not change the extent of polarity. It may be assumed that mutations which relive polarity also suppress the chromosome initiation defect in *dnaA* mutants. This suggests that the *dnaA* gene product functions in some early events of the initiation of chromosome replication which involves transcription prior to the DNA chain initiation. It is possible that the *dnaA* gene product regulates the action of RNA polymerase in the synthesis of the o-RNA primer for the chromosome initiation.

Table 4

Effect of *rpoB* mutations on the activity of histidinol
phosphate phosphohydrolase in *hisC* amber strains
of *S.typhimurium*

Strain	Enzyme Activity[*]
hisT1504 hisC2316	1.0
hisT1504 his[+]	7.2
hisT1504 hisC2316 rpoB705	2.1 ± 0.15
hisT1504 hisC2316 rpoB706	1.6 ± 0.09
hisT1504 hisC2316 rpoB711	0.8
hisT1504 hisC2316 rpoB712	0.9
hisT1504 hisC2316 rpoB713	0.8

[*]Histidinol phosphate phosphohydrolase was assayed
in toluenized cells and specific activity of the
enzyme calculated as described by Ames et al. (1963)
as modified by Roth and Ames (1966). Values
represent the enzyme specific activity as the multiplicity of the activity found in the isogenic polar
strain *hisT1504 hisC2316*.

DISCUSSION

The results of the present work confirm earlier suggestions
(Zyskind et al., 1977; Bagdasarian et al., 1977) that the function
of *dnaA* is involved in an early step of initiation of chromosome
replication which requires the action of the rifampicin-sensitive
RNA polymerase. It is possible to explain these results by assuming
several models of the initiation process:

(i) A direct interaction of RNA polymerase with the *dnaA* gene
product is necessary for the normal initiation. Mutations in the
dnaA gene alter its product in such a way that it is not recognized
by the RNA polymerase at elevated temperatures. Mutations may
alter RNA polymerase so that it recognizes the altered *dnaA* product.
This model can explain well the existence of the class of *rpoB*
mutations which suppress *dnaA*, the property of the *rpoB3012*
mutation - which suppresses *dnaA* only in the presence of rifampicin
and the apparent negative complementation of the mutation *rpoB3013*

with *dnaA46* in *E.coli* and *rpoB711, 712* and *713* with *dnaA1* in
S.typhimurium.   It is difficult however to explain by the first
model the non-specificity of suppression in respect to the
suppressor allele and the suppressed allele (Table 1). The only
way to reconcile these findings with the proposed model would be
to assume that several existing *dnaA* mutations, and particularly
those studied in the present work, result in a similar alteration
of the *dnaA* product. In view of the fact that this gene codes for
a function of primary importance to the cell life this supposition
does not seem improbable, there is however no evidence at present
to support this suggestion.

        (ii) Mutations in *dnaA* are leaky mutations which at elevated
temperatures give an extremely low frequency of initiation.
Mutations in the *rpoB* cistron might be supposed to increase the
frequency of initiation through the altered affinity for the
promoter site of the o-RNA. Alterations of promoter or terminator
recognition by RNA polymerase have been implicated as the cause of
the altered patterns of viral proteins synthesized in *rpoB* mutants
after infection with ØX174 or S13 phages (Tessman and Peterson,
1976). Model (ii) explains all of the observations presented in this
work provided that the relief of polarity caused by *rpoB* mutations
(Table 4) is interpreted as the increase in the frequency of
recognition of the main histidine operon promoter or of the weak
promoter that is presumed to be located at the start of the *hisB*
gene (Ely and Cieśla, 1974). A prediction resulting from the model
implies that strains harbouring the *rpoB* mutations which suppress
dnaA might initiate a round of DNA replication more frequently (at
lower cell mass) than wild type strains. No indication of this being
so was obtained however. Determination of the amount of residual
DNA synthesized after the inhibition of protein synthesis in several
rpoB mutants indicated that there was an equal number of replication
forks in *rpoB* and wild type strains.

        (iii) The process of the synthesis of o-RNA may be assumed
to contain regulatory elements similar to those found in tryptophan
(Lee et al., 1976; Pouwels and Pannekoek, 1976) and histidine
operon (Kasai, 1974). This model would envisage the chromosomal
origin as having a leader sequence followed by a termination signal,
at which transcription terminates with the aid of the Rho factor, or
perhaps a specific termination factor. Further transcription of
the origin would be possible through the action of a specific
antiterminator. The *dnaA* gene product is a good candidate for this
hypothetical antiterminator. Model (iii) includes both negative
and positive elements of regulation and may explain some apparently
contradictory results obtained in the studies of chromosome
initiation as, e.g., the requirement of *de novo* protein synthesis
for reinitiation yet stimulation of initiation by low doses of
chloramphenicol (Blau and Mordoh, 1972) which, according to the
model (iii), might be attributed to the partial inhibition of the

synthesis of proteins involved in the negative control.

This last model suggests several predictions most of which may be tested experimentally. Thus a class of mutations in the *rho* locus which alter the transcription termination factor might exist that would affect chromosome initiation. Mutations *dnaP* (Wada and Yura, 1974) and some suppressors of *dnaA* which map close to *ilv* locus (Wechsler and Zdzienicka, 1974) are likely candidates.

It is clear from the results of the present study that several specific classes of *rpoB* mutations may be recognized on the basis of their effect on the properties of *dnaA* mutants. It seems obvious that a functional relation exists between the *dnaA* gene product and the RNA polymerase. Further experiments are required to establish whether the alteration of RNA polymerase changes the frequency of initiation of transcription, the termination at the presumed attenuator site or at the terminus of the o-RNA. Studies of the suppression of the DNA replication defects may provide valuable data for the understanding of the mechanism of initiation and of the role played by the *dnaA* gene and RNA polymerase.

ACKNOWLEDGEMENT

This work was supported by the Polish Academy of Sciences, project No. 09.7, and by the U.S. Public Health Service Grant, No. O5-O89 N.

REFERENCES

Ames, B.N., P.E. Hartman and F. Jacob (1963) J. Mol. Biol. 7:23-42

Bagdasarian, M.M., M. Hryniewicz, M. Zdzienicka and M. Bagdasarian (1975) Molec. Gen. Genet. 139:213-231

Bagdasarian, M.M.,M. Izakowska and M. Bagdasarian (1977) J.Bacteriol. 130:577-582

Beyersmann, D., W. Messer and M. Schlicht (1974) J. Bacteriol. 118:783-789

Bird, R., M. Chandler and L. Caro (1976) J. Bacteriol. 126:1215-1223

Coppo, A., A. Manzi, J.F. Pulitzer and H. Takahashi (1975) J. Mol. Biol. 96:579-600

Coppo, A., A. Manzi, J.F. Pulitzer and H. Takahashi (1975) J. Mol. Biol. 96:601-624

Ely, B. and Z. Cieśla (1974)   J. Bacteriol. 120:984-986

Fujisawa, T. and E. Eisenstark (1973)   J. Bacteriol. 115:168-176

Gyurasits, E.B. and R.G. Wake (1973)   J. Mol. Biol. 3:156-165

Hayes, S. and W. Szybalski (1975) In: DNA Synthesis and Its
    Regulation. (eds. M. Goulian, P. Hanawalt and C.F. Fox) ICN-
    UCLA Symposium on Molecular and Cellular Biology, Vol. 3, W.A.
    Benjamin, Menlo Park, Calif. pp486-512

Helinski, R.D., M.A. Lovett, P.H. Williams, L. Katz, J. Collins,
    Y. Kupersztoch-Portnoy, S. Sato, R.W. Leavitt, R. Sparks,
    V. Hershfield, D. Guiney and D.G. Blair (1975) In: DNA
    Synthesis and Its Regulation. (eds. M. Goulian, P. Hanawalt,
    and C.F. Fox). ICN-UCLA Symposium on Molecular and Cellular
    Biology, Vol. 3, W.A. Benjamin, Menlow Park, Calif. pp514-536

Helmstetter, S., S. Cooper, P. Pierucci and R. Revelas (1968) Cold
    Spring Harbor Symp. Quant. Biol. 33:809-822

Hirota, Y., J. Mordoh and F. Jacob (1970) J. Mol. Biol. 53:369-
    387

Hong, J.S., G.R. Smith and B.N. Ames (1971) Proc. Nat. Acad. Sci.
    USA 68:2258-2262

Jacob, F., S. Brenner and F. Cuzin (1963)   Cold Spring Harbor
    Symp. Quant. Biol. 28:329-348

Kasai, T. (1974) Nature 249:523-527

Lark, K.G. (1972) J. Mol. Biol. 64:47-60

Laurent, S. (1973) J. Bacteriol. 116:141-145

Lee, F., C.L. Squires, C. Squires and C. Yanofsky (1976) J. Mol.
    Biol. 103:383-393

Louarn, J., M. Funderburgh and R. Bird (1974) J. Bacteriol. 120:
    1-5

Maaløe, O. and P.C. Hanawalt (1961) J. Mol. Biol. 3:144-155

Maaløe, O. and N.O. Kjeldgaard (1966) Control of Macromolecular
    Synthesis, W.A. Benjamin, New York pp154-187

Messer, W. (1972) J. Bacteriol. 112:7-12

Messer, W., L. Dankworth, R. Tippe-Schindler, J.E. Womack and
    G. Zahn (1975) In: DNA Synthesis and Its Regulation. (eds.
    M. Goulian, P. Hanawalt and C.F. Fox) ICN-UCLA Symposium on
    Molecular and Cellular Biology, Vol. 3 W.A. Benjamin, Menlo
    Park, Calif. pp602-617

Pouwels, P.H. and H. Pannekoek (1976) Molec. Gen. Genet. 149:
    255-265

Roth, J.R. and B.N. Ames (1966) J. Mol. Biol. 22:325-334

Sakakibara, Y. and J. Tomizawa (1974) Proc. Nat. Acad. Sci. USA
    71:802-806

Takahashi, H., A. Coppo, A. Manzi, G. Martire and J.F. Pulitzer
    (1975) J. Mol. Biol. 96:563-578

Tessman, E.S. and P.K. Peterson (1976) J. Bacteriol. 128:264-270

Wada, C. and T. Yura (1974) Genetics 77:199-220

Wechsler, J. and J.D. Gross (1971) Molec. Gen. Genet. 113:273-284

Wechsler, J.A. and M. Zdzienicka (1975) In: DNA Synthesis and Its
    Regulation (eds. M. Goulian, P. Hanawalt and C.F. Fox) ICN-
    UCLA Symposium on Molecular and Cellular Biology, Vol. 3,
    W.A. Benjamin, Menlo Park, Calif., pp624-639

Zyskind, J., L.T. Deen and D.W. Smith (1977) J. Bacteriol. 129:
    1466-1475

STABILIZED INITIATION ACTIVITY OF DNA REPLICATION IN AN Sdr^c

MUTANT OF *ESCHERICHIA COLI*

T. Kogoma*, M.J. Cunnaughton and B.A. Alizadeh

Department of Biology
The University of New Mexico
Albuquerque, New Mexico 87131, U.S.A.

INTRODUCTION

The fact that the initiation of each new round of chromosome replication in *E.coli* requires the *de novo* synthesis of protein and RNA implies that the initiation mechanism is unstable and loses activity after having been used only once. Although this is implicated in the replicon model by Jacob, Brenner and Cuzin (1963), it seemed to have escaped serious attention until several years ago when we demonstrated that under certain physiological conditions such as thymine starvation a stable replication mechanism can be synthesized and successive initiations can occur despite the absence of protein synthesis (Kogoma and Lark, 1969; 1970). This type of DNA replication in the absence of protein synthesis has been termed stable DNA replication (SDR) (Kogoma and Lark, 1975).

A similar observation was made independently by Rosenberg, Cavalieri and Unger (1969). Based on this observation, they formulated a model in which the synthesis of a repressor which occurs shortly after initiation of a round of replication prevents further initiation. After a period of thymine starvation, during which initiation activity accumulates, the addition of chloramphenicol prevents the synthesis of the repressor, resulting in continued initiations in the absence of protein synthesis (Rosenberg et al., 1969). This interpretation appears to be in conflict with the observation that after thymine starvation a period of growth in a complete medium which would have allowed the synthesis of the repressor did not abolish the ability of the cells to subsequently synthesize DNA in the presence of chloramphenicol (Kogoma and Lark, 1970).

*A.S.I. *Participant*

We have proposed an alternative hypothesis that during
thymine starvation an altered DNA replication complex is formed
which fails to destroy itself at the end of a round of replication
and therefore continues to initiate in the absence of further
protein synthesis (Kogoma and Lark, 1970; 1975). The hypothesis
assumes the existence of a system which, as part of the regulatory
mechanism of DNA replication, deliberately inactivates the
initiation ability of the cell.

To further explore the possibility of the existence of such a
regulatory system, we decided to attempt to isolate mutants which
were capable of DNA replication in the absence of protein synthesis.
Since a selective phenotype of such mutants was not (and still is not)
known, an autoradiographic technique (Wechsler, et al., 1973) was
adopted which allowed us to examine as many as 10^6 colonies at a
time. Several such mutants have been isolated (Kogoma, 1976a) and
a typical mutant has been characterized to some extent (Kogoma,
1977). Because the mutants appear to be capable of SDR without
inductive treatments, they are termed constitutive stable DNA
replication (Sdr^c) mutants. This paper summarizes the characteristics
of a particular Sdr^c mutant, describes the biological properties
of the DNA synthesized by the mutant in the absence of protein
synthesis, and discusses the implication of the findings in terms of
the regulation of chromosome replication in $E.coli$.

The materials and methods used are described elsewhere (Kogoma,
1977).

RESULTS

A.   Characteristics of Sdr^c Mutant X2

Mutant X2 was isolated after mutagenesis with N-methyl-N'-nitro-
N-nitrosoguanidine from strain AQ2, an $r_1 5 m_1 5$ derivative of 15T⁻
(555-7). The following is a summary of the characteristics of the
mutant. The details of the experiments and results are described
elsewhere (Kogoma, 1977)..

1.   The Sdr^c mutant appears to grow normally at $37^\circ C$ with a
     slightly slower growth rate than that of the parent: 1.22
     doublings/hr for X2; 1.36 doublings/hr for AQ2 in M9 salt-
     glucose medium.

2.   The mutant can continue DNA synthesis for more than 10 hours
     in concentrations of chloramphenicol (CM) that completely
     inhibit protein synthesis, resulting in the accumulation of
     DNA in the non-dividing cells.

3.   The continued DNA synthesis in the presence of CM represents

semiconservative replication of *E.coli* chromosomal DNA.

4. The replication occurs linearly at a reduced rate (40% of
    the initial rate) after 60 minutes in CM. This distinct
    slowdown in DNA replication (observed in all the Sdr^c
    mutants thus far isolated) is due to a reduced rate of DNA
    synthesis in all the cells in the population.

5. The replication also occurs in the absence of essential
    amino acids, but not in the presence of rifampin. Therefore,
    untranslated RNA synthesis appears to be required.

6. Constitutive SDR appears to require the *dnaA* and *dnaC*
    gene products.

7. The *Sdr^c* mutation has been mapped near the *pro-lac* region of
    the *E.coli* chromosome. The mutation is recessive.

8. The possibility of multi-fork replication during exponential
    growth or after the addition of CM has been ruled out.

B. Biological Properties of SDR DNA

To further assess the similarity or dissimilarity between the
systems of constitutive SDR and normal replication, we have also
examined the DNA (henceforth SDR DNA) synthesized in the absence
of protein synthesis.

First, SDR DNA was examined for the template activity upon
replication (Table 1). In this experiment, two cultures of the
mutant were labelled with (^{14}C)thymine during exponential growth
and during amino acid starvation. (^{3}H)thymine was also present only
during the exponential growth in one culture (A) and only during
the latter two-thirds of the amino acid starvation period in the
other (B). ^{14}C-labelled DNA represents the total DNA, (^{3}H)DNA
in culture A normal DNA, and (^{3}H)DNA in culture B SDR DNA. At the
end of the starvation, cells which had accumulated 2.5 times as much
DNA as normal cells were allowed to synthesize DNA in complete
medium for 90 minutes and, then for 30 minutes in medium containing
5-bromouracil (5BU) in place of thymine. DNA was isolated and
analyzed by isopycnic centrifugation in CsCl. Table 1 summarizes
the extent of conversion of radioactivities into hybrid DNA which
measures the amount of labelled DNA replicated during the 30 minutes
incubation in 5BU. The extent of the replication of ^{14}C-labelled
DNA was comparable in both cultures. On the other hand, 6% of
(^{3}H)DNA in culture B was also replicated whereas about 12% of (^{3}H)
DNA in culture A was converted into hybrid. The results indicate
that SDR DNA can be used as a template for replication although the

Table 1

Replication of Normal DNA and SDR DNA During a Period
of Growth After Amino Acid Starvation

Culture	% radioactivity in hybrid DNA	
	^{14}C	^{3}H
A	9.0	12.3
B	9.8	6.0

Two cultures of mutant X2 were labelled with (^{14}C)
thymine (5 µc/4 µg/ml) for 3 generations during
exponential growth and then for 5 hours in the absence
of essential amino acids. Culture A was also labelled
with (^3H) thymine (10 µc/4 µg/ml) only during the
exponential growth and culture B with (^3H) thymine (30
µc/4 µg/ml) only during the latter two-thirds of the
starvation period. After 90 minutes of synthesis and
growth following the starvation period, the cultures
were incubated for 30 minutes in medium containing 5BU
in place of thymine. DNA was isolated and analyzed by
isopycnic centrifugation in CsCl. Total radioactivities:
^{14}C = 29,698 cpm, ^3H = 40,433 cpm in culture A and ^{14}C =
41,648 cpm, ^3H = 185,733 cpm in culture B.

efficiency of the utilization is half as much as that of normal DNA.

To test whether SDR DNA segregates into progeny cells when
cell division is allowed upon restoration of amino acids to a
starved culture, the distribution of radioactive SDR DNA among
progeny cells was examined by autoradiography (Table 2). The mutant
and parental cells were labelled with (^3H) thymine during a 5 hour
amino acid starvation period. After the labelling, cells were
allowed to divide in non-radioactive complete medium. At every
doubling, samples were taken and fixed with 4% formalin, and autoradiograms made. The average grain number per cell among labelled
cells at 0-time (i.e., at the end of the labelling period) was
6.26 grains/mutant cell and 2.99 grains/parental cell. The mutant
cell had twice as much labelled DNA as the parent. If this extra
SDR DNA in the mutant cell segregates randomly into progeny upon
cell division, the appearance of unlabelled cells among the progeny
is expected to be delayed relative to that of the parent culture.
To test this, the percent of unlabelled cells in the populations

Table 2

Distribution of Radioactive SDR DNA in Daughter Cells
During Growth After Amino Acid Starvation

Doublings	% unlabelled cells		net increase of unlabeled cells (%)	
	X2	AQ2	X2	AQ2
0	6.0	15.1	---	---
1	26.5	36.0	20.5	20.9
2	42.4	64.0	36.4	48.9
3	50.0	---	44.0	---

Cultures of mutant X2 and parental strain AQ2 were
labelled with (^{3}H)thymine   (30 µc/4 µg/ml) during
a 5-hour amino acid starvation. Autoradiograms were
prepared according to C. Lark (1968). After 67 hours of
exposure, grains were counted and analyzed (see the
text).

was estimated using the Poisson distribution analysis (Hanawalt
et al., 1961; Lark, 1966) as shown in Table 2. The table also
includes net increases of unlabelled cells in the growing cultures.
The results clearly indicate a slight but significant delay in the
appearance of unlabelled cells in the X2 culture and suggests that
SDR DNA partially segregates into daughter cells.

This partial segregation of SDR DNA was also demonstrated in
microcolony autoradiograms. Unevenly distributed grains were
observed over several cells in a microcolony that had derived from
a single mutant cell containing radioactive SDR DNA (data not
shown).

Although SDR DNA is utilized as a template for DNA replication
and partially segregates into daughter cells upon cell division, the
DNA appears to be in such a state that it is not readily utilized
for cell division if further DNA synthesis is inhibited upon
restoration of amino acids to a starved culture. Thus, a mutant
culture starved for amino acids for several hours failed to resume
cell division upon restoration of amino acids in the presence of
nalidixic acid or exhibited little cell division in the absence of
thymine (Kogoma et al., 1977).

DISCUSSION

Although the characterization of the mode and product (DNA) of constitutive stable DNA replication has not been completed, we tentatively conclude that SDR in mutant X2 and, perhaps, other Sdr^C mutants resembles normal replication in many respects. In the discussion that follows, we assume that the replication apparatus in the mutant cell utilizes most, if not all, of the components in normal DNA replication and that the *sdr* mutation is solely responsible for the ability of the mutant cell to continue replication in the absence of protein synthesis.

Several explanations are possible for the SDR ability of the mutant. Trivial reasons such as the possibilities of the induction of SDR due to unbalanced growth and the accumulation of an excess amount of initiation capability before the addition of CM have been discussed (Kogoma, 1977). We consider these possibilities unlikely.

The mutant may fail to terminate a round of replication at the terminus and continue the replication. The mutation may occur within the termination sequence (Gefter, 1975), and the altered signal fails to be recognized by a responsible factor(s). Since the mutation maps in a different region from the terminus and is recessive, the possibility of mutant X2 having an altered termination sequence is excluded.

Another possibility is that the recognizing factor could be altered. This type of termination-defective replication is expected to be independent of the initiation function. Since the SDR in mutant X2 still requires the synthesis of untranslated RNA as well as the *dna*A and *dna*C gene products, it is unlikely that the mutant synthesizes an altered recognition factor.

Alternatively, the mutant may synthesize an inactive repressor which fails to prevent re-initiation of DNA replication. Such a mutation would give rise to multi-fork replication in the cell. However, such multi-fork replication was not detected in the mutant cell either during exponential growth or after the addition of CM.

Mutant X2 is more likely to be a regulatory mutant in which the initiation activity is somehow stabilized. This model assumes that the limited initiation in normal replication cycle is due to the instability of some of the initiation factors and that the instability is brought about by the deliberate inactivation of the factors. The mutant may be defective in this inactivation activity. Since it is known that the replicative origin becomes potentially available for re-initiation soon after the replication apparatus moves away from it (Worcel, 1970), there must occur some kind of masking of the initiation activity to prevent premature initiation.

Or, the initiation factors can be carried away from the origin as part of the replication complex (Kogoma and Lark, 1975; Kogoma, 1976b). In normal replication, the inactivation must occur sometime after the initiation and before the end of the round of replication. In the mutant, the protein factors seem to escape from the action of the altered inactivation mechanism and can be re-used for successive initiations. Whereas this particular mutant still requires RNA synthesis, we have isolated another group of Sdr^c mutants that require neither protein nor RNA synthesis for SDR. This may suggest that the inactivation reaction occurs in at least two steps.

Although the characteristics of the mutant described in this paper seem to be best explained by the last mode, it is possible, at least theoretically, that mutants which satisfy the other explanations are also found among other Sdr^c mutants. The Sdr^c mutants appear to be one of the most useful tools for the elucidation of the regulatory mechanism of chromosome replication in *E.coli*.

ACKNOWLEDGEMENT

Technical work by David Gantz and Ted Torrey is deeply appreciated. This work was supported by a grant from Public Health Service (GM22092) and a grant-in-aid for research from the University of New Mexico.

REFERENCES

Gefter, M.L. (1975) Ann. Rev. Biochem. 44:45-78

Hanawalt, P.C., Maaløe, O., Cummings, D.J. and Schaechter, M. (1961) J. Mol. Biol. 3:156-165

Jacob, F., Brenner, S. and Cuzin, F. (1963) Cold Spring Harbor Symp. Quant. Biol. 28:329-347

Kogoma, T. (1976a) Abstracts of Annual Meeting of American Society for Microbiol. p98, H15

Kogoma, T. (1976b) J. Mol. Biol. 103:191-197

Kogoma, T. (1977) Submitted to J. Mol. Biol.

Kogoma, T. and Lark, K.G. (1969) Bacteriol. Proc. GP82

Kogoma, T. and Lark, K.G. (1970) J. Mol. Biol. 52:143-164

Kogoma, T. and Lark, K.G. (1975) J. Mol. Biol. 94:243-256

Kogoma, T.,   Gantz, D. and Alizadeh, B.A. (1977) Abstracts of
     Annual Meeting of American Society for Microbiol. p173, I110

Lark, C. (1968) Methods in Enzymology XIIb, 616-625

Lark, K.G. (1966) Bacteriol. Rev. 30:3-32

Rosenberg, B.H.,   Cavalieri, L.F. and Unger, G. (1969) Proc.
     Nat. Acad. Sci. USA 63:1410-1417

Wechsler, J.A., Nüsslein, V., Otto, B., Klein, A., Bonhoeffer, F.,
     Hermann, R. Gloger, L. and Schaller, H. (1973) J. Bacteriol.
     113:1381-1388

Worcel, A. (1970) J. Mol. Biol. 52:371-386

REPLICATION OF BACTERIOPHAGE MU: DIRECTION AND POSSIBLE LOCATION

OF THE ORIGIN

Theo Goosen

Laboratory of Molecular Genetics
State University of Leiden
Leiden, The Netherlands

The temperate bacteriophage Mu-1 is distinct from other temperate phages like λ and P2 in the way it integrates in the DNA of its host E.coli. It does so by linear insertion of its DNA without any site-specificity into the bacterial chromosome, thus causing mutations (1,2,3). Also replication of Mu is very different: in contrast to λ and P2 circular DNA molecules of genome size have never been found during lytic development of the phage. However, the existence of larger covalently closed DNA molecules, containing both Mu- and host DNA sequences, has been demonstrated (4,5). As these structures are first found at 15-20 minutes after induction as well as infection (4,5), while the Mu specific DNA synthesis already starts at 8-10 minutes (6), it seems reasonable to assume these circles are not formed until at least one round of replication has been finished.

Until now, no replicative intermediates of Mu have been isolated, which makes it impossible to determine the direction and origin of replication by electron microscopy and denaturation mapping, as performed for λ, P2 and SV 40 (7,8,9).

So we determined the direction of replication of Mu in quite a different way.

Semi-conservative replication involves synthesis on both strands of short polynucleotide intermediates (Okazaki pieces), which are subsequently joined by the action of DNA ligase (10). In one strand, growing with the overall 3'-5' direction, the rate of joining of the Okazaki pieces is much slower (11), which makes it possible to isolate those fragments. In the case of P2, where by means of other methods the replication was proven to be completely

unidirectional (7), most of the isolated short fragments anneal
specifically with one strand (11,12).

Using the presence of the Okazaki pieces, the direction of
replication of Mu was determined by the following experiment.

A strain, lysogenic for Mu-1 cts 62 was induced at 43° and
after 35 minutes the DNA was labeled by a short pulse of (^{3}H)-
thymidine. The DNA was subsequently fractionated by sedimentation
through a linear 5-20% alkaline sucrose gradient on top of a
cushion of saturated CsCl. The bulk of the DNA banded in the CsCl,
while the short polynucleotides (8-12 S) were recovered as a broad
peak in the sucrose (Fig. 1). Bulk DNA and Okazaki pieces were
pooled separately, purified by isopycnic CsCl centrifugation,
sonicated and hybridized in 50% formamide to filters, loaded with
either l-strand Mu DNA, r-strand Mu DNA or denatured calf-thymus
DNA. The results are shown in Table 1A, lines 1,2 and 5. These data
show that the hybridization of the Okazaki pieces is highly
asymmetric, while the bulk of the DNA annealed equally well with
both strands. In fact the hybridization should be even more
asymmetric, but is confused by a complication: Mu possesses an
invertible segment, and so called 'G' area, which occurs in two
orientations after induction (13). Okazaki pieces, synthesized

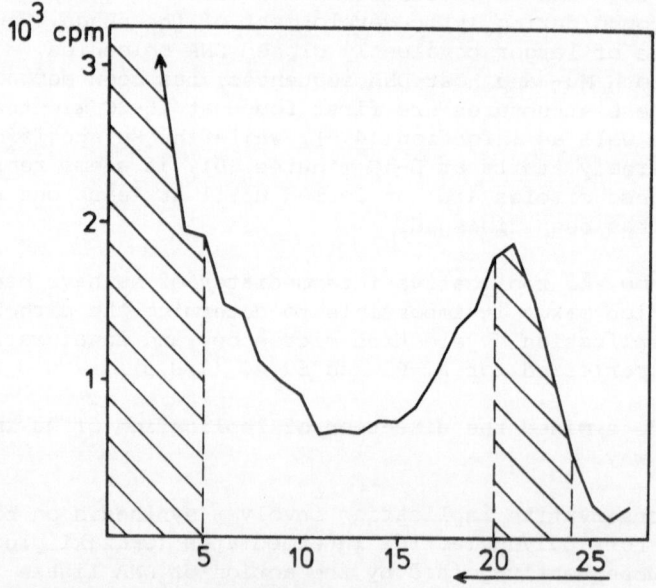

Figure 1. Isolation of Okazaki Pieces.

*A culture of a Mu cts lysogen was induced at 43°, DNA was labeled
for 30 sec with (3H)-thymidine, extracted and centrifuged in a
linear 5-20% alkaline sucrose gradient on top of a cushion of
saturated CsCl. Fractions were assayed for radioactivity and the
'long' DNA and Okazaki pieces were pooled as indicated.*

TABLE 1   Strand specificity of pulse labeled Mu DNA after induction
of a Mu lysogen.

1A.   DNA labeled at 35 minutes after induction.

DNA on filter	radioactive DNA cpm bound to filter		% strand specific Mu DNA	
	'long' DNA	Okazaki pieces	'long' DNA	Okazaki pieces
r-Mu	7.335	2.768	48,5	15,8
1-Mu	7.786	13.292	51,5	84,2
r-λMu	2.239	840	50,0	10,3
1-λMu	2.229	4.803	50,0	89,7
calf-thymus	158	328		

1B.   DNA labeled at early times after induction.

time	DNA on filter	radioactive DNA cpm bound to filter		% strand specific Mu DNA	
		'long' DNA	Okazaki pieces	'long' DNA	Okazaki pieces
8 min	r-Mu	2.223	1.739	46,7	23,4
	1-Mu	2.523	5.470	53,3	76,6
	calf-thymus	67	100		
10 "	r-Mu	3.333	1.512	50,6	20,2
	1-Mu	3.256	5.651	49,4	79,8
	calf-thymus	102	104		
12 "	r-Mu	4.407	1.517	53,2	19,7
	1-Mu	3.882	5.912	46,8	80,3
	calf-thymus	103	75		

Pooled 'long' DNA and Okazaki fragments from an alkaline sucrose gradient were purified, sonicated and hybridized to separated strands
bound to nitrocellulose filters. The percentage of strand specific Mu
DNA is given by the ratio:radioactivity bound to one strand/radio-
activity bound to both strands x 100%, after subtracting the background (DNA bound to calf-thymus DNA containing filters).

on such an inverted 'G' area, will anneal with the 'wrong' strand
of the Mu DNA, leading to hybridization with the r-strand which
at first sight is too high.

This complication can be overcome by hybridizing to the
separated strands of a plaque forming transducing λ phage, which
contains 28% of the Mu genome, but lacks the 'G' area. With those
strands, indeed the Okazaki pieces anneal more assymetrically
(Table 1A, lines 3 and 4).

In comparison to data obtained with P2, where about 90% of the
Okazaki pieces anneal with one strand (11,12), and where replication
is unidirectional (7), we conclude that the replication of Mu late
after induction is completely or almost completely unidirectional.

From the data presented here, in combination with the
previously determined direction of transcription of Mu (14), the
direction of replication can be deduced. Transcription of Mu is
from left to right on the r-strand, thus fixing the polarity of the
Mu strands (Fig. 2). Because Okazaki pieces are synthesized in the
direction opposite to the movement of the replication fork, and
which in our case mainly anneal with the l-strand, the direction of
replication has to be from left to right (Fig. 2). It is very
important to know whether Mu DNA synthesis is unidirectional in the
first round of replication. As shown in Table 1B, Okazaki pieces
isolated just after the start of the Mu replication anneal highly
asymmetric, indicating Mu replication is unidirectional at all
stages of phage development.

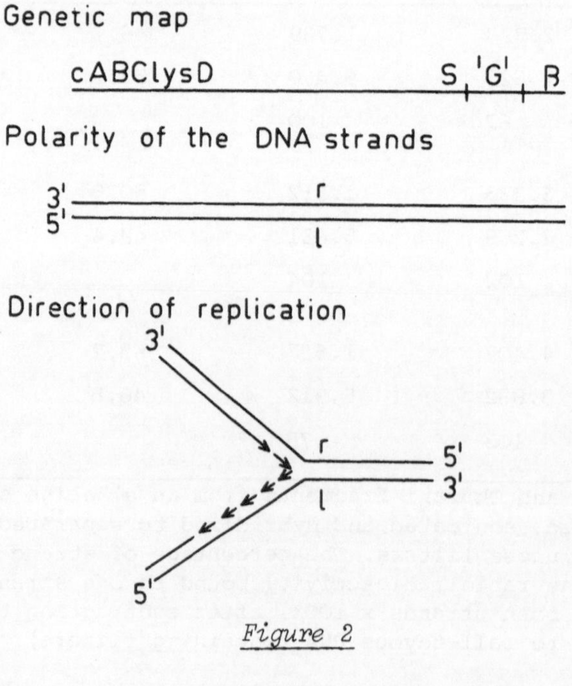

Genetic map

cABClysD   S 'G' B

Polarity of the DNA strands

3' ═══════════════════ r ═══════════════ 5'
5' ═══════════════════ l ═══════════════ 3'

Direction of replication

Figure 2

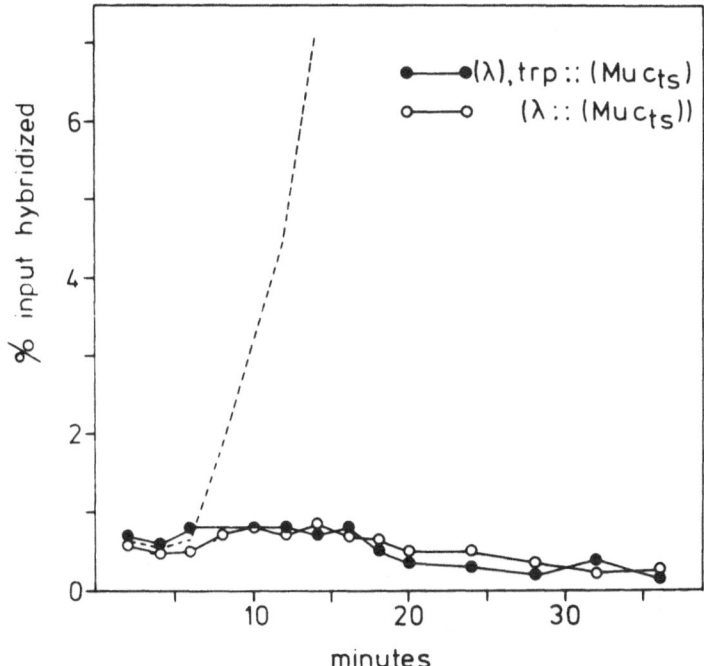

Figure 3. *Replication of Mu and λ DNA after induction of a Mu cts prophage inside a λ prophage. Mu was induced at 43° and at different times after induction DNA was pulse labeled with (³H)-thymidine during 1 minute. Cells were lysed, DNA was extracted with phenol, sonicated and denatured. The DNA was then hybridized to nitro- cellulose filters loaded with λb2 DNA. The data are given as the ratio cpm hybridized/TCA precipitable input x 100%. The dotted line reflects the Mu DNA synthesis, where samples are hybridized to Mu DNA containing filters.*

Given unidirectional replication, formation of the heterogeneous Mu-containing circles could be explained by the following
model: Mu DNA synthesis is initiated somewhere on the Mu genome,
proceeding to the right into the adjacent bacterial DNA, followed
by excision and replication of the part of Mu located to the left
of the origin. Such a model was tested by starting with a Mu cts
located within a λ prophage. After thermoinduction no Mu-induced
λ synthesis could be detected, although Mu developed normally
(Fig. 3). Therefore formation of Mu-containing circles cannot be
explained by Mu replication alone. Assuming the absence of a
normal excision after induction of a prophage, we speculate the
Mu DNA synthesis is initiated at the immunity end, with the origin
of replication possibly formed by the two extremities.

REFERENCES

(1)   Taylor, A. (1963) Proc. Nat. Acad. Sci., U.S.A. 50:1043-1051

(2)   Martuscelli, J., Taylor, A., Cummings, D., Chapman, V., DeLong,
      S. and Canedo, L. (1971) J. Virology, 8:551-563

(3)   Bukhari, A. and Zipser, D. (1972) Nature New Biol. 236:240-
      243

(4)   Waggoner, B., Gonzalez, N. and Taylor, A. (1974) Proc. Nat.
      Acad. Sci., U.S.A. 71:1255-1259

(5)   Schröder, W., Bade, E. and Delius, H. (1974) Virology 60:534-
      542

(6)   Wijffelman, C. and Lotterman, B. (1977) Molec. gen. Genet.
      151:169-174

(7)   Schnös, M. and Inman, R. (1971) J. Mol. Biol. 55:31-38

(8)   Schnös, M. and Inman, R. (1970) J. Mol. Biol. 51:61-73

(9)   Fareed, G., Garon, C. and Salzman, N. (1972) J. Virology,
      10:484-491

(10) Okazaki, R., Okazaki, T., Sakate, K., Sugimoto, K., Kainuma, R.,
      Sugino, A. and Iwatsuki, N. (1968) In: Cold Spring Harbor
      Symp. Quant. Biol. 33:129-143

(11) Kurusowa, Y. and Okazaki, R. (1975) J. Mol. Biol. 94:229-241

(12) Kainuma-Kuroda, R. and Okazaki, R. (1975) J. Mol. Biol. 94:
      213-228

(13) Daniëll, E., Boram, W. and Abelson, J. (1973) Proc. Nat. Acad.
      Sci., U.S.A. 70:2153-2156

(14) Wijffelman, C. and van de Putte, P. (1974) Mol. gen. Genet.
      135:327-337

CONTROL OF THE INITIATION OF LAMBDA REPLICATION, *oop*, *lit* AND

REPRESSOR ESTABLISHMENT RNA SYNTHESIS

Sidney Hayes

Department of Molecular Biology and Biochemistry
University of California at Irvine
Irvine, California 92717 U.S.A.

A goal of our work is to understand the mechanism for initiation of replication and stimulation of transcription from the origin (*ori*) region of bacteriophage lambda. The initiation of λ bidirectional DNA synthesis and two modes of λ transcription, (a) *oop*, *lit* RNA synthesis and (b) repressor establishment transcription, are regulated by *ori* dependent events. The requirements for *ori* dependent RNA and DNA synthesis are summarized in this report. Transcription and replication are measured from noninduced and induced prophage which are derepressed by thermally inactivating the λ repressor. This system provides for highly reproducible synchrony and permits characterization of the development of initially one phage per cell.

Requirements for Initiation of λ Replication

Lambda replicates in two stages, an early or bidirectional mode in which replication proceeds from a fixed origin (*ori*) site (Ogawa and Tomizawa, 1968; Schnös and Inman, 1970; Stevens et al., 1971; Bastia et al., 1975; Takahashi, 1975) generating circular monomers, and a late or rolling circle mode (Gilbert and Dressler, 1968; Eisen et al., 1968) in which linear DNA of greater than unit length is formed (Smith and Skalka, 1966; Carter et al., 1969; Kaiser, 1971; Skalka, 1971; Takahashi, 1975; Bastia et al., 1975; Skalka et al., 1975). The *ori* site was positioned on the λ physical map between 78.8 - 84.6 (Schnös and Inman, 1970) or 79.8 - 80.5% λ (Stevens et al., 1971). It is generally considered initiation of replication from the *ori* region does not occur at late times after infection, however, this may only be partially true (Takahashi, 1975b).

Infection experiments have shown λ will multiply in *E.coli* initiation mutants *dnaA* and *dnaC* but not in conditional lethal *E.coli dnaB, dnaE, dnaG* and *dnaZ* mutants (Hirota et al., 1968; Kohiyama, 1968; Fangman and Feiss, 1969; Gross, 1972; Shizuya and Richardson, 1974; Truitt and Walker, 1974). The bacterial gene *gro*PC or *grp*C is also required for λ DNA synthesis (Sunshine et al., 1977; Georgopoulos, 1977; Saito and Uchida, 1977). It was demonstrated that the *dnaB, dnaE* and *dnaG* products participate in the initiation of replication from an induced λ prophage (Hayes and Szybalski,1973; Hayes, submitted). In contrast, mutants in *dnaC* and *polA* reduce λ replication and phage burst by up to 95%, as does the replication inhibitor, nalidixic acid, but neither inhibits initiation of DNA synthesis (Hayes,submitted; Akiyoshi and Hayes, in preparation). The products of λ genes *O* and *P* are required for λ DNA replication (Joyner et al., 1966; Ogawa and Tomizawa, 1968) and mutants in either gene prevent the initiation of any λ replication (Stevens et al., 1971; Hayes and Szybalski, 1973; Hayes, submitted). Several λ mutants map within the region from which bidirectional replication is initiated (Eisen et al., 1966; Dove et al., 1969; Dove et al.,1971; Rambach, 1973) and have been described as *cis*-dominant *ori⁻* mutants. Those which have been characterized biochemically either reduce or prevent the initiation of λ replication (Dove et al., 1971; Nijkamp et al., 1971; Hayes and Szybalski, 1973; Hayes, submitted).

Recent attempts to sequence the *ori* region have revealed three of the *ori⁻* mutants represent small deletions within gene *O* (Denniston-Thompson et al., in press; Furth et al., in press). Two of the deletions have been shown to exert strong polar effects on rightward transcription *in vivo* (Hayes, submitted), even though the sequencing studies showed they deleted multiple triplets.

It has been suggested that genes *O* and *P* code for an endo-nucleolytic nicking activity (Shuster and Weisbach, 1969; Freifelder and Kirschner, 1971) and that rightward transcription from P_R is required for the possible introduction of a nick at the *ori* site (Inokuchi et al., 1973). The *O* and *P* products have also been found to be necessary for amplification of the synthesis of the 4S *oop* RNA and the larger *lit* RNA (about 600 nucleotides) (Hayes and Szybalski, 1973, 1975; Hayes, submitted) which are respectively transcribed from the *l*-strand *ori*-region and the distal half of gene *rex* (Hayes and Szybalski, 1973b). The *oop* RNA has been suggested to prime the initiation of λ replication (Hayes and Szybalski, 1973, 1975).

Control of *oop* and *lit* RNA Synthesis

In an uninduced prophage about 80% of the total leftward λ

transcription originates within the immunity region (cI-rex mRNA),
2% is from the ori segment, defined by phage markers imm21 and
hy42, which respectively cut near the right end of gene cII
(Oppenheim and Kapeller, 1976) and within gene O (Szpirer, 1972),
and the remainder is evenly distributed between the int and b2
regions (Hayes and Szybalski, 1973b). From 0 to 5 minutes after
thermal induction of tof^+ λcI857 lysogens the cI-rex transcription,
determined by hybridization analysis, drops from the level in the
uninduced prophage to just above the counts for a hybridization
blank in which the λ DNA strand is omitted. Between 3 to 7 minutes
after induction, a 600 nucleotide-long transcript, termed lit for
"late immunity transcription" appears in the left part of gene rex.
The lit and cI-rex transcripts both terminate at the same t_i site.
The increase in lit transcription parallels the increase (about 40-
fold) in synthesis of oop RNA from the ori segment, as if both
transcripts originated from a common promoter or were positively
regulated by a common factor at their promoters.

After the cI-rex transcription is turned off there is no
detectable leftward RNA synthesis from genes rex-cI-tof-y-cII
(refer to Tables 2,3 and Figures 3,4; Hayes and Szybalski, 1973b)
which separate the lit and oop DNA segments by about 2200 nucleotide pairs (Hayes and Szybalski, 1973b). The amplification of lit,
oop RNA synthesis, initiated between 3 to 7 minutes after prophage
induction, is regulated, since in the absence of the products of
$E.coli$ genes $dnaB$, $dnaG$, the β-subunit of RNA polymerase and λ genes
O, P and the cis-dominant ori site, *no amplification of either lit
or oop transcription occurs,* and the synthesis of oop remains at
the low level observed in either the noninduced prophage or for the
first several minutes following prophage induction in a wild type
host (Hayes and Szybalski, 1973b; Hayes and Szybalski, 1975; Hayes,
submitted). The oop RNA is transcribed from a linear λ DNA template
in vitro by RNA polymerase (Blattner and Dahlberg, 1972), and does
not require the phage and host replication proteins. Why does oop
synthesis require the above mentioned phage and host replication
proteins *in vivo* but not *in vitro* ? The *in vitro* results probably
reflect the same level of oop synthesis seen in the noninduced
prophage. The level of oop synthesis is controlled *in vivo*. As
mentioned, very little oop synthesis occurs from a noninduced
prophage where its promoter is accessible since the λ repressor
does not directly block either oop or lit transcription (Hayes and
Szybalski, 1973b).

The sequence of P_o, the promoter for oop, determined by Scherer
et al., 1977, has all the structural features exhibited by several
other promoters, i.e., it has an RNA "start site" and "strong
binding site". P_o also shares homology with the sequences of several
other promoters for a region presumed to be the polymerase
recognition site. The λ promoters P_R and P_L overlap with the serial
operator sequences in o_R1 and o_L1, whereas, the structural features

of these operators seem to be absent from the P_O promotor sequence.
However, an operator or transducer-like sequence, in exact register
with the o_R^2 and o_L^2 operators, is present in the P_O sequence. It is
possible this could be a binding site for the *tof(cro)* product,
however, *tof* does not prevent *oop* synthesis (Hayes and Szybalski,
1973b). Another possibility is that this sequence is recognized by
a complex of the above mentioned λ and *E.coli* replication activation
proteins which amplify *oop* transcription, much as the cAMP-CRP
complex potentiates transcription from the *E.coli lac* promoter.
The sequence of P_O may also overlap with the translational start
site for gene O which is transcribed from the λ *r*-strand (M. Furth,
personal communication).

As noted, *lit* synthesis requires the β-subunit of RNA polymerase, *E.coli* and λ replication gene products and the *cis*-dominant
λ *ori* site. *oop* synthesis also requires the same gene products and
DNA target site. Thus, it is possible that *lit* RNA synthesis will
occur in the absence of one or more of the replication factors if
oop participates as a cofactor for *lit* synthesis. For example,
is either *oop* RNA or an *oop* translation product required for *lit*
synthesis, rather than the λ replication proteins and *ori* site ?
In an experiment to test this, double lysogens of $\phi80imm434cI$ts
($red\beta$-N,y-*cII-ori-O-P*)λ integrated at *att*80, and combinations of
$\lambda cI857$ *ori⁻*, O^-, P^- and O^-P^- phage, integrated at *att*λ, were
constructed (Hayes and Szybalski, 1975). The *imm*434 phage
synthesizes *oop* RNA and provides the O and P products, however,
since it is deleted for the *lit* sequence, *lit* synthesis can only
occur from the *imm*λ phage. Upon thermal induction of both phage it
was found that the *imm*λ:*ori⁻*, O^-, O^-P^- and $\Delta cII-ori-O-P$ prophage
could not be complemented by the *imm*434*ori⁺O⁺P⁺* phage for *lit*
synthesis. In contrast, the *imm*λ P^-phage was complemented by the
*imm*434 phage for *lit* synthesis (Hayes and Szybalski, 1975). Since
oop was present under conditions in which *lit* synthesis did·not
occur from the *imm*λ phage, *lit* synthesis cannot depend on either
oop RNA or an *oop* translation product. The P product can be
supplied in *trans* for *lit* synthesis. The requirement of both the
O product and *ori* site can be understood if they act in *cis* for
initiation of *lit* RNA synthesis. These results can also be
explained if the O and *ori⁻* mutants employed introduce polar trans-
scriptional blocks in distal *r*-strand transcription, and the latter
block cannot be reversed by the O^+P^+ products supplied by the *imm*-
434 phage.

The inhibition of *lit* synthesis by *ori⁻* mutants (Hayes and
Szybalski, 1973; Hayes, submitted), even when the O and P products
are supplied by a heteroimmune phage, as described above, suggests
that a DNA target site 2200 nucleotides upstream from the *lit*
segment is required for *lit* RNA synthesis. Models to explain the
involvement of the *ori* site in control of *lit* RNA synthesis,

discussed in detail in Hayes (submitted), include: (1) the *ori*
site is the recognition point for formation of a modified RNA
polymerase which then recognizes the promoter for *lit* or (2) the
ori mutation inactivates initiation of transcription from P_O
which is required for *oop* and the downstream *lit* RNA synthesis. In
the latter model, the RNA synthesized between *oop* and *lit* is rapidly
and selectively degraded and thus is not detected.

Considerations for Involvement of *oop* as a
Primer RNA for Initiation of Lambda Replication

Several lines of evidence suggest *oop* RNA could prime the
initiation of λ DNA synthesis: (1) As previously noted, its
synthesis is positively regulated,i.e., greatly amplified, by the
products of phage and host replication genes. (2) The stimulation
of *oop* transcription is independent of λ DNA replication since *oop*
synthesis occurs at the wild type level (a) in induced ts*dna*E
hosts in which λ replication is inhibited (Hayes and Szybalski,
1973, 1975; Hayes, submitted), (b) in conditional lethal *pol*A
strains and (c) after the addition of 100 μg/ml nalidixic acid to
nal^s cultures. Induction of λ under the two latter conditions
limits replication to 2-4 copies but does not inhibit *oop* synthesis
(Hayes, submitted). (3) Apparent, covalently-joined *oop* RNA-DNA
fragments can be isolated from induced defective, replication⁺ λ
lysogens by hybridization and Cs_2SO_4 gradient centrifugation (Hayes
and Szybalski, 1975). (4) *oop* is the only RNA transcribed from the
l-strand between gene *c*II and a marker within gene *O*, i.e., 79.1-
81.1% of physical λ map. Another *l*-strand RNA, which we cannot
detect after induction of 594(λ*c*I857 Δ*Q-R-A-J*b), is called band 3.
This RNA, which appears to be much larger than *oop*, has been
identified after infecting cells with ϕ80*imm*λ, and is suggested to
map between 81-93% (data not provided) on the λ physical map in the
region between genes *O* and *Q* (Smith and Hedgpeth, 1975; Smith et
al., 1976). Whether this RNA is involved with either replication
or establishment transcription has yet to be determined. In summary,
there is convincing circumstantial evidence that *oop* may prime the
initiation or replication, however, an irrefutable demonstration
that it does prime has not yet appeared. The demonstration that the
Salmonella bacteriophage P22 (Roberts et al., 1976), and *E.coli*
phage λ (and presumably ϕ80) all have *oop*-RNA type molecules
transcribed adjacent to the replication genes suggests there is an
evolutionary precedent for this association, and the retention of
the *oop* molecules in divergent phage implies they serve an
important function/role for the virus.

The coupling between *oop* synthesis and the initiation of λ
replication may be explained by models in which *oop* RNA (a) primes
or (b) does not prime the initiation of λ replication. In the latter
model the observations are explained if *oop* transcription is either

(1) gene dosage dependent, (2) common factors either separately
stimulate *oop* synthesis and initiation of replication, or (3)
notification of activation of the *ori* site by these factors is
somehow telegraphed to P_o, resulting in amplification of *oop*
transcription.

A model for initiation of bidirectional DNA synthesis is
presented in Figure 1. The model is suggested from our observations
regarding the regulation of *oop* synthesis and λ replication. This
model has two unique features. First, the synthesis of an *ori*
dependent RNA primer occurs by a complex of host and phage
replication and transcription proteins termed a "repliscriptase".
It is postulated that synthesis of a primer RNA should mimic DNA,
rather than RNA synthesis in which the transcript is released from
the template. Regular transcripts synthesized by "helical strand
displacement transcription" would presumably displace the DNA
strands and destroy the template. *Displacement transcription* would
be desirable for primer RNA synthesis, however, and may require
replication proteins in addition to components of the RNA polymerase, thus the term "repliscriptase" complex was introduced.
Secondly, once strand displacement occurs, by primer synthesis and
extension of the *ori* RNA primer by DNA polymerase, bidirectional
replication becomes possible. DNA synthesis in the opposite
direction can be initiated from the displaced strand by a discontinuous mechanism, obviating the necessity for postulating the
synthesis of a second specifically initiated primer RNA, from the
displaced strand.

Consideration for Involvement of *oop* as a Leader

(1) Control of λ repressor transcription. There are two modes of
transcription for gene *c*I. In a noninduced prophage, *c*I (*maintenance*
mode) transcription occurs from a promoter *P*rm (previous
designation, *P*c) located at 78.4% on the physical λ map, and
continues downstream about 1880 nucleotides to the terminator t_i
at about 74.3% (Hayes and Szybalski, 1973b). No phage genes outside
the *c*I region are required for transcription from *P*rm (Reichardt,
1975). The repressor appears to be required for its own expression
and may either stimulate transcription from *P*rm by binding to o_R,
or activate a post-transcriptional step in *c*I polypeptide synthesis
(Reichardt, 1975). The product of gene *tof(cro)* (Pero, 1970; Calef
and Neubauer, 1968; Eisen et al., 1970), the first gene to the right
of P_R, is called the antirepressor. The *tof(cro)* product reduces
the efficiency to *c*I expression from *P*rm (Hayes and Szybalski,
1973b; Reichardt, 1975) and may act by binding at or near o_R
(Reichardt, 1975; Takeda et al., 1977).

Early after infection gene *c*I is transcribed in the
"*establishment* mode". The products of genes *c*III and *c*II plus an

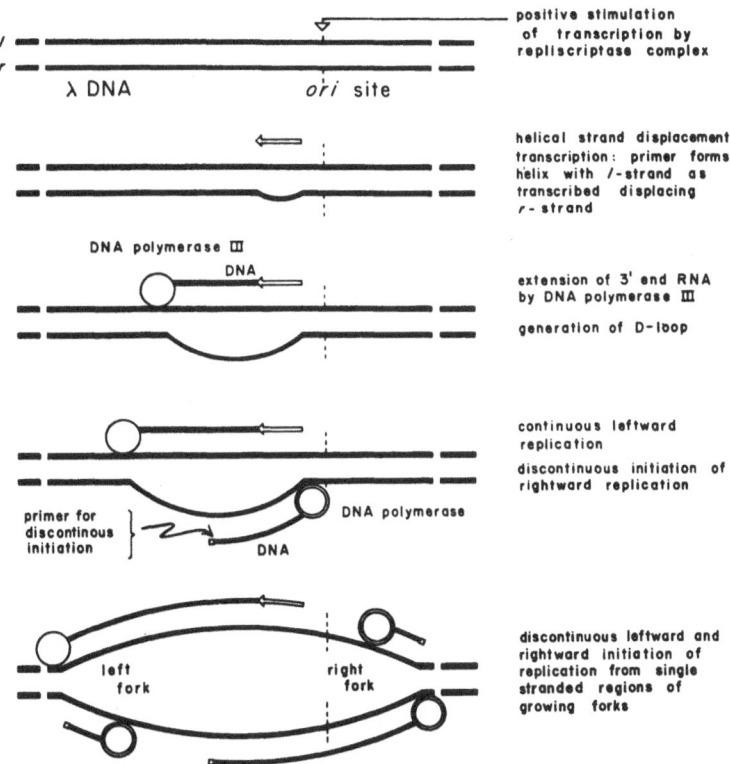

Figure 1. _Model for control of initiation of bidirectional_
replication based on the observed coupling between λ oop RNA
synthesis and initiation of λ replication. The E.coli dnaB, dnaG
and λ O, P gene products are required for both oop synthesis and
λ replication. The term repliscriptase complex is used to
collectively describe the activity of the host and phage transcrip-
tion and replication proteins. It is postulated the repliscriptase
complex synthesizes primer RNA by helical strand displacement
transcription, assumed to be more like DNA synthesis than routine
transcription. The primer displaces the r-strand as transcribed.
DNA synthesis is initiated by DNA polymerase III which extends the
3'-OH terminus of the RNA primer creating a leftward moving
replication fork which further displaces r-strand DNA. the
displaced r-strand DNA is presumably covered by the E.coli DNA
binding protein (Molineux et al., 1974) as suggested in the model
of Alberts (1973) for replication of T4. The rightward replication
fork is generated by initiation of DNA synthesis in the dis-
continuous mode as either an RNA primed Okazaki fragment or by the
knife and fork mechanism. There may be no specific site for
initiation of the rightward fork. An unspecified DNA polymerase
complex initiates discontinuous rightward and leftward replication

active y-region activate establishment of the lysogenic state
(Echols and Green, 1971; Reichardt and Kaiser, 1971; Spiegelman
et al., 1972; Reichardt, 1975b). One hypothesis for initiation of
repressor establishment transcription is that "the cII and cIII
Proteins activate cI transcription from a promoter P_{re} located
several hundred nucleotides to the right of cI" in the y-region
(Reichardt and Kaiser, 1971; Echols and Green, 1971; Spiegelman et
al., 1972). "Midway in the infective cycle", the $tof(cro)$ "product
reduces the synthesis of the cII and cIII proteins (Reichardt, 1972,
1975; Echols et al., 1973). Since the cII and cIII proteins are unstable, their intracellular levels decline, reducing the rate at
which cI is transcribed from the promoter P_{re} (Reichardt and Kaiser,
1971; Reichardt, 1972, 1975). If sufficient cI protein (λ repressor)
has been made to repress the two early phage operons, lysogeny
becomes possible" (Reichardt, 1975).

(2) _oop_ RNA as a leader for establishment transcription. Several
recent observations have supported an alternative hypothesis for
control of the initiation of repressor establishment transcription
in which oop RNA synthesis is an important first event. Honigman
et al. (1975) isolated a class of mutants called (λsar) which map
near ori and influence the synthesis of repressor transcription. It
was postulated that sar might affect the synthesis or termination
of oop RNA. Roberts et al. (1976) have suggested from an _in vitro_
analysis of the transcripts made using purified RNA polymerase and
λ, P22 and λ-P22 hybrid phage DNA templates that the oop RNA from
λ and the equivalent "b" RNA from phage P22 serve as leaders for the
repressor message. They suggested that cy mutations affect
repressor synthesis, not by inactivating P_{re} (Reichardt and Kaiser,
1971), but by somehow inhibiting elongation of the transcript from
the antiterminated "oop leader". Honigman et al. (1976) isolated a
9S RNA called "band A" from an induced dnaEts(λcI857cro27) lysogen
and have shown by the hybridization deletion mapping technique that
the startpoint for this RNA maps to the right of imm21.

We have constructed similar extrapolation curves for the
repressor establishment transcript, isolated _in vivo_, which also
suggest it is initiated to the right of the imm21 marker at 79.8%λ
(Hayes and Hayes, submitted). These observations suggest oop RNA
may be antiterminated yielding repressor establishment transcription,
however, proof consisting of a demonstration that the _in vivo_
isolated repressor establishment transcript has the sequence of oop
at its 5'-end has not yet been obtained.

_from single stranded regions of the growing forks. An alternate
model could be drawn for generation of the rightward fork in which
replication would proceed from a specific site: a terminated r-
strand transcript is extended into DNA by DNA polymerase III from
the point of termination._

A comparison of the kinetics of *imm-ori* transcription in induced λcI857 *tof*⁺ and *tof*⁻ prophage revealed that the *tof* product absolutely prevents the synthesis of repressor establishment transcription (Hayes and Hayes a,b, submitted). In contrast, uninhibited expression of repressor establishment transcription occurs after induction of the *tof*⁻ prophage (Hayes and Hayes a,b, submitted). As an example, upon induction of a *tof*⁺ λcI857 prophage, repressor transcription drops by 1/5 - 1/8 of the prophage level, whereas, in the equivalent *tof*⁻ strain repressor establishment transcription, through cI, occurs at a level 27 times greater than the noninduced prophage level. There is at least a 135-fold (27 x 5) reduction of repressor establishment transcription under *tof*⁺ conditions. This result permitted identification of the *E.coli* and λ gene products required for repressor establishment transcription. Mutants in required genes should reduce or eliminate the uninhibited expression of establishment RNA synthesis from the *l*-strand of the *y-c*II region in induced *tof*⁻ prophage. It was possible to determine that the same *E.coli* and λ genes, i.e., P, *dnaB*, *dnaG* and β-subunit of RNA polymerase, required for *oop*, *lit* RNA synthesis and the initiation of λ replication, were also required for stimulation of *any* repressor establishment transcription. In addition, the products of λ genes cIII and cII were equally required for any establishment transcription (Hayes and Hayes, b, submitted). SInce the *tof* product only partially depresses the transcription of cIII and cII after induction (Hayes and Hayes a, submitted), presumably by reducing transcription from P_L and P_R, the complete absence of detectable establishment RNA synthesis in the presence of the *tof*⁺ product suggests a specific mechanism exists for inhibition of the initiation of establishment RNA synthesis. Several such models are discussed in Hayes and Hayes (a, submitted). One idea is that the *tof* product blocks the activity rather than the expression of the cII and cIII products.

The observed parallel control in initiation of λ replication, repressor establishment transcription, and *lit*, *oop* RNA synthesis can be explained, if *oop* RNA serves a dual role, as both a primer for DNA synthesis, and as a leader for repressor establishment transcription. The cII and cIII products might function to anti-terminate *oop* RNA synthesis yielding the repressor establishment transcript in the absence of active *tof* product. There are several observations, however, which suggest this explanation for establishment transcription is not correct as stated. Although the effect of *tof* on establishment transcription has been determined, the role of *tof* is unclear. First, after induction of a *tof*⁻ λ prophage the appearance of *oop* RNA synthesis is not coordinate with the synthesis of repressor establishment mRNA, and temporally increases several minutes after the stimulation of adjacent *y-c*II transcription (Hayes and Hayes b, submitted). If the *y-c*II portion of the establishment transcript arises by anti-termination of the adjacent *oop* RNA both transcripts would be

expected to be coordinately synthesized. Second, the kinetics of
oop and establishment transcription are independently influenced by
the *tof* product. Establishment transcription is not detected after
induction of *tof*[+]prophage, whereas, vigorous *oop* synthesis occurs.
In contrast, vigorous establishment transcription occurs after
induction of *tof*[-] prophage, but, *oop* synthesis is very weak (Hayes
and Hayes b, submitted). Third, the *lit* RNA is controlled by
the same mechanism as *oop*. The simplest model for *lit* RNA is that
it represents a continuation of the *oop* RNA, and the intermediate
transcript between *oop* and *lit* is selectively degraded under *tof*[+]
conditions. However, both *oop* and *lit* synthesis occur at the normal
rate in the absence of either the *c*II or *c*III products (Hayes and
Szybalski, 1973b). If *lit* is a metabolic product of a transcript
initiated at P_o, then antitermination of *oop* would occur (but not
be detected) in the presence of *tof*, and absence of the *c*III, *c*II
products. In contrast, the establishment transcript, which occurs
in the absence of *tof*[-] and requires the *c*II and *c*III products, may
result from an inhibition of the mechanism for degrading the
intermediate transcript between *oop* and *lit*. The effect of the *tof*
product on *oop* transcription is shown in Figure 2.

The above discussion reveals the antirepressor participates
in the regulation of repressor establishment transcription and
lit oop RNA synthesis, although its mode of action is not understood. By an unknown mechanism, the *tof(cro)* product also participates in λ replication. The concept of "*cro(tof)* - lethality"
has been used to describe the inability of λ*c*I857*tof*[-] phage to
plate outside the temperature range of about 37 - 39°C. Folkmanis
et al. (1977) report a reduction in the rate of λ and *E.coli* DNA
synthesis and aberrant late λ replicative forms, measured by
[3]H-thymidine incorporation into acid precipitable counts, after
infection of both UV-killed and non-irradiated cultures of *E.coli*
with *imm*λ *tof*[-] phage at 42°C. We have titrated by using homologous ^{3}H RNA probes, the increase in DNA copies of λ*c*I857 *tof*[+]
and *tof*[-] prophage, from O-25 minutes after induction in wild type
E.coli hosts. It was found the increase in DNA copies synthesized
to the left of the origin was identical for *tof*[+] and *tof*[-] phage,
and replication to the right of the origin was 2-3X higher in the
tof[-] compared with the *tof*[+] phage. This result is supported by
phage induction experiments. When W3350(λ*c*I857cro27) is thermally
induced for 60 minutes at 42°C and the resulting lysate is plated
at 30 or 42°C, very few phage are noted. However, if the same
lysate is plated at 39°C a reasonably good phage burst is formed
(Hayes, unpublished).

The above results suggest *tof* is not directly required for λ
DNA synthesis. However, since the burst of induced *tof*[-] phage is
very low at 42°C, and, as mentioned, thymidine, as well as [3]H-
uridine incorporation (Hayes, personal communication) is reduced

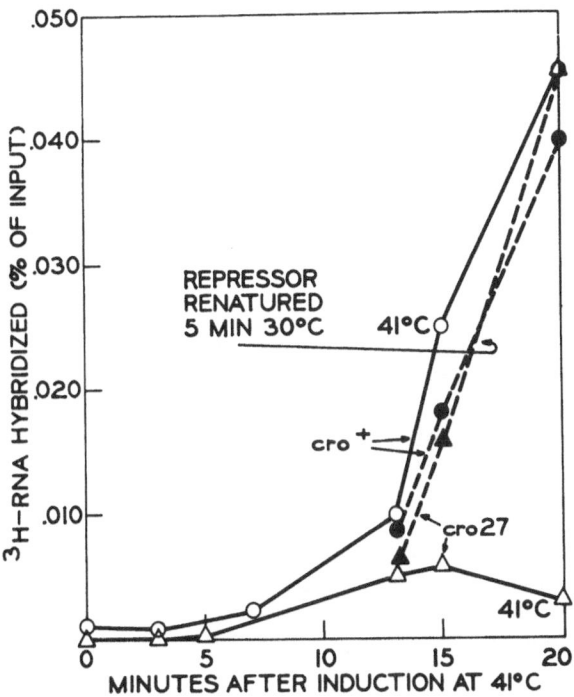

Figure 2. *Effect of repressor and* <u>tof</u> *product on the kinetics of* <u>oop</u> *RNA synthesis. Twenty ml cultures of W3350(λcI857) =* <u>tof</u>$^+$ *(*<u>cro$^+$</u>*) and W3350(λcI857cro27) =* <u>tof$^-$</u> *(*<u>cro</u>27*) were grown to* A_{575} = *.35 in labeling medium at 30°C and were thermally induced by shaking for .17 minutes at 60°C and then transferred to a 41°C shaking bath. Labeling was during the final minute of induction, at the indicated times, with 0.5 mCi of uridine-5-^{3}H. Upon termination of labeling the ^{3}H-DNA was extracted and hybridized for the presence of* <u>oop</u> *RNA as described in Hayes and Szybalski, 1973b. The solid lines reflect the level of* <u>oop</u> *RNA present in induced* <u>tof$^+$</u>*(*<u>cro$^+$</u>*) (open circles) and* <u>tof$^-$</u>*(*<u>cro</u>27*) (open triangles) cultures at the indicated times after induction. The dashed lines represent parallel* <u>tof$^+$</u>*(*<u>cro$^+$</u>*) (closed circles) and* <u>tof$^-$</u>*(cro27) (closed triangles) cultures which were similarly induced and incubated at 41°C. However, 5 minutes before pulse labeling, the cultures were shaken for .12 minutes in an ice bath and then incubated for the remaining period at 30°C in order to renature the repressor. For example, one culture was induced at 41°C for 14.5 minutes, and labeled between 14.5 - 15.5 minutes. The parallel culture in which the repressor was renatured was induced at 41°C for 10 minutes, cooled to 30°C where incubation was continued through 15.5 minutes, with labeling during the final minute. A comparison of the effect of repressor renaturation on* <u>oop</u> *and* <u>y-cII</u>*, i.e., establishment transcription is described in Hayes and Hayes (b, submitted).*

after induction, the absence of *tof* must seriously destabilize
cellular metabolism. Since the absence of *tof* can partially be
reversed by the λ repressor (Folkmanis et al., 1977; Figure 2;
Hayes and Hayes b, submitted) it is possible one activity of the
tof product is to prevent the uncontrolled expression (damaging
effect) of one or more λ gene products. It is also evident in
Figure 2, that renaturation of the repressor reverses the reduction
in either synthesis or detection of *oop* RNA observed under *tof⁻*
conditions. A major difference between induced replication
competent *tof⁺* and *tof⁻* prophage is the enormous synthesis of
repressor establishment RNA by the latter phage, and absence of
establishment transcription by the former strain. The accumulation
of the establishment transcript (30-fold increase over noninduced
*c*I prophage transcription and 100-200 fold increase over induced
tof⁺ *c*I transcription; Hayes and Hayes a,b, submitted) is suggested
to account for the *tof*-lethality observed at both 30° and 42°C. This
point has been ignored, or explained by the effect of small increases
in total *l*-strand (5 to 10-fold) or *r*-strand (up to 3-fold)
transcription (Kumar, Calef and Szybalski, 1970: Hayes, submitted)
observed after induction of *tof⁻* prophage. This hypothesis is
supported by our observation that λ*c*I857*cro*27*y*42 plates with equal
efficiency between 30 and 42°C (Hubert and Hayes, in preparation)
and the observation by Folkmanis et al., 1977, that suppressed
λ*c*Iam14*cro*27*c*II2002 plate at 32, 37 and 42°C, since both phage are
blocked in repressor establishment transcription by either the *y*
or *c*II⁻ mutations. Continued discussion of this point is beyond
the present consideration of the function and control of *oop* RNA
synthesis, but must include, however, an explanation of why λ*c*III⁻
*c*I⁻ *tof⁻* and λ*c*⁻*tof⁻*cII⁻ phage plate at 30-39° but not at 42°C.

ACKNOWLEDGEMENT

This work was supported by a grant from the National Institutes
of Health. I am very grateful to Dr. Waclaw Szybalski for his
support and kindness at the McArdle Laboratory where my investigation of these studies was initiated.

REFERENCES

1.   Alberts, B. (1973) In: Molecular Cytogenetics (eds. B. Hamkalo
     and J. Papaconstantinou) Plenum Press, New York, p233

2.   Bastia, D., Sueoka, N. and Cox, E.C. (1975) J. Mol. Biol.
     98:305

3.   Blattner, F. and Dahlberg, J. (1972) Nature New Biol. 237:
     227

4.   Calef, E. and Neubauer, Z. (1968)   Cold Spring Harbor Symp.
     Quant. Biol. 33:755

5.   Carter, B.J., Shaw, B.D. and Smith, M.G. (1969)   Biochim.
     Biophys. Acta 195:494

6.   Denniston-Thompson, K., Moore, D.D., Kruger, K.E., Furth, M.E.
     and Blattner, F.R. The physical structure of the replication
     origin of bacteriophage λ, manuscript in press

7.   Dove, W.F., Hargrove, E., Ohashi, M., Haugli, F. and Guha, A.
     (1969) Japan J. Genet. 44: (suppl.1) 11

8.   Dove, W.F., Inokuchi, H. and Stevens, W.F. (1971) In: The
     Bacteriophage Lambda (ed. A.D. Hershey) Cold Spring Harbor
     Laboratory, Cold Spring Harbor, New York. p747

9.   Echols, H. and Green, L. (1971) Proc. Nat. Acad. Sci. USA
     68:2190

10. Echols, H., Green, L., Oppenheim, A., Oppenheim, A. and
     Honigman, A. (1973) J. Mol. Biol. 80:203

11. Eisen, H.A., Fuerst, C.R., Siminovitch, L., Thomas, R.,
     Lambert, L., Pereira Da Silva, L. and Jacob, F. (1966) Virol.
     30:224

12. Eisen, H., Brachet, P., Pereira Da Silva, L. and Jacob, F.
     (1970) Proc. Nat. Acad. Sci. USA 66:855

13. Eisen, H., Pereira Da Silva, L. and Jacob, F. (1968) Cold
     Spring Harbor Symp. Quant. Biol. 33:755

14. Fangman, W. and Feiss, M. (1969) J. Mol. Biol. 44:103

15. Folkmanis, A., Maltzman, W., Mellon, P., Skalka, A.-M. and
     Echols, H. (1977) Virol. 81:352

16. Freifelder, D. and Kirschner, I. (1971) Virol. 44:223

17. Furth, M.E., Blattner, F.R., McLeester, C. and Dove, W.F.
     The genetic structure of the replication origin of phage λ,
     manuscript in press.

18. Georgopoulos, C.P. (1977) Molec. Gen. Genet. 151:35

19. Gilbert, W. and Dressler, D. (1968) Cold Spring Harbor
     Symp. Quant. Biol. 33:473

20.   Gross, J. (1972) Curr. Top. Microbiol. Immunol.   57:39

21.   Hayes, S. Positive control of *lit, oop* transcription and
      initiation of replication from coliphage lambda,   manuscript
      submitted for publication

22.   Hayes, S. and Hayes, C.   Control of λ repressor prophage and
      establishment transcription by the product of gene *tof*,
      manuscript a submitted for publication.

23.   Hayes, S. and Hayes, C.   Control of λ repressor establishment
      transcription.   Kinetics of *l*-strand transcription from the
      *y-c*II-*ori-O-P* region,   manuscript b submitted for
      publication.

24.   Hayes, S. and Szybalski, W. (1973)   In: Molecular Cytogenetics
      (eds. B. Hamkalo and J. Papaconstantinou)   Plenum Press, New
      York, p277

25.   Hayes, S. and Szybalski, W. (1973b) Molec. Gen. Genet. 126:275

26.   Hayes, S. and Szybalski, W. (1975) In: DNA Synthesis and Its
      Regulation (eds. M. Goulian and P. Hanawalt)   W.A. Benjamin,
      Menlo Park.   p486

27.   Honigman, A.A., Oppenheim, A., Oppenheim, A.B. and Stevens,
      W. (1975) Molec. Gen. Genet. 138:85

28.   Honigman, A., Hu, S.-L., Chase, R. and Szybalski, W. (1976)
      Nature   262:112

29.   Hirota, Y., Ryter, A. and Jacob, F. (1968)   Cold Spring
      Harbor Symp. Quant. Biol.   33:669

30.   Inokuchi, H., Dove, W. and Freifelder, D. (1973) J. Mol.
      Biol.   74:721

31.   Joyner, A., Isaacs, L., Echols, H. and Sly, W. (1966) J. Mol.
      Biol.   19:174

32.   Kaiser, A.D. (1971) In: The Bacteriophage Lambda (ed. A.D.
      Hershey)   Cold Spring Harbor Laboratory, Cold Spring Harbor,
      New York.   p195

33.   Kohiyama, M. (1968)   Cold Spring Harbor Symp. Quant. Biol.
      33:317

34.   Kumar, S.,   Calef, E. and Szybalski, W. (1970)   Cold Spring
      Harbor Symp. Quant. Biol.   35:331

35. Molineux, I.J., Friedman, S. and Gefter, M.L. (1974) J. Biol.
     Chem. 249:6090

36. Nijkamp, H.I.J., Szybalski, W., Ohashi, M. and Dove, W.F.
     (1971) Molec. Gen. Genet. 114:80

37. Ogawa, T. and Tomizawa, J. (1968) J. Mol. Biol. 38:217

38. Oppenheim, A.B. and Kapeller, I. (1976) Molec. Gen. Genet.
     149:121

39. Pero, J. (1970) Virol. 40:65

40. Rambach, A. (1973) Virol. 54:270

41. Reichardt, L. (1972) Thesis, Department of Biochemistry,
     Stanford University.

42. Reichardt, L. (1975) J. Mol. Biol. 93:289

43. Reichardt, L. (1975b) J. Mol. Biol. 93:267

44. Reichardt, L. and Kaiser, A. (1971) Proc. Nat. Acad. Sci.,
     USA 68:2185

45. Roberts, J.W., Roberts, C., Hilliker, S. and Botstein, D.
     (1976) In: RNA Polymerase (eds. R. Losick and M. Chamberlin)
     Cold Spring Harbor Laboratory, Cold Spring Harbor, New York.
     p707

46. Saito, H. and Uchida, H. (1977) J. Mol. Biol. 113:1

47. Scherer, G., Hobom, G. and Kossel, H. (1977) Nature 265:117

48. Schnos, M. and Inman, R.B. (1970) J. Mol. Biol. 51:61

49. Shizuya, H. and Richardson, C. (1974) Proc. Nat. Acad. Sci.,
     USA 71:1758

50. Shuster, R.C. and Weisbach, A. (1969) Nature 223:852

51. Skalka, A. (1971) In: The Bacteriophage Lambda (ed. A.D.
     Hershey) Cold Spring Harbor Laboratory, Cold Spring Harbor,
     New York. p535

52. Skalka, A., Greenstein, M. and Reuben, R. (1975) In: DNA
     Synthesis and Its Regulation (eds. M. Goulian and P. Hanawalt)
     W.A. Benjamin, Menlo Park. p460

53. Smith, G.R. and Hedgpeth, J. (1975) J. Biol. Chem. 250:4818

54. Smith, G.R., Eisen, H., Reichardt, L. and Hedgpeth, J. (1976)
     Proc. Nat. Acad. Sci. USA 73:712

55. Smith, M.G. and Skalka, A. (1966) J. Gen. Physiol. 49 (suppl.
     2) 127

56. Spiegelman, W.G., Reichardt, L.F., Yaniv, M., Heinemann, S.F.,
     Kaiser, A.D. and Eisen, H. (1972) Proc. Nat. Acad. Sci. USA
     69:3156

57. Stevens, W., Adhya, S. and Szybalski, W. (1971) In: The
     Bacteriophage Lambda (ed. A.D. Hershey) Cold Spring Harbor
     Laboratory, Cold Spring Harbor, New York p515

58. Sunshine, M., Feiss, M., Stuart, J. and Yochem, J. (1977)
     Molec. Gen. Genet. 151:27

59. Szpirer, J. (1972) Molec. Gen. Genet. 114:297

60. Takahashi, S. (1974) Biochem. Biophys. Res. Comm. 61:607

61. Takahashi, S. (1975) J. Mol. Biol. 94:385

62. Takahashi, S. (1975b) Molec. Gen. Genet. 142:137

63. Takeda, Y., Folkmanis, A. and Echols, H. (1977) J. Biol.
     Chem. 252:6177

64. Truitt, C.L. and Walker, J.R. (1974) Biochem. Biophys. Res.
     Comm. 61:1036

CONTROL OF INITIATION OF DNA REPLICATION IN MAMMALIAN CELLS

J. Herbert Taylor

Institute of Molecular Biophysics and Department of
Biological Sciences
Florida State University
Tallahassee, Florida 32306, U.S.A.

INTRODUCTION

The initiation of DNA replication in a mammalian cell is a
complex series of processes of which we have limited knowledge. We
are aware of three parameters which must be regulated. The first
and most stringent of these is the necessity to replicate every
segment of each chromosome once and only once each cell cycle.
Repair replication, which is a continuing process whether the cell
is cycling or not, can occur, in the absence of this highly
controlled semi-conservative replication. Since a chromosome
consists of a single segment of DNA which is too long to replicate
from a single origin in the time available for most cell cycles,
each chromosome must have many initiation sites. Yet, each must
function no more than once each cell cycle. If a potential site
fails to function it must be modified as the replication fork
passes over it so that it cannot function until the next cycle.
It appears likely that this type of modification or regulation will
be unique for chromosomal DNA. The DNA of animal viruses, and
perhaps organelle DNA, may be designed to initiate several to many
rounds of replication during an interval when the chromosomes
complete one round or fail to replicate at all. The basis for
this type of regulation is unknown and probably cannot be determined
until we know more about the distribution and structure of potential
origins for replication. However, a rather simple mechanism has
been proposed (Taylor, 1978) which was suggested by the methylation
patterns in *Escherichia coli*. Some years ago C. Lark (1968) found
that mutants deficient for methionine (the methyl donor in all
enzymatic DNA methylation) could grow and replicate DNA without
methionine and concurrent methylation of the DNA for one full cycle.
Ethionine would substitute for protein synthesis and allow

143

reinitiation at the origin, but could not substitute in the
enzymatic methylation of DNA. Replication ceased when the
replication fork encountered the part of the chromosome which had
replicated without methionine and therefore was only methylated
on one chain (half-methylated). Our proposal is that each
potential origin (initiation site) in a mammalian chromosome
consists of a palindromic sequence flanked at each end by identical
sequences which are methylated. After replication in the absence
of enzymatic methylation of these particular sequences, the
sequences would be half-methylated and would then fail to bind the
proteins of the replication complex. This change would have no
relation to whether the potential initiation site had actually
functioned as an origin. Once replication had been completed in
all chromosomes some regulatory signal would induce or activate
the methylating enzymes' specific for those sites and all potential
origins would be fully methylated. The cell would in that respect
be ready for another round of replication. Studies of enzymatic
replication in mammalian cells have revealed a fraction of the
total enzymatic methylation which is delayed until the G_2 phase of
the cell cycle in spite of the fact that most methylation occurs
within a few minutes of replication (Adams, 1971). However, the
distribution and sequences involved in this late methylation are
unknown. It is also well known that cells occasionally undergo a
second round of replication before division and produce another
complete set of chromosome (endoreduplication). The cell then
reaches metaphase with four chromatids per chromosome instead of
the usual two. This observation indicates that whatever
modification occurs to prepare DNA for a second round of replication
is not delayed until the G_1 phase.

The second regulatory system of which we are aware involves
a control of the sequence in which DNA is replicated. This control
of sequences is perhaps less stringent than the control of rounds
of replication, yet it was noted in the early autoradiographic
studies of DNA replication in chromosomes. Chromosomes were
labeled at many sites in pulses with ^{3}H-thymidine, but some
segments and occasionally whole chromosomes were delayed in
replication until S phase was well advanced. The patterns of
labeling observed were reproducible from cell to cell and from
cycle to cycle (Taylor, 1960). The clearest examples were the
sex chromosomes which were regulated in a way which delayed
initiation in all or some segments until S phase was at least one
half completed. Replication then extended for 20-30 minutes longer
than most other chromosomes. Variations were by no means limited
to the sex chromosomes, but involved complex patterns in all
chromosomes. This temporal regulation of replication has been
studied extensively by means of autoradiography and the use of
buoyant density differences, both natural and those induced by the
use of bromodeoxyuridine. We will consider these studies only to

the extent that mechanisms of regulation of initiation may be
indicated.

The third parameter of regulation is the control of rate or
speed of replication, the extremes of which are exemplified by
the amphibians where studies of early cleavage stages as well as
spermatogenesis and oögenesis are best known. Cycles of division
in early cleavage occur within an hour or less and during most of
this time the nuclei are in stages of mitosis. The S phase must be
completed in a fraction of the cycle although careful measurements
are difficult and do not appear to have been carried out. The S
phase preceding spermatogenesis, at the other extreme, occupies
about 10 days in *Trituris* (Callan and Taylor, 1968), while
fibroblasts in cell culture may complete the S phase in 8-12 hours.
These extreme variations could involve changes in two parameters,
rate of chain growth or frequency of operating initiation sites
along the chromosomal DNA. Evidence indicates that variations are
primarily in the density of the initiation sites as shown by Callan's
(1972) studies on the amphibians. The rate of chain growth varies
much less, but as indicated below there are measurable differences
over the cell cycle and at different stages of development.

Most of the investigations to date have been designed to
identify the units of replication, which have sometimes been called
replicons (Taylor, 1963), following the terminology originally
used by Jacob and Brenner (1963) for the bacterial chromosome.
The search has revealed smaller and smaller subunits involved in
replication. Perhaps the replicon concept is inappropriate and
attention should be centered on the nature and distribution of
origins for replication, or the potential origins of replication,
whether these are functional in a particular cell cycle or not.

Rates of Chain Growth in Replication of Chromosomal DNA in Mammalian Cells

The rate of chain growth has been measured by the use of the
density label, BrdU (bromodeoxyuridine). Following a pulse label
the DNA was sheared and banded in CsCl. The length of higher
density DNA was then estimated by two different methods (Painter
et al., 1966 and Taylor, 1968). Both were based on the ratio of
partially substituted DNA fragments to the fully substituted
fragments of sheared DNA and gave similar results, a growth rate
of 1-2 μm per minute. This translates into a rate of 3000-6000
nucleotides per minute or 50 to 100 nucleotides per second. The
estimate was based on the assumption that breakage of the DNA
occurred at random. However, our recent studies have shown that
more often the newly replicated DNA breaks near the growing fork.
This would reduce the amount of partially substituted DNA and
cause an over estimate in the rate. Our more recent estimate

indicates that one µm per minute is closer to the correct rate for
bidirectional growth, i.e., one-half µm or 1500 nucleotides per
minute for each chain (Taylor, 1977).

Estimates of chain growth rates from autoradiography were
first made by Cairns (1966). He pulse labeled HeLa cells with
^{3}H-thymidine of high specific activity, spread the DNA on a
cellulose nitrate filter and applied photographic emulsion similar
to those preparations he had already made with DNA from phage T4
and *Escherichia coli*. The labeled segments produced indicated a
rate of about one µm per minute.

Huberman and Riggs (1968) used Cairns's method and first
demonstrated bidirectional growth from origins in mammalian cells,
Chinese hamster fibroblasts in culture. They demonstrated two
forks connected by labeled chains and also showed that a ^{3}H-
thymidine pulse followed by a chase with cold thymidine produced
labeled segments with the number of grains tapering off at the ends.
The latter were correctly interpreted as indicating bidirectional
growth from origins. The chase decreased the specific activity of
the thymidine triphosphate precursor and caused the grain density
to decrease over the most recently grown parts of the new chains.
The rows of grains indicating the position of replicating DNA were
of a high density in the middle and tapered off to few grains at
each end. The initiation sites were clustered in tandem on the
fibers and varied from about 10 to 100 µm, center to center of the
labeled segments. The most common center to center distances were
between 15 and 60 µm with a mean of about 30.

These results on bidirectional growth were soon confirmed by
Callan (1972) who also showed that large variations in inter-origin
distances were observed in tissues at different stages of
development in amphibians. Early embryonic cells had the shortest
interval between initiation sites and spermatocytes had the
greatest. It was becoming clear that the differences in length of
S phase were related to the number or density of initiation sites
along the DNA, with relatively little change in the rate of chain
growth.

The Distribution of Initiation Sites in Chromosomal DNA

Initiation is a continuing process throughout the S phase in
eukaryotic cells, but it is difficult to study during S phase when
chain elongation from segments already growing, as well as from new
initiation sites, is occurring simultaneously. We have spent a
considerable amount of effort in developing a system with many
potential initiation sites and few chains which are already
initiated. Cells of Chinese hamster (CHO) are partially
synchronized by changing a log phase culture to a medium deficient

in isoleucine and glutamine. Ham's F-10 medium is prepared by omitting these two amino acids and thymidine. It is supplemented only with dialyzed calf serum. When changed to such a medium, cells accumulate in the G_1 phase or perhaps more specifically in the G_0 phase. Upon release after 20-30 hours by adding Ham's complete medium, the cells enter S phase within about 5 hrs and proceed around the cycle parasynchronously. In this way large numbers of cells can be collected at division by rolling the bottles, to which cells are attached in their normal growth, at a higher speed. At the appropriate speed in a healthy culture which is not confluent, the dividing cells detach and cannot reattach as long as the bottles are rolling. A roller bottle with 5-10 x 10^7 cells can yield 20-30 x 10^6 cells per hour at the peak of division. These cells are distributed to plastic flasks and allowed to reattach for 2 hours. Any cells not reattached are discarded with the medium, and the remaining cells are rinsed once in Ham's medium without thymidine. An aliquot of the same medium, which is supplemented only with dialyzed calf serum (10%), is added along with FdU (fluorodeoxy-uridine). The cells, in spite of these precautions, find a small amount of thymidine and can initiate at a few sites within 3-5 hours. However, if the concentration of FdU is high enough to completely inhibit thymidylate synthetase, the cells deplete this extremely small supply of thymidine and thereafter accumulate potential origins at a linear rate for many hours. Most of these origins will initiate within a minute after release by adding thymidine. The concentration of thymidine may determine the efficiency of release and the degree of synchrony, but after trying 10^{-7} M, 10^{-6} M and 10^{-5} M we have used 10^{-6} M for most experiments.

The degree of synchrony can be determined by measuring the segments produced after release with ^{3}H-thymidine, in an alkaline gradient or by autoradiography. We will consider the measurements by sedimentation in alkaline gradients first. If the cells are lysed four minutes after release the size of the newly replicated segments found in a gradient is more heterogeneous in size if the block was done with 10^{-6} M FdU rather than 10^{-5} M (Fig. 1). When the block was carried out in the presence of 10^{-5} M FdU the size distribution of segments produced in a 4 minute pulse indicated a high degree of synchrony of initiation and growth rate (Kurek and Taylor, 1977). A small fraction of the newly replicated segments are large even with 10^{-5} M FdU (Fig. 1), but the proportion of these to those of uniform size decreases with time of the block. We interpret this to mean that a few sites initiate and grow slowly until the small supply of thymidine is depleted. After that potential sites are formed, by the synthesis of the pro-tein complexes which bind at potential initiation sites, but initiation does not occur until thymidine is supplied. This conclusion is based on the observation that actinomycin D (5 µg/ml) placed in the medium within 3-5 minutes before release will almost

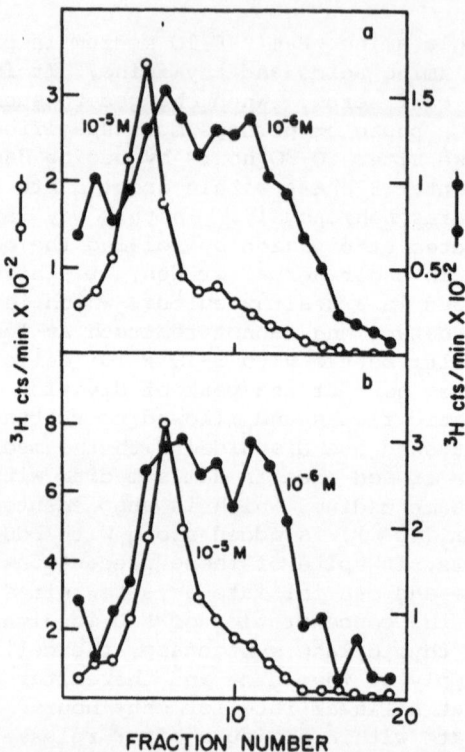

Figure 1. Replicate cultures of mitotic CHO cells were prepared in plastic flasks and incubated for 7 hours in either 10^{-5} or 10^{-6} M FdU. At the end of this incubation period (8 hours from division) the cells were pulse labeled with ^{3}H-thymidine (20 μCi/ml; 10^{-6} M) for either 4 min (a) or 8 min (b). Lysates were layered on alkaline NaI self-generating gradients and spun at 20,000 rpm for 5 hrs in the Spinco SW41 rotor. Sedimentation is from left to right. Note the homogeneous distribution of the segments produced during both 4 and 8 min pulses when the block was most effective (FdU 10^{-5} M) and the fact that the segments at 8 min are larger than those at 4 min (from Kurek and Taylor, 1977).

completely prevent the release of origins which produce the segments of uniform size (Guy, 1977). The small numbers of chains which have already been initiated grow at a normal rate up to 20 minutes or longer. However, if the actinomycin D is given 2 minutes after the ^{3}H-thymidine pulse begins, there is rapid initiation and chain growth at a rate comparable to control cells which were not treated with actinomycin D.

This experiment suggests two features of initiation. RNA transcription may be involved in initiation, but the transcript is

not completed until thymidine triphosphate is available to the cell.
Moreover, the RNA transcript is not likely to operate as an m-RNA
since the effect shows up so soon. In a mammalian cell the processing time is considerably longer than the two to three minutes it
takes for the actinomycin D to produce an effect on initiation.
Another possibility is that the RNA transcript is a primer at the
initiation site, but not in the chain elongation processes, which
proceed without interruption in the presence of actinomycin D. We
have previously presented evidence that a fraction of the larger
Okazaki pieces from mammalian cells have a segment of RNA attached
which may be as long as 180-200 nucleotides. The evidence was based
on banding the fragments in CsCl in addition to their isolation and
observation with the electron microscope. When spread by a technique
that would extend single strands of DNA but leave RNA in a tangled
bush, the segments with a higher density had bushes attached at one
end (Taylor et al., 1975). The evidence that the RNA was covalently
attached was based on the observation that the fragments retained
their higher buoyant density after heating with formaldehyde which
completely separated double stranded DNA. It appears likely that
the longer pieces of RNA (100-200 nucleotides) may be involved in
priming initiation, but that chain elongation can continue in the
presence of actinomycin D because the Okazaki fragments produced
are primed by short segments (10 nucleotides) as has been reported
for polyoma virus replication (Reichard et al., 1974).

Autoradiographic measurements of chain growth rates show that
elongation is about one half μm (1500 nucleotides) per minute at
each fork in cells that have been held at the G_1-S phase boundary
for only a few hours or are just entering S phase. This is based
on a rate measurement for the first 10-15 minutes when the
concentration of thymidine used for release was 10^{-6} M. The rate
decreased to nearly one half that amount when cells were held at
the G_1-S phase boundary for 12-14 hours (20 hours from division)
by blocking with FdU. During the same period the number of origins
operating upon release increased at least two fold and perhaps
much more. Since the origins were much closer together along the
DNA fibers, it is possible that the fraction of the DNA in which
the origins were located did not increase at all. It has not been
possible to measure accurately the fraction of DNA involved, but
it is only a minor part of the total chromosomal DNA in which the
active origins develop when cells are blocked at the beginning of
S phase.

Release of Newly Replicated Segments of Double Stranded
DNA during lysis

The release of the newly replicated DNA as double strands,
i.e., still paired with the template (parental strand) was noted
some years ago (Taylor, 1973). The mechanism of release appears
to be through the action of an enzyme bound to DNA near the growing

fork.   In the original studies we thought the segments released
might be a characteristic length not related to the pulse time, but
subsequent experiments have indicated that in well synchronized
cells at the G_1-S phase boundary, the segments released increase
linearly with time (Kurek and Taylor, 1977 and Guy, 1977). Although
the released DNA includes some forks, it appears to be mostly linear
double stranded segments which are presumably released by breakage
near the growing fork. It has not yet been possible to determine
if the break in the template chains occurs ahead of the fork, at
the fork and just behind the fork. The DNA was released most
efficiently when lysis was carried out in 0.05 M EDTA, 0.10 M
$NaHCO_3$ and .25% sarkosyl (Geigy Chemical), pH 10.8 and 37°. However, some release occurs under nearly any lysis procedure. The
release is reduced by lysis at 0° and with high salt. It also
must be reduced by lysis in dilute buffer (0.01 M Tris, 0.01 M EDTA,
pH 8.0, with 1% SDS) which is used for autoradiography. If release
of replicated segments occurred, one would not see the tandemly
linked newly replicated segments of DNA in fiber autoradiographs.
Some probably is released since the background of labeled debris
increases as cells are held longer at the G_1-S phse interface. It
will be recalled that during such blocks, the initiation sites
increase linearly with time. Upon release by pulse labeling with
3H-thymidine many newly replicated short segments are produced.
Many of these are retained in tandem linkage along the DNA fibers.

The original suggestion was that unwinding enzymes were
responsible for the breakage (Taylor, 1973). That still appears
to be the best explanation. Guy(1977) has explored various methods
for preventing the release. Lysis in high salt, urea, and at low
temperature reduces the amount, but one procedure almost completely
prevents the release. That is lysis in 2 M NaCl and 7 M urea with
.05 M EDTA and sarkosyl, pH 8.0. The Okazaki fragments are still
released but no double stranded intermediates appear. It should
be emphasized that all of the procedures used preserve the bulk of
the DNA in very long strands. The regions susceptible to breakage
have been noted only near the growing forks.

The release of the newly replicated DNA allows one to measure
the chain growth by sedimentation. The increase in highly
synchronized DNA at the G_1-S phase boundary is linear with time as
has been shown for the newly replicated chains in alkaline gradients.
The new chains released by alkaline denaturation from lysates where
the breakage is prevented are about the same length as the double
strands would be at a comparable pulse time if released by the
usual lysis procedures. However, the new chains released from the
double stranded fragments by alkaline gradients are only about one
half the length of those new chains. This indicates an average of
one nick in the new chains when release of double stranded segments
was allowed to occur during lysis. The location of the nicks, as

well as the possibility of nicks in the template chain, have not
been investigated (Guy, 1977).

The Distribution of Potential Origins in the Chromosomal DNA

When cells were held for 10 hours before pulse labeling (4 to
6 hours at the G_1-S phase) by blockage with FdU, the center to
center distances of labeled segments after pulse labeling was much
less than in those cells held only 6 hours before pulse labeling.
After 20 hours from division the number of active origins and
their density along the fibers was still greater. During such a
block the potential for ^{3}H-thymidine incorporation increased
linearly with time (Fig. 2) even though the potential for chain
elongation decreases. We interpret these changes, which coincide
with an increase of the number of labeled segments per unit length
of fibers to mean that initiation sites increase linearly with time

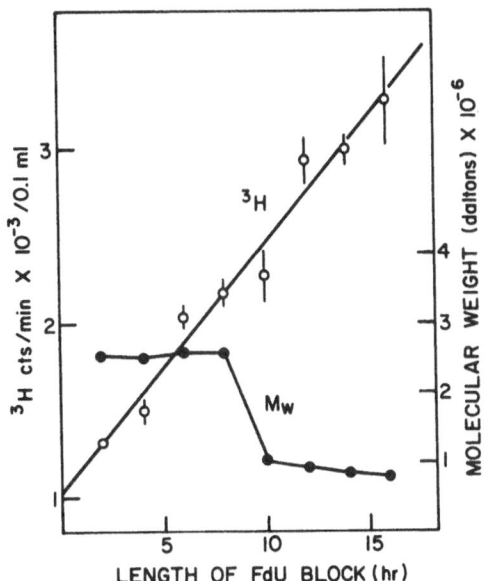

*Figure 2. Replicate cultures of mitotic CHO cells were incubated
in 10^{-5} M FdU for various lengths of time in thymidineless medium
and pulse labeled with ^{3}H-TdR (10^{-6} M) in conditioned Ham's medium
for 4 min at the times shown. Three 100 μl aliquots of the lysates
were precipitated and counted at each interval to estimate the DNA
replicated. A sample of the lysate was also analyzed in a NaI self-
generating gradient to estimate the size of replicated segments.
Since the rate of chain growth decreased with time while the
incorporation increased, the active initiation sites must have
increased at about a linear rate. (From Kurek and Taylor, 1977).*

during blockage of chain growth with FdU. More and more of the
potential sites in a portion of the DNA are activated. Upon pulse
labeling, initiation occurs at enough sites so that measurements
of center to center distances plotted as a frequency diagram fall
into a multimodal distribution in which the interval between modal
peaks is about 4 μm (Taylor and Hozier, 1976; Taylor, 1977). Fig.
3 shows autoradiographs made after pulse labeling for 4,8,24 and
40 minutes, which indicate the distribution of active origins
after holding cells at the G_1-S phase for 4-6 hours (pulse labeled
10 hours from division). Fig. 4 is a frequency diagram showing

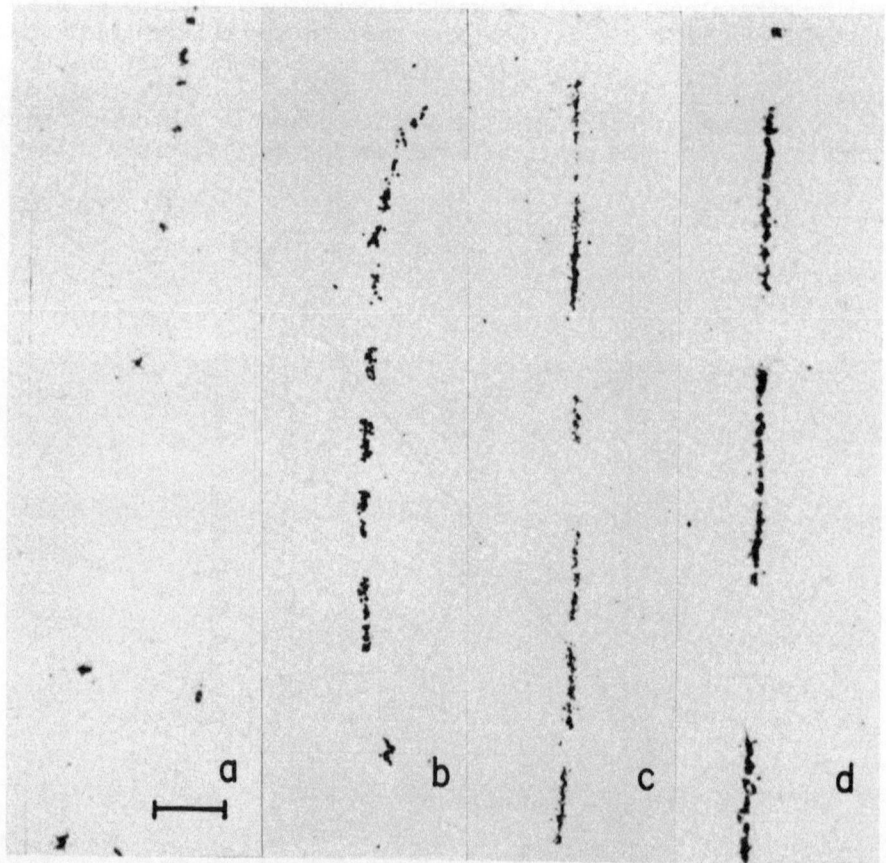

Figure 3. DNA fiber autoradiographs prepared after pulse labeling
CHO cells in ³H-thymidine (60 μCi/ml; 10⁻⁶ M) after 7 hrs in FdU
(10 hr. from division). Pulse times were 4 min (a), 8 min (b), 24
min (c) and 40 min (d). The length of the labeled segments allow
one to estimate the rate of chain elongation while the center to
center distance of segments indicate the distance between active
origins unless two segments have joined by meeting of growing forks.
The scale measures 10 μm. (From Taylor and Hozier, 1976).

the spacing between origins after several pulse times. The spacings
are so regular that we are led to the conclusion that the
chromosomes of Chinese hamsters have potential origins about every
four µm. However, only a few of these are used in fibroblasts in
culture and in most other cells of mammals which have S phases of
about the same length. The distribution of origins have been
estimated by autoradiography to be an average of 30 µm apart in
asynchronous Chinese hamster cells blocked for 12 hours with FdU
(Huberman and Riggs, 1968). In mouse L cells pulsed without blocking (Hand and Tamm, 1974; Hand, 1975a) 80% of the origins are 20-90
µm apart with a mode at a size class of 40-50 µm and a mean of 60
µm. Ockey and Saffhill (1976) have recently measured the intervals
between origins in several mammalian cell lines. The means varied
from 49.78 ± 2.13 µm for Hu P (Human embryonic fibroblast at the 4th
passage) to 84.16 ± 4.14 for mouse 3T3 cells, with CHO having a
mean distance of 73.19 ± 4.55 µm. These measurements are probably
the most accurate made to date, since they pulse labeled cells in
asynchronous growth without manipulation or drug treatment. If we
assume that the CHO cell has potential origins every four microns
only one in 18 or 19 of the potential origins are used in the
fibroblast. By holding cells at the beginning of S phase with FdU
the cell can be forced to use one in 3 or possibly more (Taylor,
1977) in a limited portion of the DNA. Presumably at some stage in
the life cycle, perhaps in early embryonic growth, most or all of
the potential origins are used during a shorter S phase.

What happens when two forks meet ? Are there structural
termini where fork movements stop ? Hori and Lark (1976) examined
autoradiographs of cells of the kangaroo rat (*Dipidomys ordii*)
which contain large quantities of late replicating satellite DNA.
They had labeled the cells with ^{3}H-thymidine from early S to near
the end of S phase. However, they found unlabeled gaps in fibers
of DNA which were otherwise labeled along most of their length.
The unlabeled DNA was shown to be satellite DNA by banding the long
segments in CsCl. The length of the gaps measured in autoradiographs
fell into modal frequencies divisible by 7 µm. They had suggested
that units of replication might have termini (Hori and Lark, 1973).
However, considering the high frequency of origins in mammalian
cells which are regularly spaced, it is reasonable to suppose that
the progress of forks is more likely to be arrested at potential
origins. Therefore, the satellite of *Dipidomys* could have origins
at intervals of 7 µm and no termini. A palindromic sequence at a
potential origin might be likely to interrupt the progress of the
polymerase. However, to explain the absence of unreplicated segments of odd multiples of the basic unit in the *Dipidomys* cells,
one may have to assume a strong but perhaps not an absolute
inhibition of firing of two adjacent origins simultaneously, i.e.,
the movement of two forks toward each other from adjacent origins.
Any unfavorable condition which interferes with replication might
then cause termination at a potential origin. When replication was

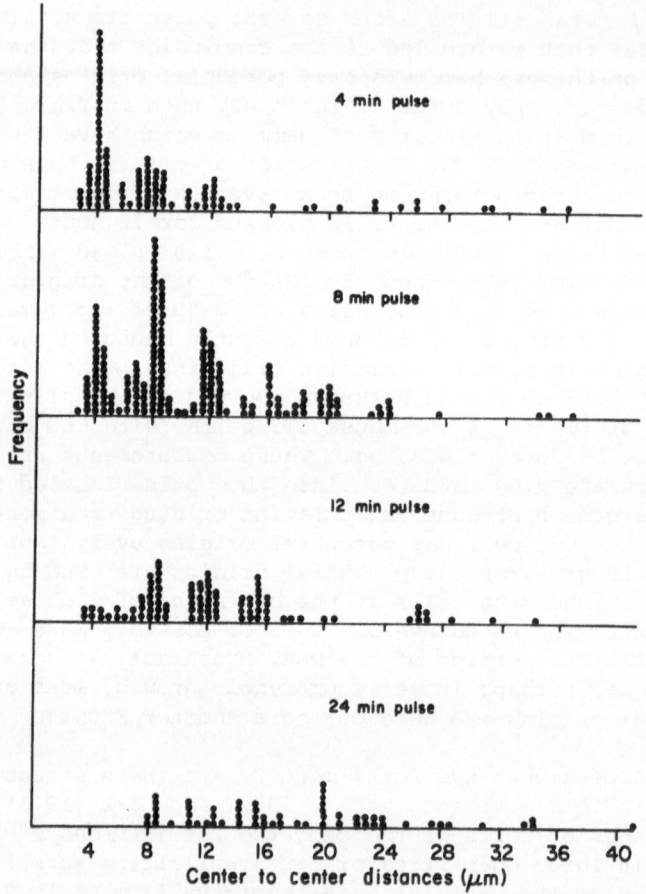

Figure 4. *Frequency diagram showing the center to center distances
of labeled segments, after the pulse times as shown when cells
were pulse labeled after 7 hrs in FdU (10 hr from the last mitosis).
Note that there are several modes in the frequency of center to
center distances with a repeating interval of about 4 μm.   (From
Taylor and Hozier, 1976).*

resumed the result could be unidirectional replication from those
sites.   This could also be the explanation for the observations
of unidirectional replication in cells where DNA synthesis was
reduced by inhibition of protein synthesis (Hand, 1975b).

    Other agents which inhibit initiation are ionizing radiation
(Painter and Young, 1975; 1976) and apparently any agent which
causes breaks in DNA (Painter, 1977; Povirk and Painter, 1976).
Moderate doses of X-rays (500 rads) have been shown to result in

a rapid decrease or cessation of replication from new origins.
Those chains already growing, however, continue at nearly the same
rate. Painter and Young detected this change by observing the change
in size of the segments of new DNA released after radiation.
Alkaline gradients of pulse labeled cells show a wide distribution
of sizes of labeled segments. Presumably the whole range of bi-
directionally growing chains would be about equally labeled by a
relatively short pulse. In a cell where origins are not produced
the size of the segments labeled in a pulse will increase with
time following the interruption. This was the method Painter and
Young used to detect the change, i.e., pulse labeling and
determination of the size of replicating chains in alkaline sucrose
gradients. Ionizing radiations, other mutagenic agents such as
methyl methane-sulfonate, and bromodeoxyuridine incorporated into
DNA and irradiated with light of 313 nm wavelength all produced
similar effects on mammalian cells in culture. Inhibition lasted
for 2-3 hours after the treatments when the damage was comparable
to that produced by 500-1000 rads of X-rays. Ultraviolet light
also produced a temporary decrease in the number of growing points,
but did not decrease the rate of cahin elongation (Povirk and
Painter, 1976).

Initiation of Replication for the Future

        Chromosomes probably consist of a single long fiber of DNA
with potential replication origins distributed in some characteristic
pattern. The rate of replication will then depend on the number of
protein molecules available to form the initiation complexes. One
initiation can replicate an indeterminate amount of the chromosome,
for the forks can move over a potential origin with only a limited
chance that an interruption will occur. Two potential origins
close together which happen to be activated will fuse to form a
single expanding loop with two forks instead of the original four
which the two expanding loops had. Perhaps there are some interactions which inhibit simultaneous firing of two adjacent origins,
but this idea is based only on the size of the unreplicated gaps
in satellite DNA of *Dipidomys*. Other explanations may prove to be
more appropriate for the observation. Otherwise the potential
origins probably fire at random, except that certain rather large
domains of the chromosomes become available for replication at
intervals in the S phase. This control of sequence could be based
in part on the base sequence of potential origins. For example,
satellite DNA's frequently replicate late in S phase during an
interval that is less than one-half of the S phase. If the delay
were due to different nucleotide sequences at the potential origins,
the satellite origins would be unable to initiate until a
particular protein, of the replication complex specific for these
sites, is synthesized in response to a regulatory signal during S
phase. Our studies of the sequence programming of DNA replication
in CHO cells (Adegoke and Taylor, 1977) indicate that the order of

replication need not be stringently controlled. A pulse label at
the beginning of S phase will label a portion of the DNA which can
replicate in the subsequent S phase at almost any time. There is
very little chance that more than a random sample of the same
segments will initiate synthesis very early in the subsequent S
phase. Likewise, the same origins need not function at the subsequent S phase since there are a great excess per unit length of
DNA.   Amaldi et al. (1973) labeled cells synchronized at the G_1-S
phase boundary with ^{3}H-thymidine during a 30 min pulse. They
resynchronized the cells at a subsequent S phase and pulse labeled
for 10 minutes with the same intermediate. They then prepared and
examined DNA fiber autoradiographs. If segments initiated during
the first cycle had been initiated again at the subsequent cycle,
they should have been labeled segments produced in 30 minutes during
the first cycle which were double labeled in the middle during the
second cycle. Very few were found. Most of the segments had not
served to initiate DNA replication the second time, but in the few
double labeled segments found the middle region was used again,
suggesting reutilization of some of the origins.

     In spite of these indications of rather non-stringent controls,
we are quite sure that reproducible autoradiographic patterns are
discernible not only for sex chromosomes, but for many portions
of autosomes as many studies have shown. We may point out that
the attempts to show reproducibilities in the replication of
sequences may have an inherent defect in that the localized radiation in a chromosome labeled with tritium may delay its initiation.
In view of the reports that low doses of radiation delay initiation,
perhaps by a few strand breakages which affect large domains of DNA,
one must entertain some reservations about the programming of
replication when the segments to be studied carry the potential for
their own breakage.

     A variation of the sequence programming appears to be necessary
to explain the modifications which are known to affect the x-
chromosomes of the females in mammals. One of the x-chromosomes
is modified in the cells of the early embryo so that each cell has
one late replicating x-chromosome (or a differential segment) and
one x-chromosome which begins replication at early S phase. The
modifications occur in individual cells of the early embryo and
once an x-chromosome has been programmed for early replication, that
cell and all of its descendants exhibit the same pattern. The early
replicating x-chromosome is the genetically active one. The
behavior of rare cells with 3 or more x-chromosomes indicates
that the modification affects the early replicating chromosomes,
for any number 1,2 or 3 of the remaining ones can be unaffected.
In the male the single x is always active. Riggs (1975) has
proposed that the modifications involve a change in methylation.
Such a change might have to involve each potential origin, unless
there is an additional level at which regulation can operate.

That some other regulatory system operates to make a particular domain of a chromosome available for replication seems likely. The fact that a small percentage of the DNA becomes involved in initiation when cells are blocked at the G_1-S phase with FdU, and the indication that this DNA need not replicate first in the next S phase, suggests that some factor other than structural properties of the origin is involved. In spite of the fact that a chromosome appears to be a single piece of DNA, it is composed of morphological domains, which are demonstrated as bands in chromosomes by various treatments and staining procedures. These bands are morphological entities, but may represent aggregates of two or more packets or domains of the size I am referring to. Lange et al. (1977) have reported that mouse chromosomes have very large subunits which were identified by centrifugation in gradients and by measurements with the viscoelastometer. We had noted such large subunits in preliminary viscoelastic measurements of hamster DNA (Taylor, 1976; Hozier, 1975). The domains contain an average of about 1.25×10^{10} daltons of DNA. When one examines the beads or incipient bands in chromosomes in CHO cells treated and stained to show bands, but stopped before the bands attain their maximum condensation, about 24-30 beads could be counted in the largest chromosomes. Since this largest chromosome contains about $30 \times 1.25 \times 10^{10}$ daltons of DNA, the number of beads correspond in size or amount of DNA to the viscoelastic subunits (Hozier, 1975). When one deals with such large packets of DNA, it is difficult to rule out various artifacts, but Lange et al. (1977) have considered a number of the possible criticisms and argues that the subunits are real and demonstratable by both sedimentation and viscoelastometry. Maybe these will prove to be the domains that are somehow regulated as units in making the DNA available for replication.

The chromosome then consists of potential initiation sites distributed at regular intervals. How these might be related to nucleosomes and transcription units is not yet indicated. The potential initiation sites are utilized nearly at random each cell cycle, but in most cells that have a long S phase only a few of the sites are used in any one cycle. There are probably several base sequences that function as initiation sites in special DNA such as satellite DNA or classes of DNA which replicate at a limited time during S phase. Otherwise the sequence regulation may be controlled at the level of domains much larger than the segment associated with each potential origin as suggested above.

REFERENCES

Adams, R.L.P. (1971) Biochim. Biophys. Acta 254:205

Adegoke, J.A. and Taylor, J.H. (1977) Exptl. Cell Res. 104:47

Amaldi, F., Buongiorno-Nordelli, M., Carnevali, F., Leoni, L.,
    Mariotti, D., and Pomponi, M. (1973) Exptl. Cell Res. 80:79

Cairns, J. (1966) J. Mol. Biol. 15:372

Callan, H.C. (1972) Proc. R. Soc. Lond. B. 181:19

Callan, H.G. and Taylor, J.H. (1968) J. Cell Sci. 3:615

Guy, Arthur (1977) Dissertation, Florida State University.

Hand, R. (1975a) J. Cell Biol. 64:89

Hand, R. (1975b) J. Cell Biol. 67:761

Hand, R. and Tamm, I. (1974) J. Mol. Biol. 82:175

Hori, T. and Lark, K.G. (1973) J. Mol. Biol. 77:391

Hori, T.A. and Lark, K.G. (1976) Nature 259:504

Hozier, J.C. (1975)   Dissertation, Florida State University.

Huberman, J.A. and Riggs, A.D. (1968) J. Mol. Biol. 32:327

Jacob, F. and Brenner, S. (1963) Compt. Rend. Acad. Sci. 256:
    298

Kurek, M.P. and Taylor, J.H. (1977) Exptl. Cell Res. 104:7

Lange, C.S., Liberman, D.F., Clark, R.W., Ferguson, P. and Sheck,
    L. Biopolymers 16:1093

Lark, C. (1968) J. Mol. Biol. 31:401

Ockey, C.H. and Saffhill, R. (1976) Exptl. Cell Res. 103:361

Painter, R.B. (1977) Mutation Res. 42:299

Painter, R.B., Jermany, D.A. and Rasmussen, R.E. (1966) J. Mol.
    Biol. 17:47

Painter, R.B. and Young, B.R. (1975) Radiation Res. 64:648

Painter, R.B. and Young, B.R. (1976) Biochim. Biophys. Acta 418:
    146

Povirk, L.F. and Painter, R.B. (1976) Biochim. Biophys. Acta
    432:267

Reichard, P., Eliasson, R., Söderman, G. (1974) Proc. Nat. Acad.
    Sci. U.S.A. 71:4901

Riggs, A.D. (1975) Cytogenet.Cell Genet. 14:9

Taylor, J.H. (1960) J. Biophys. Biochem. Cytol. 7:455

Taylor, J.H. (1963) J. Cell Comp. Physiol. 62: Suppl. 1, 73

Taylor, J.H. (1968) J. Mol. Biol. 31:579

Taylor, J.H. (1973) Proc. Nat. Acad. Sci. U.S.A. 70:1083

Taylor, J.H. (1976) Structural and functional subunits of
    chromosomes. Adv. in Pathobiol. 3 (Developmental Genetics)
    (Fenoglio, C.M., Goodman, R. and King, D.W. eds.) pp 19-26
    Stratton Intercont. Med. Book. Corp, New York.

Taylor, J.H. (1977) Chromosoma (in press)

Taylor, J.H. (1978) In: "Molecular Genetics" (J.H. Taylor, ed.)
    Academic Press, New York (in press)

Taylor, J.H., Wu, M., Erickson, L.C. and Kurek, M.P. (1975)
    Chromsoma, 53:175

Taylor, J.H. and Hozier, J.C. (1976) Chromosoma 57:341

DEVELOPMENTAL CONSIDERATIONS OF EUKARYOTIC DNA REPLICATION AND REPLICON SIZE

Peter J. Stambrook* and Douglas J. Burks

Department of Biology
Case Western Reserve University
Cleveland, Ohio 44106, U.S.A.

DNA replication in eukaryotes begins at multiple sites along the length of the chromosome (1) and proceeds bidirectionally from each point of origin (2). The factors which govern the temporal order of replication initiation remain unknown. These unresolved determinants may endow early embryonic cells with replication characteristics which distinguish them from adult cells. Among these are an altered temporal sequence of DNA replication and an extremely brief S phase. In this report I will provide evidence supporting the former contention and indirectly address the latter. It has been argued that the brevity of the early embryonic S phase is due to the activation of a number of replication initiation sites that are not available in adult cells with longer S phases (3,4). The argument proposes that the increase in the number of available replication initiation sites is accompanied by a decrease in replicon size. I wish to interject a cautionary note with respect to the second part of the argument by demonstrating that replicons as small as any seen in embryonic cells can be detected in cultured mammalian cells.

In cultured mammalian cells (5) and in vegetative cells of the slime mold *Physarum polycephalum* (6) the order in which different parts of the genome replicate appears to be under a rigid control. The DNA sequences that replicate early in one S phase replicate early in successive S phases and vice versa (5,6). In several mammalian cell lines tested, DNA that replicates early in the S phase consistently has a greater buoyant density (presumably a higher G+C content) than that which replicates late in the S phase (7-11). Likewise, satellite DNAs (7-9,12,13) and the genes which

*A.S.I. Participant

code for ribosomal RNA (14-16) replicate at defined times within
the S phase of cultured proliferating cells.

The sequence or temporal order of DNA replication appears not
to be constant in at least some cells within the life cycle of an
organism. In grasshopper spermatocytes, for example, the X
chromosome is condensed and late-replicating. In early spermato-
gonia, the X chromosome becomes euchromatic and replicates early
with the autosomes (17). A similar observation has been made in
the Chinese hamster where the long arm of the X chromosome and the
entire Y chromosome are late replicating in somatic cells but early
replicating in early cleavage stages of embryogenesis (19).

In early embryos of the frog, *Rana pipiens*, we have detected
a similar change in replication pattern (20). In these experiments
regions of mitotic chromosomes containing DNA which had replicated
during the last quarter of the S phase were identified auto-
radiographically at different stages of embryogenesis. Small
embryonic explants from different parts of the embryo and from
different developmental stages were transferred to culture dishes
containing sterile Niu-Twitty saline (21) with (^{3}H) thymidine
(25 µc/ml). The explants were incubated continuously with the
isotope for increasing intervals up to a period equivalent to the
duration of G_2 plus one quarter of the S phase. After labeling,
the explants were washed, fixed and stained with aceto-orcein,
squashed to prepare mitotic chromosomes, and processed for autoradiography. Since in amphibians the duration of both the G_2 and S
phase increases as development proceeds (22,23), the absolute
labeling period varied depending upon developmental stage.

The intent of the experiment was to determine whether any
detectable changes in late replication occurred during early
amphibian development. As the experiment is designed, cells with
labeled mitotic chromosomes were at the end or near the end of the
S phase at the time the isotope was introduced into the culture
medium. Furthermore, the position of label over the chromosomes
represents the last portion of the chromosomes to undergo replication. Regions of the chromosome that were unlabeled had replicated
prior to the introduction of isotope. In Fig. 1, the late-
replication pattern obtained from blastula stage embryos (12 hours
after fertilization) is presented. The pattern of silver grains over
the chromosomes is diffuse, with silver grains spread over much of
the length of the chromosome arms. The ends of the chromosomes are
frequently free of label. This pattern is found in both animal and
vegetal halves of the embryo. The apparent high background in these
pictures is not due to label but to the large amount of pigment
granules found in the embryos. In contrast to the diffuse labeling
pattern characteristic of blastula cells, neurulae, obtained 40 hours
later, present a very different picture (Fig. 2). The ends of the

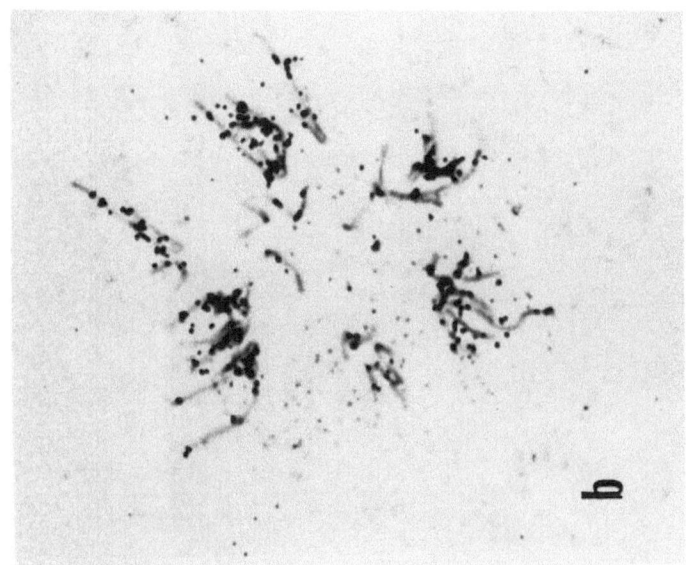

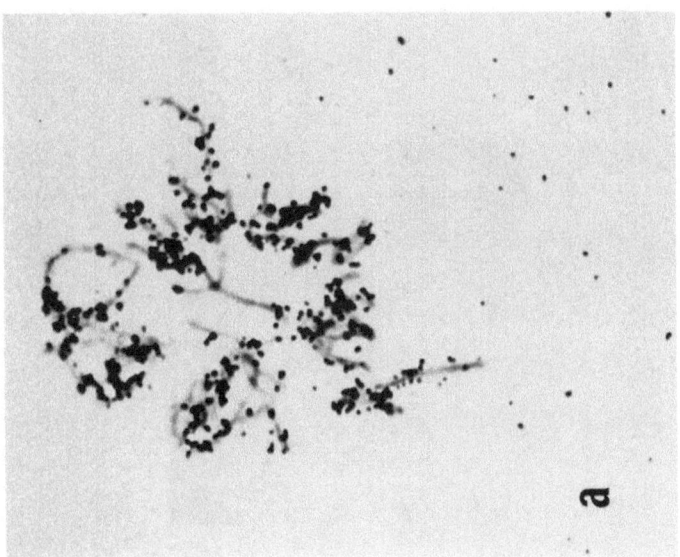

Figure 1. Late replicating patterns of two separate complements of blastula mitotic chromosomes. Blastula ectodermal explants were labeled with (³H)thymidine for a period equivalent to G₂ plus one quarter of the S phase (1 hour and 40 minutes). The explants were washed, stained, fixed, and prepared for autoradiography.

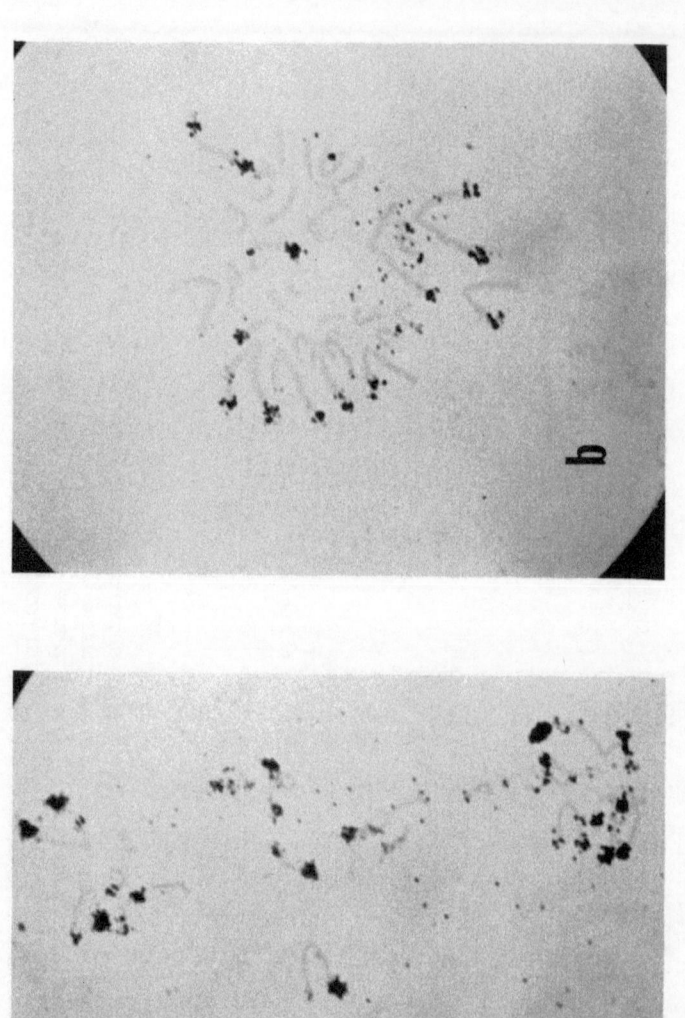

Figure 2. Late replicating patterns of two separate complements of neurula mitotic chromosomes. Neurula ectodermal-mesodermal explants were labeled with (^{3}H) thymidine for a period equivalent to G_2 plus one quarter of the S phase (four and a half hours). The explants were washed, stained, fixed, and processed for auto-radiography.

chromosomes are invariably densely labeled while most of the length
of the chromosome arms remains free of grains. Again, the pattern
was the same whether the cells were of ectoderm-mesoderm origin or
from the endoderm. *Rana pipiens* has been karyotyped (20,24), and
the replicative behavior of chromosome 1, the largest metacentric
chromosome, and chromosomes 7 and 8, the two largest submetacentric
chromosomes has been analyzed (20). In all three cases, the late
replication pattern was consistent with that of the entire mitotic
complement. During blastulation, the late replicating pattern was
diffuse with chromosome ends free of label. By the neurula stage,
late replication in each case examined was restricted to the
chromosome termini (20).

The data indicate that during early amphibian development the
temporal order of DNA replication may change, and that DNA sequences
which were relatively early replicating became late replicating and
vice versa. Although genetic inactivity has been ascribed to late
replicating DNA (25), we have no evidence that DNA sequences
contained at the chromosome termini in fact undergo a change in
transcriptive activity. The observation that the DNA replication
pattern changes during early development of *Rana pipiens* has been
independently confirmed and extended by Matsumoto and Dasgupta (27).
These investigators injected nuclei from later stages into
anucleated eggs and found that at the blastula stage the transplanted
nuclei had reverted from the terminal pattern of late replication,
characteristic of neurulae and later developmental stages, back to
the diffuse pattern characteristic of blastulae. They argued that
cytoplasmic factors play a role in governing the temporal order of
DNA replication.

The second characteristic of early embryonic cells mentioned
earlier is the brevity of the S phase which can be as short as 10
and 15 minutes (4,22,28). As development progresses, the S phase
increases to a length of 6 to 8 hours or longer. It has been argued
that variations in the duration of an S phase are due not to changes
in the rate of polymerase movement, but rather to the availability
(or lack thereof) or replication initiation sites (3,4,29). In
other words, as the S phase lengthens, fewer replication initiation
sites participate in initiation events and the effective replicon
size (i.e., the distance between adjacent replication origins)
increases. This argument is based primarily on DNA fiber autoradiography (3,29), and the comparison of embryonic replication
analyzed by electron microscopy with somatic cell replication
analyzed by DNA fiber autoradiography (4). The problem with such
a comparison is that the two techniques are measuring replication
events at very different levels. Electron microscopy is best suited
for visualizing smaller replicons, ranging from 0.1 μm up to 4 or 5
μm. In contrast, DNA fiber autoradiography will only detect larger
replicons, and is unable to resolve replicons smaller than 4 or 5 μm.

DNA fiber autoradiography also cannot unequivocally differentiate
between true large replicons and a set of clusters of small
replicons, since both would produce similar patterns of silver grain
tracks. That such clusters exist has been demonstrated by Zakian
(30) who showed that in *Drosophila virilis* embryos, two major
classes of replicating molecules could be distinguished. One class
consists of tandem replication "bubbles" or "eyes" ranging from 900
to 41,000 base pairs in size, and spaced at intervals of 4,000 base
pairs or multiples thereof. The second class of molecule is
comprised of random clusters of bubbles in which the bubbles are
smaller than 900 base pairs. The distance between the centers of
adjacent clusters again approximates 4,000 base pairs or a multiple
thereof. It is not unreasonable to hypothesize that fusion of small
replicons within a cluster gives rise to the large bubbles observed.
It is the latter type of structure that is most likely detected by
DNA fiber autoradiography.

Direct electron microscopic visualization of replicating
eukaryotic chromosomal DNA has thus far been successfully achieved
only in early insect embryos (4,28,30-32). The advantage of these
systems is that the division cycle and S phase are both very short,
providing a natural enrichment of replicating molecules. When DNA
from these cells is examined by electron microscopy about 10 percent
of the molecules visualized contain replication structures. Electron
microscopic analyses of replicating DNA from non-embryonic cells with
longer S phases has been frustrated by the very low frequency of
replicating molecules observed. As a result, replicons smaller than
4 µm have not been reported in eukaryotic cells other than from
embryonic sources.

Electron microscopic analysis of replicating DNA from non-
embryonic cells, therefore, requires a method that enriches for
replicating molecules. The technique we have developed takes
advantage of a physical property characteristic of replicating DNA,
namely the presence of single-stranded regions associated with the
replication fork (28,33). Since mercuric ion has a greater affinity
for single-stranded rather than native DNA (34), it seemed reasonable
that by interacting DNA with low levels of $HgCl_2$, molecules containing replication structures would preferentially bind with the
mercuric ion. These molecules would be more dense than the bulk of
the DNA in Cs_2SO_4 and could then be separated by equilibrium density
centrifugation.

This proposition was tested by pulse labeling culture Chinese
hamster cells with (^{3}H)thymidine (100 µc/ml) for 45 seconds. The
labeled DNA was isolated by phenol extraction, interacted with low
levels of $HgCl_2$ and centrifuged to equilibrium in Cs_2SO_4. Upon
fractionation, the gradient yielded a peak of radioactivity that
closely followed the absorbance profile (Fig. 3). Analysis of the
specific activity of the DNA in each fraction, however, revealed a

peak of specific activity on the dense side of the absorbance profile (Fig. 3). If the cells were chased with excess unlabeled thymidine for two hours after the pulse, the dense peak of specific activity was absent from the gradient (Fig. 4). This result suggests that the radioactivity is chased from the vicinity of the replication fork and its associated single-stranded DNA into wholly

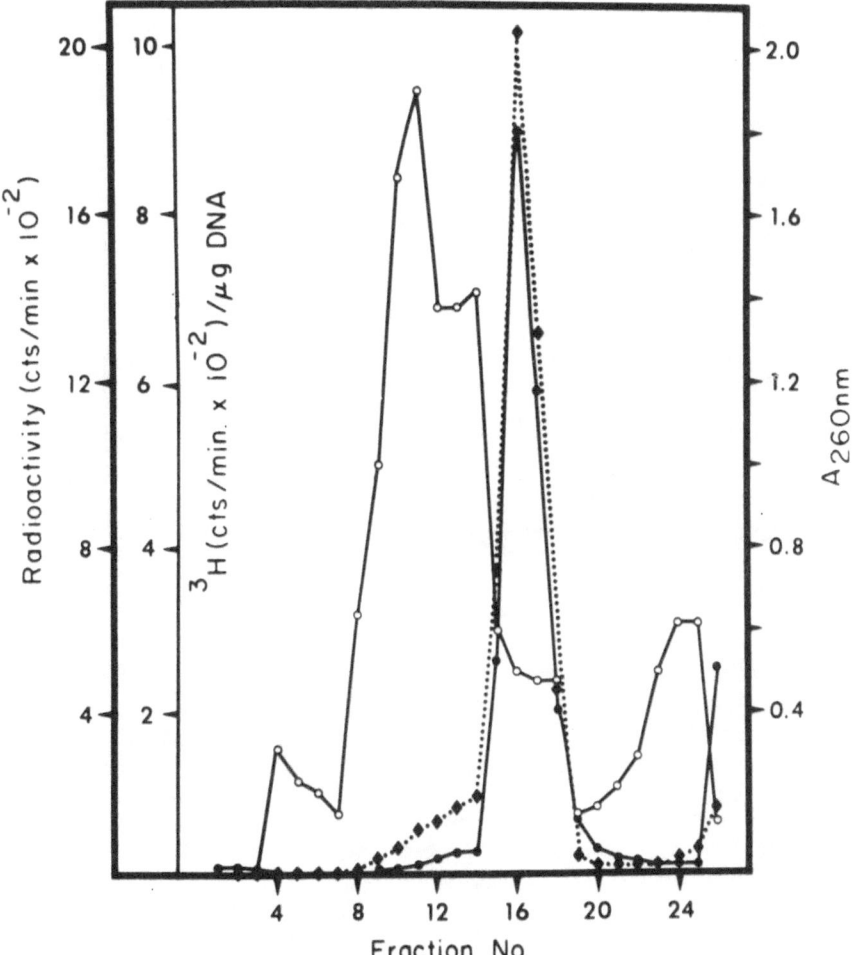

Figure 3. Centrifugation of pulse-labeled DNA in a Hg^{++}/Cs_2SO_4 gradient. Cells were labeled for 45 seconds with (^3H)thymidine, and the DNA was immediately extracted with phenol. $HgCl_2$ was added to 100 μg DNA to a Hg^{++}/DNA-phosphate ratio of 0.001 and the DNA was banded in a 9 ml Cs_2SO_4 gradient. The gradient was fractionated, the absorbance at 260 nm was determined for each fraction, and 70 μl aliquots from each fraction was precipitated with TCA and counted. ◆·····◆·····◆ , ^3H cts/min; ●—●—●, absorbance at 260 nm; O—O—O, specific activity (^3H cts/min)/μg DNA.

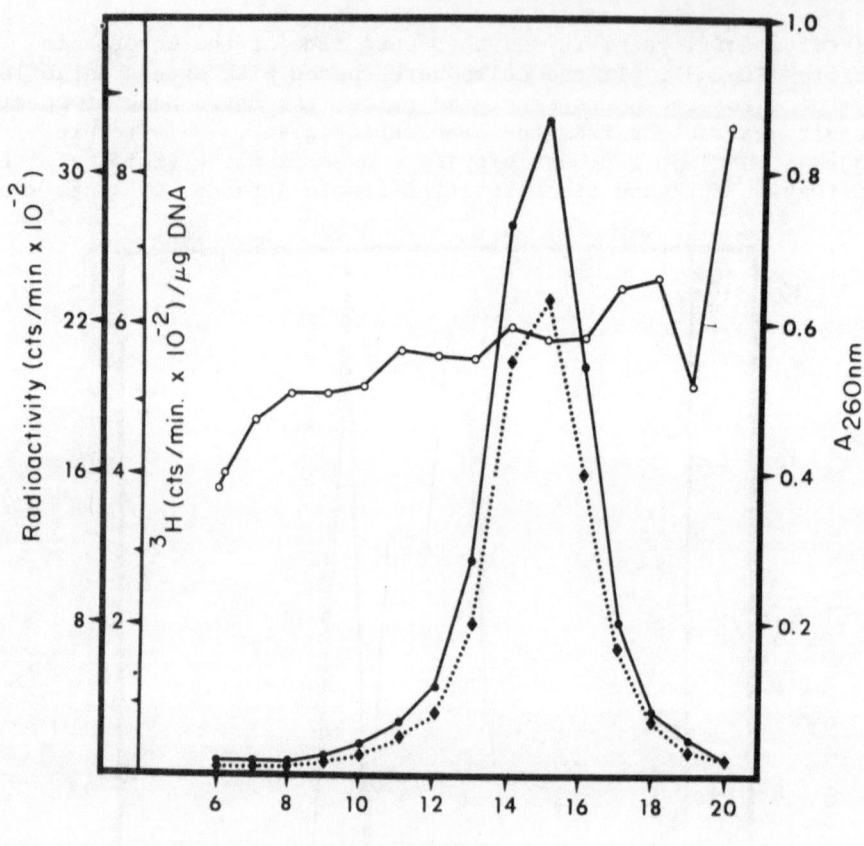

Fraction No.

Figure 4. Centrifugation of pulse-chased DNA in a Hg^{++}/Cs_2SO_4 gradient. Cells were labeled with (^3H)thymidine for 45 seconds followed by culture in medium containing 0.1 mM unlabeled thymidine for 2 hrs. The DNA was extracted and 100 µg was banded in Cs_2SO_4 after addition of $HgCl_2$ to a Hg^{++}/DNA--phosphate ratio = 0.001. The DNA was centrifuged, the gradient was fractionated, and the DNA in each fraction was quantified and counted as described in Fig. 1.
◆····◆····◆, *^3H cts/min;* ●—●—● *absorbance at 260 nm;* ○--○--○
specific activity (^3H cts/min)/µg DNA.

native replicated DNA. The dense peak of specific activity was also not apparent if the $HgCl_2$ was omitted from the gradient or if the DNA was first treated with the single-strand specific nuclease S1 before interacting with the $HgCl_2$. These two observations indicate the necessity of mercuric ion and of single-stranded DNA respectively for generating the dense peak of specific activity.

Since DNA in the density-shifted specific activity peak behaved kinetically like a replication intermediate, and since its behavior in Cs_2SO_4 gradients conformed to that predicted for a

replicating molecule, the DNA was spread for electron microscopy.
Analysis of the molecules revealed that on a per mole nucleotide
basis there was a 25 to 30 fold enrichment of structures consistent
with replication forms compared to unfractionated DNA or DNA from
other regions of the gradient. About 7 percent of the molecules
contained replication structures. These included molecules with a
single fork, or a single bubble as well as molecules with multiple
forks (Fig. 5) or with multiple bubbles (Figs. 6 and 7). Replicon
size was estimated by measuring the distance between the centers of
adjacent replication structures contained on molecules with multiple
bubbles. All replicons analyzed measured less than 5 µm and ranged
in size from 0.2 µm to 4.5 µm (Fig. 8). These sizes, which are
below the limit of resolution of DNA fiber autoradiography, are in
the same range as replicon sizes observed in early insect embryos.

   The enrichment technique described above has its limitations
since it will select for molecules with the highest single strand
to double strand ratio. These molecules would tend to be smaller
than the average, with one or more small replication forms. Such
molecules should have a high single strand to double strand ratio
and should bind relatively more mercuric ion. They should also
therefore exhibit the greatest density shift after ineraction with
$HgCl_2$ followed by centrifugation in Cs_2SO_4. Larger molecules with
large replication bubbles would have a lower single strand to double
strand ratio which would reduce their density shift and make their
detection less likely.

   This limitation of the enrichment technique restricts the
interpretation of the data. DNA fiber autoradiography is limited to

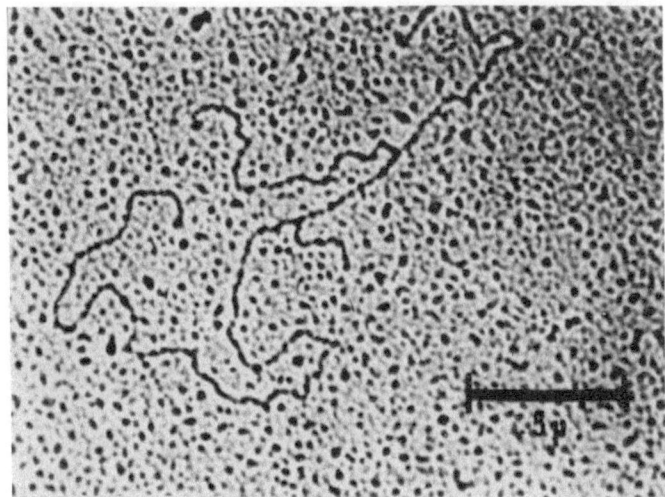

Figure 5. Molecule with multiple forks. The bar represents 0.5 µm.

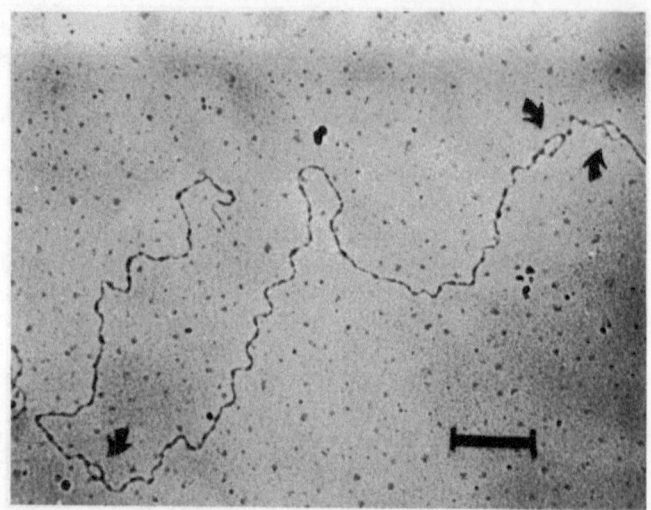

Figure 6. *Molecule with three tandem bubbles, indicated by arrows. The bar represents 0.5 μm.*

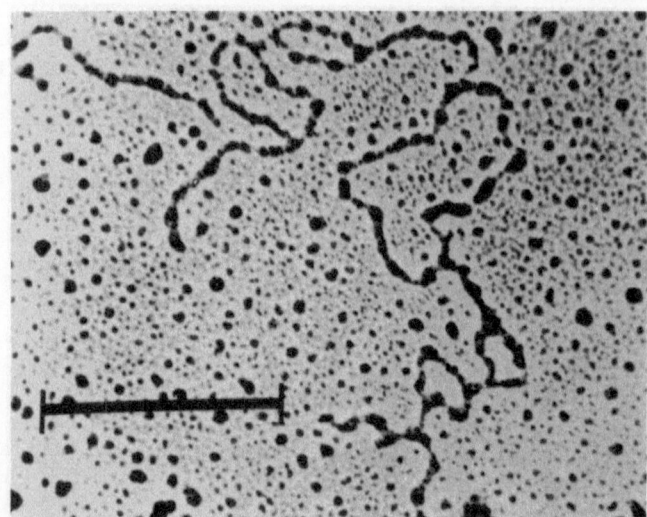

Figure 7. *Molecule with 5 tandem bubbles and a fork. The bar represents 0.5 μm.*

the analysis of large replication units while the procedure described in this report will detect predominantly smaller units. The smallest distribution of grain tracks thus far described were reported by Taylor and Hozier (35) who observed replicon sizes of 4 μm or multiples thereof. The most common replication units seen

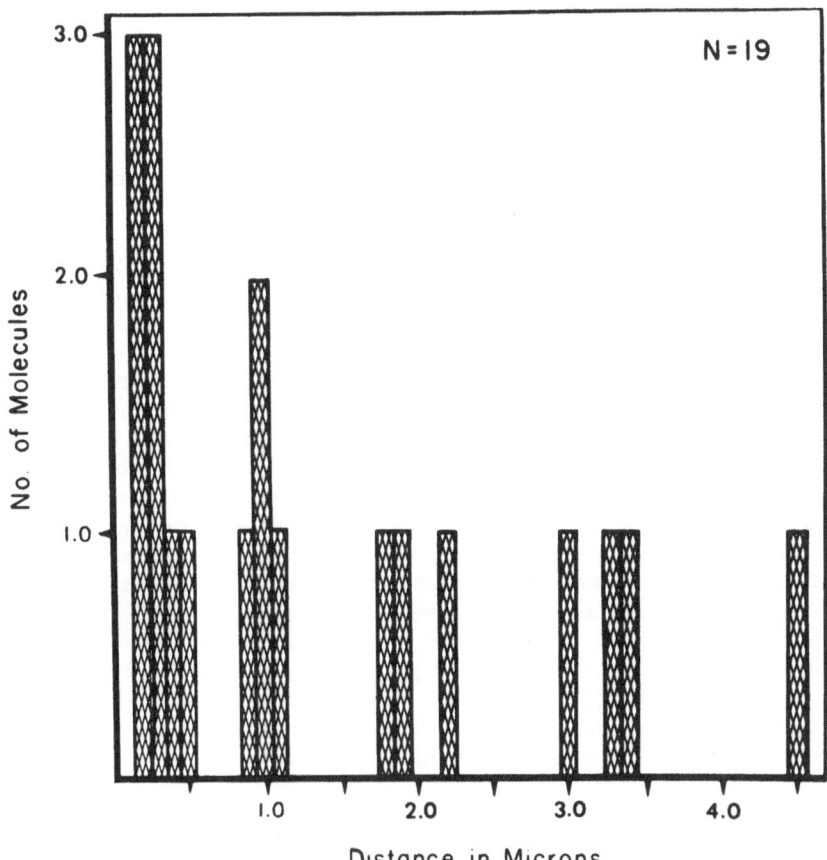

Figure 8. *Distribution of replicon sizes. The distance between centers of adjacent replication structures was measured at a 73,000 X magnification, and absolute sizes determined using circular PM 2 phage DNA as a size marker.*

by electron microscopy in this study were below 1 μm and probably represent true replicon sizes. The data unequivocally demonstrate that very small replicons exist in cultured mammalian cells and are not confined solely to embryonic systems. The data, however, cannot resolve whether such small replicons are ubiquitous within the genome since larger replicon sizes are not readily detected by this technique. The small tandem replicons may, in fact, not be representative of the majority of replicon sizes, but may reflect replication of selected regions of the genome such as tandemly repeated sequences. These would include satellite DNAs, and the DNAs coding for ribosomal RNA, 5S RNA, transfer RNA and histone messenger RNA. The tandem grain tracks observed by DNA fiber auto-radiography may also reflect true replicon sizes, but of a larger size class. Alternatively, they may represent an organization of replication superimposed upon that of the replicon, such as the distribution of clusters of small replicons that have fused to generate the grain tracks detected by fiber autoradiography.

   In summary, the data indicate that the temporal order of DNA replication for different cells of an organism need not be fixed, but may vary at least as a function of development. The late replication pattern of *Rana pipiens* blastulae has been shown to be different from that of late stages. Although the reorganization of replication sequence as a function of early development can be demonstrated in amphibians, it need not be related to the reorganization of replication units that accounts for the lengthening of the S phase. The argument that the brevity of the S phase in early embryos compared with later stages or with adult or cultured cells may be due to a greater number of active replication initiation sites, should be viewed with caution. Introduction of increased numbers of replication initiation sites would result in a reduction in replicon size. The data, however, demonstrate that very small replicons, less than 1 μm, are not unique to embryonic cells, but are also found in cultured mammalian cells. It remains possible that as the S phase lengthens, certain initiation sites are somehow inactivated. This alternative cannot be ruled out and would result in a population of larger replicons which would be far less efficiently detected by the procedures described. Nevertheless, very small replicons are retained in non-embryonic cells and must be accommodated in any models of DNA replication.

ACKNOWLEDGEMENT

This work was supported in part by U.S.P.H.S. Grant GM 22248.

REFERENCES

1.   Taylor, J.H., Woods, P.S. and Hughes, W.L. (1957) Proc. Natl. Acad. Sci. USA 43:122

2.   Huberman, J.A. and Riggs, A.D. (1968) J. Mol. Biol. 32:327

3.   Callan, H.G. (1972) Proc. Roy. Soc. London, B. 181:19

4.   Blumenthal, A.B., Kriegstein, H.J. and Hogness, D.S. (1973) Cold Spring Harbor Symp. Quant. Biol. 38:205

5.   Mueller, G.C. and Kagiwara, K. (1966) Biochim. Biophys. Acta 114:108

6.   Braun, R. and Willi, M. (1969) Biochim. Biophys. Acta 174:246

7. Tobio, A.M., Schildkraut, C.L. and Maio, J.J. (1970) J. Mol. Biol. 54:499

8. Flamm, W.G., Bernheim, J.N. and Brubaker, P.E. (1971) Exp. Cell Res. 97:64

9. Bostock, C.J. and Prescott, D.M. (1971a) Exp. Cell Res. 64:267

10. Bostock, C.J. and Prescott, D.M. (1971b) Exp. Cell Res. 64:481

11. Bostock, C.J. and Prescott, D.M. (1971c) Exp. Cell Res. 60:151

12. Tobia, A.M., Brown, W.M., Parker, R.J., Schildkraut, C.L. and Maio, J.J. (1972) Biochim. Biophys. Acta 277:256

13. Bostock, C.J., Christie, S. and Landes, I.J. (1976) J. Mol. Biol. 108:417

14. Stambrook, P.J. (1974) J. Mol. Biol. 82:303

15. Amaldi, F., Giacomoni, D. and Zito-Bignami, R. (1969) Eur. J. Biochem. 11:419

16. Balazs, I. and Schildkraut, C.L. (1976) Exp. Cell Res. 101:307

17. Nicklas, R.B. and Jaqua, R.A. (1965) Science 147:1040

18. Utakoji, T. and Hsu, T.C. (1965) Cytogenetics 4:295

19. Hill, R.N. and Yunis, J.J. (1967) Science 155:1120

20. Stambrook, P.J. and Flickinger, R.A. (1970) J. Expt. Zool. 174:101

21. Niu, M.C. and Twitty, V.C. (1953) Proc. Natl. Acad. Sci. USA 39:985

22. Graham, C.F. and Morgan, R.W. (1966) Develop. Biol. 14:439

23. Flickinger, R.A., Freedman, M.L. and Stambrook, P.J. (1967) Develop. Biol. 16:457

24. DiBerardino, M.A. (1962) Develop. Biol. 5:101

25.   Hsu, T.C., Schmid, W. and Stubblefield, E. (1964) In: The
      Role of Chromosomes in Development (ed. M. Locke) Academic
      Press, New York, pp83-112

26.   Taylor, J.H. (1964) In: Cytogenetics of Cells in Culture
      (ed. R.J.C. Harris) Academic Press, N.Y. and London

27.   Matsumoto, L.J. and Dasgupta, S. (1974) Develop. Biol. 37:
      28

28.   Kriegstein, H.J. and Hogness, D.S. (1974) Proc. Natl. Acad.
      Sci. USA 71:135

29.   Hand, R. and Tamm, I. (1974) J. Mol. Biol. 82:175

30.   Zakian, V.A. (1976) J. Mol. Biol. 108:305

31.   Wolstenholme, D.R. (1973) Chromosoma 43:1

32.   Lee, C.S. and Pavan, C. (1974) Chromosoma 47:429

33.   Habener, J.F., Bynum, B.S. and Shack, J. (1970) J. Mol. Biol.
      49:157

34.   Nandi, U.S., Wang, J.C. and Davidson, N. (1965) Biochemistry
      4:1687

35.   Taylor, J.H. and Hozier, J.C. (1976) Chromosoma 57:341

THE ORGANIZATION OF DNA REPLICATION IN A MAMMALIAN CELL LINE

Barbara R. Jasny[*], Joel E. Cohen and Igor Tamm

The Rockefeller University

New York, N.Y. 10021, U.S.A.

INTRODUCTION

The first evidence that DNA replication proceeds in an organized, nonrandom manner at the level of the chromosome was obtained by Taylor (1960) using autoradiographic techniques to examine labeled metaphase chromosomes. His work, and that of Hsu (1964), demonstrated that DNA replication is initiated at multiple sites along the chromosome and that each chromosome shows a reproducible, characteristic pattern of replication in different stages of S phase. It has also been shown that hetero-chromatin tends to be late replicating (see Lima-de-Faria and Jaworska, 1968). Even better resolution is now possible with the fluoresence labeling technique developed by Latt (1973). Both Latt (1975) and Stubble-field (1975) have suggested that chromosome bands which have been visualized as structural units by classical staining procedures are actually units of replication as observed by fluoresence labeling.

It has not yet been possible to correlate this organized pattern of replication with replication at the level of the double-stranded DNA itself. The technique of DNA fiber autoradiography, originated by Cairns (1963) and further developed by Huberman and Riggs (1968), has been used to determine the rate of fork progression, the distance between initiation sites, and the direction of replication. Hand and Tamm (1974) have suggested that there is synchrony in time of initiation of clusters of replication units, based on the occurrence of predominantly prepulse or postpulse figures in a microscopic field. Hand (1975) has obtained additional evidence for such synchrony from studies of individual fibers. The

[*]*A.S.I. Participant*

autoradiographic technique has the advantages that observations can be made on individual fibers and that many replicating fibers can be readily examined. The disadvantages are that only a small fraction of the DNA is spread well enough to be analyzed (Edenberg and Huberman, 1975), that there is subjectivity in visual observations, and that there can be wide variation in measurement of such parameters as interinitiation site distance.

To examine the organization of replication units within the genome and to determine whether initiation events occur at random along the DNA fiber, it is necessary to evaluate the processes giving rise to the autoradiograms and the replicating strands themselves in a more analytical manner than has previously been used. It is the aim of this work to provide some new insights into the organization of the initiation process.

MATERIALS AND METHODS

Cells

Monolayer cultures of uncloned L929 cells (a continuous line of mouse fibroblasts) were grown in Eagle's medium supplemented with 5% fetal calf serum. Before each experiment, the L929 cells were passaged once in 75 cm^2 Falcon flasks under conditions of exponential growth, then passaged under similar conditions into 28 cm^2 plastic dishes at approximately 1×10^5 cells per plate and allowed to double twice before use. Two dishes were used per experiment.

Radioisotopes and Chemicals

(^{3}H)thymidine (50-52 Ci/mmole) was obtained from New England Nuclear, Boston, Mass., and thymidine from Sigma Chemical Company, St. Louis, Mo. 5-Fluoro-2'-deoxyuridine (FUdR) was a gift from Hoffman-La Roche, Inc., Nutley, N.J.

Cell Labeling and DNA Spreading Procedures

The medium in culture dishes was replaced with fresh warm medium containing 2×10^{-6} M FUdR to deplete the thymidine pool and the cells were incubated at 37°C for 30-40 minutes. They were labeled for 10 minutes at 37°C with high specific activity (^{3}H)-thymidine (250 μCi/ml, 50-52 Ci/mmol, 5×10^{-6}M) in fresh medium containing 2×10^{-6} M FUdR (the "hot" pulse). Unlabeled thymidine was then added (as a 200 μM solution) so as to give a 10-fold reduction in specific activity and incubation continued for 3 hours at 37°C (the "warm" pulse). The cells were washed 2-4 times in ice-cold phosphate buffered saline deficient in Ca^{2+} and Mg^{2+} (PBS def.), removed from the plates with trypsin, and washed and re-suspended in PBS def.

One drop containing 1000-3000 cells was placed on a subbed slide adjacent to a drop of a solution of 1% SDS and 0.01 M EDTA in PBS def. The drops were mixed by touching both simultaneously with a glass rod, and were then spread down the slide. By this procedure the cells were lysed and the DNA spread out as long strands. In some experiments the labeled cells were diluted with unlabeled cells before spreading to reduce the amount of labeled DNA per slide. The slides were allowed to dry, followed by precipitation in 5% TCA and dehydration in ethanol. Slides were dipped in melted Kodak NTB2 emulsion using the red safety light, dried for 2 hours in total darkness and stored for approximately 5 months at -20°C. They were developed in Kodak D19 developer at 17°C for 4 minutes.

Analysis of Autoradiograms

Autoradiograms of (^{3}H)thymidine-labeled DNA were examined by light microscopy at magnifications of 260 x and 640 x. All measurements were made at 640 x with an ocular micrometer. Statistical analysis of the grain density was performed on DNA strands labeled lightly by (^{3}H)thymidine of low specific activity in order to distinguish breaks in strands from apparent gaps representing stretches with a low probability of grain formation. The grains over a ten division span of the ocular micrometer (17.1 μm) on 100 randomly chosen pieces of lightly labeled DNA were counted. The mean number of grains per division (which varied from 0.822 to 1.07 in different experiments) was used to determine the probability of finding a given number of strains in a given interval according to the Poisson distribution formula: $p_n = m^n e^{-n}/n!$, where m equals the mean number of grains in the interval, and n equals the number of grains actually found. If the probability of finding no grains on a stretch of lightly labeled DNA was less than 0.01, then the interval was considered a break. Typically, unlabeled stretches greater than 7 to 11 μm (with the range indicating variation between experiments) were considered significant breaks in the warm pulse-labeled DNA.

The end of a warm-pulse labeled strand is considered to be "free" if it does not terminate in a mass of nuclear material. Thus strands with two free ends have a naturally defined total length which results from breaks or termination of label at each end.

The protocol used gave rise to two types of figures: (1) Pre-pulse figures, in which initiation occurred before the hot pulse. These figures showed a central unlabeled stretch bordered by high

grain density regions (shown as rectangles in the following
diagram), followed by lightly labeled regions (shown as lines).

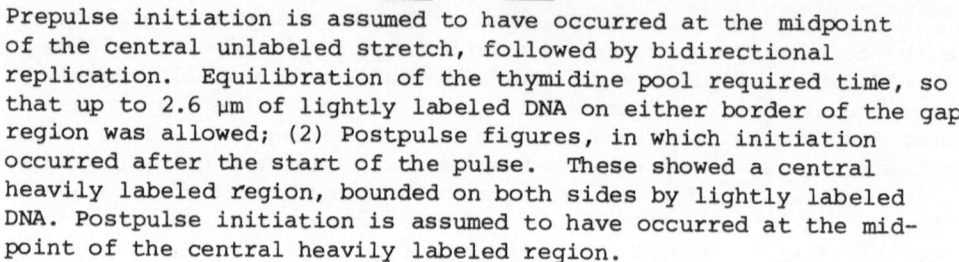

Prepulse initiation is assumed to have occurred at the midpoint
of the central unlabeled stretch, followed by bidirectional
replication. Equilibration of the thymidine pool required time, so
that up to 2.6 μm of lightly labeled DNA on either border of the gap
region was allowed; (2) Postpulse figures, in which initiation
occurred after the start of the pulse. These showed a central
heavily labeled region, bounded on both sides by lightly labeled
DNA. Postpulse initiation is assumed to have occurred at the midpoint of the central heavily labeled region.

Determinations of the occurrence of hot pulse-labeled regions
were based on clear visibility of high grain density at both the
lower and higher magnifications. Intervals of ≤ 0.87 μm (≤ 0.5
division at the high magnification) appearing either totally clear
or lightly labeled within a hot region were not considered
significant.

Huberman and Riggs (1968) used the term replication section to
refer to the stretch of DNA replicated by a single growing point.
They proposed the term replication unit to mean the basic unit of
control of the initiation of replication - presumably an adjacent
pair of diverging replication sections.

Measurements of the distances between adjacent initiation sites
provide estimates of the size of replication units. It should be
noted, however, that the stretch of DNA between two initiation sites
is not equivalent to the replication unit as defined by Huberman
and Riggs (1968), but rather comprises the neighboring halves of two
adjacent replication units. The interinitiation distance was
measured as a center to center distance for all internal figures.
A prepulse initiation figure (see diagram) was considered internal
regardless of the presence or absence of a warm pulse-labeled
stretch of DNA at the strand terminus. A postpulse initiation
figure (see diagram) was considered internal if there was a stretch
of ≥ 3.42 μm of warm pulse-labeled DNA distal to the hot pulse-
labeled region.

RESULTS

Rate of Fork Progression and Temporal Synchrony

The rate of fork progression was calculated by measuring the
lengths of the internal hot pulse-labeled halves of prepulse figures

(i.e., not at the ends of strands) and dividing such lengths by the duration of the hot pulse, i.e., 10 minutes. The frequency distribution of rate measurements is shown in Fig. 1. The mean rate of fork progression in L cells is 0.55 μm/minute, which is within the range of previous estimates from 0.4 to 1.2 μm/minute (Hand and Tamm, 1972).

To evaluate the temporal synchrony with which initiation events observed with our techniques are taking place, strands containing prepulse figures were examined, since prepulse figures potentially span a longer period of replication than postpulse figures. Very large prepulse figures may not be recognized as part of one strand because the unlabeled stretch may have been interpreted as a break rather than a prepulse gap. Only strands with lengths greater than or equal to 40 μm were used in this analysis of synchrony. For each prepulse figure on the strand, the mean distance from the assumed initiation point at the center of the figure to the ends of the hot-labeled segments was calculated (i.e., the average of the left and right halves of each prepulse figure). Then, for each possible pair of prepulse figures on a single strand, the smaller mean distance was subtracted from the greater mean distance to give a *difference* in mean distance. Fig. 2 shows the distribution of the *differences* between these mean lengths of DNA replicated from the time of initiation to the end of the 10-minute hot pulse for all of the strands. Based on the mean replication rate of 0.55 μm/minute, it is concluded that most of the observed initiations occurred within 10 minutes of each other.

The preceding results define the "window" in time through which observations are made. They indicate that there is considerable

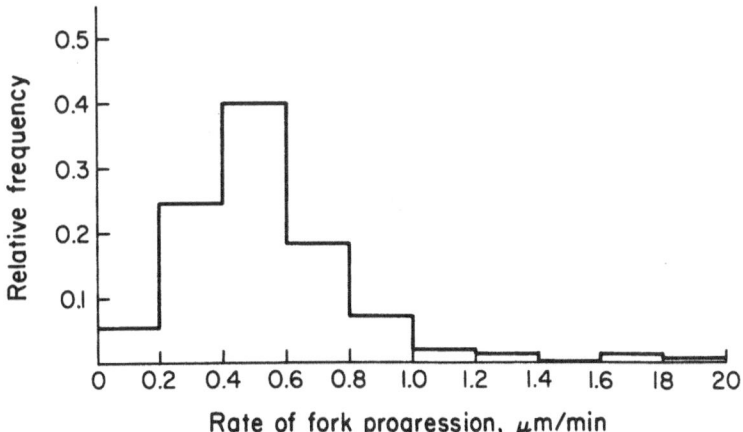

Figure 1. Frequency distribution of rates of fork progression. Mean distribution, based on the mean of three experiments (357 strands total), is shown. Strands examined did not necessarily have two free ends.

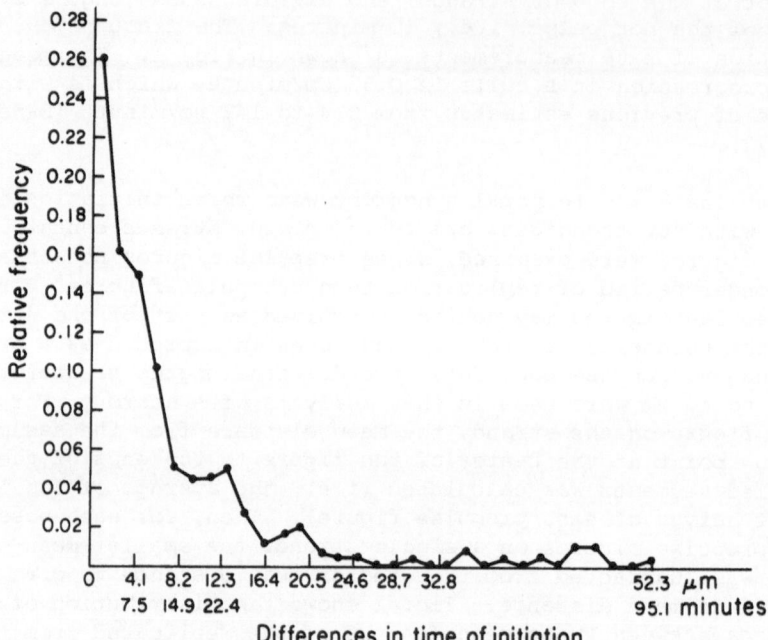

Figure 2. Frequency distribution of differences in time of initiation on individual strands. The mean distance from the assumed initiation point at the center of the figure to the ends of the two hot pulse-labeled segments was calculated for prepulse figures on each strand and for each pair of prepulse figures, the difference between these mean distances was determined. These differences were converted into time differences by multiplying the lengths by the mean rate of fork progression (0.55 μm/min = 0.33 ocular divisions/min). There were 268 differences plotted, representing data from 92 strands, not necessarily with 2 free ends, taken from three experiments.

temporal synchrony, and that a reasonable picture of the initiation pattern can be developed from the events observed on the strand using a 10-minute hot pulse followed by a 3-hour period of low specific activity labeling.

Spatial Organization

In order to determine whether initiation events occur at locations distributed at random or according to some demonstrable organization, we developed quantitatively precise predictions, based on a number of idealized assumptions, concerning what would be expected assuming randomness. First, the DNA fiber was treated as if the length of an unbroken DNA fiber greatly exceeded the average length of a strand with 2 free ends, which appears to be

true. Second, within every replication unit, initiation was
assumed to occur at a single point each located by a single
prepulse or postpulse figure. Here, the length of hot pulse-
labeled regions of prepulse or postpulse figures is irrelevant.
The model which we have compared with our data is that there are
two independent and strictly random (Poisson) processes occurring
simultaneously on a DNA fiber. One process distributes nicks at
random, with a negative exponential distribution of distances
between two adjacent nicks and with an average number λ of nicks
per unit length of DNA fiber. The nicks considered here are only
those nicks which will generate strands with 2 free ends. Thus,
we are testing the assumption that strands with prepulse and postpulse figures are broken from the DNA fiber in a random manner.
A second hypothetical process distributes initiation events, seen
as prepulse or postpulse figures, at random, again with a negative
exponential distribution of distances between adjacent events and
an average number ν of events per unit length of DNA fiber.

     For every strand with 2 free ends on two slides per
experiment, we recorded the total length of the strand, the number
of replication figures per strand, and the lengths of the external
segments. An external segment is defined as the region of warm-
labeled DNA between the end of the strand and the nearest hot-
labeled region. Only strands containing one or more hot pulse-
labeled regions were selected for observations, and all such strands
on a slide were recorded.

Nonrandom Breakage of Strands

     From the random model, it may be calculated that the
probability density function of the length y of an external
segment is:

$$g(y) = (\nu + \lambda)e^{-(\nu + \lambda)y}$$

λ = average number of nicks per unit
    length

ν = average number of initiation events
    per unit length

$y \geqslant 0$

This is a negative exponential distribution. If the external
segments y are ranked in order of increasing size and plotted as
the cumulative frequency against external segment size, the
dotted line in Fig. 3 is generated. A theoretical curve (solid
line) can be constructed by taking the integral of the probability
density function in the limits from zero to y. The data deviate

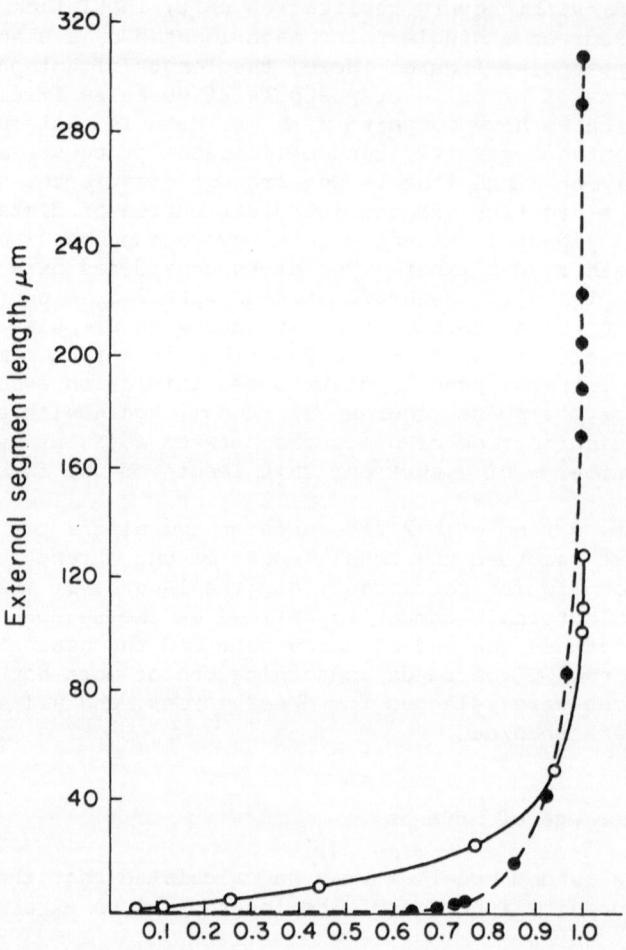

Figure 3. Frequency distribution of external segment lengths. ------
experimental data, _____ theoretical curve derived as described
in the text. All strands had 2 free ends. The data represent 3476
external segments pooled from three experiments.

greatly from the model. Far too many short external segments are
observed, and there appear to be longer segments than would have
been predicted by a random model of strand breakage.

A simple explanation for the increased probability of breakage
of a strand adjacent to a hot pulse-labeled region would be that the
high specific activity label causes excess breaks. Although this

may be a contributing factor, we do not believe that it adequately explains the nonrandom breakage. If breaks occurred only adjacent to hot regions, then the external segment lengths should have a distribution similar to that of the inter-initiation distances (the distances between adjacent initiation sites). If nonrandom breakage were occurring between adjacent hot pulse figures at single stranded regions generated during replication, then the external segment distribution should show a higher frequency of shorter lengths than the inter-initiation distance distribution. Our evidence does not support either of the above possibilities. In Fig. 4 are plotted only external segments of lengths greater than or equal to 3.42 μm which is the limit of resolution for the inter-initiation distances. All such external segments were taken from every strand with 2 free ends observed on the slides. The mean external segment length was 44.98 μm, which was 35% higher than the mean inter-initiation distance of 33.26 μm. In this comparison, each experiment was weighted according to the sample size. The difference was significant at the 0.005 probability level by t test.

To investigate breakage of the DNA fiber into strands further, we tested whether or not breakage at one end of the strand is independent of breakage at the other end. This was done by examining the number of strands having 0, 1, or both external segments longer than some given length y. If F is the fraction of all external segments which are longer than y, then, assuming that the two external segments are independent, the proportion of all strands with both external segments longer than y should be F^2, the proportion with exactly one external segment longer than y should be $2F(1 - F)$, and the proportion with zero external segments longer than y should be $(1 - F)^2$. This model is formally identical to the Hardy-Weinberg equilibrium of population genetics. There was no significant deviation from the model for any value of y in any of the three experiments as determined by χ^2 analysis (data not shown). Thus, it can be concluded that there is no interaction between the length of the external segment at one end of the strand, and the length of external segment at the other end. Possible implications of this result are discussed below.

Inter-Initiation Distance Distribution

The distances between initiation sites reveal one aspect of the organization of replication units. Although the situation is complicated by the non-random breakage of strands and by our finding that the inter-initiation distances vary with strand lengths (data not shown), it is possible to test our data for randomness in the distribution of inter-initiation distances. If the activated initiation sites are distributed randomly (i.e., with negative exponential distances between them) on the unbroken DNA fiber, then

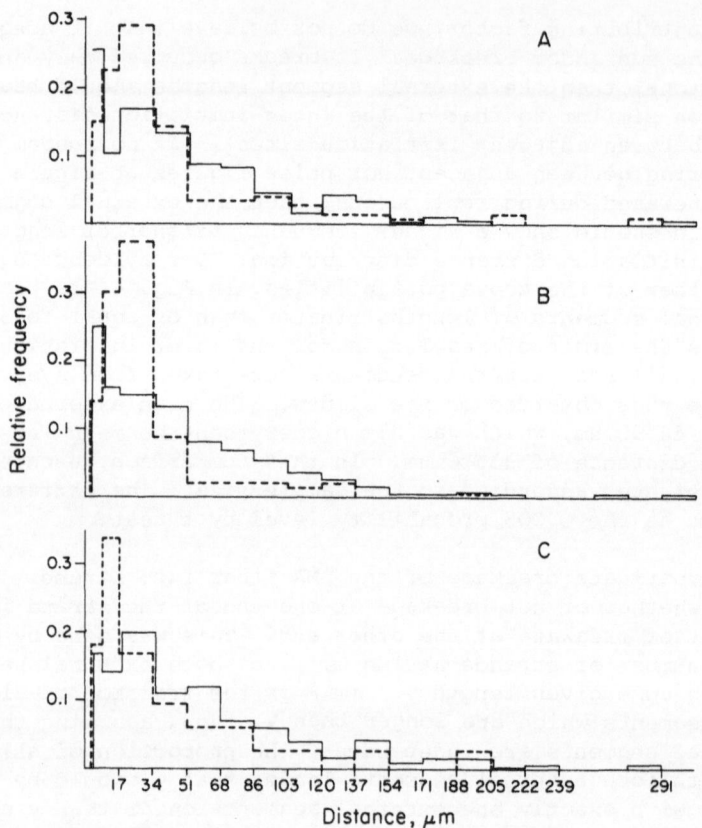

Figure 4. *Frequency distribution of external segments and inter-initiation distances ≥ 3.42 μm (2 ocular divisions) in length. Panels A-C show results from 3 separate experiments. ----- inter-initiation site distance, ——— external segment length. The data represent all strands with 2 free ends on 2 slides per experiment, plus a population of strands longer than 20-30 divisions picked from additional slides. The longer strands also had 2 free ends. For the inter-initiation distance distributions, in experiment A, n = 147; in B, n = 150; in C, n = 83. For the external segment distributions in experiment A, n = 288; in B, n = 486; in C, n = 403.*

it is possible to derive a function B such that:

$$B = 2n(\bar{x})/s,$$

n = number of observations,

$\bar{x}$ = mean inter-initiation distance,

s = standard deviation of sample mean.

As n becomes larger, B has approximately the **distribution of χ^2 with** 2n degrees of freedom. Then (Wilson-Hilferty, 1931) if $([B/(2n)]^{1/3}$ + $1/(9n) - 1)(9n)^{1/2} = z$, z has normal distribution with zero mean and unit variance. If $|z|$ is less than 2.56, then by this test the distances do not deviate significantly from exponential at the 1% level. Table 1A shows inter-initiation distances from three experiments in which all strands with 2 free ends were recorded. $|z|$ was less than 2.56 in all cases. When the mean inter-initiation distance was calculated from the 3 experiments (203 distances), z = **-2.48, which is suggestive of a random distribution. Since strands** are not equally represented by the above calculation (i.e., strands with more than two replication units can generate many inter-initiation distances), a distance distribution was calculated for hot pulse-labeled autoradiograms in strands containing only 2 activated initiation sites. For all such strands the total external segment length was subtracted from the length of the entire strand, generating a distance analogous to the inter-initiation distance (315 strands; Table 1B). A z equalling -2.13 was determined, which is also within the limits of ±2.56 set for a random distribution.

In each of the three experiments in Table 1, the mean distance between hot pulse-labeled regions on strands with two figures (Table 1B) exceeds the corresponding mean inter-initiation distances (by 6 μm in experiment 6; 11 μm in experiment 14; and 17 μm in experiment 5). The excess corresponds roughly to the mean length of a single hot pulse-labeled region. The difference arises because the two figure distances include the total lengths of the hot labeled regions in each of the two replication figures, whereas the inter-initiation distances extend from midpoint to midpoint, approximately, of adjacent replication figures.

DISCUSSION

The finding of nonrandom breakage has important implications for DNA replication. The excess breakage which may be caused by the presence of high specific activity label is not sufficient to explain the apparently nonrandom breakage at distances far removed from the hot pulse-labeled regions. If breakage were due to the presence of the single-stranded region at the replication fork, or was occurring at or near the hot pulse region , then we would not expect to observe, as we did, that the mean external segment was significantly larger than the mean inter-initiation distance.

We find that there is no interaction between the length of the external segment at one end of a strand and the length of the external segment at the other end. This finding appears to exclude the possibility that in some large stretches of the DNA fiber, containing many of the hypothesized clusters of replication units, there is a higher probability of breakage at any given distance

Table I

A. Interinitiation Distances

		Expt 5	Expt 6	Expt 14	Pooled data*
Number	n	97	39	67	203
Mean	$\bar{\chi}$ (μm)	30.44	34.7	44.5	35.76
Standard deviation	sd (μm)	31.42	48.0	56.0	42.95
	$z^{\dagger}$	-0.66	-1.82	-1.87	-2.48

B. Two Figure Distances

		Expt 5	Expt 6	Expt 14	Pooled data
Number	n	121	103	91	315
Mean	$\bar{\chi}$ (μm)	47.23	40.62	55.1	47.33
Standard deviation	sd (μm)	56.06	42.60	63.40	53.77
	$z^{\dagger}$	-1.80	-0.44	-1.27	-2.13

*In pooling the data, each distance was weighted equally, so that each experiment was weighted by n.
†Calculated from eqs. (2) and (3). If $|z| > 2.56$, we interpret the distribution of distances to be nonrandom.

from a hot pulse-labeled region than there is in other large stretches. If the DNA fiber were inhomogeneous in this way, we would have expected to observe an excess of strands with two short external segments and an excess with two long external segments, in comparison with the numbers expected using the probabilities based on strands pooled from all parts of the fiber. Thus, there do not appear to be regions, detectable by our procedures, which contain "weak spots" that cause the nonrandom breakage. It is unlikely that weak spots account for the presence of external segments which are longer than the distances between initiation sites.

Although we are not yet prepared to offer a definitive explanation of these findings, one reasonable possibility is that, at specific times during S phase, physical clusters of replication units are activated. These clusters are separated by distances longer than the measured, mean inter-initiation distance. The high degree of temporal synchrony observed (Fig. 2) indicates that the firing of initiation sites within a cluster occurs within a fairly short period of time. However, the physical site at which initiation occurs within such a cluster may be randomly determined, as suggested by the exponential distribution of inter-initiation distances and interunit distances on strands containing only 2 replication units.

Support for the theory of physical clustering of replication units can be found from studies using very different methodologies. Stubblefield (1975) and Latt (1975) have observed reproducible banding patterns when the chromosomes of cells which have been treated with BUdR during the entire S phase, except for a short period of thymidine incorporation, **are analyzed by fluorescence** labeling with the Hoechst stain 33258. These bands appear to correspond with banding patterns obtained when chromosomes are stained with quinacrine or by G banding procedures. They have theorized that entire regions of the genome, comprising lengths equivalent to hundreds of replication units, are made available for initiation at specified times within the S phase. Individual units in the cluster could initiate replication randomly. The finding of clusters of units in replicating drosophila DNA by means of electron microscopic techniques (Zakian, 1976) is also in accord with this hypothesis.

The data described in this paper do not rule out the possibility that *potential* initiation sites are located at non-random intervals. Such structures may be located at short intervals along the strand with only a small percentage of such sites being activated during any one S phase. The observed differences in inter-initiation site distances among species (Hand and Tamm, 1974) would agree with this possibility and could have evolutionary significance. There is also evidence that there are different subfractions of replicating DNA, which show different patterns of replication (Zakian, 1976; Hori and Lark, 1976). A sufficiently small percentage of the replicating DNA could have a nonrandom distribution of inter- initiation distances without being detected in our analysis. In Table I, in each experiment, the mean distance is less than the standard deviation, and z is negative. This pattern suggests the possibility of a quantitatively small deviation from randomness.

Further research in this area will entail development of more detailed structural models. We are also attempting to correlate the replication patterns observed at the fiber level with banding patterns observed on isolated chromosomes.

ACKNOWLEDGEMENTS

This investigation was supported in part by Research Grant Number CA-18608 and by Program Project Grant Number CA-18213, awarded by the National Cancer Institute, and by U.S. National Science Foundation grant BMS74-13276. B.R. Jasny is a predoctoral trainee under the Institutional National Research Service Award Number CA-09256, from the National Cancer Institute.

REFERENCES

1. Edenberg, H.J. and Huberman, J.A. (1975) Ann. Rev. Genet.
    9:245-284

2. Hand, R. (1975) J. Cell Biol. 64:89-97

3. Hand, R. and Tamm, I. (1974) J. Mol. Biol. 82:175-183

4. Hori, T.A. and Lark, K.G. (1976) Nature 259:504-505

5. Hsu, T.C. (1964) J. Cell Biol. 23:53-62

6. Huberman, J.A. and Riggs, A.D. (1968) J. Mol. Biol. 32:
    327-341

7. Latt, S.A. (1973) Proc. Nat. Acad. Sci. USA 70:3395-3399

8. Latt, S.A. (1975) Somat. Cell Genet. 1:293-321

9. Lima-de-Faria, A. and Jaworska, H. (1968) 217:138-142

10. Stubblefield, E. (1975) 53:209-221

11. Taylor, J.H. (1960) J. Biophys. Biochem. Cytol. 7:455-464

12. Wilson, E.B. and Hilferty, M.M. (1931) Proc. Nat. Acad. Sci.
    USA 17:684-686

13. Zakian, V.A. (1976) J. Mol. Biol. (1976) 108:305-331

AN UNUSUAL STRUCTURE IMPLICATED IN INITIATION OF DNA SYNTHESIS

Timothy Nilsen[*] and Corrado Baglioni

Department of Biological Sciences
State University of New York at Albany
1400 Washington Avenue
Albany, New York 12222, U.S.A.

INTRODUCTION

Despite the fact that rapid advances have been made in deciphering the detailed molecular mechanism of DNA replication in prokaryotes (1), parallel progress has not been made in eukaryotic systems. This has been due in great part to the inherent complexity of eukaryotic cells, the lack of well characterized enzymes, and the lack of specific mutations, the availability of which has greatly facilitated investigations in prokaryotes. Since the publication in 1968 of the replicon model of eukaryotic DNA replication (2), very little has been established in terms of defining the replicon at the molecular level. In particular, there is almost no knowledge concerning either possible chromosomal determinants or mechanisms involved in the initiation of DNA synthesis at the level of individual replicons. In their 1968 paper (2) Huberman and Riggs state: "It is now well established that, at the level of resolution of whole chromosome auto-radiography, regions of initiation and termination of DNA replication are reproducibly and heritably controlled. However, neither whole chromosome autoradiographic studies nor the studies presented in this paper have been able to determine whether sites of initiation (origins) are reproducible at the atomic level. The possibility remains that sites of initiation may occur anywhere within relatively long stretches of the DNA fibers". The problem, as stated above, has remained essentially unanswered. However, in the course of experiments designed to investigate the precise timing of replication of histone genes during the S phase of HeLa cells, we made an observation that may have some bearing on the question of specific initiation sites. The observation and the

*A.S.I. Participant

series of experiments that it prompted will be discussed below.
The experiments suggest and are consistent with the hypothesis
that specific nucleotide sequence(s) are involved in the
initiation of DNA synthesis at the individual replicon level.

Before discussing the experiments in detail we will briefly
review the evidence that has accumulated which suggests defined
initiation sites. This evidence can be divided into four sections:
(1) Evidence for multiple initiations in eukaryotic chromosomes.
(2) Evidence that these initiations appear to be regulated both
spatially and temporally. (3) Further evidence for a temporal
sequence of replication as determined by autoradiography, density
shift experiments, and nucleic acid hybridization. (4) Examples
of fixed specific initiation sites in bacteria, bacterial viruses,
and eukaryotic viruses.

(1) Early evidence for multiple initiation in eukaryotic DNA
replication comes from autoradiographic studies of metaphase
chromosomes (3,4,5,6). These experiments established both that
multiple initiation occurred and that initiations were not confined
to any particular region of a chromosome. However, the resolution
of this technique is not sufficient to determine either the number
or the size of replication units (replicons). Several approaches
have been used to calculate the true size of replicons. These
include fiber autoradiography (2,7) electron microscopy (7), and
sedimentation analysis on alkaline sucrose gradients (8). When
applied to many cell types, these techniques have yielded consistent
results and demonstrate that eukaryotic cells contain several
thousand replicons.

(2) The determination of the rate of chain elongation in many
eukaryotic systems (7,9) and the fact that this rate does not
change dramatically during S phase (9) has led to the conclusion
that all replicons are not active in DNA synthesis simultaneously.
For example, Blumenthal et al. (7) have estimated that if all
observed origins in Drosophila tissue culture cells were activated
at the beginning of S phase, the S phase would have a duration of
eight minutes, a value two orders of magnitude less than the
observed time. Extensive staggering of initiation has also been
postulated to extend the S phase of chicken somatic cells (10).
Although these observations are indicative, they do not demonstrate
a regulated pattern of initiation. Early evidence for selective
initiation came from Huberman and Riggs (2) who found extensive
clustering of active (replicating) replicons and suggested a model
in which adjacent replicons were under coordinate control. Since
then similar clustering has been observed in many systems (7,11,
12). Indeed the degree of synchronous initiation within clusters
has been used to assay the effect of DNA synthesis inhibitors on
initiation (12). Klevecz et al. have observed a trimodal S phase
curve in diploid cells (13). They argue that the changes in rate

of DNA synthesis can be explained by the activation of large
blocks (clusters) of replicons. Further evidence for regulation
of initiation comes from the observation that inter origin spacing
is quite different in cleavage nuclei and tissue culture cells of
Drosophila (7). It seems clear from the observations cited above
that the activation of particular replicons is regulated.
Evidence that this regulation manifests itself in a defined
temporal sequence of DNA replication is discussed below.

(3) Early observations that suggested a temporal order of
replication involved analysis of autoradiographs of metaphase
chromosomes that had been pulse labeled at different times during
S phase. This analysis indicated that certain regions of
chromosomes replicated during limited intervals of S phase (14).
Similar experiments have been used to examine the replication
pattern of polytene chromosomes in Dipteran larvae (15). These
revealed a time sequence of initiation of replication. This type
of experiment, while suggestive, lacks precision. Therefore,
several investigators have examined the timing of replication of
defined sequences of DNA in synchronized cells. It has been
shown by a variety of techniques that: (1) sequences replicated
early in one S phase are also replicated early in subsequent S
phases (16); (2) the replication of ribosomal cistrons is
confined to a particular interval in S phase (17,18,19); (3)
highly reiterated DNA in monkey cells replicates at a defined time
(20); (4) replication of integrated SV-40 occurs during a limited
portion of S phase (21); and (5) GC-rich sequences tend to
replicate before AT-rich sequences (22). All of these
observations suggest the existence of a regulated sequence of DNA
replication but, because of a lack of complete cell synchrony (23),
they are unable to reveal either the level of regulation or possible
mechanisms for it. The most direct evidence for reproducibility
of timing of replication of defined sequences comes from the
autoradiographic studies of Amaldi and co-workers (24). They
synchronized cells with vinblastine and hydroxyurea and labeled
DNA with (^{3}H) thymidine for a brief time at the beginning of S
phase. The cells were then allowed to traverse the cell cycle
and were resynchronized. At the second release the cells were
given a pulse of radioactivity for a longer duration and the DNA
was isolated and spread for autoradiography. It was found that a
significant number of DNA molecules showed double labeling and
that grain tracks generated by the first pulse were directly
centered in the tracks created by the second pulse. Although all
molecules did not show double labeling (probably because of
asynchrony; ref 25), these results demonstrate both that the timing
of replication of specific sequences is regulated and that the
initiation sites of DNA synthesis are fixed. Unique initiation
sites for DNA replication have been shown to exist in both
prokaryotic and eukaryotic systems. A few examples are presented
below.

(4)   All replicating systems studied to date have in common a
specific site for the initiation of replication (1). Using a
variety of techniques, investigators have demonstrated specific
sites in E.coli (26) phage lambda (27), T$_4$ (28), T$_7$ (29), and
B. subtilis (30). There are examples of specific initiation sites
in eukaryotes as well. Polyoma and SV-40 are small double stranded
DNA viruses that infect eukaryotic cells. They replicate in the
nucleus by employing the host replication machinery. For this
reason, it has been suggested that the replication of these viruses
could serve as a model system to study eukaryotic DNA replication
(31). Of particular interest is the fact that both viruses have
specific sites of initiation of DNA synthesis (32,33). Furthermore,
the nucleotide sequence of the SV-40 initiation site has recently
been reported (34). This sequence has many interesting features,
some of which resemble features found in sequences implicated as
protein recognition sites in other systems (34). From these
observations and the considerations presented above, we believe
that a reasonable case can be made for the hypothesis that
initiation of replication at the individual replicon level in
eukaryotes might result from the recognition, by proteins, of
specific nucleotide sequences present in each replicon. Furthermore, it seems possible that the recognition of sets or subsets
of initiator sequences could specify an orderly sequence of
replication.

                                  RESULTS

     All the experiments to be reported were performed with
synchronized HeLa cells grown in suspension. Synchrony was
achieved with a thymidine block followed by an amethopterin block
(35). Fig. 1 illustrates the pattern of DNA synthesis and mitotic
index observed after releasing the cells from the second block. In
most experiments cells were grown for one generation prior to
synchronization in the presence of trace amounts of (^{14}C)T to
label parental DNA. Newly replicated DNA was both density-
labeled with BrdU and radioactively-labeled with (^{3}H)dC. DNA was
prepared from detergent washed nuclei, phenol extracted, ethanol
precipitated, redissolved and sheared before fractionation on
neutral CsCl gradients. Fig. 2A illustrates the profile obtained
when labeled samples of phage SPO1 DNA and HeLa parental (LL) DNA
are centrifuged together on a neutral CsCl gradient. SPO1 DNA has
a density of 1.745 g/ml which is very close to the density of human
DNA fully substituted in one strand with BrdU (HL; 1.75 g/ml). SPO1
DNA was therefore used as a density marker.

     The observation that prompted the series of experiments
discussed below is illustrated in Fig. 2C. Cells were released
from the amethopterin block with 20 µM BrdU and labeled with
(^{3}H)dC for 30 min. It can be seen that there are two distinct

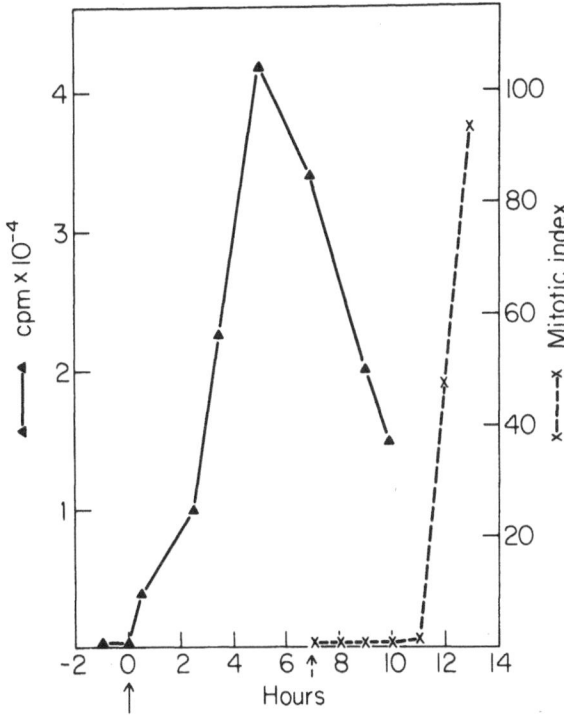

Figure 1. DNA synthesis and mitotic index following release from amethopterin block. HeLa S₃ cells grown in suspension were blocked for 12 hr with 0.5 mM thymidine. They were then released by washing and resuspending in fresh medium. After 9 hr amethopterin was added to a final concentration of 1 μM and adenosine to 0.1 mM. After 12 hr the cells were released with 80 μM T and dC. DNA synthesis was monitored by labeling aliquots of cells at the indicated times with 3 μCi/ml (³H)dC for 10 min. The cells were pelleted and resuspended in 0.5 N KOH for 1 hr. The DNA was precipitated with 10% trichloroacetic acid and collected on glass fiber filters for counting. Mitotic index was determined by direct observation of 200 cells at the indicated times.

    Solid arrow   -   *time of release from the amethopterin block.*

  Broken arrow   -   *time of addition of 6 μg/ml colcemid.*

peaks of radioactivity: HL, a peak at the position expected for DNA in which one strand is fully substituted with BrdU and HH, a peak (containing approximately 20% of the label incorporated) at the density (1.80 g/ml) predicted for DNA fully substituted in both strands with BrdU. The presence of this second peak was unexpected. Therefore, an attempt to characterize this material was made. It was found that the material banding as HH DNA: (1)

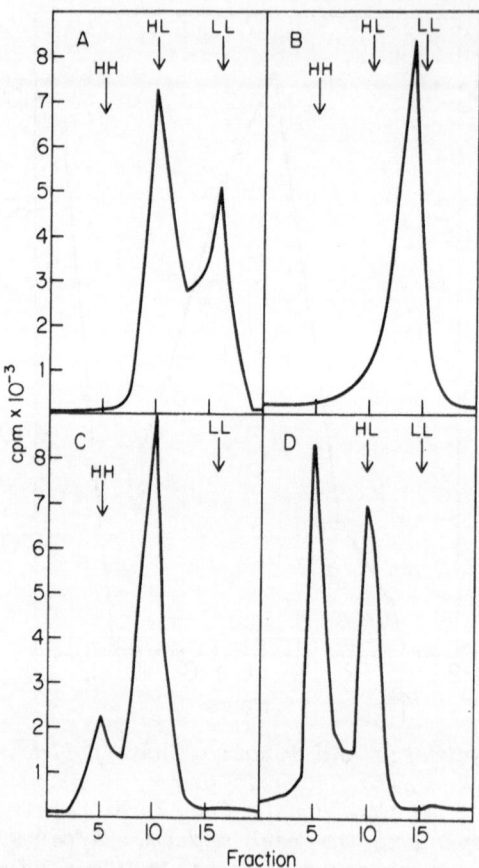

Figure 2. Neutral CsCl gradient profiles of labeled DNA. Labeling was as described in the text. 5 ml gradients with a starting density of 1.722 were centrifuged for 60 hr at 30,000 rpm in the Beckman SW 50.1 rotor. Gradient fractions were diluted with 1 ml of a solution containing 250 µg/ml unlabeled DNA, precipitated with 5% trichloroacetic acid, and collected on glass fiber filters for counting.

A. Phase SPOI DNA plus LL DNA
B. Unsheared DNA labeled in the absence of hydroxyurea
C. Sheared DNA labeled in the absence of hydroxyurea
D. Sheared DNA labeled in the presence of 1mM hydroxyurea

rebanded at the same density when isolated and rerun on CsCl gradients; (2) was insensitive to digestion by single strand specific nuclease S1; (3) rebanded at the same position after digestion with ribonuclease H, which digests the RNA in RNA/DNA hybrids; and (4) was completely hybrolyzed upon digestion with DNAse I. We therefore concluded that the radioactivity banding at the heavy density corresponded to double stranded DNA substituted in both strands with BrdU.

The possibility that HH DNA arose through extrachromosomal replication, either through an amplification event or by the replication of a virus was examined. It was expected that if HH DNA was the result of extrachromosomal replication, it would band at the heavy density even if the chromosomal DNA was not sheared. Alternatively, if HH DNA was part of the replicating chromosomal DNA, it should not be detected in a preparation of unsheared DNA because it would be connected either to HL or LL regions of DNA and band at a lower density. Labeled DNA was prepared as described above and one aliquot was sheared by passage through a 30 gauge needle. Another aliquot was left unsheared and the samples were centrifuged on parallel CsCl gradients. The results are illustrated in Fig. 2B and C. It can be seen that HH DNA is observed only in the sheared sample, indicating that it is not formed as a result of extrachromosomal replication.

We then examined the possibility that HH DNA was formed only in very early S phase. DNA was labeled as described above for 10, 20, 30 and 60 minutes following release from the amethopterin block. Also, aliquots of cells were released from the amethopterin block with thymidine and allowed to traverse either the first two hours or four hours of S phase. At these times the cells were washed and resuspended in media containing BrdU. After 30 min (^{3}H)dC was added and replicating DNA was labeled for 30 min. It was found that fully 30% of the DNA synthesized at 10 min was HH. This value declined to 20% at 20 min and 15% at 30 min in this experiment. HH DNA was also detected later in S. However, the amount of HH DNA was low, between 5 and 8%. The 30 min and 4 hr time points are shown in Fig. 3. This experiment suggested that the synthesis of HH DNA was greatest in early S but the presence of HH DNA at later times indicated that it might be a constant event. However, the possibility of asynchrony in the cell population was not excluded.

The stability of HH DNA _in vivo_ was then examined with pulse-chase experiments. DNA was labeled for twenty min after release from the amethopterin block. At this time excess cold dC was added and the chase allowed to continue for 2 hr. DNA isolated from unchased cells contained 20% HH. DNA from chased cells had no HH. The results of this experiment indicated that HH DNA may be an intermediate in DNA synthesis.

Since the percentage of HH DNA was highest in early S, when the relative proportion of initiation to elongation is greatest, it was hypothesized that HH DNA might be part of a specific structure which participates in initiation of DNA synthesis.

If HH DNA is involved in initiation, it would be expected that the proportion of HH DNA observed during a short labeling time would be considerably greater in the presence of an agent which

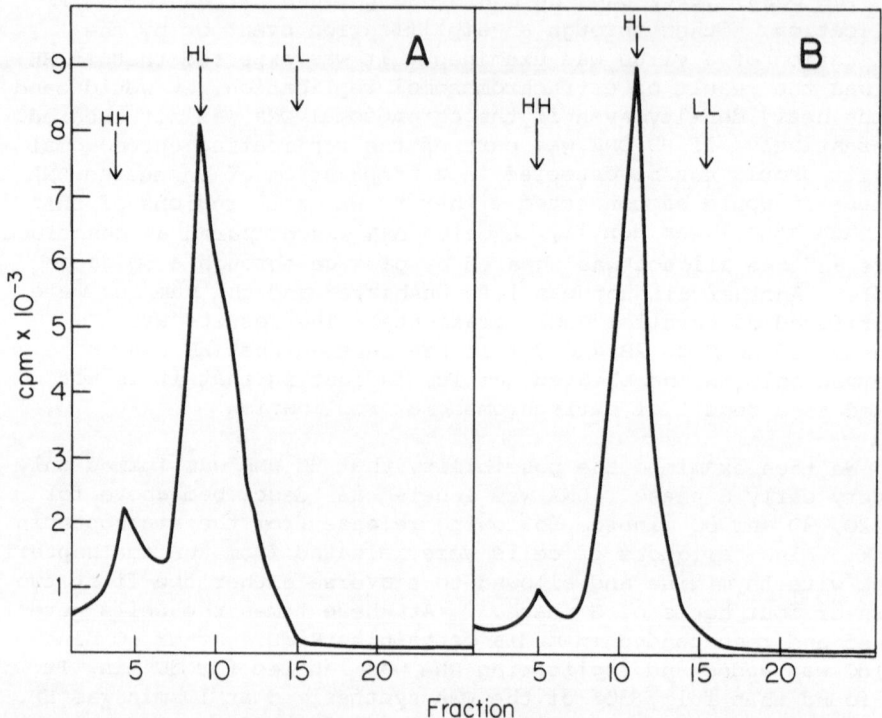

Figure 3. *Neutral CsCl gradient profiles of labeled DNA. Labeling was as described in the text. The gradients were run and fractionated as described in the legend to Fig. 2.*

    A.   *DNA labeled for the first 30 min following release from the amethopterin block.*

    B.   *DNA labeled for 30 min at four hours after release from the amethopterin block.*

preferentially inhibits elongation. It has been reported that hydroxyurea at concentrations sufficient to inhibit DNA synthesis by more than 90% does not inhibit initiation (25,36). Therefore, cells were synchronized as described above and at the time of release hydroxyurea was added with BrdU and dC. The incubation was carried out for 20 min and DNA prepared. The experimental results are illustrated in Fig. 2D. Fully 60% of the label incorporated under these conditions is found in HH DNA. Furthermore, a parallel control minus hydroxyurea revealed that total incorporation into HH was unaffected by hydroxyurea while total incorporation into HL DNA was decreased by approximately 90%.

This result allowed us to more closely examine the fate of HH

DNA using pulse chase experiments. The pulse was twenty minutes in
the presence of hydroxyurea and the chase lasted two hours either
plus or minus hydroxyurea. The results are shown in Fig. 4. While
there is slight incorporation after the inception of the chase, it
seems clear that HH DNA chases quantitatively into HL and that this
chase is inhibited by hydroxyurea. These results both strengthen
the argument that HH DNA is an intermediate in DNA synthesis and
tend to exclude the possibility that HH DNA is the result of
aberrant DNA synthesis caused by either BrdU or the synchronization
technique.

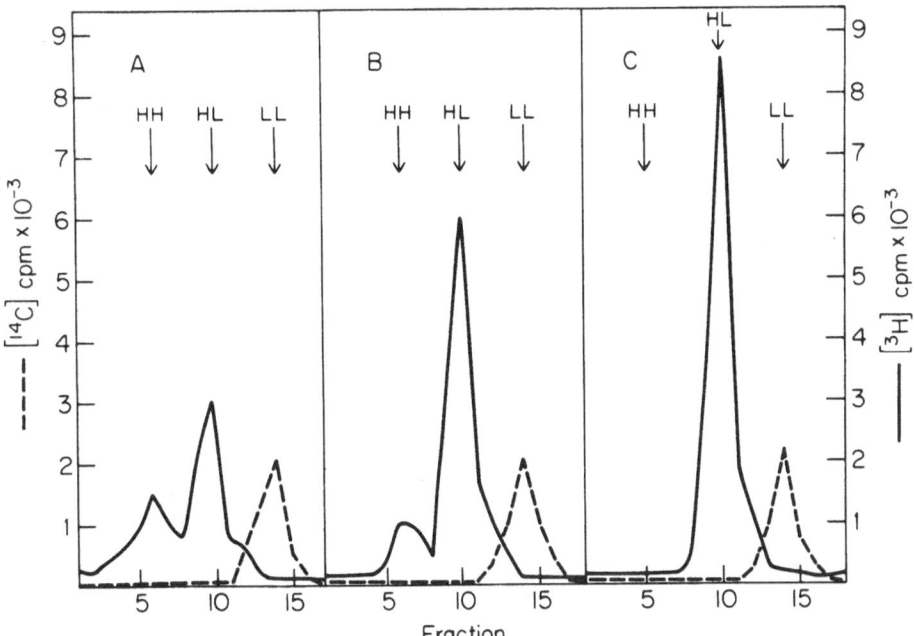

*Figure 4. Neutral CsCl gradient profiles of DNA labeled in the
presence of 1 mM hydroxyurea for the first 20 min following
release from the amethopterin block (A) and chased for 2 hrs in
the presence (B) or absence (C) of hydroxyurea. Gradients were run
and fractionated as described in the legend to Fig. 2. The ^{3}H-cpm
are normalized to a constant amount of ^{14}C-cpm in LL DNA.*

	HH	HL	Total
A.	1670	2808	4478
B.	950	4418	5368
C.	80	5470	5550

A further experiment was done to examine the possibility that amethopterin might in some way be causing the formation of HH DNA. It seemed likely that if HH DNA is formed at initiation, relatively low levels of HH DNA would be detected if the release of the cells were delayed until they had already initiated many replicons. To achieve this, we extended the release time between the thymidine block and the amethopterin block by three hours. The cells were then blocked for the usual time with amethopterin. Cells so treated could be expected to have progressed at a slow rate into S phase by the time of release, since amethopterin is not completely effective in blocking cells of the G_1-S boundary (23). The results showed a level of 8% HH DNA in cells allowed to progress into S, and a value of 20% in cells released at the beginning of S. These results tend to rule out possible artifacts due to amethopterin and strengthen our hypothesis that HH DNA is involved in initiation.

It has been reported by several investigators that agents which act by depleting nucleotide pools cause abnormal ligation of Okazaki fragments. To establish whether HH DNA might be formed by reassociation of short fragments, we have determined its size by alkaline and neutral sucrose gradient analysis of preparations sheared to different extents. The results obtained indicated that HH DNA is at least 750 nucleotides long (both single and double stranded) ruling out this possibility.

If the hypothesis that specific nucleotide sequences are needed for initiation is correct, it might be expected that these sequences would be of limited complexity. Preliminary experiments have been done to examine the complexity of HH DNA and the results are striking. HH DNA denatured in the presence of cold human placental DNA and incubated under hybridization conditions renatures with a 1/2 Cot of less than 0. Parallel experiments with LL and HL DNA show the normal renaturation kinetics representative of the entire genome. Also, HH DNA was allowed to renature by itself. No detectable reassociation occurred, indicating that HH DNA did not snap back, as would be expected if it were composed of palindromic sequences.

DISCUSSION

We have presented evidence for a structure involved in the initiation of DNA synthesis in which both newly replicated strands are base paired to each other (HH DNA). It must be stressed that, while suggestive, the experiments presented do not prove that HH DNA is a true intermediate. At present we do not know whether HH DNA exists in intact cells or whether it is formed during extraction and isolation of DNA.

If HH DNA is formed during isolation, an explanation for its relative abundance at the beginning of S phase (when elongation is slowest; ref 9) and in the presence of inhibitors of elongation is possible. The ends of replicating molecules might not be part of a stable duplex and thus may be more likely to undergo branch migration than already replicated DNA strands. If this is the case, any situation in which the rate of elongation is slowed would result in an enrichment for labeled ends. At present we cannot exclude this mechanism for the formation of HH DNA. However, the sequence complexity experiments, mentioned above, strongly suggest that branch migration, if it occurs, is restricted to a limited sequence, presumably the initiation site, common to all or many replicons.

Alternatively, HH DNA may represent a structure specifically formed during the initiation process. We plan to distinguish between these two mechanisms by performing cross-linking experiments in vivo.

Elevated levels (15%) of HH DNA in human cells have previously been reported (38). Before labeling DNA for 1 hr these unsynchronized cells were treated for 3 hr with methyl methane sulphonate, an alkylating agent. This compound inhibits DNA synthesis by more than 70% (39). The HH DNA observed under these conditions is chased and the chase is inhibited by hydroxyurea. This HH DNA has been hypothesized to result from a repair mechanism involving branch migration (38). However, if the primary effect of alkylation is to inhibit chain elongation, elevated levels of HH DNA would be expected as a result of an increased ratio of initiation to elongation (see above).

A possible explanation for the inability of agents which deplete nucleotide pools to block initiation comes from reports that among the DNA polymerases found in mammalian cells there is one (polymerase γ) that has a very low K_m for nucleotide triphosphates (37). The activity of this enzyme has been shown to rise at the onset of S phase. It is conceivable that this enzyme is involved in initiation while the other DNA polymerases are responsible for elongating a postulated "initiation complex".

Perhaps the strongest argument for a role of HH DNA in initiation is the determination of its sequence complexity. This work is a preliminary stage and we are currently in the process of making enough HH DNA to perform thorough kinetic studies.

In conclusion, we feel that the investigation of HH DNA in synchronized cells can yield information on the mechanism of DNA replication in eukaryotic cells and may shed some light on the nature of initiation at the individual replicon level.

REFERENCES

(1) Gefter, M. (1975) Ann. Rev. Biochem. 44:45

(2) Huberman, J. and Riggs, A. (1968) J. Mol. Biol. 32:327

(3) Taylor, J. (1959) In: Proc. 10th Intern. Congr. Genet.,
     Toronto. 4:63

(4) Branchi, N. (1966) Cytologia 31:276

(5) Moorhead, P. and Defendi, V. (1962) J. Cell Biol. 16:202

(6) Hsu, T. (1964) J. Cell Biol. 23:53

(7) Blumenthal, A., Kriegstein, H., and Hogness, D. (1973)
     Cold Spring Harbor Symp. Quant. Biol. 38:205

(8) Gautschi, J., Kern, R. and Painter, R. (1973) J. Mol. Biol.
     80:393

(9) Housman, D. and Huberman, J. (1975) J. Mol. Biol. 94:173

(10) McFarlane, P. and Callan, H. (1973) J. Cell Sci. 13:821

(11) Hand, R. and Tamm, I. (1974) J. Mol. Biol. 82:175

(12) Hand, R. (1975) J. Cell Biol. 67:761

(13) Klevecz, R. and Kapp, L. (1973) J. Cell Biol. 58:564

(14) Petersen, A. (1964) J. Cell Biol. 23:564

(15) Howard, E. and Plaut, W. (1968) J. Cell Biol. 39:415

(16) Mueller, G. and Kajiwara, K. (1966) Biochim. Biophys. Acta
     114:108

(17) Giacomoni, D. and Finkel, D. (1972) J. Mol. Biol. 70:725

(18) Amaldi, F., Giacomoni, D. and Zito-Bignami, R. (1969)
     Europ. J. Biochem. 11:419

(19) Zellweger, A., Ryser, U. and Braun, R. (1972) J. Mol. Biol.
     64:681

(20) Tobia, A., Brown, E., Parker, R., Schildkraut, C. and Maio, J.
     (1972) Biochim. Biophys. Acta 277:256

(21) Balazs, I., Brown, E. and Schildkraut, C. (1973) Cold
     Spring Harbor Symp. Quant. Biol. 38:239

(22) Tobia, A., Schildkraut, C. and Maio, J. (1970) J. Mol. Biol.
     54:499

(23) Amaldi, F., Carnevali, F., Leoni, L. and Mariotti, D. (1972)
     Exp. Cell Res. 74:367

(24) Amaldi, F., Buongiorno-Nardelli, M., Carnevali, F., Leoni, L.,
     Mariotti, D. and Pomponi, M. (1973) Exp. Cell Res. 80:79

(25) Walters, R., Tobey, R., and Hildebrand, C. (1976) Biochem.
     Biophys. Res. Comm. 69:212

(26) Bird, R., Louarn, J., Martuscelli, J. and Caro, L. (1972)
     J. Mol. Biol. 70:549

(27) Schnos, M. and Inman, R. (1971) J. Mol. Biol. 55:31

(28) Mosig, G. (1970) J. Mol. Biol. 53:503

(29) Wolfson, J. and Dressler, D. (1972) Proc. Nat. Acad. Sci.
     U.S.A. 69:2682

(30) Hara, H. and Yoshikawa, H. (1973) Nature New Biol. 244:200

(31) Edenberg, H., Waqar, M. and Huberman, J. (1976) Proc. Nat.
     Acad. Sci., U.S.A. 73:4392

(32) Danna, K. and Nathans, D. (1972) Proc. Nat. Acad. Sci., U.S.A.
     69:2097

(33) Fareed, G., Garon, C. and Salzman, N. (1972) J. Virol. 10:
     484

(34) Subramanian, K., Dhar, R. and Weissman, S. (1977) J. Biol.
     Chem. 252:355

(35) Butler, W. and Mueller, G. (1973) Biochim. Biophys. Acta
     294:481

(36) Hildebrand, C., Tobey, R. and Walters, R. (1976) J. Cell Biol.
     70 pt 2:58a

(37) Weissbach, A. (1975) Cell 5:101

(38) Higgins, N., Kato, K. and Strauss, B. (1976) J. Mol. Biol.
     101:417

(39) Kato, K. and Strauss, B. (1974) Proc. Nat. Acad. Sci. U.S.A.,
     71:1969

INHIBITION OF DNA SYNTHESIS BY 1-β-D-ARABINOFURANOSYLCYTOSINE: DIFFERENTIAL EFFECT ON CHAIN INITIATION AND ELONGATION IN HUMAN LYMPHOBLASTS

Arnold Fridland

Department of Biochemical and Clinical Pharmacology
St. Jude Children's Research Hospital
Memphis, Tennessee 38101, U.S.A.

INTRODUCTION

1-β-D-Arabinofuranosylcystosine (ara-c) is a potent inhibitor of DNA replication in various organisms, such as bacteria (1), DNA viruses (2) and animal cells (3). In humans, ara-c has been quite effective in the treatment of acute granulocytic leukemia (4,5). Two of the possible hypotheses for its inhibitory effect on DNA synthesis center on the replicative DNA polymerase. In the first model, the triphosphate form of ara-C inhibits competitively the utilization of deoxycytidine triphosphate during the elongation of DNA chains (6-8). In the second model, ara-CTP is incorporated into the growing DNA chain and this incorporation now prevents further chain-elongation by the polymerase (9-11). Although both of these effects have been demonstrated *in vitro* the mechanism of action for this drug has not been clearly established *in vivo* (3).

Recently, I reported on preliminary studies that implicate a new mode of action for ara-C, namely, inhibition of initiation of replication units in DNA (12). This paper presents a more detailed analysis of the inhibitory effect of the drug on the process of replicon initiation, as well as that of chain-elongation. Our results reveal that ara-c exerts a preferential effect on the process of initiation of DNA chains.

The abbreviations used are: ara-c; 1-β-D-arabinofuranosylcystosine, SSC, sodium citrate saline (pH 7.4), ara-CTP, arabinofuranosylcytosine triphosphate, FdUrd, 5-fluorodeoxyuridine.

MATERIALS AND METHODS

Cells. Human lymphoblasts (CCRF-CEM line) were grown as suspension cultures as reported earlier (12).

Sedimentation analyses in alkaline sucrose gradients. Cells were incubated at 37°, with or without ara-c, at various time-intervals, pulse-labeled with (^{3}H)dThd (10 µCi/ml, 60.7 Ci/mmol) for 5 min. The pulse was terminated by washing the cells with ice-cold SSC (0.15 M NaCl and 0.015 M sodium citrate) containing 10mM KCN and 8 mM EDTA. In pulse-chase experiments, medium containing (^{3}H)dThd was removed, the cells were washed with SSC and incubated for the desired time-interval in fresh medium containing unlabeled dThd (10 µM) with or without ara-c. After these incubations, cells were washed with ice-cold SSC, containing KCN and EDTA, resuspended in the same buffer and layered on a lysing solution on top of a 36 ml 15-30% alkaline sucrose gradient, containing 0.5 M NaCl, 0.25N-NaOH and 0.001 M EDTA. About 4×10^5 cells in 0.1 ml SSC were layered onto 0.4 ml lysing solution (0.2 M NaOH, 0.02 M EDTA, 0.1% Nonidet NP-40 (Shell oil) and allowed to lyse on the gradient in the dark at 37° for 6 hr. Gradients were centrifuged at 26,000 rev/min for 4 hr in the Beckman SW 27 rotor at 15°, and fractionated using an ISCO density gradient fractionator. The DNA of each fraction was precipitated on Whatman GF/A filters, washed 4-5 times with 5% trichloroacetic acid, once with ethanol, dried and radioactivity on each filter determined by liquid scintillation spectrometry as previously described (12).

Weight average molecular weights ($\bar{M}w$) of the labeled DNA in gradients were calculated using the relationship, $\bar{M}w = \Sigma(c_i \times M_i)/\Sigma c_i$, where C_i is dpm per min in the ith fraction and M_i is the molecular weight of molecules in the ith fraction as calculated by Studier's relationship (13). A variable amount of radioactivity (15-20%) was often found in the bottom of the gradients, which probably resulted from wall effects during sedimentation (15). These bottom fractions were always excluded from the determination of molecular weights.

Isopycnic centrifugation in alkaline CsCl gradients. For these gradients, about 4.2 g of CsCl was dissolved in 4.5 ml of sodium phosphate buffer, pH 12. About 0.1 ml of DNA solution (approximately 100 µg), isolated as described previously (14), was added to the CsCl solution in polyallomer tubes. The specific gravity of the gradient was adjusted to 1.6850 g/ml and the tubes were centrifuged for about 30 hr at 40,000 rev/min in Beckman SW-50.1 rotor at 20°C. Fractions of 0.3 ml were collected from the bottom of the gradient, and each fraction was precipitated on Whatman GF/A filters, washed 4-5 times with 5% trichloroacetic

acid, once with ethanol, dried and radioactivity on each filter
determined by liquid scintillation spectrometry.

RESULTS

(a) <u>Kinetics of inhibition of DNA synthesis in CCRF-CEM cells by
ara-c.</u> Incorporation studies with (^{3}H)-labeled thymidine
indicate that the incubation of CCRF-CEM cells with 5 nM or 50 nM
ara-c resulted in a rapid decrease of the overall rate of DNA
synthesis and for a period lasting at least 2 hr (Fig. 1). This
inhibition by ara-c was prevented by the simultaneous addition of
deoxycytidine, but neither cytidine nor uridine had a reversing
effect (data not shown). This ability of deoxycytidine to protect
the cells is presumably due to the prevention of formation of ara-
CTP from ara-c by competition for deoxycytidine kinase.

(b) <u>Analysis of pulse-labeled DNA by velocity sedimentation.</u>
Inhibition of DNA synthesis in a growing cell can be mediated by a
decrease in the rate of polymerization within individual replication
units (replicons) or a decrease in the number of replicons formed
per unit time, or both. To distinguish between these two possible
effects of ara-c, sedimentation properties of pulse-labeled DNA
were analyzed in alkaline sucrose gradients. CCRF-CEM cells were
incubated, in the absence or presence of ara-c, for 5 min with
(^{3}H)-labeled thymidine and their DNA was analyzed by velocity
sedimentation in 15-30% alkaline sucrose gradients. In control
cells, the main peak of radioactivity was associated with DNA of
molecular weight of about 2.6×10^7 daltons (49S), with more than
50% of the total radioactivity sedimenting between 10^7 to 10^8
daltons (Fig. 2). This distribution of radioactivity arises
because, in a population of reproducing cells, there are strands of
nascent DNA of all sizes between a small number of nucleotides and
the size of completed replicons. Similar radioactive profiles
have been observed in various other cell lines pulse-labeled with
(^{3}H)dThd and their origin interpreted in a similar manner (16,17).

When cells were pulse-labeled for 5 min, immediately after
addition of 5 nM ara-c, no difference in the profiles of gradients
of control and drug-treated cells was observed (data not shown).
When the time between the addition of drug and pulse-labeling was
increased, the peak of radioactivity shifted gradually toward
longer strands of DNA. As seen in the results depicted in Fig. 2A,
30 min after drug addition, the amount of radioactivity incorporated
into the fractions near the top of the gradients (fractions 2-8),
where the most recently initiated replicons would be expected, was
suppressed by more than 50%. On the other hand, there was no
detectable effect on the incorporation of radioactivity into the

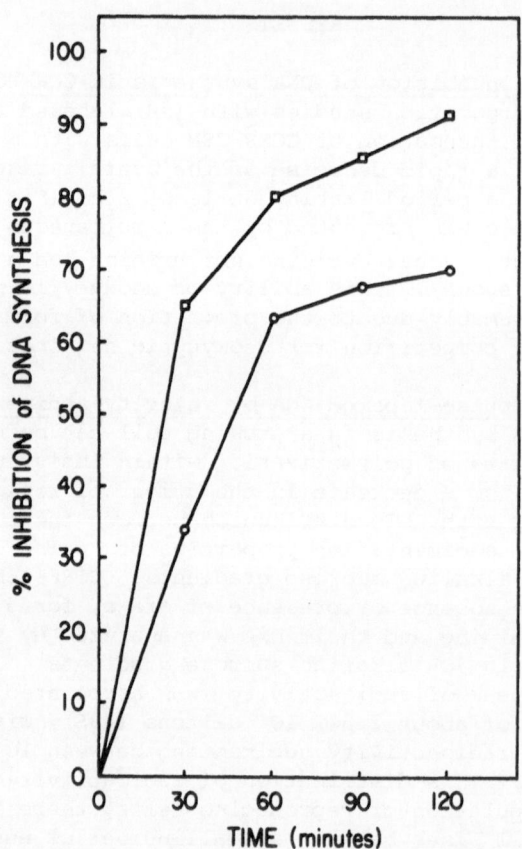

Figure 1. *Effect of ara-c on DNA synthesis of CCRF-CEM cells. Cell suspensions (6 x 10⁵ cells/ml) were incubated with ara-c at the concentrations and for the times indicated. At various times replicate 2 ml samples were withdrawn and pulse-labeled with (³H)-thymidine for 5 min at 37°C and the subsequent incorporation of radioactivity was determined as described previously (13). The data are expressed as percent inhibition of DNA synthesis, as determined by comparing the incorporation of (³H)dThd in the drug-treated cells with that of the control cells (approximately 28,000 dpm were incorporated during a 5-min pulse). Ara-c concentrations were: 5 nM,* ◯———◯ *and 50 nM* □———□ *.*

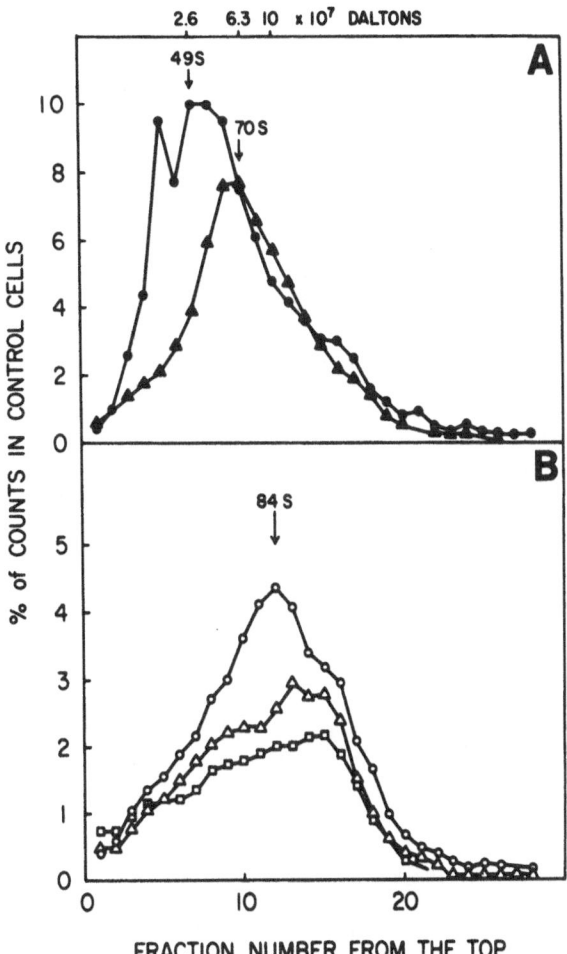

Figure 2. *Velocity sedimentation profiles of alkaline sucrose gradients of DNA from CCRF-CEM cells. Cell suspensions (6 x 10⁵ cells/ml) incubated in the absence (●━━━━●) or presence of 5 nM ara-c were pulse-labeled with 10 μCi of (³H)dThd (specific activity 60.7 Ci Mmol⁻¹) per ml at different intervals: 30 min (▲━━━━▲); 60 min (○━━━━○); 120 min (△━━━━△), and 150 min (□━━━━□). After each pulse the cells were washed with ice-cold SSC, resuspended in the same buffer and DNA analyzed in alkaline sucrose gradients as described under "Materials and Methods". The decrease of (³H)dThd-incorporation into DNA was 0.71 of control after 30 min, 0.52 after 60 min, 0.38 after 120 min, and 0.32 after 150 min. The percent of radioactivity present in each fraction of the gradient for the cells incubated at the various concentrations of ara-c multiplied by 0.71, 0.52, 0.38 and 0.32 to yield the data points shown in the profile.*

region of the gradient (fractions 10-16), containing the larger DNA
fragments. After 1 hr, there was a further suppression of radioactivity incorporated into the smaller DNA-strands and still only
a relatively small effect of the incorporation of radioactivity into
the higher molecular weight strands of DNA (Fig. 2B). By 2 hr or
2½ hr, respectively, the profile had a peak near the position
(fraction 16) where DNA of maximum size was found in these gradients
(Fig. 2B), suggesting that those replicons that were present in the
cells continued either their elongation or joining processes, or
both, to maximum size in the presence of the drug. Moreover, the
presence of a shoulder in the radioactivity profile near the
position at which the main peak occurs in control cells suggests
that either initiation of replicons is not entirely suppressed or
that the cells are beginning to recover at this time.

Similar kinetic studies were performed with cells treated with
50 nM of ara-c. Again with increasing periods of drug-treatment the
average molecular weight of newly synthesized DNA increased (Fig. 3).
At this concentration of drug, the incorporation of radioactivity
in the fractions of newly initiated replicons was suppressed almost
entirely by 30 min. In addition to inhibiting initiation, however,
50 nM ara-c also inhibited chain-elongation, as shown by the lower
amount of radioactivity in the high molecular weight region of the
gradients ((Fig. 3),fractions 10-16)).

(c) Effect of ara-c on the elongation of nascent strands of DNA.
To examine in further detail the effect of ara-c on nascent strand-
elongation, pulse-chase experiments were carried out. Exponentially
growing CCRF-CEM were pulse-labeled for 5 min with (^{3}H)thymidine
in the absence or presence of ara-c, and chased with unlabeled
thymidine. After each period of pulse or chase, the DNA was
analyzed by alkaline sucrose gradient centrifugation. The results
depicted in Fig. 4(A) to (E) demonstrate typical kinetics of the
conversion of pulse-labeled DNA into high molecular weight strands
in control cells. To facilitate the presentation of data from a
large number of gradients, the weight average molecular weight of
(^{3}H)DNA was calculated for each time-interval, as described in
Materials and Methods; this was done for both control and drug-
treated CCRF-CEM cultures. As depicted by the data in Fig. 5, the
time taken to convert pulse-labeled DNA of control cells to high
molecular weight DNA-strands of 1.7×10^8 daltons is about 60 min.
This is similar to the time required for the completion and fusion
of adjacent replicons as detected by DNA radioautography (18-20).

In the presence of 5 nM ara-c there was no detectable effect
on the size of DNA made during the pulse and during the first 30
min of chase (Fig. 5); however, as the time of incubation increased,
an inhibition in the further elongation of nascent DNA occurred.
As shown in Fig. 5, the further elongation of pulse-labeled DNA
appears to have slowed markedly after 60 min incubation. When the
concentration of ara-c was increased to 50 nM, a more profound

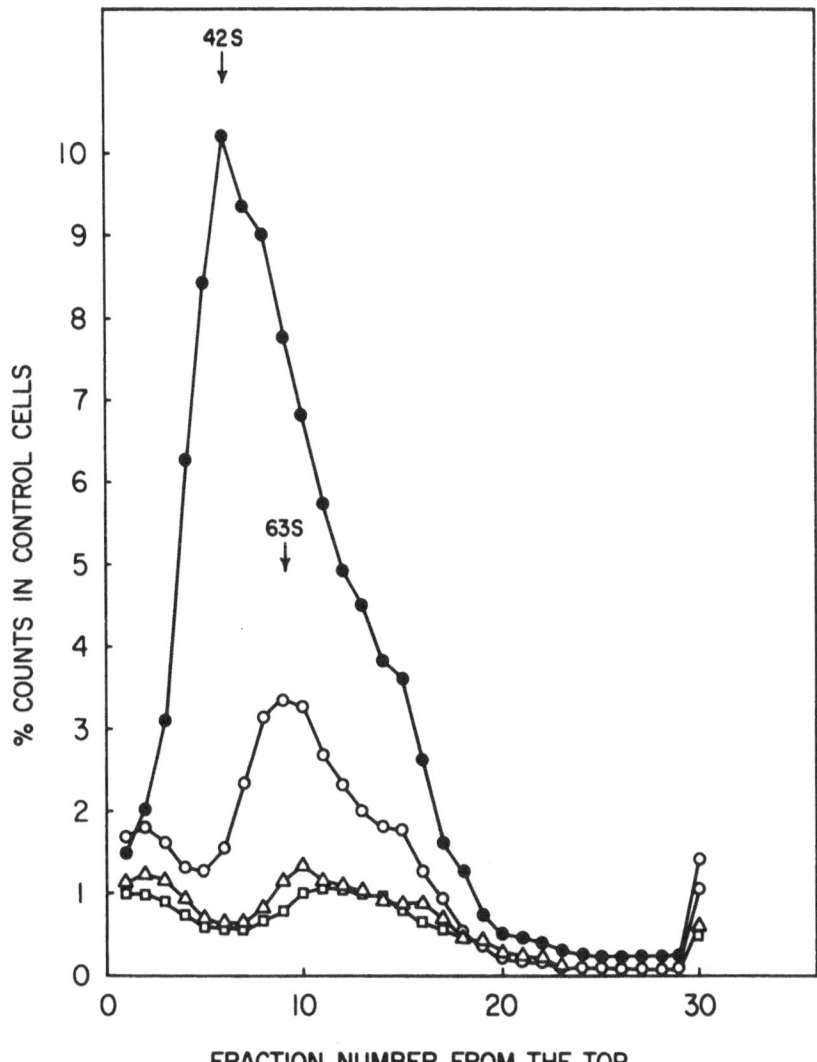

Figure 3. Velocity sedimentation profiles of alkaline sucrose gradients of pulse-labeled DNA. Cell suspensions (6 x 10⁵ cells/ml) incubated without (●——————●) or with 50 nM ara-c were pulsed-labeled with 10 μCi (³H)dThd for 5 min at different time-intervals: 30 min, O——————O; 60 min, △——————△ and 120 min, □——————□. After each pulse the DNA was analyzed in alkaline sucrose gradients as described under Fig. 2. At this concentration of the drug, (³H)-dThd-incorporation into DNA was 0.38 of control after 30 min, 0.2 after 60 min, and 0.17 after 120 min.

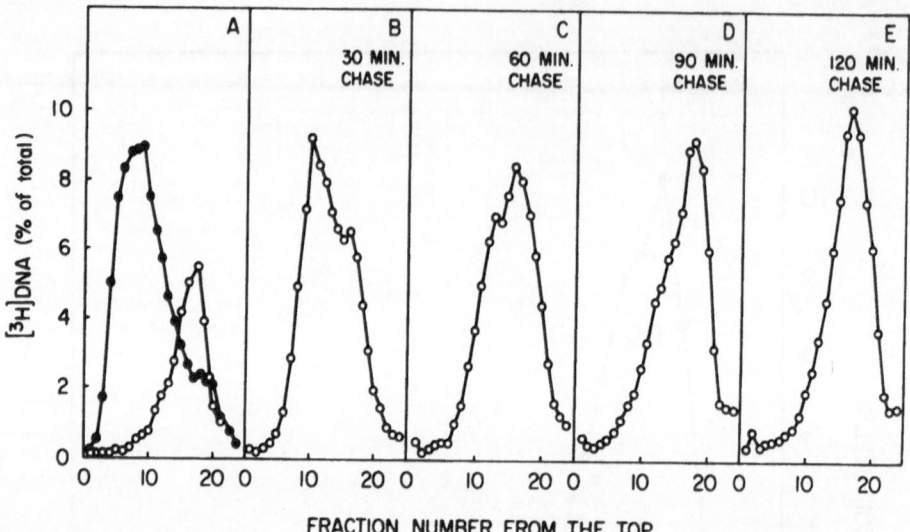

FRACTION NUMBER FROM THE TOP

Figure 4. Sedimentation profiles of (³H)dThd -- dThd pulse-chase labeled cells. Cells were pulse-labeled for 5 min (A,●━━━●) with (³H)dThd for 5 min, washed and chased with unlabeled dThd (1 μM) for 30 min (B), 60 min (C), 90 min (D), and 120 min (E). Prior to incubation with (³H)dThd, cells were grown for 16 hr in the presence of (¹⁴C) dThd (0.02 μCi/ml) to label bulk DNA (A,○━━━○). Labeled DNA was analyzed by velocity of sedimentation, as described under "Materials and Methods".

inhibition of the process of chain-elongation was obtained. As shown by the data depicted in Fig. 5, the size of DNA made after a 5-min pulse was reduced by about 1/3, and further elongation of strands occurred at a very slow rate. The residual synthesis of DNA dropped nearly to zero after 1 hr and nascent DNA could not be chased into maximum size strands.

To evaluate further the effect of ara-c on the process of chain-elongation, density-distribution of DNA molecules synthesized after incubation with (³H) labeled bromodeoxyuridine (BrdUrd) were examined. Substitution of BrdUrd for dThd during DNA synthesis in cells raises the buoyant density of newly synthesized DNA in isopycnic gradients and provides the means to determine relative rates of chain-elongation (16,21). CCRF-CEM cells were incubated for 20 min with (³H)BrdUrd, in the absence or presence of ara-c. The cells were washed free of radioactivity and then incubated for 3 hr with unlabeled thymidine. The DNA was isolated from the cells, sheared to fragments of about 2.6×10^7 daltons and analyzed by equilibrium centrifugation in alkaline CsCl gradients. As shown in the results depicted in Fig. 6, the incorporation of (³H)BrdUrd in control cells caused a marked increase in the buoyant density of newly synthesized DNA strands. When (³H)BrdUrd was added during the first 20 min of incubation with 5 nM ara-c (resulting in 20%

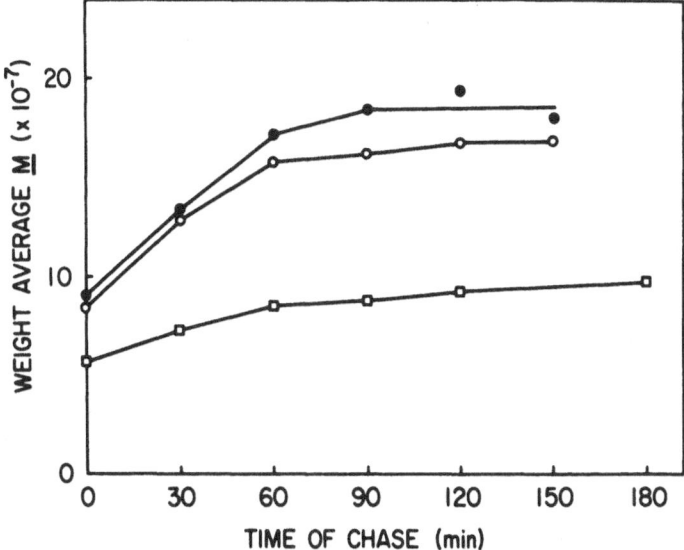

Figure 5. *Rate of formation of high molecular weight nascent DNA strands. Each symbol represents the average value of two separate experiments of (³H)dThd-dThd pulse-chase labeled cells incubated in the absence (●———●) or presence of 5 nM (○———○) or 50 nM (□———□) ara-c, as described under Fig. 4. The average molecular weights were calculated as described under "Materials and Methods".*

inhibition of DNA synthesis), the density of pulse-labeled DNA was shifted to the same position as that of the control cells (Fig. 6). These results demonstrate, therefore, that DNA synthesized during the first 20 min in the presence of 5 nM ara-c is the same size as that of the controls and confirms the previous finding (Fig. 2) that there is a step of the replication process that is more sensitive than chain-elongation to inhibition by ara-c.

In the presence of 50 nM of ara-c, the density of (³H)labeled DNA was substantially reduced (Fig. 6), indicating that the rate of chain-elongation is rapidly inhibited by this concentration of the drug. Finally, when CCRF-CEM cells were pre-treated with ara-c for 1 hr prior to labeling with (³H)BrdUrd, a more pronounced effect on the rate of chain elongation was seen at both drug concentrations. The results depicted in Fig. 7 indicate that the presence of 5 nM ara-c caused a 60% reduction of the density shift in the peak fraction, whereas with 50 nM ara-c (Fig. 7) an 85% reduction of the density shift of (³H)DNA was obtained.

DISCUSSION

In this study, DNA replication in CCRF-CEM cells was immediately inhibited when low concentrations of ara-c were added

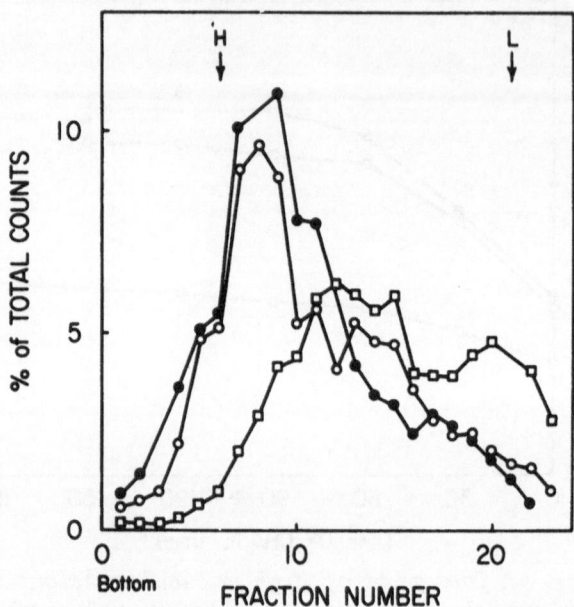

Figure 6. Density distribution of DNA in cells pulse-labeled with
(^{3}H)BrdUrd. Cells (6 x 10^5 cells ml^{-1}) were incubated without
(●——————●) or in the presence of 5 nM (O——————O) or 50 nM ara-c
(□——————□) and pulse-labeled for 20 min with 12 μCi of (^{3}H)BrdUrd
(spec. act. 14 Ci $mmol^{-1}$) per ml and unlabeled FdUrd (1 μM). After
incubation, the cells were washed and re-incubated for 5 min in
medium devoid of radioactivity. Radioactivity in DNA was chased
for 3 hr in medium containing unlabeled 1 μM dThd. Arrows indicate
the position of normal-density light strands (L) and BrdUrd-
containing DNA strands (H) obtained from cells incubated for about
8 hr with unlabeled BrdUrd (10 μM) and dCyd (10μM). DNA isolation
and centrifugation was carried out in alkaline CsCl gradients as
described under "Materials and Methods".

to the growth medium. During the first 30 min of incubation with
5 nM of drug, ara-c, the initiation of new DNA chains decreased
sharply, but the process of chain elongation was not affected (cf.
Fig. 2 with Figs. 5 or 6). Earlier investigations(7,22,23) of the
mode of action of ara-c concluded that the drug acts primarily by
inhibiting the process of chain elongation. These studies made
use of drug concentrations ranging from 5 μM to 0.1 mM, so that any

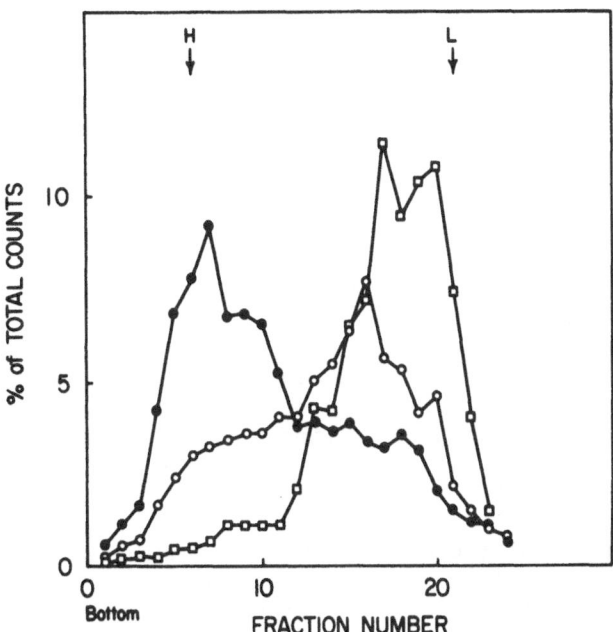

Figure 7. Density distribution of DNA in cells pulse-labeled with (^{3}H)BrdUrd. Cells (6 x 10^5 cells ml^{-1}) were incubated without (●————●) or in the presence of 5 nM (○————○) or 50 nM ara-c (□————□) and pulse-labeled 1 hr later for 20 min with (^{3}H)BrdUrd in the presence of unlabeled FdUrd. The conditions of radio-labeling, isolation, and centrifugation are described under Fig. 6.

effects on the initiation of replicons would be masked by the drastic reduction of chain elongation.

DNA replication in the nucleus of mammalian cells is thought to involve time-dependent, bidirectional synthesis of replicons which, after completion, join together to form complete DNA molecules (17-20). From autoradiographic studies, the average size of replicons in mammalian cells has been estimated to range from 30 to 60 x 10^6 daltons (19,24-26). In alkaline sucrose gradients, the average size of single-stranded pulse-labeled DNA of cells not treated with ara-c is 26 x 10^6 daltons (Fig. 2). Thus, a 5-min pulse of (^{3}H)dThd in these human lymphoblasts labels DNA segments that are somewhat smaller than replicon size. During inhibition of chain initiation by ara-c the radioactivity incorporated during a pulse is associated with DNA of much larger molecular weight than that of control cells. After a 30-min treatment with 5 nM ara-c, the peak of radioactivity was associated with DNA of about 6.3 x 10^7 daltons; at 1 hr the peak was about

1×10^8 daltons; and by 2 h, it was about 1.5×10^8 daltons
(Fig. 2). Thus, DNA strands that have been initiated before the
addition of ara-c appear to elongate to DNA chains about three
times the size of normal replicons. A similar alteration in the
size distribution of pulse-labeled DNA has been reported for HeLa
cells and Chinese hamster cells during inhibition of replicon
initiation after exposure to low doses of ionizing radiation (27).
These observations have several possible explanations. First, when
the initiation of new DNA chains is blocked by agents such as ara-c
or x-rays, the reduced number of growing DNA chains might replicate
beyond their normal termination sites. Alternatively, the
initiation or termination sites along the DNA chain might be
variable (rather than fixed according to specific base sequences),
so that drug or radiation effects on DNA could increase the distance
between points of replicon formation (24). A third possibility is
that larger DNA sizes reflect the existence of clusters of very
large replicons in the mammalian genome (27).

The biochemical route by which ara-c inhibits DNA replication
in vivo is still unclear. Some workers have proposed the
replicating DNA polymerase as a primary target of the drug (6,7).
The inhibition constant (Ki) of DNA polymerase from mammalian
cells, however, occurs at drug concentrations ranging from 1 μM to
20 μM (6-8,28), at least 200-fold higher than the level of ara-CTP
required for 50% inhibition of the overall rate of DNA synthesis
in mammalian cells (Fig. 1; also refs. 7,29). Moreover, a primary
effect on DNA polymerase would not be expected to cause a
differential effect on chain initiation and elongation.

The observations reported here are very similar to those
reported recently for DNA exposed to x-rays or methyl methane-
sulfonate and for BrdUrd-substituted DNA exposed to 313 nm of light,
in that all of these agents exerted a primary effect on replicon
initiation (27,30,31). The main lesions produced by the agents
are single-strand DNA breaks and base damage. Painter et al. (31)
have suggested that such lesions, when occurring at or near the
site of initiation of replicon clusters, can cause a change in the
supercoiled structure of DNA, thus preventing the binding of
initiation factors so that new replicons cannot be initiated.

There is now good evidence that ara-CMP is incorporated, *in
vivo*, into nascent DNA (7,22,29,32). This site of incorporation
could result in an altered helix (33) and hence prevent its further
replication. Thus, since the monophosphate derivative of ara-c is
likely present in newly initiated DNA strands, at the site of
replicon origin, it could alter the structure of the replication
complex in this region, thus blocking further growth of these short
DNA chains. The absence of an immediate drug effect on replicons
that were already formed when ara-c was added to cell culture
indicates that the processes of elongation and joining are not

appreciably affected by subtle structural changes in nascent DNA units. However, as more ara-c residues are incorporated into growing DNA chains, the structure of the DNA helix would become sufficiently altered to inhibit further polymerization, consistent with the reduced rate of chain elongation seen in these experiments after 30 min of incubation.

ACKNOWLEDGEMENTS

I am grateful to Mr. James McLellan for excellent technical assistance. This investigation was supported in part by Grant Number Ch52 from the American Cancer Society, CA18446 awarded by the National Cancer Institute, DHEW, and by ALSAC.

REFERENCES

(1) Cohen, S.S. (1966) Prog. Nucleic Acid Res. 5:1-88

(2) Ch'ien, L.T., Schabel, F.M., Jr. and Alford, C.A., Jr. (1973) In: Selective Inhibitions of Viral Functions (ed. Caster, W.A.) CRC Press, Cleveland, Ohio, pp227-256

(3) Roy-Burman, P. (1970) Rec. Res. Cancer Res. 25:1-106

(4) Gee, T.S., Yu, K-P, and Clarkson, B.D. (1969) Cancer 23:1019-1032

(5) Fleming, I., Simone, J., Jackson, R., Johnson, W., Walters, T. and Mason, C. (1974) Cancer 33:427-434

(6) Furth, J.T. and Cohen, S.S. (1968) Cancer Res. 28:2061-2067

(7) Graham, F.L. and Whitmore, G.F. (1970) Cancer Res. 30:2636-2644

(8) Rama-Reddy, G.V., Goulian, M. and Hendlers, S.S. (1971) Nature New Biology 234:286-288

(9) Momparler, R.L. (1969) Biochem. Biophys. Res. Commun. 34: 465-471

(10) Momparler, R.L. (1972) Mol. Parmacology 8:362-370

(11) Kornberg, A. (1974) DNA Synthesis. W.H. Freeman, San Francisco, Calif. pp226-228

(12) Fridland, A. (1977) Biochem. Biophys. Res. Commun. 74:72-78

(13) Studier, F.W. (1965) J. Mol. Biol. 11:373-390

(14) Fridland, A. and Brent, T.P. (1975) Eur. J. Biochem. 57: 379-385

(15) Schumaker, V.N. (1967) Adv. Biol. Med. Phys. 11:245-339

(16) Gautschi, J.R., Kern, R.M. and Painter, R.B. (1973) J. Mol. Biol. 80:393-403

(17) Huberman, J.A. and Horwitz, H. (1973) Cold Spring Harbor Symp. Quant. Biol. 38:233-238

(18) Huberman, J.A. and Tsai, A. (1973) J. Mol. Biol. 75:5-12

(19) Hori, T. and Lark, K.G. (1973) J. Mol. Biol. 77:391-404

(20) Hand, R. and Tamm, I. (1974) J. Mol. Biol. 82:175-183

(21) Bonhoeffer, F. and Gierer, A. (1963) J. Mol. Biol. 7:534-540

(22) Magnusson, G., Craig, R., Narkhammar, M., Reichard, P., Staub, M. and Warner,H. (1974) Cold Spring Harbor Symp. Quant. Biol. 39:227-233

(23) Wist, F., Krakan, H. and Prydz, H. (1974) Biochemistry, 15: 3647-3652

(24) McFarland, P.W. and Callan, H.G. (1973) J. Cell Sci. 13:821-831

(25) Hand, R. (1975) J. Cell Biol. 64:89-97

(26) Housman, D. and Huberman, J.A. (1975) J. Mol. Biol. 94: 173-181

(27) Painter, R.B. and Young, B.R. (1976) Biochim. Biophys. Acta 418:146-153

(28) Schrecker, A.W., Smith, R.G. and Gallo, R.C. (1974) Cancer Res. 34:286-291

(29) Chou, T.C., Hutchison, D.J., Schmid, F.A. and Phillips, F.S. (1975) Cancer Res. 35:225-236

(30) Painter, R.B. (1977) Mutation Res. 42:299-304

(31) Povirk, L.F. and Painter, R.B. (1976) Biochim. Biophys. Acta 432:267-272

(32) Chu, M.Y. and Fischer, G.A. (1968) Biochem. Pharmacol. 17:753-767

(33) Silagi, S. (1965) Cancer Res. 25:1446-1453

THYMIDINE METABOLISM IN BACTERIA (AND "HOW, OR HOW NOT, TO LABEL

DNA")

Gerard A. O'Donovan

Department of Biochemistry and Biophysics
Texas A & M University
College Station,
Texas 77843, U.S.A.

INTRODUCTION

Radioactive labelling techniques are of major importance in the study of DNA synthesis in bacteria. In order to label DNA in *Escherichia coli* either radioactive thymine or thymidine is generally used because these compounds are specifically incorporated into DNA. The availability of mutants, unable to synthesize thymidylate (Thy⁻), makes possible the control of the specific activity of newly synthesized DNA by adjusting the specific activity of exogenously added thymine or thymidine. In experiments which measure the *short*-term incorporation of labelled precursors other factors must be considered. These include the intracellular precursor pool sizes of bases, nucleosides and nucleotides, the rates of uptake, as well as the points of entry of the particular exogenously labelled compounds. The transport of nucleosides and bases from the medium into the cells is of considerable importance since this may result in differences between strains in the incorporation of labelled precursors into DNA. This effect is evident only when extremely low exogenous concentrations of precursors are used and it can give the appearance of altered levels of DNA synthesis.

It has frequently been shown that *the concentration of thymine* in the growth media of Thy⁻ strains has a specific effect on *the replication velocity of DNA,* that is, the rate of addition of nucleotides to a growing polynucleotide chain [for review see (1)]. In many cases this critical fact does not become obvious, even at low thymine concentrations, because one can reduce the replication velocity without simultaneously causing a detectable change in the growth rate (2). For example, between thymine concentrations of

219

0.25 and 10.0 μg per ml the replication time of *E.coli* 15T⁻ differs
by a factor of 2 to 3, without significantly affecting the growth
rate or the viability (2).

The rate of DNA synthesis by a bacterial culture growing in
steady state conditions is *independent* of the velocity of DNA
replication (3,4). The *rate* of DNA replication is determined
solely by the frequency with which successive cycles of chromosome
replication are initiated (1,5,6). This absence of coupling
between replication velocity and growth rate (the property most
frequently measured) has therefore important implications in all
studies with Thy⁻ bacteria. Thus, in Thy⁻ mutants, it would not
be valid, as is the case for other nutritional mutants, to find
the optimal concentration of thymine to add to the growth medium,
by determining the concentration which gives the maximum growth
rate and growth yield. Indeed, a concentration of thymine can be
selected that is not rate-limiting for growth but which is nevertheless rate-limiting for chain elongation (1,2).

Using a variety of techniques, Pritchard and his collaborators
(2,7,8) have been able to demonstrate that the rate of chain
elongation can be more than halved in Thy⁻ strains by reducing the
concentration of thymine in the growth medium, yet this reduction
in replication velocity does not lead to a detectable change in
growth rate. Moreover, their data strongly suggest that the
thymine concentrations usually found in growth media are *rate*-
limiting for chain elongation **in several widely used Thy⁻ strains.**

Thymine concentrations in growth media should be carefully
controlled and clearly stated for all strains. Thy⁺ wild type
strains should not be compared with Thy⁻ mutants without independent
evidence that, under the conditions of growth used, the composition
of cells of each strain is identical. This is frequently not the
case (1).

Thy⁻ mutants require different concentrations of thymine. The
minimum thymine concentration used for any strain is that compatible
with normal growth (2,7), and DNA elongation rate must be considered
among the criteria for normal growth. For example, the well-known
E.coli 15T⁻ requires only 0.25 μg thymine per ml in the growth
medium while most low thymine requiring strains (Tlr⁻) need about
2 μg per ml. High thymine requiring mutants generally require
between 20 and 50 μg thymine per ml, while other strains are known
to require still more thymine, at concentrations up to 100 μg per
ml. The reasons for these requirements are discussed in the
following pages.

During the course of this Advanced Study Institute, a great
deal has been heard about Okazaki fragments. Data from a wide variety
of experiments suggest that certain difficulties exist in observing

these short polynucleotide chains. To measure the size and the
number of these pieces one has to label the DNA with thymine or
thymidine. It is especially important to label Okazaki fragments
correctly, simply because different sizes and different numbers of
Okazaki pieces are produced *merely* by changing the method of
labelling (9). As expected, different strains (W3110 PolA[+] *versus*
W3110 PolA[-]; 15T[-] *versus* W3110T[-]) give different results.

From this brief introduction it should be clear that there are
problems in thymidine metabolism that ought to be considered. It
is therefore important to be able to distinguish those anomalous
results caused by technique from true experimental findings. A
fundamental knowledge of thymidine metabolism is required of anyone
who routinely labels DNA for any purpose. In this presentation, the
salient features of thymidine metabolism are treated in a manner
that we hope will be useful to all who work with thymidine.

BIOSYNTHESIS OF dTTP

In 1957 Friedkin and Kornberg (10) isolated the enzyme
thymidylate synthetase (EC 2.1.1.b) from extracts of *Escherichia
coli*. The enzyme catalyzed a tetrahydrofolate-dependent methylation of deoxyuridylate (dUMP) to deoxythymidylate (dTMP). The
reaction, shown below and in Fig. 1, involves the transfer

$$dUMP + N_5,N_{10}\text{-methylene-tetrahydrofolate} \xrightarrow{Mg2+}$$
$$dTMP + dihydrofolate \qquad\qquad (equation\ 1)$$

of a C_1-unit from N_5,N_{10} - methylene tetrahydrofolate to the 5-
position of dUMP. Thymidylate synthetase is unique among enzymes
that catalyze C_1 transfer reactions in that the tetrahydrofolate is
consumed (oxidized) during the reaction. Thus, tetrahydrofolate has
a dual function in the reaction, acting both as the C_1 - unit
carrier and as the reductant.

Isolation of Thy[-] mutants. Since thymine-requiring mutants
(Thy[-]) undergo thymineless death when incubated in the absence of
thymine (11), penicillin counter selection can not be used for the
isolation of thymine auxotrophs. However, Okada et al. (12,13) and
Stacey and Simson (14) discovered that Thy[-] mutants could be readily
obtained as mutants resistant to the folate analogues aminopterin
or trimethoprim in the presence of thymine. All such mutants
(*thyA*), isolated to date, lack the enzyme, thymidylate synthetase.
Mutants have been isolated in various strains of *E.coli*, *Salmonella
typhimurium* (15), *Bacillus subtilis* (16), *Enterobacter aerogenes*
(17), *Serratia marcescens* (18), *Lactobacillus fermenti* (19) and
Pseudomonas acidovorans (20).

Aminopterin and trimethoprim strongly inhibit the enzyme dihydrofolate reductase, which catalyzes the conversion of di-hydrofolate to tetrahydrofolate (Fig. 1). The reason for the success in isolating Thy$^-$ mutants lies in the fact, mentioned above, that thymidylate synthetase is unique among enzymes that transfer a C_1-unit. Since tetrahydrofolate is consumed in the thymidylate synthetase reaction, the continued action of thymi-dylate synthetase in the presence of the analogues and thymine rapidly depletes the supply of tetrahydrofolate that is obligately required for a variety of biosynthetic pathways and for the initiation of protein synthesis. The longer thymidylate synthetase functions, the less tetrahydrofolate that is available for C_1-unit transfers. Clearly, it becomes a selective advantage for the cells to lack thymidylate synthetase in the presence of the folate antagonists (e.g., trimethoprim) and thymine. That such Thy$^-$ mutants can grow on the thymine provided is an indication that both the thymine requirement and a sufficient catalytic quantity of tetrahydrofolate are supplied. For this to occur, as pointed out independently by Wilson et al. (21) and by Bertino and Stacey (22), the inhibition of dihydrofolate reductase must not be complete. Wild type Thy$^+$ cells are quickly selected against since they deplete the cell of *all* tetrahydrofolate in the course of the thymidylate synthetase reaction.

Mutants with depressed activity of thymidylate synthetase are all defective in *thyA*. The *thyA* gene encodes for the enzyme thymidylate synthetase and is located on the chromosome between *lysA* and *argA* in *E.coli* and between *lysA* and *argB* in *S.typhimurium*.

Two critical points are made here regarding *thyA* mutants in *Pseudomonas acidovorans* and *Bacillus subtilis*. To date the only *thyA* mutants (*thymidine*-requiring) reported for the genus *Pseudomonas* have been isolated by Kelln (20). The paucity of *thyA* mutants among the pseudomonads derives from two problems. The first is that most (if not all) species of *Pseudomonas* lack the enzyme, thymidine phosphorylase, which catalyzes the reaction,

Thymidine + P_i $\rightleftharpoons$ Thymine + Deoxyribose-1-P   (equation 2)

Thus it is not possible to satisfy the thymidylate requirement of *thyA* mutants in *Pseudomonas* by the addition of exogenous thymine (20). Therefore the usual selection for *thyA* mutants using the folate analogues and thymine will not work. Thymidine must be supplied instead of thymine. The second problem is that *Pseudomonas* is extremely resistant to either aminopterin or trimethoprim in the presence of thymidine. However, as demonstrated by Kelln (20), a mixture of both analogues, with a suitable concentration of thymidine, allows the isolation of thymidine-requiring mutants. Thus it should be possible to isolate *thyA* mutants in other species, such as *P. putida* or *P. aeruginosa*.

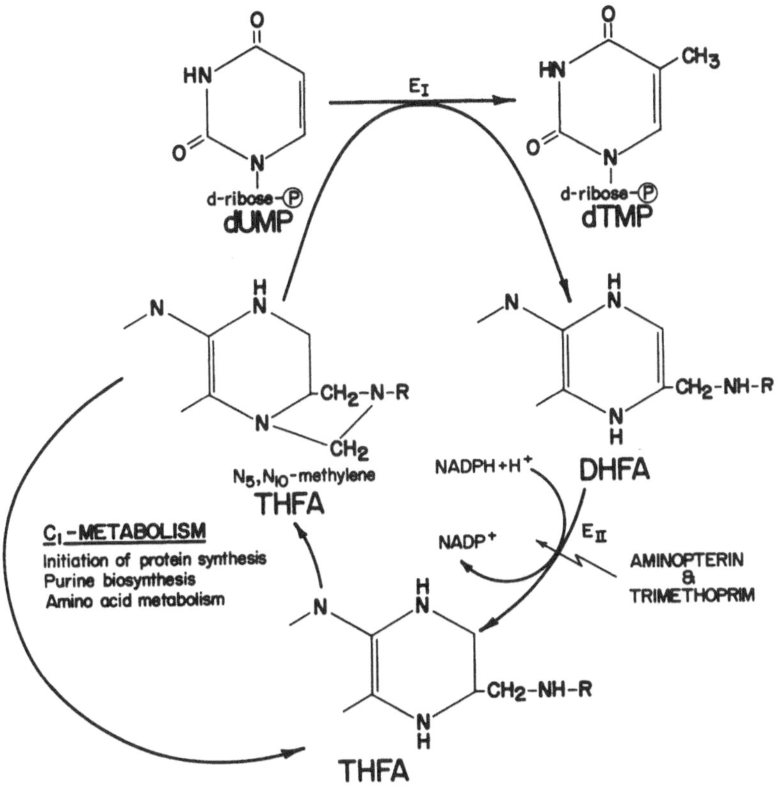

Figure 1. Thymidylate synthetase and its relation to dihydrofolate reductase and folic acid metabolism. E_I is thmidylate synthetase; E_{II} is dihydrofolate reductase.

Recently, Neuhard et al. (23) confirmed and extended an earlier discovery regarding Thy⁻ mutants in *Bacillus subtilis* (21,24). They have shown that two independent mutational events are required to isolate mutants in *B. subtilis* that are devoid of thymidylate synthetase activity. In 1966, Wilson et al. (21) and Anagnosto-poulos and Schneider-Champagne (24) suggested that two mutations in the unlinked genes, *thyA* and *thyB*, were required to create an absolute requirement for thymine in *B. subtilis*. These workers showed further that the *thyA* gene product was thymidylate synthetase but they did not identify the gene product of *thyB*. Neuhard et al. (23) and Price (25) have confirmed that the *thyA* locus codes for thymidylate synthetase and they have also identified the *thyB* gene product as a second thymidylate synthetase that can be distinguished from the enzyme encoded by *thyA* (for details see references 23,25).

Biosynthesis of dUMP from UDP. Because dUMP is the immediate precursor of thymidine nucleotide biosynthesis, it is vital for the cell to be able to maintain adequate supplies of dUMP. UDP is reduced by ribonucleoside diphosphate reductase in *E.coli* (26). The dUDP so formed, appears to be phosphorylated to dUTP which is then rapidly degraded to dUMP by dUTP pyrophosphatase (Fig. 2; references 27,28). An apparent alternative pathway for dUMP synthesis, consisting of the direct hydrolysis of dUDP to dUMP, does not seem to be of physiological significance in wild type cells. However, the direct hydrolysis of dUDP may function in *dut⁻* (29,30) and *dut⁻, ung⁻* (31) mutants of *E.coli*.

The dUTP pyrophosphatase (dUTPase) enzyme has been purified extensively from *E.coli* B by Häggmark (32). It catalyzes the reaction, given below.

$$dUTP + H_2O \longrightarrow dUMP + PPi \qquad \text{(equation 3)}$$

The enzyme is specific for dUTP and is primarily responsible in wild type cells for supplying all endogenous dUMP for thymidylate synthetase. Another function, frequently ascribed to the enzyme, is the removal of dUTP from being incorporated into DNA as uracil. Recently, *dut* mutants, defective in the enzyme dUTPase, have been isolated in *E.coli* K12 (29). Originally designated *sof* (for *s*hortened *O*kazaki *f*ragments) as well as *dnaS*, these *dut* mutants incorporate label into short (4-5*S*) DNA fragments following 5-10 second ³H-thymidine pulses, and are hyper-rec. Surprisingly the *dut⁻* phenotype does not create an absolute thymine requirement nor does it significantly expand the dUTP pool. These findings suggest that the excess of dUTP, caused by a metabolic block immediately beyond it, in the pathway to dTTP (*see* Fig. 2), is converted to dUMP by some alternate pathway, such as dUTP → dUDP → dUMP, as well as being incorporated into DNA as uracil. An interesting discussion on the *dut* genotype in different genetic backgrounds has been presented elsewhere (29,30).

Biosynthesis of dUMP from dCTP. In an attempt to study in more detail the true precursors of endogenous thymidine nucleotides, two independent series of experiments were carried out. In 1967 Karlström and Larson used a cytidine deaminase-less (*cdd*) mutant of *E.coli* B (33) to ascertain the amount of endogenous dTTP which was derived directly from a cytosine precursor. By differentially labelling cytosine and uracil in the *cdd* mutant *in vivo*, and by examining the resulting pattern of labelling in the dTMP moieties of DNA (33), they showed that about 80% of the endogenous dTTP is derived directly from a cytosine compound; only 20% is derived from a uracil compound. In 1968, Neuhard (34) used a specially constructed strain, *pyrA cdd pyrG* (see Fig. 2) which required uracil *and* cytidine for growth, in obtaining an identical result for *S. typhimurium*. After differential labelling of cytosine and uracil

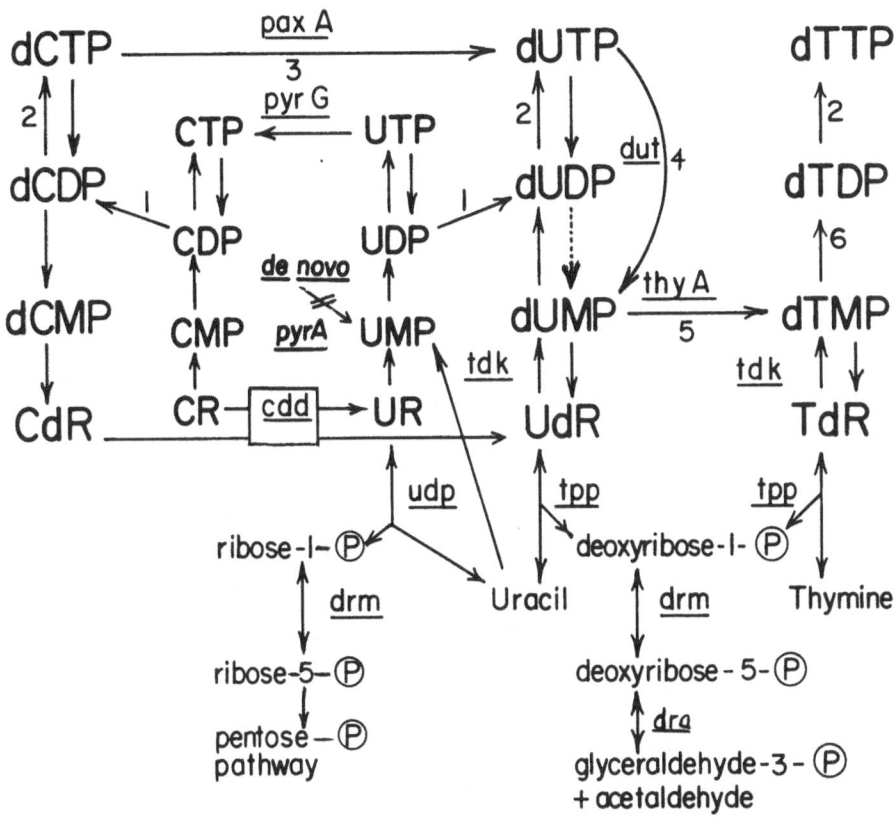

Figure 2. *Relevant pathways in the synthesis of dTTP in Escherichia coli and Salmonella typhimurium*. *The enzymes which catalyse the numbered reactions are: 1, ribonucleoside diphosphate reductase; 2, nucleoside diphosphate kinase; 3, dCTP deaminase; 4, dUTP pyrophosphatase; 5, thymidylate synthetase; 6, thymidylate kinase. The reaction dUDP* ▶*dUMP has not been reported and does not appear to be significant in vivo in wild type cells. Where two arrows pointing in different directions appear this means that two different enzymes are probably required (e.g., UdR* ⇌ *dUMP requires thymidine kinase, encoded by tdk, for the reaction UdR → dUMP and a phosphatase for the reaction dUMP → UdR.*

in vivo Neuhard (34) analyzed the acid-soluble endogenous dTTP pool from the *S. typhimurium* mutant. He found, like Karlström and Larsson (33) that 80% of the thymidine nucleotides were derived from a cytosine compound *without equilibration* with the uridine nucleotide pool of the cells. These two experiments suggest that the major pathway for the synthesis of dUMP, and hence dTMP, dTDP and dTTP, is derived directly from a cytosine compound. The logical compound, as shown in Fig. 2, is dCTP. The minor pathway for dUMP synthesis is derived from the direct reduction of UDP (Fig. 2). While examining the metabolism of cytosine compounds in *S. typhimurium* (34), Neuhard discovered an enzyme capable of specifically deaminating dCTP (35). The identification of the enzyme, dCTP deaminase, led to the elucidation of a previously unreported pathway for dUMP and subsequently dTMP biosynthesis involving the sequential action of dCTP deaminase and dUTP pyrophosphatase (Fig. 2; 27,28). This finding provided a logical explanation for the labelling data described above (33,34) and at the same time strongly suggested that the major route to dTTP synthesis involved dCTP deaminase while the minor pathway to dTTP utilized UDP reductase. Both pathways, as seen in Fig. 2, use the intermediate dUTP.

If indeed the enzyme dCTP deaminase together with dUTP pyrophosphatase, acting sequentially, is the major supplier of dUMP in wild type enteric strains, what then would be the consequences of a mutation that eliminated this enzyme ? At least four easily distinguishable characteristics would be expected for the mutant strain: (i) The endogenous pool of dTTP would be low. (ii) Likewise, the endogenous pool of dUMP would be low. (iii) The dCTP pool would be expected to be high, as a consequence of the low dTTP pool (35, 37,38). (iv) All dTTP would be derived from a uracil compound, none coming from a cytosine compound. These four characteristics were found in an *E.coli* K12 strain, HD1038, that lacked the enzyme dCTP deaminase (36). The genetic symbol, *paxA*, was designated as the genotype for the dCTP deaminase-less mutant. Recently, dCTP deaminase-less mutants have been isolated in *S. typhimurium* (39) and their genetic locus on the *E.coli* and *S. typhimurium* chromosomes has been mapped. The *paxA (dcd)* gene cotransduces with *udk* in both genera. A comment is necessary here for clarification. In recent publications (39,40) the symbol *dcd* has been ascribed to dCTP deaminase mutants. This symbol is confusing since (a) the original *E.coli* symbol *paxA* has not been superceded and is still in use. (b) The symbol *dcd* has been used for a different genotypic designation. Thus both symbols: *dcd* and *paxA*, are retained in this presentation.

In summary, two pathways for dUMP synthesis exist and dUTP pyrophosphatase plays a central role in thymidine nucleotide metabolism. In the absence of the major source of dUMP (dCTP deaminase) the other source (UDP reductase) is able to take over and provide adequate dUMP, and thus thymidine nucleotides, to ensure normal growth. The parent strain *E.coli* JC411 and its mutant derivative, strain HD1038 (*paxA*) have identical growth rates (41). This is possible only because the HD1038 cells have a low dTTP pool, which results in the derepressed synthesis of ribonucleoside diphosphate reductase (42). It should also be kept in mind that the affinity of *E.coli* ribonucleoside diphosphate reductase for UDP, as a substrate, is significantly lower than it is for CDP, ADP and GDP (43,44). Fig. 2 presents a summary of both pathways.

HOW TO LABEL DNA

Three physiological conditions should exist in *Escherichia coli* for efficient labelling of DNA with exogenous thymine. These conditions are (i) a low endogenous pool of dATP, (ii) a high endogenous pool of dUMP and (iii) a high endogenous supply of deoxyribose-1-phosphate (deoxyribose-1-P). Budman and Pardee (45) have shown that only one pathway exists for the conversion of exogenous thymine to dTTP and that this pathway requires a functional thymidine phosphorylase. Thymidine phosphorylase (encoded by *tpp*) catalyzes the reversible reaction shown below, and in Fig. 3). The enzyme has been purified to homogeneity (46) and

$$\text{Thymidine} + \text{Pi} \rightleftharpoons \text{thymine} + \text{deoxyribose-1-P} \quad \text{(equation 2)}$$

in addition to thymidine, deoxyuridine and the 5-halogen-substituted deoxyuridines are phosphorylyzed in the reaction. Thymidine phosphorylase is inhibited competitively by several ribo- and deoxyribonucleosides, notably uridine (45-48). The enzyme is found in all bacteria except *Lactobacilli* (19, 48a). In most bacteria ribosyl transfer between purines and pyrimidines is catalyzed by the coupled action of purine nucleoside phosphorylase (encoded by *pup*) and thymidine phosphorylase (49). *Lactobacilli* however contain the specific enzyme: *trans-N*-deoxyribosylase (48a) which catalyzes a phosphate *independent* transfer of deoxyribose moieties between a large number of purines and pyrimidines. The general reaction is shown below,

$$\text{Base}_1 \text{---deoxyriboside} + \text{base}_2 \longrightarrow \text{base}_1 + \text{base}_2 \text{---deoxyriboside}$$

$$\text{(equation 4)}$$

The second enzyme required for the conversion of thymine to thymidylate is thymidine kinase (encoded by *tdk*).

$$\text{Thymidine} + \text{ATP} \xrightarrow{\text{Mg}^{2+}} \text{dTMP} + \text{ADP} \quad \text{(equation 5)}$$

In addition to thymidine, deoxyuridine and the 5-halogen-substituted deoxyuridines are active as substrates (52,53). When *E.coli* cells are limited in their supply of dTTP, thymidine kinase is extremely active (54,55). The enzyme is of widespread occurrence but is singularly absent amongst the fungi. Thus to label dTTP (DNA) specifically in the fungi one has to use dTMP in a mutant strain that has been selected for its proficiency of uptake of thymidylate. Such mutants have been isolated (56) and have proved to be extremely useful.

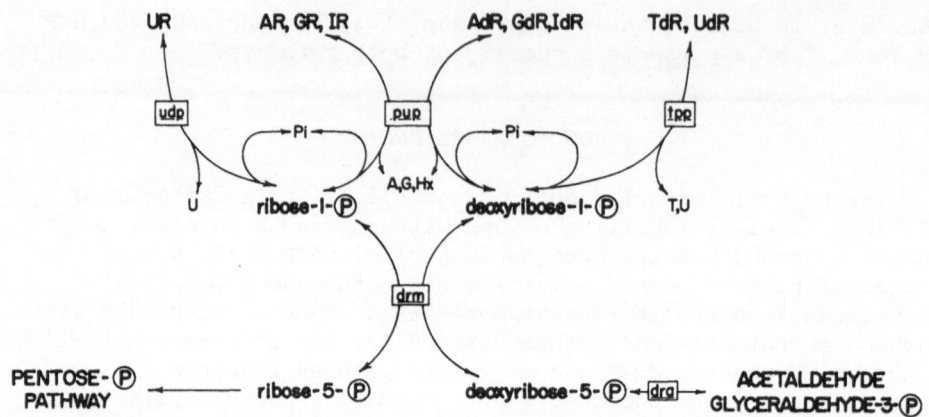

*Figure 3. Deoxyribonucleoside and ribonucleoside catabolism in
Escherichia coli and Salmonella typhimurium. Genotypic symbols
are as follows: udp specifies uridine phosphorylase; pup (deoD),
purine nucleoside phosphorylase; tpp (deoA), thymidine phosphorylase; drm (deoB), deoxyribomutase; dra (deoC), deoxyriboaldolase.
Genotypic symbols in parentheses represent alternate designations
for the same gene.*

Labelling DNA of Wild Type Cells

(a) <u>Thymine alone</u>. As can be seen from Fig. 4 exogenous
thymine is not incorporated into the DNA of wild type cells of
E.coli (57,58). None of the three criteria for successful
labelling are satisfied. However, since thymine-requiring mutants
can grow on exogenous thymine, the enzymes must exist for the
conversion of thymine to thymidylate.

(b) <u>Thymine plus a deoxyribonucleoside</u>. Although exogenous
thymine is not incorporated into the DNA of wild type cells, the
simultaneous addition of thymine and a deoxyribonucleoside (AdR,
GdR, IdR) causes excellent incorporation (44, 59-61) as can be seen
in Fig. 4. Purine deoxyribonucleosides provide much better
enhancement of incorporation than pyrimidine deoxyribonucleosides
(61). Thus, the lack of incorporation of exogenous thymine by
wild type cells is due primarily to the absence of endogenous
deoxyribosyl groups in the wild type. That exogenously supplied
deoxyadenosine enhances thymine incorporation, by supplying deoxy-
ribose-1-phosphate for the thymidine phosphorylase reaction, is
shown by the fact that no stimulation of incorporation is seen in

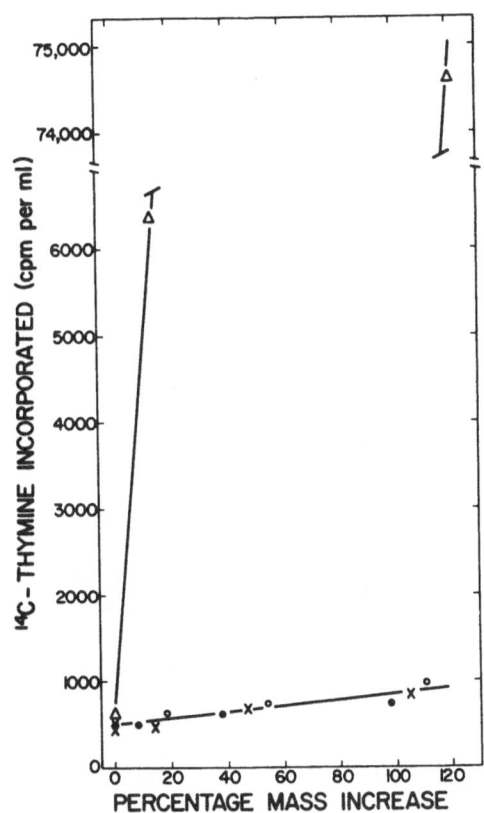

Figure 4. *Incorporation of thymine-2-^{14}C into acid insoluble material by exponentially growing cells of the following Thy$^+$ strains of Escherichia coli : HD1038 paxA (●——●); HD1038 paxA, tpp (×——×); JC411 (O——O); JC411 + deoxyadenosine (△——△). The concentrations of ^{14}C-thymine and deoxyadenosine used were 5 μg per ml and 50 μg per ml, respectively.*

pup mutants which lack the enzyme purine nucleoside phosphorylase.
The enzyme phosphorylyzes both purine ribonucleosides (AR, GR, IR)
and purine deoxyribonucleosides (AdR, GdR, IdR) with the
concomitant production of the corresponding purine base (adenine,
guanine, hypoxanthine) and either ribose-1-phosphate or deoxyri-
bose-1-phosphate as shown in Fig. 3. The deoxyribose-1-phosphate
is required by thymidine phosphorylase to convert thymine to
thymidine. One precaution should be pointed out here. In certain
strains of *E.coli* K12, deoxyadenosine (AdR) is toxic (1) so its
use as a source of deoxyribosyl groups may be limited. Deoxyguanosine (GdR) however is not toxic. Fig. 4 shows how the
addition of a purine deoxyribonucleoside with thymine enhances
incorporation.

(c) Thymidine alone. In contrast to thymine, exogenous
thymidine is rapidly incorporated into DNA of wild type cells.
However, the incorporation stops after a short time due to the
rapid degradation of thymidine to thymine by the inducible
thymidine phosphorylase (62).   Thus thymidine alone can be used
only in short pulse-labelling experiments.

(d) Thymidine in a *tpp* mutant. Accordingly, as shown by
Fangman (63), mutants of *E.coli* which lack thymidine phosphorylase
(*tpp*) incorporate thymidine for longer periods of time (63,64).
Some thymidine is still broken down in *tpp* mutants. This however is
not due to the *tpp* mutation being leaky (65) as previously
suggested, but rather is due to the presence of a functional
uridine phosphorylase (*udp⁺*) which readily degrades thymidine (65).
Indeed, the degradation of 5-bromodeoxyuridine (BUdR) by an extract
of a *tpp* mutant strain of *E.coli* (65) has been shown to be due to
the action of uridine phosphorylase. No degradation of BUdR was
seen in double mutants of the genotype *tpp udp*. An identical result
was observed by Leer et al. (66) with a purified preparation of
uridine phosphorylase from *E.coli* K12. To complete the picture it
should be pointed out that 5-fluorodeoxyuridine is also degraded
by uridine phosphorylase (65,66).

(e) Thymidine plus uridine. As mentioned above, Budman and
Pardee demonstrated that uridine inhibits thymidine phosphorylase
(45). This finding can be exploited to increase the residence time
of thymidine in the cell by the simultaneous addition of uridine
and thymidine. When radioactively labelled thymidine and uridine
(at approx. 50 μg uridine per ml) are added to an exponentially
growing culture of *E.coli* extremely efficient long term
incorporation is seen. Moreover, if one wishes to change the
labelling patterns from ^{14}C-thymine to ^{3}H-thymidine one can do so
by the simultaneous addition of uridine and ^{3}H-thymidine to a
culture previously growing exponentially in ^{14}C-thymine (Fig. 5).
This facile method should prove invaluable for workers in DNA
replication who wish to change from a long-term ^{14}C-thymine (pre-)

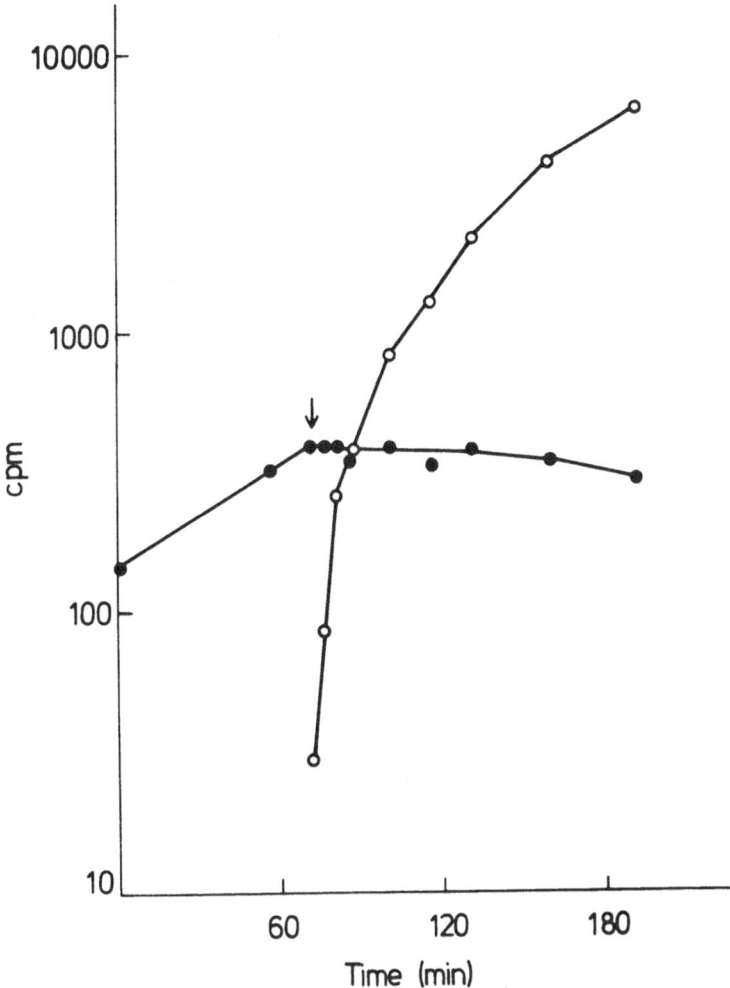

Figure 5. Incorporation of ^{14}C-thymine and ^{3}H-thymidine into acid insoluble material by exponentially growing Escherichia coli B/r. The relevant genotype of this strain is thyA drm, a low thymine requirer (67). Uridine (50 µg per ml) is added at the time indicated by the arrow. ^{14}C-thymine incorporation (●——●) was followed before and after the addition of uridine; ^{3}H-thymidine incorporation (O——O) was followed after uridine and thymidine were added simultaneously. The concentrations of thymine and thymidine were as follows: thymine, 10 µg per ml including 4 µCi ^{14}C-thymine; thymidine, 20 µg per ml, spec. act. 0.45 µCi per µg. Note that the cells began to incorporate the ^{3}H-thymidine immediately after its addition while the addition of uridine completely inhibited the incorporation of ^{14}C-thymine for 120 min. Data are from J.E.Womack (67).

label to a short or long term ^{3}H-thymidine label. No perturbation, due to filtration or centrifugation, is encountered and the addition of uridine (50 µg per ml) prolongs thymidine incorporation for up to 75 min. If longer labelling time is needed the concentration of uridine can be increased (67).

One further point seems pertinent. It is not possible to grow Thy⁻ Pyr⁻ mutants (e.g., *thyA pyrB*) in media with thymine plus *uridine* since uridine inhibits the reversible thymidine phosphorylase reaction in both directions. It is therefore necessary to use *thymine plus uracil* in order to grow a *thyA pyrB* mutant; if uridine is required, as in the case of *upp⁻* mutants, then thymine cannot be used. Thymidine (*and not thymine*) will satisfy the dTTP requirement in strains which also require uridine obligately.

(f) <u>dTMP</u>. An excellent way of promoting thymidine incorporation is to use labelled dTMP as the source of thymidine. Exogenous dTMP must first be dephosphorylated before it can be utilized (68, 69). Since the dephosphorylation is rate-limiting, the addition of dTMP results in slow-feeding of thymidine which thereby precludes the induction of thymidine phosphorylase (70).

Labelling DNA of Thy⁻ Mutants

Unlike wild type, Thy⁺, strains of *E.coli*, *Thy⁻* mutants can readily incorporate exogenous thymine into their DNA since all three criteria required for successful incorporation of thymine into DNA, exist. Thus Thy⁻ mutants have a low endogenous dTTP pool, a high endogenous dUMP pool due to a metabolic block in the conversion of dUMP to dTMP, and as a consequence of the high dUMP pool, a high endogenous pool of deoxyribose-1-phosphate (36). All three criteria are important as will be demonstrated, although the size and maintenance of deoxyribose-1-phosphate pool is most amenable to external manipulation. The addition of exogenous deoxyadenosine to a wild type Thy⁺ strain expands the deoxyribose-1-phosphate pool and thereby enhances the incorporation of thymine, as can be seen in Fig. 4. Fig. 4 also shows that no incorporation of thymine is seen in the dCTP deaminase-less mutant (*paxA*) of the *E.coli* K12 strain, HD1038, even though the mutant has been shown to have a low dTTP pool (36). In HD1038, the level of dUMP is low, and as a result the pool of deoxyribose-1-phosphate is low (41). Even though the level of deoxyribose-1-phosphate is high in Thy⁻ mutants, it is not automatically maintained but is quickly catabolized through deoxyribose-5-phosphate to glyceraldehyde-3-phosphate and acetaldehyde (see Figs 2 & 3). However, by introducing an additional mutation into a *thyA* strain that blocks further catabolism of deoxyribose-1-phosphate, one can create a low thymine

requirer. Such double mutants (*thyA drm* or *thyA dra*) maintain
a high deoxyribose-1-phosphate level and can attain normal growth
rate in media containing 1-2 µg thymine per ml. However, most
Thy⁻ strains require 1-2 µg thymine per ml since the *drm* and *dra*
mutations arise *spontaneously* in *thyA* mutants. Indeed, unless
additional thymine is added to cultures growing in rich media, such
as nutrient broth or nutrient agar, used to maintain *thyA* strains,
all such mutants acquire an additional mutation and become low
thymine requirers.

Classes of Thy⁻ Mutants

A summary of four of the classes of Thy⁻ mutants which have
been described in the literature is now presented. Particular
attention should be paid to the frequently used and poorly
understood "super low" thymine requirers, and experimenters who
use them should be aware of their unique characteristics. In the
past many such strains, such as derivatives of the *E.coli* 15
strain, first isolated by Roepke in 1947 (71,72) have been used
routinely without consideration of their physiological growth
requirements. Many times extrapolations have been made from such
experiments which do not apply to most Thy⁻ strains, e.g., *E.coli*
CR34 *thyA* (73). Differences among and between Thy⁻ strains are
the rule rather than the exception since the endogenous dTTP
concentration, alone of all growth conditions, has a profound
bearing on the DNA elongation rate (1,2,7).

(A) High thymine-requiring strains. These mutant strains
require between 20 and 50 µg thymine per ml for normal growth. All
such mutants have the genotype *thyA*, and are deficient in the
enzyme thymidylate synthetase. The *thyA* mutation causes an
accumulation of dUMP which is first degraded to deoxyuridine by a
phosphatase and then to deoxyribose-1-phosphate by thymidine
phosphorylase (69; Fig. 2). Thus the *thyA* mutation, which
prevents the *de novo* synthesis of dTTP provides the cell with the
source of deoxyribose-1-phosphate which it requires for the
incorporation of thymine (65). However, the deoxyribose-1-phosphate
so formed is readily degraded by two inducible enzymes, deoxyribomutase and deoxyriboaldolase. It is for this reason that *thyA*
mutants of *E.coli* typically require at least 20 µg thymine per ml
in the growth medium to form colonies. Such high concentrations
are necessary in order that synthesis of thymidine (and
subsequently dTTP) from deoxyribose-1-phosphate can complete
sufficiently with its degradation (1).

(B) Low thymine-requiring strains. These mutant strains can
grow in media to which 1 to 2 µg thymine per ml have been added.
All such mutants contain the *thyA* mutation and in addition, one
other mutation. The second mutation resides either in the gene

coding for deoxyribomutase (drm) or deoxyriboaldolase (dra). Thus
low thymine auxotrophs have the genotypes thyA drm or thyA dra
(see Fig. 2 & 3). Thy⁻ strains of the genotype: thyA dra can be
readily distinguished from the class with the thyA drm genotype
(69). The Thy⁻ strains which lack deoxyriboaldolase (thyA dra)
are sensitive to thymidine (75). Accordingly, they have been
shown to accumulate high intracellular levels of deoxyribose-5-
phosphate when grown on thymidine (75). This phenomenon is
discussed in more detail under "Thymidine Sensitivity". They are
often referred to by the phenotype, Tlr⁻, for thymine low requirer
(69,74). The reason for this low requirement, as seen above, is
because Tlr⁻ strains maintain a high deoxyribose-1-phosphate level.
Thus one can postulate that the more endogenous deoxyribose-1-
phosphate that is available, the lower will be the thymine
requirement.

(C) The super low thymine-requiring strains. In accordance
with the above postulate, the super low thymine auxotrophs are
those Thy⁻ mutants that can form colonies on media containing as
little as 0.2 µg thymine per ml (76). This heterogeneous Thy⁻
class appears to contain an additional mutation over the two found
in the low thymine auxotrophs. Since these super low strains are
in frequent use, it is important to understand the physiology of
such mutants. Accordingly some of the super low strains of E.coli
currently in use are now described.

One class of super low thymine requirer has been found among
Tlr⁻ strains that have the genotype, thyA drm. This class contains
an additional mutation that is unlinked to the genes in the deo
operon [for more recent information see Albrechtsen et al. (77)]
which leads to the constitutive synthesis of the enzymes involved
in nucleoside catabolism. Originally described by Ahmad and
Pritchard (76) and designated nucR, this regulatory mutation, now
called deoR (77), is known to affect the amount of thymine required
by Thy⁻ strains. In particular those strains which are thyA drm
are strikingly affected. Instead of requiring 2 µg thymine per ml,
as Tlr⁻ strains do, they can grow on media with 0.2 to 0.5 µg
thymine per ml. These thyA drm deoR strains are constitutive for
thymidine phosphorylase (and for deoxyriboaldolase; the product
of the drm locus is not made) and thus are able to rapidly convert
deoxyuridine to deoxyribose-1-phosphate (76). Then, because of
the metabolic block caused by the drm mutation, a high level of
deoxyribose-1-phosphate is maintained (see Fig. 2). By virtue of
this, minute quantities of thymine are efficiently used, thereby
allowing the thyA drm deoR cells to grow on very low quantities
of thymine (0.25 µg per ml). Mutants like this one are thus super
low thymine requirers because they are drm⁻ while still being
constitutive for thymidine phosphorylase (and deoxyriboaldolase).
One such mutant strain: E.coli Y-70-22 was originally isolated by
Okada (78) and was later described in detail by Munch-Petersen (79).

Since this single isolate (Y-70-22) appeared among Tlr⁻ mutants
similarly isolated by Okada et al. (80) it seems quite likely that
more similar mutants exist. Data by Lomax and Greenberg (81) and
Hoffee (82,83) also suggest that other super low thymine requirers
are present among *E.coli* (81) and *S. typhimurium* (82,83) Tlr⁻ strains
even though they were never screened specifically for this class.

An interesting corollary derives from the *thyA drm deoR* super
low thymine requiring strains. When the *drm⁻* mutation in strain
Y-70-22 was transduced to prototrophy by Ahmad and Pritchard (76)
they found that the *thyA drm⁺ deoR* transductant (strain SA112, ref.
76) could no longer grow on low quantities of thymine. Indeed, they
observed that the strain became *a very high thymine requirer*. This
is precisely what is anticipated. For the *deoR* renders thymidine
phosphorylase, deoxyribomutase and deoxyriboaldolase all constitutive
thereby rapidly depleting the cells of all deoxyribose-1-phosphate
pools. Under these circumstances, the reverse of what happens in
the *drm⁻* derivative of *thyA drm deoR* occurs; instead of a super low
thymine requirement, one observes a super high thymine requirement
at about 100 µg thymine per ml. Clearly then the *deoR* mutation
alters the thymine requirement in drm⁺ and drm⁻ strains. Strains
such as *thyA drm deoR* are extremely useful as super low thymine
requiring strains. In addition as shown by Ahmad and Pritchard
(76) such strains can readily be transduced to *drm⁺* and then they
are super high thymine requiring strains.

Since it is difficult to create directly a super high thymine
requirement in any strain, the method described here allows one to
make a number of high thymine requirers. These strains would be
useful in the analysis of any known super high thymine requiring
strains. One such strain is HF4704S (84) with a *dnaH* mutation
which was believed to be involved in DNA initiation (85). It must
be borne in mind, however, that mutant HF4704S (84) is a
derivative of the *E.coli* C strain, HF4704, *which is itself a Tlr⁻*
at temperatures above 37°C (86). Thus the starting strain for the
experiments of Sakai et al. (84) is a low thymine (2 µg/ml)
requiring strain at 41°C. However, after an elegant series of
experiments, Derstine and Dumas (87) consider the *dnaH* mutation to
be the result of two mutations. One mutation appears to be in *dnaA*,
a known initiation gene (88), while the second mutation appears to
be in *thyA*. Surprisingly the addition of 100 µg thymine per ml over-
comes both mutations. Since *dnaH* has been mapped proximal to *thyA*
(though not necessarily within the confines of the structural gene)
by Sakai et al. (85), it could be, as pointed out by Derstine and
Dumas, that an essential gene lies within the "*lysA - thyA - argA*
region" of *E.coli*. *dnaH* could be such a gene. This possibility
is made more feasible by the following observation. Some years ago
Karlström (O. Karlström, personal communication) was unable to
generate deletions which extended from *lysA* into *thyA*; in contrast,
Karlström readily generated deletions which extended from *argA* into

thyA. The presence of an intervening essential gene between *lysA* and *thyA* would explain such results.

More recently other initiation mutants of *E.coli* K12 have been isolated by Cyclops in Glaser's laboratory (89). Some of these new mutations have been given the genotype *dnaH* because they were mapped in the *dnaH* region and not in any other known *dna* locus, including *dnaA* (89). Presumably double mutants of the type envisioned by Derstine and Dumas would appear as *dnaA* mutants in Glaser's selection. In any case since a different strain (the parent is *E.coli* K12 strain, DG17) was used (89) it should be possible to decide the legitimacy of the *dnaH* character as described by others (84,85,87).

A strain used frequently in studies on DNA synthesis is *E.coli* 15T⁻ (90). This is another super low thymine requiring strain (2, 73). Even though this mutant and its many derivatives have the genotype *thyA drm* its super low thymine requirement is not assoc- iated with the constitutive (*deoR* or *nucR*) mutation. The low thymine requirement is due primarily to an extremely efficient method by which thymidine is converted to dTTP.

In a series of studies Pritchard and his collaborators (1,2,7, 73) have characterized in detail the unusual physiology of *E.coli* 15T⁻ (555-7). Its very low thymine requirement is readily explained by the following findings: (a) *E.coli* 15T⁻ (555-7) has a higher endogenous pool of dTTP than other Tlr⁻ strains, such as the *E.coli* K12 derivative CR34 Thy⁻, at all levels of exogenously supplied thymine. (b) Thymine and thymidine are converted to dTTP with greater efficiency than in other Tlr⁻ strains. Since it has been previously observed by Beacham et al.(73) and by Cairns (J. Cairns, personal communication) that thymidine inhibits its own conversion to dTTP (*see* below, under "Thymidine Inhibition"), it is reasonable to assume that some relief from thymidine inhibition is found in the 15T⁻ strain (555-7). (c) The strain probably contains a higher level of deoxyribose-1-phosphate than other Tlr⁻ strains, such as CR34 Thy⁻, based on the following experiment. The addition of low concentrations of thymine plus deoxyguanosine (1m*M*) to strain CR34Thy⁻(a *thyA drm* mutant) increased the endogenous dTTP pool to a level comparable to that found in 15T⁻ (555-7) grown in a medium with the same concentration of thymine *without* deoxyguanosine (73).

In summary it seems that *E.coli* 15T⁻ (555-7) which is a descendant of the first ever Thy⁻ strain isolated (71,72), may be unique among other Thy⁻ strains of *E.coli*. However the strain can still be used for economic labelling of DNA since it requires but 0.25 µg thymine per ml for normal growth. As demonstrated many times the strain is ideal for computing the rates of DNA synthesis. However when comparisons are made between *E.coli* 15T⁻ and other *E.coli* Thy⁻ derivatives, such as CR34 Thy⁻ or W3110 Thy⁻, caution must be exercised, for the latter strains require ten times more thymine for *normal growth* than 15T⁻.

THYMIDINE INHIBITION

In certain strains of *E.coli* the steady state level of endogenous dTTP found in Thy⁺ derivatives cannot be reached when isogenic Thy⁻ derivatives are grown under optimal conditions with thymine or thymidine. In this regard, strain 15T⁻ achieves approximately 50% of the dTTP found in 15T⁺ strains which is considerably better than *E.coli* CR34 Thy⁻ which achieves only 20% of the dTTP of its Thy⁺ isogenic strain. The probable reason for the low maximal dTTP pools in Thy⁻ strains is the phenomenon of thymidine inhibition (*cf.* thymidine sensitivity of Tlr⁻ strains of the genotype *thyA dra*). When different concentrations of thymidine are added to a culture of a Thy⁻ derivative of *E.coli* CR34 an unusual result is seen (73). At very low thymidine concentrations (5 µM) wild type Thy⁺ levels of dTTP are approached. As the thymidine concentration is progressively increased only about 20% of the wild type Thy⁺ level of dTTP is reached. This level is similar to what can be obtained using thymidine or thymine plus deoxyguanosine (1mM) at various concentrations (from 20 to 200 µM). It seems very clear that somehow thymidine is inhibiting its own entry into the synthesis of dTTP. As we pointed out above this inhibitory effect of thymidine is far more pronounced in *E.coli* CR34Thy⁻ than in *E.coli* 15T⁻ which probably accounts for the fact that *E.coli* 15T⁻ can grow on minute quantities of thymine (0.25 µg/ml). Indeed Beacham et al. (73) have suggested that a mutation causing the above difference in the inhibitory effect underlies the super low thymine requirement of *E.coli* 15T⁻.

Thymidine inhibition has been observed many times by Beacham et al. (1,73) for the *E.coli* strains 15T⁻, 15T⁺ and CR34Thy⁻, by Cairns for 15T⁻ (J. Cairns, personal communication) and by Hosono and Kuno (91) in CR34Thy⁻ and in 2 Y strains (76,78,79,91).

Three findings suggested to Beacham et al. (73) that the maximal attainable level of dTTP in Thy⁻ cells was limited by some inhibitory effect of thymidine on one of the subsequent steps on the pathway to dTTP.

(1)   The addition of deoxyguanosine to the growth
        medium of *E.coli* 15T⁻ (555-7) in order to provide
        an additional supply of deoxyribose-1-phosphate
        reduced rather than increased the dTTP
        concentration, as thymine concentrations were
        increased.

(2)   The addition of thymidine itself to a Thy⁻ strain

gave an identical result in *E.coli* CR34 Thy⁻;
as thymidine concentration was increased the
lower became the pool of dTTP.

(3) The addition of thymidine to a Thy⁺ strain
likewise inhibited dTTP synthesis and reduced
the endogenous concentration of dTTP.

As of now it is not known for sure if the inhibition exerted
by thymidine on its own conversion to dTTP occurs in all strains
of *E.coli* and *S. typhimurium* or if it is unique to the *E.coli*
strains 15T⁻ and CR34Thy⁻. Since wild type (Thy⁺) levels of dTTP
in these strains can not be achieved in their Thy⁻ counterparts
it is unlikely that the phenomenon is universal. For it is
possible to swell the dTTP pool to the level found in Thy⁺ strains
in many Thy⁻ strains of *S. typhimurium* (35) and *E.coli* (92) by the
addition of exogenous thymine or thymidine. In any event, much
data regarding thymidine inhibition derived from 15T⁻ experiments
remain to be explained. The inhibited step in dTTP synthesis must
involve both the *de novo* synthesis of dTTP, derived from dCTP
(the major pathway) and from dUDP (the minor pathway) as well as
the dTTP synthesis from exogenously supplied thymidine, since *both*
syntheses are inhibited by thymidine. The inhibited step should
therefore be common to both pathways. It can be seen from Fig. 2
that only one of two steps, namely dTMP → dTDP or dTDP → dTTP
(steps No. 6 and No. 2, respectively in Fig. 2) are likely to be
involved. Beacham et al. (73) intimated, but did not prove, that
the step involving the conversion of dTDP to dTTP was the likely
one to be inhibited. Recently however Hosono and Kuno (91) have
strongly suggested that the inhibited step was the conversion of
dTMP to dTDP. Indeed thymidine-5' monophosphate kinase appears
to be a likely enzyme for inhibition by thymidine. Clearly
thymidine inhibition is real and it occurs to a certain extent
every time thymidine is added to Thy⁻ or Thy⁺ strains at
concentrations commonly used for growth. In four different Thy⁻
mutants of *E.coli*, this thymidine inhibition has been observed
(91). These widely differing Thy⁻ strains include a derivative
of *E.coli* CR34 *thyA dra* (CR34 is synonymous with C600); *E.coli*
15T⁻ derivatives and two derivatives of *E.coli* strain Y-70-2089
(*thyA dra*). If thymidine inhibits its own conversion to dTTP at
concentrations that are commonly used for normal growth then it is
easy to use excess thymidine. Indeed it has been shown (1) that
when one labels Thy⁺ strains with thymidine (or with thymine, plus
deoxyadenosine or deoxyguanosine) a point of diminishing returns
is frequently reached. For the *higher* concentration of thymidine
(or thymine plus deoxyadenosine) the smaller is the contribution
of the *exogenously* added compound to the dTTP pool, and the smaller
the total pool. Thus for maximum incorporation of label the *lowest
possible* concentration of thymidine (or thymine) should be used.

THYMIDINE SENSITIVITY

All Thy⁻ mutants are defective in the enzyme thymidylate synthetase and have the genotype *thyA*. Such mutants typically require 20 to 50 µg thymine per ml and are generally known as high thymine requirers. When grown under conditions where thymine becomes limiting (less than 20 µg per ml) spontaneous secondary mutations arise which permit growth at low thymine concentrations. These double mutants have the genotype *thyA drm* or *thyA dra*, are frequently known as Tlr⁻ mutants and because they can grow at 1 to 2 µg thymine per ml are known as low thymine requirers. Both high and low thymine requiring strains have been described above under "Classes of Thy⁻ Mutants".

The two genotypic classes of Tlr⁻ mutants can also be distinguished phenotypically. Tlr⁻ mutants of the genotype, *thyA drm*, are insensitive to thymidine toxicity while Tlr⁻ mutants of the genotype, *thyA dra* are *sensitive to thymidine* (74,75).

It can be seen from Figs. 2 & 3 that the only difference in the two Tlr⁻ classes is that *thyA drm* strains accumulate deoxyribose-1-phosphate while the *thyA dra* strains accumulate deoxyribose-5-phosphate. Clearly then thymidine sensitivity in deoxyriboaldolase-minus strains has something to do with the accumulation of deoxyribose-5-phosphate. Initial work on thymidine sensitivity was carried out with *S. typhimurium* (75). This genus has an obvious advantage over *E.coli* in that it can use deoxyribose as its sole carbon and energy source (75). The pathway by which this occurs is given below:

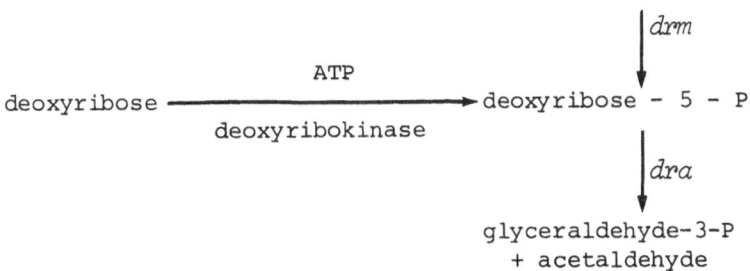

Thus one can distinguish between the two different classes of Tlr⁻ mutants using deoxyribose as the carbon source in *S. typhimurium*. Since wild type *S. typhimurium* can grow on deoxyribose, Tlr⁻ mutants which retain the ability to grow on deoxyribose must be *dra⁺* and thus *drm⁻*, while Tlr⁻ mutants which have lost the ability to grow on deoxyribose must be *dra⁻* and thus *drm⁺*. This test cannot be used for *E.coli* due to its inability to grow on deoxyribose as sole carbon and energy source.

Though we always seem to associate thymidine sensitivity with
Tlr$^-$ mutants of the genotype *thyA dra* this is not necessarily so.
Indeed, it makes no difference if the strains in question are
thyA$^-$ or *thyA*$^+$; for all that is needed for thymidine sensitivity
is the *dra*$^-$ mutation. No amount of thymidine can inhibit the
growth of Tlr$^+$ cells. It should be pointed out that *all Tlr*$^-$
strains exhibit some inhibition by deoxyribonucleotides. The
degree of inhibition differs greatly among various strains and is
greatest in *dra*$^-$ strains where the growth rate is profoundly
affected.

When thymidine is added to a growing culture in liquid medium
of genotype *thyA*$^-$ *dra*$^-$ or *thyA*$^+$ *dra*$^-$, a typical result is seen.
There is an initial lag followed by a reduction in growth rate and
then by recovery. The amount of thymidine required for maximal
inhibition is about $2 \times 10^{-5}M$. Data by Beacham et al. (75) show
that inhibition is not absolute even at thymidine concentrations of
400 µM, 20-fold above the maximal required. The typical response
found for a number of sensitive strains of *S. typhimurium* is a
reduction in growth rate rather than a complete cessation of growth
in liquid cultures (75). All deoxyribosides have similar inhibitory
properties to thymidine, but purine deoxyribonucleosides are less
inhibitory than their pyrimidine counterparts.

Thymidine sensitivity is partially reversed by the addition of
equimolar concentrations of other ribonucleosides. By adding
thymidine (30 µM) and uridine (30 µM) one sees partial inhibition;
with 30 µM thymidine and 150 µM uridine, no inhibition is seen.
Here again pyrimidine ribosides are effective at lower
concentrations than purine ribosides.

The finding that thymidine sensitivity is due to the lack of
deoxyriboaldolase (Dra$^-$) is true for Thy$^-$ and Thy$^+$ strains of
S. typhimurium and *E.coli*. This strongly suggests that the sensitivity is due, directly or indirectly, to the accumulation of
deoxyribose-5-phosphate derived from the added thymidine (and other
deoxyribonucleosides). Beyond that however it is difficult to say
with certainly what the nature of the toxicity is. The regulation
of the *deo* operon by the *deoR* repressor, the *cytR* repressor, and
the cyclic AMP - cAMP receptor protein control system has recently
been demonstrated *in vitro* (94). It has now been shown
conclusively that the synthesis of thymidine phosphorylase is
induced by deoxyribose-5-phosphate; the role of deoxyribose-5-
phosphate as effector for the *deoR* repressor has also been shown.
This explains why lower levels of thymidine phosphorylase are
found in *drm* than in *dra* mutants.

The following key features can be stated about thymidine
toxicity:

(i)   All Thy$^-$ mutants with the Tlr$^-$ phenotype show
      some degree of sensitivity to exogenously supplied
      thymidine (and other deoxyribonucleosides).

(ii) The sensitivity is exhibited by Thy$^-$ and Thy$^+$
      strains, e.g., *thyA$^+$ dra$^-$* and *thyA$^-$ dra$^-$* are
      sensitive.

(iii) Deoxyribose-5-phosphate is the key compound. It
      accumulates in *dra$^-$* mutants, it induces the synthesis
      of thymidine phosphorylase as the effector for the
      deoR repressor, and it probably reaches toxic levels
      in *dra$^-$* strains.

(iv) That *thyA dra* is a low thymine requirer is due to
      the fact that it contains a high endogenous pool
      of deoxyribose-1-phosphate. Thus *thyA dra* mutants
      have high levels of *both* deoxyribose-5-phosphate
      and deoxyribose-1-phosphate. It is therefore
      possible that the extreme sensitivity of *dra* strains
      is due to the total concentration of deoxyribophosphate
      rather than deoxyribose-5-phosphate alone.

(v)   When ^{3}H-thymidine is added to non-sensitive Thy$^+$
      strains (no mutation in *dra* or *drm*) in the presence
      and absence of uridine, no effect of uridine on the
      rate of thymidine incorporation is seen (75). Thus
      competition for uptake is ruled out. In addition,
      as we pointed out earlier under "How to Label DNA",
      the presence of uridine increases the residence
      time of thymidine in the cell. This is due to the
      fact that uridine inhibits the induction of thymidine
      phosphorylase (45,94). The presence of uridine
      therefore increases the plateau level of
      incorporation of ^{3}H-thymidine.

One final point seems pertinent. The toxic effect of
deoxyribose phosphates, seen here, might well be analogous to many
other cases of carbohydrate toxicity in bacteria. The toxic effect
is generally caused by the accumulation, in mutant strains, of
phosphorylated intermediates. Typically, as for *L*—arabinose (95),
glycerol (96), glucose (97), galactose (98), fructose (99), rhamnose
(100), galactitol (101) and mannitol (102), the cause of the
toxicity is a metabolic block beyond a phosphorylation reaction in
the *normal* catabolic pathway for that carbohydrate. Just as feeding
the nucleoside sugar, thymidine renders *thyA dra* cells sensitive
to thymidine due to the accumulation of the toxic deoxyribose-5-
phosphate, so too is feeding galactose to cells which lack the
transferase, epimerase or pyrophosphorylase but retain the kinase.

The accumulation of galactose-1-phosphate, known to be toxic for
E.coli mutants (103) as well as in human galactosemia (104), occurs.

Reiner (105) recently discovered a further novel means of
achieving toxicity in certain *E.coli* mutants. He found that a
certain *E.coli* C mutant was sensitivie to xylitol while other
E.coli K12 mutants were inhibited by *D*-arabitol. Unlike the more
general case, described above, these mutants *lacked* the catabolic
pathways for pentitol (xylitol and arabitol) catabolism. The
phosphotransferase system (106) is involved in the toxic syntheses
caused by xylitol and arabitol addition (105). These phosphotransferases seem to be purposefully non specific (101, 106, 107).
For example, sorbitol phosphotransferase acts on fructose (108)
mannitol (109), and galactitol (101) as well as sorbitol. These
alternate substrates are now directly phosphorylated and no toxic
intermediates occur (105). Xylitol and D-arabitol are normally
either oxidized to the appropriate pentulose before phosphorylation
in the case of *E.coli* C, or not catabolized, in the case of *E.coli*
K12 and B strains (110). Therefore if phosphotransferases were to
act on xylitol or *D*-arabitol, one might anticipate some phosphorylated intermediates with no means of further metabolism. Indeed,
Reiner has shown that fructose-phosphotransferase results in toxic
synthesis from xylitol, and that either galactitol - or sorbitol-
phosphotransferase results in toxic synthesis from D-arabitol (105).

These two toxicity-producing mechanisms may well have ubiquitous
relevance.

The *nup* and *tsx* Genes

Recently Hanawalt and coworkers (111) have shown that a widely
used strain of *E.coli* K12 is deficient in the transport of a number
of nucleosides including thymidine. A low-thymine requiring
derivative of strain AB1157 (*thyA, drm*) exhibited strikingly lower
thymidine pulse-label incorporation at low thymidine concentrations
(less than 1.0 μg per ml) than many other strains. No difference
was observed for the base thymine. A single gene, designated *nup*
(for *n*ucleoside *u*ptake) has been localized at about 10 min on the
E.coli chromosome. In *nup*[+] strains such as W3110 *thyA, drm*, the
maximal level of ^{3}H-thymidine incorporation is reached at thymidine
concentrations as low as 0.2 to 0.3 μg per ml. In *nup*[-] strains
such as AB1157, a concentration of 1 μg thymidine is required for
maximal incorporation. However since thymidine is taken up and
incorporated into wild type Thy[+] cells it was important for McKeown
et al. (111) to check for thymidine uptake in Thy[+] derivatives of
W3110 (*nup*[+]) and AB1157 (*nup*[-]). They found that the deficiency seen
in Thy[-] derivatives of AB1157 was also seen in the isogenic Thy[+]
strains. The outer membrane of *E.coli* is known to exclude hydro-

philic compounds with molecular weights above 700 daltons (112).
Smaller molecules, such as nucleotides, are transported poorly
through the outer membrane, at a rate generally sufficient for
cell growth. However for effective uptake of nucleosides the
presence of certain group-specific outer membrane proteins is
required. Many such transport-facilitating membrane proteins have
now been detected as receptors for various phages and colicins.
The λ receptor protein enhances maltose transport (114,115). It
appears therefore that many additional outer membrane proteins,
detected originally as phage and colicin receptors, are likely to
be involved in the uptake of critical substrate molecules to which
the outer membrane would otherwise be impermeable.

Accordingly, in an independent study Hantke (116) has examined
a series of *tsx* mutants in *E.coli* Kl2 selected for their resistance
to phage T6 and to colicin K (113). He has shown that the receptor
protein for phage T6 and colicin K is a constituent of a "channel"
through the outer membrane; in *tsx* mutants, which lack a single
outer membrane protein of molecular weight 25,000 the uptake of
nucleosides is severely impaired. The *tsx* mutation has been mapped
at 9 min (*nup* was shown to map at about 10 min) on the *E.coli*
chromosome and it affects the uptake of ribo- and deocyribonucleo-
sides. Thus it appears that the *nup* and *tsx* genotypes are identical.
Moreover it is known that strain AB1157 and all its derivatives are
tsx (117). However Hantke (116) argues that the *nup* gene product
may involve a "pore" function rather than a permease as suggested
by Hanawalt (111) in the light of its low specificity.

The important feature of the *nup* and *tsx* mutations is the
frequency of their occurrence. Many widely used strains of
E.coli contain the *nup* or *tsx* mutation. These same strains are
frequently used in thymidine pulse-labelling experiments. Strain
AB1157, for example, is the parent strain for a number of commonly
used recombination - and repair-deficient strains (consult
Bachmann, ref. 117). More often than not Thy[+] strains (such as
AB1157) are pulse-labelled with thymidine at concentrations of
thymidine far lower than 1.0 µg per ml. A typical thymidine
concentration is 0.05 µg per ml. For this reason it is important
to realize that some obvious difficulties in the interpretation
of pulse-labelling experiments with thymidine occur. Pulse-
labelling differences are expected between thymine and thymidine-
labelling of nascent DNA chains (see below under "Labelling
Okazaki Fragments"). In *nup*[-] and *tsx* strains these differences are
accentuated since thymidine (at concentrations less than 1.0 µg/ml)
uptake is severely impaired. However thymidine is more rapidly
accumulated in *nup*[+] cells than is thymine, and pulses of thymine
give a slower approach to dTTP concentrations that are required for
normal DNA synthesis. In addition to the problem of thymine and
thymidine metabolism, the presence of a thymidine uptake deficiency

(nup^- or tsx in all AB1157 derivatives, whether Thy^+ or Thy^-) in
many commonly used strains makes interpretation of radioactive
labelling procedures extremely difficult. It should be pointed out
that a large number of tsx strains are in common use.

From the studies of Hanawalt (111) we wish to emphasize some
difficulties in the interpretation of pulse-labelling experiments.
Different strains (for example nup^+ and nup^-) show strikingly
different levels of transport of exogenously added nucleosides.
As we pointed out above, AB1157 (which is tsx, hence nup^-) has
been used extensively as the starting strain to construct a large
number of commonly used repair - and recombination-deficient
strains (111). Rather than reconstruct a new strain, experimenters
frequently request previously described mutant strains from their
colleagues. This is especially common in the case of $E.coli$ 15T$^-$
which has been extremely useful for a wide variety of experiments
on nucleic acid synthesis and its regulation. The problem with
$E.coli$ 15T$^-$ is its incomparable ability of growing in media with
as little as 0.2 μg thymine per ml. This is discussed in detail
above under "Thymine-requiring mutants". Exogenously added nucleosides and bases are found at different concentrations in the cell
than in the medium. These intracellular concentrations are known
to fluctuate during the early periods of pulse labelling. Such
concentration differences may help explain the extreme differences
for thymine and thymidine, during pulse labelling of nascent DNA
chains (118-120). One further point regarding differences between
thymine and thymidine is important. In most strains thymidine is
efficiently incorporated into cells, so that an internal
concentration of thymidine (hence dTTP) sufficient to permit normal
DNA synthesis rates is obtained quite rapidly. Thymine however is
not accumulated fast enough to allow normal levels of DNA synthesis
at low thymine concentrations; neither is thymine metabolized for
incorporation into DNA nearly as efficiently as thymidine. Thus
thymine pulses give a slower approach to dTTP concentrations
sufficient to support normal DNA synthesis. This should be an
important consideration in all DNA labelling and especially in the
case of labelling Okazaki fragments, as seen below.

LABELLING OKAZAKI FRAGMENTS

Okazaki fragments have been the subject of intensive
investigations every since Okazaki et al. (119) showed that nascent
bacterial DNA, pulse labelled with ^{3}H-thymidine, was first detectable
as small polynucleotide chains (approximately 4S) which then grow
to a full length of about 10S (approximately 2000 nucleotides)
before being joined together by the action of DNA ligase (121). These
small polynucleotide chains vary in number and length and are
generally regarded as $bona$ $fida$ intermediates of DNA replication. In

order to find out the length and number of Okazaki pieces one has to label DNA. Here, as elsewhere, the method of labelling is very important because it appears that one can change *both the number and the size* of Okazaki pieces by the way one labels them (9).

The most common means of labelling Okazaki fragments is to use ^{3}H-thymidine since thymidine as we have seen is rapidly incorporated into DNA. Another method used to label Okazaki pieces is to use thymine label in a thymine auxotroph. This method is not nearly as effective as using thymidine because a large precursor pool has to be negotiated before DNA is specifically labelled. To obviate this problem Thy⁻ cells are often starved for thymine, which depletes the intracellular nucleotide pools. This method generally creates another problem in that many more smaller pieces are seen after thymine starvation (118,122,123). Likewise, the advantage offered by using thymidine may be offset by its creation of far more Okazaki pieces than would be found if thymine was used to label (120). The difficulty of labelling and identifying Okazaki pieces is compounded further by the fact that only about 1% of the total DNA is in the form of Okazaki fragments.

Confusing results have frequently been observed. We do not consider here any of the methods of analysis of Okazaki fragments *after* labelling but the reader should be aware that some problems exist. One school of thought, the more popular one in 1977, believes that Okazaki fragments are regular intermediates of DNA replication, consisting of short chains of about 2000 bases in length on one or both sides of the replication fork. The less popular view suggests that Okazaki fragments are not *bona fide* intermediates of DNA replication but rather are the result of experimental techniques. This view may well be incorrect but it is valid until some purposeful methodology, modelled after that presented by Anderson (9) can accurately compute the exact number of size of Okazaki fragments at different times *in several different strains* of bacteria. Different results are obtained from strain to strain. For example when a Tlr⁻ *polA*⁺ derivative of *E.coli* strain W3110 is compared with a Tlr⁻ *polA* W3110 considerable differences are observed. When the super low thymine requirer: *E.coli* 15T⁻ (*thyA dra arg met trp*), cured of its *E.coli* 15 plasmid is compared with the two W3110 strains considerable differences are seen (9).

To label short chain intermediates in DNA replication the following problems must be addressed:

(1) Labelling Okazaki fragments is often achieved by using short pulses of ^{3}H-thymidine or ^{3}H-thymine (+ deoxyadenosine) in Thy⁺ bacteria. In this way a rough estimate of the *size* of the short chain intermediates is obtained. However, since the specific

activity of the DNA precursor pool can fluctuate during the pulse
labelling one cannot compute accurately the absolute amount of DNA
from the number of counts per minute. Thus one cannot get an
estimate of the *number* of pieces.

(2) Labelling Okazaki pieces with thymine or thymidine in Thy⁻
strains which have been depleted of their precursor pools by thymine
starvation obviates the above problem of specific activity. But as
was mentioned earlier, starvation causes the formation of a very
large number of very small pieces so that again the true *size* and
indeed the number of different pieces remain obscured. A useful
approach to a related problem but applicable to this one has been
presented by Brewin (123). In Brewin's method Thy⁻ bacteria are
starved for 10 min at 25°C before being pulsed with minute
quantities (0.5 pmol ^{3}H-thymidine per 10^8 bacteria) of thymidine.
The ^{3}H-thymidine so administered is completely and quickly
incorporated, leaving *no residue of labelled material* in the TCA
soluble precursor pool so that any subsequent chase can be carried
out precisely.

(3) The number of chains of *any* given size in a population that
makes up only about 1% of the total DNA population must be
measured.

This is precisely the problem to which Anderson addressed
herself (for details consult Anderson, these proceedings). She
analyzed the number of molecules and size distribution of the DNA
chains *from uniformly labelled (with thymine) bacteria* growing under
steady state conditions. She clearly established that Thy⁻ mutants
could be used, and that thymine should be used for unform labelling
of cells. One final point is important. As pointed out by Anderson
care should be taken that the radioactive thymine used to label the
cells is freshly purified. Radiation-induced products often arise
when radioactively labelled thymine is stored for long periods.
These products are known to cause bacteria to accumulate up to
25% of their DNA as low molecular weight molecules (9). Thus it is
difficult to estimate the number and the size of Okazaki fragments.
It is also difficult to estimate how many Okazaki fragments are
chaseable (i.e., are intermediates of replication). Anderson (9)
found that only three, from a total of 20, pieces could be chased
in the *polA⁺* W3110 strain. This figure could well represent zero
which would suggest that *none* of the 20 Okazaki pieces between 700
and 9000 bases in size is a *bona fide* intermediate of DNA replication. Proving or disproving these findings is critical for a large
number of old as well as new experiments.

ACKNOWLEDGEMENTS

I express thanks to those who so generously contributed to us their unpublished results, to those who read earlier drafts of the manuscript for their many helpful suggestions and criticisms, and to Dr. Ian Molineux for his patience, without which this manuscript would not have materialized.

Research in G.A.O'D's laboratory was supported by a NATO Collaborative International Grant (with Jan Neuhard, Copenhagen) and by a grant from the Robert A. Welch Foundation, Houston, Texas.

REFERENCES

1. Pritchard, R.H. (1974) Phil. Trans. Royal Soc. London 267: 303-336

2. Pritchard, R.H. and A. Zaritsky (1970) Nature 226:126-131

3. Cooper, S. and C.E. Helmstetter (1968) J. Molec. Biol. 31:519-540

4. Helmstetter, C.E. and S. Cooper (1968) J. Molec. Biol. 31: 507-518

5. Maaloe, O. (1976) In: Alfred Benzon Symposium IX on Control of Ribosome Synthesis. Munksgaard, Copenhagen (eds. N.O. Kjeldgaard and O. Maaloe) Academic Press, New York pp.15-21

6. Pritchard, R.H., P.T. Barth and J. Collins (1969) XIX Microbial Growth, Symp. Soc. Gen. Microbiol. pp-263-297

7. Zaritsky, A. and R.H. Pritchard (1971) J. Molec. Biol. 60: 65-74

8. Zaritsky, A. and R.H. Pritchard (1973) J. Bacteriol. 114: 824-837

9. Anderson, M.L.M., these proceedings

10. Friedkin, M. and A. Kornberg (1957) In: A Symposium on the Chemical Basis of Heredity. (eds; W.D. McElroy and B. Glass) Johns Hopkins Press, Baltimore. pp.609-613

11. Cohen, S.S. and H.D. Barner (1954) Proc. Nat. Acad. Sci. USA 40:885-893

12. Okada, T., J. Homma and H. Sonohara (-962) J. Bacteriol. 84: 602-603

13.   Okada, T., K. Yanagisawa and F.J. Ryan (1961) Z. Vererb. 92:
      403-412

14.   Stacey, K.A. and E. Simson (1965) J. Bacteriol. 90:554-555

15.   Eisenstark, A., R. Eisenstark and S. Cunningham (1968)
      Genetics 58:493-506

16.   Farmer, J.L. and F. Rothman (1965) J. Bacteriol. 89:262-263

17.   Harrison, A.P. (1965) J. Gen. Microbiol. 41:321-333

18.   Ishibashi, M., Y. Sugino and Y. Hirota (1964) J. Bacteriol.
      87:554-567

19.   Marouf, B.A. (1973) Ph.D. Thesis, Texas A & M University

20.   Kelln, R.A. and R.A.J. Warren (1973) J. Bacteriol. 113:510-
      511

21.   Wilson, M.C., J.L. Farmer and F. Rothman (1976) J. Bacteriol.
      92:186-196

22.   Bertino, J.B. and K.A. Stacey (1966) Biochem. J. 101:32-33c

23.   Neuhard, J., A.R. Price, L. Schack and E. Thomassen (1977)
      Proc. Nat. Acad. Sci. USA (in press)

24.   Anagnostopoulos, C. and A. Schneider-Champagne (1966) C.R.
      Acad. Sci., Paris 262:1311-1314

25.   Price, A.R., these proceedings

26.   Larsson, A. and P. Reichard (1967) Progr. Nucl. Acid Res.
      7:303-347

27.   Bertani, L.E., A. Häggmark and P. Reichard (1963) J. Biol.
      Chem. 238:3407-3413

28.   Greenberg, G.R. and R.L. Somerville (1962) Proc. Nat. Acad.
      Sci. USA 48:249-257

29.   Tye, B.-K., P.-O. Nyman, I.R. Lehman, S. Hochhauser and
      B. Weiss (1977) Proc. Nat. Acad. Sci. USA 74:154-157

30.   Lehman, I.R., these proceedings

31.   Warner, H., these proceedings

32. Häggmark, A. (1967) 4th FEBS Meeting, Oslo, p296

33. Karlström, O. and A. Larsson (1967) Eur. J. Biochem. 3: 164-170

34. Neuhard, J. (1968) J. Bacteriol. 96:1519-1527

35. Neuhard, J. and E. Thomassen (1971) J. Bacteriol. 105:657-665

36. O'Donovan, G.A., G. Edlin, J.A. Fuchs, J. Neuhard and E. Thomassen (1971) J. Bacteriol. 105:657-665

37. Munch-Petersen, A. (1970) Eur. J. Biochem. 15:191-202

38. Neuhard, J. and A. Munch-Petersen (1966) Biochim. Biophys. Acta 114:61-71

39. Neuhard, J. and E. Thomassen (1976) J. Bacteriol. 126:999-1001

40. Beck, C.D., J. Neuhard and E. Thomassen (1977) J. Bacteriol. 129:305-318

41. O'Donovan, G.A. (1970) Biochim. Biophys. Acta 209:589-591

42. Biswas, C., J. Hardy and W.S. Beck (1965) J. Biol. Chem. 240:3631-3639

43. Larsson, A. and P. Reichard (1966) J. Biol. Chem. 241:2533-2539

44. Larsson, A. and P. Reichard (1966) J. Biol. Chem. 241:2540-2549

45. Budman, D.R. and A.B. Pardee (1967) J. Bacteriol. 94: 1546-1550

46. Schwartz, M. (1971) Eur. J. Biochem. 21:191-198

47. Beacham, I.R., P.T. Barth and R.H. Pritchard (1968) Biochim. Biophys. Acta 166:589-592

48. Munch-Petersen, A. (1968) Eur. J. Biochem. 6:432-442

48a. Minghetti, A. (1959) Ital. J. Biochem. 8:224-230

49. Imada, A. and S. Igarasi (1967) J. Bacteriol. 94:1551-1559

50. Saunders, P.P., B.A. Wilson and G.F. Saunders (1969) J. Biol.
     Chem. 244:3691-3697

51. Wang, T.P. and J.O. Lampen (1951) J. Biol. Chem. 192:
     339-347

52. Okazaki, R. and A. Kornberg (1964) J. Biol. Chem. 239:269-
     274

53. Okazaki, R. and A. Kornberg (1964) J. Biol. Chem. 239:
     275-284

54. Neuhard, J. (1967) Biochim. Biophys. Acta 145:1-6

55. Neuhard, J. (1966) Biochim. Biophys. Acta 129:104-115

56. Fäth, W.W. and M. Brendel (1974) Molec. Gen. Genet. 131:
     57-67

57. Crawford, L.W. (1958) Biochim. Biophys. Acta 30:428-429

58. Siminovitch, L. and A.F. Graham (1955) Can. J. Microbiol.
     1:721-732

59. Boyce, R.P. and R.B. Setlow (1962) Biochim. Biophys. Acta
     61:618-620

60. Kammen, H.O. (1967) Biochim. Biophys. Acta 134:301-311

61. Munch-Peterson, A. (1967) Biochim. Biophys. Acta 142:
     228-237

62. Rachmeler, J., J. Gerhart and J. Rosner (1961) Biochim.
     Biophys. Acta 49:222-225

63. Fangman, W.L. (1969) J. Bacteriol. 99:681-687

64. Fangman, W.L. and A. Novick (1966) J. Bacteriol. 91:
     2390-2394

65. Beacham, I.R. and R.H. Pritchard (1971) Molec. Gen. Genet.
     110:289-298

66. Leer, J.C., K. Hammer-Jespersen and M. Schwartz (1977) 75:
     217-224

67. Womack, J.E. (1977) Molec. Gen. Genet. (in press)

68. Lichtenstein, J., H.D. Barner and S.S. Cohen (1960) J. Biol.
     Chem. 235:457-465

69. O'Donovan, G.A. and J. Neuhard (1970) Bacteriol. Rev. 34: 278-343

70. Breitman, T.R., R.M. Bradford and W.D. Cannon, Jr. (1967) J. Bacteriol. 93:1471-1472

71. Roepke, R.R. (1967) J. Bacteriol. 93:1188-1189

72. Roepke, R.R. and F.E. Mercer (1947) J. Bacteriol. 54:731-743

73. Beacham, I.R., K. Beacham, A. Zaritsky and R.H. Pritchard (1971) J. Molec. Biol. 60:75-86

74. Alikhanian, S.E.I., T.S. Iljina, E.S. Kaliaeva, S.V. Kameneva and U.V. Sukhodolec (1966) Genet. Res. 8:83-100

75. Beacham, I.R., A. Eisenstark, P.T. Barth and R.H. Pritchard (1968) Molec. Gen. Genet. 102:112-127

76. Ahmad, S.I. and R.H. Pritchard (1971) Molec. Gen. Genet. 111:77-83

77. Albrechtsen, H., K. Hammer-Jespersen and A. Munch-Petersen (1976) Molec. Gen. Genet. 146:139-145

78. Okada, T. (1966) Genetics 54:1329-1336

79. Munch-Petersen, A. (1968) Biochim. Biophys. Acta 161:279-282

80. Okada, T., H. Torii and S. Kuno (1969) Jap. J. Genet. 44: 193-200

81. Lomax, M.S. and G.R. Greenberg (1968) J. Bacteriol. 96: 501-514

82. Hoffee, P.A. (1968) J. Bacteriol. 95:449-457

83. Hoffee, P.A. and B.C. Robertson (1969) J. Bacteriol. 97: 1386-1396

84. Sakai, H., S. Hashimoto and T. Komano (1974) J. Bacteriol. 119:811-820

85. Sakai, H. and T. Komano (1975) Biochim. Biophys. Acta 395: 433-445

86. Bertani, G. (1951) J. Bacteriol. 62:293-300

87. Derstine, P.L. and L.B. Dumas (1976) J. Bacteriol. 128: 801-809

88. Blau, S. and J. Murdoh (1972) Proc. Nat. Acad. Sci. USA 69:2895-2898

89. Sevastopoulos, G.G., C.T. Wehr and D.A. Glaser (1977) Proc. Nat. Acad. Sci. USA 74:3485-3489

90. Lark, K.G. (1966) Bacteriol. Rev. 30:3-32

91. Hosono, H. and S. Kuno (1975) Eur. J. Biochem. 57:177-179

92. Munch-Petersen, A. (1970) Eur. J. Biochem. 15:191-202

93. Breitman, T.R. and R.M. Bradford (1964) Biochem. Biophys. Res. Commun. 17:786-791

94. Valentin-Hansen, P., B.A. Svenningsen, A. Munch-Petersen and K. Hammer-Jesperen (1977) Molec. Gen. Genet., in press

95. Englesberg, E., R.L. Anderson, R. Weisberg, N. Lee, P. Hoffee, A. Hottenhauer and H. Boyer (1962) J. Bacteriol. 84:137-146

96. Cozzarelli, N.R., J.P. Koch and E.C.C. Lin (1965) J. Bacteriol. 90:1325-1329

97. Böck, A. and F.C. Neidhardt (1966) J. Bacteriol. 92:470-476

98. Kurahashi, K. and A.J. Wahba (1958) Biochim. Biophys. Acta 30:298-302

99. Ferenci, T. and H.L. Kornberg (1973) Biochem. J. 132:341-347

100. Englesberg, E. and L.S. Baron (1959) J. Bacteriol. 78:675-686

101. Lengeler, L. (1975) J. Bacteriol. 124:26-38

102. Solomon, E. and E.C.C. Lin (1972) J. Bacteriol. 111:566-574

103. Yarmolinsky, M.B., H. Wiesmeyer, H.M. Kalckar and E. Jordon (1959) Proc. Nat. Acad. Sci. USA 45:1786-1791

104. Schwartz, V., L. Golberg, G.M. Komrower and A. Holzel (1956) Biochem. J. 62:34-40

105. Reiner, A.M. (1977) J. Bacteriol. 132:166-173

106. Postma, P.W. and S. Roseman (1976) Biochem. Biophys. Acta 457:213-257

107. Curtis, S.J. and W. Epstein (1975) J. Bacteriol. 122:1189-1199

108.Joyce-Mortimer, M.C. and H.L. Kornberg (1976) J. Gen.
    Microbiol. 96:383-392

109.Legeler, J. and E.C.C. Lin (1972) J. Bacteriol. 112:840-848

110.Reiner, A.M. (1975) J. Bacteriol. 123:530-536

111.McKeown, M., M. Kahn and P. Hanawalt (1976) J. Bacteriol.
    126:814-822

112.Nakae, T. and H. Nikaido (1975) J. Biol. Chem. 250:7359-7365

113.Smeleman, S. and M. Hofnung (1975) J. Bacteriol. 124:112-118

114.Michael, J.G. (1968) Proc. Soc. Exp. Biol. Med. 128:434-438

115.Weltzien, H.U. and M.A. Hesaitis (1971) J. Exp. Med. 133:
    534-553

116.Hantke, K. (1976) FEBS Lett. 70:109-112

117.Bachmann, B.J. (1972) Bacteriol. Rev. 36:525-557

118.Diaz, A.T., D. Wiener and R. Werner (1975) J. Molec. Biol.
    95:45-61

119.Okazaki, R., T. Okazaki, K. Sakabe, K. Sugimoto, R. Kainuma,
    A. Sugino and N. Iwatsuki (1968) Cold Spring Harbor Symp.
    Quant. Biol. 33:129-143

120.Werner, R. (1971) Nature 230:570-572

121.Olivera, B.M. and I.R. Lehman (1967) Proc. Nat. Acad. Sci.
    USA 57:1426-1433

122.Dingman, C.W., M.P. Fisher and M. Ishizawa (1974) J. Molec.
    Biol. 84:275-295

123.Brewin, N. (1977) J. Molec. Biol. 111:343-352

DEOXYRIBONUCLEOSIDE TRIPHOSPHATES AND DNA POLYMERASE IN BACTERIO-
PHAGE PBS1-INFECTED *BACILLUS SUBTILIS*

Ronald A. Hitzeman and Alan R. Price[*][Δ]

Department of Biological Chemistry
University of Michigan
Ann Arbor, Michigan, 48109, U.S.A.

Jan Neuhard and Henrik Møllgaard

Enzyme Division, Institute of Biological Chemistry B
University of Copenhagen, Denmark

INTRODUCTION

Bacteriophage PBS1 and its clear-plaque variant PBS2 are unique viruses in that their DNA contains uracil instead of thymine (1,2). One of the changes affecting deoxyribonucleotide metabolism in *Bacillus subtilis* following PBS1 or PBS2 infection (see Fig. 1) is an increase in the level of DNA polymerase activity. We predicted that the phage-induced DNA polymerase would be a new enzyme, whose substrate in vivo would be dUTP for the synthesis of uracil-DNA. This report describes the discovery of dUTP in vivo, evidence for its biosynthetic origin, and characteristics of the purified phage DNA polymerase.

RESULTS

<u>Metabolism of dUTP and dUMP in PBS1-infected *B.subtilis*</u>. We devised a new chromatographic system to resolve dUTP from dTTP on polyethyleneimine-coated cellulose plates (Fig. 2), since the commonly used method (4,5) gave little or no separation of these compounds. An acidic organic solvent was used in the first dimension to resolve uracil from thymine compounds (6), followed

---

[*]*A.S.I. Participant*
[Δ]*To whom correspondence should be addressed*

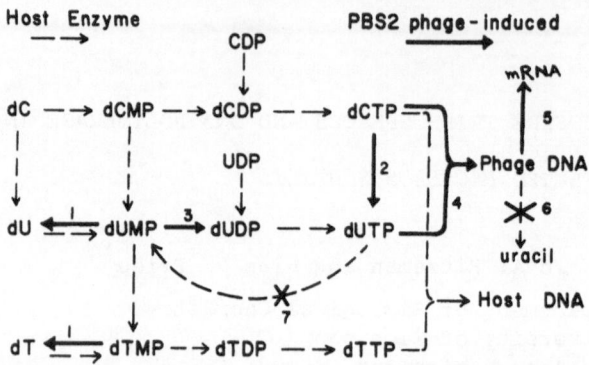

<u>Figure 1</u>.   *Proposed scheme for deoxyribonucleotide metabolism in
B.subtilis as altered by PBS1 and PBS2 phage infection. Known
phage-induced functions (heavy lines) are indicated: (1) dTMP
(dUMP) 5'-phosphatase; (2) dCTP deaminase; (3) dUMP kinase; (4)
DNA polymerase; (5) RNA polymerase; (6) inhibitor of host's N-
glycosidase for uracil-containing DNA; and (7) inhibitor of host's
dUTPase. Original citations on these proteins may be found in
Ref 3.*

by a salt elution in the second dimension to separate on the basis
of charge. This sytem gave excellent separation of dUTP from dTTP
and other pyrimidine nucleotides, all of which can be labelled
in vivo by radioactive uracil.

This new thin-layer chromatography system was employed to
quantitate the levels of pyrimidine nucleoside triphosphates in
uninfected and PBS1-phage infected *B.subtilis* (Fig. 3). The
concentration of dTTP and the other triphosphates dropped
dramatically within 3 to 5 min after infection, followed by a
rapid rise in the levels of UTP,CTP and dCTP. the dTTP concentration, however, remained low (less than 10% of its original value),
and dUTP appeared and accumulated from 10 to 20 min after infection,
simultaneously with the onset of phage DNA synthesis (Fig. 3).
Our demonstration (confirmed with another system (5) as well)
represents the first time that dUTP has ever been detected in vivo,
and it confirms our prediction that dUTP is the probable substrate
for the replication of uracil-containing PBS1 DNA.

To investigate the metabolic origin of dUTP in PBS1-infected
B.subtilis, a similar experiment was performed using motile strain
SB19E *lys pyr-2 cdd*-1 (from I. Takahashi, this strain lacks
cytidine/deoxycytidine deaminase, so it cannot degrade these
compounds; see Ref.7) and our isolate of a derivative strain, SB19E

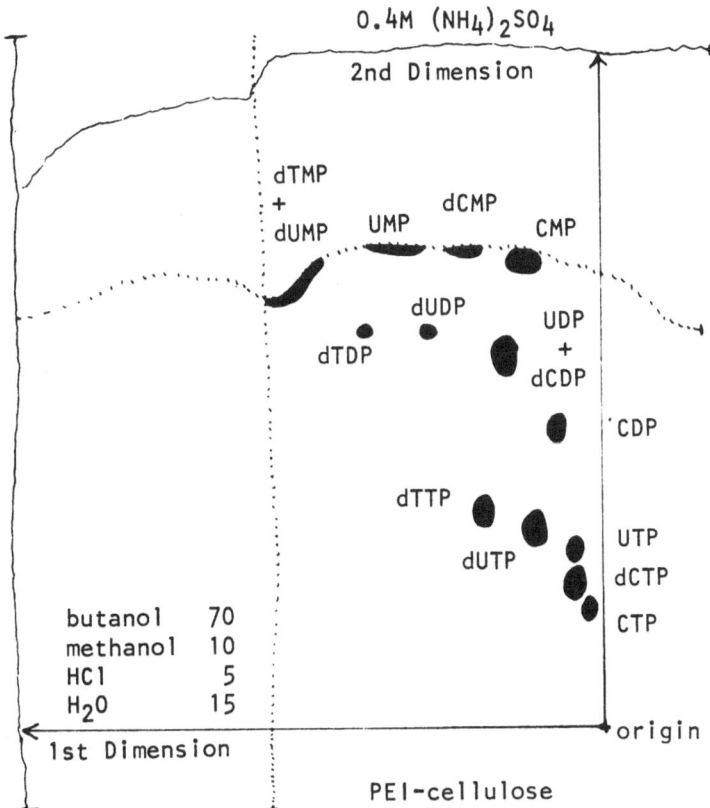

Figure 2.   *Two-dimensional chromatographic separation of dUTP from dTTP and other pyrimidine nucleotides.   Thin layers of polyethyleneimine-coated cellulose on plastic sheets (18 cm square) are prepared and washed (4) and pre-run in 65% methanol.   Aqueous 100-μl samples are applied by automatic syringe to the "origin" spot with 11 nmoles of each marker nucleotide. After washing in methanol to remove salts, the dried chromatogram is hung for 30 min in the chamber to equilibrate and then lowered into the solvent to develop about 4 h in the first dimension.   Then the sheet is dried under a cool air stream to remove the HC1, followed by washing in 10mM Tris in methanol and then in methanol to neutralize and remove residual acid (4).   The dried sheet is then developed about 40 min in the second dimension solvent, dried, washed in methanol, and examined under a UV-light to locate the markers.*

lys pyr-2 cdd-1 dcd-1 (which in addition lacks dCMP deaminase, and thus has no pathways for converting cytosine compounds to uracil compounds).   Cultures were labelled as in Fig. 3, except that (^{3}H-5) uracil was employed to label uracil-nucleotides plus $^{32}P_i$

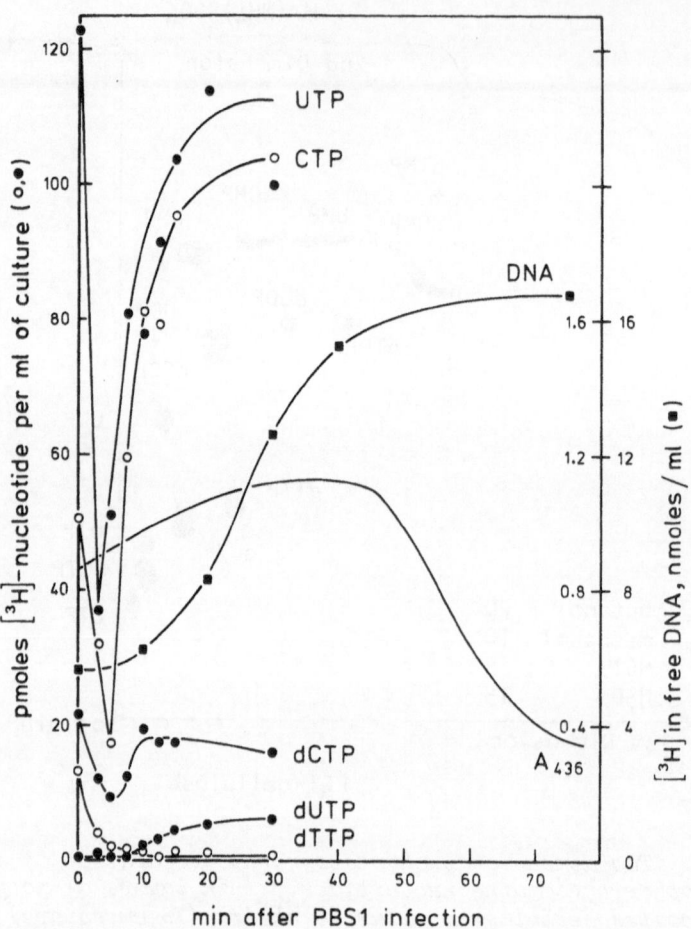

Figure 3. Amounts of pyrimidine nucleoside triphosphates labelled
by (³H-6)uracil in uninfected and PBS1-infected *B.subtilis.* Strain
SB19E *lys pyr*-2 (from I. Takahashi, Ref 7) was grown at 37° in Bis-
Tris minimal medium (8) plus 0.2% glucose, 0.2% casamino acids, and
uracil at 10 µg/ml. Log phase cells (with a generation time of 43
min) were labelled with (³H-6)uracil at 10 µCi/ml for 2 h to uniformly labelled nucleotide pools in this strain, which cannot
synthesize pyrimidines de novo. At A_{436} = 1.0 (about 10^8 cells/ml,
for about 200 µg dry weight per ml), an uninfected 5-ml cell sample
was removed, and the rest was infected with PBS1 at 10 phage per
cell (which killed 95% of the cells within 15 min). Samples were
taken at the indicated times for rapid filtration and extraction of
pools at 0° in 2.5 ml of 0.33 N HCOOH; after centrifugation, lyo-
philization of 2-ml portions, and resuspension in 200 µl of water
(5), samples were applied to chromatograms and developed as in Fig.
2. Exposure to X-ray film confirmed that the tritium spots coincided

(10 μC_i/ml at 10 μC_i/μmole) to label all nucleotides, and unlabelled cytidine (30 μg/ml) to reduce the amount of tritium entering cytosine-containing compounds. Uninfected and 30 min PBS1-infected samples were extracted as above for resolution of all 8 common ribo- and deoxyribonucleoside triphosphates (4,5) and double-label counting of both isotopes in each nucleotide. In the uninfected dcd^+ cell, dTTP as well as the 4 purine compounds contained no tritium, as expected since the 5-tritium from uracil in dUMP is lost on methylation to dTMP (10). From the ^{3}H/^{32}P ratio in dUTP (0.51) relative to that in UTP (0.88) and in CTP or dCTP (0.25), one can calculate that (^{3}H) dUTP in infected cells was derived 42% from reduction of uridine nucleotides versus 58% from deamination of cytidine nucleotides in the dcd^+ strain (which contains host dCMP deaminase and phage dCTP deaminase activities; see Fig. 1). In the experiment using the dcd^- strain, which lacks the host dCMP deaminase, similar results (46% versus 54% respectively) were obtained. This suggests that the phage-induced dCTP deaminase is sufficient to make the necessary dUTP and that the host dCMP deaminase may not contribute much to dUTP production (or that activity of the phage enzyme increases to compensate for the absence of the host enzyme). A parallel experiment using uninfected cells labelled with (^{3}H-6)uracil and ^{32}P$_i$ showed that dTTP was derived 55% from uridine nucleotides and 45% from deoxy-cytidine compounds in the dcd^+ strain, but 100% from uridine nucleo-tide reduction in the dcd^- strain is expected. This finding proves that the dcd^- strain has no deoxycytidine nucleotide deaminase activity *in vivo*. Similar conclusions have been drawn by Rima (11) from the labelling of phage DNA or host DNA in such mutants.

We next considered whether the *B.subtilis* dTMP synthetase activity was inhibited after PBS1 infection, since infected cells contained little or no dTTP (Fig. 3) and the uracil-DNA phage would presumably have no need for dTMP. However, we found about 80% of normal dTMP synthetase activity (assayed by the release of tritium

---

with the pyrimidine nucleoside triphosphate markers; however, additional radioactive areas were observed near the origin, around the diphosphate markers, and along the secondary front in the second dimension (making quantitation of diphosphates and monophosphates unreliable). The triphosphate spots were cut out, eluted with 250 μl of 2N NH_3, and counted at 4° in Beckman BioSolv scintillation fluid; the data were converted to pmoles of nucleotide per milli-liter of original uninfected culture, based on the determined specific activity of the (^{3}H)uracil added to the medium. The amount of free DNA (not inside phage heads) was measured as before (9). The A_{436} (l cm path-length in an Eppendorf photometer) was measured to follow lysis of the culture, which was complete at 75 min and yielded a typical burst of 2.5 x 10^9 phage per ml.

from (^{3}H-5) dUMP (10,12) in either sonic extracts (0.5 nmole/min/
mg protein) or toluenized cells (the same activity per ml as in
sonic extracts). This confirms the spectral assay results of
Tomita and Takahashi (13). Furthermore, we discovered that
uninfected wild-type *B. subtilis* extracts contained *two* dTMP
synthetases resolvable on a DEAE-cellulose column (8). Both of
these peaks were also present in PBS1-infected cells (Fig. 4).
The major peak A is dependent on and probably coded by the *thy*A
gene, the minor peak B by the *thy*B gene, and either enzyme is
sufficient to support normal growth rates without exogenous thymine
(8). (Preliminary data by Rima (11) using unfractionated mutant
cell extracts point to the same conclusion). The two dTMP
synthetases differ in their heat-stability (B is very labile),
their K_M for dUMP (4 μM for A versus 20 μM for B), and the
inhibition caused by dTMP and 5-F-dUMP (A is 10-fold more sensitive).

The dTMP synthetase A and B activities are also measurable
in vivo, by growth in minimal or rich medium in the presence of
(^{3}H-5)uracil, whose tritium is lost on conversion of (^{3}H) dUMP
to dTMP in the cells (15). We demonstrated that the growth rate
and the amount of tritium released per cell was the same in *thyA*$^+$
thyB$^+$, *thyA*$^-$ *thyB*$^+$, and *thyA*$^+$ *thyB*$^-$ cells at 37° (*thyA*$^-$*thyB*$^-$cells
have very little activity and require thymine for growth). Further-
more, the rate of tritium release remained unchanged (60 to 100%
of uninfected cells) for at least 30 min after PBS1 infection of
each strain.

Different evidence to support this point was derived by
labelling in broth cultures with (^{3}H-5) deoxycytidine using the
above strain SB19E *lys pyr*-2 *cdd*-1 *dcd*-1. This strain can only
convert the label to deoxycytidine compounds, since it lacks
deoxycytidine deaminase (7) and dCMP deaminase; thus no deoxyuridine nucleotides can be formed so no tritium can be released by
its dTMP synthetase reaction (less than 3% of the rate of tritium
release of its parent *dcd*$^+$ strain). However, PBS1 infection of
labelled *dcd*$^-$ cells gave a rapid rate of tritium release from
(^{3}H-5) deoxycytidine for over 20 min, presumably because the phage-
induced dCTP deaminase produced (^{3}H) dUTP which is converted to
(^{3}H) dUMP to form dTMP and (^{3}H) H$_2$O (see Fig. 1). Thus, the
production of dTMP by dTMP synthetases A and B appears to continue
after phage infection; the unneeded dTMP is presumably degraded by
the phage-induced dTMPase to thymidine (4).

PBS2 phage-induced DNA polymerase. The dUTP and other deoxyribonucleoside triphosphates found in phage-infected cells are
presumably polymerized enzymatically to form phage uracil-DNA. A
large increase in the specific activity of DNA polymerase was
observed after PBS2 infection of *B. subtilis* (2). This enzyme has
now been purified 900-fold in 22% yield from lysozyme extracts by

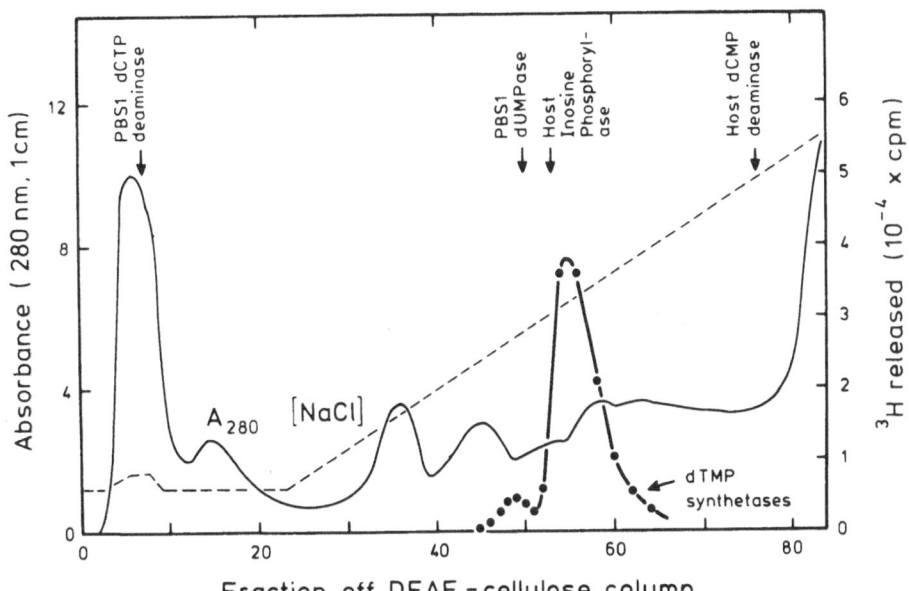

Figure 4. Two dTMP synthetase activities separable by DEAE-cellulose column chromatography of PBS1-infected cell extracts. A 2-liter culture of B. subtilis 168 ilvA1 sacA78 in Penassay Broth was grown at 37° to A 436 = 1.0 for infection by PBS1 phage at a multiplicity of 5 for 30 min. The cells were chilled, harvested by centrifugation, washed in 0.9% NaCl, and frozen. Extracts were prepared by sonic oscillation and centrifugation in 50 mM TrisHCl buffer (pH 7.5) containing 5 mM 2-mercaptoethanol. The dTMP synthetase specific activity (see Ref. 8) was 0.4 nmole/min/mg protein. The activity of one PBS1-induced enzyme tested was normal (260 nmole/min/mg for dUMPase (see Ref. 14)). The 10 ml extract (277 mg protein) was applied to a 1.5 x 15 cm column of DEAE-cellulose and washed in the above buffer, followed by elution at 14 ml/h using a linear 200-ml gradient from 0 to 0.4 M NaCl in buffer with collection of 2.4 ml fractions. The A280, relative conductivity (NaCl), and peak elution positions of some PBS1 and B. subtilis enzymes as markers are shown. The dTMP synthetase activity (●) was measured on 50-μl aliquots incubated for 30 min at 30° in 200 μl assays with (³H-5) dUMP at 10,000 CPM/nmole to measure tritium release (8). Parallel chromatography of uninfected cell extracts gave 2 dTMP synthetase peaks at identical positions, although the first peak (B enzyme) was generally twice as large; apparently the phage-induced dUMPase (14) which elutes here causes a reduction in dTMP synthetase B activity by hydrolyzing the substrate.

precipitation with streptomycin sulfate, enzymatic digestion of
contaminating nucleic acids, ammonium sulfate fractionation, and
Sephadex G200, DEAE-cellulose, phosphocellulose, and DNA-cellulose
column chromatography (16). The protein appears to form unstable
aggregates (and even precipitates) in low salt solutions, but in
high salt it behaves on Sephadex columns or sucrose gradients as a
stable enzyme of 142,000 molecular weight. When subjected to polyacrylamide gel electrophoresis in sodium dodecyl sulfate, the
preparation shows equimolar amounts of 2 subunits of 68,000 and
76,000 daltons. Studies of suppressor-sensitive PBS2 mutants have
also indicated that 2 genes control the PBS2 DNA polymerase (17).

The purified phage DNA polymerase requires a high ionic
strength for optimal activity (300 mM NaCl optimum at 14 mM $MgCl_2$,
or 60 mM $MgCl_2$ optimum at 60 mM NaCl). The activity is completely
dependent upon added DNA, uses denatured DNA (uracil or thymine-
containing DNA) much better than native DNA, and being most active
on "activated" DNA (DNase-digested to fragments with single
stranded gaps). The apparent K_M for DNA decreases with decreasing
DNA size, from 500 µg/ml for 10^7-dalton DNA to 20 µg/ml for 10^5-
dalton denatured PBS2 DNA.

The PBS2 DNA polymerase requires all 4 common deoxyribonucleoside triphosphates for maximal activity on natural DNA templates.
The enzyme displays competitive kinetics for dUTP ($K_M=K_i=16$µM, with
V_{max} = 12.7 µmoles/min/mg protein) versus dTTP ($K_M=K_i=7$µM, with
V_{max} = 9.9) on single-stranded PBS2 uracil-DNA (Fig. 5). Likewise
the K_M for dUTP (4 µM) and dTTP (3 µM) were low with thymine-
containing salmon sperm DNA as the template. Thus the PBS2 DNA
polymerase, like many other known polymerases, is rather non-
selective in its substrate specificity, using uracil or thymine-
containing DNAs and triphosphates at similar rates.

A nuclease activity is associated with the PBS2 DNA polymerase, as suggested by parallel purification, stabilization by
albumin or salt solutions, effects of high NaCl concentrations on
activity and molecular size, heat inactivation, inhibition by
dextran sulfate (50% at 0.5 µg/ml), and preference for denatured
DNA. Preliminary studies using ^{32}P-5'-labelled DNA containing (^{3}H)
thymidine residues suggests that the attack is exonucleolytic from
the 3'-end to produce 5'-monophosphate products. The nuclease is
inhibited by the addition of the 4 triphosphates (polymerizing
conditions); likewise, the production of mononucleotides from the
triphosphates during polymerase assays is dependent upon added DNA.
Thus these activities all appear to reside in the PBS2 DNA polymerase protein. The enzyme differs in many respects from the 3
known *B. subtilis* DNA polymerases (Table 1).

In conclusion, we have purified and characterized the PBS2
DNA polymerase, proving it is a unique enzyme different from the

Table I

PBS2 phage DNA polymerase has different properties than *B. subtilis* DNA polymerases I, II and III[a]

DNA polymerase	Molecular Weight daltons	Subunit Size daltons	Optimal NaCl[b] mM	Template Preference	Associated Exonuclease	Percent Inhibition By araCTP at 50 μM[c]	NEM at 2 mM[d]	HPUra at 100 μM[e]
PBS2	142,000	68,000 + 76,000	350	denatured or activated	Yes	73	75	0
Host I	115,000	one chain	250	denatured or activated	No	0	0	0
Host II	170,000	unknown	100	activated	No	52	0	0
Host III	166,000	one chain	15	activated	Yes	15	90	99

[a] *B. subtilis* DNA polymerases data from Gass and Cozzarelli (18,19) and others (20, 21).

[b] With $MgCl_2$ at 7 mM in the assay.

[c] With dCTP at 40 μM, araCTP was used as an inhibitor.

[d] Without added 2-mercaptoethanol, N-ethylmaleimide was added as an inhibitor.

[e] With dGTP at 40 μM, 6-(p-hydroxyphenylhydrazino)-uracil was tested as an inhibitor.

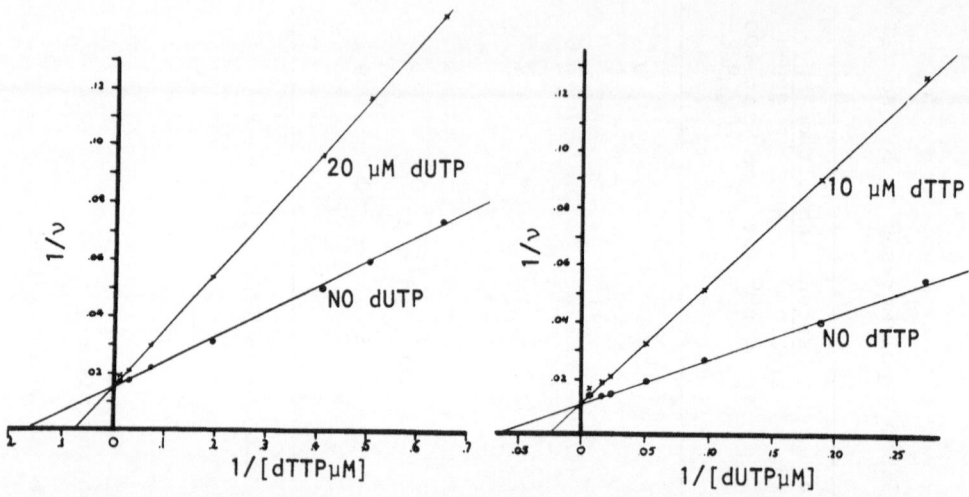

Figure 5. *Double-reciprocal plot of PBS2 DNA polymerase activity with limiting concentrations of dTTP or dUTP as substrates or competitive inhibitors.* Assays (110 μl) contained the indicated concentrations of (methyl-³H) dTTP or (5-³H) dUTP as substrate (220,000 CPM), with or without the indicated concentrations of unlabelled inhibitor. In addition to other assay reagents (16), denatured PBS2 DNA was present as template at 290 μg/ml; dCTP, dGTP, and dATP at 37 μM; MgCl₂ at 14 mM; NaCl at 300 μM; and 1 μl purified enzyme (7.3 ng) per assay. Aliquots of 50 μl were removed after 5 and 10 min at 37° to determine tritium incorporated into acid-insoluble product (v in nmole per min per ml of enzyme)

known host DNA polymerases. Although this enzyme probably synthesizes only uracil-containing DNA in vivo, inside a cell containing undegraded host thymine-DNA (22), the phage polymerase does not display any particular specificity for uracil- versus thymine-containing DNA nor for dUTP versus dTTP in vitro. Apparently the triphosphate pools are altered by phage-induced proteins (Fig. 1), so that dTTP is excluded, even though dTMP synthesis continues. Then dUTP accumulates in the infected cell, to permit only uracil-DNA synthesis. As yet undiscovered factors must prevent the replication of thymine-containing host DNA.

ACKNOWLEDGEMENTS

This work was supported in part by a U.S. Energy Research and Development Administration grant (Report No. COO-2101-33) to A.R.P.,

an NIH traineeship (USPHS 5-T01-GM00187-17) to R.A.H., and a
Carlsberg Foundation grant to J.N.   A.R.P. expresses his warm
thanks to the Josiah Macy Jr. Foundation for a Faculty Scholar
Award and to the Enzyme Division and Jan Neuhard for their
generous hospitality during his sabbatical year.

REFERENCES

1.   Takahashi, I. and Marmur, J. (1963) Biochem. Biophys. Res.
      Commun. 10:289-292

2.   Price, A.R. and Cook, S.J. (1972) J. Virol. 9:602-610

3.   Price, A.R. (1976) In: Microbiology - 1976 (ed. D. Schless-
      inger) American Society for Microbiology, Washington. pp290-
      294

4.   Randerath, K. and Randerath, E. (1967) Methods Enzymol.
      12A:323-347

5.   Neuhard, J. and Thomassen, E. (1971) Eur. J. Biochem. 20:
      36-43

6.   Flanegan, J.B. and Greenberg, G.R. (1977) J. Biol. Chem.
      252:3019-3027

7.   Rima, B.K. and Takahashi, I. (1977) J. Bacteriol. 129:
      574-579

8.   Neuhard, J., Price, A.R., Schack, L. and Thomassen, E.,
      submitted for publication

9.   Post, L. and Price, A.R. (1975) J. Virol. 15:363-371

10.   Lomax, M.I.S. and Greenberg, G.R. (1967) J. BIol. Chem.
      242:109-113

11.   Rima, B.K. (1974) Ph.D. thesis, McMaster University, Hamilton,
      Ontario, Canada

12.   Roodman, S.T. and Greenberg, G.R. (1971) J. Biol. Chem.
      246:2609-2617

13.   Tomita, F. and Takahashi, I. (1976) In: Microbiology - 1976
      (ed. D. Schlessinger) American Society for Microbiology,
      Washington. pp315-318

14.   Price, A.R. and Fogt, S.M. (1973) J. Biol. Chem. 248:1372-
      1380

15. Tomich, P.K., Chiu, C.S., Wovcha, M.G. and Greenberg, G.R.
     (1974) J. Biol. Chem. 249:7613-7622

16. Hitzeman, R.A. (1977) Ph.D. thesis, University of Michigan,
     Ann Arbor, Michigan.

17. Levine, M. (1975) Ph.D. thesis, McMaster University, Hamilton,
     Ontario, Canada.

18. Gass, K.B. and Cozzarelli, N.R. (1973) J. Biol. Chem. 248:
     7688-7700

19. Low, R.L., Rashbaum, S.A. and Cozzarelli, N.R. (1976) J. Biol.
     Chem. 251:1311-1325

20. Clements, J.E., D'Ambrosio, J. and Brown, N.C. (1975) J. Biol.
     Chem. 250:522-526

21. Ganesan, A.T., Laipis, P.J. and Yehle, C.O. (1973) In:
     DNA Synthesis In Vitro (eds. R. Wells and R. Inman) University
     Park Press, Baltimore. pp405-433

22. Tomita, F. and Takahashi, I. (1975) J. Virol. 15:1073-1080

IN VIVO SYNTHESIS AND PROPERTIES OF URACIL-CONTAINING DNA

Huber R. Warner[*] and Bruce K. Duncan

Department of Biochemistry
College of Biological Sciences
University of Minnesota
St. Paul, Minnesota 55108, U.S.A.

INTRODUCTION

All known living cells contain thymine in their DNA, but it is not clear why uracil, which base pairs adenine as does thymine, would not suffice. A considerable expenditure of energy is required to synthesize dTTP from dUTP, and the pathway shown in Fig. 1 suggests that dUTP is an obligate precursor of dTTP in wild type *Escherichia coli*, and probably most other organisms. An alternate route could be the salvage of thymine or thymidine, but it is not known how much dTTP is synthesized by this pathway in various organisms. In wild type *E.coli* probably very little, if any, dTTP is synthesized in this manner (1).

There are two known ways in which uracil can be incorporated into DNA. One is by use of dUTP in place of dTTP by DNA polymerase during replication or repair of DNA. *E.coli* DNA polymerase I·can use dUTP in place of dTTP (2), and presumably other DNA polymerases can also. Incorporation of dUTP into DNA probably occurs rather infrequently, because the presence of deoxyuridine 5'-triphosphatase (dUTPase) activity in *E.coli* and other cells (3,4) keeps the intracellular dUTP pool low. However, when the dUTPase activity is greatly reduced by mutation, dUTP incorporation into DNA probably becomes very significant (5). The uracil-DNA glycosidase activity in the cell serves as a backup system to remove this uracil (6). The second way that uracil can occur in DNA is by deamination of cytosine. This deamination occurs more readily than the deamination of adenine or guanine (7), and poses a mutagenic challenge to the cell unless the uracil can be removed before replication occurs. Specific repair of this damaged region of the DNA can be initiated

A.S.I. Participant

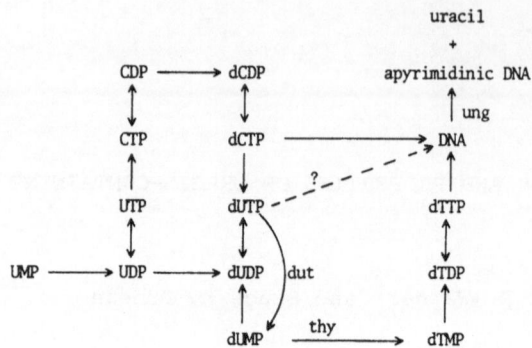

Figure 1. *Pyrimidine nucleotide metabolism in* E.coli

by uracil-DNA glycosidase activity in *E.coli* and other organisms
(8).

The mutagenic potential of cytosine deamination would
certainly be deleterious to cells, but it is not known whether the
presence of uracil in DNA would otherwise seriously affect the
biological activity of the DNA. In this paper we present direct
evidence for the incorporation of uracil into DNA, and we describe
some of the properties of this DNA.

RESULTS AND DISCUSSION

To facilitate *in vivo* synthesis of uracil-containing DNA, we
felt it would be necessary to isolate mutants deficient in uracil-
DNA glycosidase activity. An *E.coli* W3110 suspension was treated
with N-methyl-N'-nitro-N-nitrosoguanidine, and the mutagenized
surviving cells were segregated, grown up in liquid medium, and
plated as described by Weiss and Milcarek (9). Three thousand
colonies were picked and grown to stationary phase in liquid medium.
These were assayed for uracil-DNA glycosidase, and eight enzyme-
deficient isolates were detected; these mutants are referred to as
ung mutants. The properties of four of these are shown in Table I.
Two of the mutants have barely detectable activity *in vitro*, and
the other two have reduced, but heat-sensitive activity. Since the
mutants which are the most deficient grow as well as the *ung*[+]
parent, the uracil-DNA glycosidase activity is not essential for
normal growth.

Some other properties of the *ung*-1 mutant are listed in Table
II. The mutant supports the growth of all wild type bacteriophage
tested so far. It is not sensitive to mutagenic agents such as
methyl methane sulfonate and ultraviolet light. The *ung* mutation
reverses the formation of very short Okazaki fragments during DNA

Table I

Uracil-DNA glycosidase activity in extracts of *E.coli ung* mutants

Strain	Genotype	Generation time at 42°	Enzyme Activity[*] 25°	42°	$Q42^{\circ}/25^{\circ}$
W3110	*ung*$^+$	27	3.1	12.8 (100)	4.1
BD10	*ung*-1	27	<0.003	0.003(0.02)	-
BD13	*ung*-2	31	0.003	0.08 (0.7)	-
BD15	*ung*-3	73	0.31	0.88 (6.9)	2.8
BD21	*ung*-6	69	0.04	0.05 (0.4)	1.3

[*] nmol uracil released per min per mg protein. The values in ( ) are the % residual activity at 42°.

Table II

Properties of *ung E.coli* mutants

1. Growth rate not related to *ung* genotype

2. Plates T4, T5, T7, λ, P1 and M13 bacteriophages

3. Normal sensitivity to methyl methane sulfonate, mitomycin C, nalidixic acid, and uv light

4. Prevents synthesis of short Okazaki fragments in presence of dUTP during DNA synthesis on cellophane disks (10)

5. The *ung* mutation maps at about 56 minutes between *nadB* and *tyrA*

6. Permits replication of U-containing T4 and T5 phage

synthesis in the presence of dUTP on cellophane disks (10), suggesting that short Okazaki fragment formation in this replication system is due to a nicking process initiated by uracil-DNA glycosidase. Uracil-containing bacteriophage can also replicate in these *E.coli* as will be described below.

The formation of short Okazaki fragments has also been observed *in vivo* in *sof* or *dnaS* mutants (11). These mutants have the same phenotype as *dut* mutants which are deficient in dUTPase activity, and the *dut* and *sof* mutations apparently both occur in the same gene (5,12). The *dut-11 ung-1* double mutant has been compared with the properties of *dut (sof)* mutants in Table III. The *ung* mutation reverts the *sof* phenotype *in vivo*, again indicating that these short DNA pieces arise from the incorporation of dUTP into DNA followed by uracil-DNA glycosidase-initiated repair. Surprisingly, we did not find that the hyper-rec phenotype of *sof* mutants (11) was also reverted by the *ung* mutation. The *dut* mutants are sensitive to high concentrations of uracil on plates, possibly due to expansion of the dUTP pool resulting in such massive incorporation of uracil into DNA that repair cannot keep pace. If repair is blocked by the *ung* mutation, the uracil-sensitivity is lost. Furthermore, dUTPase-deficient bacteriophage fail to make plaques on *dut E.coli* (unpublished observation), but do replicate on *dut ung* hosts as described below.

The above experiments suggest that dUTP is incorporated into DNA in *dut E.coli* mutants, and that this uracil remains in the DNA if the *ung* mutation is also present. To directly measure the amount of uracil in the DNA we have grown cells in the presence of $(6-^{3}H)$-uridine, and have examined the base composition of this DNA. The results shown in Table IV show that considerable thymine is replaced by uracil in the *dut ung* cells, but not in wild type or *ung* cells. Thymine is not labeled in this experiment since the strains are all *thy^{-}*, but some thymine is available in the medium, which contains 0.1% yeast extract. The ratio of radioactivity

Table III

Effect of *E.coli ung* mutation on *E.coli dut* phenotype

dut mutant	*dut ung* mutant
1. synthesizes short Okazaki fragments (5)	reverts *sof* phenotype (Tye and Lehman, unpublished)
2. hyper-rec phenotype (11)	hyper-rec phenotype
3. uracil-sensitive in spot test (Weiss, unpublished)	reverts uracil sensitivity
4. restricts replication of *dut* T4 and T5 phage	replicates *dut* T4 and T5 phage

Table IV

Incorporation of $(6-^3H)$-uridine into *E.coli* DNA

E.coli genotype		% Radioactivity		
		C	U	T
W3110	(dut^+ ung^+ thy^-)	98	<1	1
BD10	(dut^+ ung-1 thy^-)	99	<1	<1
BW212	(dut-11 ung-1 thy^-)	69	30	<1

The *E.coli* were labeled for 60 minutes in casamino acid-salts medium containing 0.1% yeast extract, 4 μCi and 0.001 μmol $(6-^3H)$-uridine per ml. The DNA was isolated and analyzed for base composition (19). When cytosine was subjected to the procedure used to determine the base composition of DNA about 0.5% of the cytosine was deaminated to uracil. Hence, we consider any values up to 1% to be rather unreliable.

incorporated into cytosine and uracil in DNA during a 30 minute labelling period of *E.coli* BW212 is rather constant (Fig. 2A), and not much of the uracil is removed from the DNA when the cells are washed free of $(6-^3H)$-uridine and incubated in growth medium containing cold uridine (Fig. 2B). These data indicate that the uridine is continuously incorporated into the DNA, and that once there it is removed slowly, if at all. Thus, uracil-DNA glycosidase must be the major enzyme responsible for removal of uracil from DNA, and the activity of endonuclease V, which is somewhat uracil-specific, on this uracil-containing DNA must be minimal (13).

We have also attempted to synthesize uracil-containing bacteriophage DNA. In these experiments we have made use of the multiple T4 phage mutant 56⁻*den*A *den*B *alc*10. Gene 56 codes for a phage-induced dCTPase-dUTPase activity (14), and must be mutated for the purpose of this experiment to delete the dUTPase activity. The *den*A and *den*B genes code for phage-induced endonuclease II and IV, and must be mutated to prevent degradation of host DNA and cytosine-containing phage DNA (15-17). The *alc* gene codes for a phage-induced protein which prevents transcription of late genes in cytosine-containing phage DNA (18). These phage and wild type T4 were grown in *E.coli* W3110 (dut^+ ung^+) and BW212 (dut-11 ung-1) in the presence of $(6-^3H)$-uridine; the progeny phage were purified by CsCl gradient centrifugation, and the pyrimidine composition

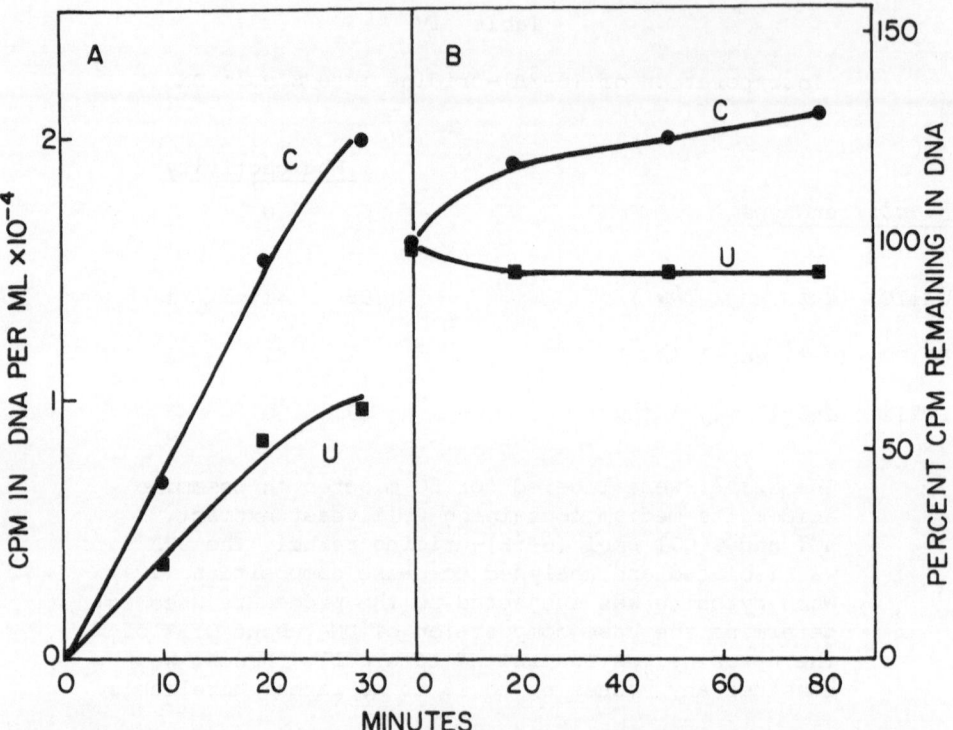

Figure 2. *Incorporation of (6-³H)-uridine into E.coli DNA. A.*
E.coli BW212 was grown in the presence of (6-³H)-uridine using the
conditions described in Table IV. At various times after addition
of the (6-³H)-uridine the DNA was isolated and analyzed for the
amount of radioactivity incorporated into its pyrimidine bases.
B. E.coli BW212 was labeled with (6-³H)-uridine for 30 minutes,
and then the cells were harvested, washed with non-radioactive
medium, and growth was continued in casamino acid-salts medium
containing 0.1% yeast extract, 20 μg thymidine, 100 μg deoxy-
adenosine, and 0.1 μmol uridine per ml. At various times the DNA
was isolated and analyzed for the amount of radioactivity remaining
in its pyrimidine bases.

of the DNA was determined (19). The mutant phage contain uracil in
their DNA when grown in *E.coli* BW212, but not when grown in the
wild type host (Table V), while wild type phage do not contain
uracil after replication in either host. Thus, it is possible to
synthesize uracil-containing T4 phage DNA in which about one-third
of the thymine has been replaced by uracil. This DNA is apparently
suitable for transcription of the late T4 genes, and can also be
packaged to make mature phage.

To determine whether these uracil-containing phge can replicate

Table V

Incorporation of $(6-^3H)$-uridine into T4 phage

E.coli genotype	Phage	% Radioactivity			
		H	C	T	U
W3110 $(dut^+ ung^+)$	Wild type T4	33	<1	66	1
BW212 $(dut-11\ ung-1)$	Wild type T4	38	<1	61	1
W3110 $(dut^+ ung^+)$	T4(56⁻ denA denB alc10)	9	24	66	<1
BW212 $(dut-11\ ung-1)$	T4(56⁻ denA denB alc10)	9	30	42	19

The infected E.coli were labeled from 8 to 90 minutes
after infection using the conditions of Table IV, and
the phage produced were purified by CsCl gradient
centrifugation. Since the base ratio of radioactivity
in H and T reflects the known base composition of T4 DNA,
the % radioactivity is probably a reliable estimate of
the DNA in the phage.

as well as comparable thymine-containing phage, E.coli W3110 (ung^+)
and BD10 $(ung-1)$ were infected with each, and the production of
progeny phage was followed (Fig. 3). The results indicate that
thymine-containing phage replicate equally well in the wild type
and ung hosts, and that the uracil-containing phage also
replicate in the ung host. When wild type cells are infected with
uracil-containing phage, however, less than one progeny phage is
produced per infected cell by 80 minutes after infection. Thus,
uracil-containing T4 phage can replicate, but only in an ung host.

The failure of uracil-containing T4 phage to replicate in a
wild type host suggests that the parental phage DNA must be so
rapidly attacked by uracil-DNA glycosidase that repair cannot keep
up with the rate of uracil excision. Thus, net degradation of the
parental DNA may be occurring. When wild type E.coli is infected
with radioactive uracil-containing phage, the parental DNA was
rapidly acid-solubilized (Fig. 4). In a number of experiments we
have obtained 50-65% acid-solubilization of the uracil-containing
DNA by five minutes after infection. Much less acid-solubilization
occurs in ung cells, and even less occurs when thymine-containing
phage are used. Acid-solubilization would not be expected to reach
100% since any uracil produced by the action of uracil-DNA glycosidase can be reincorporated into RNA, and continued incubation of

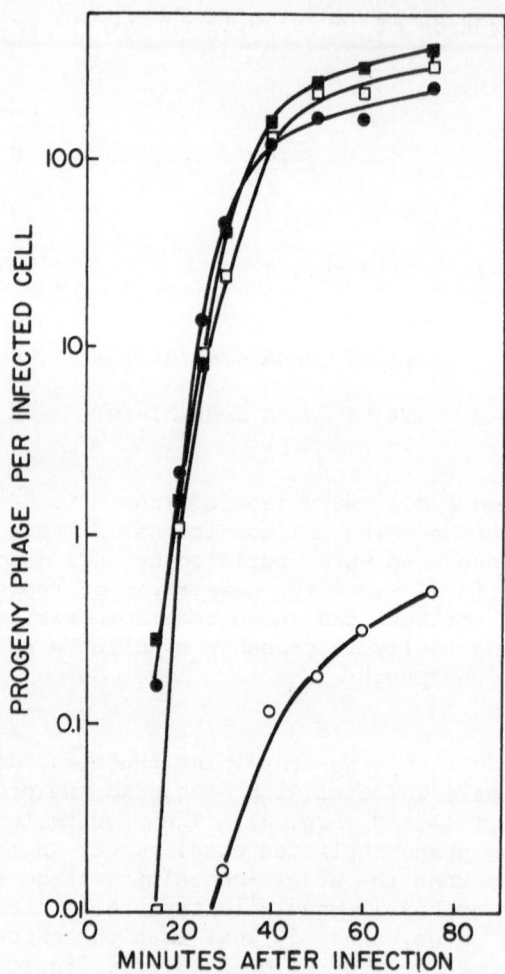

Figure 3. *Replication of uracil-containing phage.* E.coli *W3110 and* BD10 *were infected with uracil-containing (U-) and uracil-free (T-) phage. The infected cultures were shaken vigorously at 37°, and were diluted 2.5 x 10³ at 25 minutes after infection to prevent reinfection of cells by progeny phage. The infected cells were lysed with chloroform at various times and the phage were titered on* E.coli *CR63 (ung⁺) , so only uracil-free phage would produce plaques. The multiplicities of infection were 7 for* E.coli *W3110 infected with T-phage (●), 8 for* E.coli *W3110 infected with U-phage (○), 15 for* E.coli *BD10 infected with T-phage (■), and 17 for* E.coli *BD10 infected with U-phage (□), respectively. The titer of infected cells was assumed to be equal to the initial cell titer.*

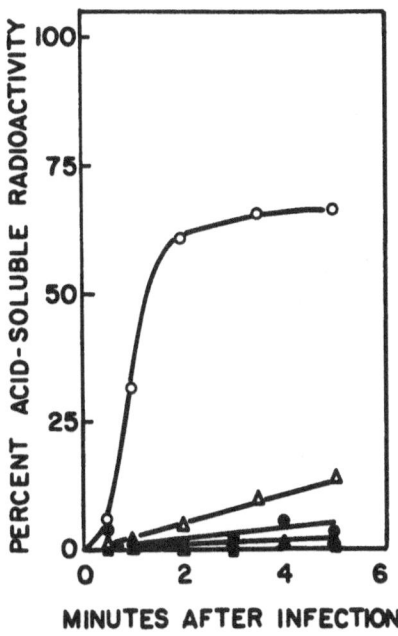

Figure 4. *Acid-solubilization of parental phage DNA.* E.coli *W3110
and BD10 were infected with wild type T4, T-, and U- phage at low
multiplicities (0.4 - 0.7) to maximize the amount of parental
phage DNA injected. These phage had been synthesized in the
presence of (6-³H)-uridine and the distributions of radioactivity
in the phage were comparable to those shown in Table V. At various
times samples were removed and the amount of acid-soluble radioactivity was determined. The results shown above correspond to
T- or U-phage with no cells present (▪), T4 with* E.coli *W3110 (▲),
T-phage with* E.coli *W3110 (●), and U-phage with* E.coli *BD10
(△) and W3110 (○).*

the culture beyond five minutes does lead to partial reincorporation
of the acid-soluble products into acid-insoluble form.

    The above results with *E.coli* and phage T4 indicate that
uracil-containing DNA can be made in *dut ung E.coli*, and that this
DNA must be reasonably functional. We do not yet know, however,
whether DNA in which all thymine has been replaced by uracil would
also be functional. The uracil-containing DNA is rapidly degraded
if introduced into a cell containing uracil-DNA glycosidase
activity, but we do not yet know how much uracil replacement can
be tolerated for the DNA to be successfully repaired.

    In our experiments the uracil-DNA glycosidase is presumably

Table VI

Viability and uracil content of phage grown on *dut ung E.coli*

Parental Phage	Primary E.coli Host	% Pyrimidine as Uracil	Progeny Titers ung^+/ung
Wild type T4	W3110[*]	1	1.19
	BW212[*]	1	1.25
T4(56⁻*den*A *den*B *alc*10)	W3110	<1	1.05
	BW212	19	0.02
Wild type T5	W3110	<1	1.45
	BW212	<1	1.53
Wild type λ	W3110	N.D.[≠]	1.09
	BW212	N.D.	0.03
Plvir	W3110	N.D.	1.19
	BW212	N.D.	0.03

[*]W3110 is dut^+ ung^+: BW212 is *dut*-11 *ung*-1

[≠]Not determined

attacking A-U base pairs in native DNA. Lindahl et al. (20) have
shown that *in vitro* the enzyme releases uracil from native phage
PBS1 DNA and from deaminated *E.coli* DNA at 40% and 70% the rate
of release of uracil from denatured PBS1 DNA, respectively. This
indicates that although the uracil-DNA glycosidase may have a
slight preference for G-U mismatches over A-U base pairs, the
enzyme does not possess sufficient specificity to selectively
repair G-U mismatches in DNA containing A-U base pairs rather than
A-T base pairs.

   We have also screened phages T5, P1 and λ for their ability
to form plaques on wild-type *E.coli* and the *ung-1* mutant after
being plated on *dut-11 ung-1* (Table VI). The results suggest that
uracil-containing P1 and λ phage, but not T5 phage, are produced
in *dut ung E.coli*. From this we concluded that neither P1 nor λ
induces a dUTPase, but that T5 does. We have confirmed this latter
conclusion. Other phages known to induce dUTPase activities
include T2 and T6 (19), *Bacillus subtilis* phage Φe (21), *Rhizobium
leguminosarum* 317 (22). All of these dUTPase-inducing phages,
as well as T4 and T5, have relatively large genomes ($\geq 40 \times 10^6$),
and in most cases are known to induce the synthesis of numerous
enzymes to achieve a high rate of DNA synthesis. The dUTPase is
probably important both for the exclusion of uracil from DNA, and
for synthesis of the substrates for dTMP synthetase and dUMP
hydroxymethylase.

   Since *E.coli* and T4 DNA containing substantial amounts of
uracil can function biologically, the occasional misincorporation
of dUTP into DNA in place of dTTP in normal cells would not be
expected to result in serious consequences for those cells. This
conclusion suggests that the removal of uracil produced by
cytosine deamination may be a more critical function for uracil-
DNA glycosidase than is the removal of misincorporated uracil.
These results strengthen the argument that the use of dTTP for
DNA synthesis, rather than dUTP, permits efficient recognition
of any uracil produced by cytosine deamination so that such DNA
can be immediately repaired to avoid mutations.

ACKNOWLEDGEMENTS

   This investigation was supported by U.S. Public Health
Service Research Grant GM 21464 and Postdoctoral Fellowship GM
05401 from the National Institute of General Medical Sciences.
We thank Elizabeth Kutter and Larry Snyder for the gift of the
alc mutant T4 phage, and Bernard Weiss for constructing strain
E.coli BW212 (*dut-11 ung-1*).

REFERENCES

1. Kammen, H.O. (1967) Biochim. Biophys. Acta 134:301-311

2. Bessman, M.J., I.R. Lehman, J. Adler, S.B. Zimmerman, E.S. Simms and A. Kornberg (1958) Proc. Natl. Acad. Sci. USA 44:633-640

3. Bertani, L.E., A. Häggmark and P. Reichard (1963) J. Biol. Chem. 238:3407-3413

4. Greenberg, G.R. and R.L. Somerville (1963) Proc. Nat. Acad. Sci. USA 48:247-257

5. Tye, B., P. Nyman, I.R. Lehman, S. Hochhauser and B. Weiss (1977) Proc. Nat. Acad. Sci. USA 74:154-157

6. Lindahl, T. (1974) Proc. Nat. Acad. Sci. USA 71:3649-3653

7. Lindahl, T. and B. Nyberg (1974) Biochemistry 13:3405-3410

8. Lindahl, T. (1976) Nature 259:64-66

9. Weiss, B. and C. Milcarek (1974) In: Methods in Enzymology (eds. L. Grossman and K. Moldave) Vol XXIX, Academic Press, Inc. New York., p180-193

10. Olivera, B.M., B. Tye and I.R. Lehman (1977) Federation Proceed. 36:2020

11. Konrad, E.B. and I.R. Lehman (1975) Proc. Nat. Acad. Sci. USA 72:2150-2154

12. Hochhauser, S.J. and B. Weiss (1976) Federation Proceed. 35:1492

13. Gates, F.I. and S. Linn (1977) J. Biol. Chem. 252:1647-1653

14. Warner, H.R. and J.E. Barnes (1966) Proc. Nat. Acad. Sci. USA 56:1233-1240

15. Sadowski, P.D., H.R. Warner, K. Hercules, J.L. Munro, S. Mendelsohn, and J.S. Wiberg (1971) J. Biol. Chem. 246:3431-3433

16. Sadowski, P.D. and D. Vetter (1973) Virology 54:544-546

17.   Kutter, E., A. Beug, R. Sluss, L. Jensen and D. Bradley
       (1975) J. Mol. Biol. 99:591-607

18.   Snyder, L., L. Gold and E. Kutter (1976) Proc. Nat. Acad.
       Sci. USA 73:3098-3102

19.   Price, A.R. and H.R. Warner (1969) Virology, 37:882-892

20.   Lindahl, T., S. Ljungquist, W. Siegert, B. Nyberg and
       B. Sperens (1977) J. Biol. Chem. (in press)

21.   Dunham, L.F. and A.R. Price (1974) Biochemistry 13:2667-
       2672

22.   Ley, A.N., H.R. Warner and P. Kahn (1972) Can. J. Microbiol.
       18:375-384

EXCISION REPAIR OF URACIL IN DNA AND ITS CONTRIBUTION TO THE POOL

OF OKAZAKI FRAGMENTS

I.R. Lehman** and Bik-Kwoon Tye*

Department of Biochemistry
Stanford University School of Medicine
Stanford, California 94305, U.S.A.

INTRODUCTION

Among mutants that were initially identified as "hyper rec", i.e., showing an abnormally high frequency of recombination between intrachromosomal duplications, one group was found in which the hyper rec phenotype coincided with the abnormal persistence of nascent DNA (Okazaki) fragments (1). The increased frequency of nicks and gaps in the chromosomes of such mutants might be expected to provide a corresponding increase in the number of sites at which recombination can occur and in this manner, generate the hyper rec phenotype (2). These hyper rec mutants fell into three classes. One consisted of mutants defective in DNA ligase (lig^-), and the second consisted of mutants with a defect in DNA polymerase I ($polA^-$). Both DNA ligase and DNA polymerase I have been implicated in the discontinuous replication of the $E.coli$ chromosome (3,4), hence, isolation of mutants with defects in these enzymes was to be anticipated. However, the third class, that we have termed $dnaS$ or sof, was unanticipated and contained neither DNA ligase nor DNA polymerase I defectives. They could, however, be identified by the transient appearance of abnormally small (4-6S) Okazaki fragments (5) following brief pulses with (^{3}H) thymidine. A comparison of the pulse labeling patterns observed in wild type, $polA$ and sof mutants (designated as dut, see below) is shown in Fig. 1. Like the nascent fragments seen in wild type and $polA$ strains, the small fragments observed in sof mutants appeared transiently and could be chased

* *Present address: Section of Biochemistry, Cell and Molecular Biology, Cornell University, Ithaca, New York, 14853, U.S.A.*

** *A.S.I. Participant*

efficiently into high molecular DNA by the addition of an excess
of unlabeled thymidine. Thus, with the exception of their
relatively small size, Sof fragments were indistinguishable from
Okazaki fragments. The *sof* mutants were subsequently shown to be
identical to mutants (*dut*) defective in deoxyuridine triphosphate
hydrolase (dUTPase) (6,7,8). This enzyme serves two crucial
functions in pyrimidine nucleotide metabolism, (i) it removes dUTP
as a substrate for DNA synthesis and thus prevents incorporation of
uracil into DNA, and (ii) it provides the only known enzymatic
route for the synthesis *de novo* of dTMP and hence dTTP.

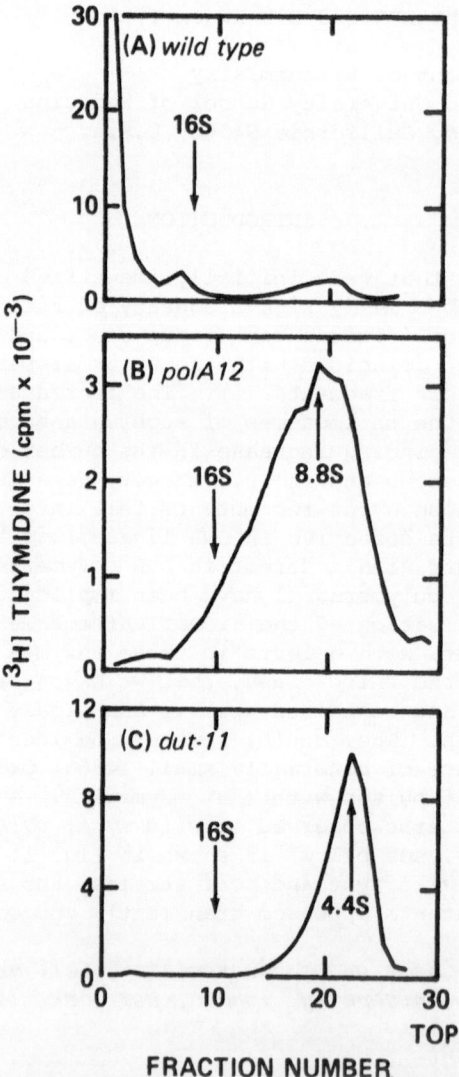

Figure 1. Role of dUTPase in the *de novo* synthesis of dTTP and in
the elimination of dUTP for DNA synthesis.

How can a defect in dUTPase produce the Sof phenotype ?   A
review of the functions of dUTPase (Fig. 2) provides a plausible
explanation for the transient accumulation of small DNA fragments
in mutants with reduced dUTPase activity. Under these conditions,
available dUTP should be increased markedly relative to dTTP, and
the frequency of uracil incorporation into DNA during replication
correspondingly increased. The subsequent action of uracil N-glycosidase (9) and appropriate endonucleases (10,11), which recognize
and excise the uracil from DNA, produces a nick or gap which can
be repaired by the action of DNA polymerase I and DNA ligase (3,4).
An excision-repair process of this kind (Fig. 3) would be expected
to result in the transient appearance of small fragments in the newly
replicated DNA.

In this paper we shall consider (i) the evidence in support of
the identity of the *sof* and *dut* mutations, (ii) the effect of
additional mutations on the Sof phenotype, (iii) the implications
of the *sof* mutation for the model of discontinuous replication of
the *E.coli* chromosome.

Identity of *sof* and *dut* Mutations

Although the *sof* and *dut* mutations were isolated independently
and, in fact, by different screening procedures (8), it is clear
that they represent mutations in the same genetic locus. The
evidence in favor of their identity is the following. (i) Both

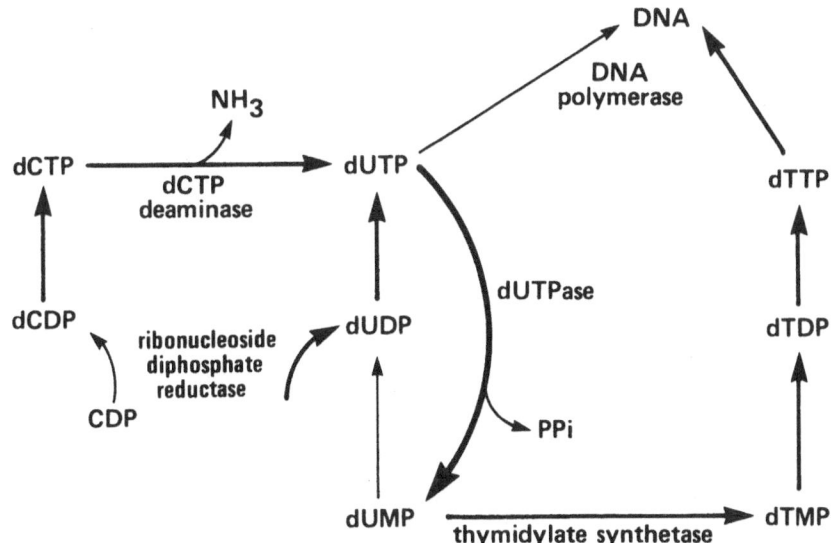

*Figure 2. Sedimentation profiles in alkaline sucrose density
gradients of (³H)thymidine pulse-labeled DNA from wild type, polA12
and dut 11 (sof1) mutants of E.coli.*

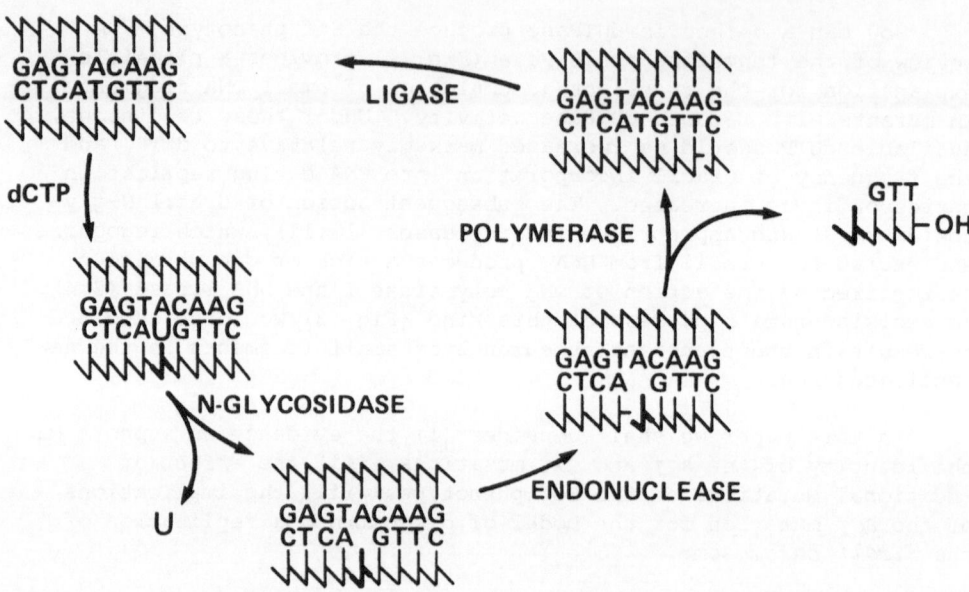

Figure 3. *Diagrammatic representation of the probable steps in the post replication excision-repair of uracil incorporated into DNA in dUTPase defective strains.*

sof and *dut* are closely linked to *pyr*E at 81 minutes on the *E.coli* map, (ii) both *sof* and *dut* mutants display the hyper rec phenotype, (iii) mutants isolated by their Sof phenotype and mutants isolated initially as defective in dUTPase, all have reduced levels of dUTPase and accumulate short DNA fragments after brief pulses with (^{3}H)thymidine (Fig. 4). In fact, the size of the labeled fragments and the extent to which they accumulate correlate well with dUTPase activity in extracts, (iv) a revertant of one of the *sof* mutants has both a normal dUTPase activity and no longer accumulates DNA fragments.

Effect of Additional Mutations on the Sof Phenotype
DNA Polymerase I and DNA Ligase

Sof mutants have no demonstrable phenotype other than their hyper rec character and the transient accumulation of DNA fragments; their growth characteristics are indistinguishable from otherwise isogenic wild type strains of *E.coli*. However, introduction into *sof* mutants of mutations in DNA polymerase I (*polA12*) (12) or in DNA ligase (*lig4*) (13) which, by themselves, produce only abnormal temperature sensitivity of DNA repair, results in the inability of the double mutants to grow at elevated temperatures. This temperature-sensitive conditional lethality is a consequence of the inhibition of DNA replication due, in turn, to the inability of the

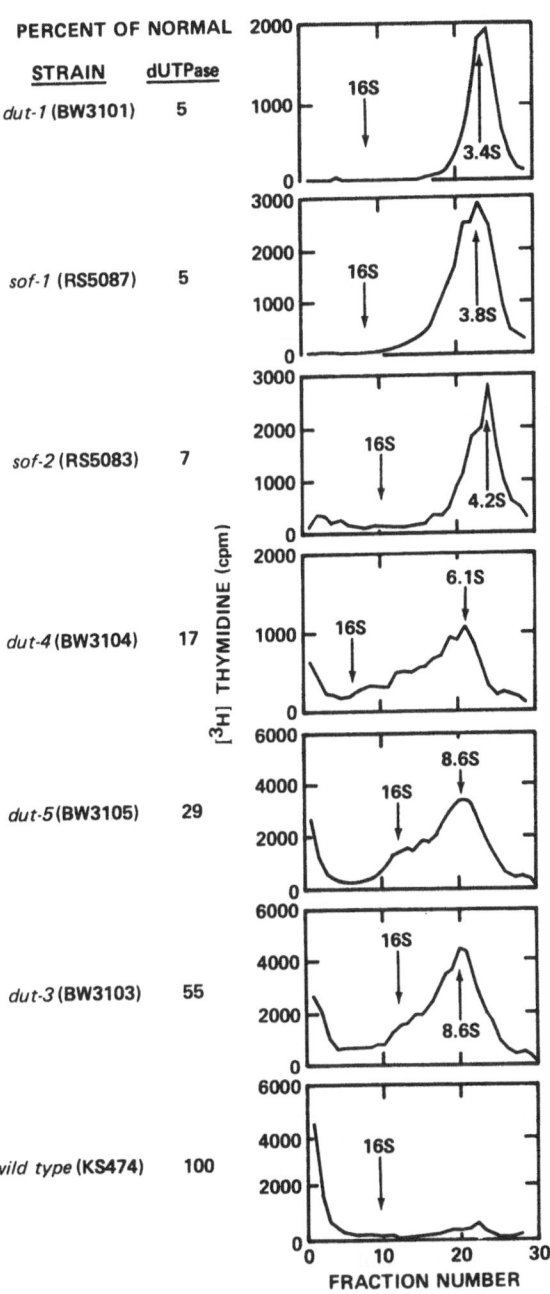

<u>Figure 4</u>. Accumulation of short DNA fragments in <u>sof</u> and <u>dut</u> mutants. Pulse-labeling of cells with (³H)thymidine was carried out for 10 seconds at 30°. Values of dUTPase activity are relative to that of an extract of the wild type strain (KS474).

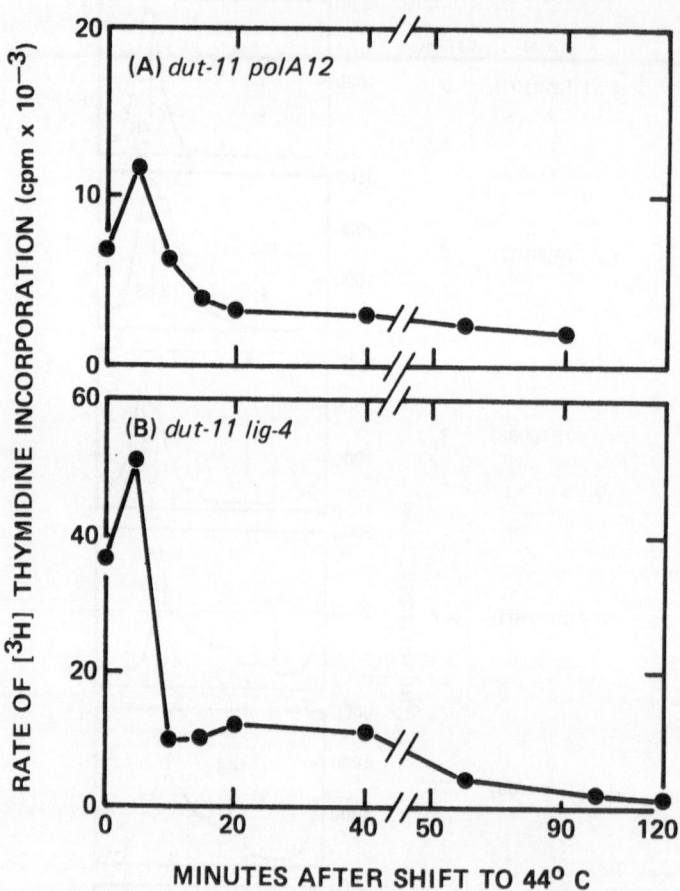

Figure 5. Shut off of DNA replication in *dut11 polA12* and *dut11 lig4* double mutants at 44°C. Cultures were pulsed with (³H) thymidine for 30-second periods. Acid-insoluble radioactivity was measured at the times indicate. (a) *dut11 polA12*, (b) *dut11 lig4*.

fragments to be joined. As shown in Fig. 5, shift of a culture of the *dut⁻polA⁻* double mutant from 30° to 44° caused cessation of DNA synthesis within 10 minutes. Thus, combination of the *dut⁻* and *polA⁻* mutations results in the inability of the double mutant to grow at 44° and blocks DNA synthesis at this temperature. The same is true for the *dut⁻lig⁻* double mutant.

The defect in DNA replication in the *dut⁻polA⁻* double mutant was further examined in pulse-labeling experiments. A culture of the double mutant was grown at 30°, then shifted to 44° and pulsed with (³H) thymidine for increasing periods of time. In contrast to the *dut⁻* mutant, the 4-6S fragments that appeared in the double mutant persisted for pulse periods ranging from 10 seconds to 5

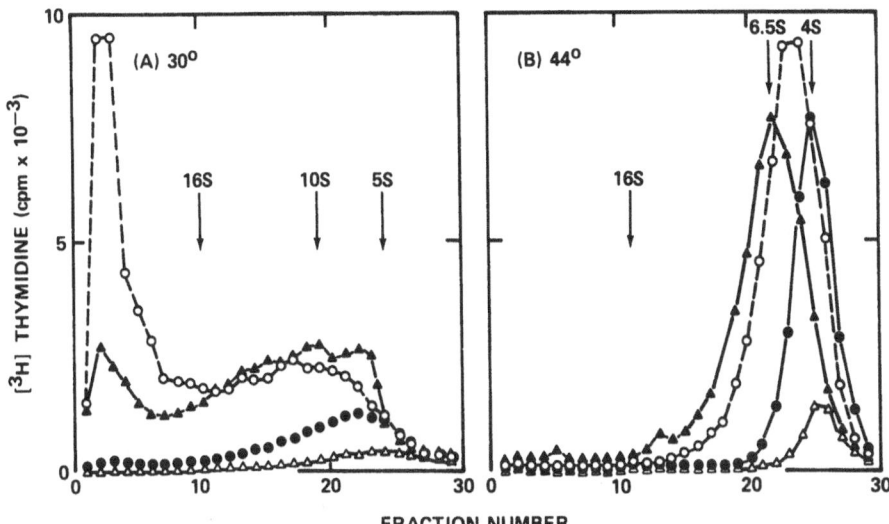

Figure 6. *Sedimentation profiles in alkaline sucrose density gradients of (³H)thymidine pulse-labeled DNA from the dut11 polA12 double mutant. Cultures were (a) pulsed with (³H)thymidine at 30°C for 10 sec (△——△), 30 sec (●——●), 1 min (▲——▲), and 2 min (○——○); (b) shifted to 44°C for 5 min before being pulsed with (³H)thymidine for 10 sec (△——△), 30 sec (●——●), 2 min (○---○), and 5 min (▲——▲).*

minutes (Fig. 6). Thus, at 44°, the *polA⁻* mutation appears to block conversion of the 4-5S DNA fragments to high molecular weight DNA. In contrast, the sedimentation profile of DNA isolated from the *dut⁻polA⁻* double mutant after pulsing with (³H)thymidine at 30° was similar to that of the *dut⁻* mutant at this temperature, i.e., (³H)thymidine was incorporated into DNA of increasing molecular weight as length of the pulse period increased.

Uracil N-glycosidase

Uracil N-glycosidase cleaves the N-glycosidic bond linking uracil to deoxyribose in DNA and may, therefore, be involved in the removal of uracil residues in DNA (9). Mutants are known that are defective in uracil N-glycosidase (*ung*) (14), however, they have no apparent phenotype. To determine whether a defect in uracil N-glycosidase can decrease the level of 4-5S fragments that appear in *dut* mutants, the double mutant *dut⁻ung⁻* was constructed and pulsed with (³H)thymidine for 10 seconds at 30°. The sedimentation profile of the isolated DNA was similar to that of the wild type (Fig. 7). Thus, an *ung⁻* mutation suppresses the Sof phenotype, presumably by preventing cleavage of phosphodiester bonds at apyrimidic sites generated by the action of uracil N-glycosidase.

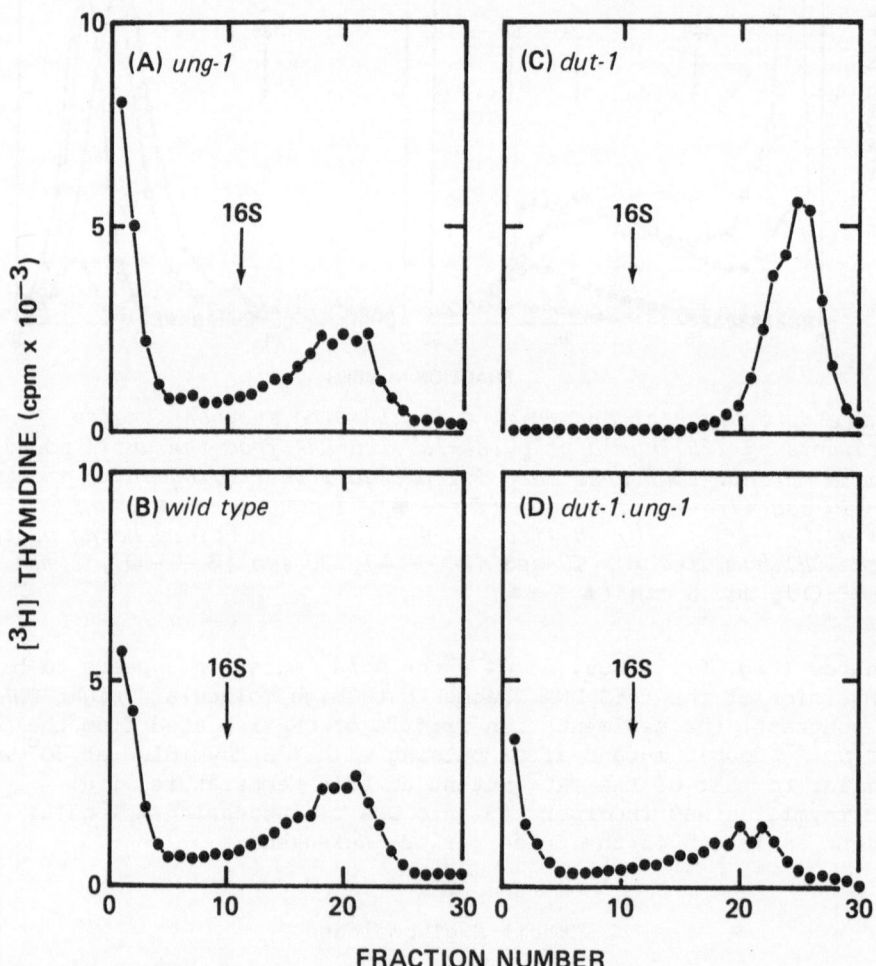

Figure 7. Sedimentation profiles in alkaline sucrose density gradients of (^{3}H)thymidine pulse-labeled DNA from wild type and ung1, dut1, and dut1 ung1 mutants of E.coli. Cultures were pulse-labeled with (^{3}H)thymidine at 30°C for 10 sec. (a) ung1, (b) wild type, (c) dut1, (d) dut1 ung1.

dCTP Deaminase

Introduction of a deletion in the dCTP deaminase gene *(dcd)* (15) restores the viability of the double mutant *dut⁻polA⁻* at 43°. It also suppresses the Sof phenotype. Thus, the alkaline sucrose sedimentation profile of the pulse-labeled DNA of the triple mutant *dut⁻polA⁻dcd⁻* showed predominantly 8-10S fragments, unlike the 4-5S fragments seen in the *dut⁻polA⁻* double mutant (16). Inasmuch as 75-80% of the dUTP synthesized in *E.coli* results from the demination of dCTP (17), some suppression of the *dut* mutation by deletion of the *dcd* locus is to be anticipated.

Implications of the Sof Mutation for Discontinuous DNA Replication

An obvious inference from the findings described here is that DNA fragments that are labeled following short pulses with (³H) thymidine and can be subsequently chased into high molecular weight DNA need not be replication intermediates. Inasmuch as dUTP is a normal product of pyrimidine nucleotide metabolism, low levels of uracil could be incorporated into DNA despite the presence of dUTPase. The size of the fragments generated as a result of uracil incorporation would then depend largely upon the frequency with which such incorporation occurs. Spontaneous deamination of cytosine in DNA, which is believed to occur under physiological conditions, is yet another means for the generation of uracil residues in DNA (18). Thus, it is possible that some portion of the Okazaki fragments that are seen even in wild type cells may result from excision-repair processes.

To explore the possibility that uracil is normally incorporated and excised from DNA during its replication, pulse labeled DNA from an *ung⁻* strain was isolated and treated with purified uracil N-gly-cosidase *in vitro*. As noted above, excision of uracil results in the generation of an apyrimidic site which is sensitive to alkaline cleavage. Consequently, alkaline sucrose density gradient sedimentation of the N-glycosidase-treated DNA should provide an estimate of the amount of uracil present in the pulse labeled DNA. As shown in Fig. 8, DNA isolated from the *ung⁻* mutant following a 10 second pulse with (³H)thymidine displays an alkaline sucrose density gradient profile very similar to that observed for *ung⁺* (wild type) cells pulsed under the same conditions (Compare Figs. 8 and 9). Approximately 30% of the label sedimented very rapidly (>30S) and the remainder had an average sedimentation coefficient of 8-14S. Upon treatment of the DNA with uracil N-glycosidase followed by alkaline sucrose density gradient sedimentation, essentially all of the rapidly sedimenting DNA was reduced to fragments with sedimentation coefficients in the range of 8 to 16S. A considerably less marked effect was observed with DNA from the

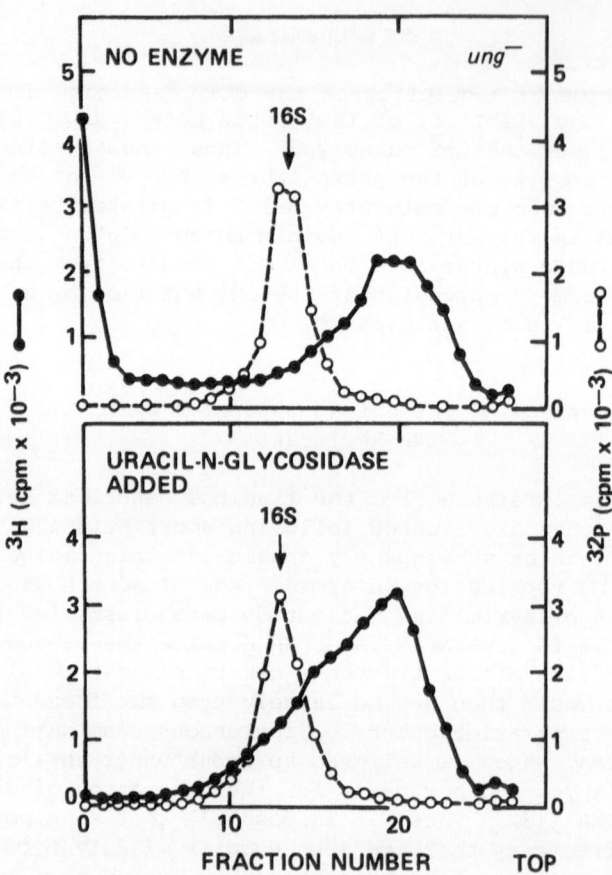

Figure 8. *Persistence of uracil in pulse-labeled DNA from the* *ung1 mutant.* *The* *ung1* *mutant was pulse-labeled with (³H)thymidine at 37°C for 10 sec; the cells were harvested and ³²P-labeled ØX174 phage were added to the cell pellet. The DNA was isolated and treated with uracil N-glycosidase, then sedimented in an alkaline sucrose density gradient. The reaction mixture without uracil N-glycosidase was incubated and sedimented the same way.*

wild type strain (ung^+) (Fig. 9). (The sedimentation properties of ³²P-labeled ØX174 DNA, which lacks uracil and which served as a control in these experiments, remained unchanged after treatment with uracil N-glycosidase). Thus, uracil is incorporated into DNA even in the presence of normal levels of dUTPase, but is rapidly excised by the action of uracil N-glycosidase.

Since treatment of the DNA with uracil N-glycosidase *in vitro* results in a change in sedimentation coefficient from approximately 30S to 8-16S, uracil incorporation probably occurs on an average

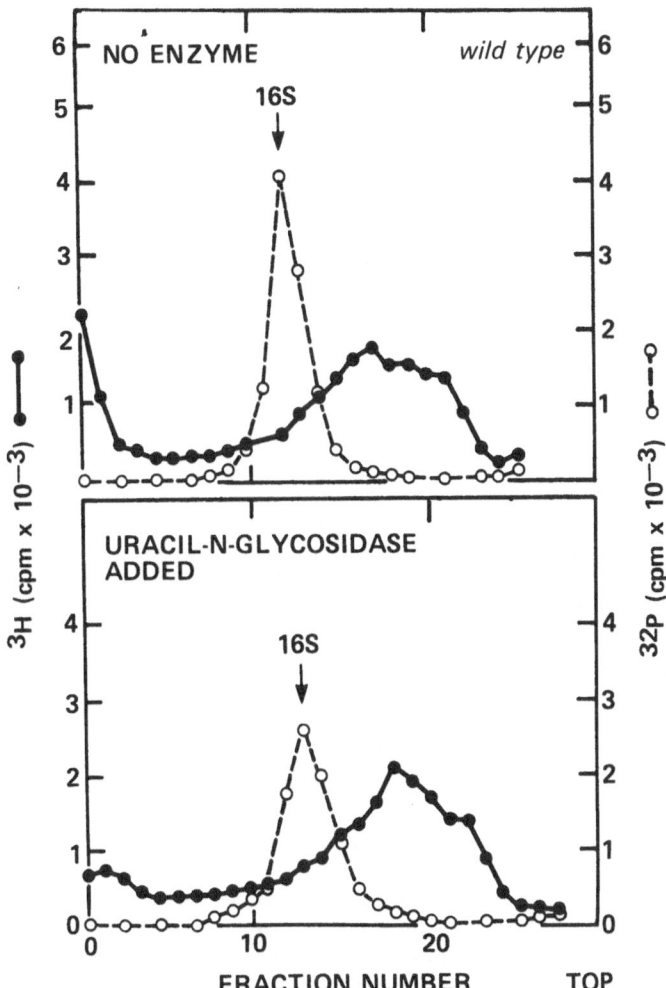

Figure 9. *Absence of uracil from DNA of wild type cells (ung⁺).*
Ung⁺ cells were pulse-labeled with (³H)thymidine at 37°C for 10 sec.
The cells were harvested and ³²P-ØX174 phage were added to the cell
pellet. The DNA was isolated and treated with uracil N-glycosidase
then sedimented in an alkaline sucrose density gradient.

of one per 2000-3000 nucleotides. A recent analysis of several of
the kinetic parameters of a homogeneous dUTPase from *E.coli* by
Shlomai and Kornberg (19) indicates that an intracellular dUTP concentration of 0.5 µM is required to generate sufficient dTTP (from
dUMP) to sustain a normal rate of DNA synthesis. Inasmuch as DNA
polymerase III holoenzyme (20) has nearly the same K_m for dUTP and
dTTP, this level of dUTP in the presence of an intracellular dTTP
concentration of 150 µM would be sufficient to permit uracil

incorporation into DNA at a frequency of 1 per 300 thymidines or
1 per 1200 nucleotides. It is therefore clear that at least some
fraction of the Okazaki fragments normally seen in (^{3}H)thymidine
pulse labeling experiments does indeed originate from post
replication repair of uracil-induced damage to DNA.

Essentially all newly synthesized DNA in *polA* mutants appears
in the form of 8-10S Okazaki fragments. Since the Sof phenotype of
dut mutants can be suppressed by an *ung⁻* mutation, we wished to
determine whether the appearance of Okazaki fragments following
comparable pulses of *polA* mutants can be similarly suppressed by the
ung⁻ mutation. Two different *polA⁻ung⁻* double mutants were
constructed using *E.coli polAex2* (2) and *E.coli polA4113* (21), both
of which are temperature sensitive, conditionally lethal mutants.
The *polA⁻* strains were pulse-labeled at 44° with (^{3}H)thymidine and
the corresponding *polA⁻ung⁻* strains were pulse-labeled with (C^{14})thy-
midine. DNA from the two strains was mixed and sedimented in an
alkaline sucrose gradient. As shown in Fig. 10, the sedimentation
profiles of the *polA⁻* and *polA⁻ung⁻* pulse-labeled DNAs were
similar, with the *polA⁻ung⁻* DNA sedimenting somewhat more rapidly
than that from the polA⁻ strain. Because of the relatively weak
dependence of sedimentation coefficient on molecular weight (22), a
more sensitive method is required to determine the actual difference
in size between the two populations of DNA fragments. However, it is
apparent that in contrast to Sof fragments, the Okazaki fragments
that accumulate in *polA⁻* mutants are not eliminated by a mutation
in uracil N-glycosidase.

Olivera (23) has found that replication of the *E.coli* chromosome in an *in vitro* system, i.e., concentrated cell lysates on
cellophane discs (24), proceeds discontinuously on only one of the
two strands at the replication fork. The rather minimal difference
in pulse labeling patterns between the *polA⁻* and *polA⁻ung⁻* mutants
would suggest that if DNA replication *in vivo* is also discontinuous
on only one strand, then repair mechanisms other than that
following uracil incorporation fragment the continuously synthesized
strand. Alternatively, DNA replication *in vivo* may occur discontinuously on both strands (25).

Persistence of Uracil in DNA of the *dut⁻ung⁻* Double Mutant

Uracil is incorporated into pulse-labeled DNA of the *dut⁻ung⁻*
double mutant and persists for at least the length of the pulse
period (10 sec). Thus, when pulse-labeled DNA from the *dut⁻ung⁻*
mutant was treated with uracil N-glycosidase and then sedimented in
an alkaline sucrose density gradient. The labeled DNA, most of which
was initially of high molecular weight (>30S), was degraded to 4S
fragments (Fig. 11). To determine whether uracil is retained for
periods longer than 10 seconds, the *dut⁻ung⁻* double mutant was

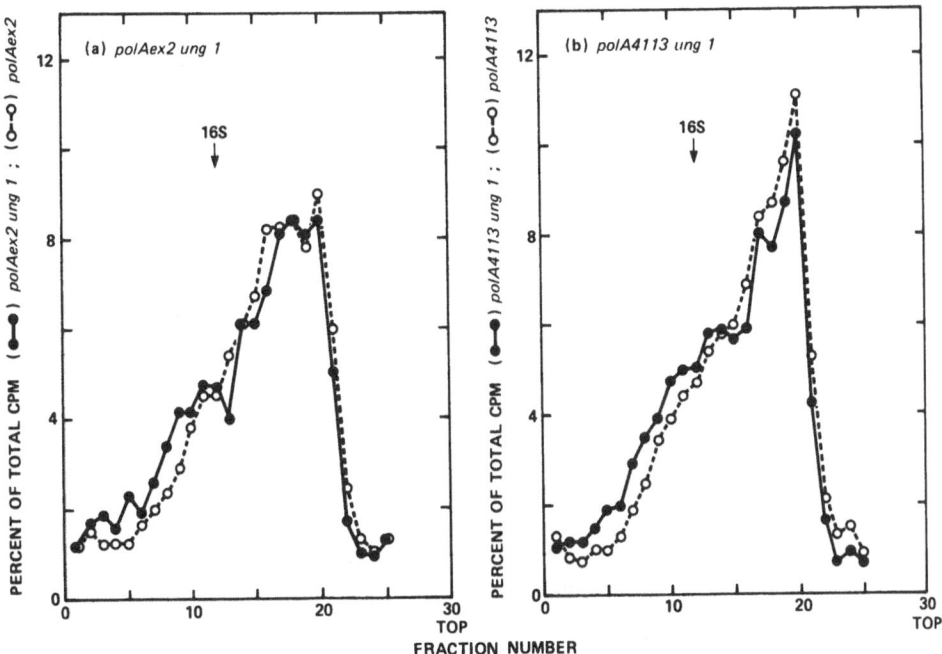

Figure 10. Comparison of sedimentation profiles of pulse-labeled DNA from polA⁻ and polA⁻ ung⁻ mutants. The polA⁻ung⁻ cultures, i.e., (polAex2 ung1) and (polA4113 ung1), were pulsed at 44°C for 10 sec with (¹⁴C)thymidine. The polA⁻ cultures, i.e., (polAex2) and (polA4113), were pulsed at 44°C for 10 sec with (³H)thymidine. Following the pulses, the cells were centrifuged and lysed. The polA⁻ and polA⁻ung⁻ lysates were mixed and sedimented in an alkaline sucrose density gradient. ³²P-labeled ØX174 DNA served as a 16S sedimentation marker. Sedimentation profiles are plotted as percent of total counts in each fraction.

labeled with (³H)thymidine during 4 hours of growth. Treatment of the labeled DNA with uracil N-glycosidase followed by sedimentation in an alkaline sucrose density gradient resulted in a sedimentation profile very similar to that seen after comparable treatment of the pulse-labeled DNA (Fig. 12). Thus, uracil may persist in DNA through several generations in the dut⁻ung⁻ double mutant.

The finding that a mutant defective in both dUTPase and uracil N-glycosidase retains uracil in its DNA through several generations suggests that uracil N-glycosidase is the major enzyme responsible for the removal of uracil residues from DNA. Endonuclease V, also known to recognize uracil in DNA and to cleave phosphodiester bonds at or near these residues (26), is probably of lesser significance in the removal of uracil at A:U base pairs. Possibly, it participates in the excision of uracil residues at G:U mismatches that result from the deamination of cytosine in DNA. The *dut⁻ung⁻*

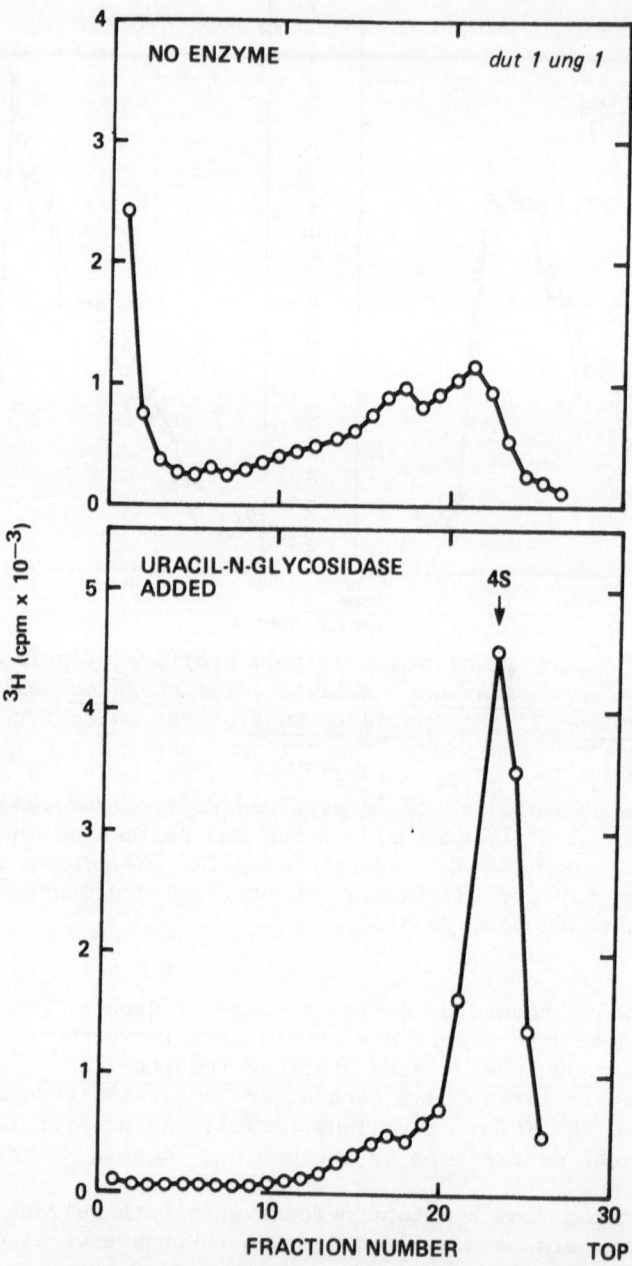

Figure 11. Persistence of uracil in pulse-labeled DNA from the
dut1 ung1 double mutant was pulse-labeled with (³H)thymidine at 37°C
for 10 sec. The isolated DNA was treated with uracil N-glycosidase
then sedimented in an alkaline sucrose density gradient.

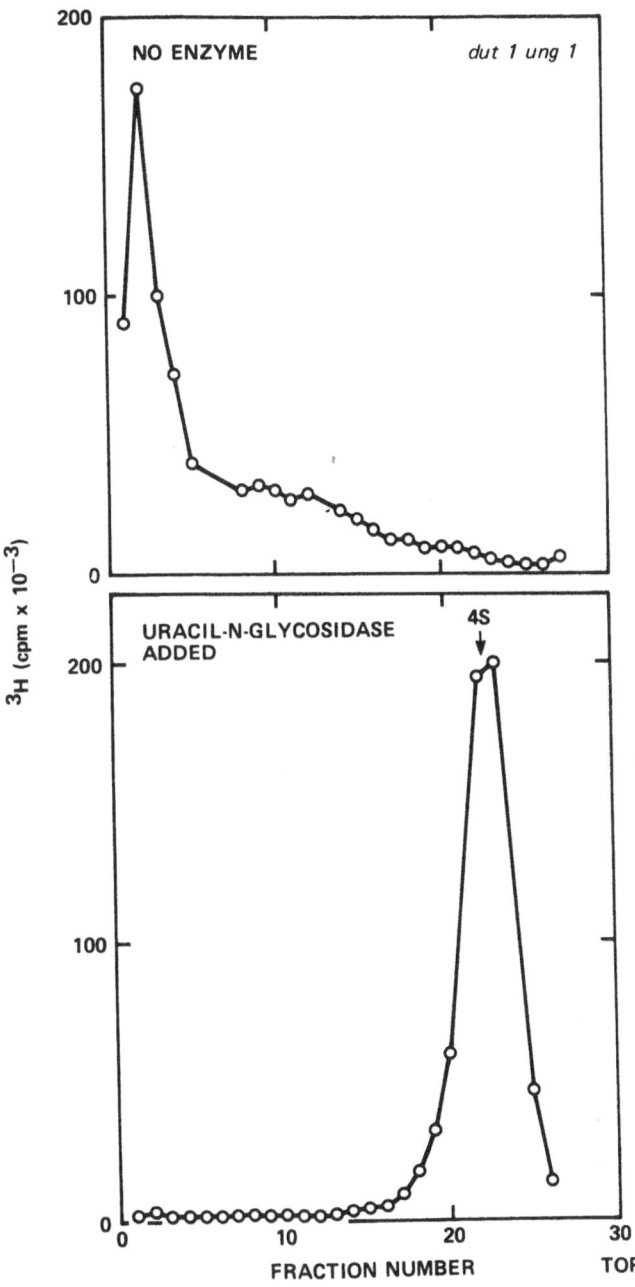

Figure 12. Persistence of uracil in DNA after long term labeling of the *dut1 ung1* double mutant. The *dut1 ung1* double mutant was labeled at 37°C with (³H)thymidine over a period of 4 hr. The isolated DNA was treated with uracil N-glycosidase, then sedimented in an alkaline sucrose density gradient.

double mutant appears to be largely unaffected by levels of uracil in its DNA up to one per hundred nucleotides. A similar result has been reported by Warner and Duncan (14). We have not yet determined the effects of more extensive substitution of uracil for thymine in *E.coli*.

CONCLUSIONS

Mutants of *Escherichia coli* defective in the enzyme dUTPase incorporate uracil into newly synthesized DNA. Excision-repair of the incorporated uracil then results in the generation of labeled DNA fragments observed following brief pulses with (^{3}H)thymidine. This process requires the action of four enzymes, uracil N-glyco-sidase, an endonuclease, DNA polymerase I and DNA ligase. Since uracil is also incorporated into the newly synthesized DNA of strains of *E.coli* that contain normal levels of dUTPase, DNA frag-ments generated by the post replication excision-repair of uracil may contribute to the pool of Okazaki fragments that normally appear in wild type strains.

Mutants that are defective both in dUTPase and in uracil N-gly-cosidase incorporate uracil into their DNA with a high frequency (up to one per hundred nucleotides). These uracil residues once incorporated, persist on the DNA without any obviously adverse effect on the growth of the cells.

ACKNOWLEDGEMENTS

This investigation was supported by grants from the United States Public Health Service (GM 06196) and the National Science Foundation (PCM 74-00865).

B.T. was a Post Doctoral Fellow of the Helen Hay Whitney Foundation.

REFERENCES

1. Okazaki, R., Okazaki, T., Sakabe, K., Sugimoto, K. and Sugino, A. (1968) Proc. Nat. Acad. Sci. USA 59:598-605

2. Konrad, E.B. (1977) J. Bact. 130:167-172

3. Lehman, I.R. (1974) Science 186:790-797

4. Lehman, I.R. and Uyemura, D.G. (1976) Science 193:963-969

5. Konrad, E.B. and Lehman, I.R. (1975) Proc. Nat. Acad. Sci. USA 72:2150-2154

6. Bertani, L.E., Häggmark, A. and Reichard, P. (1963) J. Biol. Chem. 238:3407-3413

7. Greenberg, G.R. and Somerville, R.L. (1962) Proc. Nat. Acad. Sci. USA 48:247-257

8. Tye, B.-K., Nyman, P.-O., Lehman, I.R., Hochhauser, S. and Weiss, B. (1977) Proc. Nat. Acad. Sci. USA 74:154-157

9. Lindahl, T., Ljungquist, S., Siegert, W., Nyberg, B. and Sperens, B. (1977) J. Biol. Chem. 252:3286-3294

10. Hadi, S.J. and Goldthwait, D.A. (1971) Biochemistry 10:4986-4993

11. Verly, W.G. and Rassart, E. (1975) J. Biol. Chem. 250:8214-8219

12. Monk, M. and Kinross, J. (1972) J. Bact. 109:971-978

13. Gottesman, M.M., Hicks, M. and Gellert, M. (1973) J. Mol. Biol. 77:531-547

14. Warner, H.R. and Duncan, B.K. (1977) Proceedings of NATO ASI "DNA Synthesis: Present and Future", Plenum Press

15. Beck, C.F., Eisenhardt, A.R. and Neuhard, J. (1975) J. Biol. Chem. 250:609-616

16. Tye, B.-K. and Lehman, I.R. (1977) J. Mol. Biol. in press

17. Karlström, O. and Larsson, A. (1967) Eur. J. Biochem. 3:164-170

18. Shapiro, R. and Klein, R.S. (1966) Biochemistry 5:2358-2362

19. Shlomai, J. and Kornberg, A. J. Biol. Chem., submitted

20. Wickner, W. and Kornberg, A. (1974) J. Biol. Chem. 249:
     6244-6249

21. Olivera, B.M. and Bonhoeffer, F. (1974) Nature (Lond.) 250:
     513-514

22. Studier, F.W. (1965) J. Mol. Biol. 11:373-390

23. Olivera, B.M. (1977) Proc. Nat. Acad. Sci. USA, in press

24. Olivera, B.M. and Bonhoeffer, F. (1972) Nature New Biol.
     240:233-235

25. Alberts, B.M. and Sternglanz, R. (1977) Nature, in press

26. Gates, F.T. and Linn, S. (1977) J. Biol. Chem. 252:1647-1653

DNA REPLICATION INTERMEDIATES IN *ESCHERICHIA COLI*

M. Raggenbass and L. Caro

Université de Genève
Département de Biologie Moléculaire
30 quai Ernest-Ansermet
CH-1211, Genève, Switzerland

INTRODUCTION

Bacterial DNA replication is, at least in part, discontinuous. An important fraction of the newly synthesized DNA can be isolated in the form of short molecules (Okazaki fragments). In the process of chain elongation these fragments are subsequently joined to long molecules (Sakabe and Okazaki, 1966; Okazaki et al., 1973). In *Escherichia coli* Okazaki fragments are preferentially synthesized in the 3' → 5' direction of replication (Louarn and Bird, 1974).

In 1968, a particular subclass of fragments was demonstrated in *E.coli*: after a short pulse of ^{3}H-thymidine, a significant amount of newly synthesized DNA extracted under *non-denaturing* conditions was isolated as single stranded DNA by hydroxyapatite chromatography (Oishi, 1968a; Okazaki et al., 1968). Analogous fragments were also detected in phage infected bacterial cells (Oishi, 1968b; Paetkau et al., 1975) and in eukaryotic cells (Habener et al., 1970). We shall, henceforth, call these single stranded newly synthesized molecules "neutral fragments" since they are extracted under neutral conditions. The work reported here is part of a study undertaken to further characterize these neutral fragments. In these particular experiments we have studied the size distribution and the strand specificity of neutral and Okazaki fragments in cultures of *E.coli* whose DNA replication had been synchronized. The strain used carried the prophage Mu-1, inserted in the chromosome near the origin of replication, and the prophage λ at its normal attachment site. The fraction of the fragment population produced by the replication of these two markers was identified and analyzed by DNA-DNA hybridization to filters loaded with either Mu or λ DNA.

The experimental approach was essentially the same as that used by Louarn and Bird (1974) in their study of the polarity of Okazaki fragments. These authors used an *E.coli* strain lysogenic for λ. After a short ^{3}H-thymidine pulse, the fragments were extracted and hybridized with the separated strands of λ DNA. In an exponentially growing culture, the fraction of newly synthesized DNA which can hybridize with the prophage DNA is low (about 2%). In order to increase the sensitivity of the method, and also in order to get information on the specificity, with respect to the replication fork, of the two classes of fragments (Okazaki fragments and neutral fragments), we used synchronized cultures. It should be noted that the Okazaki fragments always include the neutral fragments.

Fig. 1 shows the map of the *E.coli* strain used. LC 560 is K12 W1485 *leu⁻ thy⁻* and is wild type for DNA replication. It carries the RTF-Tc plasmid pAR132 (Chandler et al., 1977). Mu-1 is integrated 2 to 5 μm away from the origin of chromosome replication at the *ilv* locus, λ *ind⁻* is integrated at its normal attachment site, *att* λ.

RESULTS

In a first experiment, cells were prelabeled with ^{14}C-thymine, chromosome replication was terminated by a period of amino acid starvation and synchronized by a period of thymine starvation equivalent to one generation, in the presence of amino acids. The replication was initiated by adding 4 μg/ml thymine and, simultaneously, ^{3}H-thymidine. This pulse labeling was terminated after 15 sec at 30°. Under these conditions, DNA replication starts

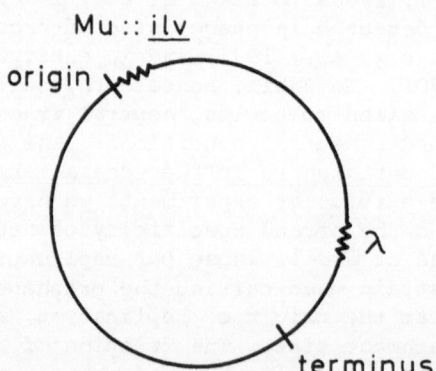

Figure 1. *The relative position of Mu-1 and λ ind⁻ on the genetic map of strain LC 560. The relative positions of origin and terminus of replication (Bird et al., 1976) are also indicated.*

synchronously from the origin as was shown by DNA-DNA hybridization studies (Louarn et al., 1974; Bird et al., 1976). An aliquot of the cells was lysed in alkaline medium and the lysate sedimented on an alkaline sucrose gradient in the presence of two internal markers: λ ^{32}P - DNA and ØX174 ^{32}P - DNA. A second aliquot was lysed in neutral medium; DNA was extracted and analyzed on a hydroxyapatite column (Oishi, 1968a).

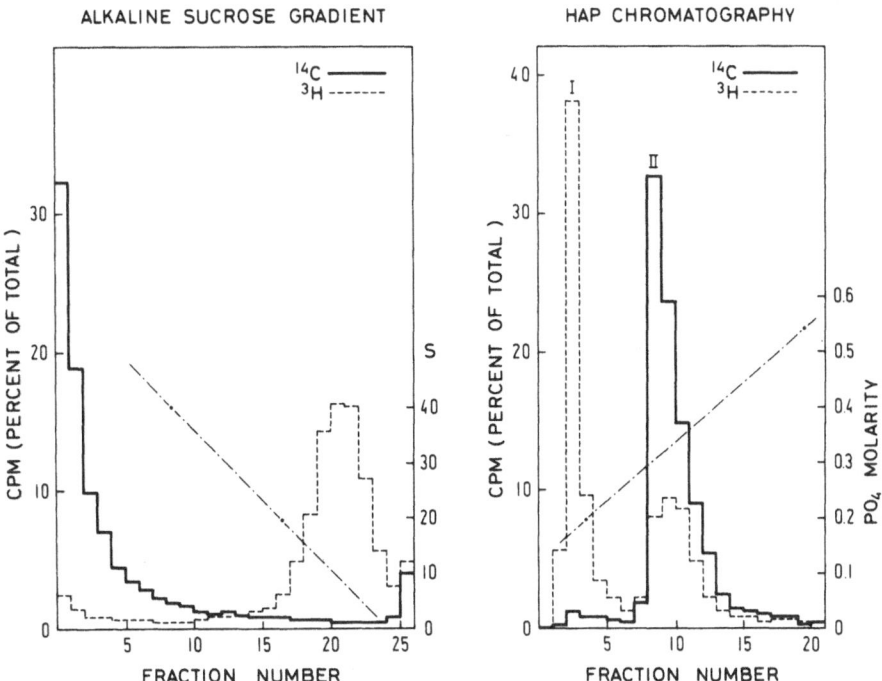

Figure 2. Size distribution in alkaline sucrose gradient and hydroxy-apatite chromatography of DNA labeled during a short pulse at the time of chromosome initiation. A culture of LC 560, uniformly labeled with ^{14}C-thymine in glucose Casamino Acids medium, was synchronized as described in the text and pulse labeled with ^{3}H-thymidine during 15 sec at 30°.
Left: An aliquot of the cells was lysed in alkaline medium (0.2 N NaOH) and the lysate sedimented on an alkaline 5-20% sucrose gradient (0.1 N NaOH, 0.9 M NaCl). Two internal markers were present : λ ^{32}P-DNA and ØX174 ^{32}P-DNA. Their sedimentation constants are 40S and and 16,4S respectively (Studier, 1965).
Right: A second aliquot of the cells was lysed in neutral medium by lysozyme, Sarkosyl, and Proteinase K treatment. DNA was analyzed on an hydroxyapatite column at 60°. Elution was done by a linear Na-PO$_4$ gradient. In these conditions single stranded and double stranded DNA elute at 0.12 and 0.25 M Na-PO$_4$ respectively. Recoveries were always greater than 80%.

The results are shown in Fig. 2. In the alkaline sucrose
gradient essentially all the newly synthesized DNA sediments slowly,
at sedimentation constants ranging from 5 to 20. This is in
contradiction to results obtained with pulses during exponential
growth (Louarn and Bird, 1974) where some newly synthesized DNA
always appeared linked to high molecular weight DNA. It seems, thus,
that at initiation of chromosome replication, none of the newly
synthesized molecules are covalently linked to pre-existing DNA;
because of the short duration of the pulse, these molecules are
still of low molecular weight and sediment relatively slowly.

In the hydroxyapatite chromatography about 60% of the newly
synthesized DNA elutes as single stranded molecules (peak I) while
[14]C-prelabeled DNA is almost entirely double stranded (peak II).
Only about 4% of this DNA elutes in the single stranded peak.
Therefore, neutral fragments constitute a large fraction of the
newly synthesized DNA. Since this experiment shows that the
synchronized culture produces both types of fragments (Okazaki and
neutral fragments), just as an exponentially growing culture does,
we can use this system to study the strand specificity of both types.

In the second experiment, DNA replication was synchronized and
the cells were pulse labeled with ^{3}H-thymidine for 40 sec, instead
of 15 sec, at 30°. The culture was divided in two aliquots which
were treated as described in the preceding experiment.

Fig. 3 (top) shows the size distribution of the newly
synthesized DNA in an alkaline sucrose gradient. Markers (covalently
twisted circles and nicked circles of polyoma ^{14}C-DNA) were run in
parallel. Most of the newly synthesized DNA sediments slowly, with
sedimentation constants between 5 and 30S. About 20% of the newly
made molecules have a sedimentation constant greater than 40S.

The gradient was fractionated in size-classes. Each class was
hybridized with λ DNA, with Mu DNA, and with each of the separated
strands of Mu DNA.

Fig. 3 (bottom) gives the fraction of DNA of each size class
hybridizing with the H strand of Mu DNA. The 5 to 20S fragments
hybridize preferentially with the H strand, the shortest ones
showing the highest specificity. Molecules longer than 20S have
an opposite bias or no bias at all; this opposite bias disappears
if the length of the pulse is increased from 40 sec to 60 sec).
About 8% of each class of labeled DNA hybridizes with Mu DNA; less
than 0.1% hybridizes with λ DNA.

Fig. 4 shows the results of hydroxyapatite chromatography of
the newly synthesized DNA: 27% of the label elutes as single
stranded DNA. These neutral fragments show some strand specificity,

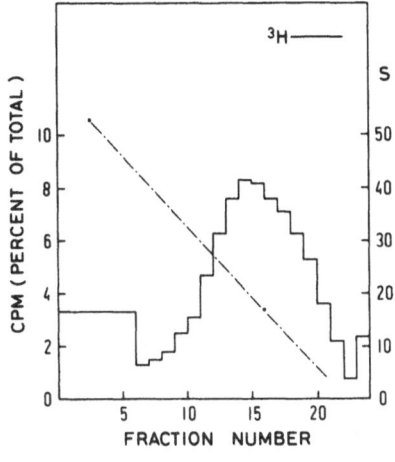

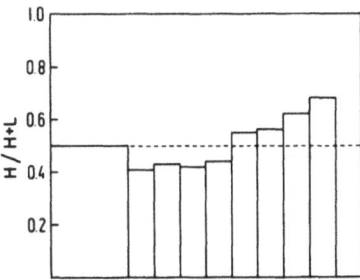

Figure 3. Size distribution and molecular polarity of Okazaki fragments. A culture of LC 560 (not prelabeled with ^{14}C-thymine) was grown and synchronized as described under Fig. 2 and pulse labeled with ^{3}H-thymidine during 40 sec at 30°.

Top: An aliquot of the culture was lysed in alkaline medium and the lysate sedimented as described under Fig. 2. Markers (covalently twisted and nicked circles of polyoma ^{14}C-DNA) were run in parallel. Their sedimentation constants are 53S and 16-18S respectively (Weil and Vinograd, 1963).

Bottom: The alkaline sucrose gradient was subdivided in size-classes. Each of them was hybridized to filters loaded with each of the separated strands of Mu DNA. DNA-DNA hybridizations were done according to the method of Kourilsky et al., (1971). Phage Mu-1 DNA strands were separated by isopycnic centrifugation of denatured DNA in presence of poly (U, G) according to the method of Szybalski et al., (1971). In the diagram, H and L indicate the amount of DNA hybridizing with H and L strand of Mu DNA respectively. All the data, including those concerning filters loaded with Mu or λ DNA, were corrected for background binding (amount of label binding to calf-thymus DNA filters) and for non-specific hybridization (amount of non-lysogenic E.coli DNA hybridizing, under the same conditions, with Mu or λ DNA filters).

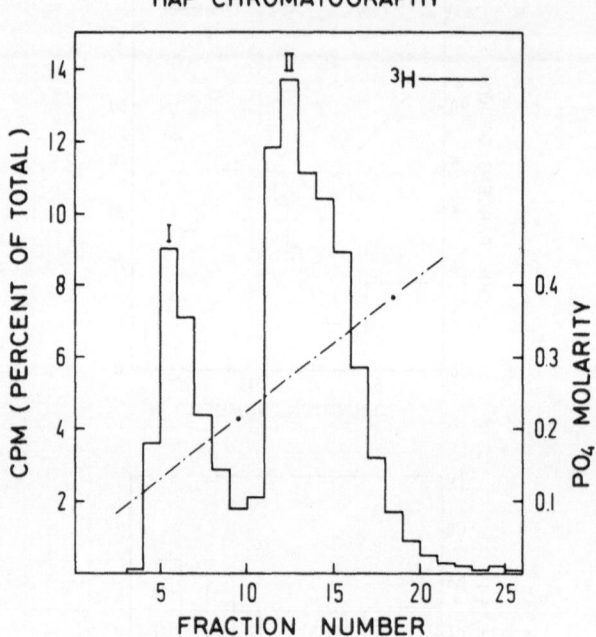

Figure 4. Molecular polarity of neutral fragments.
Top: A second aliquot of the synchronized culture pulse labeled
during 40 sec at 30° was lysed in neutral medium and DNA was
submitted to hydroxyapatite chromatography as described under Fig.
2.
Bottom: Peaks I and II from hydroxyapatite chromatography were
hybridized with filters loaded with each of the separated strands
of Mu DNA. DNA-DNA hybridizations and corrections of the data
were done as described under Fig. 3.

the hybridization with the H strand of Mu DNA being more pronounced than that with the L strand. Again, about 8% of the neutral fragments hybridize with Mu DNA, while no significant hybridization with λ DNA was detected.

DISCUSSION

The DNA-DNA hybridization results show that, at the beginning of chromosomal replication in a synchronized system, no class of fragments hybridizes significantly with prophage λ DNA. It was shown by Louarn et al. (1974) that at 37°, in glucose Casamino Acids medium, this marker is duplicated only after 15 min of synchronous DNA replication. A significant hybridization with λ DNA would have indicated that some newly synthesized DNA could be produced randomly on the chromosome, perhaps as a result of repair. Our results mean that both Okazaki and neutral fragments are preferentially synthesized at or near the replication fork.

About 8% of the pulse labeled DNA hybridize with Mu DNA. LC 560 strain carries a resistance transfer factor (RTF), which replicates mostly unidirectionally and starts replication at the same time as the chromosome under the conditions used (Silver et al., this volume). Chromosomal replication being bidirectional (Bird et al., 1972), we estimate that forks traveling clockwise from the chromosome origin contribute one third to the total DNA synthesis. The maximal efficiency of the DNA-DNA hybridization method used is 50%. We conclude that at least half of the chromosomal DNA synthesized in the clockwise direction during a 40 sec pulse, at 30°, is Mu specific.

Fig. 3 shows that Okazaki fragments, extracted from a culture of *E.coli* whose DNA replication had been synchronized, hybridize preferentially to the H strand of Mu-1. This is in agreement with our results (data not shown) and those of Louarn and Bird (1974) for cultures of *E.coli* growing exponentially. The strand specificities of Okazaki fragments appear to be less pronounced in the synchronized system than in the exponentially growing one. This difference could be due to the treatment used to achieve complete chromosome synchronization.

Fig. 4 shows that neutral fragments produced by a synchronized culture exhibit a slight strand specificity. They hybridize preferentially with the H strand of Mu DNA, as Okazaki fragments do. This, also, is confirmed by our results for exponentially growing cultures (data not shown).

More than 80% of newly synthesized DNA after a 15 sec pulse appear as Okazaki fragments; 60% of this same newly synthesized

DNA appear as neutral fragments.   Evidently, neutral fragments
constitute a significant fraction of Okazaki fragments. Considering
the strand specificity of neutral fragments and the fact that they
represent, in the 40 sec pulse, 27% of the newly synthesized DNA
(Fig. 4), we can estimate their minimal size distribution on the
basis of the data for Okazaki fragments (Fig. 3). The calculations
give a mean size corresponding to a sedimentation constant of
10 ± 2S.   Direct estimation of their size distribution on an
alkaline sucrose gradient gave a mean value of 7 ± 2S (data not
shown).   Thus, neutral fragments appear among the shorter Okazaki
fragments.

In experiments not reported here, we observed that the fraction
of newly synthesized DNA isolated as neutral fragments decreases
progressively as the pulse length is increased from 5 sec to 30 sec.
for a 5 sec pulse, in an exponential culture, 70% of newly
synthesized DNA appeared as neutral fragments; this value decreased
to 55% for a 15 sec pulse and 40% for 30 sec.   In the experiments
reported here the corresponding values are 60% and 27% for 15 and
40 sec pulses,   respectively.   Thus, neutral fragments behave like
precursors in the process of DNA chain elongation. If they were
mainly products of some post-replication repair synthesis we would
expect, in a series of very short pulses, an opposite behavior.

The precursor-like behavior of neutral fragments, their slight
molecular polarity and their relatively low mean size lead us to
believe that neutral fragments are intermediates in DNA chain
elongation in both the 3' → 5' and in the 5' → 3' directions of
replication.

Finally, our data give a rough estimate of the number of
Okazaki and neutral fragments per replication fork. The time needed
to replicate the chromosome at 30° is estimated at 65 min (L. Silver
and M. Chandler, unpublished data). It follows that during a 40 sec
pulse approximately 11 μm of single stranded DNA will be synthesized
per replication fork (or 17 μm during a 60 sec pulse). From the
sedimentation profile we can estimate that this new DNA will be
distributed as follows: eleven 2 S pieces (0.06μm), three 9 S pieces
(0.4μm), 1.3 14 S pieces (1.3μm), and 0.7 20 S pieces (3μm) for the
40 seconds pulse. The contribution of higher molecular weight
material is very small. At 60 sec, the same distribution is
obtained with the difference that there is considerably more material
at molecular weights higher than 20 S. Approximately 80% of each of
these classes is constituted by neutral fragments.

In summary, we conclude that the neutral fragments constitute
a subclass of Okazaki fragments, comparable to them in size, that
they are produced at the replication fork in both directions of
replication with, however, a slight preference for the 3' → 5'
direction.   They thus behave like intermediates in DNA replication
in both directions.

ACKNOWLEDGEMENTS

We thank Dr. R.E. Bird, Mrs. Grete Kellenberger-Gujer and Dr. J.-M. Louarn for helpful advice and discussions. This work was supported by grant no. 3.679.75 from the Swiss National Science Foundation.

REFERENCES

Bird, R.E., J. Louarn, J. Martuscelli and L. Caro (1972) J. Mol. Biol. 70:549-566

Bird, R.E., M. Chandler and L. Caro (1976) J. Bacteriol. 126: 1215-1223

Chandler, M., R.E. Bird and L. Caro (1975) J. Mol. Biol. 94: 127-132

Chandler, M., B. Allet, E. Gallay, E. Boy de la Tour and L. Caro (1977) Molec. Gen. Genet. 153:289-295

Habener, J.F., B.S. Bynum and J. Shack (1970) J. Mol. Biol. 49: 157-170

Kourilsky, Ph., J. Leidner and G.Y. Tremblay (1971) Biochimie 53:111-114

Louarn, J.-M., and R.E. Bird (1974) Proc. Nat. Acad. Sci. USA 71:329-333

Louarn, J.-M., Funderburgh and R.E. Bird (1974) J. Bacteriol. 120:1-5

Oishi, M. (1968a) Proc. Nat. Acad. Sci. USA 60:329-336

Oishi, M. (1968b) Proc. Nat. Acad. Sci. USA 60:1000-1006

Okazaki, R., T. Okazaki, K. Sakabe, K. Sugimoto and A. Sugino (1968) Proc. Nat. Acad. Sci. USA 59:598-605

Okazaki, R., A. Sugino, S. Hirose, T. Okazaki, Y. Imae, R. Kainuma-Kuroda, T. Ogawa, M. Arisawa and Y. Kurosawa (1973) In: DNA Synthesis in Vitro. (eds. R.D. Wells and R.B. Inman) University Park Press, Baltimore, Md. pp 83-106

Paetkau, V., L. Langman and R.C. Miller, Jr. (1975) J. Mol. Biol. 98:719-737

Sakabe, K. and R. Okazaki (1966)   Biochim. Biophys. Acta   129:
    651-654

Studier, F.W. (1965)   J. Mol. Biol.   11:373-390

Szybalski, W., H. Kubinski, Z. Hradecna and W.C. Summers (1971)
    In: Methods in Enzymology, Vol. XXI D (eds. L. Grossmann and
    K. Moldave)   Academic Press, New York-London, pp383-413

Weil, R. and J. Vinograd (1963)   Proc. Nat. Acad. Sci. USA   50:
    731-738

SIZE DISTRIBUTION OF SHORT CHAIN DNA IN TWO STRAINS OF

ESCHERICHIA COLI

Margaret L.M. Anderson

Imperial Cancer Research Fund
Mill Hill Laboratories
Burtonhole Lane, London, NW7 1AD, England

INTRODUCTION

It is now generally accepted that DNA replication in *E. coli* proceeds by a discontinuous mechanism, i.e., nascent DNA on one, if not on both, sides of the replication fork is synthesized in short stretches called Okazaki pieces which are subsequently sealed together to form large molecular weight DNA. The work reported here was undertaken in order to determine the size distribution and number of Okazaki pieces per cell in *E. coli* growing under steady state conditions.

In order to detect Okazaki pieces, it is necessary to label them, and this usually involves pulse-labelling thymine auxotrophs with radioactive thymine or thymidine. Problems arise because the distribution of label in different size classes of DNA after such pulses seems to depend on how the pulse labelling is carried out. Thymidine tends to be the label of choice because it is incorporated into DNA with little or no lag. However it has been shown that there are profound changes in the intracellular thymine nucleotide pools during thymidine pulses (Werner 1971a), so the cells are not under steady state conditions. During a pulse with radioactive thymine, there is a lag in the appearance of label in DNA which is caused by thymine mixing with the large intracellular nucleotide precursor pools. Label is initially incorporated into DNA at low but continuously changing specific activity. It is common practice, therefore, to precede a thymine pulse by starving auxotrophs for thymine in order to empty precursor pools. These cells are clearly not under steady state conditions. In general, more short chain DNA molecules are seen after a thymidine pulse than a thymine pulse (Werner, 1971b), and

where thymine starvation is used before labelling, very many short
DNA chains are observed (Dingman, Fisher and Ishizawa, 1974; Diaz
et al., 1975; Brewin, 1977). The steady state distribution of
chain lengths of DNA has been determined in unlabelled cells by
labelling 5' ends *in vitro* (Jacobson and Lark, 1973). However
with this procedure it is difficult to know what proportion of
molecules are replication intermediates. Ogawa et al. (1977) have
measured the number of RNA-terminated DNA chains in several strains
of *E. coli*. But here only molecules smaller than 1800 nucleotides
were examined and it is not known what fraction of the total
nascent molecules these formed.

In the present study it was decided to determine the size
distribution of DNA which had been uniformly labelled by growing
thymine auxotrophs for several generations in medium containing
radioactive thymine. Analysis of how this distribution changed
when cells were transferred to unlabelled medium should show which
pieces were, in a general sense, replication intermediates.

The commonly used technique of alkaline sucrose gradient
centrifugation is not suitable for analysing the size distribution
of short molecules of uniformly labelled DNA. This is because only
about 1% of the total DNA is in the form of short chains (Cairns,
unpublished results) and there is always contamination of small
molecular weight species, by the excess of faster sedimenting,
higher molecular weight DNA. Instead, a technique must be used
whereby small molecules move faster than large so that the former
can be separated free from larger molecules even though the latter
constitute the bulk of the DNA. In this work separation of DNA
according to size was achieved by gel electrophoresis.

Gel electrophoresis of DNA commonly requires that the DNA be
first purified from other cell constituents which interfere with
DNA migration. The disadvantage of purifying DNA is that recovery
is never complete and there is a danger that losses may not be
random with respect to size class. Further, the more the DNA is
handled the higher is the probability that the DNA will be broken.
In the present work, a denaturing gel system was used which had
been developed specifically to allow separation of DNA from
unfractionated cell lysates. This ensured that the recovery of
DNA from cells was maximal while the breakage was minimal.

METHODS

The procedure used to label cells, prepare lysates and run
alkaline gels is described in detail in another paper (Anderson,
submitted for publication). Briefly, labelled cells were killed
in a 1% phenol, 37% ethanol-containing buffer, pH 7.9, harvested
and washed carefully to remove unincorporated radioactive thymine.

Lysates were prepared in low ionic strength buffer by treating
cells first with lysozyme, then with pronase and sarkosyl. DNA
was denatured by treatment with alkali. The ionic strength of
lysates was reduced by diluting with 5% sucrose containing the
dye bromophenol blue. Lysates were then applied to wells in
cylindrical, 19 cm long 1% agarose gels at pH 12.7. Electrophoresis was carried out until the dye had migrated about half
way down the tube. Gels were sliced with stacked razor blades
and the radioactivity in slices was determined in an alkaline
scintillant.

Calibration gels containing standard DNAs and unlabelled
cell extract were run in parallel in every experiment. The
relationship between electrophoretic mobility and molecular
weight of DNA is shown in Fig. 1. Under the conditions used,
resolution of DNA molecules with chain lengths greater than 9000
nucleotides is poor. Standards shorter than 700 nucleotides
were not used, so all calculations are restricted to the range
700 - 9000 nucleotides.

Two strains of E.coli were used: W3110 a wild type, low
thymine requirer and p3478, a *pol*A1 derivative of W3110. Here,
p3478 is referred to as *pol*A1.

RESULTS

Fig. 2(a) compares the distribution of counts observed on gels
for *pol*A1 cells uniformly labelled with ^{3}H-thymine for 5
generations with that of *pol*A1 cells pulse labelled with ^{3}H-
thymidine for 15 sec. For uniformly labelled DNA, 97.69% of the
DNA had a size greater than 9000 nucleotides, 2.25% fell in the
range 700 - 9000 nucleotides while 0.06% was smaller than 700
nucleotides. In contrast, for pulse labelled DNA, the corresponding
values were 24.7%, 73.83% and 2.43%. Fig. 2(b) shows the lower
molecular weight end of the distribution of counts after uniform
labelling. The proportion of total counts is very low in this
region of the gel, but the counts are significant since over 1.3
x 10^6 counts per min were applied to the gel.

Although a rough estimate of the size range of nascent DNA
can be obtained from the gel of pulse labelled DNA, a conversion
of counts per minute into absolute quantity of DNA cannot be made
because the specific activity of the DNA in each size class is not
known.

Absolute values can be obtained for the uniformly labelled
DNA. The total number of nucleotides in DNA per cell under the
conditions of growth used here is 1.5 x 10^7 (Anderson, submitted
for publication). Since the percentage of total counts in each gel

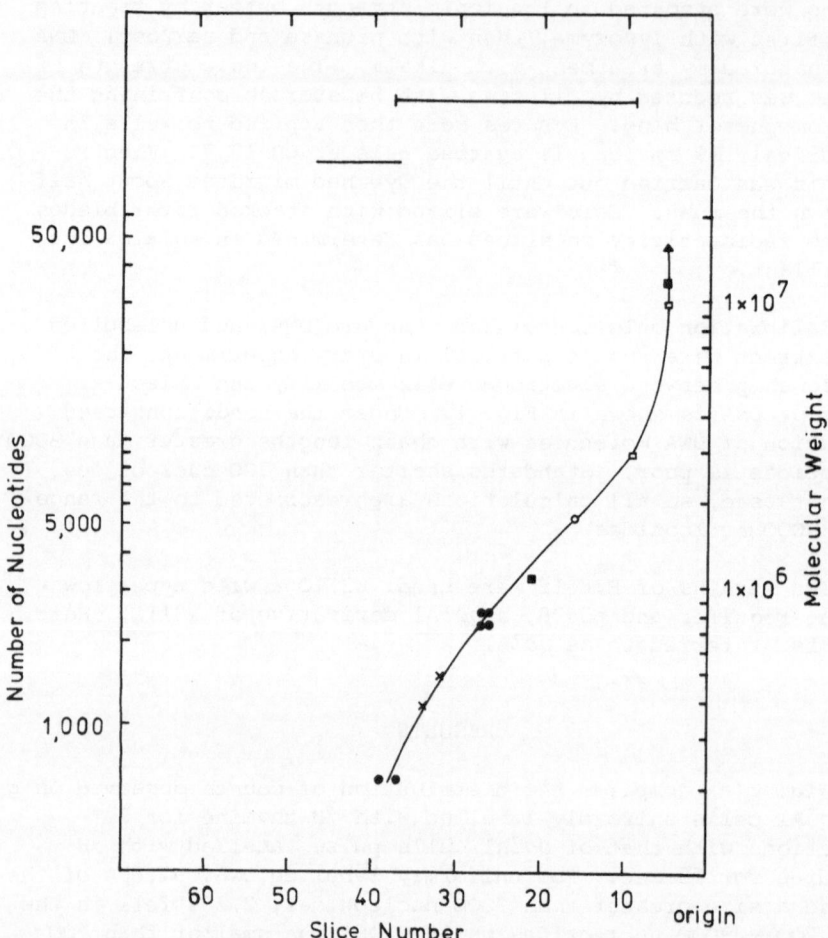

Figure 1. *Relative electrophoretic mobility of DNA standards on alkaline agarose gels. Radioactive standard DNAs were electrophoresed on gels at 20v in the presence of unlabelled lysate from 3 x 10⁸ cells. The position of most of the markers shown here came from one gel but, some additional markers have been added from other gels by expressing their position relative to EcoR1 digested polyoma DNA, which was included in every gel. The following denatured DNA sizes were used:* ▲, *λDNA, 15.4 x 10⁶ daltons;* ■, *EcoR1 digested λ variant X DNA (Murray and Murray (1974), 11.73 x 10⁶, 1.06 x 10⁶ daltons;* □, *EcoR1 digested λ variant IX DNA, 9.86 x 10⁶, 2.93 x 10⁶ daltons;* ○, *EcoR1 digested polyoma DNA, 1.75 x 10⁶ daltons;* ●, *HhaI digested polyoma DNA, 8.05 x 10⁵, 7.35 x 10⁵, 2.1 x 10⁵ daltons;* ✗, *HpaII digested polyoma DNA, 4.78 x 10⁵, 3.74 x 10⁵ daltons. The long horizontal bar denotes the region covered by chain lengths 700-9000 nucleotides. The short bar denotes the position of the tracking dye.*

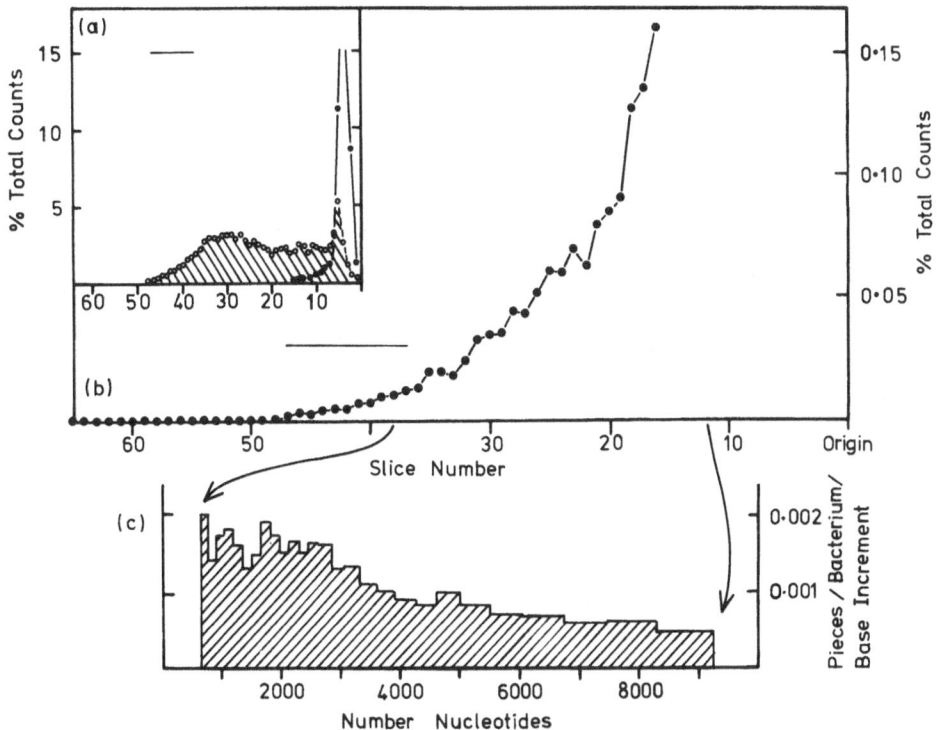

*Figure 2. Gel electrophoresis of labelled DNA from E.coli pol A1.
(a) The distribution of label following 15 sec exposure to
³H-thymidine (cross hatched area) compared to the distribution after
uniform labelling with ³H-thymine observed in a separate gel. Total
cts/min after pulse labelling, 21,000.   (b) The low molecular
weight end of the distribution after uniform labelling with ³H-
thymine for 5 generations. Total cts/min 1,390,000.   (c) The same
distribution as in (b), expressed as pieces per bacterium for each
base increment in size from 700-9000 bases.*

slice is known, the equivalent number of nucleotides in DNA can be
easily calculated. The size range of DNA falling into each gel
slice is known by referring to the calibration gels run in
parallel. The number of molecules of DNA in any gel slice is
obtained by dividing the total number of nucleotides present by the
average chain length of DNA in that slice.

By summing the numbers obtained across the range 700-9000
nucleotides (slices 12-38 in Fig. 2) a value of 80 molecules per

cell was obtained. A distribution of lengths is shown in Fig. 2(c).
It takes into account the fact that gel slices do not cover an equal
increment of nucleotides. Results are expressed as number of
molecules of DNA per bacterium per unit increment in length (i.e.,
per nucleotide). It is clear that not all sizes are equally
common and that there is an excess of small pieces.

In order to determine which molecules were chaseable, the
change in distribution of label was studied when cells were
transferred to medium lacking radioactive label. Fig. 3 shows a
comparison of the distribution of counts in *pol*A1 and W3110 before
and after a 10 min chase.

For *pol*A1, the percentage of counts in the range 700-9000
nucleotides fell from 1.9% to 1.2%. This represented a drop from
about 70 molecules to 50. For W3110, Fig. 3(b), a similar chase
produced a drop from 0.66% to 0.58% of the total counts and this
was equivalent to a fall from 20 to 17 molecules per cell. For
both *pol*A1 and W3110 *pol*$^+$ repeated experiments gave qualitatively
similar results although the absolute number of molecules varied
between 65 and 85 for *pol*A1 and between 20 and 40 for *pol*$^+$. This
will be discussed later.

In the case of *pol*$^+$ the 10 min chase was complete. No further
reduction in the number of short chain molecules was observed when
the time of chasing was prolonged to 90 min. The same result was
obtained whether the chasing was carried out by filtration and
suspension of cells in unlabelled medium, or by adding a twenty
fold excess of unlabelled thymine or by diluting the cells twenty
fold into unlabelled conditioned medium. Similarly, unchaseable
pieces persisted even when cells were grown into stationary phase
when it was expected that the amount of ongoing DNA synthesis
would be minimal.

These results suggest that in W3110 the total number of
cheaseable DNA molecules per cell is very small. The value of 3
obtained from the experiment shown in Fig. 3(b) represents the sum
of fractions of molecules spread over 33 gel slices. Hence this
estimate is not very reliable. The correct value probably lies
between 3 and 6.

The results of the experiments in Fig. 3 suggest that *pol*A1
continues more short chain molecules of DNA per cell, than W3110.
This was confirmed by running on the same gel a lysate from a
mixture of ^{14}C-labelled *pol*A1 cells and ^{3}H-labelled W3110 cells
(Fig. 4). The same result was obtained when the labels in *pol*A1
and W3110 were reversed. On chasing, the *pol*A1 profile slowly
approached that of W3110 until by 60 min chasing, the two were
superimposable. This suggests that sealing of new molecules of DNA
occurs rather slowly in *pol*A1 and occurs over a large stretch of the
chromosome.

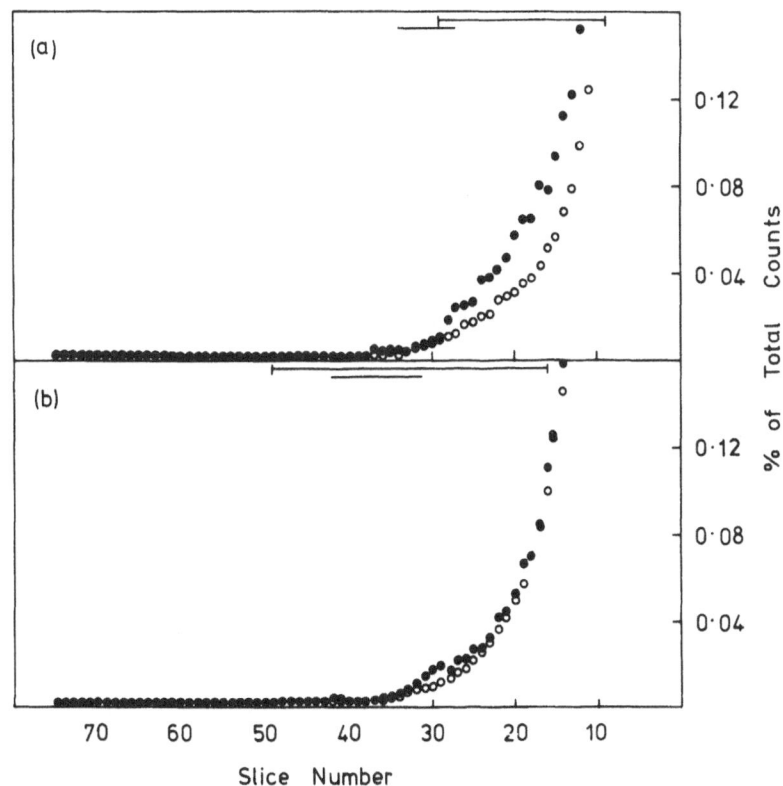

Figure 3. *Electrophoresis of uniformly labelled and chased DNA on alkaline gels. Cells uniformly labelled with ^{14}C-thymine were killed at a cell count of 3 x 10^8 cells/ml. Cells uniformly labelled with ^3H-thymine to a cell count of 3 x 10^8 cells/ml were collected on Millipore filters, washed and resuspended in conditioned medium. The chase in conditioned medium was carried out for 10 min and the cells were killed. Killed cells were mixed to give three times more ^3H cpm than ^{14}C cpm. Preparation of lysates and electrophoresis was by the standard procedure. The short horizontal bar shows the position of the tracking dye. The long horizontal bar shows the region covered by chain lengths of 700-9000 bases.* ● ● ● ^{14}C; ○ ○ ○ ^3H. *(a)* E.coli polA1, *^3H-432,700 cts/min; ^{14}C-161,600 cts/min; gel run for 17 hr at 20v. (b)* W3110 pol$^+$, *^3H-641,700 cts/min; ^{14}C-237,146 cts/min; gel run for 19.5 hr at 25v.*

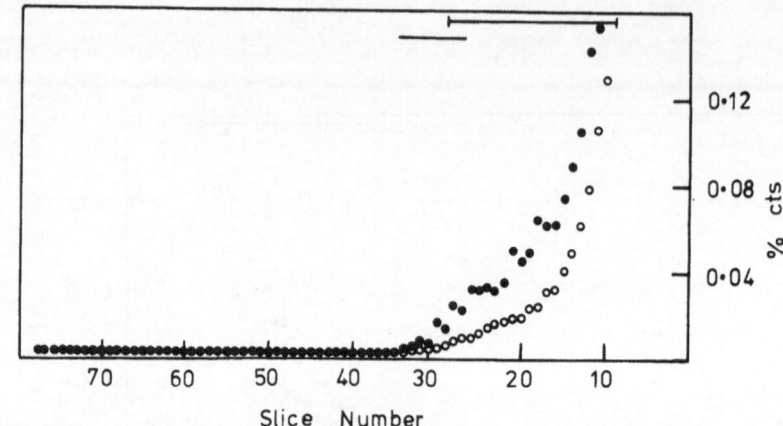

Figure 4. Electrophoresis on the same gel of DNA from E.coli polA1 uniformly labelled ^{14}C-thymine, and W3110 uniformly labelled with ^{3}H-thymine. ● ● ● ^{14}C; ○ ○ ○ ^{3}H. ^{14}C-142,915 *cts/min.* ^{3}H- 741,553 *cts/min.*

DISCUSSION

The steady state size distribution and number of molecules of short chain DNA has been determined in two strains of *E.coli*. The approach used was designed primarily to avoid problems of transient changes in the levels of nucleotide precursor pools and hence changes in the specific activity of radioactive nucleotides being incorporated into DNA, which accompany the usual method of pulse labelling DNA. In practice, DNA was uniformly labelled by growing cells for several generations in medium containing radio- active thymine and the size of the DNA was measured by electro- phoresis of unfractionated lysates on alkaline agarose gels. Chaseable molecules of DNA were determined by observing how the size distribution of DNA changed when cells were transferred to unlabelled medium.

Repeated analysis of a given cell lysate showed some variation between one gel and another and therefore, there is some error in aligning gel slices from assay and calibration gels. This means that there is some uncertainty about the absolute number of pieces per cell. For pol^{+} values found varied between 20 and 40 while for *pol*A1 the equivalent values were 65 and 85.

For *E.coli* W3110 about 20-40 molecules of DNA with a size range

between 700 and 9000 nucleotides were found per cell. The number
of chaseable molecules is so small that it is difficult to give a
reliable figure; the true value probably lies between 3 and 6.
Most of the short chain molecules did not chase under any of the
conditions tested. The size distribution of these unchaseable
molecules is very close to that expected for a random distribution
of breaks throughout the chromosome. If the probability of there
being a break at each internucleotide link has a constant value, $\underline{k}$,
then the average length of the chains will be $1/k$ and the proportion
of all chains (P_c) and of the total DNA mass (P_m), present in
chains with length between L_1 and L_2, are given by the following
equations:

$$P_c = e^{-kL_1} - e^{-kL_2} \qquad (1)$$

$$P_m = (1 + kL_1)\, e^{-kL_1} - (1 + kL_2)\, e^{-kL_2} \qquad (2)$$

The tail end of the experiment distribution for Fig. 3 and
the theoretical random distribution calculated according to
equation (2) are shown in Fig. 5(a). It can be seen that the
distribution of label in unchaseable DNA molecules is very close to
the theoretical distribution for random breakpoints distributed
at an average of somewhere between 100,000 and 66,000 nucleotides
(i.e., $\underline{k}$, lying between 1×10^{-5} and 1.5×10^{-5}).

Given such a distribution, it can be calculated according to
equation (1) that there should be a total of between 12 and 26
molecules in the size range 700-9000 nucleotides. This is
compatible with the value of 17 found experimentally.

The breaks giving rise to this distribution may exist as
such in the chromosome. Alternatively, unchaseable pieces might be
created by the action of alkali on sensitive sites such as those
created by depurination and depyrimidation. It is unlikely that
they arise through radiation damage because exactly the same
distribution of label is found when cells have been grown in the
presence of high specific activity ^{3}H-thymine and in the presence
of low specific activity ^{14}C-thymine (data not shown). Jacobson
et al. (1973) also found unchaseable pieces in cells by using the
method of end labelling DNA *in vitro* and they suggested that they
might be located at sites remote from the replication fork.

The distribution of short chain DNA molecules in *polAl* is
clearly not random even after a 10 min chase (Fig. 5b). There are
too many short molecules. However, by 60 min the distribution of
counts for *polAl* is superimposable on that of W3110. For unchased
cells, *polAl* contains roughly twice as many molecules in the range

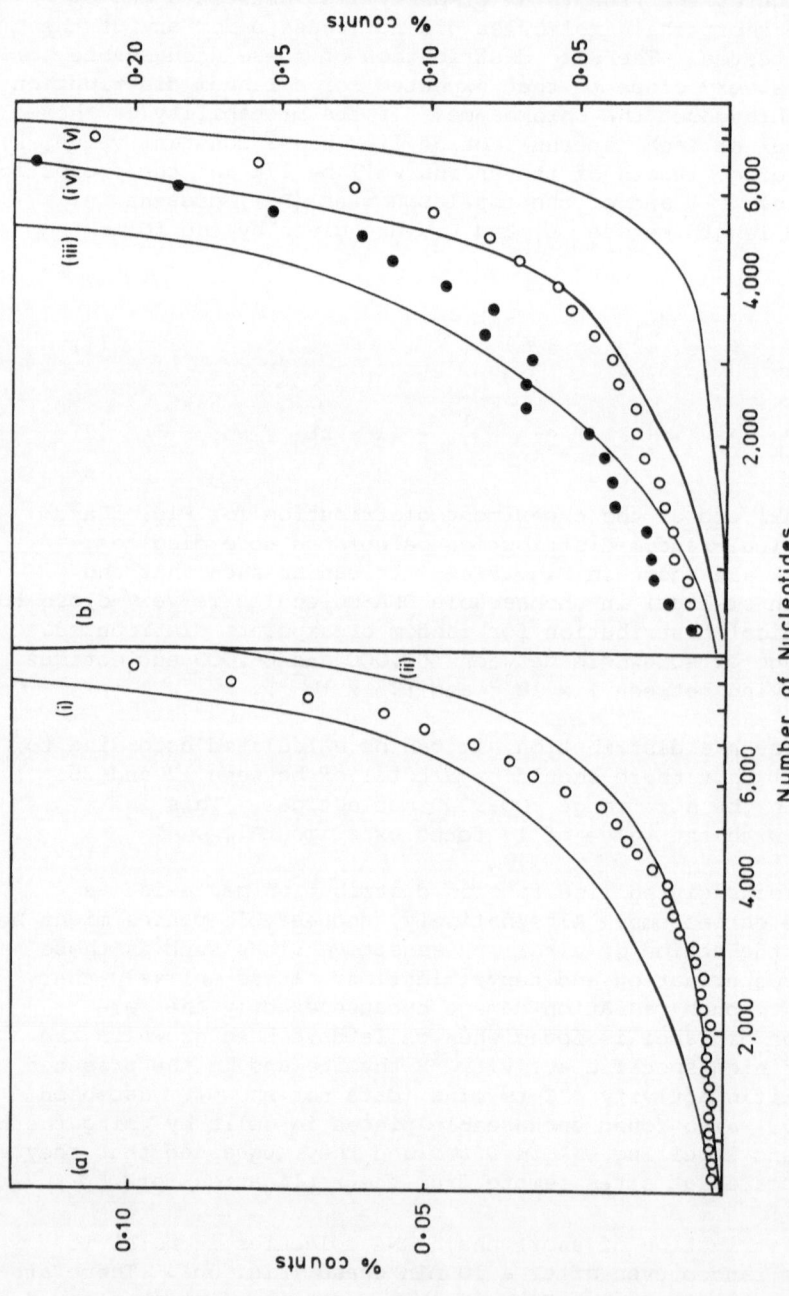

Figure 5. Comparison of the observed distribution of label (a) in W3110 *pol⁺* after a 10 min chase and (b) in *polA1* before and after a chase (taken from Fig. 3) with the expectation for a random distribution of breaks, given the following rate constants for breakage: (i) 1.5×10^{-5}, (ii) 1×10^{-5}, (iii) 3×10^{-5}, (iv) 2×10^{-5} and (v) 1×10^{-5} base^{-1}.

700-9000 nucleotides as W3110 (Fig. 4). However, this excess stretches right back to the fourth gel slice which is equivalent to a size of 45000 nucleotides. This is what would be expected if newly synthesized DNA in *pol*A1 contains breaks which are only slowly sealed such that the process of sealing is spread over a rather long stretch of the chromosome. A detailed set of chase experiments should give the length of this stretch and should show if sealing occurs in a random or programmed manner.

ACKNOWLEDGEMENTS

I should like to acknowledge the help and support of J. Cairns. I am grateful to C. Smith and M. Legate for technical assistance.

REFERENCES

Brewin, N.J. (1977) J. Mol. Biol. 111:343-352

Dingman, C.W., Fisher, M.P. and Ishizawa, M. (1974) J. Mol. Biol. 84:275-295

Diaz, A.T., Weiner, D. and Werner, R. (1975) J. Mol. Biol. 95:45-61

Jacobson, M.K. and Lark, K.G. (1973) J. Mol. Biol. 73:371-396

Murray, N.E. and Murray, K. (1974) Nature 251:476-481

Ogawa, T., Hirose, S., Okazaki, T. and Okazaki, R. (1977) J. Mol. Biol. 112:121-141

Werner, R. (1971a) Nature New Biol. 233:99-103

Werner, R. (1971b) Nature 230:570-572

EVIDENCE FOR THE ABSENCE OF TRIPHOSPHATE TERMINI FROM THE 5' ENDS

OF NEWLY SYNTHESIZED *E. COLI* DNA MOLECULES

David T. Denhardt

Department of Biochemistry
McIntrye Medical Sciences Building
3655 Drummond Street
McGill University, Montreal, Quebec, Canada

INTRODUCTION

A picture of how *E. coli* DNA replicates is slowly coming into
focus. There appears to be a specific origin of replication at
around 82 minutes on the genetic map. From this origin replication
is initiated once every generation and proceeds in both directions
around the circular *E. coli* DNA; termination is presumed to occur
when the two "growing forks" meet. The term "growing fork" is used
to refer to that region of a replicating DNA molecule ahead of
which the DNA has not been replicated and behind which the DNA has
been replicated. A variety of proteins required for DNA replication
have been identified by virtue of mutations in the genes coding for
them (Wechsler, this volume), and several of these proteins have
been highly purified and some of their enzymatic activities
elucidated (Kornberg, this volume; Wickner, ibid). The bacterial
chromosome is a circular double-stranded DNA molecule in a superhelical state (Wang, this volume) most likely in association with
one or more "histone-like" proteins. During replication the two
parental DNA strands must be unwound and separated, as it is known
that DNA replication is semi-conservative. This requires repeated
"nicking" and repair of the parental strands.

It is well established that DNA polymerase III (product of the
dnaE gene) is the replicative polymerase. This implies that one of
the newly synthesized strands (that whose 5' → 3' polarity is
contrary to the direction of growing fork progression) is synthesized in a discontinuous fashion. This leads inexorably to a
requirement for repeated initiations of synthesis of DNA molecules
in addition to the primary initiation event that occurs at the

321

origin (Okazaki et al. 1968). Although the complementary strand, which is synthesized in the 5' → 3' direction as the growing fork moves, could in principle be synthesized continuously, the weight of the evidence favors, in my opinion, discontinuous synthesis of this strand also. It does, however, appear to be synthesized either in pieces of longer length or in pieces that are more rapidly joined together, in contrast to its complement.

One question under investigation in my laboratory for the past few years is how synthesis of these intermediate chains, or Okazaki fragments, is initiated. Some years ago I theorized that a protein, a π protein, allowed synthesis to initiate by positioning a deoxynucleotide appropriately so that the 3' hydroxyl was available to a DNA polymerase (Denhardt, 1972). At about the same time, evidence was obtained suggesting that what actually occurred was the synthesis of an RNA oligo-nucleotide (Brutlag et al., 1971; Sugino et al., 1972); this ribonucleotide polymer was then in turn used as a primer by the appropriate DNA polymerase. Except for a few minor problems this is an attractive mechanism in so far as it requires no new concepts or unprecedented protein functions. One problem is that the RNA must be removed after it has served its priming function, and in terms of maximizing the metabolic efficiency of the cell this could be wasteful. [It has been argued, in contrast, that this is useful because it might reduce the probability of an error (D. Botstein, cited in Dressler, 1975)]. A second possible problem is that the RNA/DNA duplex may be in a different conformation from the DNA/DNA duplex, and it is conceivable that the DNA polymerase adapted to use the RNA primer on a DNA template may not work as well on a DNA-primed DNA template. This consideration has been discussed elsewhere (Denhardt et al., 1975; Denhardt, 1977). However these speculations are not sufficient reason to dismiss RNA priming, especially since it is known to be possible *in vitro* and there is evidence that it occurs *in vivo*.

The *in vitro* work has been discussed by others elsewhere in this volume, and it will not be considered further here. The most recent *in vivo* work on *E. coli* is that published by Ogawa et al. (1977) and by Miyamoto and Denhardt (1977). Both of these reports concur that DNA molecules terminated with one or a few 5' terminal ribonucleotides can be identified in *E. coli polAex* strains (see Lehman, this volume, for discussion of the properties of these strains). The evidence seems good therefore that some DNA molecules in *E. coli polAex* are initiated with an RNA, or at least ribonucleotide, primer. Ogawa et al. (1977) reported evidence for RNA-terminated DNA molecules in other strains but at a somewhat lower frequency. Miyamoto and Denhardt (1977) were unable to detect a significant number (i.e., amounts that could be shown not to result from a small amount of RNA contamination) of DNA molecules with 5' terminal ribonucleotides in "wild-type" strains.

A logical interpretation of the story thus far is that all DNA molecules are initiated *in vivo* with an RNA primer and that the RNA primers are rapidly removed by the 5' → 3' exonuclease activity of *E. coli* DNA polymerase I. This is indeed a widely accepted view. There are, however, several disturbing facts - facts which I have been unable so far to reconcile with this view. One fact is that Miyamoto and Denhardt (1977) used a method for isolating DNA (bringing the culture to 100°C very quickly in the presence of 0.7% sodium dodecyl sulphate and 1% phenol) that should have allowed at least some of the nascent DNA molecules in wild-type strains to retain their RNA. A second fact is that the proportion of the molecules with 5' terminal ribonucleotides in *E. coli*Δex2 was greater in the longer molecules than in the shorter molecules (Miyamoto and Denhardt, 1977). If RNA priming were the universal mechanism, followed by removal of the RNA at some constant rate, independent of the size of the DNA, then the naive expectation is that the shorter the molecule the more likely it will be to possess a 5' terminal ribonucleotide. But this was not what was found. Nevertheless, these facts could be reconciled with RNA priming if the shortest molecules were not true intermediates in DNA replication; for instance, they could be products of a repair process, or even of abortive replication events in the cell. However, as of this writing the evidence favors the involvement of these shortest DNA molecules (about 20 nucleotides long, the "class C" DNA of Miyamoto and Denhardt, 1977) as intermediates in DNA replication, but the case is not yet firm. Anderson (this volume) has discussed the fact that there are a number of short DNA molecules in *E. coli* that do not chase very rapidly into high molecular weight DNA.

Because the arguments in favor of universal RNA priming are still weak, it is worth considering other alternatives, and it is one of these alternatives that I wish to develop here. Although DNA polymerase III appears *in vitro* to be incapable of, or at least very inefficient at, initiating synthesis of a DNA molecule *de novo*, it cannot be excluded that *in vivo* in collaboration with other proteins *de novo* initiation occurs. If initiation occurred with a deoxynucleoside 5'-triphosphate then nascent DNA chains should possess, at least briefly, a 5' terminal triphosphate. One might also expect that any DNA chain begun with an RNA primer would retain transiently a ribonucleoside triphosphate terminus. The experiments described below were designed to detect a very small number of 5' triphosphate termini on DNA molecules.

METHODS

An exponentially growing culture of cells (e.g., *E. coli pol*Δex *rep*) was labeled with ^{32}P (20 mCi carrier-free phosphate) for

several minutes at 32-34° in 400 ml of a medium (TPG) containing
low phosphate (0.17 mM KH$_2$PO$_4$). Note that this is the permissive
temperature for this strain;[4] the exonuclease is still impaired
however (Hours and Denhardt, unpublished experiments). Labeling was
terminated and the DNA purified as described by Miyamoto and
Denhardt, 1977. Since the details are published I will simply
summarize the highlights here.

In order to inactivate cellular degradative enzymes rapidly,
the culture of cells is quickly combined with a half-volume of a
solution of SDS (2%) and phenol (3%) at 100°. The DNA is denatured and small molecules (less than 10^7 daltons) released from
the cell. High molecular weight DNA (presumably greater than 10^7
daltons) remains trapped in the cell wall macromolecular sacs and is
removed by a low speed centrifugation. The small DNA molecules in
the supernatant are purified by extraction with phenol-chloroform,
precipitation with isopropanol, extraction with methoxyethanol,
precipitation with cetyltrimethyl ammonium bromide, (CTAB)
chromatography on nitrocellulose (twice), and finally equilibrium
banding in a cesium chloride density gradient. The methoxyethanol-
CTAB step removes contaminating carbohydrate and polysaccharide
materials as well as residual denatured proteins. The nitrocellulose chromatography separates out the RNA since DNA, but not RNA,
adsorbs to nitrocellulose in 0.5 M potassium chloride. The DNA is
then eluted by a low salt buffer.

In the work described here the DNA was labeled for 5 minutes
with ^{32}P. This represents about 10% of the generation time. This
long labeling period was used to insure that the phosphate in the
pool of nucleotides in the cell was reasonably well equilibrated.
The (^{32}P)DNA was concentrated by alcohol precipitation after the
CsCl banding and dissolved in 20-40 μl of H$_2$O. Roughly 10^7 cpm and
10-20 μg of DNA were recovered from 400 ml of cells at about 4 x 10^8
per ml. This specific activity was approximately that expected if
most of the DNA recovered was labeled with the specific activity of
the medium. Brief heating at 100° at several stages during the
purification was performed to facilitate dissolving of the DNA
after alcohol precipitation and to minimize contamination with
free RNA molecules resulting from the formation of RNA-DNA
complexes.

This procedure yields a reasonably pure preparation of DNA,
and we estimate the recovery of small DNA molecules to be somewhat
less than 25%. It is representative of about 10-20% of the mass of
DNA in the cell. The remaining 80-90% of the high molecular weight
cellular DNA is removed with the cell wall ghosts in the first
centrifugation. If the cells are labeled for only a few seconds
with (^{3}H)thymidine, then most, if not all, (the quantitation is
difficult) of the radioactivity is found in the population of short

DNA molecules. However, it appears that much of the DNA in the preparation existed in the cell for at least some minutes before extraction. In other words, in the preparation of molecules under investigation only a fraction can be considered truly nascent. This aspect will be reported in detail elsewhere.

Purified snake venom phosphodiesterase (Worthington VPH, 3.5 mg/ml, treated according to Sulkowski and Laskowski, 1971, to inhibit nucleotidase) and DNase I (2 mg/ml, DNase A in 50% glycerol stored at -20°) were generously provided by Dr. J.H. Spencer. Nuclease digestions were performed for 2-3 hours at 36°C in 50 mM Tris-HCl, pH 9, and 5 mM $MgCl_2$. The DNA concentration was about 100 µg/ml, and the phosphodiesterase and deoxyribonuclease were respectively 0.7 and 0.2 mg/ml in a 10 µl reaction volume.

Electrophoresis of the entire digest on 3 x 47.5 cm Cellogel strips (obtained from Mandel Scientific Co., Montreal) was performed as described by Harbers et al. (1976) except that the voltage was 2-2.5 kilovolts and the pH of the buffer was adjusted to 3.5 with pyridine. A mixture of dyes (xylene cyanol FF, acid fuchsin, and methyl orange) was used to provide markers. Electrophoresis was continued until the leading pink dye had moved to within 5-10 cm of the end of the strip. The strip was then cut into 1 cm pieces and the radioactivity in each piece determined in a liquid scintillation counter with a triton-toluene-fluor cocktail.

RESULTS

Evidence for a 5' triphosphate terminus was sought from the presence of a pyrophosphate residue that would be liberated from such a terminus during hydrolysis by snake venom phosphodiesterase. Pyrophosphate can be efficiently detected by high voltage electrophoresis on Cellogel strips in a pH 3.5 buffer. Two advantages of this methodology are that all triphosphate termini contribute to the pyrophosphate peak (thus enhancing the sensitivity) and that the mononucleotides, which constitute over 99% of the radioactivity, electrophorese more slowly (thus there is no background trail that could mask miniscule amounts of pyrophosphate).

Fig. 1 shows the distribution of radioactivity along a Cellogel strip after electrophoresis of $(\gamma^{32}P)$ATP and (^{33}P)phosphate, and also of $(\gamma^{32}P)$ATP after digestion with snake venom phosphodiesterase. The rapidly migrating component between the faster red and pink dye markers is pyrophosphate. Evidence for this assignment was obtained by the fact that treatment with pyrophosphatase converted it to phosphate; chromatography on PEI cellulose confirmed the ability of snake venom phosphodiesterase to cleave triphosphates into monophosphates plus pyrophosphate (data not shown).

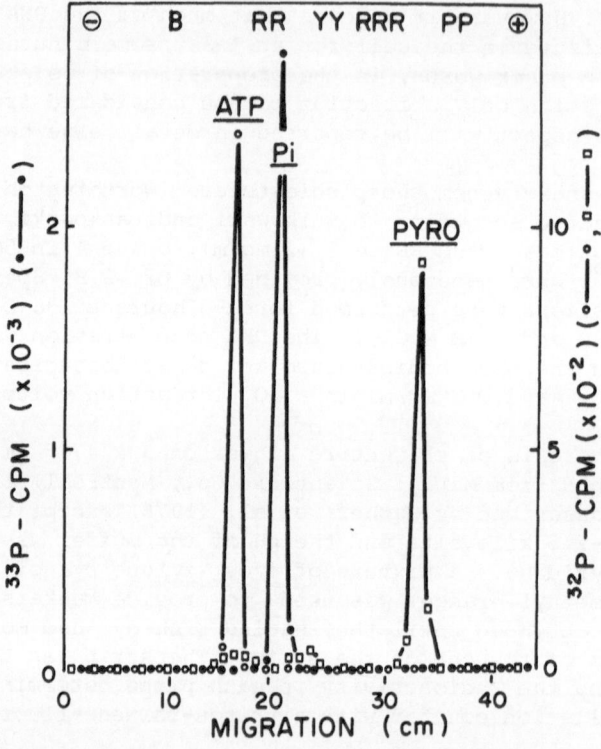

Figure 1.   *Electropherogram of* $(\gamma^{32}P)ATP$ *before (O - O) and after*
(□ - □) digestion with snake venom phosphodiesterase (+ DNase I).
The (^{33}P)*phosphate (● - ●) was run together with the* $(\gamma^{32}P)ATP.$
Nuclease digestion and high voltage electrophoresis were performed
as described in Methods. The locations of the blue (B), red
(R), yellow (Y) and pink (P) dye bands are indicated along the top
of the graph.

Fig. 2 shows a similar experiment but with RNA labeled *in vivo*
for 5 minutes with [32]P. The RNA was purified from the flow-through
of the first nitrocellulose column (see Methods) by equilibrium
banding in a cesium sulphate-formamide density gradient and
subsequent gel filtration on Sephadex G-25. Hydrolysis of the
purified RNA with snake venom phosphodiesterase generated a peak
of radioactivity in the position of pyrophosphate. Although as
plotted the peak appears small, it contained 20,000 cpm and
represented about 0.7% of the total radioactivity.

In a number of experiments with DNA from several different
E.coli strains labeled *in vivo* with ^{32}P for 4-6 minutes, no
evidence for the presence of pyrophosphate after digestion of the
(^{32}P)DNA with DNase I plus snake venom phosphodiesterase was

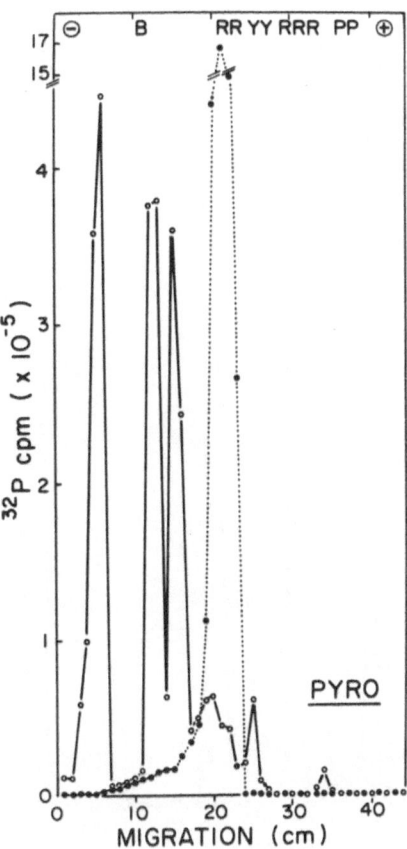

*Figure 2.   Electropherogram of (^{32}P)RNA before (● – ●) and after
(O – O) digestion with snake venom phosphodiesterase (+ DNase I).
The conditions were as in Fig. 1.*

obtained. (DNase I was included in the RNA experiment shown in
Fig. 2 so one cannot argue that the DNase destroyed the pyrophosphate.) A typical experiment is shown in Fig. 3. C, A, G, and T
indicate the positions of the deoxymononucleotides, and Pi indicates
phosphate. Because there were over a million counts per minute
in the hydrolysate electrophoresed, I calculate that less than
0.002% of the ^{32}P could be in pyrophosphate. (Note the hundred-fold
change in the scale at fraction 20). This is about 300 x less
than was observed in the RNA digestion described in Fig. 2. Since
the average size of the DNA chains was around 1,000 nucleotides,
this means that fewer than 1/100 of the molecules terminated in a
triphosphate.

Given that 10% of the nascent DNA from 10^{11} cells labeled with

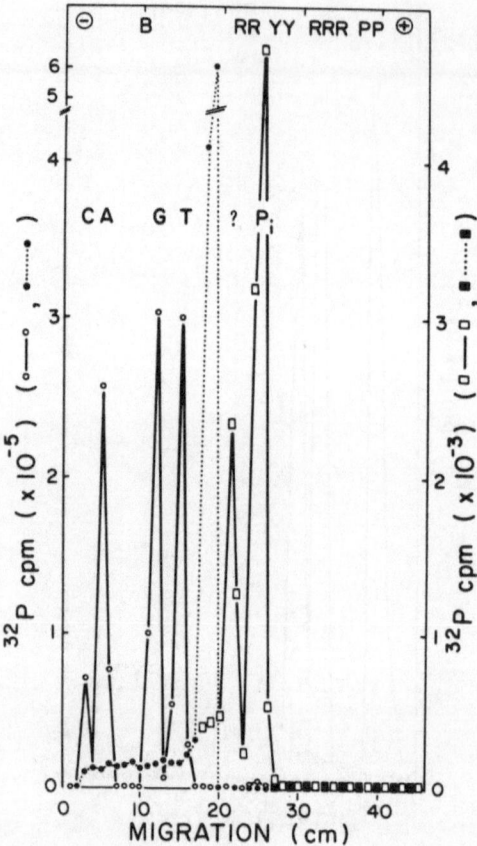

Figure 3. *Electropherogram of (^{32}P)DNA before (●···●;■···■)*
and after (○-○;□-□) digestion with snake venom phosphodiesterase
and DNase I. Note that the scale on the ordinate changes by a factor
of 100 around fraction 20. The conditions were as in Fig. 1.

300 mCi/mM ^{32}P-phosphate was analyzed, then there is less than one
molecule terminated with a triphosphate per cell.

I have also considered the possibility that the nascent 5'
terminus could be a diphosphate. Since ribonucleotides are
reduced to deoxyribonucleotides at the diphosphate level in *E. coli*
this possibility is intriguing. In this case, snake venom
phosphodiesterase would release phosphate during the degradation,
and indeed an inspection of Fig. 3 reveals a peak of ^{32}P in the
position of phosphate. Between 0.4 and 1.2% of the cpm were found
here (the rest were in the C, A, G and T peaks). However, I have
several reasons for attributing the origin of this phosphate to
something other than diphosphate termini. One reason is that

similar amounts of phosphate were generated during digestion of long-term ^{32}P-labeled DNA, and also ^{32}P-labeled ϕX174 viral DNA (a circular single-stranded molecule). Secondly, DNA previously treated with alkaline phosphatase, and thus lacking any terminal phosphates, yielded phosphate after digestion. Thirdly, phosphate was present in digests of DNA possessing ^{32}P at the 5' end only as the result of the preparation of DNA molecules having been end-labeled with polynucleotide kinase and (γ[32]P)ATP. I suspect that the phosphate is generated during the hydrolysis by the nucleases because of a low level of nucleotidase activity. The origin of the smaller amount of an unidentified component migrating a little more slowly than the phosphate is unknown. Both of these compounds were detected in roughly similar amounts in every digestion performed. The unidentified material was not sensitive to alkali (0.3 M NaOH, 36°C, 6 hr) or phosphatase (indeed phosphatase treatment of a snake venom phosphodiesterase + DNase I digestion increased the amount some five-fold). Both components were liberated from DNA previously exposed to alkaline phosphatase or to alkali. It may be significant that there was always close to a 2 to 1 ratio, phosphate in excess, of these two compounds.

If the 5' terminus is not a di- or triphosphate, what is it ? A previous paper (Miyamoto and Denhardt, 1977) showed that in the total population of DNA molecules isolated from *E. coli* pol*A*ex2 by these procedures only a few percent of the molecules were terminated with a ribonucleotide. From the data in that paper one can estimate[*] that about 0.03% of the ^{32}P would be expected to be found in pNp residues after alkaline hydrolysis of the DNA preparation. This was approximately what was found when the preparation of (^{32}P)DNA was examined for alkali-labile [32]P.

---

[*]*From Tables 3 and 6 of Miyamoto and Denhardt (1977) one can make the following calculations for* E. coli pol*A*ex2 *DNA:*

Size Class	Percent with ribonucleotide terminus	Average number of nucleotides	Fraction of phosphate in a pNp	Percent of total DNA mass
A	25	2000	1/4000	70
B	20	800	1/2000	30
C	2	20	1/500	2

Thus if we <u>assume</u> *uniform labeling of the DNA, and a 5' terminal monophosphate, we expect the percent of the* [32]P *found in pNps after alkaline hydrolysis to be*

$$\frac{0.7 + 0.3 + 0.02}{(0.7)(4000) + (0.3)(2000) + (0.02)(500)} = 0.03\%$$

Fig. 4 shows the amount of ^{32}P released from the (^{32}P)DNA
after exposure to alkali or to phosphatase. In this experiment
the digests were chromatographed on PEI-cellulose thin layer plates
with 3 M sodium formate, pH 3.5. The amount of ^{32}P released by
alkaline hydrolysis was quite low and close to the 0.03% expected.
However the quantitation of this small fraction of radioactivity
is not precise. (0.01% of the radioactivity equals 100 cpm; a
background of 10 cpm has been subtracted.) The amount of ^{32}P in

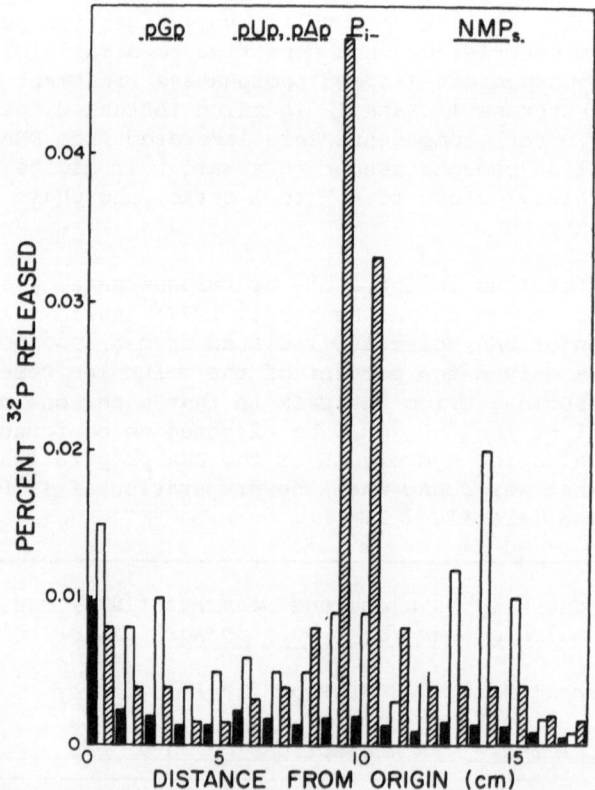

Figure 4. Distribution of the radioactivity on a PEI-cellulose plate
after chromatography with 3 M sodium formate, pH 3.5. (^{32}P)DNA was
prepared as described in the text and chromatographed after exposure
to alkali (open bars) or to alkaline phosphatase (striped bars); the
solid bars indicate unreacted DNA. In all three cases over 99% of
the radioactivity remained at the origin, which is not shown. The
percent of the total radioactivity (about 10⁶ cpm) is plotted for
each fraction. A more detailed description of the chromatography,
and of the hydrolyses with alkali and with alkaline phosphatase,
has been published (Miyamoto and Denhardt, 1977).

ribonucleotides was only about 3 x more than the apparent amount
in pNps, and this suggests that the average length of the ribonucleotide primer was only 3-4 nucleotides. This is a maximum
estimate since contamination with RNA would contribute
predominantly to the rNMP peak.

Exposure to bacterial alkaline phosphatase released about 0.1%
of the ^{32}P from the DNA as phosphate. This is what would be
expected if the average size of the labeled molecules in the
population was 1000 nucleotides and they were all terminated with
5' monophosphates. This estimate is crude however since the DNA
was not uniformly labeled, and because there was such a wide
distribution in sizes of the molecules that an independent estimate
of the "average" size is not easily made. In different experiments
the percent of phosphate released ranged between 0.07% and 0.44%.
About 10% of this phosphate (1/2 of 0.03%) presumably comes from 5'
ribonucleotide termini. Control experiments indicated that only a
small proportion of the total number of termini present were
generated during the purification of the DNA. That the DNA was not
uniformly labeled is evident from the low molar ratio of (^{32}P)dCMP
present in the DNA (see Fig. 3); uniformly labeled DNA yielded the
expected molar ratios.

DISCUSSION

In conclusion, the population of DNA molecules that was studied
here and which was shown to include essentially all of the nascent
DNA appears to consist largely (90%) of molecules terminated with a
2'-deoxyribonucleoside-5'-monophosphate. A minority of the DNA
molecules from *E. coli polAex2* terminate in one, or a few, ribonucleotides. Less than 1% of the molecules, fewer than one per
cell, terminate in a triphosphate. All four bases are found at
the 5' end (data not shown; but were obtained by labeling other
preparations with polynucleotide kinase and (γ^{32}P)ATP) of the
smallest, class C, as well as the largest, class A, molecules. One
cannot conclude however that these DNA molecules were initiated with
a deoxymononucleotide. I have not excluded that the 5' monophosphate ends represent the product of a processing event - for
example the removal of an RNA or protein primer; experiments
investigating these possibilities are in progress.

The amount of short DNA molecules recovered from 400 ml of ,
cells in these experiments was about 20 µg with a recovery of 20%.
If we make our very crude approximation that the average size of
the molecules was 1000 nucleotides this suggests that there were
900 molecules per cell. Miyamoto and Denhardt (1977) calculated
(from end-labeling experiments) that there were about 300 molecules
of this size per cell, with a smaller number of larger molecules
and a larger number of smaller ones. Without further information it

is unfortunately not possible to determine to what extent the set of molecules labeled during the 5 minute period *in vivo* with ^{32}P-orthophosphate corresponds to the total set of molecules isolated from the cell and labeled *in vitro* with ^{32}P using polynucleotide kinase. What proportion of the molecules are growing at any one moment in time is not evident.

ACKNOWLEDGEMENTS

I thank Ken Hastings, Stewart Millward, Klaus Harbers and John Spencer for advice, and John Spencer for gifts of materials and for the use of his high voltage electrophoresis apparatus. This research was supported by the Medical Research Council of Canada and the National Cancer Institute of Canada. I thank Linda Robinson for excellent secretarial assistance.

REFERENCES

Brutlag, D., Schekman, R. and Kornberg, A. (1971) Proc. Natl.
    Acad. Sci. 68:2826-2829

Denhardt, D.T. (1972) J. Theor. Biol. 34:487-508

Denhardt, D.T. (1977) In: Comprehensive Biochemistry, Volume 24,
    A. Neuberger (ed.) - Biological Information Transfer. Elsevier/
    North Holland Biomedical Press, pp 1-53

Denhardt, D.T., Eisenberg, S., Harbers, B., Lane, H.E.D. and
    McFadden, G. (1975) In: DNA Synthesis and Its Regulation
    (M. Goulian, P. Hanawalt and C.F. Fox, eds.) W.A. Benjamin
    Inc. Menlo Park, Calif. pp398-422

Dressler, D. (1975) Ann. Rev. Microbiology 29:525-559

Harbers, B., Delaney, A.D., Harbers, K. and Spencer, J.H. (1976)
    Biochemistry 15:407-414

Miyamoto, C. and Denhardt, D.T. (1977) **J. Mol. Biol. 116:681-707**

Ogawa, T., Hirose, S., Okazaki, T. and Okazaki, R. (1977) J. Mol.
    Biol. 112:121-140

Okazaki, R., Okazaki, T., Sakabe, K., Sugimoto, K. and Sugino, A.
    (1968) Proc. Natl. Acad. Sci. USA 59:598-605

Sugino, A., Hirose, S. and Okazaki, R. (1972) Proc. Natl. Acad.
    Sci. 69:1863-1867

Sulkowski, E. and Laskowski, M. (1971) Biochim. Biophys. Acta
    240:443-447

ON THE POSSIBLE DIRECT CONVERSION OF RNA TO DNA

S. Barlati[*] and A. Brega

Istituto di Biologia Generale
Facoltà di Medicina e Chirurgia
Università degli Studi di Milano (Italy) and
Laboratorio di Genetica ed Evoluzionistica del
C.N.R. di Pavia (Italy)

INTRODUCTION

In the present study we have investigated the possibility
that an RNA molecule can be directly converted in DNA by sequential
action of ribonucleotide reductase- and thymidylate synthetase-like
activities. The reasons for advancing such an unorthodox
hypothesis arose from some considerations on the nuclear RNA
(hnRNA) from eukaryotic cells.

The role and the fate of the hnRNA which is synthesized and
degraded inside the eukaryotic cell by more than 90-95% while
migrating from the nucleus to the cytoplasm is not yet clear (1).
The possibility has not been considered however that at least a
part of this RNA, rather than being degraded, might be chased in
DNA without leaving the nucleus. Indeed, when using 5-(^{3}H)-uridine
for labelling RNA, methylation in the 5 position of the uridine
moiety of this RNA would be accompanied by a loss of radioactivity
which might be confused with degradation of RNA.

Furthermore, the synthesis of polynucleotides is a process
which needs a very large input of energy and one may sometimes
wonder if there are sufficient reasons for the evolution of such
energetically unfavourable processes like the one which involves
continuous synthesis and degradation of RNA molecules like that
observed for hnRNA.

We decided therefore to test the possibility that hnRNA might
chase into DNA, in in vitro cell cultures, under the conditions
*A.S.I. Participant

which are generally used for the preparation of hnRNA.

After approaching this problem in eukaryotic cells : secondary chick embryo fibroblasts (CEF) and HeLa ćells we also applied the same experimental approach to the prokaryote <u>Bacillus</u> <u>subtilis</u>.

The experimental scheme was to pulse-label the cells with 5 or 6-(^{3}H)-uridine and then follow, in a chase, the fate of the ^{3}H-labeled RNA into total TCA precipitable material and into TCA precipitable, alkali resistant material under conditions where there is complete degradation of RNA molecules.

As is known, and briefly summarized in the scheme of Fig. 1, the label of 5-(^{3}H)-uridine can be incorporated in DNA mainly as dCTP. The amount of cytidine incorporated into DNA in such a pathway can be monitored by following the incorporation of 5-(^{3}H)-uridine into DNA. The use of this label will not permit detection of the uridine transformed in thymidine since, as already mentioned, methylation of 2' deoxyuridine monophosphate by thymidylate synthetase will displace the 5-(^{3}H) label and give rise to cold thymidine monophosphate. Only by using 6-(^{3}H)-uridine is it possible to obtain labeled TTP which can then give rise to (^{3}H)-thymine labelled DNA.

Therefore while 5-(^{3}H)-uridine can account for most of the incorporation of ^{3}H labelled cytidine into DNA (2,3), 6-(^{3}H)-uridine will account for the incorporation of ^{3}H labelled cytidine and thymidine into DNA. The amount of uridine incorporated in DNA as

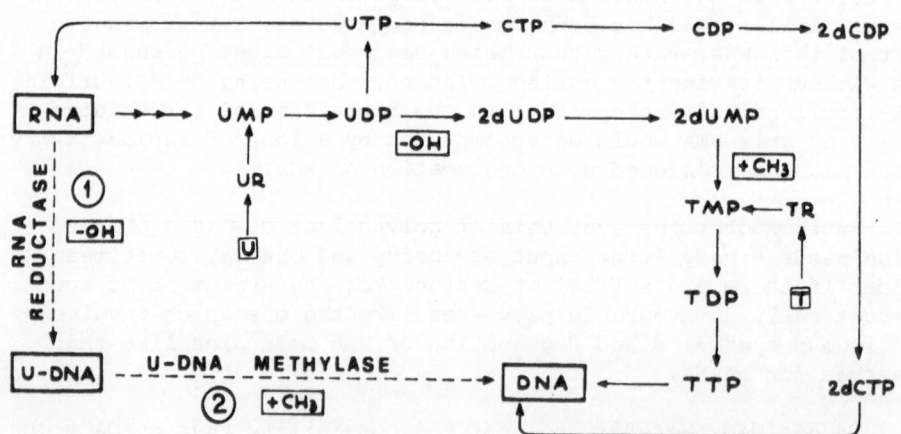

Figure 1. Scheme of uracil metabolism. The dashed arrows illustrate the proposed pathway for conversion of RNA into DNA with the formation of an intermediate uracil containing DNA (U-DNA).

thymidine can therefore be followed by the differential incorporation between 6- and 5-(^{3}H)-uridine into DNA.

Under chase conditions however, if the chase is efficient, the incorporation of cytidine and thymidine, deriving from uridine, into DNA should be blocked. In order to test whether this was the case, parallel experiments were performed in which a pulse of (^{3}H)-thymidine was followed under the same chase conditions as for cells pulsed with (^{3}H)-uridine. In such a way we could verify whether the (^{3}H)-TTP pool could be chased under our experimental conditions. If we therefore make the assumption that (^{3}H)-TTP deriving from (^{3}H)-thymidine has to be chased in the same way as the (^{3}H)-TTP deriving from 6-(^{3}H)-labelled uridine, we can control and have an estimate of the utilization of the labelled TTP under conditions of chase whether derived from thymidine or uridine. Under these conditions we could also take into account the possible contribution of nucleotides deriving from the breakdown of RNA, since, in order to be re-utilized into DNA, they should follow the same pathway for the conversion to deoxyribonucleotides (Fig. 1).

In conclusion, during a chase, one should not expect an increase of the incorporation of 5- or 6-(^{3}H)-uridine into DNA. On the contrary, in the case that macromolecular RNA could be directly reduced and methylated, one would expect an increase of label into DNA during chase conditions. This is summarized in the dashed pathway reported in Fig. 1, in which the step (1) corresponds to the reduction of ribose and step (2) to the methylation of uracil. After the first reduction step there would be the transient formation of uracil containing DNA (U-DNA).

The experiments presented in this work are in agreement with the possible existence of such a pathway, although they are also amenable to different interpretations.

MATERIALS AND METHODS

Reagents: Methyl (^{3}H)-thymidine (19 ci/mmole), 6-(^{3}H)-uridine (16.3 Ci/mmole), and 5-(^{3}H)-uridine (28.1 Ci/mmole) were purchased from Amersham (England).

Actinomycin D was furnished by Serva Feinbiochemica and HPUra was obtained from Dr. B.W. Langley (Imperial Chemical Industries, London).

Cells and Media

Chick embryo fibroblasts (CEF) at the 2nd passage were seeded in 5 cm Petri dishes and used for the experiments in the exponential growth phase. Eagle Medium with a double concentration of amino-

acids and vitamins (EM 2X) supplemented with 5% calf serum and
10% Tryptonephosphate broth (TPB) was used for cells' growth.
HeLa cells were also grown as monolayer cultures and the medium
was EM 2X supplemented with 5% calf serum. Throughout the
experiments the serum concentration was lowered to 2% and TPB
omitted.

    B.subtilis SB25 (his$^-$, trp$^-$), SB168 (trp$^-$) and the ts mutant
in RNA synthesis PB1653 (4), were gorwn in Davis minimal medium
supplemented with 0.5% glucose 0.1% casaminoacid hydrolysate and
20 µg/ml of required aminoacids (C medium).

    The experimental conditions are reported under the figure
legends.

DNA preparation

    B. subtilis 168 growing exponentially at 35° in C medium
was pulsed with 1 µc/ml of 6-(^{3}H)-uridine in the presence of 5 µg/
ml thymidine and 2.5 µg/ml cytidine. After 5 min at 35°, 20 µg/ml
uridine and 5 µg/ml thymidine were added. After 20 min at 35° the
cells were centrifuged and the nucleic acids extracted following
the Marmur procedure (5).

                              RESULTS

Eukaryotic Cells

    Fig. 2 and 3 report results obtained in pulse-chase
experiments using secondary cultures of growing CEF.

    In Fig. 2 the cells were pulsed for 1 hr with 5- or 6-(^{3}H)-
uridine at the time of addition of Actinomycin D (0.04 µg/ml) to
the culture medium. This concentration of actinoymycin is
sufficient for blocking rRNA but not hnRNA or DNA synthesis.

    It can be seen that, under these conditions, during the
chase there is extensive degradation of the prelabelled pulsed
RNA which is evidenced by the decrease of total TCA precipitable
counts; similar patterns are obtained for 5- and 6-(^{3}H)-pulsed
RNA. At the same time however, an increase can be observed of
the alkali-resistant counts in the case the cells were pulsed
with 6-(^{3}H)-uridine. No such increase was observed in the case of
5-(^{3}H)-uridine labelled cells.

    In order to avoid the extensive RNA breakdown observed under
the above experimental conditions, the pulse was performed after
a pretreatment of the cells for 45 min with 0.04 µg/ml Actinomycin
D. Under these conditions no degradation of prelabelled RNA is

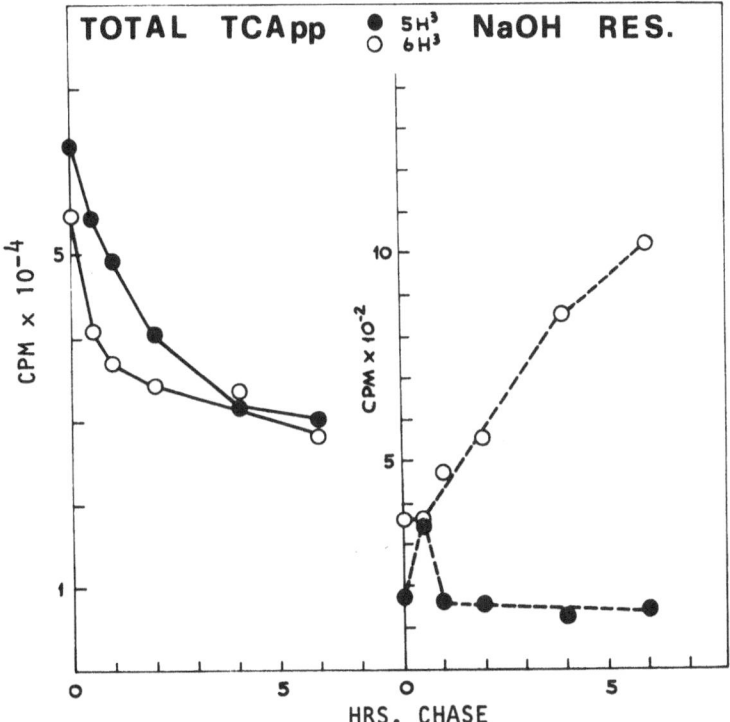

Figure 2. *Monolayer CEF in exponential growth were pulsed for 1 hr with 5 μc/ml of 5-(³H)-uridine (●) or 2 μc/ml of 6-(³H)-uridine (○) in the preparation of 0.04 μg/ml of Actinomycin D, 5 μg/ml of thymidine and 2.5 μg/ml of cytidine. After the pulse the cells were washed twice with cold medium containing in addition 5 μg/ml of uridine. At the indicated times cells were removed from the dish with 2 ml of 0.5% SDS; a 0.2 ml aliquot of the homogenate was TCA precipitated and the remaining 1.8 ml was treated for 25 min at 100°C in 0.1 M NaOH and TCA precipitated. Samples were then filtered on millipore filters and the radioactivity determined. In the left panel is reported the total TCA precipitable radioactivity and in the right panel the TCA precipitable alkali-resistant counts.*

observed either in the case of 5- or 6-(³H)-uridine label. However, as reported in Fig. 3, although there has been a drastic change in the pattern of the fate of the prelabelled total TCA precipitable radioactivity, no change is observable for the chase of prelabelled RNA into alkali-resistant-TCA-precipitable material as compared with the results reported in Fig. 2; in either case after 4 hrs of chase approximately 5% of the total RNA has been made alkali-resistant.

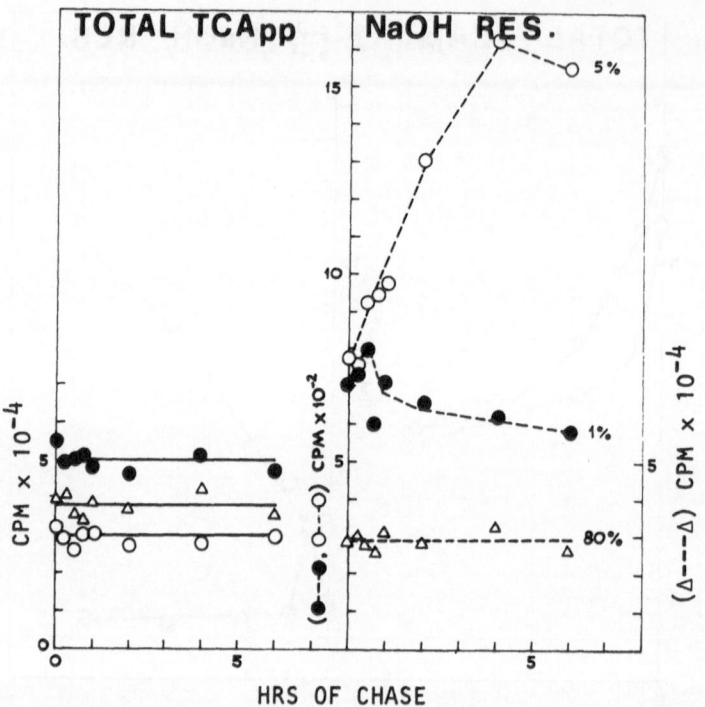

Figure 3. *Monolayer CEF in logarithmic growth were pretreated for 45 min with 0.04 µg/ml of Actinomycin D and then pulsed with 10 µc/ml of 5-(^{3}H)-uridine (●), 4 µc/ml of 6-(^{3}H)-uridine (○) or 0.5 µc/ml of (^{3}H)-thymidine (△). Experimental conditions were as those reported in Fig. 2 except that cells were removed from the dish with 0.1 M NaOH and, for alkali digestion, 1.8 ml of the cells homogenate was incubated at $37^\circ C$ for 18 hr in 0.3 M NaOH. In the dishes pulsed with (^{3}H)-thymidine, cold thymidine was omitted during the pulse and added throughout the chase at the concentration of 5 µg/ml. Left panel: total TCA precipitable counts; right panel: TCA precipitable alkali-resistant counts.*

In Fig. 3 the chase of (^{3}H)-thymidine pulsed cells is also reported; it is clear that under these conditions the (^{3}H)-TTP prelabelled pool, which is derived from the exogenous (^{3}H)-thymidine, is very efficiently chased so that the increase of label, during the chase of 6-(^{3}H)-uridine into DNA, should not derive from free (^{3}H)-TTP.

The recovery of (^{3}H)-thymidine labeled DNA after alkali treatment was constant and approximately 80% of the total TCA precipitable radioactivity. The percentage of labeled alkali-resistant material after 4 hrs of chase was approximately 1% for

5-(^{3}H) and 5% for 6-(^{3}H)-uridine.

It should be noted that, in the experiments reported in Fig. 2 and 3, a transient increase of alkali-resistant radioactivity is observed after 30 min of chase when the cells were pulsed with 5-(^{3}H)-uridine.

This might suggest that in the proposed model of conversion of RNA into DNA the reduction of ribose to deoxyribose (alkali-resistance) preceeds the methylation of uracil to thymine (removal of the 5-(^{3}H) label).

The utilization of synchronized CEF has also shown that the chase of 6-(^{3}H)-uridine prelabeled RNA in DNA is highest in the S phase of the cell cycle (data not reported).

Fig. 4 reports the results obtained when HeLa cells, in place of CEF, are utilized under experimental conditions analogous to those of the experiment reported in Fig. 3. Qualitatively the results with HeLa cells are very similar to those obtained with CEF. Quantitatively however, the percentage of alkali-resistant; TCA-precipitable material is considerably increased and after 4 hrs the chase is approximately 5% for 5-(^{3}H) and 20% for 6-(^{3}H)-uridine labeled cells.

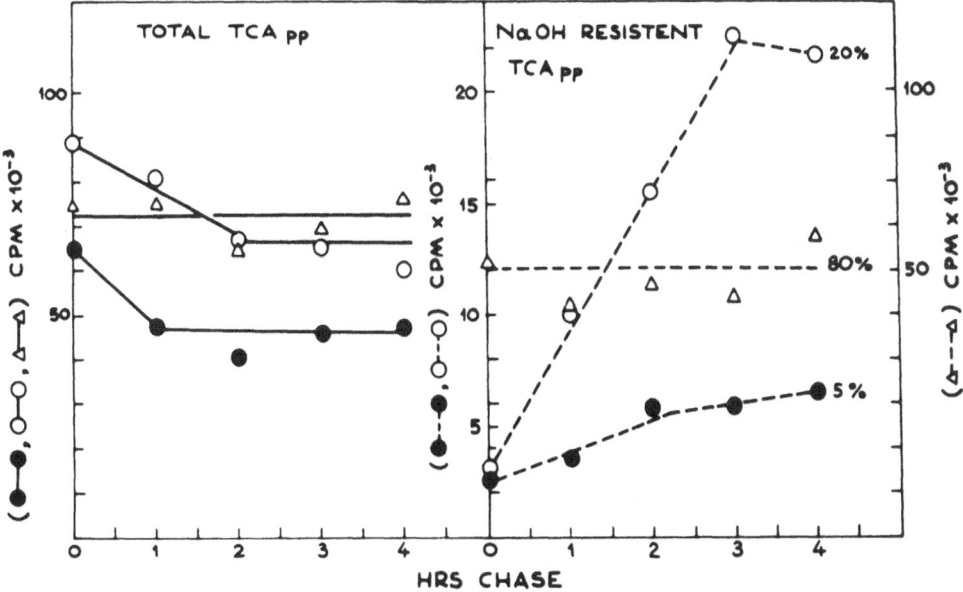

Figure 4. Chase of 5 (●), 6 (○) (^{3}H)-uridine and (△) and (^{3}H)-thymidine in monolayer HeLa cells. The experimental conditions are those of the experiment reported in Fig. 3.

Prokaryotic Cells

The same kind of experiment but omitting Actinomycin D was also performed on B. subtilis. In this case the pulse of labeled precursors was for 5 min for cells growing in minimal medium. This pulse corresponds roughly to 1/20 of the generation time of these bacteria under the experimental conditions employed. This is approximately the same fraction of the generation time used for CEF or HeLa cells (1 hr pulse over 20-24 hrs generation time).

In Fig. 5 are reported the results of the chase of 6-(^{3}H)-uridine and (^{3}H)-thymidine into total and alkali-resistant, TCA precipitable material. With prokaryotic cells also one can see the

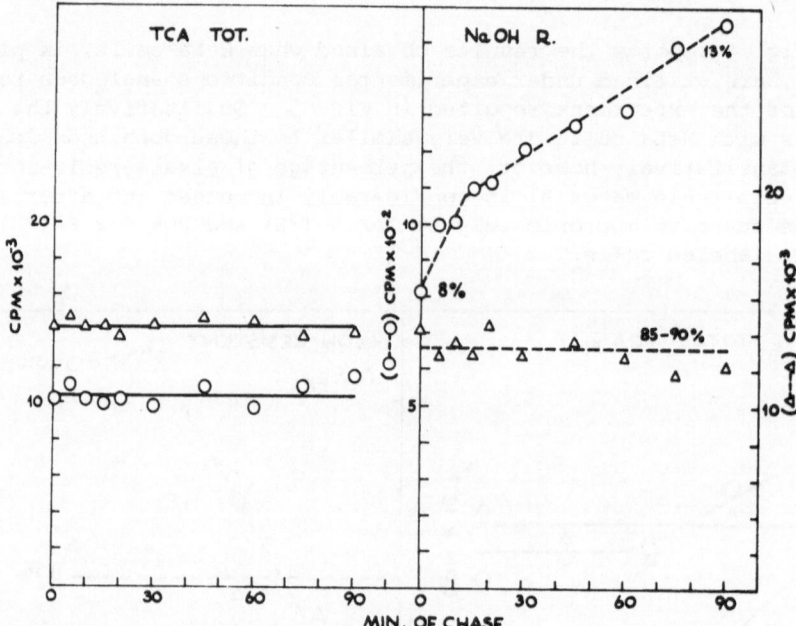

Figure 5. *B.subtilis SB25 in exponential growth was pulsed for 5 min with 1 µc/ml of 6(^{3}H)-uridine (O) or 0.4 µc/ml of (^{3}H)-thymidine (△) in C medium plus 2.5 µg/ml of cytidine. After the pulse the radio-activity was removed by centrifugation, the cells washed once and fresh C medium containing 5 µg/ml of thymidine, 5 µg/ml of uridine and 2.5 µg/ml cytidine added. At the indicated times 2 ml samples were removed and divided in a 0.5 ml aliquot which was TCA precipitated and the remaining 1.5 ml aliquot was treated with 0.1 M NaOH for 25 min at 100°C and then TCA precipitated. Samples were filtered on millipore filters and the radioactivity determined. Left panel: total TCA precipitable counts; right panel: TCA precipitable alkali-resistant counts.*

same pattern previously observed. Again an increase of radioactivity deriving from 6-(^{3}H)-uridine prelabeled cells is evident,
during the chase, in the alkali-resistant fraction.

   Again, this increase should not derive from free (^{3}H)-TTP
labeled DNA precursors deriving from 6-(^{3}H)-uridine since, under
these conditions, the (^{3}H)-TTP deriving from (^{3}H)-thymidine is
efficiently chased as evidenced by the stability of the (^{3}H)-
thymidine labeled material during the chase.

   The alkali-resistant fraction of 6-(^{3}H)-uridine labeled
material goes from 8 to 13% after a 60 min chase.

   In order to be sure that the uridine incorporation monitored
in the alkali-resistant, TCA precipitable fraction was indeed DNA,
the nucleic acids from cells pulsed for 5 min with 6-(^{3}H)-uridine
and chased for 20 min, were extracted. Fig. 6 reports the results
of the sensitivity of nucleic acids to pronase, RNase and DNase
after the chase.

   From the data obtained it is seen that approximately 20% of
the nucleic acid is sensitive to DNase but not to pronase (Fig. 6A)

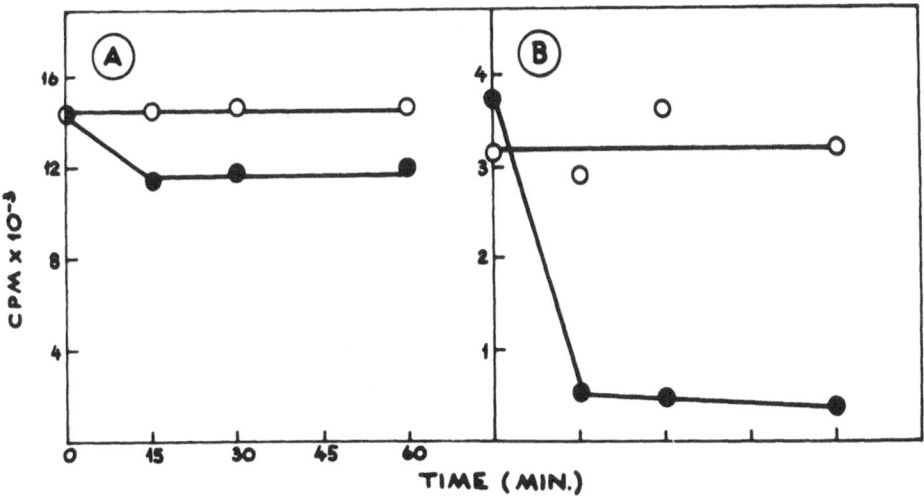

Figure 6. Sensitivity to 8 μg/ml of DNase (●) and 200 μg/ml of
pronase (O) of the nucleic acid extracted from B.subtilis SB168
pulsed for 5 min with 6(^{3}H)-uridine and chased for 20 min. The
response to the enzymatic treatments before (Fig. 6A) and after
(Fig. 6B) digestion of the total nucleic acid for 30 min at 37°C
with a mixture of pancreatic RNase (10 μg/ml) and T_1-RNase (10
units/ml), are reported.

and that after digestion with RNases or alkali, the residual radioactivity is almost completely (95%) sensitive to DNase but not to
pronase (Fig. 6B). So, although unlikely, the possibility that the
label from 6-(^{3}H)-uridine could go in some cellular structures,
other than DNA, was excluded.

We then decided to study the possible dependence of RNA and/or
DNA synthesis for the chase of 5- or 6-(^{3}H)-uridine into DNA.

For testing the dependence from RNA synthesis we took
advantage of a B.subtilis mutant (PB1653) temperature sensitive for
RNA synthesis (4). This mutant has its RNA synthesis completely
blocked after 5 minutes from the shift from the permissive (33°C) to
the non permissive (45°C) temperature.

In Fig. 7 are reported the results of the chase of 5- and
6-(^{3}H)-uridine prelabeled cells after pulsing the cells for 5 min
with the 33°C and then chasing at 33° or 45°C.

It is clear that very similar results were obtained if the
chase is performed at either 33 or 45°C; one can therefore conclude
that the chase of prelabeled RNA into DNA does not depend on
RNA synthesis.

A slight increase of the alkali-resistant counts is also
evident in the case of 5-(^{3}H)-uridine labeled cells, but it is
not comparable however with that observed for 6-(^{3}H)-uridine
labeled cells.

For testing the dependence of the chase from DNA synthesis, the
compound 6-(p-Hydroxyphenylazo)-uracil (HPUra) was utilized which
is known to specifically inhibit DNA polymerase III in B.subtilis
(5,6). The results of the chase of 6-(^{3}H)-uridine in the presence
or in the absence of HPUra are reported in Fig. 8.

It can be seen that HPUra completely prevents the chase of
radioactivity into DNA. It can therefore be concluded that the
chase of label into DNA depends on DNA synthesis. The results of
this experiment give rise to two different interpretations, as
will be discussed; one compatible with the model proposed and one
which disproves it.

DISCUSSION

In the present study we have tried to develop an experimental
scheme in order to verify the possibility of direct conversion of
RNA in DNA.

The results of the pulse-chase experiments described, either

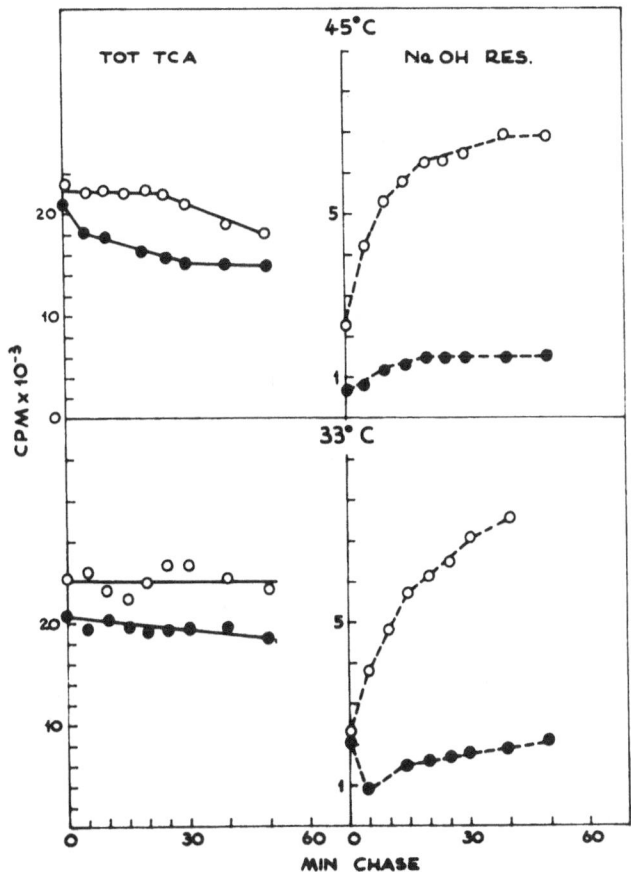

Figure 7. PB1653 cells were grown at the permissive temperature 33°C and pulsed for 5 min with 1 µc/ml of 5-(^{3}H)-uridine (●) or 0.4 µc/ml of 6-(^{3}H)-uridine (○) in the presence of 5 µg/ml of cold thymidine and 2.5 µg/ml of cytidine. Each culture was then split in two parts for the incubation at 33° or 45°C and the chase started by adding 20 µg/ml of uridine and 5 µg/ml of cold thymidine. At each time 1 ml sample was removed: a 0.3 ml aliquot was TCA precipitated, the remaining 0.7 ml incubated at 37°C for 18 hrs in the presence of 0.3 M NaOH and then TCA precipitated. Samples were filtered on millipore filters and radioactivity determined. Left panel: total TCA precipitable counts; right panel: TCA precipitable alkali-resistant counts.

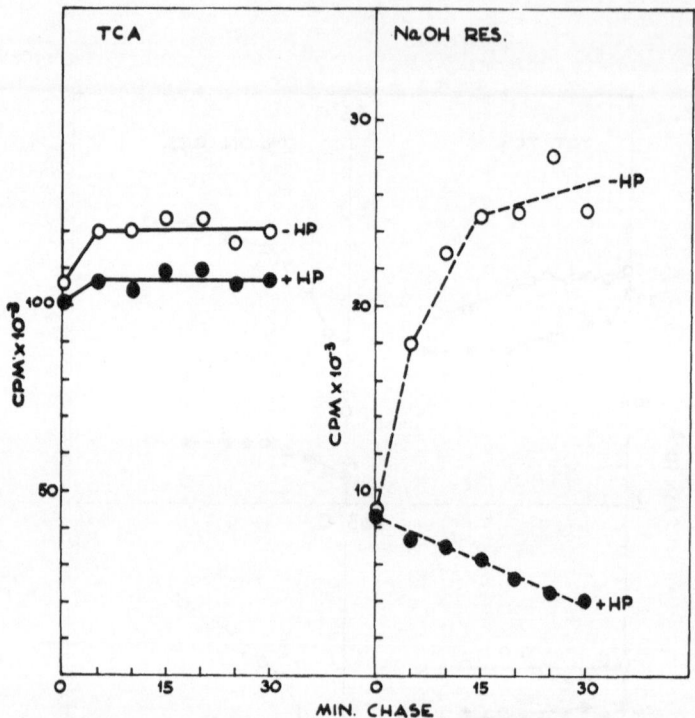

Figure 8. *B.subtilis SB168 in logarithmic growth were pulsed with 1 µc/ml of 6-(^{3}H)-uridine. After the pulse the culture was split in two parts for the incubation in the presence (●) or in the absence (○) of 30 µg/ml of HPUra. Experimental conditions were as those reported under Fig. 7. Left panel: total TCA precipitable counts; right panel: TCA precipitable alkali-resistant counts.*

in eukaryotic or prokaryotic cells, clearly show that prelabeled RNA can chase into DNA when using, for labeling RNA, 6-(^{3}H)-uridine.

A major criticism of this kind of pulse-chase experiments is that it is always difficult to exclude that the labeled nucleotides chasing into DNA are derived from free nucleotides or from degraded RNA molecules which are incorporated into DNA after being reduced to deoxyribonucleotides.

However, since the experimental conditions were such that the prelabeled (^{3}H)-TTP pool deriving from exogenous (^{3}H)-thymidine was efficiently chased, we consider it unlikely that the (^{3}H)-TTP pool deriving from 6-(^{3}H)-uridine, could not be, under the same experimental conditions, also efficiently chased.

The possibility still exists (though difficult to envisage) that the pool of TTP arising from uridine and that deriving from thymidine are in two different compartments, do not mix and only the TTP pool from exogenous thymidine can be efficiently chased.

If this is not the case, as seems more likely, one could therefore exclude that the chase of prelabeled RNA in DNA is due to free labeled nucleotides or derives from degradation products of RNA. We would therefore favour the interpretation that the chase of prelabeled RNA into DNA is a measure of a direct conversion of a part of the prelabeled RNA into DNA by subsequent action, on the RNA molecules, of RNA reductases and specific (U-DNA)-methylases (i.e., thymidylate synthetase). That reductases might act before methylases is suggested by the experiments reported in Fig. 2 and 3 in which, during the chase of pulse-labeled 5-(^{3}H)-uridine RNA, the transient increase of alkali-resistant radioactivity observed (reduction of ribose to deoxyribose) is followed by a decrease of radioactivity (methylation of 5-(^{3}H)-uridine in the 5 position with removal of the (^{3}H)-label).

That the increased radioactivity chasing into DNA does not derive from degradation of RNA is also suggested by the fact that the amount of radioactivity made alkali-resistant does not depend on the extent of RNA degradation (compare total TCA precipitable counts from Fig. 2 and 3).

The experiments performed with bacteria indicate that the chase of prelabeled RNA in DNA depends on DNA synthesis but not on RNA synthesis.

The results of these experiments may be also interpreted to disprove our working hypothesis. Indeed, since HPUra seems to block directly the polymerization of deoxyribonucleotide triphosphates, the inhibition of the chase of label into DNA might depend on the block of incorporation of free labeled deoxynucleotides arising from 6-(^{3}H)-uridine or from RNA breakdown and subsequent conversion according to a known pathway. One would expect that the direct conversion of an RNA molecule into a DNA molecule should not be sensitive to the HPUra block. On the other hand, our interpretation would be that RNA conversion is strictly coupled with DNA synthesis.

Whether or not the chase of RNA in DNA depends on direct RNA modification or from the utilization of the prelabeled nucleotides' pool is still an open question. Several problems remain to be clarified in order to be able to understand the extensive chase of 6-(^{3}H)-uridine-labeled RNA into DNA observed

under conditions in which the (^3H)-TTP pool deriving from
exogenous thymidine is efficiently competed. This direct
conversion, if proven, might have a role in handling RNA
primers for the Okazaki pieces.

ACKNOWLEDGEMENTS

We thank Drs G. Mazza and A. Galizzi for providing the
facilities for bacterial growth and Dr S. Riva for the B.subtilis
mutant 1653. We also thank Drs A. Falaschi and S. Spadari for
helpful and critical discussion of the data.

REFERENCES

(1) Weinberg, R.A. (1973) Ann. Rev. Biochem. 329-354

(2) Barlati, S. (1970) J. Bacteriol. 101:330-332

(3) Kelley, W.N. (1972) In: Growth, Nutrition and Metabolism
     of Cells in Culture. No. 7 pp.212-256. Academic Press, New
     York and London.

(4) Riva, S., Villani, G., Mastromei, G. and Mazza, G. (1976)
     J. Bacteriol. 127:679-690

(5) Marmur, J. (1961) J. Mol. Biol. 3:208-218

(6) Brown, N.C. (1970) P.N.A.S. 67:1454-1461

(7) Low, R.L., Rashbaum, S.A. and Cozzarelli, N.R. (1974) P.N.A.S.
     71:2973-2977

SOME ASPECTS OF DNA STRAND SEPARATION

James C. Wang

Department of Biochemistry and Molecular Biology
Harvard University
Cambridge, Massachusetts, 02138, U.S.A.

Ever since the double helix structure was first proposed for
DNA, the problem of strand separation during replication has
attracted much attention. When I was asked by the organizers of
this meeting to give a general survey on the physical chemistry of
DNA, I decided to focus my discussions on strand separation. At
the molecular level however, even the simplest *in vitro* DNA
replication system is rather complicated, and little quantitative
information is available. Therefore, the discussions to follow are
qualitative and conceptual in nature.

The Energetics of Strand Separation

For a purified DNA, the energetics of strand separation have
been measured. In a given medium, there is a temperature, T_m,
usually referred to as the "melting temperature" or the "transition
temperature" of the DNA, at which paired and unpaired bases are at
equilibrium. At this temperature the standard free energy change
$\Delta G^\circ (T_m)$ for the disruption of a base pair is zero. This zero ΔG° is
composed of an unfavorable ΔH° term of 7-9 kcal per mole of base
pairs disrupted depending on whether it is an AT or a GC pair, and
a favorable ΔS° term of approximately 23 cal per degree per mole
of base pairs disrupted[1]. From the standard thermodynamic relation

---

[1]
*There is considerable spread in the ΔS° values reported in the
literature, from 20 to 27 cal deg^{-1} $mole^{-1}$ base pair at neutral
pH (Shiao and Sturtevant, 1973; Klump and Ackermann, 1971; Privalov
et al., 1969). Within experimental error, ΔS° is independent of
the base composition (Klump and Ackerman, 1971), the ionic
strength of the medium (Privalov et al., 1969), and whether Mg(II)
or monovalent counterions are present (Baba and Kagemoto, 1974).*

$\partial \Delta G^{\circ} / \partial T = -\Delta S^{\circ}$, and assuming that ΔS° is temperature independent, it follows immediately that at a temperature T in the same medium, the standard free energy change $\Delta G^{\circ}(T)$ for the disruption of a base pair is $(-\Delta S^{\circ})$ $(T-T_m)$ or ΔS° (T_m-T).

In a typical "physiological" salt medium containing a dilute buffer, a couple of tenth molar monovalent cations and a few milli-molar of Mg^{++}, T_m of a 100% (G+C) DNA is about $110^{\circ}C$ and that of a 100% (A+T) DNA is about $70^{\circ}C$. Therefore in such a medium the standard free energy change at $37^{\circ}C$ for the disruption of a GC pair is approximately 23 x (110-37) cal or 1.7 kcal, and that for the disruption of an AT pair is approximately 23 x (70-37) cal or 0.8 kcal.

The rough estimates above are the *average* quantities for the separation of two fairly long complementary strands. In some cases we wish to know the free energy change for the disruption of one or a few base pairs in the middle of an originally intact helix. This introduces an additional free energy term: the free energy of interior loop formation. The additional unfavorable free energy of generating an interior loop containing 2-6 unpaired bases is about 2 kcal; for a larger loop containing 7-20 bases, this term is about 3 kcal (Tinoco et al., 1973). As an example, suppose two AT and four GC pairs are disrupted in the middle of an originally intact duplex, forming an interior loop containing 12 unpaired bases. The ΔG° under physiological conditions at $37^{\circ}C$ is estimated to be 0.8 x 2 (two AT pairs) + 1.7 x 4 (four GC pairs) + 3 (interior loop containing 12 bases) or 11 kcal.[2]

The physical significance of the free energy of loop formation can be thought of as follows. In a completely denatured DNA chain, each monomer unit can assume many configurations. Not all these configurations can be assumed by an unpaired base in an interior loop, since the closing of a loop by base pairs at the two ends imposes constraints on the configurations of the unpaired units in between. This gives rise to an unfavorable free energy change. For loops containing more than 20 bases, theoretical consideration indicates that the free energy of loop formation increases with the logarithms of the number of the unpaired bases (Zimm, 1960). Since the free energy for disrupting the base pairs increases linearly with the total number of base pairs disrupted, for the denaturation of a long helical segment the free energy of interior loop formation is but a small part and is therefore unimportant.

---

[2] *To be more rigorous, the free energy change is not only base composition dependent but also sequence dependent (Tinoco et al., 1973). At the time of writing not enough data for DNA are available to allow more precise estimates.*

The Energetics of Strand Separation in the Presence of a Protein
Which Binds Specifically to Single-Stranded DNA

In the presence of a protein which binds specifically to
single-stranded DNA, the energetics of strand separation is affected.
Qualitatively, the presence of such a protein lowers T_m and the free
energy of strand separation. Alberts and Frey (1970) have shown
that in the presence of 100 µg/ml of the T4 gene 32 protein, poly
dAT but not T4 DNA (which has a higher melting temperature) can be
denatured at 37° under physiological conditions.

For the cooperative binding of the *E. coli* DNA binding protein
to single-stranded DNA, the free energy change is about -15 kcal
per mole of tetrameric binding protein (Ruyechan and Wetmur, 1975),
each of which covers about 32 nucleotides. In the presence of 1
µM (80 µg/ml) of the tetrameric binding protein, it can be
calculated that the specific binding of the protein to single-
stranded DNA contributes a favorable free energy of about 0.4 kcal
per mole of cooperatively denatured base pairs. This is still a
bit below what is required to compensate the unfavorable free
energy of disrupting an AT pair and is considerably below that for
a GC pair. Sigal et al. (1972) estimated that there are 800
monomer units of the binding protein in an *E. coli* cell. The total
concentration is therefore of the order of µ*M*. But since most of
the protein molecules are likely to be bound, the free protein
concentration is probably considerably lower than 1 µ*M*.

It should be pointed out that the above estimate is for the
cooperative binding of the protein molecules. The free energy
gain for the binding of a single tetrameric protein to denatured
DNA is about one half the value for the cooperative binding of
each tetramer (Ruyechan and Wetmur, 1976). This lower gain, plus
the unfavorable free energy of loop formation, make the disruption
of a short stretch of base pairs more difficult.

The Effect of Superhelical Turns on Strand Separation

A large number of double-stranded DNA rings from natural
sources are *covalently closed* in the sense that there is no single-
chain scission anywhere in either strand. A basic parameter of
such a covalently closed duplex ring is the *linking number* or the
topological winding number $\underline{\alpha}$. This quantity is an invariant unless
a single-chain scission is transiently introduced. Suppose $<\underline{\alpha}>$
is the average linking number of a population of such molecules.
If one or a few backbone bonds of each of the molecules are subject
to breaking and rejoining, $<\underline{\alpha}>$ will change to a value $<\underline{\alpha}_{-m}>$ The
difference $<\underline{\alpha}> - <\underline{\alpha}_{-m}> \equiv i$ will be referred to as the *linking
difference*. If i is negative, the DNA is said to be underwound or

negatively twisted or negatively superhelical. Covalently closed
duplex DNA rings from natural sources are usually underwound. The
definition of $\underline{\alpha}$ and an introduction to the energetics of a
covalently closed circular DNA are presented in the Appendix.

It has been shown that relative to the nicked circular or a
linear DNA, it is easier to unwind an underwound duplex DNA ring
(Bauer and Vinograd, 1970; Hsieh and Wang, 1975). A combination
of single-strand specific binding protein and negative superhelical turns is likely to achieve the unpairing of AT rich
segments under physiological conditions. Delius et al. (1972)
observed that the disruption of a DNA segment by T4 gene 32 binding
protein occurred more readily with negatively twisted SV40 DNA
than with nicked circular SV40 DNA.

The linking difference of a covalently closed double-stranded
DNA has a strong effect on any process which involves the unwinding
or winding of the DNA helix (Davidson, 1972; Bauer and Vinograd,
1970; Hsieh and Wang, 1975). Conversely, any process which is
strongly affected by the linking difference probably involves a
step which requires the unwinding of the DNA helix, either by
disrupting base pairing or by changing the helix configuration.
Examples are transcription by *E. coli* RNA polymerase (Botchan et al.
1973; Wang, 1974), reduction of the linking difference by *E. coli*
and *M. luteus* ω proteins (Wang, 1971; Kung and Wang, 1977), binding
of *lac* repressor to *lac* operator (Wang et al., 1974), nicking of
DNA duplex by single-strand specific nucleases (Kato et al., 1973;
Beard et al., 1974; Wang, 1974). λ *int* protein promoted
recombination (Mizuuchi and Nash, 1976), incorporation of single-
stranded DNA into an underwound DNA (Holloman et al., 1975), and
the endonucleolytic cleavage of double stranded øX DNA by the *cis*
A protein (Ikeda et al., 1976; Scott et al., 1977). The linking
difference of a DNA might affect both the initiation step and the
propagation step of DNA replication (to be discussed in a later
section).

Strand Separation During DNA Synthesis

Though strand separation under physiological conditions is
energetically unfavorable, when strand separation is accompanied
by DNA synthesis using one or both of the strands as the template,
the overall reaction is the conversion of deoxynucleotide triphosphates to single-stranded or double-stranded DNA, which is
clearly favorable energetically. Nevertheless, it appears that
strand separation imposes a formidable barrier to DNA synthesis.

Consider first the case with a single-stranded DNA as the
template. Under physiological conditions a typical natural single-
stranded DNA exhibits a fair amount of secondary structure due to
intrastrand base pair formation. Because of the short size of the

helical runs and sequence mismatching, the stability of the helical
segments in a single-stranded DNA is usually lower than that of a
perfect double helix. The single-stranded DNA from phage fd for
example, exhibits a braod melting profile with a melting temperature
much lower than that of a double-stranded DNA of similar base
composition (Marvin and Hohn, 1969). Among purified DNA polymerases,
E. coli polymerase I is capable of extended copying of a single-
stranded template, but *E. coli* DNA polymerase II and III and T4 DNA
polymerase are incapable of such synthesis (cf. the review by
Gefter, 1975). The addition of a particular DNA binding protein
and or other factors frequently stimulates synthesis on a single-
stranded template: examples are the effect of T4 gene 32 protein
on T4 DNA polymerase (Huberman et al., 1971), the effect of *E. coli*
DNA binding protein on *E. coli* DNA polymerase II (Sigal et al., 1972;
Gefter et al., 1975), and the combined effect of *E. coli* DNA
binding protein and other factors on *E. coli* DNA polymerase III
(Hurwitz and Wickner, 1974; Livingston et al., 1975). For the
E. coli protein-DNA polymerase II combination, the destabilization
of the intrastrand helical structure by the binding protein
contributes to the extent of template copying (Sherman and Gefter,
1976). The rate of synthesis is also enhanced by the formation of
a specific complex between the binding protein and the polymerase.
This complex in turn forms a ternary complex with the DNA template.
The low dissociation rate of the ternary complex causes a processive mode of synthesis, which increases the rate of synthesis
(Gefter et al., 1975). In the absence of the DNA binding protein,
the mode of synthesis by DNA polymerase II is distributive, and the
rate of synthesis is limited by the rate of polymerase binding to
DNA (Gefter et al., 1975).

For DNA synthesis involving strand displacement, purified
E. coli DNA polymerases II and III and T4 DNA polymerase are
unable to carry out the process. Although *E. coli* DNA polymerase
I is capable of strand displacement, this process appears to be
rather inefficient. Once strand displacement occurred, the
polymerase may switch template to copy the displaced strand. Or
branch migration might occur, resulting in the displacement of
the newly synthesized strand containing the 3' hydroxyl terminus
by the originally displaced strand (Masamune and Richardson, 1971;
Kelly et al., 1970; Heijneker et al., 1973; Schildkraut et al.,
1964). Similar results were observed with a combination of T4 DNA
polymerase and T4 gene 32 protein (Nossal, 1974).

Use of Underwound DNA Templates

The above discussion indicates that in order to facilitate
DNA synthesis involving strand displacement, it might be necessary
to destabilize the pairing between the *parental* strands. Though
the addition of a single-strand specific binding protein appears

to offer a solution, it also imposes a dilemma. Any agent which
destabilizes pairing between parental strands must at the same
time destabilize pairing between the newly synthesized strand and
the template strand, which may lead to template switching or
branch migration.

One solution to the dilemma is the use of a negatively
twisted DNA. When short RNA or DNA fragments containing 3' OH
termini are annealed to underwound DNA molecules to serve as primers,
E. coli DNA polymerase I can efficiently lengthen the priming
fragments and separate the parental strands, until the DNA is
relaxed, (Liu and Wang, 1975). Note that the negative linking
difference specifically destabilizes pairing between parental
strands and has no effect on the pairing between the newly
synthesized strand and its template.

Active Unwinding of the parental Strands

The idea that strand separation might be actively driven by
a structure in a replicating DNA was raised many years ago by
Cairns (1963b). In the last several years, enzymes which can
actively unwind the DNA duplex have been found in a number of
cases, and their biological functions have been receiving
increasing attention. Friedman and Smith (1973) found that an ATP
requiring exonuclease from *H. influenza* yielded, during the early
stage of digestion of linear DNA molecules, high molecular weight
products with long single-stranded tails with 3' OH termini. The
DNA fragments released from the original duplex were also large,
and only a fraction of which was acid soluble. They suggested
that several enzyme molecules might act cooperatively to melt out
a region of the DNA, using ATP hydrolysis to supply the necessary
free energy. A single-strand specific endonucleolytic activity
associated with the enzyme could release the DNA fragments by
cleaving the melted portion of the duplex.

Independently, MacKay and Linn (1974) found that during the
early stage of digestion, the *rec*BC ATPase-nuclease from *E. coli*
gave duplex DNA molecules with single-stranded tails several
thousand nucleotides long and single-stranded fragments several
hundred nucleotides long. These intermediates were formed even
in synchronized reactions with DNA in molar excess over the enzyme.
For the duplex molecules with single-stranded ends, the single-
stranded ends were found to possess 3' or 5' ends with equal
frequency. MacKay and Linn suggested that the enzyme could unwind
the DNA duplex from an end. The enzyme would act as a barrier to
the reassociation of the strands. The long single-stranded ends
were formed by a mechanism in which the enzyme would track down
one strand, cleave it every few hundred nucleotides, while

remaining bound to the terminus of the other strand, which could either have 3' or 5' termini.

The active unwinding of the duplex by *rec*BC nuclease has been demonstrated by the addition of the *E. coli* DNA binding protein to the reaction mixture (MacKay and Linn, 1976). The binding protein presumably binds to the unwound strands and blocks the nucleolytic action of the *rec*BC enzyme. The net reaction then becomes the active separation of the DNA duplex to protein covered single-strands. It is interesting to point out that the final product in the above reaction is thermodynamically unstable with respect to renaturation. Renaturation under physiological conditions is slow, however, even in the presence of the *E. coli* DNA binding protein, unless a polyamine is present (Christiansen and Baldwin, 1977).

Hoffmann-Berling and his coworkers (Abdel-Monem and Hoffmann-Berling, 1976; Abdel-Monem et al., 1976,1977a) isolated another DNA-dependent ATPase capable of unwinding the DNA duplex. This enzyme is a fibrous protein of a peptide weight of 180,000. It can unwind a DNA duplex if the duplex is covalently linked to a single-stranded region. Unwinding of the DNA required dephosphorylation of ATP, and the unwinding of each DNA molecule appear to require many enzyme molecules.

Two groups have found that the *rep* gene product of *E. coli*, which is required for phage ϕX replication, is a DNA-dependent ATPase. Scott et al. (1977) have found that the *rep* protein exhibits a single-stranded DNA dependent ATPase activity. It has been demonstrated that with covalently closed double-stranded ϕX DNA as the substrate, upon cleavage of the plus-strand by the ϕX *cis* A protein, separation of the plus- and minus-strand can occur by the action of the *rep* protein in the presence of ATP and the *E. coli* DNA binding protein. The binding protein presumably facilitates the reaction by binding to the unwound strands and therefore prevents (kinetically) renaturation. Abdel-Monem et al. (1977b,c) have shown that the *rep* gene product is a DNA unwinding ATP-α-phosphohydrolase with a molecular weight of 75,000. It is probably identical to the ATPase described by Richet and Kohiyama (1976). In the system used by Scott et al. (1977), the action of the *rep* protein is catalytic, and strand separation is achieved by a few *rep* protein molecules. In the system used by Abdel-Monem et al. (1977b,c), strand separation is achieved only by a large number of *rep* protein molecules.

Kolodner and Richardson (1977) have found that the T7 gene 4 protein is a single-stranded DNA-dependent nucleoside 5'-triphosphate phosphohydrolase (NTPase). It has no effect on DNA synthesis by T7 DNA polymerase on a single-stranded template. For synthesis initiating at single-strand scissions in duplex DNA

molecules, however, the elongation of the strand with the 3' OH
terminus by T7 DNA polymerase, which involves strand displacement,
can occur only if the gene 4 enzyme is functioning. The joint
action is specific for the pair: neither T4 DNA polymerase nor
E.coli DNA polymerase I,II and III are effective in the presence of
the T7 gene 4 product. Approximately 10 gene 4 protein molecules
are required per T7 DNA polymerase, and 4-6 triphosphates are
hydrolyzed per base pair unwound.

The Enzyme Gyrase

A unique ATP requiring *E.coli* enzyme gyrase has been
discovered by Gellert et al. (1976a). *In vitro* this enzyme converts
a covalently closed circular DNA duplex to a highly negatively
twisted supercoil. The enzyme requires ATP, which is hydrolyzed
during the reaction. It is shown that an antibiotic, novobiocin,
which inhibits DNA replication *in vivo*, inhibits gyrase *in vitro*.
Gyrase isolated from a novobiocin resistant strain is not
inhibited by the antibiotic (Gellert et al., 1976b). These results
strongly suggest that gyrase is required for DNA replication *in vivo*.
If the major role of gyrase is to facilitate strand separation, it
could achieve this indirectly by constantly maintaining the DNA in a
negatively twisted state, or more directly by unwinding the duplex
close to the advancing replicating fork.

The Requirement of a Swivel in Strand Separation

The possible requirement of a swivel for the separation of two
intertwined parental strands has been discussed extensively by a
number of authors (Delbrück and Stent, 1957; Cairns, 1963a,b; Cairns
and Davern, 1967). The requirement of a swivel at certain if not
all times during the replication cycle is best seen for the
replication of a covalently closed circular DNA duplex: without a
swivel, complete separation of the strands is topologically
impossible (Cairns, 1963a,b; Vinograd and Lebowitz, 1966).

The simplest swivel would be a single-strand scission in one
of the parental strands. It would be necessary to seal the scission
before the replication fork passes through this point, otherwise
the continuity of the progeny molecule could not be maintained.
Presumably, in the presence of excess ligase, an endonuclease which
makes single-chain scissions could fulfill the role of a swivel.

Another class of enzymes which could fulfill the role of a
swivel are enzymes which can break and rejoin, in a contiguous way,
DNA backbone bonds. Those found in the prokaryotic organisms have
been referred to as ω proteins (Wang, 1971, 1973; Küng and Wang,

1977); those found in eukaryotic organisms have been referred to as untwisting activity (Champoux and Dulbecco, 1972), relaxation activity (Keller, 1975a), ω protein (Baase and Wang, 1974; Pulley-blank and Morgan, 1975) and nicking-closing enzyme (Vosberg and Vinograd, 1976). Since all these enzymes are detected by their ability to isomerize topological isomers of DNA perhaps they might be called DNA *topoisomerases*. It was postulated that the breaking of a DNA backbone by such an enzyme might be associated with the simultaneous formation of a covalent DNA-protein bond; the dissociation of the protein from the DNA could occur only by reforming the DNA backbone bond (Wang, 1971). Recent evidence support such a mechanism (Depew et al., 1976; Champoux, 1976; Keller, W., personal communcation). Studies on the ϕX *cis* A protein indicate that this protein can also break and rejoin DNA backbone bonds by a similar mechanism (see Kornberg, this volume). These enzymes will be reviewed in detail elsewhere.

Since the enzyme gyrase can alter the superhelicity of a covalently closed circular DNA duplex, it must also contain an activity which can break and rejoin DNA backbone bonds. Thus it could fulfill the dual roles necessary for strand separation of a closed circular DNA; active unwinding and providing a swivel in the meantime. The possible relation between the two mutually antagonistic activities gyrase and ω protein, and how the two activities are regulated *in vivo*, remain to be elucidated.

The Segregation of Progeny DNA Rings

For replicating circular DNA duplex, the final segregation of progeny DNA rings requires the complete unwinding of one parental strand around the other. In the meantime, the circularity of both progeny molecules is maintained. How is the complete unwinding of the parental strands achieved without interrupting the circularity of the progeny molecules has been a puzzling question to molecular biologists. A model has been proposed, in which circularity is maintained by base sequence redundancy straddling the point at which a scission is introduced to allow the complete separation of the parental strands (Gefter, 1975).

Recently, it has been shown that the *E.coli* ω protein can catalyze the conversion of a single-strand ring to a knotted ring in which topological knots have been introduced into the ring. Furthermore, the role of the protein ω is a catalytic one: under conditions the simple ring without knots is thermodynamically favored, the protein can convert the knotted form back to the simple ring form (Liu et al., 1976). Topologically, the untying of a knot has the same characteristics as the segregation of two intertwined rings: the DNA strand must be broken, the knot untied, and at the same time the circularity of the product is preserved.

Though the *E.coli* ω protein is the only enzyme studied to date for the interconversion between simple and knotted rings, it is expected that other topisomerases can also carry out such a reaction. It is tempting to think that the segregation of the progeny rings can simply be achieved by such an enzyme. The question then becomes, how does such an enzyme do it ?

It is perhaps appropriate to end this brief review with a question mark. Though many problems related to strand separation have been realized since the double-helix structure of DNA was first proposed, it is only in recent years that the possible roles of special enzymes necessary for strand separation have received much attention. Many questions remain to be answered.

REFERENCES

1. Abdel-Monem, M., Dürwald, H. and Hoffman-Berling, H. (1976) Eur. J. Biochem. 65:441

2. Abdel-Monem, M., Lauppe, H.-F., Kartenbeck, J., Dürwald, H. and Hoffmann-Berling, H. (1977a). J. Mol. Biol. 110:667

3. Abdel-Monem, M., Chanal, M-C. and Hoffmann-Berling, H. (1977b) Eur. J. Biochem., in press

4. Abdel-Monem, M., Dürwald, H. and Hoffmann-Berling, H. (1977c) Eur. J. Biochem, in press

5. Abdel-Monem, M. and Hoffmann-Berling, H. (1976) Eur. J. Biochem. 65:431

6. Baase, W.A. and Wang, J.C. (1974) Biochemistry 13:4299

7. Baba, Y. and Kagemoto, A. (1974) Biopolymers 13:339

8. Bauer, W.R., Ressner, E.C., Kates, J. and Patzke, J.V. (1977) Proc. Natl. Acad. Sci. USA 74:1841

9. Bauer, W.R. and Vinograd, J. (1970) J. Mol. Biol. 47:419

10. Beard, P., Morrow, J.F. and Berg, P. (1973) J. Virol. 12:1303

11. Botchan, P., Wang, J.C. and Echols, H. (1973) Proc. Natl. Acad. Sci. USA 70:3077

12. Cairns, J. (1963a) J. Mol. Biol. 6:208

13. Cairns, J. (1963b) Cold Spring Harbor Symp. Quant. Biol. 28:43

14. Cairns, J. and Davern, C.I. (1967)   J. Cell Physiol.   70: Suppl. 1, 65

15. Champoux, J.J. (1976) Proc. Natl. Acad. Sci. USA 73:8488

16. Champoux, J.J. and Dulbecco, R. (1972) Proc. Natl. Acad. Sci. USA 69:143

17. Christiansen, C. and Baldwin, R.L. (1977) J. Mol. Biol., in press

18. Davidson, N. (1972) J. Mol. Biol. 66:307

19. Delbrück, M. and Stent, G.S. (1957) In: The Chemical Basis of Heredity (W.D. McElroy and B. Glass, eds.) The Johns Hopkins Press, Baltimore, pp699

20. Delius, H., Mantell, N.J. and Alberts, B. (1972) J. Mol. Biol. 67:341

21. Depew, R.E., Liu, L.F. and Wang, J.C. (1976) Fed. Proc. Amer. Soc. Exp. Biol. 35:1493 (Abstr. No. 684).

22. Depew, R.E. and Wang, J.C. (1975) Proc. Natl. Acad. Sci. USA 72:4280

23. Friedman, E.A. and Smith, H.O. (1973) Nature New Biol. 241:54

24. Gefter, M.L. (1975) Ann. Rev. Biochem. 44:45

25. Gefter, M.L., Molineux, I.J., Pauli, A. and Sherman, L. (1976) In: DNA Synthesis and Its Regulation. (M. Goulian and P. Hanawalt, eds.) W.A. Benjamin Inc., Menlow Park, Calif. p2

26. Gellert, M., Mizuuchi, K., O'Dea, M.H. and Nash, H.A. (1976a) Proc. Natl. Acad. Sci. USA 73:3872

27. Gellert, M., O'Dea, M.H., Itoh, T. and Tomizawa, J. (1976b) Proc. Natl. Acad. Sci. USA 73:4474

28. Heijneker, H.L., Ellens, D.J., Tjeerde, R.H., Glickman, B., van Dorp, V., and Pouwels, P.H. (1973) Mol. Gen. Genet. 124:83

29. Holloman, W.K., Wiegand, R., Hoessli, C. and Radding, C.M. (1975) Proc. Natl. Acad. Sci. USA 72:2394

30. Hsieh, T.-S. and Wang, J.C. (1975) Biochemistry 14:527

31. Huberman, J.A., Kornberg, A. and Alberts, B.M. (1971) J. Mol. Biol. 62:39

32. Hurwitz, J. and Wickner, S. (1974) Proc. Nat. Acad. Sci. USA 71:6

33. Ikeda, J., Yudelevich, W. and Hurwitz, J. (1976) Proc. Natl. Acad. Sci. USA 73:2669

34. Kato, A.C., Bartok, K., Fraser, M.J. and Denhardt, D.T. (1973) Biochim. Biophys. Acta 308:68

35. Keller, W. (1975a) Proc. Natl. Acad. Sci. USA 72:2550

36. Keller, W. (1975b) Proc. Natl. Acad. Sci. USA 72:4876

37. Kelly, R.B., Cozzarelli, N.R., Deutscher, M.P., Lehman, I.R. and Kornberg, A. (1970). J. Biol. Chem. 245:39

38. Klump, H. and Ackermann, T. (1971) Biopolymers 10:513

39. Kolodner, R. and Richardson, C.C. (1977) Proc. Natl. Acad. Sci. USA 74:1525

40. Kung, V.W. and Wang, J.C. (1977) J. Biol. Chem., in press

41. Liu, L.F., Depew, R.E. and Wang, J.C. (1976) J. Mol. Biol. 106:439

42. Liu, L.F. and Wang, J.C. (1976) In: DNA Synthesis and Its Regulation (M. Goulian and P. Hanawalt, eds.) W.A. Benjamin, Inc., Menlo Park, Calif. p38

43. Livingston, D.M., Hinckle, D.C. and Richardson, C.C. (1974) J. Biol. Chem. 250:461

44. MacKay, V. and Linn, S.(1974) J. Biol. Chem. 249:4286

45. MacKay, V. and Linn, S. (1976) J. Biol. Chem. 251:3716

46. Marvin, D.A. and Hohn, B. (1969) Bacteriol. Rev. 33:172

47. Masamune, Y. and Richardson, C.C. (1971) J. Biol. Chem. 246:2692

48. Mizuuchi, K. and Nash, H.A. (1976) Proc. Natl. Acad. Sci. USA 73:3524

49. Nossal, N.G. (1974) J. Biol. Chem., 249:5668

50. Privalov, P.L. and Ptitsyn, O.B. (1969) Biopolymers 8:559

51. Pulleyblank, D.E. and Morgan, A.R. (1975) Biochemistry, 23: 5205

52. Pulleyblank, D.E., Shure, M., Tang, D., Vinograd, J. and Vosberg, H. P. (1975) Proc. Natl. Acad. Sci. USA 72:4280

53. Richet, E. and Kohiyama, M. (1976) J. Biol. Chem. 251:808

54. Ruyechan, W.T. and Wetmur, J.G. (1975) Biochemistry 14:5529

55. Ruyechan, W.T. and Wetmur, J.G. (1976) Biochemistry 15:5057

56. Schildkraut, C.L., Richardson, C.C. and Kornberg, A. (1964) J. Mol. Biol. 9:24

57. Scott, J.F., Eisenberg, S., Bertsch, L.L. and Kornberg, A. (1977) Proc. Natl. Acad. Sci. USA 74:193

58. Sherman, L.A. and Gefter, M.L. (1976) J. Mol. Biol. 103:61

59. Sigal, N., Delius, H., Kornberg, T., Gefter, M.L. and Alberts, B. (1972) Proc. Natl. Acad. Sci. USA 69:3537

60. Shiao, D.D.F. and Sturtevant, J.M. (1973) Biopolymers 12:1829

61. Tinoco, I. Jr., Borer, P.N., Dengler, B., Levine, M.D., Uhlenbeck, O.C., Crothers, D.M. and Gralla, J. (1973) Nature New Biol. 246:40

62. Vinograd, J. and Lebowitz, J. (1966) J. Gen. Physiol. 49: Suppl., 103

63. Vosberg, H.-P. and Vinograd, J. (1976) In: DNA Synthesis and Its Regulation (M. Goulian and P. Hanawalt, eds.) W.A. Benjamin, Inc., Menlo Park, Calif. p94

64. Wang, J.C. (1971) J. Mol. Biol. 55:523

65. Wang, J.C. (1973) In: DNA Synthesis *in Vitro*. (R.D. Wells and R.B. Inman, eds), University Park Press, Baltimore, Md, p153

66. Wang, J.C. (1974) J. Mol. Biol.   87:797

67. Wang, J.C., Barkley, M.D. and Bourgeois, S. (1974) Nature, 251:249

APPENDIX

An Introduction to the Energetics of
Covalently Closed Double-Stranded DNA Rings

Before discussing the energetics of covalently closed double-stranded DNA rings, it is necessary to have some intuitive feeling about the quantity called the *linking number*, and its relation to the double helical structure of DNA[1].

We shall start by performing the following experiment. Take a long strip of paper, $\sim\frac{1}{2}$ inches by 1 foot. Draw two lines parallel to the edges, labelled + and -, representing two strands of complementary sequences. The + strand goes from Z to A, with the arrows indicating the polarity of the strand. The - strand goes from A' to Z', and is antiparallel to the + strand. These are illustrated in Fig. 1(a). There is no helical turn between the two

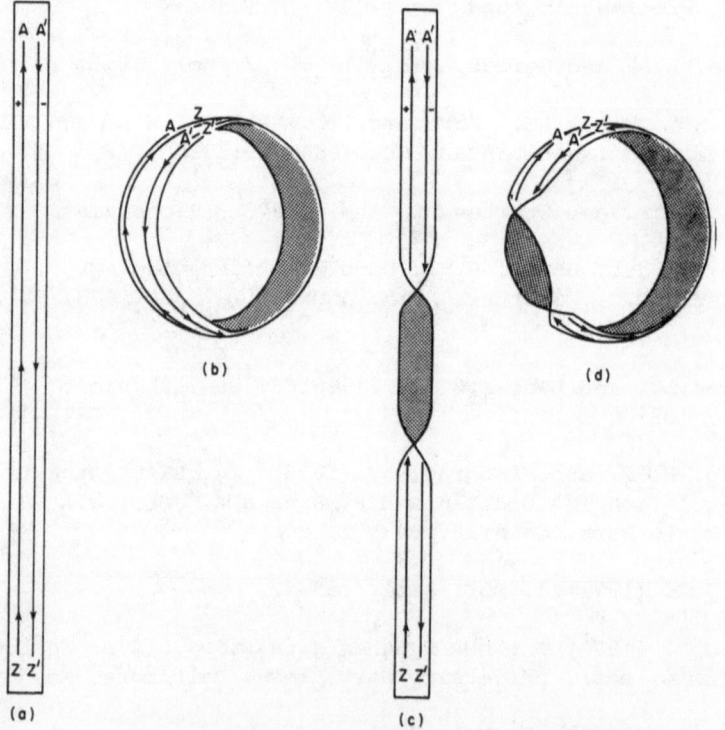

Figure 1. See text for descriptions.

---

[1] *For further discussions, cf. F.H.C. Crick (1976), Proc. Natl. Acad. Sci. USA 73:2639, and references therein.*

linear strands shown in Fig. 1(a). We can cyclize the linear
strands by joining Z to A and Z' to A', as shown in Fig. 1(b). It
is easy to see that the two circular strands shown in Fig. 1(b) are
not linked. If we take a pair of scissors and cut the strip into
two along its center, we end up with two separate rings.

We now form one helical turn between the two linear strands
shown in Fig. 1(a), by holding the top of the strip with one hand
and rotate the bottom part along the longitudinal axis a full turn,
as illustrated in Fig. 1(c). We could either form a right-handed
turn, as shown in Fig. 1(c), or form a left-handed turn by
rotating in the other way. Again we cyclize the structure shown
in 1(c) to give the one shown in 1(d).[2] Now the two circular
strands are linked with a linking number 1: this can be seen by
cutting the strip into two along the center. Had we started with
a left-handed helical turn, we would have obtained two linked
rings distinguishable from the two linked rings resulting from a
right-handed helical turn. Therefore, the linking number can be
either plus or minus. By convention the two rings shown in Fig.
1(d) have a linking number $\underline{\alpha}=+1$. Since we are almost always dealing
with right-handed turns in DNA, α is almost always positive.

We continue our experimentation by starting with the structure
shown in Fig. 1(a), form two right-handed turns between the strands,
cyclize, and split the strip. The resulting two rings are linked
by a linking number $\underline{\alpha}=2$.

It should be emphasized that although the linking number and the
helical turns are related, there is a fundamental difference
between the two quantities. The linking number is a topological
quantity independent of the structural details of the component
rings. Disruption of all base pairs or the formation of many more
base pairs will not change $\underline{\alpha}$ from 1 for the structure shown in
Fig. 1(d). The quantity $\underline{\alpha}$ is an invariant provided that the
continuity of the strands is not disrupted, even transiently. We
shall not concern ourselves here about the rigorous relation
between the linking number, the helical turns, and the spatial
configuration of the strip[1].

We now consider an experiment in which a population of identical
circular DNA molecules, each containing a single-chain scission
per molecule, are converted to the covalently closed form by ligase.
What would be the average value of α for the product molecules ?
Intuitively, it is clear that the average should be around N/k,

---

[2] *Note that the + strand must join to the + strand and the − strand
must join to the − strand when cyclization is performed. Therefore
we are not allowed to rotate the strip by half a turn to form
a Mobius strip.*

where N is the number of base pairs per DNA molecule and k is the number of base pairs per DNA helical turn under the conditions that the ligase reaction is carried out. We also expect, that since molecules with different values of $\underline{\alpha}$ differ in free energy, there might be a distribution in $\underline{\alpha}$ values for the product molecules if the free energy differences are comparable to the thermal energy kT per molecule or RT per mole of molecules (k and R are the Boltzmann and the gas constant respectively).

When such an experiment is carried out, a distribution in α is clearly seen. This is illustrated in Fig. 2. We let α_m be the $\underline{\alpha}$ value for the most abundant species, then the $\underline{\alpha}$ values of all other species can be assigned as $\alpha_m \pm 1$, $\alpha_m \pm 2$........etc (cf. Depew and Wang, 1975). We can easily calculate the free energy difference between any two species from their relative amounts in the equilibrium mixture. The free energy difference ΔG^o_i between a species with $\underline{\alpha} = \alpha_m + i$ and the species $\underline{\alpha} = \alpha_m$ is, from the relation between the standard free energy change and the equilibrium constant,

$$\Delta G^o_i \equiv G^o(\alpha_m + i) - G^o(\alpha_m)$$

$$= -RT \ln \{ \{\alpha_m + i\} / \{\alpha_m\} \}$$

where $G^o(\alpha_m + i)$ and $G^o(\alpha_m)$ and $(\alpha_m + i)$ and (α_m) are the standard free energies and the concentrations of the species with $\underline{\alpha} = \alpha_m + i$ and $\underline{\alpha} = \alpha_m$ respectively. Experimentally, it is found that when the absolute value of i is reasonably large (say $|i| > 3$),

$$\Delta G^o_i = Ki^2$$

where K is roughly a constant inversely proportional to the size of the DNA molecule. In a physiological medium, $K = 1000\ RT/N$, where N is the number of base pairs per DNA molecule (Depew and Wang, 1975; Pulleyblank et a., 1975). Note that as defined above, ΔG^o_i is the standard free energy change for the conversion of a closed circular DNA with $\underline{\alpha} = \alpha_m$ to $\underline{\alpha} = \alpha_m + i$; it is always positive. We refer the quantity i as the *linking difference*[3].

So far we have only considered individual species with integral $\underline{\alpha}$ values. Similar equations can be written for populations of molecules, with quantities for the population averages

_____

[3]
The term linking difference corresponds roughly to the term super-helical turns used in earlier literatures (cf. refs. 62 and 9). The new term is rigorously defined here in terms of the topological parameters.

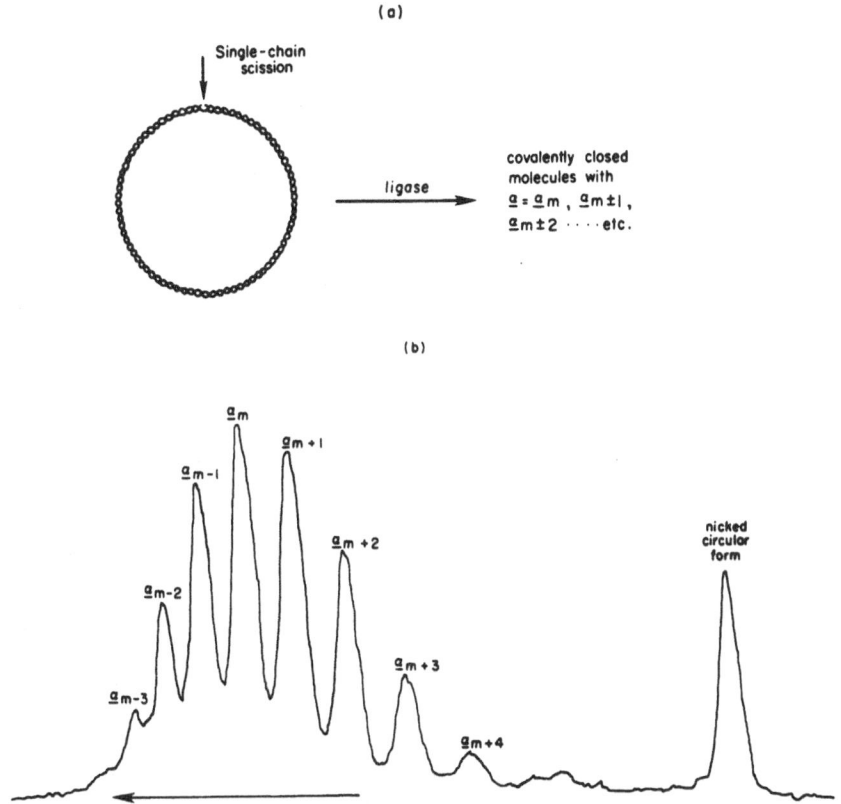

Figure 2. *(a) The covalent closure of circular double-stranded DNA molecules, each containing a single-chain scission, to give a population of molecules with $<\alpha> = \alpha_m, \alpha_m \pm 1, \alpha_m \pm 2$ ---- etc. (b) Electrophoretic pattern in agarose gel of the DNA sample after covalent closure by ligase. The slowest band is nicked circular DNA not sealed by ligase. The values of α for the group of covalently closed species are indicated in the figure. For details cf. refs. 21 and 52.*

substituting for the quantities for the individual species. We define $<\alpha_m>$ as the population average of α for a population of molecules which are "relaxed", in the sense that $<\alpha_m>$ will not be changed by a cycle of breaking and rejoining of backbone bonds. Note that although each individual species has an integral value of α, the population average $<\alpha_m> \equiv \Sigma \, x_j \alpha_j$, where x_j is the mole fraction of the species with $\alpha = \alpha_j$, is in general not an integer. For a population of molecules with $<\alpha> = <\alpha_m> + <i>$, the molecules are referred to as *overwound* if $<i> > 0$ and *underwound* if $<i> < 0$. The terms positively twisted or positively superhelical have also been used for the case $<i> > 0$, and the terms negatively twisted or negatively superhelical for the care $<i> < 0$.

We can now consider the energetics of strand separation for a covalently closed DNA. Suppose we have an *underwound* DNA with

$$<\underline{\alpha}> = <\underline{\alpha}_m> + <i>$$

$$= <\underline{\alpha}_m> - p$$

where $p = -<i>$ is a positive number. The DNA can be "relaxed" by the denaturation of $p\,k$ base pairs. This can be seen by considering a simple numerical example. Consider a DNA molecule with $N=1000$ base pairs and $<\underline{\alpha}> = 90$. Assume that $k = 10$. We expect $<\underline{\alpha}_m> = 100$, and $k = -<i> = 10$. Now we denature $p\,k = 100$ base pairs, leaving 900 base pairs in the helical region. The quantity $<\underline{\alpha}>$ is unchanged and is still 90. But 90 is the expected $<\underline{\alpha}_m>$ value for a helical DNA with 900 base pairs. Therefore the original DNA with 100 base pairs *kept in an unpaired state* is "relaxed" in the sense that breaking and rejoining of a backbone bond will not change $<\underline{\alpha}>$.

Fig. 3 illustrates a path for the conversion of an underwound DNA with a linking difference $<i>$ to one with $p\,k$ base pairs denatured but is otherwise "relaxed". The first step is the relaxation of the DNA, the standard free energy change is $-\Delta G_i^o$. The second step is the introduction of a single-chain scission; the standard free energy change is designated as ΔG_1^o. The third step is the disruption of $p\,k(p = -<i>)$ base pairs; the standard free energy change is ΔG_d^o. The final step is the rejoining of the nick introduced in the second step; the standard free energy change is $-\Delta G_1^o$. Note that the overall free energy change, $-\Delta G_i^o + \Delta G_d^o$, which differs from the free energy change ΔG_d^o for the disruption of the same number of base pairs in a nicked circular (or linear) DNA by the term $-\Delta G_i^o$. Note that $-\Delta G_i^o = -Ki^2$ is always negative; therefore the overall free energy change for the process illustrated in Fig. 3 is less than that for the second step. As a numerical example, we consider SV40 DNA. The linking difference $<i>$ is -26 (Keller, 1975b) and $N = 5300$. Thus $-\Delta G_i^o = 1000\ RT\ (-26)^2/5300$ or 80 kcal. This is to be compared with a ΔG_d^o of $\sim$310 kcal for the denaturation of 260 base pairs.

For a small *unwinding* of the DNA helix by ∂p turns, it can be shown that a DNA with a linking difference of $<i>$ turns is associated with an additional free energy term

$$-\partial \Delta G_i^o = \left(\frac{d\Delta G\ i}{di}\right)\delta p$$

$$= 2K<i>\delta p$$

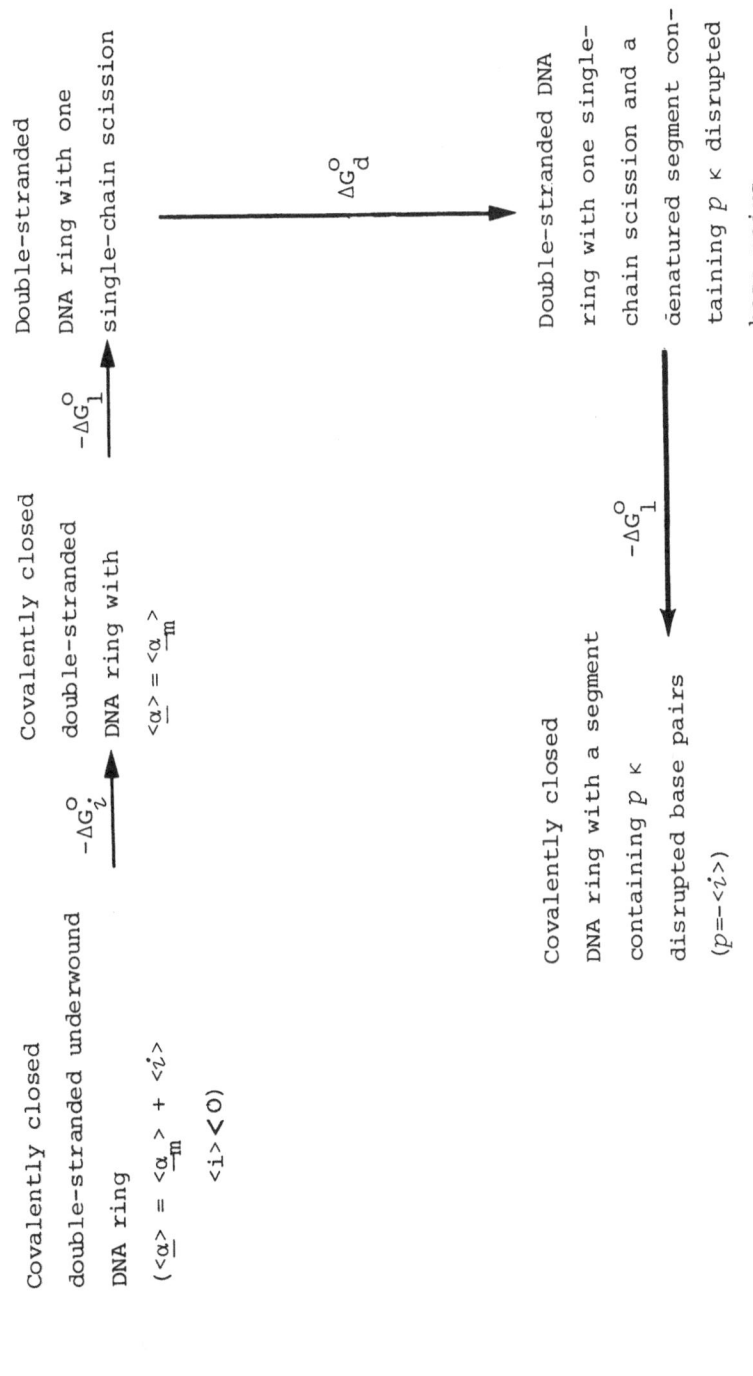

Figure 3. The disruption of p κ base pairs in an underwound covalently closed DNA with a linking difference $\langle i \rangle = -p$. The overall free energy change is $-\Delta G_i^o + \Delta G_d^o$. The free energy change for the disruption of the same number of base pairs in a nicked DNA is ΔG_d^o.

relative to the same DNA with a linking difference of zero. Again
taking SV40 DNA as an example, the disruption of one base pair or
the unwinding of 0.1 turns is associated with a free energy of
2 x 1000 RT x(-26)x 0.1/5300 or - 0.6 kcal.

MULTIPLE INTERACTIONS OF A DNA BINDING PROTEIN, GENE-32 PROTEIN OF PHAGE T4, DURING DNA REPLICATION AND RECOMBINATION

G. Mosig[*], A. Breschkin, R. Dannenberg and S. Bock

Department of Molecular Biology
Vanderbilt University
Nashville, Tenn. 37235, U.S.A.

INTRODUCTION

It is now generally accepted that proteins involved in DNA synthesis function as complexes (Kornberg, 1974; Alberts et al., 1975). Genetic and biochemical studies have shown that, in addition to other proteins, DNA binding proteins, DNA polymerases, ligases and nucleases are required in various steps of DNA replication, repair of radiation damages and genetic recombination. While some of these proteins are essential in all processes, others appear to be important in only some, e.g., certain nucleases function in recombination, but not in DNA elongation. How do these proteins interact in the various steps of DNA metabolism ? Our results indicate that they function in different ways in DNA replication or recombination, forming different complexes. They interact with a DNA-binding protein which coordinates their different actions on DNA.

Phage T4 provides an excellent model system for analyzing such interactions because many conditional lethal mutations in several genes affecting DNA metabolism are known and readily manipulated to facilitate combined biochemical and genetic experiments. In T4, certain mutations affecting recombination also affect DNA replication and progeny production, but the causal relationship of these defects is not known (for reviews see Mosig, 1970; Miller, 1975; Broker and Doermann, 1975). It should be recalled that molecular recombination is required for progeny production of T4 because of its packaging mechanism: mature DNA molecules in virions represent circular permutations of the genetic map and have

A.S.I. Participant

approximately 3% of their sequences repeated at both ends as
"terminal redundancies". After infection, recombination between
terminal redundancies or between terminal regions of different
molecules (Figure 1) has to generate DNA molecules which are longer

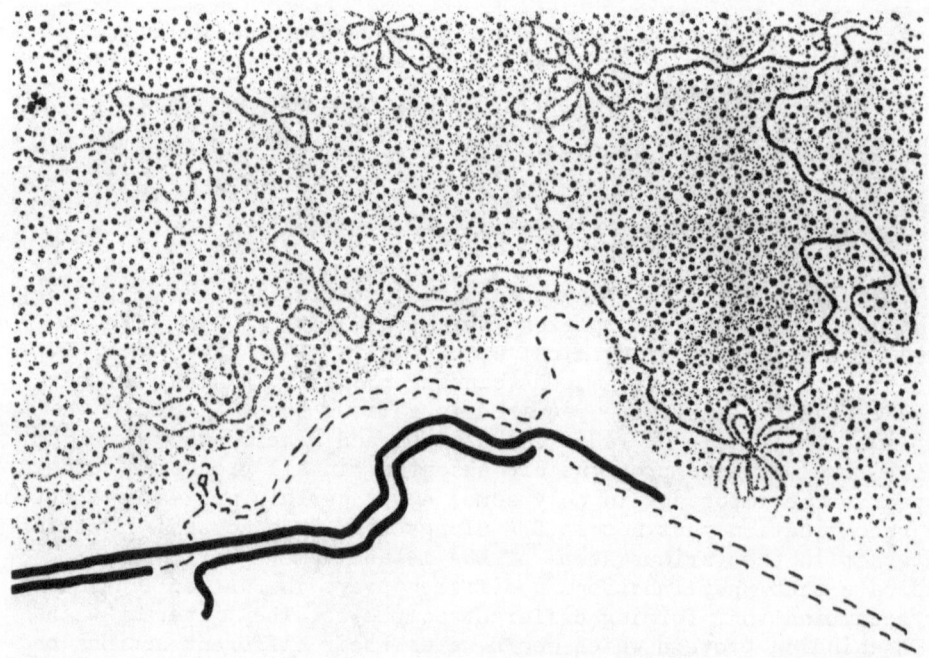

Figure 1. *A looped recombinational intermediate, isolated early
after infection with replication proficient T4. Two ends at
different genetic locations have become single-stranded and are
invading homologous regions as postulated on the basis of genetic
experiments (Mosig et al., 1971). Many of the recombination forks
but few of the replicative forks have long single-stranded whiskers
(Mosig, Dannenberg and Breschkin, 1975; Dannenberg and Mosig,
manuscript in preparation).*

than unit length ("concatemers") because mature molecules are
packaged and cut to "head-full" sizes from such concatemers
(Streisinger, Emrich and Stahl, 1967). Thus, in T4, mutations are
lethal when they abolish all recombination as well as when they
interfere with DNA replication.

We have focussed our analysis on a DNA binding protein of
phage T4, gene-32 protein, because it has a key role in various
steps of DNA metabolism (for review see Broker and Doermann, 1975).
It has been postulated that gene-32 protein functions in DNA
replication and in recombination by (i) unwinding duplex DNA, (ii)
facilitating reannealing and/or (iii) protecting single-stranded
DNA from nucleolytic degradation. To what extent are these functions
a consequence of the binding of gene-32 protein to DNA ? There is no
evidence that pure gene-32 protein unwinds DNA in vivo or in vitro
(Wang, this volume). Our recent experiments, summarized below,
have shown that gene-32 protein has multiple interaction sites
which result in multiple functions. While the N-terminal end is
essential to binding to DNA, other specific N-terminal regions
interact with DNA ligase, DNA polymerase, and with membrane
components, and the C-terminal domain is important in moderating
recombination nucleases. These interactions can be differentially
inactivated by certain mutations as shown in Figure 2. We conclude

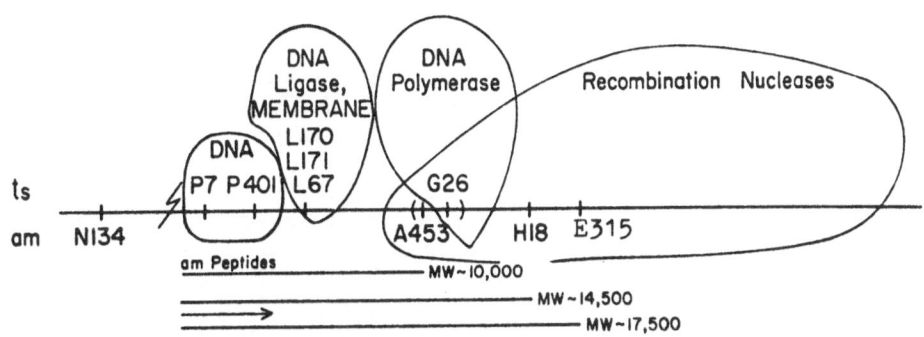

Figure 2. Map of gene-32 mutations (Mosig et al., 1977) and
interactions which are affected by these mutations. The apparent
molecular weights of the am peptides have been determined in
polyacrylamide gels (Krisch et al., 1974; Gold, O'Farrell and
Russel, 1976). Molecular weight of the wild type protein has been
reported to be 35,000 (Alberts and Frey, 1970).

from these and other results that these interactions are important
in forming a complex that coordinates various steps in
recombination (Figure 3) as well as in initiation of DNA replication.
To some extent, these interactions must be of a different nature
than interactions in replicative DNA elongation. Most remarkably,
the short am peptide of the am mutant A453 must contain all
interaction sites that are essential for DNA replication. This

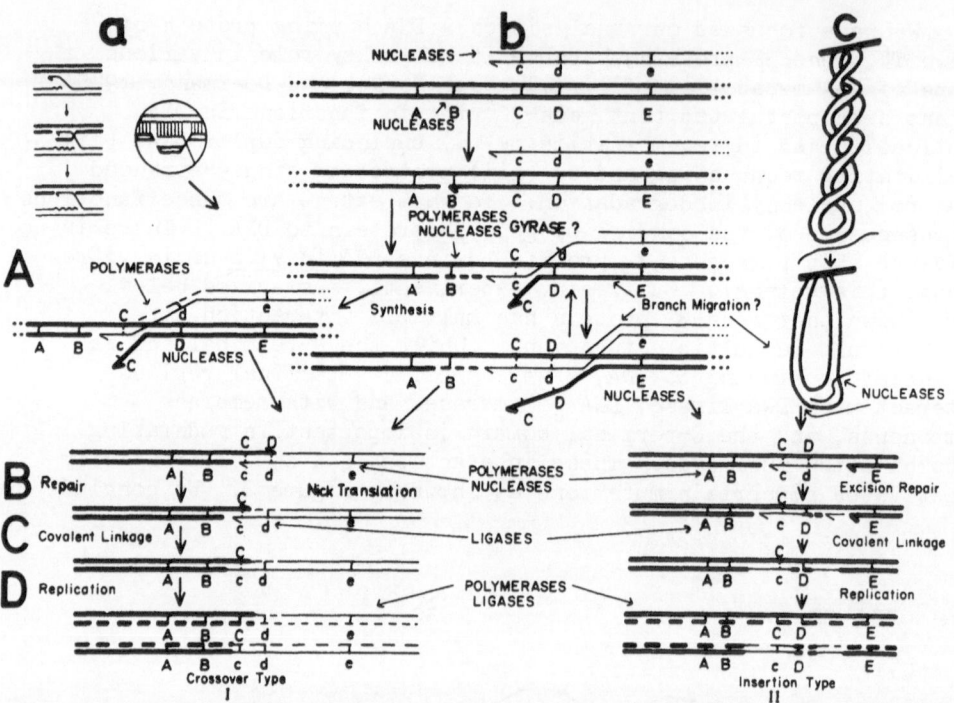

*Figure 3. A diagram of some current models of general genetic recom-
bination modified from Whitehouse (1963) and Holliday (1964) to show
the postulated roles of various recombination enzymes. Branched intermediates (Broker and Lehman, 1971) are formed first (step A). The
top part summarizes three plausible and mutually compatible models
for initiation of base pairing in heteroduplex regions of branched
intermediates: (a) displacement synthesis generates single-stranded
segments which invade hmologous DNA (Hotchkiss, 1974; Meselson and
Radding, 1975), (b) nucleolytic degradation from blunt ends generates
invading single-stranded segments (Broker and Lehman, 1971; Mosig et
al., 1971), (c) supercoiled DNA takes up single-stranded* in vitro
*(Holloman et al., 1976). Depending on how branched intermediates are
resolved, crossover-type (mode I, lower left panel) or insertion-type
(mode II, lower right panel) recombinants are generated. Hetero-du-
plex regions may be partially repaired (step B) before or after exchanged strands are covalently linked (step C). Finally DNA replication (step D) resolves unrepaired heteroduplexes to true recombinants.
We propose that most, if not all, of these steps are coordinated by
binding of the responsible enzymes to a DNA binding protein (Figure
2). Missing or defective enzymes will often result in aberrant intermediates. Parental DNA: thick or thin solid lines. Newly synthesized DNA: broken lines. 5' ends are marked with arrows. Genetic alleles of different parental origin are shown as capital or lower case
letters respectively. They are marked on both DNA strands only when
heteroduplexes contain different alleles.*

perhaps explains why all available gene-32 ts mutations map within this relatively short promoter proximal segment of the gene 32. On the other hand, the C-terminal domain is essential in modulating recombination nucleases while the very C-terminal end appears to be dispensable.

We have used a genetic approach, outlined in Figure 4, to help understand the biologically important functions of specific interactions of gene-32 protein with DNA and with other proteins. This approach is particularly useful when a protein participates in different biological functions and when it has no enzymatic activity of its own. Furthermore, this approach is unaffected by potential proteolytic degradation of any of the interacting components.

If gene-32 protein participates in a different manner in DNA replication than in recombination by interacting with DNA and with different sets of proteins (as shown in Figure 4 center and right panels), we expect that mutations which abolish binding of gene-32 protein to DNA affect most or all recombination and replication steps. On the other hand, mutations that specifically inactivate certain protein-protein interactions might affect only certain steps. Thus, we have asked whether certain steps in recombination as well as in DNA replication are blocked under restrictive conditions for progeny production in each of the different gene-32 mutants shown in Figure 2. To do this we have differentially labeled parental and progeny DNA with a combination or radioactive and density isotopes and investigated the intracellular DNA in each of the mutants after infection (under conditions that are restrictive for progeny production) by density and sedimentation analyses and by electron microscopy. We have also measured incorporation of precursors into DNA and DNA degradation by various methods. This allows us to distinguish replication of parental DNA (primary replication) from subsequent rounds (secondary replication) and to detect recombinational intermediates under conditions which do not allow progeny production.

In addition, we have characterized by these methods, secondary mutations in other phage or host genes which influence phenotypic expression of certain "primary" gene-32 mutations. This analysis detects two different kinds of interactions: (i) host functions that may partially substitute for and thus mask specific defects in certain gene-32 mutants (cf. Figure 4, left panel) and (ii) secondary mutations in another gene that may restore interactions with the corresponding protein gene product (Figure 4, center panel). (Of course, combinations of certain mutations might completely abolish any residual interaction of the corresponding proteins and thus show "negative suppression"). The model shown in Figure 2 provides the most consistent summary of our results, which are described in detail elsewhere (see references below). Here we

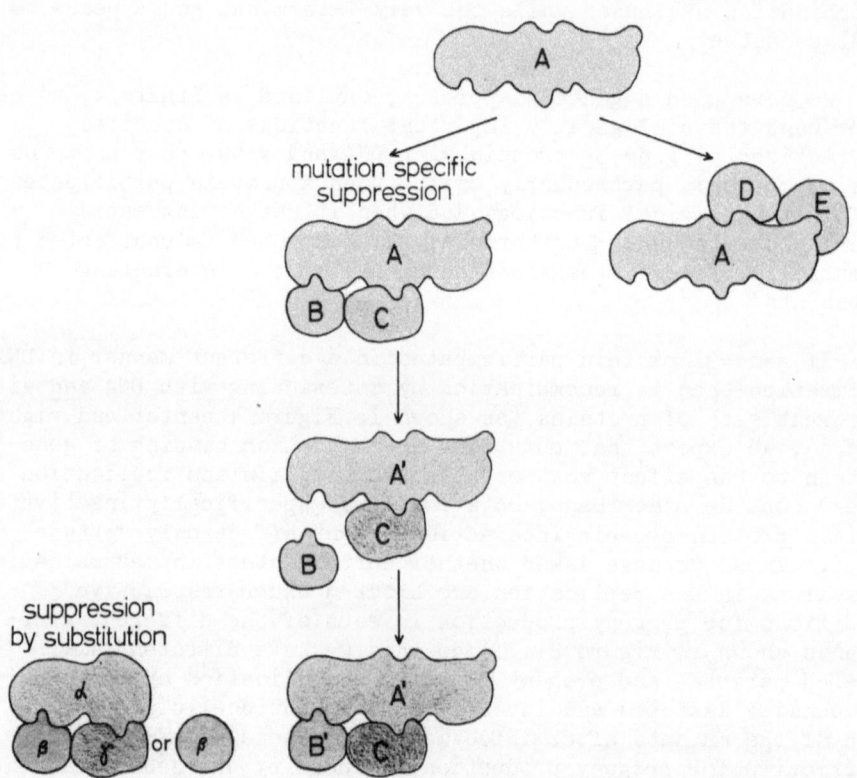

Figure 4. _A genetic approach to analyze interactions of gene_
products by suppressor studies. _If protein A participates in two_
different processes by interacting with different cellular
components, we expect that certain mutations differentially
inactivate interactions with B, thus inactivating B function,
without necessarily affecting other functions. _In this situation_
we expect that certain site specific mutations in gene B partially
restore interaction of B' with the altered A' protein. _Other_
mutations in gene B may have the opposite or not effect.
Alternatively a protein β or a protein complex containing β may
substitute for the defective interaction of A with B. _These types_
of interactions predict that suppression is mutation specific
(at least in one of the genes). _Steric and allosteric interactions_
cannot be distinguished. _Genetic suppression by more indirect_
mechanisms (e.g., affecting precursor pools, rates of synthesis,
etc) is expected to be gene specific but not mutation specific.

emphasize only selected aspects which led us to formulate this model and relate them to other findings:

(1) All gene-32 am mutants, including am A453 which produces the shortest am peptide, replicate as much of their parental DNA as gene-32$^+$ phage (Figure 5), using the am peptides, and not substituting host functions, for this replication (Breschkin and Mosig, 1977a,b).

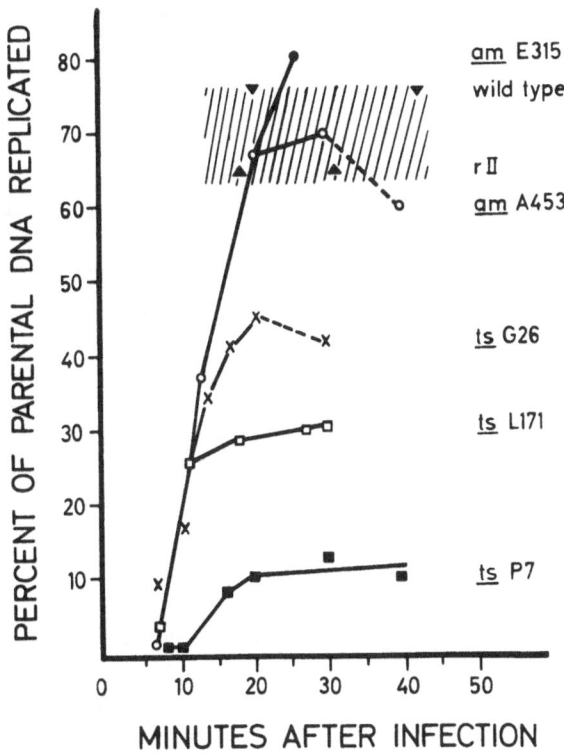

Figure 5. Time course of replication of parental DNA of different gene-32 mutants under restrictive conditions for progeny production. The shaded area indicates the range of wild-type DNA replication (Breschkin and Mosig, 1977b).

(2)   In the am mutants as well as in ts G26, but not in the other ts mutants, DNA is excessively degraded by recombination nucleases (Mosig and Bock, 1976), i.e., certain mutant peptides are unable to modulate recombination nucleases.

(3)   In contrast to the am mutations, all gene-32 ts mutations affect DNA replication albeit to a different extent. The closer they map to the beginning of the gene the more they interfere with primary replication. In fact, only the most promoter proximal mutations tsP7 and tsP401 are completely defective in all gene-32 functions, presumably because their peptides do not bind to DNA at restrictive temperatures in vivo (Breschkin and Mosig, 1977a) as well as in vitro (Curtis and Alberts, 1976). In this respect the gene-32 protein resembles another DNA binding protein: the lac repressor (Müller-Hill, 1975) whose binding to DNA is mainly affected by promoter proximal mutations.

(4)   Residual DNA replication of tsP7, but not of the other gene-32 mutants, depends on $dnaC^+$ and $dnaG^+$ host genes. It is important to recall that tsP7 whose peptide does not bind to DNA, utilizes the same host genes for its replication as incomplete T4 chromosomes, most of which lack gene-32 and thus do not produce gene-32 protein (Mosig et al., 1972). Functions of E.coli dnaC and dnaG protein have been extensively discussed during this A.S.I. Since there is no evidence that either of these proteins functions in a similar manner as T4 gene-32 protein it is plausible to assume that one (or more) host protein complex(es) involving dnaC and dnaG protein can substitute in vivo for a defective T4 complex containing gene-32 protein. Apparently this substitution occurs only if mutant gene-32 protein does not block access of host proteins to the T4 DNA. It has been shown by Huberman et al. (1971) that DNA binding proteins of E.coli and T4 specifically stimulate their "homologous" DNA polymerases in vitro. Thus, it is noteworthy that tsP7 DNA replication in vivo requires functional T4 DNA polymerase and not E.coli dnaE product (Breschkin and Mosig, 1977b), suggesting that subcomplexes of heterologous replication proteins can interact in vivo to a certain extent.

(5)   Specific interaction of T4 DNA polymerase with T4 gene-32 protein is, however, essential in initiation of DNA replication and recombination, and it must be of a different nature than the interaction for which dnaC and dnaG products can substitute. When this interaction is defective (in tsG26), E.coli DNA polymerase I can partially substitute for the (T4 DNA polymerase-gene-32 protein) complex. We believe that DNA polymerase I can substitute for the tsG26 defect because it can catalyze displacement synthesis (Kornberg, 1974) while T4 DNA polymerase requires additional gene-32 protein to perform this function (Nossal, 1974). Both initiation of DNA replication (Robberson et al., 1972; Tomizawa, 1975) and recombination (Hotchkiss, 1974; Meselson and Radding,1975;

LeClerc and Setlow, 1975) are thought to involve displacement synthesis.

(6) Coinfecting amA453 particles (whose DNA is unlabeled) help replication of labeled tsP7 DNA. This does not reflect intergenic complementation nor classical intragenic complementation; there is no progeny production under these conditions. In fact, while the amA453 peptide (which presumably binds to DNA) promotes replication of tsP7 DNA (whose peptide does not bind), it also renders tsP7 DNA susceptible to excessive degradation. (Recall that DNA of tsP7 alone is not degraded). This and other results (see below) suggest that recombination nucleases require modification by gene-32 protein to promote nucleolytic action and to prevent excessive degradation during recombination. Presumably, after infection with tsP7 under restrictive conditions, its DNA is not degraded because a substituting host complex is bound to DNA and this host complex has no affinity for, nor allows action of, the T4 recombination nucleases (which are controlled or coded for by genes 46 and 47). The situation, as well as its consequence is probably different when bacteria are infected with tsP7 under permissive conditions and then are shifted to restrictive temperatures (Curtis and Alberts, 1976). Under the latter conditions, tsP7 protein at first binds to DNA and thus might facilitate excessive nuclease activity after the temperature shift.

(7) Under permissive conditions for progeny production, certain gene-32 mutants are deficient in insertion-type recombination, but none are deficient in crossover-type recombination (Mosig et al., 1977), (see Figure 3). Under these conditions heteroduplex regions are not shorter in any gene-32 mutants, but are in fact longer in some of them. This indicates that formation of branched recombinational intermediates (see Figures 1 and 3) is normal under these conditions. In certain gene-32 mutants, heteroduplex lengths are increased under these conditions. We believe that these mutants are also deficient in heteroduplex repair for two reasons: (i) the gene-32 mutants are more sensitive to UV irradiation than wild type under these conditions (Mosig et al., in preparation) and (ii) many recombinational forks contain single-stranded whiskers (see Figure 1). These whiskers most likely represent single strands which were displaced during heteroduplex formation. They measure 200 to 2000 nucleotide residues or more. Thus, they are larger than genetic estimates of heteroduplex lengths in mature particles (for reviews see Mosig, 1970; Broker and Doermann, 1975), supporting the view that in T4 there is heteroduplex repair in vivo (Berger and Pardoll, 1976).

(8) Gene-32 mutations affect the rate of gene-32 protein synthesis. This autoregulation occurs at the level of translation (Krisch et al., 1974; Gold et al., 1976; Krisch and van Houwe,

1976; Krisch et al., 1977). It has been postulated that this autoregulation is related to defective binding of mutant proteins to DNA. Results of Wood and Bernstein (1977), together with our results, imply that the effects of these mutations on auto-regulation are more or less independent of their effects on binding to DNA. Note also that over-production of certain mutant peptides apparently does not greatly affect residual activity in DNA replication. Although peptides of all am mutants are over-produced (as seen in pulse-label experiments), the short amA453 peptide does not accumulate (i.e., it is rapidly degraded), while the long amE315 peptide does persist (our unpublished experiments). Nevertheless, amA453 and amE315 replication are similar (Figure 5). Apparently, the E315 peptide can fold into a stable (protease resistant) form even though its C-terminal domain is missing, while the A453 peptide shows imperfect folding.

Our model of multiple gene-32 interactions is further supported by analysis of secondary suppressor mutations (cf. Figure 4). One example is the mutual suppression of one gene-32 mutation (tsL171) and rII mutations (Mosig and Breschkin, 1975; Mosig et al., 1977). The rII proteins are membrane proteins (Weintraub and Frankel, 1972; Ennis and Kievitt, 1973). The mutual suppression of rII and the gene-32 mutation L171 (which involves other membrane proteins and phage and host ligase) suggests that ligation of recombining DNA, if not all recombination, occurs at the membrane. Membrane attachment would provide topological constraints and thus permit unwinding of linear and nicked DNA by DNA gyrase (Gellert et al., 1976) and/or by other unwinding proteins (Abdel-Monem et al., 1976).

Another example of our analysis is the suppression of a T4 ligase (gene 30) mutation (tsB20) by two "silent" alterations of gene-32. The ligase mutant B20 synthesizes DNA after infection but this DNA is later degraded in a similar pattern as in gene-32 am mutants, even though functional gene-32 protein is present. DNA degradation in the ligase mutant B20 is suppressed by a certain point mutation at the promoter distal portion of gene-32 (isolated by us) or by a deletion del 9 isolated by Homyk and Weil (1974). In del 9 the C-terminal end of 32 protein is probably substituted with a segment of another protein (our unpublished results). We conclude that the interaction of ligase with gene-32 protein is important in moderation of recombination nucleases. All of these results indicate that the entire recombination complex is involved in coordinating different recombination steps and in preventing excessive DNA degradation in vivo. Since gene-32 protein interacts with many components of this recombination complex, gene-32 mutations may differentially affect various recombination steps.

One of the major conclusions from our analysis is that several proteins which participate in DNA replication and recombination (as well as in repair) (e.g., DNA polymerase and DNA ligase) are modified and channeled into recombination complexes by associating in a different manner at specific regions with a DNA binding protein. We want to speculate on two possible implications:

(1) It is intriguing to note that mutations in several recombination genes (gene 32, gene 30 (DNA ligase), genes 46 and 47 recombination nuclease) show premature cessation of DNA replication (i.e., have a "DNA arrest" phenotype). It is also noteworthy that all gene-32 mutations affect recombination and initiation of DNA replication coordinately, while most other gene-32 functions are differentially affected (Breschkin and Mosig, 1977b). Early T4 recombination generates forks in DNA molecules when ends of molecules invade the homologous regions in the interior regions of other molecules (see Figure 1). These forks differ from replication forks by the presence of single-stranded whiskers shown in Figure 1 (Broker and Lehman, 1971; Mosig et al., 1975; Reed et al., 1976; Dannenberg and Mosig, manuscript in preparation). Perhaps such recombinational forks can be converted to replication forks and likewise recombination complexes are converted to replication complexes after the whiskers have been eliminated and nucleases dissociate from the complex. This process would provide an alternate way of initiating DNA replication which does not depend on recognition of unique origin DNA sequences by specific proteins. If these whiskers are not eliminated, recombination forks might act as roadblocks for replicative forks initiated at the origin. Thus, DNA replication ceases in 46/47 mutants because the recombination nuclease is defective and in the tsP7 gene-32 mutant because the nuclease is not activated. In the gene-32 am mutants (as well as in certain gene-30 ligase) mutants recombination nucleases work excessively at the recombination forks and thus eventually destroy the templates.

(2) The integration of tumour virus DNA sequences into host DNA by recombination may involve modification of host enzymes by virus-coded DNA binding proteins in a similar manner as the modifications by gene-32 protein that we have described. This might be one of the multiple functions of tumour antigens (Weil et al., 1974; van der Vliet, this volume; Levinson et al., 1976; Weinberg, 1977).

ACKNOWLEDGEMENTS

Research in our laboratory has been supported by NIH grant GM 13221. G.M. thanks the Alexander-von Humboldt-Foundation for generous support during sabbatical leave at the Max-Planck-Institut für Züchtungsforschung, Köln-Vogelsang.

REFERENCES

(1) ABDEL-MONEM, M., DÜRWALD, H. and HOFFMANN-BERLING, H. (1976) Eur. J. Biochem. 65,441

(2) ALBERTS, B. and FREY, L. (1970) Nature 227,1313

(3) ALBERTS, B., MORRIS, C.F., MACE, D., SINHA, N., BITTNER, M. and MORAN, L. (1976) In: DNA Synthesis and Its Regulation (Goulian, M. and Hanawalt, P., Eds) pp 241-295 W.A. Benjamin Inc. Menlo Park, Calif.

(4) BERGER, H. and PARDOLL, D. (1976) J. Virol. 20,441

(5) BRESCHKIN, A.M. and MOSIG, G. (1977a) J. Mol. Biol. 112,279

(6) BRESCHKIN, A.M. and MOSIG, G. (1977b) J. Mol. Biol. 112,295

(7) BROKER, T.R. and LEHMAN, I.R. (1971) J. Mol. Biol. 60,131

(8) BROKER, T.R. and DOERMANN, A.H. (1975) Ann. Rev. Genet. 9,213

(9) CURTIS, M.J. and ALBERTS, B. (1976) J. Mol. Biol. 102,793

(10) ENNIS, H.L. and KIEVITT, K.D. (1973) Proc. Nat. Acad. Sci., U.S.A. 70,1468

(11) GELLERT, M., O'DEA, M.H., ITOH, T., and TOMIZAWA, J.-I. (1976) Proc. Nat. Acad. Sci., U.S.A. 73,4474

(12) GOLD, L., O'FARRELL, P.Z. and RUSSEL, M. (1976) J. Biol. Chem. 251,7251

(13) HOLLIDAY, R. (1964) Genet. Res. 5,282

(14) HOLLOMAN, W.K. and RADDING, C. (1976) Proc. Nat. Acad. Sci., U.S.A. 73,3910

(15) HOMYK, T. Jr., and WEIL, J. (1974) Virology 61,505

(16) HOTCHKISS, R.D.(1974) Ann. Rev. Microbiol. 28,445

(17) HUANG, W.M. and LEHMAN, I.R. (1972) J. Biol. Chem. 247,3139

(18) HUBERMAN, J., KORNBERG, A. and ALBERTS, B. (1971) J. Mol. Biol. 62,39

(19) KORNBERG, A. (1974) DNA Synthesis, W.H. Freeman and Co., San Francisco.

(20) KRISCH, H.M., BOLLE, A. and EPSTEIN, R.H. (1974) J. Mol. Biol. 88,89

(21) KRISCH, H.M. and VAN HOUWE, G. (1976) J. Mol. Biol. 108,67

(22) KRISCH, H.M., VAN HOUWE, G., BELIN, D., GIBBS, W. and EPSTEIN, R.H. (1977) Virology, 78,87

(23) LECLERC, E.J. and SETLOW, J.K. (1975) J. Bacteriol. 122,1091

(24) LEVINSON, A., LEVINE, A.J., ANDERSON, S., OSBORN, M., ROSENWIRTH, B. and WEBER, K. (1976) Cell 7,575

(25) MESELSON, M.S. and RADDING, C.M. (1975) Proc. Nat. Acad. Sci. U.S.A. 72,358

(26) MILLER, R.C. Jr. (1975) Ann. Rev. Microbiol. 29,355

(27) MOSIG, G. (1970) In: Advances in Genetics Vol. 15, Academic Press, New York.

(28) MOSIG, G. and BOCK, S. (1976) J. Virol. 17,756

(29) MOSIG, G. and BRESCHKIN, A. (1975) Proc. Nat. Acad. Sci., U.S.A. 72,1226

(30) MOSIG, G., BOWDEN, D.W. and BOCK, S. (1972) Nature New Biol. 240,12

(31) MOSIG, G., BERQUIST, W. and BOCK, S. (1977) Genetics 86,5

(32) MOSIG, G., EHRING, R., SCHLIEWEN, W. and BOCK, S. (1971) Molec. gen. Genetic. 113, 51

(33) NOSSAL, N.C. (1974) J. Biol. Chem. 249,5668

(34) REED, S.I., STARK, G.R. and ALWINE, J.C. (1976) Proc. Nat. Acad. Sci., U.S.A. 73, 3083

(35) ROBBERSON, D.L., KASAMATSU, H. and VINOGRAD, J. (1972) Proc. Nat. Acad. Sci., U.S.A. 69,737

(36) STREISINGER, G., EMRICH, J. and STAHL, M.M. (1967) Proc. Nat. Acad. Sci., U.S.A. 57,292

(37) WEIL, R., SALOMON, C., MAY, E. and MAY, P. (1974) Cold Spring Harbor Symp. Quant. Biol. 39,381

(38) WEINBERG, R.A. (1977) Cell 11,243

(39) WEINTRAUB, S.B. and FRANKEL, F.R. (1972) J. Mol. Biol. 70,589

(40) WHITEHOUSE, H.L.L. (1963) Nature 199, 1034

(41) WOOD, W.J. Jr., and BERNSTEIN, H. (1977) J. Virol. 21,619

INTERACTION OF PROTEIN HD FROM E.COLI WITH NUCLEIC ACIDS

Verena Berthold[*] and Klaus Geider

Max-Planck-Institut fur Medizinische Forschung, Abteilung
Molekulare Biologie
Jahnst. 29, 6900 Heidelberg, F.R.G.

INTRODUCTION

DNA binding proteins play an important role in the regulation
of nucleic acid metabolism in cells and in the life cycle of
bacteriophages or animal viruses. They can influence the structure
of nucleic acids and interfere with enzymatic activities by their
binding or can directly associate with other proteins.

RESULTS AND DISCUSSION

We have isolated a low molecular weight DNA binding protein
from E.coli (1) and called it protein HD for its heat resistance and
binding to double-stranded DNA. It also binds to single-stranded
DNA and to RNA. An E.coli cell may contain at least 10,000 copies
of the protein which was isolated according to the scheme of Table 1.

Physical Properties of HD Protein

Amino acid analysis reveals 85 residues. The molecular weight
of the monomer is 9,000 in agreement with its migration on polyacrylamide dodecylsulphate gels. In the native state it occurs as
tetramer shown by cross-linking with dimethylsuberimidate. Clusters
of lysine and arginine in its amino acid sequence (2) may give
histone-like properties, although its isoelectric point is around
7.5. Met-Ala-Lys at the N-terminus of the protein is shared with
the end of other DNA binding proteins.

[*] A.S.I. Participant

Table 1.   Isolation Scheme for Protein HD

E.coli cells,

mechanical disruption,

2 M NaCl, 10% PEG: removal of nucleic acids,

dialysis

↓

DNA cellulose column (native or denatured)

0.4 M salt eluate, dialysis

↓

DEAE cellulose pass through

↓

carboxylmethylcellulose,

0.3 M salt elution

Binding to Nucleic Acids

Complexes of protein HD with single-stranded phage fd DNA,
double-stranded phage T7 DNA or phage fr RNA show an increase of
the sedimentation rate compared to the free nucleic acids. The
gradual shift in sedimentation of the DNAs for increasing amounts
of protein HD indicate non-cooperative binding which was also
demonstrated for the RNA in a filter binding assay. Competition
experiments showed no preference for one or the other class of the
nucleic acids checked. Therefore we assume there might be a roughly
equal distribution of the protein among cellular RNA and double- and
single-stranded DNA. Preparation of washed ribosomes are also
enriched for protein HD (unpublished results). Subramanian and
co-workers (personal communication) find the association correlated
to the growth cycle of E.coli cells.

Interaction with Single-Stranded DNA·Protein Complexes

Single-stranded phage fd-DNA requires DNA binding protein I
(3,4) for its conversion to double-stranded replicative form (5).This
step of DNA replication can be inhibited by the addition of protein
HD (1) or phage coded gene 5 protein (5). The ring like structure of
the fd DNA·DNA binding protein I complex is converted into
characteristic structures upon the addition of one of the two
proteins (6). Transfection experiments with fd-DNA indicated that
gene 5 protein can displace all of the DNA binding protein I, but
protein HD can only substitute some of the binding protein I (7).

Experiments with ^{125}I-labelled DNA binding protein I confirmed these results, showing a complete displacement of the label comigrating with the fd DNA in the presence of gene 5 protein (8). Protein HD can displace not more than half of the binding protein I, no matter which of the proteins is added first.

Stability of the Protein Complexes with fd DNA

The complex formation of single-stranded DNA with DNA binding proteins is fast, even at 0°C. Protein HD, gene 5 protein and DNA binding protein I protect the DNA against nucleolytic attack. Kinetics show a long term protection for the two latter proteins, but protection with HD protein was expressed only within the first minutes of incubation and the DNA finally completely degraded by the endonuclease (8). This reflects a fast dissociation of HD protein according to its non-cooperative binding. The cooperative binding of DNA binding protein I and gene 5 protein cause very little exposure of the DNA to nucleolytic degradation. On the other hand, DNA binding protein I is exchanged fast in its complex with DNA.

Effects of Protein HD on Nucleic Acid Metabolism

The DNA binding protein HD inhibits DNA synthesis with DNA polymerase I or DNA polymerase III holoenzyme. The binding to the template may stop chain elongation by displacement of the polymerase or - for DNA polymerase III holoenzyme - by removing a part of DNA binding protein I. Protein HD may also disturb the 3' end of the priming DNA which is then no longer suitable for chain elongation. A similar mechanism can be assumed for the inhibition of exonuclease III by HD protein. Degradation of linear single-stranded DNA by exonuclease I is not effected by the protein, and transcription of double-stranded DNA by RNA polymerase is slightly stimulated (1).

The properties of HD protein are compared in Table 2 with DNA binding protein I and phage fd coded gene 5 protein.

CONCLUDING REMARKS

Protein HD is similar to DNA binding proteins described from other laboratories (9,10,11,12). They may play a role in the organization of the DNA in the cell like the histones in chromatin.

Furthermore, these proteins may have a regulatory function in nucleic acid metabolism like enhancing transcription at specific sites (9) or slowing down DNA replication (1). They could also protect RNA against fast degradation or influence translation by their binding to ribosomes. As protein HD has no strong affinity to one type of nucleic acids, an equal distribution may be rendered more specific by its possible interaction with other cellular proteins.

Table 2.   Properties of DNA Binding Proteins

	HD Protein	DNA Binding Protein I	Phage fd Gene 5 Protein
binding to SS-DNA	+	+	+
cooperative binding to SS-DNA	-	+	+
binding to DS-DNA	+	-	-
binding to RNA	+	±	-
inhibiton of DNA synthesis	+	-	+
inhibition of transfection	-	+	-

REFERENCES

(1) BERTHOLD,V. and GEIDER, K. (1976) Eur. J. Biochem. 71, 443

(2) TSUGITA, A. manuscript in preparation

(3) SIGAL, N., DELIUS, H., KORNBERG, T., GEFTER, M.L. and
     ALBERTS, B. (1972) Proc. Natl. Acad. Sci. USA, 69, 3537

(4) WEINER, J.H., BERTSCH, L.L. and KORNBERG, A. (1975) J. Biol.
     Chem. 250, 1972

(5) GEIDER, K. and KORNBERG, A. (1974) J. Biol. Chem. 249, 3999

(6) ZENTGRAF, H., BERTHOLD, V. and GEIDER, K. (1977) Biochim.
     Biophys. Acta 474, 629

(7) UHLMANN, A. and GEIDER, K. (1977) Biochim. Biophys. Acta 474, 639

(8) GEIDER, K. manuscript in preparation

(9) CREPIN, M., CUKIER-KAHN, R. and GROS, F. (1975) Proc. Natl. Acad. Sci. USA 72, 333

(10) ROUVIÈRE-YANIV, J. and GROS, F. (1975) Proc. Natl. Acad. Sci. USA 72, 3428

(11) GOSH, S. and ECHOLS, H. (1972) Proc. Natl. Acad. Sci. USA 69, 3660

(12) KOHIYAMA, M., KOLLEK, R., GOEBEL, W. and KEPES, A. (1977) J. Bacteriol. 129, 658

A PUTATIVE DNA GYRASE MUTANT OF *E. Coli*

Steven J. Projan and James A. Wechsler*

Department of Biological Sciences
Columbia University
New York, New York, 10027, U.S.A.

INTRODUCTION

The *E. coli* chromosome, as well as plasmid DNA, many viral genomes, and mitochondrial DNA can be isolated as supercoiled molecules. The chromosome is presumably supercoiled in the native state as it can be quantitatively isolated as a folded structure even when replication is not occurring and in the absence of protein synthesis (1,2). In addition, *in vitro* RNA synthesis increases with increased supercoiling of template DNA (3). Proteins which relax supercoiled DNA have been isolated from both prokaryotes and eukaryotes (4,5).

Any biological process which unwinds or overwinds the duplex of a covalently closed DNA molecule will generate superhelical twists (6). Addition of superhelical twists in parental DNA during replication is supported by analysis of replicating SV40 and mitochondrial DNA molecules (7,8). Recently, a DNA gyrase activity which supercoils relaxed cccDNA has been isolated from *E. coli* (9,10). This report discusses preliminary experiments on a temperature-sensitive lethal mutant that is apparently defective in this activity.

RESULTS AND DISCUSSION

The *dna-517* mutation in strain E517 was originally classified as a *dnaA* allele though it was noted at that time that this designation was questionable because *dna-517* could not be integratively suppressed (11). Several results led to our

A.S.I. Participant

postulating that this mutant was defective in DNA gyrase: (1) coumermycin Al-resistant cells (Cou^r) produce a coumermycin-resistant DNA gyrase and the *cou^r* locus is 95% cotransducible with *dnaA* (12 and 2) though the *dna-517* mutant synthesizes substantial amounts of DNA after the shift to restrictive temperature, the rate of protein synthesis, and presumably RNA synthesis, decreases within an hour of the first observable decline in the rate of DNA synthesis (Wechsler, unpublished). Among dna mutants this is an unusual phenotype but might be an expected consequence of a gross conformational change due to the loss of supercoiling.

Genetic evidence that the *dna-517* mutation affects the same gene product as do *cou^r* mutations was obtained by isolating ten independent sets of spontaneous temperature-resistant revertants from E517. Among each set of temperature-resistant colonies between 20% and 80% were Cou^r. Because it is impossible to determine how many of the revertants in each set arose from separate mutational events, the frequency of co-mutation to temperature-resistance and *cou^r* cannot be stated with certainty but is probably between 20% and 50%. Some revertants to temperature-resistance were also more coumermycin-sensitive than E517 or CR34, its wild-type parent. The converse experiment was done by isolating spontaneous Cou^r (to 50 µg/ml) derivatives of E517 at 30C. The number of Cou^r that were temperature-resistant ranged from approximately 5% to approximately 30% in this experiment. We, therefore, conclude that the *dna-517* and the *cou^r* mutations are probably allelic. These results do not, however, exclude the possibility that the *dna-517* and *cou^r* mutations affect different genes whose products interact.

Assuming that *dna-517* and *cou^r* both affect the same gene, cotransduction of the two phenotypes should be close to 100%. These transductions have not yet been done because of technical problems in selecting for *cou^r* transductants and because E517 reverts to temperature-resistance at too high a frequency to place confidence in such an experiment.

If E517 is defective in gyrase, covalently closed circular DNA from the mutant incubated at 42° should have a reduced number of supercoil twists. To test this, E517 and CR34 were transformed with a hybrid plasmid derived from ColEl (13). The plasmid pMB9 (supplied by Dr. H. Boyer) carries the genes for tetracycline-resistance and contains one EcoRl site (H. Boyer, personal communication). A purified transformant of each strain was grown into logarithmic phase at 30°, shifted to 42° for six hours and harvested. The plasmid was isolated from cleared lysates on CsCl-ethidum bromide density gradients (14), and the relative degree of supercoiling was measured by agarose gel electrophoresis (15).

As manipulation of the cells during harvesting and lysis could not be done at high temperature and the mutant product might quickly renature on return to low temperature, sodium azide was added to the cells at the termination of the 43° incubation. Because the effect of azide on ATP production might well reduce wild-type gyrase activity, gel electrophoresis was done in the presence of ethidium bromide (15). In the absence of ethidium bromide or in its presence when azide is omitted from the procedure, only two bands are seen with pMB9 from CR34/pMB9: One at the relaxed position near the top of the gel and one further down at the fully supercoiled position (data not shown).

In Fig. 1, the bulk of the pMB9 DNA from CR34/pMB9 forms a broad band that is slightly less than fully supercoiled. The pattern of supercoiling of the plasmid from E517/pMB9 is significantly shifted toward the relaxed position (Fig. 1C). Only four species, three of similar intensity, can be detected: one at the relaxed position, one slightly less than fully supercoiled, and two intermediate to these. *In vitro* action of gyrase on relaxed molecules produces a series of supercoiled species dependent upon the enzyme concentration (9). The reason why only four species are observed when plasmid DNA is isolated from the mutant is unclear.

Since the plasmid band from E517/pMB9 was concentrated several-fold prior to electrophoresis, it is also evident that far less cccDNA was obtained from the mutant strain. This is, in part, due to a deficiency in replication as will be shown below. Also, the method of lysis using sodium dodecyl sulfate in the cleared lysate procedure followed by CsCl gradient centrifugation results in the loss of relaxation complexes from the cccDNA band (16,17). If the plasmid were preferentially converted to relaxation complexes or otherwise not covalently closed in the mutant, the yield of plasmid obtained by this procedure would be markedly reduced. Nevertheless, the distribution of supercoiled molecules from E517/pMB9 is clearly shifted toward the relaxed position compared to plasmid from CR34/pMB9.

Examination of pMB9 replication in the mutant was done utilizing the same methodology except that the DNA was labeled with ^{3}H-thymine and the CsCl-ethidium bromide gradients were fractionated and counted. The extent of plasmid synthesis measured as cccDNA when ^{3}H-thymine is added simultaneously with the shift to 42C or 2 hours after the shift is shown in Fig. 2. Two hours was chosen, as by that time incorporation of ^{3}H-thymine into E517 chromosomal DNA has virtually ceased (Wechsler, unpublished).

Results with E517/pMB9 at 30° and CR34/pMB9 at 30° or 42° are indistinguishable (data not shown).

It is obvious that cccDNA is not synthesized after

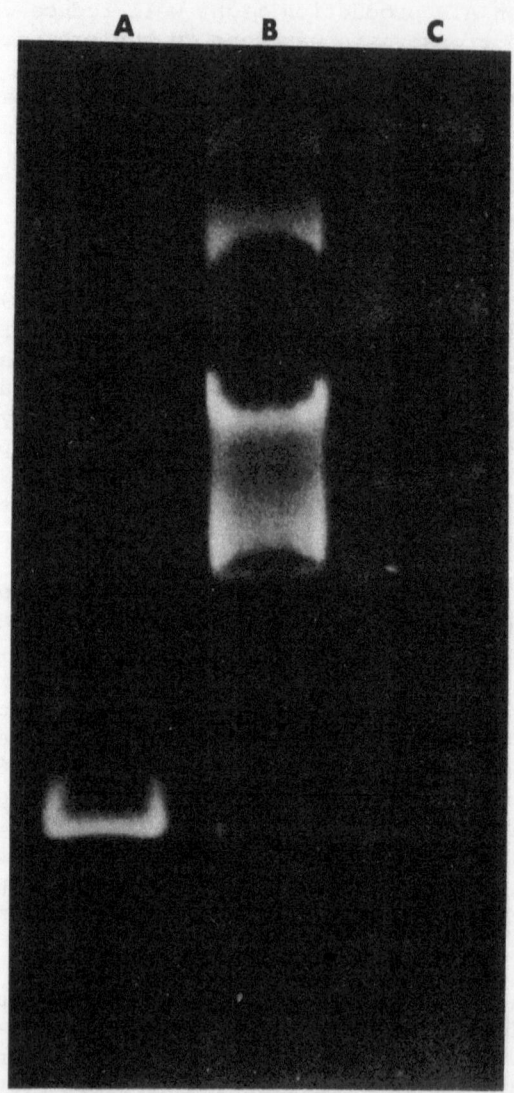

Figure 1. *Agarose gel electrophoresis of pMB9. Electrophoresis was according to Keller (15).   Ethidium bromide, 0.024 µg/ml was added to the agarose and the buffer solutions (15). (a) pMB9 treated with EcoR1; (b) pMB9 from CR34/pMB9; (c) pMB9 from E517/pMB9.*

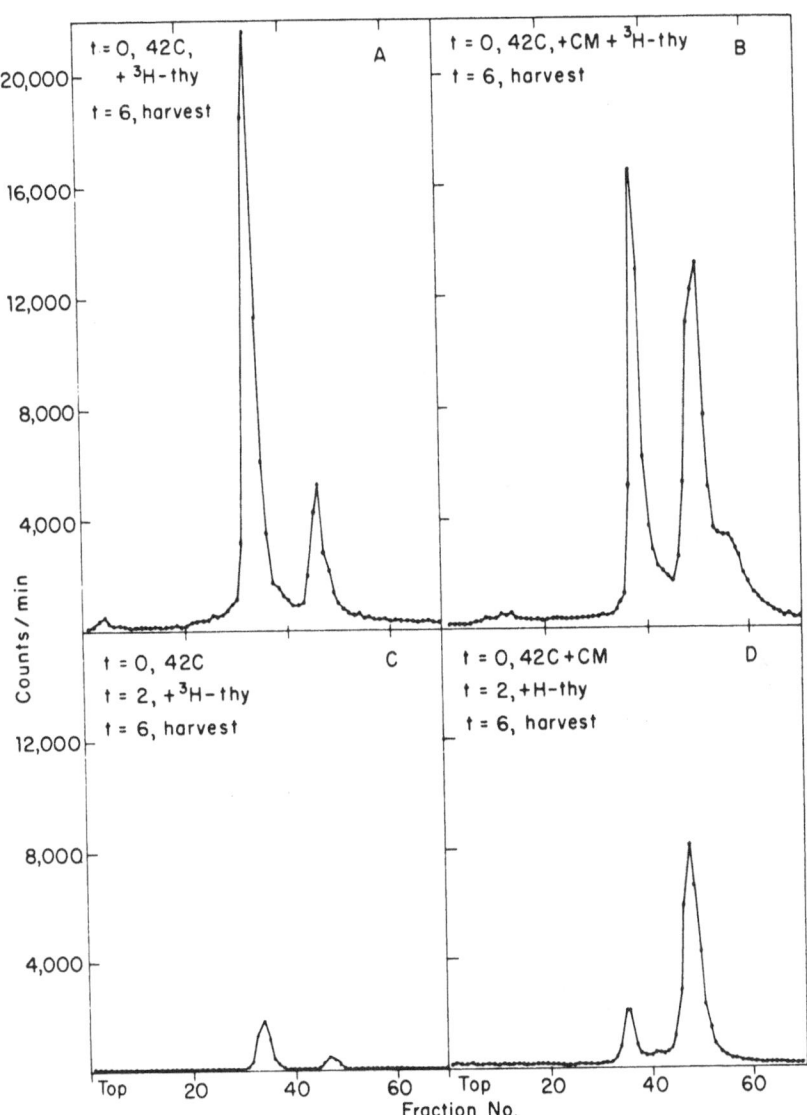

Figure 2. *CsCl-ethidium bromide gradients of cleared lysates from E517/pMB9. Cells were grown at 30° in m9-casamino acids medium containing 0.2% glucose and thymine 10 µg/ml. Logarithmic 30° cultures were transferred to 42° at 0 time. Preparation of lysates for centrifugation was as described in text.*

inactivation of the *dna-517* gene product (Fig. 2B). If the plasmid
were replicated but not converted to cccDNA, it would band at the
less dense position along with chromosomal DNA. The virtual
absence of counts at both positions shows that pMB9 is not
replicated in E517/pMB9 after 2 hours at restrictive temperature.

In the absence of protein synthesis, ColEl replication
increases (16). To test the effect of *dna-517* on the amplification
of ColEl, the experiment was done identically, except chloramphenicol (CM) was added at the time of the shift to 42°. Though
there is a decrease in the amount of radioactivity in the more
dense band in 2D relative to 2C, part of this may be explained by
the decreased labeling time (4 hrs versus 6 hrs). The proportion
of relaxation complexes among the total number of plasmid molecules
decreases rapidly in the presence of CM and is not a consideration
in this experiment (16). It would appear that amplification of
pMB9 is substantial even when the *dna-517* gene product is
inactivated.

In the experiment in Fig. 2, CM was added and amplification
synthesis begun prior to the inactivation of the *dna-517* gene
product. The experiment in Fig. 3 shows that if CM is not added
until after 2 hours of incubation at 42°, no significant plasmid
replication is observed.

In summary the *dna-517* mutation appears to be in the same
gene that can mutate to *cou*^{r}. The alternative explanation is that
the product of dna-517 physically interacts with the *cou* gene
product. Plasmid supercoiling from E517/pMB9 grown at restrictive
temperature is consistent with this strain having a defective DNA
gyrase activity. The replication of pMB9 is eliminated at 42°
approximately in parallel with the cessation of chromosome
replication. If, however, CM is added prior to inactivation of the
dna-517 product, amplification of the plasmid appears to occur
though it may be partially inhibited. If CM is not added until
after the dna-517 product is inactivated, amplification does not
occur.

The *dna-517* mutation is presumed to be leaky as there is
approximately a six-fold increase in total DNA following a shift
to restrictive temperature (Wechsler, unpublished). This amount
of residual DNA synthesis is greater than would be expected with
an initiation defective mutant. It is, however, not clear what
the expected kinetics of gyrase inactivation would be. Supercoils
will be generated by any process that adds or removes twists from
the double helix and, thus, a lack of gyrase is likely to result
in only a relative decrease in the number of supertwists.

It is possible that DNA replication could continue for long
times in spite of this reduction in supertwist density.

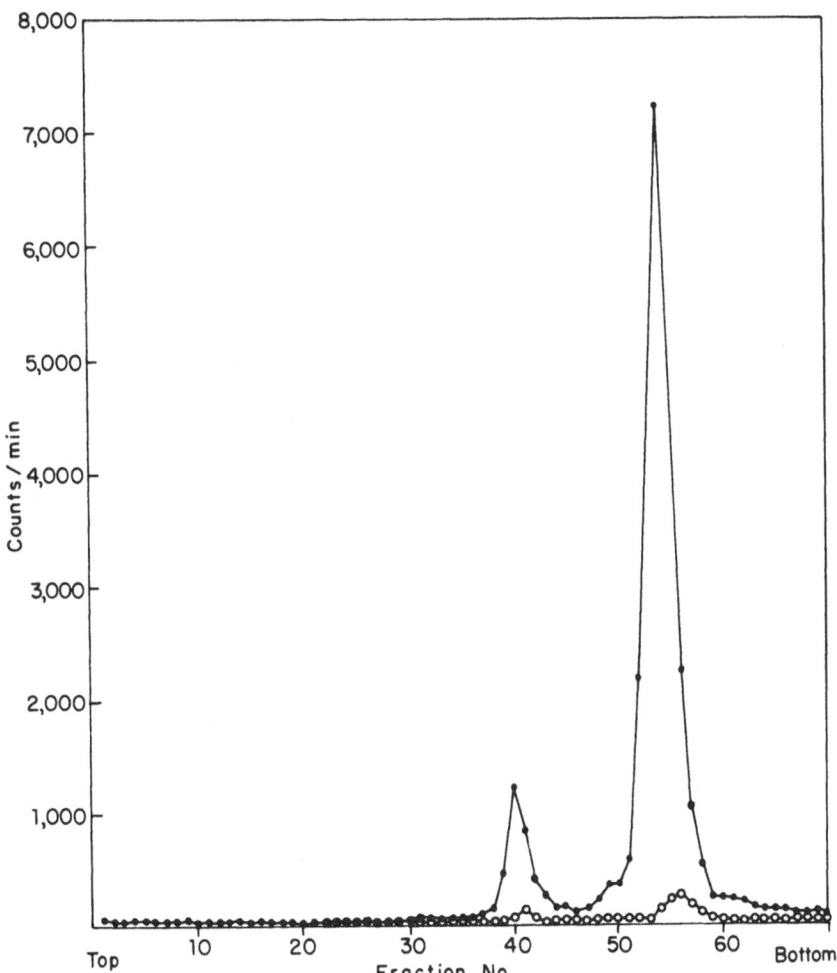

Figure 3. *CsCl-ethidium bromide gradients of cleared lysates from E517/pMB9. Methods as in Fig. 2.* (●) *CM added after 15 min. at 42°;* (○) *CM added after 2 hrs at 42°.* ³*H-thymine was added to both cultures after 2 hrs at 42°.*

The response of gyrase activity to antibiotics does not offer a convincing answer to this question. The DNA gyrase activity is complex as Cou^r cells produce a Cou^r gyrase and nalidixic acid resistant cells with a mutation in *nalA* produce a Nal^r gyrase (N. Cozzarelli, personal communication). The genes which can mutate to *cou^r* and *nalA* map far apart in the *E. coli* chromosome (18). The DNA gyrase activity that introduces supercoil twists in relaxed circular DNA is known to be sensitive to coumermycin (10). There are three measurable, nalidixic-acid sensitive activities of DNA gyrase: (1) an activity that introduces supercoil twists; (2) an SDS-activated nuclease that results in double-stranded DNA fragments; and (3) a nicking-closing activity (N. Cozzarelli, personal communication).

The response of cells to coumermycin and nalidixic acid is not identical. Coumermycin affects DNA synthesis immediately and continued incorporation of thymine is reduced to a very low level. In addition, RNA synthesis is also quickly inhibited though inhibition is much less severe (12). Nalidixic acid inhibits DNA replication almost completely and does not affect RNA or protein synthesis at low concentrations (19,20,21). There is often some DNA degradation in the presence of Nal (19) and there is also a low but continuing level of Nal^r synthesis (21). If both antibiotics stop DNA replication by blocking gyrase activity only, these differences are difficult to rationalize. It is possible that the response to the particular antibiotic may depend upon the pattern of inhibition of the various DNA gyrase-associated activities and the ratios of these activities *in vivo*.

We are currently attempting to isolate mutants in the *cou, dna-517* gene that are more temperature-sensitive than E517.

ACKNOWLEDGEMENTS

We are indebted to N. Cozzarelli for stimulating discussions and for communication of results prior to publication. This investigation was supported by grant number CA-12590, awarded by the National Cancer Institute DHEW and by research grant BMS75-02300 from the National Science Foundation.

REFERENCES

(1)   Jones, N.C. and Donachie, W.D. (1974)   Nature   251:252-253

(2)   Worcel, A. and Burgi, E. (1974)   J. Mol. Biol.   82:91-105

(3)   Wang, J.C. (1974)   J. Mol. Biol.   87:797-816

(4) Wang, J.C. (1971) J. Mol. Biol. 55:523-533

(5) Champoux, J.C. and Dulbecco, R. (1972) Proc. Nat. Acad. Sci.
     U.S.A. 69:143-146

(6) Bauer, W. and Vinograd, J. (1968) J. Mol. Biol. 33:141-171

(7) Salzman, N.P., Sebring, E.D. and Radanovich, M. (1973) J.
     Virol. 12:669-676

(8) Robbersson, D.L. and Clayton, D.A. (1972) Proc. Nat. Acad.
     Sci., U.S.A. 69:3810-3814

(9) Gellert, M., Mizuuchi, K., O'Dea, M.H. and Nash, H.A. (1976)
     Proc. Nat. Acad. Sci., U.S.A. 73:3872-3876

(10) Gellert, M., O'Dea, M.H., Itoh, T. and Tomizawa, J.-I. (1976)
     Proc. Nat. Acad. Sci., U.S.A. 73:4474-4478

(11) Wechsler, J.A. and Gross, J.D. (1971) Molec. Gen. Genet.
     113:273

(12) Ryan, M.J. (1976) Biochemistry 15:3769-3777

(13) Lederberg, E.M. and Cohen, S.N. (1974) J. Bacteriol. 119:
     1072-1074

(14) Clewell, D. and Helsinki, D. (1969) Proc. Nat. Acad. Sci.,
     U.S.A. 62:1159-1166

(15) Keller, W. (1975) Proc. Nat. Acad. Sci., U.S.A. 72:4876-4880

(16) Clewell, D.B. and Helsinki, D.R. (1972) J. Bacteriol. 110:
     1135-1146

(17) Blair, D.G., Clewell, D.B., Sherratt, D.J. and Helsinki, D.R.
     (1971) Proc. Nat. Acad. Sci., U.S.A. 68:210-214

(18) Bachmann, B.J., Low, K.B. and Taylor, A.L. (1976) Bacteriol.
     Revs. 40:116-167

(19) Dietz, W.H., Cook, T.M. and Gross, W.A. (1966) J. Bacteriol.
     91: 768-779

(20) Cook, T.M., Dietz, W.H. and Gross, W.A. (1966) J. Bacteriol.
     91:774-783

(21) Pedrini, A.M., Geroldi, D., Siccardi, A. and Falaschi, A.
     (1972) Eur. J. Biochem. 25:359-365

THE ROLE OF ATP IN *E.COLI* CHROMOSOME REPLICATION

Patrick Forterre*, Evelyne Richet* and Masamichi
Kohiyama*

Institut de Recherche en Biologie Moléculaire du CNRS
et de l'Université Paris VII
75221 Paris Cedex 05, France.

For a long time it has been thought that the only energy-requiring event occurring in DNA replication was the breakage of the dNTP phosphodiester bond during the process of polymerization. In 1968, however, Cairns and Denhardt noticed that carbon monoxide or cyanide inhibited the replication of the chromosome *E.coli* and that the inhibition could not be explained by depletion of the intracellular pool of dNTP (1). This result suggested the existence of another energy-consuming step in DNA replication and lead the authors to postulate the existence of an energy-dependent unwinding of the DNA duplex. The discovery of *in vitro* replication systems soon confirmed the requirement of ATP for DNA replication; for example, toluene-treated bacteria require ATP in addition to the four dNTP for continuation of chromosome replication (2). Since then, many investigators have concentrated their efforts on discovering the mechanism of this energy dependence. It is now clear that not only one, but several, ATP-dependent reactions may take place in the process which leads to the duplication of the chromosome. In this article, we will focus on some enzymes potentially capable of unwinding the chromosome in the presence of ATP, as well as on the role of ATP in chromosome replication *in situ*.

1.   <u>DNA-Dependent ATPases of *E.coli*</u>. Whatever hypothesis one may choose, the basic biochemical reaction which gives rise to the ATP requirement is the hydrolysis by some enzymes of ATP in the presence of DNA. Thus, the isolation of DNA-dependent ATPases is one of the approaches by which one may elucidate the mechanism of ATP requirement in chromosome replication.

A.S.I. Participants

Table 1

E. coli DNA-dependent ATPases

Enzyme	Structural Gene	Involvement in *E. coli* replication	Involvement in *in vitro* ØX174 DNA replication			
			SS	RF	RF	RF
Rep protein	*Rep*	+	−		+	
DNA gyrase	$COUM^R$	+	−		+	
dnaB protein	*dnaB*	+	+		+	
Holopoly-merase III (a)	*dnaE* *dnaZ* (b)	+	+		+	
ATPase I						
ATPase II						
DNA-unwinding enzyme						
Factor Y				+		

(a) Holopolymerase III (Kornberg's terminology) is equivalent to protein and the elongation factors I and III.

(b) The structural genes of factors I and III are unknown.

Table 1/ contd.

E. coli DNA-dependent ATPases

Action on DNA		Specificity for poly-nucleotide as cofactor	Specificity for nucleo-side tri-phosphates as substrates	Molecular weight	NEM sensi-tivity	Refer-ences
Bind-ing	unwind-ing					
+	+	single-stranded DNA	(d)ATP			3
		circular relaxed double-stranded DNA	ATP	140,000	+	4,5
–		single-stranded DNA	rNTP	250,000	–	6
		primed single-stranded DNA	(d)ATP		+	7
+ (single-stranded DNA)		single-stranded DNA	(d)ATP	74,000	+	8
+ (single-stranded DNA)		double-stranded polynucleo-tide	(d)ATP	86,000	–	9
+ (single-stranded DNA)	+	single-stranded DNA	(d)ATP GTP CTP UTP	180,000	+	10
+ (single-stranded DNA)		single-stranded DNA	(d)ATP	55,000	+	11

Hurwitz's mixture of these four proteins; DNA polymerase III, *dna*Z

Table I summarizes the properties of *E.coli* enzymes capable
of hydrolyzing ATP in the presence of DNA. At least four of them
have been shown to participate in DNA replication: the ATPase
associated with the *dna* B gene product, the ATPase associated
with the holopolymerase III, DNA gyrase and the ATPase found in the
preparation of the *rep* gene product. Besides these enzymes, four
other DNA-dependent ATPases have been reported. The most important
characteristics of these DNA-dependent ATPases are: first, their
ability to bind DNA; and secondly, at least for one of them, its
capacity to denature double-stranded DNA in the presence of ATP.
It has been shown by Abdel-Monem and Hoffmann-Berling that their
180,000 molecular weight DNA-unwinding enzyme could denature DNA.
RNA hybrid duplexes or DNA.DNA partial duplexes of fd DNA, and that
this process required simultaneous (d)ATP or rNTP dephosphorylation
(12,13). The same observation has been made by Kornberg with the
DNA-dependent ATPase associated with the *rep* gene product. In this
case, for complete separation of complementary strands of the ØX174
RF I DNA, the *E.coli* DNA binding protein and the ØX174 *cis* A
protein are required in addition to the presence of ATP (3).

Although such an important function of the DNA-dependent ATPase
has been clarified only very recently, the first finding of this
kind of enzyme dates from the year 1974 when Moss and Paoletti
isolated the nucleoside triphosphate phosphohydrolase I from
vaccinia virus cores (145,15). At that time, only two particular
properties of the enzyme drew attention: its ability to bind single-
stranded DNA, its specificity for rATP or dATP and its requirement
for single-stranded DNA as a cofactor. Another enzyme (called
ATPase I) having almost the same characteristics as Moss's enzyme
(molecular weight of 74,000, single-stranded DNA dependency, DNA
binding ability, and adenine specificity) was isolated from *E.coli*
several years later in our laboratory (8). Moreover, because this
enzyme has the same characteristics as the ATPase found in the
purified preparation of *rep* gene product, Scott and Kornberg have
suggested that this enzyme is identical to the *rep* gene product (3).
An intriguing aspect of those ATPases is that most of them require
single-stranded DNA as a cofactor for ATP hydrolysis (to ADP and
Pi) but not as substrate. They can attach firmly to DNA even in the
absence of ATP. We can imagine that the enzymes when attached to
DNA change their configuration so as to reveal an active site for
ATP hydrolysis and that the hydrolysis of ATP further changes the
configuration which may result in a movement of enzyme molecule.
Such a model has already been presented in the contraction of
muscle in which myosin polymer hydrolyses ATP moving along the actin
polymer which has a polarity with respect to the molecule. Thus
we can further imagine that the ATPase can move along the DNA chain
which also has a polarity, hydrolyzing ATP. This movement may be
the driving force for denaturation of double-stranded DNA.

Another type of indication that a DNA-dependent ATPase can

denature a double-stranded DNA has come from our studies on the
ATPase II which was originally found as a contaminating enzyme of
a crude preparation of the ATPase I (9). Like ATPase I, ATPase II
behaves as a DNA-binding protein and hydrolyses only in the
presence of DNA, ATP (dATP) but not other base types of
nucleoside-triphosphate. The main differences between ATPases I
and II are: ATPase II is not adsorbed to DEAE cellulose even at low
ionic strength, and it requires specifically a double-stranded
polynucleotide as cofactor of ATP hydrolysis. We have examined
several polynucleotides for their ability to act as cofactor and
have found that there is a correlation between cofactor efficiency
and their melting temperature - the lower the melting temperature
of a polynucleotide, the greater the stimulatory effect on ATP
hydrolysis (poly d(A.T)>poly d(BrU-A)>poly (dA-Iu)>poly(rA)-(dT)>
Calf thymus DNA). Our interpretation of this phenomenon is that
ATPase II can insert into a double-stranded structure of polynucleotide hydrolyzing ATP and that it can slide faster (thus
hydrolyzing ATP faster) when a polynucleotide has a weak base
pairing force (9).

   The common characteristic of these DNA-dependent ATPases is
that they can degrade ATP as well as dATP. If one of the DNA-
dependent ATPases is responsible for the ATP requirement in
chromosome replication, dATP should be able to replace ATP. Thus
the answer to the question can be obtained from an *in vitro*
replication system.

II. Specificity of ATP for chromosome replication in toluene-
treated bacteria. Using nucleoside triphosphates, (purified on a
column of Dowex-I (16)) we have examined whether toluene-treated
bacteria can proceed with DNA replication in the presence of any
base type of nucleoside triphosphate (10^{-4} - 2.10^{-3}M) when supplied
with the four dNTP (3.10^{-5}M). Although several factors modify the
effect of a NTP on DNA synthesis (e.g., duration of toluene
treatment, concentrations of dNTP and incubation period for DNA
synthesis), ATP is the best of the eight NTP or dNTP with respect
to the stimulation of DNA synthesis (17). Thus comparing the
plateau (Fig. 1), the level of ^{3}H-TTP incorporated in the presence
of ATP is about ten times greater than that observed in the presence
of any other type of (d)NTP.dATP cannot replace the function of ATP.
However, this does not exclude the possible participation of DNA-
dependent ATPases in chromosome replication for the reason that
in the presence of dATP or GTP, for example, there is a small
amount of DNA synthesis which is not observed in their absence.
However, if this minor DNA synthesis (called abortive synthesis)
is under the control of DNA-dependent ATPases, one may find a
selective stimulatory effect of deoxyadenine nucleoside triphosphate.
But any preference of dATP compared to other NTPs (except ATP) has
not been shown in this minor type of DNA synthesis.

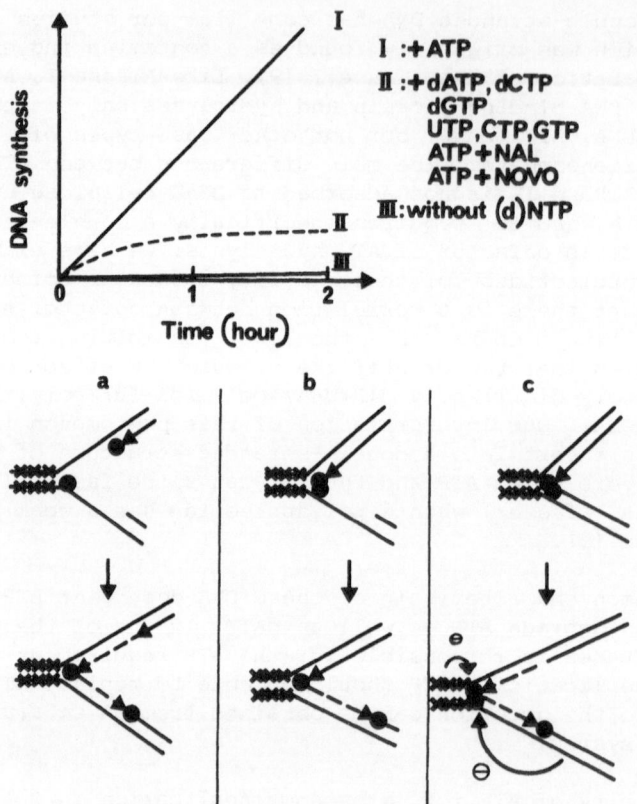

Figure 1. _Kinetic of DNA replication in toluene-treated bacteria_
and models for the progression of the replication fork.
The upper figures of the models represent the state of the
replication fork at the time of toluene treatment. The lower
figures represent the state of the replication fork after abortive
DNA synthesis _in vitro_.   _Inhibitory effect,_ ⎯⎯→ _; abortive_
DNA synthesis ------→_; chromosomal superstructures,_ ⬤⬤⬤⬤ _;_
replication complex, ⬤

     Although one could envisage other minor mechanisms (a complexed
form of a DNA-dependent ATPase degrading solely ATP) it is difficult
to imagine any of the known DNA-dependent ATPase as the only enzyme
giving rise to ATP requirement.

III. Characterization of abortive DNA synthesis. In order to
understand the ATP requirement, we have compared the characteristics
of DNA synthesis occurring in the presence of ATP with those of

synthesis occurring in the presence of one of the other NTPs. The
kinetic experiments have shown that DNA synthesis in toluene-
treated *E.coli* cells stops prematurely after 10-15 minutes at
30° when ATP is replaced by CTP for example (18). This DNA
synthesis, called abortive synthesis is believed to be the
continuation of chromosome replication occurring *in vivo* prior to
toluene treatment for the following reasons: first, *in vivo*
labeled growing points are connected with DNA synthesized in the
presence of CTP after toluenization, second, at least the *dna* B
and *dna* E gene products are involved in the abortive synthesis,
and third, the abortive synthesis is inhibited by inhibitors of
DNA replication, as araATP, mitomycin C and sodium azide (18).

We have found that this abortive DNA synthesis is not
influenced by either novobiocin or by nalidixic acid (18). However,
DNA synthesis dependent on ATP is completely inhibited by these
inhibitors under the same conditions. Thus, the kinetic of dTTP
incorporation in the presence of ATP becomes identical to that
obtained in the presence of CTP. Therefore, we have concluded that
ATP is required principally in the novobiocin and nalidixic acid
sensitive step and that the non ATP specific polymerization step is
indifferent to these inhibitors.

IV.   Possible role of (d)NTPs in the abortive synthesis. We have
pulse-labeled toluene-treated bacteria with (^{3}H)-TTP in the presence
of CTP, or ATP+novobiocin, or ATP+nalidixic acid, and analyzed the
DNA produced in the "abortive synthesis" by alkaline sucrose
gradient centrifugation. We obtained 9S and 16S pieces after short
term incubation and long pieces (30-100S) after sixty minutes
incubation. Therefore, the initiation, elongation and sealing of
Okazaki fragments can occur in the presence of any base type of
nucleoside triphosphate or in the presence of both ATP and nalidixic acid (or novobiocin) (18). This DNA synthesis resembles *in
vitro* ØX174 RF synthesis. According to results obtained with this
phage, the possible steps in which NTP is required at the following:
firstly, the formation of an initiation complex between the DNA,
DNA-binding protein, *dna* B and *dna* C proteins (ATP specific) (19);
secondly, the synthesis of the primers by *dna* G protein; and thirdly,
the formation of an elongation complex between primed DNA, factor I
and the polymerase III in which the *dna* Z gene product and factor
III participate (ATP or dATP dependent) (7). The *dna* B gene
product is associated with an ATPase which degrades any type of
rNTP (6). Therefore, because of this non-specificity, one
possibility is that the nonspecific NTP requirement in abortive
synthesis is due to the *dna* G gene product. If the primer is an
RNA (20), the four rNTPs may be required for abortive synthesis.
It is possible that one of any rNTPs could be sufficient for the
primer synthesis carried out by *dna* G protein. However, we do not
yet know whether *dna* G protein participates in the abortive

synthesis. The binding of polymerase III to a primer requires either
ATP or dATP but its Km value is about ten times lower than the Km
found for ATP in DNA synthesis produced by toluene-treated cells
(7). This low requirement could be supported by the dATP
present as precursor of DNA.

V.   Possible identity of ATP specific reaction. In ØX174 DNA *in
vitro* replication, novobiocin or nalidixic acid inhibits only RF
to RF replication but not viral to RF (21). This suggests that the
novobiocin or nalidixic acid sensitive step which is identical
to the ATP specific reaction defined by us is required only for
double-stranded DNA replication.

    At this time, no published data is available concerning the
characterization of the nalidixic acid sensitive protein. None
of the known DNA-dependent ATPases is sensitive to nalidixic acid
or novobiocin (except DNA gyrase). However, it has been shown that
the DNA-dependent ATPase coded for by *rep* gene is involved in DNA
replication. Thus, the *rep* mutant replicates the chromosome about
three times slower than the wild type (23). We have found in
toluene-treated *rep*$_3$ mutants that ATP stimulated DNA synthesis is
reduced by a factor of three compared to the wild type but, on the
contrary, abortive DNA synthesis is not affected (18).

VI.   Models for the progression of the replication fork. The *E. coli*
chromosome is not a naked molecule randomly coiled in the cytoplasm
but a very ordered structure. Several independent domains of
supercoiling have been observed in isolated nucleoids (24).
Nucleoids prepared in 0.1 M NaCl contain a histone like protein,
the HU protein (25). It is likely that the association of HU and
DNA results in the formation of complexes similar to the
nucleosomes of Eukaryotes. Such structures have been observed in
E. coli by electron microscopy (26). Moreover, we have noticed that
in toluene-treated bacteria 40% of the DNA is resistant to micro-
ccocal nuclease (27) as it has been observed in eukaryotic isolated
nuclei. These data lead us to propose the following model (Fig. 2).
Step 1: elimination of chromosomal superstructures to form a
suitable template for the polymerization complex (this step
includes the unwinding of DNA and perhaps the removal of HU
protein). Step 2: polymerization. Step 3: reformation of
chromosomal superstructures behind the replication fork. When
steps 1 and/or 3 are blocked by nalidixic acid or novobiocin or by
the absence of ATP *in vitro*, a limited portion of the chromosome
is nevertheless replicated which corresponds to the abortive
synthesis. In contrast, the inhibitors of step 2, azide, mitomycin C
and araATP completely shut-off DNA synthesis.

    Three hypotheses may be envisaged considering the existence of
the portion of chromosome ready to be replicated in the absence of
the ATP specific reaction (Fig. 1). (a) In step 1, a large portion

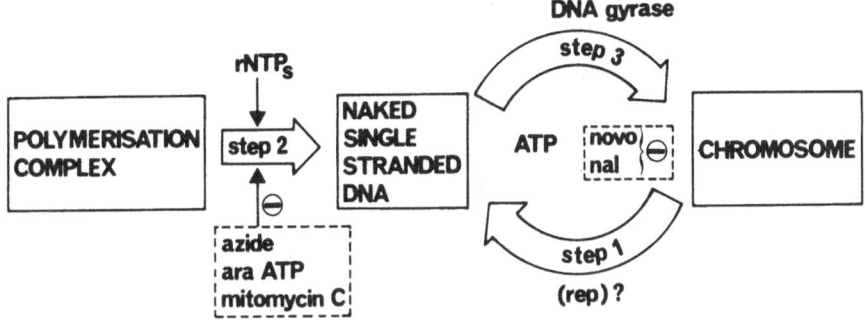

Figure 2. *Model of the events occurring at the replication fork.*

of the chromosome is unwound *in vivo* before being replicated, thus
providing a suitable template for abortive synthesis *in vitro*.

In that case we should notice that the amount of DNA
synthesized in the abortive DNA synthesis (1-2% of the chromosome)
fits well with the size of one domain of supercoiling according to
the work of Worcel (24). This suggests, if this hypothesis is
correct, that the domains of supercoiling are replicated in a
sequential manner. (b) Steps 1 and 2 occur simultaneously but one
of the two strands, probably the lagging strand, is replicated with
some delay and serves as a template for abortive DNA synthesis once
the unwinding stops. (c) Both strands are replicated at the point
where the double helix is unwound. In that case, we would have to
imagine that some superstructure should be maintained in front of
or behind the growing point to allow its progression. For example,
if relaxation of the DNA should occur in coordination with the
unwinding process, premature relaxation in the absence of gyrase
could stop DNA replication when the growing point reaches this
relaxed part of the chromosome. The same situation will occur if
accumulation of relaxed DNA behind the replication fork prevents
its movement.

In order to choose between these hypotheses, we must wait for
further characterization of the enzymes involved in steps 1 and 3
and for a thorough knowledge of the topology of the replication
fork.

REFERENCES

(1) Cairns, J. and Denhardt, D.T. (1968) J. Mol. Biol. 36:335-
      342

(2)   Moses, R.E. and Richardson, C.C. (1970) Proc. Nat. Acad. Sci.
      U.S.A. 67:674-681

(3)   Scott, J.F., Eisenberg, S., Bertsch, L.L. and Kornberg, A.
      (1977) Proc. Nat. Acad. Sci. U.S.A.   74:193-197

(4)   Gellert, M., Mizuuchi, K., O'Dea, M.H. and Nash, H.A. (1976)
      Proc. Nat. Acad. Sci. U.S.A. 73:3872-3876

(5)   Gellert, M., O'Dea, M.H., Itoh, T. and Tomizawa, J.I. (1976)
      Proc. Nat. Acad. Sci. U.S.A. 73:4474-4478

(6)   Wickner, S., Wright, M. and Hurwitz, J. (1974) Proc. Nat.
      Acad. Sci. U.S.A.   71:783-787

(7)   Wickner, S. (1976) Proc. Nat. Acad. Sci., U.S.A. 73:3511-3515

(8)   Richet, E. and Kohiyama, M. (1976) J. Biol. Chem.   251:808-812

(9)   Richet, E. and Kohiyama, M. (manuscript in preparation).

(10) Abdel-Monem, M. and Hoffmann-Berling, H. (1976) Eur. J.
      Biochem. 65:431-440

(11) Wickner, S. and Hurwitz, J. (1975) Proc. Nat. Acad. Sci. U.S.A.
      72:3342-3346

(12) Abdel-Monem, M., Dürwald, H. and Hoffmann-Berling, H. (1976)
      Eur. J. Biochem. 65:441-449

(13) Abdel-Monem, M., Lauppe, H.F., Kartenbeck, J., Dürwald, H.
      and Hoffmann-Berling, H. (1977) J. Mol. Biol. 110:667-685

(14) Paoletti, E.,   Rosemond-Hornbeak, H. and Moss, B. (1974)
      J. Biol. Chem. 249:3273-3280

(15) Paoletti, E. and Moss, B. (1974) J. Biol. Chem. 249:3281-3286

(16) Hurlbert, R.B., Schmitz, H., Brumm, A.F. and Potter, V.R. (1954)
      J. Biol. Chem. 209:23

(17) Forterre, P. and Kohiyama, M. (1972) In: Mechanism and
      Regulation of DNA Replication (Kolbert, A.R. and Kohiyama, M.,
      eds.) Plenum Press

(18) Forterre, P. and Kohiyama, M. (manuscript in preparation).

(19) Weiner, J.H., McMacken, R. and Kornberg, A. (1976) Proc. Nat.
      Acad. Sci. U.S.A. 73:752-756

(20) Sugino, A., Hirose, S. and Okazaki, R. (1972) Proc. Nat. Acad.
     Sci. U.S.A. 69:1863-1867

(21) Sumida-Yasumoto, C., Yudelevich, A. and Hurwitz, J. (1976)
     Proc. Nat. Acad. Sci. U.S.A. 73:1887-1891

(22) Marians, K.J., Ikeda, J.E., Schlagman, S., and Hurwitz, J.
     (1977) Proc. Nat. Acad. Sci. U.S.A. 74:1965-1968

(23) David-Lane, H.E. and Denhardt, D.T. (1975) J. Mol. Biol.
     97:99-112

(24) Worcel, A. and Burgi, E. (1972) J. Mol. Biol. 71:127-147

(25) Rouviere-Yaniv, J. XLII Cold Spring Harbor Symp. Quant. Biol.
     (1977) in press

(26) Griffith, J.D. (1976) Proc. Nat. Acad. Sci. U.S.A. 73:563-567

(27) Forterre, P. and Kohiyama, M. (manuscript in preparation).

NUCLEIC ACID BINDING GLYCOPROTEINS WHICH SOLUBILIZE DEOXYRIBONUCLEIC ACID IN DILUTE ACID: SPECIES DISTRIBUTION AND POSSIBLE ROLE IN DNA CONDENSATION

D.R. Champ, P. Unrau[*],C.E. Grant and J.L. Young

Biology and Health Physics Division
Atomic Energy of Canada Limited
Chalk River Nuclear Laboratories
Chalk River, Ontario KOJ 1JO, Canada

INTRODUCTION

Holloman (1975) first described the isolation, from the smut fungus *Ustilago maydis*, of a glycoprotein which paradoxically made DNA soluble in 5% TCA without extensive template degradation. He suggested that such a glycoprotein might influence DNA condensation in the cell, and thus function to control nuclear DNA structure. This acid solubilization of DNA has important implications for studies on DNA replication, recombination and repair. A widespread practice in studies of DNA metabolism utilizing radioactive tracers is the meticulous acid washing of reaction products (Bollum, 1966), gradient fractions (Curtis and Alberts, 1976) or DNA samples in general (e.g., Unrau, Wheatcroft and Cox, 1972). Washing in acid is used to remove DNA precursor pools in DNA polymerase studies (Kornberg and Gefter, 1971).   The release of DNA bases and photoproducts in an acid soluble form is interpreted as a consequence of DNA repair because it is currently thought that DNA damage which is acid soluble has been degraded (Setlow and Carrier, 1964).

If, however, some high molecular weight DNA were to be complexed in such a way that it was soluble in cold acid without being degraded (e.g., Unrau, 1975), then the direct interpretation of the acid solubility of incorporated label as indicating nucleolytic DNA breakdown might have to be re-evaluated. The presence of glycoproteins or other compounds having the ability to complex with DNA and render it acid soluble, in organisms used for studies on replication and repair, might lead to a misinterpretation of some aspects of such studies. For example, the presence of higher molecular weight complexes in the acid soluble replicative precursor

[*]*A.S.I. Participant*

pools might lead to an underestimate of the extent of replication.

The implications of Holloman's work led us to re-examine his observations in the smut fungus *Ustilago maydis* (Unrau, Champ, Young and Grant, in preparation). As we were interested in the broader implications of acid-soluble large DNA fragments in relation to replication and repair we looked for evidence of such complexing activities in phage-infected bacteria, bacteria, fungi, algae, chick embryos and human fibroblasts. The presence of proteins (ASP) which bind to DNA and render the complex soluble in dilute acid could be demonstrated either by an assay for acid soluble ^{3}H-thymidine-labelled DNA fragments which became acid precipitable after alkaline sucrose gradients or alkaline treatment because of the dissociation of the NDA-protein complex (Holloman, 1975) or by isolation of ASP from cellular extracts.

The anti-condensation properties of ASP have been incorporated into a simple model for DNA condensation control. To examine some of the effects of ASP on DNA condensation, a spermidine-induced DNA condensation assay was developed. The assay utilized T7 DNA, *Ustilago* ASP, spermidine and other factors expected to affect DNA condensation (Lezzi, 1970; Flink and Pettijohn, 1975). This paper presents some preliminary observations, hypotheses, and models, in an attempt to demonstrate that a particular class of DNA binding proteins may be important in studies of DNA conformation and metabolism.

MATERIALS AND METHODS

Preparation of ASP. The extraction and partial purification of ASP from *U. maydis* is a modification of the method described by Holloman (1975) and will be described in detail elsewhere. Here we emphasize that reducing conditions are mandatory for stability and the optimum DNA binding activity of ASP. Affinity chromatography on Concanavalin-A-Sepharose (Pharmacia) results in the partial purification of a Con-A bound fraction (ASP I) as well as a non-binding fraction (ASP II). ASP have also been isolated from *Micrococcus radiodurans, Saccharomyces cerevisiae, Ankistrodesmus braunii,* and *Gallus gallus* (chick embryos).

Assay of solubilizing proteins. The assay described by Holloman (1975) has been modified by eliminating a carrier-DNA precipitation step at the suggestion of G.R. Banks, and by the addition of DTT to the reaction mix. T7 DNA, prepared as described previously (Champ, 1975) and labelled with ^{3}H-thymidine to a specific activity of 2.24 kBq/nmol of nucleotide was used as substrate in an assay mix (50µl) containing 2 nmol of DNA nucleotide equivalents, 10 mM MOPS buffer, pH 7.5, 1 mM DTT, and ASP protein. The mixture was incubated at 37°C for 10 minutes and the reaction terminated by the addition of

60µl of ice cold 10% (w/v). The mixture was held on ice for 10
minutes, spun 2 minutes at 12,000 X g and a 50µl sample was spotted
on to a Whatman 4 MM paper rectangle. The rectangles were washed
twice in 96% ethanol, air dried and counted. Although the DNA-ASP
complex is soluble in dilute acid, it is insoluble in 96% ethanol.
One unit is defined as that amount of protein which renders 50% of
the DNA in a standard reaction acid-soluble but ethanol-insoluble.
It was advantageous to use the 50% solubility point as the unit
criterion because the effect of promoters or inhibitors of the
reaction could be detected easily.

Condensation assay. 100 mM Spermidine-HCl was added to a
buffered reaction mix, final volume 50µl containing 2 nmol nucleotide
equivalent of phage T7 DNA. The resulting mix was incubated at 22°C
for 120 minutes, spun for 2 minutes at 12,000 X g and 40µl were
spotted on to filter paper, washed as above, and counted.

Gradients. Acid-soluble reaction products were sedimented in
5-20% alkaline sucrose gradients, 4.2 ml containing 0.3 M NaOH and
0.7 M NaCl. Sedimentation was at 50,000 rpm and 20°C for 240
minutes in a Beckman SW-56 rotor. Alkaline gradients were either
collected on strips, washed once in acid (10% TCA), twice in 96%
EtOH, dried and then counted, or into centrifuge tubes for acidif-
ication, precipitation and counting.

Preformed 5-20% neutral sucrose gradients (4.2 ml) containing
10 mM Tris-HCl buffer, pH 7.6 and a 0.3 ml 50% sucrose shelf were
run at 50,000 rpm and 20°C for 90 minutes in an SW-56 rotor.
Gradient fractions were collected onto paper strips and counted.

RESULTS

Table 1 summarises some of the evidence for the presence of
ASP in various organisms. As will be noted, ASP has been isolated
directly from *M.radiodurans*, *S.cerevisiae*, *U. maydis*, *A. braunii*
and *G. gallus*. The presence of ASP in phages T4 and T7-infected
cells, *E.coli*, and *H.sapiens* was demonstrated by isolating acid-
soluble ethanol-insoluble DNA complexes and dissociating these
complexes in alkaline to yield acid-precipitable DNA fragments.
Acid-soluble fragments were analyzed by alkaline gradients in the
case of phage T7, *M.radiodurans* and *U. maydis*. In Fig. 1 we show that
material sedimenting with $S_{20,w}$ values ranging from 4 to 60 is
present in acid extracts of T7 lysates. Acid precipitation of the
alkaline gradient fractions demonstrated that the previously acid-
soluble DNA could in fact be acid precipitated. The dissociation by
alkali of DNA-protein complexes (Holloman, 1975) results in
previously soluble DNA becoming acid-precipitable. The size range
observed in Fig. 1 reflects both that of so-called replicative
intermediates (Okazaki et al., 1968) and of high molecular weight

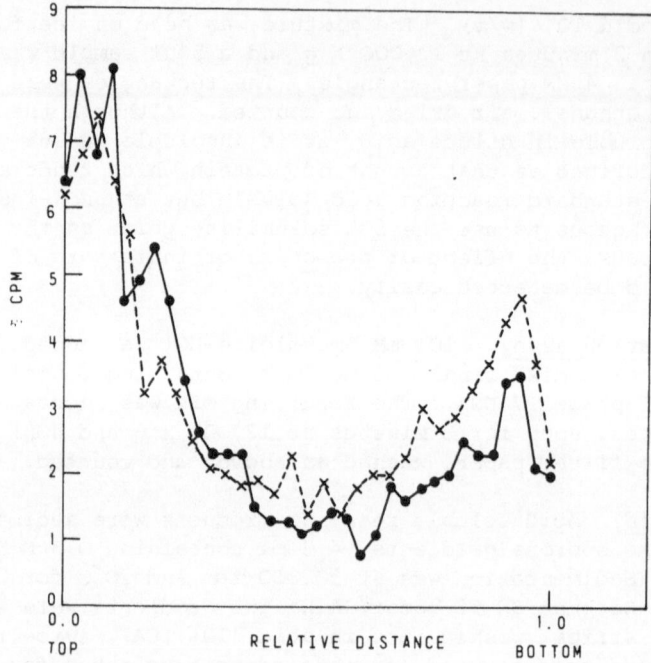

Figure 1. *Alkaline sucrose sedimentation of acid soluble* ³H-Thy
labelled fragments from phage T7 pysates. *Phage T7 infected E.coli*
lysates were acid precipitated and the acid soluble material was
sedimented in alkaline sucrose. *Fractions were washed in acid and*
ethanol on strips (x - - - - x) or were precipitated with acid and
carrier DNA (o ———— c) and the pellet counted.

acid-soluble complexes.

In Fig. 2 we show the sedimentation of 4 nmol nucleotide
equivalents of T7 DNA in neutral sucrose after saturation with a
partially purified preparation of *Ustilago* protein (ASP I). The
amount of material sedimenting in front of the unit length DNA
position was proportional to the amount of protein added. On
addition of 1.2 µg of protein, (20 X the amount required to make the
DNA totally soluble in dilute acid), 47% of the DNA sedimented
ahead of T7 unit length position. This rapidly sedimenting DNA
presumably is due to the formation of a DNA-protein complex.

The ASP I fraction bound to Con-A on isolation. However, on
prolonged storage at -20°C the amount of material binding to Con-A
decreased. We have demonstrated that the ASP II-like subfraction
of ASP I (i.e., that which no longer binds to Con-A) interferes
with spermidine condensation of DNA, while ASP I does not interfere
at the spermidine level tested. Fig. 3A shows the acid solubilizing
activity present on a second passage of ASP I over Con-A after 6

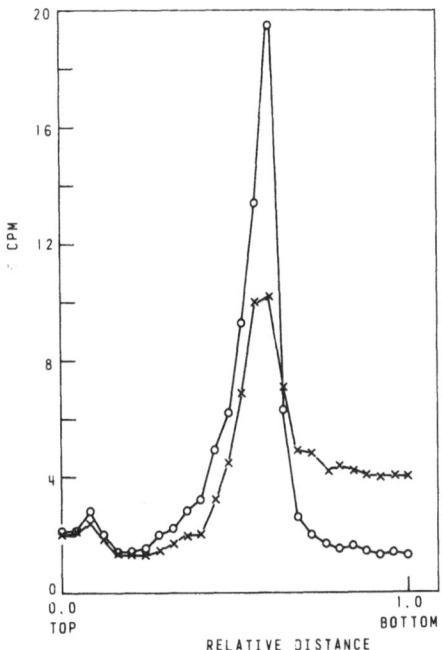

Figure 2. *Sedimentation in neutral sucrose of phage T7 DNA without and with ASP I of U.maydis. Without ASP I (o———o); with ASP I (x———x).*

months storage at -20°C, while Fig. 3B shows the anti-condensation activity assayed on the same fractions. Fig. 3 shows that ASP I and ASP II both cause DNA to become soluble in dilute acid, while only ASP II prevents DNA condensation in spermidine.

Fig. 4 shows the time course of spermidine-induced T7 DNA condensation. The initial reaction is very rapid and was complete within a few minutes; the final level of condensation was constant for at least 480 minutes, so a standard time of 120 minutes was chosen for subsequent tests of DNA condensation (see Methods).

Fig. 5 shows that spermidine-induced condensation follows cooperative kinetics; that over a range of less than .2 mM spermidine the DNA may condense entirely from solution and that condensation is sensitive to pH changes. Fig. 6 shows the effects of varying the concentration of the divalent cation Mg^{2+} on spermidine-induced DNA condensation. Concentrations at and above 2 mM inhibit condensation while concentrations below 0.5 mM have little effect. Data for 0.2 and 5 mM Mg^{2+} concentrations are shown.

The amount of spermidine required to condense DNA when it is

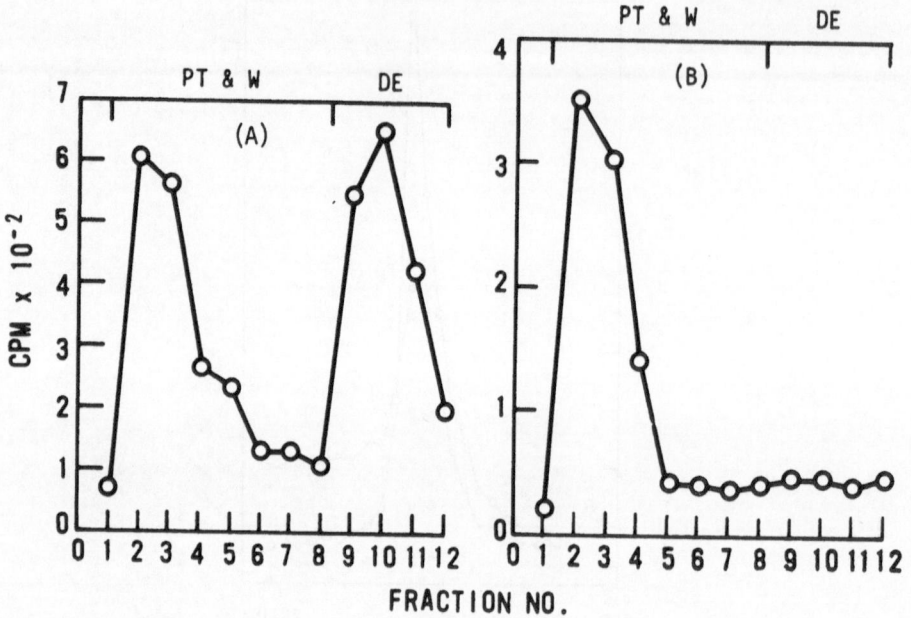

Figure 3A. Activity profile of ASP I rechromatographed on Con-A.
100 units of ASP I, after storage for 6 months, were passed over
a washed Con-A-sepharose column in 50 mM MOPS buffer, pH 7.5,
containing 1 mM DTT, 1 mM Ca^{2+}, 1 mM Mg^{2+}, and 1 mM Mn^{2+}. The
pass through and wash (PT & W) were collected and the bound
material was removed with a buffered 20% dextrose solution
starting at fraction 8 (DE), and all the fractions were assayed
for solubilizing activity.

Figure 3B: The column fractions were also assayed in the presence
of 2 mM spermidine for interference with phage T7 DNA condensation.

complexed with ASP I increases with the amount of ASP I added, as
shown in Fig. 7. The biphasic nature of the condensation curve
presumably reflects the presence of both ASP I and II in the binding
mixture. A critical feature of the reaction between ASP and DNA
is shown in Fig. 8, where anticondensation activity was stimulated
at 0.2 mM Mg^{2+} and inhibited by concentrations above 0.5 mM (5 mM
shown).

                          DISCUSSION

Several unexpected observations were made during the course of

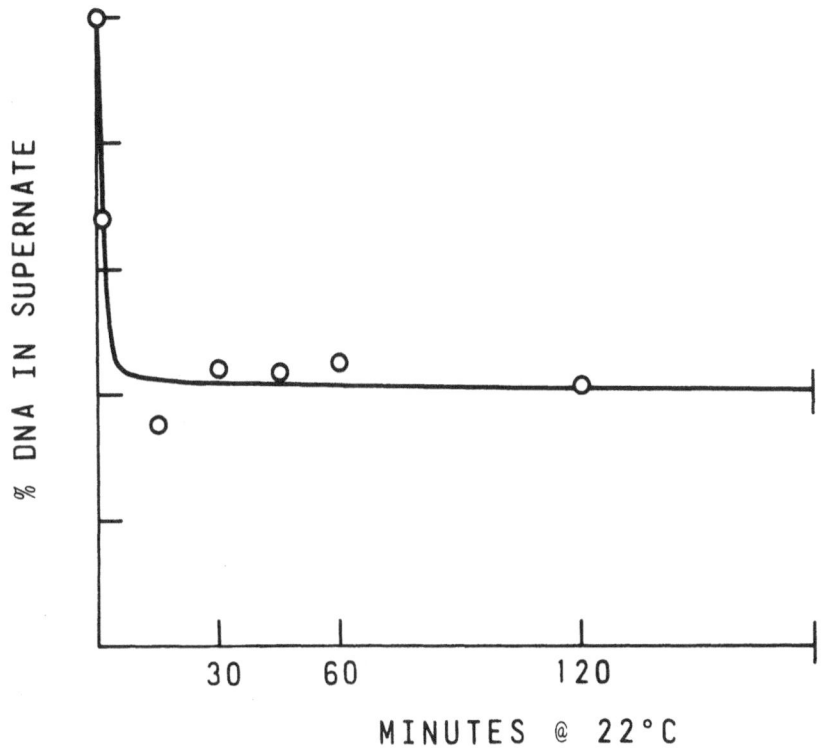

Figure 4. *Time course of condensation of phage T7 DNA with spermidine. 0.55 mM spermidine was added to DNA in pH 7.5 MOPS buffer.*

this study. The presence in diverse organisms of material with the character first reported by Holloman, as shown in Table 1, suggests that in these diverse organisms such DNA-protein complexes fulfill a common and perhaps critical function. We believe that our data supports Holloman's original contention, that the function of such glycoproteins is to control the state of condensation of the DNA. We originally isolated ASP from *Ustilago maydis* to test a simple model for the control of DNA condensation by DNA binding proteins. The basis of the model is the interaction of binding proteins with DNA to prevent DNA condensation. Certain parts of the genome would be uncondensed and therefore amenable to replication and transcription only while complexed with the relevant binding protein. A change in nuclear environment might dissociate the binding proteins leading to DNA condensation and consequent termination of replication or transcription of the affected genes. Organization of more complex genomes might require the presence of more than one of this class of binding proteins.

The model stemmed from Lerman's observations of the ψ-DNA

Table 1

Direct isolation of ASP or of acid soluble radioactive DNA fragments from a variety of organisms

Organism	ASP[1]	Acid Soluble[2] Fragments	Akaline Gradients
T$_4$ phage	nt[3]	+	nt
T$_7$ phage	nt	+	+
E.coli	nt	+	nt
M. radiodurans	+	+	+
S. cerevisiae	+	nt	nt
U. maydis	+	+[4]	+[4]
A. braunii	+	nt	nt
G. gallus	+	nt	nt
H. sapiens	nt	+	nt

[1] ASP were extracted and assayed by the methods previously described.

[2] Fragments labelled with ^{3}H-TdR were soluble in 5% TCA; after being made alkaline they precipitated in 5% TCA.

[3] nt - not tested

[4] ^{14}C-uracil-labelled fragments penetrated alkaline sucrose gradients and were acid precipitable after overnight RNA digestion in alkali.

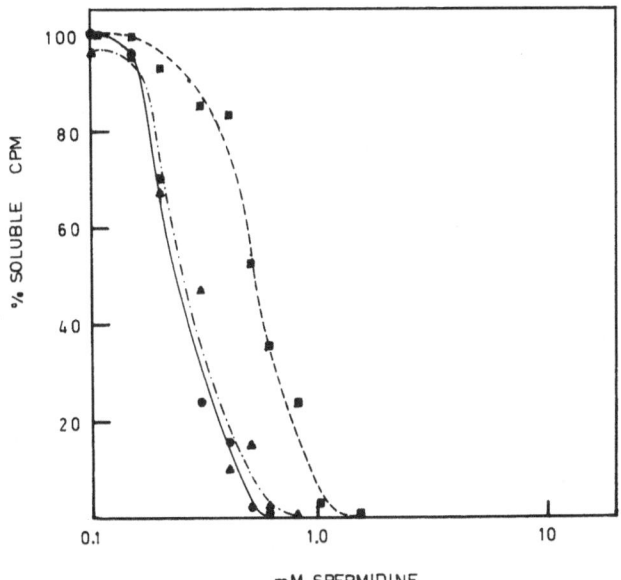

Figure 5. Phage T7 DNA condensation in buffered spermidine as a function of pH. (●————●) Na Acetate, pH 6.0; (▲————▲) MOPS pH 7.5; (■ ——— ■) Bicine pH 9.0.

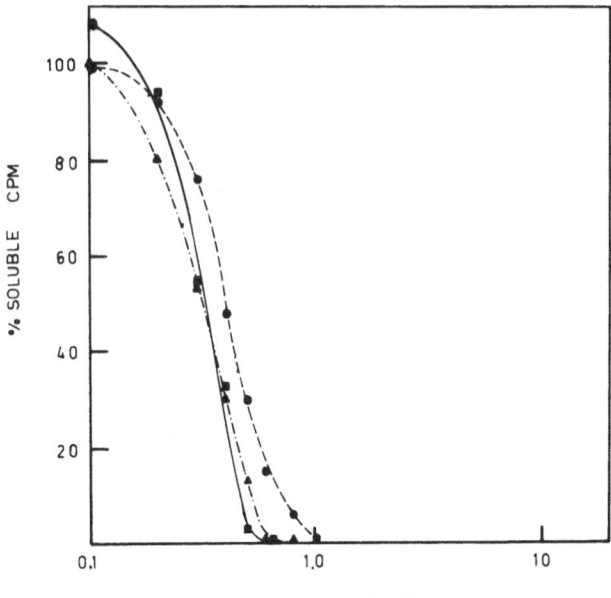

Figure 6. The effect of Mg²⁺ ions on DNA condensation. Condensation was at pH 7.5 in MOPS buffered spermidine; added Mg²⁺ - 0 (▲———▲); .2 mM (■————■); 5 mM (●———●).

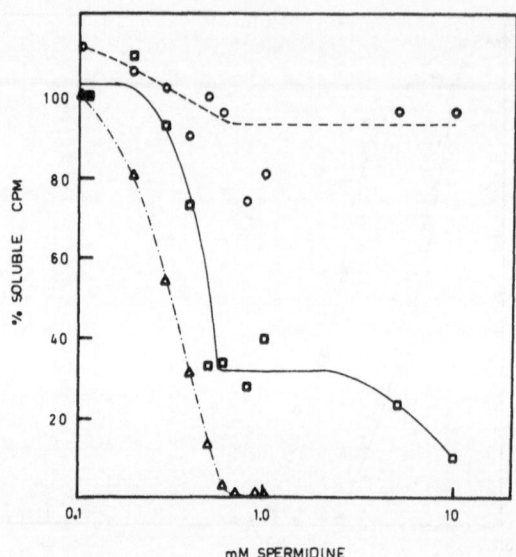

Figure 7. *The effect of ASP I from <u>U. maydis</u> on phage T7 DNA condensation. Condensation was at pH 7.5 in MOPS buffered spermidine. (▲————▲) control; (■————■) 100 % saturation; (○——————○) 10 X saturation. Saturation was with respect to solubilization in dilute acid.*

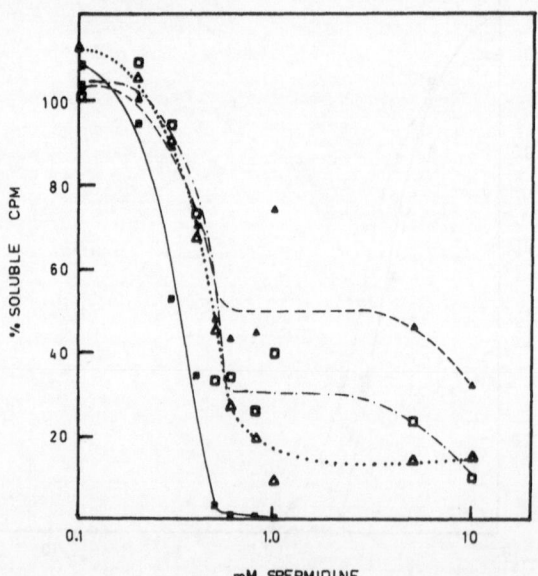

Figure 8. *Phage T7 DNA condensation in MOPS buffered spermidine. pH 7.5 at saturating ASP I levels with varying Mg^{2+} concentration. (■————■), 0.2 mM Mg^{2+} control; (□——————□), no Mg^{2+}; (▲——————▲), 0.2 mM Mg^{2+}; (△......△), 5 mM Mg^{2+}.*

transition (Lerman, 1971; Lerman, 1973) in which it was shown that the collapse of DNA into fast-sedimenting molecules was a cooperative process dependent on biophysical interactions between DNA molecules and compounds such as polyethylene oxide (PEG). He pointed out that the process was non-linear so that a small change in effector concentration could lead to a large change in sedimentation velocity. Furthermore, the ψ-DNA transition was very sensitive to ion effects. Lerman proposed that ψ-DNA-transition reflects the process of condensation of the DNA beginning from sites of nucleation. We suggest that such nucleation sites become available when ASP-like binding proteins dissociate from the DNA. Gosule and Schullman (1976) have also interpreted the condensation of DNA in spermidine as reflecting a ψ-DNA-like transition. Our DNA condensation system demonstrates that a small change in spermidine concentration leads to a marked change in DNA condensation. Similar effects of polyamines have also been reported by Osland and Kleppe (1977).

If the process of DNA condensation in the cell in essence mimics the effects shown in Figs. 4-8, we suggest that its degree of condensation may be critically dependent on pH, the concentrations of DNA, binding proteins, metal ions and other factors such as spermidine or F_1 histones (Johns, 1966). Some of these components such as polyamines and histones will promote condensation, while others such as ASP would prevent condensation. Our data shows that changes in the concentration of critical ions, proteins, or pH may have an effect on the degree of condensation of the DNA. This suggestion is in agreement with the ideas of Lerman (1973) and Lezzi (1970). A correlation between condensation and gene inactivation has been observed in chromosome puffs (Beerman, 1952; Clever and Karlson, 1960; Lezzi, 1970) and has been interpreted to mean that transcription depends on DNA being uncondensed. We propose, then, that selective condensation of DNA, dependent on interactions between anti-condensation proteins, DNA and the nuclear milieu, may prove to be an economical system for transcriptional control.

In Fig. 9 we summarize some of the features of a model based on the above speculations. An idealized representation of the condensation kinetics of DNA or chromatin is shown in Fig. 9A, where a set of genes which are coordinately controlled by virtue of similar ASP binding sites condense as shown by line A. When anti-condensation proteins complex with this particular set of genes, condensation kinetics may change to those shown by line B because the proteins strongly inhibit condensation. Finally, if a third effector influences the binding coefficient of the anticondensation proteins, so that they are less tightly bound, they will be displaced by relatively small increases in the level of the primary condensation factor, as shown by line C. Particular gene sets will

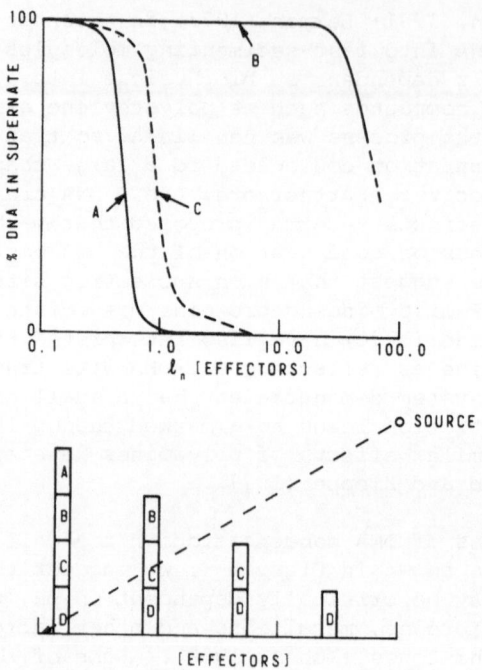

Figure 9A. *A model for the control of condensation by binding-protein-DNA interactions. Line A – condensation of DNA or chromatin as a function of increasing effector levels. Line B – condensation of DNA after complexing with ASP-like binding proteins. Line C – condensation of DNA + binding proteins when a third variable such as metal ions influences the protein-DNA binding coefficient.*

Figure 9B. Gradient effects on DNA-binding protein complexes and chromatin condensation. Blocks A, B, C and D of the genome represent DNA-binding-protein complexes. As effector levels rise, in cells nearer the source, blocks of DNA condense because the binding coefficient changes and the complexes fall apart.

behave uniformly by condensing or uncondensing in response to small changes in effectors, ions, or other factors in the nuclear environment.

In Fig. 9B we have extended this speculation to show how a gradient of effectors that affect condensation might influence gene action (and thus differentiation). Blocks of genes and their autocondensation proteins are presumed to be sensitive to an effector of condensation diffusing from a source. Those cells furthest from the source will contain the lowest level of effector which would allow a greater degree of protein binding leading to

DNA uncondensation and transcription. Those cells nearest the source will have uncondensed only those genes which are bound to the most tightly complexed anticondensation proteins. A gradient where the effector promotes binding of the anticondensation proteins can also be envisaged and would lead to enhanced cellular activity nearer the effector source.

At present we have tested only one minute matrix involving a known genome, a partially characterized binding protein, and a partially characterized condensation system. It is clear that a substantial effort will be required before the complete role of the ASP proteins in cells is elucidated.

ACKNOWLEDGEMENTS

We have appreciated helpful discussions with N.E. Gentner, R.E.J. Mitchel and members of the Biology Branch, CRNL, who have also provided some of the diverse organisms for the species survey. We also appreciate discussions with L.S. Lerman, W.K. Holloman, and with members of the Genetics Division, N.I.M.R., Mill Hill. We acknowledge the collaboration of G.R. Banks and R. Holliday of N.I.M.R.

REFERENCES

Holloman, W.K. (1975) J. Biol. Chem. 250:2993-3000

Bollum, F.J. (1966) In: Procedures in Nucleic Acid Research.
    (eds. G.L. Cantoni and D.R. Davies) Harper and Row, New York,
    pp296-300

Curtis, M.J. and Alberts, B. (1976) J. Mol. Biol. 102:793-816

Unrau, P., Wheatcroft, R. and Cox, B.S. (1972) B.B.A. 269:311-321

Kornberg, T. and Gefter, M.L. (1971) Proc. Nat. Acad. Sci. USA
    68:761-764

Setlow, R.B. and Carrier, W.L. (1964) Proc. Nat. Acad. Sci. USA
    51:226-231

Unrau, P. (1975) Mutation Res. 29:53-65

Lezzi, M. (1970) In: International Review of Cytology. (eds.
    G.H. Bourne and J.F. Danielli) Academic Press, New York and
    London. pp127-167

Flink, I. and Pettijohn, D.E. (1975) Nature 253:62-63

Champ, D.R. (1975) Ph.D. thesis, University of Toronto.

Okazaki, R., Okazaki, T., Sakabe, K., Sugimoto, K. and Sugino, A.,
    (1968) Proc. Nat. Acad. Sci. USA   59:598-605

Lerman, L.S. (1971) Proc. Nat. Acad. Sci. USA   68:1886-1890

Lerman, L.S. (1973) Cold Spring Harbor Symp. Quant. Biol.   38:
    59-73

Gosule, L.C. and Schullman, J.A. (1976)   Nature   259:333-335

Osland, A. and Kleppe, K. (1977) Nucleic Acids Res. 4(3):685-695

Johns, E.W. (1966) Biochem. J.   106:No.3, 23

Beerman, W. (1952) Chromosoma   5:139-198

Clever, U. and Karlson, P. (1960)   Exptl. Cell Res.   20:623-626

ATP DEPENDENT DEOXYRIBONUCLEASE OF *BACILLUS SUBTILIS*; SUBUNIT STRUCTURE AND SUBSTRATE SPECIFICITY OF DNA DEPENDENT -ATPase ACTIVITY

Janine Doly

Centre de Génétique Moléculaire
C.N.R.S.
91190 Gif sur Yvette, France

ATP dependent deoxyribonucleases have been found in several bacterial species: *Micrococcus luteus* (1), *Escherichia coli* (2,5), *Diplococcus pneumoniae* (6), *Haemophilus influenzae* (7.10), *Mycobacterium smegmatis* (II), *Bacillus laterosporus* (12) and *Bacillus subtilis* 13-15). Evidence exists that this enzyme may be involved in genetic recombination. *E.coli* K12 *recB* and *recC* mutants which lack activity of the enzyme are defective in genetic recombination and DNA repair. Absence of the *recB recC* DNase was also found to be important in determining transformability (16). The level of genetic transformation is decreased in ATP dependent nuclease mutants of *Diplococcus pneumoniae* (6), but in *H.influenzae*, the role of the ATP dependent nuclease in recombination is not clear as mutants lacking this enzyme activity still keep their ability to recombine. The ATP dependent DNases purified from *E.coli*, *M.luteus* and *H.influenzae* have both exonucleolytic and endonucleolytic activity. They can degrade both single stranded and double stranded DNA from either the 3' or the 5' end and also possess a DNA dependent ATPase activity.

Work in this laboratory on *B. subtilis rec* mutants prompted us to purify this enzyme and study its subunit structure and enzymic properties. The enzyme has been purified from *B.subtilis* 168M by ammonium sulphate fractionation, Sephadex gel filtration, DEAE cellulose chromatography and gel electrophoresis on discontinuous polyacrylamide gradient as previously described (17). The molecular weight of the enzyme was estimated to be 270,000 by electrophoresis under non denaturing conditions. SDS gel electrophoresis showed it to be composed of five different subunits of molecular weights 81,000, 70,000, 62,000, 52,500 and 42,500.

An X-ray sensitive mutant (GSY 1290, $recE_5$ $ilvC_I$ $trpE_{26}$) of $B.subtilis$ was isolated after treatment of strain GSY 1269 ($ilvC_I$ $trpE_{26}$) with N-methyl-N'-nitro-N-nitrosoguanidine (14). The mutant is also mitomycin C (0.05 µg/ml) sensitive, defective in transformation and transduction (frequencies are only 5-10% of the parental strain) and lacks an ATP-dependent DNase activity. Extracts of the mutant were subjected to the same purification procedure as used for the wild type enzyme ($B.subtilis$ 168M). At the last purification step on a polyacrylamide gel, an inactive protein was eluted with a similar migration to the wild-type ATP dependent DNase. SDS gel electrophoresis of the purified inactive protein revealed five subunits. Four of the subunits were of identical molecular weight to those of the wild type enzyme. The 70,000 MW subunit, however, was replaced by a smaller polypeptide of MW 56,500. From these results it was concluded that the $recE_5$ mutation affects the structural gene coding the 70,000 MW subunit. The $recE_5$ mutation may be either a nonsense mutation, a deletion or a mutation resulting in specific degradation of the modified subunit (17). Revertants which have recovered both the enzyme activity and wild type levels of resistance to mitomycin C were isolated. None of the revertants support growth of a nonsense mutant of phage SPOI. The nonsense suppressor (su-3$^+$) (18) does not restore either enzyme activity or mitomycin C resistance to strain GSY 1290. The $recE_5$ mutation is therefore unlikely to be a nonsense mutation.

The molecular weight of the ATP dependent DNase from $B.subtilis$ 168M (308,000) estimated by SDS gel electrophoresis (17) is comparable to that of the $H.influenzae$ (300,000) (19) and $E.coli$ (270,000) (5) enzymes. Important differences however exist when comparing the number and size of the enzyme subunits. We have found that the $B.subtilis$ DNase is composed of detected five polypeptides of different molecular weight. The $H.influenzae$ enzyme however is composed of three subunits (117,000, 115,000, 68,000) (19) and the $E.coli$ enzyme of only two (α : 60,000 and β : 170,000 (20)). Mutations in either the $recB$ or $recC$ locus affect the β subunit. The $B.subtilis$ $recE_5$ mutant lacks both exonuclease and ATPase activity. Exonuclease activity of the $B.subtilis$ purified enzyme was assayed using $B.subtilis$ and T_7 DNA as substrates as for the $E.coli$ and $D.pneumoniae$ enzymes (5,6) : Its pH optimum is broad (pH 9.6 to 11.0). The enzyme activity requires Mg^{2+} with a sharp optimum at 50 mM. This is higher than the optimum Mg^{2+} requirement (10-20 mM) reported for DNases from $H.influenzae$ (8), $D.pneumoniae$ (6) and $E.coli$ (5). As is the case for these other enzymes, the $B.subtilis$ exonuclease activity has an absolute requirement for ATP with an optimal concentration of 70µM - the ATP is split into ADP and inorganic phosphate.

Studies on the substrate specificity of the ATP dependent DNases may yield some information on the role of the enzyme in genetic recombination. We initiated such studies on the purified

Table 1

Some properties of ATP dependent from several bacterial species

Bacteria	*D.pneumoniae* 6	*M. luteus* I
Requirements		
Optimum pH	9.25 to 9.8	9.2 to 10
Mg^{++}	$7.10^{-3}M$ to $2.10^{-2}M$	$1.5 \cdot 10^{-2}M$
ATP	$1.10^{-5}M$	$5.10^{-5}M$
Other nucleoside triphosphates	dATP	dATP
Structure		
M.W. of enzyme		
No. of subunits (M.W.'s)		
Substrate specificity		
Linear DS	+ATP exonuclease	
Linear SS	+ATP less effective	endonuclease activity only
Linear DS	no hydrolysis	
Circular SS	no hydrolysis	
Rate of ADP formed to phosphodiester bonds split		3

SS: single stranded DNA;   DS: double stranded DNA

Table 1 (Contd)

Some properties of ATP dependent from several bacterial species

E.coli recBC 5.3	H. influenzae 7-10, 19	B. subtilis 17
exonuclease activity : pH 9.0 endonculease activity : pH 7.0	7.5 to 9.0	9.6 to 11.0
$2.10^{-5}M$	1 to $2.10^{-2}M$	$5.10^{-2}M$
$2.10^{-5}M$ (DS DNA)	$4.10^{-5}M$	$7.10^{-5}M$
$1.10^{-4}M$ (SS DNA)		
any ribo or deoxy- ribonucleoside triphosphate	any 5' triphosphate	dATP
	270,000 17	260,000 15
270,000	300,000 19	308,000 17
		81,000
140,000	117,000	70,000
2 128,000	3 115,000	5 62,000
	19 68,000	52,500
		42,500
+ATP exonuclease activity	+ATP exo-endo	+ATP exonuclease
+ATP exo and endo nuclease activity	+ATP 1/10 rate of duplex hydrolysis	no hydrolysis
	no hydrolysis	no hydrolysis
	no hydrolysis	no hydrolysis
20	31 to 39	5

B.subtilis enzyme.   This DNase readily degrades linear double stranded DNA molecules of various origins.   Experiments with denatured *B.subtilis* DNA had previously suggested a low ATP dependent activity on single stranded DNA. In order to exclude the possibility that a limited amount of double stranded DNA is formed in these preparations, the two strands of the *B.subtilis* DNA (heavy and light) were separated on a MAK column. No activity was detected with this separated strand with or without ATP. The following DNAs were not attacked by the enzyme in the presence or absence of ATP; phage fd circular single stranded DNA, SV40 form I (double stranded circular DNA) and SV40 form II (double stranded circular DNA with single strand breaks produced by pancreatic DNase), fd single stranded linear DNA (after pancreatic DNase treatment).   Studies on substrate specificity have so far been reported for *H.influenzae* (10) and *E.coli* (21) purified ATP dependent DNAse. Both of them preferentially hydrolyse linear duplex DNA molecules from either DNA terminus and do not degrade circular single stranded or double stranded DNA.   Unlike the *B.subtilis* enzyme they degrade linear single stranded DNA. Table I is a summary of the enzyme properties which have been studied in different organisms.   Recent data have shown that the only products of DNA hydrolysis by the purified *B.subtilis* ATP dependent DNAse are mononucleotides.   Also double stranded DNA is hydrolyzed equally from either end. This suggests that this enzyme is, as are those from other bacteria, an ATP dependent exonuclease which can progressively hydrolyse DNA at the 5' or the 3' end. The consumption of ATP during the cleavage of a DNA phosphodiester bond is under investigation.   Preliminary data suggest that an average of 5 ATP molecules are degraded for each phosphodiester bond cleaved. The DNAses from *H.influenzae* and *E.coli* have been reported to hydrolyse greater amounts of ATP;40 ATP molecules in the case of the *H.influenzae* enzyme and more than 20 in the case of the *E.coli* enzyme. The ATP requirement of those DNAses and the fact that they possess both exo- and endonuclease activity led several authors to propose elaborate models for their mode of action (3, 21-25).

DNA unwinding by a DNA dependent ATPase from *E.coli* described by Abdel Monem et al. (26) also consumes a high amount of ATP. In that case, ATP is dephosphorylated at a rate of several thousand molecules per enzyme monomer.   The discrepancy between the amount of ATP degraded and that of DNA hydrolysed during the action of *H.influenzae* (9) and *E.coli* (5) ATP dependent DNAses may therefore be attributed to the unwinding activity of these enzymes on duplex DNA molecules.   ATP cleavage to ADP and Pi may be necessary for the binding of ATP dependent DNase to DNA; the formation of an enzyme-DNA complex has been suggested by the results of Shemyakin et al. with *B.subtilis* DNase (27).   Ohi and Sueoka (15) has shown that this enzyme brings DNA molecules into "close proximity" but the functional meaning of this "pairing" is not yet clear.

REFERENCES

1. Anai, M ., Hirahaschi, T. and Takagi, Y. (1970) J. Biol. Chem.
    245:767-774

2. Buttin, G. and Wright, M. (1968) Cold Spring Harbor Symp.
    Quant. Biol. 33:259-269

3. Wright, M., Buttin, G. and Hurwitz, J. (1971) J. Biol. Chem.
    246:6453-6555

4. Nobrega, F.G., Rola, F.M., Pasetto-Nobrega, M. and Oishi, M.
    (1972) Proc. Nat. Acad. Sci. USA 69:15-19

5. Goldmark, P.J. and Linn, S. (1972) J. Biol. Chem. 247:1849-
    1860

6. Vovis, G.F. and Buttin, G. (1970) Biochim. Biophys. Acta
    224:42-54

7. Friedman, E.A. and Smith, H.O. (1971) Fed. Proc. 30:595-

8. Friedman, E.A. and Smith, H.O. (1972) J. Biol. Chem. 247:
    2846-2853

9. Smith, H.O. and Friedman, E.A. (1972) J. Biol. Chem. 247:
    2854-2858

10. Friedman, E.A. and Smith, H.O. (1972) J. Biol. Chem. 247:
    2859-2865

11. Winder, F.G. and Lavin, M.F. (1971) Biochim. Biophys. Acta
    247:542

12. Anai, M.J. (1967) Japan Biochem. Soc. 39:167

13. Chestukin, A.V., Shemyakin, M.F., Kalinina, N.A. and
    Prozorov, A.A. (1972) FEBS Letters 24:121-125

14. Doly, J., Sasarman, E. and Anagnostopoulos, C. (1974) Mutat.
    Res. 22:15-23

15. Ohi, S. and Sueoka, N.J. (1973) J. Biol. Chem. 248:7336-7344

16. Cosloy, S.D. and Oishi, M. (1973) Molec. Gen. Genet. 124:
    1-10

17. Doly, J. and Anagnostopoulos, C. (1976) Eur. J. Biochem.
    71:309-316

18. Georgopoulos, C.P. (1969) J. Bacteriol. 97:1397-1402

19. Wilcox, K.W., Orlosky, M., Friedman, E.A. and Smith, H.O.
     (1975) Fed. Proc. 34:515

20. Lieberman, R.P. and Oishi, M. (1974) Proc. Natl. Acad. Sci.
     USA 71:4815-4820

21. Karu, A.E., Mackay, V., Goldmark, P.J. and Linn, S. (1973)
     248:4874-4884

22. Friedman, E.A. and Smith, H.O. (1973) Nature New Biol. 241:
     54-58

23. Ferdinand, F.J. and Knippers, R. (1975) Eur. J. Biochem.
     52:291-299

24. Wilcox, K.W. and Smith, H.O. (1976) 251:6127-6134

25. Eichler, D.C. and Lehman, I.R. (1977) J. Biol. Chem. 252:
     499-503

26. Abdel-Monem, M., Durwald, H. and Hoffmann-Berling, H. (1976)
     Eur. J. Biochem. 441-449

27. Shemyakin, M.F., Chestukin, A.V., Kalinina, N.A. and Prozorov,
     A.A. (1973) FEBS Letters 31:31-34

P1 TRANSDUCTION FREQUENCIES: A CLUE TO CHROMOSOME STRUCTURE ?

Millicent Masters

Department of Molecular Biology
Edinburgh University
Edinburgh, Scotland

The generalized transducing phage P1 is able to transduce any gene on the chromosome of *Escherichia coli* (1). Although the frequency with which each gene is transduced is constant and characteristic for that gene, different genes are transduced with different frequencies (2). One of the factors determining the frequency of transduction of a gene is its concentration relative to other genes. Thus the relative gene frequencies in cultures of *E.coli* can be determined by comparing the frequency with which phage lysates prepared from the cultures transduce a selection of markers. This method has been used to advantage in studying the replication of the *E.coli* chromosome (2,3,4,5).

When using individual lysates for transduction we noted that there appeared to be a gradient in transduction frequency from the origin to the terminus of replication (2). We therefore decided to measure the transduction frequencies of a large number of genes located at different points on the chromosome to see if we could find a correlation between the position of a gene and its transduction frequency.

Cultures of the F^- strain W3110 growing on broth with a generation time of twenty minutes were used to prepare P1 lysates. Under these conditions of growth replication is expected to be dichotomous and gene frequencies thus to vary from 4 at the origin of replication to 1 at the terminus. These lysates were used to transduce a multiply auxotrophic strain, generally AB2147 or one of its derivatives as previously described (2). Transductants were counted and compared by setting the number of arg H^+ = 1. The results are shown in Table 1 and Fig. 1. Also included are data from Wall and Harriman (6). The transduction frequency distribution

Table 1

Transduction frequencies of *E.coli* genes

Marker	Map Location	Mean Transduction Frequency: F⁻ Donor	Calculated Gene Frequency	Transduction Frequency / Calculated Gene Frequency
malA	73.9	1.8 ± 0.5	1.08	1.67
xyl	78.8	1.29 ± 0.12	1.25	1.03
cysE	80.0	0.45 ± 0.18	1.28	0.35
pyrE	80.7	0.32 ± 0.07	1.3	0.25
uhp	81.0	5.2 ± 1.5	1.3	4.0
tna	82.2	5.2 ± 2.4	1.42	3.7
bglB	82.3	6.6 ± 2.5	1.45	4.6
rbs	83.0	6.8 ± 2.0	1.33	5.1
ilv	83.2	5.6 ± 1.7	1.35	4.1
metE	84.0	1.25	1.30	0.96
rha	86.1	0.36± 0.04	1.22	0.30
metB	87.1	0.79± 0.15	1.2	0.66
arg	87.8	1.0	1.16	0.86
purD/H	89.0	0.66± 0.08	1.14	0.58
metA	89.2	0.7 ± 0.04	1.12	0.49
mel	91.8	0.7 ± 0.05	1.07	0.65
purA	93.4	0.66	1.0	0.66
pyrB	94.7	0.41± 0.07	0.96	0.43
leu	1.6	1.94± 0.5	0.8	2.4
lac	7.9	0.83± 0.2	0.68	1.22
purB	25.0	0.32	0.42	0.76
trp	27.3	0.45± 0.14	0.40	1.12
his	44.1	0.23± 0.07	0.48	0.48
purF	49.6	0.31± 0.05	0.57	0.54
purC	52.8	1.1 ± 0.4	0.62	1.77
argG	68.0	0.67	0.93	0.72
nalA	8.0	0.22	0.54	0.41
gua	53.5	0.65	0.63	1.0
nad	55.2	0.54	0.66	0.82
tyr	56.0	0.59	0.67	0.88
thy	60.0	1.61	0.74	2.1
lys	61.0	1.73	0.76	2.3
proA	6.0	0.72	0.72	1.0
gal	7.0	0.34	0.52	0.65

Gene frequencies were calculated (column 4) from an exponential line
varying over a four fold range from origin to terminus. This line
was placed so that equal numbers of the observations from column 3
would fall above and below it. Data for markers from *nalA* to *gal*
are from (2).

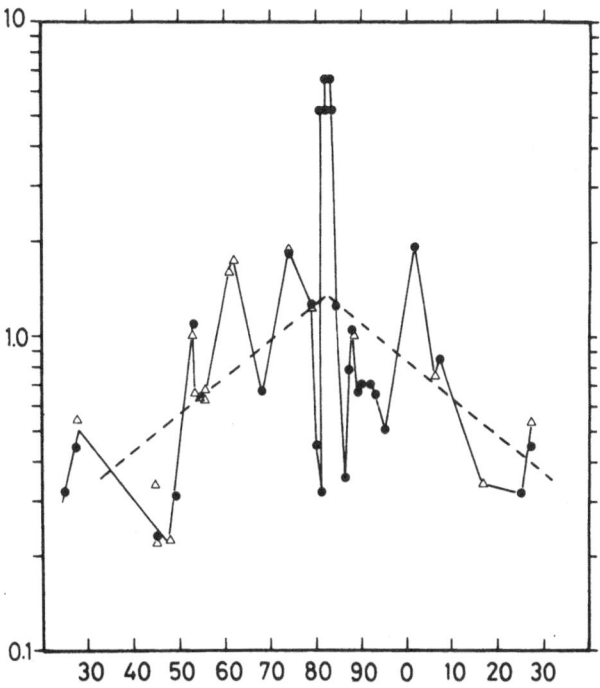

Figure 1. *Transduction frequencies as a function of map position. Closed circles are new data (malA - argG in Table 1). Triangles are from (2) and Table 1 (nalA - gal). Open circles are from (6) normalized to the above data (x 0.75). Map positions are from (7). The dashed line represents the gene frequency distribution of an exponential culture similar to those used to prepare the phage lysate used for frequency distributions. It is placed as described in the text. It was used to calculate the values in Table 1, column 4.*

differs from the gene frequency distribution in the following respects:

(1)   Markers near the origin are transduced with very high frequency (*tna, uhp, rib, ilv, bgl*). These markers cover a span of less than 2% of the chromosome length and are cotransducible.

(2)   Markers flanking this peak are transduced with a much lower frequency resulting in troughs on either side of the origin peak.

(3)   The majority of other markers measured are transduced with intermediate frequencies. There is a gradient of frequency from origin to terminus.

(4)   Superimposed on this gradient are additional peaks and
       troughs, none, however, so marked as that at the origin.

Troughs are not the Result of Poor Expression
of Particular Markers

It would be hypothesized that the low frequencies of transduction obtained for markers such as pyr E and rha are the result
of poor expression under the selective conditions employed, that
is, cells which have integrated these markers do not necessarily
form colonies on plates.

That this is not the case can be shown by an analysis of the
frequencies of cotransduction between neighboring markers with
differing frequencies of transduction. If poor expression accounted
for the small numbers of rha^+ or pyr E^+ transductants recovered, then
an increased number might be expected if they were scored as the
unselected marker in a transductional cross. This is not the case.
The pyr E^+ represent only a small fraction of transductants
selected as uhp^+ although a large fraction of pyr E^+ transductants
are uhp^-. In fact if one knows the frequencies with which two
neighboring markers are transduced it is possible to predict the
degree of asymmetry which will occur in their cotransduction
frequencies. Thus, if there are two neighboring markers A and B,
and if A is transduced with a frequency a, and B with a frequency
b, then:

$$\frac{B^+ \text{ cotransduced with } A^+/A^+}{A^+ \text{ cotransduced with } B^+/B^+} = \frac{b}{a}$$

This is true both at the origin where the gradient in transduction frequency is steep and elsewhere where it is less steep
(Table 2).

It has frequently been noted that the cotransduction frequencies
between two markers can depend on which of the pair is selected in a
cross. Bachmann et al. (7) discuss the implications of this fact for
transductional mapping studies. The data presented here suggest
that when a lack of reciprocality in cotransduction is noted, a
corresponding difference in transduction frequency will be found for
the markers concerned.

Peaks in Transductional Frequency Occur in Regions
of High Gene Density

About 700 $E.coli$ genes have been described and mapped (7).
Although these genes probably represent about a third of the coding

Table 2

Cotransduction of some pairs of linked markers
with differing transduction frequencies

Selected Marker	% Cotransduction		Ratio of Cotransduction Frequencies	Frequency of Transduction of the Selected Marker	Ratio of Transduction Frequencies
pyrE uhp	uhp 58	pyrE 6.6(±2.8)	8.8 (6.2 - 15)	0.32 5.2	16.2
rha arg	arg 12	rha 5	2.4	0.36 1	2.8
rha metB	metB 24	rha 8	3.0	0.27 0.8	3.0
ilv metE	metE 36.7	ilv 7.5	4.9	5.6 1.25	4.5

Cotransduction frequencies were determined by
scoring at least 100 transductants for the unselected marker. The cotransductional data for
metE-ilv are from Bachmann et al. (1976).

capacity of the *E.coli* chromosome they are very unevenly distributed
along the physical length of the chromosome. Some regions are
densely packed with genes; other regions have very few. Bachmann
et al. (7) have analyzed and commented upon the distribution of the

positions of genes. The faint line in Fig. 2 shows the number of
genes per minute of length on the *E.coli* chromosome. Also shown in
this figure are the transduction frequencies taken from Table 1
and adjusted to represent transduction from an unreplicated
chromosome. Each of the seven distinct peaks in transduction
frequency matches a peak in gene density. There are two further
peaks in gene density, one not matched by a peak in transduction
frequency (at 17 min.). There is insufficient transductional data
to determine whether there is a peak matching the gene density peak
at 45 min. The relative broadness of many of the transductional
peaks results from the large gaps between some of the markers
measured. Measurement of transduction frequency for additional
markers could be expected to result in the sharpening or sub-
division of some of the peaks.

Speculations Regarding Chromosome Structure

Bachmann et al. (7) suggest that gene density may reflect the
three dimensional structure of the chromosome. Worcel and Burgi (8)
have suggested that the *E.coli* chromosome is composed of non-freely
rotating loops of DNA, perhaps held together at their bases to
form a rosette. Bachmann et al. (7) speculate that perhaps gene
dense regions are on the outside of this structure and hence more
accessible to transcription and translation than regions buried
within it. They note that genes required for protein and DNA
synthesis are located in gene dense regions.

If the genome has such a structure it would not be surprising
to find that gene dense regions are particularly amenable to transduction, either because their DNA is accessible to phage enzymes
during DNA packaging in the donor cell, or because it is accessible
to the recombination system in the recipient cell.

The high frequency with which markers near the origin of
replication are transduced is particularly striking. If the origin
were attached to a membrane site during most of the cell cycle as
has been suggested (9) it might be thus rendered even more
accessible than other exposed regions of the chromosome.

Other explanations have been offered to explain why all genes
are not transduced with the same frequency (10,11). These
generally postulate a preferred base sequence or a hierarchy of
such sequences from which encapsulation into phage heads proceeds.
Genes near these sequences would be transduced with higher frequency
than those far from them. A prediction of this sort of model is
th-t the transducing DNA in phage heads would have a frequency
distribution corresponding to the transductant frequency
distribution.

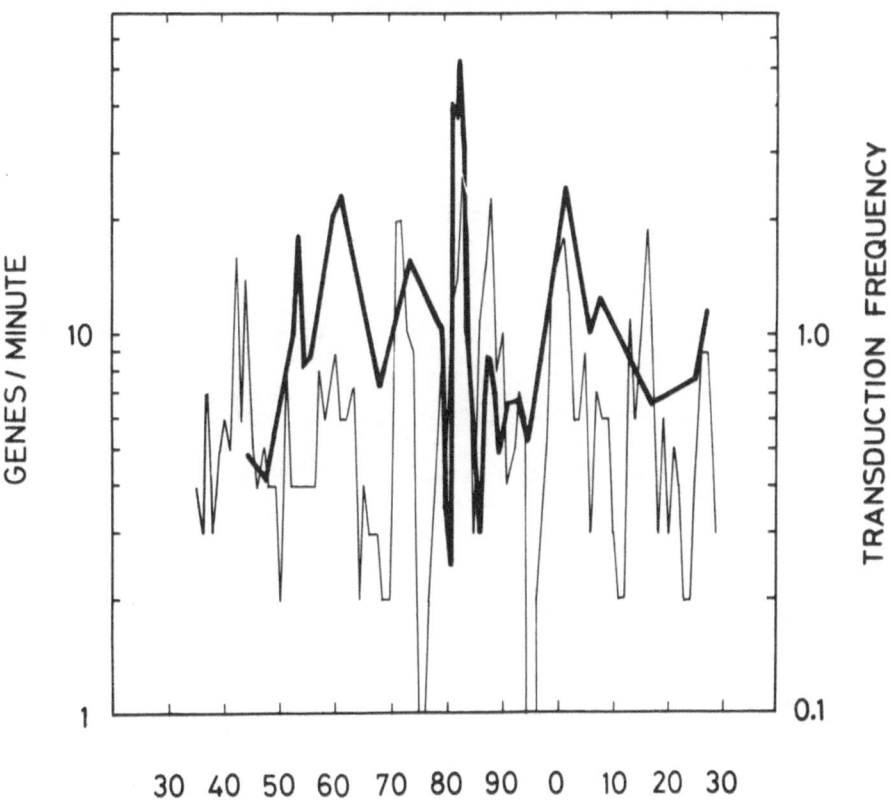

Figure 2. *Comparison of gene density and transduction frequency as a function of map position.*

The faint line shows genes per minute and is derived by counting the genes on the most recent E.coli map (7). All loci shown other than phage attachment sites, operators and promoters were counted. The heavy line shows transduction frequencies normalized to represent an unreplicated donor chromosome.

An alternative explanation is that all DNA is equally likely
to be incorporated into transducing phage but differs in its
probability of being integrated into the recipient chromosome.
The unintegrated DNA could give rise to the high number of
abortive transductants that have been reported for some markers
(12). Recent work indicates that 85% of the transducing DNA
which is injected into the recipient cell is not incorporated into
the recipient chromosome (13). Experiments are in progress in our
laboratory to determine the gene frequency distribution of the DNA
in transducing phage and hence to help to distinguish amongst the
above alternatives.

References

1.   Lennox, E.S.(1955) Virology 1:190

2.   Masters, M., and Broda, P. (1971) Nature New Biol. 232:137-
     140

3.   Caro, L. and Berg, C.M. (1969) J. Mol. Biol. 45:325-336

4.   Wolf, B., Newman, A. and Glaser, D.A. (1968) J. Mol. Biol.
     32:611-629

5.   Masters, M. (1970) Proc. Nat. Acad. Sci. (Wash.) 65:601-608

6.   Wall, J.D., and Harriman, P.D. (1974) Virology, 59:532-544

7.   Bachmann, B.J., Low, K.B. and Taylor, A.L. (1976) Bact. Rev.
     40:116-166

8.   Worcel, A. and Burgi, E. (1972) J. Mol. Biol. 71:127-147

9.   Parker, D.L. and Glaser, D.A. (1974) J. Mol. Biol. 87:153-
     168

10. Harriman, P.D. (1972) Virology 48:595-600

11. Chelala, C.A. and Margolin, P. (1974) Mol. Gen. Genet. 131:
     97-112

12. Arber, W. (1960) Virology, 11:273

13. Sandri, R.M. and Berger, H. (1977) In: Abstracts of the Annual
     Meeting of the American Society for Microbiology. Washington,
     p136

BACTERIAL NUCLEOID STRUCTURE AFTER INHIBITION OF DNA REPLICATION:

THE ROLE OF RNA SYNTHESIS

Steinar Øvrebø* and Christopher Korch

Department of Biochemistry
University of Bergen
Arstadveien 19, 5000 Bergen, Norway

In 1971 Stonington and Pettijohn (7) devised a procedure for the isolation of the E.coli genome in a compact, non-viscous structure. They showed the existence of two different structures and named them the membrane-associated and the membrane-free folded chromosome, Studies of the membrane-free structure have shown that in size and compactness it grossly resembles the in vivo DNA structure seen with the electron microscope, which is usually termed the nucleoid (1). Because of this resemblance, the isolated folded chromosome has been called the nucleoid (5) a term which also will be used here.

One controversy in the study of the nucleoid has been its envelope (membrane) association as a function of DNA replication (2,6,10). To thoroughly study this question it is of great importance to understand the nucleoid structure during DNA replication and in its absence. The present study deals with the nucleoid structure during a block in DNA replication upon amino acid starvation and chloramphenicol treatment. Results indicating that a change in the nucleoid structure is dependent upon a concomitant block in RNA synthesis will be described.

It was shown earlier (9) that the nucleoid structure changes during amino acid starvation in E.coli. The results showed a decrease in sedimentation value of the nucleoid during amino acid starvation resulting in a discrete sized species. The authors (9) concluded that the observed decrease in sedimentation value was due to the fact that the DNA initiation was blocked in cells undergoing amino acid starvation, and for this reason the chromosomes ended up in one genome size. This in contrast to the genome size in an exponential culture, which has an average size of approximately 1.5. Later,

*A.S.I. Participant

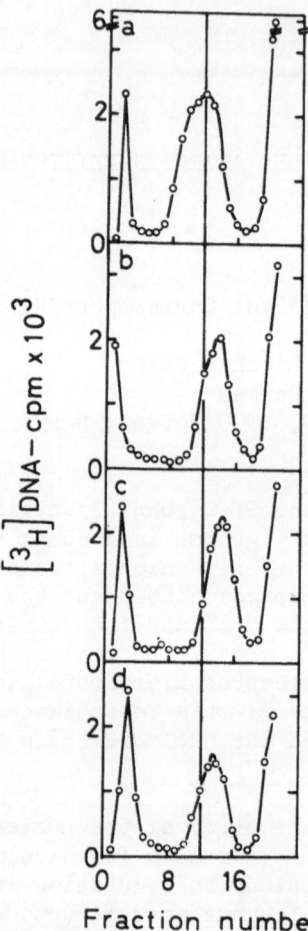

Figure 1. *Sedimentation profile of nucleoids isolated from E.coli*
CP 78 *(B1, leu, arg, and rel⁺). The DNA was labelled with*
³H-*thymidine. Harvested cells were resuspended in a solution*
containing 0.01 M Tris-HCl pH 8.0, 0.01 M sodium azide, 20% w/v
sucrose and 0.1 M NaCl. 0.2 ml of the resuspended cells were mixed
with 0.05 ml of a solution containing 4 mg egg-white lysozyme/ml,
0.12 M Tris-HCl pH 8.0 and 0.05 M Na₂EDTA. After 5 mins incubation
at 0°C the cells were lysed by addition of 0.25 ml of a solution
containing 1% Brij-58, 0.4% deoxycholate, 1% sarcosyl, 2 M NaCl and
0.01 M Na₂EDTA. This final solution was incubated for 10 minutes at
18°C. The lysate so obtained was applied to a 10 to 30% w/v sucrose
gradient and centrifuged for 45 minutes at 17,000 rpm with a SW 40
Ti rotor in a Beckman ultracentrifuge. Sedimentation was from right
to left. Nucleoids isolated from (a) exponential growing cells,
(b, c and d) cells starved for leucine in 10 , 20 and 30 minutes.

however, it was reported (3) that blocking of DNA initiation by
amino acid starvation gives chromosomes with genome size of 2; this
means it is not initiation but termination which is blocked by amino
acid starvation.

Using the method described by Stonington and Pettijohn (7) with
minor modifications (see legend, Fig. 1) we have studied the nucleoid
structure during amino acid starvation. Figure 1 shows sucrose
gradient centrifugation of nucleoids isolated from exponentially
growing E.coli cells and cells starved of leucine for 10,20 and 30
minutes. As can be seen from Figure 1, already after 10 minutes of
amino acid starvation there is a reduction in the sedimentation rate
of the nucleoid. The sedimentation rate is not further changed with
longer starvation times. Since 10 minutes is only about 15% of the
generation time it is unlikely that the change in sedimentation rate
seen is due to completion of DNA replication cycles, since this would
only effect a minor part of the chromosomes. Pulse labelling
experiments also support the view that the change in nucleoid
structure is independent of a block in DNA replication (data not
shown).

Pettijohn and Hecht have shown that cells treated with
rifampicin gave a completely unfolded nucleoid (5) in accordance
with in vitro experiments which showed that the nucleoid structure
is extremely sensitive to RNase treatment. Since the nucleoid is
very dependent on RNA, changes in RNA synthesis rate is also likely
to result in a concomitant change in the nucleoid. A decrease in
the rate of RNA synthesis is one of the consequences of amino acid
starvation in a stringent rel^+ E.coli cell. Figure 2 shows
incorporation of uracil in stringent rel^+ and relaxed rel^- strains.
This incorporation shows the accumulation of RNA in a cell. It has
been shown that it is primarily stable RNA (8) which is made in
lesser amounts during amino acid starvation in rel^+ cells, whereas
the rate of synthesis of several specific mRNAs, for example
β-galactosidase mRNA, is not changed during amino acid starvation
in rel^+ cells (4). In contrast to this, amino acid starvation in
rel^- cells causes no immediate reductions in RNA synthesis and this
resembles the RNA synthesis seen in rel^+ cells treated with
chloramphenicol. To see if this lower rate of RNA synthesis in rel^+
strains during amino acid starvation can explain part of, or all of,
the structural changes in the nucleoid, attempts were made to
isolate the nucleoid from a rel^- strain during amino acid starvation
as shown in Figure 3. No significant transition to a lighter complex
could be demonstrated.

Amino acid starvation in both rel^+ and rel^- cells blocks
initiation/termination of DNA replication. Another way to block
DNA replication is to treat a culture of E.coli cells with
chloramphenicol. The block in DNA initiation/termination in this
method is also thought to be mediated through inhibition of protein

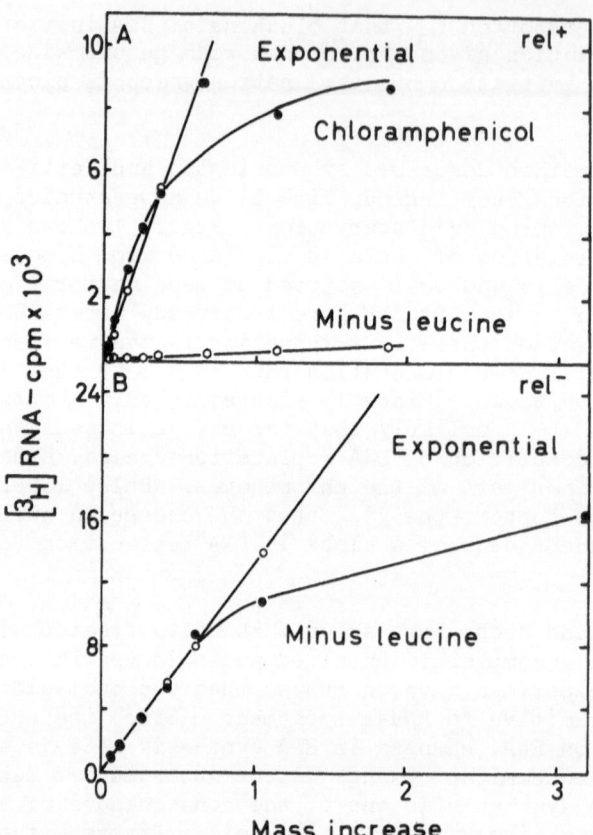

Figure 2. *Incorporation of ³H-uracil into RNA. Cells were labelled with uracil and samples withdrawn at indicated times and TCA precipitated (DNA + RNA) and NaOH treated before TCA precipitation (DNA). The radioactivity was measured and the DNA radioactivity was subtracted before plotting. (A) E.coli CP 78 (rel⁺). Exponential, chloramphenicol treated and amino acid starved cultures. (B) E.coli CP 79 (B1, leu, thr, his, arg, and relA). Exponential and amino acid starved cultures.*

synthesis. Figure 4 shows that chloramphenicol also does not change the sedimentation rate of the nucleoid.

The nucleoid with the lower sedimentation value, isolated from rel⁺ cells, has a greater sedimentation rate than the nucleoids isolated from rifampicin treated cells (5). Rifampicin gives a completely unfolded nucleoid while amino acid starvation results only in a partially unfolded nucleoid. Pettijohn and Hecht (5) have described a model for the structure of the nucleoid. In this model they describe a core of RNA which holds the DNA loops together. These

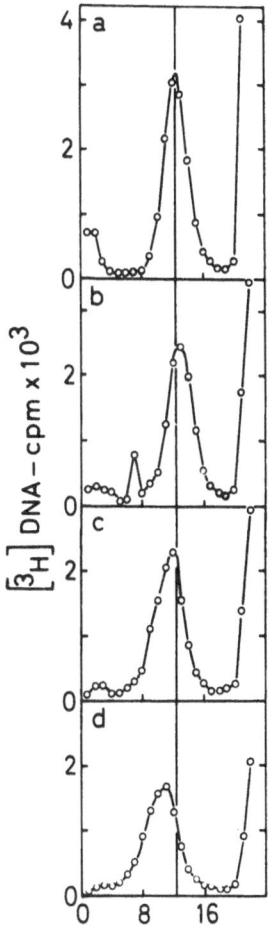

Figure 3. Sedimentation profile of nucleoids isolated from E.coli CP 79 (rel⁻) (a) exponential gorwing cells, (b, c and d) cells starved for leucine in 10, 20 and 30 minutes. Sedimentation from right to left.

DNA loops are thought to be the main structural elements of the nucleoid. Inhibiting RNA synthesis with rifampicin results in this RNA not being made, and the nucleoid will thus be found in a fully extended form, termed the unfolded nucleoid (5). The nucleoid isolated from amino acid starved rel⁺ cells are partially unfolded nucleoids as judged by their sedimentation rates (5). This suggests that some of the core RNA is removed or not synthesized in these cells. From the above mentioned model it is reasonable that amino acid starvation in a rel⁺ strain inhibits synthesis of only part of the core RNA which holds the DNA loops together. Is this core RNA

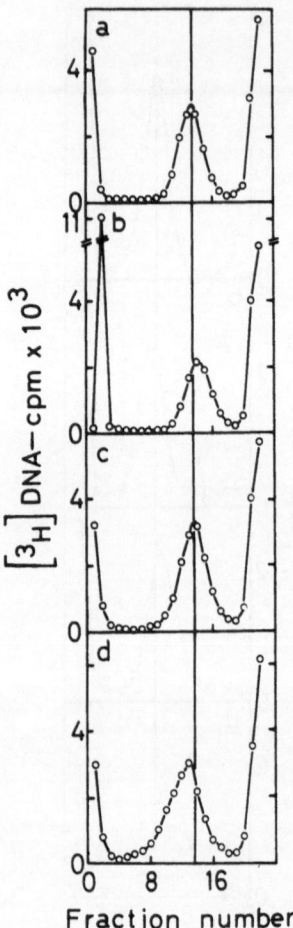

Figure 4. Sedimentation profile of nucleoids isolated from E.coli CP 78 (rel⁺) (a) exponential growing cells, (b, c and d) cells treated with chloramphenicol for 10, 20 and 30 minutes. Sedimentation from right to left.

a specific type of RNA ? If so, the synthesis of this RNA must in part be regulated through the mechanism of stringent control.

ACKNOWLEDGEMENT

We would like to thank Professor Kjell Kleppe for encouragement and interest shown throughout this study. The research reported here was supported by grants from the Norwegian Research Council for Science and Humanities.

REFERENCES

(1) JACOB, F., BRENNER, S. and CUZIN, F. (1963) Cold Spring
     Harbor Symp. Quant. Biol. 28,323

(2) KORCH, C., ØVREBØ, S. and KLEPPE, K. (1976) J. Bacteriol.
     127,904

(3) MARUNOUCHI, T. and MESSER, W. (1973) J. Mol. Biol., 78,221

(4) MORRIS, D.W. and KJELDGAARD, N.O. (1968) J. Mol. Biol. 31,145

(5) PETTIJOHN, D.E. and HECHT, R.M. (1973) Cold Spring Harbor
     Symp. Quant. Biol. 39,3141

(6) RYDER, O.A. and SMITH, D.W. (1974) J. Bacteriol. 120,1356

(7) STONINGTON, O.G. and PETTIJOHN, D.E. (1971) Proc. Nat. Acad.
     Sci., U.S.A. 68,6

(8) TRAVERS, A.(1974) In:Control of Transcription (Biswas, B.B.,
     Mandal, R.K., Stevens, A. and Cohn W.E., eds.) pp 67-80,
     Plenum Press, New York.

(9) WORCEL, A. and BURGI, E. (1972) J. Mol. Biol. 71,127

(10) WORCEL, A. and BURGI, E. (1974) J. Mol. Biol. 82,91

SPECIFIC PROPERTIES OF THE END GROUPS OF rDNA FROM *TETRAHYMENA*

H. Klenow

Biochemical Institute B
University of Copenhagen
Copenhagen, Denmark

In some eukaryotic cells the DNA coding for ribosomal RNA exists as extrachromosomal molecules. This has been demonstrated for oocytes of various animals (1) and for some primitive organisms like *Tetrahymena* and *Physarum*. From the latter two organisms this kind of discrete DNA molecules (rDNA) has been isolated and characterized. They have in both cases been found to contain an axis of two-fold rotational symmetry; i.e., the individual strands are self complementary (2,3,4). DNA molecules with this special arrangement of base sequence is often referred to as palindromic molecules. Heating and rapid cooling of solutions of palindromic DNA molecules results in formation of fold-back or smap-back forms of hairpin-like structures with molecular weights which are half of that of the native molecules. The special arrangement of bases in palindromic molecules also means that cleavage with site specific endonucleases results in odd numbers of fragments. The rDNA of *Tetrahymena* strain BVII which has a molecular weight of about 13.2×10^6 has thus been found to give rise to three fragments by treatment with restriction enzyme EcoRI, two identical fragments (molecular weight 1.6×10^6) from the ends of the molecule and one fragment containing the center part of the molecule (molecular weight 10.0×10^6) (3). The snap-back form of this rDNA molecule would be expected to give rise to one each of these two fragments by treatment with EcoRI, i.e., the center fragment in hairpin form (molecular weight 5.0×10^6) and the end group fragment.

In order to be able to study free palindromic DNA molecules in further detail a simple procedure has been developed for the isolation of them in the snap-back form. Total cellular DNA is isolated, heated above the melting temperature and rapidly cooled.

The bulk of the DNA will be present in single stranded form while
free palindromic molecules are present as completely double
stranded fold-back structures. Such DNA structures may be isolated
free of single stranded DNA by the aqueous two-phase system of
Albertsson (5).   Density equilibrium centrifugation of a preparation
of this kind of molecules from *Tetrahymena* has shown the presence
of two density classes of DNA molecules in the snap-back form. The
heavier one has a density identical with rDNA and is according
to hybridization data more than 95 per cent pure rDNA (6).

Analysis by agarose gel electrophoresis of the heavy component
shows, however, the presence of not only the expected 6.6×10^6
molecular weight band but also of a 13.2×10^6 band and in addition
up to four fainter and slower moving but distinct bands. Digestion
with the restriction endonuclease EcoRI gives rise to the expected
end group fragment (molecular weight 1.6×10^6) and the hairpin
fragment (molecular weight 5.0×10^6) and in addition to three
fainter bands (molecular weights about 3.2×10^6, 6.5×10^6 and
9.5×10^6 respectively) (see Fig. 1). These bands correspond to
di-tetra- and hexamers of the 1.6×10^6 molecular weight fragment.
Partial digestion with restriction enzyme EcoRI gives rise to at
least two additional bands, a 8×10^6 molecular weight band and a
more slowly moving band. These findings suggest that the snap-back
form of rDNA is present not only as monomers but also as di- and
oligomers held together by hydrogen bonds localized near the end
group of the monomer molecules.

Treatment of the snap-back preparation of rDNA with the large
fragment of DNA polymerase I from E.coli in the presence of all 4
deoxyribonucleoside triphosphates gives rise to conversion of the
dimers and the oligomers to monomers. This suggests that extension
of 3'-end groups displaces strands that a responsible for holding
together of the dimer and oligomer structures. Similarly, almost
complete conversion to monomers is obtained in the presence of
dGTP and dTTP but not in the presence of either of these two triph-
phosphates alone or in the presence of dATP and dCTP. This suggests
that the strands displaced during synthesis contain only guanine
and thymine.

Treatment of the snap-back preparations with the single
stranded specific nuclease S_1 leads also to the conversion of both
dimers and oligomers to monomers suggesting the presence of non-base
paired regions near the part of the molecules responsible for the
structure of the dimers and the oligomers. Finally, digestion with
the double stranded-specific 3' → 5' exonuclease III from E.coli
gives rise to conversion of 25-30 per cent of the DNA to an acid-
soluble product, while exonuclease III treatment after EcoRI
digestion gives rise to degradation of about 50 per cent of the DNA.
This may suggest that only about half of the 3'-end groups in the
rDNA snap-back preparation are susceptible to degradation by exonuclease III.

MW ×10^6

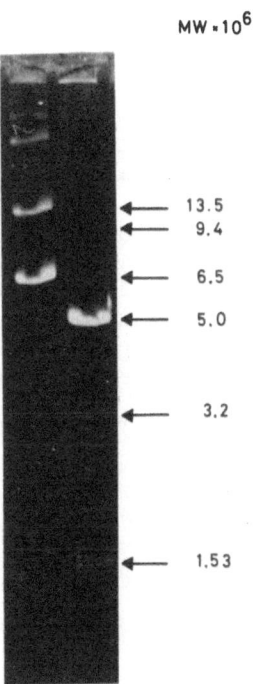

← 13.5
← 9.4

← 6.5
← 5.0

← 3.2

← 1.53

Figure 1. *Gel-electrophoresis analysis of a snap-back preparation of rDNA from Tetrahymena. Right lane: untreated preparation; left lane: preparation digested completely with EcoRI.*

The following two molecular models for rDNA which may account for the formation of dimers and oligomers of the snap-back form have been considered.

Model A postulates the presence of cohesive 5'-end groups and of a non-palindromic region at one end of the molecule the rest of which is a perfect palindrome. After heating and rapid cooling two different types of snap-back molecules are formed. One of them has a free 3'-end group corresponding to the non-palindromic region and a free 5'-end group corresponding to one of the 5'-end groups of the native molecule. The other type of snap-back molecules has no free 3'-end group but a free 5'-end group corresponding for the inner part to the non-palindromic region and for the outer part to the other free 5'-end group of the native molecule. Dimers may be formed from one of each of these molecules. Dimer formation may be due to the non-palindromic region or due to the cohesive 5'-end groups. Either kind of dimer contain single-stranded stretches

allowing formation of oligomers (see Fig. 2). The dimers and oligomers contain double-stranded 3'-end groups the extension of which
under concomitant displacement of strands in front of synthesis
would cause formation of monomer molecules. The extended strands
contain only guanine and thymine and only deoxyGTP and deoxyTTP are,
therefore, required for the reaction. The displaced strands which
also contain only guanine and thymine consist for one part of the
strand of the non-palindromic region which has a 3'-end group and
for the other part of the 5'-end group of the same strand.
Conversely, both the strand of the non-palindromic region which has
no end-group and its extension forming a free 3'-end group in the
native form of the molecule contain only adenine and cytostine. At
the centers of the cruciforms of the dimers and oligomers some nonhydrogen bonded bases would for steric reasons be expected to be

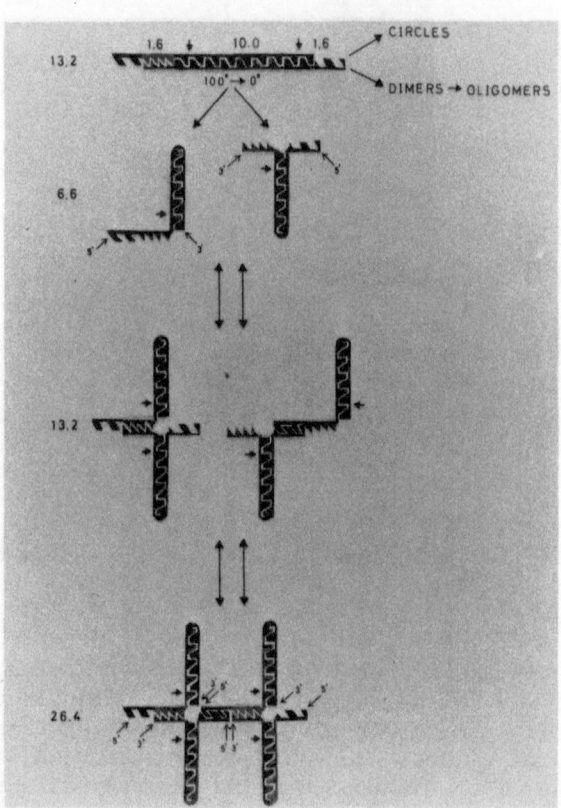

Figure 2. Model A for the structure of rDNA from Tetrahymena. The
short arrows indicate the cleavage site for restriction enzyme
EcoRI. The numbers indicate molecular weights in millions of
daltons. The model postulates the presence of cohesive 5'-end
groups and a non-palindromic region at one end of the molecule the
rest of which is a perfect palindrome. The snap-back form may give
rise to formation of dimers and oligomers.

present. The single-stranded specific endonuclease S$_1$ would, therefore, cause the oligomer structures to fall apart giving rise to the snap-back form of the palindromic region and to double stranded structures of the non-palindromic region and of the cohesive ends of the native molecule. The structure at the center of the cruciform would prevent exonuclease III to degrade the double stranded 3'-end group of the non-palindromic region beyond this point. The degradation by exonuclease III from double stranded 3'-end group of paindromic part would, therefore, cause degradation of only about 25 per cent of the molecules. It should be emphasized that the non-palindromic region and the cohesive 5'-end groups may be quite small compared to the rest of the molecule.

A molecule with the structure of this model may have arisen from a closed circular molecule consisting of a giant palindrome, the ends of which are joined to a non-palindromic region. One of the strands of this region would contain only adenine and cytosine while the other would contain only thymine and guanine. The open form of the circle may be formed by introduction of nicks in the non-palindromic region. For the strand containing only adenine and cytosine the nicks would be introduced primarily at the 5'-end of the non-palindromic region. The snap-back form of this strand of the molecule would have a free 5'-end group containing adenine and cytosine and a double stranded hairpin structure with a 3'-end group. For the strand of the circle containing only guanine and thymine in the non-palindromic region nicks would be introduced internally in this region. The snap-back form of this strand would contain both a free 5'-end group and a free 3'-end group. It is possible that the non-palindromic region and the cohesive 5'-end groups of model A contain the same base sequences. In that case the non-palindromic region of the proposed circular structure would contain repeated base sequences. Model A is supported by the electron microscopic observations by Karrer et al. (3) of the presence of a circular form of rDNA.

Model B postulates that rDNA is a perfect palindromic molecule from one end to the other and that the molecule contains a number of site specific nicks near the ends. These nicks are present in both strands but in different numbers in different molecules. When such molecules are heated and rapidly cooled different types of snap-back molecules with varying lengths of free 3'- and free 5'-end groups are formed. Some of the free 3'-end groups will be partially complementary to some of the free 5'-end groups and they will, therefore, allow the formation of aggregates held together by hydrogen bonds (see Fig. 3). These aggregates correspond to the designed dimers and oligomers observed.

Nicks introduced in the two strands of a molecule may never be present precisely opposite to each other. Short stretches close to where two arms in the aggregate of the model meet will,therefore,

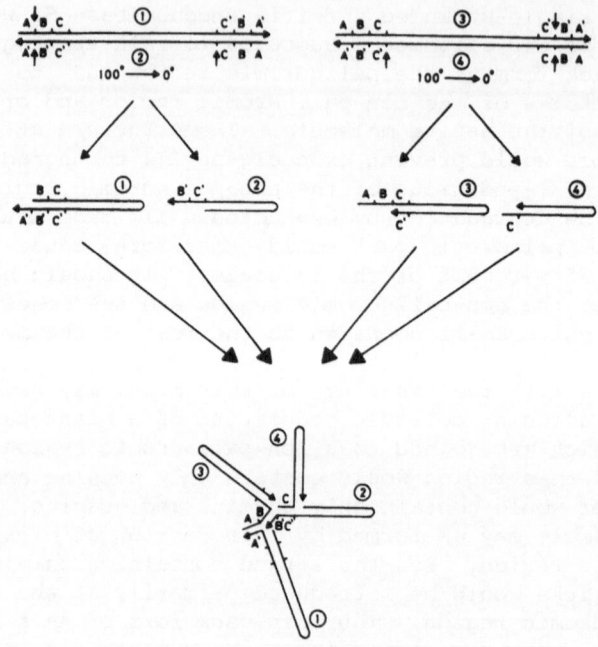

Figure 3. *Model B for the structure of rDNA from Tetrahymena. The short arrows indicate sites of single-stranded nicks. The numbers identify the different DNA strands. The model postulates that the molecule is a perfect palindrome from one end to the other and that it contains a number of site specific nicks near the ends. The nicks are present in both strands but in different numbers in different molecules. The snap-back form may give rise to formation of aggregates. In the example shown here a tetramer-like aggregate may be formed.*

be single-stranded. These stretches will be susceptible to cleavage by endonuclease S_1. The presence of these stretches will also mean that in the average half of the 3'-end groups of the aggregates will not be susceptible to degradation by exonuclease III. Also for this model treatment with exonuclease III will, therefore, give rise to only 25 per cent degradation of the snap-back form and of its aggregates. Extension of 3'-end groups under concomitant strand displacement will also for this model convert the aggregates of the snap-back form to a monomer type of molecule. The model further postulates that the nicked 3'-end group region contains only guanine and thymine. Only dGTP and dTTP are, therefore, required for extension of strands resulting in conversion of the aggregate to monomer type of molecules. The nicked end group regions may be quite small compared to the rest of the molecule.

Experiments designed to distinguish between the two models are in progress.

ACKNOWLEDGEMENTS

The author wants to thank Drs Jan Engberg and Ed. Orias for fruitful discussions and Birthe Juul Hansen for expert assistance.

REFERENCES

1.   Tobler, H. (1975) In: Biochemistry of Animal Development, Vol. 3, Academic Press, New York. p91

2.   Engberg, J., Andersson, P., Leick, V. and Collins, J. (1976) J. Mol. Biol. 104:455

3.   Karrer, K. and Gall, J. (1976) J. Mol. Biol. 104:421

4.   Vogt, V.M. and Braun, R. (1976) J. Mol. Biol. 106:567

5.   Öberg, B., Albertsson, P.A. and Philipson, L. (1965) Biochim. Biophys. Acta 173:

RECOMBINATIONAL MAPPING OF THE *polA* LOCUS OF *ESCHERICHIA COLI* K12:

GENETIC FINE STRUCTURE

W.S. Kelley[*], J.A. Reehl and N.C. LeDonne, Jr.

Department of Biological Sciences
Mellon Institute of Science
Carnegie-Mellon University
4400 Fifth Avenue, Pittsburgh, Pa. 15213, U.S.A.

Recently we described a series of experiments by which the orientation of the *E. coli polA* cistron was established with respect to the linked genetic loci *metE* and *rha* (Kelley and Grindley, 1967A). It was possible to orient the cistron unambiguously because of the unique enzymic properties of the wild type *polA* gene product, DNA polymerase I, and the properties of the products of three allelic *polA* mutations, *polA1, polA6* and *polA107*.

Continued interest in the role of DNA polymerase I in recombination, replication and repair processes has stimulated further experiments with *polA⁻* mutants and the enzymatic defects in several other allelic forms of DNA polymerase have been documented (for review see Lehman and Uyemura, 1976). All told, approximately thirty different *polA⁻* mutations have been isolated and reported in the literature, although extensive enzymic characterization has been limited to a half dozen with isolable activities. Because the enzyme is a multifunctional molecule whose several functions appear to work in concert, the effects of different alleles on cellular processes are not identical (e.g., see Grindley and Kelley, 1976; Konrad, 1977).

In order to confirm our genetic analysis and to extend our knowledge of several characterized *polA⁻* mutations, we have applied our genetic mapping techniques to several further alleles. Not all of the mutations tested thus far appear equally amenable to this type of genetic analysis and interpretations of the mapping experiments are facilitated by some knowledge of the characteristics of the different mutant enzymes. However, the results which we are reporting here confirm our previously published gene orientation.

*A.S.I. Participant

455

MATERIALS AND METHODS

Transductional techniques with phage PICM (Rosner, 1972)
and media are as described previously (Grindley and Kelley, 1976;
Kelley and Grindley, 1976a). All crosses were carried out at 32°
and scored as described in those experiments. The *polA* alleles
utilized in this study were obtained from the individuals indicated
in Table 1 whom we thank for their generosity. The original allele
designations have been altered to fit the numbering scheme proposed
by the *E. coli* Genetic Stock Center (Bachmann, personal communication.

Enzymological techniques and assay systems are as described in
Kelley and Grindley (1976b).

Table 1

Bacterial Strains

JG108   = *E. coli* K12 W3110 F$^-$ *thyA$^-$ lac$^-$ rha$^-$ metE$^-$ str^r*, from
     J.D. Gross via J.A. Wechsler

1R713   = *E. coli* K12 F$^-$ prototroph, from E.S. Anderson

DS602   = *E. coli* K12 F$^-$ *thyA$^-$ lac$^-$ rha$^-$ polA214 str^r*, from D.J.
     Sherratt (*polA214* originally was designated *polAts214*)

DS694   = *E. coli* K12 F$^-$ *thyA$^-$ lac$^-$ rha$^-$ polA1 str^r*, from D.J.
     Sherratt

BT4113   = *E. coli* K12 F$^-$ *thy$^-$ polA4113*, from B.M. Olivera
     (*polA4113* originally was designated *polABT4113 or 4113*)

RS5064   = *E. coli* K12 F$^-$ *trpA$^-$ polA480*, from I.R. Lehman (*polA480*
     or *polA480*ex) originally was designated *polAexl*)

KMBL1789 = *E. coli* K12 F$^-$ *thyA$^-$ argA$^-$ bio$^-$ pheA$^-$ endA$^-$ polA107*,
     from H.L. Heijneker (*polA107* originally was designated
     polA'107)

KMBL1791 = *E. coli* K12 F$^-$ *thyA$^-$ argA$^-$ bio$^-$ pheA$^-$ endA$^-$ resA1*, from
     H.L. Heijneker

JG111   = *E. coli* K12 F$^-$ *thyA$^-$ metE$^-$ polA6*, from J.D. Gross

These original strains were used to construct *metE$^-$ polA$^-$
rha$^+$*, *metE$^+$ polA$^-$ rha$^-$* and *metE$^-$ polA$^-$ rha$^+$* derivatives of
JG108 via phage P1CM using the techniques described in Kelley
and Grindley (1976a). The transductants, isogenic except
for the *polA* allele, were used for growth experiments, for
recombinational mapping and as a source of starting material
for enzyme preparation.

Table 2

Plating Efficiencies of $polA^-$ Strains

Strain	Genotype polA allele	Medium	Plating Efficiency (per cent of 32° value)			
			32°	37°	42°	45°
CM1062	$met^-\ pol^+$	Broth	100%	100%	100%	60%
CM3760	$met^+\ pol^+$	Broth	100	60	82	49
CM1069	$met^+\ polA1$	Broth	100	73	100	30
CM3819	$met^+\ polA107$	Broth	100	100	100	100
CM3593	$met^+\ polA6$	Broth	100	73	67	80
CM4127	$met^+\ polA480$	Broth	100	60*	0.08*	0.10*
		Minimal	100	100	100	5
CM3844	$met^+\ polA4113$	Broth	100	100*	0.05*	0.03*
		Minimal	100	20	0.03	<0.001
CM3109	$met^+\ polA214$	Broth	100	100*	0.5*	0.08*
		Minimal	100	100	100	0.07

Cultures of each strain indicated were grown from a single clone in L Broth and growth **followed turbidometrically using** the Klett-Summerson colorimeter. All cultures doubled in 70-71 minutes at 32°. At Klett readings of 70-100 (at least two doublings before stationary phase) samples were removed, diluted and spread plated onto nutrient plates of the same composition or onto M9 minimal glucose plates (with thymine and streptomycin). Plates were incubated for 20.24 hours (nutrient) or 40-50 hours (minimal) and then scored.

*One hundred survivors of growth at each of the three elevated temperatures (37°, 42°, 45°) were picked and patched onto an L.B. master plate. After 12 hours growth they were replica plated onto plates containing 0.025% MMS at 45°; no pol^+ revertants were detected in any case.

RESULTS

(a) The *polA* conditional lethal phenotype. Although a number
of reports of the growth behavior of temperature sensitive
mutations exist, we began our study by comparing the growth
characteristics of isogenic strains carrying the six *polA⁻* alleles
shown in Table 2. These experiments indicated that the three
alleles *polA480*, *polA4113* and *polA214* have observably different
plating behavior. Similar observations have been reported recently
by Cooper (1977) and by Chase and Masker (1977) and it is clear that
"conditional lethality" is a functional definition rather than an
absolute physiological standard when used to describe the growth
of these strains. Interestingly, the *polA214* allele displays a
conditional lethal phenotype although this facet of its behavior was
not noted when it was originally described in the literature.

(b) Backcrosses. The traditional definition of a *polA⁻*
mutation has been, in part, that it is cotransducible with the
metabolic markers *metE* and *rha*, located at 85 and 86 minutes,
respectively, on the *E. coli* genetic map (Bachmann, Low and Taylor,
1976). Complementation data have been cited for the alleles *polA1*,
polA6 and *polA107* (Kelley and Grindley, 1976a), but no complementation data exist for other *polA* mutations. We have confirmed that
the three conditional lethal *E. coli* K12 alleles cotransduce with
metE and *rha* by first constructing either *metE⁻ polA⁻* or *rha⁻ polA⁻*
strains by transduction and then backcrossing these strains to met⁺
pol⁺ or rha⁺ pol⁺. In addition, we have carried out the same
experiment with the mutation *resA1* originally isolated in *E. coli* B
(Kato and Kondo, 1970) confirming that it too is a *polA⁻* mutation.
Our data (Table 3) indicate that the linkage to the outside
markers *metE* and *rha* is similar to that which we previously
reported for *polA1*, *polA6* and *polA107* (Kelley and Grindley, 1976a).
Further, these results show that transductions of this type do not
exert any apparent selective pressure to favor the growth of *pol⁺*
revertants or transductants.

(c) Control crosses to estimate phage-induced reversion. In
order to eliminate the possibility that a transductional event
itself generates *pol⁺* character in a transductant, either through
reversion or by "selfing" (Demerec, 1963), we have examined a
series of phage P1CM transducing lysates for their ability to
generate *pol⁺* clones of *E. coli* when a homo- or self-cross was
performed. Our data (Table 4) indicate that such self-transduction
is very rare and, as will be seen, is in all cases well below the
levels seen in most recombinational crosses.

(d) Three factor transductional crosses. Using the methods
previously described (Kelley and Grindley, 1976a) we have carried
out a series of three factor transductional crosses, utilizing

Table   3

Backcrosses of $polA^-$ Alleles

Donor	Recipient	% met^+ pol^+	% rha^+ pol^+
met^+ $polA^+$ rha^+	met^+ $polA480$ rha^-	---	71/1261=5.6%
	$metE^-$ $polA480$ rha^+	74/738=10.0%	--
met^+ pol^+ rha^+	met^+ $polA4113$ rha^-	---	38/591=6.4%
	$metE^-$ $polA4113$ rha^+	50/856=5.8%	--
met^+ $polA^+$ rha^+	met^+ $polA214$ rha^-	---	37/536=6.9%
	$metE^-$ $polA214$ rha^+	42/385=10.9%	--
met^+ $polA^+$ rha^+	met^+ $resA1$ rha^-	--	not determined
	$metE^-$ $resA1$ rha^+	158/1380=11.4%	--

Transductions were carried out and scored as described in
Kelley and Grindley (1976a). Frequencies for $polA1$, $polA6$
and $polA107$ backcrosses were published there.

Table   4

Tests for Selfing

Donor	Recipient	Number of transductions	Percent met^+ pol^+
met^+ $polA1$ rha^+	x $metE^-$ $polA1$ rha^+	10	1/6284 = 0.016%
met^+ $polA6$ rha^+	x $metE^-$ $polA6$ rha^+	7	1/9416 = 0.011%
met^+ $polA107$ rha^+	x $metE^-$ $polA107$ rha^+	6	0/10001 < 0.01%
met^+ $polA480$ rha^+	x $metE^-$ $polA480$ rha^+	4	0/7400 < 0.014%
met^+ $polA4113$ rha^+	x $metE^-$ $polA4113$ rha^+	2	0/2928 < 0.03%
met^+ $polA214$ rha^+	x $metE^-$ $polA214$ rha^+	5	0/8605 < 0.01%
met^+ $resA1$ rha^+	x $metE^-$ $resA1$ rha^+	1	0/2280 < 0.04%

To test for reversion in phage stocks or for selfing, crosses were
performed and scored as described in Kelley and Grindley (1976a).

Table 5

Transductional Data

	Donor		Recipient	met$^+$ pol$^+$	pol$^+$	%pol$^+$	Gene Order				
1a	met$^+$ polA107	x	metE$^-$ polA214	4417	0	0.00%	metE---A107 A214				
1b	met$^+$ polA214	x	metE$^-$ polA107	6566	6	0.09%				A214	
2a	met$^+$ polA107	x	metE$^-$ polA480	3536	0	0.00%	metE---A107 A480				
2b	met$^+$ polA480	x	metE$^-$ polA107	6855	3	0.04%					
3a	met$^+$ polA107	x	metE$^-$ polA4113	4058	0	0.00%		A4113			
3b	met$^+$ polA4113	x	metE$^-$ polA107	10328	3	0.03%	metE---A107				
4a	met$^+$ polA1	x	metE$^-$ polA214	5112	5	0.10%	metE---		A1		
4b	met$^+$ polA214	x	metE$^-$ polA1	7265	13	0.18%				A214	
5a	met$^+$ polA1	x	metE$^-$ polA480	4050	4	0.10%			A1		
5b	met$^+$ polA480	x	metE$^-$ polA1	4538	1	0.02%	metE--- A480				
6a	met$^+$ polA1	x	metE$^-$ polA4113	6538	17	0.26%	metE---	A4113	A1		
6b	met$^+$ polA4113	x	metE$^-$ polA1	3800	1	0.03%					
7a	met$^+$ polA214	x	metE$^-$ polA4113	8172	12	0.15%	metE---	A4113		A214	
7b	met$^+$ polA4113	x	metE$^-$ polA214	6307	5	0.08%					
8a	met$^+$ polA214	x	metE$^-$ polA480	9287	2	0.02%	metE--- A480				
8b	met$^+$ polA480	x	metE$^-$ polA214	10923	1	0.01%	metE--- A214				
9a	met$^+$ polA480	x	metE$^-$ polA4113	6722	2	0.03%	metE--- A480	A4113			
9b	met$^+$ polA4113	x	metE$^-$ polA480	6327	8	0.13%					
10a	met$^+$ polA6	x	metE$^-$ polA214	4991	14	0.28%	metE--- A214				A6
10b	met$^+$ polA214	x	metE$^-$ polA6	9612	18	0.19%					
11a	met$^+$ polA6	x	metE$^-$ polA480	10191	20	0.20%	metE--- A480				A6
11b	met$^+$ polA480	x	metE$^-$ polA6	4245	1	0.02%					

Table 5/contd

	Donor	Recipient	met⁺	pol⁺	%pol⁺		Gene Order
12a	met⁺ polA6	x metE⁻ polA4113	8029	9	0.11%	metE--	A6 A4113
12b	met⁺ polA4113	x metE⁻ polA6	10414	19	0.18%		
13a	met⁺ polA1	x metE⁻ polA6	3527	2	0.06%	metE--	A1 A6
13b	met⁺ polA6	x metE⁻ polA1	4171	12	0.28%		
14a	met⁺ polA107	x metE⁻ polA1	5699	3	0.05%	metE--A107	A1
14b	met⁺ polA1	x metE⁻ polA107	3691	6	0.16%		
15a	met⁺ polA6	x metE⁻ polA107	4630	18	0.39%	metE--A107	A6
15b	met⁺ polA107	x metE⁻ polA6	3176	0	0.00%		
16a	met⁺ polA1	x metE⁻ resA1	10829	9	0.08%	metE--	resA1 A1
16b	met⁺ resA1	x metE⁻ polA1	11352	2	0.02%		
17a	met⁺ polA6	x metE⁻ resA1	4905	4	0.08%	metE--	resA1 A6
17b	met⁺ resA1	x metE⁻ polA6	3518	0	0.00%		
18a	met⁺ resA1	x metE⁻ polA107	6592	9	0.14%	metE--A107	resA1
18b	met⁺ polA107	x metE⁻ resA1	8160	8	0.10%		
19a	met⁺ polA4113	x metE⁻ resA1	5708	8	0.14%	metE--	resA1 A4113
19b	met⁺ resA1	x metE⁻ polA4113	9697	9	0.09%		
20a	met⁺ resA1	x metE⁻ polA480	3251	4	0.12%	metE--	A480 resA1
20b	met⁺ polA480	x metE⁻ resA1	621	0	0.00%		

metE ————————————————— polA ————————————————— rha

A107 A480   resA1 A214   A4113 A1 A6

Crosses were performed and scored as in Kelley and Grindley (1976a). Data shown for crosses 13a and 13b; 14a and 14b; 15a and 15b are from that reference.

$metE^+$ selection in the hope of positioning the four further $polA$
alleles with respect to each other and to $polA1$, $polA6$ and $polA107$
in the intracistronic map. As we have previously discussed, the
crosses of any pair of alleles with respect to the outside $metE$
marker must be compared to establish relative marker order;
frequencies of recombination may or may not indicate the genetic
distance separating two markers and should be interpreted with
extreme caution. The data for the crosses completed thus far
indicate an internally consistent genetic fine structure map with
three exceptions. Our data and the suggested intragenic map are
presented in Table 5.

    (e) <u>The $polA214$ polymerase</u>. Since the conditional lethal
phenotype has been associated with the presence of a defective
$5' \rightarrow 3'$ exonuclease function in the case of $polA480$ and $polA4113$
(Konrad and Lehman, 1974; Olivera and Bonhoeffer, 1974; respectively), we anticipated a similar phenomenon to be the case with $polA214$.
To investigate this possibility we carried out a purification of
the polymerase from this cell line using the procedure developed by
us for isolating the $polA6$ polymerase (Kelley and Grindley, 1976b).
Beginning with 800 grams of cells wer purified the polymerase
through DNA agarose, phosphocellulose, hydroxylapatite and Bio-Gel
column chromatography steps to obtain a final preparation amounting
to 13% of the initial polymerizing activity found in the crude
extract. This polymerase showed extremely low levels of $5' \rightarrow 3'$
exonuclease activity when tested. When the temperature dependence
profile of the $polA214$ $5' \rightarrow 3'$ exonuclease was compared with that
of either whole wild type DNA polymerase I or of the small subtilisin fragment, prepared by the cleavage method of Jacobsen,
Klenow and Overgaard-Hansen (1974), it was definitely more thermolabile (Fig. 1). Thus, this polymerase has a temperature sensitive
$5' \rightarrow 3'$ exonuclease which accounts for its properties in part.

    Subtilisin cleavage or natural fragmentation during isolation
produces a polymerase fragment from the $polA214$ enzyme which is
analogous to the 76,000 dalton "large fragment" produced by the
subtilisin cleavage of wild type DNA polymerase I (Klenow and
Hennigsen, 1970; Jacobsen et al. 1974). When the two proteolytic
fragments are tested for polymerization in identical assay
conditions the pH optimum for the $polA214$ polymerase is a half a
pH unit lower than that for wild type polymerase (Fig. 2). In
addition to its lowered pH optimum, the $polA214$ polymerase has a
lower specific activity than the wild type enzyme and is more easily
fragmented during the isolation procedure. Polymerization activity
does not appear to be significantly more thermo-labile than that of
wild type enzyme nor do there appear to be any differences between
the two enzymes in their primer/template requirements.

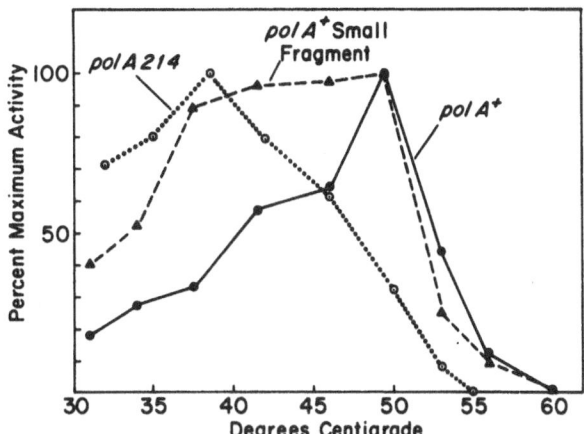

Figure 1. Temperature sensitivity of 5' → 3' exonuclease from
            polA214, polA⁺ whole enzyme and polA⁺ small fragment.
Assays were conducted as described in Kelley and Grindley (1976b),
testing the enzymes' capacities to solubilize micrococcal nuclease-
nicked ³H-dAT copolymer. The activities are expressed here as
percent of the maximum value for each enzyme. Specific activity
levels for the three enzymes were not the same.

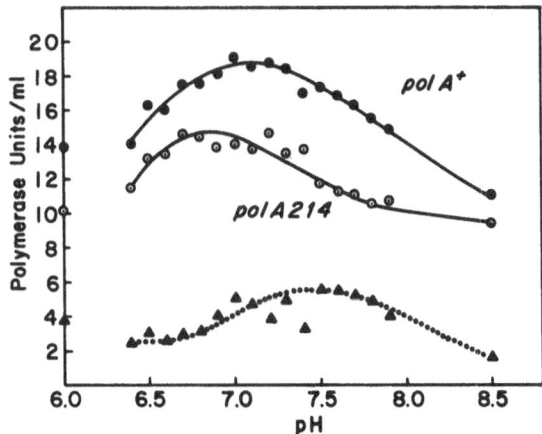

Figure 2. pH optima for the polA214 and polA⁺ DNA polymerase I
            large fragment.
The polA⁺ and polA214 large proteolytic fragments were isolated from
the whole enzyme by Bio-Gel P-150 column chromatography as described
in Kelley and Grindley (1976b) and were diluted to give similar
activity levels. Samples were assayed in triplicate using a series
of 0.066 M tris buffers prepared and titrated immediately before
use. The data are present as units of polymerase per ml for each
enzyme. A difference curve was plotted by subtracting the values of
the polA214 curve from those of the polA⁺ curve. Individual assay

DISCUSSION

We have carried out careful transductional analysis of the genetic recombination between several different *polA⁻* alleles in an attempt to establish a more complete fine structure map than that previously published. With the exception of three pairs of crosses, those between *polA214* and *polA1; polA214* and *polA4113;* and *polA4113* and *polA6,* the gene order is consistent with the fine structure map suggested in Table 5. We believe these exceptions may be rationalized on the basis of information which is available describing the enzymatic defects found in the purified mutant polymerases. Assignment of genetic map distances between the allelic *polA⁻* mutations is not possible on the basis of this data.

In considering recombinational mapping experiments involving two *polA⁻* mutations there are at least two important potential sources of difficulty in interpreting experimental data. The first is the assumption, implicit in our previous arguments, that a double mutant recombinant will display the *polA⁻* phenotype (or be totally non-viable). The second is the assumption that there is a reasonable chance of recombination between two linked mutations which is proportional to their genetic separation - i.e., that a phenomenon such as negative interference will not play a significant role in the recombinational event. At the present time any consideration of our recombinational data must be made in the light of enzymological evidence since very little is known about the primary structure of the DNA polymerase I molecule and there is reason to believe that several, if not all, of the known *polA⁻* mutations confer the cellular phenotype characterized by Konrad (1977) as "hyper-rec". Any elevation of recombinational frequencies resulting from the hyper-rec phenotype will certainly make it impossible to interpret data to give genetic linkage distances in traditional map units. Viable double mutants in which a combination of two mutations would yield a *polA⁺* phenotype would not only bias calculation of genetic distances but could also result in incorrect inference of map positions.

In considering the three anomalous crosses one must bear in mind the enzymological evidence which is available as well as the history of the particular mutant. Two of the pairs of crosses, 4a and 4b and 7a and 7b of Table 5 involving *polA214* indicate a map position for this mutation which lies toward the *rha* proximal end of the gene. Our enzymological evidence shows that the *polA214* polymerase certainly has a thermo-labile $5' \rightarrow 3'$ exonuclease, implying that the predominant genetic lesion is in the amino terminal portion of the polypeptide. These data are consistent with the results of crosses 8a and 8b which indicate that the *polA214* lesion is virtually identical, in genetic terms, with that

difference values are shown as open triangles along this line.

of *polA480* which is also known to affect the enzyme's 5' → 3'
exonuclease (Uyemura, Eichler and Lehman, 1976). However, the low
specific activity of the purified *polA214* polymerizing fragment
and its altered pH optimum also argue for there being a mutational
effect within the carboxy terminal portion of the molecule. The
polA214 mutation was originally isolated as the result of a complex
screening procedure involving nitrosoguanidine mutagenesis followed
by bromouracil treatment (Kingsbury and Helinski, 1973). It is
therefore possible that this mutation is actually not a single
point mutation but rather a double mutation itself. Since the
recombinational data are not clearcut, we have not designated an
actual position for the *polA214* mutation in our fine structure map
but rather have indicated that it lies between *polA107* and *polA6*.

The *polA4113* mutation of Olivera and Bonhoeffer (1974) also
results in anomalous recombinational data both in the crosses with
polA214 described above and when crossed with *polA6* (crosses 7a and
7b and 12a and 12b, respectively). At present there is no evidence
in print which describes the enzymological properties of the
polA4113 polymerase molecule. However, unpublished data of Olivera
and Lundquist (personal communciation) indicate that this polymerase
is defective in its 5' → 3' exonucleolytic function and possesses
a normal polymerizing capacity, although deficient in its ability
to carry out coordinated nick transplation synthesis. Our results
from crosses 12a and 12b are therefore much more difficult to
rationalize unless the recombinational events in these crosses
generate viable double mutants.

The product of the *resA1* mutation has not been characterized,
but our mapping data indicate that it is clearly a *polA⁻* mutation
and it will therefore have to be assigned a *polA* allele number.
This mutation is useful for physiological studies because it is
fully sensitive to methylmethane sulfonate at temperatures from
32° to 45° and reverts at a very low frequency.

ACKNOWLEDGEMENTS

This work was supported by National Science Foundation Grant
PCM76-14715 and by the Allegheny County (Pennsylvania) Health
Research and Services Foundation Grant R-85 to WSK.

We wish to thank Mrs Charlene F. Rizzo and Mrs Arax T. Leney
for their technical assistance in carrying out this project.

REFERENCES

Bachmann, B.J., Low, K.B. and Taylor, A.L. (1976) Bact. Revs.
40:116-167

Chase, J.W. and Masker, W.E. (1977) J. Bact. 130:667-675

Cooper, P. (1977) Molec. Gen. Genet. 150:1-12

Demerec, M. (1963) Genetics 48:1519-1531

Grindley, N.D.F. and Kelley, W.S. (1976) Molec. Gen. Genet.
    143:311-318

Jacobsen, H., Klenow, H. and Overgaard-Hansen, K. (1974) Eur. J.
    Biochem. 45:623-627

Kato, T. and Kondo, S. (1970) J. Bact. 104:871-881

Kelley, W.S. and Grindley, N.D.F. (1976a) Molec. Gen. Genet.
    147:307-314

Kelley, W.S. and Grindley, N.D.F. (1976b) Nucleic Acid Research
    3:2971-2984

Kingsbury, D.T. and Helinski, D.R. (1973b) Genetics 74:17-31

Klenow, H. and Henningsen, I. (1970) Proc. Nat. Acad. Sci. (Wash.)
    65:168-175

Konrad, E.B. (1977) J. Bact. 130:167-172

Konrad, E.B. and Lehman, I.R. (1974) Proc. Nat. Acad. Sci. (Wash.)
    71:2048-2051

Lehman, I.R., Uyemura, E.G. : DNA Polymerase I: Science 193:963-969
    (1976).

Olivera, B.M. and Bonhoeffer, F. (1974) Nature (Lond.) 250:513-514

Rosner, J.L. (1972) Virology 48:679-689

Uyemura, D., Eichler, D.C. and Lehman, I.R. (1976) J. Biol. Chem.
    251:4085-4089

DNA POLYMERASE III OF *B.subtilis:* CHARACTERIZATION OF THE BINDING

SITE FOR ARYLHYDRAZINOPYRIMIDINE INHIBITORS

Neal C. Brown* and George E. Wright

Department of Pharmacology
University of Massachusetts Medical School
Worcester, Massachusetts, 01605, U.S.A.

A distinct class of 6-substituted pyrimidines, exemplified by the 6-(arylazo)pyrimidine, 6-(*p*-hydroxyphenylazo)uracil (HPUra)) selectively inhibits the replicative synthesis of DNA of Gram-positive bacteria (1). The site of drug action is the replication-specific enzyme, DNA polymerase III (2,3,4) (Pol III). The mechanism of pol III inhibition by HPUra and analogous 6-substituted uracils - a mechanism which has been studied in detail in our laboratory and that of Cozzarelli (3,4,5,6,7,8,9) - involves the following three processes:- (1) reduction of the drug to an active, hydrazino (H_2) form (i.e., $-N=N- \rightarrow -\overset{H}{N}-\overset{H}{N}-$, (2) hydrogen bonding, in competition with dGTP, to an unpaired cytosine residue in the template strand immediately distal to the 3' hydroxyl primer terminus, i.e. ⬚⬚⬚⬚__3'OH ◄ and, (3) reversible reaction of the

template-bound drug with enzyme, sequestering the latter from productive activity as a component of a relatively stable, ternary complex. The novel mechanism by which $H_2 \cdot$HPUra pairs with cytosine (4) and our current concept of the drug:DNA:enzyme complex (8) are depicted, respectively, in parts A and B of Fig. 1.

Arylhydrazinopyrimidine sensitivity is apparently a property unique to the replication-specific, type III polymerases derived from Gram-positive bacteria. We propose that these Gr+ polymerases are uniquely drug-sensitive because they possess a common, structurally unique "aryl" binding site which firmly engages the aryl moiety of the inhibitor. We have begun a comprehensive examination of the structure of the "aryl" site of Gr+ polymerases, exploiting *B.subtilis* pol III as a prototype. The results of our

*A.S.I. Participant

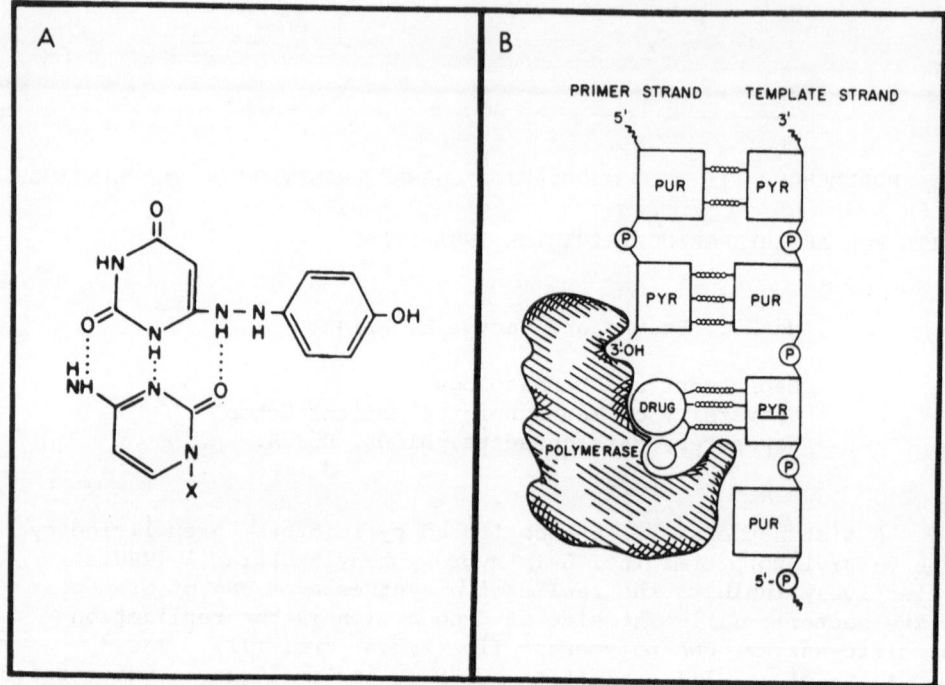

Figure 1. (A) Mechanism by which H$_2$·HPUra hydrogen bonds to template cytosine; (B) Proposed model for the arylhydrazino-pyrimidine:DNA:pol III complex.

initial efforts, which have exploited a mutant, H$_2$·HPUra-resistant enzyme and structure-activity relationships involving a variety of 6-substituted uracils, have permitted construction of a working model of this site; this paper describes this model and summarizes, in part, its experimental basis.

Is There a Discrete Aryl Site in pol III ?

The existence of a specific aryl site in *B.subtilis* pol III and, conversely, the importance of the aryl group as a drug moiety involved in complex formation, are supported by two substantial pieces of evidence. The first is the observation (8) that an unsaturated aryl moiety, i.e., a phenyl or naphthyl group, is required for inhibitory activity; replacement of the aryl group with saturated rings or simple alkyl groups destroys inhibitor activity

Table 1[a]

Effects of inhibitors on wild-type pol III and pol III/azp-12

Inhibitor	Ki (μM)	
	wild type pol III	pol III/azp-12
U-NH-NH-⬡-OH (H$_2$·HPUra)	0.4	20
U-NH-NH-⬡ (H$_2$·PUra)	1.4	2.1
U-NH-NH-⬡-CH$_3$ (H$_2$·TUra)	0.5	0.4
U-NH-NH-⬡-NH$_2$	7.1	76
U-NH-CH$_2$-⬡	2.3	2.5
U-NH-CH$_2$-⬡-NH$_2$	18	130
U-NH-CH$_2$-CH$_2$-⬡	32	30

[a] from ref. 12, with permission

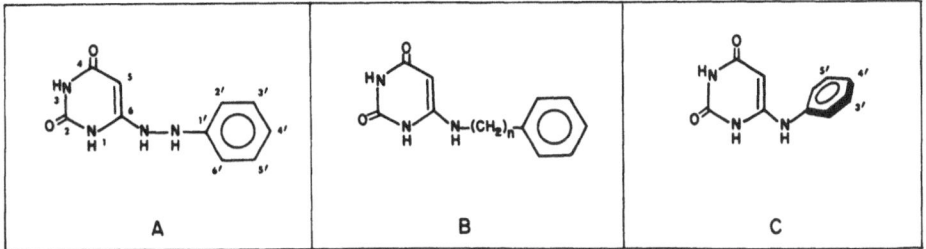

Figure 2. 6-substituted uracils: (A), 6-phenylhydrazino)uracil (H$_2$·PUra); (B), 6-(phenylalkylamino)uracils, n = 1-4; (C), 6-anilino-uracil.

completely. The second, and most convincing fragment of evidence
for the existence of an aryl site is based on the observation (8)
that the arylhydrazinopyrimidine resistance of a mutant enzyme is
determined by mutational alteration of a site which specifically
recognizes and reacts with a substituent of the aryl group. The
effect of mutational alteration is indicated by the data of Table
1, which summarizes an experiment examining the influence of
specific aryl substituents on the inhibitor sensitivity of wild-
type pol III and pol III/*azp*-12, a mutant enzyme derived from a
spontaneous mutant of *B.subtilis* selected for resistance to HPUra
(6); the data indicate that the resistance to an arylhydrazino or
arylalkylaminopyrimidine is dictated specifically by the presence
of a polar, hydrophilic *para*-substituent such as -OH or -NH$_2$ in its
aryl ring.

Properties of the Aryl Site

Hydrophobicity and probable source of binding energy. The
specific capacity of the aryl site to accommodate planar, unsaturated rings indicates that it consists of distinctly hydrophobic
space. We have not identified the amino acid moiety or moieties
which constitute the major source of aryl:enzyme reaction;
consideration of the chemistry of the phenyl group and the strength
of the aryl:enzyme binding suggests that binding energy may be
derived from a combination of hydrophobic binding and a specific
interaction such as that provided by van der Waal's or charge-
transfer forces.

Location relative to the substrate binding site. The specific
capacity of dGTP to reverse the inhibition induced by arylhydrazino-
uracils strongly suggests that the aryl site closely approximates
the triphosphate binding site. The observations (5,6,8) that mutant
enzymes may be altered profoundly with respect to K_i for aryl-
hydrazinopyrimidines but remain unaffected with respect to K_m for
substrate strongly implies that the aryl and substrate binding
sites are not strictly identical.

Location relative to the site of template binding. The distance
between the aryl site and the H-bonded 6-NH moiety (cf. Fig. 2,
for numbering system) of the drug:cytosine complex has been
measured by examining the relationship between inhibitory potency
and the number of bonds separating the 6-NH and aryl group (10,11).
Optimal inhibitory capacity occurs when the phenyl and pyrimidine
rings are separated by 3 bonds, the number found in 6-(phenylhydrazino)uracil (H$_2$·PUra, Fig. 2A) and 6-(benzylamino)uracil (Fig. 2B,
n=1). 6-(phenylethylamino)uracil (Fig. 2B, n=2) retains moderate
activity, while the 6-(phenylpropyl)- and 6-(phenylbutyl)homologs
(Fig. 2B, n=3 and 4, respectively) are essentially inactive.
Contraction of the aryl-pyrimidine distance *via* deletion of the

l'-NH yields 6-anilinouracil (Fig. 2C), which *per se* has no
inhibitory activity; however, substitution of 6-anilinouracil with
hydrophobic halo or alkyl groups in the *para* and/or *meta* positions
yields compounds of unusually high potency (i.e., *m,p*-dimethyl or
m,p-dichloro-anilinouracil, K_i's, 1 µM; or *m,p*-trimethylene
$(-(CH_2)_3-)$, K_i, 0.4 µM).

Presence of a secondary, polar binding site, Y. The most
potent pol III inhibitor among the 6-substituted uracils is the
p-OH derivative, H_2·HPUra. We have proposed (8) that the unique
capacity of the *p*-OH group to enhance inhibitory potency resides
in its capacity to engage a polar, hydrogen bonding moiety in the
"*para*" domain of the aryl site, and we refer to this secondary
hydrophilic site as "Y". We base the existence of the hydrophilic
Y site on the results of the experiments with pol III/*azp*-12
summarized in Table I and discussed above. The behavior of the
azp-12 enzyme, which is *resistant* to the polar *p*-OH and *p*-NH$_2$
compounds and *normally sensitive* to the *p*-H- and *p*-CH$_3$- inhibitors,
is most simply explained by assuming that mutational alteration has
occurred in a site which specifically accommodates the polar, *para*
group.

Shape. The aryl site has not only restricted space but
apparently a rather strict conformation which accommodates only
those arylhydrazino or arylalkylamino compounds in which the aryl
group is conformationally mobile (cf. Fig. 3A) and capable of
rotating freely about the 6-NH moiety. Substituents which
restrict mobility about the 6-NH severely diminish inhibitory
potency (10).

We have no precise information on the position of the aryl site
relative to the plane of the base pair formed between the uracil
moiety and the template cytosine: the expression of inhibitory
activity by the substituted 6-anilinouracils in which the orientation of the phenyl ring is restricted to that depicted in Figs.
2(C) and 3(B) strongly suggests that a similar conformation (cf.
Fig. 3A and B) must occur in the case of the arylhydrazino and
benzylamino analogues.

Of the possible configurations of phenylhydrazino- and benzyl-
aminouracils which place the phenyl ring in a plane identical to
that of the 6-anilino analogues, we favor the specific *cis*
conformation depicted in the left portion of Fig. 3(B). Part (C)
cf Fig.3 melds the concepts of part (B) and compares, in composite
diagram, how we specifically envision the positions assumed by the
respective phenyl groups of the 6-anilinouracil and H_2·PUra as each
projects into the aryl site.

We believe that the aryl site is shaped such that it binds the
maximal number of ring carbons when the phenyl ring is presented

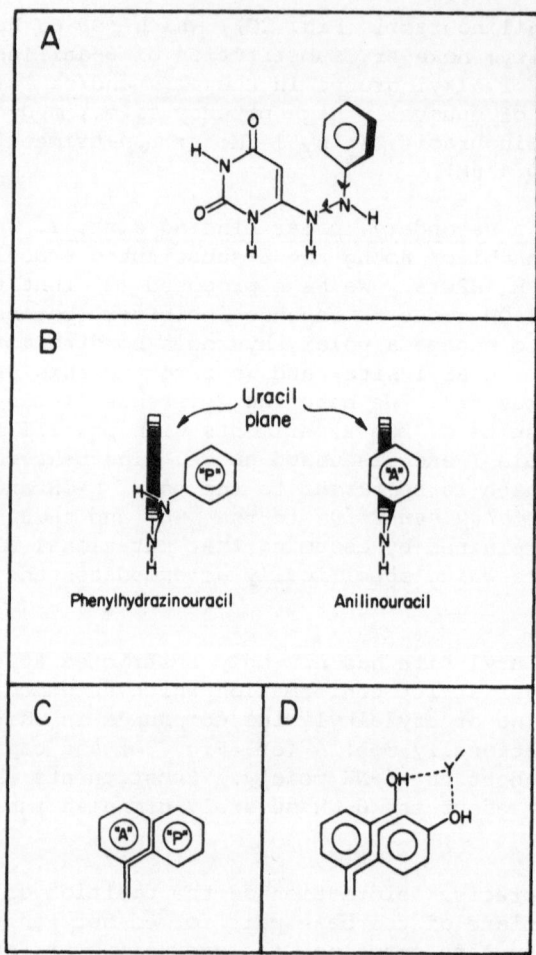

<u>Figure 3</u>. *Active conformations of phenylhydrazinouracils and 6-anilinouracil. (A), $H_2 \cdot$PUra drawn to stress requirement for conformational mobility about the 6- and the 1'-NH groups; (B) display of putative "active" conformations of phenylhydrazino- and anilinouracils; (C) composite depiction of the positions of the aryl groups of anilinouracil ("A") and phenylhydrazinouracil ("P") during their engagement with the aryl site; (D) probable spatial relationship of domain Y with the <u>meta</u> position of <u>m</u>-hydroxymethylanilinouracil and the <u>para</u>-position of <u>p</u>-hydroxyphenyl-hydrazinouracil.*

in the phenylhydrazino or "P" mode. The "P" preference explains why 6-anilinouracil, which presents fewer carbons to the aryl site,

is essentially inactive as an inhibitor. The preference for the "P"
configuration also explains how *meta* and *para* substitution of 6-
anilinouracil creates inhibitors with high potency; the substituents
simply extend the anilino moiety into the "P" domain, and thus
present more surface to engage in hydrophobic bonding.

The results of recent, unpublished experiments (Wright and
Brown, to be published) with pol III/*azp*-12 and 6-anilinouracils
substituted in either the *meta* or *para* position with the polar
hydroxymethyl (-CH$_2$OH) group lends further support to the presence
of a "P" conformation of the aryl site. The experiments indicate
that pol III/*azp*-12 displays resistance specifically to the *meta*
isomer; presentation of the -CH$_2$OH group in the *para* position does
not distinguish pol III/*azp*-12 from wild type. Fig. 3(D) depicts
the probable basis for the experimental result; the *para* -CH$_2$OH
group is simply too far from the *para* -specific "Y" domain of the
aryl site to be affected significantly by the Y-specific mutation.
Conversely, the *meta*-CH$_2$OH group is considerably closer to the "Y"
domain of the "P" oriented site and, therefore, is strongly
influenced by changes in it.

A Working Model of the Aryl Site of Wild Type
pol III and pol III/*azp*-12

Fig. 4 summarizes our current concept of the structure of the
aryl site. The essential features of the site in the wild type
enzyme include: (1) the *primary* site of reaction with the aryl
group, designated Θ; (2) the presence of the secondary, polar site,
Y, and (3) an "up" (cis) uracil/aryl conformation.

Fig. 5 depicts the mechanism by which the aryl site reacts
with the phenyl ring of 3 different inhibitors; the left panels,
A, B and C, designate the mechanism by which the wild type aryl
site accommodates, respectively, the unsubstituted H$_2$·PUra, the
p-methyl derivative 6-(*p*-tolylhydrazino)uracil, H$_2$ TUra) and the
p-hydroxylated, H$_2$·PUra. In each of the three cases Θ provides
the *primary* source of energy involved in the binding of the aryl
moiety; in the case of H$_2$·PUra the polar "Y" site augments the
aryl: Θ reaction, possibly by serving as a source of H bonding.

The right panels, D, E and F depict the aryl site of pol
III/*azp*-12 and its reaction with H$_2$·PUra, H$_2$·TUra, and H$_2$·HPUra.
We do not specify in our scheme for pol III/*azp*-12 any change in
the basic Θ : aryl configuration or interaction, primarily because
the affinity of enzyme for an unsubstituted ring, e.g., that of
H$_2$·PUra, is identical to that of wild type. The specific effect of
the *azp*-12 mutation, as we envision it, is to neutralize the polar
status of amino acid moiety "Y", either by effecting its
replacement with a hydrophobic amino acid moiety or by effecting a

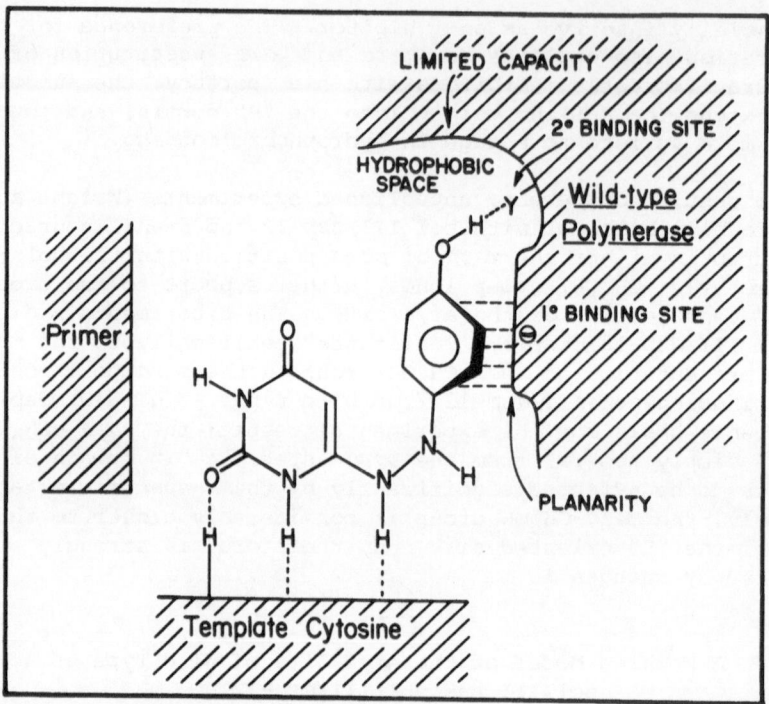

Figure 4. *Summary of the features of the aryl binding site of pol III and its reaction with* $H_2 \cdot HPUra$.

conformational change in its environment – a change which results in its submersion in hydrophobic space.

Either mechanism of the hydrophobic conversion of Y can explain why pol III/*azp*-12 is normosensitive to the unsubstituted $H_2 \cdot$PUra, slightly hypersensitive to the p-methylated, $H_2 \cdot$TUra, and resistant to phenylhydrazinouracils carrying polar groups in the p-position. The aryl group of $H_2 \cdot$PUra (panel D), whose binding is not strongly affected by Y (panel A), fits normally in the aryl site. The θ:aryl reaction (panel E) of $H_2 \cdot$TUra, a binding reaction which in wild type pol III (panel B) may be influenced slightly *negatively* by Y, may be facilitated slightly by loss of the latter *via* the *azp*-12 mutation. In the case of $H_2 \cdot$HPUra (panel F) the conversion of Y to a hydrophobic environment results in the *repulsion* of the *para* polar substituent.

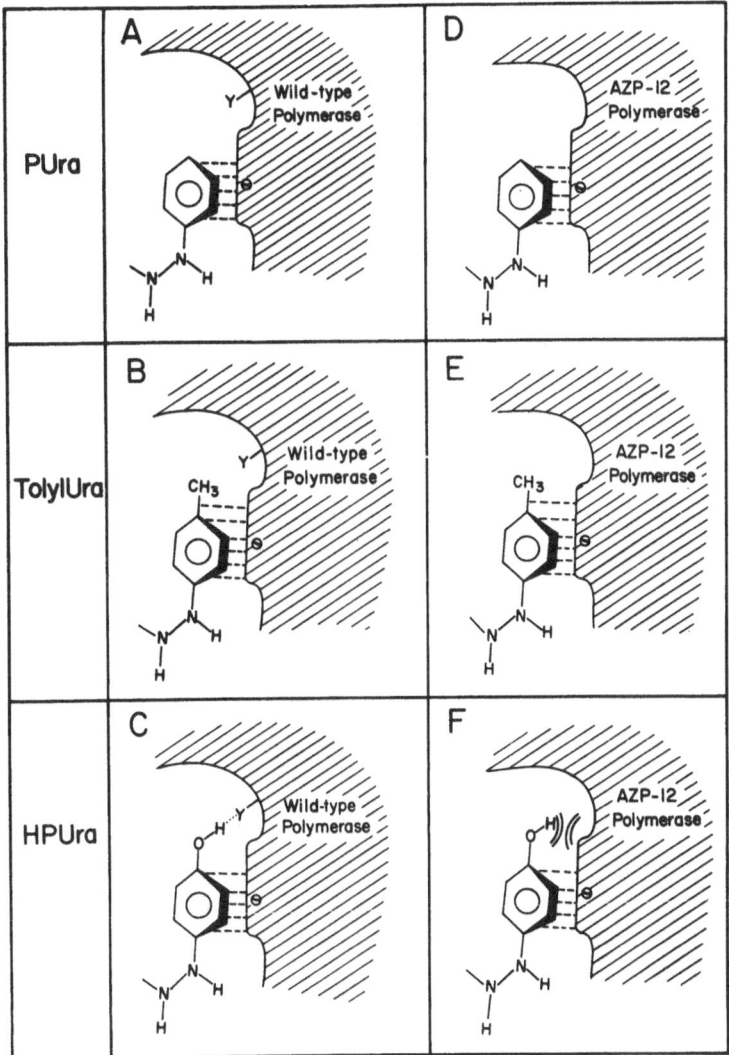

Figure 5. Comparison of the binding sites of wild-type pol III and pol III/azp-12 and their reaction with different aryl moieties.

Future Directions

Aryl site labeling. The model of Fig. 4, a concept of the aryl site based on a broad interpretation of structure-activity relationships, is clearly speculative. However, certain features of the model, in particular those concerning the nature of site Y and its mutation, can be examined if the active site regions of pol III and pol III/*azp*-12 can be "tagged" covalently with

appropriate drug and substrate derivatives, and analyzed for amino
acid and sequence analysis. We are attempting to exploit this
approach using an anilino derivative as a basis for constructing
an alkylating inhibitor.

Development of inhibitors for naturally resistant DNA
polymerases. The capacity of 6-arylamino-, 6-arylhydrazino- and
6-arylalkylaminouracils to mimic dGTP should accord these
compounds specific facile access to the substrate site of a *wide*
spectrum of DNA polymerases - including those which do *not* possess
the characteristic aryl-specific site of Gram-psotive DNA polymerases III. We believe that the dGTP mimicry can be exploited
to construct 6-substituted pyrimidine(s) with 6-substituents which
will penetrate and engage reversibly, an "aryl" site or its
equivalent in enzymes which are normally drug-resistant, i.e.,
E.coli pol III, polymerases I and II, and eukaryotic polymerases.
Therefore, we are constructing a screen involving a large number of
6-substituted pyrimidines and a variety of DNA polymerases.

Acknowledgements

This work was supported by U.S.P.H.S./N.I.H. research grants
CA15915 (from the Nat. Cancer Institute to N.C.B.) and GM21747
(from the National Institute of General Medical Sciences, to G.E.W.).

References

1.   Brown, N.C. and R.E. Handschumacher (1966) J. Biol. Chem.
     241:3033

2.   Bazill, G. and J. Gross (1973) Nature New Biol. 243:241

3.   Cozzarelli, N. and R. Low (1973) Biochem. Biophys. Res.
     Commun. 57:151

4.   Mackenzie, J., M. Neville, G.E. Wright and N.C. Brown (1973)
     Proc. Nat. Acad. Sci. USA 71:512

5.   Low, R., R.Rashbaum and N.R. Cozzarelli (1974) Proc. Nat.
     Acad. Sci. USA 71:2973

6.   Clements, J., J. D'Ambrosio and N.C. Brown (1975) J. Biol.
     Chem. 250:522

7.   Cozzarelli, N.R. (1977) Ann. Rev. Biochem., in press

8.   Brown, N.C. and G.E. Wright (1977) In: Proceedings of the
     Symposium on Drug Action at the Molecular Level. Macmillan,
     London.

9.   Brown, N.C. (1972) Biochim. Biophys. Acta 281:202

10. Wright, G.E. and N.C. Brown (1977) J. Med. Chem., in press

11. Brown, N.C., J. Gambino and G.E. Wright (1977) J. Med. Chem.,
     in press

12. Brown, N.C. and G.E. Wright (1977) International Encyclopedia
     of Pharmacology and Therapeutics, Pergamon Press, Ltd.

DNA PROOF-READING BY A EUKARYOTIC DNA POLYMERASE

G.T. Yarranton[*] and G.R. Banks[**]

National Institute for Medical Research
Mill Hill
London, NW7 1AA, England

Replication of DNA, as reflected in the low mutation frequency of the cell must occur with a high degree of fidelity. In agreement is the observation that when directed by synthetic template-primers, many prokaryotic and eukaryotic DNA polymerases misincorporate nucleotides not complementary to the template at an extremely low frequency (1-3). This fidelity may, in fact, be greater than can be accounted for by the measure difference in free energy between complementary and noncomplementary base pairs (4). It is likely, therefore, that DNA polymerases and/or other proteins increase fidelity above that determined by base specificity alone. The 3' → 5' exonuclease activity associated with prokaryotic polymerases (5) is a prime candidate to accomplish this by excising a noncomplementary base misincorporated at the 3'-terminus of a growing DNA chain by the polymerase activity. Indeed, where specifically tested, this exonuclease will excise such a preformed primer nucleotide before primer extension can occur (6). In addition, there is a correlation between the polymerase to exonuclease specific activity ratios of DNA polymerases from phage T4 strains harbouring mutations in their structural polymerase genes and the mutation frequencies of these strains (7,8).

In contrast, mammalian DNA polymerases, with the possible exception of polymerase -δ (9), do not possess a 3' → 5' exonuclease activity, yet do possess a high degree of fidelity and in one case, can extend a primer from a non-complementary 3'-terminus (3). This

---

[*] *Present address: Department of Biology, Massachusetts Institute of Technology, Cambridge, Mass. 02139, U.S.A.*

[**] *A.S.I. Participant*

has promoted a belief that its presence or absence characterizes
prokaryotes or eukaryotes respectively and that in the latter, it
may have evolved into a separate gene function. The DNA polymerase
of the lower eukaryote *Ustilago maydis* does, however, possess
an associated $3' \rightarrow 5'$ exonuclease activity and the experiments
reported in this paper establish that it can proof-read *in vitro*
(10). The recent demonstration that a polymerase from *Saccharomyces
cerevisiae* can also proof-read suggests that this class of lower
eukaryotes may have retained exonuclease associated DNA polymerases
(11).

The exonuclease associated with the *U.maydis* DNA polymerase, in
the absence of polymerase substrates, hydrolysed the non-
complementary (^3H)dAMP primer termini in poly(dA)·(dT)$_{\overline{320}}$-(^3H)
(dA)$_{\overline{0.97}}$ about 4 times faster than complementary (^3H)dTMP in
poly(dA)·(dT)$_{\overline{320}}$-(^3H)(dT)$_{\overline{1.7}}$ at $37°$ and at $25°C$, 12 times faster
(Fig. 1). Similar results obtained for the hydrolysis of non-
complementary (^3H)dCMP in poly(dA)·(dT)$_{\overline{320}}$-(^3H)(dC)$_{\overline{0.82}}$. Thus the
exonuclease shows a preference for non-complementary $3'$-terminal
nucleotides, a preference enhanced at low temperatures. It is
probable that "fraying" or terminal denaturation of poly(dA)·(dT)$_n$
polymers is responsible for the considerable hydrolysis of the
(^3H)dTMP termini at $37°C$, resulting in the larger temperature coefficient than found for the non-complementary termini.

When poly(dA)·(dT)$_{\overline{320}}$-(^3H)(dT)$_{\overline{1.7}}$ was incubated with the
enzyme in the presence of $(\alpha-^{32}P)$dTTP, (^{32}P)poly(dT) synthesis was
observed accompanied by an initial hydrolysis of about 10% of the
(^3H)dTMP nucleotides, the remainder being resistant (Fig. 2). This
resistance required substrates (dTTP or dUTP) complementary to the
template, suggesting a requirement for primer extension by the
polymerase activity. dATP, dGTP or dCTP, which cannot extend the
primer with the above template-primer produced no resistance
(Table 1). In contrast, under identical reaction conditions,
complete and rapid hydrolysis of the (^3H)dAMP termini in poly(dA)·
(dT)$_{\overline{320}}$-(^3H)(dA)$_{\overline{0.97}}$ occurred as (^{32}P)poly(dT) synthesis
proceeded (Table 1 and Fig. 3). Again identical observations were
made with poly(dA)·(dT)$_{\overline{320}}$-(^3H)(dC)$_{\overline{0.82}}$. These results suggest
that when extension of a complementary $3'$-terminus occurs, it is
protected from hydrolysis, whereas extension of a non-complementary
terminus is blocked until complementarity is restored by the exonuclease activity.

Incubation of poly(dA)·(dT)$_{\overline{320}}$-(^3H)(dC)$_{\overline{6.5}}$ with the polymerase
in the absence or presence of $(\alpha-^{32}P)$dTTP resulted in the hydrolysis
of only 10% of the total (^3H)dCMP and in its presence, the rate of
(^{32}P)poly(dT) synthesis was only about 10% of that observed with
poly(dA)·(dT)$_{\overline{320}}$-(^3H)(dA)$_{\overline{0.97}}$ in which all the (^3H)dAMP was
hydrolysed (Fig. 4). Because the enzyme was in molar excess over

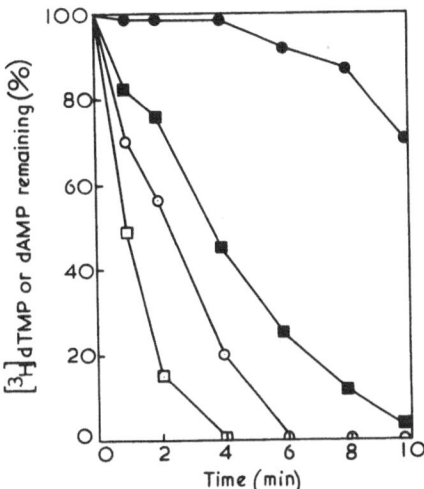

Figure 1. *Hydrolysis of complementary and noncomplementary terminal primer (^{3}H)-nucleotides in the absence of polymerase substrates. 3.0 nmoles nucleotides (in a 1:1 template:primer ratio) poly(dA)·(dT)$\overline{\frac{320}{}}$ -(^{3}H) (dT)$\overline{\frac{}{7.7}}$ ($\bullet$——$\bullet$,$\blacksquare$——$\blacksquare$) or poly(dA)·(dT) $\overline{\frac{320}{}}$ -(^{3}H)(dA)$\overline{\frac{320}{0.97}}$ (O——O, □$\overline{\frac{1.7}{}}$□) was incubated with 10 units of U. maydis DNA polymerase at 37°C ($\blacksquare$——$\blacksquare$,□——□) or 25°C ($\bullet$——$\bullet$,O——O). At the times indicated, aliquots of the reaction mixtures were removed and (^{3}H)-label remaining in the primer determined (6).*

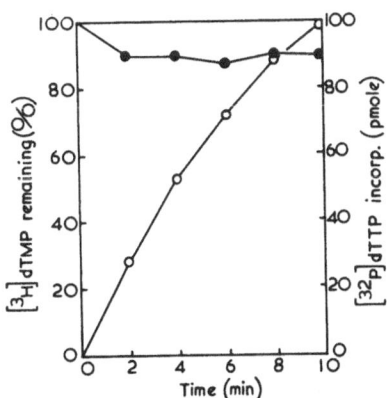

Figure 2. *Hydrolysis of complementary terminal primer (^{3}H)dTMP and simultaneous (^{32}P)poly(dT) synthesis. 3.0 nmoles (in a 1:1 template: primer ratio) poly(dA)·(dT)$\overline{\frac{320}{}}$ -(^{3}H)(dT)$\overline{\frac{}{7.7}}$ was incubated with 10 units U.maydis DNA polymerase in the presence of 13.3µM (α-^{32}P)dTTP at 37°C. (^{3}H)dTMP remaining in the primer ($\bullet$——$\bullet$) and (^{32}P)poly (dT) synthesized (O——O) was determined at the times indicated (6).*

Table 1

The influence of deoxyribonucleoside triphosphates on the
hydrolysis of primer terminal (^{3}H)-nucleotides

Primer	dNTP	(^{3}H)dNMP hydrolysis
$(dT)\overline{320} - (^3H)(dT)\overline{1.7}$	None	+
	dTTP or dUTP	−
	dATP, dGTP or dCTP	+
$(dT)\overline{320} - (^3H)(dA)\overline{0.97}$	None	+
	dTTP or dUTP	+
	dATP, dGTP or dCTP	+

Reaction conditions were as given for Fig. 1 with
triphosphates, when present, at 13.3 μM. The
template was poly(dA) in all cases. (+) hydrolysis
detected; (−) hydrolysis not detected.

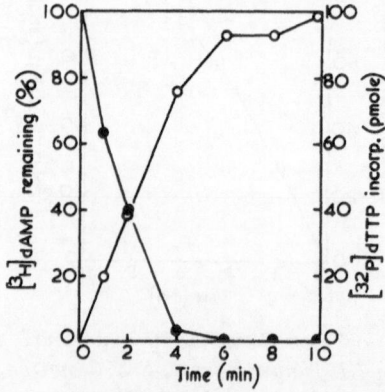

Figure 3. *Hydrolysis of noncomplementary terminal primer (^{3}H)dAMP
and simultaneous (^{32}P)poly(dT) synthesis. Details were as given for
Fig. 2 but with 3.0 nmoles poly(dA)·(dT)$\overline{320}$ −(^{3}H)(dA)$\overline{0.97}$. (●——●)
(^{3}H)dAMP hydrolysis; (○——○) (^{32}P)poly(dT) synthesis.*

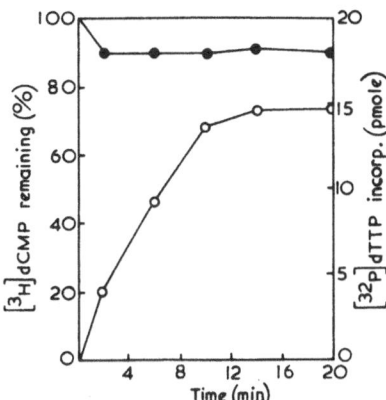

Figure 4.   Hydrolysis of noncomplementary terminal primer (^{3}H)dCMP and simultaneous (^{32}P)poly(dT) synthesis. Details were as given for Fig. 2 but with 3.0 nmoles poly(dA)·(dT)$\overline{320}$ -(^{3}H)(dC)$\overline{6.5}$. (●———●) (^{3}H)dCMP hydrolysis; (O———O)(^{32}P)poly(dT) synthesis.

primer termini, the rate of poly(dT) synthesis was proportional to
the number of functional primer sites, suggesting that with the
former template-primer, only 10% of the primer molecules were
capable of supporting synthesis. The distribution of labelled
terminal nucleotides added onto (dT)$\overline{320}$ by terminal addase during
synthesis of these primers follows a Poisson distribution (12,13)
and calculations reveal that about 10% of the primer molecules in
poly(dA)·(dT)$\overline{320}$ -(^{3}H)(dC)$\overline{6.5}$ would contain 1-3 (^{3}H)dCMP nucleotides per primer. We suggest that it is this class which is
responsible for the observed limited hydrolysis and which can then
serve as functional primers for polynucleotide synthesis.

In the light of the above observations, the homopolymer substrate conformation specificity of the exonuclease activity was
determined (Fig. 5). (dT)$\overline{320}$ -(^{3}H)$\overline{1.7}$ alone was resistant to
hydrolysis, whereas when complexed with poly(dA) it was susceptible
and thus the exonuclease activity was specific for the apparently
double stranded conformation (a reason for this was discussed
above). The conflict between the observations of the resistance of
the termini of the single stranded homopolymer and of 90% of the
(^{3}H)dCMP in poly(dA)·(dT)$\overline{320}$ -(^{3}H)(dC)$\overline{6.5}$, and the susceptibility
of (^{3}H)dAMP and (^{3}H)dCMP in (dT)$\overline{320}$ -(^{3}H)(dA)$\overline{0.97}$ and (dT)$\overline{320}$ -(^{3}H)
(dC)$\overline{0.82}$ complexed with poly(dA), may be resolved if the enzyme
binds to double stranded DNA at a 3'-terminus on one strand and if
the exonuclease site then accommodates 1-3 single stranded nucleotides only.

Additional evidence that the polymerase activity requires a
base paired primer 3'-terminus has come from studies of the template:
primer specificity of pyrophosphorolysis and nearest neighbour base

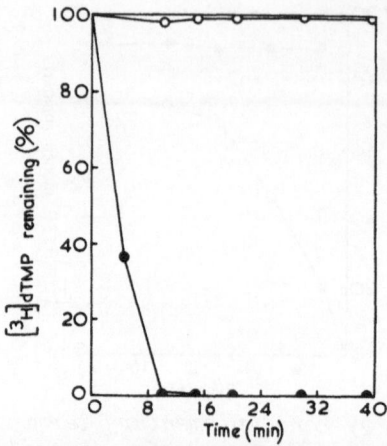

Figure 5. *Hydrolysis of (^{3}H)dTMP termini in single and double stranded homopolymers. 1.0 nmole (dT)$\overline{320}$ -(^{3}H)(dT)$\overline{1.7}$ alone (O——O) or complexed with 1.0 nmole poly(dA) (●$\overline{1.7}$●) was incubated with 10 units of U.maydis DNA polymerase at 37°C. (^{3}H) dTMP remaining in the polynucleotide was determined at the times indicated (6).*

analysis of the reaction products. Pyrophosphorolysis (formation of dNTP from DNA and inorganic pyrophosphate) is a direct reversal of polymerisation and should have identical template primer specificity. Table 2 shows that this was the case, in that the conversion of non-complementary primer nucleotides to their free triphosphates was not

Table 2

Template-primer specificity of the pyrophosphorolysis reaction

Primer	PPi	% total primer label as	
		dNMP	dNTP
(dT)$\overline{320}$ -(^{3}H)(dT)$\overline{1.7}$	−	100	<0.01
	+	47	53
(dT)$\overline{320}$ -(^{3}H)(dA)$\overline{0.97}$	−	99.5	0.5
	+	100	<0.01

Reaction conditions were as given for Fig. 1 but with 3.3 μM tetrasodium pyrophosphate present (+) or absent (−). Poly(dA) was the template in all cases. Reaction products were identified by thin layer chromatography.

Table 3

Nearest neighbour base analyses of the products of DNA
synthesis directed by homopolymer template-primers.

Primer	(^{32}P) 3'-dNMP	Amount (pmole)
$(dT)\overline{320} - (^{3}H)(dT)\overline{1.7}$	dTMP	131
	dAMP	0.07
	dCMP	0.06
$(dT)\overline{320} - (^{3}H)(dA)\overline{0.97}$	dTMP	130
	dAMP	0.08
	dCMP	0.05
$(dT)\overline{320} - (^{3}H)(dC)\overline{0.82}$	dTMP	126
	dAMP	0.06
	dCMP	0.05
$(dT)\overline{320} - (^{3}H)(dC)\overline{6.5}$	dTMP	20
	dAMP	0.05
	dCMP	0.04

Reaction conditions were as given for Fig. 2 with
$(\alpha-^{32}P)dTTP$ at 13.3 µM. The template was poly(dA)
in each case. A standard nearest neighbour bases
analysis of the reaction products was made and the
3'-dNMP's identified by paper chromatography.

detected. Table 3 shows that addition of (^{32}P)dTMP onto non-complementary termini was also not detected when transfer of (^{32}P) from the (α-^{32}P)dTTP polymerase substrate was examined by nearest neighbour base analysis. Only transfer to the internal complementary dTMP was observed.

The above experiments clearly establish that a *U. maydis* DNA polymerase can proof-read *in vitro*, the associated exonuclease removing noncomplementary primer nucleotides, thus restoring complementarity and providing functional primer termini for the polymerase activity.

REFERENCES

1.   Hall, Z.W. and Lehman, I.R. (1968) J. Mol. Biol. 36:321-333

2.   Battula, N. and Loeb, L.A. (1974) J. Biol. Chem. 249:4086-4093

3.   Chang, L.M.S. (1973) J. Biol. Chem. 248:6983-6992

4.   Loeb, L.A., Springgate, C.F. and Battula, N. (1974) Cancer Research 34:2311-2321

5.   Kornberg, A. (1974) DNA Synthesis, W.H. Freeman & Co., San Francisco.

6.   Brutlag, D. and Kornberg, A. (1972) J. Biol. Chem. 247: 241-248

7.   Muzyczka, N., Poland, R.L. and Bessman, M.J. (1972) J. Biol. Chem. 247:7116-7122

8.   Hershfield, M.S. (1973) J. Biol. Chem. 248:1417-1423

9.   Byrnes, J.J., Downey, K.M., Black, V.L. and So, A.G. (1976) Biochemistry, 15:2817-2823

10. Banks, G.R. and Yarranton, G.T. (1976) Eur. J. Biochem. 62: 143-150

11. Chang, L.M.S. (1977) J. Biol. Chem. 252:1873-1880

12. Hayes, F.M., Mithcell, V.E., Radcliff, R.L. and Williams, D.L. (1967) Biochemistry 6:2488-2492

13. Yarranton, G.T. and Banks, G.R. (1977) Eur. J. Biochem, in press.

THE THREE DNA POLYMERASES OF ANIMAL CELLS: PROPERTIES AND FUNCTIONS

Arturo Falaschi and Silvio Spadari

Laboratorio di Genetica Biochimica ed Evoluzionistica
del Consiglio Nazionale delle Ricerche
Via S. Epifanio, 14
27100 Pavia,   Italy

INTRODUCTION

For the understanding of the events leading to the
advancement of the DNA growing fork in animal cells, a description
of the molecules involved in this process is essential. The
search for these molecules is oriented by the wealth of information
obtained in prokaryotic systems, where the cross-evidence from
the biochemical data and the analysis of conditional lethal mutants
altered in DNA replication has allowed the beginning of a
description of the complex series of reactions occurring at the
growing point. From these systems it is apparent that the basic
polymerization reaction is assured by a DNA dependent DNA polymerase
(1).

Enzymes with properties similar to the bacterial and phage
induced DNA polymerase (1) have been described for a long time in
animal cells (2), and their properties have been reviewed in the
last years by several authors (3-7). Here we shall try and
summarize in a comprehensive fashion the properties of the three
DNA polymerases that exist, in different proportions but with
analogous properties, in all higher animals. The characteristic
features of each of the three enzymes are similar enough in the
different animal cells to allow an attempt at a generalized common
description. We shall also review the evidence (mainly still
indirect) that permits some tentative conclusion on their *in vivo*
function.

All cells from vertebrate animals contain (in different
relative proportions) three distinguishable DNA polymerases; each
of them has properties so similar in different organisms that a

common nomenclature has been agreed among the investigators active
in this field (8). Table 1 reports the basic properties that allow
a preliminary recognition and classification of the three enzymes,
called DNA polymerase α, β and γ in order of discovery.

These three enzymes are present in all animal cells tested,
with the possible exception of protozoa, which lack the β-enzyme (9).
No other different template-dependent DNA polymerase, so far, has
been demonstrated convincingly in animal cells, except in the
presence of some parasitic organism. We shall now look in detail
at the general properties of the three enzymes. These are also
summarized in Table II.

DNA Polymerase α

Localization. This has been the first DNA polymerase described
in animal cells, and the most abundant one present in proliferating
tissues. The enzyme is found mainly in the "cytoplasmic" fraction
of the cells following aqueous extraction (though a significant
proportion, in the order of 20%, was always found also in the
nuclear fraction). This has brought to some uncertainty and
controversy in the past as to the exact location *in vivo* of the
enzyme (4,5). Several independent lines of evidence have recently
demonstrated that the enzyme is found in the cytoplasmic fraction

Table 1

The Three DNA Polymerases of Animal Cells

α	- high M.W. ($>1 \times 10^5$); found in cytoplasm, following aqueous extraction; uses activated DNA, not poly rA-oligo dT; inhibited by N.E.M.; acidic iso-electric point.
β	- low M.W. ($<5 \times 10^4$); found in nucleus; can use also poly rA-oligo dT; insensitive to N.E.M.; basic isoelectric point.
γ	- high M.W.; found in nucleus, cytoplasm and mito-chondria; prefers poly rA-oligo dT; inhibited by N.E.M. acidic isoelectric point.

Table II

Properties of Animal DNA Polymerases α, β and γ

Properties	α	β	γ	References
Physical and chemical properties				
Molecular weight: native	130-180,000	45,000	>140,000[a]	3-7
Molecular weight: denatured (SDS)	70-90,000	45,000		15,18,28 Korn (this vol.)
S value	6-8	3.5	7-9[a]	4,5
Homogeneity	yes	yes	no	15,18,28 Korn (this vol.)
Isoelectric point	5.6-6	9-9.4	5.4-6.1	13,35
Tendency to aggregation	yes	yes	yes	4,5
Chromatographic behaviour (order of elution with phosphate buffer, pH 7.5:				
DEAE - cellulose	III	I[b]	II	4,5,35
Phosphocellulose	I,II[c]		III	
Hydroxylapatite	I,II[c]		III	
Native DNA	I	III	II	

Table II/contd.

Properties	α	β	γ	References
Functional and catalytical properties				
In vivo localization	nucleus	nucleus	nucleus mitochondria	10,11,12,37
% of total activity: growing cells	>85	10-15	2-5	⎰ 4,5,35,37,38
" " : resting cells	0-5	90-95	5-10	⎱
Specific activity (units/mg)	800,000	200,000	25,000[d]	4,32, Korn (this meeting)
Number of molecules/cell (estimated)	6×10^4	7×10^4		
Polymerization rate (nucleotide x molecule^{-1} x sec^{-1}) (estimated)	30	2.5		
Effect of DNA extending proteins	stimulation	no effect	no effect	23-26
Effect of histones (f_1, f_{2a2}, f_3, total histones)	inhibition	inhibition	inhibition[e]	24
dNTP degradation	no	no		⎫
Pyrophosphorolysis	yes	no		⎬ 4,18,32
Pyrophosphate exchange	yes	no	no	⎭
Associated DNAses 3'→5'	no	no		
5'→3'	no	no		
Associated RNAse H	no	no	no	Spadari, unpubl.

Table II/contd.

Properties	α	β	γ	References
Inhibition by antibodies against α	yes	no	no	16,73,74
Inhibition by antibodies against β	no	yes	no	38,73
Inhibition by antibodies against γ	no	no	yes	88
Inhibition by antibodies against Reverse Transcriptase	no	no	no	16,34,88
Misincorporation frequency	$<10^{-4}$	$<10^{-4}$	$<10^{-4}$	39
Ability to copy DNA across pirimidine dimers	yes	yes	yes	39
K_m for dNTP's (μM)	2-12	8-12	0.2-0.6	4,5,31
Optimal pH	7.2^f	8.5-9	8	4,5,35,37
Preferred divalent ion: activated DNA	Mg^{++}	Mg^{++}	Mg^{++}	} 4
synthetic polymers	MN^{++}	Mn	Mn	
High ionic strength effect	inhibition	stimulation[g]	stimulation[h]	4,5,32
Effect of pCMB	inhibition	inhibition	inhibition	} 4,5,14,35,38
Effect of N.E.M.	inhibition	no effect	inhibition	
Presence of Zn^{++} in molecule	yes	yes	yes	3

Table II/contd.

Properties	α	β	γ	References
Template : primer-complexes				
(a) Homopolymers				
deoxytemplate : deoxyprimer	yes	yes	yes	⎫
deoxytemplate : riboprimer	yes	poor	poor	⎬ 4,5
ribotemplate : deoxyprimer	no	yes	yes	⎭
(b) Natural DNA				
Activated native double-stranded gapped DNA	yes	yes	yes	4,5
Nicked native double-stranded DNA (T5 phage DNA)	no	no	no	Spadari, unpubl.
Denatured	no	no	no	⎫ 4,5
Native DNA	no	no	no	⎭
RNA - primed DNA	yes	no	no	20

Footnotes to Table II

a. Exception: liver mitochondrial polymerase
b. Not adsorbed at 0.02 M K-phosphate buffer, pH 7.5
c. Poor separation
d. Best partial purification (32)
e. Activated DNA was used as primer-template. Recently (32) stimulation of γ-polymerase by histone f1 has been reported using $dT_{12-18} \cdot rA$.
f. K-phosphate, 20 mM, is the preferred buffer
g. Maximal stimulation in 50 mM Tris HCl buffer (pH 8.5) is obtained at 0.1-0.13 M KCl
h. For combinations of different KCl and K-phosphate concentrations see Ref. 32. Highest enzyme activity is obtained in the presence of 50 mM K-phosphate and 100 to 200 mM KCl.

only because it leaks out of the nuclear membrane after cell
disruption. Herrick et al. (10) have enucleated mouse cells with
cytochalasin B, and have thus shown that the α-enzyme is found in
fact in the nuclear fraction. Martini et al. (11) have enucleated
by micromanipulation *Xenopus laevis* oocytes and shown that the
α-activity was found in the nucleus. Foster and Gurney (12) have
utilized a non-aqueous extraction procedure for mouse and human
cells, by extracting lyophilized cultures with glycerol, with the
same results. Thus, the DNA polymerase α is located in the nucleus,
where it most probably exerts its action, but easily leaks out
during the common aqueous extraction procedures.

Preferred template. The most efficient template-primer complex
for this enzyme is any natural DNA containing single stranded
portions interspersed between double stranded regions containing
3'OH termini (4); the 3'OH terminated stretches of DNA serve as
primers, which the enzyme prolongs with its polymerizing activity.
The need for a 3'OH terminated primer is absolute for this and for
the other two animal DNA polymerases, and seems a common feature
of all template dependent DNA polymerizing enzyme, from any source
(1). Among the synthetic homopolymer structures, the α-enzyme can
copy all deoxypolymer templates, whether primed by a deoxy- or by a
ribo-oligonucleotide (4,5). It cannot use ribopolynucleotides as
templates, neither ribo- nor deoxy-primed; this is an important
feature to distinguish it from the β- and γ-enzymes. Also natural
RNAs do not serve as templates at all for the α-polymerase.

Purification and main properties. The enzyme, first described
by Bollum (2) has been extensively purified from calf thymus by the
same author (4) as well as from *in vitro* cultured HeLa cells by
Weissbach (5), and KB cells by Sedwick et al. (13). The most
purified fractions are probably homogeneous and have a specific
activity of the order of 800,000 u/mg (communicated by D. Korn at
this meeting) (where a unit is the amount of enzyme incorporating
one nmole of total dNMP per hour).

The pure enzyme has an acidic isoelectric point (between 5.6
and 6) (13); for the reaction, the preferred buffer is phosphate
at pH 7.2 (a pH lower than for the other two enzymes) (4). One
important property of the enzyme is the sensitivity to high ionic
strength and this must be borne in mind when assaying high salt
extracts or fractions. A very useful property of this enzyme is
the high sensitivity to NEM: preincubation of the enzyme with this
reagent at 5 mM for 30 min at 0° inhibits completely and
irreversibly the activity. This, as will be shown later, is
extremely practical for distinguishing the three different DNA
polymerases in the same extract.

The M.W. of the pure enzyme ranges in different systems between

130,000 and 180,000 (4,5,14). The 130,000 form can be disaggregat-
ed in SDS-polyacrylamide gels into two proteins with a similar M.W.
(of the order of 70,000) (15 and D. Korn, this meeting). The enzyme
has a high tendency to form aggregates in low ionic strength: thus,
e.g., Spadari et al. (16), lowering the ammonium sulfate
concentration of a preparation of HeLa α-polymerase from 0.125 M
to 0.01 M could show the appearance, besides the usual 7 S peak, of
two other discrete peaks at 10 and 12 S; the physiological
significance of these aggregates is dubious and the observation
invites to a great caution in the study of high M.W. complexes
containing α-polymerase.

Associated enzyme activities. The pure enzyme molecule does
not have any nucleolytic activity,neither endo- nor exo-, neither
from 3' nor from 5' end of DNA, neither for ribo- nor for deoxy-
structures (17,18). This important feature is common also to the
β-enzyme, and distinguishes the animal DNA polymerases from those
of the prokaryotes, as well as from those of some lower eukaryotes
like yeast and *Ustilago* (7,19). The role of the associated exonucleases in bacteria seems well established, either for error-
correcting processes (3' to 5' exonuclease) or for removal of
damaged structures (5' to 3' exonuclease) (1); these processes are
certainly no less important in animal cells, but they will probably
rely on appropriate specialized enzymes, rather than on the molecule used for the polymerization reaction.

The lack of associated exonuclease explains also the lack of
capacity of α-polymerase to degrade dNTPs to dNMPs (4). Instead,
pyrophosphorolysis of the DNA can be observed, as well as exchange
of pyrophosphate into dNTPs in the presence of DNA (4); in other
words, the conditions can be found to observe microreversion of the
polymerizing reaction.

Complex forms of the α-polymerase. Several authors observed
a multiplicity of molecular forms having the same fundamental
properties as the α-enzyme, but different M.W.; while some heterogeneity may be ascribed to aggregation artefacts in low ionic
strength, the evidence for the permanent association of a simpler,
catalytically active "core" to other proteins seems convincingly
demonstrated. The most thorough study in this context was made by
I.R. Johnston and collaborators (14); they were able to
fractionate on DEAE cellulose four different forms of α-polymerase,
two of which (called A and C) have been more characterized. The A
form has a M.W. ranging between 200,000 and 230,000, whereas the
C form has a M.W. ranging between 155,000 and 170,000. The A form
is distinguished from the C one by its lower sensitivity to NEM
and to heat inactivation. Treatment of the A form with 2.4 M urea
transforms the A enzyme into the C one; after this treatment a
peptide with a M.W. of approximately 70,000 can be fractionated

from A enzyme preparations; this peptide can be added to the C
enzyme reforming a structure having all the properties of the A
enzyme, including the relative NEM and heat resistance. Thus, the
C enzyme, which is probably analogous to the simple structure
isolated also by the other groups quoted previously, can be
associated *in vivo* to at least one specific protein which changes
somewhat its properties.

A. de Recondo and her collaborators (this meeting) have shown
evidence for the tight association of a DNA binding factor with the
catalytical core of the α-polymerase of rat liver; whether this DNA
binding protein is the same as the protein described by
Dr. Johnston's group remains to be established; certainly the
possible interactions of α-polymerase with other cell proteins,
considering also the lack of associated exonucleases typical of
animal polymerases, is worth further experimental effort.

Relevant specific properties. Two aspects of the catalytic
activity of the α-polymerase which are specific for this enzyme
deserve a special consideration; in the first place, this enzyme
is the only animal polymerase which can use as template-primer
complex a native natural DNA partially transcribed by the *E.coli*
RNA polymerase (20); in view of the presence of ribo-oligonucleotide
primers for synthesis of the Okazaki pieces in animal systems (21,
22) this observation points to a preferential role of the
α-polymerase for such a synthesis. Also, among the animal
polymerases, the α-enzyme is the only one whose polymerization
rate *in vitro* is accelerated by the presence of the denatured DNA
binding proteins; this was shown by Herricks et al. (23) who, using
α-polymerase and single stranded DNA binding protein, both isolated
from calf thymus, could show an approximately 10 fold stimulation
of the polymerization rate on gapped λ-DNA; this stimulation was
specific for the α-enzyme (the β-polymerase was insensitive) and
for the binding protein (neither a different binding protein from
calf thymus nor the gene-32 protein of T_4 phage could stimulate
the α-enzyme). Similar results were obtained by Weissbach et al.
(24) with the HeLa enzyme and calf thymus binding protein; they
could further show that also the γ-polymerase is not stimulated.
Finally, Otto et al. (25) from Ehrlich ascites cells, and Carrara
et al. (26) from *Xenopus laevis* oocytes, have purified a binding
protein which stimulates specifically the α-polymerase from the
same source. This effect and its specificity, are reminiscent
of the same observation of the specific stimulation of T_4 polymerase by the gene 32 protein, which seems a good reproduction
in vitro of the likely role of this sort of protein in DNA
replication (1).

Thus, these two observations point to a role of the DNA polymerase α in DNA replication.

DNA Polymerase β

DNA polymerase β is less abundant than α in proliferating tissues (see later sections). It does not leak out of the nuclear membrane during the extraction procedure and has thus always been found associated with the nuclei (5). Its most distinctive features from the α-enzyme are the small size (3.5 S) and the insensitivity to NEM (4,5). The enzyme has been extensively purified by Weissbach group (27), Bollum's group (4) and by Korn's group (13). The pure enzyme consists of a single polypeptide chain in SDS gels, with a M.W. of 45,000 (18,28), identical to that of the native enzyme. The enzyme is clearly distinguished in crude extracts from the α-polymerase, thanks to the NEM insensitivity, to a different pH optimum (greater than 8.5) and to the stimulatory effect of high ionic strength. It can utilize the same template-primer complexes as the α-enzyme, with one addition and one exception: it can also copy poly rA - oligo dT, contrary to α; and it cannot utilize ribo-primed DNA templates. The activity on poly rA-oligo dT is inhibited by phosphate buffer, and this, as will be shown later, is important for distinguishing it from the γ-enzyme. Further distinctive features from the β-polymerase are a basic isoelectric point (13) (greater than 9), which accounts for its inability to adsorb to DEAE cellulose, a property which eases the fractionation from the α-enzyme.

Like the α-enzyme, it does not have any associated exonuclease (17,18); also (and contrary to α) neither pyrophosphate exchange nor pyrophosphorolysis are demonstrable with this enzyme, pointing to the difficulty in showing the microreversion of its catalytic activity. As mentioned above, the polymerization rate by this enzyme is not stimulated by DNA binding proteins.

DNA polymerase γ

This is, chronologically, the last of the three DNA polymerases to be described in animal cells; its discovery was spurred by the diffusion of the investigations on the RNA virus-induced reverse transcriptase; when it was demonstrated that one could use poly rA-oligo dT as template-primer complexes for the viral reverse transcriptase (29), rather than the less convenient viral RNAs, a number of reports appeared of "reverse transcriptase" in a variety of uninfected animal cells. Weissbach and Spadari purified from HeLa cells the enzyme responsible for this activity (30,31); curiously enough their enzyme preparation contained enough phosphate buffer to inhibit the action of the β-enzyme on this template (see above). They were able in any case to purify away this activity from the other known polymerases, and obtain thus a novel polymerizing activity, clearly distinguishable from the α- and β-enzymes. It was soon clear that this enzyme was totally inactive on any natural RNA as template, whether primed or unprimed, whether viral or messenger RNA (5). The enzyme prefers homo- polymer structures as templates, both ribo- and deoxy-, preferentially with a deoxy-primer. The poly rA templates, are copied better than poly dA (see Table II) whereas poly dG is copied better than poly rG. Furthermore, also the standard gapped DNA is a good template, only about 1/4 to 1/10 as good as the poly rA - oligo dT. In any case the enzyme does not appear to be in any respect a reverse transcriptase - like polymerase.

The activity of the γ-polymerase on activated DNA (as well as the activity on other templates: see below) is not inhibited by the antibodies against α (32,89). This demonstrates that the utilization of natural DNA templates is not due to a contamination with the α-enzyme, but is a peculiar property of the γ-polymerase. Also the antibodies against different reverse transcriptases do not affect the γ-activity (31,33); furthermore, the latter enzyme was purified away from the reverse transcriptase in extracts of human leukemic cells (34).

The enzyme fractionates partially from the α-polymerase in DEAE cellulose gradients; the best separation from the other two polymerases is given by native DNA cellulose chromatography.

The γ-polymerase is found in all mammalian cells, usually in a small proportion with respect to the other two enzymes. The action on poly rA - oligo dT in the presence of phosphate assures a highly specific assay (32); the enzyme is sensitive to NEM and has an acidic isoelectric point (35). The M.W. is rather large though somewhat variable in different systems; the values most commonly found are between 140,000 and 180,000 but molecules of 300,000 (36) and of 60,000 (37) daltons have been reported. The presence or absence of associated nucleases has not yet been established, since no group as yet has obtained a highly purified fraction.

In the last year it has become apparent, thanks to the work of Bolden et al. (37), that the γ-polymerase is indistinguishable from the mitochondrial polymerase, an enzyme which has been described in animal cells before the discovery of the γ-enzyme. Other papers in this volume describe the fact in detail. Table III summarizes the work of different groups leaving little doubt as to the identical properties of the two enzymes. It must still be borne in mind that the enzyme is found associated to the nuclei in appreciable proportions (close to 40%) (5,38), whereas the fraction of γ-polymerase found in "cytoplasmic" extracts is probably due to leakage either from the nuclei or from the mitochondria.

Table III

Identify of DNA Polymerase γ and Mitochondrial DNA Polymerase[a]

Properties	γ-polymerase	mt-polymerase
Molecular weight	140,000–180,000	140,000–180,000
S value in high ionic strength	7–9	7–9
Isoelectric point	5.4	5.4
pH optimum	8	8
Preferred divalent cation	Mn^{++}	Mn^{++}
Optimal (Mn^{++}) (mM)	0.1	0.1
Optimal (Mg^{++}) (mM)	5	5
Effect of ionic strength	stimulation	stimulation
Effect of K-phosphate	stimulation	stimulation
Effect of N.E.M.	inhibition	inhibition
Effect of ethidium bromide	inhibition	inhibition
Chromatographic behaviour on DEAE cellulose, phospho-cellulose, hydroxyl-lapatite and native DNA cellulose	identical	
Heat inactivation	identical	
Template: primer (relative to activated DNA)		
Activated DNA	100	
poly(rA).dT_{12-18}	400–1,000	
poly(dA).dT_{12-18}	20–30	20–150
poly(rC).dG_{12-18}	10–30[b]	
poly(dG).dG_{12-18}	150–500	400–1,200
poly(dA)	<0.1	
dT_{12-18}	<0.1	
mRNA:dT_{12-18}	<0.1	

[a]References are No. 32,35,37,38. Reference 37 refers to HeLa and rat liver cytoplasmic γ- and mt-polymerases; References 35 and 38 to nuclear γ- and mt-polymerases from rat neurons and chick embryos, respectively.
[b]Better reading for poly(rC) template is obtained at low salt conditions (32).

General Comments on the Properties of the Three Polymerases

Table II gives a comprehensive (if schematic) summary of the properties of the animal DNA polymerases; most of the data have been reviewed briefly above. The property common to all three (and to all known DNA polymerases) is the ability to copy in complementary fashion a single stranded DNA template, primed by a 3'-OH terminated sequence; another common property (at least to α and β) but that distinguishes them from the prokaryote polymerases, is the absence of associated DNAses.

The three enzymes are certainly distinct molecules; antibodies against each polymerase do not cross-react with the others.

The fidelity of copying of the three enzymes seems good, as compared to that of the bacterial enzymes: the reported figures, measured on homopolymer templates, of less than one error of incorporation vs 10^4 polymerized nucleotides, must not be taken as absolute values, but relative to the similar misincorporation measurement in the same assay conditions by bona fide "error-free" enzymes like the DNA polymerase I of *E.coli* (39). It is interesting to note that, contrary to the bacterial enzymes, the animal DNA polymerases can copy across a UV-induced thymine dimer on the template (39); if this were the case *in vivo*, one would not expect to observe the strong inhibition of DNA replication caused by UV-irradiation, and would expect a high number of mutations induced by this agent in animal cells, which also does not seem to be the case. Considering also the absence of error-correcting or dimer excising activities in the animal polymerases, one has to think that probably *in vivo* the DNA polymerases advance as part of a more complex integrated system, containing probably also the error-correcting DNAses; this system will probably respond in a more sensitive way to this type of damage.

The K_ms for the triphosphates are similar for the α- and β-polymerases, and of the order of 10 μM; they are lower by a factor of 10 for the γ-enzyme; the affinity for DNA seems instead lower in the α-polymerase than in β or γ (27,39).

The relative abundance of each polymerase in the cell is variable: the α-polymerase is always the most abundant one in rapidly proliferating tissues, and the one that is more prone to important variations in absolute terms; indeed in some resting tissues its level may be so low as to be almost non detectable, whereas the other enzymes are always present. The γ is present always in lower amounts than the β-enzyme, varying between 1/10 and 1/4 of the β activity (each assayed with its optimal reaction mixture).

With the qualification that the γ-polymerase is present also

in the mitochondria, all these enzymes can be safely assumed to
reside in the nucleus *in vivo*.

The catalytic properties of the three enzymes are different
enough to allow their simultaneous qualitative assay without
appreciable interference, and even if the relative proportions vary
in a wide range; the β-polymerase assay is absolutely specific,
thanks to the insensitivity to NEM, the high pH optimum and the
preference for high ionic strength; the γ-assay, using polyrA -
oligo dT in the presence of phosphate is also absolutely specific.
The α-assay measures also the β-activity, though only at 30%
efficiency; if in the same preparation an independent assay for
β-polymerase is run, its contribution to the α-assay can be
rigorously estimated, and the α-activity corrected accordingly. That
this is the case can be easily demonstrated by fractionating the
enzyme mixture in a sucrose gradient, and assaying all the fractions
with the three specific assays; Fig. 1 reports the results of an
idealized experiment, in which the three enzymes are presented in
proportions of 70%, 28% and 2% for α, β and γ respectively. The
absolute specificity for the β and γ assay is well demonstrated
thanks to the large difference in S; also the α and γ-polymerase,
though their sedimentation properties are close, are clearly
distinguished in these conditions, and the lack of correspondence
of the two symmetrical peaks shows the lack of interference of
the two activities; finally, it is also evident that the β-
polymerase responds in part to the α-assay, and the conditions for
correcting this contribution are evident. The detailed assay
conditions are given in the legend to Fig. 1.

Viral Induced Polymerases

The infection of animal cells with DNA viruses leads to the
increase of DNA polymerase activities (and also of other DNA enzymes).
In some cases it can be shown that the enzyme activity that
increases is a new polymerase, different from the three described
above. Thus, Citarella et al. (40) were able to fractionate the
activity that raises after vaccinia infection of HeLa cells from
the other known polymerases, and showed that it was an enzyme
with some different properties from the α- and β-enzymes. It has
a M.W. of 120,000, is immunologically distinct from α and β , and
is distinguished also by other features. No increase of host
polymerase activity is observed. Several other investigators, as
summarized in Table IV, have reported similar results following
infection of different cells with Herpes simplex and Herpes-related
viruses; in all cases, a new DNA polymerase is induced after
infection; this can be recognized different from the three host
polymerases by a number of criteria, among which, the immunological
specificity, the chromatographic properties, and the sensitivity to

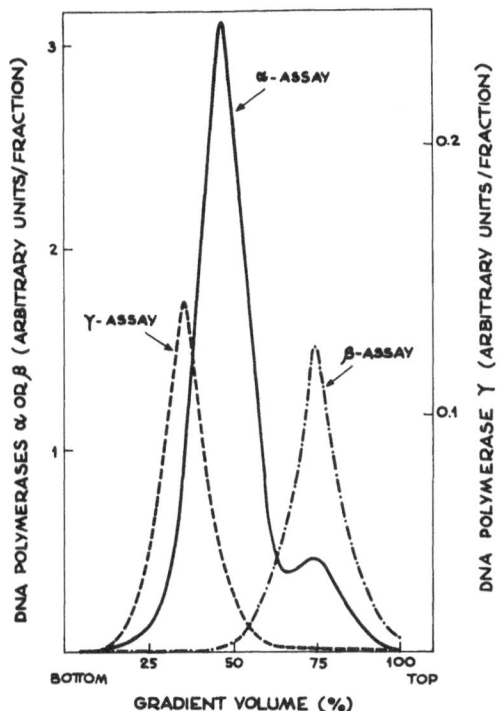

Figure 1. *An idealized representation of the independent assays of the three polymerases in sucrose gradients.*

Preparation of crude extracts and sucrose gradient centrifugation: Cells are suspended in 5 volumes of 0.1 M KCl, 0.1 M K-phosphate (pH 7.2), 0.5% Triton X-100, 5 mM DTT, 1.5 mM MgCl$_2$ and homogenized in a Potter; the suspension is centrifuged for 1 hr at 100,000 x g; 0.2 ml of the supernatant are applied on a 5-20% (w/v) sucrose gradient (5 ml) in 0.1 M K-phosphate (pH 7.2), 0.1 M KCl, 1 mM DTT; centrifugation is at 0°C for 16 hrs at 42,000 rpm in a Spinco 50, 1 rotor; 0.25 ml fractions are collected from the bottom.

α-polymerase reaction: 20 mM K-phosphate (pH 7.2), 0.1 mM EDTA, 0.5 mM DTT, 10 mM MgCl$_2$, 250 µg/ml BSA, 200 µg/ml of activated DNA, all four deoxynucleoside triphosphates at 50 µM each, with ³H-TTP. Under these conditions, β-polymerase responds with a 30% efficiency to the α-polymerase assay; γ-polymerase is inactive under these conditions (32).

(β-polymerase reaction: 50 mM Tris-HCl (pH 8.5), 0.1 M KCl, 10 mM MgCl$_2$, 1 mM DTT, 250 µg/ml BSA, 200 µg/ml of activated DNA, all four deoxyribonucleoside triphosphates at 50 µM each, with ³H-TTP. When enzymes samples are preincubated with 5 mM NEM at 0°C for 30 min (to inactivate α- and γ-polymerases) the β-assay is strictly specific for the β-polymerase.

γ-polymerase reaction: 50 mM Tris-HCl (pH 8.5), 50 mM K-phosphate (pH 8.5), 0.13 M KCl, 0.5 mM McCl$_2$, 1 mM DTT, 250 µg/ml BSA, phosphonacetic acid (an antiviral agent much less effective on the host DNA polymerases (41)). In several cases there is good evidence that the enzyme is coded by a viral gene: temperature-sensitive mutants of Herpes virus have been obtained, unable to synthesize viral DNA at non permissive temperature (42,43): these produce a temperature sensitive DNA polymerase. Also, mutants of Herpes virus resistant to phosphonacetic acid produce a DNA polymerase resistant to the same drug (44). Thus, it seems evident that (a) the Herpes viruses (and vaccinia too ?) produce a viral DNA-coded DNA polymerase; (b) this is essential for viral DNA synthesis.

In the case of the much smaller SV40 and polyoma DNA viruses instead, no evidence of production of a new polymerase is found (45, 46,47); in both these cases, the host α-polymerase increases its level; the β-polymerase remains constant. Contrasting data are available for the γ-polymerase which increases after polyoma infection of mouse fibroblasts, but not after SV40 infection of monkey cells (47). The relevance of these data to the in vivo function of the animal polymerases is discussed below.

In Vivo Function of the Animal DNA Polymerases

General considerations. The lack of conditional mutants, as emphasized in the introduction, makes it difficult to ascribe unambiguously a function to a molecule; even in prokaryotes, where mutants are available and the properties of proteins involved in DNA metabolism are much more known than in eukaryotes, the attribution or exclusion of a function to a molecule is not without doubts and errors. The answers, as of now, will needs be tentative, and based on correlations and inferences, rather than on direct evidence. While it seems unreasonable to assume that the DNA polymerases are geared for anything else than some form of DNA synthesis, we know again from the prokaryote work that at least two rather distinct types of DNA synthesis can exist in living organisms; thus, one particular DNA polymerase, (the polymerase III of *E. coli* and *B. subtilis*) (1) is mainly involved in, and responsible for, the chromosome replicative synthesis, whereas the other two polymerases are mainly involved in, and responsible for, repair type synthesis. Also this is in fact not an absolute distinction, since the need for a small fraction of the DNA polymerase I molecules in one aspect of chromosome replication (the removal of the RNA primers to the Okazaki pieces and resynthesis of removed sequences) seems well established (1), and, conversely, an involvement of DNA

50 µg/ml poly (rA)- oligo dT, 50 µM ^{3}H-TTP. Under these conditions the assay is strictly specific for γ-polymerase as β-polymerase activity on poly(rA)- oligo dT is inhibited by phosphate (32) and α-polymerase does not utilize this template-primer (4,5).

Table IV

DNA Polymerases Induced by Virus Infection[a]

Host Cells	Virus	Viral DNA Polymerase	Host Polymerase			References
			α	β	γ	
HeLa S3	Vaccinia	+	-			40
L-cells, chick cells	"	+				75
BHK, HEp-2	Herpes Simplex Cirus (HSV)	+				76
HeLa F	"	+				77
Rabbit kidney cells	"	+				78
Duck cells	Marek's disease Herpes virus	+				79
BHK 21/C13	HSV-2 (temperature sensitive mutants)	+ (at permissive temperature)				42
Hep-2	"	+ (at permissive temperature)				43
BHK 21/C13	HSV-1 HSV-2 (mutants resistant to phosphonacetic acid)	+ (resistant to phosphonacetic acid)				44
BHK 21/C13	Pseudorabies	+				80
CV-1 monkey cells	SV-40	-	+	-	-	45
Mouse kidney	Polyoma	-		+		46
3T3 mouse fibroblasts	"	-	+	-	+	47
KB	Adenovirus	-				81

[a] + and - represent the presence and absence, respectively, of induced polymerase.

polymerase III in some form of repair-type synthesis (including
that necessary for recombination) seems quite likely (48,49). Still,
in bacteria a gross separation of main functions remains, and we
can wonder which of the animal DNA polymerases is more involved
in either process.

A first indirect hint may come from the measure of relative
affinity of the three enzymes to dNTPs. As reported in Table III,
whereas the α and β-enzymes have roughly the same K_m for precursors,
that of the γ-enzyme is lower by a factor of 10. Now, it is known
that a lowering of the intracellular pool of dNTPs (such as obtained
by hydroxyurea, a drug inhibiting the conversion of rNDPs to dNDPs
by interaction with the specific reductase) totally inhibits
replicative synthesis, whereas repair synthesis still proceeds at
a good rate (50). This observation can be rationalized in
evolutionary terms (in conditions where DNA precursors are scarce,
it is advisable to stop replication but maintain a well protected
and repaired DNA until better times come) and could be explained by
the lower K_m of the polymerase involved in repair synthesis. This
would fit with the idea that DNA polymerase γ is mainly geared, in
the nucleus, for repair synthesis. The argument is attractive, but
weakened by the likelihood that the *in vivo* interaction of the DNA
polymerases with other cofactors might affect drastically the
affinity for dNTPs observed *in vitro*. This last possibility is
made even more likely by the fact that the measured pools of dNTPs
in animal cells in proliferative conditions are lower by at least an
order of magnitude than the K_m of even the γ-polymerase (51). We
must then look for other correlations to get hints on the function
of each polymerase.

α-polymerase. The greatest (and most convincing) wealth of
data comes from the study of the **correlation of enzyme levels with**
DNA replication rates. The first indication came from the work of
Chang and Bollum; whether they studied cultures of L cells in
different growth conditions (52), or regenerating rat liver (53),
or phytohaemoagglutinin-stimulated lymphocytes (54), they always
found a striking correlation between the level of α-polymerase and
that of DNA replication rate; the β-enzyme instead, in terms of
units/cell, did not undergo significant variations in these
different physiological conditions. Similar results were obtained
in BHK cells by Craig et al. (55); in synchronized HeLa cells,
though the quantitative variations are less spectacular than in
unsynchronized cultures going from stationary to exponential growth
and back, the α-polymerase rises with the S phase (56). (Also the
γ-polymerase increases in this case, but this observation is
weakened by the considerations that will be made below). Cells
from different other organisms have been studied in different
physiological conditions causing drastic changes in cell proliferation and DNA synthesis rate (see Table V). In all cases, it is

the α-enzyme that rises in parallel to DNA replication rates, the
other two enzymes remaining relatively constant in terms of units/
cell. These variations, in some cases, are quite spectacular; thus
in stimulated lymphocytes (54,57,58), the replication rate and the
α-polymerase level may go up in parallel by a factor of 50 or
more; in rat neurons, during perinatal development, in coincidence
with the arrest of cell proliferation and DNA replication, the DNA
polymerase α level drops within two days to 1/6 of its original
level (59), etc.

The only apparent exception to the rule that only the α-
polymerase responds to proliferation stimuli, are the synchronized
HeLa cells and the PHA stimulated lymphocytes (see Table V). For
the synchronized cells, we must consider that the γ-assay was made
in total extracts, and no fractionation of the mitochondria was
attempted: thus, it is conceivable that this increase may be due to
mitochondria proliferation. For the stimulated lymphocytes, again
the same criticism can be made concerning the γ-increase; more
important, the quantitative variations of β and γ are much less
dramatic than those of α; whereas this enzyme may increase by a
factor of 50 or more, the other two enzymes raise by a factor of 5
at most in concomitance with the peak in DNA replication rate.
Furthermore, as will be discussed below, the increase of β-polymerase may not be correlated with DNA replication rate, but with other
cellular phenomena.

We may also recall here that, of the three polymerases, only
the α-enzyme can elongate RNA primers on natural DNAs, and is
stimulated by the specific animal DNA extending proteins, both
features considered typical of replicative synthesis. Also, in
the case of the DNA viruses not coding for their own polymerase,
the main or only polymerizing enzyme which increases after infection
(45) and which is found associated with replicating particle is
the α-enzyme (Fanning et al., this meeting).

In conclusion, it seems difficult to escape the contention
that the α-polymerase is involved in (and is the main enzyme
responsible for) the polymerization necessary for DNA replication.

β-polymerase. The poor or nil correlation of the level of this
enzyme with DNA replication rate, and its constance in different
physiological conditions has invited the suggestion that it may be
related to repair type synthesis. A more direct indication in this
sense comes from the work with the DNA stimulated lymphocytes:
these, besides the increase of α-polymerase in correspondence to
the peak in DNA synthesis rate, show a second increase of DNA
polymerase activity (60) occurring at later times, when DNA
replication rate is at a minimum. This second wave of DNA polymerase is parallel to a peak in the level of other enzymes acting

Table V

Correlation of DNA Polymerases α, β and γ
with Cellular DNA Replication Rate[a]

System	α	β	γ	References
Cell cultures				
Mouse L-cells	+	-		52
Synchronized HeLa cells	+	-	+	56,82
BHK-21/Cl3 cells	+	-		55
PHA stimulated lymphocytes	+	+[b]	+[b]	54,57,58
Organs				
Regenerating rat liver	+	-		53,83,84
Antigen stimulated spleen	+	-		4
Neurons during perinatal development	+	-	-	59,85
Spleen during perinatal development	+	-	-	59
Heart muscle during perinatal development	+	-	-	59,86
Erythropoietin induced spleen cells	+	-		87

[a] + and - represent the presence and absence, respectively, of positive correlation.

[b] Increases much less than α

on DNA, like DNA ligase and a DNase acting on single stranded DNA
(60). The second polymerase peak is due mainly to polymerase β,
which reaches only at this late time (2 or 3 days after the peak in
DNA synthesis rate) its maximum level in this system, and is indeed
the prevalent polymerase in the lymphocytes at this stage (58).
This late peak of DNA enzymes corresponds to a peak in the capacity
of the stimulated lymphocytes to perform repair synthesis following
UV irradiation. Though the exact significance of this increase in

DNA enzymes at this stage is not clear (perhaps an involvement with immunoglobulin production), it seems well established that the β-polymerase level is correlated with the peak in DNA repair capacity. The inference that the β-enzyme is involved mainly in repair synthesis seems reasonable. Further circumstantial evidence in this sense comes from the fact that, at the late stages of maturation of spermocytes, when the meiotic process involves only recombination steps and not overall chromosome replication, only the β-polymerase is found associated with the cells, and is actually increasing in absolute terms (61); again, the inference for its involvement in the repair synthesis necessary for recombination comes to the mind.

γ-polymerase. The location of this enzyme, in part at least, in the mitochondria, where it is the only polymerase present, strongly suggests that one of its roles is the replication and the repair of mitochondrial DNA. This need not be the exclusive (or even the main) role of this enzyme, considering the appreciable proportion found in the nucleus; it is unlikely that it should concentrate there to be stored for mitochondrial needs ! An involvement in repair synthesis may be suggested by the observation of the low K_m, whereas a main role in replication seems unlikely from its scarce response to proliferative stimuli, and its low polymerization capacity (see below).

CONCLUDING REMARKS

The DNA polymerases are not the only putative piece of machinery for DNA replication which has been described in animal cells; it is well established that also in these cells the synthesis at the growing fork is discontinuous (62); it seems also well demonstrated that short RNA sequences (synthesized by a still undefined RNA polymerase) prime the synthesis of the Okazaki pieces (21,22); the template could be maintained in an extended position by one or more of the several DNA binding proteins described in animal cells (23-26) (some of these have the properties in fact of DNA extending factors, according to the definition proposed by Hoffmann-Berling (63)). The Okazaki pieces are then probably synthesized by the α-polymerase; RNase H activities have also been described that could remove the primers (64); we do not know whether the same polymerase that synthesizes the Okazaki pieces also fills the gaps left by the degraded primers, but there is no evidence against it. A polynucleotide ligase has been described in animal cells having the properties necessary for joining the short intermediates (65) (and a different ligase could do the same job in mitochondrial DNA).

Can the α-polymerase alone justify the *in vivo* rate of synth-

esis ? It is not easy to answer rigorously this question, in view
of the uncertainties and variability of the measures of the parameters necessary to this calculation, but we may boldly try and
see whether we are far away from the expectations estimable from
the most frequently obtained data. The most popular estimate of
the rate of advancement of each growing fork in mammalian cells is
of 1 μm/min, that means an overall rate of polymerization of 50
nucleotides x sec^{-1} x $helix^{-1}$ (62). The specific activity of the
best preparation of α-polymerase is of the order of 800,000 U/mg
(Korn, this meeting); given a M.W. of 130,000 this corresponds to
approximately 30 nucleotides x sec^{-1} x molecule of $enzyme^{-1}$. Thus
we are not too far from the *in vivo* synthesis rate; the situation
is made even better if one considers that, if the discontinuous
mode of synthesis applies to both helices, more than one polymerase
molecule may be active at the same time on one helix, and that the
DNA extending protein may specifically accelerate even more the
polymerization rate of the α-enzyme.

   We can look at the problem from another point of view, and
wonder if the polymerizing capacity of the α-polymerase molecules
present in the cell is adequate to assure the chromosome replication
rate. From the data published by several authors, there are
approximately 1,000 units of α-polymerase/10^8 cells, in human
established cell lines (4,56) (like those used for the calculations
of the previous paragraph), i.e., 1 x 10^{-5} units/cell. Since one
unit can incorporate 0.28 pmol of·nucleotide x sec^{-1}, i.e., 1.7 x
10^{11} nucleotides/sec, and one polymerase molecule incorporates
30 nucleotides/sec, we get 5.6 x 10^9 molecules of polymerase/unit,
i.e., 5.6 x 10^4 molecules /cell. How much synthesis goes on in the
cell at any time ? In an asynchronous population, in order to
synthesize all the DNA of the cell in 24 hrs (generation time of an
in vitro established cell line) the amount of synthesis averages
out 1.5 x 10^5 nucleotide/sec. Thus, the minimum number of α-
polymerase molecules needed is 1.5 x 10^5 nucleotide x sec^{-1}/30
nucleotides x sec^{-1} x $molecule^{-1}$), i.e., 5 x 10^3 molecules. Thus,
we have an excess of 10 fold in number of molecules of α-polymerase
over the minimum needed. The values refer to non-synchronous
cultures, and one could argue that all the polymerase molecules
present in the cells out of S phase are "useless". On the other
hand, this would increase the need for α molecules only by a factor
of 4 (assuming an S phase of 6 hrs over a 24 hr cell cycle) and
thus still below the 10 fold excess observed; furthermore, it is
known that S phase cells contain approximately twice as much α-
polymerase than the other cells.

   A similar calculation made for the β-polymerase gives a
polymerization rate of 2.5 nucleotides x sec^{-1} x $molecule^{-1}$; this
value is 1/10 of that of α, and makes it hard to justify a chain
growth rate of 50 nucleotides x sec^{-1} x $helix^{-1}$; even the discontinuous mode of replication seems hard to reconcile with the need

of at least 20 molecules per helix active at the same time. The
overall polymerization capacity of the cell due to the β-enzyme,
calculated as above, gives a need of 6 x 10^4 molecules/cell; the
calculated number of molecules present (based on a value of 100 U/
10^8 cells (4)) is of 6.7 x 10^4: barely enough to the purpose. This
is another argument against the β-polymerase capacity to be
responsible for replicative synthesis.

Finally, for the γ-enzyme, the data available taken at face
value tend to rule out even more that it could sustain the overall
rate of synthesis of the cells: we said that we have an excess of
10 fold polymerization capacity for α, vs replication rate; now,
the γ-enzyme is lower than α (as units per nucleus) by a factor of
approximately 50; another factor of 10 is lost if we do not use
the poly rA-oligo dT template, but natural DNA. Thus, with all the
limitations implicit in this sort of argument, the γ-polymerase
is unlikely to sustain all the cell replication rate.

In summary, the evidence in favour of the α-polymerase as the
main enzyme for replicative synthesis are the folliwng:

1.   Its polymerizing rate per molecule is faster than that of
      the β-enzyme and close to the known chain growth rate *in
      vivo*.

2.   It is the most abundant polymerase in the cell, measured
      as overall polymerizing capacity.

3.   It is the only polymerase whose level correlates consistently
      and strongly with the DNA replication rate.

4.   It is the only polymerase that can use the RNA primed-
      natural DNA as template.

5.   It is the only polymerase stimulated by DNA binding
      proteins.

Each of these arguments is not a satisfactory proof of its own, but
the overall weight of the evidence is becoming overwhelming.

Undoubtedly many other molecules and factors will be
necessary to build up the whole replicative machinery of the growing
point: from the bacterial data again, we can assume that some
unwinding enzyme or enzyme system will be necessary, similar to
the *rep* protein (66) and to the DNA-dependent ATPase described by
Hoffmann-Berling (63) and by Richet and Kohiyama (67). The activity
of a ω-like protein (68) in relaxing excess supercoiling (described
in animal cells by several groups (69)) is probably also essential.
In view of the fact that also the eukaryote chromosome is a highly
netatively supercoiled structure (70), we can assume that some

activity like the bacterial gyrase (71) will be present in the
cell and by the topological constrains imposed will cooperate in
the active unwinding of the advancing growing fork. Finally, we
have no idea of the initiation signals for replication synthesis
start.Perhaps the simpler viral systems (21) may help to unravel
the process much like the ØX and similar systems are doing in
E.coli. It must also be borne in mind that the animal
chromosome is always found associated with histones, and that
these molecules all reduce the rate of action of the animal
polymerase *in vitro* (24). This may reduce the apparent excess of
polymerizing activity present in the cell; for α, the difficulty
may be balanced by the specific stimulation of polymerization due
to the DNA binding protein.

As for the function of the three enzymes in repair synthesis
we can add little to the circumstantial evidence summarized above;
the low K_m may indicate the γ-enzyme; the correlation of the
β-enzyme with repair capacity in lymphocytes, and with late meiotic
stages, points to this molecule as well. A number of repair
mutants (as described elsewhere in this volume (72)) are beginning
to be available in human cells: some of these have been assayed for
the three DNA polymerases, and found normal in this respect.
Nonetheless, the extension of the search for repair mutants and the
desirable future availability of replication mutants might one day
allow a more direct recognition of the role of the three
polymerases in animal cells.

At the present, the best working hypothesis, in our opinion,
gives a main role in DNA replication to α-polymerase, and in
repair-type synthesis (including recombination) to the β-enzyme,
and at least a role in mitochondrial DNA replication and repair to
DNA polymerase γ.

REFERENCES

(1)   Kornberg, A. (1974) DNA Synthesis. W.H. Freeman and Co.,
       San Francisco, U.S.A.

(2)   Bollum, F.J. and Potter, V.R. (1957)   JACS   79:3603

(3)   Loeb, L.A. (1974)   In: "The Enzymes" (Boyer, P.D. ed.) vol.
       10, 173-209

(4)   Bollum, F.J. (1975) Prog. Nucl. Acid Res. and Mol. Biol.
       15:109-144

(5)   Weissbach, A. (1975)   Cell 5:101-108

(6)   Holmes, A.M. and Johnston, I.R. (1975) FEBS Letters 60:233-243

(7) Wintersberger, E. (1977) TIBS   58-61

(8) Weissbach, A., Baltimore, D., Bollum, F.J., Gallo, R.C. and
     Korn, D. (1975) Science 190:401-402

(9) Chang, L.M.S. (1976) Science 191:1183-1185

(10) Herrick, G., Spear, B.B. and Veomett, G. (1976) Proc. Natl.
     Acad. Sci. U.S.A. 73:1136-1139

(11) Martini, G., Tató, F., Gandini Attardi, D. and Tocchini-
     Valentini, G.P. (1976) Biochem. Biophys. Res. Commun. 72:
     875-879

(12) Forster, D.N. and Gurney, T. (1976) J. Biol. Chem. 251:7893-
     7898

(13) Sedwick, W.D., Wang, T.S.-F. and Korn,D. (1972) J. Biol. Chem.
     247:5026-5033

(14) Holmes, A.M., Hesslewood, I.P., Wickremasinghe, R.G. and
     Johnston, I.R. (1977) "The Biochemistry of the Cell Nucleus",
     Biochem. Soc. Symp. Vol. 42, in press

(15) Sedwick, W., Wang, T.S.-F. and Korn, D. (1975) J. Biol. Chem.
     250:7045-7056

(16) Spadari, S., Muller, R. and Weissbach, A. (1974) J. Biol.
     Chem. 249:2991-2992

(17) Chang, L.M.S. and Bollum, F.J. (1973) J. Biol. Chem. 248:
     3398-3404

(18) Wang, T.S., Sedwick, W.D. and Korn, D. (1974) J. Biol. Chem.
     249:841-850

(19) Banks, G.R. and Yarranton, G.T. (1976) Eur. J. Biochem. 62:
     143-150

(20) Spadari, S. and Weissbach, A. (1975) Proc. Natl. Acad. Sci.
     U.S.A. 72:503-507

(21) Reichard, P., Eliasson, R. and Söderman, G. (1974) Proc.
     Natl. Acad. Sci., U.S.A. 71:4901-4905

(22) Tseng, B.Y. and Goulian, M. (1975) J. Mol. Biol. 99:339-346

(23) Herrick, G., Delius, H. and Alberts, B. (1976) J. Biol. Chem.
     251:2142-2146

(24) Weissbach, A.,   Spadari, S. and Knopf, K.W. (1975) In:
     "DNA Synthesis and Its Regulation", Goulian, M. and Hanawalt,
     P. (eds.) Benjamin Press, Inc. 64-81

(25) Otto, B., Baynes, M. and Knippers, R. (1977) Eur. J. Biochem.
     73:17-24

(26) Carrara, G., Gattoni, S., Mercanti, D. and Tocchini-Valentini,
     G. (1977) Nucleic Acids Res., in press

(27) Weissbach, A., Schlabach, A., Fridlender, B. and Bolden, A.
     (1971) Nature New Biology 231:167-170

(28) Chang, L.M.S. (1973) J. Biol. Chem. 248:3789-3795

(29) Spiegelman, S., Burny, A., Das, M.R., Keydar, J., Schlom, I.,
     Travnicek, M. and Watson, K. (1970) Nature 228:430-432

(30) Fridlender, B., Fry, M., Bolden, A. and Weissbach, A. (1972)
     Proc. Natl.   Acad. Sci., U.S.A. 69:452-455

(31) Spadari, S. and Weissbach, A. (1974) J. Biol. Chem. 249:
     5809-5815

(32) Knopf, K.W., Yamada, M. and Weissbach, A. (1976) Biochemistry,
     15:4540-4548

(33) Lewis, B.J., Abrell, J.W., Smith, R.G. and Gallo, R.C. (1974)
     Science, 183:867-868

(34) Lewis, B.J., Abrell, J.W., S-ith, R.G. and Gallo, R.C. (1974)
     Biochim. Biophys. Acta 349:148-160

(35) Hübscher, U., Kuenzle, C.C. and Spadari, S. (1977) Eur. J.
     Biochem., in press

(36) Matsukage, A., Bohn, E.W. and Wilson, S.H. (1975) Biochemistry,
     14:1006-1020

(37) Bolden, A., Pedrali-Noy, G. and Weissbach, A. (1977) J. Biol.
     Chem. 252:3351-3356

(38) Bertazzoni, U., Scovassi, A. and Brun, G.M. (1977) Eur. J.
     Biochem., in press

(39) Villani, G., Defais, M., Spadari, S., Caillet-Fauquet, P.,
     Boiteux, S. and Radman, M. (1977) "Proceedings of the 7th
     Int. Congress of Photobiology", Plenum Press Co., New York,
     in press

(40) Citarella, R.V., Muller, R., Schlabach, A. and Weissbach, A. (1972) J. Virol. 10:721-729

(41) Mao, T.C.-H., Robishaw, E.E. and Overby, L.R. (1975) J. Virol. 15:1281-1283

(42) Hay, J., Moss, H., Jamieson, A.T. and Timbury, M.C. (1976) J. Gen. Virol. 31:65-73

(43) Purifoy, D.J.M. and Benyesh-Melnick, M. (1975) Virology, 68:374-386

(44) Hay, J. and Subak-Sharpe, J.H.(1976) J. Gen. Virol. 31:145-148

(45) Riva,S., Cline, M.A. and Laipis, P.J. (1977), this volume.

(46) Wintersberger, U. and Wintersberger, E. (1975) J. Virol. 16:1095-1100

(47) Närkhammar, M. and Magnusson, G. (1976) J. Virol. 18:1-6

(48) Youngs, D.A. and Smith, K.C. (1973) Nature New Biol. 244:240-241

(49) Canosi, U., Siccardi, A.G., Falaschi, A. and Mazza, G. (1976) J. Bact. 126:108-121

(50) Reichard, P. (1972) Adv. Enz. Regul. 10:3-16

(51) Bray, G. and Brent, T.P. (1972) Biochim. Biophys. Acta 269: 184-191

(52) Chang, L.M.S. and Bollum, H.J. (1973) J. Mol. Biol. 74:1-8

(53) Chang, L.M.S. and Bollum, F.J. (1972) J. Biol. Chem. 247: 7948-7950

(54) Coleman, M.S., Hutton, I.J. and Bollum, F.J. (1974) Nature, 248:407-409

(55) Craig, R.K., Costello, P.A. and Keir, H.M. (1975) Biochem. J. 145:233-240

(56) Spadari, S. and Weissbach, A. (1974) J. Mol. Biol. 86:11-20

(57) Mayer, R.J., Smith, R.G. and Gallo, R.C. (1975) Blood 46:509-518

(58) Bertazzoni, U., Stefanini, M., Pedrali-Boy, G., Giulotto, E., Nuzzo, F., Falaschi, A. and Spadari, S. (1976) Proc. Natl. Acad. Sci., U.S.A.   73:785-789

(59) Hübscher, U., Kuenzle, C.C. and Spadari, S. (1977) Nucleic Acid Res., in press

(60) Pedrali-Noy, G., Dalpra', L., Pedrini, M.A., Ciarrocchi, G., Giulotto, E., Nuzzo, F. and Falaschi, A. (1974)   Nucleic Acids Res. 1:1183-1199

(61) Hecht, N.B.,   Farrell, D. and Davidson, D. (1976) Dev. Biol. 48:56-66

(62) Edemberg, H.J. and Huberman, J.A. (1975) Ann. Rev. Genetics 9:245-284

(63) Abdel-Monem, M., Lauppe, H.F., Kartenbeck, J., Dürwald, H. and Hoffmann-Berling, H. (1977) J. Mol. Biol. 110:667-685

(64) Stein, H. and Hausen, P. (1969)   Science 166:393-395

(65) Söderhäll, S., and Lindahl, T. (1976)   FEBS Letters 67:1-8

(66) Scott, J.F., Eisenberg, S., Bertsch, L.L. and Kornberg, A. (1977)   Proc. Natl. Acad. Sci., U.S.A. 74:193-197

(67) Richet, E. and Kohiyama, M. (1976)   J. Biol. Chem. 251:808-812

(68) Wang, J.C. (1971)   J. Mol. Biol. 55:523-533

(69) Champoux, J.J. and Dulbecco, R. (1972) Proc. Natl. Acad. Sci., U.S.A. 69:143-146

(70) Benyajati, C. and Worcel, A. (1976)   Cell 9:393-407

(71) Gellert, M., Mizuuchi, K., O'Dea, M.H. and Nash, H.A. (1976) Proc. Natl. Acad. Sci., U.S.A. 73:3872-3876

(72) Falaschi, A. and Pedrini, A.M. (1977), this volume.

(73) Brun, G.M., Assairi, L.M. and Chapeville, F. (1975) J. Biol. Chem. 250:7320-7323

(74) Smith, R.G., Abrell, J.W., Lewis, B.J. and Gallo, R.C. (1975) J. Biol. Chem. 250:1702-1709

(75) Lazier, C. and Helleiner, C.W. (1970) J. Gen. Virol. 8:149-152

(76) Keir, H.M., Hay, J., Morrison, J.M. and Subak-Sharpe, J.
     (1966) Nature 210:369-371

(77) Weissbach, A., Hong, S.-C.L., Aucker, J. and Muller, R. (1973)
     J. Biol. Chem. 248:6270-6277

(78) Müller, W.E.G., Falke, D. and Zahn, R.K. (1973) Arch.
     Gesamte Virusforsch 42:278-284

(79) Boezi, J.A., Lee, L.F., Blakesley, R.W., Koenig, M. and Towle,
     H.C. (1974) 14:1209-1219

(80) Halliburton, I.W. and Andrew, J.C. (1976) J. Gen. Virol.
     30:145-148

(81) Ito, K., Arens, M. and Green, M. (1975) J. Virol. 15:
     1507-1510

(82) Chiu, R.W. and Baril, E.F. (1975) J. Biol. Chem. 250:
     7951-7957

(83) Baril, E.F., Jenkins, M.D., Brown, D.E., Laszlo, J. and
     Morris, H.P. (1973) Cancer Res. 33:1187-1193

(84) William, L.E., Surrey, S. and Lieberman, I. (1975) J. Biol.
     Chem. 250:8179-8183

(85) Chiu, F.J. and Sung, S.C. (1972) Biochim. Biophys. Acta 269:
     364-369

(86) Claycomb, W.C. (1975) J. Biol. Chem. 250:3229-3235

(87) Roodman, G.D., Hutton, J.J. and Bollum, F.J. (1976) Biochim.
     Biophys. Acta 425:478-491

(88) Robert-Guroff, M. and Gallo, R.C. (1977) Biochemistry, 16:
     2874-2880

(89) Robert-Guroff, M., Schrecker, A.W., Brinkman, B.J. and Gallo,
     R.C. (1977) Biochemistry 16:2866-2873

STRUCTURE AND CATALYTIC PROPERTIES OF HUMAN DNA POLYMERASES α

AND β

David Korn, Duane C. Eichler, Paul A. Fisher and
Teresa Shu-Fong Wang

Laboratory of Experimental Oncology, Department of
Pathology
Stanford University School of Medicine, Stanford,
California 94305, U.S.A.

INTRODUCTION

In this paper we shall describe some of our recent studies on
the structure and catalytic properties of DNA polymerases α and β
(1) that we have purified from cultured human KB cells and from
adult human livers. The principal topics of discussion will be
the definition of the polypeptide composition of near homogeneous
KB cell DNA polymerase α, the problem of the intracellular
localization of polymerase α, and the results of our initial
studies on the specific *in vitro* replication of human D-loop
mitochondrial DNA (mtDNA) by near homogeneous human DNA polymerase
β.

The Structure of Human DNA Polymerase α (11)

The high molecular weight class of eukaryotic DNA polymerase,
currently designated DNA polymerase α, was first identified in calf
thymus extracts about 20 years ago (2) and has subsequently been
shown to be ubiquitous among eukaryotes and to represent the
predominant polymerase species in actively dividing cells (3-7).
Despite considerable recent effort, attempts to achieve the complete
purification and characterization of this enzyme have thus far met
with only limited success. Several groups have obtained
polymerase α preparations free of significant contaminating
activities and of sufficient purity to describe some of its
distinctive catalytic properties (8-10). However, this enzyme has
been unusually difficult to purify completely (4,5,10), and thus its
molecular structure has not yet been convincingly demonstrated.

Table 1.

PURIFICATION OF DNA POLYMERASE-α (11)

STEP	FRACTION	VOLUME[a]	PROTEIN[b]	ACTIVITY[a]	SPECIFIC ACTIVITY	YIELD
		ml	mg	$units$	$units/mg$	%
Crude Extract	I	9.9	43	880	20	(100)
pH 5.5 precipitation	II (resolubilized precipitate)	2.2	19	890	47	108
	II'(supernatant)	-	-	60	-	-
Ultracentrifugation	III (supernatant)	2.8	9	790	88	97
	III'(pellet)	-	-	50	-	-
First DEAE	IV (adsorbed)	2.9	9×10^{-1}	490	550	67
	IV'(flow through)	-	-	100	-	-
Second DEAE	V	1.1	4×10^{-1}	280	700	32
Phosphocellulose	VI	3.8×10^{-1}	4×10^{-2}	240	6,000	28
Hydroxylapatite	VII	6.6×10^{-2}	1×10^{-2}	130	13,000	15
DNA-cellulose	VIII	6.5×10^{-2}	1×10^{-3}	33	33,000	4
Gel electrophoresis[b]	IX	-	1.6×10^{-4}	-	206,000	-

a. Quantities are expressed per g wet wt. of KB cells

b. Aliquots, 400µl, of Fraction VIII were used for non-denaturing gel electrophoresis. The protein value was derived by densitometry of a stained gel. Recovery of DNA polymerase activity by elution of slices of parallel unstained gels varried between 50% and 95%. The specific activity is based on units of loaded activity.

Purification of DNA Polymerase α

We have succeeded in developing a highly reproducible purification scheme (Table 1) by which we can obtain essentially homogeneous preparations of DNA polymerase α from cultured human KB cells. Several explanatory comments are in order regarding this protocol. It should be noted that the data presented in Table I are expressed per gram wet weight of freshly harvested KB cells and that the final yield of homogeneous enzyme (Fraction IX) is on the order of 0.1 to 0.2 µg per gram of starting cell pellet. This yield is of the order of magnitude (about 1 mg per kg of tissue) predicted earlier by Dr. F.J. Bollum (cited in (5)).

At each of the initial four steps of the purification procedure, minor amounts of DNA polymerase activity are fractionated away from the bulk activity and subsequently ignored. Thus far we have examined only the first DEAE cellulose flow through activity, which can represent up to 30% of the yield at this step and is observed both with fresh extracts and in those stored for several months at -70°. When this activity is chromatographed on phosphocellulose, hydroxylapatite and DNA cellulose, it behaves similarly to the main Fraction IV material, and it also appears to be similar in gross enzymatic properties. However, if this flow-through activity is re-chromatographed on DEAE cellulose in 10 mM NaPO$_4$, it adsorbs to the resin and can be eluted in two peaks at ionic strength roughly equivalent to 65 mM and 95 mM KPO$_4$. In this respect the fraction is very similar to the "Al" and "A2" species of polymerase α that Holmes et al. (12-14) observe as the predominant form of this activity in calf thymus extracts.

The behavior of polymerase α on phosphocellulose (Fig. 1A) is critically dependent on the immediately antecedent step. Fraction IV is loaded directly onto phosphocellulose and eluted as described for Fraction V, the complex chromatographic profile in Fig. 1B is reproducibly generated, with peaks of activity at 0.17 M, 0.21 M and 0.29 M KPO$_4$. The total recovery of polymerase activity relative to the crude extract is identical to that obtained with Fraction V, but the specific activities of the peak fractions are on order of magnitude lower than that of Fraction VI. Thus the second DEAE step not only simplifies the subsequent phosphocellulose profile, but it enhances the specific activity of the post-phosphocellulose fraction substantially.

The complex chromatographic behavior exhibited by DNA polymerase α preparations on DEAE cellulose and phosphocellulose, together with the heterodisperse activity profiles that are encountered in density gradient sedimentation and gel filtration (12-14), have been universally noted and have confounded past efforts to achieve the complete purification of this enzyme (2-5). As yet there is no entirely satisfactory explanation for this

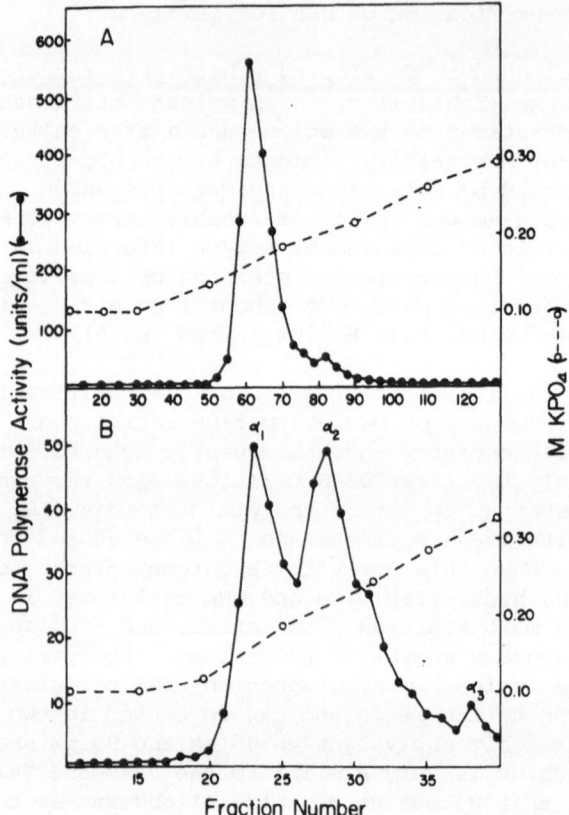

Figure 1. *Phosphocellulose chromatography of KB cell DNA*
polymerase α (11). *(A): Fraction V enzyme.* *(B): Fraction IV enzyme*

physical heterogeneity, nor for the important simplification that
is afforded by the seemingly redundant second DEAE cellulose step
in our procedure. It is worth noting, however, that similar, but
not identical, chromatographic complexity can also be observed
with nuclear preparations of polymerase α (15), and that the
chromatographic behavior of both the nuclear and the cytoplasmic
fractions appears to be modulated by (among other things) the
growth state of the cell culture.

In the experiment illustrated in Fig. 2, nuclear and cytoplasmic extracts were prepared (11,15) from KB cells harvested in
early log (Panels A,C) and late log (Panels B,D), respectively,
and were chromatographed on phosphocellulose after only a single
prior DEAE step. The cytoplasmic polymerase profiles (Panels C,D)
from early and late log cells are rather similar and demonstrate
the 0.17 M, 0.21 M and 0.29 M KPO_4 peaks that were noted in Fig 1B.

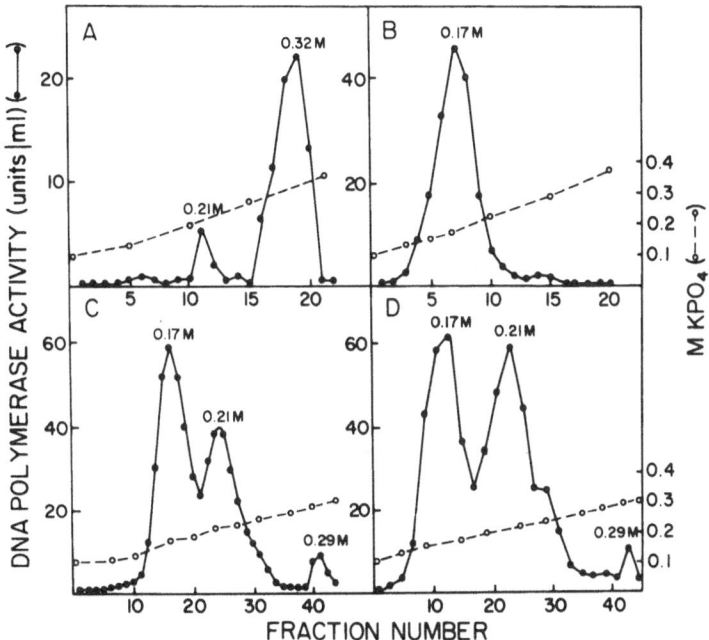

Figure 2. Phosphocellulose chromatography of KB cell nuclear and cytoplasmic fractions of DNA polymerase α. Cells were harvested at densities of 9.4 x 10⁴/ml (early log) or 2.8 x 10⁵/ml (late log), and nuclear and cytoplasmic crude extracts were prepared (11,15). The respective polymerase α activities were partially purified through only a single DEAE step before being chromatographed on phosphocellulose. *Panel A:* Nuclear α, early log. *Panel B:* Nuclear α, late log. *Panel C:* Cytoplasmic α, early log. *Panel D:* Cytoplasmic α, late log. Total polymerase α activity/g of cells was identical in each crude extract, and the recoveries of polymerase activity at prior steps of purification were the same in all four preparations.

The corresponding nuclear polymerase profiles (Panels A,C) differ dramatically, however. At high cell density there is essentially a single peak of nuclear α activity eluting at 0.17 M KPO_4, while at low cell density the 0.17 M peak is barely detectable and the predominant form of the enzyme activity is recovered at a unique elution position, 0.32 M KPO_4, that is not observed with the cytoplasmic fraction at all. We do not understand the basis of these intriguing changes in chromatographic profiles, but we feel that caution is appropriate in evaluating impure polymerase α chromatograms and in attempting to reach conclusions from them about novel or altered polymerase α species.

Physical Properties of DNA Polymerase α

When polymerase Fractions VI, VII and VIII were analyzed
electrophoretically on non-denaturing 5.0% and 8.0% polyacrylamide
gels (Fig. 3), the R_f values of the single DNA polymerase activity
peak in all three fractions were identical. Further gel analysis
of Fraction VIII (Fig. 4) revealed comigration of the enzyme activity
with a single discernible protein band, from densitometric
quantitation of which (relative to bovine serum albumin) it could be
concluded that the electrophoretically isolated polymerase Fraction
IX had a specific activity of >200,000 (nmol of labeled dTMP
incorporated per hour per mg of protein). Similar gels were run
at six additional concentrations of acrylamide, and in each instance
there was exact coincidence of the polymerase activity with the
protein band (Fig. 5), at a constant specific activity. No
additional protein species were resolved. The linear relationship
depicted in Fig. 5 establishes the electrophoretic homogeneity of
the Fraction VIII polymerase and permits the calculation of the

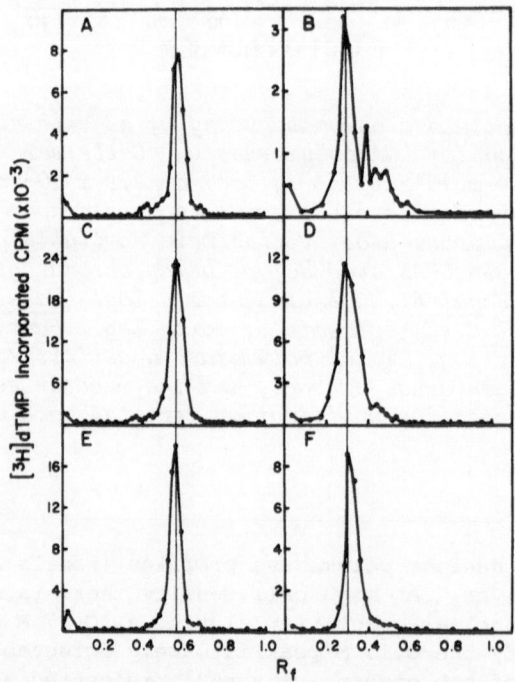

Figure 3. Non-denaturing polyacrylamide gel electrophoresis of DNA
polymerase α Fractions VI, VII and VIII (11). Panel A: Fraction VI,
5% gel. Panel B: Fraction VI, 8% gel. Panel C: Fraction VII, 5% gel.
Panel D: Fraction VII, 8% gel. Panel E: Fraction VIII, 5% gel.
Panel F: Fraction VIII, 8% gel.

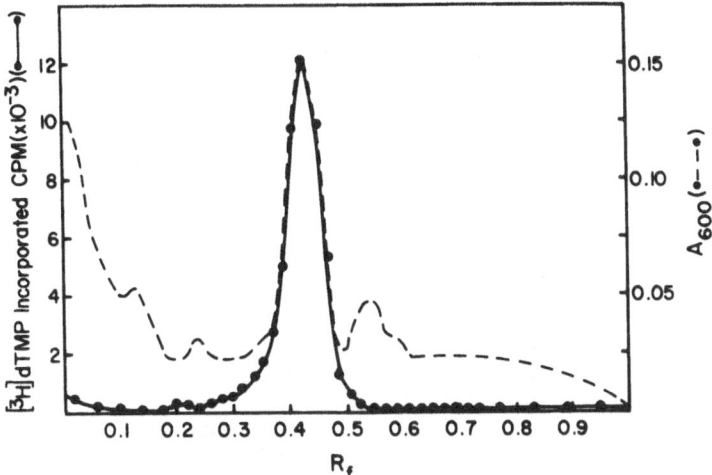

Figure 4. Non-denaturing polyacrylamide gel analysis of Fraction VIII polymerase α activity and protein (11).

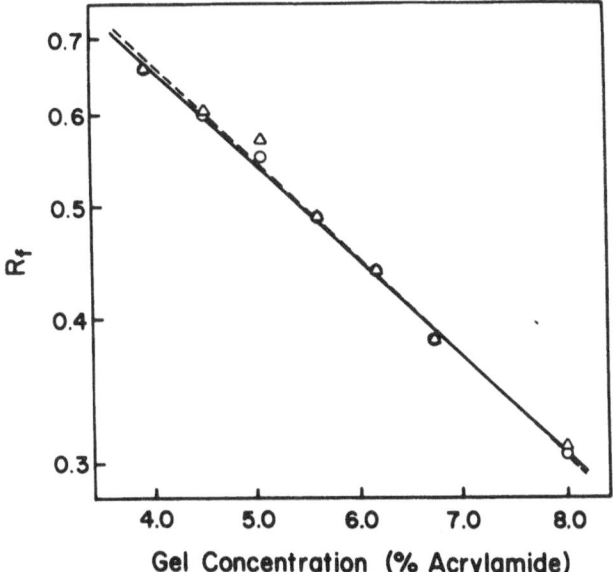

Gel Concentration (% Acrylamide)

Figure 5. Plot of mobility versus gel concentration for DNA polymerase α, Fraction VIII (11). Pairs of 3 mm polyacrylamide gels were formulated at acrylamide concentrations of 3.92%, 4.48%, 5.04%, 5.60%, 6.16%, 6.72% and 8%. One gel of each pair was stained for protein while the parallel gel was used to recover polymerase activity. The Figure shows the R_f values of the single discernible protein band on each stained gel (△----△) and the R_f values of the single peak of polymerase activity detected on the corresponding sister gel (○----○).

molecular radius of polymerase α (4.3 nm) (Fig. 6) and the
estimation of its molecular weight (265,000 to 280,000, assuming V̄
of 0.71 to 0.75).

Fractions VI and VIII behaved identically on Sephacryl S-200
gel filtration (Fig. 7) and eluted at positions consistent with an
apparent molecular weight of 140,000, a value of one-half of that
estimated from gel electrophoresis. Similarly, Fractions VI, VII,
VIII and IX sedimented identically in linear glycerol gradients at
low (Fig. 8) and high (100 mM KPO$_4$) ionic strength. The calculated
S values of 7.1 to 7.2 are in close agreement with the molecular
weight estimates obtained by gel filtration, and taken together,
these data suggest that polymerase α migrates as a homodimer under
conditions of gel electrophoresis. Isoelectric focusing of Fraction
VIII (Fig. 9) in a glycerol-stabilized pH 4 to 5 gradient reveals
a single sharp peak of polymerase activity at pI 5.0 to 5.2.
Polymerase α appears to be only marginally soluble at its isoelectric point, since focusing of large amounts of enzyme leads to
the appearance of a precipitate in the peak region.

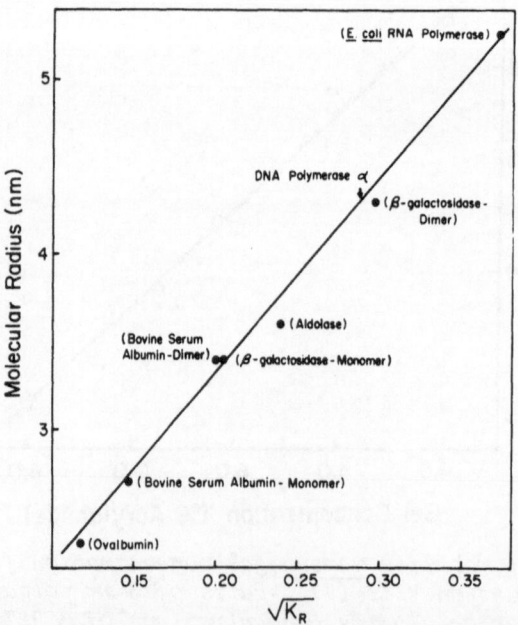

Figure 6. *Determination of molecular radius of DNA polymerase α
from non-denaturing gel electrophoresis (11). Analyses of standard
proteins were performed as in Fig. 5 and the negative slopes of the
straight lines thus determined (Kr). The molecular radii of the
standards were computed from published values of Mr and V̄.*

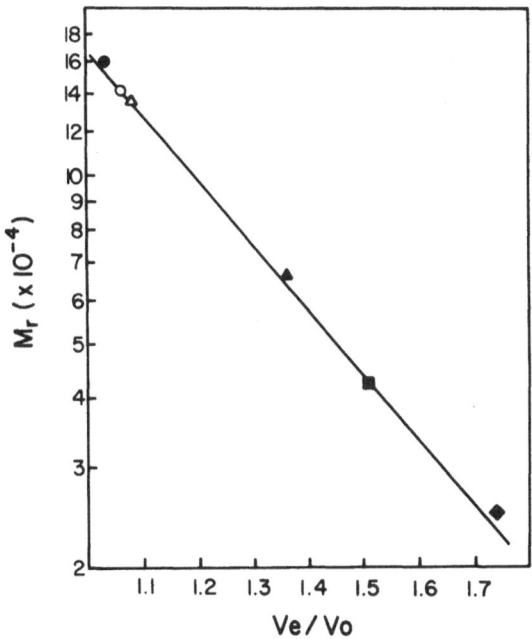

Figure 7. *Sephacryl S-200 gel filtration of polymerase α Fractions VI (△) and VIII (○) in 100 mM KPO₄. The standard proteins were bovine γ-globulin (●), bovine serum albumin (▲), ovalbumin (■), and chymotrypsinogen (◆). Vo was determined with blue dextran.*

When Fraction VIII was analyzed by denaturing gel electro-phoresis in the <u>absence</u> of a stacking gel, considerable variability was observed on different runs of the same material. Two predominant patterns of protein bands were encountered, with evidence of a transition between them. In one case there was a tight cluster of bands in the region from 135,000 to 150,000 daltons, while in the other there was a two-band pattern, with polypeptides in approximately equimolar ratios at 76,000 and 66,000 daltons. An intermediate pattern is illustrated in Fig. 10. The single discernible protein band at 142,000 daltons is

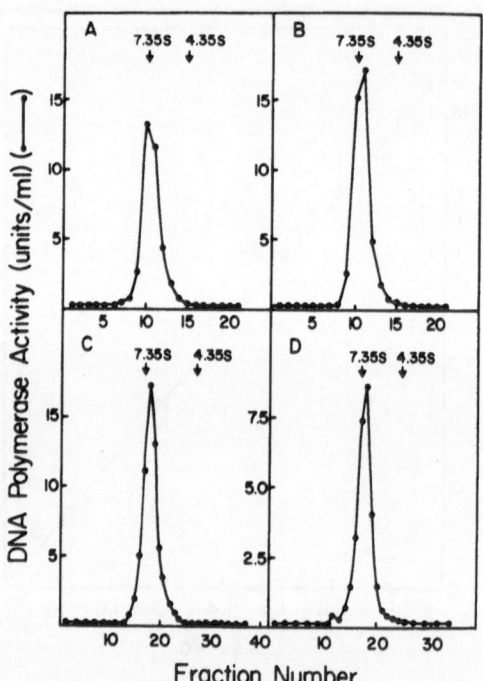

Figure 8. Glycerol gradient sedimentation of polymerase α in 10 mM KPO_4. The marker proteins were aldolase (7.35S) and bovine serum albumin (4.35S). *Panel A:* Fraction VI. *Panel B:* Fraction VII. *Panel C:* Fraction VIII. *Panel D:* Fraction IX.

accompanied by a smear of Coomassie blue staining material in the middle portion of the gel.

To resolve this problem of gel irreproducibility, we repeated these analyses in the presence of a stacking gel. By this method, examination of Fraction VIII revealed only the small band pattern (Fig. 11A) with two resolvable polypeptides at 76,000 and 66,000 daltons. Given the estimated molecular weight of the Fraction VIII polymerase activity of 140,000, and its electrophoretic homogeneity by native gel analysis, it was a reasonable interpretation that the two bands in Fig. 11A comprised the principal subunits of polymerase α.

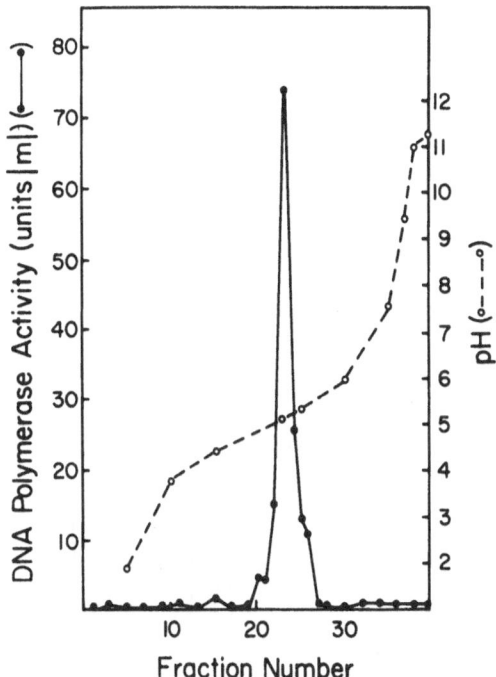

Figure 9. *Isoelectric focusing of polymerase α Fraction VIII (11).*

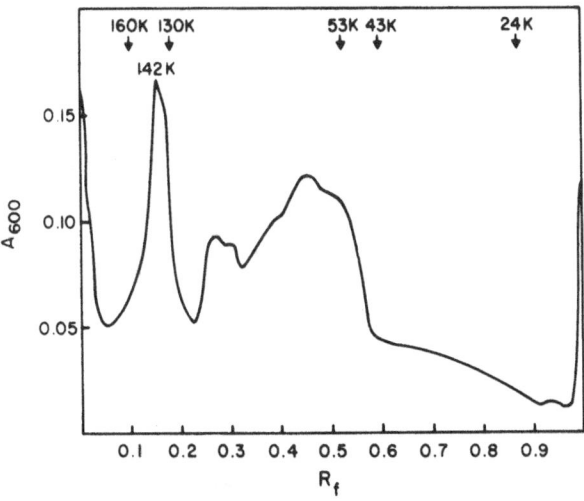

Figure 10. *Sodium dodecylsulfate gel analysis of polymerase α Fraction VIII. Electrophoresis was performed on an 8% polyacrylamide gel without stacking gel. The gels were stained with Coomassie blue and scanned at 600 nm. The standard proteins (↓) were bovine γ- globulin (16,000), β-galactosidase (130,000), bovine γ-globulin heavy chain (53,000), ovalbumin (43,000) and chymotrypsinogen (24,000).*

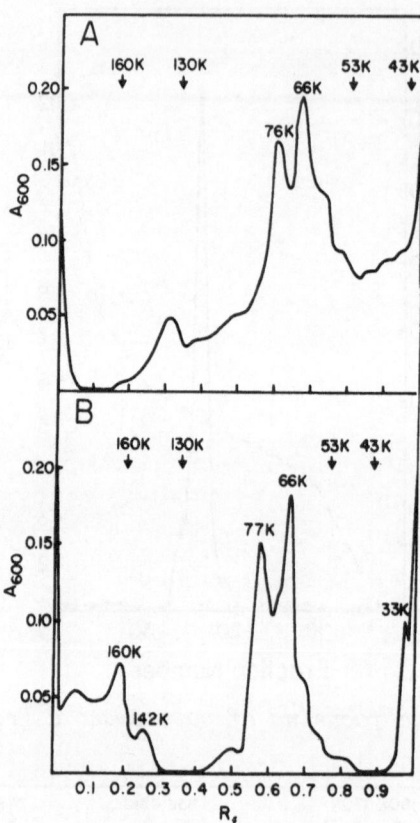

Figure 11. *Sodium dodecylsulfate analysis of DNA polymerase α (11). Panel A: Electrophoresis of Fraction VIII on an 8% polyacrylamide gel system that included a stacking gel. The Mr markers were as described in Fig. 10. The peaks labeled 76K and 66K were the only stained bands discernible by eye. Panel B: Fraction IX enzyme was analyzed as above. Only those stained bands that were detectable by eye are designated with their assigned molecular weights.*

        To confirm this interpretation, polymerase Fraction IX, generated at two different native gel concentrations, was transferred directly onto a sodium dodecylsulfate gel and subjected to electrophoresis (Fig. 11B). In each instance, the single band of protein associated with the polymerase activity gave rise to two bands on the denaturing gel, at 77,000 and 66,000 daltons, in approximately equimolar quantities. Moreover, the gel pattern was in excellent agreement with that obtained with Fraction VIII, except that most of the diffuse Coomassie blue staining background shown in Fig. 11A had been eliminated. The gel scan in Fig. 11B

displays a faint cluster of higher molecular weight bands in the
140,000 to 160,000 region, as well as a doublet near the dye front.
Since these components are not present in Fraction VIII, and given
both the denaturing gel variability that we have addressed and the
technical complexity of the gel transfer experiments, we believe
that these quantitatively insignificant components most likely
represent handling artifacts. Finally, when native gel slices
immediately flanking the polymerase band were transferred and
analyzed on denaturing gels, no stainable bands at all could be
detected.

Comment:

Attempts to deduce the structure of polymerase α have heretofore been based on two different experimental approaches, both of
them indirect and limited by the absence of homogeneous enzyme
fractions. In one approach, attention has focused on the remarkable
physical heterogeneity that is exhibited by impure polymerase
preparations and has led to the suggestions that polymerase α
exists in a multiplicity of partially interconvertible forms that
may have subtle differences in catalytic properties (4,5,10,12-14).
The second approach has involved denaturing polyacrylamide gel
analysis of partially purified polymerase fractions, and although
not definitive, has led to the suggestion that this enzyme may
be comprised of subunits of 70,000 to 90,000 molecular weight
(6,10,16,17).

In the purification scheme described here, the
serendipitous step of rechromatography on DEAE cellulose has
allowed us to follow over 80% of the polymerase α activity as a
single chromatographic entity of constant charge to mass ratio
through the remainder of the purification. This fact obviates
concerns about possible differential loss of polymerase sub-species
and permits confidence that our Fraction IX enzyme is in fact the
catalytically active core component of KB cell DNA polymerase α.
The native gel analyses demonstrate that our Fraction VIII enzyme
is electrophoretically homogeneous, and although the enzyme band
represents only 15 to 50% (in different gel runs) of the total
Coomassie blue stainable material on the gels, no other discrete
proteins can be detected in this fraction. The denaturing gel
studies indicate that the purified polymerase, with an estimated
Mr of 140,000, contains two subunits of 76,000 and 66,000 daltons,
respectively. It should be recognized that the difference between
this result and that reported by Holmes et al. (14) with the calf
thymus polymerase α may be more apparent than real, since, as
shown in Fig. 10, denaturing gel electrophoresis of our Fraction
VIII in the absence of a stacking gel yields a protein pattern
quite consistent with that obtained by the British group with a
similar procedure. The effect of the stacking gel is dramatic and

suggests that in its absence, the reduced polymerase subunits may reassociate during electrophoresis. We do not yet know whether either or both of the polymerase α subunits possesses independent catalytic activity.

Intracellular Localization of DNA Polymerase α (21)

It has been a consistent and somewhat perplexing observation that the majority of DNA polymerase α activity is usually recovered in the cytoplasmic fraction of aqueously extracted cells and tissues (4-6). During the past year, three alternative methods of tissue fractionation have been described (18-20) that purport to lead to the nearly quantitative recovery of α activity in the nuclear compartment. These techniques include the use of cyto-chalasin B to separate cultured cells into karyoplasts and cytoplasts (19); nonaqueous fractionation of lyophilized cells in 100% glyercol (20); and tissue extraction in unbuffered isotonic sucrose solutions containing 4 mM Ca^{2+} (18). Of these methods, the last appeared to be most convenient and suited to large scale biochemical applications.

When we attempted to apply the Ca^{2+}-sucrose method to the isolation of DNA polymerases from cultured human KB and lympho-blastoid cells, we obtained results substantially different from those reported by Lynch et al. (18). When freshly harvested KB cells are extracted under standard MG^{2+}-hypotonic conditions, 75 to 85% of the total DNA polymerase activity is typically recovered in the cytoplasmic fraction (Table II). If the extraction medium was made isotonic by addition of sucrose to 0.3 M, the distribution of polymerase activity was unchanged. However, extraction of the cells in the Ca^{2+}-sucrose medium of Lynch et al., or in 4 mM Ca^{2+} alone, reduced the recovery of cytoplasmic α activity by 80 to 95% without affecting the yield of nuclear activity. Further examin-ation of this phenomenon (Table III) demonstrated precisely similar findings with cultured WI-L2 lymphoblastoid cells, grown in a completely different culture medium, and indicated strongly that the loss of enzyme activity was almost certainly due to the presence of the calcium cation. Several control experiments were performed that demonstrated that the loss of cytoplasmic polymerase α could not be explained by Ca^{2+} inhibition of activity in the assays but rather was caused by the apparently irreversible inactivation of the enzyme.

Although the data in Table II indicated that Ca^{2+}-sucrose extraction did not affect the recovery of nuclear DNA polymerase activity, it seemed desirable to ensure that the nuclear crude extract assays were reliable indicators of total nuclear polymer-ase α activity and that we were not overlooking some qualitative alterations of nuclear enzyme activity that might be of interest.

Table II

Recovery of DNA Polymerase-α Activity in Nuclear and Cytoplasmic
Fractions of KB Cells (21)

Experiment	Extraction Medium	DNA Polymerase Activity	
		Nuclear (units/10^9nuclei)	Cytoplasmic (units/10^9cells)[a]
1	Mg^{++}-hypotonic	730	3600
2	Mg^{++}-sucrose	1010	3400
3	Ca^{++}-sucrose	870	100
4	Ca^{++}-hypotonic	1000	150

a. 2.1×10^8 cells per g (wet wt.)

KB cells were grown, harvested and disrupted as described (11). The
composition of the extraction media were: Mg^{++}-hypotonic - 5 mM
KPO_4, pH 7.5, 1 mM EDTA, 2 mM $MgCl_2$, 1 mM β-mercaptoethanol; Mg^{++}-
sucrose - identical to preceding, except for addition of 0.3 M
sucrose; Ca^{++}-sucrose - 0.3 M sucrose plus 4 mM $CaCl_2$ (18); Ca^{++}-
hypotonic - 4 mM $CaCl_2$ in H_2O. All extraction media contained in
addition 1 mM p-toluenesulfonylfluoride and 1% isopropanol.

Thus, nuclear extracts prepared by conventional and Ca^{2+}-sucrose
procedures were partially purified through the step of
phosphocellulose chromatography (15). The recoveries of polymerase
α activity at each purification step were identical for the two
extracts and as shown in Fig. 12, the activity profiles and yields
were also essentially the same at the final chromatography step.
This result serves to validate the crude extract assays and to
corroborate the conclusion that Ca^{2+}-sucrose extraction of KB cells
is without effect on the nuclear fraction of polymerase α.

Comment:

It is generally assumed that the recovery of the majority of
the DNA polymerase α activity of dividing cells in the cytoplasmic

Table III

Effects of extraction media on recovery of cytoplasmic DNA polymerase-α from KB and WI-L2 cells (21)

Experiment	Extraction Medium	Cytoplasmic DNA polymerase-α Activity			
		KB $(Units/10^9 cells)^a$		WI-L2 $(Units/10^9 cells)a$	
1	Mg^{++}-hypotonic	7950	(100%)	1940	(100%)
2	Mg^{++}-hypotonic minus KPO_4	-	-	2560	(132%)
3	2 mM $MgCl_2$	5740	(72%)	-	-
4	H_2O	7040	(89%)	-	-
5	4 mM $CaCl_2$	1420	(18%)	610	(31%)

a.   KB: 1.2×10^8 cells per g (wet wt.). WI-L2: 5.4×10^8 cells
per g (wet wt.). The discrepancy between Tables II and III in
yield of cytoplasmic activity per 10^9 KB cells is within our
normal range of variation and apparently reflects differences
in average cell size in different cultures. If computed on a
per g wet wt. basis, the yields of KB cytoplasmic polymerase
activity in the two Tables are roughly equivalent.

fraction is an artifact of nuclear leakage during aqueous tissue
fractionation. Yet it is noteworthy that with the very same
techniques, a number of investigators (15,22,23) have found that a
significant fraction of polymerase α activity remains firmly
associated with purified nuclei and can only be extracted at high
ionic strength. The basis for this nuclear-cytoplasmic partitioning
of enzyme activity is not clear, but it may reflect potentially

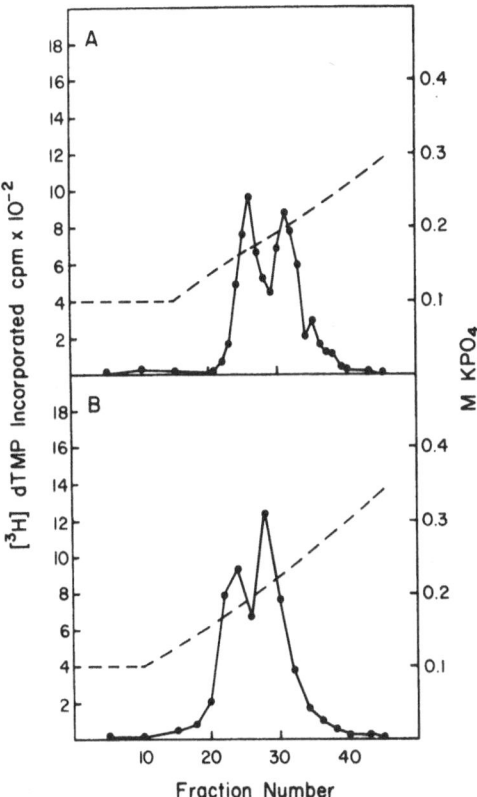

Figure 12. *Phosphocellulose column chromatography of DNA polymerase*
α activity purified from nuclei prepared in (A) Mg^{2+}-sucrose or (B)
Ca^{2+}-sucrose extraction media (as defined in Table II) (21). The
cells were harvested in the mid-log phase of the growth curve.

important distinctions among functionally and/or structurally
different polymerase α sub-classes. Certainly, for an enzyme that
is synthesized in the cytoplasm and which must migrate to the
nucleus to participate in DNA replication, the modulation of intracellular polymerase translocation could be an important component
in the regulation of DNA polymerase activity. For a variety of
reasons, the development of reliable methods with which to examine
the *in situ* compartmentalization of DNA polymerase α would be of
considerable importance. Of the three methods recently proposed
(18-20), the Ca^{2+}-sucrose procedure was particularly appealing
since it was based on a fractionation technique that is well
established in the cell biology literature (24,25).

        In our application of this method to cultured human cells, we
have obtained results that provide an alternative, and
disappointing, explanation for the apparent restriction of polymerase

α to the nuclear compartment, i.e., irreversible inactivation of
the majority, cytoplasmic fraction of the enzyme. Although our
findings are not strictly comparable to those obtained with
regenerating rat liver (18), the uncertainties of DNA polymerase
assays with crude extracts are such that we do not believe the data
reported in the rat liver study exclude the possibility that an
identical phenomenon was encountered. In any event it is a
reasonable conclusion that the Ca^{2+}-sucrose method of tissue
fractionation may be of limited applicability, and its use in the
study of DNA polymerase quantity and subcellular distribution can
lead to erroneous results.

The Effect of Mn^{2+} on the *In Vitro* Activity of
Human DNA Polymerase β (26)

The standard assay of DNA polymerase activity requires the
presence of a divalent cation, generally Mg^{2+} or Mn^{2+}, and an
appropriate primer-template, usually a suitable "activated" DNA
preparation or a variety of oligomer-initiated homopolymer systems
(3,4,6,27). Such assay conditions, although convenient, are of
limited interpretability and are unsuited to studies concerned
with the reconstruction of *in vitro* DNA replication systems in
which the use of defined natural DNA primer-template molecules is
essential. In previous reports (15,28,29) we have described the
structure and certain enzymatic properties of near homogeneous
DNA polymerase β obtained from cultured human KB cells. To obtain
further insight into the catalytic capabilities of this enzyme we
have attempted to extend these studies to the use of defined DNA
molecules. In our initial investigations of specifically gapped
Col El DNA, the cohesive ends of bacteriophage Ø80 DNA, and KB cell
closed circular D-loop mt DNA, we could not detect reactivity of
polymerase β under standard assay conditions with Mg^{2+} (15), but
we found that the enzyme was highly active with these primer-
templates in the presence of Mn^{2+}. More detailed analysis of this
phenomenon revealed that the substitution of Mn^{2+} for Mg^{2+} in the
standard assay has a profound effect on the kinetic parameters of
the polymerase β reaction. Of particular importance is the
demonstration that in incubations with low concentrations of defined
DNA primer-templates that contain relatively few primer-sites per
molecule, Mn^{2+} is the obligatory metal activator for this enzyme.

Purification of DNA Polymerase β from Adult Human Liver

In pursuing these catalytic studies, we have developed a
modified protocol for the purification of DNA polymerase β from
normal adult human livers, removed within 60 to 90 minutes of
"death" from cardiac transplant donors. Our purpose in using livers
was both to avail ourselves of an additional tissue source and to
determine whether the detailed characterization of human polymerase

β which had heretofore been passed exclusively on enzyme prepared from cultured KB cells (28,29), would prove true for the "normal" enzyme. Our initial efforts to apply our previously published procedure (15,28) to the isolation of polymerase β from purified human hepatic nuclei (26) were unsuccessful because the enzyme was found to chromatograph anomalously on DEAE-cellulose. Specifically a substantial fraction of polymerase β activity (∿50%) adsorbed to the resin and eluted coincidentially with DNA polymerase α in a 0.3 M KPO$_4$ (pH 8.2) step. It is of interest that in our experience with these normal livers, we find that ∿50% of the polymerase activity in purified nuclei, and ≥70% of the activity in whole liver homogenates, appears to be polymerase α. This result is at odds with reports that the normal liver (rat) lacks significant (or detectable) levels of this enzyme (18,30). Because the splitting of polymerase β activity proved disadvantageous to its further purification, we developed a new procedure that is modified from those earlier described by ourselves (28) and Chang (31).

The new purification scheme, summarized in Table IV, is equally applicable to KB cell and human hepatic nuclei, and it provides in the final fraction from either tissue an enzyme product in comparable yield and of comparable specific activity. The polymerase β activities from both sources behave identically at each purification step, and the Fraction VII preparations are essentially indistinguishable in their structural and enzymatic properties. The polymerases have identical sedimentation values in low salt glycerol gradients (∿3.5S) and elution positions from Sephacryl S-200 gel columns (coincident with ovalbumin), and they isoelectric focus identically at a pI of 9.2. The Fraction VII polymerases appear to be 50 to 70% pure, as estimated from Coomassie blue staining of SDS-polyacrylamide gel electropherograms (Fig. 13). In each instance, the mobility of the major protein band is the same within experimental error and is consistent with a size of 35,000 daltons. This size value is smaller than that of 43,000 to 45,000 previously determined for the KB (28) and calf thymus chromatin (31) polymerases, but very similar to the 32,000 value reported for the Novikoff hepatoma enzyme (32). In both of the earlier KB and calf thymus studies, polymerase β was found to migrate close to ovalbumin in denaturing polyacrylamide gels. In the present study, the R$_f$ values of these two proteins are again very similar, but the standard protein (a glycoprotein) migrates anomalously and does not fall on the calibration curve. The principal conclusion we wish to establish here is that the behavior of the KB and hepatic enzymes is identical.

With respect to enzymatic properties, the two polymerases show identical primer-template preferences, pH and divalent cation optima, response to salt, and resistance to sulfhydryl-blocking reagents (Table V). Both preparations are similarly free of contaminating DNase activities: 3' → 5'-exonuclease, <0.04% of the polymerase

Table IV

Purification of Human DNA Polymerase β

STEP	FRACTION	VOLUME[a] (ml)		ACTIVITY[a] (units)		PROTEIN[a] (mg)		SPECIFIC ACTIVITY (units/mg)		YIELD (%)	
		KB	liver	KB	liver	KB	liver	KB	liver	KB	liver
Crude Extract	I	140	350	3540[b]	4640[b]	350	490	10	9.5	-	-
First Phospho-cellulose	II	110	170	4050[b]	3860[b]	104	85	40	45	-	-
DEAE	III	85	120	280	1420	22	20	13	75	(100)	(100)
Second Phospho-cellulose	IV	2.5	3	140	700	3.6	10	40	70	50	50
Sephacryl S-200	V	2.6	13	85	660	1.5	3.4	60	190	30	46
DNA-Cellulose	VI	2.7	8.3	45	270	c	c	c	c	16	19
Isoelectric focusing	VII	1.8	10.8	40	270			~8000[d]	~8000[d]	14	19

a. 7.3×10^{10} nuclei from 84 g of KB cells and 10.6×10^{10} nuclei from 2160 g of human liver were used.

b. Fractions I and II contain substantial amounts of nuclear polymerase-α activity which is separated from polymerase β at the DEAE step. Thus the yield of polymerase β in Fraction III is set at 100%.

c. The polymerase activity in Fraction VI is extremely unstable and must be collected into bovine serum albumin. Therefore protein could not be determined at this step.

d. Estimated from Coomassie blue staining of SDS polyacrylamide gels.

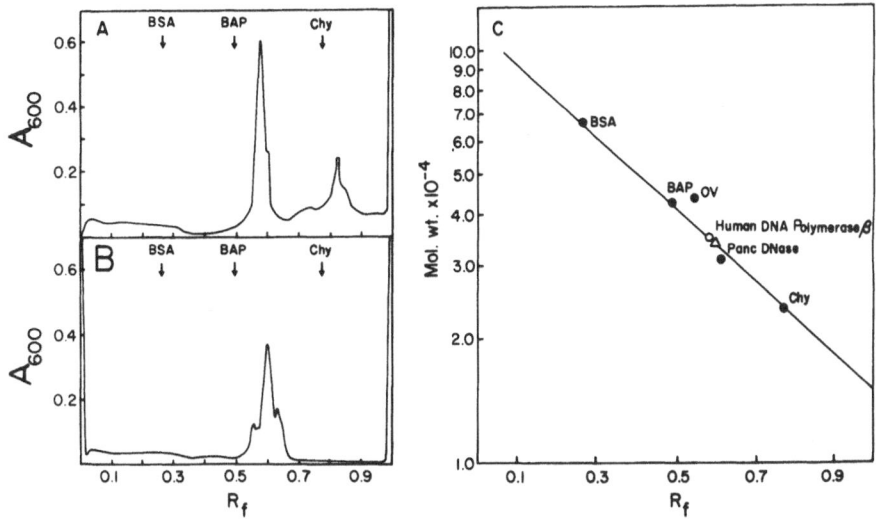

Figure 13.   *Sodium dodecylsulfate polyacrylamide gel electrophoresis*
of human DNA polymerase β (26). Panel A: Hepatic Fraction VII, 34
units.   Panel B: KB Fraction VII, 26 units.   Panel C: Standard semi-
log plot of Mr versus Rf.   Hepatic polymerase β (○);   KB polymerase
β (△).   The standard proteins are bovine serum albumin (BSA 67,000);
bacterial alkaline phosphatase (BAP, 42,500); ovalbumin (Ov, 43,000);
pancreatic DNase (Panc DNase, 31,000); and chymotrypsinogen (Chy,
24,000).

activity; $5' \rightarrow 3'$-exonuclease, $<6 \times 10^{-4}$%; and endonuclease activity,
$<10^{-5}$%. On the basis of these analyses, we conclude that the two
polymerase β preparations are indistinguishable, and we have used
both Fraction VII enzymes interchangeably in the catalytic studies
that follow. In a recent study of the phylogeny of the DNA
polymerases, Chang (33) found that enzymes of the polymerase α
class were ubiquitous among eukaryotes, but activity of the β class
appeared to have evolved only at the time of development of the
metazoa. Moreover, although the *in vivo* functions of the eukaryotic
DNA polymerases have not been clearly established, considerable
circumstantial evidence suggests that polymerase α is involved in
DNA replication, while polymerase β has been proposed to be a
"repair" enzyme (3,4,6). These considerations suggest that there
may have been less evolutionary pressure for the stringent
conservation of polymerase β than of α and led us to wonder whether
the enzyme that had been extensively characterized from the highly
aneuploid transformed KB cell line was prototypic of the normal
human polymerase. The results of this comparative examination
indicate that polymerase β has in fact been highly conserved in a
cell line that has been in continuous culture for over 20 years.

Table V

Comparison of Properties of KB Cell and Human
Hepatic DNA Polymerase-β

Primer-Template	Addition		Polymerase β Activity (%)	
			KB	Liver
Activated Salmon	None		(100)	(100)
Sperm DNA	KCl	20 mM	120	140
		40 mM	150	150
		80 mM	160	120
		120 mM	135	110
	NEM	2 mM	130	90
		4 mM	120	100
		10 mM	110	90
	pCMB	20 μM	60	50
		50 μM	20	50
	Ethanol	0.2%	140	100
		5%	110	100
		10%	100	90
		20%	50	30
	Urea	0.2 M	130	80
		0.4 M	80	100
		0.8 M	60	60
$(dA)n.(dT)\overline{16}$	KCl	20 mM	110	120
		100 mM	180	140
		200 mM	200	160
		500 mM	40	15
$(A)n.(dT)\overline{16}$	KCl	20 mM	120	120
		100 mM	310	280
		200 mM	300	180
		500 mM	60	20

0.10 unit of Fraction VI DNA polymerase β from KB cells or human
liver was assayed as described (15,29).

Effect of Mn^{2+} on Kinetic Parameters of DNA Polymerase β

At the relatively high concentrations of activated DNA (>100 µM in nucleotide) that are routinely used in the standard assay for polymerase β (15), Mg^{2+} is the preferred cation (Fig. 14), but at low DNA concentrations Mn^{2+} is preferred; at very low concentrations ($\leq$10 µM), this preference is striking (Fig. 14, insert). In the presence of Mn^{2+}, the apparent Km for polymerase β with activated DNA is <10 µM (Table VI), while with Mg^{2+}, this value is $\sim$300 µM. Table VI also shows that the two cations have comparable, although less dramatic, effects on the affinity of this enzyme for substrate dNTPs. The results of action mixing experiments are complex and are illustrated in Fig. 15. If the polymerization reaction is optimized for one divalent cation, the addition of increasing amounts of the second cation leads to substantial inhibition of the reaction. It is of interest that in Fig. 15A, 50% inhibition is produced by adding 1 mM Mg^{2+} to a reaction containing 1 mM Mg^{2+}, while in Fig. 15B, comparable inhibition is achieved by the introduction of 0.5 mM Mn^{2+} to an incubation containing 20 mM Mg^{2+}.

It has been generally recognized that in reactions of polymerase β with synthetic homopolymer primer-templates Mn^{2+} is the preferred

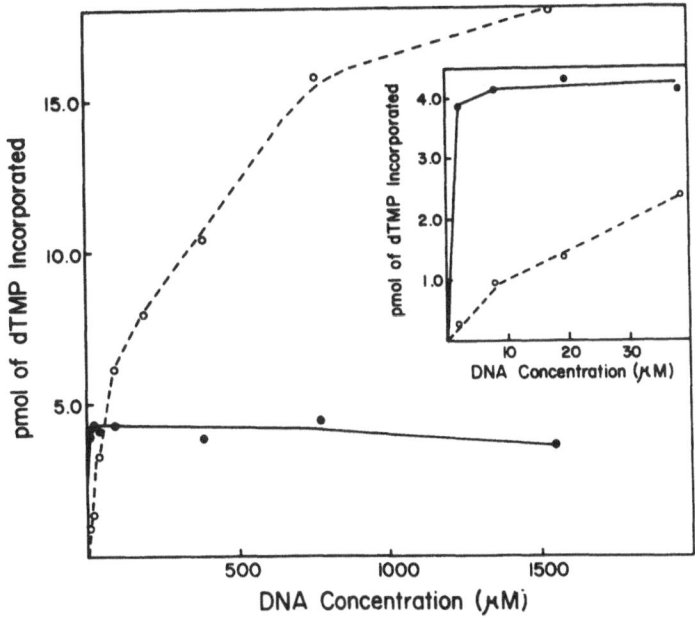

Figure 14. *The effect of divalent cation on utilization of activated DNA by DNA polymerase β (26). Assays were carried out in the presence of 20 mM $MgCl_2$ (O----O) or 5 mM $MnCl_2$ (●----●). The insert shows the results obtained at very low DNA concentrations.*

Table VI

Effect of Divalent Cation on Michaelis-Menton Constants (26)

Primer-Template	dNTP	Divalent Cation	Km	
			Primer Template (μM of nucleotide)	dNTP (μM of Each)
"Activated" Salmon Sperm DNA	4 dNTP	Mg^{+2}	300	160
		Mn^{+2}	<10	20
$(dA)n.(dT)\overline{16}$	dTTP	Mg^{+2}	7	(a)
		Mn^{+2}	2	10
$(A)n.(dT)\overline{16}$	dTTP	Mg^{+2}	(b)	(b)
		Mn^{+2}	53	500

(a) In the presence of Mg^{+2} and poly(dA).oligo(dT)$\overline{16}$, saturating levels of dTTP are not reached up to 500μM of the triphosphate.

(b) Human DNA polymerase β can not copy poly(A).oligo(dT)$\overline{16}$ in the presence of Mg^{+2}.

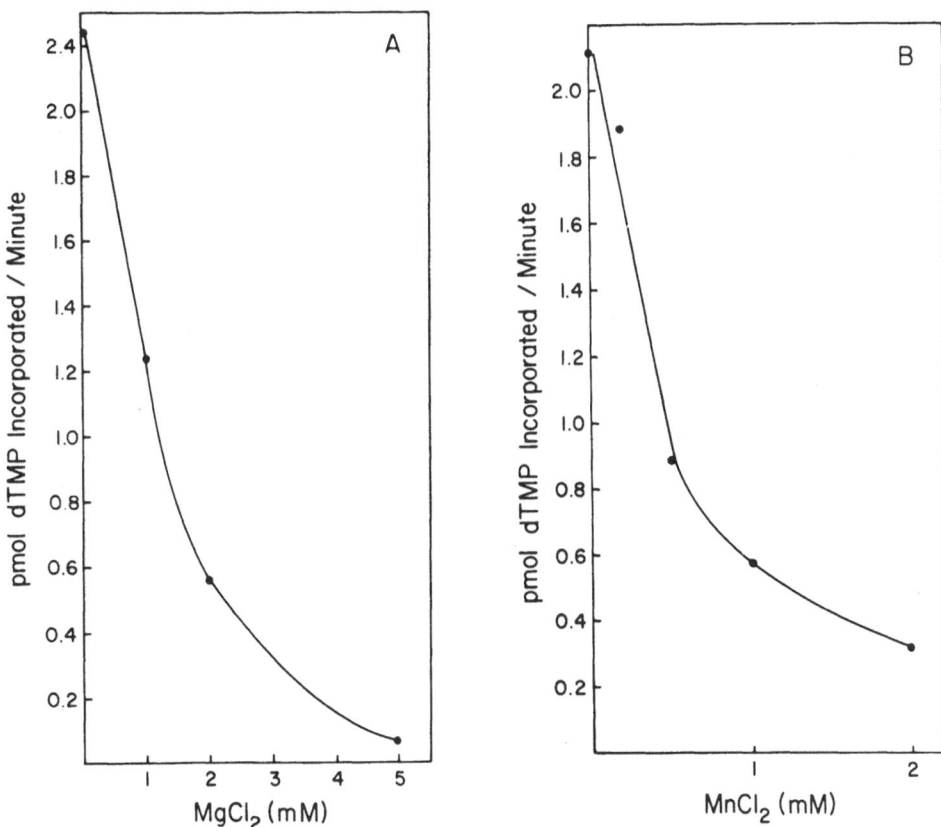

Figure 15. *Inhibition of the polymerization reaction by one divalent cation in the presence of the other (26). The reaction mixtures were set up identically, except that that in Panel A contained activated DNA, 50 μM, and 1 mM MnCl$_2$; while that in Panel B contained activated DNA, 500 μM, and 20 mM MgCl$_2$. The concentration of the other divalent cation was varied as indicated.*

cation. As demonstrated in Fig. 16, this preference is maintained over a wide range of concentration of the deoxyhomopolymer template although the relative advantage of Mn^{2+} is substantially decreased at very high primer-template concentrations. In the presence of Mn^{2+}, the apparent Km of the polymerase for the primer-template is about 3-fold lower than the value determined with Mg^{2+} (Table VI). It is curious that in reactions containing (dA)n.(dT)$\overline{16}$ a Km value for dTTP is readily measured in the presence of Mn^{2+}, while with Mg^{2+} as cation, saturating levels of dTTP are not

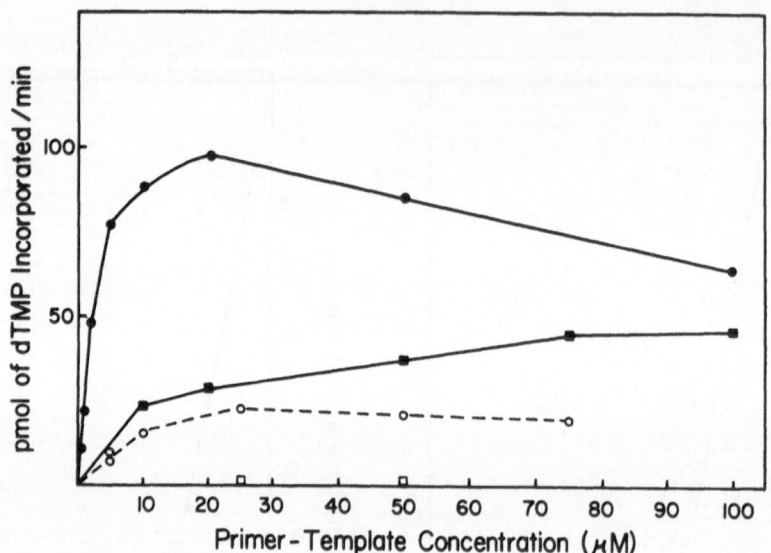

Figure 16. *The effect of divalent cation on utilization of synthetic primer-template by DNA polymerase β (26). Reactions were performed with poly(dA).oligo(dT)$\overline{16}$ at the indicated concentrations (as nucleotide) and either 0.5 mM MnCl$_2$ (●----●) or 5 mM MgCl$_2$ (○----○); or with poly(A).oligo(dT)$\overline{16}$ and either 1.5 mM MnCl$_2$ (■----■) or 5 mM MgCl$_2$ (□).*

attained at substrate levels up to 500 μM. In reactions with the initiated homoribopolymer, (A)n-(dT)$\overline{16}$, Mn^{2+} is absolutely required; polymerase β is completely unreactive with this primer-template in the presence of Mg^{2+} (Fig. 16). As shown in Table VI, in the reaction with the ribopolymer and Mn^{2+}, the apparent affinity of the enzyme for both primer-template and substrate dTTP is lower by 25- to 50-fold than that measured with the homologous deoxypolymer. Whether this reflects an intrinsic difference in the manner in which the polymerase protein interacts with these two synthetic primer-templates, or whether it is artifactual and explicable, for example, by differences in the conformation of the two polymers under incubation conditions is not clear.

The effect of divalent cation on the reactivity of polymerase β with DNA is most dramatically observed in assays employing defined DNA primer-templates that contain only a few primer sites per molecule (Fig. 17). The experiment illustrated in Fig. 17A is a control with activated salmon sperm DNA and demonstrates the strong preference of polymerase β for Mn^{2+} at low DNA concentrations. Assays using defined primer-template molecules are shown in Fig. 17B and 17C. With low concentrations of Ø80 DNA

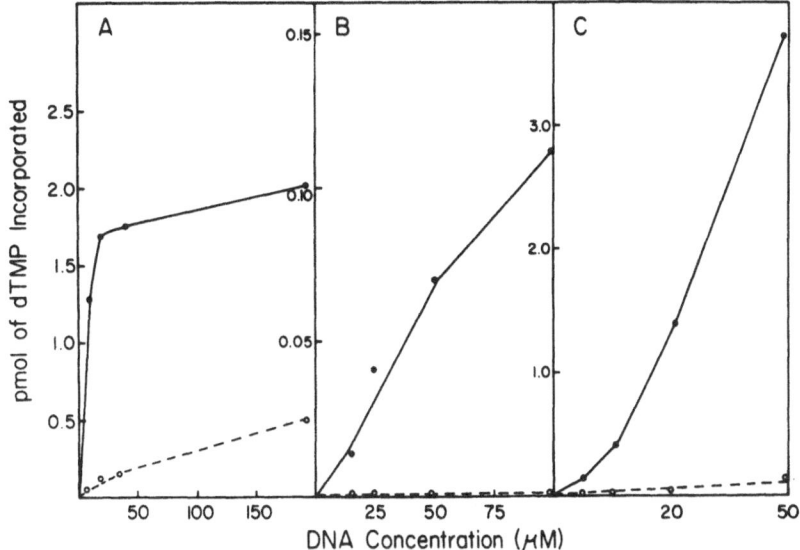

Figure 17. The effect of divalent cation on utilization of specific DNA primer-templates by DNA polymerase β (26). Panel A: activated DNA; Panel B: φ80 DNA; Panel C: "gapped" Col E1 DNA. $MnCl_2$ (●----●); $MgCl_2$ (○----○).

(Fig. 17B), in which the template consists of the two 12-nucleotide long cohesive ends per molecule, or of exonuclease III-gapped Col E1 DNA (Fig. 17C), the polymerase was inactive with Mg^{2+} but fully capable of promoting DNA synthesis in the presence of Mn^{2+}.

Comment:

On the basis of current models of the role of divalent cations in the DNA polymerization reaction, a satisfactory explanation for the results presented here is not readily apparent. Two different metals are recognized as obligatory participants in this enzymatic reaction, zinc, which is thought to function in the binding of polymerase to the 3'-terminus of the primer chain (34,35); and Mg^{2+} or Mn^{2+}, which appear to coordinate the binding and conformational alteration of the incoming dNTP within the active site of the polymerase molecule (35-37). In a recent magnetic resonance study of the interaction of Mn^{2+} with *E. coli* DNA polymerase I, Slater et al. (35) demonstrated the presence of multiple metal binding sites (1 tight, 4 ± 1 intermediate), all of which were postulated to be involved with the binding of deoxynucleotides but not with the binding of the polymerase to the primer-template. An additional class of weak binding sites (20 ± 4 per molecules) were thought to be possible sites of metal inhibition.

Despite the important insights that have been provided by
these studies, they clearly do not offer a complete explanation for
the diverse effects of divalent cations on the complex DNA polymerase reaction. For example, they do not address the important role
of these metals in determining the ability of different polymerases
to respond to a variety of natural and synthetic primer-templates.
Many studies with eukaryotic DNA polymerases have been concerned
with differential primer-template utilization (4,6) and this
parameter has been incorporated into the current system of nomenclature of these enzymes (1). It is generally accepted that Mg^{2+}
is the preferred cation with DNA, while Mn^{2+} may be preferred or
even required for reactivity with different homoribo- and homo-
deoxyribopolymers (4). However, the designation of Mg^{2+} as the
cation of choice in DNA templated reactions is based entirely on
the results of assays that employ "activated" (gapped) DNA, in
which the concentration of DNA nucleotides is very high and the
actual number and disposition of functional primer sites are
undefined (15,38). Although the mechanisms that are at play in the
phenomena that we have described here are obscure, our results are
of considerable practical significance. They specify for the first
time, feasible assay systems that will permit the use of defined
DNA primer-template molecules for more sophisticated catalytic
studies with polymerase β.

In Vitro Replication of KB Cell Closed Circular Mitochondrial
DNA by Human DNA Polymerase β (39)

Recognition of the profound effect of Mn^{2+} on the apparent
affinity of DNA polymerase β for DNA provides novel opportunities
for the detailed study of the catalytic properties of this enzyme
with defined DNA molecules of known structure. In following up
the observations described in the previous section of this paper,
we discovered that in the presence of Mn^{2+} polymerase β was capable
of performing substantial synthesis on KB cell closed circular
mtDNA. This reaction was of particular interest for at least two
reasons. First, closed circular mtDNA can be readily obtained in
reasonable yield at >95% purity from cultured eukaryotic cells.
Second, preparations of closed circular mammalian mtDNA contain a
significant fraction of D-loop species (40-42). These comprise a
unique population of naturally primed duplex circular DNA molecules
whose mechanism of *in vivo* replication has been described in great
detail (40,43,44).

In this section we shall describe the initial results of our
study of the partial *in vitro* replication of mtDNA by polymerase
β. We demonstrate that in the presence of Mn^{2+} this enzyme, in the
absence of any additional protein factors, carries out a highly

selective synthetic reaction consisting of the specific elongation
of the D-loop primer fragment, followed by the replication of the
displaced parental template strand. The point at which the enzyme
switches template strands is most likely that at which all negative
superhelical turns have been removed and an energetically unfavorable
introduction of positive superhelical turns would be required for
further synthesis on the initial parental template. The product
of the reaction is an enlarged D-loop that has been converted
predominantly to a fully duplex structure. The highly purified
polymerase is free of contaminating or associated DNase activities,
and most likely for this reason the enzyme neither recognizes nor
generates any additional "illegitimate" sites of deoxynucleotide
inorporation in the primer-template population.

Utilization of KB Cells mtDNA by DNA Polymerase β

The reaction of DNA polymerase β with twice-purified closed
circular mtDNA in the presence of Mn^{2+} is illustrated in Fig. 18A.
By contrast, if Mg^{2+} is substituted for Mn^{2+} under otherwise
identical conditions, no polymerization is detectable. Examination
of the kinetics of the reaction at high (100 μM) and low (3 μM)
dNTP concentrations (Fig. 18B) reveals an initial interval of rapid
synthesis that is followed by a plateau, during which there is no
loss of incorporated radioactivity. Addition of fresh enzyme to
the reaction after 120 minutes leads to no further incorporation;
thus the limited extent of replication is not due to enzyme
inactivation. Electron microscopic examination of the unreacted
mtDNA preparation (Table VII), purified by two cycles of dye-
buoyant density centrifugation, revealed a population of closed
circular monomeric (79 ± 6%) and catenated (21 ± 6%) DNA molecules,
estimated to be >95% pure by mass. 14 ± 2.7% of all monomers and
monomeric subunits of catenanes contained D-loops. On the
assumption (to be proved below) that only D-loop molecules are
utilized in the enzymatic reaction, and given a Mr of human
monomeric mtDNA of 10.9×10^6 (42,45), the data in Fig. 18B indicate
that the limits of incorporation at low and high dNTP concentrations
are 350 and 1400 nucleotides per D-loop molecule, respectively.

Analysis of the Synthetic Product by Buoyant Density and Velocity Sedimentation

The label that is incorporated in the *in vitro* polymerization
reaction at low or high dNTP concentration is distributed
identically in three regions of the dye-buoyant density gradient
(Fig. 19), designated as lower, middle and upper bands,
respectively. Studies of *in vivo* replication of mouse mtDNA (40,
43) have established that the lower band contains closed circular
molecules with and without standard D-loops; the middle band is
enriched for replicating (expanded D-loop) forms; and the upper band

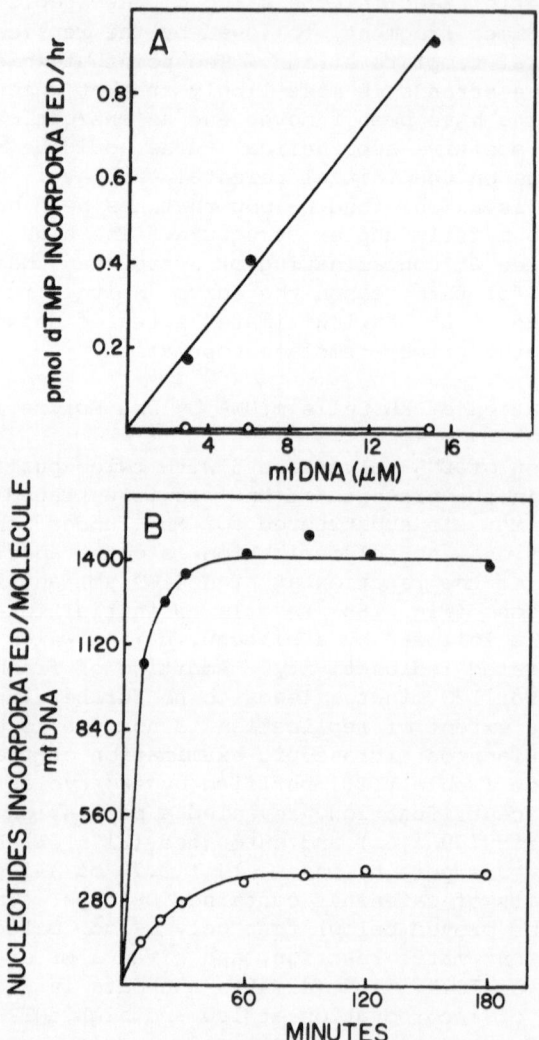

Figure 18. *Utilization of KB cell closed circular mtDNA by human DNA polymerase β (39). The DNA was purified by two cycles of centrifugation in ethidium bromide-CsCl gradients. Panel A: Reactions contained either 5 mM MnCl$_2$ (●----●) or 20 mM MgCl$_2$ (○----○). Incubations were for 10 minutes at 37°. Panel B: Reactions contained 20 μM mtDNA, 2.8 units of DNA polymerase β and either 3 μM dNTPs (○----○) or 100 μM dNTPs (●----●). For purposes of computation, the mass of the KB cell mtDNA monomer was taken as 10.9 x 10^6 daltons (42,45) and polymerization was assumed to occur only on D-loop molecules, which comprised 14% of the initial DNA population (Table VII).*

Table VII

Frequency, Size and Morphology of D-loops in KB Cell
Lower Band mtDNA[a] (39)

Type of D-loop	Unreacted mtDNA[b]	Reacted mtDNA[b] (100 μM dNTPs)
Standard[c] D-loop	14 ± 2.7%	1.5 ± 0.8%
Enlarged duplex[d] D-loop	<0.1%	10 ± 2%
Enlarged duplex D-loop[c] with extruded tails	<0.1%	6 ± 1.7%
Normalized D-loop length	1.0	1.5_5

(a) *In vitro* replicated mtDNA was prepared and purified by re-banding in an ethidium bromide-CsCl density gradient as described in Fig. 19.

(b) All frequencies have been derived by grouping monomer mtDNA and monomeric subunits of catenated mtDNA. 618 molecules in the unreacted DNA sample and 800 molecules in the reacted DNA preparation were examined by electron microscopy.

(c) The standard D-loop is a duplex region comprised of a primer fragment hydrogen bonded to a parental template strand. The opposite parental strand is displaced as a single strand. In KB cell mtDNA the naturally occurring D-loop is ∿600 nucleotides long (Gillum, A., and Clayton, D.A., unpublished observations). The size and standard deviation of unreacted mtDNA D-loops was 2.62 ± 0.32 arbitrary length units. (See Fig. 22A).

(d) Enlarged duplex D-loops are double-stranded over ≥80% of the entire D-loop. The size and standard deviation of reacted mtDNA D-loops was 4.05 ± 0.54 arbitrary length units. (See Fig. 22B).

(e) Illustrated in Fig. 22C.

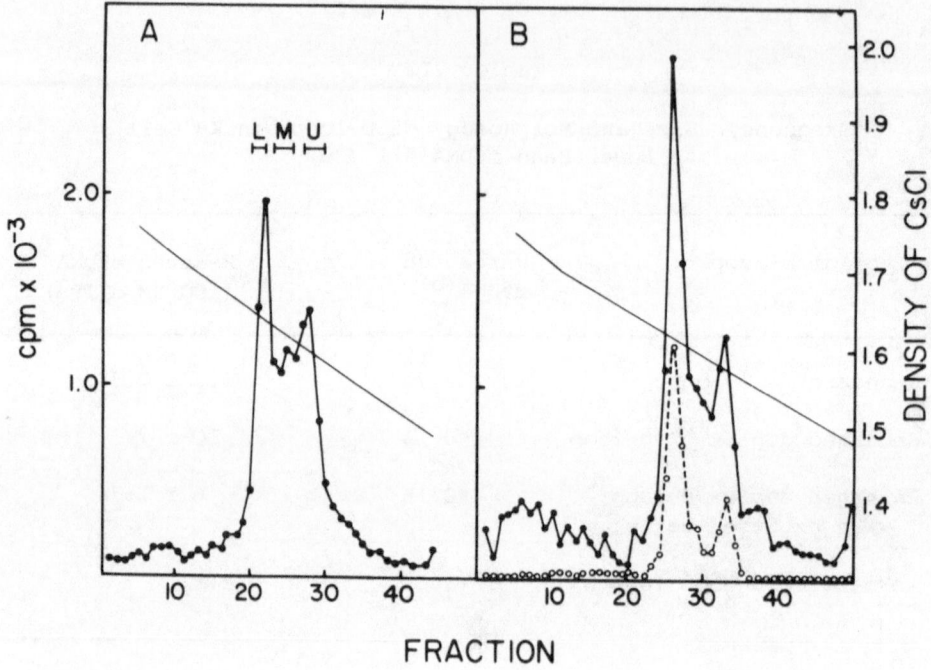

Figure 19. Ethidium bromide-CsCl buoyant density centrifugation of in vitro replicated mtDNA (39). The reactions contained (A) 50 μM mtDNA, 3 μM dNTPs, (α-^{32}P)dTTP (3.8 x 10^4 cpm/pmol); (B) 21 μM (^{3}H)mtDNA, 100 μM dNTPs, (α-^{32}P)dTTP (758 cpm/pmol). Incubations were for 60 minutes at 37°. The bars in Panel A indicate the fractions that were pooled as lower (L), middle (M), and upper (U) bands for further study. Panel A: ^{32}P, ●----●; Panel B: ^{32}P, ●----● ^{3}H, ○----○.

contains predominantly nicked and gapped circles. The pattern of radioactivity in Fig. 19A is highly reproducible and generally reveals ∼2/3 of the label in lower band and ∼1/3 of the label in upper band. The resolution in these gradients is not sufficient to permit confident assessment of the middle band region.

Since upper band is known to contain open circles, it was important to determine whether the polymerase was nicking the primer-template during the course of incubation. Several control experiments with *in vivo* labeled KB cell mtDNA were performed, including mock incubations of primer-template alone and with enzyme in the absence of dNTPs. The results (data not shown) indicate that with different preparations, handling of the mtDNA alone could result in conversion of 15 to 30% of the initially lower band material to upper band material (presumably due to nicking), but

this shift was not enhanced by the polymerase. In an experiment
with (^{3}H)mtDNA primer-template and (α-^{32}P)dTTP (Fig. 19B) it was
possible to determine the ratio of ^{32}P to ^{3}H across the buoyant
density gradient and demonstrate that there was only a small
($\leq$2-fold) preference of polymerase β to utilize the nicked molecules.
We conclude from these analyses that at least the majority of upper
band product in Fig. 19 results from polymerization on nicked DNA
molecules that are generated spontaneously by storage and handling.

     The lower, middle and upper band fractions (Fig. 19A) were
separately analyzed by velocity sedimentation, before and after
brief heat denaturation to cause the selective melting of the D-loop
primer fragment (40) (Fig. 20). As a whole, the patterns shown in

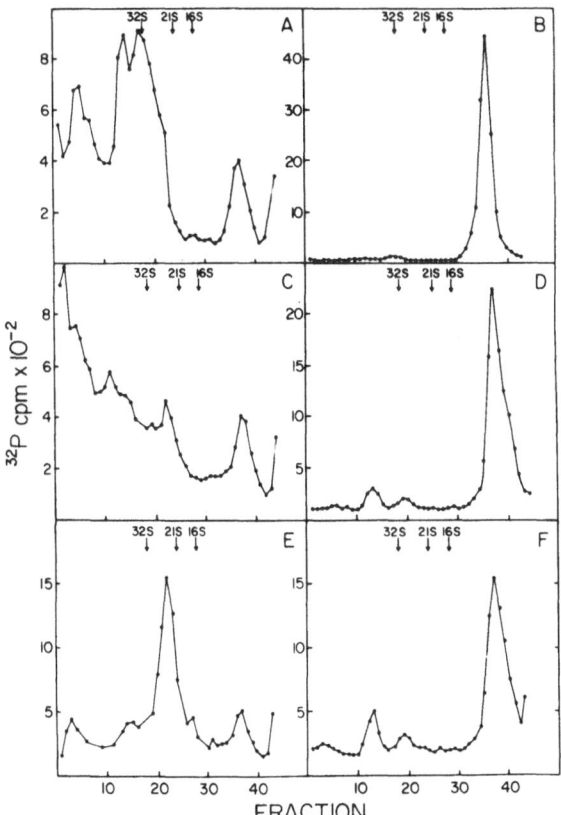

*Figure 20. Velocity sedimentation of in vitro replicated mtDNA (39).
The isolated lower, middle and upper band fractions in Fig. 19A were
divided in half, and one sample of each pair was heated at 100° for
60 seconds and rapidly quenched. The sedimentation markers were
phage T7 DNA (32S) and SV40 Form I (21S) and Form II (16S) DNA. Lower
band mtDNA: Panel A, native; Panel B, melted. Middle band mtDNA:
Panel C, native; Panel D, melted. Upper band mtDNA: Panel E, native;
Panel F, melted.*

Fig. 20 are remarkably similar to those reported by Berk and
Clayton (40) in their study of *in vivo* pulse-labeled mouse mtDNA.
Most of the label recovered from lower band (Fig. 20A) sediments
at >30S, a rate consistent with that of full length mtDNA monomers
and catenated oligomers; a small amount of slowly sedimenting, 9S,
material is present. After melting, however, (Fig. 20B) all of the
label migrates as 9S species. Thus, most of the radioactivity that
sediments at >30S before heating must be in short D-loop fragments.

The middle band fraction sediments vary heterogeneously in
sucrose gradients (Fig. 20C), with the bulk of the label in large
structures of >26S. This fraction is not well resolved in the
ethidium bromide-CsCl gradient, and the heterodisperse sedimentation pattern is consistent with the presence of an admixture of
open circles, closed circular monomers and catenanes of variable
complexity. After heat denaturation (Fig. 20D), however, almost
all of the label is again recovered as 9S species. Finally,
identical analysis of the upper band fraction shows that most of
the label sediments at 27S (Fig. 20E), the position expected for
open circles, while after melting (Fig. 20F), the radioactivity
is again displayed in 9S species. From these results it can be
concluded that all of the incorporated label that is recovered
from the buoyant density gradient is in discrete forms of low
molecular weight, 9S, DNA that, prior to heating, are associated
with intact monomeric and oligomeric mtDNA molecules. The data
permit the interpretation that polymerase β specifically recognizes
and elongates the primer strand of the D-loop. No other sites
of nucleotide incorporation are recognized or generated by the
enzyme in the closed circular primer-template population.

Analysis of Replicated mtDNA in Alkaline CsCl Gradients

The 7S primer sequence in mouse D-loop mtDNA has been shown
to be heavy (H) strand (40,41). From *in vivo* labeling studies it
is known that this strand is elongated unidirectionally to at least
0.6 genome unit before light (L) strand synthesis begins on the
displaced parental H-strand (43). Thus, *in vivo* pulse-labeling
of mouse mtDNA (40) leads to incorporation of label exclusively
into H-strand products. The mechanism of human mtDNA replication
has not been elucidated in comparable detail, and in fact, the
density of the human D-loop primer sequence has not yet been
demonstrated. The results of analysis of the *in vitro* replicated
mtDNA in alkaline CsCl gradients are shown in Fig. 21. *In vivo*
labeled KB cell (^{3}H)mtDNA served as a reference marker and was
resolved into H- and L-strands, as expected (46). The labeled
replication products, in which synthesis had occurred to an average
of <50, 350 and 1400 nucleotides per primer chain (Panels A, B and
C), respectively, all banded heterogeneously over both the heavy
and light density regions of the gradient. With increased extent

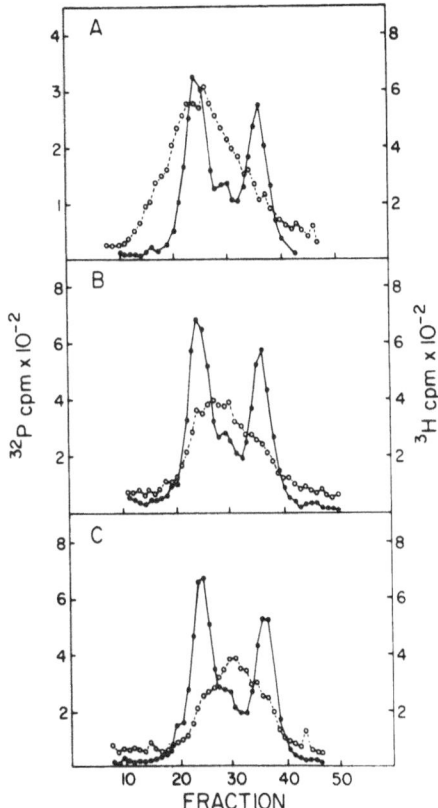

Figure 21. Alkaline CsCl equilibrium of *in vitro* replicated mtDNA
(39). mtDNA was incubated with DNA polymerase β under conditions
that led to the incorporation of an average of <50, 350 and 1400
nucleotides per primer chain (Panels A, B and C). KB cell (^{3}H)mtDNA,
labeled *in vivo*, was used as a marker (●----●); ^{32}P-labeled mtDNA,
replicated *in vitro* (○----○).

of synthesis, however, the average density of the product species
appeared to be skewed progressively toward the L-strand position.
This result suggests that the KB cell D-loop primer fragment is
comprised of both heavy and light density sequences and the sequences
that are synthesized in the limited polymerization reaction are
predominantly of light density. The data are also consistent with
the interpretation that either copying of the displaced parental
strand, or fold-back copying of the primer strand itself from its
3'-terminus (which is formally indistinguishable), occurs to a
significant degree during the *in vitro* reaction. The latter
possibilities are supported by the finding that the melted out
reaction products, isolated by velocity sedimentation, are from 50

to 70% resistant to digestion by S1, the single-strand specific
nuclease. This indicates that the elongated primer chains that
are synthesized by polymerase β in D-loop mtDNA contain rapidly
annealing regions of self-complementarity.

Electron Microscopic Analysis of KB Cell mtDNA
Before and After *In Vivo* Replication

D-loop mtDNA can be characterized by its formal appearance
when displayed for electron microscopy. One arm of the D-loop
region is duplex due to the presence of the complementary primer
chain (H-strand in the mouse), and the other arm is the
corresponding displaced parental strand (Fig. 22A). Since the
analyses presented above indicated that polymerase β was
specifically extending the primer fragment, we examined the reaction
products for alteration in the D-loop region. As shown in Table
VII and Fig. 22, two major differences were detected between control
and replicated mtDNA preparations. First, almost all of the D-loops
in the reacted molecules appeared to be fully duplex (Fig. 22B), and
a significant number showed extruded tails at one end of the D-loop
(Fig. 22C). These tails had both single- and double-stranded
character. Second, the size of the D-loop in reacted mtDNA is
approximately 1.5 times that of naturally occurring D-loops. The
simplest explanation for these observations is that the enzyme
extends the primer fragment by ∿300 nucleotides and then switches
to the displaced template strand and proceeds to polymerase an
additional ∿900 nucleotides in the opposite direction. The extent
of synthesis suggested by the ultrastructural analysis is in
excellent agreement with the limit of dNMP incorporation that was
illustrated in Fig. 18B. A plausible mechanism for the formation
of tails is branch migration of the 5'-end of the original primer
fragment as a free end, which could serve as a template for the
polymerase upon its postulated return to the origin of the D-loop
and support the incorporation of an additional small number of
nucleotides. The net effect of branch migration would be to
maintain the enlarged D-loop at a constant size, but the position
of the D-loop on the closed circular DNA molecule would be
displaced in the 5' → 3' direction.

Comment:

In the absence of any added protein factors, highly purified
human DNA polymerase β can carry out a specific, partial
replication of closed circular D-loop mtDNA molecules. The data
presented from combined biochemical and electron microscopic
analyses of the reaction product indicate that the polymerase
catalyzes a limited elongation of the 9S primer chain and then
switches strands to copy the displaced parental template strand.

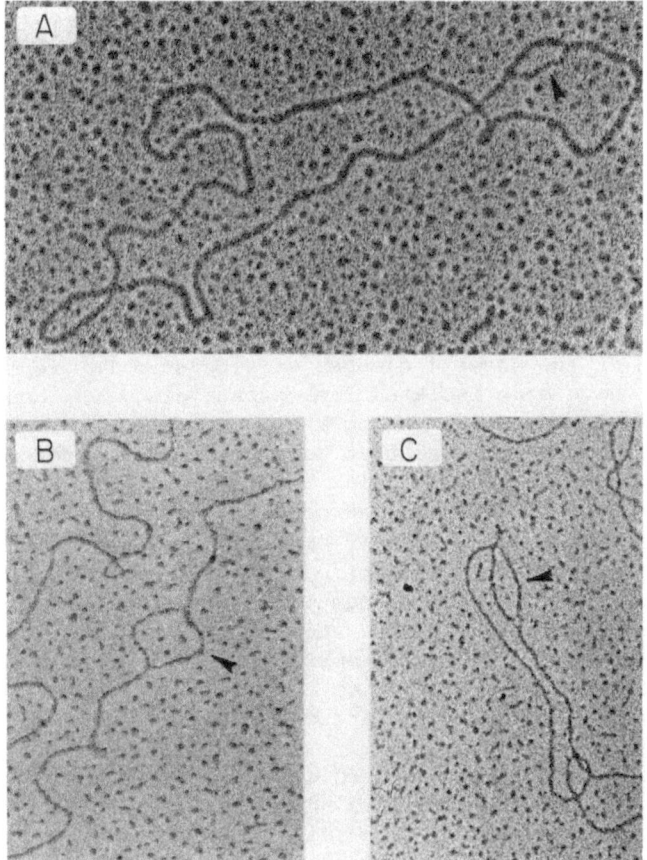

Figure 22. Electron micrographs of unreacted and in vitro replicated D-loops. (A): Unreacted KB cell mtDNA. The arrow designates the single-stranded arm of a standard D-loop, the size of which is ∿600 nucleotides. (B): Reacted mtDNA. The arrow designates a fully duplex D-loop, approximately 1.5 times larger than the standard D-loop. (C): Reacted mtDNA. The arrow designates an enlarged, fully duplex D-loop with a small extruded tail. DNA samples were prepared for electron microscopy by the formamide technique (43).

The product of the reaction is a fully duplex D-loop that has been enlarged approximately 1.5-fold.

From the known superhelix density of human (HeLa) mtDNA, -0.056 (47), it can be calculated that the closed circular DNA molecule of 16.6 kilobase pairs contains ∿90 negative superhelical turns, and thus the unreacted D-loop molecule has the equivalent of ∿30 negative superhelical turns. From the extent of the

polymerization reaction, it is likely that the 9S, ∿600
nucleotide long primer fragment is lengthened by about 50% (∿300
nucleotides) before duplication of the displaced template strand
begins. This degree of elongation would be expected to be
accompanied by the removal of ∿30 superhelical turns (48),
suggesting that the primer-elongation phase of the reaction ceases
when the superhelix density of the DNA molecule has been reduced
to zero. Continuation of synthesis on the original template
strand beyond this point would require the introduction of
energetically unfavorable positive superhelical turns.

It will clearly be of interest to determine the effects on
this reaction of the several classes of accessory DNA replication
proteins that have been isolated from prokaryotic (49) and
eukaryotic (50-54) sources. Among these, the effect of the ATP-
dependent DNA gyrase activity that has recently been described in
E. coli (55) would be of particular interest, since that enzyme
has been shown specifically to introduce negative superhelical
turns into relaxed closed circular DNA. The re-introduction of
such turns into the partially replicated mtDNA might be expected
to drive the primer-elongation reaction, in a stoichiometry of 10
nucleotides incorporated for each negative turn that is
subsequently removed, and in the appropriately coupled reaction it
may be possible for polymerase β to accomplish the complete
replication of the initial template strand.

In recent years there has been considerable effort to develop
appropriate *in vitro* systems with which to employ purified
soluble proteins to explore the detailed biochemistry of DNA
replication (49). Significant progress has thus far been realized
only with prokaryotic systems in which the availability of defined
phage DNA primer-templates and a large number of *E. coli* mutants
thermosensitive for DNA replication has been of inestimable benefit.
Substantial insight has been gained into the mechanisms of
replication of relatively simple, single-stranded circular DNA
molecules (56,57), but the replication of duplex circular (58,59)
and linear (60) phage genomes is less completely understood. With
respect to eukaryotes, the biochemical basis of DNA replication
is much less clear, and the design of useful *in vitro* systems
has been largely limited to purified intact and permeabilized
nuclei from normal or virus-infected cells (61-65). The highly
specific and delimited reaction that is carried out by polymerase
β with D-loop mtDNA suggests that this might be a particularly
useful model system with which to study the *in vitro* replication of
duplex circular DNA with purified eukaryotic (or prokaryotic)
components.

REFERENCES

(1) Weissbach, A., Baltimore, D., Bollum, F.J., Gallo, R. and Korn, D. (1975) Science 190:401-402; Eur. J. Biochem. 50:1-2

(2) Bollum, F.J. (1960) J. Biol. Chem. 235:2399-2403

(3) Loeb, L.A. (1974) In: The Enzymes (ed. P.D. Boyer) 10:173-209, Academic Press, New York.

(4) Bollum, F.J. (1975) In: Prog. Nuc. Acid Res. and Molec. Biol. (ed. W.E. Cohn) 15:109-144, Academic Press, New York.

(5) Holmes, A.M. and Johnston, I.R. (1975) FEBS Letters 60:233-243

(6) Weissbach, A. (1975) Cell 5:101-108

(7) Weissbach, A. (1977) Ann. Rev. Biochem. in press

(8) Chang, L.M.S. and Bollum, F.J. (1973) J. Biol. Chem. 248: 3398-3404

(9) Brun, G., Rougeon, F., Lauber, M. and Chapeville, F. (1974) Eur. J. Biochem. 41:241-251

(10) Sedwick, W.D., Wang, T.S-F. and Korn, D. (1975) J. Biol. Chem. 250:7045-7056

(11) Fisher, P.A. and Korn, D. (1977) J. Biol. Chem. 252:6528-6535

(12) Holmes, A.M., Hesslewood, I.P. and Johnston, I.R. (1974) Eur. J. Biochem. 43:487-499

(13) Holmes, A.M., Hesslewood, I.P. and Johnston, I.R. (1975) Nature 255:420-422

(14) Holmes, A.M., Hesslewood, I.P. and Johnston, I.R. (1976) Eur. J. Biochem. 62:229-235

(15) Sedwick, W.D., Wang, T.S-F. and Korn, D. (1972) J. Biol. Chem. 247:5026-5033

(16) Smith, R.G., Abrell, J.W., Lewis, B.J. and Gallo, R.C. (1975) J. Biol. Chem. 250:1702-1709

(17) Chang, L.M.S. (1977) J. Biol. Chem. 252:1873-1880

(18) Lynch, W.E., Surrey, S. and Lieberman, I. (1975) J. Biol. Chem. 250:8179-8183

(19) Herrick, G., Spear, B.B. and Veomett, G. (1976) Proc. Nat. Acad. Sci. U.S.A. 73:1136-1139

(20) Foster, D.N. and Gurney, T. Jr. (1976) J. Biol. Chem. 251: 7893-7898

(21) Eichler, D.C., Fisher, P.A. and Korn, D. (1977) J. Biol. Chem. 252:4011-4014

(22) Weissbach, A., Schlabach, A., Fridlender, B. and Bolden, A. (1971) Nature New Biol. 231:167-170

(23) Chiu, R.W. and Baril, E.F. (1975) J. Biol. Chem. 250:7951-7957

(24) Wang, T.Y. (1967) Methods Enzymol. 12A:417-421

(25) Busch, H. (1967) MEthods Enzymol. 12A:421-448

(26) Wang, T.S-F., Eichler, D.C. and Korn, D. Biochemistry, in press

(27) Kornberg, T. and Kornberg, A. (1974) In: The Enzymes (ed. P.D. Boyer) Vol. 10 pp. 173-209, Academic Press, New York.

(28) Wang, T. S-F., Sedwick, W.D. and Korn, D. (1974) J. Biol. Chem. 249:841-850

(29) Wang, T.S-F., Sedwick, W.D. and Korn, D. (1975) J. Biol. Chem. 250:7040-7044

(30) Baril, E.F., Jenkins, M.D., Brown, O.E., Laszlo, J. and Morris, H.P. (1973) Cancer Res. 33:1187-1193

(31) Chang, L.M.S. (1973) J. Biol. Chem. 248:3789-3795

(32) Stalker, D.M., Mosbaugh, D.W. and Meyer, R.R. (1976) Biochem. 15:3114-3121

(33) Chang, L.M.S. (1976) Science, 191:1183-1185

(34) Slater, J.P., Mildvan, A.S. and Loeb, L.A. (1971) Biochem. Biophys. Res. Commun. 44:37-43

(35) Slater, J.P., Tamir, I., Loeb, L.A. and Mildvan, A.S. (1972) J. Biol. Chem. 247:6784-6794

(36) Englund, P.T., Huberman, J.A., Jovin, T.M. and Kornberg, A. (1969) J. Biol. Chem. 244:3038-3044

(37) Sloan, D.L., Loeb, L.A., Mildvan, A.S. and Feldman, R.J. (1975) J. Biol. Chem. 250:8913-8920

(38) Wickner, R.B., Ginsberg, B. and Hurwitz, J. (1972) J. Biol. Chem. 247:498-504

(39) Eichler, D.C., Wang, T. S-F., Clayton, D.A. and Korn, D. J. Biol. Chem., in press - November 1977

(40) Berk, A.J. and Clayton, D.A. (1974) J. Mol. Biol. 86: 801-824

(41) Kasamatsu, H., Robberson, D.L. and Vinograd, J. (1971) Proc. Nat. Acad. Sci. U.S.A. 68:2252-2257

(42) Brown, W.M. and Vinograd, J. (1974) Proc. Natn. Acad. Sci. U.S.A. 71:4617-4621

(43) Robberson, D.L. and Clayton, D.A. (1972) Proc. Natn. Acad. Sci. U.S.A. 69:3810-3814

(44) Berk, A.J. and Clayton, D.A. (1976) J. Mol. Biol. 100:85-102

(45) Robberson, D.L., Clayton, D.A. and Morrow, J.F. (1974) Proc. Natn. Acad. Sci. U.S.A. 71:4447-4451

(46) Clayton, D.A., Davis, R.W. and Venograd, J. (1970) J. Mol. Biol. 47:137-153

(47) Smith, C.A., Jordan, J.M. and Vinograd, J. (1971) J. Mol. Biol. 59:255-272

(48) Bauer, W. and Vinograd, J. (1968) J. Mol. Biol. 33:141-172

(49) Kornberg, A. (1974) "DNA Synthesis", chap. 7. W.H. Freeman & Co., San Francisco.

(50) Herrick, G. and Alberts, B. (1976) J. Biol. Chem. 251:2124- 2132; 2133-2141

(51) Otto, B., Baynes, M. and Knippers, R. (1977) Eur. J. Biochem. 73:17-24

(52) Vosberg, H-P., Grossman, L.I. and Vinograd, J. (1975) Eur. J. Biochem. 55:79-93

(53) Champoux, J.J. and Dulbecco, R. (1972) Proc. Natn. Acad. Sci. U.S.A. 69:143-146

(54) Keller, W. (1975) Proc. Natn. Acad. Sci. U.S.A. 72:2550-2554

(55)   Gellert, M., Mizuuchi, K., O'Dea, M.H. and Nash, H.A. (1976)
       Proc. Natn. Acad. Sci. U.S.A.   73:3872-3876

(56)   Schekman, R., Weiner, A. and Kornberg, A. (1974) Science,
       186:987-993

(57)   Wickner, S., Wright, M., Berkower, I. and Hurwitz, J. (1974)
       In: "DNA Replication" (Wickner, R.B. ed.) p.195-215, Marcel
       Dekker, Inc., New York.

(58)   Eisenberg, S., Scott, J.F. and Kornberg, A. (1976) Proc.
       Natn. Acad. Sci. U.S.A.   73:3151-3155

(59)   Scott, J.F., Eisenberg, S., Bertsch, L.L. and Kornberg, A.
       (1977) Proc. Natn. Acad. Sci. U.S.A.   74:193-197

(60)   Hinckle, D.C. and Richardson, C.C. (1974)   J. Biol. Chem.
       249:2974-2984

(61)   Laipis, P.J. and Levine, A.J. (1973) Virology, 56:580-594

(62)   Winnacker, E.L., Magnusson, G. and Reichard, P. (1972)
       J. Mol. Biol. 72:523-537

(63)   Bolden, A., Aucker, J. and Weissbach, A. (1975) J. Virol.
       16:1584-1592

(64)   Hershey, H.V., Stieber, J.F. and Mueller, G.C. (1973) Eur.
       J. Biochem. 34:383-394

(65)   Tseng, B.Y. and Goulian, M. (1975) J. Mol. Biol. 99:317-337

ADDENDUM

A refining of the molecular weight calibration curves of standard
proteins has led us to recompute the apparent size of the β
polymerase. Our current estimate is that β has a molecular weight
of 39,000. The difference between the text value of 35,000 and
this new estimate is probably not significant and is due to
differing gel methodologies and the behaviour of the particular
standard proteins that are used to generate the calibration curves.
The principal conclusion that we established remains: that the
behaviour of the KB and hepatic enzymes is identical in our hands.

PROPERTIES AND INTERACTIONS OF DNA POLYMERASE α, DNA POLYMERASE β

AND A DNA BINDING PROTEIN OF REGENERATING RAT LIVER

A.M. de Recondo[**], J.M. Rossignol, M. Méchali and
M. Duguet

Unité d'Enzymologie, Institut de Recherches Scientifiques
sur la Cancer
94800, Villejuif, France

During the last years, considerable progress has been made
towards the understanding of the mechanism and control of DNA
replication. The results obtained with prokaryotes (1) have greatly
favoured studies of eukaryotic DNA replication; in both cases, DNA
replication involves the concerted action of several enzymes and
non enzymic factors integrated in a replication complex. However,
the genetic approach of this problem is particularly difficult in
eukaryotes, and to determine the roles of enzymes and factors
required for DNA synthesis *in vivo*, reconstructed systems involving
purified proteins must be used as well as the kinetic complementation
method described by L. Chang (2). In this attempt, we have purified
DNA polymerase-α, DNA polymerase-β and a DNA binding protein* from
rat liver to study their properties and their interactions in
several *in vitro* DNA synthesis systems.

Purification of DNA Polymerase α, DNA Polymerase β and DNA
Binding Protein 25K from Regenerating Rat Liver

The purification procedure of DNA polymerase-α is shown in
Table 1. It involves the preparation of a postmicrosomal supernatant, precipitation by ammonium sulfate, chromatography on DEAE
cellulose, on hydroxyapatite in the presence of 0.5 M NaCl on
phosphocellulose, and on denatured DNA cellulose.

*To conform to the nomenclature adopted in this meeting, we have
called DNA-binding protien (BP) the protein of 25K previously named
DNA-unwinding protein (15,22) by analogy with the calf thymus DNA-
unwinding protein (UP1) of Herrick and Alberts (6,19,20).
**A.S.I. Participant

Table I

Purification of DNA polymerase-α

Fraction	Total protein	Activities in the presence of poly(dC).(dG)$_{12-18}$	
	mg	units[1]	units/mg
I Supernatant	4,500	31,350	7
II $(NH_4)_2SO_4$	514	84,000	165
III DEAE cellulose	30	14,050	450
IV Hydroxyapatite	4.75	11,490	2,420
V Phosphocellulose	0.69	2,185	3,165
VI DNA cellulose	0.030	557	16,560

Purification of DNA polymerase-β

Fraction	Total protein	Activities in the presence of poly(dC).(dG)$_{12-18}$	
	mg	units	units/mg
I Supernatant	2131.5	2278	1.07
II PEG precipitation	1483	2319	1.56
III DNA cellulose (peak 0.8 M)	0.15	548	3653
IV Phosphocellulose (peak 0.6 M)	0.0058	119.7	20450

Purification of DNA-binding protein (BP)

Fraction	Total protein	Binding Activity	Proportion of 25 K
	μg	units[2]	%
III single stranded DNA cellulose (peak 0.4 M)	1250	16,000	65
IV phosphocellulose (peak 0.5 M)	500	9,000	97

(1) 1 unit = 1 nmole of total nucleotide incorporated per hour in initial velocity conditions.

(2) A binding unit is defined as the amount of protein able to retain 10^{-2} μg SV40 DNA I in the conditions of the binding assay.

The DNA polymerase obtained by this method has a specific
activity of 10,000 to 18,000 nmoles of deoxynucleotides
incorporated per hour and per mg of protein, in the presence of
poly(dC).(dG)$_{12-18}$ or activated calf thymus DNA as a template.

Electrophoresis of the purified DNA polymerase-α in poly-
acrylaide gel in non denaturing conditions, indicates that one
protein band is present, in addition to some protein(s) which is
not able to migrate in this gel even after treatment of the
preparation with DNase I. In sodium dodecyl sulfate acrylamide
gels, all the preparations studied exhibit two major bands in the
molecular weight range of 6 to 7 x 10^4 and 1.6 x 10^5. The first
represents approximately 68% of total protein, and the second 25%.
Occasionally a band in the molecular weight range of 9 x 10^4 appears
(less than 10% of total proteins). Similar results have been
reported by Bollum (3), Sedwick, Wang and Korn (4), and Holmes,
Hesslewood and Johnston (5).

DNA polymerase-β and DNA-binding protein of 25,000 daltons
have been purified from the same material, using differential DNA
cellulose affinity chromatographies (Table II). The purification
procedure was the following: the postmicrosomal supernatant
(Step 1) was brought to 2 M NaCl and to 10% (w/v) polyethylene
glycol (PEG) in order to remove nucleic acids. After centrifugation,
DNA polymerase-α stayed selectively in the PEG precipitate, whereas
DNA polymerase-β and DNA-binding protein remained in the supernatant (Step II). The supernatant was dialyzed and loaded onto
two DNA cellulose columns connected in series according to the
procedure of Herrick and Alberts (6). The first column was packed
with native DNA cellulose, the second with denatured DNA cellulose.
After adsorption of the extract, the columns were washed, then
uncoupled. DNA polymerase-β binds tightly to double stranded DNA
and was eluted from native DNA cellulose (Step III). Then it was
purified on a phosphocellulose column (Step IV). The fraction IV
exhibits a specific activity of 20,000 nmoles of deoxynucleotides
incorporated per hour and per mg of protein (Table I).

Proteins specifically adsorbed on single stranded DNA
cellulose were eluted by three elution steps (0.15 M NaCl, 0.40
M NaCl and 0.8 M NaCl). The fractions eluted at 0.4 M NaCl (DNA-
binding protein or BP, Step III) were pooled and applied to a
phosphocellulose P11 column. A DNA-binding protein was eluted
at 0.5 M NaCl (Step IV) and was purified to apparent homogeneity,
as shown by SDS polyacrylamide gel electrophoresis of fractions
III and IV. Fraction IV exhibits a single protein peak (> 97% of
total proteins) at 25,000 daltons and possesses a high binding
activity (detected by retention of SV40 DNA I on nitrocellulose
filters), a very slight endonuclease activity, but neither
detectable DNA polymerase nor ATPase activities (Table II).

Table II

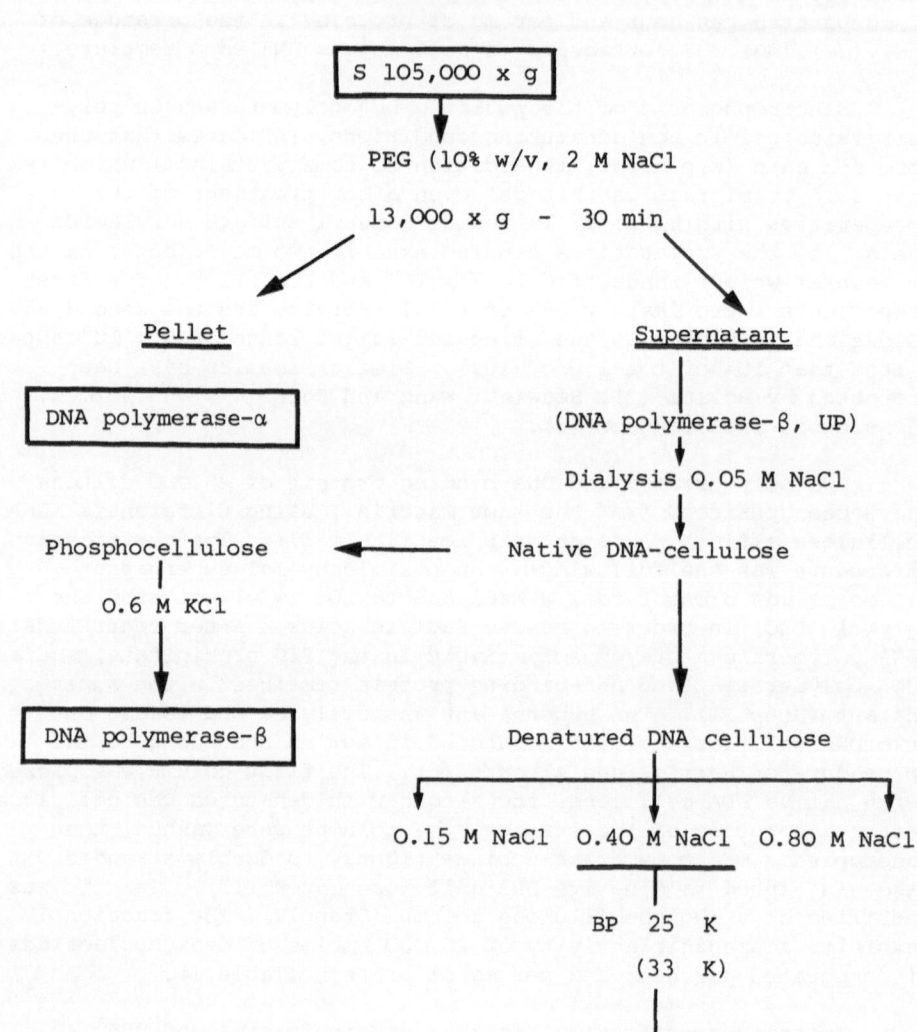

Purification procedure of DNA polymerase-β and BP 25 K

DNA polymerase-α can be purified from the PEG precipitate and the procedure described above provides a simple method to isolate α-, β-polymerase and DNA-binding protein from a same extract in a single preparation. Moreover, when applied to nuclear extracts, another DNA-binding protein of 33,000 daltons can be purified.

Properties of DNA Polymerase α

This enzyme has all the properties of the other mammalian DNA polymerases-α: neutral pH optimum, sensitivity to sulfhydryl-blocking agents, heparin and high salt concentrations. After chromatography on denatured DNA cellulose, the α-polymerase is completely devoid of exo- and endo-nuclease activities.

We have previously described in detail the template requirements of DNA polymerase-α (7). The purified enzyme cannot initiate new chains, heat-denatured DNA and native DNA are poor templates, and single-stranded polydeoxynucleotides cannot be used as templates. Activated calf thymus DNA or polydeoxyribonucleotides initiated with complementary oligodeoxyribonucleotides are excellent templates for the enzyme.

This enzyme does not copy the ribostrand of synthetic hybrids poly(A).(dT)$_{12-18}$ or poly(C).(dG)$_{12-18}$ even in the presence of Mn^{++} but can use a ribostrand as an initiator. Moreover as described by Spadari and Weissbach (8), the DNA polymerase-α activity is highly stimulated by concomitant RNA synthesis (Table III).

Two types of template competition experiments were also performed to study the first step of replication, i.e., the binding of the enzyme to the template molecule.

(a) In the presence of the four deoxyribonucleoside triphosphates, we studied the effect of double- or single-stranded templates on the replication of a partially single-stranded template.initiator (Fig. 1). When both templates were mixed together in equal amounts before addition of the enzyme, poly(dC).(dG)$_{12-18}$ inhibited the copying of poly [d(A-T)] whereas poly [d(A-T)] had no effect on poly(dC).(dG)$_{12-18}$ directed synthesis (Fig. 1A and 1B). The single-stranded poly(dG), which was not used as template by the enzyme under our experimental conditions, strongly decreased the copying of poly [d(A-T)] and the copying of poly(dC).(dG)$_{12-18}$ F9gi. 1A and 1B). The effect of poly(dG) on the copying of poly [d(A-T)] implies that the DNA polymerase-α can bind to poly(dG) but is unable to dissociate from a polymer that is devoid of a primer and which has a very compact structure in solution. Thus, the binding of the enzyme to the template must be distinguished from the polymerization reaction.

Table III

Response of regenerating rat liver DNA polymerase-α
to an RNA-primed DNA template formed in situ

	dNMP incorporated (pmol)
Complete system	1,650
- rATP, rCTP, rGTP, rUTP	315
- RNA polymerase	396
- RNA polymerase and 4 rNTPs	270
- DNA polymerase-α	0
- dTTP, dCTP, dATP, dGTP	0

The complete system (150 µl) contained 50 mM
Tris HCl (pH 7.9) 10 mM $MgCl_2$, 0.75 mM dithiothreitol, 80 µM EDTA, 2 mM KPO_4 (pH 7.5), 100 µM
ribonucleotide triphosphates, [(^{3}H)rGTP with a
specific activity of 500 counts/min per pmol],
2 units of *E.coli* RNA polymerase, 40 µM of each
deoxyribonucleoside triphosphate [(^{14}C)dTTP with a
specific activity of 80 counts/min per pmol), 16.5
µg of calf thymus denatured DNA, 0.5 units of DNA
polymerase-α and 30 µg of SAB. Incubation was
carried out at 37° for 60 min. Under these
conditions, 5.10 percent of total DNA were
transcribed by *E.coli* RNA polymerase.

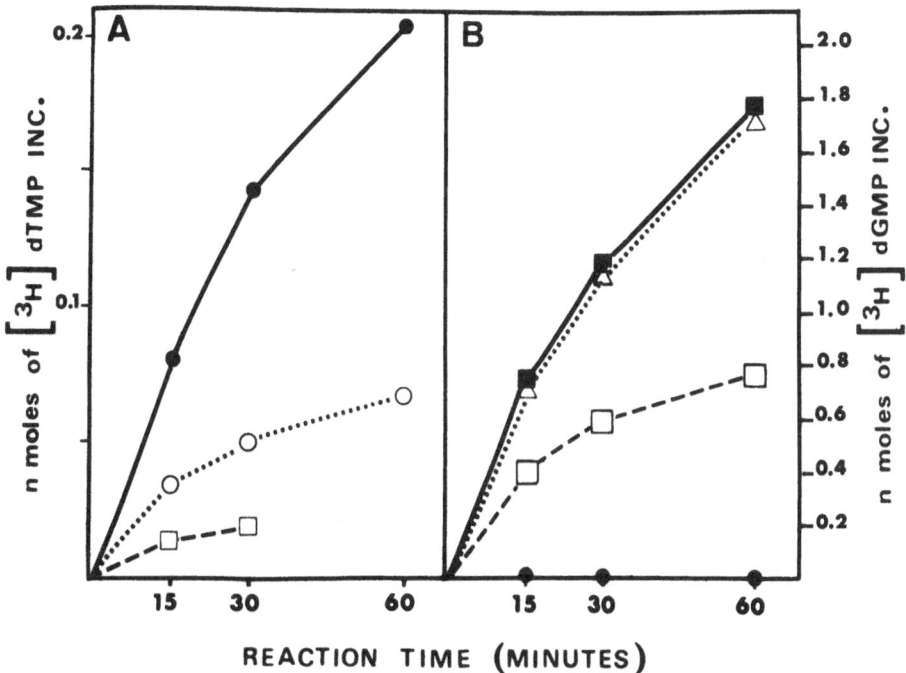

Figure 1. <u>Template competitions.</u> *The reaction was performed in a total volume of 100 μl containing 2.6 units of Fraction VI, the various templates at a concentration of 22 μM, 200 μM concentrations of each of the four dNTPs, optimal concentrations of MgCl₂ and KC1, 100 mM 2-mercaptoethanol, 60 mM Tris HC1 at pH 7 for A, and pH 7.6 for B. Activity was expressed as nmoles of labelled deoxynucleotide incorporated per assay.*

A. Template capacity of poly [d(A-T)] in the presence or absence of other template : (●——●) 22 μM poly [d(A-T)] (○---○) plus 22 μM poly(dC).poly(dG)12-18, (□----□) plus 22 μM poly(dG).

B. Template capacity of poly(dC).(dG)12-18 in the presence or absence of other templates: (■——■) 22 μM poly(dC).(dG)12-18, (△——△) plus 22 μM poly [d(A-T)], (□——□) plus 22 μM poly(dG). Control: 22 μM poly(dG) alone (●——●).

(b) We have also studied the processive or the distributive nature of enzymatic DNA synthesis catalyzed *in vitro* by DNA polymerase-α as Chang did for calf thymus DNA polymerase-α and -β, and *E.coli* DNA polymerase I (2). Using a high molecular ratio of template to enzyme, we examined the effect of the addition of initiated template on the ongoing replication reaction of another template. Although the addition of poly(dC) template to ongoing poly(dA) directed synthesis catalyzed by the α-polymerase resulted in an immediate inhibition of dTMP incorporation (Fig. 2A) as observed by Chang, the copying of poly(dC) was very low, suggesting that the enzyme molecules in the process of copying poly(dA) were not

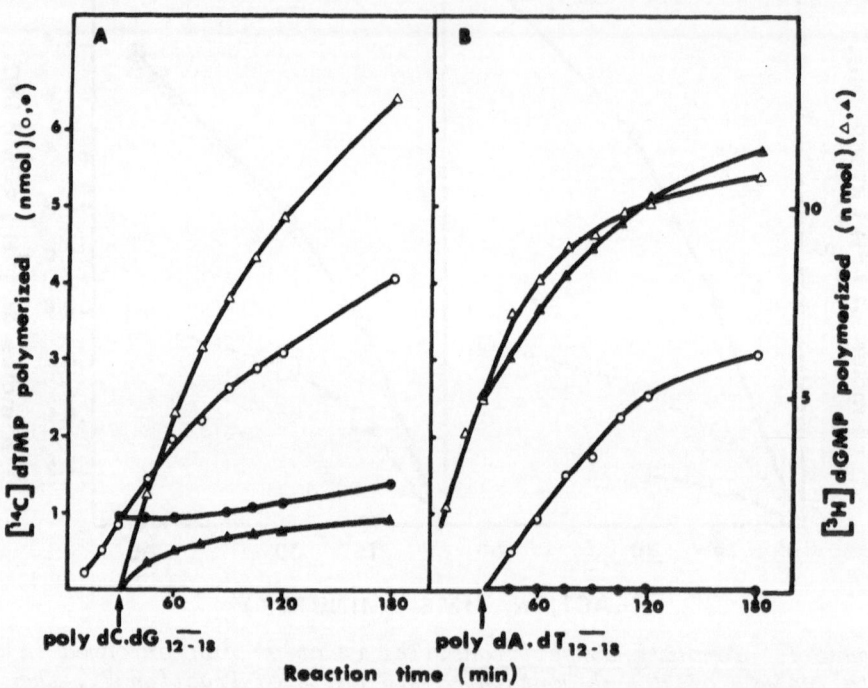

Figure 2. _Nature of the reaction catalyzed by DNA polymerase-α._
This figure shows the effect of the addition of initiated poly(dC)
template to an ongoing poly(dA) directed synthesis (A) and
reciprocally (B). The reactions were carried out at 35°C in a
total volume of 0.25 ml containing 50 μl of Fraction VI (8 units).

A. _Reactions 1 and 2 involving 100 μM poly(dA) and 10 μM dT 12-18_
were started by the addition of the enzyme. After 30 min of
incubation, a mixture containing 100 μM poly(dC) and 10 μM dG 12-18
was added to reaction 2. Reaction 3 involving 100 μM poly(dG) and
10 μM dG 12-18 alone started by the addition of the enzyme at the
same time as reaction 2.

B. _Reactions 1 and 2 involving 100 μM poly(dC) and 10 μM dG 12-18_
were started by the addition of the enzyme. After 30 min of
incubation, a mixture containing 100 μM poly(dA) and 10 μM dT 12-18
was added to reaction 2. Reaction 3 involving 100 μM poly(dA) and
10 μM dT 12-18 alone was started by the addition of the enzyme at
the same time as reaction 2.

The concentration of other components in the reaction mixture were
200 μM (14C) dTTP and 200 μM (3H) dGTP (30 cts/min per pmol), 3 mM
MgCl2, 1.2 mM KC1, 100 μM 2-mercaptoethanol, 50 μM Tris-HC1 pH 7

completely dissociated from this template at the time when the
competing template poly(dC) was added. Moreover, Fig. 2B shows
that under our conditions, synthesis on primed poly(dC) by DNA
polymerase-α was processive and was not inhibited by the addition
of poly(dA). The added poly(dA) was not utilized as a template,
indicating no dissociation of the enzyme from the poly(dC) template
had occurred.

Different hypotheses can be advanced : the processive or
distributive mechanism of DNA polymerase in $vitro$ depends on the
experimental conditions, or on the kinds of template-initiator used
if the enzyme binds more tightly with one of them. But the Km
values of regnerating rat liver DNA polymerase-α for various template systems were of the same order. On the other hand, our
results could be explained if we assume that regenerating rat liver
α-polymerase is composed of a catalytic core associated with a DNA
binding factor which differs from the polymerase core in template
affinity and which could act as an initiating factor. After
initiation of synthesis on poly(dA), the addition of poly(dC)
caused a release of the catalytic core which reassociated with
poly(dC), whereas the "initiating" factor remained on the poly(dA).
In this case, poly(dC) would be copied but with a low efficiency
whereas poly(dA) directed synthesis would stop. On the contrary,
after initiation of synthesis on poly(dC) the addition of poly(dA)
would result in no change in the ongoing poly(dC) directed synthesis
because only the "initiating" factor dissociated from poly(dC)
to poly(dA). In this case, the copying of this homopolymer
template was processive. These results imply that under our
experimental conditions the "initiating" factor has a higher
affinity constant for poly(dA) than for poly(dC) as opposed to
that found with the α-polymerase core.

In order to dissociate the catalytic and the binding properties
of DNA polymerase-α, we have studied the binding of this enzyme to
a DNA molecule containing no free ends: SV40 DNA I. When DNA
polymerase-α was preincubated with labelled SV40 DNA I, a nucleoprotein complex was formed, as indicated by the retention of the DNA
on a nitrocellulose filter. As shown in Fig. 3, the amount of SV40
DNA I retained by the filter was linearly dependent on the amount
of DNA polymerase-α added to the reaction medium in the range from
0 to 60-95% of the added DNA. The linear part of the binding curve
indicates that 1 μg of DNA can be retained to the filter by 1.8 μg
of protein. It is difficult to extrapolate the corresponding

_____

*for A and pH 7.6 for B, and 50 μg of bovine serum albumin per assay.
For each reaction, 20 μl aliquots were removed at various times and
analyzed. The open symbols represent the control reactions for
poly(dA) (O——O) and poly(dC) (Δ——Δ) replication conditions
(reactions 1 and 3). The closed symbols represent the results of
the competition experiments (reactions 2, ●——● , ▲——▲).*

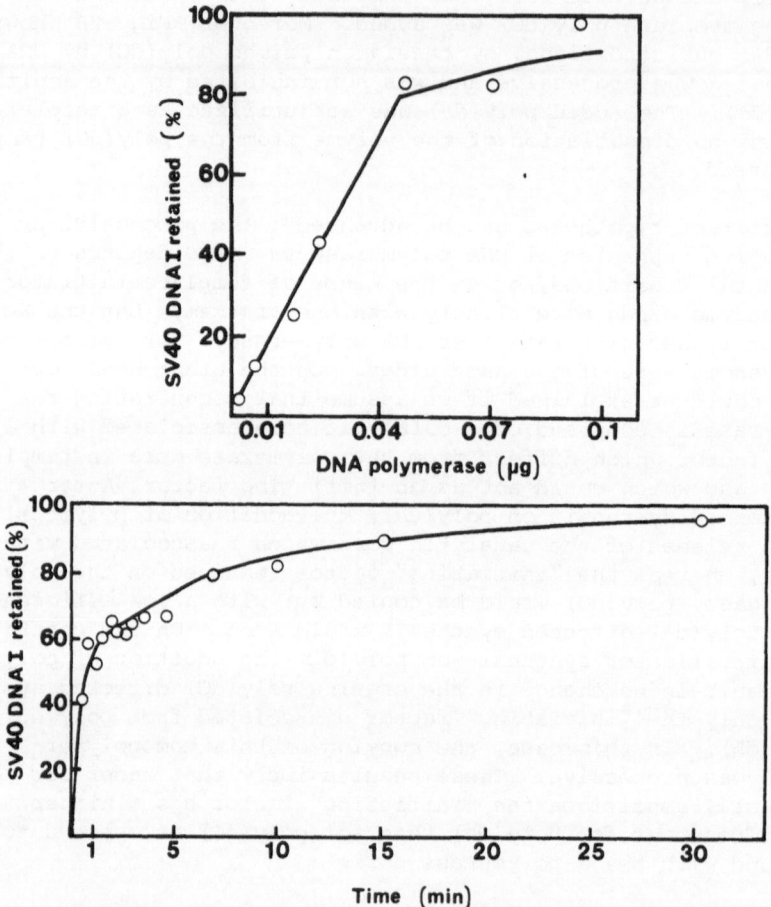

<u>*Figure 3. Binding of DNA polymerase-α to SV40 (³H) DNA I.*</u>

The binding activity was determined using a nitrocellulose filter assay. Reaction media contained 0.028 µg SV40 (³H) DNA I (31,000 cpm/µg) and increased quantities of enzyme. The amount of DNA retained in the absence of enzyme was 1.5%.

<u>*Kinetic of formation of the nucleoprotein complex.*</u>

The reaction media contained 0.745 µg (³H)SV40 DNA I (34,000 cpm/µg) and 1.14 µg DNA polymerase-α in 2.4 ml buffer containing 250 µg/ml bovine serum albumin. The solution was preincubated for 10 min at 37°C without the enzyme and the reaction was initiated by the addition of DNA polymerase-α. After the indicated incubation periods, 0.1 - ml samples were removed, diluted in 1 ml of buffer and filtered immediately. In this experiment, the time required between dilution and complete filtration was about 10 to 15 sec.

enzyme/DNA molar ratio since the molecular weight of DNA polymerase-α can range from 130,000 to at least 450,000, depending on the conditions used for the measurement of the molecular weight. Moreover the enzyme may aggregate under the low ionic strength conditions used in the binding assay.

The kinetics of formation of the complex (Fig. 3) is relatively rapid with a half-time of reaction less than 50 sec. Formation of the nucleoprotein complex does not require divalent cations and at $MgCl_2$ concentrations above 1 mM the formation of the complex is reduced.

The presence of this binding activity in highly purified DNA polymerase-α led us to try to dissociate the catalytic and the binding activities during the purification of the DNA polymerase-α. All the described chromatographic procedures used to purify the enzyme failed to separate the two activities. Moreover, in the presence of 10 mM or 200 mM KCl, they cosediment in sucrose gradients and thus they cannot be separated on the criteria of size and shape of the molecules. It was not possible to determine if the binding and DNA polymerase-α activities increased simultaneously through the first steps of purification because of the presence of other proteins which could bind to SV40 DNA I.

Sucrose gradients containing SV40 DNA I, or DNA polymerase-α, or nucleoprotein complexes formed at different protein/DNA ratios, were then made. Fig. 4A shows the position of SV40 DNA I and polymerase-α sedimented separately. At 38% retention on the filter of the (^{3}H) DNA I involved in a nucleoprotein complex, the major part of DNA I sediments at 21-22 S, but about 27% of SV40 DNA sediments at 24-26 S (Fig. 4B). At 72% retention on the filter of DNA involved in a nucleoprotein complex (Fig. 4C) the (^{3}H) DNA I sediments near the bottom of the gradient, and a broad peak between 25 and 33 S is obtained. In all cases, even when all the SV40 DNA I molecules are involved in a nucleoprotein complex (not shown), the DNA polymerase-α activity determined with poly(dC).(dG)$_{12-18}$ or activated DNA as template, does not co-sediment with SV40 DNA I. However, a trailing edge of DNA polymerase-α activity towards the heavy region of the gradient can always be observed when a nucleoprotein complex has been formed. The fractions containing DNA polymerase activity and SV40 DNA I were pooled separately (Pools P and D of Fig. 4C). DNA polymerase contained in the pool P has lost its binding activity whereas SV40 DNA I contained in the pool D is retained on nitrocellulose filter. As judged by electron microscopy and alkaline sucrose gradients, the SV40 DNA I of the pool D remained supercoiled DNA I.

One would expect an alteration in the sedimentation behaviour of α-polymerase activity if the binding factor is associated with

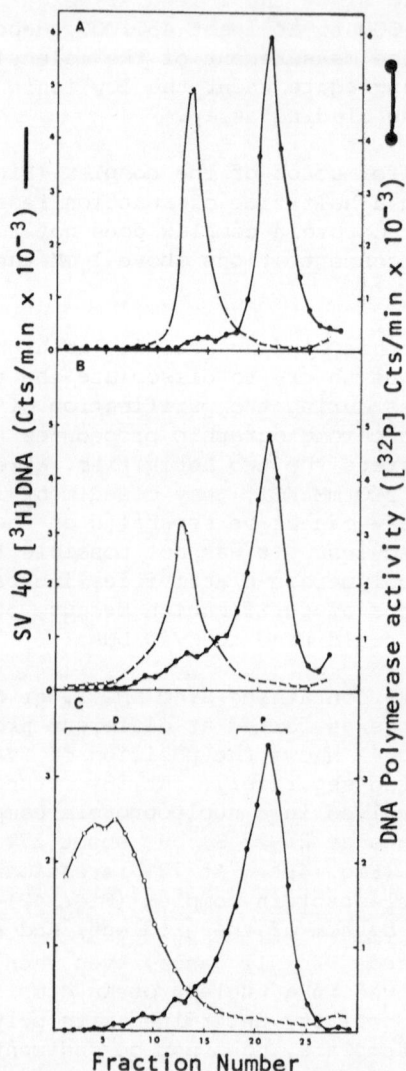

<u>*Figure 4*</u>.   *Separation of the SV40 DNA I – DNA-binding factor*
complex from the catalytic activity of DNA polymerase-α.

Centrifugation of SV40 (^{3}H) DNA I.   DNA polymerase-α, or mixtures,
were preformed at 4°C in 4.8 ml of 5 to 20% sucrose gradients in
50 mM Tris-HC1 pH 7.6, 10 mM KC1, 5 mM 2-mercaptoethanol, 0.2 mM
MgCl$_2$, which have been layered on top of a 0.1 -ml cushion of 50%
sucrose. After 225 min at 40,000 rpm in a Spinco SW 50.1 rotor, 180
μl – fractions were collected from the bottom of the tubes. <u>Panel A</u>:
DNA polymerase-α and SV40 DNA I sedimented separately. <u>Panel B</u>: DNA
polymerase-α and SV40 DNA I preincubated at 37°C in conditions
allowing to retain 38% SV40 DNA I on nitrocellulose filter.

DNA polymerase-α prior to binding to DNA. The centrifugation conditions of Fig. 4 do not however allow to observe such an alteration. It is important to note that the recovery of α-polymerase activity is always less than in usual sedimentation experiments and the α-polymerase activity, when dissociated from the binding factor, became very unstable.

On the other hand, the SV40 DNA I of the nucleoprotein complex could not be cleaved by S_1 nuclease, suggesting that the protein bound to SV40 DNA I on partially denatured regions of genome, in particular A-T rich regions. So, we believe that centrifugation can dissociate the binding factor, which remained bound to SV40 DNA I, from the catalytic unit DNA polymerase-α which then became unstable. These results agree with those of template challenge experiments described above.

The possible presence of a "conversion factor" in DNA polymerase-α preparations has been pointed out by Probst et al. (9) and Holmes et al. (10), and similar studies in progress in different laboratories lead to question the exact nature of the molecular subunit structure of DNA polymerase-α. We are attempting now to find conditions permitting the study of the dissociated binding and catalytic activities.

Properties of DNA Polymerase β

The rat liver enzyme presents the main properties of other mammalian DNA polymerases-β: resistance to N-ethylmaleimide, stimulation by high ionic strength (0.1 M KCl), alkaline pH optimum (pH 8.8) and a sedimentation coefficient of 3.1 S in high salt concentrations. The activities of various nucleic acids as templates are given in Table IV.

RNA- and DNA-dependent activities were copurified throughout the different steps of the purification procedure of DNA polymerase-β. Moreover the R-dependent activity associated with the β-enzyme differs from the γ-polymerase activity which could be detected in rat tissue at very low levels and sediments at 5 - 6 S. We have studied the specific inhibition of RNA- DNA-dependent activities of DNA polymerase-β to determine if the mechanism of the template recognition by this enzyme implies only one or two catalytic centers.

Panel C: DNA polymerase-α and SV40 DNA I preincubated at 37°C in conditions allowing retention of 68% SV40 DNA I on nitrocellulose filter. (●——●) DNA polymerase activity as cts/min (^{32}P)dGMP incorporated with poly(dC).(dG)12-18 as template. (O——O), position of SV40 DNA I as (^{3}H)cts/min. Fractions containing SV40 (^{3}H) DNA I (1 to 11) and DNA polymerase activity (18 to 23) were pooled separately (pools D and P).

Table IV

Template requirements of DNA polymerase-β

Template	Total nucleotides incorporated
	nmol
Native DNA	0.993
Denatured DNA	0.496
Activated DNA	4.35

	dAMP	dTMP	dGMP
	nmol	nmol	nmol
poly [d(A-T)]	2.7	2.7	
poly(dC).(dG)$_{12-18}$			47.1
poly(dA).(dT)$_{12-18}$		25.6	
poly(A)-(dT)$_{12-18}$		23.9	
poly(A).poly(dT)	0.27		

Fig. 5 shows the effect of several inhibitors or effectors on RNA-dependent and DNA-dependent activities of DNA polymerase-β. Firstly, high ionic strength (100 mM KCl) increases the rate of the RNA-dependent reaction more than the DNA-dependent reaction, and the optimal concentrations of KCl are different according to the template used (Fig. 5A). In the experiment shown in Fig. 5B we have tested the effect of various concentrations of potassium phosphate on the template activity of poly(A) or poly(dC) by the DNA polymerase-β in the presence of Mn^{++}. The use of potassium phosphate to distinguish poly(A) replication catalyzed by β-polymerase from that catalyzed by γ-polymerase was recently described by Knopf et al. (11). At a concentration of 50 mM of potassium phosphate the RNA-dependent activity of β-polymerase is completely abolished whereas the activity of γ-polymerase is stimulated. The same concentration of potassium phosphate depresses the DNA-dependent activity of DNA polymerases α, β and γ (11). In our experiments RNA-dependent and DNA-dependent activities of β-enzyme are inhibited by high potassium phosphate concentrations but the shape of the inhibition curves is different and the concentration necessary to obtain a total inhibition of reactions also differs (40 mM for RNA-dependent activity and 80 mM for DNA-dependent activity). As noted by Chang and Bollum (12) the effect of phosphate on a DNA polymerase reaction is likely to be a complex function of the divalent cation, the type of template used and the pH of the reaction. It is also important to note that this compound which acts as dNTPs competitor, more strongly inhibits the RNA-dependent activity of DNA polymerase-β than its DNA-dependent activity.

Hecht (13), Sedwick et al. (14), have previously mentioned that ethidium bromide affects DNA polymerase-α but not the β-polymerase. As shown in Fig. 5C we have obtained similar results. In the presence of a DNA template the rat liver β-polymerase is more resistant than α-polymerase, in a concentration range of ethidium bromide from 2 to 20 μM. However the RNA-dependent activity of β-polymerase is rapidly abolished. Other experiments (not shown) indicate that the inhibition by ethidium bromide also depends on the template used. The replication of single stranded polydeoxynucleotide template initiated by oligomers is more sensitive than the copying of activated DNA, but the greatest inhibition is always observed with a hybrid template. These results can be explained by the greater affinity of ethidium bromide for RNA than hybrid and than DNA, and by the fact that this compound could also bind to single stranded structures.

Therefore all these results do not allow to determine if the RNA-dependent and DNA-dependent activities of β-polymerase are carried by different catalytic centers because all the effectors studied could not only act on the enzyme but also on templates or substrates. However in all cases, the R-dependent activity of

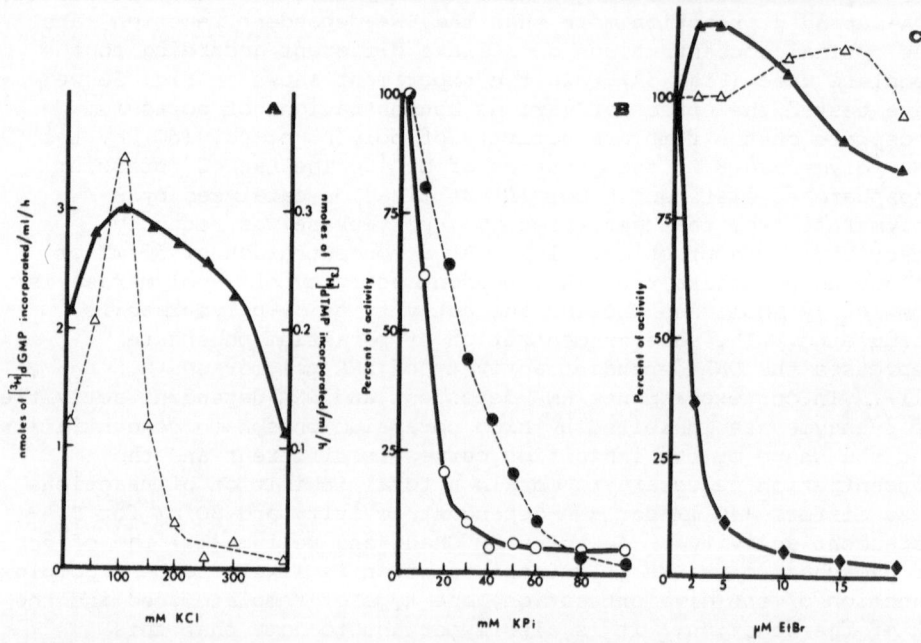

Figure 5. Effect of different effectors on DNA polymerase-β.

A. *Effect of KC1 concentration. Reaction mixture contained in a total volume of 50 μl : 50 mM Ammediol-HC1, pH 8.8, 1 mM dithiothreitol, 0.1 mM $MnCl_2$, 2 μg of bovine serum albumin, 125 μM(^{3}H) dGTP or (^{3}H)dTTP (1 μCi/assay)and 25 μM poly(dC).(dG)12-18 (▲——▲) or poly(A).(dT)12-18 (△---△). Reaction was performed during 30 min at 37°C for poly(dC).(dG)12-18 or 27°C for poly(A). (dT)12-18.*

B. *Effect of potassium phosphate. Reaction was the same as described in A except that potassium phosphate was added at the concentrations indicated in the figure, and that KC1 was added at a concentration of 0.1 M. At 0 mM potassium phosphate, DNA polymerase-β activity was 75 nmoles/ml/h (100% activity) in the presence of poly(dC).(dG)12-18 (●---●) and 3.5 nmoles/ml/h (100% activity) in the presence of poly(A).(dT)12-18 (○——○).*

C. *Effect of ethidium bromide. Reaction mixture for poly(A).(dT) 12-18 was the same as described in A, except that KC1 was added at the concentration of 0.1 M and ethidium bromide was present at the concentration indicated in the figure. Reaction mixture for activated calf thymus DNA contained in a total volume of 50 μl :*

β-polymerase is more sensitive than DNA-dependent activity and this fact emphasizes the differences between β- and γ-enzymes.

Properties of DNA-Binding Protein

As indicated above, the Fraction IV (BP) contains a single polypeptide chain of 25,000 daltons as determined by gel electrophoresis in denaturing conditions (Fig. 6A).

After sucrose velocity sedimentation at low ionic strength the protein, detected by nitrocellulose filter binding assay, is resolved in two peaks. The main peak appears at 5.9 S and the second at 2.3 S. At a higher ionic strength, the binding activity is detected in a single peak at 2.3 S (Fig. 6B).

The diffusion coefficient of BP was determined by gel filtration through Biogel A 0.5 M in the same ionic strength conditions. The insertion of S and D in the Svedberg equation gives a molecular weight of 104,000 ± 7,000 consistent with a tetrameric form in low ionic strength conditions, and of 24,000 ± 1,500 (consistent with an SDS polyacrylamide gel) in high ionic strength conditions. The physical properties of rat liver BP are summarized in Table V.

When poly [d(A-T)] is incubated at 13,5°C with various amounts of binding protein, an increase in absorbance at 260 nm of the nucleic acid which corresponds to the denaturation of the polymer, is observed (15). The rat liver DNA-binding protein also depresses the helix melting temperature (Tm) of synthetic hybrid poly(A).poly(dT) and double stranded poly(A).poly(U), indicating that both DNA and RNA are recognized by this protein.

Despite its poor affinity for double stranded DNA, the protein binds to supercoiled SV40 DNA I. This finding may be facilitated by the melting of A-T rich regions on this DNA (20). The formation of the nucleoprotein complex at 37°C is followed by a slight increase in the absorbance at 260 nm and is very sensitive

---

50 mM Ammediol-HC1 buffer, pH 8.8, 5 mM MgCl$_2$, 1 mM 2-mercapto-ethanol, 125 µM of (^{3}H)dATP (1 µCi/assay), 125 µM each dCTP, dGTP and dTTP, activated DNA 200 µg/ml, 2 µg of bovine serum albumin and ethidium bromide at the concentrations indicated in the figure. At 0 mM of ethidium bromide DNA polymerase-α activity (100% activity) was 37,36 nmoles of labelled nucleotide ml/h in the presence of activated DNA (▲——▲), DNA polymerase-β activity (100% activity) was 4.37 nmoles of labelled nucleotide/ml/h in the presence of activated DNA (△---△), and 10-80 nmoles of labelled nucleotide/ml/h in the presence of poly(A).(dT)12-18 (◆——◆).

Table V

Physical properties of rat liver BP

	Subunit	Native
SDS molecular weight	25,000	
$S_{20,w}$	2.3 S	5.9 S
$D_{20,w}$	8.7×10^{-7} cm/s	5.3×10^{-7} cm/s
Molecular weight	24,800	104,000
Stokes radius	24.7 Å	40.5 Å
Frictional coefficient	1.27	1.29

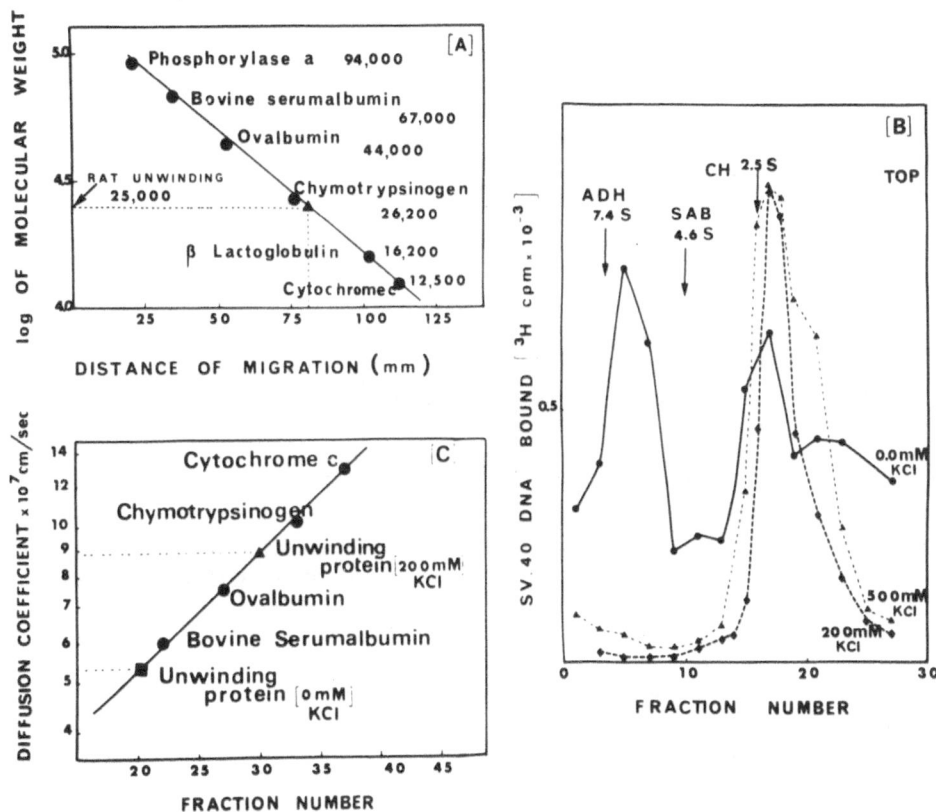

Figure 6. *Molecular weight determination of native and dissociated binding protein.*

A. *SDS-polyacrylamide-slab gel electrophoresis in 15% acrylamide. Four micrograms of each protein were denatured in the presence of SDS and electrophoresis was performed.*

B. *Sucrose velocity sedimentation of binding protein at various ionic strengths. Protein samples (5 μg binding protein, 200 μg alcohol dehydrogenase, 500 μg bovine serum albumin and 500 μg chymotrypsinogen) were subjected to a 16 hr run at 45,000 rpm in a SW 50.1 rotor and assayed in specific conditions for each protein. Fractions were collected from the bottom to the top. (●——●). 0 mM KC1, (♦ — — ♦) 200 mM KC1, (▲——▲) 500 mM KC1. Absolute corresponding ionic strengths are 0.05, 0.25 and 0.55 M.*

C. *Gel filtration on Biogel A 0.5 M. Fractions containing binding protein were dialyzed against buffer before assayed by nitrocellulose filter binding. (■) 0 mM KC1 (μ = 0.05), (▲) 200 mM KC1 (μ = 0.25).*

to ionic strength and to magnesium ions (15). The saturated
complex has a sedimentation coefficient of about 30 S (Fig. 7) and
contains a protein to DNA weight ratio of 4/1 and an average
number of 1 protein subunit per 7 nucleotides. In these
conditions all the A-T rich regions are probably melted and
covered with protein. Since the origin of SV40 replication seems
to be related to a short G-C rich sequence surrounded by A-T rich
regions, the binding of protein to these regions could provide a
good recognition signal for proteins that initiate replication.

Interactions Between DNA Polymerase α or β
and Homologous DNA Binding Protein BP 25 K

Stimulation of DNA polymerases by homologous DNA-binding
proteins is one of the strongest arguments to assign a role to
these proteins but also as a function of the length of single
stranded regions of this template. The stimulatory effect of DNA-
binding protein is more detectable when the initiator/template ratio
is low (1/10). A similar observation has been made for DNA
polymerase-α and a DNA binding protein from mouse cells, in the
presence of activated and denatured DNA as templates (21).

Stimulation of DNA Polymerase-β

The effect of rat liver DNA-binding protein on the activity
of DNA polymerase-β in the presence of poly(dC).oligo(dG) as a
template is very similar to that observed with DNA polymerase-α
(22). Nevertheless, unlike the above results, the stimulatory
effect of BP is higher when a template with short single stranded
regions is used (initiator/template ratio $\geq \frac{1}{2}$).

We have therefore also tested the stimulation of DNA polymerase-β on a double stranded template with few 3'OH ends: the
synthetic copolymer poly [d(A-T)]. This template readily melted
(at 37°C) by rat liver BP at a protein/DNA weight ratio to 10.
In this case a 2- to 5-fold stimulation of activity is observed
depending on the template concentration (22). At 20 μM poly [d(A-T)]
the stimulation is maximum for a BP/DNA weight ratio of 3/1 (i.e.,
smaller than the BP/DNA ratio necessary to unwind totally the
poly [d(A-T)]).

Since the Mg^{++} concentration affects both DNA polymerase-β
activity and binding of the BP to DNA, the influence of this ion
in the stimulation process has been tested. Optimum Mg^{++}
concentration is 2 mM with poly [d(A-T)] alone and 4 mM in the
presence of binding protein. This result is consistent with the
stabilization of the double-helix by Mg^{++} which is partially
prevented by BP (22).

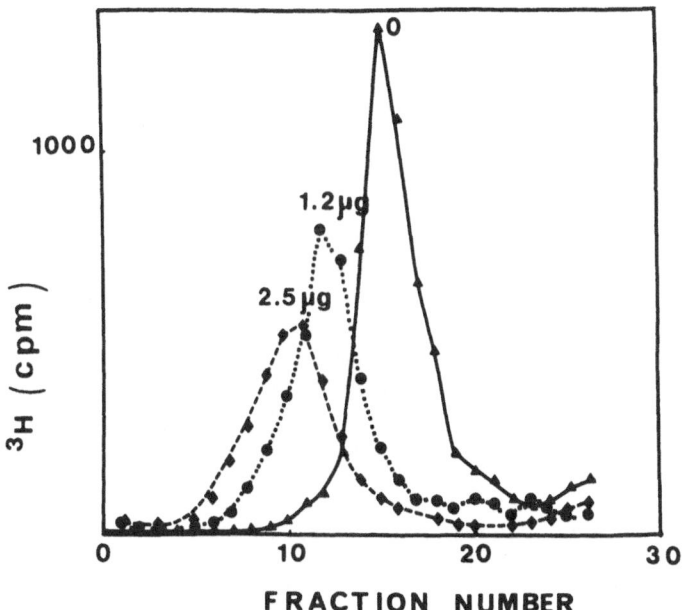

Figure 7. Sedimentation of rat liver BP-DNA complexes. 0.305 µg of supercoiled (³H) SV40 DNA was incubated at 37°C for 30 min without (▲) or with 1.2 µg (●) or 2.5 µg (♦) of rat liver binding protein in buffer in a final volume of 100 µl. Then the three samples were loaded onto a 5 to 20% sucrose gradient in 50 mM Tris-HCl, pH 7.6, 10 mM KCl, and 5 mM 2-mercaptoethanol, which had been layered on top of a 0.1 ml cushion of 50% sucrose in the same buffer. Centrifugation was performed for 3 hrs at 4°C and 40.000 rpm in a SW 50.1 Spinco rotor. 26 equal fractions were collected from the bottom. The radioactivity in each fraction was determined by liquid scintillation counting. The arrow indicated the position of supercoiled SV40 DNA (21 S).

Stimulation of a DNA polymerase by an homologous binding protein can be interpreted in two ways. In the first hypothesis, the protein may act on the structure of DNA, suppressing thermodynamic barriers to the progression of polymerase molecules along the DNA. This hypothesis is consistent with an optimal BP/DNA ratio smaller than that necessary for complete coating of single stranded portions of DNA or smaller than the BP/DNA ratio required to totally unwind the template.

In the second hypothesis, binding protein may form binary complexes with DNA polymerases by direct protein-protein inter-actions, or ternary complexes with the DNA (24). This hypothesis implies specific interactions between DNA polymerases and binding proteins. Banks and Spanos (25,26) had reported that DNA synthesis of repair type, catalyzed by the unique DNA polymerase of the primitive eukaryote *Ustilago maydis*, was stimulated by homologous binding protein, but attempts to detect protein-protein interaction failed. Herrick and Alberts (20) with UP1 of calf thymus (25,000 daltons), Otto et al. (21) with a binding protein from mouse cells (33,000 daltons), had observed a specific stimulation for DNA polymerase-α and not for DNA polymerase-β from the same source. The phosphorylation of the mouse cells binding protein drastically reduced its effect on the polymerizing activity of DNA polymerase-α although it did not interfere with the binding of the protein to single stranded DNA. In contrast, our results show a stimulatory effect of rat DNA-binding protein (25,000 daltons) on the two homologous DNA polymerases (α and β). This difference might be explained by the extreme sharpness of the stimulation range in regard to the binding protein/DNA ratio. Another possibility is that stimulation by rat liver DNA-binding protein is not strictly specific to one DNA polymerase, but due to the action of this protein on DNA structure as well as to direct interaction with the enzymes. This last possibility is supported by the slight stimulation of *E.coli* DNA polymerase I by rat liver DNA-binding protein observed on poly [d(A-T)] template (not shown). Further experiments are required before a conclusion concerning the specific interaction between mammalian DNA-binding protein and homologous DNA polymerase(s).

CONCLUSION

We have purified three proteins of a eukaryotic replication complex and studied their properties.

The template specificity of DNA polymerases α and β has been determined. Template competition experiments performed with DNA polymerase-α indicate that the binding of the enzyme to the template can be distinguished from the polymerization itself and that the *in vitro* synthesis catalyzed by this α-polymerase is not distributive in a classical sense. Moreover a DNA-binding factor able to bind to a double stranded DNA containing no free ends (SV40 DNA I) has been reproducibly detected in highly purified DNA polymerase-α. This factor could not be separated from the catalytic unit by any of the conventional separation procedures used, but its dissociation from the catalytic core may be observed after formation of a nucleoprotein complex between DNA polymerase-α and SV40 DNA I. This factor could be a subunit of DNA polymerase-α, which appeared to be composed of two polypeptides of 160,000 and 60,000 daltons, as well as the β subunit of the AMV reverse transcriptase (27) or copol III* (28). Further studies have to be made to elucidate the nature of this DNA-binding factor and its relationship with the catalytic unit of DNA polymerase-α.

The effect of a homologous DNA-binding protein of 25,000 daltons on the reaction catalyzed by the α- and β-enzymes of rat liver in the presence of different synthetic templates has been studied. Rat liver BP 25 K increases notably the activity of DNA polymerase-α but also that of DNA polymerase-β depending on the experimental conditions used. That is not very surprising for a protein which acts on template too.

A large variety of nuclear proteins are phosphorylated in rapidly proliferating cells and the phosphorylation seems to play a role in the regulation of transcription and replication processes. Therefore we try to know if the rat liver DNA-binding protein could be phosphorylated *in vitro* and if phosphorylation affects its b inding to DNA or its stimulatory effect on DNA polymerases as it was recently made for a mouse cell's DNA-binding protein (21). A comparative study of the effect of protein 32 and BP on the *in vitro* replication of several templates by DNA polymerase α and β is also undertaken.

We believe that reconstructed systems using these purified proteins and natural DNA templates (e.g., ØX174, fd or M13 phages, SV40 DNA, or linear viral DNA), which could be complemented by viral proteins or proteins from permissive cells, may allow us to understand better the DNA replication in eukaryotic cells.

REFERENCES

1.   Schekman, R., Weiner, A. and Kornberg, A. (1974) Science 186: 987-993

2.   Chang, L.M.S. (1975) J. Mol. Biol. 93:219-235

3.   Bollum, F.J. (1975) Progress in Nucleic Acid Research 15: 109-144

4.   Sedwick, W.D.,   Wang, T.S.F. and Korn, D. (1975) J. Biol.
      Chem. 250:7045-7056

5.   Holmes, A.M., Hesslewood, I.P. and Johnston, I.R. (1976)
      Eur. J. Biochem. 62:229-235

6.   Herrick, G. and Alberts, B.M. (1976) J. Biol. Chem. 251:2124-
      2132

7.   De Recondo, A.M., Lepesant, J.A., Fichot, O., Grasset, L.,
      Rossignol, J.M. and Cazillis, M. (1973) J. Biol. Chem. 248:
      131-137

8.   Spadari, S. and Weissbach, A. (1975) Proc. Nat. Acad. Sci. USA
      72:503-517

9.   Probst, G.S., Stalker, M., Mosbaugh, D.W. and Meyer, R.R.
      (1975) Proc. Nat. Acad. Sci. USA 72:1171-1174

10.   Holmes, A.M., Hesslewood, I.P. and Johnston, I.R. (1975)
      Nature 255:420-422

11.   Knopf, K.W., Yamada, M. and Weissbach, A. (1976) Biochemistry
      15:4540-4548

12.   Chang, L.M.S. and Bollum, F.J. (1973) J. Biol. Chem. 248:
      3398-3404

13.   Hecht, N.B. (1974) J. Reprod. Fert. 41:345-354

14.   Sedwick, W.D., Wang, T.S.F. and Korn, D. (1972) J. Biol. Chem.
      247:5026-5033

15.   Duguet, M. and de Recondo, A.M. (1977) submitted to J. Biol.
      Chem.

16.   Huberman, J.A., Kornberg, A. and Alberts, B.M. (1971) J. Mol.
      Biol. 62:39-52

17.   Molineux, I.J., Friedman, S. and Gefter, M.L. (1974) J. Biol.
      Chem. 249:6090-6098

18.   Weiner, J.H., Bertsch, L.L. and Kornberg, A. (1975) J. Biol.
      Chem. 250:1972-1980

19.   Herrick, G. and Alberts, B.M. (1976) J. Biol. Chem. 251:
      2133-2141

20.   Herrick, G., Delius, H. and Alberts, B.M. (1976) J. Biol.
      Chem. 251:2142-2146

21. Otto, B., Baynes, M. and Knippers, R. (1977) Eur. J.
     Biochem. 73:17-24

22. Duguet, M., Soussi, T., Rossignol, J.M., Méchali, M. and
     de Recondo, A.M. (1977) FEBS Letters 79:160-164

23. Molineux, I.J. and Gefter, M.L. (1975) J. Mol. Biol. 98:
     811-825

24. Cavalieri, S.J., Neet, K.E. and Goldthwait, D.A. (1976)
     J. Mol. Biol. 102:697-711

25. Banks, G.R. and Spanos, A. (1975) J. Mol. Biol. 93:63-67

26. Yarranton, G.T., Moore, P.D. and Spanos, Â. (1976) Molec.
     Gen. Genet. 145:215-218

27. Grandgenett, D.P. (1976) J. Virol. 17:950-961

28. Wickner, W. and Kornberg, A. (1974) J. Biol. Chem. 249:
     6244-6249

DNA POLYMERASES OF SEA URCHIN EMBRYOS

Benita De Petrocellis[*], Elio Parisi[*], Silvana Filosa[**],
Antonio Capasso, Fabrizia Giovinazzi

Laboratory of Molecular Embryology
Arco Felice
Naples, Italy

It is generally accepted that DNA synthesis followed by cell
division is one of the leading events on which the formation of a
multicellular organism from the unicellular zygote depends. Sea
urchin embryos represent a suitable system for the study of the
molecular events connected with the process of cell division. During
oogenesis the sea urchin oocyte differentiates into a mature egg
which loses the ability to divide. BY removing the block to cell
division, fertilization turns the mitotically inert egg into a
rapidly proliferating cell. During cleavage, i.e., during the
first 12 hr of development, the rate of DNA synthesis in the growing
sea urchin embryo is close to that of a bacterial cell, which is
considered as being one of the fastest dividing organisms, Sometimes
around the blastula stage, however, the mitotic activity and hence
the rate of DNA synthesis drops to a very low level.

Several lines of evidence indicate that the activity level of
DNA polymerase α (high molecular weight, N-ethylmaleimide sensitive)
is positively correlated with the rate of cell proliferation in
regenerating rat liver (Iwamura et al., 1968; Baril et al., 1973),
in tissue culture cells (Chang et al., 1973; Spadari and Weissbach,
1974; Craig et al., 1975; Chiu and Baril, 1975) and in phytohema-
gglutinin-stimulated lymphocytes (Fridlender et al., 1974).

Conversely, the activity of DNA polymerase β (low molecular
weight, N-ethyl-maleimide insensitive) appears to be relatively
constant during the cell cycle. In this paper we present evidence

[*] *A.S.I. Participants*

[**] *Institute of Histology and Embryology, University of Naples, Italy.*

that also in the sea urchin *Paracentrotus lividus* two major forms
of DNA polymerases are present (De Petrocellis et al., 1976). In
many respects they resemble the enzymes found in other eukaryotic
cells. Unlike what has been observed in other rapidly proliferating
tissues, sea urchin embryos have a constant level of DNA polymerase
activity throughout development; in particular no increase in DNA
polymerase α has been found during the period of intensive cell
division. We show that mitotically inert egg contains a high level
of DNA polymerase, 70% of which is represented by polymerase α.
This suggests that the egg contains a large store of enzyme which
is called into action following fertilization.

Another DNA-dependent DNA polymerase has been solubilized
from mitochondria. The activity of this enzyme increases during
development and reaches a maximum at the blastula stage. At the
onset of gastrulation the activity of the mitochondrial DNA
polymerase decreases again to the level found in mitochondria of
unfertilized eggs. This pattern of activity closely parallels the
endogenous incorporation measured with intact mitochondria (Gadaleta
et al., 1977).

(1)   Multiple Forms of DNA Polymerase in *Paracentrotus* Eggs

Paracentrotus eggs contain multiple forms of DNA polymerase.
Cell-free extracts prepared from unfertilized eggs and chromatographed on DEAE-cellulose were resolved in three peaks of activity
(Fig. 1). The first peak (DNA polymerase I) eluted at 0.075 M
NaCl, the second peak (DNA polymerase II) at 0.15 - 0.20 M NaCl
and the third peak at 0.40 M NaCl. Because of the low level and
high instability the third activity was not further characterized.
The two major peaks of activity could be differentiated on the
basis of their sensitivity to NEM and to high ionic strength.

DNA polymerase I activity was not affected by NEM or by high
ionic strength, while polymerase II was 97% inhibited by 10 mM NEM
and 70% by 200 mM KCl (Table I). DNA polymerase activity isolated
from nuclei of embryos at the blastula stage exhibited an elution
profile from DEAE-cellulose which was indistinguishable from that
of polymerase I (Fig. 2). In addition, the nuclear enzyme has
characteristics of resistance to NEM and salts similar to those of
polymerase I (Table I). We therefore concluded that the two enzymes
are identical or closely related forms.

(2)   Purification and Properties of DNA Polymerases I and II

The biochemical and biophysical properties of the two major DNA
polymerases present in sea urchin eggs have been studied on

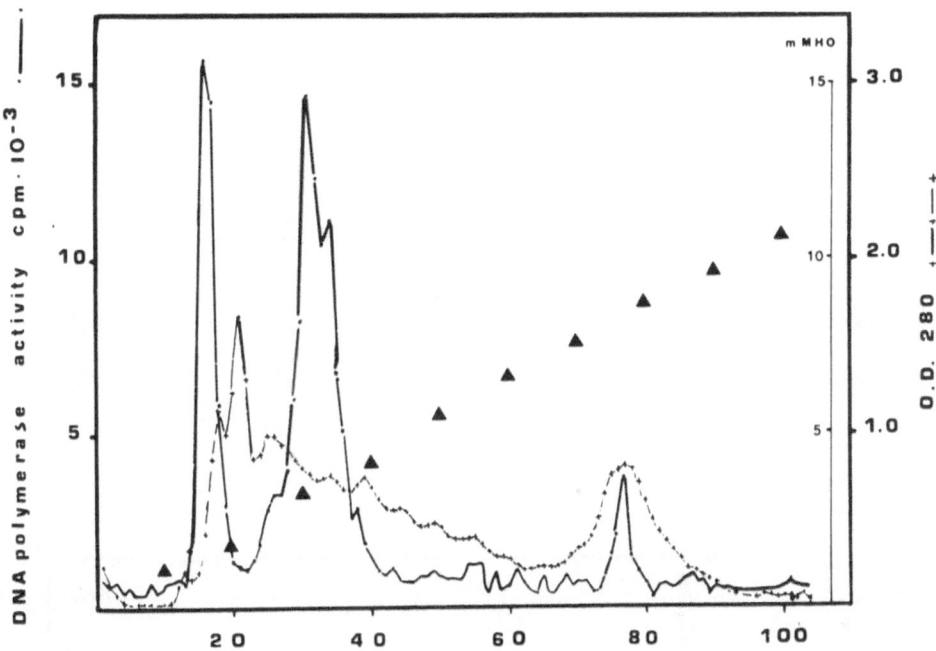

Figure 1. *DEAE-cellulose chromatography of DNA polymerase from unfertilized eggs of* Paracentrotus lividus. *Eggs were intermittently sonicated in 20 mM Tris-HCl pH 8.5 containing 25% glycerol, 1 mM EDTA, 1 mM 2-mercaptoethanol, 0.3 M NaCl. After high speed centrifugation followed by dialysis against the same buffer without NaCl, the extract was chromatographed on a DEAE-cellulose column which was eluted with a linear gradient of NaCl. Recovery of DNA polymerase activity ranged from 70 to 100%. Enzyme activity was measured with activated DNA and the four deoxynucleotide triphosphates one of which was* ^{3}H-labelled.

partially purified enzyme preparations. Embryonic nuclei were used as source for polymerase I while polymerase II was purified from unfertilized eggs. For DNA polymerase I, nuclei prepared by the method of Hinegardner (1962) from embryos at the mesenchyme blastula stage were disrupted by sonication in high ionic strength buffer. The nuclear extract was then purified by a DEAE-cellulose batch followed by chromatography on DEAE-cellulose. Final purification was about 150-fold with respect to a homogenate prepared from whole embryos. For DNA polymerase II, the homogenate of unfertilized eggs was first fractionated by streptomycin sulfate and ammonium sulfate and then chromatographed on DEAE-cellulose. The successive steps included chromatography on phosphocellulose and affinity chromatography on DNA-cellulose. Final purification was 500-fold.

Table I

Characteristics of the Two Major DNA Polymerases from
Unfertilized Eggs and of the DNA Polymerase from Embryonic Nuclei

Enzyme	Elution from DEAE-cellulose (mMHO)	% Inhibition by 10 mM NEM	% Inhibition by 200 mM KCl
DNA polymerase I	2	23	20
DNA polymerase II	5	97	70
Nuclear DNA Polymerase	2	25	22

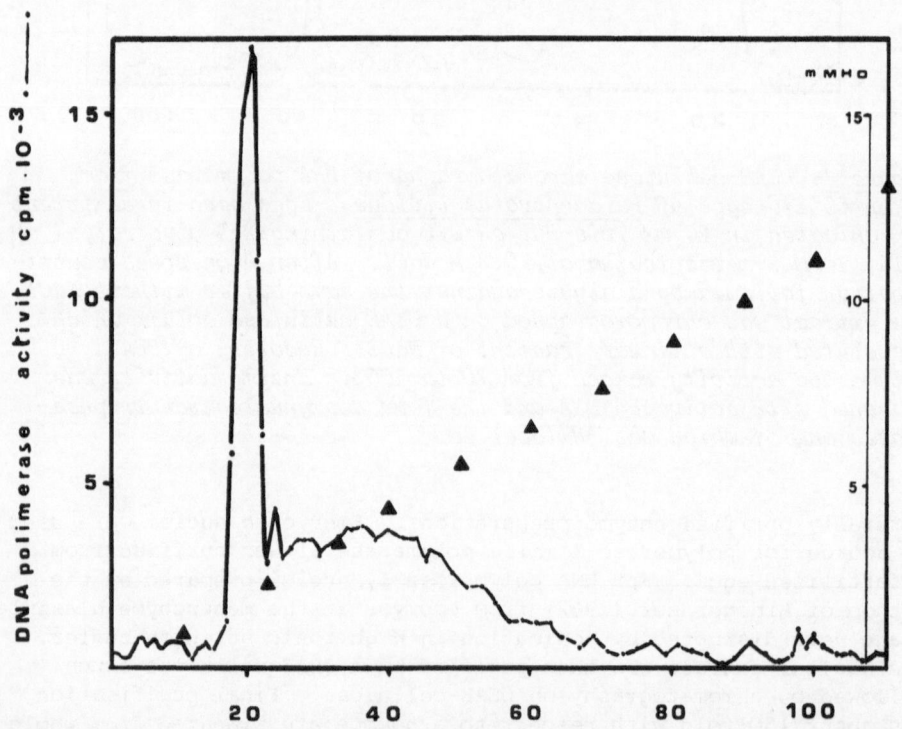

Figure 2. DEAE-cellulose chromatography of DNA polymerase prepared
from nuclei of Paracentrotus embryos. Nuclei were prepared by the
method of Hinegradner (1962). The preparation of the nuclear
extract and the conditions of chromatography were similar to those
reported in the legend of Fig. 1.

The biophysical characteristics of the purified enzymes are summarized in Table II. By gel filtration on Sephadex G-200 the apparent molecular weight of polymerase I was estimated to be 68,000 and that of polymerase II 150,000. Nevertheless, the two polymerases had identical sedimentation properties. The sedimentation coefficients determined by glycerol density gradient sedimentation was 3.6S for both enzymes. An estimate of the "true" molecular weights was obtained by this value of the sedimentation coefficient and the values of the Stokes radius derived from gel filtration data (Laurent and Killander, 1964). The partial specific volume was assumed to be 0.725 ml g^{-1}. The computed molecular weights were 50,000 for DNA polymerase I and 70,000 for DNA polymerase II. This latter value is considerably lower than that obtained by gel filtration alone and is indicative of anomalous behaviour on gel filtration probably due to molecular assymetry (Sedwick et al., 1975; Holmes et al., 1973; Barry et al., 1972; Craig et al., 1975).

According to the previous observations the purified enzymes exhibited differential sensitivity to sulphydryl-group reagents. Increasing concentrations of NEM and HO-MB decreased DNA polymerase I activity by only 25%, while DNA polymerase II was almost completely inhibited by both reagents (Fig. 3A and B). Finally, the two enzymes were also differentiated for their ability to utilize different primer-templates (Table III). Poly (dA·dT) was in general utilized by both enzymes better than activated DNA. Incorporation was template-directed because there was little incorporation when (^{3}H)-dTTP was substituted with (^{3}H)-dCTP. The homopolymer $A_n:(dT)_{\overline{12}}$ could be utilized by polymerase I but only with Mn^{++} as divalent cation. Neither enzyme was able to copy the double-stranded polymer Poly (dA).Poly (dT).

Table II

	Stokes Radius ($^{\circ}$A)	Sedimentation Coefficient (S)	Molecular Weight	
			Gel Filtration	"True"
DNA polymerase I	35	3.6	68,000	50,000
DNA polymerase II	46	3.6	150,000	70,000

Biophysical parameters of purified DNA polymerase I and II

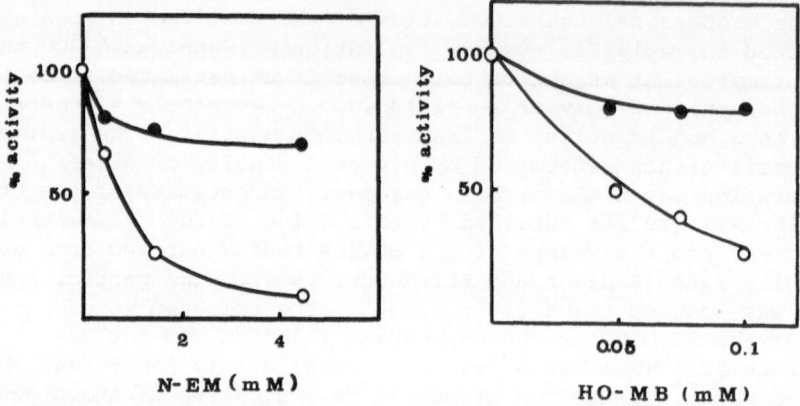

Figure 3. *Effect of N-ethylmaleimide (A) and p-hydroxymercuri-benzoate (B) on DNA polymerase I and II. Purified preparations of DNA polymerase I and II were preincubated for 10 min. at 0°C with the indicated concentrations of NEM. HO-MB was directly added to reaction mixture.* ●——● *DNA polymerase I;* O——O *DNA polymerase II.*

Table III

Effect of different primer-templates on the DNA
polymerase I and II from *Paracentrotus* 1

Primer-template	divalent cation	dNTP	DNA poly I pMoles/ 2 hrs.	DNA poly II pMoles/ 2 hrs.
Activated DNA	12 mM Mg^{++}	3 dNTP+(^{3}H)-dTTP	80	61.0
$(rA)_n(dT)_{\overline{12}}$	12 mM Mg^{++}	3 dNTP+(^{3}H)-dTTP	4	6.9
$(rA)_n(dT)_{\overline{12}}$	0.8 mM Mn^{++}	3 dNTP+(^{3}H)-dTTP	6	7.5
$(rA)_n(dT)_{\overline{12}}$	2.0 mM Mn^{++}	3 dNTP+(^{3}H)-dTTP	37	5.5
Poly(dA.dT)	12 mM Mg^{++}	dATP+(^{3}H)-dTTP	233	1095.0
Poly(dA).Poly(dT)	12 mM Mg^{++}	dATP+(^{3}H)-dTTP	2	5.1
Poly(dA.dT)	12 mM Mg^{++}	dATP+(^{3}H)-dCTP	4	13.8

(3) Activity level of DNA polymerase in embryos at
Different Stages of Development

During the early development of *Paracentrotus* DNA polymerase
activity is almost constant and independent of the rate of DNA
synthesis in the growing embryo (Fig. 4). The presence of
inhibitors was ruled out by experiments carried out with mixed
extracts.

The activity levels of DNA polymerase I and DNA polymerase II
present in eggs and embryos at different stages were measured on
extracts preincubated for 10 minutes at 0°C with and without 10 mM
NEM. The fraction of activity not inhibited by NEM was considered
to be polymerase I, while the enzyme activity suppressed by NEM
was attributed to DNA polymerase II. No significant changes of the
activity of the two polymerases were observed during cleavage
(Table IV). Similar results were obtained with experiments in

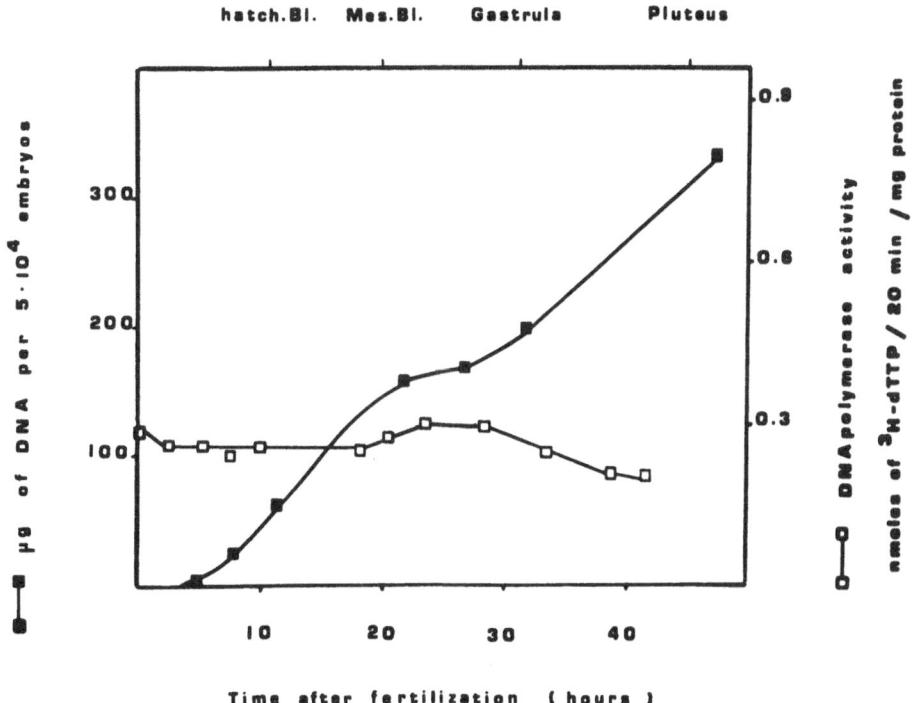

Figure 4. DNA polymerase activity in embryos at different stages
of development.

Table IV

Percent of NEM-sensitive and NEM-insensitive DNA polymerase
activity in homogenates from *Paracentrotus* embryos
at different stages of development

Extract source	DNA polymerase activity	
	NEM-sensitive %	NEM-insensitive %
Unfert. eggs	81	19
2-Cells	66	34
Morula	57	43
Blastula	77	23
Mes. blastula	72	28

Table V

Percent of DNA Polymerase I and II in Extracts
of *Paracentrotus* embryos

Extract source	DNA polymerase activity	
	DNA polym. II %	DNA polym. I %
Unfert. eggs	76	24
Morula	63	37
Blastula	74	26
Mes. blastula	86	14

12 mg protein of the extracts from embryos at the
indicated stages were fractionated on DEAE-cellulose
columns (0.5 x 6 cm). DNA polymerase I was eluted
with 0.05 M NaCl, DNA polymerase II with 0.2 M NaCl.

which the two enzymes were separately assayed after fractionation
of extracts on DEAE-cellulose columns (Table V).

These data indicate that no strict correlation exists in sea
urchin embryos between activity of DNA polymerase I and II and DNA
synthesis.

(4) Mitochondrial DNA Polymerase During Development of *Paracentrotus*

The data reported in section 3 do not take into account mitochondrial DNA polymerase activity. Mitchondria isolated from
Paracentrotus eggs incorporate deoxynucleoside triphosphates into
DNA. The rate of incorporation was dependent on the stage of
development and was maximal in mitochondria from blastulae
(Gadaleta et al., 1977). We have solubilized a DNA polymerase
activity from a pure fraction of mitochondria isolated from
Paraentrotus eggs. The enzyme is tightly bound and could only be
extracted by means of non-ionic detergents or IM KCl. The
solubilized enzyme was unable to utilize $A_n:(dT)_{\overline{12}}$ as template
under conditions in which this polymer was copied by polymerase I.
In addition, the behaviour on DEAE-cellulose was substantially
different from that of the two polymerases already described. The
activity level of the mitochondrial polymerase has been measured
under conditions which allowed complete solubilization and high
recovery from purified mitochondria. The results reported in Fig. 5
show that the mitochondrial DNA polymerase activity assayed with
exogenous template began to increase from the morula stage, reached
a maximum in the blastulae and subsequently decreased again to
the level of the unfertilized eggs. This pattern of activity closely
followed that of the endogenous incorporation also reported in
Fig. 5.

DISCUSSION

The two major forms of DNA polymerase present in *Paracentrotus*
eggs exhibit features that are typical of the eukaryotic DNA
polymerases.   DNA polymerase I is also present in the embryonic
nuclei and, like other DNA polymerase β, is relatively insensitive
to thiol-group blocking reagents and is able to copy the synthetic
template $A_n:(dT)_{\overline{12}}$. DNA polymerase II, on the other hand, exhibits
many characteristics of a DNA polymerase α. A purified fraction of
this enzyme has a sedimentation coefficient of 3.6S which is rather
unusual for α polymerase. Less purified fractions of the same
enzyme, however, sediment with a peak at 5.8S and this raises the
possibility that the 3.6S form might be a subunit of the enzyme.
The characteristics of DNA polymerase II appear clearly different
from those of polymerase I.   The main distinctive property of DNA

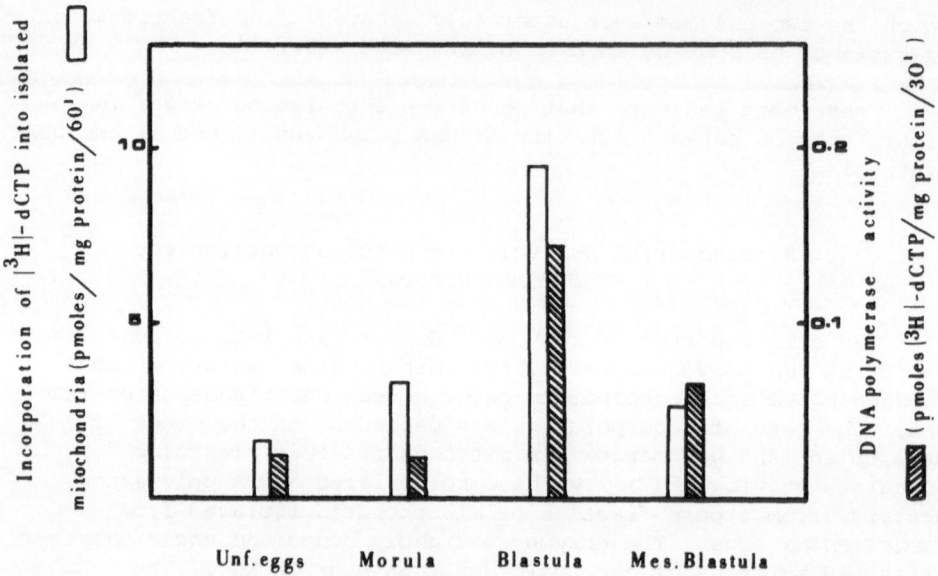

<u>Figure 5</u>. *DNA polymerase activity in mitochondria from embryos at
different stages of development. Mitchondria were prepared from
embryos at the indicated stages and purified on sucrose gradient.
DNA polymerase activity was solubilized by extraction with 1 M
KC1 and after extensive dialysis, assayed with exogenous DNA
template. The endogenous activity was measured in intact mitochondria with no added template.*

polymerase II is its pronounced sensitivity to the reagents of
the sulphydryl groups. We have exploited this peculiarity for
the quantitative assay of the two enzymes in homogenates prepared
from embryos at different stages of development. In particular,
we have analyzed the period between fertilization and the onset of
gastrulation because this coincides with the time of intensive
synthesis of DNA in the embryos. We have found no change in total
DNA polymerase activity during this period and in particular
no increase in activity of the α-like DNA polymerase II. Similar
results have been recently reported by Morris and Rutter (1976)
in *Strongylocentrotus franciscanus*.

Since DNA polymerase α activity is in general markedly
increased in proliferating tissues, the fact that such an increase
does not occur in the rapidly dividing sea urchin embryo suggests
that the unfertilized egg is equipped with sufficient enzyme to
cope with the high level of DNA synthesis that begins after
fertilization.   Indeed, the egg contains a large store of DNA

polymerase of type α and β, the former being the predominant form. It has been reported that during the early development of the sea urchin embryos DNA polymerase activity is progressively transferred from the cytoplasm to the nucleus (Fansler and Loeb, 1972).

This might suggest that the enzymes which are stockpiled in the egg are compartmentalized and gradually called into action as development proceeds.

Less information is available on sea urchin mitochondrial DNA polymerase. The data concerning the characteristics of this enzyme are still preliminary. Nevertheless they suggest that the mito-chondrial polymerase is a distinct enzymic entity. The activity of this enzyme undergoes a transient increase during development. Since data concerning *in vivo* synthesis of mitochondrial DNA are not available for sea urchin we do not know whether the increase in mitochondrial DNA polymerase activity can be related to the bio-genesis of new mitochondria.

REFERENCES

Baril, E.F., Jenkins, M.D., Brown, O.E., Laszo, J., and Morris, H.P. (1973) Cancer Res. 33:1187-1193

Barry, J. and Alberts, B. (1972) Proc. Natl. Acad. Sci. USA 69: 2717-2721

Chang, L.M.S., Brown, M. and Bollum, F.J. (1973) J. Mol. Biol. 74:1-8

Chiu, R.W. and Baril, E.F. (1975) J. Biol. Chem. 250:7951-7957

Craig, R.K., Costello, P.A. and Keir, H.M. (1975) Biochem. J. 145:233-240

De Petrocellis, B., Parisi, E., Filosa, S. and Capasso, A. (1976) Biochem. Biophys. Res. Commun. 68:954-960

Fansler, B. and Loeb, L.A. (1972) Exptl. Cell Res. 75:433-441

Fridlender, B., Virasoro, S. and Mordoh, J. (1974) Fed. Proc. 33:1277

Gadaleta, M.N., Nicotra, A., Del PRete, M.G. and Saccone, C. (1977) Cell Differentiation (in press).

Hinegardner, R.T. (1962) J. Cell Biol. 15:503-508

Holmes, A.M. and Johnstone, I.R. (1973) FEBS Letters 29:-1-6

Iwamura, Y., Ono, T. and Morris, H.P. (1968) Cancer Res. 28:
    2466-2476

Laurent, T.C. and Killander, J. (1964) J. Chromatog. 14:317-322

Morris, P.W. and Rutter, W.J. (1976) Biochemistry 14:3106-3113

Sedwick, W.D., Shu-Fong Wang, T. and Korn, D. (1975) J. Biol. Chem.
    250:7045-7056

Spadari, S. and Weissbach, A. (1974) J. Mol. Biol. 86:11-20

IDENTITY BETWEEN NUCLEAR AND MITOCHONDRIAL DNA POLYMERASES γ FROM CHICK EMBRYO

Gilbert M. Brun[1], Anna I. Scovassi[1] and
Umberto Bertazzoni*[2]

[1]Laboratoire de Biochimie-Fondation Curie-Section de
Biologie
26, rue d'Ulm 75231 Paris Cedex 05, France
[2]Laboratorio di Genetica Biochimica ed Evoluzionistica
del C.N.R.
Via S. Epifanio 14, 27100 Pavia, Italy

The availability of specific tests for DNA polymerases α and β has made possible to obtain quantitative determinations of these enzymes in crude extracts and to study their physiological roles (1). A new specific assay for DNA polymerase γ has been described recently by the Weissbach group (2) who showed that while γ-polymerase is active on poly(A)·oligò(dT) in the presence of phosphate, β-polymerase is completely inhibited. This has opened up the possibility to distinguish this enzyme from α and β-polymerases and thus to study its localization and function.

We have studied the DNA polymerase γ in chick embryo since the information about this enzyme in this organism is incomplete and some confusion with β-polymerase and Reverse Transcriptase has been claimed. The very recent finding by Bolden et al. (3) that HeLa cell and rat liver mitochondrial DNA polymerase have the same properties of the cytoplasmic γ-polymerase prompted us to study the subcellular localization of this enzyme in chick embryo.

We have found that chick embryo mitochondria contain a single DNA polymerase whose properties are not distinguishable from the DNA polymerase γ found in the nuclei of the same cells (4).

RESULTS AND DISCUSSION

We have studied the subcellular distribution of the DNA poly-

*A.S.I. *Participant*

merase γ in chick embryo by fractionating the cells into the three
main fractions, namely nuclei, mitochondria and cytosol.   The
relative amounts of the α, β and γ -polymerases have been determined
in crude extracts and after sucrose gradient sedimentation.   The
results are shown in Fig. 1.   The γ-polymerase is present in the
nuclear fraction as a minor component whereas in the mitochondrial
fraction it becomes the prevalent enzyme.   In more extensively
purified preparations of mitochondria it is evident that these

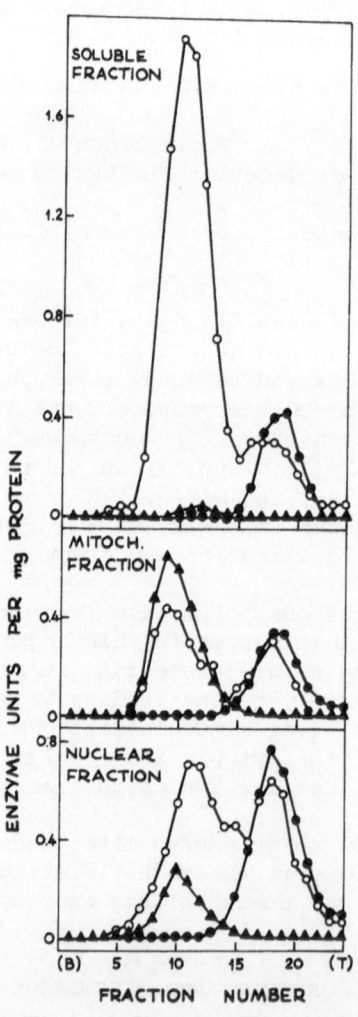

Figure 1. Sucrose gradient sedimentation analysis of the chick
embryo soluble, mitochondrial and nuclear fractions.
O -- O α assay, ● -- ● β assay, ▲ -- ▲ γ assay.

organelles contain a single DNA polymerase whose activity responds only to the γ-polymerase specific assay and is not unspecifically associated with their external membranes (4).

In order to ascertain the identity between the nuclear and the mitochondrial γ-polymerases we have purified these enzymes several hundred folds from chick embryo nuclei and mitchondria. Table 1 summarizes the properties of the purified γ-polymerases and Figs. 2 and 3 give details of K_m and heat inactivation curves.

Table 1

Properties	γ-polymerases	
	Nuclear	Mitochondrial
Sedimentation coefficient	7-7.5S	7-7.5S
K (Fig. 2)	0.8-1.2μM	0.8-1.2μM
pH optimum	7.8-8	7.8-8
Preferred divalent cation	Mn^{++}	Mn^{++}
Optimal K phosphate	70mM	70mM
Optimal KCl	100mM	100mM
Chromatographic behaviour	identical	
Heat inactivation (Fig. 3)	identical	
Antibodies against DNA polymerase α	no cross reactivity	
" " " " β	no cross reactivity	
NEM inhibition (1mM)	12%	15%
(5mM)	59%	62%
Ethidium bromide inhibition (1μg/ml)	68%	75%
Template specificity (% of maximal act.)		
poly(A)·(dT)$_{15}$	100	100
poly(dA)·(dT)$_{15}$	17	14
poly(C)·(dG)$_{15}$ (-KP)	28	22
poly(dC)·(dG)$_{15}$	7	5
activated DNA	12	11
globin mRNA·oligo(dT)	0.3	0.3

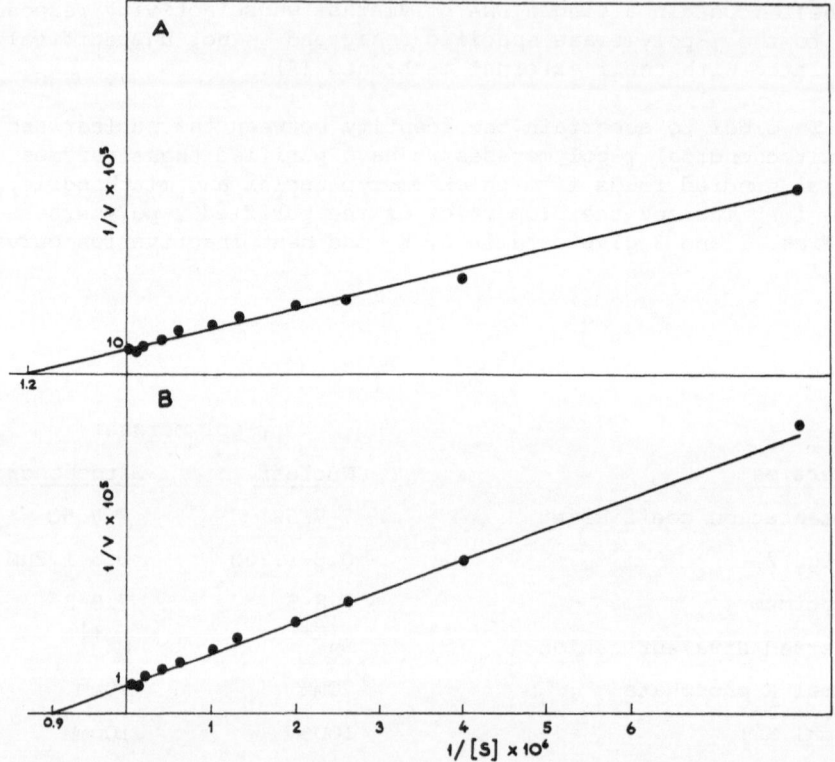

Figure 2. _Determination of K_m for the nuclear (panel A) and mitochondrial (panel B) DNA polymerases._

The two enzymes show the same physicochemical properties (sedimentation, optimal pH, K_m values, stimulation by K phosphate and
chloride ions) and are inhibited at the same extent by NEM and
ethidium bromide. They are not inhibited by antibodies against DNA
polymerases α and β from chick embryo. The response to different
template-primers systems is very similar if not identical. The
properties of the purified enzymes are very closed to those
reported recently by Weissbach and collaborators for the HeLa cell
DNA polymerase γ (2). and HeLa mitochondrial DNA polymerase γ (3).

We have also determined the variations of the DNA polymerase
γ during the chick embryo development. The levels of the γ-enzyme
at different days of egg incubation, as compared to the α- and β-
polymerase, are given in Table 2.

The changes of the three DNA polymerases are not significant
since their amounts remain very much the same during the stages of
development considered here. It seems therefore that all three

Table 2

Levels of DNA Polymerase α, β and γ in
Different Day Old Chick Embryo

Days of incubation	Enzyme units/g of embryo		
	α	β	γ
5	608	92	42
8	490	100	28
10	516	106	35
12	465	90	34
14	510	100	36

The assays were performed as described in ref. 4. A
unit of enzyme cprresponds to the nmol of dTMP
(multiplied by 3.5 in the case of α and β-assay)
incorporated in 1 hour per gram of embryo.

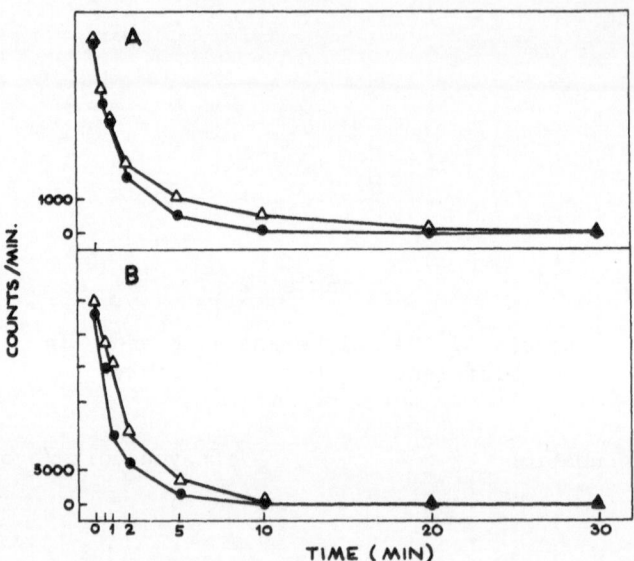

Figure 3. *Heat inactivation curves of the nuclear (panel A) and*
mitochondrial (panel B) γ-polymerases.
△ — △ *with added poly(A)·(dT)*$_{15}$, ● — ● *without poly(A)·(dT)*$_{15}$

enzymes are needed during this early period of chick development.

In conclusion the results presented above suggest that the
same γ-enzyme is present in two different cellular compartments,
namely nuclei and mitochondria. As to the function of this
polymerase very little can be said about the nuclear activity
since no direct correlation with the DNA metabolism has yet been
found, while the mitochondrial γ-activity is very probably
involved in the replication and repair of the mitochondrial DNA,
being the only DNA polymerase species found in mitochondria
purified from HeLa cell and rat liver (3), rat neurons (5);
Xenopus laevis (6) and chick embryo (this paper and 4). It is
possible that all mitochondria from different species contain a
DNA polymerase with the properties of the -polymerase. We are
presently studying the phylogeny of this enzyme and we have
already evidence that it can be found also in lower eukaryotes
as yeast and sea urchin (our unpublished observations).

ACKNOWLEDGEMENTS

U.B. is a EURATOM scientific agent and this publication is
contribution No1473 of the Biology Division of European
Communities.

REFERENCES

1. Bollum, F.J. (1975) Progr. Nucleic Acid Res. Mol. Biol. 15:109-144

2. Knopf, K.W., Yamada, M. and Weissbach, A. (1976) Biochemistry 15:4540-4548

3. Bolden, A., Pedrali Noy, G. and Weissbach, A. (1977) J. Biol. Chem. 252:3351-3356

4. Bertazzoni, U., Scovassi, A.I. and Brun, G.M. (1977) Eur. J. Biochem., in press

5. Hübscher, U., Kuenzle, C.C. and Spadari, S. (1977) Eur. J. Biochem., in press

6. Bazzigalupo, P. and Baldi, M.I., paper in preparation.

NEURONS AS A MODEL SYSTEM FOR *IN VIVO* STUDIES OF THE POSSIBLE FUNCTION OF DNA POLYMERASES IN DNA REPLICATION AND REPAIR SYNTHESIS DURING DEVELOPMENT

Ulrich Hübscher[1]*, Clive C. Kuenzle[1] and Silvio Spadari[2]

[1]Department of Pharmacology and Biochemistry
School of Veterinary Medicine, University of Zurich
Zurich, Switzerland

[2]Laboratorio di Genetica Biochimica ed Evoluzionistica
Consiglio Nazionale delle Ricerche
Pavia, Italy

INTRODUCTION

The identification of three classes of DNA polymerases (α, β and γ) in mammalian cells (1) has naturally led to the question of the role of these DNA polymerases in DNA replication and repair. This paper will present studies on rat neurons DNA polymerases which are pertinent to this question. Forebrain neurons were selected for several reasons:

(a) they are essentially unable to replicate their DNA, undergo mitosis and proliferate after birth (2-4).

(b) they can be isolated from glial cells (5,6) and represent an *in vivo* intact unicellular system which can be used to correlate development with the capacity to perform DNA replication and DNA repair synthesis.

(c) synaptic end knobs contain mitochondria completely unassoc-iated with nuclei and these structures can be isolated as distinct particles (synaptosomes) surrounded by a closed membrane (7), thus allowing the isolation of mitochondrial DNA polymerase and *in vitro* studies on the role of mt-polymerase in the replication and repair of mitochondrial DNA.

Our studies can be summarized as follows:
A.S.I. Participant

605

1.   Correlation between DNA polymerase α, β and γ activities and
      the *in vivo* rate of mitotic activity of the neurons.

      The results are reported in Fig. 1.   DNA polymerase α shows its
highest specific activity before birth and then declines rapidly
reaching a level of almost zero by the 23rd day after birth (1A).
The enzymes decrease approximately by 80% between days -4 and birth
and correlates temporally well with the decline in mitotic activity
(Fig. 1B) which drops to zero at birth.   The enzyme seems to be
synthesized and degraded at a rapid rate in order to keep up with
the requirements of the cell. Mixing experiments gave no evidence
for the presence of soluble dissociable inhibitors or activators of
DNA polymerases at different times during development.

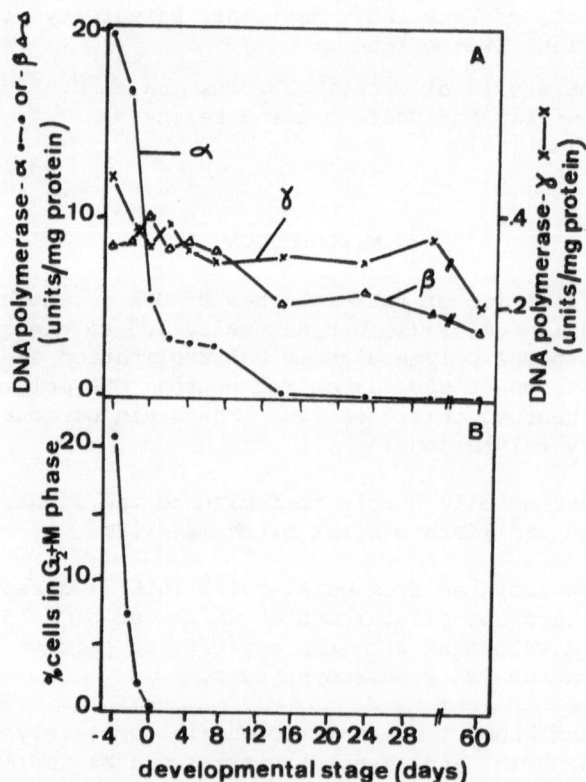

Figure 1 (A).   *Developmental profiles of DNA polymerases* α, β *and*
γ *in forebrain neurons*.

DNA polymerases α, β *and* γ *activities have been determined using
assay conditions which permitted direct estimation of each DNA
polymerase in crude extracts (8,9)*.

Figure 1 (B).   *Percentage of cells in G₂ and M phase
as determined by cell-cycle analysis (10)*.

In contrast the activities of DNA polymerase β and γ are constant during proliferation of neurons and do not change with the arrest of mitotic activity, thus remaining independent of the *in vivo* rate of cell multiplication. Therefore since mitosis no longer occurs during the remaining lifetime of the neurons, one or both enzymes may serve a repair rather than a replicative-type function in these cells.

This can be checked *in vivo* by looking, by autoradiography, if injection of adult rats (whose neurons no longer contain α-polymerase) with mutagens able to cause bulky DNA lesions, results in positive unscheduled DNA synthesis of neuronal nuclei. One could then eventually attempt to correlate *in vivo* development (differentiation and ageing) with the capacity to perform both DNA replication and DNA repair synthesis in neurons. These studies are in progress in our laboratory.

2.   Identity of DNA polymerase γ from rat synaptosomal mito-
     chondria and cortical nuclei.

Until recently mt-polymerase was thought to be a distinct enzyme species, but recent work by Bolden et al. (11) has shown that the mitochondrial DNA polymerase in HeLa cells and rat liver resembles, and may even be identical with, the γ-polymerase found in the cytoplasm. We had thus the idea of approaching this problem by using a novel system particularly suited for separating nuclear and mitochondrial compartments. Taking advantage of the fact that in the brain synaptic end knobs contain mitochondria completely unassociated with nuclei and that these structures can be isolated as distinct particles (synaptosomes) surrounded by a closed membrane (7) we have decided to isolate the γ-polymerase from cortical nuclei on one hand and the mt-polymerase from synaptosomes on the other (Fig. 2).

Glycerol gradient centrifugation of a crude intrasynaptosomal mitochondria extract and the results of several chromatographic steps (DEAE-cellulose, phosphocellulose and native DNA-cellulose) have provided evidence for the presence of a single DNA polymerase activity in synaptosomal mitochondria. In addition isoelectric focusing of the purified polymerase revealed a single sharp band of polymerase activity.

A great similarity between the purified mitochondrial DNA polymerase and nuclear γ-polymerase was found by several criteria reported in Table I.

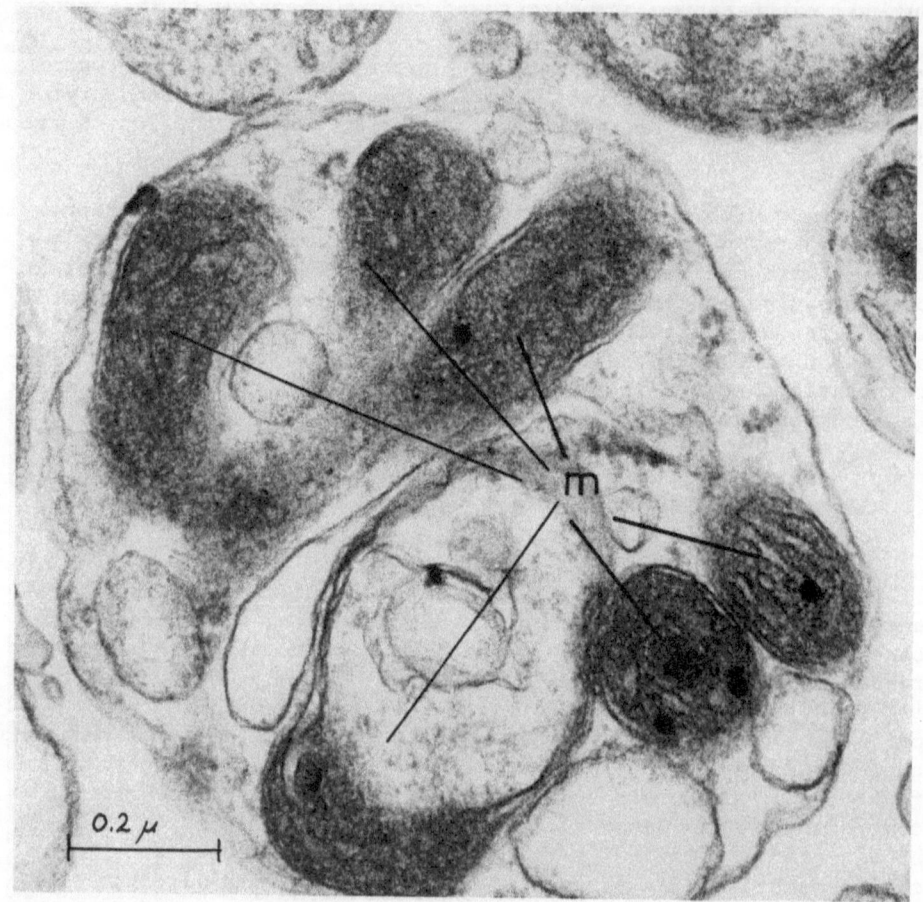

0.2 μ

Figure 2. Electromicrograph of a synaptosome with intrasynaptosomal mitochondria (m).

(3)   Deoxyribonucleic acid synthesis dependent on exogenous
       triphosphates in synaptosomes.

       Considering that synaptosomes may offer significant protection
of mitochondria during isolation, we have planned to use them for
studying the role of mitochondrial DNA polymerase in the
replication and/or repair of mitochondrial DNA. The first step was
to elucidate the optimal conditions for incorporation of deoxyribonucleoside triphosphates into mitochondrial DNA of synaptosomes.
These are: preincubation of synaptosomes at 37°C for 30 min in
50 mM Tris-HCl (pH 8.5), 0.5 mM ATP, 120 mM KCl, 2 mM $MnCl_2$, 1 mM
DTT, 0.1 mg/ml BSA followed by addition of the four deoxynucleotides

Table I

Similarity between purified nuclear DNA polymerase γ and
synaptosomal mitochondrial DNA polymerase

Properties	Nuclear γ-polymerase	Synaptosomal mt-polymerase
Molecular weight	180,000	180,000
Isolelectric point	5.4	5.4
Optimum reaction pH	8.0	8.0
Preferred divalent cation	Mn^{++}	Mn^{++}
Optimal Mn^{++} (mM)	0.1	0.1
Optimal Mg^{++} (mM)	5	5
Effect of NEM	inhibition	inhibition
Effect of ionic strength	stimulation	stimulation
Effect of K-phosphate	stimulation	stimulation
Chromatographic behaviour	identical	
Heat inactivation	identical	
Template: $rA_n \cdot dT_{12-18}$	400	500
activated DNA	100	100
$dA_n \cdot dT_{12-18}$	26	80
polyA	<0.1	<0.1
dT_{12-18}	<0.1	<0.1
$mRNA \cdot dT_{12-18}$	<0.1	<0.1

triphosphates (50 μM each) with (^{3}H)-dTTP and (^{3}H)-dGTP (1,000 cpm/pmol).

Requirements for optimal incorporation of deoxynucleosides triphosphates are shown in Table II.

Table II

Requirements for optimal incorporation of deoxynucleoside triphosphates into mitochondrial DNA of synaptosomes

Conditions	cpm/25 μl synaptosomes (0.3 mg protein)	%
complete with preincubation	4,800	100
complete without preincubation	3,000	63
pH 7.4	1,352	28
minus KCl	2,328	48
minus MnCl$_2$, plus MgCl$_2$ (5 mM)	1,000	21
plus K-phosphate (50 mM, pH 8.5)	802	17

On neutral CsCl gradient (Fig. 3) the in vitro labelled mt-DNA of synaptosomes bands at a density of 1.704 g/cm^3, while neuronal nuclear DNA banded at 1.697 g/cm^3 on a parallel gradient. We have no data yet to conclude that *in vitro* DNA synthesis in synaptosomes is a replicative rather than a repair process. However, increasing doses of novobiocin, known to inhibit replicative but not repair DNA synthesis in bacteria (13,14), progressively inhibits incorporation of deoxynucleosides triphosphates into mitochondrial DNA of synaptosomes. Since we have shown (9) that synaptosomal mitochondria contain a single DNA polymerase activity, we are now trying to elucidate its role in DNA replication and/or repair of mitochondrial DNA.

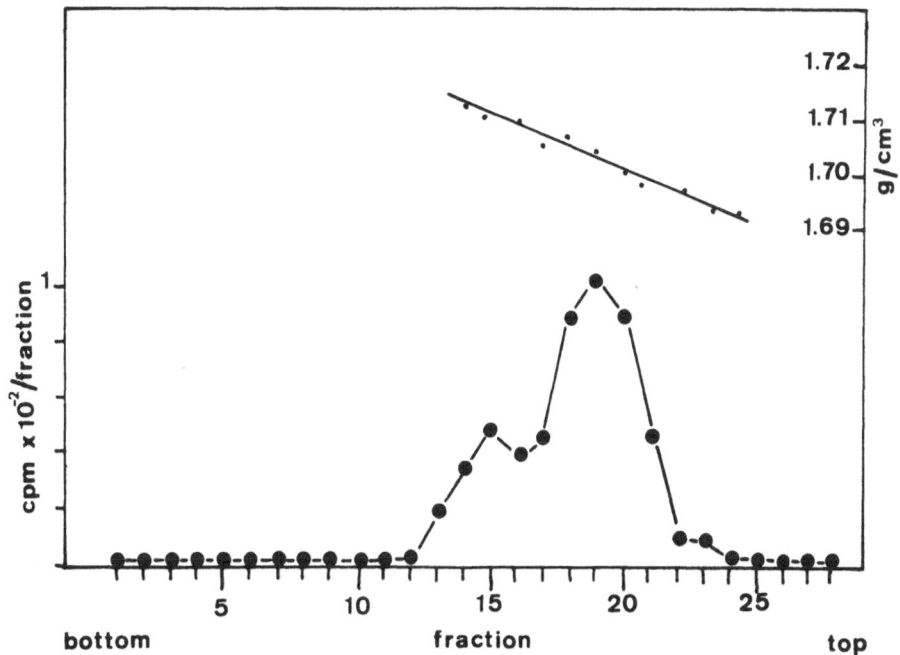

Figure 3. *Neutral CsCl centrifugation of mitochondrial DNA from synaptosomes.*
Incubation was performed as described in the text and the mt-DNA was extracted and centrifuged in CsCl as described by Flamm et al.
(12).

REFERENCES

(1)   Weissbach, A., Baltimore, D., Bollum, F.J., Gallo, R.C. and
      Korn, D.   (1975)   Science 190:401

(2)   Altman, J. In: "Handbook of Neurochemistry", Lajtha, J. (ed.)
      Vol 2, p. 137, Plenum Press, New York (1969)

(3)   Jacobson, M. In:"Developmental Neurobiology", Jacobson, M.
      (ed.) p.1 Rinehard and Winston Inc., New York (1970)

(4)   Haas, R.J., Werner, J. and Fliender, T.M.   (1970)   J. Anat.
      107:421

(5)   Sellinger, O.Z., Azcurra, J.M., Johnson, D.E., Ohlsson, W.G.
      and Lodin, Z. (1971) Nature New Biol. 230:253

(6) Kuenzle, C.C., Knüsel, A., and Schümperli, D. In: "Methods in Cell Biology", Prescott, D.M. (ed.) Vol XV, p.89, Academic Press, New York (1977)

(7) Morgan, I.G., Wolfe, L.S., Mandel, P. and Gombos, G. (1971) Biochim. Biophys. Acta 241:737

(8) Hübscher, U., Kuenzle, C.C. and Spadari, S. ((1977) Nuc. Acid Res. in press

(9) Hübscher, U., Kuenzle, C.C. and Spadari, S. (1977) Eur. J. Biochem. in press

(10) Crissman, H.A. and Tobey, R.A. (1974) Science 184:1297

(11) Bolden, A., Pedrali, Noy, G. and Weissbach, A. (1977) J. Biol. Chem. 252:3351

(12) Flamm, W.G., Birnstiel, M.L. and Walker, P.M.B. In: "Sub-cellular Components" Birnie, G.B. (ed.) p.279, Butterworth, London (1972)

(13) Staudenbauer, W.L. (1975) J. Mol. Biol. 96:201

(14) Ryan, M.J. and Wells, R.D. (1976) Biochemistry 15:3778

DNA POLYMERASE IN CHROMATIN

Ellen Fanning[*], Karl-Heinz Klempnauer, Bernd Otto,
Ernst-Jürgen Schlaeger and Rolf Knippers

Fachbereich Biologie
University of Konstanz
D-7750 Konstanz, Germany

INTRODUCTION

Two major types of DNA polymerase from mammalian cells have
been described. The DNA polymerase α (6-8S) is the main activity
in growing cells. The smaller DNA polymerase β (3-4S) constitutes
about 5-15% of the total polymerase activity in growing cells. The
roles of each of these polymerase activities in the replication of
mammalian chromatin are not yet clear. However, the evidence
available indicates that the α polymerase activity is closely
correlated with DNA replication in growing L cells (Chang and
Bollum, 1973), regenerating rat liver (Baril et al., 1973), in HeLa
cells (Spadari and Weissbach, 1974; Chiu and Baril, 1975) and BHK21
cells (Craig et al., 1975). The characteristics of these enzymes
have been described in some detail (for review, see Weissbach, 1975;
Wintersberg, 1977).

Most studies of mammalian DNA polymerases have employed isolated
enzymes to copy templates selected by the investigators, but perhaps
a better understanding of eukaryotic DNA replication could be gained
by studying mammalian DNA polymerases on their natural template
chromatin.

We have employed two systems in our study of DNA polymerases
in chromatin. The first, bovine lymphocytes prepared from lymph
nodes, offers the possibility of studying polymerase activity in
resting lymphocytes or in lymphocytes stimulated to divide by
various plant lectins. Bovine lymphocytes are composed of about
60% T-cells, which can be stimulated by Concanavalin A (Con A). At
24-36 h after addition of Con A to lymphocytes in culture, DNA
A.S.I. Participant

replication begins, reaching its maximum at about 48 h after
stimulation (Fig. 1). We have begun to exploit a second system,
simian virus 40 (SV40) DNA nucleoprotein complexes, to study
enzymes involved in DNA replication. These SV40 DNA-protein
complexes can be eluted from nuclei prepared from monkey cells
productively infected with SV40 (Su and DePamphilis, 1976). The
ability to separate replicating SV40 chromatin from nonreplicating
complexes by zone velocity sedimentation (Su and DePamphilis, 1976)
offers the opportunity to study DNA polymerases, as well as other
replication enzymes, on a defined natural template.

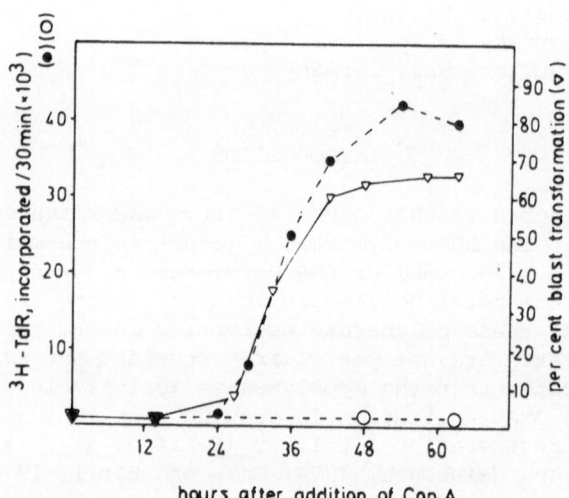

hours after addition of Con A

Figure 1. ³H-thymidine incorporation and blast transformation in Con
A-activated lymphocytes (Sons, 1976).
Bovine lymphocytes (5 x 10⁶ cells/ml) were labeled with 2.5 µCi/ml
³H-thymidine (spec.act.: 17 Ci/mmole) at different times after
stimulation by Concanavalin A. Incorporation was stopped after 30
min at 37°C by adding 5 volumes of ice cold phosphate-buffered
saline. The cells were pelleted and resuspended in 5 ml cold 5%
trichloroacetic acid (TCA). The precipitates were collected on
glass fiber filters and washed with TCA. All determinations were
performed in duplicate.

Con A-activated cells (●)

Untreated control cells (○)

The percentage of blast formation was followed in the Con A-
treated cell culture by staining methanol-fixed cells with Giemsa.
Blast cells were identified by their size and by their high plasma-
to nucleus ratio.

MATERIALS AND METHODS

Cell culture. Lymphocytes prepared from bovine retro-
pharyngeal lymph nodes were cultured as described earlier (Peters,
1975). Lymphocytes were stimulated after incubation for about 20 h
at 37°C in a humid 5% CO_2 atmosphere by adding 5 µg/ml of
Concanavalin A (Boehringer, Mannheim) to the culture.

Mouse ascites cells were propagated in adult Balb/c mice
and harvested 10 days after intraperitoneal injection of 10^6 cells
per animal.

CV-1 cells were grown in monolayer cultures in Dulbecco's
modified Eagle medium (DMEM) (Gibco) and 10% calf serum (Flow
Laboratories).

Virus. Simian virus 40 (SV40), strain Rh 911, was a gift from
G. Sauer, Heidelberg. The virus was propagated in CV-1 cells
infected with plaque-purified SV40 at a multiplicity of 0.01 plaque-
forming units per cell and harvested 10-12 days later.

Micrococcal nuclease digestion. Lymphocyte nuclei or chromatin
prepared from mouse ascites cells according to Hancock (1974) was
digested with 20-100 units/ml of micrococcal nuclease (Boehringer,
Mannheim) in 10mM Tris-HCl, pH 7.5, 10mM 2-mercaptoethanol, 1mM
PMSF, 1mM $CaCl_2$ for 15-20 min at 37°C. The digestion was stopped
by adding EDTA to a final concentration of 5 to 10 mM, and in some
experiments Triton X-100 (final concentration, 1%) and sodium
deoxycholate (final concentration, 0.5%).

Zone velocity sedimentation of micrococcal nuclease digestion
products. Insoluble material was removed by low speed centrifugation and the supernatant was layered on a linear 5 to 20% sucrose
gradient containing 2mM EDTA, pH 7.5 and 10mM 2-mercaptoethanol.
The conditions of sedimentation are given in the figure legends.
The gradients were collected through a flow-through cuvette to
record absorption at 260 nm.

DNA polymerase assay. The assay mixture contained 50mM Tris-
HCl, pH 7.6, 20mM $MgCl_2$, 10mM 2-mercaptoethanol, 500 µg/ml bovine
serum albumin, 100 µg/ml activated calf thymus DNA and 20 µM dATP,
dGTP, and dTTP. ^{3}H-dCTP (2 µM) was used as the radioactive tracer
at a specific activity of 600 cpm/pmole. The sample to be assayed
(0.05 ml) was incubated in 0.2 ml of assay mixture at 37°C for 60
min. Radioactivity incorporated was determined by TCA precipitation.

SV40 DNA-protein complexes. Confluent CV-1 cells were infected
with 1-5 plaque-forming units per cell of SV40. At 36-38 h post-
infection, the infected cultures were washed 3 times with isotonic
saline, the cells were scraped off the plates in Buffer A (10mM

Hepes, pH 7.8,   5mM KCl, 0.5mM $MgCl_2$ and 1mM dithiothreitol).
Nuclei were preapred and SV40 complexes were eluted from them in
the same buffer, as described previously (Su and DePamphilis, 1976).

RESULTS AND DISCUSSION

1. Expression of DNA Replication Enzymes in Lymphocytes

During the DNA replication phase in Con A-stimulated lymphocytes several different enzyme activities have been found in
significantly greater amounts than in resting lymphocytes. These
enzymes could possibly be involved in DNA replication. For example,
dormant lymphocytes contain the β polymerase and very low levels of
the α polymerase. In S-phase lymphocytes, however, the activity
of the β polymerase is increased by a factor of about 2 (relative
to cell number, number of nuclei, or amount of DNA), while the
activity of the α polymerase is increased more than 5 to 10-fold
(Loeb et al., 1968; Hauser et al., 1976).

The following enzyme activities have also been detected in
increased amounts in S-phase lymphocytes; single-strand-specific
endonuclease (Otto and Knippers, 1976), RNase H (Büsen et al., 1977),
and three DNA-dependent ATPases (Otto, 1977).

If any of these enzyme activities do, in fact, have a role in
DNA replication in stimulated lymphocytes, then one might expect
to detect them in chromatin, and preferentially in chromatin from
S-phase lymphocytes compared to that from unstimulated resting
lymphocytes.

3. DNA Replication in Isolated Nuclei
Prepared from Lymphocytes

Nuclei were prepared by published methods (Benz and Strominger,
1975) from resting lymphocytes or from activated lymphocytes at
different times after addition of Con A to the cultures. As *in
vivo*, the capacity of the isolated nuclei to synthesize DNA increases
with time of exposure to Con A (Fig. 2).

In other characteristics as well, DNA synthesis in isolated
lymphocyte nuclei resembles DNA replication. For example, it is
dependent on ATP (Schlaeger, manuscript in preparation). Furthermore, the first detectable products of synthesis are 4S fragments
which are rapidly assembled into longer (>30S in alkali) pieces of
DNA. The DNA synthesis observed is semiconservative; when dTTP is
replaced by BrdUTP in the reaction mixture, and the DNA, sheared
to 10-20S fragments, is analyzed by equilibrium density centrifugation in neutral CsCl gradients, the newly synthesized DNA bands at

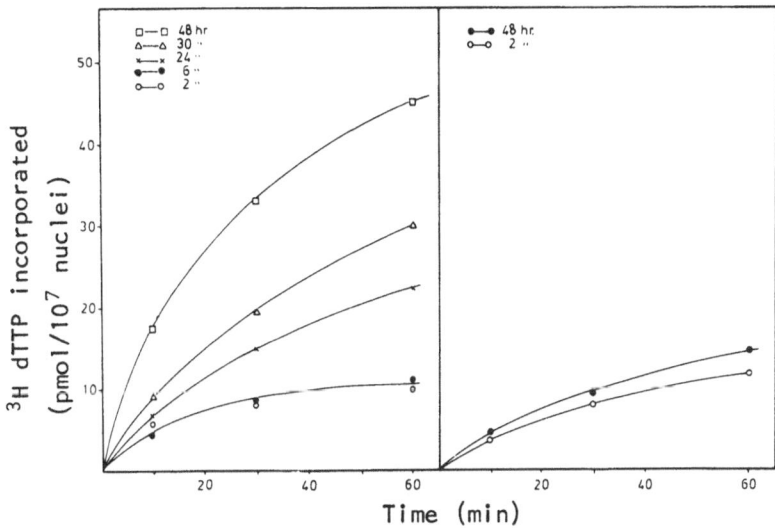

Figure 2. Incorporation of ³H-dTTP into nuclei prepared from Concanavalin A-stimulated lymphocytes at different times after stimulation.
Nuclei were prepared from bovine lymphocytes at 2, 6, 24, 30 and 48 h after Con A stimulation and incubated in vitro as described previously (Benz and Strominger, 1975). In the control, nuclei were prepared from unstimulated lymphocytes cultured for 48 h. At different times the reactions were terminated by adding 10% TCA (Hauser et al. (1976).

Left: Nuclei from Con A-stimulated lymphocytes

Right: Nuclei from unstimulated control cultures

the position expected for fragments of the hybrid density (Schlaeger, in preparation). Thus, most of the DNA synthesis observed in isolated nuclei from lymphocytes corresponds to DNA replication.

Some features of DNA replication in isolated nuclei, however, may differ from replication *in vivo*. We have investigated whether DNA replicated *in vitro* is associated with chromatin proteins. Chromatin proteins protect the segments of DNA with which they are associated from nuclease digestion (Clark and Felsenfeld, 1971). Thus, if newly replicated DNA is associated with nucleoproteins, it should be no more vulnerable to nuclease action than bulk DNA.

Nuclei were prepared from S-phase lymphocytes which had been prelabeled *in vivo* for 16 h with ^{14}C-thymidine and from an unlabeled parallel culture.

The unlabeled nuclei were incubated *in vitro* under standard
conditions (see legend to Fig. 2). After various times of
incubation, the nuclei were treated with micrococcal nuclease.
About 50% of the *in vivo* ^{14}C-label and 25% of the *in vitro* ^{3}H-label
remained nuclease-resistant after 30 min of digestion (Fig. 3). This
result demonstrates that only about half of the DNA synthesized
in vitro is protected from nuclease attack, possibly because too
little chromatin protein is available in the isolated nuclei.
Experiments in which nuclei from cells which had been exposed to
cycloheximide *in vivo* were treated with nuclease yielded similar
results (Weintraub, 1976).

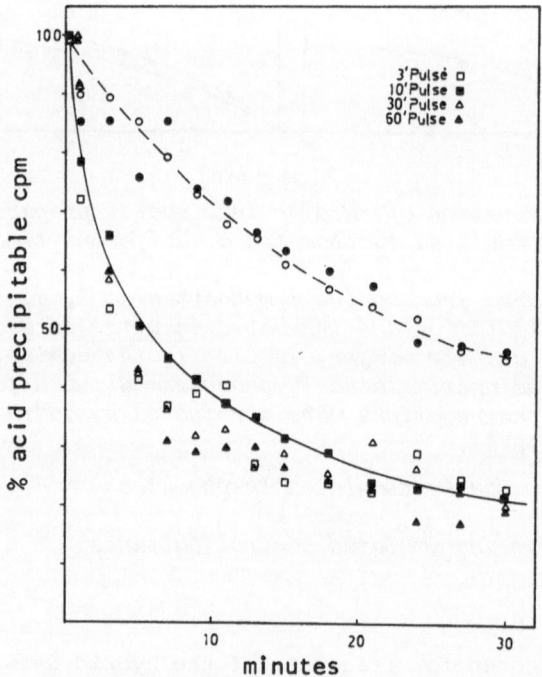

Figure 3. *Degradation of lymphocyte chromatin by micrococcal
nuclease.*
*Nuclei were prepared from Con A-activated lymphocytes labeled with
[14]C-thymidine for 16 hours during the DNA replication phase. Nuclei
were also prepared from a parallel culture of unlabeled cells and
incubated under standard conditions (Benz and Strominger, 1975) with
a mixture of deoxyribonucleoside triphosphates, including ^{3}H-dTTP,
for 3, 10, 30 and 60 min. The nuclei (5 x 10[6]/ml) were then digested
with 20 units/ml of micrococcal nuclease, as described in Materials
and Methods, except that the digestion mixture also contained 1mM
MgCl$_2$, 0.25 M sucrose and 2% dextran. At the indicated times,
aliquots of 0.05 ml were removed and the DNA was precipitated with*

10% TCA. (Klempnauer and Schlaeger, in preparation).

(,   )[14]C-labeled nuclei.

The nuclease-resistant chromatin from lymphocyte nuclei was analyzed by zone velocity sedimentation (Fig. 4). Not only the bulk chromatin, but also the *in vitro* ^{3}H-labeled chromatin is organized into nucleosomes, as shown by the characteristic sedimentation pattern.

We tentatively conclude that nucleosomes remain largely intact during the progress of the replication fork. These data are compatible with results obtained from *in vivo* experiments (Seale, 1976a,b; Jackson et al., 1976).

3. Association of DNA Polymerase with Nucleosomes

Since newly synthesized DNA in isolated nuclei from lympho-cytes is observed in nucleosomes, it is possible that DNA replication may occur on the nucleosomes themselves and that replication enzymes may then be associated with the nucleosomes.

We have looked for DNA polymerase activity associated with nucleosomes from resting and proliferating lymphocytes. Nucleo-somes were generated by digestion of nuclei from dormant or S-phase lymphocytes with micrococcal nuclease and the digestion products were analyzed by zone velocity sedimentation in neutral sucrose density gradients (Fig. 5). Each fraction of the gradient was assayed for DNA polymerase activity using "activated" calf thymus DNA as primer-template (Otto et al., 1977). Most of the enzymic activity was recovered in those fractions of the gradient which contained monomeric nucleosomes. Moreover, monosomes from S-phase lymphocytes contained more polymerase activity than those from resting lymphocytes.

However, mammalian DNA polymerases are known to form aggregates in solutions of low ionic strength, and thus, the presence of polymerase activity apparently associated with monosomes could be the result of simple cosedimentation with the monosomes. Control experiments make this possibility unlikely. First, the DNA polymerase activity coprecipitated with nucleosomes in the presence of 10-20 mM $MgSO_4$ (Honda et al., 1975), but exogenous preparations of α or β polymerase, added to fraction 28-30 of the gradient, did not coprecipitate in the presence of $MgSO_4$. DNA polymerase activity can be dissociated from Mg-precipitated monosomes by increasing the ionic strength or by preincubating the nucleosomes with "activated" calf thymus DNA

(R. Knippers, unpublished observations). These data suggest that the observed association of DNA polymerase activity with monosomes (Fig. 5) may be specific.

Nucleosomes prepared from mouse ascites cell chromatin were also examined for DNA polymerase activity. The α polymerase

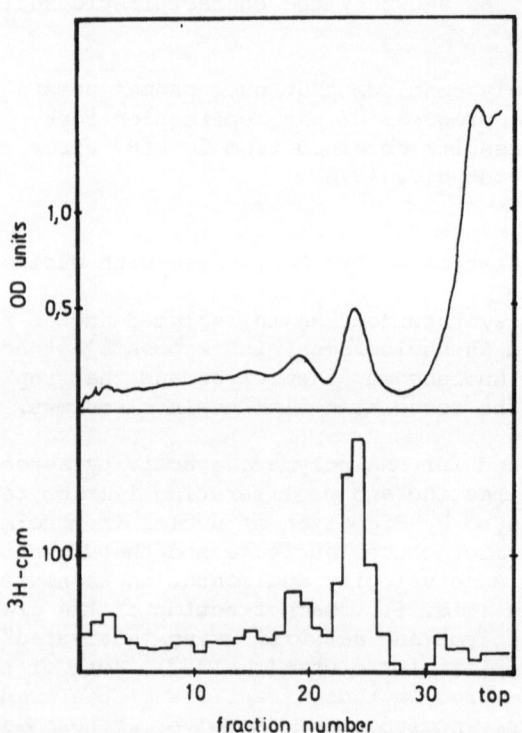

Figure 4.   In vitro replicated DNA is associated with nucleosomes.
Nuclei prepared from S-phase lymphocytes were incubated for 10 min
at 37°C in the presence of deoxyribonucleoside triphosphates,
including ³H-dTTP.   Chromatin prepared 5 x 10⁶ ³H-labeled nuclei
(Hancock, 1974) was mixed with twice the amount of chromatin from
unlabeled nuclei before micrococcal nuclease treatment (100 units/
ml).   The digestion was stopped by adding EDTA, Triton X-100 and
sodium deoxycholate.   The digestion products were analyzed by zone
velocity sedimentation for 25 h at 4°C and 34,000 rpm in the
Beckman SW41 rotor.   Qualitatively similar results were obtained
when the isolated nuclei had been incubated for 60 min (Klempnauer,
unpublished).

Top: OD₂₆₀

Bottom: TCA-precipitable radioactivity

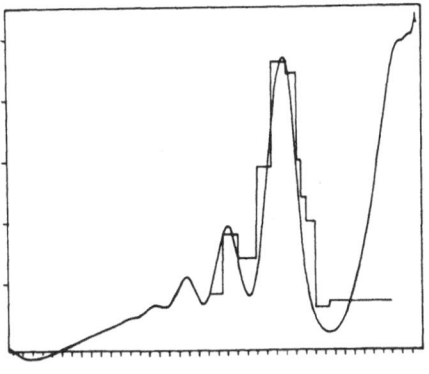

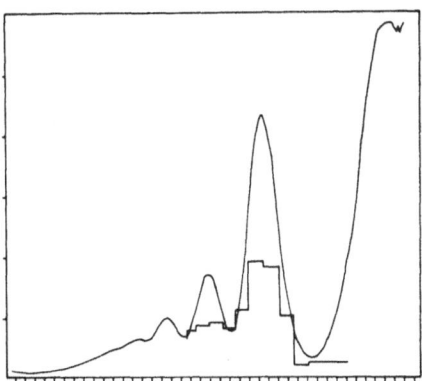

*Figure 5.   DNA polymerase activity in chromatin from lymphocytes.
Nuclei were prepared from resting (bottom) and from S-phase
lymphocytes, 48 h after addition of Con A(top).   Nuclei (10⁸) were
digested with 100 units/ml of micrococcal nuclease.   The reaction
was stopped by adding EDTA, and the digestion products were
analyzed by zone velocity sedimentation for 17 h at 24,000 rpm and
4°C in the Beckman SW27 rotor.   DNA polymerase activity in 0.05 ml
aliquots of each fraction was assayed as described previously (Otto
et al. 1977). The monomeric nucleosomes prepared by this procedure
are "core particles", as defined by Noll and Kornberg (1977),
which contain DNA of about 140 base pairs in length.*

(Ω)   OD₂₆₀

(⊓)   DNA polymerase activity

Abscissa: fraction number.   Sedimentation was from right to left.

*Ordinate: each unit corresponds to 300 cpm ³H (top) or 250 cpm ³H
               (bottom) in acid-precipitable material, or 0.1 OD₂₆₀ units.*

activity was not associated with chromatin or with nucleosomes. Only β polymerase activity was associated with chromatin, and of the chromatin-bound polymerase activity, only 50% remained associated with nucleosomes after digestion with micrococcal nuclease (Fig. 6). We do not yet understand how all of the polymerase activity in lymphocyte nuclei can be associated with nucleosomes, so that no soluble enzyme activity is detected after nuclease digestion whereas in mouse ascites cell nuclei, only half of the chromatin-bound polymerase activity is found on nucleosomes. Further experiments will be necessary to solve this problem.

4.   Characterization of Soluble and Chromatin-Bound DNA Polymerase

Nucleosomal DNA polymerase activity was dissociated from Mg-precipitated nucleosomes from dormant and S-phase lymphocytes in 0.4M NaCl. Soluble DNA polymerase is that polymerase activity

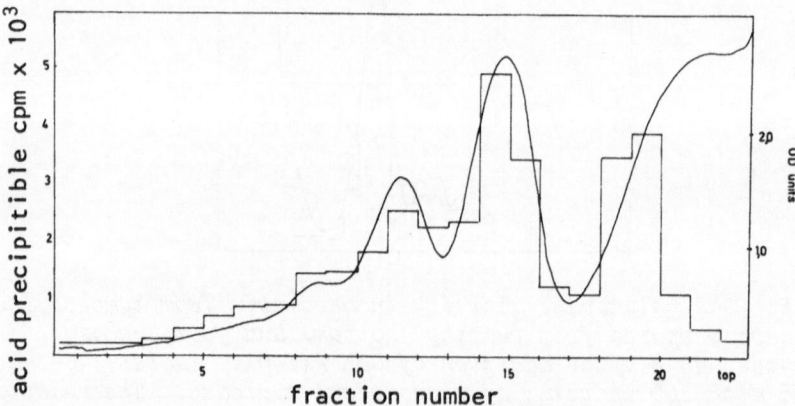

Figure 6. *DNA polymerase activity in chromatin from mouse ascites cells.*
Chromatin from mouse ascites cells (containing about 4 mg DNA) was digested with 100 units/ml of micrococcal nuclease. Digestion was terminated with EDTA and the products were analyzed by zone velocity sedimentation on linear 6-20% sucrose density gradients in a Beckman SW27 rotor for 16 h at 0°C and 25,000 rpm. Samples of 0.05 ml from each fraction were assayed for DNA polymerase activity as described in Materials and Methods. All DNA polymerase activity recovered in this gradient corresponds to β polymerase, as judged by its sedimentation in sucrose gradients containing 0.4 M NaCl and by its salt optimum (Schlaeger, van Telgen, Klempnauer and Knippers, in prep.).

(⌒) OD_{260}

(⊓)   DNA polymerase activity

which remains in the supernatant after the preparation of nuclei from resting or S-phase lymphocytes. Each of these enzyme fractions was analyzed by zone velocity sedimentation in sucrose density gradients (Fig. 7).

The nucleosomal DNA polymerase from dormant lymphocytes sediments with 3-4S (relative to beef hemoglobin, 4.4S, as sedimentation marker). The fraction of soluble DNA polymerase from dormant cells also contains a 3-4S species and a faster sedimenting 6-7S species. Based on the sedimentation behavior and on the different salt sensitivities of these activities (Weissbach, 1975), we have identified the slower sedimenting form as β- and the faster sedimenting form as α-polymerase, respectively. In S-phase lymphocytes, the major activity on the nucleosomes and soluble corresponds to α-polymerase, although the β-enzyme is also present (Fig. 7).

Thus, nucleosomes from S-phase lymphocytes contain polymerase activity, whereas nucleosomes from dormant lymphocytes do not. This result need not mean, however, that the nucleosomes which contain α-polymerase activity are undergoing replication. S-phase lymphocytes contain much more soluble α-polymerase activity, which is presumably available to associate with the nucleosomes. However, this explanation for the preferential binding of α-polymerase activity to S-phase nucleosomes is unlikely on the basis of the control experiments discussed above (cf. section 3). Moreover, the following data indicate that the α-polymerase activity is, in fact, associated with replicating chromatin and probably not associated with nonreplicating chromatin.

5. DNA Polymerase in SV40 DNA-Protein Complexes

In nuclei of SV40-infected cells, at least two types of protein-SV40 DNA complexes ("minichromsomes") are observed (Su and DePamphilis, 1976). Most of the viral DNA is covalently closed circular DNA (Form I) associated with nucleosomal subunits. This form sediments at about 75S through neutral sucrose density gradients (Fig. 8). A second less abundant DNA-protein complex sediments at about 95S (Fig. 8). This form contains viral DNA in the process of replication, as well as histones organized into nucleosomes and a variety of nonhistone chromatin proteins.

SV40 DNA-protein complexes of both types were labeled *in vivo* for 15 min with ^{3}H-thymidine, eluted from the nuclei and separated by zone velocity sedimentation in neutral sucrose density gradients according to published procedures (Su and DePamphilis, 1976). Each fraction of the g radient was assayed for DNA polymerase activity (Otto et al., 1977) and the results are shown in Fig. 8. DNA polymerase activity was found associated with the 95S replicating

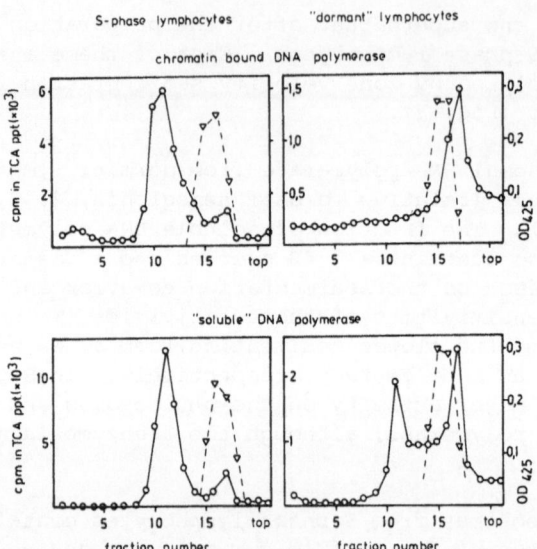

*Figure 7.   Analysis of DNA polymerase activity in lymphocytes.
(a) "chromatin-bound" DNA polymerase.   Nuclei were prepared from
dormant and from activated lymphocytes 48 h after addition of
Concanavalin A.   About 10^8 nuclei were pelleted, resuspended in
2 ml of digestion buffer and treated with 20 units/ml of micrococcal
nuclease, as described in Materials and Methods.   Monomeric nucleosomes prepared on sucrose gradients were precipitated with 20mM
$MgSO_4$ and resuspended in Buffer B (50mM Tris-HCl, pH 7.5, 1 mM EDTA,
10mM 2-mercaptoethanol and 10% glycerol) containing 0.4 M NaCl to
dissociate DNA polymerase activity from chromatin.*

*(b) "soluble" DNA polymerase.   The supernatants, recovered after the
preparation of nuclei, were centrifuged at 100,000 x g for 60 min.
The supernatant was passed through a DNA-cellulose column (2 cm x
5 cm).   The adsorbed DNA polymerase activity was eluted with 0.5 M
NaCl in Buffer B. More than 90% of the input activity was recovered
in a single peak.*

*Samples of 0.4 ml of each preparation, (a) and (b), were layered on
linear 5-20% sucrose gradient containing 0.4 M NaCl in Buffer B and
centrifuged in the Beckman SW41 rotor at $2^{\circ}C$ at 38,000 rpm for 40 h.
Beef hemoglobin (4.4S) served as sedimentation marker ($-- \nabla --$) and
was identified by its characteristic color (OD_{425}).   The DNA polymerase activity in each fraction of the gradient was determined as
described earlier (Otto et al., 1977).*

SV40 chromatin and soluble at the top of the gradient, but little
or no polymerase activity was detectable in 75S SV40 complexes. DNA
polymerase activity could be coprecipitated with SV40 DNA-protein
complexes in the presence of 20 mM $MgSO_4$.   The polymerase activity

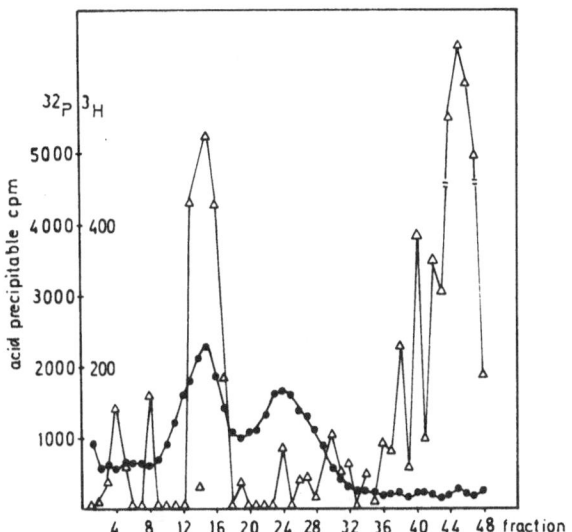

Figure 8. DNA polymerase in SV40 chromatin. CV-1 cells infected with SV40 were labeled for 15 min with 40 μCi/ml of ^{3}H-thymidine at 38 h post-infection. SV40 DNA-protein complexes were eluted from nuclei them, according to the method of Su and DePamphilis (1976). The eluate containing the complexes was analyzed by zone velocity sedimentation in 5-30% sucrose density gradients made up in 10mM Hepes, pH 7.8 (SW50.1, 0°, 45,000 rpm). An aliquot of each fraction was precipitated in 10% TCA at 0° and another aliquot was tested for DNA polymerase activity using "activated" calf thymus DNA as primer-template and α-[32]P dATP, as described previously (Otto et al., 1977) with minor modifications. The sedimentation co-efficients of the faster and slower sedimenting SV40 DNA-protein complexes were estimated to be 95S and 75S, respectively, by comparison with the sedimentation behavior of bacteriophage T7 DNA (34S) (data not shown).

●-●   *[3]H cpm*

△-△   *[32]P cpm*

from the Mg-precipitate was further characterized by its salt dependence and by zone velocity sedimentation (Fig. 9). All detectable polymerase activity sedimented faster than 4.4S bovine hemoglobin marker. Since DNA polymerase β would·sediment more slowly (3-4S) (Fig. 7), the polymerase activity found in replicating SV40 chromatin appears to be predominantly DNA polymerase α.

The SV40 DNA-protein complexes have also been tested for other enzyme activities thoughtto be involved in DNA replication. For

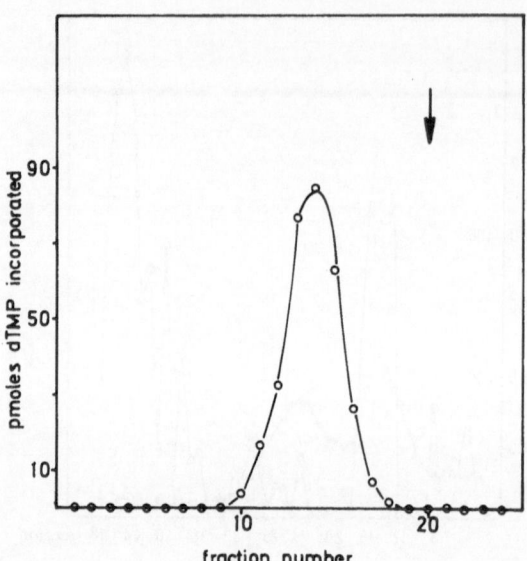

Figure 9. *Sedimentation of DNA polymerase from SV40 nucleo-protein complexes.*
CV-1 cells infected with SV40 were labeled from 24-36 h post-infection with 2 μCi/ml of ^{3}H-thymidine. SV40 DNA-protein complexes were eluted from 4 x 10[7] nuclei as described previously (Su and DePamphilis, 1976). The complexes were precipitated from the eluate in 20 mM MgCl$_2$ for 30 min at 0° and pelleted by centrifugation at 12,000 x g for 10 min. The pellet was washed once with Buffer A containing 20mM MgCl$_2$, and then resuspended in 0.5 ml of Buffer A containing 0.5mM MgCl$_2$. A portion (0.2 ml) of the resuspended material, made 0.5 M in NaCl to dissociate the DNA polymerase activity from the complexes, was analyzed by zone velocity sedimentation in a sucrose density gradient. The gradient consisted of 0-15% sucrose in 50mM Tris-HCl, pH 7.5, 2mM EDTA, 250mM $(NH_4)_2SO_4$, 5mM DTT, 0.5 mg/ml bovine serum albumin and 20% glycerol. Bovine hemoglobin (4.4S) was included in a parallel gradient as a sedimentation marker, shown by the arrow. The gradients were centrifuged in the SW50.1 rotor at 45,000 rpm and 0° for 20 h. Aliquots (10 μl) of each fraction were assayed for DNA polymerase activity, using activated calf thymus DNA and ^{3}H-dTTP (800 cpm/pmole), as described earlier (Otto et al., 1977).

example, we have detected two different DNA-dependent ATPase activities, one sedimenting slightly slower than the replicating SV40 chromatin and one, slightly faster than 95S. Thus, the association of replication enzymes, and of the α polymerase in particular, with SV40 chromatin appears to be specific.

Supported by the Deutsche Forschungsgemeinschaft (SFB 138-B4)

REFERENCES

Baril, E.F., M.D. Jenkins, O.E. Brown, J. Laszlo and H.P. Morris (1973) Cancer Research 33:1187-1193

Benz, W.S. and J.L. Strominger (1975) Proc. Nat. Acad. Sci. USA, 72:2413-2417

Büsen, W., H.J. Peters and P. Hausen (1977) Eur. J. Biochem. 74: 203-208

Chang, L.M.S. and F.J. Bollum (1973) J. Biol. Chem. 248:3398-3404

Chiu, R.W. and E.F. Baril (1975) J. Biol. Chem. 250:7951-7957

Clark, R.J. and G. Felsenfeld (1971) Nature New Biol. 229:101-106

Craig, R.K., P.A. Costello and H.M. Keir (1975) Biochem. J. 145: 233-240

Hancock, R. (1974) J. Mol. Biol. 86:649-663

Hauser, H., R. Knippers, K.P. Schäfer, W. Sons and H.-J. Unsöld (1976) Exp. Cell Res. 107:79-84

Honda, B.M., D.L. Baillie and E.P.M. Candido (1975) J. Biol. Chem. 250:4643-4647

Jackson, V., D. Grauner and R. Chalkley (1976) Proc. Nat. Acad. Sci. USA 73:2266-2269

Loeb, L.A., S.S. Agarwal and A.M. Woodsick (1968) Proc. Nat. Acad. Sci. USA 61:827-834

Noll, M. and R.D. Kornberg (1977) J. Mol. Biol. 109:393-404

Otto, B. and R. Knippers (1976) Eur. J. Biochem. 71:617-622

Otto, B. (1977) FEBS Lett., in press

Otto, B., M Baynes and R. Knippers (1977) Eur. J. Biochem. 73: 17-24

Peters, J.H. (1975) In: Methods in Cell Biology (ed. D.M. Prescott) Vol IX, Academic Press, New York, pp1-11

Seale, R.L. (1976a) Proc. Nat. Acad. Sci. USA 73:2270-2274

Seale, R.L. (1976b) Cell 9:423-429

Sons, W. (1976) Dissertation, University of Konstanz

Spadari, S. and A. Weissbach (1974) J. Mol. Biol. 86:11-20

Su, R.T., and M.L. DePamphilis (1976) Proc. Nat. Acad. Sci. USA
      73:3466-3470

Weintraub, H. (1976) Cell 9:419-422

Weissbach, A. (1975) Cell 5:101-108

Wintersberg, E. (1977) TIBS 2:58-61

DNA POLYMERASE AND THE ONSET OF DNA SYNTHESIS IN WHEAT EMBRYOS

S. Litvak[*], B. Mocquot[**] and M. Castroviejo

Laboratoire de Biochimie
UER-BBC, Université de Bordeaux II
351 cours de la Libération
33405, Talence, France

INTRODUCTION

The existence of multiple DNA polymerases in animal cells is well established (see review by A. Falaschi, this volume). DNA polymerases α, β and γ have been found in all mammalian cells studied up to now. Recently it has been shown by Weissbach (1) and Bertazzoni et al. (this volume) that polymerase γ and the mitochondrial DNA polymerase are identical.

In the case of lower eukaryotes, like fungus or algae, multiple DNA polymerases have also been described (2-5).

Although much less information is available in higher plants, there is good evidence pointing to the presence of multiple DNA polymerases in these organisms. We have described the purification of two DNA polymerases from ungerminated wheat (6), one of those enzymes having properties very similar to γ polymerase (7).

Seed germination is characterized by the activation of metabolism in the quiescent embryo. Protein synthesis starts 30 min after germination is initiated by water imbibition (8). DNA synthesis cannot be observed before 15 hours of germination. This synthesis seems to be dependent on protein synthesized during the first 9 hours of germination. The sequence of these events renders

[**]*Département Physiologie Végétale, I.N.R.A., Domaine de la Grande Ferrade, 33 Villenave-d'Ornon, France.*

---

[*]*A.S.I. Participant*

the germinating embryo a useful system for the study of the
initial steps of DNA synthesis and for the search of the protein
factors involved in the onset of DNA synthesis.

The characterization of three DNA polymerases from ungerminated wheat embryos, the changes during wheat germination and the
role of the dNTP pool in the onset of DNA synthesis are presented
and discussed.

RESULTS

1. Purification of DNA Polymerases A, B and C

Fig. 1 shows schematically the procedure used for purifying
3 distinct DNA polymerases from ungerminated wheat embryos. (A
detailed procedure will be published elsewhere). The starting

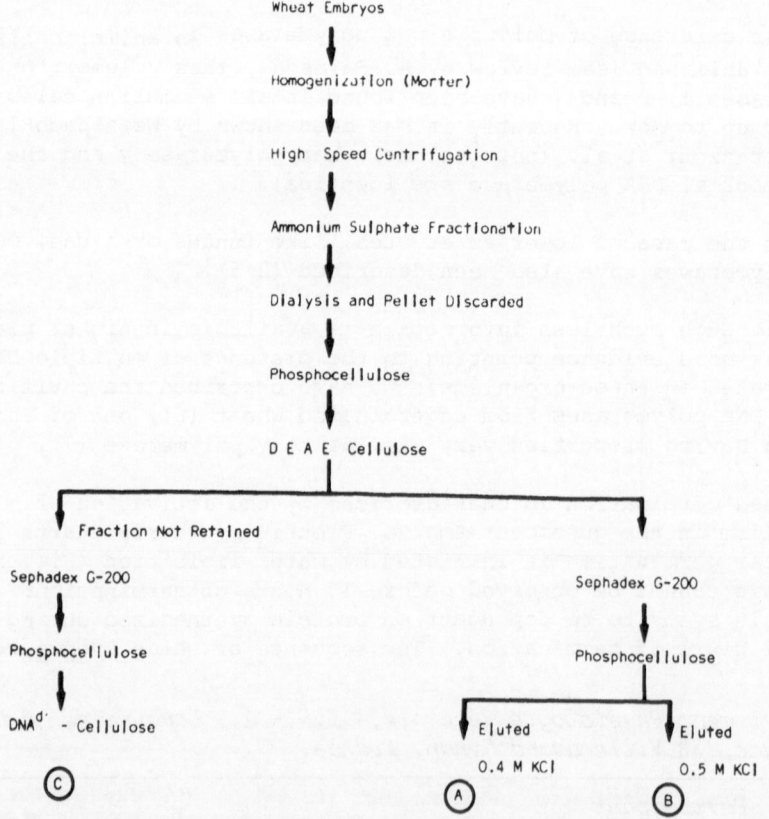

*Figure 1. Scheme of purification of DNA polymerases A, B and C
from ungerminated wheat embryos.*

material for the purification procedure was freshly prepared
commerical wheat germ. Enzyme C is separated from A and B by DEAE-
cellulose, while enzymes A and B are separated on phosphocellulose.
The separation of A and B can be followed with great detail, since
the template requirements for both are different.

The specific activity of enzyme C is of 200-500 units/mg
protein, while enzymes A and B have specific activities ten times
lower. Enzyme C gives two main bands in SDS acrylamide gel electrophoresis, one having a molecular weight of 65,000 and the other
50,000.

2. Properties of DNA Polymerases A, B and C

Some properties of enzymes A, B and C are shown in Table 1.
The three DNA polymerases are purified from the cytoplasmic high
speed supernatants. No low molecular weight DNA polymerase (β type)
has been found in wheat. All three enzymes are strongly inhibited
by N-ethyl maleimide.

In Table II the template requirements of wheat DNA polymerases
is shown. Mn^{++} stimulates all incorporation involving synthetic
templates, while this cation inhibits TMP incorporation when
activated DNA or poly d(AT) is used. Enzyme A is the only one to
utilize the template poly(A)- dT_{12} in the presence of Mg^{2+}. This
property is very useful for separating enzymes A and B in a phosphocellulose column. None of the wheat enzymes are able to
recognize poly(C)- dG_{12}, a very good template for the oncornavirus
reverse transcriptases.

3. Mitochondrial DNA Polymerase

When mitochondria, purified from ungerminated wheat embryos as
described by Cunningham (9), are lysed with Triton X - 100, sonicated
in the presence of 0.2 M KCl and centrifuged at 18.000 x g, the
supernatant shows DNA polymerase activity. This activity is
retained in DEAE-cellulose and phosphocellulose and elutes at 0.5 M
KCl from the latter (not shown).

4. DNA Synthesis During Wheat Embryo Germination

A - DNA synthesis. Only after 10-15 hours after water
imbibition, a dramatic increase in DNA synthesis is observed (Fig.
2). The experiment was carried out by germination of 100-150
embryos obtained by the Johnson and Stern procedure (10), for
different lengths of time as described in the abscissa, and then

Table 1

Properties of Enzymes A, B, C

	A	B	C
Localization	cytoplasm	cytoplasm	cytoplasm
Sedimentation coefficient	7.5	6.3	5.9
Molecular weight	140.000	110.000	100.000
Isoelectric point	6.8	5.2	7
Optimum pH	8	7	8
KCl 0.2 M inhibition	not	yes	yes
Optimum Mg^{+2}	5 mM	5 mM	5 mM
Optimum Mn^{+2}	inhibition	0.1 mM	inhibition
NEM inhibition	yes	yes	yes

Some general properties of DNA polymerases A, B and
C from ungerminated wheat embryos

incubating all samples for 1 hour in the presence of (^{3}H)thymidine.

B - DNA polymerases changes during germination. Four samples
of wheat embryos were germinated for 0, 12, 24 and 48 hours and
then DNA polymerases were purified up to the DEAE-step (see Fig. 1).
The non retained material corresponds to enzyme C, the retained
activity, to enzyme A plus B. It can be seen in Fig. 3 that the
activity corresponding presumably to enzyme C decreases dramatically
during the germination period studied by us; while the total
activity of enzyme A and B seems not to change.

5. Changes in the dNTP Pool During Germination

Wheat embryos were germinated for different lengths of time
and the size of each pool was determined by the use of *E.coli* DNA
polymerase I as described previously (11). In the case of the study
of the ATP pool a fluorometric method was used. As seen in Fig. 4,
ATP increases to a maximum only 2-5 hours after water imbibition,

Table II

Template	Primer	Precursor	A		B		C		AMV R.T.	
			Mg	Mn	Mg	Mn	Mg	Mn	Mg	Mn
Activated DNA		TTP	1.23	0.87	1.05	0.52	12.65	5.52	7.60	3.55
Poly d(AT)		TTP	1.77	1.55	1.95	1.05	17.76	9.22	20.33	5.80
Poly A	dT_{12}	TTP	8.83	12.94	<0.1	5.85	<0.1	<0.1	137	249
Poly dA	dT_{12}	TTP	4.03	18.20	1.90	6.86	6.46	7.51	4.0	28.6
Poly dC	dG_{12}	dGTP	8.36	18.14	4.49	4.77	12.12	14.31	127	112
Poly C	dG_{12}	dGTP	<0.1	<0.1	<0.1	<0.1	<0.1	<0.1	65	22

Template requirement of DNA polymerases A, B and C from ungerminated wheat embryos. The incubation conditions are found in Reference 6.

Activity corresponds to nmoles nucleotide incorporated per mg protein per minute.

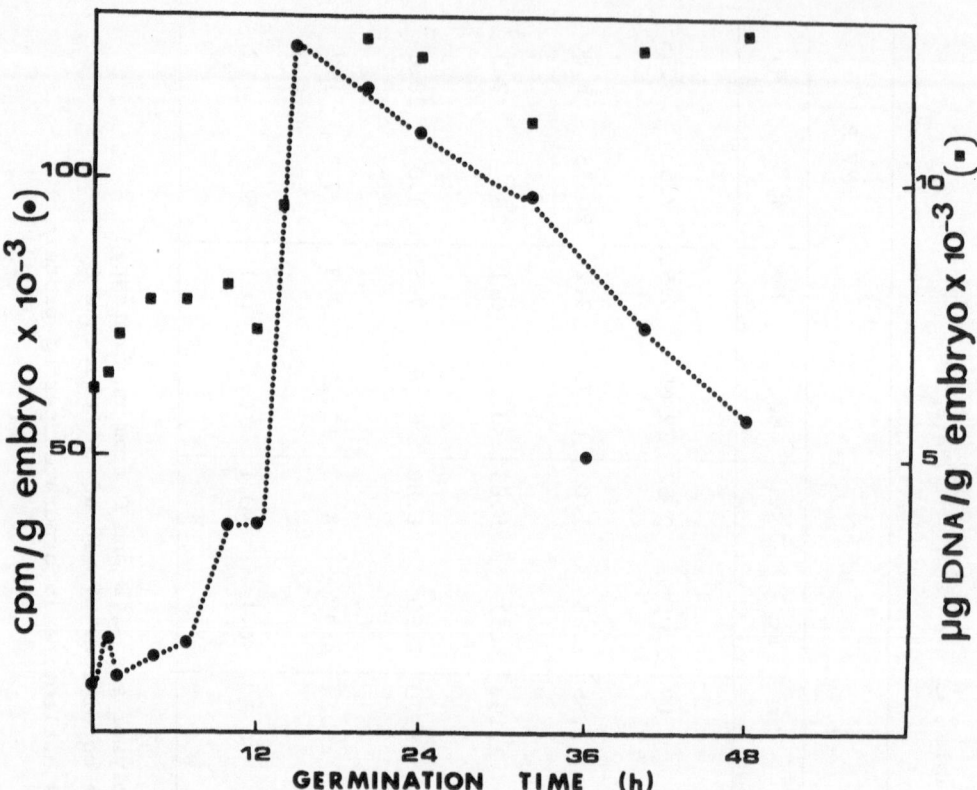

Figure 2. *Incorporation of (³H)thymidine and DNA content in wheat embryos during germination. The radioactive incubation mixture contained in 1 ml : 20 mM Tris-HC1 pH 7.5; 100 μg chloramphenicol and 1 μC (³H)thymidine (Amersham) 18 C/mmole. All samples were incubated for one hour at 25° C and nucleic acids isolated with perchloric acid. More details in the text.*

of water imbibition.

DISCUSSION

As a first step in the study of the mechanism involved in the onset of DNA synthesis once the quiescent embryo is imbibed with water, we have purified and characterized the enzymes having DNA polymerase activity found in the ungerminated wheat embryo.

After extensive purification, three DNA polymerases, whose properties are shown in Tables I and II, were found, which have some properties in common with mammalian DNA polymerases. DNA while dCTP and dTTP attained a maximal plateau only after 15 hours

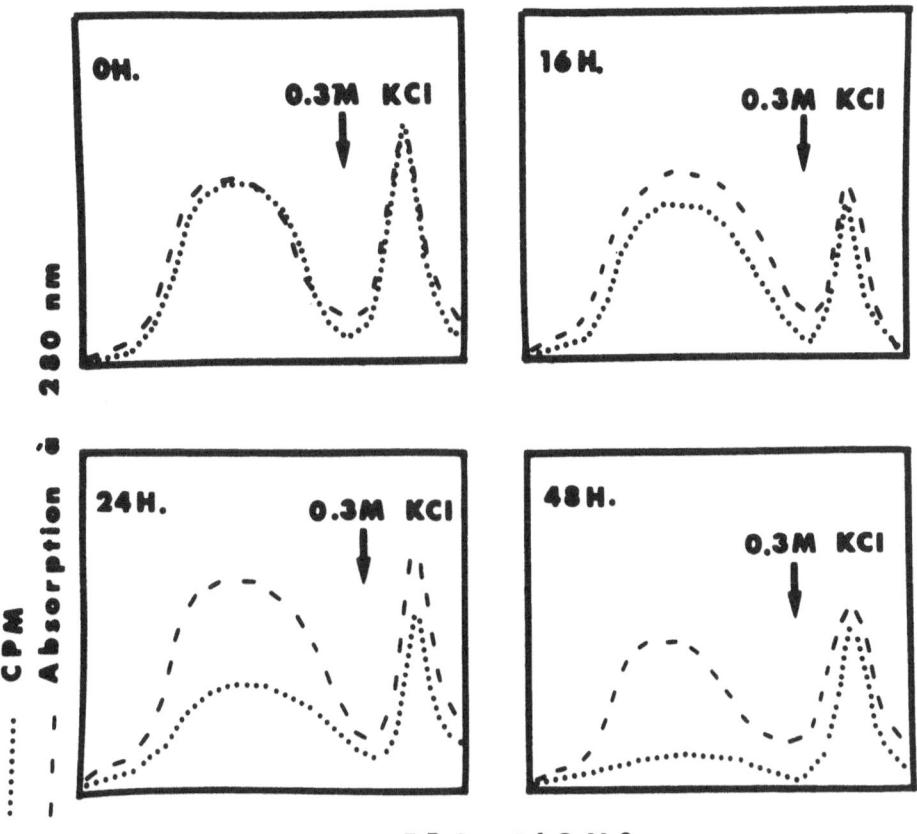

Figure 3. DNA polymerases changes during wheat embryo germination.

polymerase C resembles polymerase β in its chromatographic properties on DEAE-cellulose, but it has a relatively high molecular weight; it is not able to utilize polyrA - dT_{12} in the presence of Mn^{++} as does the polymerase β, and it is sensitive to NEM.

DNA polymerase A has a striking similarity with polymerase γ, since it is able to utilize poly(A)- dT_{12} in the presence of Mg^{++} or Mn^{++}. DNA polymerase B has some chromatographic and physico chemical similarities with polymerase α, but it recognizes poly(A)- dT_{12} in the presence of Mg^{++} while polymerase α does not.

It is interesting to note that the only activity we have found in wheat mitochondria resembles DNA polymerase B from the soluble cytoplasm in the following parameters: retention on DEAE and phosphocellulose, elution from the latter at 0.5 M KCl and recognition of poly(A) - dT_{12} only in the presence of Mn^{++}. This may

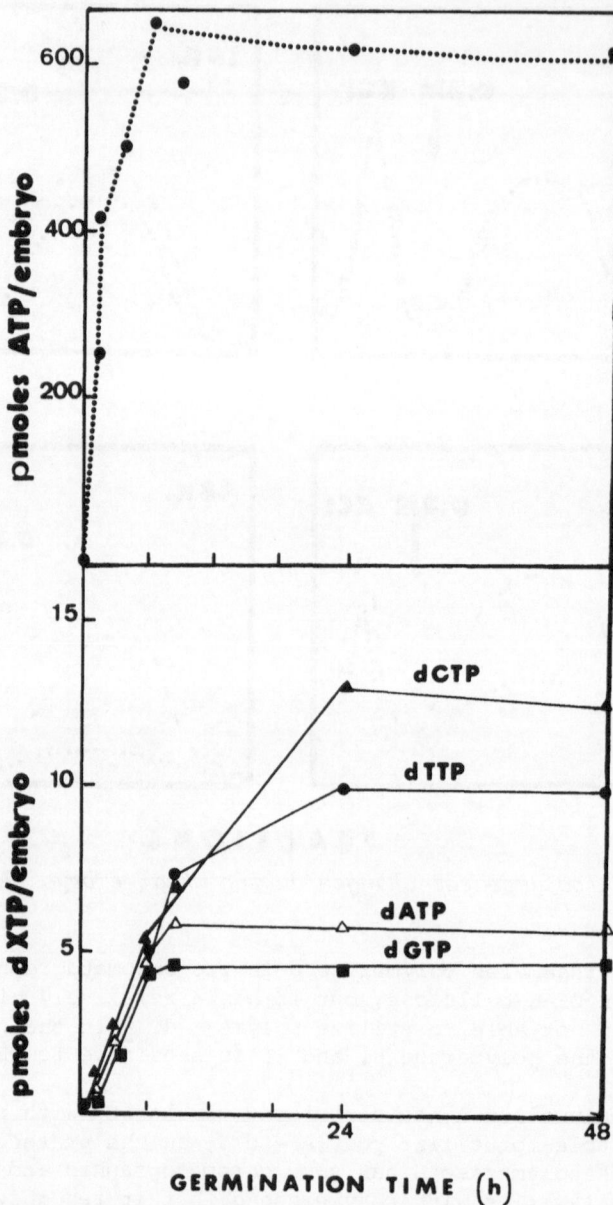

Figure 4. *Changes in the dNTP pool during wheat embryo germination.
For more details see the text.*

indicate an important difference with other systems, e.g., HeLa
cells (1), chick embryos (Bertazzoni, this volume) or mice EMT-6
cells (our unpublished results), where the main activity of the

mitochondrial fraction is identical to DNA polymerase γ. However, caution must be taken when using the so called "γ assay" that appears to be very specific for the mammalian DNA polymerase γ activity, as in lower eukaryotes or higher plants this specificity has not been proven. In wheat, for instance, two enzymes are able to recognize poly(A)- dT_{12} with Mn^{++} and the "γ assay" in this case would lead to error.

As pointed out in the introduction the onset of DNA synthesis takes place only 10-15 hours after the embryos have been imbibed with water. Protein synthesis and RNA synthesis seem to start quite earlier. Which may be the limiting steps for the entrance of the wheat embryo into S phase of active DNA synthesis ? (12). There are several possibilities:

I   - The DNA template in the chromatin may be unavailable

II   - Decreased amount of enzymes involved in the biosynthesis of
       DNA precursors and hence a shortage of DNA precursor.

III - Decreased amount of enzymes involved directly in DNA synthesis
       like DNA polymerase - DNA dependent, RNA polymerase - DNA
       dependent, ligases, repair enzymes, etc.

IV - The possibility of the presence of inhibitors of the enzymes
       mentioned before cannot be eliminated.

Most probably the answer to the problem of onset of DNA synthesis in germination will involve several of these parameters. We have described in this article some of the factors that may be involved in this phenomenon.

A very high level of DNA polymerase activity is found in ungerminated wheat. The possibility of DNA polymerase induction in germinating enzymes seems improbable (13). No increase was observed when the activity was followed during germination of whole embryos. Moreover, DNA polymerase C decreased markedly during the first 40 hours of germination. Although no variation in total A plus B activity was observed, we cannot eliminate the possibility of an internal rearrangement of both activities. We can speculate that the polymerase C was an enzyme involved in DNA synthesis during embryogenesis and that disappears during germination.

The DNA precursor levels are extremely low in ungerminated wheat (Fig. 4) and the increase during germination is very slow. This may be the reason why the onset of DNA synthesis takes so much time. Several enzymes involved in DNA precursor synthesis may be involved in this shortage. We are currently studying two of them, nucleotide reductase and thymidine kinase, since these enzymes are known to increase dramatically in proliferating systems (14).

Concerning DNA template availability (point I above) in chromatin some preliminary experiments performed in our laboratory show that chromatin from ungerminated embryos has a better template capacity than the chromatin from 40 hours germinated embryos.

ACKNOWLEDGEMENTS

This research was supported by C.N.R.S. (A.T.P. 1897) and D.G.R.S.T.

Reverse Transcriptase from Avian Myeloblastose Virus was a kind gift of Dr. J.W. Beard (Life Science Inc., Florida, U.S.A.)

REFERENCES

1.   Bolden, A., Pedrali Noy, G. and Weissbach, A. (1977)
     J. Biol. Chem. 252:3351-3356

2.   Wintersberger, E. (1974) Eur. J. Biochem. 50:41-47

3.   McLennan, A.G. and Keir, H.M. (1975) Nucleic Acid Research
     2:223-237

4.   Loomis, L.W., Rossamando, E.F. and Chang, L.M.S. (1976)
     Biochim. Biophys. Acta 425:469-477

5.   Chang, L.M.S. (1977) J. Biol. Chem. 252:1873-1880

6.   Castroviejo, M., Tarragó-Litvak, L. and Litvak, S. (1975)
     Nucleic Acid Research 2:2077-2090

7.   Tarragó -Litvak, L., Castroviejo, M. and Litvak, S. (1975)
     FEBS Letter 59:125-130

8.   Marcus, A., Feeley, J. and Volcani, T. (1966) Plant Physiology
     41:1167-1172

9.   Cunningham, R.S. and Gray, M.W. (1977) Biochim. Biophys. Acta
     475:476-491

10. Johnson, F.B. and Stern, H. (1957) Nature 179:160-161

11. Mocquot, B., Pradet, A. and Litvak, S. (1977) Plant Science
     Letters, in press

12. Mory, Y.Y., Chen, D. and Sarid, S. (1977) Plant Physiology
     49:20-33

13. Mory, Y.Y., Chen, D. and Sarid, S. (1975) Plant Physiology
     55:437-442

14. Berglund, O. (1972) J. Biol. Chem. 247:7270-7276

EFFECT OF SV40 INFECTION ON DNA POLYMERASE ACTIVITY IN

MONKEY CELLS

Silvano Riva[Δ][*], Martha A. Cline and Philip J. Laipis

Department of Biochemistry
University of Florida
Gainesville, Florida, 32610, U.S.A.

INTRODUCTION

Infection of permissive host cells by either simian virus 40 (SV40) or polyoma virus causes a wide spectrum of changes in cell metabolism. The activities of a number of enzymes concerned with DNA metabolism increase; host cell DNA synthesis is usually initiated and one round of DNA replication appears to occur; cell histone and non-histone chromosomal proteins are synthesized at increased rates. Changes in the cell surface and in cell morphology occur, along with the expression of viral specific non-capsid antigens, such as T, U, or TSTA (references 1-5 are recent reviews of SV40 and polyoma virus molecular biology). We have carried out experiments examining one aspect of this virus-host cell interaction more closely: the effect of SV40 infection on monkey cell DNA polymerase activity.

SV40 or polyoma virus induce an increase in DNA polymerase activity after infection (6-8). We have partially purified the three major DNA polymerase species from monkey cells, and have compared the activities present in uninfected resting confluent cells to SV40-infected or uninfected growing cells. The enzyme fractions have been characterized by several techniques. We have carried out a number of studies *in vivo* and *in vitro* to determine the role of the DNA polymerases in SV40 DNA replication.

[Δ]*Present address*: *Laboratorio di Genetica Biochimica ed Evoluzionistica, Consiglio Nazionale delle Richerche, Pavia, Italy.*

[*]*A.S.I. Participant*

RESULTS

The kinetics of cellular and viral DNA synthesis and the increase in total DNA polymerase activity that occurs after SV40 infection of resting confluent cells are shown in Figs. 1A and 1B. DNA polymerase activity begins to increase 24-36 hours after infection by SV40 virus, and has reached its maximum level by about 60 hours post-infection. In this experiment total activity increased 4.5 fold. As a positive control, uninfected confluent cells were stimulated to divide by the addition of media containing 50% calf serum. Approximately 45% of the cells did divide, as judged by the increase in cell number, and total DNA polymerase activity increased two-fold. The maximum rate of DNA synthesis in both serum-stimulated and SV40-infected cells precedes the maximum increase in DNA polymerase activity, and polymerase activity remains high after DNA synthesis has ceased. Similar observations have been reported by others (9, 25).

These experiments measured the total DNA polymerase activity in crude cell lysates prepared by sonication. In order to identify the enzyme (or enzymes) which increased after infection, and determine if any new or modified enzymes had appeared, it was necessary to purify the three major DNA polymerases and study their properties when separated from each other. The polymerases were purified using two DEAE cellulose columns followed by a glycerol gradient.

Confluent CV-1 cells were placed in 2% media and allowed to rest four days at 37°. The cells were scraped from the plates, washed, and either frozen at -70° or used immediately. They were lysed by sonication, the debris pelleted, and the supernatant applied to a DEAE column. Over 80% of the protein eluted from the column with the wash, including 80-100% of the applied polymerase activity.

This run-through peak was dialyzed and applied to a second DEAE cellulose column. One DNA polymerase activity, later identified as β polymerase, ran right through the column again. The remaining α and γ activity eluted between $0.1 - 0.14$ M PO_4. These activities were further purified by sedimentation in 10-30% glycerol gradients in 0.2 M PO_4. The S values obtained for the three enzymes in different preparations are reported in Table 1. The gradient fractions were pooled, concentrated against Carbowax PEG 6000 and the resulting enzymes stored in 50% glycerol at -20°. β was the most unstable, with 50% of the activity being lost after six months storage. These glycerol gradient enzymes were used for the drug inhibition studies reported below.

In this purification scheme, the three major polymerase species were first identified on the basis of column binding behavior,

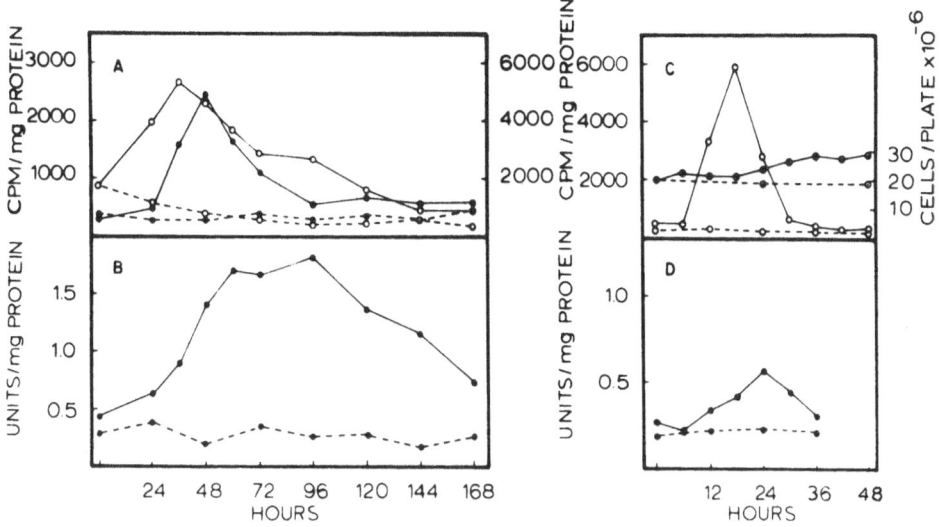

Figure 1. CV-1 cells were allowed to become confluent in Dulbecco's Modified Eagles medium (DMEM) with 10% calf serum, then the media was changed to DMEM plus 2% calf serum and the cells were allowed to remain four days at 37°. They were either used directly (confluent uninfected control cells), virus-infected or serum-stimulated with the addition of DMEM plus 50% calf serum. DNA synthesis was measured by the incorporation of (³H)thymidine into either SV40 DNA (extracted by the method of Hirt (36)) or high molecular weight cell DNA (the Hirt pellet). Cells from two plates at each time point were harvested, sonicated, and DNA polymerase activity measured (9). Fig. 1A. (³H)thymidine incorporated into the Hirt extract (SV40 DNA) (●—●) and cell DNA (O—O) in infected cells. (³H)thymidine incorporated into the Hirt extract (●--●) and cell DNA (O--O) in uninfected cells. Fig. 1B. DNA polymerase activity in infected (●—●) and uninfected cells (●--●). Fig. 1C. (³H)thymidine incorporated into cell DNA in serum-stimulated (O—O) and control cells (O--O). Cell number in serum-stimulated (◉—◉) and control cells (◉--◉). Fig. 1D. DNA polymerase activity in serum-stimulated (●—●) and control cells (●--●).

sensitivity to N-ethylmaleimide, and ability to use an oligo dT:poly r A primer-template complex as a substrate for synthesis. Once the enzymes had been separated by glycerol gradients, we confirmed that these properties were sufficient to differentiate the three monkey cell enzymes. This allowed us to quantitate each enzyme in the presence of the other two. DNA polymerase α (or total DNA polymerase activity) was measured in an assay containing activated calf thymus

Table I

Physical Properties of Monkey Cell DNA Polymerases

	Confluent, Uninfected			Serum-Stimulated			SV40-Infected		
	α	β	γ	α	β	γ	α	β	γ
Total amount units/3G cells	420	197	41	1701	189	48	2282	225	74
Specific activity units/mg protein (DEAE I)	3.8	1.8	0.43	12.1	1.3	0.27	32	3.2	1.06
Elution from DEAE II (PO$_4$)	0.14M	0.02M	0.12M	0.13M	0.02M	0.1M	0.13M	0.02M	0.11M
S Value	6.6	3.9	9.7	8.1	4.1	10	7.1	3.7	10

Physical properties of CV-1 monkey cell DNA polymerases. Yield and specific activity data are from three separate experiments. S values were determined from 10-30% glycerol gradients using human hemoglobin (4.6S) as marker.

DNA substrate, with salt and triphosphate concentrations identical to those previously used for HeLa cells (9). β polymerase was measured in the same assay plus 2.5 mM N-ethylmaleimide. γ polymerase was measured using an oligo dT:poly rA substrate (9); α and β polymerases showed no activity in this assay. γ polymerase was about 10-20% as active in the activated calf thymus assay as it was in the dT:rA assay. No differences in activity between confluent, serum-stimulated, or SV40-infected cell enzymes were noted when KCl or substrate concentrations were varied in the assay for each of the three types of polymerase.

Similar fractionations were carried out using SV40-infected, and confluent cells stimulated to divide by the addition of media containing 50% calf serum. A summary of the yield and physical properties of three polymerases from the three different cell populations is shown in Table I.

We conclude from this purification that DNA polymerase α is the only polymerase to show major changes in activity levels, with increases of five to ten-fold in total units or specific activity. Polymerase γ, in several experiments, has increased by up to 100%. There was no obvious change in physical characteristics such as relative sedimentation coefficient (determined by glycerol gradient sedimentation of the first DEAE column fractions), elution from the second DEAE column, or response to salt and substrate level changes in the assay. We have further characterized polymerase α by determining its relative mobility on non-denaturing acrylamide gels. There was no difference in mobility between the enzyme activity from uninfected or SV40 infected cells; however, peaks of activity were broad, suggesting aggregation of the enzyme was occurring, and these experiments are being repeated, using 0.05% NP40 to reduce aggregation and improve the resolution.

As noted, there was no marked increase in DNA polymerase γ activity. However, Spadari and Weissbach (9) have reported that DNA polymerase γ increases many-fold early in the S phase of synchronized HeLa cells, and there is one report (10) that polyoma virus infection increased DNA polymerase γ activity. We have examined the polymerases at 72-96 hours post infection, or 24 hours after serum stimulation, when total activity was highest. It is possible that γ increases rapidly, and then decreases. The experiment shown in Fig. 2 is designed to detect a transient increase in DNA polymerase γ after SV40 infection.

No increase early in the infective cycle is observed. A preliminary experiment further suggests γ polymerase does not increase in confluent CV-1 cells when they are serum-stimulated.

When enzyme preparations for all three polymerases from each of the three cell populations were available, we examined the

inhibition of the three polymerase activities by the inhibitors
N-ethylmaleimide (NEM) and phosphonoacetic acid (PAA).

There is no significant difference in the inhibition profiles
of either α, β or γ polymerase from infected or uninfected cells.
We have also used the inhibitor 1-β-D arabinofuranosylcytosine
triphosphate (AraCTP) to test DNA polymerases α and β and obtained
the same result.

DNA polymerase α is drastically inhibited by NEM, while
polymerase γ is also sensitive to the compound. As is consistent
with other reports, polymerase β is not affected by NEM (11). All
three monkey cell DNA polymerases are inhibited by PAA, although
α is the most sensitive. DNA polymerase α is also inhibited by
AraCTP, consistent with previous experiments (12,13).

With this inhibitor data in hand, we have examined *in vivo* and
in vitro SV40 DNA synthesis in an attempt to determine the function
of the three polymerases in SV40 replication. We had previously
reported that AraC inhibited both the synthesis of Okazaki fragments
and the gap-filling step required before Okazaki fragments could
be joined into longer SV40 progeny strands *in vivo* (14). We

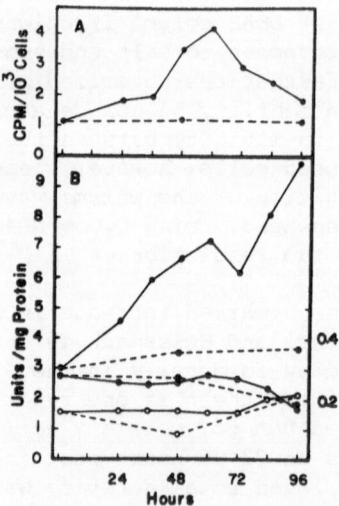

Figure 2. CV-1 cells were treated as described in Fig. 1. At
intervals DNA synthesis was measured by the incorporation of (^3H)
thymidine and two plates were harvested to measure polymerase
activity. Different assays were used to measure each polymerase
in association with the other two (see text). Figure 2A. (^3H)
thymidine incorporated into the Hirt extract. (●—●) infected
cells; (●--●) uninfected cells. Figure 2B. α DNA polymerase
activity in infected (●—●) or uninfected cells (●--●); (O)
β polymerase; (◉) γ polymerase.

attempted to examine the effect of PAA on SV40 replication *in vivo*. However, PAA is not an efficient inhibitor of DNA synthesis in CV-1 cells. This compound has been reported to inhibit Herpes virus DNA polymerase, and to block Herpes virus replication in Wi-38 and HeLa cells (15--17). Four-fold higher doses and times of inhibition (4 mg/ml culture media, 4 hours at 37°) were required to inhibit either cell or SV40 DNA synthesis by 50% in CV-1 cells. Following this treatment, there was obvious cytotoxicity, with many rounded and detached cells present. This *in vivo* resistance to the drug is in contrast to the facile inhibition of purified polymerase or SV40 DNA synthesis in a nuclei system.

We have developed a nuclear system from SV40 infected CV-1 cells, similar to the polyoma-infected 3T6 cell nuclei system of Winnacker et al. (18), which will incorporate (^{3}H)thymidine triphosphate into SV40 DNA (Fig. 3). Figure 3 demonstrates an unfortunate property of these systems; the more effort put into nuclei preparation, the less complete DNA replication becomes.

Figure 3A shows alkaline sucrose gradients of SV40 DNA labeled *in vitro* with (^{3}H)thymidine triphosphate in "dirty" nuclei (18). Such nuclei have DNA/protein ratios of 10-15 and are contaminated with many membrane fragments, etc., when examined by light microscopy. When these nuclei are prepared with a final wash of either Triton X-100 or NP-40, nonionic detergents, there is a signficant improvement in the visual appearance and DNA/protein ratio of the preparation. This is balanced by a decrease in the efficiency of joining of short fragments into longer, progeny length SV40 DNA molecules, and the absence of form I daughter molecules, Figure 3B. These "clean" nuclei have been used for subsequent assays, because they are sufficiently permeable to allow relatively large proteins, such as DNA polymerase, to enter freely, a property lacking in "dirty" nuclei, and one which was necessary if reconstruction experiments using isolated enzymes are to be carried out.

We have used this nuclei system from SV40 infected CV-1 cells to examine the viral DNA product synthesized when NEM and PAA were used to inhibit DNA synthesis.

While NEM inhibits drastically the rate of incorporation of (^{3}H)thymidine triphosphate into Okazaki fragments of SV40 DNA (data not shown), a more complex picture is presented by PAA (Fig. 4). In the presence of EGTA, Figure 4B, which is a constituent of the nuclear assay, PAA did not significantly inhibit the total incorporation of (^{3}H)thymidine triphosphate into SV40 DNA. There does appear to be a decrease in the amount of Okazaki fragments joined into longer length progeny strands. However, Bolden et al. (17) have reported that PAA strongly inhibits DNA synthesis in isolated nuclei. We could not observe such inhibition until either

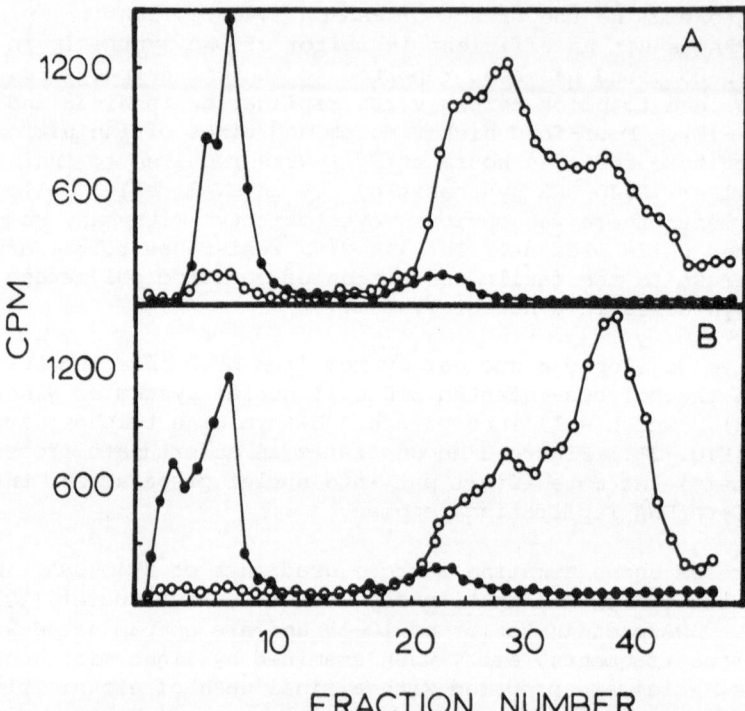

Figure 3. _Nuclei were prepared by the methods of Winnacker et al._
(18), Figure 3A, or given a second wash with 0.5% Triton X-100 in
nonionic buffer, Figure 3B. An assay mixture containing all four
ribotriphosphates, an ATP generating system, EGTA, and three
deoxtriphosphates plus (^{3}H)-labeled thymidine triphosphate was
added, the reaction incubated at 30°, and SV40 DNA extracted by a
modified Hirt procedure. The DNA was sedimented in an alkaline
sucrose gradient₃(5-20%, 60% cushion) with ^{32}P-SV40 marker DNA
(●—●); in vitro (^{3}H)-labeled SV40 (○—○).

EGTA was removed from the assay, or 15 mM $CaCl_2$ was added, a three-
fold excess over the EGTA. Under these conditions, DNA synthesis,
as measured by the incorporation of (^{3}H)thymidine triphosphate into
acid precipitable material, was significantly inhibited. When this
DNA was examined in alkaline sucrose gradient, Figure 4C, only
short 2-3 S fragments of labeled DNA are present, distinctly smaller
than the 4-6 S Okazaki fragments present in the uninhibited,
Figure 4A, or PAA plus EGTA, Figure 4B, sucrose gradients.

    Similar experiments have been carried out using AraCTP and the
adenosine analog, AraATP, to inhibit SV40 DNA replication in the
nuclei system. The results are qualitatively and quantitatively
similar to those obtained earlier, when these compounds were used

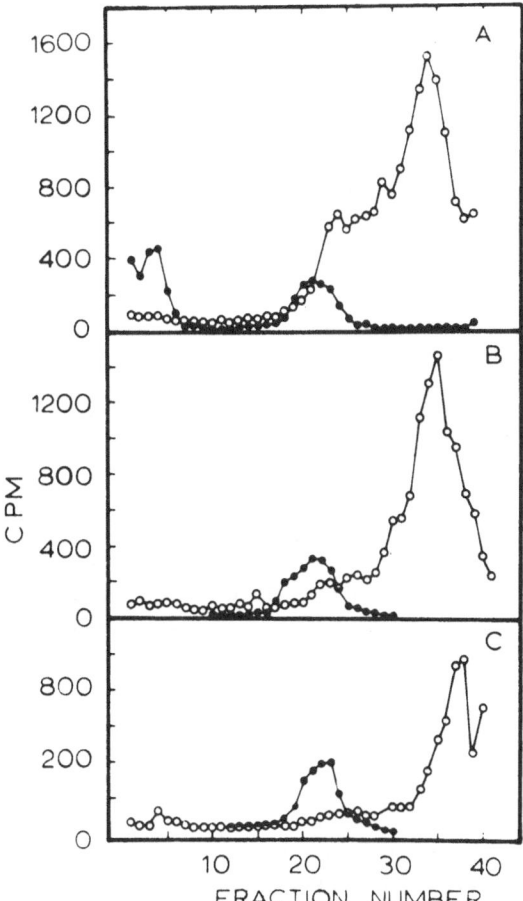

Figure 4. *Alkaline sucrose gradients of in vitro* (^{3}H)*-labeled SV40 DNA synthesized in nuclei inhibited with PAA. Figure 4A. Control, no inhibitor; Figure 4B. 100 µM PAA (EGTA present at 5 mM); Figure 4C. 100 µM PAA (EGTA absent from the reaction; similar results were obtained if a three-fold excess of CaCl$_2$ was added in the presence of EGTA).* ^{32}P*-SV40 marker DNA (*●—●*); in vitro* (^{3}H)*-labeled SV40 (*○—○*).*

to inhibit replication *in vivo* (14). Total incorporation of (^{3}H)-thymidine triphosphate was greatly reduced, but there appears to be a preferential inhibition of gap-filling similar to that observed *in vivo*, since Okazaki fragments are present, albeit at reduced levels.

DISCUSSION

Our results indicate that after SV40-infection or serum-stimulation of confluent CV-1 cells, a five to ten-fold increase in DNA polymerase α activity occurs. DNA polymerases β and γ do not increase in activity by striking amounts, when measured late in infection or in late S phase. Similar results have been obtained for polyoma virus infected mouse cells (11, 19) and SV40 infected monkey cells (20,21). In addition, Evans et al. (22) have recently observed an activity resembling DNA polymerase α in SV40 nucleoprotein complexes. We have further characterized the DNA polymerases from these three cell types, by a number of physical characteristics and by their response to various inhibitors. It appears that there are no obvious differences between the polymerases of uninfected, serum-stimulated, or SV40-infected cells, other than amount.

A number of investigators have examined the changes in DNA polymerase activities between rapidly growing or nongrowing cells in culture and in regenerating rat liver (9, 24-26). Our results in both srerum-stimulated and SV40-infected cells are in agreement with their main conclusions, that DNA polymerase α increases in growing cells, compared to resting cells. We do not see the increase in DNA polymerase γ that Spadari and Weissbach (9) observed in HeLa cells. Whether this failure of polymerase γ to increase represents a real difference between SV40-infected monkey and growing HeLa cells remains to be determined.

A common theme to all of the enzymes of DNA synthesis in the mammalian cell seems to be the presence of multiple forms of the same enzyme activity. Perhaps the reported regulation of the DNA polymerases, and thymidine kinase (27, 28) is typical of many of the enzymes of DNA metabolism. Cells may possess two sets of the enzymes of DNA synthesis - one set adapted to preserving the status quo in the resting cell, another, different class of enzymes better adapted to the milieu of the actively growing cell. Further research into the regulation of the enzymes of DNA metabolism, and especially into the interaction occurring when SV40 virus infects a resting host cell, may clarify this issue.

Finally, we have carried out experiments designed to indicate what role the DNA polymerases were playing in SV40 replication. Replication is discontinuous on SV40 or polyoma template molecules (14,29). It has been suggested that two DNA polymerases are involved in replication, one to synthesize the fragment, and a second enzyme to fill in the gaps remaining between fragments or when an RNA primer is removed (14, 29-33). Studies on isolated DNA polymerases (34,35) with RNA:DNA primer template complexes have shown that DNA polymerase α is able to use an RNA primer, while polymerase β does not initiate efficiently on such primer molecules.

These results would indicate that DNA polymerase α could serve to initiate Okazaki fragment DNA synthesis *in vivo*. We have used an *in vitro* replication system employing SV40-infected cell nuclei and several inhibitors of DNA polymerase α to attack this question.

We conclude from the drug inhibition data that DNA polymerase α is involved in the synthesis of the Okazaki fragment itself, since NEM and the arabinoside analogs drastically reduced incorporation into these 4-6S fragments of DNA. However, the results of *in vivo* and *in vitro* (23) inhibition by AraA and AraC and their triphosphates, and the results of PAA inhibition shown in Figure 4B, suggest DNA polymerase α also plays a role in the gap-filling step, between Okazaki fragments. These conclusions are tempered by (1) the poor specificity of the inhibitors for various enzymes, and (2) the realization that drug-inhibited cells and nuclear systems might not faithfully reflect the true mechanism of SV40 replication. NEM seems relatively specific for α polymerase; however, if β or γ polymerases are complexed, or otherwise associated with a sulfhydryl-containing protein as part of a DNA synthetic complex, they too might be NEM sensitive during SV40 DNA replication. The specificity of inhibition by PAA of the isolated polymerases is not striking, although DNA polymerase α is relatively more inhibited than polymerases β and γ. The effect of calcium on PAA inhibition of DNA synthesis in nuclei is unexplained. In order to observe inhibition of DNA synthesis *in vitro* in nuclei, it was necessary to omit EGTA from the reaction, or add an excess of calcium. Since EGTA has no effect on the inhibition of DNA polymerase α by PAA in the normal polymerase assay, we would suggest at least part of the effect of PAA on (^{3}H)thymidine incorporation into DNA in nuclei is due to a calcium-dependent nuclease degradation of newly synthesized DNA. The results of Figure 4B would thus tend to suggest that gap-filling was preferentially inhibited by PAA, and that gross inhibition of synthesis observed in Figure 4C was cuased by degradation of newly synthesized DNA.

The results of inhibition by AraA and AraC *in vivo* (14) and their triphosphates *in vitro* could indeed indicate that DNA polymerase α plays a role in the gap-filling step, or could also reflect (1) a slower rate of fragment synthesis, as suggested by Hunter and Franke (23), (2) an inability of ligase to seal Okazaki fragments together if they are terminted with an arabinoside on the 3'-OH side of a nick, or (3) preferential endonucleolytic degradation of AraC-containing regions of DNA.

So, while such drug inhibition studies are indicative, they are by no means conclusive. Thus we are now attempting to isolate CV-1 cell mutants which are temperature sensitive or otherwise restricted in SV40 or cell DNA replication, so that we might ascertain by other criteria whether DNA polymerase α is indeed involved in both functions of DNA replication.

ACKNOWLEDGEMENTS

    We would like to thank S.E. Hillier and G.S. Michaels for expert technical assistance. S.R. was the recipient of a fellowship from the Italo-American Medical Education Foundation. This work was supported in part by Grant Number CA17662 from the National Cancer Institute, DHEW and a Basil O'Connor Starter Research Grant from the National Foundation - March of Dimes.

REFERENCES

1. Benyesh-Melnick, M. and Butel, J. (1974) The Molecular Biology of Cancer (ed. H. Busch) Academic Press, New York, p403

2. Levine, A. (1974) Progr. Med. Virol. 20:1

3. Butel, J. (1975) Progr. Med. Virol. 21:88

4. Levine, A. (1976) Biochim. Biophys. Acta 458:213

5. Kelly, T. and Nathans, D. (1977) Adv. Virus Res. 21:85

6. Kara, J. and Weil, R. (1967) Proc. Nat. Acad. Sci. USA 57:63

7. Kit, S., Piekarski, J. and Dubbs, D. (1967) J. Gen. Virol. 1:163

8. Sambrook, J. and Shatkin, A. (1969) J. Virol. 4:719

9. Spadari, S. and Weissbach, A. (1974) J. Mol. Biol. 86:11

10. Närkhammar, M. and Magnusson, G. (1976) J. Virol. 18:1

11. Weissbach, A. (1975) Cell 5:101

12. Momparler, R., Rossi, J. and Labitan, A. (1973) J. Biol. Chem. 248:285

13. Sartiano, G., Lynch, W., Boggs, S. and Neil, G. (1975) Proc. Soc. Exp. Med. 150:718

14. Laipis, P. and Levine, A. (1973) Virology 56:580

15. Mao, J., Robishaw, E. and Overby, L. (1975) J. Virol. 15:1281

16. Huang, H. (1975) J. Virol. 16:1560

17. Bolden, A., Aucker, J. and Weissbach, A. (1975) J. Virol. 16:1584

18. Winnacker, E., Magnusson, G. and Reichard, P. (1972) J. Mol. Biol. 72:523

19. Wintersberger, U. and Wintersberger, E. (1975) J. Virol. 16:1095

20½ Lazarus, L. and Kitron, N. (1976) Arch. Virol. 52:113

21. McLenaghan, I., Adams, R., Eason, R. and Lindsay, J. (1976) Biochem. Soc. Trans. 4:800

22. Evans, M., Wagar, M. and Huberman, J. (1977) Abstracts, 77th Annual Meeting Amer. Soc. Microbiol. 8-13 May. H-98

23. Hunter, T. and Franke, B. (1975) J. Virol. 15:759

24. Chang, L., Brown, M. and Bollum, F. (1973) J. Mol. Biol. 74:1

25. Baril, E., Jenkins, M., Brown, O., Lazlo, J. and Morris, H. (1973) Cancer Res. 33:1187

26. Craig, R., Costello, P. and Keir, H. (1975) Biochem. J. 145:233

27. Postel, E. and Levine, A. (1975) Virology 63:404

28. Kit, S., Leung, W., Trkula, S. and Jorgensen, G. (1974) Int. J. Cancer 13:203

29. Fareed, G. and Salzman, N. (1972) Nature New Biol. 238:274

30. Magnusson, G., Winnacker, E., Eliasson, R. and Reichard, P. (1972) J. Mol. Biol. 72:539

31. Magnusson, G. (1973) J. Virol. 12:600

32. Franke, B. and Hunter, A. (1974) J. Mol. Biol. 83:99

33. Reichard, P., Eliasson, R. and Söderman, G. (1974) Proc. Nat. Acad. Sci. USA 71:4901

34. Chang, L. and Bollum, F. (1972) Biochem. Biophys. Res. Commun. 46:1354

35. Spadari, S. and Weissbach, A. (1975) Proc. Nat. Acad. Sci.
     USA 72:503

36. Hirt, B. (1967) J. Mol. Biol. 26:365

CHARACTERIZATION OF BACTERIOPHAGE T7-INDUCED DNA PRIMASE

Eberhard Scherzinger*, Günter Hillenbrand, Atsushi Yuki
and Erich Lanka*

Max-Planck-Institut für Molekulare Genetik
1000 Berlin 33, Ihnestrasse 63-73
Germany

Bacteriophage T7, although relatively small in size (∿30 genes), relies largely on its own replicative apparatus for multiplication. Extensive genetic and biochemical analyses (for review see ref. 1) have demonstrated that the products of genes 1 to 6 are all involved in phage DNA synthesis. Of these, the two proteins encoded by genes 4 and 5 play a major role in replication. The gene-5 protein combines with *E. coli* thioredoxin to become the major sub-unit of the phage-induced DNA polymerase (2). The gene-4 protein has been shown by electron microscopy (3) to participate in the events at the replication fork that involve synthesis of DNA as short fragments. Recent studies with the purified gene-4 protein, which we call T7 DNA primase, strongly support a role of the enzyme both in the initiation *de novo* and extension of polynucleotide chains (4,5,6). The multiple enzymic operations which are carried out by T7 DNA primase (either alone or in conjunction with T7 DNA polymerase) are schematically represented in Fig. 1.

Complementation in Extracts of T7- and T3- Infected Cells

In vitro studies of T7 DNA replication with a soluble enzyme fraction from T7-infected *E. coli* have revealed a dependence of the system on the gene-4 protein (7,8,9). We have, therefore, initially used a complementation assay to isolate this protein from T7-infected cells. An analysis of the purified T7 gene-4 protein (hydroxyapatite fraction of ref.5) by NaDodSO$_4$-gel electro-phoresis shows only the 66,000- and the 58,000-dalton polypeptides were eluted from the colum coincident with the gene-4 complementing activity, the T7 DNA polymerase stimulating activity, the priming activity, and the NTPase activity. Gel electrophoresis of
A.S.I. Participants

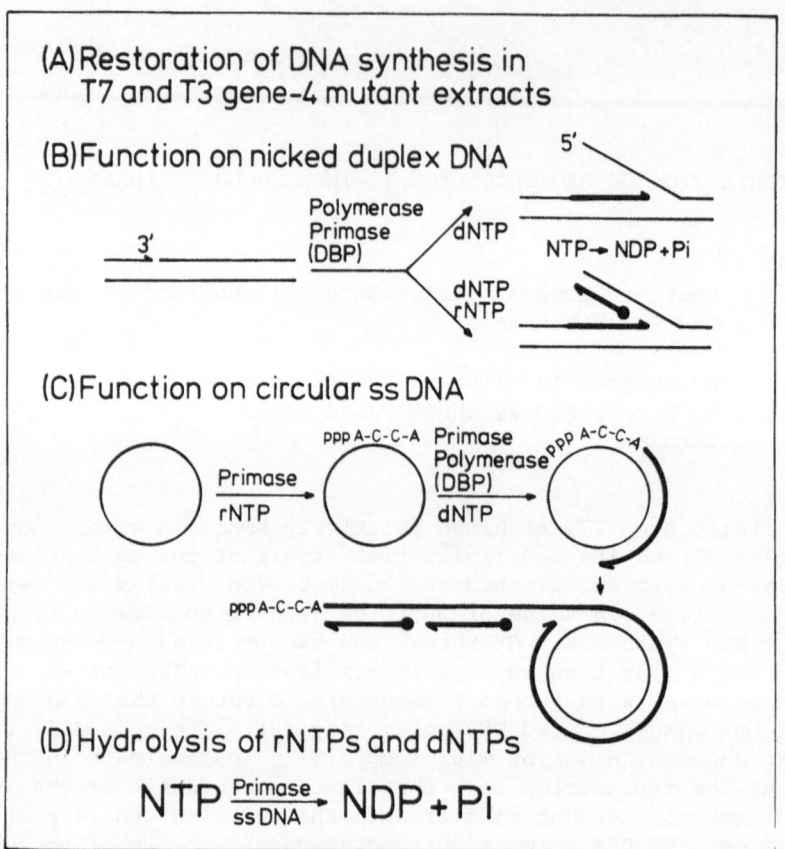

Figure 1. _Schematic representation of the multiple activities associated with T7 DNA primase._

(^{14}C)-labeled proteins synthesized by cells infected with T7$^+$ and T7 gene-4 amber mutants indicates that two polypeptides of the same size are eliminated in the mutant infections (5). We assume, therefore, that these two polypeptides are the product of the T7 gene-4, one being perhaps a proteolytic cleavage product of the other.

Phage T3, which is closely related to T7, induces an analogous enzyme, which we have also purified to near homogeneity. As shown in Fig. 2, the T3 gene-4 protein can restore T7 DNA synthesis in a T7 gene-4 mutant extract equally well as its T7 counterpart. Similarly, the T7 gene-4 protein can substitute for the corresponding T3 enzyme in the complementation of a T3 gene-4 mutant extract. This observation indicates that the T3 and T7 gene-4 proteins do not discriminate between their homologous DNA substrates. On the other hand, it is also evident that the complete

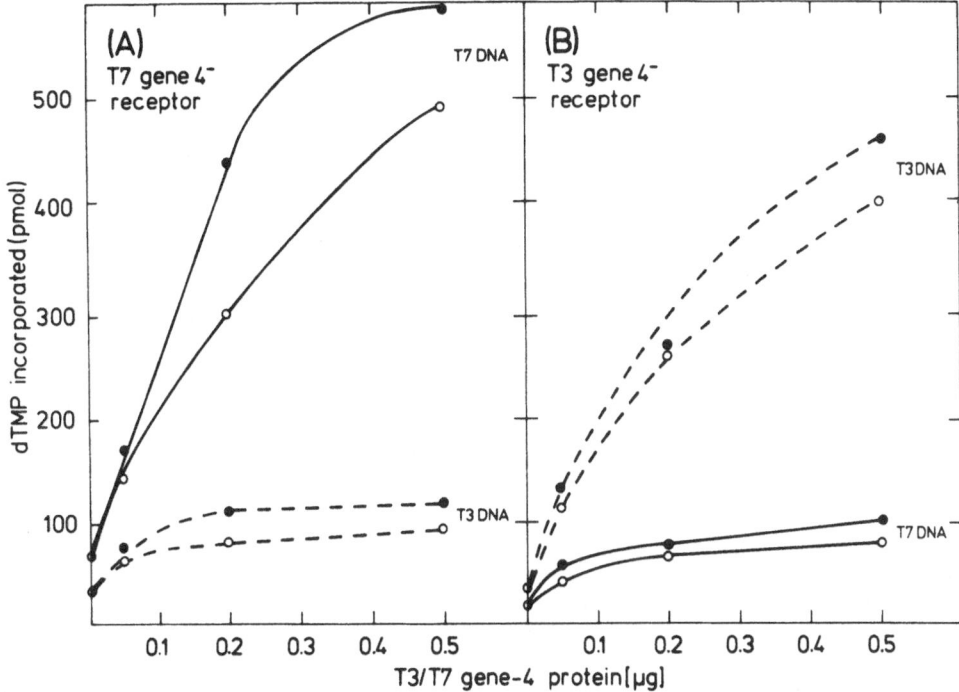

Figure 2 (A and B).   Complementation in vitro of T7 and T3 gene-4 mutant extracts by purified T7 (O) or T3 (●) gene-4 proteins. Incubations were carried out as described in ref. 5, using receptor extracts of gently lysed T7$_{3,4,6}$-or T3$_{3,4,6}$-infected cells (8); native T7 or T3 DNA served as a template, as indicated.

T3 and T7 enzyme systems are highly selective for their homologous DNA *in vitro* (Fig. 2). The agent responsible for this template specific replication has been identified as the viral RNA polymerase, an enzyme known to transcribe preferentially its homologous DNA (10). These findings suggest a dual role of the viral RNA polymerase in both transcription and replication. It may be that a brief transcriptional operation of the RNA polymerase near a promoter site signals the start of replication from a specific origin.

Function of T7 DNA Primase on Nicked Duplex DNA
(strand displacement)

Whereas T7 DNA polymerase by itself cannot synthesize DNA on

nicked duplex templates it can, in conjunction with T7 primase,
catalyze extensive synthesis. This reaction is stimulated by, but
is not dependent on, the presence of rNTPs. An earlier analysis (5)
of the reaction products synthesized on T7 DNA has shown that, in
the absence of rNTPs, synthesis starts from single strand
interruptions within duplex molecules and proceeds by a mechanism
involving the displacement of the 5' end of the strand which is
annealed to the template strand. Later in the reaction, by
processes called "template switching" and "branch migration",
rapidly renaturable product molecules are generated. To obtain
maximum stimulation, T7 primase is required in approximately molar
equivalence to the DNA polymerase. Furthermore, the reaction is
specific for T7 DNA polymerase; neither T4 DNA polymerase nor
E.coli DNA polymerases I,II or III can cooperate with T7 primase
(5,9). Thus, it seems likely that T7 primase acts catalytically
in conjunction with T7 DNA polymerase to promote the strand
displacement.

The apparent function of T7 primase in the extension of
polynucleotide chains, however, only partially explains its
essential role in the replication of duplex DNA. Alkaline sucrose
and CsCl gradient centrifugation of the products made on T7 DNA
in the presence of rNTPs has revealed that under these conditions
of synthesis much of the product DNA is not covalently linked to
the template (5). This observation suggested that, in addition to
synthesis via strand displacement (as expected for synthesis on
the "leading side" of a replication fork), also RNA-primed
initiation of new DNA chains is occurring in vitro (as expected
for synthesis on the "lagging side" of a replication fork). The
rNTP-dependent initiation reaction is not restricted to T7 DNA
(5). Therefore, we studied this type of reaction in more detail
using circular single strands as templates (e.g., fd and øX174 DNA).

Function of T7 DNA Primase on a Circular Single-Stranded Template
(DNA Chain Initiation de novo)

The conversion of øX174 (or fd) DNA to duplex forms requires
T7 DNA primase, T7 DNA polymerase, ATP and CTP, all 4 dNTPs and
Mg^{++} Omission of either one of these components reduces DNA
synthesis to less than 2%. Addition of T7 DNA binding protein in
amounts sufficient to saturate the template results in a 2- to 20-
fold stimulation of DNA synthesis, depending on the temperature
(Table I). The polymerization reaction can be separated into 3
stages which will be discussed in turn below (see also Fig. 1C):
(i) ssDNA-dependent synthesis of an oligoribonucleotide primer by
T7 primase; (ii) primer extension by T7 DNA polymerase and primase;
(iii) displacement of the newly made strand from the viral strand
and new initiations on the displaced strand.

Table 1

Temperature ($^{\circ}$C)	dTMP incorporated		Ratio +/- T7 DBP
	+T7 DBP	- T7 DBP	
	pmol/5 min		
5	8	< 1	< 20
10	18	1	18
20	163	36	4.5
30	302	147	2.0
35	393	195	2.0
40	328	172	1.9

Stimulation of ⌀X174 DNA-directed synthesis by T7 DNA
binding protein. Conditions of synthesis were as described
in the legend to Table 2, except that ATP (0.5 mM) and CTP
(0.1 mM) were used in place of synthetic primers.

(i) Primer synthesis: When homogeneous preparations of T7 (or T3)
primase are incubated with ATP and (α-^{32}P)CTP in the presence of
⌀X174 DNA and Mg^{++}, a limited number of oligoribonucleotides is
produced in high yield. As shown in Fig. 3a, fractionation of the
dephosphorylated products by gel electrophoresis resolves mainly 3
radioactive bands. These were identified by re-electrophoresis on
DEAE-paper (at pH 3.5) as A-C (I), a mixture of A-C-C and A-C-A (II),
and A-C-C-A (III), Qualitatively the same products are synthesized
on ⌀X174 DNA when all 4 rNTPs are present (f) and also under
conditions of active DNA synthesis (b). If ⌀X174 DNA is replaced
by native T7 DNA, efficient oligomer formation is only observed
under conditions of replication (c,d). In this case, synthesis
is likely to occur on single strands generated via strand displacement during replication (as expected for RNA-priming during "lagging
strand" synthesis). No oligomer formation takes place in the absence
of DNA (e), or with homo-deoxypolymers as template (not shown).
Thus, T7 primase is a novel ssDNA-dependent oligoribonucleotide
synthetase, which appears to generate the same products on all
natural templates tested.

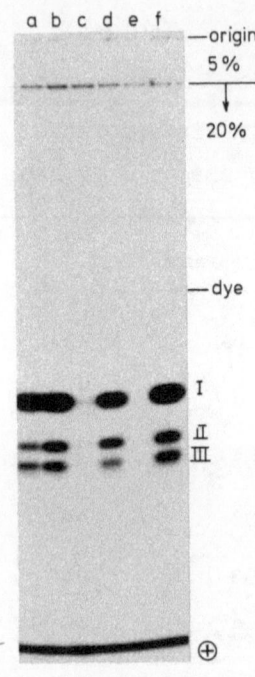

<u>*Figure 3*</u>. *Autoradiograph of fractionated oligoribonucleotides synthesized by T7 DNA primase. The complete reaction mixture (20 µl) contained 40 mM Tris-HCl (pH 7.6), 10 mM MgCl₂, 5 mM dithiothreitol, 0.2 mg/ml bovine serum albumin, 100 µM ATP, 10 µM (α-³²P)CTP (10⁴cpm/pmol), 0.2 µg ØX174 DNA, and 1.5 units of T7 DNA primase. Incubation was for 5 min at 30°C. The reaction was stopped by heating (2 min at 90°C), and after treatment with alkaline phosphatase for 60 min at 37°C the samples were analyzed by electrophoresis at 25 V/cm on a 20% polyacrylamide - 7 M urea slab gel (11). Using the above conditions, 1 unit of T7 DNA primase permits the incorporation of 1 pmol CMP into oligoribonucleotides. (a) complete reaction, (b) T7 DNA polymerase (0.15 µg), and all 4 dNTPs (50 µM each) added, (c) ØX174 DNA replaced by T7 DNA, (d) T7 DNA; polymerase and dNTPs added, (e) ØX174 DNA omitted, and (f) GTP and UTP (10 µM each) added. Dye is xylene cyanol FF.*

(ii) <u>Primer extension</u>: Inspection of the 5' end of the poly-nucleotide chains synthesized by the T7 enzymes on ØX174 (or fd) DNA has revealed that pppA-C-C-A is primarily (>70%) used for the covalent extension by the DNA polymerase (5). This led us to ask whether T7 DNA polymerase alone or in combination with T7 primase might also be able to extend short oligomers other than A-C-C-A. Therefore, the primer activity of a number of synthetic oligo-nucleotides was tested in reactions containing ØX174 DNA covered with DNA binding protein as a template (Table 2). The data show

Table 2

Oligonucleotide Additions	dTMP incorporated	
	− T7 primase	+ T7 primase
	pmol / 5 min	
None	6	17
A−C	5	17
C−A	5	18
A−C−C	4	73
C−C−A	17	112
A−C−C−A	80	234
A−C−C−G	14	132
A−C−C−U	7	102
A−C−C−C	15	123
A−C−A−A	8	160
A−A−C−A	11	185
C−C−C−A	44	178
C−C−A−A	45	159
C−C−A−A−A	37	146
A−C−C−A−A	82	179
U−U−U−U	4	34
C−C−C−C	4	91
A−A−A−A	6	188
dA−dA−dA−dA	6	147

Effect of T7 DNA primase on oligonucleotide-primed ØX174
DNA synthesis. Complete reaction mixtures (50 µl) contained
40 mM Tris-HCl (pH 7.6), 10 mM $MgCl_2$, 20 mM NaCl, 5 mM dithiothreitol, 0.2 mg/ml bovine serum albumin, 50 µM each
of dATP, dGTP, dCTP, and (^{3}H)TTP (400 cpm/pmol), 50 µM
oligonucleotide primers, 0.2 µg ØX174 template DNA, 1.6 µg
T7 DNA binding protein, 0.15 µg T7 DNA polymerase, and 0.75
unit of T7 DNA primase. After incubation for 5 min at 30°C,
(^{3}H)dTMP incorporated was determined.

that, at the relatively high primer concentration used (50 μM),
T7 DNA polymerase alone is able to extend certain oligomers,
particularly those containing C-C-A sequences. Whether this
selectivity by T7 DNA polymerase simply reflects an increased
annealing efficiency of these oligomers to complementary sequences
on the template or is due to a special affinity of T7 DNA
polymerase to these oligomers is not yet known. The data further
show that the efficiency of priming with all tri-, tetra-, and
pentanucleotides tested is drastically increased in the presence
of T7 primase. Thus, we conclude that T7 primase is also required
for primer extension.

In contrast to T7 DNA polymerase, T7 primase appears not to
be selective for oligomers of a special base composition. However,
the chain length of the primer seems to be important for the
recognition by T7 primase as shown by the following experiment.
ØX174 DNA-dependent synthesis was measured using either oligo(rA)
or oligo(rC) nucleotides of various chain length to prime DNA
synthesis. In reactions containing T7 primase, a distinct preference for tetranucleotide primers is observed with both sets of
oligoribonucleotides, whereas T7 DNA polymerase alone prefers
primers of chain lengths >4 (Fig. 4). The relatively poor primer
activity of $(rC)_{5-8}$ compared to $(rA)_{5-8}$ may be explained by the
fact that the longest complementary sequences in ØX174 DNA are $(dG)_4$
and $(dT)_7$, respectively (12).

The stimulatory effect of T7 primase on oligomer-primed
synthesis is markedly influenced by the temperature (Table 3). A
10-fold stimulation of A-C-C-A-primed synthesis is observed at $40^{\circ}C$,
while at temperatures below $10^{\circ}C$ stimulation is barely detectable.
At high temperature annealing of the primer to the template and
consequently also the efficiency of chain initiation by the DNA
polymerase is impaired. Therefore, it seems likely that T7 primase
acts by stabilizing primer and polymerase on the template DNA.

This stabilization presumably occurs at base-complementary
regions on the template, as suggested by the following experiment.
ØX174 DNA was converted to the duplex form using either ATP and CTP,
or oligoribonucleotides of various base composition for initiating
DNA synthesis. The ^{32}P-labeled DNA products were cleaved by the
restriction enzyme HpaI. This enzyme produces 3 fragments which
were first fractionated under neutral conditions on a 1.6% agarose
gel. HpaI fragment 2 (1264 nucleotides) was isolated from the gel
and denatured in order to separate the newly made strands from the
unlabeled template. Subsequent gel electrophoresis on poly-
acrylamide-urea at $45^{\circ}C$ resolved a series of bands characteristic
for each primer species (Fig. 5). This is the result expected if
synthesis with an individual primer starts from distinctive sites
within the fragment examined. It is interesting to note that the
pattern produced by the use of A-C-C-A is very similar to that

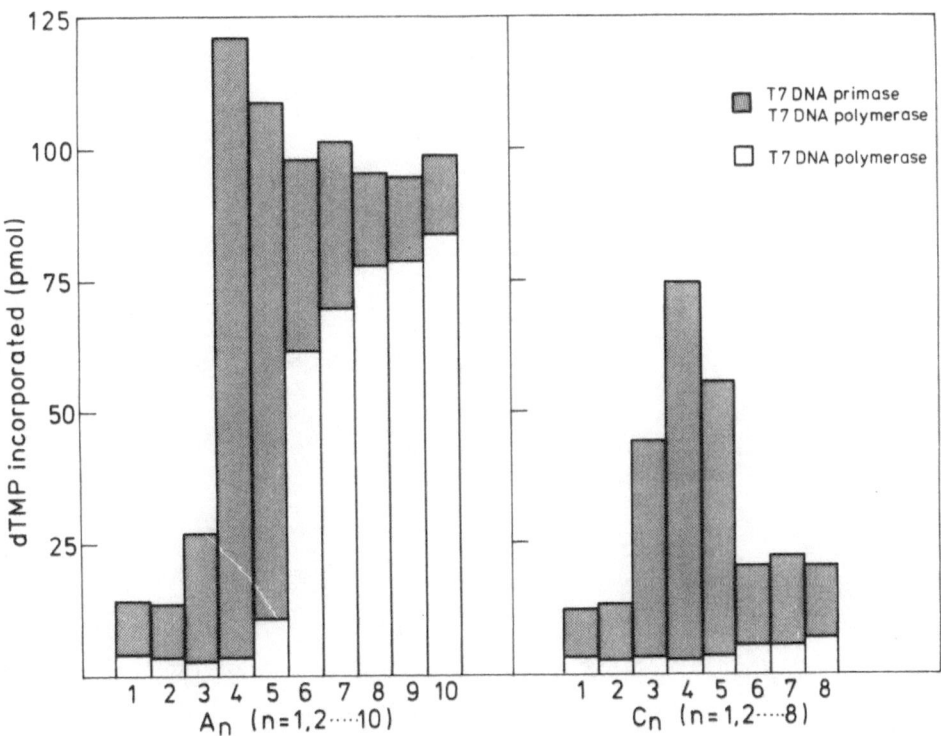

Figure 4. *Effect of T7 DNA primase on priming activities of oligo-ribonucleotides of various chain lengths. Incubations were performed as described in the legend to Table 2, using either 5 μM oligo(rA)ₙ or 50 μM oligo(rC)ₙ to prime DNA synthesis on ØX174 DNA.*

Table   3

Temperature (°C)	dTMP incorporated + T7 primase	- T7 primase	Ratio +/- T7 primase
	pmol / 5 min		
5	3	3	1
10	17	11	1.5
20	88	49	1.8
30	222	66	3.4
35	258	54	4.8
40	220	21	10.5

Influence of temperature on A-C-C-A-primed ØX174 DNA synthesis. Conditions of synthesis were as described in the legend to Table 2.

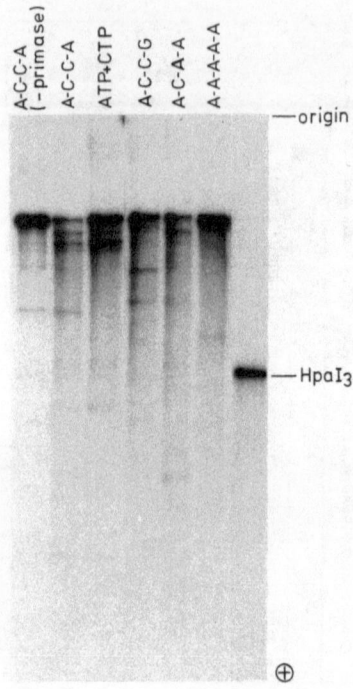

Figure 5.   *Autoradiograph of a 7.5% polyacrylamide - 7 M urea*
slab gel used to fractionate isolated HpaI-fragments 2 of øX174
duplex form DNA under denaturing conditions.   HpaI-fragment 3
(392 nucleotides) was run in parallel as a marker. Experimental
procedures are outlined in the text.

obtained from samples containing ATP and CTP.   This result suggests
that site selection in principle is the same whether the primer is
generated *in situ*, or whether exogenous A-C-C-A is employed for
priming DNA synthesis.

(iii) <u>Strand displacement</u>: Once synthesis has started on a circular
template, polymerization proceeds around the circle to generate a
circular double-helix with one or more single strand interruptions,
depending on the number of starts.   Continuation of synthesis
induces strand displacement and generates circular molecules with
double-stranded tails plus short linear fragments.   Some of the
product molecules contain single-stranded regions at the tail-circle
junction, as visualized by electron microscopy (Fig. 6). Moreover,
single- and double-stranded regions are occasionally seen to
alternate along the tail.   Therefore, we conclude that new chains
can also be initiated on the displaced complementary strand (as
expected from a rolling circle model).

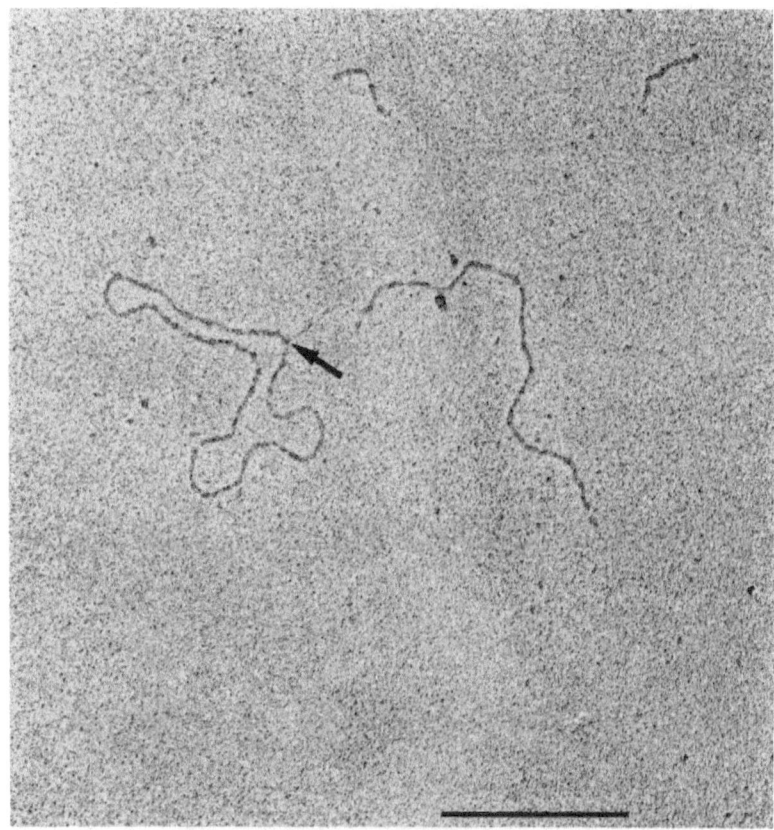

Figure 6. Electron micrograph of a product molecule synthesized on øX174 DNA in the presence of T7 DNA binding protein, polymerase and primase. Experimental conditions are described in ref. 5. The arrow points to a single-stranded region at the tail-circle junction. Bar equals 0.5 µM.

NTP Hydrolysis by T7 DNA Primase

Purified T7 primase displays an activity which hydrolyzes both ribo- and deoxyribonucleotide triphosphates (except rGTP) to the corresponding diphosphates and inorganic phosphate (5,6). Dephosphorylation of NTPs by T7 primase is stimulated by single-stranded but not by duplex DNA. Recently, Kolodner and Richardson (6) have reported that DNA synthesis catalyzed by T7 DNA polymerase and primase on duplex DNA is closely coupled to the hydrolysis of NTPs. It has been suggested that the cleavage of NTPs by T7 primase might provide the energy for opening the DNA strands at the replication fork. The mechanism by which T7 primase utilizes the

energy to facilitate strand displacement is not known. By analogy
with the presumed role of the T4 gene 44-62 ATPase in the elongation
of polynucleotide chains (13), one might speculate that T7 primase
acts by enhancing the affinity of the DNA polymerase to the template
strand; its NTP hydrolysis may serve to generate changes in the
protein conformation which would help drive the polymerase along
one of the base-paired DNA chains, thereby displacing the
complementary chain.

ACKNOWLEDGEMENT

    Stimulating and critical comments by H. Schuster are highly
acknowledged. We thank J. Reeve for valuable advice in preparation
of the manuscript and B. Geschke and D. Seiffert for expert
technical assistance.

REFERENCES

(1)   Hausmann, R. (1976) Curr. Top. Microbiol. Immunol. 75:77-110

(2)   Mark, D.F. and Richardson, C.C. (1976) Proc. Natn. Acad. Sci.
      U.S.A. 73:780-784

(3)   Wolfson, J. and Dressler, D. (1972) Proc. Natn. Acad. Sci.
      U.S.A. 69:2682-2686

(4)   Scherzinger, E. and Klotz, G. (1975) Mol. Gen. Genet. 147:
      233-249

(5)   Scherzinger, E., Lanka, E., Morelli, G., Seiffert, D. and
      Yuki, A. (1977) Eur. J. Biochem. 72:543-558

(6)   Kolodner, R. and Richardson, C.C. (1977) Proc. Natn. Acad.
      Sci. U.S.A. 74:1525-1529

(7)   Hinkle, D.C. and Richardson, C.C. (1974) J. Biol. Chem. 249:
      2974-2984

(8)   Scherzinger, E. and Seiffert, D. (1975) Mol. Gen. Genet. 141:
      213-232

(9)   Hinkle, D.C. and Richardson, C.C. (1975) J. Biol. Chem. 250:
      5523-5529

(10) Dunn, J.J., Bautz, F.A. and Bautz, E.K.F. (1971) Nature (New
      Biol.) 230:94-96

(11) Maxam, A.M. and Gilbert, W. (1977) Proc. Natn. Acad. Sci.
     U.S.A.   74:560-564

(12) Sanger, F., Air, G.M., Barrell, B.G., Brown, N.L., Coulson,
     A.R., Fiddes, J.C., Hutchinson III, C.A., Slocombe, P.M. and
     Smith, M. (1977) Nature (Lond.) 256:687-695

(13) Alberts, B., Morris, C.F., Mace, D., Sinha, N., Bittner, M.
     and Moran, L. (1975) In: "DNA Synthesis and Its Regulation"
     (eds. Goulian, M., Hanawalt, P. and Fox, C.F.) W.A. Benjamin,
     Menlo Park, Ca) 241-269

SUPPRESSION OF *E.coli dnaB* MUTANTS BY PROPHAGE P1*bac:*

A BIOCHEMICAL APPROACH

E. Lanka[*], M. Schlicht, M. Mikolajczyk, B. Geschke,
C. Edelbluth and H. Schuster[*]

Max-Planck-Institut für Molekulare Genetik
1000 Berlin 33
Ihnestrasse 63-73, Germany

INTRODUCTION

Bacteriophage P1 codes for a *dnaB* analog *(ban)* protein, which is repressed in wild-type P1 prophage and expressed constitutively in P1*bac (dnaB* analog control) mutants (3,6). *E.coli dnaB*(P1*bac)* lysogens are able to grow at temperatures that arrest DNA synthesis in the non-lysogen (3,6). In addition, p1*bac ban* mutants have been isolated in which the expression of the viral *dnaB* analog is prevented. These phage mutants do not suppress the *dnaB* character (3).

An attempt has been made to imitate the suppressing action of p1*ban* protein (7) in an *in vitro* system for DNA replication (8). A soluble enzyme fraction that catalyzes the conversion of ϕX174 single-stranded DNA to duplex DNA was isolated from p1*bac*- and P1*bac ban* lysogens of *E.coli* wild type and *dnaB* ts mutants. It was found that DNA synthesizing activity of *ban* protein containing fractions from wild type or *dnaB*(P1*bac)* lysogens was more temperature-resistant than that from *dnaB* or *dnaB*(P1*bac ban)* mutants. It was suggested that the temperature-resistant DNA synthesis with fractions from P1*bac* lysogens is mediated by the P1*ban* protein (7).

In this paper the purification of the *dnaB* complementation activity from P1*bac* lysogens of two types of *E.coli dnaB* mutants is described: These mutants are firstly three nearly isogenic strains carrying the *dnaB*266 amber mutation and differing only with respect to the *dnaB* alleles and amber suppressors (6,3), and secondly a novel mutant, *dnaB*252, defective in DNA initiation (11,1) and suppressible by prophage P1*bac* (E. Lanka, B. Geschke, and

[*]*A.S.I. Participants*

H. Schuster, in preparation). Purification is achieved by
following an assay as described for the isolation of *E.coli dnaB*
wild type protein (10).

We begin by describing a model for the suppression of *dnaB*
mutants by prophage P1*bac*, which is entirely based on results
obtained by *in vivo* experiments. A brief description of the
experimental strategy for the isolation of P1*ban* protein follows.
Then we demonstrate that in P1 lysogens a protein is synthesized
which is under the control of P1*bac* and which copurifies with the
dnaB complementation activity. Finally, the biochemical data
obtained are compared with the model.

A Model for Suppression of *E.coli dnaB* Mutants
by P1*bac* Prophage

The model (Fig. 1) is based on the assumption that *dnaB* and
ban protein monomers are rather *similar*, in order to be capable of
associating with each other at random. Depending on the amount and
the ratio of both proteins in a *dnaB* ts (P1*bac*) lysogen, multimers
consisting of *dnaB*, *dnaB + ban*, and *ban* monomers can exist at 30°C
or below. Following a shift to temperatures non-permissive for the
dnaB function (≥ 40°C), *ban* either stabilizes or substitutes for
the defective *dnaB* protein. Stabilization is effective during
continuous growth at 40°C. Substitution may occur either during
the process of DNA elongation or only after completion of a round of
DNA replication following the temperature shift. Whether
stabilization or substitution takes place may depend on the
particular *dnaB* mutant suppressed by P1*bac*. In any case, the
average composition of the multimers in any cell should be
relatively more rich in *ban* monomers (up to 100%) than *dnaB* monomers
at 40°C as compared to 30°C, in order to account for the inability,
or reduced ability, to replicate phage λ at temperatures of 40°C
or above (3).

The model is based on the following results obtained by *in vivo*
experiments. Both *dnaB* and *ban* are able to replace each other in
bacterial and P1 DNA replication (3,6,2), indicating *similarities*
between both proteins. The existence of functional multimeric
aggregates of *dnaB* subunits that are at least tetramers was suggested
by results obtained with *E.coli dnaB* heteroallelic diploids (5). An
interaction of *dnaB* and *ban* proteins was deduced from the ability
of P1*bac* prophage to suppress *groP* (*dnaB*) mutants, thus restoring
λ growth at least at low temperatures (3). Inspite of their
similarities, *ban* and *dnaB* protein must at the same time differ
distinctly from each other, as revealed by the properties of a
dnaB amber (P1*bac*) lysogen. In this mutant strain, *E.coli* DNA
replication is restricted to temperatures above 32°C and λ does not
replicate at all (3,6).

TEMPERATURE   FUNCTIONAL MULTIMERIC AGGREGATES OF <u>dnaB</u> (●) AND/OR <u>ban</u> (■) MOLECULES

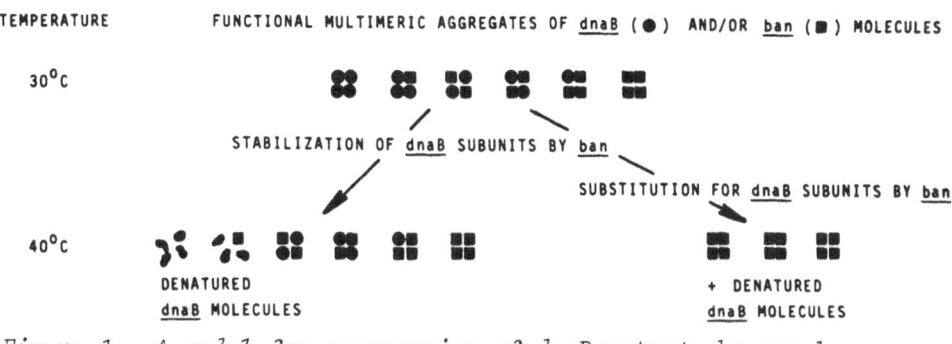

Figure 1. A model for suppression of <u>dnaB</u> mutants by prophage
P1<u>bac</u>.

Experimental Strategy

For the isolation of *dnaB* complementation activity[†] the
*dnaB*266 amber mutant and its P1 lysogens were used. Activity was
purified from strains in which the amber mutation is either
unsuppressed (su⁻, viable only as P1*bac* lysogen), or suppressed
by *supE* to temperature sensitivity, or by *supF* to temperature
resistance (Table 1). All P1*bac* lysogens were grown at 40°C, a
temperature at which suppression of *dnaB* by P1*bac* is effective,
whereas for all strains in which the *dnaB* mutant character is
not suppressed by P1*bac*, a growth temperature of 25°C was used
(Table 1). Complementing activity was purified as outlined in
Fig. 2.

In preliminary experiments using P1*bac* lysogens of *dnaB* ts
mutants, it was found that the DNA synthesizing activity did not
retain its temperature resistance upon further purification. This
temperature resistance had been found to be characteristic for
ammonium sulfate fractions containing *ban* protein (7). As a
consequence, the complementing activity not only became nearly
indistinguishable from that of the corresponding *dnaB* non-lysogen,
but also was obtained in rather low yield (Table 2). In addition,
the activity in crude extracts of the (su⁻)P1*bac* lysogen was very
often rapidly lost. These findings raised the question as to
whether *ban* activity could be followed at all by the assay used,

---

[†]in the following abbreviated to "activity" or "complementing
activity.

Table 1

E.coli K12 *dnaB* mutants

Strain (relevant characters)			*dnaB* Gene Product	Growth Temperature ($^\circ$C)
dnaB 266 amber mutant	*supF*	non-lysogen (P1) (P1*bac*)	*dnaB* tr	40
	supE	non-lysogen (P1*bac*) (P1*bac ban*)	*dnaB* ts	25 40 25
	su⁻	(P1*bac*)	*dnaB* am	40
HfrH	252 252 (P1*bac*)		*dnaB* ts	25 40

tr = temperature resistant; ts = temperature-sensitive;

am = amber

because the P1 protein might only be able to support øX174 DNA-dependent DNA synthesis by stabilizing, but not by substituting, *dnaB* protein. Therefore another mutant, *viz. dnaB*252, was included in these studies for the following reason. No *dnaB* complementation activity can be isolated from this mutant (S. Wickner, cited in 11) unless it contains the P1*bac* prophage (E.Lanka, B. Geschke, and H. Schuster, in preparation), thus ensuring that any activity purified from the 252(P1*bac*) lysogen must at least contain *ban* protein.

In view of the instability of the *dnaB* ts protein, it was very important to avoid time-consuming purification procedures, and therefore affinity chromatography was used as a quick and efficient purification step, as described in the next section.

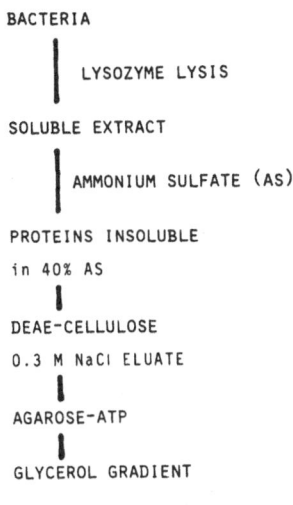

BACTERIA

    LYSOZYME LYSIS

SOLUBLE EXTRACT

    AMMONIUM SULFATE (AS)

PROTEINS INSOLUBLE

in 40% AS

DEAE-CELLULOSE

0.3 M NaCl ELUATE

AGAROSE-ATP

GLYCEROL GRADIENT

FRACTIONATION SCHEME

(dnaB COMPLEMENTATION ACTIVITY)

Figure 2. *Fractionation scheme for the purification of* *dnaB*
complementation activity.

Use of Affinity Chromatography for the Purification
of *dnaB(ban)* protein

A ribonucleoside triphosphate activity has been found to be
associated with the *dnaB* protein of *E.coli* (9). This suggested that
it would be possible to make use of affinity chromatography on
immobilized ATP for the purification of *dnaB* complementation
activity. Two types of agarose-ATP were used, firstly N6(6-amino-
hexyl)-ATP coupled to agarose (Type 2, P·L biochemicals), and
secondly ATP which had been reacted with periodate to couple the
ribose moiety to agarose via adipic acid dihydrazide as a spacer
(Type 4, P·L biochemicals). Complementing activity is adsorbed to
both types of agarose ATP and can be subsequently eluted by ATP
(C. Edelbuth, E. Lanka and H. Schuster, in preparation). An
example is given in the next section.

Two Proteins Copurify with the *dnaB* Complementation
Activity from *E.coli* Pl*bac* Lysogens

The DEAE-cellulose fraction of *supE*(Pl*bac*) was adsorbed to
agarose-ATP. Elution with ATP yielded four major proteins as shown
by SDS gel electrophoresis (Fig. 3 and Fig. 4, right half). The

Table 2

Purification of *dnaB* complementation activity from *E.coli supE dnaB* ts (P1*bac*)

Fraction	Total Units	Total Protein (mg)	Specific Activity (units/mg)	Yield (%)
Extract	-	13,700	-	-
Ammonium sulfate	540	1,530	0.35	100
DEAE-cellulose	76	7	11	14
Agarose-ATP[a]	62	0.09	62	11
Glycerol gradient[a]	18	0.008	2,380	3

One unit of *dnaB* complementation activity incorporates 1 nmol of dTMP in 30 min at 25°C

[a]These steps were performed with only a part of the fraction from the preceding purification step. The values reported assume that the yield and purification would be the same if the entire DEAE-cellulose- and agarose-ATP fractions were subjected to the following procedure(s).

approximate molecular weights of these proteins are A = 70,000, B = 62,000, C = 56,000, and D = 38,000. In the corresponding aga-rose-ATP fractions of *supE* and of *supE(P1bac ban)*, protein C is absent (Fig. 3 and Fig. 4, right half). When the agarose-ATP fractions were subjected to glycerol gradient centrifugation, protein B and protein B + protein C cosediment with the complementing activity of *supE*, *supE(P1bac ban)* and *supE(P1bac)*, respectively, whereas the bulk of proteins A and D sediments faster. The native molecular weight of the complementing activity of *supE(P1bac)* is about 350,000 (data not shown).

When the DEAE-cellulose fractions of *supF* and its P1 lysogens, and of *su⁻(P1bac)* were subjected to the same procedure, the following

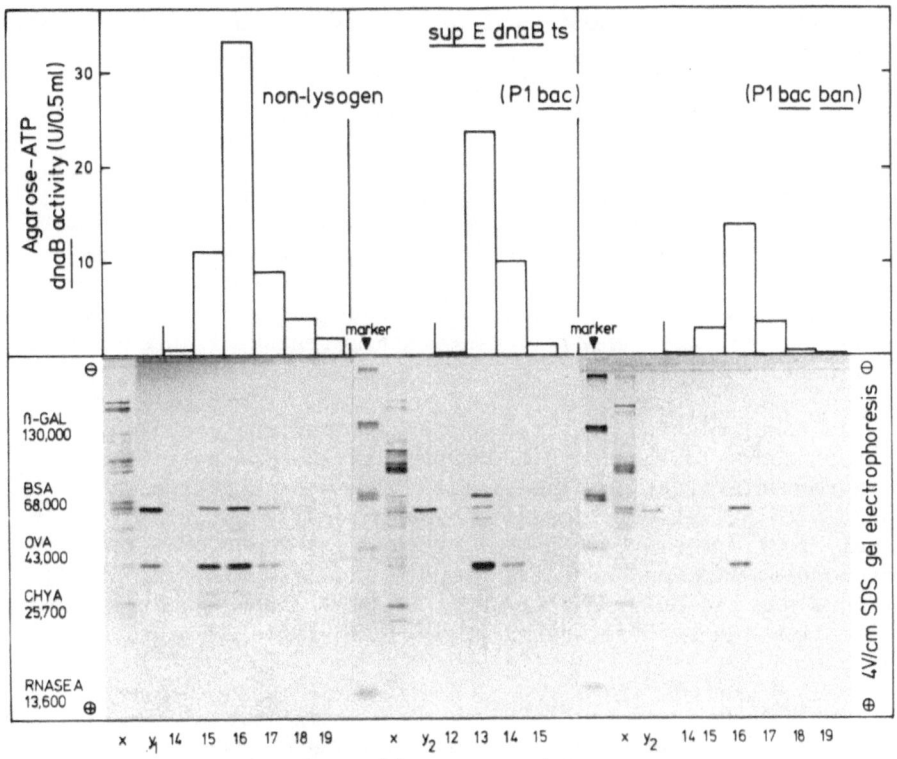

Figure 3. Agarose-ATP affinity chromatography and SDS gel electro-phoresis of dnaB *complementation activity. DEAE-cellulose fractions (4 mg protein each) were adsorbed to agarose-ATP type 4 (1 ml bed volume) and eluted by 5 mM ATP. Fractions (0.1 ml) were subjected to SDS electrophoresis on 15% polyacrylamide gels (4). x = loading material; agarose-ATP fractions of* supF *(y₁) and* supF- *(P1bac) (y₂) were run in parallel.*

results were obtained. Protein C is found only in the *supF*(P1*bac*) lysogen but not in the corresponding P1 wild type lysogen or non-lysogen (Fig. 4, left half). In the case of *su⁻*(P1*bac*), protein B is absent, but in contrast protein C is visible (Fig. 4, right half). In addition, all fractions contain variable amounts of proteins A and D. Whereas the molecular weights of proteins A, C and D of all strains tested are identical, as judged by SDS gel electro-phoresis, protein B of *supF* strains is slightly smaller (MW = 61,000) than protein B of *supE* strains. Glycerol gradient centrifugation of the agarose-ATP fractions again resulted in a cosedimentation of complementing activity with protein B and B + C, respectively (Fig. 5), with native molecular weights in the range of 300,000 to 350,000. (Sedimentation analysis of the *su⁻*(P1*bac*) fraction did not give unequivocal results because of the low amounts available). These results indicate that proteins B and C may be subunits of *dnaB* and *ban*, respectively. The apparent difference in size between the *dnaB* monomers of *supE* and *supF* is not understood, since these proteins should differ only by the amino acids glutamine and tyrosine.

Although the complementing activities of *supF*(P1*bac*) and of *supE*(P1*bac*), both contain proteins B and C, they differ from each other in two respects. (1) The ratios of proteins B to C are about 2.7 for *supF*(P1*bac*) and about 1.4 for *supE*(P1*bac*). (2) The complementing activity of *supE*(P1*bac*) is temperature-sensitive whereas that of *supF*(P1*bac*) is temperature-resistant (DEAE-cellulose fractions).

The complementing activity of 252(P1*bac*) was isolated as described above for the *supE* and *supF* strains and their lysogens. The agarose-ATP fraction also yields four major proteins when subjected to gel electrophoresis. The molecular weights of three of them are identical to those of proteins A, C and D of the corresponding *supE* and *supF* fractions. The molecular weight of protein B was slightly smaller (MW = 60,000) than the corresponding protein from the *supF* series (Fig. 6). The ratio of protein B to C is about 1.6.

The isolation of *dnaB* protein from strain 252 is more difficult since no *dnaB* complementation activity can be extracted (S. Wickner, cited in 11). *In vivo*, this protein is temperature-sensitive for DNA initiation but not for DNA elongation (11,1). Assuming that for this reason the protein might retain the ATPase activity which has been found associated with wild type *dnaB* protein (9), the DEAE-cellulose fractions were tested for N-ethyl-maleimide resistant ATPase activity. A peak fraction eluting at about 0.3 M NaCl, beyond the bulk of ATPase activity, was pooled and subjected to agarose-ATP chromatography. The activity is adsorbed to and eluted from agarose-ATP by appropriate concentrations of ATP. When subjected to SDS gel electrophoresis, three major proteins are

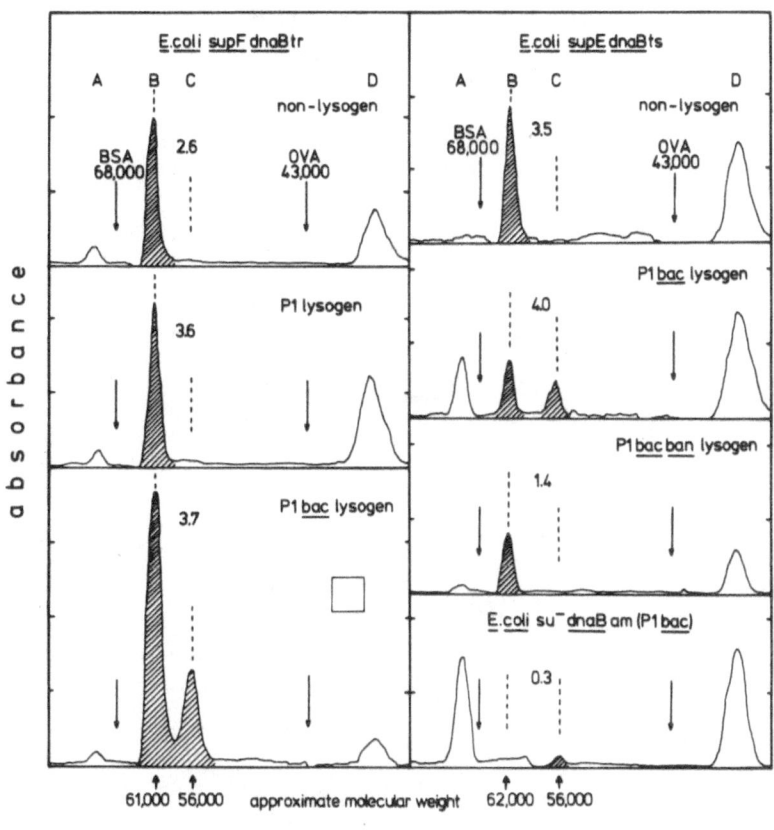

Figure 4. Densitometer tracings of proteins (agarose-ATP fractions) after SDS gel electrophoresis. Numbers refer to *dnaB* units applied to the gel. Hatched area = proteins cosedimenting with *dnaB* complementation activity during glycerol gradient centrifugation. Area of inset square corresponds to 1 µg BSA traced under identical conditions.

found, identical to proteins A, B and D of 252*bac* (Fig. 6). The ATPase activity can be inactivated by antibody directed against *dnaB*, as is shown in the next section (E. Lanka, B. Geschke, and H. Schuster, in preparation).

Inactivation of *dnaB* Complementation and ATPase Activity by Antibody Directed Against *dnaB* Protein

Preparations of *E.coli dnaB* gene product contain ribonucleoside triphosphatase activity that is stimulated 10-fold by DNA and is

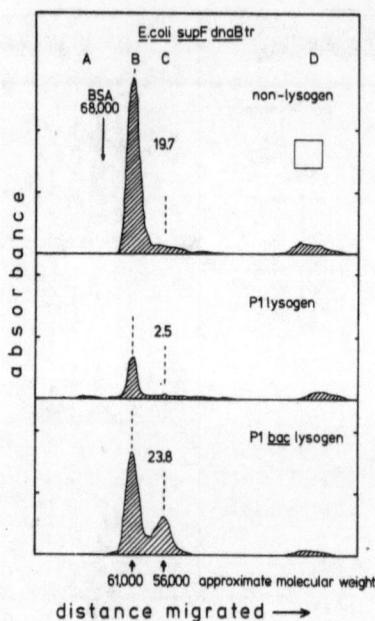

Figure 5. Densitometer tracings of proteins (glycerol gradient fraction) after SDS gel electrophoresis. For explanation see legend to Fig. 4).

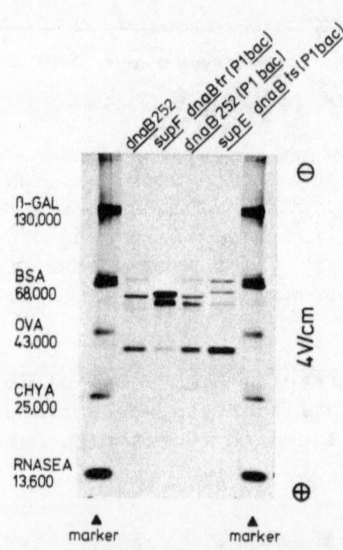

Figure 6. SDS gel electrophoresis of dnaB complementation or ATPase activity of different strains (agarose - ATP fractions). Agarose-ATP fractions were prepared and subjected to SDS gel electrophoresis as described in the legend to Fig. 3. The ATPase activity of strain dnaB252 was purified as described in the text.

resistant to N-ethylmaleimide (9). An N-ethylmaleimide resistant
ATPase activity is also found associated with the *dnaB* complementation activity of all strains tested (ATP-agarose- and glycerol
gradient fractions). The ATPase activities of Pl*bac* lysogens and
non-lysogens differ from each other with regard to their stimulation
by DNA. The activity from Pl*bac* lysogens is stimulated to a lesser
extent by øX174 DNA than that from the non-lysogens. The results
obtained with *supF* and 252 and their Pl*bac* lysogens are shown in
Table 3. The agarose-ATP fraction of strain 252, although
completely inactive in complementation, retains its ATPase activity
(Table 3).

Antibody directed against *dnaB* inactivates the complementing
activities of both *supF* and *supF*(Pl*bac*) to the same extent. A
similar inactivation is observed with 252(Pl*bac*) (Table 3).
Parallel with the complementing activity, the øX174 DNA dependent
ATPase but not the DNA independent ATPase activity of all strains
is inactivated by *dnaB* antibody (Table 3). The fact that the ATPase
activity of strain 252 can be inactivated by *dnaB* antibody confirms
the association of this ATPase with the *dnaB* protein.

Do the Biochemical Dats Fit with the Model ?

Two proteins copurify with the *dnaB* complementation activities
from Pl*bac* lysogens of strains *supF*, *supE* and 252. It is assumed
that these proteins represent *dnaB* and Pl*ban*. Whereas the
approximate molecular weight of the *ban* subunit is 56,000 in all
three strains, the molecular weight of the *dnaB* subunit differs
slightly among the mutants, being about 61,000, 62,000 and 60,000
for *supF*, *supE* and 252, respectively. The molar ratios of *dnaB* and
ban subunit molecules may be calculated on the basis of the relative
amounts of *dnaB* and *ban* protein in the complementing activities,
taking into account the molecular weight differences of *dnaB* subunits among the mutants. In this way, values of 5 : 2, 3 : 2, and
3 : 2 are obtained for the ratio of the number of *dnaB* and *ban*
molecules in the lysogens of *supF*, *supE* and 252.

These calculations are based on the assumption that the *dnaB*
complementation assay used in the purification does not discriminate
between *dnaB* and *ban* protein. Discrimination might however occur
because multimers composed predominantly or totally of *ban* protein
subunits might complement with lower efficiency or not at all. The
low recovery of a complementing activity from *su* (Pl*bac*) and the loss
of temperature resistance of DNA synthesis - a property of crude
protein fractions containing *ban* (7) - upon further purification
may indicate that the assay used selects against *ban* protein.

Nevertheless, the results obtained so far are compatible with

Table 3

Inactivation of *dnaB* complementation and ATPase activity
by antibody directed against *dnaB* protein

Assay	*dnaB* Antibody	ØX174 DNA	Agarose-ATP fraction of strain			
			supF dnaB tr non-lysogen (P1*bac*)	252 *dnaB* ts non-lysogen (P1*bac*)	252 *dnaB* ts non-lysogen (P1*bac*)	252 *dnaB* ts non-lysogen
dnaB complementation	–	+	100	100	100	inactive
	+	+	1.3	1.8	6.2	–
	–	+	100	100	100	100
	+	+	18.8	26.6	19.9	16.4
ATPase	–	–	14.4	23.7	27.4	17.4
	+	–	15.2	21.4	27.1	16.9

ATPase activity was measured as described by Wickner et al. (9). One unit of ATPase activity catalyzed the production of 1 µmol of $^{32}P_i$ in 30 min at 25°C.

a model in which suppression is due at least partially to a stabilization of *dnaB* ts by Pl*ban* monomers at temperatures normally restrictive for *dnaB* mutants: Firstly, the presence of two proteins in the complementing activities of Pl*bac* lysogens from three mutants indicates the existence of multimers composed of a host and a prophage coded protein. Secondly, the presence of both *dnaB* and *ban* protein in the complementing activity from Pl*bac* lysogen of two *dnaB* mutants grown at 40°C indicates preservation of *dnaB* ts molecules under conditions of effective Pl*bac* suppression. Thirdly, the temperature sensitivity of the complementing activity from *supE*(Pl*bac*), as compared to that from *supF*(Pl*bac*), supports the assumption that *dnaB* ts molecules are preserved.

ACKNOWLEDGEMENTS

We are much indebted to Danièle Touati-Schwartz and Aline Jaffé-Brachet for supplying us with bacterial and phage strains. We also thank them, R. d'Ari, M.B. Yarmolinsky and D.W. Smith for making results available to us prior to publication. The helpful suggestions provided by E. Scherzinger are highly appreciated. We are grateful to A. Kornberg and R. McMacken for the gift of a *dnaB* antibody preparation. We thank R. Brimacombe for correcting the English.

REFERENCES

1. Beyersman, D., M. Schlicht and H. Schuster (1971) Molec. Gen. Genet. 111:145

2. Beyersman, D. and H. Schuster (1971) Molec. Gen. Genet. 114:173

3. D'Ari, R., A. Jaffé-Brachet, D. Touati-Schwartz and M.B. Yarmolinsky (1975) J. Mol. Biol. 94:341

4. Laemmli, U.K. (1970) Nature 227:680

5. Lark, K.G. and J.A. Wechsler (1975) J. Mol. Biol. 92:145

6. Ogawa, T. (1975) J. Mol. Biol. 94:327

7. Schuster, H., M. Mikolajczyk, J. Rohrschneider and B. Geschke (1975) Proc. Natl. Acad. Sci. 72:3907

8. Wickner, W., D. Brutlag, R. Schekman and A. Kornberg (1972) Proc. Natl. Acad. Sci. 69:965

9.   Wickner, S., M. Wright and J. Hurwitz (1974) Proc. Natl. Acad.
     Sci. 71:783

10. Wright, M., S. Wickner and J. Hurwitz (1973) Proc. Natl. Acad.
     Sci. 70:3120

11. Zyskind, J.W. and D.W. Smith (1977) J. Bacteriol. 129:1476

TWO REPLICATION FUNCTIONS IN PHAGE P1: *ban*, AN ANALOG OF *dnaB*, AND

bof, INVOLVED IN THE CONTROL OF REPLICATION

Danièle Touati-Schwartz[*]

Institut de Recherche en Biologie Moléculaire
4 Place Jussieu
75005, Paris, France

INTRODUCTION

Coliphage P1 is a temperate phage which can either enter a lytic cycle, or establish itself as a prophage in a plasmid state. The plasmid is a covalently closed circular DNA of about 60×10^6 daltons, present in one or two copies per bacterial chromosome (Ikeda and Tomizawa, 1968; Prentki et al., 1977). Prophage loss occurs at a frequency less than 10^{-4} per cell per generation (Rosner, 1972).

The fact that P1 is both a plasmid and a phage makes it a very interesting and convenient tool for the study of the regulation of plasmid replication and its relation with cell division. Our first approach was to look for P1 functions involved in its replication. We could show (d'Ari et al., 1975) that P1 determines a product (ban) which is analog to the *dnaB* bacterial function, and which allows this phage to grow lytically in *dnaB* mutants (Lanka and Schuster, 1970). When P1 infects a thermosensitive *dnaB* mutant at the non permissive temperature, a transient synthesis of bacterial DNA occurs until lysis (Beyersman and Schuster, 1971).

In the prophage state the *ban* gene is repressed by the product of gene CI, the repressor of P1 lytic functions (d'Ari, 1977). Mutants in which *ban* is expressed constitutively have been found (d'Ari et al., 1975; Ogawa, 1975). In some of them *ban* expression is due to a mutation of the CI repressor (*dan* mutants), in others the operator which controls *ban* expression is modified (*bac* mutants).

[*] *Present address: Institut de Biologie Physico-Chimique, Rue Pierre et Marie Curie, 75005, Paris, France.*

Both types of mutants allow survival of *dnaB* ts lysogens at a non permissive temperature. Genetic evidence will be presented here suggesting that the *ban* product synthesized in P1 *bac* lysogens interacts with the *dnaB* product of the host.

The stable maintenance of a low number of plasmid copies implies a strict control of its replication. In 1969, Pritchard proposed a model where an inhibitor of replication, synthesized very early at each new replication cycle, would prevent further initiation until its concentration in the growing cells falls below a critical value. According to this model anything which would affect the amount of activity of the replication inhibitor would change both the number of copies per cell and the compatibility properties of the plasmid. We describe here a new mutant of P1, called *bof*, which is apparently present at a higher number of copies per cell and is impaired in its compatibility properties. It may be affected in the production of the replication inhibitor postulated by Pritchard. Similar mutants have been described for R factors (Uhlin and Nordström, 1975).

RESULTS

Interactions Between the *ban* and *dnaB* Products

Two tests have been used to study the complementation between *ban* and *dnaB*. The first one is the viability of various *dnaB* mutants when lysogenized by P1 mutants (Table 1). The second, which requires some comments, is the ability of the P1 lysogens to allow the growth of phage lambda (Table 2). This phage requires the *dnaB* product for growth. Not only does it fail to grow on thermosensitive *dnaB* mutant at a restrictive temperature, but it also fails to grow on some of them (called *gro*P) at the permissive temperature. Non thermosensitive bacterial mutations have been found which block the replication of lambda DNA. They have been called *grp*, and one class, *grpA* maps in *dnaB* while three others (grpC, D and E) map elsewhere. Mutants of phage lambda have been found which can grow on *gro*P or *grp* strains. They are called Lambda π (Georgopoulos and Herskowitz, 1971) or lambda *reg* (H. Saito and H. Uchida, 1977), and are mutated in cistron P. Growth of phage lambda in the P1 lysogens therefore appears as a more stringent test for *dnaB* activity than mere growth of the lysogens themselves.

The results in Table 1 and 2 can be summarized as follows:

Positive complementation is seen in three cases:

- wild type *ban* product can complement the growth deficiency in

Table 1

Growth of lysogens depending on the state of dnaB and ban products in the cell

state of ban product / of dnaB product	repressed	expressed at high level (1)		expressed at low level (2)	
	wild type or altered	wild type	altered (3)	wild type	altered (3)
wild type	+	+	+	+	+
altered (a) { directly : dnaB ts	+	+	-	+	+
altered (a) { indirectly : grpE	+	(c)	± (g)	+	+
destroyed: dnaB ts (b) { class I	-	+	-	+	-
destroyed: dnaB ts (b) { class II	-	+	-	-	-
dnaB252 (f)	-	+	+	+	-
no product (dnaB266 amb)	(c)	+ (e)	(c)	(d)	(d)

Legend Table 1

(a)   .dnaB ts at a temperature at the limit of the thermo-
      sensitivity. Cells are still viable and able to form
      colonies but start to lightly elongate .grPE at any
      temperature.

(b)   at non persmissive temperature

(c)   the strain does not exist

(d)   not done

(e)   the lysogen is cryosensitive and does not allow λ to grow
      (Table 2)

(f)   a dnaB mutant which behaves as an initiation mutant
      (Zyskind and Smith, 1977)

(g)   the lysogen exist, but grow very poorly

(1)   in bac mutants

(2)   in bof bac mutants (see last paragraph of the results)

(3) ban_1 mutation. (Pl ban_1 phage does not grow in dnaBts at non
      permissive temperature)

Legend Table 2

(a)   at permissive temperature for bacterial growth

(b)   at non permissive temperature for the non lysogen strain

(c)   does not exist

(d)   not done

(e)   depending on the temperature or/and the allele

(f)   the plaques are still very small

(g)   the lysogen is growing very poorly

Table 2

Growth of λ^+ on lysogens as a function of dnaB and of ban products in the cell

state of the ban product / of dnaB product	repressed wild or altered	expressed at high level		in Cr mutants	expressed at low level	
		wild	altered		wild	altered
wild type	+	+	+	+	+	+
altered (a) groP	−	+	−	+	± (e)	−
altered (a) grpA	−	−	−	+	−	−
altered (a) grpC	−	+ (f)	−	(g)	+ (f)	−
altered (a) grpD	−	−	−	+	−	−
destroyed : dnaBts (b)	(c)	± (e)	(c)	+	(d)	(d)
no product : dnaB amb 266	(c)	−	(c)	+	(d)	(c)

dnaB thermosensitive mutants at high temperature as well as in a *dnaB* amber mutant.   (Table 1, lines 4,5,6,7).

- altered *ban* product can complement the growth deficiency of a peculair thermosensitive *dnaB* mutant (*dnaB* 252) which behaves as an initiation mutant (Table 1, line 6).

- wild type *ban* product restores the ability to support lambda growth in *groP*, and some *grpC* strains (Table 2, line 2,4).

Instances of negative complementation are the following:

- altered *ban* product in several thermosensitive *dnaB* mutants leads to inviable lysogens at a temperature where the non lysogens are still able to form colonies (Table 1, line 2).

- wild type *ban* product complements negatively a *grpE* mutation (Table 1, line 3).

The last columns in the tables show the effect of a decrease in the relative amount of *ban* product with respect to *dnaB* product. In a thermosensitive *dnaB* strain lysogenized with P1 *bac* the quantity of *ban* product is roughly the same as that of *dnaB* product (Schuster, personal communication). A lowering of the level of ban product in the cell results in the disappearance of negative complementation with *dnaB* ts at permissive temperature and in the impossibility to replicate some dnaB ts mutants (class II in the table).   These results suggest that the *ban* product can randomly substitute with the *dnaB* product to form hybrid molecules, the mean composition of which depends on the relative amounts of *ban* and *dnaB* products in the cell.   Lanka and Schuster performed biochemical studies on this system and reached similar conclusions (E. Lanka and H. Schuster, this volume).

The *ban* product cannot totally substitute for *dnaB*. Indeed an amber *dnaB* strain (*dnaB* 266) lysogenized with P1 *bac* is cryo-sensitive and does not allow phage lambda to grow at the permissive temperature.   This may reflect either a requirement for more *ban* product or a functional difference between *dnaB* and *ban*. A mutant has been obtained from P1 *bac* which allows the growth of a *dnaB* 266 strain even at low temperatures and renders the strain able to support lambda growth.   Very little is known of the additional mutation which has been provisionally called *Cr*. P1 *bac Cr* also restores capacity for lambda growth in grp A and grp D strains where P1 *bac* does not.

P1*bof*, a Copy Mutant of Phage P1

In an attempt to elucidate the role of the *ban* product in the

life cycle of phage P1 we looked for an amber mutation in this
gene. Our selection was for amber mutants eliminating the negative
interaction between *ban* and *dnaB* altered products. Instead of a
ban mutation we found an amber mutation reducing the expression of
ban in *bac* prophages. The mutation, called *bof*, has a pleiotropic
phenotypic expression and will be shown below to be a copy mutation.

Genetic Evidence That P1 bof is a Copy Mutant

PlCm and PlKm are derivatives of phage P1 carrying drug
resistance genes obtained from an R factor (Kondo et al., 1970;
Takano, 1976). Bacteria lysogenic for these phages are resistant
to certain doses of chloramphenicol or kanamycin. The presence of
the *bof* mutation on this phage increases the level of drug
resistance which they confer to bacteria (Table 3). The relative
increase due to the *bof* mutation is the same whatever the other
mutations in the phage or the host. This increased drug resistance
of *bof* lysogens suggests that the *bof* mutation leads to an
increased synthesis of the drug destroying enzymes coded by the
prophage, itself resulting from a possible increase in the number
of plasmid copies in the cell. Unfortunately, however, the
relationship between the number of copies and the degree of drug
resistance is unknown.

Table 3

strain with	drug resistance (a)	bof^+ (b)	bof^- amb
no suppressor	chloramphenicol resistance	1	2,1
amber suppressor		1	1,1
no suppressor	kanamycin resistance	1	2,6

(a) The resistance of a lysogen is determined as
being the maximum drug concentration allowing
a 100% survival.

(b) Results are normalized to wild type. The basic
level of drug resistance varying with the strain
being used.

The immunity system of P1 is bipartite (Wandersman and Yarmolinsky, 1977). Virulence can be obtained by the loss of sensitivity to the CI repressor or by the constitutive expression of the "ant" product which antagonizes repression (d'Ari, 1977). Plvir, which belongs to the second class, is not able to form plaques on Pl*bof* lysogens. In liquid medium it can grow on such lysogens, but there is a delay in the lytic cycle. This result would be consistent with an increased synthesis of CI product by Pl*bof* lysogens as could be expected from the presence of an increased number of plasmid copies.

An increased copy number, resulting in an increased concentration of CI product in the cell, could account for the property which leads to the isolation of Pl*bof*, i.e., a decreased expression of *ban* product. Indeed *ban* expression in wild type P1 is under CI control. The *bac* mutation suppresses this control presumably creating a *bac* site with a lower affinity for the CI product. In the presence of higher concentrations of CI product in the cell, the remaining affinity of the *bac* site for this product could result in the partial repression of the *ban* gene (S. Bourgeois and M. Pfahl, 1976).

Compatibility Properties of the bof Mutant

Two P1 prophages cannot be maintained together in a bacterial strain, even transiently. Superinfection of a lysogen leads either to prophage displacement or to loss of the superinfecting phage, at comparable frequencies. Table 4 shows preliminary results obtained with *bof* mutants, using a *recA* host to decrease the recombination between the resident prophage and the superinfecting phage. Double lysogens can be obtained and a symmetrical segregation of the two prophages is observed, as expected if the segregation of the plasmid copies at division is random (Uhlin and Nordström, 1975).

All *bof* phenotypes were found to reverse simultaneously. The *bof* gene was shown by deletion mapping to be located far from both *ban* and CI.

The results summarized above suggest that the *bof* product is involved in the control of plasmid replication. However the absence of *bof* product, due to the amber mutation, does not lead to a "relaxed" replication as would be expected in the absence of the replication inhibitor, according to the model of Pritchard. This observation can be explained in several ways:

(a)   the amber mutation may be at the end of the *bof* gene and may only incompletely inactivate it.

Table 4

resident prophage	superinfecting phage	% of double lysogens/ phage displacement (a)	lysogens repartition after six generations without drug			
			$CM^R KM^R$		CM^R	KM^R
CM bof$^+$	KM bof$^+$	10^{-4} to 10^{-3} (b)	100%		0%	0%
CM bof	KM bof	10^{-2} to 1	53%		24%	23%

(a) displacement in recA strain is low, as well as lysogeny (Rosner, 1972)

(b) those are probably illegitimate recombinants

(b) the *bof* product is only one of several elements controlling the inhibition replication

(c) the replication of Pl *bof* is limited by the availability of some phage coded product, since the phage is in the repressed state.

We somewhat favour the last hypothesis because the presence of a mutation decreasing the immunity (CIts) leads to the induction of Pl *bof* at temperatures where neither this phage nor PlCIts is normally induced (unpublished results). Work is in progress to determine whether the limiting product might not be that of gene *ban*.

REFERENCES

Beyersman, D. and Schuster, H. (1971) Mol. Gen. Genet. 114:173-176

Bourgeois, S. and Pfahl, M. (1976) Advances in Protein Chemistry, Vol. 30, 1-93

d'Ari, R., Jaffé-Brachet, A., Touati-Schwartz, D. and Yarmolinsky, M. (1975) J. Mol. Biol. 94:341-366 d'Ari, R. (1977) J. Virol. (in press)

Georgopoulos and Herskowitz (1971). In: Bacteriophage Lambda
     (ed. Hershey) Cold Spring Harbor, New York, p553-564

Ikeda, H. and Tomizawa, J.I. (1968) Cold Spring Harbor Symp. Quant.
     Biol. 33:791-798

Kondo, E., Haapala, D.K. and S. Falkow (1970) Virol. 40:431-440

Lanka, E. and Schuster, H. (1970) Molec. Gen. Genet. 106:274-285

Ogawa, T. (1975) J. Mol. Biol. 94:327-340

Prentki, P., Chandler, M. and Caro, L. (1977) Molec. Gen. Genet.
     152:71-76

Pritchard, R.H., Barth, P.T. and Collins, J. (1969) Symp. Soc. Gen.
     Microb. 19:263-267

Rosner, J.L. (1972) Virol. 48:679-689

Saito, H. and Uchida, H. (1977) J. Molec. Biol. 113:1-26

Takano, T. and Ikeda, S. (1976) Virol. 70:198-200

Uhlin, B.E. and Nördström,   (1975) J. Bact. 124:641-649

Wandersman and Yarmolinsky (1977) Virol. 77:386-400

Zyskind, J. and Siith, D. (1977) J. Bact. 129:1476-1486

THE PRESENT STATUS OF ØX174 DNA REPLICATION IN VIVO

David T. Denhardt[*] and Christian Hours

Department of Biochemistry
McIntyre Medical Sciences Building
McGill University
3655 Drummond Street
Montreal, Quebec, Canada H3G 1Y6

INTRODUCTION

We have a more detailed understanding of the mechanism of replication of the DNA of bacteriophage ØX174 than of any other organism, largely because of its small size and the concentrated efforts of a number of laboratories. The viral DNA is a circular, single-stranded DNA molecule of 5,375 nucleotides, and the sequence of these is known (Sanger et al., 1977). After infection of *E.coli*, the first event is the synthesis of a complementary strand to form the duplex parental replicative form (RF). This RF is transcribed and then replicated several times to form progeny RF molecules. Finally, synthesis of viral single-stranded DNA occurs and progeny virus particles are produced. A more extensive discussion of these and other events in the virus life cycle can be found in reviews by Denhardt (1975,1977).

In this Chapter, we attempt to analyze some of the recent work on the synthesis, primarily *in vivo*, of the complementary and viral strands. An important point to recognize is that there are significant differences in how these two DNA strands are synthesized. There is a profound asymmetry in the replication of the viral DNA, imposed perhaps because of the fact that only one DNA strand is found in the virus.

There are also requirements for different DNA polymerases adapted to fit specific tasks during the synthesis of the individual strands.

[*]*A.S.I. Participant*

Complementary Strand Synthesis

The evidence is compelling that initiation of synthesis of
the complementary strand *in vivo* can occur at many locations on
the viral strand template. The number of initiation events taking
place on any one viral strand template probably varies from 2 to
more than 10, depending upon conditions. The sites at which
initiation happens do not seem to be randomly located, and this
may be because there is an orderly progression of initiation events
starting at a preferred location in the molecule and proceeding
in a semiregular fashion around the DNA. One argument that multiple
initiation events occur is that attempts to halt synthesis of a
complementary strand with a chain-terminating compound (2',3'-dide-
oxyadenosine, McFadden and Denhardt, 1977) or by introducing lesions
into the viral DNA by ultraviolet radiation (Hourcade and Dressler,
personal communication) were not as effective as would have been
expected if there had been a single specific origin of synthesis.

Independent evidence for multiple origins of ØX174 complementary strand synthesis comes from studies on the gaps in nascent
parental RF and on the size of the complementary strand DNA
synthesized under conditions of ligase deficiency. There are
multiple gaps (stretches of missing nucleotides, comprising, on the
average, about 15 nucleotides) in the newly synthesized complementary strand *in vivo*. Analysis of the pyrimidine tracts present
in the gaps (after filling in the gaps with $[\alpha^{32}P]$dNTPs using T4
DNA polymerase) revealed that in the population of complementary
strands studied the gaps occurred at many places in the genome,
although not at random (Eisenberg et al., 1975). When synthesis
of the complementary strand was allowed to occur under conditions
of ligase deficiency (*E.coli ts7* at the restrictive temperature)
the complementary strand was found in pieces that averaged some
1/5 - 1/10 the size of the complete strand (McFadden and Denhardt,
1975). It is unlikely that this was the result of degradation
because under similar conditions the viral strand was usually
found as a unit-length molecule.

Zuccarelli et al. (1976) obtained evidence that there was a
single preferred initiation site near gene A. Curiously, however,
the nascent complementary strand still contained several discontinuities, and they speculated that the DNA polymerase "stutters"
during synthesis. In apparent contrast to ØX174, G4 appears to have
one specific site for the initiation of complementary strand
synthesis *in vivo* (Hourcade, Sims and Dressler, personal communication).

Problems that remain unresolved mostly concern the mechanism
of initiation of synthesis of the complementary strand DNA molecules
and the joining of these nascent molecules. Good evidence of the
requirement *in vivo* for the *dnaB, dnaG,* and *dnaZ* functions exists

(Dumas and Miller, 1974; McFadden and Denhardt, 1974; Haldenwang and Walker, 1976). It seems that these proteins (or a subset) have the capacity, probably in conjunction with other proteins also, to initiate, or cause the initiation of, DNA chains at almost any location in a single-stranded DNA molecule. However, it is notable that something about fd (also M13, f1) and G4 (also St-1) single-stranded DNA molecules prevents them (in appropriate extracts containing discriminatory factors), at least *in vitro*, from utilizing this mechanism and instead they have their own mechanisms (Vicuna et al., 1977; Kornberg, this volume, Wickner, ibid). Dumas et al. (1975) have reported that in *dnaC* mutants at the restrictive temperature (but not at the permissive temperature) formation of ØX174 parental RF is inhibited *in vivo* by rifampicin; this work suggests that under certain conditions RNA polymerase, or at least the β subunit, is capable of substituting for the *dnaC* function in the primer-generating reaction. Also, there is evidence that *in vivo* some of the nascent complementary strand DNA molecules contain one or a few ribonucleotides at the 5' end (Machida et al., 1977). This is consistent with the *in vitro* studies already alluded to. The possibility that some chains are initiated by a mechanism not using the 3' hydroxyl of a ribonucleotide has not been excluded, but there is no evidence for it either. Be that as it may, given initiation, DNA polymerase III and associated elongation factors are usually responsible for the synthesis of the complementary strand segments. Under conditions of reduced DNA polymerase III activity other DNA polymerases appear able to substitute however (Denhardt et al., 1973).

Completion of the complementary strand involves the removal (when present) of the RNA (or other) primer, filling in of the gaps separating the nascent DNA chains, and ligation. Although DNA polymerase I normally is the preferred enzyme for gap-filling (Schekman et al., 1971) there is no evidence of a requirement *in vivo* for the 5' → 3' exonuclease activity. As shown in Table 1, the joining of the nascent segments *in vivo* to form RF I seems almost normal in *E.coli polAex2* (defective in the 5' → 3' exo-nuclease activity) at the restrictive temperature. One could use this as an argument that the gaps are not the record of an RNA primer; it may also be relevant that the average size of the gap (10-15 nucleotides) is larger than the size of the apparent RNA primer in *E.coli* (Denhardt, this volume).

It seems to us that we are still missing some pieces of this puzzle and the hypothesis that the gaps reflect the transient presence of a π protein primer is still tenable (Denhardt, 1972). (The π protein was envisioned to be a protein that formed a complex with the DNA template and a deoxymononucleotide so that the 3' OH was available to be used as a primer for a DNA polymerase). The *dnaG* protein is a good candidate for a π-like protein. Although instead of positioning a single deoxyribonucleotide at specific

Table 1

Effect of the *polAex2* mutation on RF I formation

E.coli Strain	Time of Incubation (min)	Temperature	RF I/II Ratio
rep pol[+]	10	34°C	1.32
	10	43°C	1.56
	20	43°C	1.63
	30	43°C	1.38
rep polAex2	10	34°C	0.92
	10	43°C	1.04
	20	43°C	1.04
	30	43°C	0.69

The appropriate *E.coli* strain was grown in 100 ml TPGA at the permissive temperature (34°C). Ten, twenty or thirty minutes before infection (indicated in the second column) chloramphenicol was added to a final concentration of 150 μg per ml of culture, and the temperature was either maintained at 34°C or shifted to 43°C. Cultures at a density of about 3×10^8 cells/ml were infected with ^{32}P-labeled ØX*am*3 at a multiplicity of infection of 10. The cultures were incubated for an additional 20 min at the pre-incubation temperature and then poured into a cold ethanol-phenol solution. The DNA was extracted as described by Eisenberg et al. (1975).

The amount of RF I and RF II from each sample was determined by sedimentation velocity centrifugation in alkaline sucrose gradients (5-20% sucrose, containing 1 M NaCl, 2 mM EDTA and 0.05% sarcosyl, centrifuged for 80 min at 15°C and 49,000 rpm in the SW56 rotor in a Beckman L2-65B ultracentrifuge). This research will be described in more detail elsewhere.

locations, as originally suggested, it may act as a primer-
independent, sugar-unspecific DNA polymerase synthesizing short
ribo, deoxyribo, or mixed, oligonucleotides at any of a potentially
large number of locations.

To summarize ØX174 complementary strand synthesis: There are
three easily distinguishable stages, each apparently preferring *in
vivo* to make use of the services of a different DNA polymerase.
These are outlined in Fig. 1. First, repeated initiations, at nonrandom but relatively non-specific sites, by a process requiring
at least the *dnaB* and *dnaG* functions. If there were an "entry"
site in the region of gene A for an enzyme complex that then
migrated along the DNA, generating priming elements at intervals,
then the results of Zuccarelli et al. (1976) could be explained.
Second, elongation , usually by DNA polymerase III although there
is evidence that DNA polymerase I or II can substitute. For some
reason elongation does not proceed right up to the 5' end of the
adjacent chain, and because of this there is a requirement for a
"gap-filling" event. We do not understand the necessity for this
third polymerization event, but whatever it is, DNA polymerase I
would normally seem to be the preferred polymerase. Ligation of the
juxtaposed DNA molecules by polynucleotide ligase and the subsequent
introduction of superhelical turns by gyrase completes the process.
It is not an unreasonable extrapolation that an identical set of
events occurs during the replication of *E.coli* DNA after a stretch
of single-stranded *E.coli* DNA has been exposed.

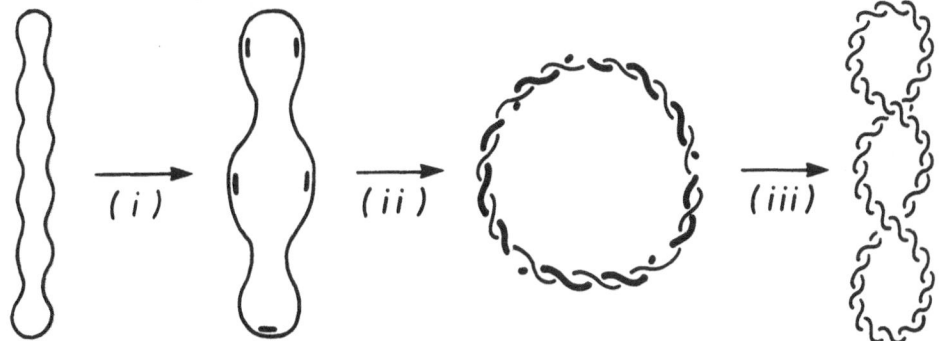

Figure 1.   Synthesis of a ØX174 complementary strand on a viral
single-stranded DNA template. (i) Synthesis at relatively non-
specific locations of 2-20 short oligonucleotide (ribo, deoxyribo
or mixed) primers; at the limit these may consist of one nucleotide
appropriately positioned. This event seems to involve the activity
of the dnaB, dnaC, dnaG, and other as yet poorly characterized
proteins. (ii) Elongation of the primers, requiring the activity
of the dnaE, dnaZ, and probably other, proteins. (iii) Joining of
the complementary strand segments, normally making use of the
activities of DNA polymerase I and polynucleotide ligase. Superhelical turns are introduced by gyrase.

Viral Strand Synthesis

In contrast to complementary strand DNA synthesis, which is initiated on a single-stranded template molecule, viral strand DNA synthesis is initiated on a double-stranded DNA molecule. This leads to distinct requirements for the process, and also makes available another mechanism for initiating synthesis of DNA chains.

One of the novel requirements is the *rep* function. This function was first identified by Denhardt et al. (1967) as an *E.coli* function required for the replication of ØX RF but not of *E.coli* DNA or the synthesis of ØX174 complementary strand DNA. Lane and Denhardt (1975) found that the bacterial DNA in the *rep* mutant was replicated at one-half the normal rate. Takahashi et al. (1977) identified the *rep* protein as a 70,000 dalton polypeptide in SDS gels. Scott et al. (1977) partially purified the *rep* protein and presented evidence that it was capable of unwinding duplex DNA in an ATP-dependent reaction. The evidence thus points to a role for the *rep* protein in separating the strands of double-stranded DNA.

A second requirement for viral strand synthesis is the ØX174 gene A function. There is evidence that *in vivo* this protein is a site-specific endonuclease (Francke and Ray, 1971; Henry and Knippers, 1974) that puts a nick into a specific site in ØX174 RF I (Baas et al., 1976). The nick is quite possibly a "swivel" for the unwinding of the two parental DNA strands, and it provides a 3' hydroxyl that potentially can be used as a primer for the initiation of synthesis of a new viral strand. The gene A protein may well also have other functions in initiation , elongation, or termination of viral strand synthesis. *In vitro* studies suggest that it remains covalently bound to the 5' end of the nicked viral strand and moves around the complementary strand DNA template with the *rep* protein, reforming an intact viral strand circle on completion of the circuit (Ikeda et al., 1976; Eisenberg et al., 1977). Thus it seems to have a ligating as well as a nicking capability. Evidence from *in vivo* studies exists for the interaction of *rep* with gene A and gene F proteins, and for the role of gene A in generating circular unit-length progeny DNA *in vivo* (Tessman and Peterson, 1976; Fujisawa and Hayashi, 1976).

The possibility that the 3' hydroxyl terminus of a (at that time) hypothetical gene A-nicked viral strand was used *in vivo* as a primer was inferred from the experiments of Dressler and Denhardt (1968) and Sinsheimer et al. (1968). Both groups obtained evidence for the existence of nascent viral strands longer than unit-length during viral strand synthesis. The presence of viral strands longer than unit-length in association with a circular template led Gilbert and Dressler (1968) to develop their Rolling Circle Model of ØX174 DNA replication. Essential facets of the model were the utilization of a nicked parental strand as a primer and the copying of a

circular template in an endless fashion. The validity of the
Rolling Circle Model applied to other systems has been variously
supported (e.g., ribosomal gene amplification (Hourcade et al.,
1973; Rochaix et al., 1974) and not supported (e.g., T4 DNA
replication [Doermann, 1973]). Late replication in lambda seems
to occur in a rolling circle mode (Takahashi, 1975). In the case
of ØX174 the existence of strands longer than unit-length has been
confirmed in a number of laboratories and in so far as this aspect
of the model is concerned it appears true (Schröder et al., 1973).

     Although it can be argued that the presence of viral strands
longer than unit-length does not *per se* demand parental strand
priming, those arguments are not compelling and priming by nicked
viral strands seems most likely. Perhaps the strongest evidence
comes from the *in vitro* work of Eisenberg et al. (1977), where, in
the absence of any other known priming mechanism (e.g., RNA synthesis), synthesis of longer-than-unit-length strands occurs.
However, another fundamental precept of the Rolling Circle Model,
that is that once initiated there is a continuous peeling off of
the viral strand, has no support. There is no good evidence for
linear intermediates or viral strand tails longer than unit-length,
while there is evidence that termination and a new initiation event
are required at each round of replication (Fujisawa and Hayashi,
1976). Thus the Rolling Circle Model must be modified (it no
longer "rolls" very far) if it is to be retained as a true accounting
of ØX174 viral strand replication.

     However, two sorts of observations make one wary of accepting
that parental strand priming and a modified Rolling Circle Model
mechanism are established. One is the existence of short pieces
of viral strand DNA. Several investigators have reported viral
strands shorter than unit-length during viral DNA synthesis *in vivo*
(Yokahama et al., 1971; Machida et al., 1977), and we are looking
into this question also. As illustrated in Fig. 2, viral strand
sequences shorter than unit-length can be easily demonstrated.
Although at first blush this result would appear to be incompatible
with parental strand priming, it can be accommodated if the gene A
protein were to nick at the origin where progeny and parental viral
strand DNA were joined prior to completion of synthesis of the
progeny viral strand. Note, however, that if the RF molecule must
be superhelical in order for gene A protein to act (Marians et al.,
1977), then it is not clear how a second nicking event could occur
prior to completion of the first round of synthesis since the
replicating molecule is not superhelical (Eisenberg et al., 1977).
Alternatively, the viral strand could be transiently nicked, by ω
or gyrase for example, and then "trapped" in the nicked form by
appropriate "stopping" methods. Thus the existence of short viral
strand DNA is subject to various interpretations, including possibly
being the result of a repair process induced by the incorporation
of uracil, (see also Lehman and Tye, this volume; Warner and Duncan,
ibid), and more information is needed.

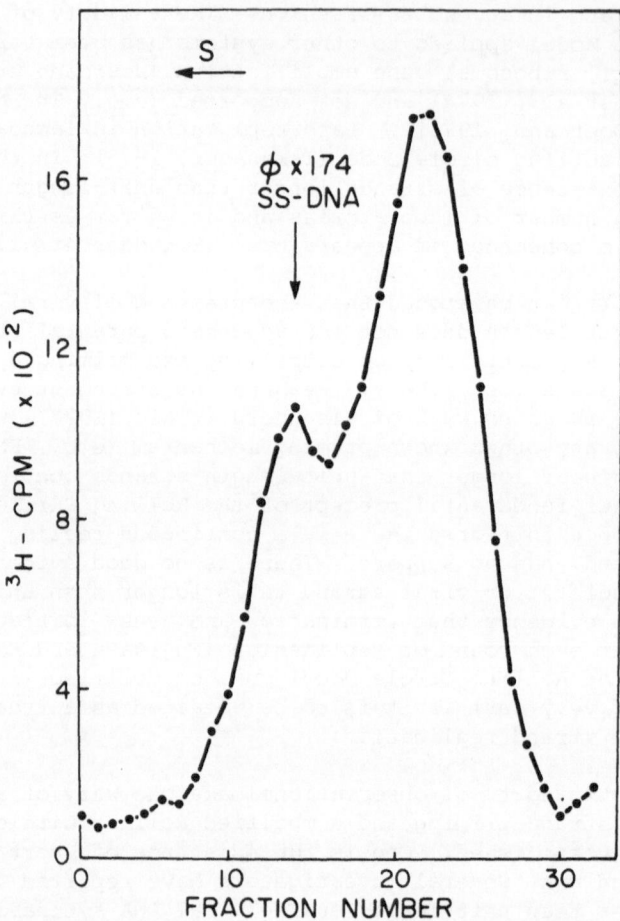

Figure 2. *E.coli C cells were grown in 200 ml tryptone-KCl medium at 30°C. When the cells reached the density of about 2 x 10⁸ cells per ml, they were infected with ØXam3 (lysis-defective phage) at a multiplicity of infection of 5. The culture was incubated for an additional 30 min at 30°C, then pulse-labeled for 3 sec with 1 mCi of (³H)thymidine. The incorporation was stopped by pouring cells into an equal volume of a boiling solution containing 2% SDS, 3% phenol and 10 mM EDTA. After phenol-chloroform extraction, the DNA was purified through a nitrocellulose column and centrifuged on an alkaline sucrose gradient (5-20% sucrose containing 1 M NaCl, 2 mM EDTA, and 0.05% sarcosyl), for 270 min at 15°C and 45,000 rpm in the SW 50.1 rotor of a Beckman L2-65B ultracentrifuge. Control experiments, to be reported elsewhere, indicate that the short pieces of DNA are not the result of an artifactual breakdown of larger molecular weight DNA. DNA-DNA hybridization studies revealed that 80-90% of the radioactivity was in ØX174 viral strand sequences.*

et al. (1977) of viral strand DNA molecules terminated at the 5'
end with a ribonucleotide. If this observation is confirmed, and
if it is true that the ribonucleotides are acting as primers, then
it brings into question the idea that the 3' hydroxyl end of the
gene A-induced nick in the viral strand is used as a primer.
Denhardt (1977) has suggested that one way of accommodating the
existence of viral strands longer than unit-length with a priming
mechanism that does not use the gene A nick is by invoking a
requirement for molecules of greater than one genome in length in
termination (the "Reciprocating Strand Model"). A somewhat similar
situation obtains in lambda, where two *cos* sites are required for
generating and packaging unit-length lambda DNA (Emmons, 1974).

    In our opinion, the weight of the evidence favors the idea
that the 3' hydroxyl of the nick introduced by the gene A protein
into the viral strand of RF I is used as a primer. A diagram of the
events occurring during viral strand synthesis is shown in Fig. 3.
Synthesis of a new viral strand is *continuous, but for only one
round of replication*. Repeated nicking-closing events, at the origin
or elsewhere, occur, possibly to allow unwinding. Depending on how
the viral DNA is isolated from infected cells, viral strands longer
or shorter than unit-length may be observed. Gene A protein
function is required again at termination to reform the original

A potentially more serious problem is the report by Machida

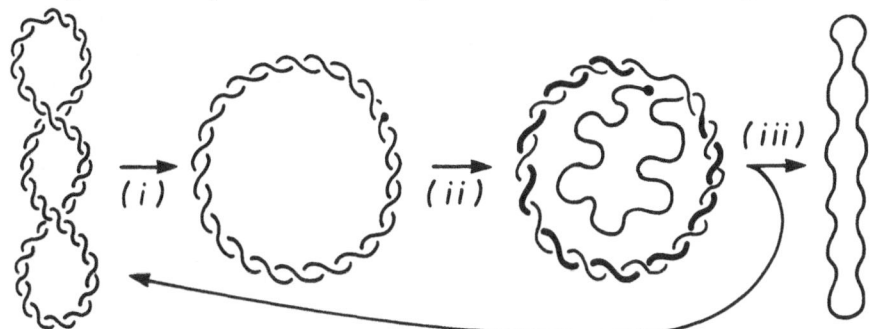

*Figure 3. Synthesis of a ØX174 viral strand on the complementary
strand of a double-stranded molecule. (i) Nicking of superhelical
RF I by gene A protein at a specific site. (ii) Separation of the
parental viral strand from the parental complementary strand,
probably by the action of the rep protein, and synthesis of a new
viral strand on the complementary strand template, by DNA polymerase
III and associated factors. (iii) Regeneration of the circularity
of the displaced parental viral strand. Reformation of a new RF I
structure by the combined actions of a DNA polymerase, polynucleotide ligase, and gyrase. DNA polymerase I normally seems to
be the most efficient enzyme for filling in the gap in the newly
synthesized progeny viral strand. This occurs after the viral
strand has been displaced and circularized.*

circular viral strand, now freed from its association with the
complementary strand. The newly synthesized viral strand, left in
association with the parental complementary strand, has a small gap
in the region of the origin-terminus, the same region as the gene
A nick. It is not known whether or not this unique viral strand
gap exists for the same reason as the multiple gaps in the
complementary strand (Eisenberg et al., 1975). A new RF I molecule
is produced when the gap is filled in, the nick closed by ligase,
and an appropriate superhelix density conferred on the molecule
by gyrase. Meanwhile the displaced parental viral strand either can
be incorporated into a progeny phage particle or can have a new
complementary strand synthesized on it; either of these processes
could of course start before displacement of the viral strand is
completed.

Future research on the replication of ØX174 DNA will be directed
predominantly toward the study of *in vitro* systems and purified
enzymes. A variety of fascinating questions concerning the mode of
action of the many proteins identified as playing a role in DNA
replication remain to be delved into. The mechanism of priming
is not entirely clear, particularly with regard to why viral
single-stranded DNA molecules from the different phages have
different requirements for complementary strand synthesis. Why
are there gaps in the nascent DNA ? The control of the initiation
of replication and the function of superhelicity are largely
unexplored. Although we have discussed viral and complementary
strand synthesis as totally separate events, it may be that *in vivo*,
during RF replication, the two types of syntheses are somehow
coupled. Fortunately, it cannot yet be said that we completely
understand how ØX174 DNA is replicated.

ACKNOWLEDGEMENTS

This work was supported by the National Cancer Institute of
Canada, the Medical Research Council of Canada, and the Ministére
de l'Education du Québec.

We thank Neil Miyamoto for assistance with some of the
experimental work reported herein.

REFERENCES

Baas, P.D., Jansz, H.S. and Sinsheimer, R.L. (1976) J. Mol. Biol.
    102:633-656

Denhardt, D.T. (1972) J. Theor. Biol. 34:487-508

Denhardt, D.T. (1975) CRC Critical Reviews in Microbiology 4:161-223

Denhardt, D.T. (1977) In: Comprehensive Virology (ed. H. Fraenkel-Conrat and R.R. Wagner) Vol. 7, Plenum Press, pp1-104

Denhardt, D.T., Dressler, D.H. and Hathaway, A. (1967) Proc. Natl. Acad. Sci. USA 57:813-820

Denhardt, D.T., Iwaya, M., McFadden, G. and Schochetman, G. (1973) Can. J. Biochem. 51:1588-1597

Doermann, A.H. (1973) Ann. Rev. Genetics 7:325-341

Dressler, D.H. and Denhardt, D.T. (1968) Nature 219:346-351

Dumas, L. and Miller, C.A. (1974) J. Virol. 14:1369-1379

Dumas, L.B., Miller, C.A. and Bayne, M.L. (1975) J. Virol. 16:575-580

Eisenberg, S., Harbers, B., Hours, C. and Denhardt, D.T. (1975) J. Mol. Biol. 99:107-123

Eisenberg, S., Griffith, J. and Kornberg, A. (1977) Proc. Natl. Acad. Sci. USA 74:3198-3202

Emmons, S.W. (1974) J. Mol. Biol. 83:511-525

Francke, B. and Ray, D.S. (1971) J. Mol. Biol. 61:565-586

Fujisawa, H. and Hayashi, M. (1976) J. Virol. 19:416-424

Gilbert, W. and Dressler, D.H. (1968) Cold Spring Harbor Symp. Quant. Biol. 33:473-484

Haldenwang, W.G. and Walker, J.R. (1976) Biochem. Biophys. Res. Commun. 70:932-938

Henry, T.J. and Knippers, R. (1974) Proc. Natl. Acad. Sci. USA 71:1549-1553

Hourcade, D., Dressler, D. and Wolfson, J. (1973) Proc. Natl. Acad. Sci. USA 70:2926-2930

Ikeda, J.-E., Yudelevich, A. and Hurwitz, J. (1976) Proc. Natl. Acad. Sci. USA 73:2669-2673

Lane, H.E.D. and Denhardt, D.T. (1975) J. Mol. Biol. 97:99-112

Machida, Y., Okazaki, T. and Okazaki, R. (1977) Proc. Natl. Acad. Sci. USA 74:2776-2779

Marians, K.J., Ikeda, J.-E., Schlagman, S. and Hurwitz, J. (1977) Proc. Natl. Acad. Sci. USA 74:1965-1968

McFadden, G., and Denhardt, D.T. (1974) J. Virol. 14:1070-1075

McFadden, G. and Denhardt, D.T. (1975) J. Mol. Biol. 99:125-142

McFadden, G. and Denhardt, D.T. (1977) Virology 76:870-875

Rochaix, J.-D., Bird, A. and Bakken, A. (1974) J. Mol. Biol. 87:473-487

Sanger, F., Air, G.M., Barrell, B.G., Brown, N.L., Coulson, A.R., Fiddes, J.C., Hutchison, C.A. III, Slocombe, P.M. and Smith, M. (1977) Nature 265:687-695

Schekman, R.W., Iwaya, M., Bromstrup, K. and Denhardt, D.T. (1971) J. Mol. Biol. 57:177-199

Schröder, C.H., Erben, E. and Kaerner, H.-C. (1973) J. Mol. Biol. 79:599-613

Scott, J.F., Eisenberg, S., Bertsch, L.L. and Kornberg, A. (1977) Proc. Natl. Acad. Sci. USA 74:193-197

Sinsheimer, R.L., Knippers, R. and Komano, T. (1968)   Cold Spring Harbor Symp. Quant. Biol. 33:443-447

Takahashi, S. (1975) Molec. Gen. Genet. 142:137-153

Takahashi, S., Hours, C. and Denhardt, D.T. (1977) FEMS Letters (in press).

Tessman, E.S. and Peterson, P.K. (1976) J. Virol. 20:400-412

Vicuna, R., Ikeda, J.-E. and Hurwitz, J. (1977) J. Biol. Chem. 252:2534-2544

Yokoyama, Y., Komano, T. and Onodera, K. (1971) Agr. Biol. Chem. 35:1353-1362

Zuccarelli, A.L., Benbow, R.M. and Sinsheimer, R.L. (1976) J. Mol. Biol. 106:375-402

ENZYMATIC REPLICATION OF DNA IN *E.COLI* PROBED BY SMALL PHAGES

Arthur Kornberg

Department of Biochemistry
Stanford University School of Medicine
Stanford, California, 94305, U.S.A.

Many of us here either have a vested interest in studying DNA synthesis or are considering how seriously to undertake research in the subject. I think DNA synthesis is important and attractive to pursue for several reasons. (1) Interest in biological sciences is in the ascendency, with even wider appeal now than the physical sciences. (2) DNA replication is one of the fundamental processes in all biological systems. (3) Major questions about DNA replication remain to be answered; information is increasing at a rapid and exponential rate and opportunities for resolving these major questions are great. For these reasons I am pleased and grateful that Masamichi Kohiyama and Ian Molineux had the vision and made the effort to organize this school and that NATO has sponsored it.

As background, I would like to enumerate some significant advances of the last few years that invite further studies in DNA replication and make progress in our understanding of its mechanism and control likely:

(1) Sequence analysis of DNA (25,36): Analyses are now being performed in many laboratories, including ours, with remarkable ease (1000 nucleotides/month). The total sequence of øX174 DNA is now known; those of M13 (fd) and G4 are almost complete. Structure of the origins of replication, promoters, and terminators will soon be in hand.

(2) Dissection of a genome by restriction nucleases and amplification of genes by cloning in *E.coli* using a plasmid or λ lysogen as a vehicle (31,

50, 51):   Identification of genes and their
amplification are now commonplace and the tempo
and range of this work will increase. Knowing
the arrangement of genes in the chromosome will
facilitate understanding how their expression is
regulated.

(3) Organization of eukaryotic DNA as a chain of
nucleosomes (23):   Understanding the structure of
chromatin will surely lead to better analyses of
control of replication and gene expression.

(4) Resolution and reconstitution of replication proteins
(20,22,61): Identification and isolation of the parts
of the replication machinery of *E.coli* has partially
clarified mechanisms of initiating and elongating DNA
chains and should lead to a detailed molecular
description of the whole intricate process. Important
progress has been made in purifying and
characterizing eukaryotic polymerases (11,19).

(5) Proteins (enzymes) for binding, unwinding, twisting
and untwisting DNA: Recognition of these proteins
and advances in analyses and separation of various
conformations of DNA. These proteins include: ω
protein for untwisting (7,54); gyrase for twisting
(13); gene 32 (T4) and *E.coli* DNA binding proteins
for helix destabilization (2) and for annealing of
single strands (8); *rep* protein for ATP-driven
unwinding of a replication fork (42,43); other ATP-
driven unwinding enzymes (1,33).

(6) Greater variety and abundance of cellular and viral
mutants, specific drug inhibitors, and nutritional
factors: Analyses of coded complex pathways and their
regulation have been and will be aided, both *in vivo*
and *in vitro*, by an enlarged reservoir of these tools
and approaches.

The magnitude of advances in some of these areas has been
enormous and, because they were not anticipated, startling.
Progress in DNA replication has not been disappointing; it has been
steady and encouraging. We have succeeded in obtaining many
components of the replicating machinery of prokaryotic systems in
soluble form and in resolving and reconstituting these systems.
We have learned a great deal about the eukaryotic polymerases. But
progress has not been electrifying ! Why not ? (1) Replicating
material is scarce ! Transcription and translation are ongoing
major metabolic processes and RNA polymerase and ribosomes are
major constituents of cells, but replication on the other hand,

happens only once in a cell's lifetime. (2) the "replisome" falls apart very easily ! Lysates prepared by the gentlest procedures are unstable; missing bits and pieces in the debris of cell lysates remain to be recovered. (3) The work is too difficult to attract large numbers of investigators; startling advances are not anticipated. Perhaps someone here will soon make the strike that will send everyone digging for discoveries in DNA replication.

I will not attempt a comprehensive review of DNA synthesis and replication but will focus on recent progress in the enzymolgoy of DNA replication as seen from the *E. coli* systems employed by the small DNA phages for their replication.

Small DNA Phages as Probes of Replication of the *E. coli* Chromosome

Although important advances in characterizing replication mechanisms of the coliphages T7 (19,41) and T4 (2) have been made, these mechanisms are not as pertinent to host cell chromosome replication as those of phages M13, G4 and øX174. The latter rely virtually entirely on the cellular machinery for their replication enzymes. Each of these three phages has proved to be a remarkably effective tool for illuminating a distinctive piece of the machinery that the cell uses for replication of its own chromosome or of its extrachromosomal plasmids.

These phage chromosomes contain only about 5000 nucleotides in a single-stranded circle (SS). They have several attractive features. They are relatively easy to isolate, characterize and use as templates. They are converted to a double-stranded circular replicative form (RF) and they require initiation of a new strand, perhaps at a unique signal. And finally, they afford by their apparent simplicity, a strong psychological impetus; surely, one should be able to obtain a soluble-enzyme preparation from cell extracts capable of replicating so simple a chromosome.

Because of the instability of the *E. coli* replication enzyme system, early studies had to rely on cells made permeable to substrate molecules (30,53) or on lysates immobilized on cellophane discs (37) and permeable to small and large molecules. Such systems are not amenable to refined analysis of molecular events. We therefore attempted by classical enzyme techniques using small phage DNAs as templates, to resolve and reconstitute the enzymes of DNA replication. We have learned how to obtain at least part of this replication apparatus in molecularly dispersed form, and to resolve it into at least a dozen discrete components.

Priming of DNA Synthesis on Single-Stranded Viral DNA In Vitro

Among the many viral, bacterial, and animal DNA polymerases thus far isolated, none can start a chain *in vitro* (20). Each requires, in addition to a template strand, a 3'-hydroxyl-terminated chain (primer terminus). Because they lack such termini, the single-stranded, circular coliphage DNA chromosomes have proved ideal templates for studying the mechanisms used by *E.coli* to initiate new DNA chains (38,61).

Since RNA polymerases are known to initiate new RNA chains, it appeared that an RNA transcript might prime DNA synthesis (5,21) and rifampicin, a specific inhibitor of *E.coli* RNA polymerase, should inhibit DNA synthesis. RNA polymerase was shown to be essential in the conversion of phage M13 single-stranded DNA (SS) to the duplex replicative form (RF). Following this, soluble extracts from gently lysed *E.coli* cells were demonstrated to support the conversion of various single-stranded phage chromosomes to their replicative forms (38,64). Fractionation and identification of the proteins involved in this apparently simple replication showed not only that synthesis of the complementary DNA chain is surprisingly complex, but also that there are at least three different DNA strand initiation systems in *E.coli* (22,38-40, 59,61). The chromosomes of filamentous phage M13 and icosahedral phages G4 and øX174 are each replicated *in vitro* by a different set of bacterial replication proteins that differ primarily in the mechanism by which primer transcripts are generated (Table 1).

Table 1

Protein Requirement for Replication of
Single-Stranded Viral DNAs

	Template		
Stage	M13	G4	øX174
Prepriming	none	none	DBP, *dna*B, *dna*C, proteins i and n
Priming	RNA polymerase DBP	Primase DBP	Primase *dna*B, DBP
Elongation	Holoenzyme DBP	Holoenzyme DBP	Holoenzyme DBP
Termination	DNA polymerase I DNA ligase	DNA polymerase I DNA ligase	DNA polymerase I DNA ligase

Replication of M13 Single-Stranded DNA

Resolution and purification of the components needed for the conversion of M13 SS to RF I disclosed the requirements for the five proteins shown in Fig. 1 (12). DNA binding protein (DBP) masks the single strand except at a single region, and primer synthesis can take place only in this position. In the absence of DNA polymerase I, the gap in RF II is found at this "promoter" region (12,47).

Evidence for RNA priming of M13 DNA replication (21,56,64), besides the effects of RNA polymerase inhibitors, is: (*i*) all four rNTPs are required; (*ii*) conversion of SS and RF takes place in two stages: the initial stage of RNA synthesis produces a primed SS that could be isolated and converted to RF in a second stage in the absence of rNTPs and in the presence of rifampicin; (*iii*) a phosphod-iester linkage of a deoxyribonucleotide to a ribonucleotide is found in the isolated RF in stoichiometric equivalence to the number of RF molecules formed; and (*iv*) persistence of an RNA fragment at the 5' end of the newly synthesized complementary strand can be inferred from the behavior of DNA polymerases and DNA ligases in filling and sealing the small gap in the RF II product. The size and nature of the primer transcript has recently been determined.[*] The DNA sequence at the RNA-polymerase-binding site has shown (15, 32) two "hairpin" structures about 60 nucleotides long. The significance of the strongly duplex character of the SS DNA near the origin of M13 complementary strand synthesis is not yet clear.

Replication of G4 Single-Stranded DNA

The rifampicin-resistant replication of phage G4 DNA (14) to form parental duplex (RF II) requires only three proteins: DBP, primase (*dnaG* protein), and holoenzyme (65) Table 1, Fig. 2).

---

[*]*Geider, K., Beck, E., and Schaller, H. personal communication.*

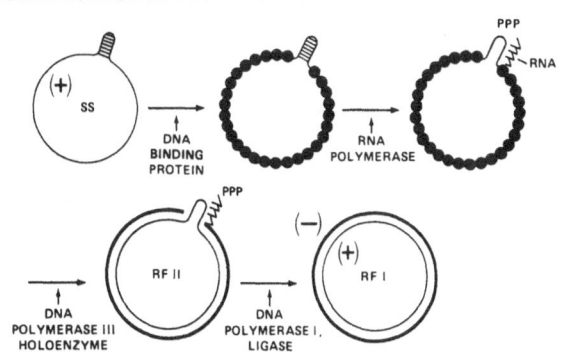

Figure 1. Scheme for conversion of M13 single-stranded DNA to RF.

Purification of each of these proteins was greatly facilitated by
the use of G4 conversion to RF as an assay.

With these purified proteins in hand, some significant features
of primer synthesis were found. A unique RNA transcript results
when primase acts on G4 DNA, in the presence of DBP and the four
rNTPs; rifampicin has no effect (4). This RNA fragment is found at
or near the sole site on the template strand used for initiation
of the complementary strand; it is covalently linked at the 5' end
of this chain after replication by holoenzyme.

The sequence of the RNA primer covalently attached to the
synthetic complementary strand (Fig. 3) (3) contains a GC-rich
duplex region. The complementary region in the SS DNA template,
left uncoated by DBP, might serve as a recognition site for primase.
Analysis of the covalent linkage formed between RNA and DNA when
holoenzyme was incubated with G4 SS (previously primed with RNA)
shows that DNA synthesis can begin after primer residue 26,27,28 or
29. This variation in primer size may be due to premature
termination of RNA synthesis or to removal of residues from the 3'
end by RNase. The sequence of the G4 chromosome has recently been
ascertained (14); it contains a sequence complementary to the primer

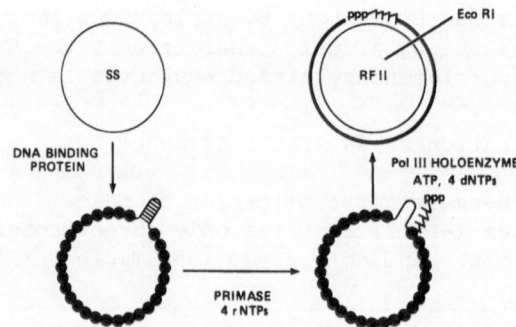

Figure 2. *Scheme for conversion of G4 single-stranded DNA to RF II*.

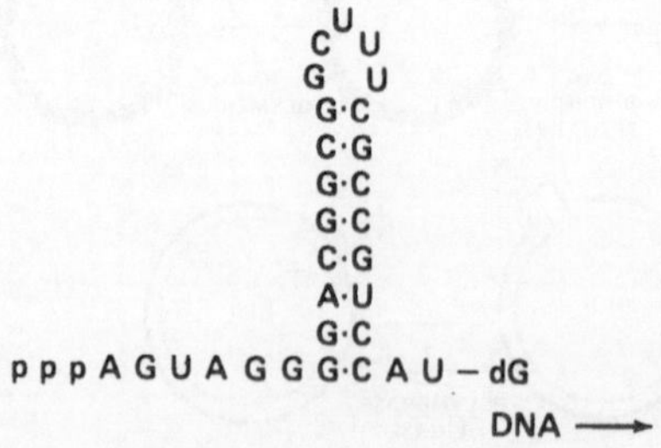

```
               C   U
                G   U
                G·C
                C·G
                G·C
                G·C
                C·G
                A·U
                G·C
  pppA G U A G G G·C A U – dG
                      DNA ⟶
```

Figure 3. *Sequence of G4 primer RNA*.

sequence. The primer site is located in the intercistronic region
between genes F and G; as anticipated from the more complex protein
requirements for replicating øX DNA, no such sequence is found in
its chromosome (36).

Properties of Primase in Priming G4 DNA Synthesis

Using conversion of G4 viral DNA to RF as an *in vitro* assay
for primase, this 60,000-dalton protein has been purified to near
homogeneity (34). Three activities copurify with primase at each
stage: (*i*) RNA synthesis on G4 DNA, (*ii*) complementation of an
extract from a temperature-sensitive *dnaG* mutant for replication of
G4 DNA, and (*iii*) priming of G4 DNA replication carried out by
purified holoenzyme. Moreover, the role of primase in primer
synthesis is corroborated by the thermolability of both nucleotide
incorporation into primer (58) and DNA replication when primase from
temperature-sensitive *dnaG* mutants is employed (4).

Primase, in the presence of DBP, is remarkably specific for
G4 DNA as a template for primer synthesis (34,58). Other unrelated
single- and double-stranded DNAs are inactive. This specificity
appears to reside in a nucleotide sequence in G4 DNA which enables
primase to bind only to this DBP-coated template (58; D. Bates and
A. Kornberg, unpublished observations). Primase action is not
affected by rifampicin or streptolydigin, two potent inhibitors
of *E.coli* RNA polymerase; primer synthesis is, however, inhibited
by actinomycin D (34).

Although ATP is the initiating nucleotide in primer synthesis
on G4 DNA, several ATP analogs can be utilized by primase as
substitutes, such as adenosine 5'-tetraphosphate, ADP, adenosine
5'-O-(3-thiotriphosphate), and adenyl-imidodiphosphate (35). In.
addition, as discovered in studies of primase-dependent synthesis
of primers for øX DNA replication (27), primase can incorporate
dNTPs as well as rNTPs (22,35,58). Primase adds either a ribonucleotide or a deoxyribonucleotide to the 3'-OH of a ribo- or a
deoxyribonucleotide residue of a primer terminus. Although dNTPs
are incorporated they profoundly inhibit rNTP incorporation at low
concentrations (5-50 μM), causing shorter transcripts to be
synthesized (35).

Synthesis of the complete, 29-residue RNA transcript is not
essential for effective priming of G4 replication, but synthesis
of at least a dinucleotide is required (35). The amount of primer
actually synthesized in the complete system (i.e., coupled priming-
replication) depends on the concentration of dNTPs (35); at 20 μM
rNTPs and 50 μM dNTPs, only 2 to 6 residues of RNA are found in the
replicated product, whereas at 5 μM dNTPs, up to full-length RNA
primer is synthesized before extension by holoenzyme occurs. These

data suggest that primase transcribes G4 DNA until it reaches a
termination site or is stopped prematurely by dNTPs at inhibitory
levels. At any point after a dinucleotide has been formed, DNA
polymerase may displace primase and extend the primer terminus.

Replication of ØX Single-Stranded DNA
and the Role of *dna*B Protein

Conversion of the ØX chromosome to RF occurs by a rifampicin-
resistant reaction, *in vivo* and *in vitro* (64). Fractionation and
identification of the required proteins indicates that at least nine
proteins participate in synthesis of the complementary strand,
including each of those required for G4 replication and four
additional proteins - *dna*B, *dna*C, and i and n proteins (Table 1,
Fig. 4) (38,39,59). Partially purified, the latter four proteins
convert DBP-coated ØX DNA to a nucleoprotein complex which is
rapidly replicated upon the addition of primase and holoenzyme
(55,59). We have termed this complex the ØX replication intermediate.

Because of its remarkable stability in the presence of ATP, the
replication intermediate could readily be isolated (55). Although
the exact composition of the intermediate is not yet determined,
the available evidence suggests that proteins i and n are not
essential components. Not only do these two proteins act
catalytically during intermediate formation, but the isolated
intermediate is not affected by antibodies directed against them
(52). On the other hand, the isolated intermediate contains a stoichiometric amount (approximately 150 molecules) of DBP (55; J. Weiner
and A. Kornberg, unpublished data). *E. coli dna*B protein is judged
to be a key constituent of ØX replication intermediate (27,29,52)
for these reasons: (*i*) replication activity of the isolated intermediate is sensitive to anti-*dna*B γ-globulin, (*ii*) DNA-dependent
ATPase activity of *dna*B protein (63) is present in the isolated
intermediate; and (*iii*) labeled *dna*B protein is incorporated into
the complex.

From the stoichiometry of *dna*B protein used during intermediate
formation and from the level of *dna*B ATPase activity associated with
the intermediate, it is concluded that a single *dna*B molecule is
present in the intermediate (29,55). The precise functions of
protein i, protein n, and *dna*C protein in the transfer of *dna*B
protein to the DBP-coated ØX chromosome remain to be elucidated.
It is known, however, that stoichiometric quantities of *dna*C protein
are required for the reaction, perhaps because *dna*B protein must
first form a complex with *dna*C protein (60) before it is transferred to the ØX DNA. Unfortunately, lack of an antibody directed
against *dna*C protein has prevented a more rigorous analysis of *dna*C
protein function in ØX SS replication.

The multienzyme system for priming *in vitro* replication of

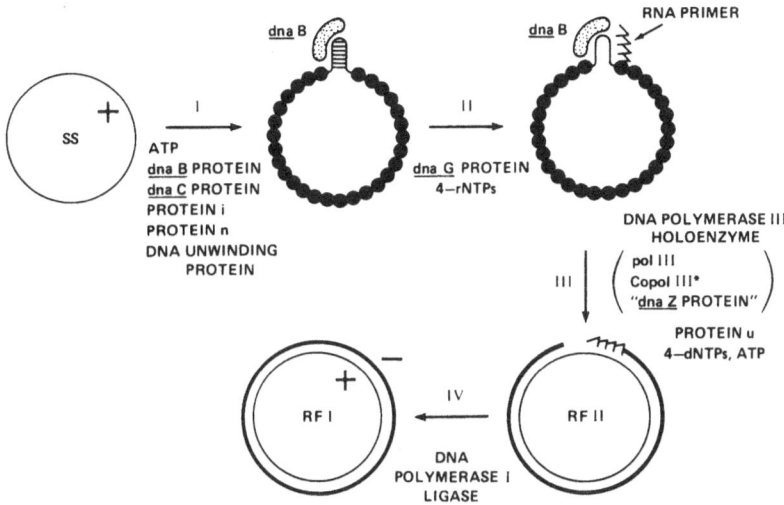

Figure 4. Scheme for conversion of ⌀X174 single-stranded DNA to RF.

⌀X DNA (22,27,29,55) consists of primase and each of the five proteins required for formation of the ⌀X replication intermediate (Table 1, Fig. 4), in addition to ATP, Mg^{2+} and the nucleoside triphosphates. Both rNTPs and dNTPs can serve as substrates; hybrid primers are formed when both are present (22,28). When the concentration of ribonucleotides is high relative to deoxyribonucleotides, as *in vivo*, both rate and extent of ribonucleotide incorporation are significantly higher than those for deoxyribonucleotide. Both RNA and DNA primers appear to be initiated solely with ATP; neither GTP nor dATP can substitute as the initiating nucleotide (3,28). Because ATP is required during DNA primer synthesis to stabilize the replication intermediate (containing template-bound *dnaB* protein and its ATPase activity), AMP residues are interspersed in the DNA transcripts. Greater than 98% of both the DNA and RNA transcripts stably hybridized to ⌀X template are extended by holoenzyme into complementary strand DNA (27,28).

As judged by polyacrylamide gel electrophoresis in 7 M urea, RNA primers ranged from 14 to 50 nucleotides in length, whereas DNA primers were reported to be substantially shorter, 4 to 20 residues (27,28). Primase will transcribe only those ⌀X circles which have been converted to intermediate form. With antibodies against the proteins required for primer synthesis, it was shown that in addition to primase, both proteins known to be in the intermediate, *dnaB* protein and DBP, participate in primer synthesis.

The magnitude of RNA synthesis by the reconstituted system in the absence of coupled replication is far in excess of that required

for priming complementary strand synthesis. As much as 300 ribonucleotide residues (i.e., transcription of nearly 6% of the øX
template) or 80 deoxyribonucleotide residues can be incorporated
per input circle (28). The complexity of the fingerprint pattern
of the oligoribonucleotides produced after T1 RNase digestion of
øX RNA primers indicates that transcripts are synthesized from
multiple (probably random) sites on øX DNA (28). Furthermore,
using $(\gamma^{32}P)$ATP incorporation as a measure of chain initiation,
it was discovered that multiple RNA primers (usually 5 to 8) are
made on each circle (27). The single *dnaB* protein molecule bound
in the replication intermediate was shown to participate in *de novo*
initiation of each of these primer transcripts. From the size of
the DNA chains synthesized when the gaps between the primers are
subsequently filled by polymerase action, it appears that at high
concentrations of primase the multiple primers are spaced at
intervals, in the range of 70 to 500 nucleotides, along the
template strand.

These data suggest (27) that *dnaB* protein, once fixed in the
intermediate, migrates processively along the DBP-coated DNA
strand, perhaps utilizing ATP hydrolysis to energize its movement,
and that recognition by primase of bound *dnaB* protein or of an
induced secondary structure in the template leads to initiation of
a primer transcript at or near that locus. The *dnaB* protein,
remaining on the DNA, apparently moves along the strand to a nearby
site where synthesis of another primer is initiated by primase.
The distance between neighboring primer transcripts appears to be an
inverse function of the primase concentration. Polarity of movement
is presumably in a direction opposite to growth of the primer and
therefore 5' → 3' *on the single-stranded template*. Uncoupled from
replication, this sequence is repeated many times over. With
holoenzyme and dNTPs present, the first primer is rapidly elongated
to a full-length, linear, complementary strand. These events are
schematized in Fig. 5.

Thus *dnaB* protein acting on a øX viral strand circle appears to
function as a "mobile promoter" for primer synthesis by primase to
initiate the growth of a DNA strand. Similarly, in replication of
the bacterial chromosome, we propose (27) that *dnaB* protein migrates
processively along the newly exposed lagging strand, behind the
replication fork in the direction of fork movement (Fig. 6). *dnaB*
protein (or a secondary structure it generates in the template) is
recognized by primase to initiate primer synthesis for nascent
(Okazaki) fragments of the *E. coli* chromosome. The *dnaB* protein-
template complex would be a signal to primase that this single-
stranded DNA is engaged in replication rather than prepared for
recombination, repair, or other metabolic processes involving
single-stranded DNA. *dnaB* protein, once bound, might travel with
the replication fork without dissociating until the round of
replication is completed. However, premature dissociation of the

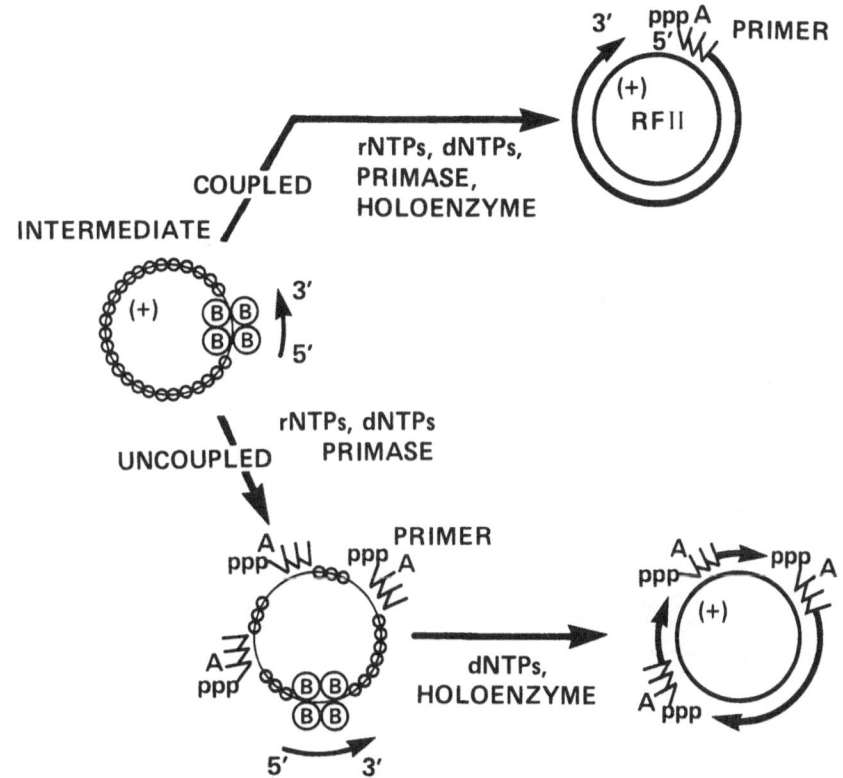

Figure 5. *Scheme for* *dnaB* *protein action as a "mobile promoter"*
in initiating DNA replication.

dnaB protein from the replicating DNA might be lethal if there were
no means for rapidly restoring *dnaB* at or near the replication fork;
dnaC protein and proteins i and n could participate in such a
process.

     The mechanism proposed for *dnaB* protein action may also
apply to initiation of bidirectional replication at the chromosomal
origin (Fig. 7) (27). A *dnaB* protein molecule is bound on each
strand at the replication origin and moves in the 5' → 3' direction
of that strand. Thus each *dnaB* molecule would direct priming both
of the continuously synthesized strand (the single initiating event
for the leading strand) and of nascent fragments of the
discontinuously synthesized strand (multiple events for the lagging
strand).

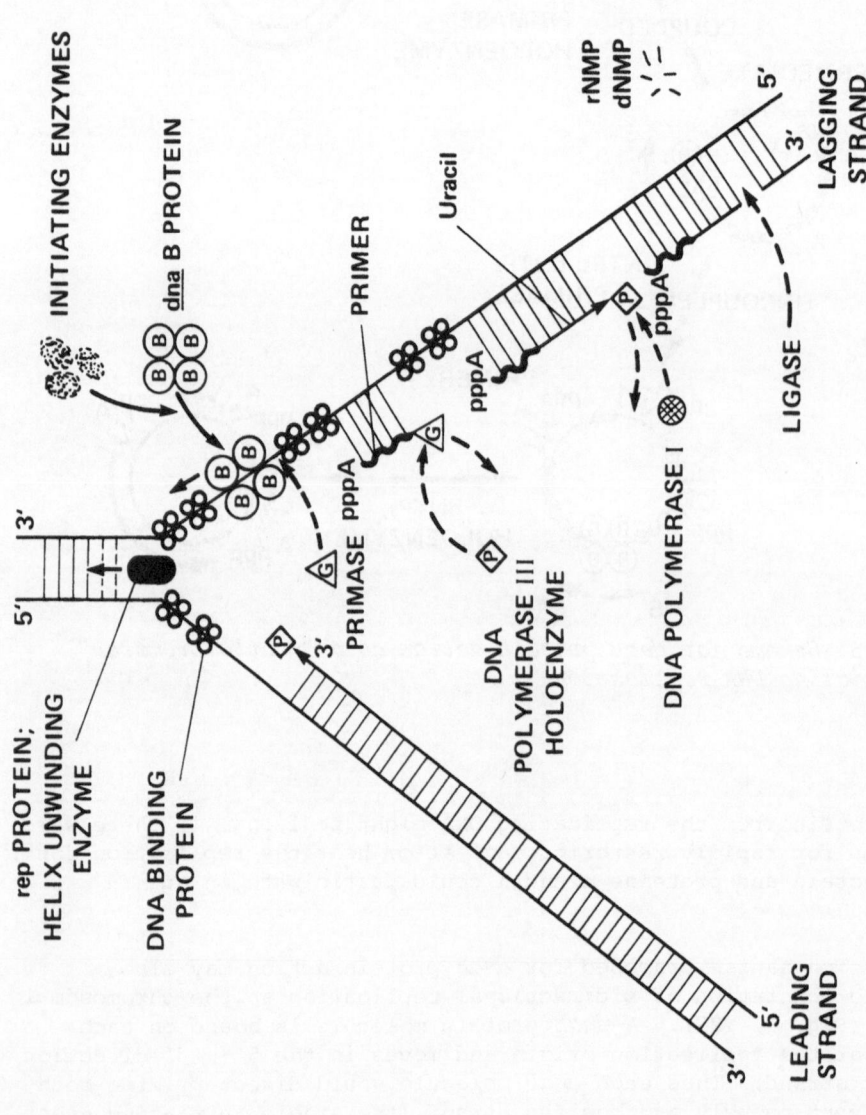

Figure 6. Scheme for dnaB protein action at a replication fork in the E.coli chromosome.

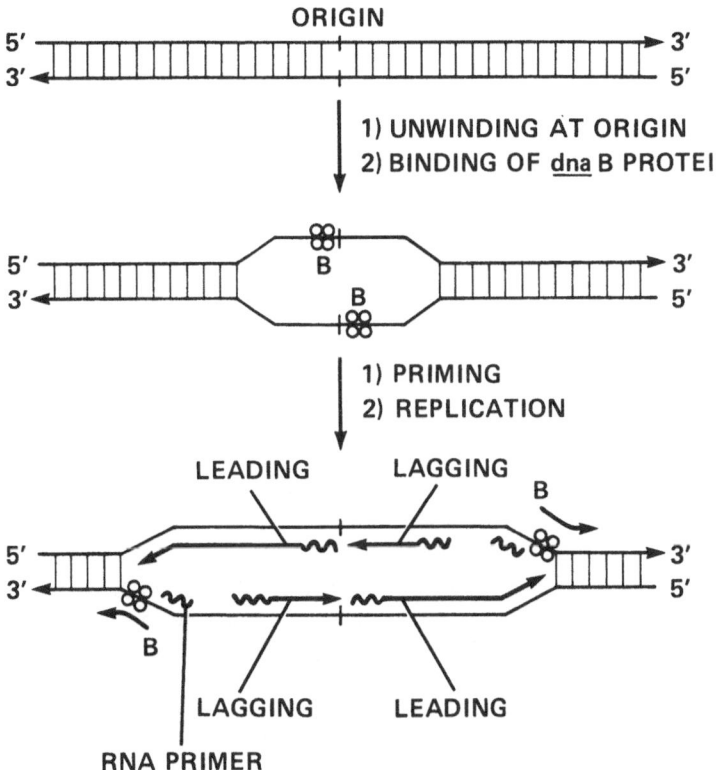

Figure 7. Scheme for <u>dnaB</u> protein action in priming bidirectional replication at the chromosomal origin; B represents <u>dnaB</u> protein.

Elongation of DNA Chains

The mechansim of elongation of SS phage chromosomes, primed and coated with DBP, appears to be universal, regardless of the template and primer (Table 1) (12,26,27,38,57,62). In all *in vitro* studies, the multisubunit holoenzyme (Fig. 8) catalyzes covalent extension of the primer terminus into complementary strand DNA. Although DBP in this step can be replaced by spermidine, both for rate and extent of DNA synthesis, binding protein is required to make polymerase action maximally processive (12,44).

Multiplication of the øX Duplex:
Role of *cistron* A and *rep* Proteins

Following conversion of single-stranded viral DNA to replicative form (SS → RF), two other stages in the life cycle of

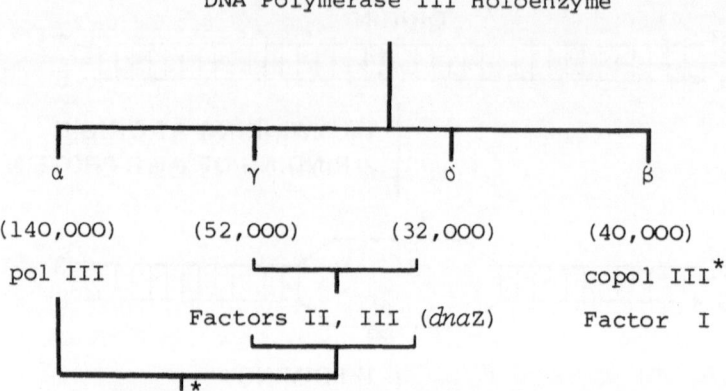

Figure 8. Subunits of holoenzyme

φX are recognized: first, multplication of the double-stranded RF;
and second, synthesis of viral strands using the complementary (-)
strand of RF as template (45).   RF replication *in vitro* is sustained
by crude enzyme systems (9,46), thus providing assays by which
cistron A protein, induced by φX, and *rep* protein of the host cell
(9,17) were purified. From genetic studies these proteins are
known to be required for RF replication and single-stranded DNA
synthesis *in vivo* (48,49).

To attempt to reconstitute RF replication led to the discovery
that four proteins, namely *cis*A protein, *rep* protein, DBP, and
holoenzyme can use supertwisted φX RF I DNA to synthesize viral (+)
circles in quantities that far exceed the input template (10,42).
In the absence of holoenzyme, the other three proteins were shown
to catalyze separation of the two strands of RF I DNA (12) (Fig. 9).

Phage induced *cis*A protein nicks φX RF I DNA (16,17,42). Pure
*cis*A protein and RF I incubated together (with Mg^{2+}) yield a
*cis*A-RF II complex, which has been purified and examined in the
electron microscope (10a) (Fig. 10A); location of the *cis*A protein
is at a unique site as judged by cleavage by Pst 1 nuclease (Fig.
10B,C).

Schematically the multiple functions of *cis*A protein are
shown in Fig. 11. The unique nucleotide sequence in an SS region,
exposed in superhelical RF I DNA[*], and nicked by *cis*A protein is
located in the viral (+) strand, between nucleotides 4290 and 4330

---

[*]*Covalently closed, relaxed φX duplex DNA is not nicked by cisA
protein (16,17).*

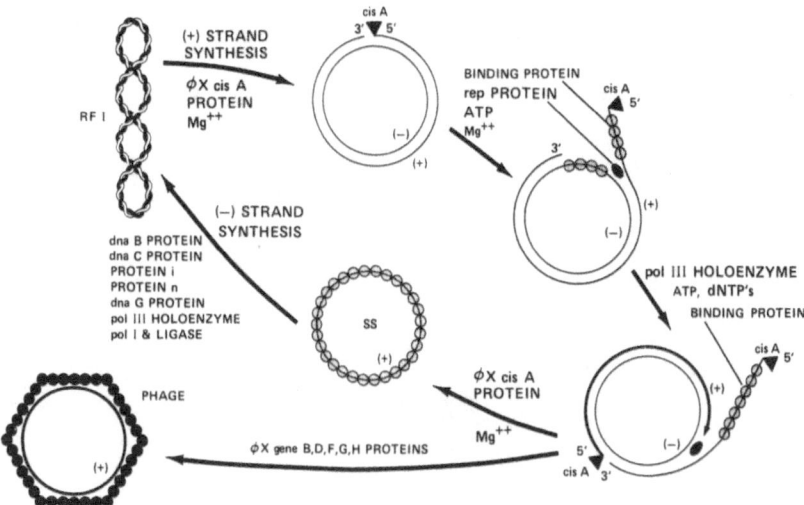

Figure 9. *Scheme for ØX174 RF replication*

(10a) on the ØX sequence map (36). The nicking results in what we
presume to be a covalent attachment of the *cis*A protein to the 5'
end of the viral strand. It seems plausible that this linkage, as
in the case of the untwisting enzymes (which carry out repeated
nicking and ligating steps without an energy requirement (6,54),
conserves the energy of the cleaved phosphodiester bond. To
explain the nicking and ligating activity of bound *cis*A protein,
we propose two active sites or subunits in the bound *cis*A protein,
only one of which (open circle in Fig. 11) is bound covalently to
the 5' end in a given complex.

Electron microscopic (Fig. 12) and biochemical studies (10a)
suggest that the *rep* protein recognizes the origin by its interaction with the bound *cis*A protein. Unwinding of the two strands,
sustained by the ATPase activity of *rep* protein and by DBP, then
ensues. Movement of the 5' end into the duplex would displace
the viral (+) strand from its complementary (-) circular template.
The 3' OH terminus, at the origin, serves as a primer and is
extended by holoenzyme to regenerate the origin. When the *cis*A
protein-*rep* protein complex tranverses the full length of the
template circle, a unit-length ØX DNA would be excised by the
unoccupied active site of the *cis*A protein (solid circle in Fig. 11)
by displacing and attacking the regenerated origin in the viral (+)
strand. Concurrent with the nucleolytic scission carried out by the
second *cis*A subunit (solid circle in Fig. 11), the first subunit
of the *cis*A protein (open circle) ligates the newly created 3' OH
with the bound 5' end, using the conserved energy of the covalent

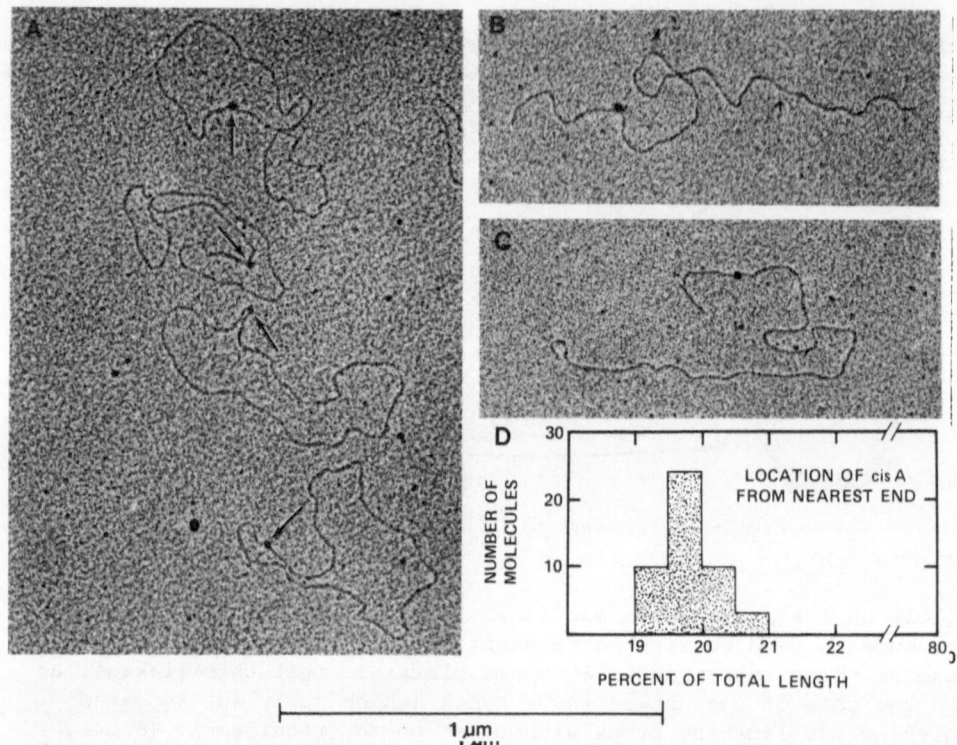

Figure 10. A. *CisA protein-RF II complex*. B. *and C. Complexes cleaved by* Pst *1 nuclease*.

protein-DNA bond. In this mechanism using two active sites or subunits of the *cis*A protein alternately, the *cis*A protein-RF II complex is restored with the completion and release of each viral circle and the catalytic role of the complex is assured. This mechanism ensures efficient use of *cis*A protein and its proper orientation for the continuous synthesis of viral (+) DNA on a single template.

Initiation of complementary (-) strand synthesis by host proteins (including *dnaB*, *C* and *G* proteins, proteins i and n, holoenzyme and other proteins) takes place either on the displaced viral loop of the replicating intermediate, or on the released viral (+) circle. Further studies in which the purified (+) strand synthesizing system is coupled with that for the (-) strand are now required to answer this question.

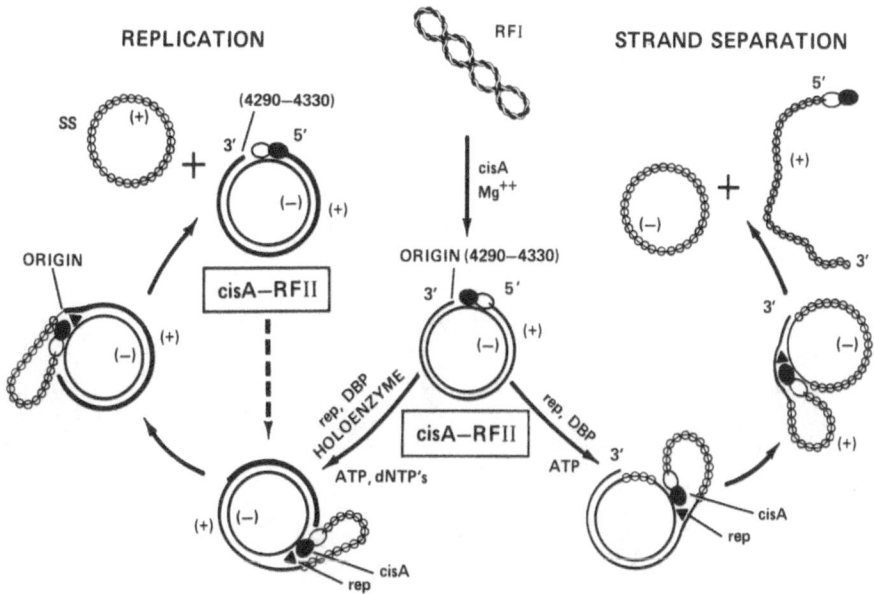

Figure 11. Scheme illustrating multiple functions of *cisA* protein.

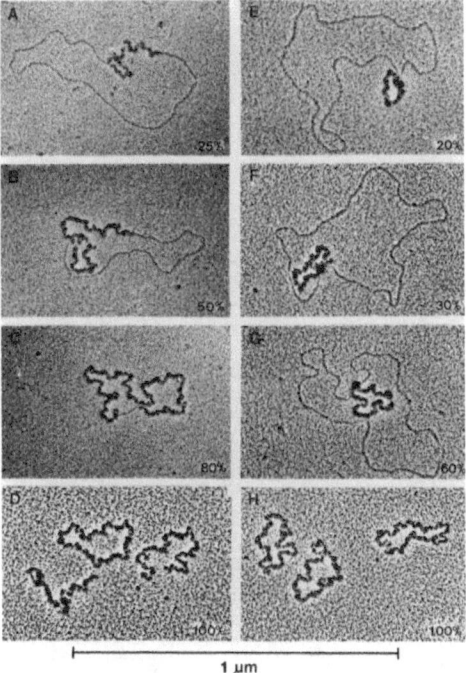

Figure 12. Intermediates in strand separation and replication.

Table 2

E.coli Replication Proteins

Polypeptide	Size (K)	Subunits	Function	Unamplified yield molecules/cell	Unamplified yield mg/kg*	Amplification
DBP	74	4	binding DNA	300	20	–
dnaB	250	4	mobile promoter	20	0.3	5
*dnaC***	29	1	prepriming	200	–	–
protein i	80	4	prepriming	150	0.5	–
protein n**	–	–	prepriming	–	–	–
primase	60	1	primer formation	100	0.2	–
pol III holo	300	4		20	0.2	–
pol III (α)	140	1		–	–	–
copol III (β)	40	1		–	–	–
52K (γ)	53	1	DNA synthesis	–	–	10
32K (δ)	32	1		–	–	–
protein u**	–	–		–	–	–
pol I	109	1		250	10	70
ligase	74	1	ligation	300	11	500
gyrase	140	–	supertwisting	–	–	–
rep	65	1	strand separation	30	0.6	10
*cis*A (ØX174)	60	1	initiation	3000	20	–
dUTPase	64	4	dUTPase	350	3	–

*wet weight;   **dnaC preparation 30% pure;   proteins n and u less pure

A mechanism analogous to that for øX RF replication may also operate during the replication of other chromosomes. Initiation of synthesis of one of the strands by extension of a nick introduced by *cis*A-like protein would represent an origin of replication. (In chromosomes that replicate bidirectionally, nicking at the origin would be introduced in both strands). This leading strand, synthesized continuously, is analogous to the øX viral (+) DNA. The other lagging strand, analogous to the øX complementary (-) DNA, would be initiated repeatedly on the exposed parental (+) template.

Replication Proteins and Their Significance

The list of *E. coli* replication proteins (Table 2) is already formidable and impressive. Further insights into the mechanism of DNA replication may emerge from future efforts to reconstitute these enzyme systems for the concerted, rapid and efficient replication of RF, as well as the replication stages that precede and follow it.

A great deal has been learned from *in vivo* studies of øX replication. Inevitably there are advantages and limitations of that approach as compared to the enzymologic approach. When making a choice between them, one must place bets and take chances, as has been the case throughout the history of biochemistry.

In vivo studies involve the least perturbation of the system, although starvation, pulse-chase, and various extraction manipulations are not without some peril to the normal course of cellular events. No *in vivo* procedure can describe molecular events with any confidence. One learns, for example, that *cis*A and *rep* proteins are important in RF replication but not what they are doing.

In delineating a metabolic pathway, the *in vivo* approach cannot distinguish a true intermediate from a spur off the main track. In the sequence:

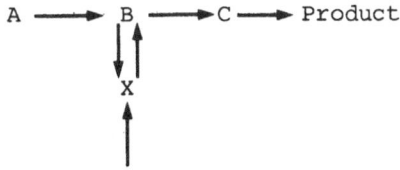

a labeled or visible intermediate may be B or it may be X. Often the concentration of B is too small to be detected, for the very reason that it is rapidly converted to the product. On the other hand X may be prominent, and therefore the only molecule observed; it may be chased effectively into the product even though it may be only a detour (for repair or other diverse purposes) and not a major stage in the biosynthetic pathway.

Enzymological studies, as I have emphasized have serious limitations. We violently disrupt the cell, destroy its organization, impose unphysiological demands that the molecular components by hydrophilic (soluble enzymes) and select a combination of components, conditions, and assays that strongly influence our measurements.

With the use of isolated phage DNAs as template, we ignore that the injection of the viral DNA into the cell is not a random process but is tightly coupled to replication. Certain capsid or pilot proteins which may be crucial to the first event of replication are ignored as are the cell membranes. Instead we use phenol to extract the circles from the virus and pipet 10^{10} of them into a test tube.

However if we observe the clear and unequivocal action of an enzyme, if its specificity and rates are plausible for an *in vivo* role, and if its appearance and disappearance from the cell by genetic manipulation or drug inhibition match expectations for its functions, then our confidence is enhanced.

To conclude, we must recognize that reconstitution assays do not assure us that we have reproduced the physiological replication event. We may be missing important parts of the replication machinery or we may have substituted foreign parts derived from other cell machinery. It is therefore essential to test the validity of isolated components and reaction sequences and mechanisms by introducing them into cruder and more organized systems such as the gently solubilized lysate (38), the cellophane-disc lysate (37) or the folded chromosome (24).

ACKNOWLEDGEMENTS

This work was supported in part by grants from the National Institutes of Health and the National Science Foundation.

REFERENCES

1. Abdel-Monem, M., Lauppe, H.-F., Kartenbeck, J., Durwald, H., and Hoffmann-Berling, H. (1977) J. Mol. Biol. 110:667

2. Alberts, B. et al. (1977) In: Nucleic Acid - Protein Recognition (H.J. Vogel, ed.) Academic Press, N.Y., p31

3. Bouché, J.-P., Rowen, L. and Kornberg, A. (1977) J. Biol. Chem. in press

4. Bouché, J.-P., Zechel, K. and Kornberg, A. (1975) J. Biol.
    Chem. 250:5995

5. Brutlag, D., Schekman, R. and Kornberg, A. (1971) Proc. Nat.
    Acad. Sci. USA 68:2826

6. Champoux, J.J. (1976) Proc. Nat. Acad. Sci., U.S.A. 73:3488

7. Champoux, J.J. and McConaughy, B.L. (1976) Biochemistry
    15:4638

8. Christiansen, C.H. and Baldwin, R.L. (1977) J. Mol. Biol.
    in press

9. Eisenberg, S., Scott, J.F. and Kornberg, A. (1976) Proc. Nat.
    Acad. Sci. USA 73:1594

10. Eisenberg, S., Scott, J.F. and Kornberg, A. (1976) Proc. Nat.
    Acad. Sci. USA 73:3151

10a. Eisenberg, S., Griffith, J. and Kornberg, A. (1977) Proc. Nat.
    Acad. Sci. USA 74: in press

11. Falaschi, A. (1977), this volume

12. Geider, G. and Kornberg, A. (1974) J. Biol. Chem. 249:3999

13. Gellert, M., Mizuuchi, K., O'Dea, M.H. and Nash, H.A. (1976)
    Proc. Nat. Acad. Sci. USA 73:3872

14. Godson, G.N. Cold Spring Harbor Symp. Quant. Biol., in press

15. Gray, C.P., Sommer, R., Pole, C., Beck, E. and Schaller, H.
    (1977) Proc. Nat. Acad. Sci. USA, in press

16. Henry, J.T. and Knippers, E. (1974) Proc. Nat. Acad. Sci.
    USA 71:1549

17. Ikeda, J.,Yudelevich, A. and Hurwitz, J. (1976) Proc. Nat.
    Acad. Sci. USA 73:2699

18. Kolodner, R. and Richardson, C.C. (1977) Proc. Nat. Acad.
    Sci. USA 74:1525

19. Korn, D. (1977) this volume

20. Kornberg, A. (1974) DNA Synthesis. W.A. Freeman and Co.,
    San Francisco.

21. Kornberg, A. (1976) In: RNA Polymerases, Cold Spring
     Harbor Symp. Quant. Biol. pp331

22. Kornberg, A. (1977) Biochem. Soc. Trans. 5:359

23. Kornberg, R.D. (1977) Ann. Rev. Biochem. 46:931

24. Kornberg, T.B., Lockwood, A. and Worcel, A. (1974) Proc. Nat.
     Acad. Sci. USA 71:3189

25. Maxam, A.M. and Gilbert, W. (1977) Proc. Nat. Acad. Sci. USA
     74:560

26. McHenry, C. and Kornberg, A. (1977) J. Biol. Chem., in press

27. McMacken, R., Bouché, J.-P., Rowen, S.L., Weiner, J.H., Ueda,
     K., Thelander, L., McHenry, C. and Kornberg, A. (1977) In:
     Nucleic Acid-Protein Recognition (Vogel, H.J., ed.) Academic
     Press, N.Y. pp15-29

28. McMacken, R. and Kornberg, A. (1977) J. Biol. Chem., in press

29. McMacken, R., Ueda, K. and Kornberg, A. (1977) Proc. Nat.
     Acad. Sci., USA 74: in press

30. Moses, R.E. and Richardson, C.C. (1970) Proc. Nat. Acad. Sci.,
     USA 67:674

31. Murray, N.E., Brammar, W.J. and Murray, K. (1977) Mol. Gen.
     Genetics 150:53

32. Ravetch, J.V., Horiichi, K. and Zinder, N.D. (1977) Proc.
     Nat. Acad. Sci. USA, in press

33. Richet, E. and Kohiyama, M. (1976) J. Biol. Chem. 251:808

34. Rowen, L. and Kornberg, A. (1977) J. Biol. Chem., in press

35. Rowen, L. and Kornberg, A. (1977) J. Biol. Chem., in press

36. Sanger, F., Air, G.M., Barrel, B.G., Brown, N.L., Coulson,,
     A.R., Fiddes, J.C., Hitchison, C.A. III, Slocomb, P.M.Y. and
     Smith, M. (1977) Nature 265:687

37. Schaller, H., Otto, B., Nusslein, V., Huf, J., Hermann, R.,
     and Bonhoeffer, F. (1972) J. Mol. Biol. 63:183

38. Schekman, R., Weiner, A. and Kornberg, A. (1974) Science,
     186:987

39. Schekman, R., Weiner, J.H., Weiner, A. and Kornberg, A. (1975)
     J. Biol. Chem. 250:5859

40. Schekman, R., Wickner, W., Westergaard, O., Brutlag, D.,
     Geider, K., Bertsch, L.L. and Kornberg, A. (1972) Proc. Nat.
     Acad. Sci. USA 69:2691

41. Scherzinger, E., Lanka, E., Morelli, G., Seiffert, D. and
     Yuki, A. (1977) Eur. J. Biochem. 72:543

42. Scott, J.F., Eisenberg, S., Bertsch, L.L. and Kornberg, A.
     (1977) Proc. Nat. Acad. Sci. USA 74:193

43. Scott, J.F. and Kornberg, A. (1977) J. Biol. Chem., in press

44. Sherman, L.A. and Gefter, M.L. (1976) J. Mol. Biol. 103:61

45. Sinsheimer, R.L. (1968) Prog. Nuc. Acid. Res. Mol. Biol.
     8:115

46. Sumida-Yasumoto, C.,Yudelevich, A. and Hurwitz, J. (1976)
     Proc. Nat. Acad. Sci. USA 73:1887

47. Tabak, H., Griffith, J., Geider, K., Schaller, H. and
     Kornberg, A. (1974) J. Biol. Chem. 249:3049

48. Tessman, E.S. (1966) J. Mol. Biol. 17:218

49. Tessman, E.S. and Peterson, P.K. (1976) J. Virol. 19:416

50. Thomas, M., Cameron, J.R. and Davis, R.W. (1974) Proc. Nat.
     Acad. Sci. USA 71:4579

51. Timmis, K., Cabello, F. and Cohen, S. (1975) Proc. Nat. Acad.
     Sci. USA 72:2242

52. Ueda, K., McMacken, R. and Kornberg, A. (1977) J. Biol. Chem.,
     in press

53. Vosberg, H.-P. and Hoffmann-Berling, H. (1971) J. Mol. Biol.
     58:739

54. Wang, J. (1971) J. Mol. Biol. 55:523

55. Weiner, J.H., McMacken, R. and Kornberg, A. (1976) Proc. Nat.
     Acad. Sci. USA 73:752

56. Westergaard, O., Brutlag, D. and Kornberg, A. (1973) J. Biol.
     Chem. 248:1361

57.   Wickner, S. (1976) Proc. Nat. Acad. Sci. USA   73:3511

58.   Wickner, S. (1977) Proc. Nat. Acad. Sci. USA   74:2815

59.   Wickner, S. and Hurwitz, J. (1974) Proc. Nat. Acad. Sci.
      USA   71:4120

60.   Wickner, S. and Hurwitz, J. (1975)   Proc. Nat. Acad. Sci.,
      USA 72:921

61.   Wickner, S. and Hurwitz, J. (1975)   In: DNA Synthesis and Its
      Regulation (Goulian, M. and Hanawalt, P., ed.) W.A. Benjamin,
      Menlo Park, Calif.   pp227

62.   Wickner, S. and Hurwitz, J. (1976) Proc. Nat. Acad. Sci. USA
      73:1053

63.   Wickner, S., Wright, M. and Hurwitz, J. (1974) Proc. Nat.
      Acad. Sci. USA   71:783

64.   Wickner, W., Brutlag, D., Schekman, R. and Kornberg, A. (1972)
      Proc. Nat. Acad. Sci. USA 69:965

65.   Zechel, K., Bouché, J.-P. and Kornberg, A. (1975)   J. Biol.
      Chem. 250:4684

THE *dnaG* GENE PRODUCT OF *ESCHERICHIA COLI*: CHARACTERIZATION OF

THE CATALYTIC ACTIVITY

Daniel Capon and Malcolm Gefter*

Department of Biology
Massachusetts Institute of Technology
Cambridge, Massachusetts, 02139, U.S.A.

INTRODUCTION

The initiation of ∅X174 replicative form synthesis from viral single-stranded DNA *in vitro*, occurs in at least two distinct stages (1-3). The first step requires the products of the *dnaB* and *dnaC* gene, the DNA-binding protein, and additional protein factors[1] to "activate" ∅X174 single-stranded DNA in an ATP-dependent reaction. This activated DNA may be isolated by gel filtration after incubation of DNA with ATP and a partially purified cell extract (1) or resolved protein fractions (2). Furthermore "activated" DNA promoted immediate complementary strand synthesis, requiring the *dnaG* gene product in addition to DNA polymerase and factors as opposed to the kinetic lag observed for unreacted DNA. The absolute requirement for the *dnaG* protein in this latter stage suggests that it plays an essential role in DNA chain initiation. It has been reported that the *dnaG* protein functions, (phage G4 replicative form synthesis), by synthesizing an oligoribonucleotide primer in a rifampicin-resistant reaction (4). It was suggested that the template for this reaction: binding protein complexed with single-stranded DNA, is the functional equivalent of "activated" ∅X174 DNA; and that *dnaG* protein catalyzes the same reaction on both DNAs.

The experiments described here deal with our attempt to analyze *dnaG* protein reaction in the absence of net DNA synthesis on both ∅X174 DNA and G4 DNA. We will describe experiments indicating that

---

[1] *Replication factors x, y and z (3), or i and n (2)*

A.S.I. Participant

the *dnaG* protein catalyzes the synthesis of a "primer" employing
deoxyribonucleotide substrates as well as ribonucleotide substrates.
These results are similar to those recently described (10).

METHODS AND MATERIALS

Conditions for growing cells and preparing partially purified
cell extracts have been previously described (1). Homogeneous DNA-
binding protein was prepared by the method of Molineux et al. (5).
dnaG protein, DNA polymerase III* and copolymerase III* were
prepared from the same cells (*E.coli* H562) by modifications of
existing procedures (see below). Cells were gently lysed, and a
high speed supernatant was passed through DEAE cellulose at 0.4M
$(NH_4)_2SO_4$. The flow-through was adjusted to 45% saturation with
ammonium sulfate and the pellet was washed progressively with
buffer containing 40%, 30% and 20% ammonium sulfate as described
(1). The supernatant of the 20% wash which contained at least 50%
of each of the above protein activities was brought to 50%
saturation with ammonium sulfate, and the pellet dissolved in Buffer
A (50mM Tris HCl pH 7.5, 1mM EDTA 2mM DTT and 20% Glycerol) and
dialyzed against 3 changes of 100 volumes each of buffer A for 2-3
hours each. Phosphocellulose chromatography on this fraction
yielded separate activities. The material which flowed through
contained copolymerase III* activity and was further purified by
DEAE-cellulose and Sephadex G-150 chromatography (6). Material
eluting between 0.075 and 0.15 M NaCl contained the *dnaG* and DNA
polymerase III* activities, which were resolved by DEAE-Sephadex
chromatography. The *dnaG* activity eluted at approximately 0.11 M
NaCl and polymerase at 0.15 M NaCl. Each was further purified by
an additional phosphocellulose column. The same procedure was
used to prepare the temperature-sensitive *dnaG* protein. The latter
was prepared from strain NY73.

CONDITIONS FOR FORMATION OF BACTERIOPHAGE G4
OR ØX174 "COMPLEX" ACTIVATED DNA

Reaction mixtures (50 μl) were 0.02M Tris.HCl pH 7.5, 0.012M
$MgCl_2$, 0.004M DTT, 0.002M KH_2PO_4 0.0005M ATP and ØX174 DNA, (100
pmoles nucleotide); 1.5 μg rifampicin and partially purified cell
extract (approximately 50 μg protein). Generally, a twentyfold
scale reaction mixture was incubated at 30°C for 10 minutes and
filtered through a biogel A5-M column (0.9 x 30 cm) in 0.02M Tris.
HCl pH 7.5; 0.001 M EDTA, 0.012M $MgCl_2$, 0.0005M ATP, and 2.5%
sucrose. Active fractions (0.5 ml) which eluted at the void volume,
would support efficient RF II synthesis when the following components
were added to 50 μl aliquots: 2.0 nmoles each dATP, dCTP and dGTP:
0.2 nmoles (^{3}H) TTP (6.0 x 10^3cpm/pmole); *dnaG* protein [0.05 - 0.1

units (9)], DNA polymerase III[*] [.1 - .5 units (6)] and copolymerase III[*] [.1 - .5 units (6)]. Incubation was at 30°C for 20 minutes.

PHAGE DNA

Bacteriophage ϕX174 am3 phage was grown on *E.coli* C and purified according to Sinsheimer (7) except that a discontinuous CsCl gradient was substituted for the phase partition step. G4 phage (isolate provided by Dr. G.N. Godson) was prepared by infecting a culture of *E.coli* C at 5 x 10[8] cells/ml at a *moi* = 0.1 at 37°C, allowing lysis to occur (about 2 hours later), precipitating the phage from the lysate with NaCl and polyethylene glycol (8) and then a discontinuous CsCl gradient. Each phage preparation was further purified on a continuous CsCl gradient and DNA was prepared from the final preparation. The preparation of dialyzed phage was brought to 0.5% in sarkosyl and heated to 70°C and extracted three times with an equal volume of redistilled phenol that was saturated with 0.01M Tris.HCl pH 7.5, 0.001M EDTA. The aqueous phase was dialyzed extensively against 0.01M Tris.HCl pH 7.5, 0.001M EDTA. An additional 5-20% neutral sucrose gradient in 70% formamide was necessary for the G4 DNA preparation to make its conversion to RFII dependent upon added *dnaG* protein. Ribotriphosphates were freed of detectable deoxyribotriphosphates by ascending chromatography on Whatman #3MM paper in iso-propanol (30): conc.NH$_4$OH (10): EDTA 0.001M in water (60).

CONDITIONS FOR SYNTHESIS OF G4 RFII

Reaction mixtures (50 μl) were 0.02M Tris.HCl pH 7.5, 0.012M MgCl$_2$, 0.004M DTT 2mM KH$_2$PO4 0.0002M ATP and contained G4 DNA, (50 nmoles nucleotide); 2 nmoles each of dATP, dCTP and dGTP; 0.2 nmoles (^{3}H) TTP; 5 nmoles of GTP, UTP and CTP (when added) DNA binding protein (1-5 μg), *dnaG* protein (.05 1 .2 units), DNA polymerase III[*] (0.1 - 0.5 units), copolymerase III[*] (0.1 - 0.5u) incubations were at 30°C.

RESULTS

The basic strategy employed in these experiments is to utilize a thermolabile *dnaG* protein in order to determine if the product catalyzed by the *dnaG* protein had been formed and the protein activity no longer required. The assay for the product catalyzed by the *dnaG* protein is primer-independent DNA synthesis. The systems employed in this study are the conversion of bacteriophages G4 and ϕX174 single stranded DNAs to the double stranded replicative forms. These reactions require in addition to other proteins, the *dnaG* protein for the SS to RF conversion.

The demonstration that the systems employed require the activity of the *dnaG* protein is shown in Table I. In the case of ØX174 DNA, the substrate for RF synthesis is an "activated" form synthesized by incubating ØX174 DNA and ATP in a partially purified cell extract. The "activated" form is isolated by gel filtration [see Methods, and (1)]. The conversion of the "activated" form to RF requires DNA polymerase III and elongation factors (DNA polymerase III* and copolymerase III*) and the *dnaG* protein. Although the ØX174 DNA is incubated in an extract capable of forming RF (provided all four deoxynucleoside triphosphates are added), i.e., it contains the *dnaG* protein and the DNA polymerase and factors, they are not able to form a complex with the DNA that is stable to filtration.

In contrast, bacteriophage G4 DNA incubated under the same conditions and then isolated does not show an absolute requirement for the addition of *dnaG* protein or DNA polymerase III and factors. Purified G4 DNA, however, requires (in addition to DNA binding protein) absolutely the *dnaG* protein and DNA polymerase III and factors. Thus the ØX174 "DNA complex" and purified G4 DNA can be used to assay for the product of the *dnaG* protein.

The strategy employed in analyzing the *dnaG* protein product is to incubate the substrate DNA with the thermolabile *dnaG* protein (*dnaG*^{ts} protein) and ribonucleoside triphosphates [as suggested by previous studies (4)] along with other necessary proteins at 30°C. The temperature is then raised to 38°C (which inactivates the *dnaG* protein) and deoxyribonucleoside triphosphates are added. DNA synthesis is monitored by incorporation of one radioactive deoxyribonucleoside triphosphate. If a primer has been synthesized at 30°C and the *dnaG* protein is no longer needed, then DNA synthesis should commence immediately on addition of deoxytriphosphates.

In confirmation of previous findings using G4 DNA (4) the *dnaG* protein and all four ribonucleoside trophosphates allow for synthesis of DNA by synthesizing on RNA primer (Table II). That the *dnaG* is required for DNA synthesis is shown by the failure to observe DNA synthesis if the reaction is run at 38°C (permissive for *dnaG* protein but not for *dnaG*^{ts} protein).

To investigate whether or not deoxyribonucleoside triphosphates can replace ribonucleoside triphosphates in the *dnaG* reaction, the 30°C incubation is carried out with a mixture of 3 deoxytriphosphates and ATP, the temperature was raised to 38°C and the fourth triphosphate is added. (The overall reaction is completely dependent on the addition of all four deoxytriphosphates - see Table IV, line 6). As seen in Table II, although not as effective as ribonucleosides in supporting subsequent DNA synthesis, the addition of ATP,

Table I
Bacteriophage øX174 "DNA Complex" SS→RF

System	Percent Reaction
complete	100 (2.1p moles)
-*dnaG*	9
-polymerase - factors	10
-*dnaG* - polymerase - factors	3

Bacteriophage G4 "DNA complex" SS→RF

complete	100 (1.3p moles)
-*dnaG* - polymerase - factors	26
-polymerase - factors	75
-*dnaG*	28

Bacteriophage G4 DNA SS→RF

complete	100 (4.2p moles)
-DNA B.P.	3
-polymerase - factors	<1
-*dnaG*	<1

TTP, dGTP and dCTP (line 7) at 30°C is sufficient to allow for primer synthesis. As seen in line 5 and 6 dGTP and dCTP are absolutely required during the first incubation. However, if GTP is included in the first reaction, then the requirement for dGTP is abolished (line 9). The requirement for dCTP is still observed (line 10). It should be noted, that in the presence of ATP and GTP the overall reaction is more efficient as assayed by subsequent DNA synthesis, suggesting that GTP is preferred to dGTP for primer synthesis.

We conclude from these studies that *dnaG* protein·dependent primer synthesis proceeds in the presence of either deoxy or ribo-triphosphates and that GTP is preferred to dGTP. These results however are not observed when øX174 DNA complex replaces G4 DNA in this reaction.

As seen in Table III, ribonucleoside triphosphates do not satisfy a substrate requirement for *dnaG*-dependent primer synthesis. Incubation of the "complex" with ATP followed by the addition of all four deoxytriphosphates leads to the synthesis of RF (product

Table II

Bacteriophage G4 SS→RF

Ribonucleoside triphosphates or deoxyribonucleoside triphosphates
satisfy primer requirement

	5' at 30°C	20' at 30°C	20' at 30°C or 20' at 38°C	Percent Reaction
1.	ATP,GTP,CTP,UTP	4dNTP	-	100 (4p moles)
2.	ATP,GTP,CTP,UTP	-	4dNTP	40
3.	ATP,GTP,CTP,UTP at 38°C	-	4dNTP	<4
4.	ATP	-	4dNTP	<4
5.	ATP + TTP,dCTP,dATP	-	dGTP	<4
6.	ATP + TTP,dGTP,dATP	-	dCTP	<4
7.	ATP + TTP,dGTP,dCTP	-	dATP	13
8.	ATP,GTP + TTP,dGTP,dCTP	-	dATP	63
9.	ATP,GTP + TTP,dCTP,dATP	-	dGTP	38
10.	ATP,GTP + TTP,dGTP,dATP	-	dCTP	8

Reaction mixture contains G4 DNA, *dnaC*ts protein, DNA binding protein, DNA polymerase III
+ factors and rifampicin.

Table III

Bacteriophage φX174 SS→RF

Ribonucleoside triphosphates *do not* satisfy primer requirement

	5′ at 30°C	20′ at 30°C	20′ at 38°C	Percent Reaction
1.	ATP	4dNTP	–	(100 (0.5p moles)
2.	ATP	–	4dNTP	<4
3.	ATP	–	4dNTP + *dnaG*+	210
4.	ATP 2′ at 38°C	–	4dNTP	<4
5.	ATP,GTP,CTP,UTP	4dNTP	–	100
6.	ATP,GTP,CTP,UTP	–	4dNTP	<4
7.	ATP,GTP,CTP,UTP	–	4dNTP + *dnaG*+	112
8.	ATP,GTP,CTP,UTP 2′ at 38°C	–	4dNTP	<4

Reaction mixture contains φX174 "complex" *dnaG*ts protein, DNA polymerase III + factors, rifampicin, PO_4^{-3}

Table IV

Bacteriophage ØX174 "DNA complex" SS→RF

Deoxyribonucleoside triphosphates satisfy primer requirement

	5´ at 30°C	20´ at 30°C	or 20´ at 38°C	Percent Reaction
1.	-	4dNTP	-	100 (0.5p moles)
2.	-	-	4dNTP	<5
3.	2´ at 38°C	-	4dNTP	<5
4.	ATP,TTP,dATP,dGTP	dCTP	-	72
5.	ATP,TTP,dATP,dGTP	-	dCTP	<5
6.	ATP,TTP,dATP,dGTP	-	-	<5
7.	ATP,TTP,dATP,dCTP	dGTP	-	79
8.	ATP,TTP,dATP,dCTP	-	dGTP	13
9.	ATP,TTP,dCTP,dGTP	dATP	-	100
10.	ATP,TTP,dCTP,dGTP	-	dATP	45
11.	ATP,dATP,dCTP,dGTP	TTP	-	100
12.	ATP,dATP,dCTP,dGTP	-	TTP	13

Reaction mixture contains ØX174 complex, $dnaG^{ts}$ protein, DNA polymerase III + factors, rifampicin, PO_4^{-3}.

analysis is by alkaline sucrose gradient) as shown in line 1. The reaction does not work at 38°C (line 2) but the addition of wild-type *dnaG* protein to the reaction allows synthesis to take place at 38°C. Thus the reaction in *dnaG* protein dependent and temperature-sensitive when the *dnaG*ts protein is used. Incubation of the DNA complex with all four ribotriphosphates at 30°C does not lead to the synthesis of a primer (line 6). Inclusion of the four ribotriphosphates does not inhibit the reaction using the *dnaG*ts protein (line 5) or does it prevent the wild-type protein from functioning at 38°C. Thus in contrast to the G4 RF reaction, ribonucleoside triphosphates do not appear to satisfy the primer requirement for ∅X174 RF synthesis.

As shown in Table IV, deoxynucleoside triphosphates do satisfy the primer requirement when incubated in the presence of ∅X174 "DNA complex", ATP and the *dnaG*ts protein. The deoxynucleotide that is absolutely required is dCTP (line 5), dGTP and dTTP are stimulatory and dATP is essentially not required (line 10). Thus the deoxy-nucleotide requirement for primer synthesis is similar to what was observed for the G4 reaction.

We conclude that for ∅X174 RF formation a stable primer can be formed with deoxy but not ribotriphosphates.

DISCUSSION

We have confirmed earlier studies (4) that the *dnaG* protein is able to prime DNA synthesis by employing ribonucleoside triphos-phates as substrates. In addition, we show that deoxynucleotides can substitute for ribonucleotides in the case of G4 phage DNA and are essential in the case of ∅X174 DNA. One difference between these two systems is that the G4 reaction is carried out with pure components whereas the ∅X174 reaction utilizes a "complex". The latter may contain activities that destroy an RNA-DNA hybrid and not a DNA-DNA hybrid. We cannot rule out this possibility although we have no evidence for the existence of such an activity. Since we are demanding in our studies a *stable* primer, it is possible that the *dnaG* protein does catalyze the synthesis of an RNA primer but is naturally unstable whereas the DNA primer is not. The DNA primer may in fact be the "natural" primer. A system similar to the discrimination reaction for fd and ∅X174 RF synthesis may be operating here (11) in that RNA synthesized on ∅X174 DNA may not stable whereas that synthesized on G4 DNA is. Recent studies (10) have shown that a ribonucleotide is only needed for the 5´ terminal residue of an otherwise deoxynucleotide-containing polymer.

Our results are in general agreement with both those of Kornberg and colleagues as well as S. Wickner (10). We have found, however, in contrast to the latter results, a requirement of dCTP

for primer synthesis in addition to ATP and dGTP and dTTP. This
discrepancy may be due to the demands of our experimental system,
i.e., stability to 38°C as opposed to stability at low temperature
to filtration.

The major questions that are raised by this study are: The role
of ribonucleotides and deoxyribonucleotides in primer synthesis,
the nature of the product (sequence, length, origin and terminus)
and the nature of the "natural" reaction product when primer
synthesis is coupled to DNA synthesis. These questions are
currently being addressed.

REFERENCES

1.   Ray, R., Capon, D. and Gefter, M. (1976) Biochem. Biophys.
      Res. Comm. 70:506-512

2.   Weiner, J., McMacken, R. and Kornberg, A. (1976) PNAS 73:
      752-756

3.   Wickner, S. and Hurwitz, J. (1974) PNAS 71:4120-4124

4.   Bouché, J.P., Zechel, K. and Kornberg, A. (1975) J. Biol.
      Chem. 250:5995-6001

5.   Molineux, I., Friedman, S. and Gefter, M. (1974) J. Biol.
      Chem.   :6090-6098

6.   Wickner, W., Shekman, R., Geider, K. and Kornberg, A. (1973)
      PNAS 70:1764-1767

7.   Sinsheimer, R. (1966) Procedures in Nucleic Acid Research
      (eds. G. Cantoin and D. Davice) Harper and Row. pp569-576

8.   Yamamato, K., Alberts, B., Benzinger, R., Lanthorne, L. and
      Reiber, G. (1970) Virology 40:734-744

9.   Wickner, S., Wright, M. and Hurwitz, S. (1973) PNAS 70:
      1613-1618

10.   Wickner, S. (1977) PNAS, in press

11.   Vicuna, R., Hurwitz, J., Wallace, S. and Girard, M. (1977)
      J. Biol. Chem. 252:2524-2533

G4 AND ST-1 DNA SYNTHESIS *IN VITRO*

Sue Wickner

Laboratory of Molecular Biology
National Cancer Institute
National Institues of Health
Bethesda, Maryland 20014, U.S.A.

INTRODUCTION

The purpose of studying DNA replication *in vitro* has been to learn about the proteins and the biochemical mechanisms involved in this process. The advantage of using *Escherichia coli* is that genetic studies of others have defined many genes whose functions are essential for chromosome replication. The genes required for *E. coli* DNA replication include those needed for initiation of rounds of replication (*dnaA, dnaB, dnaC(D), dnaI, rpoB and dnaP*) and those which have a role in continuing a round of replication (*dnaB, dnaC(D), dnaG, dnaZ, polC, polA, nalA and cou*) (reviewed in ref. 1). Some of the *E. coli* DNA replication proteins are also required for the conversion of single-stranded circular phage DNA to duplex DNA catalyzed by crude extracts of uninfected *E. coli* (2-4). Three distinct enzymic pathways exist by which viral single-stranded DNA is converted to double-stranded DNA. Each of these pathways has been reconstituted with purified *dna* proteins plus additional proteins, not yet defined by temperature sensitive mutants. Each pathway is distinct in its nucleotide and protein requirements and its DNA specificity. These three enzyme systems differ primarily in their mechanism of chain initiation. The fd DNA-dependent system uses RNA polymerase to make an RNA primer (2, 5). The primer can be elongated *in vitro* by DNA polymerase I alone, DNA polymerase II and DNA binding protein (6), or by DNA polymerase II or III in combination with *dnaZ* protein, and two other factors defined by their requirement in this reaction and referred to as DNA elongation factors I and III (DNA EF I and III) (7). The øX174 DNA-dependent system requires *dnaB, dnaC(D) and dnaG* proteins, DNA binding protein, other *E. coli* proteins (referred to as replication factors X, Y and Z (8) or factors i and n (9)) only ATP of the rNTPs (8), 4 dNTPs and DNA elongation components (*dnaZ* protein, DNA polymerase III, DNA EF I and DNA EF III). Most likely, the first seven proteins are involved in primer synthesis. The G4 or ST-1 DNA-dependent system uses the same elongation systems as do the fd and ØX174 systems, but needs only the *dnaG* protein and DNA binding protein to begin new chains (10,11). Thus some structure of G4 and ST-1 DNA allows *dnaG* protein to function independent of *dnaB* and *dnaC(D)* proteins and replication factors X,Y and Z. As first discovered by Schekman et al. (10), and studied by others (11-14) this relatively simple system facilitates the study of the function of the *dnaG* protein.

The results presented below suggest the following mechanism of initiation of G4 DNA synthesis (Fig. 1; ref. 14). DNA binding protein covers G4 DNA and promotes the binding of *dnaG* protein to the DNA. An oligonucleotide is synthesized by this protein DNA complex in a reaction requiring ADP, dTTP (or UTP), and dGTP (or GTP). The oligonucleotide is elongated by DNA polymerase I or the combination of DNA polymerase III, *dnaZ* protein, DNA EF I and DNA EF III.

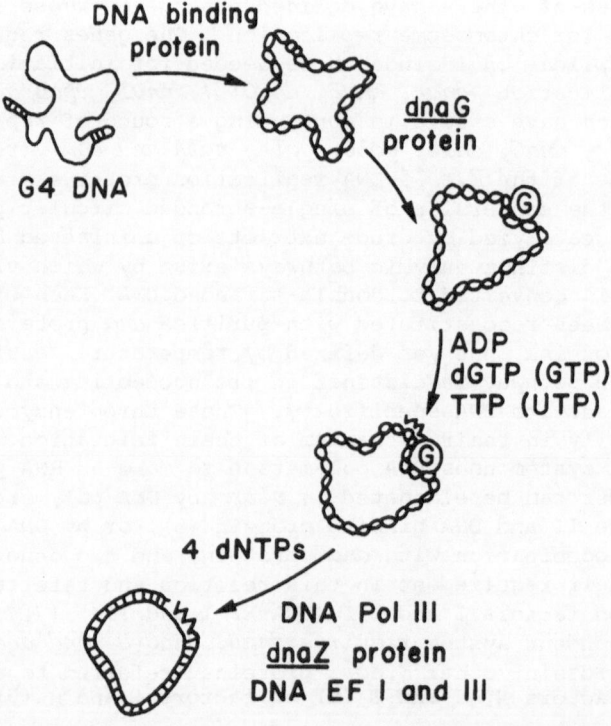

Figure 1. Mechanism of priming of G4 DNA-dependent DNA synthesis.

Site specific binding of *dnaG* protein to G4 and ST-1 DNA

The *dnaG* protein binds specifically to G4 and ST-1 in a reaction requiring DNA binding protein. This was shown by the physical isolation of *dnaG* protein bound to G4 and ST-1 DNA (Fig. 2, A and C). The binding of *dnaG* protein to these DNAs required that the DNA be covered with DNA binding protein (Fig. 2, B); nucleotide cofactors were not required. The binding reaction was reversible as demonstrated by the dissociation of *dnaG* protein from the DNA by high salt. With excess *dnaG* protein relative to G4 DNA, about one molecule of *dnaG* protein bound per circle of DNA. In contrast to G4 and ST-1 DNA, fd and ØX174 DNA did not bind *dnaG* protein (Fig. 2, D and E). It is reasonable to assume that fd DNA has no binding site for *dnaG* protein since this protein is not required for fd DNA synthesis. However, *dnaG* protein is required for ØX174 DNA synthesis. If its function in ØX174 DNA synthesis is similar to its function in G4 DNA synthesis, then perhaps some of the other proteins required for ØX174 DNA synthesis but not for G4 synthesis (*dnaB* and *dnaC(D)*) proteins and replication factors X, Y and Z) are needed to form a site on the ØX174 DNA which can be recognized by *dnaG* protein.

Primer synthesis catalyzed by *dnaG* protein

DNA synthesis by the isolated complex of G4 or ST-1 DNA, *dnaG* protein and DNA binding protein required 4 dNTPs and DNA elongation proteins. In contrast to reports of others (10,12,13), rNTPs were not required. However, ADP stimulated DNA synthesis with G4 about 15-fold and with ST-1 about 4-fold. The K_m for ADP was about 2 µM. Other ribo- or deoxynucleoside diphosphates or triphosphates or AMP did not stimulate the G4 reaction; all at 20 µM. ATP and dADP stimulated G4 DNA synthesis to the same extent but only a higher concentrations; their K_m was about 50 µM.

The G4 DNA synthesizing reaction was separated into an ADP-dependent priming reaction (first reaction) and an ADP-dependent DNA elongation reaction (second reaction). It was found that ADP, dTTP, dGTP, *dnaG* protein, G4 DNA and DNA binding protein were required in the first reaction for incorporation of label from $(\alpha^{32}P)$dGTP or $(\alpha^{32}P)$dTTP into a protein-DNA complex and for ADP-independent DNA synthesis by the complex in a second reaction Fig. 3). In contrast, dATP and dCTP were not required in the first reaction. However, when they were included, label from them was found in a protein-DNA complex. In the priming reaction, UTP satisfied the dTTP requirement; dideoxyTTP and TDP did not satisfy the requirement. GTP substituted for dGTP but dGDP did not. There was no preference for utilization of GTP over dGTP in the first reaction.

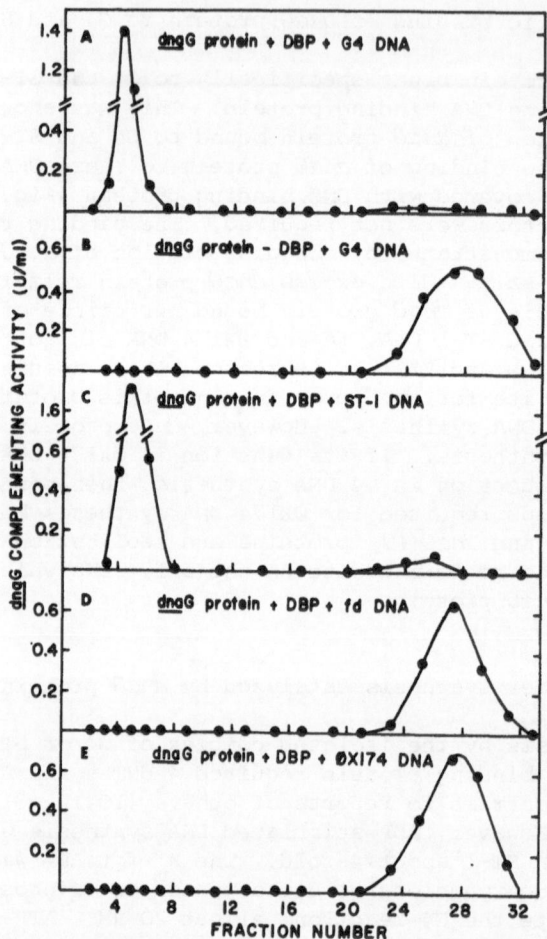

<u>Figure 2</u>. *Specific binding of <u>dnaG</u> protein to G4 and ST-1 DNA.*
(A) The first reaction (60 μl) contained: reaction buffer (1 mM
dithiothreitol, 20 μg/ml rifampicin, 50 mM Tris·HCl (pH 7.5), 1 mg/
ml bovine serum albumin, 8 mM MgCl₂), 6 nmol of G4 DNA (as
nucleotide), 8 μg of DNA binding protein (6) and 1.7 U of <u>dnaG</u>
protein (15). After 10 min at 30°, the mixture was applied at
room temperature to a 22 x 0.5 cm column of Sepharose 6B equilibrated
with 50 mM Tris·HCl (pH 7.5), 1 mM dithiothreitol, 0.1 mg/ml bovine
serum albumin, 5% glycerol and 5 mM MgCl₂. The column was eluted
with the same buffer and 0.11 ml fractions were collected. The
excluded volume (where DNA eluted) was fraction 4; the included
volume was fraction 36. Fractions were assayed for <u>dnaG</u> activity
(15). (B) The reaction was as in (A) but DNA binding protein was
omitted. (C,D,E) The reaction mixtures were as in (A) but G4 DNA
was omitted and 6 nmol of ST-1, fd and ØX174 DNA were included in
experiments C, D and E, respectively.

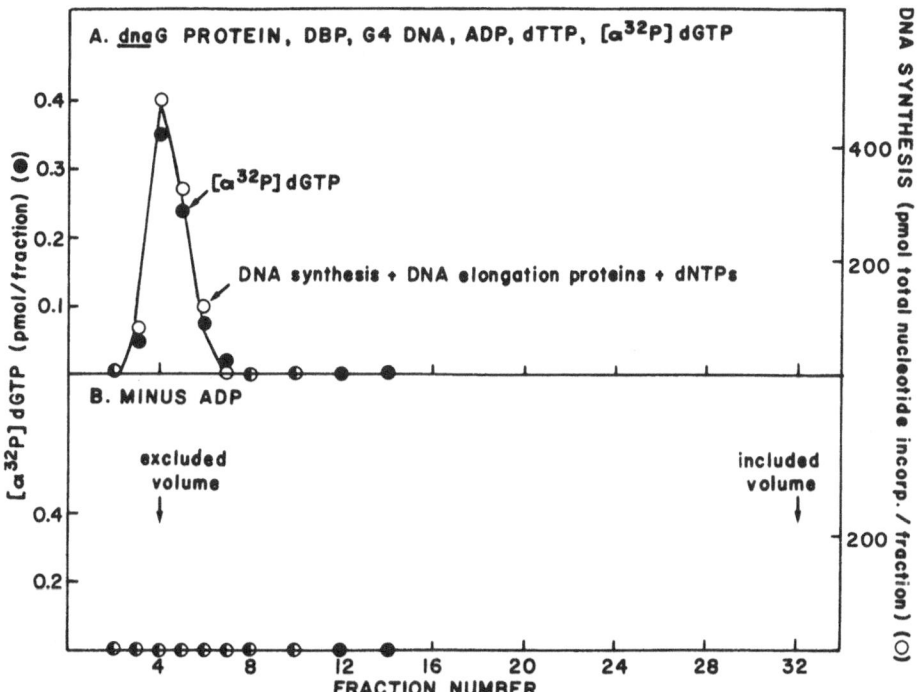

Figure 3. Priming of G4 DNA.
(A) First reaction mixtures (60 µl) contained: Reaction buffer described in Fig. 2, 6 nmol of G4 DNA, 2.1 U of dnaG protein, 8.0 µg of DNA binding protein and 16 µM ADP, dTTP and (α³²P) dGTP (50,000 cpm/pmol). After 15 min at 30°, mixtures were subjected to gel filtration as in Fig. 2. The excluded volume fractions were measured for ³²P (●). They were also assayed for DNA synthesis in a second reaction (60 µl) containing: 10-50 µl of isolated protein-DNA complex, reaction buffer described in Fig. 2, 50 µM each of dATP, dCTP, dGTP and (³H)dTTP (800 cpm(pmol), 0.4 U of dnaZ protein (7), 0.4 U of DNA EF I (16), 0.4 U of DNA EF III (16) and 0.6 U of DNA polymerase III (17). After 10 min at 30°, acid insoluble (³H)dTTP was measured (○). The dTMP incorporation measured was the yield. (B) The experiment was as in (A) but ADP was omitted from the first reaction.

    The results with ST-1 DNA were similar but not identical to those with G4 DNA. Incorporation of label from (α³²P)dGTP or GTP into a protein-DNA complex required only ADP, ATP, dADP or dATP. When fd or ØX174 DNA was used with ADP and 4 dNTPs, label from (α³²P)dGTP or dATP was not associated with the DNA. Thus only those DNAs which bind dnaG protein and whose DNA synthesis is initiated by dnaG protein alone incorporate dNTPs (rNTPs) into a protein-DNA complex.

Several lines of evidence demonstrate that label from $(\alpha^{32}P)$-dGTP associated with the protein-DNA complex was in a primer molecule: (a) The same components were required to incorporate label from $(\alpha^{32}P)$dGTP into the complex and to form a complex active for DNA synthesis. (b) After DNA elongation of the isolated complex, the label from $(\alpha^{32}P)$dGTP sedimented in an alkaline sucrose gradient with nearly full-length linear DNA (Fig. 4). (c) The label from $(\alpha^{32}P)$dGTP was recovered from the complex as a phenol extractable oligonucleotide. While the oligonucleotides synthesized from G4 and St-1 DNA differed in base composition, both were 10 to 30 nucleotides long. In addition, the primer molecules could be ribo-, deoxy-, or mixed ribo- and deoxyoligonucleotides depending on the nucleotides added to the priming reaction (D. Capon, M. Gefter and S. Wickner, unpublished results). Although *dnaG* protein can utilize dNTPs or rNTPs, the overall DNA synthesizing reaction only requires dNTPs since dNTPs are specifically required for DNA polymerase activity.

As proof that the *dnaG* protein is involved in primer synthesis, oligonucleotide synthesis was shown to be thermolabile when thermolabile *dnaG* protein was used (Fig. 5). Oligonucleotide primers synthesized by thermolabile *dnaG* protein at low temperature (30°) could be elongated by any one of the DNA elongation systems at high temperature (37°).

DNA elongation of primed DNA

Once a primer is synthesized DNA elongation can occur *in vitro* by any one of several mechanisms: (A) DNA polymerase I alone (B) DNA polymerase II plus DNA binding protein (6), or (C) DNA polymerase II or III in combination with *dnaZ* protein, DNA EF I and DNA EF III (7). The reaction catalyzed by DNA polymerase III has been studied extensively recently (18-22). Since *dnaE* (DNA polymerase III) and *dnaZ* proteins are involved in *E. coli* chromosome replication, mechanism (C) probably shares the most similarities with the *in vivo* mechanism. The DNA elongation reaction catalyzed by DNA polymerase III, *dnaZ* protein, DNA EF I and DNA EF III has been studied by the physical isolation of protein-protein and protein-DNA complexes preceeding dNMP incorporation (Fig. 6; ref. 16). (A) DNA EF III and *dnaZ* protein form a protein complex independent of DNA polymerase III, DNA EF I, primed template or triphosphates. When separately subjected to Sephadex gel filgration, *dnaZ* protein eluted as expected for a protein of 125,000 daltons and DNA EF III eluted as expected for a protein of 65,000 daltons. When the two proteins were mixed prior to gel filtration, they eluted as a larger complex. (2) Together *dnaZ* protein and DNA EF III catalyze the transfer of DNA EF I to a primed DNA template. When *dnaZ* protein, DNA EF I, DNA EF III, primed-template and ATP were incubated and the reaction mixture subjected to agarose

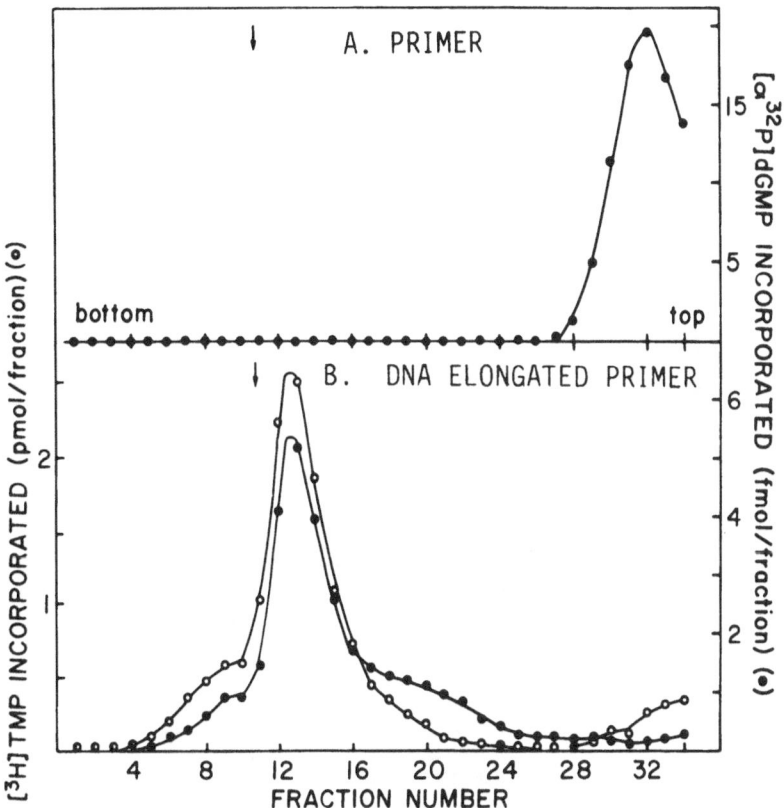

Figure 4.   *Alkaline sucrose sedimentation of primer and DNA*
           elongated primer.
(A) Products of the priming reaction were prepared and isolated as
described in Fig. 3A by gel filtration.   One portion of the isolated
primed-DNA complex was adjusted to 20 mM EDTA, 0.8 M NaCl and 0.2 M
NaOH in 100 μl.   The sample was sedimented in a 10 to 25% alkaline
sucrose gradient containing 2 mM EDTA, 0.8 M NaCl and 0.2 M NaOH
for 6 hrs in a SW50.1 Spinco rotor at 50,000 rpm at 10°.   34
fractions were collected and radioactivity from (α³²P)dGTP was
measured directly (●). (B) Another portion of the isolated DNA
complex was incubated in a second reaction (60 μl) in the absence
of ADP with dATP, dCTP, dGTP and (³H)dTTP as described in Fig. 3
but with 0.6 μg of T4 DNA polymerase.   After 10 min at 30°, the
reaction was adjusted to 20 mM EDTA, 0.8 NaCl and 0.2 M NaOH in
100 μl and sedimented as in part A.   34 fractions were collected
and acid insoluble radioactivity from (α³²P)dGTP (●) and from (³H)-
dTTP (○) was measured. The arrow refers to the position of (¹⁴C)
ØX174 DNA.   The primer was also found to be elongated by DNA
polymerase I or by the combination of DNA polymerase III, dnaZ
protein, DNA EF I and DNA EF III (results not shown).

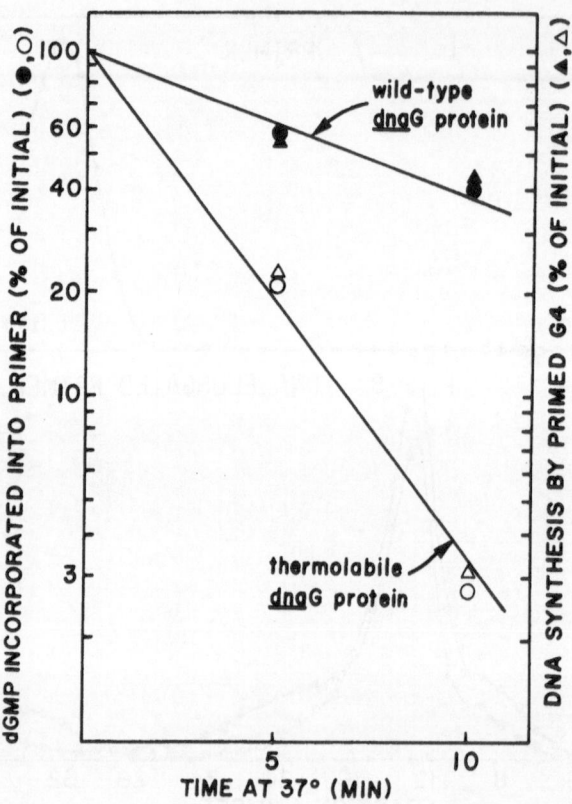

TIME AT 37° (MIN)

*Figure 5. Thermolability of primer synthesis catalyzed by thermo-
labile dnaG protein.   dnaG proteins (wild-type and thermolabile)
were heated (diluted to 300 U/ml in 10% glycerol, 50 mM Tris-HCl
(pH 8.0), 1 mM dithiothreitol and 1 mg/ml bovine serum albumin) for
0, 5 or 10 min at 37°.   The dnaG proteins (10 μl) were then
incubated in first reactions with reaction buffer, G4 DNA, DNA
binding protein, ADP, dTTP and (α³²P)dGTP (as described in Fig. 3)
and subjected to gel filtration (as described in Fig. 3).   ³²P was
measured in the DNA complexes isolated from reactions containing
wild-type dnaG protein (●) and from reactions containing thermolabile
dnaG protein (○).   100% represented 1.02 and 0.80 pmol of dGMP
incorporated with wild-type and thermolabile dnaG protein-DNA
complexes, respectively.   ADP-independent dTMP incorporation in a
second reaction was measured as described in Fig. 3 with DNA
complexes formed using wild-type dnaG protein (▲) and thermolabile
dnaG protein (△).   100% represented 1040 and 712 pmol of total
nucleotide incorporated with wild-type and thermolabile dnaG protein-
DNA complexes, respectively.   When the first reaction contained a
mixture of wild-type and thermolabile dnaG proteins heated 10 min
at 37°, both dGMP incorporation into primer and ADP-independent
DNA elongation were the expected sums.*

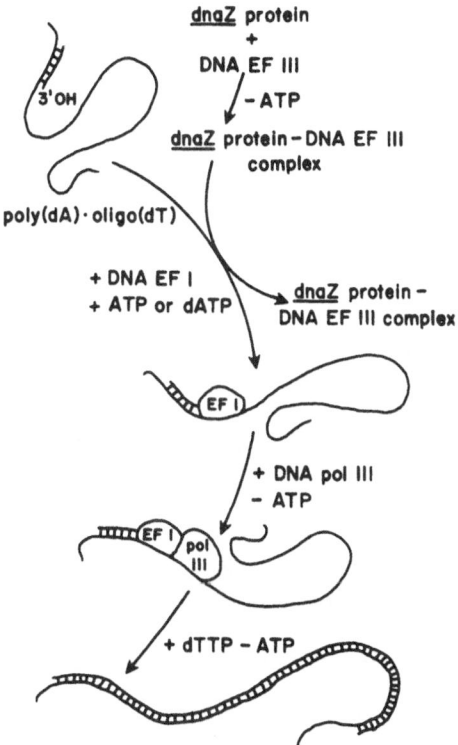

Figure 6. Mechanism of DNA elongation catalyzed by DNA polymerase III in combination with dnaZ protein, DNA EF I and DNA EF III.

gel filtration, DNA EF I was recovered associated with the primed-template. *dnaZ* protein, DNA EF III and ATP were not associated with the DNA. Formation of the DNA EF I-primed-template complex required ATP (or dATP), DNA EF I, DNA EF III, *dnaZ* protein and primed DNA. The function of ATP in this reaction is still unknown. dTMP incorporation by the isolated complex required DNA polymerase III and dNTPs. When the primed template was poly(A) oligo(dT), dTMP incorporation no longer required ATP or dATP as a cofactor. (3) DNA polymerase III binds to the complex of primed template and DNA EF I but not to primed template alone and catalyzes DNA synthesis upon addition of dNTPs.

A reasonable interpretation of the data is that the DNA EF I-primed DNA complex facilitates or stabilizes the binding of DNA polymerase III to the primer. It is not known yet if DNA EF I is associated with DNA polymerase III during chain propagation. In addition it is not known if, under these conditions, DNA binding protein is required to melt secondary structure and allow synthesis of long DNA chains.

DISCUSSION

These multicomponent reactions by which *E. coli* proteins
catalyze single-stranded phage DNA-dependent DNA synthesis are
excellent model systems for studying the biochemical basis of
replication. With purified proteins and defined DNA templates
it has been possible to dissect the overall process into partial
reactions. Since it is likely that the replication proteins
function in the same way to replicate *E. coli* DNA as they do to
replicate phage DNA, information gained from the study of these
biologically important proteins in phage replication will provide
insight into the more complicated process of *E. coli* chromosome
replication.

REFERENCES

(1)   Wechsler, J.A., this volume

(2)   Wickner, W.T., Brutlag, D., Schekman, R. and Kornberg, A.
      (1972) Proc. Nat. Acad. Sci. U.S.A. 69:965-969

(3)   Schekman, R., Wickner, W.T., Westergaard, O., Brutlag, D.,
      Geider, K., Bertsch, L.L. and Kornberg, A. (1972) Proc.
      Nat. Acad. Sci. U.S.A. 69:2691-2695

(4)   Wickner, R.B., Wright, M., Wickner, S. and Hurwitz, J. (1972)
      Proc. Nat. Acad. Sci. U.S.A. 69:3233-3237

(5)   Geider, K. and Kornberg, A. (1974) J. Biol. Chem. 249:3999-4005

(6)   Sigal, N., Delius, H., Kornberg, T., Gefter, M.L. and Alberts,
      B. (1972) Proc. Nat. Acad. Sci. U.S.A. 69:3537-3541

(7)   Wickner, S. and Hurwitz, J. (1976) Proc. Nat. Acad. Sci. U.S.A.
      73:1053-1057

(8)   Wickner, S. and Hurwitz, J. (1974) Proc. Nat. Acad. Sci.
      U.S.A. 71:4120-4124

(9)   Schekman, R., Weiner, J.H., Weiner, A. and Kornberg, A. (1975)
      J. Biol. Chem. 250:5859-5865

(10) Schekman, R., Weiner, A. and Kornberg, A. (1974) Science
      186:987-993

(11) Wickner, S. and Hurwitz, J. (1975) In: DNA Synthesis and Its
      Regulation, eds. Goulian M., Hanawalt, P. and Fox, C.F.
      (W.H. Benjamin, Menlo Park, Ca.) pp. 227-238

(12) Bouche, J.-P., Zechel, K. and Kornberg, A. (1975) J. Biol. Chem. 250:5995-6001

(13) Zechel, K., Bouche, J.-P. and Kornberg, A. (1975) J. Biol. Chem. 250:4684-4689

(14) Wickner, S. (1977) Proc. Nat. Acad. Sci. U.S.A. 74:2815-2819

(15) Wickner, S., Wright, M. and Hurwitz, J. (1973) Proc. Nat. Acad. Sci. U.S.A. 70:1613-1618

(16) Wickner, S. (1976) Proc. Nat. Acad. Sci. U.S.A. 73:3511-3515

(17) Livingston, D.M., Hinckle, D.C. and Richardson, C.C. (1975) J. Biol. Chem. 250:461-469

(18) Wickner, W., Schekman, R., Geider, K. and Kornberg, A. (1973) Proc. Nat. Acad. Sci. U.S.A. 70:1764-1767

(19) Wickner, W. and Kornberg, A. (1974) J. Biol. Chem. 249:6244-6249

(20) Hurwitz, J., Wickner, S. and Wright, M. (1973) Biochem. Biophys. Res. Commun. 51:257-267

(21) Hurwitz, J. and Wickner, S. (1974) Proc. Nat. Acad. Sci. U.S.A. 71:6-10

(22) Wickner, W. and Kornberg, A. (1972) Proc. Nat. Acad. Sci. U.S.A. 70:3679-3683

PLASMID REPLICATION

Robert H. Rownd

Laboratory of Molecular Biology and
Department of Biochemistry
University of Wisconsin
1525 Linden Drive, Madison, Wisconsin 53706, U.S.A.

INTRODUCTION

Bacterial plasmids offer many advantages for the study of
DNA replication and its control. These extrachromosomal genetic
elements have been discovered in over 40 different bacterial genera
and they display a wide variety of phenotypic characteristics (1-4).
Moreover, within an individual taxonomic grouping of bacteria,
such as the Enterobacteriaceae, many different types of plasmids
have been identified and characterized. These plasmids may differ
in their molecular size, DNA base composition, and replication
properties (1-7). As would be expected, larger plasmids are more
diverse in their functions, since they carry a larger number of
genes. There is recent evidence, however, that only a relatively
small region of a plasmid specifies the genetic information for its
autonomous replication (8-11).

The three major classes of plasmids which have been
characterized most extensively are F plasmids, colicinogenic (Col)
plasmids, and drug resistance (R) plasmids (1-7). F plasmids are
sex factors which are capable of promoting their own transfer, the
transfer of other plasmids, or the transfer of the host bacterial
chromosome during bacterial mating. The classical F plasmid which
was originally isolated in the 1950's has a molecular weight of
about 65×10^6 and the same DNA base composition as *Escherichia
coli* (50% G + C), the host in which it was isolated. There is
approximately one copy of an F plasmid per chromosome equivalent
of DNA. Derivatives of F plasmids which have incorporated a
segment of the bacterial chromosome (called F' plasmids) have also
been isolated and extensively characterized. Col plasmids specify
the synthesis of proteins (colicins) which kill certain species of

751

bacteria. Some Col plasmids are also sex factors since they can convert host cells to be genetic donors. This type of Col plasmid usually has a relatively large molecular weight (*ca*. 60-80 x 10^6), a base composition of approximately 50% G + C, and there is usually one copy per chromosome equivalent of DNA. Other Col plasmids are non-transmissible and usually have molecular weights in the range 5-10 x 10^6 and are present in multiple copies per chromosome equivalent of DNA. R plasmids confer multiple drug resistance to host cells and have a number of other phenotypic traits. Many R plasmids are sex factors which are transmissible to other bacteria. These R plasmids typically have molecular weights in the range to 50-70 x 10^6 and there are 1-2 copies per chromosome. They usually confer resistance to several antibiotics (typically 3 to 6). Resistance to each of the antibiotics is determined by specific genes on the R plasmid. Non-transmissible R plasmids have also been isolated which confer resistance to only one of two anti- biotics, have molecular weights in the range 5-10 x 10^6, and are present in multiple copies (10-20) per chromosome equivalent of DNA. In addition to these 3 classes, bacterial plasmids have been isolated in many genera of bacteria which confer on their host such diverse properties as pathogenicity (Ent plasmids), hemolysin production (Hly plasmids), specific surface structures and antigens, and catabolic systems (Cam, Sal, and Oct plasmids). Plasmids have also been implicated in the production of antibiotics, plant pathogenicity, and in nitrogen fixation in the genus *Rhizobium* (3).

There is considerable diversity among the plasmids which are included in the classes mentioned above. Genetic experiments have shown that closely related plasmids are unable to coexist in the same host, a phenomenon which is referred to as "incompatibility" (1,12). Unrelated plasmids can be harbored by the same host simultaneously and thus are "compatible". The mechanism of incompatibility between related plasmids is presently unknown, although it seems very likely that it is related to the control of their replication. Approximately 20 "incompatibility groups" have been identified among plasmids which confer drug resistance, that is, that would be classified as R plasmids (12). Plasmid incompatibility studies have been of fundamental importance in the study of the epidemiology and classification of plasmids. The existence of many different incompatibility groups indicate that closely related plasmids may share many features in the mechanism which control their replication. Conversely the large number of incompatibility groups also indicates that different plasmids utilize a variety of mechanisms for the control of their replication.

The R plasmids of the FII incompatibility group have a number of characteristics which make them particularly interesting subjects for studying the control and mechanism of DNA replication in bacteria (5-7, 13, 14). These R plasmids are composite structures consisting of a resistance transfer factor (RTF) which mediates the

transfer of the plasmid during bacterial mating and an r-determinants component which harbors the majority of the drug resistance
genes. Each of these components appears to be capable of autonomous replication under certain growth conditions and thus each
must contain at least one origin of replication (13-15). When the
host cells are cultured in medium containing appropriate drugs,
the r-determinants component can undergo extensive gene amplification and recombination to form R plasmids consisting of one RTF
component which is attached to multiple, tandem copies of the r-
determinants component (13, 16, 17). This phenomenon is interesting
both as an example of gene amplification and as a model system for
chromosomes containing multiple origins of replication, which may
be analogous in structure to eukaryotic chromosomes.

Replication Properties of Representative
Bacterial Plasmids

Molecular characteristics of plasmid DNA. As mentioned
previously, different bacterial plasmids have been found to have
different replication properties. In general, plasmids which are
included within the same incompatibility group usually have similar
replication properties. When a number of replication properties
are considered, different bacterial plasmids may broadly be
classified into two groups (Table I), although a particular plasmid
may not necessarily share all of the characteristics of these two
groups. The most extensively characterized representative of
group I plasmids are ColEl plasmid (18), Clo DF13 plasmid (19), and
a number of non-transmissible R plasmids (3). F plasmids and a
number of transmissible R plasmids of the FII incompatibility group
are representative of group II plasmids. Group I plasmids are not
sex factors and have molecular weights in the range 4-8 x 10^6 and
thus contain about 6-12 genes. There are usually multiple copies
of these plasmids per host chromosome which are selected at random
for replication during the duplication of the multicopy plasmid
pool (3, 18, 19, 20). Group II plasmids are usually transmissible
and hence are sex factors. They are considerably larger in size,
and typically have molecular weights in the range of 40-60 x 10^6
(5). As much as 20% of the plasmid DNA may carry genes involved
with transfer functions. In the case of the F plasmid and FII
incompatibility group R plasmids, the transfer genes are contiguous
and form a single operon (21). There are usually only 1-2 copies
of group II plasmids per host chromosome; however, there may be
multiple copies of these plasmids per cell since bacterial cells
may harbor more than one chromosomal equivalent of DNA. Although the
copy number of these plasmids is more stringently controlled,
individual copies of transmissible R plasmids are also selected at
random during the duplication of the multi-copy pool of plasmids
during the cell division cycle (22, 23).

Table I

Replication Properties of Bacterial Plasmids

	Group I	Group II
Representative Plasmids	ColE1	F Plasmids
	Non-transmissible R plasmids	Transmissible R plasmids
Sex Factors	No	Yes
Molecular Size (daltons)	$5\text{-}10 \times 10^6$	$25\text{-}75 \times 10^6$
Copies per Chromosome	10-40	1-2
Selection for Replication	Random	Random (R plasmids)
Required for Replication		
Protein Synthesis	No	Yes
RNA Synthesis	Yes	No
dna A	No	No
dna B	Partial dependence	Yes
dna C/D	Yes	Yes
dna E (Pol III)	Partial dependence	Yes
dna F	Yes	Yes
dna G	Yes	Yes
pol I	Yes	No
Association with Folded Chromosome	Majority of plasmid DNA unassociated	Majority of plasmid DNA associated

Macromolecular synthesis required for plasmid replication.
Studies on bacterial chromosome replication have shown that protein
synthesis is required for the initiation of a round of replication,
but not for the completion of rounds of replication which are in
progress (24). These observations suggest that one or more proteins
are required for the initiation of chromosome replication. At the
present time, very little is known about the function of these
"initiator proteins".

When protein synthesis is inhibited in E.coli harboring the
plasmid ColEl (7, 18) [or other plasmids which would be included in
group I (3, 26)] by addition of chloramphenicol to the growth
medium, plasmid replication continues for about 10 hours whereas
the host chromosome only completes the round of replication which
is in progress. As a result, the number of plasmid copies per
chromosome may be increased 40-60 fold. Rifampicin, an inhibitor
of RNA polymerase, inhibits the replication of these plasmids in
the absence of protein synthesis (18, 25, 26), suggesting that RNA
synthesis is required as an essential feature of the replication
process.

It is interesting to note that when segments of ColEl plasmid
are deleted to form smaller (mini) plasmids or when additional DNA
segments are inserted into ColEl using restriction endonucleases
and subsequent ligation, the amount of plasmid DNA per cell after
replication in the presence of chloramphenicol is approximately
the same irrespective of the size of the plasmid (45% of the total
DNA of the cell) (18, 27). Since the plasmid derivatives of
ColEl have different sizes, there must be a larger number of copies
of the smaller plasmids after replication in the presence of
chloramphenicol. Moreover, there is a larger number of copies of
mini ColEl per cell than of ColEl plasmid from which mini ColEl
was derived. Thus, reduction of the size of the plasmid by deletion
does not result in a corresponding reduction in the amount of
plasmid DNA per cell. These findings suggest that the host cells
may regulate the total mass of plasmid DNA per cell rather than
just the number of plasmid copies per cell.

When protein synthesis is inhibited in bacterial cells
harboring either F plasmids (28) or transmissible R plasmids (14,
29, 30), there is only a 20-50% increase in the amount of plasmid
DNA. Under these conditions there is a 40-50% increase in the
amount of chromosomal DNA, owing to the completion of rounds of
chromosome replication in progress at the time protein synthesis
was arrested. As a specific example, there is a 25% increase in
the amount of R plasmid NR1 DNA during inhibition of protein
synthesis in either E.coli or Proteus mirabilis (14, 29, 31). The
residual chromosomal replication during inhibition of protein
synthesis usually requires 60-90 minutes for completion, whereas
the residual replication of the R plasmid DNA is usually complete

within 10-15 minutes (32). Since only a small fraction of the R
plasmids in the cells should be in the process of replication at
the time protein synthesis is arrested, the 25% increase in the
amount of R plasmid DNA indicates that new rounds of R plasmid
replication are initiated in the absence of protein synthesis.
Taken together, these findings suggest that one or more proteins
are required for the initiation of R plasmid replication and that
a small pool of "initiator protein" is present in the cells during
exponential growth (14, 29, 31). The utilization of this initiator
protein would account for the residual plasmid replication which
takes place after inhibition of protein synthesis. The residual
replication of the R plasmid *NR1* in the absence of protein synthesis
is not inhibited by rifampicin in *P.mirabilis* (32). Thus,
rifampicin-sensitive RNA synthesis is not required for R plasmid
NR1 replication. Either this R plasmid does not require RNA
synthesis for replication, or uses a rifampicin-insensitive RNA
polymerase for its replication.

The R plasmid *NR1* DNA which is replicated in the absence of
protein synthesis in *E.coli* is not converted to the covalently
closed circular (CCC) form unless protein synthesis is allowed to
begin again, whereas *NR1* DNA in *P.mirabilis* or F' lac DNA in
E.coli which is replicated in the absence of protein synthesis is
converted to the CCC form (32). The non-CCC R plasmid DNA
molecules which are formed in *E.coli* may represent molecules whose
replication is arrested just prior to the completion of
replication.

Plasmid replication in *E.coli* DNA replication mutants. As
discussed in more detail in the article by Wechsler in this volume,
many mutants have been isolated which affect the replication of the
host bacterial chromosome (33, 34). These mutants are generally
temperature-sensitive mutants in which the mutant gene product is
functional when the cells are cultured at the permissive temperature,
but is inactivated when the growth temperature is increased to the
restrictive temperature. Two types of temperature-sensitive
mutants which effect host chromosomal replication have been
isolated (33). One type seems to affect only the initiation of
chromosome replication; at the restrictive temperature, the rounds
of chromosome replication which were in progress are completed,
but the initiation of the new rounds of replication is inhibited.
The second type of host mutants are elongation mutants in which DNA
chain elongation is arrested immediately when the cells are cultured
at the restrictive temperature. Of the initiation mutants in
E.coli, dnaA is not required for the replication of the small non-
transmissible plasmids like ColE1 (35) or the larger transmissible
F plasmids and R plasmids (32, 36). However, dnaC is required for
the initiation of the replication of all of these plasmids (32,
35, 36). Essentially all of the plasmids which have been examined
require the dnaB gene product for chain elongation, although ColE1

shows only a partial dependence on dnaB gene function. The replication of all plasmids which have been examined have been found to require the dnaF and dnaG gene products for chain elongation (Table I). Several other types of initiation mutants (dnaI and P) and elongation mutants (dnaZ) have been isolated more recently but plasmid replication has not as yet been examined in these mutants except for Clo DF13 which requires all of these gene products for replication (19).

Like the host bacterial chromosome, F plasmids and transmissible R plasmids require DNA polymerase III (which is temperature-sensitive in dnaE mutants) for replication, but have no dependence on DNA polymerase I. However, ColEl and a number of the small, non-transmissible plasmids require DNA polymerase I for replication and usually have a partial dependence on DNA polymerase III since their replication is inhibited to some degree in dnaE mutants (Table I).

The study of the replication of a variety of bacterial plasmids in mutants which affect the replication of the host bacterial chromosome has shown that much of the replication machinery of bacterial cells is shared by both the bacterial chromosome and various plasmids (Table I).

Plasmid DNA-protein relaxation complexes. A number of plasmids, including Col plasmids, R plasmids, and F plasmids have been isolated from *E.coli* in the form of a covalently closed circular (CCC) DNA-protein relaxation complex which can be converted to a nicked circular (relaxed) DNA form by treatment with agents which denature or degrade proteins (7, 37). The relaxed molecules contain a single nick in a unique strand and at a unique site of the DNA duplex. At one time it was hypothesized that the nick involved in the relaxation of the DNA may be involved in the initiation of plasmid replication. However, more recent experiments have indicated that the relaxation nick does not occur at the plasmid origin of replication (38), but probably occurs at the origin of transfer of the single strand of plasmid DNA which is transferred to recipient cells during bacterial mating (39, 40).

Association of plasmids with the folded chromosome. Under special lysis conditions, the *E.coli* chromosome can be isolated in a condensed or folded state which consists of 20-50 domains of DNA within which both strands of the duplex are covalently continuous (41, 42). RNA may be involved in the stabilization of this "folded chromosome". The folded chromosome appears to be attached to the bacterial membrane. Bacterial plasmids which are present in low copy number in the cells, such as F plasmids and transmissible R plasmids, appear to be associated with the folded chromosome when it is sedimented in a sucrose gradient, whereas only a small fraction of the DNA of plasmids present in high copy number such as ColEl or non-transmissible R plasmids co-sediment with the folded chromosome (43).

Mutants of Plasmid Replication

In addition to the host mutants which affect both chromosome and plasmid replication, other mutants have been isolated which appear to undergo normal chromosome replication, but in which the replication of specific plasmids or groups of plasmids is altered (Table II). One example of this type of mutant is Plasmid copy mutants in which there is an increase in the number of copies of the plasmid per cell. F plasmid (44), R plasmid (31, 45), and Colicin plasmid (Clo DF13) (46) copy mutants have been isolated anc characterized. The majority of copy mutants are mutants of the plasmid itself, since the increased number of plasmid copies remains after the transfer to another host (31, 45, 46), although host mutants which effect plasmid copy number have also been isolated (44). Two types of temperature-sensitive mutants of plasmid replication have been isolated. One type is a mutant of the plasmid itself, since the replication of the plasmid remains temperature-sensitive after transfer to a new host strain (47). Temperature-sensitive host mutants which effect the replication of specific plasmids or groups of plasmids have also been isolated (48). In a detailed study of temperature-sensitive host mutants for plasmid replication three groups of mutants were isolated which were identified by their inability to replicate all of the plasmids which were examined (group I), their inability to replicate ColEl and F-like plasmids (group II), or their inability to replicate ColEl (group III). A host mutant which falls into group III is a mutant which is temperature-sensitive for DNA polymerase I since this enzyme is required for the replication of Col-El (Table I). It is expected that the biochemical characterization of these various types of mutants will eventually help to elucidate the role of each of these gene products in plasmid replication.

Isolation and Characterization of
Replicating Plasmid DNA

The small size of plasmid DNA is advantageous for the characterization of replicating molecules since it minimizes the artifactual breakage of the DNA during isolation and handling. However, it also introduces the problem that only a small fraction of the plasmid molecules will be in the process of replication at any instant in a random culture of bacterial cells. This is because it takes only a small fraction of the cell division cycle to duplicate a replicon which is only a few percent (or less) of the size of the host bacterial chromosome (20, 22). Two methods have

Table II

Mutants of Plasmid Replication

1.   Copy mutants: F plasmids, R plasmids, Colicin plasmids

2.   ts Plasmid mutants: F plasmids, R plasmids, Colicin plasmids

3.   ts Host mutants: Group I   - unable to maintain all
                                         plasmids tested (ColEl, F'lac,
                                         ColV2, ColIb, Rl, R64)

                            Group II   - unable to maintain ColEl and
                                         F-type plasmids (F'lac, ColV2,
                                         Rl), but can maintain I-type
                                         plasmids (ColIB and R64)

                            Group III - unable to maintain ColEl

been used to increase the fraction of replicating plasmid molecules in bacterial cells. There is a substantial increase in the fraction of replicating plasmid DNA in *Proteus mirabilis* and *Escherichia coli* when the rate of DNA chain elongation is decreased by limiting the concentration of DNA precursors (13, 14, 49). Substrate limitation can be achieved by growth of a thymine auxotroph in medium containing limiting concentrations of thymine or by the addition of hydroxyurea (an inhibitor of ribonucleoside diphosphate reductase) to the growth medium. Under these conditions, there is an increase in the time required for a DNA replication fork to traverse the p lasmid DNA from the origin to the terminus of replication (S period). Since it requires a larger fraction of the division cycle to replicate the plasmid DNA, more plasmids are in the process of replication during substrate limitation. Bacterial mass synthesis is not significantly affected during the first few hours of substrate limitation (49, 50). It appears that the initiation of plasmid replication continues at near the normal rate. Another method which has been used to increase the fraction of replicating molecules is to first subject a thymine auxotroph to a period of thymine starvation, and then to restore a low concentration of thymine to the growth medium (which should be equivalent to thymine limitation) for a brief period (51, 52). Presumably under these conditions, inhibitor proteins continue to be synthesized during thymine starvation, so that a substantial amount of initiation plasmid replication occurs as soon as thymine is restored to the growth medium. It should be noted that both methods used to increase the fraction of replicating plasmid molecules utilize non-physiological growth conditions which may affect the pattern of subsequent plasmid replication.

The majority of replicating plasmid molecules have a buoyant
density intermediate between the covalently closed circular (CCC)
and the nicked circular form of the plasmid DNA in an ethidium
bromide-cesium chloride density gradient. It should be noted,
however, that there have usually been no attempts to evaluate
the recovery of replicating plasmid DNA in these gradients, so
that the intermediate density, CCC replicating plasmid DNA
represents an undetermined fraction of the plasmid DNA in the cells
(51-53). In any event, these observations indicate that plasmid
molecules replicate as CCC or supercoiled structures, as previously
observed for replicating molecules of mitochondrial and SV40 and
polyoma DNA in eukaryotes (54). The newly synthesized DNA in pulse-
labeled replicating plasmid molecules is considerably smaller in
size than the entire plasmid DNA and increases in size with
increasing duration of the pulse (55). These results indicate that
the newly synthesized (daughter) strands of replicating plasmid
molecules are not attached by covalent bonds to the parental DNA.
Although a large fraction of non-replicating ColEl and R plasmid
molecules exist in the form of DNA-protein relaxation complexes,
the replicating form of these plasmid molecules are not susceptible
to relaxation (nicking) under the same conditions (53, 55). This
suggests that either the replicating plasmid molecules do not exist
in the form of a DNA-protein relaxation complex as do the non-
replicating forms, or, if they do, the replicating molecules are
not susceptible to relaxation under these conditions.

Several novel forms of replicating plasmid molecules have
been observed which may represent plasmid replication intermediates.
In *E.coli*, recently replicated molecules of several plasmids appear
to have a lower number of superhelical turns than the non-
replicating form of the DNA (56, 57). In *P. mirabilis* a significant
fraction of replicating molecules of a transmissible R plasmid and
ColEl plasmid have a density in an ethidium bromide-cesium chloride
gradient which is lighter than the nicked form of the plasmid DNA
(53, 55). Moreover, almost all of the replicating molecules of
these two plasmids appear as "rosette-like" structures in which the
DNA strands are extensively intertwined when viewed in an electron
microscope. The molecular basis for these unusual properties is
presently unknown. Since these unusual forms are observed only for
replicating molecules or recently replicated molecules, they may
represent intermediates in plasmid replication.

Replicating molecules of plasmid DNA have been found to be
θ-type (Cairns) circular structures in which both daughter branches
are joined to the non-replicated duplex DNA at the two branch
points. Analysis of the position of the two branch points in
replicating molecules using either denaturation mapping or specific
cleavage of the plasmid DNA at specific sites using various
restriction endonucleases has made it possible to determine the
origin and direction of replication of plasmid molecules. A number

of plasmids have been found to undergo unidirectional replication
from a fixed origin at which replication is initiated, including
ColEl (18) and a small non-transmissible R plasmid pSC101 (which
shares most of the characteristics of group II plasmids rather
than group I) (57). These two plasmids have also been joined
together at their EcoRI cleavage sites by ligation. Depending on
the conditions of cell culture, this composite plasmid (pSC134)
is able to initiate replication at either of the origins of the
component plasmids and to undergo unidirectional replication (57).
Several plasmids have been found to undergo bidirectional replication from a fixed origin, including the cloned replicator fragment
of the F plasmid (mini F plasmid) (58).

Several plasmids have been found to contain more than one
origin of replication. Although the R plasmid R6K uses a single
origin (59), a deletion mutant of R6K has been found to initiate
replication from two distinct origins (51, 60). Usually only one
origin is used to initiate replication, although in about 5% of
the replicating molecules both origins were used simultaneously.
More recent experiments indicate that there may also be a third
origin on R6K (60). In the case of the mini F plasmid the origin
of replication is not coincident with a region of the plasmid which
has been identified as the incompatibility region (61).

Composite R plasmids which consist of RTF and r-determinants
components also contain multiple origins of replication (14, 15,
62, 63). Denaturation mapping studies of the R plasmid *NR1*
revealed that replicating molecules contained at least two origins
of replication; one in the RTF component and the other in the r-
determinants component (13-15). More recently, analysis of
replicating *NR1* molecules having smaller extents of replication
using EcoRI and Sal I restriction endonucleases has revealed two
additional origins of replication on the composite R plasmid. Two
of these four origins are located in the r-determinants component
on EcoRI fragments M and L. The other two origins are located in
Sal I fragments D and E, respectively, which are located within
EcoRI fragment B (Fig. 1) (62, 63). In cloning experiments it has
been found that EcoRI fragments B of FII incompatibility group R
plasmids is capable of independent replication in the absence of
any of the other EcoRI fragments (8). It is also possible to
isolate small plasmids (miniplasmids) derived from copy mutants of
FII composite R plasmids which are capable of autonomous replication (10, 11). The smallest of these miniplasmids contain only
about 2.5×10^6 daltons of EcoRI fragment B (the region between the
EcoRI cleavage site between fragments B and H and the cleavage
site between Sal I fragments D and E) (Fig. 1) (64). Other small
miniplasmids have been isolated which contain only the segment of
the RTF component just described and all or part of the r-determinants component (64).

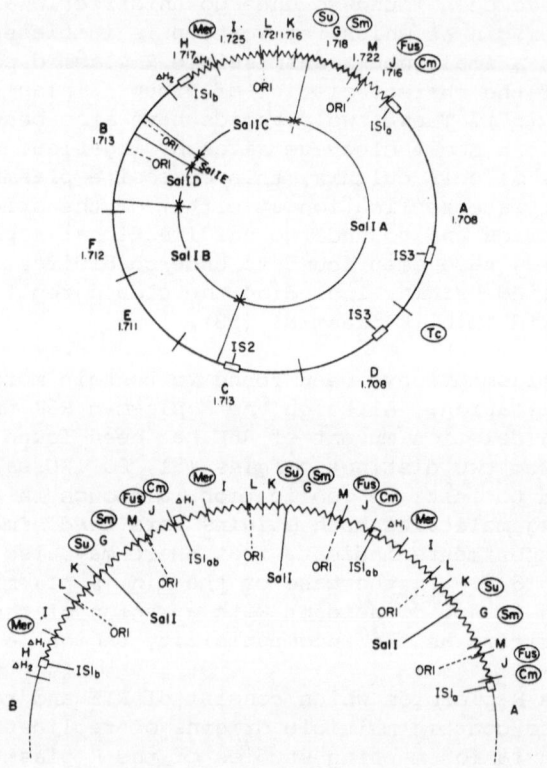

Figure 1. EcoRI and Sal I restriction endonuclease map of R
plasmid NR1 DNA. (Top) Composite NR1 DNA; (Bottom) Poly r-determinant NR1 DNA molecule containing three tandem copies of the
re-determinants component (the RTF region of the molecule is not
shown). The 13 EcoRI fragments of NR1 DNA (designated A – M) are
indicated on the outside of the circular DNA molecule (67, 68).
The numbers below the letters refer to the buoyant densities (in
g/ml) of the individual EcoRI fragments in a CsCl gradient (68,
70). The location of the drug resistance genes was determined
in cloning experiments with the EcoRI fragments of NR1 DNA (75).
The four origins of replication which have been identified on a
copy mutant of NR1 called pRR12 are designated ORI (62,63). The IS
elements of FII composite R plasmids were located on the EcoRI
fragments as described previously (67). The location of the Sal I
cleavage sites was determined by C. Barton in this laboratory.

Thus, the larger transmissible plasmids may contain multiple
origins of replication, some of which may not be normally used for
the initiation of replication. However, these "silent" origins
may be activated when other origins are deleted. It is also
possible that silent origins may be activated under different

conditions of cell growth. Deletion mutants and miniplasmids of
the FII incompatibility group R plasmid *NR1* are presently available
which contain one or more of the four origins of replication which
have been identified on the R plasmid (64). The characterization
of these plasmids should make possible an analysis of the interaction between various origins of replication in the initiation of
replication.

R Plasmid Drug Resistance Gene Amplification

The R plasmids of the FII incompatibility group have several
characteristics which make them particularly interesting for
studying the control and mechanism of DNA replication in bacteria.
These R plasmids are usually composite structures consisting of
RTF and r-determinants components. As mentioned previously, each
of these components appears to have origins of replication and each
component is capable of autonomous replication under certain
growth conditions. For example, when *P. mirabilis* harboring the
composite R plasmid *NR1* is cultured in drug-free medium, the
composite R plasmid DNA appears as a single satellite band of
density 1.712 g/ml to the *P. mirabilis* chromosomal DNA (1.700 g/ml)
(Fig. 2A). However, if the cells are cultured in medium containing
drugs to which the re-determinants component confers resistance,
there is an increase in the density and the amount of the R plasmid
DNA, as illustrated in Fig. 2. Under these conditions R plasmids
are formed which consist of a single copy of the RTF and multiple,
tandem copies of the r-determinants component (poly-r-determinant
R plasmids) (Fig. 3). Autonomous poly-r-determinants consisting of
multiple copies of the r-determinants component are also formed.
The recombination of multiple copies of r-determinants (1.718 g/ml)
with an RTF (1.710 g/ml) increases the density and the size of the
R plasmid DNA. The density of poly-r-determinant R plasmids
containing a large number of copies of r-determinants approaches
1.718 g/ml as a limiting value, since most of the mass of the R
plasmid DNA is due to r-determinants (Fig. 2C). The extra copies of
r-determinants required for the formation of these poly-r-determinant molecules appear to be formed by their autonomous replication
during growth of the cells in medium containing appropriate drugs
(6, 13, 16, 17, 29).

The amplification of r-determinants in *P. mirabilis* is a
reversible process. Subsequent growth of cells harboring poly-r-
determinant molecules in drug-free medium results in a decrease in
the percentage of the amplified (1.718 g/ml) r-determinants DNA
(Fig. 2). After prolonged growth in drug-free medium, the R plasmid
DNA returns to the basic composite form which consists of a single
copy of the RTF and a single copy of the r-determinants component.
These observations suggest that poly-r-determinant molecules are
unstable, perhaps owing to the extensive homology of the repeated

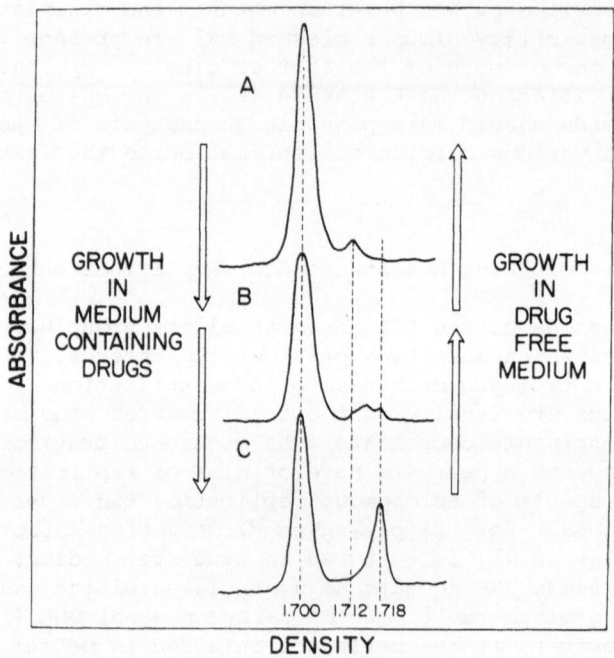

Figure 2. *Systematic changes in the DNA density profiles of the R plasmid NR1 in P. mirabilis, depending on whether the host cells are cultured in drug-free medium or medium containing appropriate drugs. More detailed analyses of the changes in the R plasmid DNA density profiles are described elsewhere (13, 16, 17, 29, 65, 66).*

r-determinants sequences. Since there is no difference in the growth rates of cells harboring R plasmids in the 1.712 g/ml form (Fig. 2A) and cells harboring R plasmids in the 1.718 form (Fig. 2C) (65), it appears that r-determinants undergo only a limited amount of replication after dissociation from poly-r-determinant molecules and that the extra copies of r-determinants are diluted passively by cell division after dissociation (6, 13, 16, 17, 29).

The reversible amplification of the r-determinants component of composite R plasmids indicates a remarkable plasticity in the control of the replication of this genetic element (16, 17, 66). During growth in medium containing appropriate drugs, r-determinants apparently must be able to replicate, at least transiently, more frequently than the RTF component in order to obtain amplification of the drug resistance genes. During subsequent growth of cells harboring amplified r-determinants in drug-free medium r-determinants are diluted from the cells and thus must replicate less

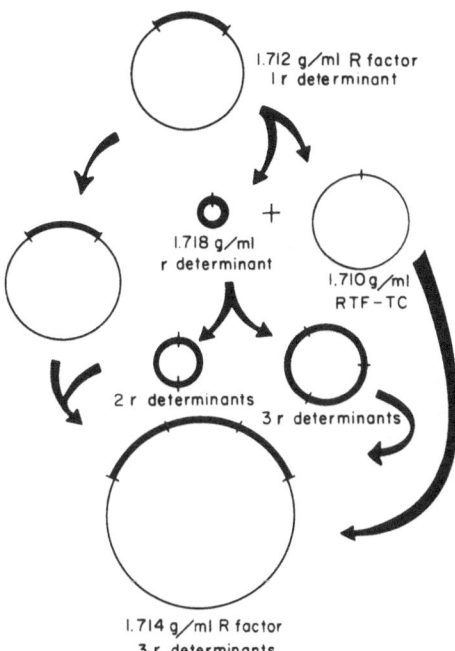

1.712 g/ml R factor
1 r determinant

1.718 g/ml
r determinant

+

1.710 g/ml
RTF-TC

2 r determinants

3 r determinants

1.714 g/ml R factor
3 r determinants

Figure 3. *Schematic diagram illustrating the dissociation and reassociation of the RTF and r-determinants in P. mirabilis and the density increase that accompanies the incorporation of multiple copies of r-determinants into individual R plasmids to form poly-r-determinant R plasmids. After the incorporation of a large number of copies of r-determinants, the poly-r-determinant R plasmid DNA would have essentially the same density as r-determinants DNA (1.718 g/ml), since most of the mass of the DNA is due to r-determinants.*

frequently than the RTF. Thus, mechanisms must exist for "turning on" and "turning off" the r-determinants replication system. Moreover, poly-r-determinant molecules must contain multiple origins of replication in the RTF component and the tandem sequences of r-determinants (Fig. 1). Thus, the dissociation and reassociation of composite R plasmids is interesting both as an example of gene amplification and as a model system for chromosomes containing multiple origins of replication, which may be analogous in structure to eukaryotic chromosomes.

In all of the experiments with the composite R plasmid *NR1* in *P. mirabilis* the entire r-determinants component is excised, amplified, and incorporated in poly-r-determinant molecules (17, 67, 68). However, with another FII incompatibility group R plasmid

(NR84) amplification and formation of tandem sequences of at least
five different segments of the R plasmid have been observed (17,
66, 68). This indicates that there may be multiple dissociation or
excision sites on NR84 which mediate drug resistance gene
amplification. At least two of the segments which can be amplified
do not contain any of the four origins of replication which have
been identified on the closely related composite R plasmid *NR1*. If
replication is necessary for gene amplification, NR84 may have
additional "silent" origins of replication which may be activated
under particular growth conditions in order to obtain gene
amplification.

Although the r-determinants component of FII composite R
plasmids can be amplified in *P. mirabilis* when the host cells are
cultured in the presence of appropriate drugs, r-determinants do
not appear to be able to replicate efficiently, if at all, in
Salmondella typhimurium which is another member of the Enterobacteriaceae. In *S. typhimurium* there is a relatively high frequency
of loss of the drug resistances conferred by r-determinants,
presumably due to R plasmid dissociation and the loss of r-determinants from the cells (68, 69). Thus, FII incompatibility group R
plasmids exhibit remarkably different behavior in different genera
of Enterobacteriaceae especially when the control of replication
of the r-determinants component is considered.

The base compositions of the individual EcoRI fragments of
NR1 DNA have been determined by measuring the buoyant density of
these fragments in cesium chloride gradients. There is considerable
heterogeneity in the density and, hence, in the base composition,
of the various fragments (70). It is interesting to note that there
is an origin of replication on EcoRI fragment L in the r-determinants which is in a high density (high G + C) region of the r-
determinants (Fig. 1). There is also an origin of replication on
EcoRI fragment M which is in a lower density region of the r-
determinants. Other experiments in this laboratory have indicated
that transcription occurs in opposite directions in these two
regions of the r-determinants component. These considerations
suggest that complex plasmids such as the R plasmid *NR1* may have
resulted from the fusion of smaller plasmids which, at least at
one time, were independent replicons. These considerations may
also explain why different segments of these plasmids may dissociate,
amplify, and reassociate to form plasmids containing repeated tandem
sequences of specific segments. The amplified segments may
utilize the replication functions of the progenitor plasmids which
may be activated during growth in medium containing appropriate
drugs.

Most studies on drug resistance genes amplification have been
with FII incompatibility group R plasmids in *P. mirabilis*. Recently,
however, similar and perhaps related observations on gene

amplification have been observed in *Streptococcus* (71),
Pseudomonas (72) and *E.coli* (73).   It will be interesting to see
whether plasmid gene amplification is of relatively widespread
occurrence. Gene duplications and the formation of tandemly
repeated sequences have also been observed for segments of bacterial
chromosomes (74).   It is possible that these chromosomal rearrangements may share similar mechanisms with the amplification
of drug resistance genes.

ACKNOWLEDGEMENTS

I thank Dr. David D. Womble for his discussions and critical
comments during the preparation of this manuscript.   The research
carried out in my laboratory was supported (in part) by a U.S.
Public Health Service Research Grant (GM 14398).

REFERENCES

1.   Novick, R. (1969) Bacteriol. Rev. 33:210-264

2.   Meynell, G. (1973) Bacterial Plasmids, M.I.T. Press,
     Cambridge, Mass.

3.   Falkow, S. (1975) Infectious Multiple Drug Resistance.
     Pion, Ltd., London.

4.   Lewin, B. (1977) Gene Expression-3. Plasmids and Phages.
     John Wiley & Sons, New York.

5.   Clowes, R. (1972) Bacteriol. Rev.   36:361-405

6.   Davies, J.E. and Rownd, R.H. (1972) Science 176:758-768

7.   Helinski, D.R. (1973) Ann. Rev. Microbiol.   27:437-470

8.   Timmis, K., Cabello, F. and Cohen, S.N. (1975) Proc. Natl.
     Acad. Sci. USA 72:2242-2246

9.   Lovett, M.A. and Helinski, D.R. (1976) J. Bacteriol.   127:
     982-987

10. Mickel, S. and Bauer, W. (1976) J. Bacteriol. 127:644-655

11. Taylor, D.P., Greenberg, J. and Rownd, R.H. (1977) J.
     Bacteriol. 132:

12. Datta, N. (1975) In: Microbiology - 1974 (ed. D. Schlessinger)
     American Society for Microbiology, Washington, D.C. pp9-15

13. Rownd, R.H., Perlman, D. and Goto, N. (1975) In:
     Microbiology - 1974 (ed. D. Schlessinger) American Society
     for Microbiology, Washington, D.C. pp76-94

14. Rownd, R.H., Perlman, D., Womble, D.D., Taylor, D.P., Morris,
     C.F. and Hill, W.E. (1975) In: Microbial Drug Resistance
     (ed. S. Mitsuhashi and H. Hashimoto) University of Tokyo
     Press, Tokyo. pp27-50

15. Perlman, D. and Rownd, R.H. (1976) Nature 259:281-284

16. Rownd, R.H., Perlman, D., Hashimoto, H., Mickel, S.,
     Appelbaum, E. and Taylor, D. (1973) In: Cellular Modification
     and Genetic Transformation by Exogenous Nucleic Acids (eds.
     R.F. Beers, Jr. and R.C. Tilgham) The Johns Hopkins University
     Press, Baltimore. pp115-128

17. Rownd, R.H., Goto, N., Appelbaum, E.R. and Perlman, D. (1975)
     In: Microbial Drug Resistance (eds. S. Mitsuhashi and H.
     Hashimoto) University of Tokyo Press, Tokyo. pp3-25

18. Helinski, D., Lovett, M., Williams, P., Katz, L., Collins, J.,
     Kupersztoch-Portnoy, Y., Sato, S. Leavitt, R., Sparks, R.,
     Hershfield, V., Gainey, D. and Blair, D. (1975) In: DNA
     Synthesis and Its Regulation. (eds. M. Goulian and P.
     Hanawalt) W.A. Benjamin, Inc., Menlo Park, California. pp514-
     536

19. Veltkamp, E. and Nijkamp, H.J. (1977) In: Plasmids: Medical
     and Theoretical Aspects. Springer-Verlag, Berlin. pp203-
     212

20. Bazaral, M. and Helinski, D.R. (1970) Biochemistry 9:399-406

21. Achtman, M. and Helmuth, R. (1975) In: Microbiology - 1974
     (ed. D. Schlessinger) American SOciety for Microbiology,
     Washington, D.C. pp95-103

22. Rownd, R. (1969) J. Mol. Biol. 44:387-402

23. Terawaki, Y. and Rownd, R. (1972) J. Bacteriol. 109:492-
     498

24. Lark, K.G. (1969) Ann. Rev. Biochem. 38:569-604

25. Clewell, D.B., Evenchik, B. and Cranston, J.W. (1972)
     Nature New Biol. 237:29-31

26. Veltkamp, E., Barendsen, W. and Nijkamp, H.J. (1974) J.
     Bacteriol. 118:165-174

27. Hershfield, V., Boyer, H.W., Chow, L. and Helinski, D.R.
     (1976) J. Bacteriol. 126:447-453

28. Kline, B.C. and Helinski, D.R. (1971) Biochemistry 10:
     4975-4980

29. Rownd, R., Kasamatsu, H. and Mickel, S. (1971) Ann. N.Y.
     Acad. Sci. 182:188-206

30. Terawaki, Y., Nakaya, R. and Rownd, R. (1974) J. Bacteriol.
     117:687-695

31. Morris, C., Hashimoto, H., Mickel, S. and Rownd, R. (1974)
     J. Bacteriol. 118:855-866

32. Womble, D.D. and Rownd, R.H. (1978) Submitted to Plasmid

33. Hirota, Y., Mordoh, J., Scheffler, I. and Jacob, F. (1972)
     Fed. Proc. 31:1422-1427

34. Gefter, M. (1975) Ann. Rev. Biochem. 44:45-78

35. Collins, J., Williams, P. and Helinski, D.R. (1975) Molec.
     Gen. Genet. 136:273-289

36. Arai, T. and Clowes, R. (1975) In: Microbiology - 1974
     (ed. D. Schlessinger) American Society for Microbiology,
     Washington, D.C. pp141-155

37. Helinski, D.R. and Clewell, D.B. (1971) Ann. Rev. Biochem.
     40:899-942

38. Tomizawa, J., Ohmori, H. and Bird, R.E. (1977) Proc. Natl.
     Acad. Sci. USA 74:1865-1869

39. Dougan, G. and Sherratt,D. (1977) Molec. Gen. Genet. 151:
     151-160

40. Inselburg, J. (1977) J. Bacteriol. 132:332-340

41. Stonington, O. and Pettijohn, D. (1971) Proc. Natl. Acad.
     Sci. USA 68:6-9

42. Worcel, A. and Burgi, E. (1972) J. Mol. Biol. 71:127-147

43. Kline, B.C. and Miller, J.R. (1975) J. Bacteriol. 121:165-
     172

44. Cress, D.E. and Kline, B.C. (1976) J. Bacteriol. 125:635-642

45. Nordström, K., Ingram, L. and Landbäck, A. (1972) J. Bacteriol.
     110:562-569

46. Kool, A.J. and Nijkamp, H.J. (1974) J. Bacteriol. 120:
     569-578

47. Jacob, F., Brenner, S. and Cuzin, F. (1963) Cold Spring
     Harbor Symp. Quant. Biol. 28:329-348

48. Kingsbury, D.T. and Helinski, D.R. (1973) J. Bacteriol. 114:
     1116-1124

49. Perlamn, D. and Rownd, R.H. (1975) Molec Gen. Genet. 138:
     281-291

50. Barnes, M.H. and Rownd, R. (1972) J. Bacteriol. 111:750-757

51. Crosa, J.H., Luttropp, L.K., Heffron, F. and Falkow, S. (1975)
     Molec. Gen. Genet. 140:39-50

52. Lovett, M.A., Katz, L. and Helinski, D.R. (1974) Nature
     251:337-340

53. Womble, D.D., Warren, R.L. and Rownd, R.H. (1976) Nature
     264:676-678

54. Kasamatsu, H. and Vinograd, J. (1974) Ann. Rev. Biochem.
     43:695-719

55. Womble, D.D., Warren, R.L. and Rownd, R.H. (1978) Submitted
     to J. Bacteriol.

56. Crosa, J.H., Luttropp, L.K. and Falkow, S. (1976) Nature
     261:516-519

57. Cabello, F., Timmis, K. and Cohen, S.N. (1976) Nature 259:
     285-290

58. Eichenlaub, R., Figurski, D. and Helinski, D. (1977) Proc.
     Natl. Acad. Sci. USA 74:1138-1141

59. Lovett, M.A., Sparks, R.B. and Helinski, D.R. (1975) Proc.
     Natl. Acad. Sci. USA 72:2905-2909

60. Crosa, J.H., Luttropp, L.K., Thomas, M. and Falkow, S. (1978)
     In: Microbiology - 1978. (eds. J. Davies and R. Novick)
     American Society for Microbiology, Washington, D.C.

61. Palchaudhuri, S. and Maas, W. (1978) In: Microbiology - 1978 (eds. J. Davies and R. Novick) American Society for Microbiology, Washington, D.C.

62. Warren, R.L., Womble, D.D., Tanaka, N., Barton, C.R.,Scallon, S., Cramer, J.H. and Rownd, R.H. (1977) In: Plasmids: Medical and Theoretical Aspects. Springer-Verlag, Berlin. pp253-264

63. Warren, R.L., Womble, D.D., Barton, C.R., Easton, A.E. and Rownd, R.H. (1978) In: Microbiology - 1978 (eds. J. Davies and R. Novick) American Society for Microbiology, Washington, D.C.

64. Greenberg, J. and Rownd, R.H. (1978) Submitted to J. Bacteriol.

65. Hashimoto, H. and Rownd, R. (1975) J. Bacteriol. 123:56-58

66. Rownd, R.H., Perlman, D., Goto, N. and Appelbaum, E.R. (1975) In: DNA Synthesis and Its Regulation (eds. M. Goulian and P. Hanawalt). W.A. Benjamin, Inc., Menlo Park, Calfornia. pp537-559

67. Tanaka, N., Cramer, J.H. and Rownd, R.H. (1976) J. Bacteriol. 127:619-636

68. Rownd, R.H., Miki, T., Appelbaum, E.R., Miller, J.R., Finkelstein, M. and Barton, C.R. (1978) In: Microbiology - 1978 (eds. J. Davies and R. Novick) American Society for Microbiology, Washington, D.C.

69. Watanabe, T. and Ogata, Y. (1970) J. Bacteriol. 102:363-368

70. Finkelstein, M. and Rownd, R.H. (1978) Submitted for publication.

71. Clewell, D.B., Yagi, Y. and Bauer, B. (1975) Proc. Natl. Acad. Sci. USA 72:1720-1724

72. Broda, P., Bayley, S.A., Duggleby, C.J., Worsey, M.J. and Williams, P.A. (1978) In: Microbiology - 1978. (eds. J. Davies and R. Novick) American Society for Microbiology, Washington, D.C.

73. Huffman, G., Miki, T. and Rownd, R.H. Manuscript in preparation.

74. Anderson, R.A. and Roth, J.R. (1977) Ann. Rev. Microbiol. 31:473-505

75. Miki, T., Easton, A.E. and Rownd, R.H. (1977) Molec. Gen.
     Genet. (in press)

COPY NUMBER CONTROL MUTANTS OF THE R PLASMID R1 IN

ESCHERICHIA COLI

Bernt Eric Uhlin[*], Petter Gustafsson, Søren Molin and
Kurt Nordström

Unit of Experimental Microbiology
Department of Molecular Biology
University of Odense, Niels Bohrs Alle
DK-5000 Odense, Denmark

INTRODUCTION

DNA replication is carefully regulated in bacteria which
results in a constant average number of genomes per cell in an
exponentially growing population. This applies to the bacterial
chromosome as well as to plasmids carried by the cells. Despite
extensive studies on DNA synthesis there is very little knowledge
available so far about the mechanisms involved in the regulation of
DNA replication. It is generally believed that the control is
exerted at the level of initiation (Pritchard, 1974). Mutant plasmids
which show an increased number of copies compared to the wild type
plasmid, so called copy mutants, have been isolated from several
plasmids (Nordström et al., 1972; Timmis and Winkler, 1973; Kool and
Nijkamp, 1974; Morris et al., 1974; Matsubara and Takeda, 1975).
Such plasmid mutants may be used as a tool in studies of control of
DNA replication. In this paper we will report on studies with copy
number control mutants of the plasmid R1 in *Escherichia coli* K-12.

THE SYSTEM USED

Escherichia coli K-12 was used throughout this work. The
plasmid investigated is the R plasmid R1*drd-19*, which is a
derivative of plasmid R1 derepressed with respect to transfer
(Meynell and Datta, 1967). The relevant strains and plasmids are
listed in Table 1.

Plasmid copy number is defined as plasmid copies per chromosome
[*]A.S.I. Participant

equivalent. Plasmid-borne mutations affecting the copy number are
referred to as *copy mutations* (phenotype Cop, genotype *cop*). The
ratio between the copy number of a copy mutant and that of the
wild type has been denoted *copy effect*. The amount of plasmid DNA
was determined as covalently closed circles by ethidium bromide-
cesium chloride gradient centriguation or by alkaline sucrose
gradient centrifugation.

ISOLATION OF COPY MUTANTS

Mutants of the plasmid Rl*drd-19*, showing a several-fold
increase in the copy number, have been isolated by selection on
solid media for increased resistance to ampicillin (Nordström et
al., 1972; Uhlin and Nordström, 1975). By transfer to a second
host it was confirmed that the increase in copy effect is due to
a plasmid-borne mutation. Such mutants have been obtained showing
copy numbers in the range from that of the wild type plasmid to a
ten-fold higher level. The ampicillin resistance is mediated by a
degrading enzyme, β-lactamase, which is coded for by the plasmid.
Both the specific activity of the enzyme and the resistance to the
drug have been shown to be proportional to the copy effect (Uhlin
and Nordström, 1977) as illustrated by the equations:

$$\beta\text{-lactamase activity } = k_1 \times \text{(copy effect)}$$

$$\text{Ampicillin resistance} = k_2 \times \text{(copy effect)}$$

β-lactamase and ampicillin resistance can therefore be used to
titrate the copy effect.

Escherichia coli K-12 carrying the R plasmid Rl*drd-19* is able
to form colonies on rich media plates containing D-ampicillin at
concentrations up to 100 µg/ml. At a 50% higher drug concentration
the colony count is reduced about 100-fold. The colonies found on
plates containing more than 200 µg of ampicillin/ml are all mutant
clones. These mutant clones seem all to carry copy mutants of the
plasmid. The higher the selecting drug concentration, the higher
the copy number of the mutant plasmids. Therefore, it is possible
to directly select for a desired copy effect. We have not found
any mutants of the plasmid Rl*drd-19* analogous to the type
described by Odakura et al. (1974). From a different R plasmid
than Rl they isolated a mutant in which the increased resistance
to ampicillin was due to an increased number of the β-lactamase
gene within the plasmid genome. In that case the copy number was
similar to that of the wild type plasmid. The copy mutants
obtained from Rl*drd-19* can be further mutated to higher resistance
levels as shown in Fig. 1. It is not known whether such additional
increases in resistance are due to mutational alterations in the
same or in different functions involved in the copy number control.

Table 1

Characteristics of bacterial strains and R plasmids used

Denotion	Genotype[a]	Reference
Bacterial strains		
1005	*met, nal, rel*	Grinsted et al. (1972)
D11	*his, pro, trp, strA*	Boman et al. (1968)
R plasmids[b]		
R1*drd-19*	*bla$^+$, cat$^+$, aphA$^+$,* *aadA$^+$, sul$^+$, tra$^+$, fin*	Meydell & Datta (1967)
pKN103	*cop*	Nordström et al. (1972)
pKN142	*cop*	Uhlin & Nordström (1975)
pKN301	*cop(td)*	
pKN303	*cop(am)*	This laboratory, unpublished.
pKN401	*cop(td)*	
R100	*cat$^+$, aadA$^+$, sul$^+$, tet$^+$* *tra$^+$, fin$^+$*	Egawa & Hirota (1962)

(a) Symbols used for bacteria and plasmids are as proposed by Demerec et al. (1966) and Novick et al. (1976), respectively. The symbols *cop(td)* and *cop(am)* denote temperature dependent copy mutants and amber suppressible copy mutants, respectively.

(b) The listed pKN-plasmids are all derived from plasmid R1*drd-19*.

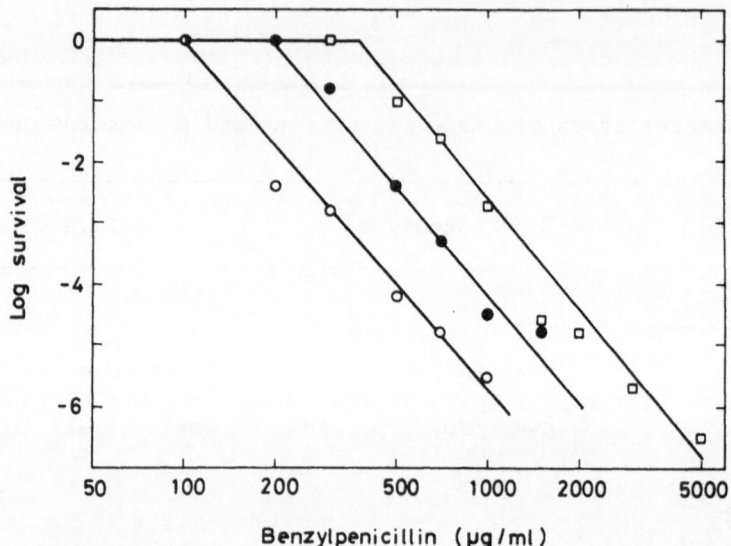

Figure 1. *Survival on benzylpenicillin plates after mutagenic*
treatment of cells carrying wild type or copy mutants of the
R plasmid R1drd-19. Logarithmically growing cultures of strain
1005 carrying the plasmid R1drd-19 (○), the copy mutant pKN103
(●), and the copy mutant pKN142 (□), in LB medium [Bertani,
(1951)] were treated with ethyl methane sulphonate at 37C. After
phenotypic expression had been allowed the cells were diluted and
plated on LA plates (LB solidifed with agar) containing different
amounts of benzylpenicillin. The survival was calculated from the
colony count after the plates had been incubated overnight at 37C.

     None of the copy mutations have so far been mapped. A
preliminary analysis of the fragment pattern after cleavage with
the restriction endonuclease EcoRl indicates that some of the copy
mutants carry a deletion, but it is not known if those deletions
are responsible for the altered copy number. A formal complementation analysis is made difficult by the incompatibility (inability
to coexist stably) exerted between related plasmids. Therefore, at
present it is not known how many plasmid genes there are in the
copy number control. But, as will be discussed later, there are
indications that two or more functions may be involved.

                      COPY NUMBER CONTROL FUNCTIONS

Two types of copy mutants with a conditional copy effect have

been isolated (Table 1) (Gustavsson and Nordström, in
preparation). One type has a copy effect that is suppressed when
the host cell carries an amber suppressor activity. The amber
copy mutant pKN303 has a copy number of 1-2 times that of the wild
type plasmid in the presence of a strong amber suppressor, while in
the absence of any amber suppressor the mutant has a 4-6 times
higher copy number. The behaviour of the amber copy mutant strongly
indicates that a protein with a repressor function is involved in
the copy number control of plasmid Rldrd-19.

The other type of conditional mutants has a temperature-
dependent copy effect. The plasmid mutant pKN301 shows the wild
type copy number when the host cells are grown at 30C, while when
grown at 40C the plasmid copy effect is 4-6 times higher. So, both
types of conditional copy mutants (plasmids pKN301 and pKN303) can
shift between two different copy levels, the wild type level and a
4-6 times higher level. This may indicate, though it has so far
not been proven, that the two mutants are altered in the same
control function.

A RUNAWAY COPY MUTANT

For both types of conditional copy mutants described above it
is quite obvious that when the mutation is expressed the copy
number is carefully controlled at an increased, but still defined,
level. This finding lead to the search for mutants which have no
control of the copy number at a condition where the mutation is
expressed. After mutagenic treatment, with ethyl methane sulphonate,
of a strain carrying the temperature-dependent mutant pKN301
selection was made for an increased ampicillin reistance at 30C.
The clones so obtained were tested for their behaviour at higher
temperatures and some of the isolates showed a drastically reduced
viability at 37C and 42C. The mutant denoted pKN401 is such a
temperature-dependent mutant which has been further studied (Uhlin
and Nordström, in preparation). The plasmid pKN401 was transferred
to an unmutagenized host strain and some of its properties are
illustrated in Fig. 2. The strain carrying the mutant plasmid,
D11-pKN401, was grown at different temperatures. There was a
drastic increase in the amount of plasmid DNA at temperatures above
34C (Fig. 2a). A similar increase was shown for the total amount of
DNA in the cells (Fig. 2b), and there was a corresponding decrease
in the growth rate of the cells (Fig. 2c). The bacteria did not
grow at a measurable rate at temperatures above 35C.

The inhibitory effect of plasmid pKN401 on cell growth at
higher temperatures was also demonstrated by shifting a culture
growing at 30C to a higher temperature (Fig. 3). Immediately after
the shift to 40C strain D11-pKN401 grew at about the same rate as
the control strain, D11-Rldrd-19, but after about 4 doublings in

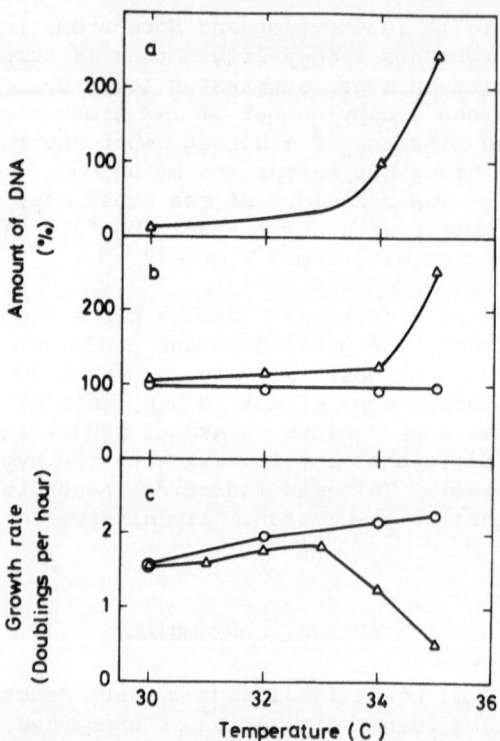

Figure 2. *Properties of the temperature-dependent mutant pKN401.*
Strain D11 carrying plasmid R1drd-19 (O) or pKN401 (Δ) was
grown logarithmically in LB medium at different temperatures. (a)
The amount of covalently closed circular (CCC) DNA was analyzed
by ethidium bromide-cesium chloride gradient centrifugation after
labelling with ³H-thymidine. The plasmid DNA content was calculated
as % of chromosomal DNA. (b) The total amount of DNA was determined
by the diphenylamine method as described by Giles and Myers (1965)
and the protein content was measured according to Lowry et al. (1951).
The ratio DNA/protein is shown in a relative scale with the value
for strain D11-R1drd-19 at 30C set to 100%. (c) Growth rate
(doublings/hour) was determined by measuring the optical density in
a Klett-Summerson colorimeter.

mass at the post-shift temperature the growth of strain D11-pKN401
slowed down and after about one more doubling in cell mass there was
no further increase in optical density.   There was a drastic
decrease in viability of the cells after about 4 mass doublings
even when they were plated at 30C.   The situation is obviously
lethal to the cell growth, though there was no detectable lysis of
the cells as judged from optical density measurements. The decrease
in viability shown by the viable count is not due only to a
sensitivity to plating, since also liquid cultures were found to be

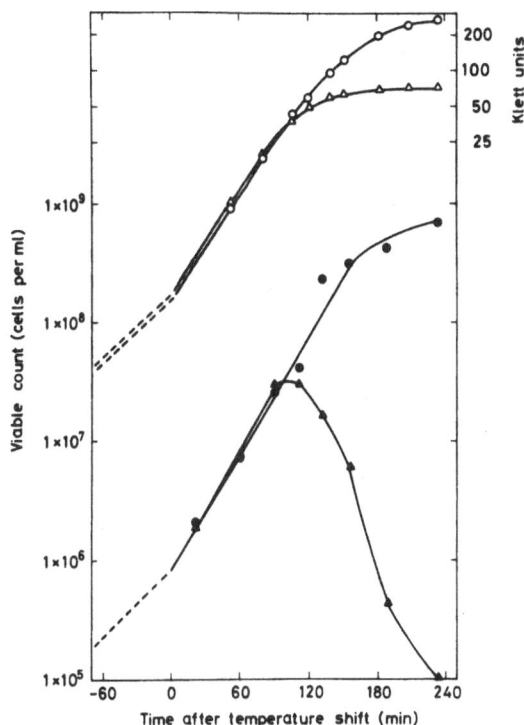

*Figure 3. Effect of plasmid pKN401 on cell growth after a
temperature shift from 30C to 40C. Strain D11 carrying plasmid
R1drd-19 (O) or pKN401 (Δ) was grown logarithmically in LB medium
at 30C (dotted lines). Growth was followed as optical density
using a Klett-Summerson colorimeter and by determination of viable
count. The samples for viable count were immediately diluted in
0.9% NaCl and plated on LA plates which were then incubated overnight at 30C. The temperature shift was carried out by
transferring (at time zero) part of the culture to prewarmed medium
and growth was allowed to continue at 40C. Optical density (open
symbols) and viability (assayed at 30C) (closed symbols) were
followed as before the temperature shift.*

unable to restore growth when shifted back to 30C.

The rate of plasmid synthesis was studied during the initial
cell doublings after cells carrying plasmid mutants pKN301 or
pKN401, respectively, had been shifted from 30C to 40C. The
differential plot in Fig. 4 shows that the synthesis of plasmid
pKN301 adjusted rapidly to a new, constant rate, and the resulting
steady state copy effect was 4-6 times higher than at 30C. The
mutant pKN401, however, showed an accelerating rate of synthesis

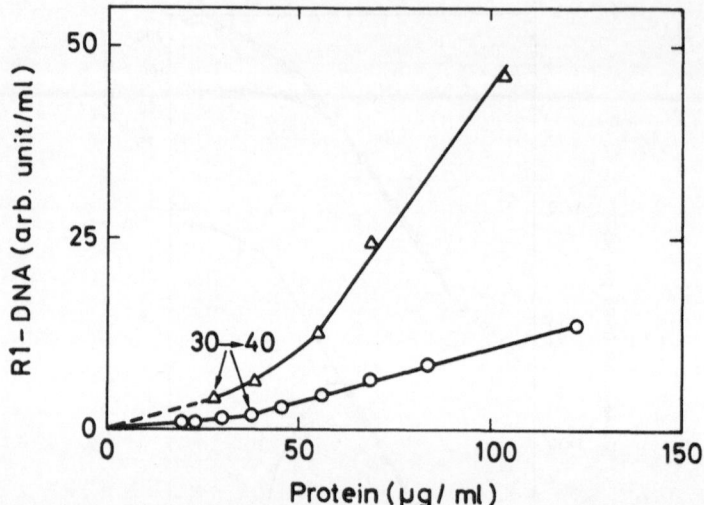

Figure 4. *Differential rate of plasmid DNA synthesis during a temperature shift from 30C to 40C. Cultures of strain 1005-pKN301 (O) and strain D11-pKN401 (Δ), growing logarithmically at 30C, were shifted to 40C at the times indicated by the arrows. The DNA was labelled with ³H-thymidine and samples were taken for analysis of CCC-DNA by alkaline sucrose gradient centrifugation and for determination of the protein content.*

as judged from the analysis of the CCC-DNA. The data for plasmid pKN401 at times after the shift were replotted as shown in Fig. 5; the amount of pKN401-DNA increased logarithmically after the temperature shift with a doubling time of about 12 minutes.

Thermo-resistant clones of strain D11-pKN401 can be isolated at a frequency of about 10^{-5}-10^{-6} (table 2). The plasmid-carrying clones isolated at 42C formed two classes as judged by determination of resistance to ampicillin. One class showed a temperature-dependent copy effect similar to the mutant pKN301, while the other class had a copy number very similar to wild type at both high and low temperature.

INCOMPATIBILITY

Plasmid incompatibility has been suggested to be linked to the control of plasmid replication (Pritchard et al., 1969). The copy mutants of plasmid R1*drd-19* have been tested for their incompatibility properties by the use of the related R plasmid R100 as a test plasmid (Uhlin and Nordström, 1975). From such tests the mutants could be grouped into two classes, defined by showing

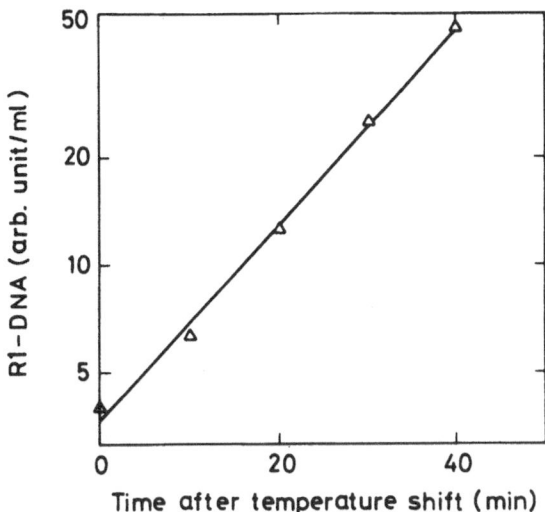

Figure 5. *The increase in plasmid pKN401 DNA after a temperature shift from 30C to 40C. The data for strain D11-pKN401 from the experiment in Fig. 4 has been replotted on a logarithmic scale versus the time.*

Table 2

Temperature-resistant derivatives obtained
from strain D11-pKN401[a]

Class	Frequency (%)	Copy Effect	
		30C	42C
A	70	1-2	4
B	20	1-2	1-2
C	10	0	0

(a)   Strain D11-pKN401 was grown exponentially in LB medium at 30C. At a cell density of about 10^8 cells/ml samples were spread on LA plates and these were incubated overnight at 30C and 42C. About 10^{-5}-10^{-6} of the cells plated gave rise to colonies at 42C. These were tested for copy effect in a preliminary test by determination of resistance to ampicillin at 30C and 42C.

an increased or a decreased incompatibility, respectively, towards
the test plasmid when they were compared to the wild type plasmid
R1*drd-19*. All the copy mutants isolated were found to be altered
with respect to incompatibility properties, indicating that there
may be a link between incompatibility and copy number control.

It has been suggested that the function exerting
incompatibility could be an inhibitor acting in the replication
control (Pritchard et al., 1969). If such an inhibitor, by
mutational alteration, lost its repressor activity, the plasmid
should show a decreased incompatibility towards a related plasmid.
The temperature-dependent copy mutant pKN301 was tested with R100
as test plasmid at 30C and 40C. The result found was that the
mutant pKN301 did not show a decreased incompatibility towards
plasmid R100 at 40C, but there was an increased expression of this
property as compared to the situation found at 30C. At the latter
temperature plasmid pKN301 was similar to wild type R1*drd-19* with
respect to incompatibility as well as copy number. The temperature-
sensitive activity involved in the control of the copy effect for
mutant pKN301 is apparently not mediating plasmid incompatibility.

DISCUSSION

Plasmids are dispensible genetic elements and therefore it
should be possible to obtain mutants, altered also in essential
control functions, which might give information about the
replication control. The copy number mutants of plasmid R1 mediate
readily distinguished phenotypic properties. The increased gene
dosage of the copy mutants results in increased amounts of the
antibiotic degrading enzymes, and that increase is proportional to
the increase in plasmid copy number (Uhlin and Nordström, 1977).
Mutations affedting the copy number have been described for a
number of different plasmids. In many cases these mutations are
located on the plasmid (Nordström et al., 1972; Timmis and Winkler,
1973; Kool and Nijkamp, 1974; Morris et al, 1974; Matsubara and
Takeda, 1975). Mutations located on the bacterial chromosome giving
increase in the copy number of plasmids have also been reported
(Macrina et al., 1974; Cress and Kline, 1976; Davis and Vapnek,
1976).

None of the copy mutations have so far been properly mapped
but some indications have come out from a preliminary analysis of
the fragment pattern after digestion of the plasmid DNA with
restriction endonuclease. It is not known whether the deletions
observed in some cases are directly responsible for the copy
effect. Copy mutants may spontaneously give rise to smaller
plasmids, miniplasmids, that are able to replicate autonomously
(Goebel and Bonewald, 1975; Mickel and Bauer, 1976; Molin and Uhlin,
unpublished results). Such miniplasmids, being the result of very

large deletions, may prove useful for genetic analysis and mapping. A preliminary analysis of miniplasmids derived from plasmid Rl suggests that the genes for regulation of the copy number level are located close to the origin region where the initiation of replication occurs (Luibrand et al., 1977; Molin and Uhlin, unpublished results).

The number of control elements involved in copy number regulation is not known but it may be argued that the temperature-dependent mutants pKN301 and pKN401 represent mutational alterations in two different functions. The one element setting the copy number at wild type level and the other element keeping the level about 4 times higher. However, the results can also be interpreted as if two mutations are affecting, to a different extent, the same control element. Further analysis and mapping of the mutants and revertants might distinguish between these possibilities.

A basic question for an understanding of the control is the mode of function of the elements involved. The results with the conditional copy mutants clearly demonstrate that there is at least one negatively acting element involved. A control element with a repressor function has also been shown to act in the replication control of plasmid λdv (Berg, 1974; Matsubara, 1976). Models for control of DNA replication involving repressors have been proposed by several groups (Pritchard et al., 1969; Rosenberg et al., 1969; Sompayrac and Maaløe, 1973). The plasmid λdv system seems also to contain one more copy number control element, the RNA synthesis terminator site t_{R1}. The frequency of replication may be affected by the phage λ N gene product, acting as an anti-rho substance (Berg and Kellenberg-Gujer, 1974).

The runaway behaviour of the mutant pKN401 suggests that at high temperature there is no control of the copy number which may result in an increase of the amount of plasmid DNA to levels that become lethal to the host cell. Experiments with miniplasmids derived from the runaway mutant have shown that it is not the number of plasmid copies that eventually inhibits the growth. A miniplasmid only about 1/10 the size of pKN401, but retaining the runaway-behaviour, is synthesized in equally high total amounts of DNA as pKN401 after a temperature shift to 40C. In this case the actual number of copies per chromosome equivalent is about 10 times higher. It appears as if the cells can tolerate quite large amounts of extra chromosomal DNA (Fig. 2). What actually sets the upper limit of plasmid DNA that can be tolerated by the host cell is not known, but this type of plasmid mutants offers a system for studies on how cells are affected by the presence of large amounts of non-chromosomal DNA.

The synthesis of plasmid pKN401 increased logarithmically after

the shift to 40C (Figs. 4 and 5), and the doubling time of plasmid
DNA was about 12 minutes. This is a much higher value than
expected if the rate of elongation was the only limitation on the
synthesis. Plasmid Rl is replicated by DNA polymerase III (Nordström
et al., 1974). From the rate of elongation of the bacterial
chromosome it can be calculated that replication of plasmid Rl
should take 1 or 2 minutes (assuming bi- or uni-directional
replication, respectively). The actual generation time of
replication covalently closed circular molecules has been estimated
for phage λ, which has about half the molecular weight of plasmid
Rl. During the early time after infection phage λ replicates as
circular molecules with a generation time of 2-3 minutes (Mackinlay
and Kaiser, 1969; Carter and Smith, 1970).

The relatively long cycling time found for plasmid pKN401
suggests that there is a rate-limiting step in plasmid Rl
replication. Sheehy and Novick (1975) studied the replication
cycle of a plasmid in *Staphylococcus aureus*. They found evidence
for a rate-limiting multi-step process after the newly replicated
molecules had been fully polymerized. The conversion of the
dimeric terminal molecules to mature closed circular monomers was
estimated to take 3-4 minutes at 37C. Pulse-chase experiments have
shown that it takes 3-4 minutes for plasmid Rl to completely
transfer replicating DNA molecules into CCC DNA (Gustavsson and
Nordström, in preparation). Furthermore, the majority of the
plasmid pKN401 DNA was present as CCC DNA, which suggests that a
step before the onset of the next replication is rather slow. It
should be stressed that plasmid pKN401 offers a unique
possibility for studies of the various stages in plasmid replication,
since this plasmid has no copy number control at temperatures above
35C, so it is possible to isolate large quantities of the molecules
involved.

ACKNOWLEDGEMENTS

The skillful technical assistance of Eva Andersen, Dorthe
Dolleris, Eva Heyn Olsen, and Marianne Hald Rasmusen is gratefully
acknowledged. The work was supported by the Danish Medical Research
Council (project No. 6866) and the Danish Natural Science Council
(project Nos. 6532 and 7077).

REFERENCES

Berg, D.E. (1974) Virology 62:224-233

Berg, D.E. and G. Kellenberger-Gujer (1974) Virology 62:234-241

Bertani, G. (1951) J. Bacteriol. 62:293-300

Boman, H.G., K.G. Eriksson-Grennberg, S. Normark and E. Matsson (1968) Genet. Res. 12:169-185

Carter, B.J. and M.G. Smith (1970) J. Mol. Biol. 50:713-718

Cress, D.E. and B.C. Kline (1976) J. Bacteriol. 125:635-642

Davis, R. and D. Vapnek (1976) J. Bacteriol. 125:1148-1155

Demerec, M., E.A. Adelberg, A.J. Clark and P.E. Hartman (1966) Genetics 54:61-76

Egawa, R. and Y. Hirota (1962) J. Genet. 37:66-69

Giles, K.W. and A. Myers (1965) Nature 206:93

Goebel, W. and R. Bonewald (1975) J. Bacteriol. 123:658-665

Grindsted, J., J.R. Saunders, L.C. Ingram, R.B. Sykes and M.H. Richmond (1972) J. Bacteriol. 110:529-537

Kool, A.J. and H.J.J. Nijkamp (1974) J. Bacteriol. 120:569-578

Lowry, O.M., N.J. Rosebrough, A.L. Farr and R.J. Randall (1951) J. Biol. Chem. 193:265-275

Luibrand, G., D. Blohm, H. Mayer and W. Goebel (1977) Molec. Gen. Genet. 152:43-51

Mackinlay, A.G. and A.D. Kaiser (1969) J. Mol. Biol. 39:679-683

Macrina, F.L., G.G. Weatherly, and R. Curtiss III (1974) J. Bacteriol. 120:1387-1400

Natsubara, K. (1976) J. Mol. Biol. 102:427-439

Matsubara, K. and Y. Takeda (1975) Molec. Gen. Genet. 142:225-230

Meynell, E. and N. Datta (1967) Nature 214:885-887

Mickel, S. and W. Bauer (1976) J. Bacteriol. 127:644-655

Morris, C.F., H. Hashimoto, S. Mickel and R. Rownd (1974) J. Bacteriol. 118:855-866

Nordström, K., L.C. Ingram and A. Lundbäck (1972) J. Bacteriol. 110:562-569

Nordström, U.M., B. Engberg and K. Nordström (1974) Molec. Gen.
    Genet. 135:185-190

Novick, R.P., R.C. Clowes. S.N. Cohen, R. Curtiss III, N. Datta
    and S. Falkow (1976) Bacteriol. Rev. 40:168-189

Odakura, Y., H. Hashimoto and S. Mitsuhashi (1974) J. Bacteriol.
    120:1260-1267

Pritchard, R.H. (1974) Philos. Trans. R. Soc. London 264:303-336

Pritchard, R.H., P.T. Barth and J. Collins (1969) Symp. Soc. Gen.
    Microbiol. 19:263-297

Rosenberg, B.H., L.F. Cavalieri and G. Ungers (1969) Proc. Natl.
    Acad. Sci. USA 63:1410-1417

Sheehy, R.J. and R.P. Novick (1975) J. Mol. Biol. 93:237-253

Sompayrac, L. and O. Maaløe (1973) Nature New Biol. 241:133-135

Timmis, K. and U. Winkler (1973) Molec. Gen. Genet. 124:207-217

Uhlin, B.E. and K. Nordström (1975) J. Bacteriol. 124:641-649

Uhlin, B.E. and J. Nordström (1977) in press

DETERMINATION OF ORIGIN AND DIRECTION OF REPLICATION OF LARGE

PLASMIDS OF <u>E.COLI</u> BY TWO ELECTRON MICROSCOPIC TECHNIQUES

L. Silver[*], M. Chandler, E. Boy de la Tour and L. Caro

Université de Genève
Département de Biologie Moléculaire
30 quai Ernest-Ansermet
CH-1211, Geneve 4, Switzerland

We have undertaken a study of the origin and direction of replication of the large drug resistance factor, R100.1 (Egawa and Hiroto, 1962) and of its resistance transfer factor (RTF) derivative, pAR132, i.e., an R100-1 which is deleted for the region carrying most of the drug resistance markers or r-determinant region.

The origin and direction of replication of several small, naturally occurring and artificially constructed plasmids in <u>E.coli</u> have been determined by two main methods. One consists of the electron microscopic analysis of restriction-endonuclease treated replicative intermediates (Inselburg, 1974; Lovett et al., 1974, 1975; Crosa et al., 1975, 1976; Cabello et al., 1976; Eichenlaub et al., 1977). In such studies, restriction fragments are inspected for the occurrence of "eye" or "Y" forms, enabling the localization of a replication start-point and the determination of the direction of replication. A second electron microscopic technique, partial denaturation mapping of replicative inter- mediates, has been used successfully to determine the starting point and mode of replication of the bacteriophage λ, P2 and T7 (Schnös and Inman, 1970, 1971; Dressler et al., 1972). Partial denaturation mapping of circular molecules, such as λ, P2 and plasmids, has the inherent disadvantage of the lack of a fixed point of reference for the alignment of the molecules. Both of these electron microscopic techniques require the selective isolation of replicative intermediates or, at least, their enrichment. Such an enrichment is more difficult to achieve for plasmids than for bacteriophage because while bacteriophage replication is more or less synchronized in a population of infected cells, replication of a plasmid occurs during only a short portion of the cell cycle; thus

[*]*A.S.I. Participant*

examination of plasmids isolated from an exponentially growing
culture might show only a very small proportion of molecules in the
act of replication. In practice, with large plasmids like R100.1
and pAR132, DNA species isolated by the cleared lysate procedure
of Clewell and Helsinki (1969) and found at densities of CCC DNA
or lighter in EtBr-CsCl gradients yield about 0.1% replicative
intermediates (by inspection in the EM) of the Cairns' type
(theta form).

     In our experiments on the replication of R100.1 we have
sought an enrichment for replicative intermediates by synchronizing
the cultures for initiation of replication by sequential amino
acid and thymine starvation, followed by re-addition of thymine
to start DNA replication anew. This regime has been shown, by
DNA-DNA hybridization studies, to synchronize initiation of
replication for the E.coli chromosome and for R100.1 (Louarn et al.,
1974; Bird et al., 1976). We have also tried to reduce some of the
uncertainties inherent to the identification of an origin of
replication of a circular molecule by using two different methods
of analysis. One method, already mentioned, was the determination
of the position of the forks with respect to the partial
denaturation map of replicating molecules. The other method was
the study, by electron microscopic autoradiography, of molecules
pulse labeled, during a synchronized round of replication, with
(^{3}H)-thymidine.

     Both methods gave the same result : DNA replication of both
R100.1, pAR132 after the synchronization procedure is predominantly
unidirectional in a single sense from a unique origin at Kb 8.8 on
the R100 map. The data presented here are those of experiments
with the RTF, pAR132. Parallel experiments with R100.1 gave
entirely similar results. A more detailed account of this work
appears in Silver et al.,(1977).

                              AUTORADIOGRAPHY

     The method of autoradiography of partially denatured DNA
species has been used by Rochaix et al. (1974) to show that
ribosomal RNA genes of Xenopus laevis are amplified by rolling
circle replication. We applied this technique to plasmids isolated
as CCC species from cultures which had been synchronized, pulsed
for a short time with (^{3}H)-thymidine, then chased with cold thymine
and thymidine. The first step was to obtain a partial denaturation
map of the molecule. Purified plasmids were prepared either by a
modification of the method of Inman and Schnös (1970), with
denaturation at pH 10.7, or by the method of Davis and Hyman (1971)
with denaturation at pH 8.3 in 80% formamide. Projections of
electron micrographs or the partially denatured plasmid molecules
were traced with the aid of a Numonics digitizer (Numonics Corp.),

interfaced for recording purposes with a Hewlett Packard HP9830A
desk top calculator. The coordinates of denaturation loops and
total molecule length were thus recorded and processed on the
HP9830A calculator, and a Hewlett Packard HP plotter, with the
help of several programs. A denaturation loop histogram was
constructed from the collected data from each collection of
molecules. For this each molecule was divided into 200 and 250
segments (bins) and each occurrence of a denatured region was
tabulated for each bin. A standard partial denaturation map was
thus prepared for each plasmid; such histograms showed that the
partial denaturation pattern of a single species of molecule is very
reproducible.

For autoradiography, the purified, pulse labeled, plasmids
were prepared as for partial denaturation. Grids supporting these
plasmids were stained with uranyl acetate, shadowed with platinum-
palladium and overlayed with a thin carbon film. A monomolecular
layer of Ilford L4 photographic emulsion was applied to the
prepared grids and these were stored in light-tight boxes for 6
to 8 weeks. Development was with Physical Developer (Caro and van
Tubergen, 1962). Grids were scanned in a Philips EM300 electron
microscope and photographs were taken of all molecules which had
associated photographic grains. Under optimal conditions of
synchrony, spreading and exposure, approximately 40% of molecules
were associated with grains.

Fig. 1 shows one such molecule of pAR132, pulse labeled for
20 seconds at the time of initiation. Molecules such as these were
aligned by comparison with the standard partial denaturation map
prepared from exponentially grown plasmid molecules. The partial
denaturation histogram generated from the 54 molecules analyzed is
shown in Fig. 2a. The position of the grains, thus localized with
respect to the plasmid map, was plotted and a histogram of grain
distribution generated, which is shown in Fig. 2b. It can clearly
be seen that the radioactive thymidine administered during the 20
second pulse period has been incorporated preferentially in a
region of the molecule between Kb8 and Kb12 on the map, and that
label incorporation, i.e., grain distribution, is discontinuous,
following a gradient from the putative origin toward the right,
at its lowest to the left of the origin. (It should be remembered
that this is a linear representation of a circular molecule).

Partial Denaturation Mapping of Replicative Intermediates

When plasmid DNA, isolated from synchronized cultures by the
cleared lysate method was run on EtBr-CsCl gradients no replicative
intermediates were observed at any density. This led us to think
that replicative intermediates of this plasmid might be entrained

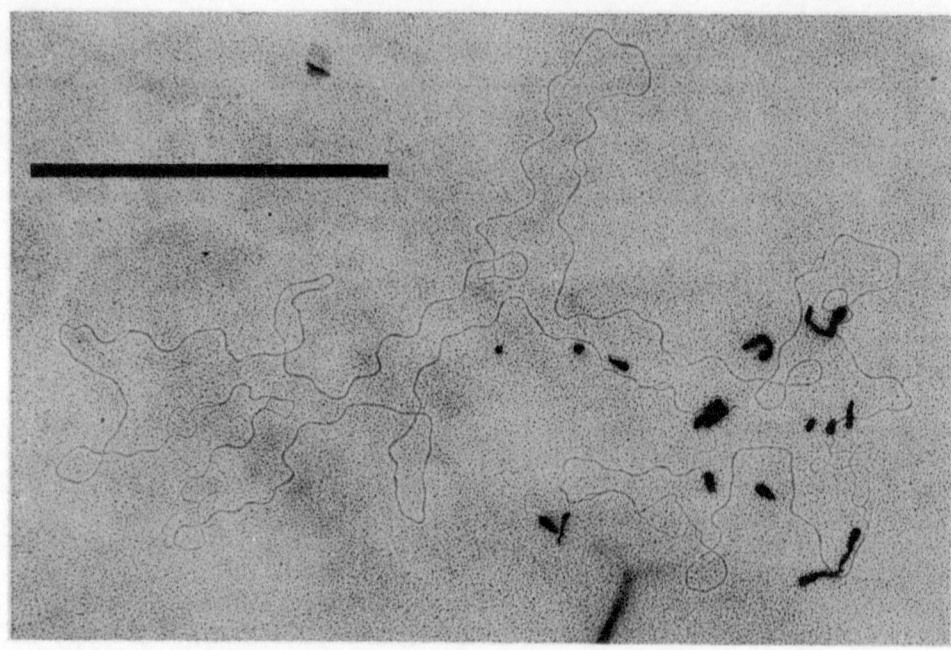

Figure 1. *Electron microscope-autoradiograph of pAR132, synchronized and pulse labelled for 20 seconds with (^{3}H)-thymidine and chased for 30 minutes; experimental details of isolation and microscopy are as described in Silver et al., 1977. The bar is 5 Kb.*

with or attached to cellular material which was either lost in the clearing-centrifugation of the Clewell-Helsinki procedure or which might float on EtBr-CsCl gradients (which have an average density of 1.56g/cc). In fact, we found that plasmid replicative intermediates had a density <u>lighter</u> than that of bulk chromosomal DNA, and we presume that this is due to the attachment or association of replicating plasmid DNA to a cellular component such as membrane. We also found that such an association seemed to be more stable when the replication took place in the presence of bromouracil.

Replicative intermediates were thus easily isolated from synchronized cultures initiated in the presence of 5-BU. They were isolated from lysozyme-sarkosyl-proteinase-K lysed cells, the whole lysate having been run on a single CsCl gradient. Replicative intermediates were found in the gradient at densities between 1.68 and 1.70, hence lighter than bulk <u>E.coli</u> chromosomal DNA (1.71), and they were analyzed by partial denaturation mapping. Fifty-seven molecules of pAR132 thus analyzed are represented in Fig. 3. For

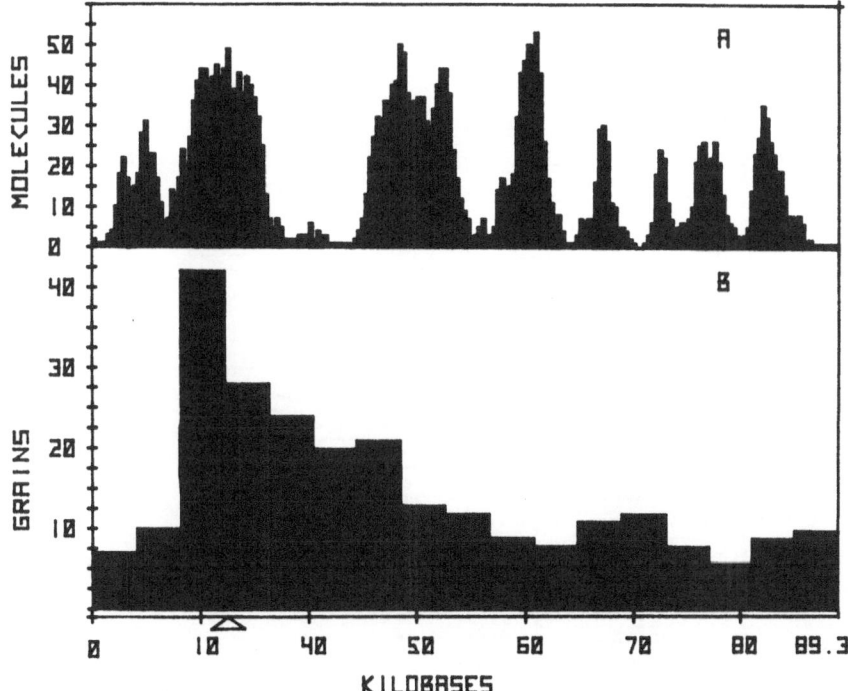

Figure 2. (a) Denaturation loop histogram of 54 molecules prepared as in Fig. 1.

(b) Photographic-grain distribution of these molecules. The kilobase map units refer to the R100 map (Ohtsubo and Ohtsubo, 1976; Silver et al., 1977). The small triangle at Kb 12.5/32.5 denoted the position of the deletion of the r-det in pAR132. The grain bin size is 4.1 Kb.

this mapping, molecules one through fifty were aligned so that one of the replication forks, chosen after visual inspection, was treated as a fixed point; the histogram thus generated was then aligned, for Kilobase-map position, with the standard denaturation map; molecules fifty-one through fifty-seven, which did not fit this pattern were aligned by eye. The generated histogram is presented in Fig. 4. A comparison of this histogram with that obtained from the autoradiography experiment (Fig. 2a) shows that the histogram obtained from the replicative intermediates which were, for the most part, aligned "by fork" is more precise than that aligned by eye, with more definition in the denatured regions. The replicative intermediate data show that replication is predominantly unidirectional in a single sense from an origin at Kb8.8. Seven molecules, fifty-one through fifty-six, appear to replicate bi-directionally, while molecule fifty-seven replicates

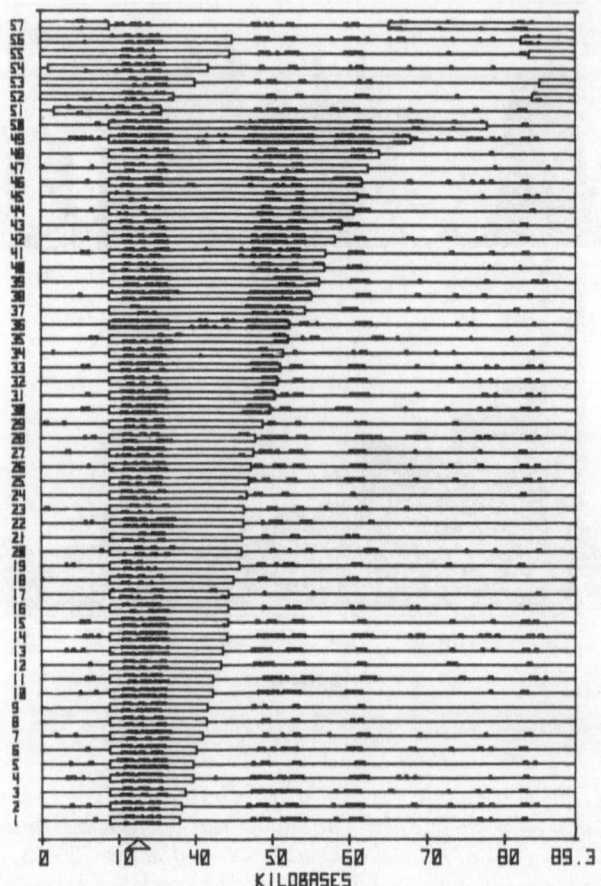

Figure 3. Representation of 57 partially denatured replicative intermediates of pAR132, prepared and aligned as described in the text.

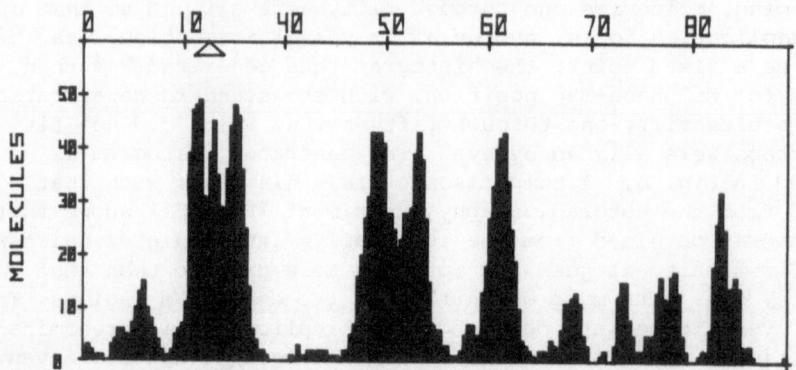

Figure 4. Denaturation loop histogram generated from the 57 molecules
represented in Fig. 3.

unidirectionally from the Kb8.8 origin, but in the sense opposite to
that of the majority class. This result, then, agrees very well
with the results of the autoradiography experiments.

While the design of the replicative intermediate experiment
included the use of 5-BU as a density label, the results indicate
the replicative intermediates are found at light-densities,
probably due to their association with a cellular component, possibly
membrane; hence, it was reasonable to repeat the experiment with
initiation in thymine rather than 5-BU. Although this has been done,
using pulse lengths of 10 to 30 seconds, replicative intermediates
of thymine initiated plasmid DNA have not been isolated under
these conditions. The results with 5-BU, however, are extremely
reproducible. Since the lysis procedure includes a heat
denaturation step (10 minutes at $70^{\circ}C$), we thought that this
differential recovery of 5-BU over thymine replicative intermediates might be due to the relative resistance of 5-BU containing
DNA to heat denaturation (Kit and Hsu, 1961). However, repetition
of the procedure with heat-treatment at 60° for 20 minutes, under
which conditions BU-intermediates were found as usual, did not
give thymine-initiated replicative intermediates. Nevertheless, the
fact that the replicative form data agrees with the autoradiography
data, indicates that the mode of replication under conditions of
initiation in thymine (+ thymidine) and 5-BU are the same. We
believe the failure to find thymine-containing intermediates to be
due in some way to greater stability of the BU-containing DNA,
although this remains unproven.

In summary, then, electron microscope autoradiography provides
a technique for the analysis of replication origin and direction
in DNA species which are not amenable to the isolation of
replicative intermediates. The resolution of the technique is
good, giving localization of the origin region to ± 2Kb, or 0.6
microns. Secondly, in the case of unidirectionally replicating
molecules, a partial denaturation map can be established for a
circular molecule, which uses the replication origin as a fixed
reference point.

ACKNOWLEDGEMENTS

We thank E. Gallay for excellent electron microscopy. This
work was supported by grant No. 3.679.75 from the Swiss National
Science Foundation and by a grant from the Fondation Ernst et
Lucie Schmidheiny. L.S. was the recipient of Public Health Service
Postdoctoral Fellowship 5F22 AIO2248-02 from the National
Institute of Allergy and Infectious Diseases. M.C. was a
recipient of a NATO/Royal Society European Fellowship for part of
this study.

REFERENCES

(1) Bird, R.E., Chandler, M. and Caro, L. (1976) J. Bacteriol. 126:1215-1223

(2) Cabello, F., Timmis, K. and Cohen, S.N. (1976) Nature 259: 285-290

(3) Caro, L. and van Tubergen, R.P. (1962) J. Cell Biol. 15, 173-188

(4) Clewell, D.B. and Helinski, D.R. (1969) Proc. Nat. Acad. Sci. U.S.A. 62:1159-1166

(5) Crosa, J.H., Luttropp, L.K., Heffron, F. and Falkow, S. (1975) Molec. gen. Genet. 140:39-50

(6) Crosa, J.H., Luttropp, L.K. and Falkow, S. (1976) J. Bacteriol. 126: 454-466

(7) Davis, R.W. and Hyman, R.W. (1971) J. Mol. Biol. 62:287-302

(8) Dressler, D., Wolfson, J. and Magazin, M. (1972) Proc. Nat. Acad. Sci., U.S.A. 69:998-1002

(9) Egawa, R. and Hirota, Y. (1962) Jap. J. Gen. 37:66-69

(10) Eichenlaub, R., Figurski, D. and Helinski, D.R. (1977) Proc. Nat. Acad. Sci., U.S.A. 74:1138-1141

(11) Inman, R.B. and Schnös, M. (1970) J. Mol. Biol. 49:93-98

(12) Inselburg, J. (1974) Proc. Nat. Acad. Sci., U.S.A. 71:2250-2259

(13) Kit, S. and Hsu, T.C. (1961) B.B.R.C. 5:120-124

(14) Louarn, J., Funderburgh, M. and Bird, R.E. (1974) J. Bacteriol. 120:1-5

(15) Lovett, M.A., Katz, L. and Helinski, D.R. (1974) Nature, 251:337-340

(16) Lovett, M.A., Sparks, R.B. and Helinski, D.R. (1975) Proc. Nat. Acad. Sci., U.S.A. 72:2905-2909

(17) Ohtsubo, H. and Ohtsubo, E. (1976) Proc. Nat. Acad. Sci., U.S.A. 73:2316-2320

(18) Rochaix, J-D., Bird, A. and Bakken, A. (1974) J. Mol. Biol. 87:473-487

(19) Schnös, M. and Inman, R.B. (1970) J. Mol. Biol. 51:61-73

(20) Schnös, M. and Inman, R.B. (1971) J. Mol. Biol. 55:31-38

(21) Silver, L., Chandler, M., Boy de la Tour, E. and Caro, L. (1977) J. Bacteriol. (in press).

REPLICATION OF COLICIN El PLASMID DNA *IN VITRO*

Jun-ichi Tomizawa

Laboratory of Molecular Biology
National Institute of Arthritis, Metabolism and
Digestive Diseases
National Institutes of Health
Bethesda, Maryland 20014, U.S.A.

Several methods for *in vitro* synthesis of DNA have been developed in recent years (cf. Gefter, 1975; Geider, 1976). These methods have been applied to the synthesis of DNA fragments on double-stranded template DNA and to the synthesis of the strands complementary to the viral strands of single-stranded phages. However, complete *in vitro* replication of double-stranded DNA had not been successful at the time we started the work to be described. Considering the importance of circular double-stranded DNA and the inherent limitation of *in vivo* experiments as means to elucidate mechanisms of DNA replication, we attempted to construct an *in vitro* system to replicate circular double-stranded DNA. After selecting colicin El plasmid (Col El) DNA as a model material, an *in vitro* system capable of carrying out a complete cycle of replication of the plasmid DNA was established (Sakakibara and Tomizawa, 1974; Tomizawa et al., 1975).

In this essay, properties of the *in vitro* replication of Col El DNA are described after a brief introductory description of certain characteristics of the plasmid DNA and of its *in vivo* replication.

1. Structure of Col El DNA

Col El DNA extracted from *E.coli* has a supercoiled closed circular duplex structure (Bazaral and Helinski, 1968). The molecular weight of a monomeric molecule is approximately 4.4 million (H. Ohmori and J. Tomizawa, unpublished result). Cleavage maps of Col El DNA have been constructed by using restriction endo-

nucleases (Oka and Takanami, 1976; Tomizawa et al., 1977). The
single site cleaved by the restriction endonuclease *Eco*Rl provides
a convenient reference position (Tomizawa et al., 1974; Inselburg,
1974; Lovett et al., 1974). No genetic map by recombinational
analysis has been made. A large number of deletion and insertion
variants of Col El have been isolated and have permitted partial
identification of regions that affect plasmid functions (Inselburg,
1977).

Two types of Col El DNA molecules having unusual chemical
composition have been found. One of them is a DNA-protein complex
called a relaxation complex (Clewell and Helinski, 1969). This
complex, which has a supercoiled structure, is converted to an
open-circular structure by treatment with a protease or sodium
dodecyl sulfate. The resulting open-circular molecule has a
strand-break in an H-strand which is heavier than the complementary
L-strand in a CsCl solution containing poly(U,G) (Clewell and
Helinski, 1970). When the open-circular molecules are digested
to various extents by exonuclease III and then by endonuclease
*Eco*Rl, the length of the longer double-stranded region of the
treated molecules decreases as the digestion by exonuclease III
progresses but the length of the shorter double-stranded region
is conserved at approximately 20% of unit length (Sugino et al.,
1975). This result provides information on the polarity of the
DNA strands as well as on the position of the strand-break, as
illustrated in Fig. 1. The location of the strand-break is very
close to but separated from the site of initiation of DNA
replication (see later). At least three kinds of proteins have been
identified as the components of a relaxation complex (Lovett and
Helinski, 1975). When a relaxation complex is disrupted by sodium
dodecyl sulfate, the largest protein is attached to the 5'-end of
the H-strand (Guiny et al., 1975).

Another type of Col El DNA which has unusual chemical
composition is closed-circular DNA with an RNA sequence in a DNA
strand (Blair et al., 1972). The fraction of such plasmid
molecules in normally growing colicinogenic bacteria is negligible,
but rises to about 50% when the bacteria are cultured overnight
under conditions where protein synthesis is prevented. One, or less
frequently two, RNA sequences exist in a molecule (Blair et al.,
1972; Sugino et al., 1975) and are distributed approximately equally
among H- and L-strands (Blair et al.,1972). The composition of
nucleotides and the partial nucleotide sequences of the RNAs have
been reported (Helinski, 1975). The idea that the RNA sequence in
the L-strand exists at one unique site (Blair et al., 1972;
Helinski et al., 1975) in the plasmid molecule is not supported by
the observation of a nearly random localization of the RNA
sequences (Sugino et al., 1975), of which about half are in the
L-strands. The biological significance of these molecules of
unusual chemical composition will be discussed later.

2. *In Vivo* Replication

Each colicinogenic bacterium carries 20 to 30 Col El DNA molecules (Clewell and Helinski, 1972). Results of electron microscopic examination and sedimentation analysis of replicating molecules of Col El DNA show that most of the molecules replicate semiconservatively in a Cairns type structure and some in a rolling circle type structure (Inselburg, 1970; Inselburg and Fuke, 1971). One of the branch points of replicating molecules is located at a fixed region: approximately 17% of unit length from the single *Eco*Rl cleavage site (Inselburg, 1974; Lovett et al. 1974). The location of the other branch point varies depending on the size of the replication loop. These results suggest that the plasmid replication initiates at a fixed region and proceeds unidirectionally (Reservation: see Section 3).

A unique characteristic of the replication of Col El DNA is its insensitivity to inhibition of protein synthesis in bacteria (Blair et al., 1972). Although replication of the bacterial chromosome gradually ceases in the absence of protein synthesis, plasmid replication continues for more than ten hours and thus the number of the plasmid molecules increases to over 1000 per bacterial cell. The overall rate of replication of the plasmid in the presence or absence of concomitant protein synthesis is about the same.

Many experiments have been done to elucidate the role of various bacterial functions in Col El DNA replication. One approach has been to study the effect of limitation of the activity of a given bacterial protein on replication of the plasmid DNA. The analysis of the results of this type of experiments is generally not simple because of the presence of residual activity of the protein in question at the restrictive condition and because of the variation of expression of gene function due to different genetic backgrounds and culture conditions. Therefore, it is not too surprising to find apparent discrepancies in the reports dealing with the role of gene functions for plasmid replication. Even when one accepts the notion that a certain gene function is involved in plasmid DNA replication, the possibility of elucidating the mechanism of action of the gene function by *in vivo* experiments is severely limited.

Col El DNA replication in colicinogenic bacteria is inhibited by rifampicin or streptolydigin, which inhibits the action of *E. coli* DNA-dependent RNA polymerase (Clewell et al., 1972; Clewell and Evenchik, 1973). Col   El DNA synthesis in rifampicin-resistant bacteria is resistant to the drug (Clewell et al., 1972). These results indicate that the RNA polymerase, or at least its β-subunit, is involved in plasmid replication. Since the plasmid

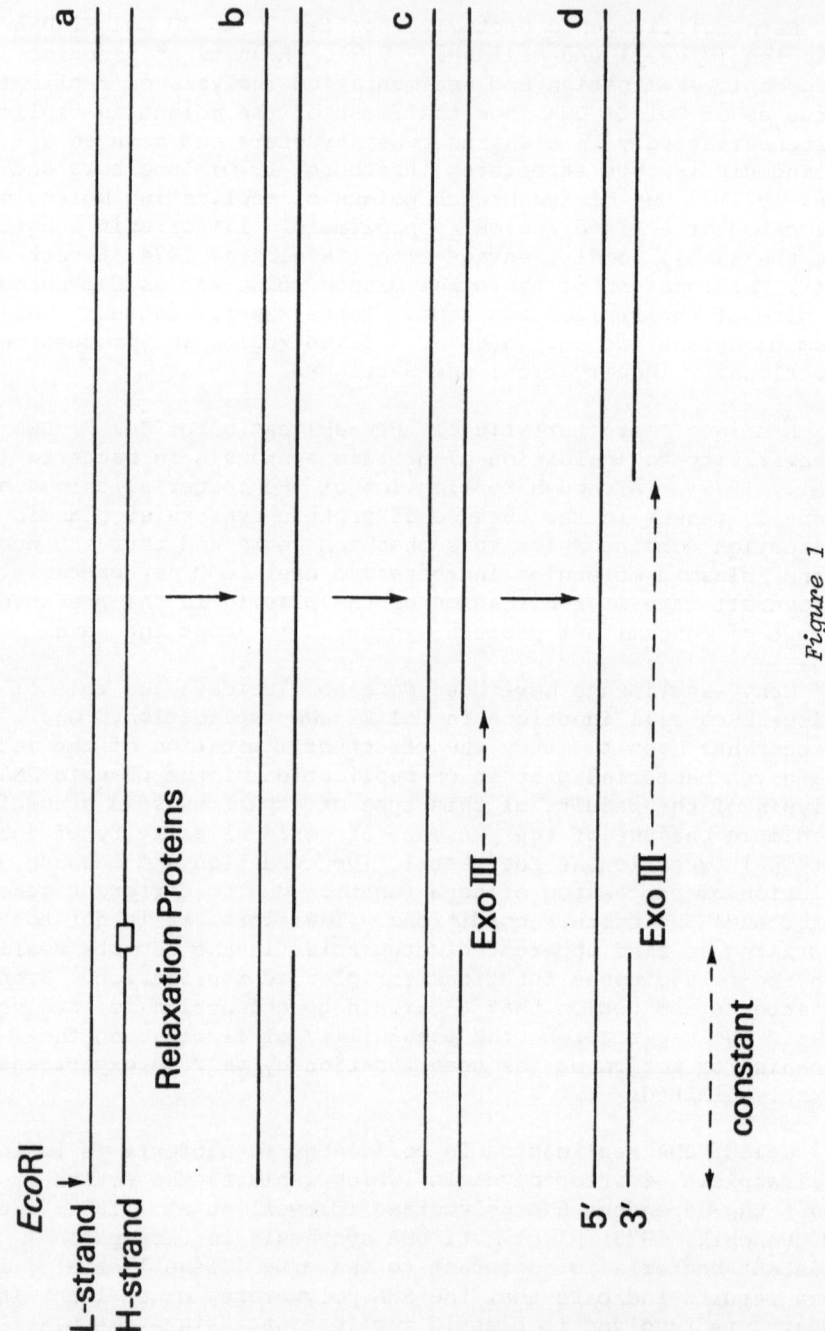

Figure 1

Figure 1.   *Polarity of the strands of Col E1 DNA.*

The principle of electron microscopic determination of the location of the strand-break that is created by removal of the relaxation proteins is presented (Sugino et al., 1975). The polarity of the strands of Col E1 DNA is deduced from the results obtained.   The molecules a and b represent those obtained by digestion of a relaxation complex with EcoR1 before and after digestion of protein by pronase, respectively.   The proteins or the strand-break could not be visualized by the electron microscopic method used. For other samples, the molecules were digested with exonuclease III which hydrolyses the DNA strands in the 3' to 5' direction. When the molecules digested with exonuclease III to different extents were examined, molecules with a "bush" caused by single stranded DNA were observed.   The position of the "bush" is determined with the molecules further digested by EcoR1 (c and d).   The length from the "bush" to one of the ends is constant and about 20% of unit length.   The strand-break formed after removal of the relaxation proteins is known to be in an H-strand.   The 3' terminus of the H-strand of a linear molecule formed by cleavage by EcoR1 is located closer to the strand-break formed by the removal of the relaxation proteins than the 5' terminus of the strand.   The end of the molecule with the 3' terminus of the H-strand is defined as the left end of the molecule.

← ───

continues to replicate in the absence of concomitant protein synthesis, RNA made by RNA polymerase may be utilized without translation. The role of RNA as the primer for DNA synthesis at some stage of replication is thus suggested.

Col El DNA cannot be maintained stably in *polA*1 bacteria, indicating that the polymerase activity and/or 3' → 5' exonuclease activity of DNA polymerase I are required for plasmid maintenance (Kingsbury and Helinski, 1970). The interpretation that the 5' → 3' exonuclease activity of DNA polymerase I is dispensable for Col El DNA replication (Grindley and Kelley, 1975; Tacon and Sherratt, 1976) can be accepted with assumption that the residual enzyme activity of the *polAex* bacteria was not sufficient for the plasmid replication under the conditions used.

The ability of various bacteria carrying temperature-sensitive mutations for the synthesis of chromosomal DNA to support Col El DNA synthesis has been studied. Col El DNA synthesis continues after cessation of chromosomal DNA synthesis in a *dnaB* strain at a high temperature (Goebel, 1970; Goebel and Schrempf, 1972). Synthesis of the plasmid DNA is partially inhibited in *dnaC* and *dnaD* strains (Goebel, 1973; Goebel, 1974; Collins et al., 1975) and is strongly inhibited in a *dnaG* strain (Collins et al., 1975) at high temperature. Apparently contradictory results have been reported for the requirement for the *dnaA* function (Goebel, 1970; Collins et al., 1975) and also for the *dnaE* function (Goebel, 1972; Collins et al., 1975). In all these cases, there is no evidence that the functions in question are directly involved in plasmid replication.

Bacterial mutants in which Col El shows temperature-sensitive replication have been isolated (Kingsbury and Helinski, 1973a; Kingsbury et al., 1973). Some of them form a temperature-sensitive DNA polymerase I (Kingsbury et al., 1973b). There is a report that may suggest the isolation of Col El mutants which cannot replicate at high temperature (Kingsbury and Helinski, 1973a).

3. The Process of *in vitro* Replication

We developed a method for *in vitro* replication of Col El DNA by modifying the method used for *in vitro* synthesis of the complementary strands of single-stranded phage DNAs (Wickner et al., 1972). Col El DNA is synthesized in a reaction mixture that contains Col El DNA, the four deoxyribonucleoside triphosphates, the four ribonucleoside triphosphates, spermidine, salts and a cell extract (Sakakibara and Tomizawa, 1974a). The cell extract is prepared by gentle lysis of a concentrated cell suspension. On the premise that some structure of Col El DNA which would be destroyed by purification and some functions which would be expressed by the DNA might be essential for its replication, we first studied DNA synthesis that depended on the endogenous Col El DNA in an extract made from colicinogenic bacteria. Later it was found that Col El DNA added to the reaction mixture can replicate as does endogenous DNA. It was further shown that an extract made from bacteria which do not carry Col El can support replication of plasmid DNA added to the extract (Tomizawa et al., 1975).[*]

---

[*] *Col El DNA was selected as a model material to study* in vitro

One of the major products of *in vitro* synthesis of Col El DNA
is a class of completely replicated molecules synthesized semiconservatively (Sakakibara and Tomizawa, 1974a). These are circular
monomeric molecules with normal supercoiling. They are active in
transfection of non-colicinogenic bacteria and can be used as
templates for further *in vitro* replication. Another major product
of *in vitro* replication is a class of molecules which have a small
replication loop in a unique region (Tomizawa et al., 1974). The
loop has a size of approximately 1/15 of unit length and contains a
newly synthesized DNA fragment(s) having an average sedimentation
constant of approximately 6S (Sakakibara and Tomizawa, 1974a).
These fragments are referred to as 6S fragments. The replication
loop consists of either one double-stranded branch and one single-
stranded branch or two double-stranded branches (Tomizawa et al.,
1974). The 6S fragments present in the former type of replication
loops are predominantly L-DNA fragments that hybridize to the H
strands of Col El DNA (Tomizawa, 1975). One of the branch points of
a replication loop is located at approximately 17% of unit length
of Col El DNA from the *EcoR*l site (Fig. 2). The size of these
replication loops and the length of 6S fragments are not homogeneous
(Tomizawa et al., 1974; also see Section 4a).

The rate of synthesis of these molecules with a small replication loop is enhanced by adding 10% (vol/vol) glycerol to the
reaction mixture (Sakakibara and Tomizawa, 1974b). On the other
hand, the addition of glycerol suppresses synthesis of completely
replicated molecules. These molecules with a small replication
loop that have been accumulated in the presence of glycerol can

---

*replication of double-stranded circular DNA because of the suitable
size and ease of preparation of the DNA and because of the various
properties expressed by in vivo replication that would have
facilitated experimentation. For example, one can enrich the
concentration of endogenous template by growing colicinogenic cells
in the presence of chloramphenicol before harvesting. This was
considered to be important, since the system had to be developed by
examining DNA synthesis that is dependent on endogenous template.
Relaxation complexes and RNA-containing molecules of Col E1 DNA can
be used as exogenous templates as well as those without detectable
non-DNA components. The proteins or the RNA of these molecules are
removed in cell extracts. Open-circular molecules of Col E1 DNA can
also be used. They are converted to closed-circular twisted
molecules in cell extracts. The activity of cell extracts did not
significantly decrease after storage at −20°C for more than two years
when DNA synthesis was measured in the presence of spermidine and
glycerol. Gradual loss of activity was noticed for DNA synthesis in
the absence of glycerol. A modified method of preparation of cell
extracts has been reported (Staudenbauer, 1976a).*

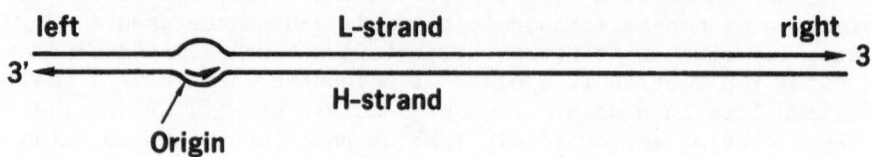

Figure 2.   *Schematic representation of an early replicative inter-*
mediate containing a 6S L-fragment after cleavage by EcoR1
(Tomizawa, 1975). The location of the loop is determined electron
microscopically.   The 5'-end of a 6S L-fragment is tentatively
assigned as the origin of replication.

complete replication when incubated with a cell extract in the
absence of glycerol (Tomizawa, 1975) or simply by reduction of the
concentration of glycerol in a reaction mixture by dilution
(Sakakibara and Tomizawa, 1974b).   Therefore these molecules consist
of a class of replicative intermediates.   They are named early
replicative intermediates.   The first product of synthesis is a 6S
L-fragment (Tomizawa, 1975). Dimethyl sulfoxide (1 to 10% vol/vol)
has an effect similar to glycerol (Y. Sakakibara, personal
communcation; T. Itoh, unpublished result).   The mechanism of
action of these substances is unknown.

The results described above indicate that a rate limiting
step separates the formation of the early replicative intermediates
from their further replication and that this separation is
accentuated by the presence of glycerol.   By analysing the processes
that occur in the presence of glycerol we can study the early events
in Col El DNA replication separately from those that occur later.

Replication of the early replicative intermediates continues by
unidirectional expansion of the replication loop as shown by
electron microscopic examination of molecules replicated to various
extents (Tomizawa et al., 1974)[*].   Replication occurs more or less
simultaneously on both parental strands so that both branches of the
replication loop are mostly double-stranded.   Before completion of
replication of the molecule, two daughter molecules are formed. They

---

[*]*Unidirectional expansion of a replication loop, as was found in*
electron microscopic studies, does not mean that initiation of
replication occurred from the fixed branch point.   The method, in
principle, does not reveal the site of initiation of replication and
direction of replication within the smallest replication loop
observed (Tomizawa, 1975).   In addition, there is no reason to assume

are open-circular monomeric molecules with a gap at the origin-terminus region of the newly synthesized L- or H-strand (Sakakibara and Tomizawa, 1974c and Section 7). The gaps are then filled and completely replicated twisted circular molecules are formed.

In addition to completely replicated monomeric molecules, dimeric catenanes which consist of two interlocked circular mono-meric molecules are formed in the reaction mixture as a minor product (Sakakibara et al., 1976). Catenanes consisting of two open-circular monomeric units with a gap in the newly synthesized strand are first synthesized. The sealing of the open-circular unit with a gap in the L-strand is followed by the sealing of the open-circular unit with a gap in the H-strand. Thus sequential sealing results in the formation of a dimeric catenane consisting of two monomeric twisted circular units.

The processes of formation of monomeric molecules or a catenane by replication of a closed-circular molecule are schematically presented in Fig. 3. These processes are subjected to further discussion in Sections 6 and 7.

4. Analysis of the Functions that are Required for *in vitro* Col E1 DNA Synthesis

Col E1 DNA added to a reaction mixture containing a cell extract made from non-colicinogenic bacteria can complete a round of replication in the absence of *de novo* protein synthesis (Tomizawa et al., 1975). This result indicates that the functions coded in the plasmid DNA are not essential for Col E1 DNA replication. In other words, the plasmid DNA can replicate entirely by functions provided by the bacteria.

The rate of synthesis of 6S L-fragments by a cell extract (prepared by centrifugation for 20 min at 60,000 x g) is essentially unchanged in the supernatant fluid obtained by further centrifugation

---

that a replication loop expands in the direction of replication of the first DNA fragment involved in initiation of DNA replication. To assign the site of initiation of replication by electron microscopy, it is essential to at least know the polarity of DNA strands and the strand on which the first fragment is synthesized. In this way the site of initiation (origin) of DNA replication, as shown in Fig. 2, is tentatively assigned. However, the site of origin thus assigned simply means the 5'-end of the fragment in the molecule under observation. The fragment might have been subjected to nuclease action and the end might thus represent a structure generated by its resistance to the nuclease action.

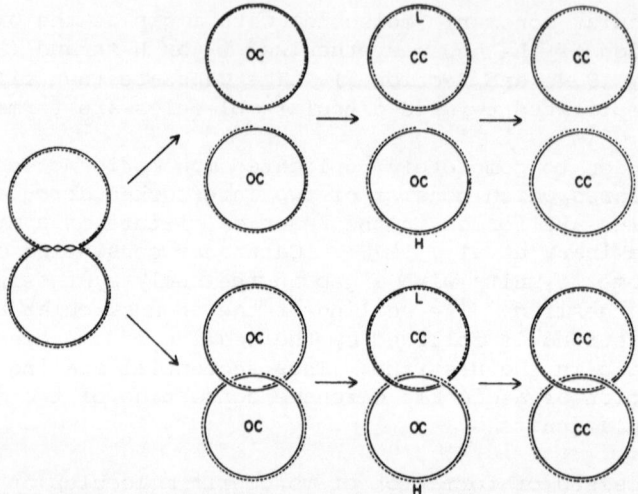

Figure 3. *Schematic representation of the processes of formation of monomeric molecules or a catenane by replication of a closed-circular DNA molecule (Sakakibara et al., 1976). Continuous and dotted lines indicate parental and newly formed strands, respectively. An outer circle shows an H-strand and an inner circle shows an L-strand. OC and CC represent open-circular and closed-circular units, respectively.*

for 3 hours at 150,000 x g. This latter supernatant fluid also is able to carry out complete replication of Col El DNA, although the rate is reduced to about half (J. Tomizawa, unpublished result). Therefore, particulate materials do not seem to be essential components for Col El DNA replication.

In vitro replication of Col El DNA consists of various consecutive steps. In this and the following sections, the functions that are required for each step are discussed.

a. The First Phase of DNA Synthesis: Synthesis of 6S L-Fragments

The requirement for the four ribonucleoside triphosphates for synthesis of 6S fragments suggests the involvement of RNA synthesis in synthesis of the DNA (Sakakibara and Tomizawa, 1974a). Inhibition of synthesis of 6S fragments by rifampicin (Sakakibara and Tomizawa, 1974a and b) or by antibody to *E.coli* DNA-dependent RNA polymerase (J. Tomizawa, unpublished result) indicates participation of RNA polymerase, or at least its β subunit, in DNA synthesis.

Rifampicin added to a reaction mixture at the beginning of

incubation completely inhibits Col El DNA synthesis. However, when
Col El DNA is pre-incubated with the ribonucleoside triphosphates in
a cell extract, from which small molecular weight components have
been removed by gel filtration or dialysis, and then rifampicin and
the deoxyribonucleoside triphosphates are added, Col El DNA is
synthesized (Sakakibara and Tomizawa, 1974a and b). The cell
extract used for the preincubation can be replaced by purified RNA
polymerase when a cell extract and the deoxyribonucleoside triphosphates as well as rifampicin are subsequently added (J. Tomizawa,
unpublished result). In either case, molecules with a small
replication loop containing 6S fragment(s) at the normal position
are formed in the presence of glycerol. This result indicates that
the process that involves RNA synthesis precedes synthesis of 6S
L-DNA.

When an extract of *polAl* bacteria is used, synthesis of Col El
DNA is not detected (J. Tomizawa and T. Kakefuda, unpublished
result). By the addition of purified DNA polymerase I, the system
acquires the ability to synthesize DNA. The synthesis is sensitive
to rifampicin and early replicative intermediates which are
indistinguishable from the normal ones are formed in the presence of
glycerol. The *polAl* extract contains the 5' → 3' exonuclease
activity of DNA polymerase I indicating that the polymerase activity
and/or 3' → 5' exonuclease activity of DNA polymerase I are involved
in Col El DNA replication. The absence of detectable Col El DNA
synthesis in the *polA* extract suggests the participation of the
polymerase activity in 6S L-DNA synthesis. A requirement for DNA
polymerase I was also observed for Col El DNA synthesis in a cell
extract prepared by a modified method (Staudenbauer, 1976b).

Synthesis of early replicative intermediates in an extract is
inhibited by novobiocin and the related antibiotic coumermycin
(Gellert et al., 1976b). These drugs have been shown to be
preferential inhibitors of the action of DNA gyrase. Synthesis
of 6S fragments in an extract made from coumermycin-resistant
bacteria is as resistant to the drug as DNA gyrase from the
resistant bacteria. Inhibition of synthesis of 6S fragments by
coumermycin in an extract made from sensitive bacteria is relieved
by the addition of DNA gyrase purified from the resistant bacteria.
In vitro Col El DNA synthesis is inhibited by nalidixic acid and
the related drug oxolinic acid which also inhibit the function of
DNA gyrase (Gellert et al., 1977). These drugs do not inhibit Col
El DNA synthesis in extracts made from *nalA* bacteria that are
resistant to these drugs and which have drug-resistant DNA gyrase.
Addition of DNA gyrase prepared from the *nalA* bacteria permits Col
El DNA synthesis in the presence of oxolinic acid in an extract
prepared from the sensitive bacteria.

In the case of *in vitro* replication of øX174-RFI DNA, a
supercoiled DNA substrate is required to permit the endonucleolytic

action of ∅X174-A protein that is necessary for initiation of DNA
synthesis (Ikeda et al., 1976). On the other hand, there is no
evidence of previous endonucleolytic action for the initiation of
replication of Col El DNA. Although supercoiling of Col El DNA
is required for synthesis of 6S fragments, it is not known for
which early step of DNA synthesis the function of DNA gyrase is
required.

    A mixture of RNA polymerase, DNA polymerase I and DNA gyrase
with or without DNA binding protein cannot replace the cell extract
in synthesis of 6S fragments. An active enzyme mixture can be
obtained by adding to the above mixture an enzyme fraction obtained
by ammonium sulfate fractionation (40 to 55% saturation) of the
supernatant fluid of a streptomycin-treated cell extract. Most of
the DNA synthesized in the mixture are L-strands longer than a third
of unit length. An ammonium sulfate fraction (0 to 40%
saturation) contains a component that limits the length of the
strands to approximately the size of 6S fragments. In either case,
DNA synthesis is sensitive to rifampicin and initiates from unique
nucleotides (see Section 5) (T. Itoh and J. Tomizawa, unpublished
result).

    6S L-fragments occur in a number of discrete sizes as analyzed
by acrylamide gel electrophoresis (H. Ohmori, R.E. Bird and J.
Tomizawa, unpublished result). The average length varies depending
on the kind of cell extract used and the conditions of incubation.
Since the synthesis of these fragments is initiated from a unique
region (see Section 5), heterogeneity of the length of the fragments
is probably a result of the heterogeneity of the termination points.
Heterogeneous but discrete points of termination of synthesis of
6S L-fragments may be a reflection of intermittent polymerization
of deoxyribonucleotides in DNA synthesis. Such phenomena as
detachment of proteins from DNA and dissolution of a cruciform made
by a palindrome ahead of a replication fork, to provide a single
stranded region to be copied, are not likely to occur at a constant
rate.

    Why is chain elongation arrested after synthesis of DNA
fragments having a certain size range ? The possibility that
reduction of negative superhelical density of a molecule by
replication is the cause of termination is unlikely since the
average sizes of early fragments made on derivatives of Col El DNA
of various sizes are roughly the same (J. Tomizawa and T. Kakefuda,
unpublished result). In view of the existence of a cellular
component(s) that limits the length of L-strands as described
above, it seems likely that the length is determined by the
relative activities of components that participate in DNA synthesis.

b. The second phase of DNA synthesis: Chain elongation

Col El DNA replicates by unidirectional expansion of a
replication loop of an early replicative intermediate (Tomizawa
et al., 1974). Usually DNA synthesis proceeds more or less
simultaneously on both parental strands. However, under some
conditions (see later) only the L-strand extends by forming an
expanded D-loop structure.

If one assumes that nucleotides are polymerized only by
5' → 3' extension of the polynucleotides, the presence of molecules
with replication loops of various sizes which carry two double-
stranded branches can be taken to mean that at least the H-strand
is synthesized by a discontinuous mechanism (Okazaki et al., 1968).
In fact, DNA fragments which are presumed to be synthesized discontinuously have been obtained by pulse-labelling of Col El DNA
replicating in intact (Oka and Inselburg, 1975) or toluene-treated
bacteria (Staudenbauer, 1974). The interesting question is whether
the L-strand is synthesized by a discontinuous mechanism or not.
Experiments carried out with an extract made from *lig ts7* bacteria
with thermolabile DNA ligase show that early replicative intermediates are synthesized but further replication of the intermediates to completion is depressed at a restrictive temperature
for the DNA ligase (Y. Sakakibara, personal communication). Instead
of synthesis of elongated strands, DNA fragments of approximately
7S are synthesized at that temperature. They are hybridizable
to various portions of the L- and H-strands of Col El DNA. The
DNA fragments extend to longer DNA chains upon chasing at a low
temperature. These results imply that both strands of Col El DNA
are synthesized by a discontinuous mechanism. However, the
possibility that these fragments are formed as a result of incomplete
repair of some kinds of mismatched nucleotides in the process of
replication (Tye et al., 1977) has not been completely ruled out.

It has been shown that nicotinamide mononucleotide (NMN), an
inhibitor of the action of *E.coli* DNA ligase, does not affect
continuation of replication and formation of daughter open-
circular molecules but blocks the final sealing of the newly
synthesized strands for circularization (Sakakibara and Tomizawa,
1974c). Under these conditions, NMN blocks sealing of DNase I-
generated nicks in Col El DNA. If *E.coli* DNA ligase is involved
in the joining process during discontinuous replication, it must be
in some way protected from the inhibitory action of NMN.
Alternatively the joining process might be carried out by another
DNA ligase, the action of which is not inhibited by NMN.

E.coli DNA-dependent RNA polymerase is involved in the
initiation of synthesis of the first fragment made on Col El DNA.
Once this process is completed, the RNA polymerase is no longer
needed for completion of a round of replication of the intermediate.

On the other hand, some functions that do not participate in the synthesis of 6S L-fragments are involved in the progression of replication beyond this point. One such function is the *dnaC* (and *dnaD*) function (T. Itoh and J. Tomizawa, unpublished result). In an extract made from *dnaC*28 (and *dnaD*7) bacteria only 6S L-fragments are synthesized. The defective extract can be complemented by purified *dnaC* protein for synthesis of completely replicated molecules. An ammonium sulfate fraction (40 to 70% saturation) of the supernatant fluid of a streptomycin-treated *dnaC*28 extract synthesizes L-strands which are elongated to various extents. When another fraction (0 to 40% saturation) is added to the above fraction, the length of L-strands synthesized are shortened. For synthesis of completely replicated molecules, both of the ammonium sulfate fractions and the *dnaC* protein are required.

Extracts made from *dnaB*1029 bacteria can support synthesis of 6S L-DNA but cannot carry out complete replication. The defective extract can be complemented with purified *dnaB* protein for completion of replication (T. Itoh and J. Tomizawa, unpublished result). Relative resistance of *in vivo* Col El DNA replication in *dnaC*ts and *dnaB*ts bacteria by incubation at high temperature is probably due to incomplete inactivation of these proteins.

DNA polymerase III may be involved in the second phase of DNA synthesis (Y. Sakakibara, personal communication; Staudenbauer, 1976b).

Coumermycin, which inhibits 6S DNA synthesis (Gellert et al., 1976), also prevents further replication in the presence of rifampicin of previously formed replicative intermediates (J. Tomizawa and T. Itoh, unpublished result). This result indicates that elongation of DNA chains requires a supercoiled DNA template. DNA gyrase may act to relieve the positive supercoiling strain imposed by replication or more actively, to facilitate separation of the parental strands by imposing a negative supercoiling strain. Relaxed covalently closed circular DNA has previously been observed *in vivo* as a transient, newly replicated, species of certain plasmids (Timmis et al., 1976; Crosa et al., 1976). These molecules may represent a situation where newly sealed covalently circular DNA can be transiently captured in a relaxed state before DNA gyrase can act on it.

The mode of synthesis of DNA fragments on Col El DNA is schematically presented in Fig. 4 and summarized as follows: (1) Synthesis of the L-fragment from the origin of replication requires a unique set of proteins. A different set(s) of proteins is required for synthesis of later fragments; (2) The H-strand is synthesized discontinuously; (3) The L-strand is extended continuously (Fig. 4A) or synthesized discontinuously (Fig. 4B); (4) DNA is synthesized more or less simultaneously on both strands

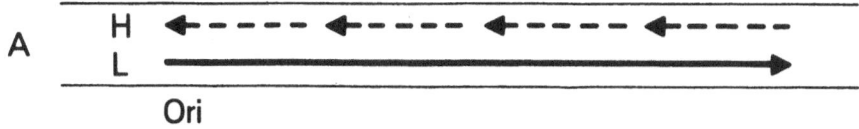

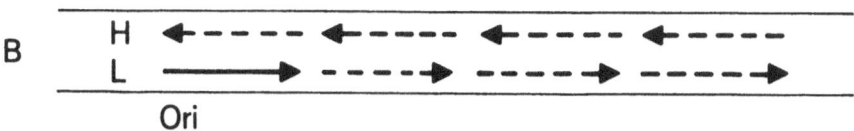

Figure 4. *Schematic representation of DNA fragments synthesized on the H- and L-strands. The continuous and discontinuous arrows represent DNA fragments synthesized by different mechanisms. Separate strands are joined eventually.*

in normal replication. Synthesis of H-strand fragments is required for continuous synthesis of the L-strand (Fig. 4A) or discontinuous synthesis of L-fragments beyond the first one (Fig. 4B).

In any case, the mechanisms of synthesis of the first fragment from a specific origin of replication is different from that of the fragments subsequently formed. If one type of synthesis is dependent upon the other, utilization of two distinct mechanisms for synthesis of DNA fragments by a single replicon may be advantageous by reducing the chance of aberrant initiation of replication of the replicon (Tomizawa, 1975).

5. Origin of DNA Replication

In the following discussion on the location of the origin of DNA replication in the nucleotide sequence, the origin is defined as the deoxyribonucleotide residue of the DNA molecule from which polymerization of deoxyribonucleotides is initiated.

The direction of replication of a 6S L-fragment and the location of the replication loop in an early replicative inter-mediate indicate that the 5'-end of a 6S L-fragment is located approximately 17% of the length from the left end of Col E1 DNA cleaved by *Eco*R1 (see Fig. 2). The position is in the *Hae*II-C segment of pNT1, a smaller derivative of Col E1 (Fig. 5) (Tomizawa et al., 1977). Examination of DNA fragments formed by digestion with *Hae*II and *Hae*III of the molecules prepared by hybridization of uniformly labelled 6S L-fragments to H-strands of Col E1 DNA shows that the 5'-end of the 6S L-fragment, treated, or not treated with alkali to hydrolyze RNA, is in the δ segment or near

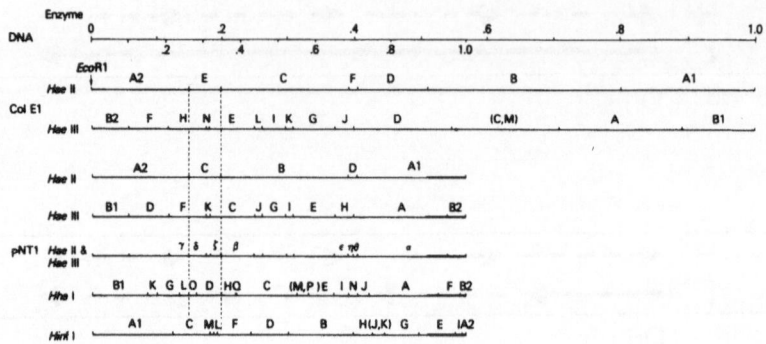

Figure 5. *Cleavage maps for Col E1 and pNT1 (Tomizawa et al.,*
1977).

the left end in the *Hae*III-K segment of pNT1.   If the origin of
DNA replication is conserved in the isolated 6S L-fragments, the
site of the origin must be in the region described above.
Experiments to determine the site are carried out by assuming
that some 6S L-fragments may retain presumed primer RNA.   Some
fragments, in fact, were found to contain ribonucleotide residues
at their 5'-ends and this allowed us to determine the site of
origin of replication as follows.   6S L-fragments were treated
with alkali to hydrolyze the primer RNA and the hydroxyl groups
at the 5'-ends uncovered by alkali treatment were labelled with
$(\gamma-^{32}P)$ATP and polynucleotide kinase.   The deoxyribonucleotide
residue thus labelled was found to be the dA, dC or dC residue
which is located at 14, 15 or 16 residues from the right end of
the δ segment, respectively (Fig. 6).   Analysis of 6S L-fragments
labelled at the 5'-ends show that one, or a few rA residues
are present on some 6S L-fragments.   Examination of the 5'-ends of
the 6S L-fragments which have been treated with alkaline phosphatase and alkali shows that the 5'-end is almost exclusively
one of the three residues described above (Tomizawa et al., 1977).

A search for 2'(3')-ribonucleoside monophosphates formed by
alkali treatment of 6S L-fragments labelled with one $(\alpha^{32}P)$
deoxyribonucleoside monophosphate has shown that ribonucleotide-
deoxyribonucleotide linkages in the 6S L-fragments are
predominantly rA-dA and rA-dC (Bird and Tomizawa, 1977) in agreement
with the result described above. It was also found that most of
the 6S L-fragments that contain ribonucleotides have a single rA
residue and a large fraction of this residue is phosphorylated at
the 5'-end.   From an estimate of the frequency of transfer of
phosphorus atoms in DNA to ribonucleotides by alkali hydrolysis
and the length of 6S L-DNA, it is calculated that between 20 and
50%, depending on the preparation, of 6S L-fragments contain at
least one ribonucleotide residue. This result is confirmed by

```
              -100        -90         -80         -70         -60         -50
L-strand : 5'-GGGAAACGCCTGGTATCTTTATAGTCCTGTCGGGTTTCGCCACCTCTGACT-
                        *                      *

H-strand : 3'-CCCTTTGCGGACCATAGAAATATCAGGACAGCCCAAAGCGGTGGAGACTGA-
                         *
```

```
           -40         -30        -20        -10          1          10
TGAGCGTCTATTTTTGTGATGCTCGTCAGGGGGGCGGAGCCTATGGAAAAACGCCTGCTACGTGG-

ACTCGCAGATAAAAACACTACGAGCAGTCCCCCCGCCTCGGATACCTTTTTGCGGACGATGCACC-
```

Figure 6. The nucleotide sequence of the right-hand two-thirds of the δ segment (Tomizawa et al., 1977). Č indicates a modified cytosine residue. The number indicates the position from the left-most RNA-DNA transition point.

determining the sensitivity to bovine spleen phosphodiesterase (Kurosawa et al., 1975) of some preparations of uniformly labelled 6S L-fragments which have been phosphorylated at the 5'-ends and then treated with alkali. Spleen phosphodiesterase is known to hydrolyze 5'-hydroxyl terminated but not 5'-phosphoryl terminated polynucleotides (Razzell and Khorana, 1961).

Since synthesis of 6S L-fragments requires prior RNA synthesis by *E. coli* DNA-dependent RNA polymerase with all four ribonucleoside triphosphates, the synthesis probably depends on an RNA primer which contains all four ribonucleoside monophosphates. In that case the presence of DNA molecules with a few or no ribonucleotide residues means that most or all of the residues of the RNA primer have been removed from the DNA product. Although the RNA primers seem to be removed efficiently, most of the deoxyribonucleotide residues from which polymerization of deoxyribonucleotides is initiated are preserved.

There are at least two enzymes in the cell extracts that possess activities capable of degrading an RNA primer if it remains hybridized to the template and leaving a phosphate group at the 5'-end on the polynucleotide. One of them is the 5' → 3' exonuclease of DNA polymerase I (Kornberg, 1969). This nuclease activity, however, does not necessarily stop at the end of the RNA primer. We have shown that most of the deoxyribonucleotide residues of 6S L-fragments from which polymerization of deoxyribonucleotides is initiated are preserved. Therefore this enzyme is not a likely candidate for the removal of the RNA primer. The other enzyme, RNase H, digests RNA hybridized to DNA (Berkower et al., 1973; Miller et al., 1973; Henry et al., 1973). It has been reported that when RNase H digests the RNA primer, the ribonucleotide adjacent to the first deoxyribonucleotide remains. On the other hand, some preparations of RNase H have been shown to

remove this last ribonucleotide (Darlix, 1975). Whatever the
mechanism by which the RNA is removed, the bulk of the RNA primer
and sometimes even the last ribonucleotide are removed, while
leaving the first deoxyribonucleotide in place.

The transition from the primer RNA to 6S L-DNA occurs in a
region consisting of a stretch of 5 A residues. The 5'-side of the
stretch, one or two turns of the helix away, is extremely rich in
G and C. The 3'-side is also rich in G and C. It has been shown
that mRNA tends to terminate in a region rich in A/T pairs and that
one turn before that region is rich in G/C pairs (cf. Gilbert, 1976;
Roberts, 1976). Such a structural similarity of the terminal region
of the presumed primer RNA to that of some mRNA species suggests
that the signal for termination of synthesis of the primer RNA may
reside in the nucleotide sequence of this region. If so, some
heterogeneity of the site of origin of replication may reflect
heterogeneity of the site of termination of synthesis of the primer
RNA molecules. Some heterogeneity of termination site for synthesis
of mRNA molecules has been known (M. Rosenberg, personal
communication). However, it is also possible that the heterogeneity of the site of RNA-DNA junctions may be a result of
processing of the RNA molecules that have been synthesized over
the region. The site from which synthesis of the primer RNA
initiates has not been identified.

The primer for DNA replication is utilized only once in a
generation of each molecule. It is not known whether the RNA with
potential priming function is synthesized once in a generation
or many times throughout a generation.

Although the structure of the region of the RNA-DNA
transition described above seems to be quite unusual, the structural
requirements for proper functioning of the region cannot be
obtained from a single example of the sequence at the RNA-DNA
transition point. Future comparative study of various replicons,
including mutant and "recombinant" replicons, may reveal which
features of the structure are essential.

In this connection, the nucleotide sequence in the region
215 to 241 residues from the origin of replication, in the
direction of overall replication attracts attention. This region
is extremely similar to the region of the transition (Fig. 7) and
faces in the opposite direction. Since we have been unable to find
DNA fragments whose synthesis is initiated in this region, the
region is not utilized, or is utilized inefficiently as the site
of initiation of synthesis of a DNA fragment. Either the structure
of the region itself is defective in initiating DNA synthesis,
or the region is not transcribed or is transcribed infrequently.
The region might be an auxiliary origin that would operate when the

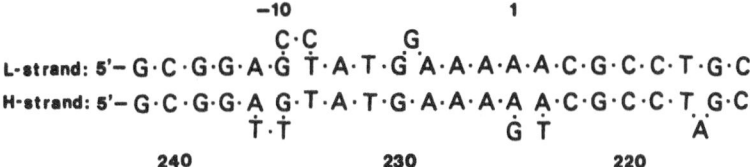

Figure 7.   Comparison of the nucleotide sequence of the region -16 to +9 (top) of the L-strand and that of the region +215 to +241 of the H-strand (bottom). The sequences (Tomizawa et al., 1977; H. Ohmori and J. Tomizawa, unpublished result) are presented in a way to show maximum matching.

primary origin does not operate. Even if the region is totally inactive as an origin of replication of the plasmid, some evolutionary significance seems to be indicated. Studies of transcription of the region, of a plasmid containing an active promoter near the region, of molecules which delete the primary origin and of various other replicons should provide answers to these questions.

The site and the mechanism of initiation of *in vitro* synthesis of Col El DNA are probably the same as those of *in vivo* synthesis. The site of one of the branch-points of replication of Col El DNA molecules as determined by electron microscopic observation of *in vitro* replicating molecules (Tomizawa et al., 1974) agrees with that of *in vivo* replicating ones (Inselburg, 1974; Lovett et al., 1974). Rifampicin, which blocks initiation but does not block continuation of *in vitro* replication (Sakakibara and Tomizawa, 1974), also inhibits *in vivo* replication of Col El (Clewell et al., 1972). We therefore speculate that both *in vivo* and *in vitro* replication initiate at the same site and that both are primed by RNA synthesized by *E.coli* DNA-dependent RNA polymerase.

It has been proposed that DNA replication initiates from the site of the strand-break formed by detachment of the proteins from the relaxation complex (Helinski et al., 1975) and the site is located close to the origin of replication (Lovett et al., 1974; Sugino et al., 1975). However, the site is in the *Hha*-H segment (of pNT1) separated from the origin of DNA replication by approximately 300 nucleotides in the direction of overall DNA replication (D. Bastia and S.M. Weissman, personal communication). The site is thus unlikely to serve as the normal origin of replication. A possible function of the relaxation complex will be discussed later.

Some Col El DNA molecules formed *in vivo* in the absence of protein synthesis contain a stretch of ribonucleotides for which a partial nucleotide sequence has been reported (Helinski et al.,

1975). We do not find any such sequence in the region of 1,200
base pairs around the origin of replication. The RNA sequences
might be remnants of RNA fragments formed during the process of
chain elongation or they might be products which are formed under
the special condition of prolonged block of protein synthesis.

6. Separation (Segregation)

Two daughter molecules are formed by separation from a
replicating molecule toward the end of a replication cycle. Col El
DNA molecules formed *in vitro* upon separation have a gap in the
newly synthesized strand at the origin-terminus region (Sakakibara
and Tomizawa, 1974). The gap is sealed and a round of Col El DNA
replication is completed by formation of two twisted circular
molecules. In addition to monomeric molecules, dimeric catenanes
are formed as a minor product (Sakakibara et al., 1976). In *in
vivo* replication of simian virus 40 DNA, where replication initiates
in a fixed region and proceeds bidirectionally to a termination
point opposite the origin of replication (Fareed et al., 1972),
molecules with an interruption in the newly synthesized strand at
the termination point have also been observed (Fareed et al., 1973).

It has been proposed that catenanes are obligatory or
preferred intermediates in the formation of monomeric molecules by
replication (Kupersztoch and Helinski, 1973; Novick et al., 1973;
Jaenisch and Levine, 1973). The experimental basis for this
proposal is that the radioactive label in the catenanes consisting
of an open-circular unit and a closed-circular unit was chased into
closed-circular molecules. The closed-circular molecules formed
by the chase were assumed to be monomeric. However, this assumption
has not been tested and, in fact, is almost impossible to test by an
in vivo experiment. In contrast, in *in vitro* Col El DNA replication,
the fate of any species of molecule can be examined by reincubating
the purified molecules in a reaction mixture. It has been
demonstrated by such an experiment that closed-circular molecules
formed from catenanes consisting of an open-circular unit and a
closed-circular unit are not monomeric molecules but catenanes
consisting of two closed-circular units (Sakakibara and Tomizawa,
1976). In this system, two monomeric molecules or a catenane are
formed as the alternative products by replication of a monomeric
molecule. Like the daughter monomeric molecules formed upon
separation, both of the monomeric units of the catenane first formed
are open-circular molecules with a gap in the newly synthesized
strands. The parental strands are intact and circular.*

---

*Catenanes once formed are maintained stably in the reaction mixture.
Formation of catenanes by interaction of monomeric molecules is insignificant during incubation in the reaction mixture. These results*

An essential event in separation of daughter molecules is complete unwinding by rotation of the parental DNA strands. For a circular molecule, this necessitates breakage of one or both of the parental strands. A model of separation has been proposed which states that the breakage of the parental strand(s) results in the temporary loss of circularity of at least one of the daughter molecules. The final circularization of the molecule in this model is ensured by terminal redundancy created by synthesis and strand-displacement (Gefter, 1975). However, breakage of a DNA strand does not necessarily result in separation of the two ends of the strand. The action of DNA unwinding proteins, such as ω·protein (Wang, 1971) or DNA gyrase (Gellert et al., 1976), shows that DNA strands can wind around each other without necessarily exposing the free ends of the s trand at the strand break. Both ends of the strand at the break are probably joined while they are held together by the enzyme. A model of separation by continuation of the replication process is presented in Fig. 8. The parental strands in the unreplicated portion of a circular replicating molecule would rotate by the action of an unwinding enzyme until they could no longer be held together. Since the unwinding requires rotation to create negative superhelicity, the enzyme used to carry out this rotation is probably DNA gyrase. If the parental strands separate when they are completely unwound, two daughter molecules will be formed. If the parental strands separate while they are still interlocked, a

---

suggest that the occurrence of genetic recombination among Col E1 DNA molecules in the in vitro system is rare. In colicinogenic bacteria, Col E1 DNA molecules usually exist as monomeric molecules. In contrast, for some derivatives of Col E1 DNA, such as Col E1-kan, mini-Col E1 (Hershfield et al., 1976) (including pNT1), Col E1d6-6 (Inselburg, 1977), a large fraction of molecules exists in polymeric forms of various extents of multiplicities (Tomizawa, unpublished result). The extent of multiplicity varies in cells of different clones and cannot be maintained at a constant value. The multiplicity is independent of recA function. These results suggest that Col E1 DNA is a poor substrate for recA mediated genetic recombination. This made it possible to study the formation of catenanes of Col E1 DNA by replication without disturbance by genetic recombination. A significant fraction of various species of circular DNA is known to exist as catenanes (cf. Sakakibara and Tomizawa, 1976). It is quite possible that some of them are formed by recombination between the component units and separate again into those units by recombination. In connection with the presence of polymeric molecules, it is worth mentioning that the copy number of a plasmid replicon has been frequently estimated by dividing the total amount of the plasmid DNA per cell by the DNA content of a monomeric molecule. The multiplicity of polymerization of a plasmid is an essential parameter for estimation of the number of copies of a plasmid replicon.

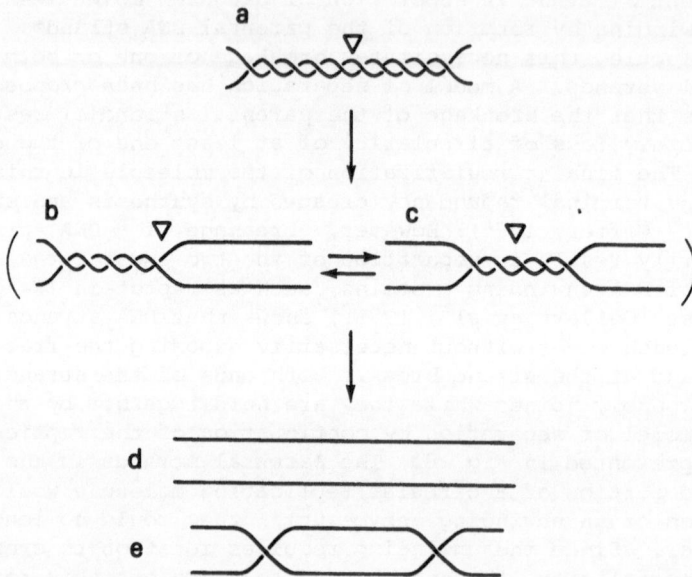

Figure 8. *A model for separation of daughter circular molecules
and formation of a catenane. In (a), the unreplicated portion of
a circular replicating molecule near the end of replication is
depicted. The lines show the parental strands. The triangle
indicates the site where DNA gyrase acts if the enzyme has site-
specificity. The parental strands unwind as the enzyme acts (b, c).
As long as the parental strands stay together, the site remains (c)
or can be generated (b to c). When the strands are completely
unwound, two daughter monomeric molecules are formed (d). When they
separate while they are still interlocked, a catenane is formed (e).*

catenane will be formed. Thus separation of daughter molecules may
require no special processes, but simply a continuation to completion
of the process of the elongation phase of replication.

7. Completion of Replication

The daughter molecules separate while they have a gap in the
newly synthesized strand at the origin-terminus region. The gap is
filled and a closed-circular molecule is formed. At least DNA
polymerase I and DNA ligase are involved in this process.* The

---

** Daughter open-circular molecules obtained by pulse labelling for
3 min from 27 min of incubation at 30°C were purified and reincub-*

process of filling the gap is one of the rate limiting processes in the *in vitro* replication of Col El DNA. However, the difference in the rates of filling of the gap in the L-strand and in the H-strand, which is most clearly demonstrated in the process of conversion of a catenane as discussed above, suggests that the rate limiting step is not simply the final ligation which may be common for the sealing of both newly synthesized strands. For synthesis of the L-strand, which is sealed faster, this step could be extension of the 3'-end existing at the time of separation of daughter molecules. For the H-strand, extension of the 3'-end which must have been prevented by some mechanism before separation, could now operate after separation. In addition, it might be necessary to initiate synthesis of an Okazaki fragment before final gap-filling of the H-strand could occur. The slower sealing of the H-strand may be a reflection of such a complex event.

A round of replication is completed when supercoiling is then imposed on the daughter molecules by DNA gyrase.

8. Direction of Replication

Replication of Col El DNA initiates from a fixed region and procedes unidirectionally. Synthesis of the leading L-fragment from the origin of replication is followed by synthesis of the lagging H-fragment. The overall direction of synthesis of the H-strand is opposite to the direction of polymerization of nucleotides. Extension of the H-strand over the origin of replication in its direction of polymerization does not occur. This inhibition of extension of the H-strand could be the crucial event that causes unidirectional replication. Without this inhibition the molecule could replicate bidirectionally.

This inhibition of extension of the H-strand could be caused by a pre-existing DNA structure. The presence of such a structure near the origin of replication could be accidental or, more likely, the structure would have an essential role for synthesis of the leading fragment at the origin of replication. Alternatively, synthesis of the leading fragment would itself create a structure that blocks extension of the lagging H-strand in the direction discussed above.

---

ated in an extract of polA1 bacteria for 30 min in the presence of rifampicin. Approximately 70% of the molecules were converted to closed-circular molecules when the extract was supplemented with purified DNA polymerase I. The efficiency was approximately 10% when polymerase I was not added. These results show that daughter molecules have a gap in a DNA strand immediately after separation and DNA polymerase I is involved in the process of filling the gap.

The mechanisms of this inhibition are not known. However,
inhibition of polymerization of nucleotides on the lagging strand
or of synthesis of a new Okazaki fragment for elongation of the
strand in the direction in question because of the presence of the
leading strand could be imagined. It should be noted that the
H-strand extends in the direction in question after formation of
open-circular daughter molecules (see Section 7) and probably
in transfer replication (see Section 9). Proteins which are
different from those used in the chain elongation phase of DNA
replication would participate in the synthesis after separation of
daughter molecules or the event of separation itself would dissolve
the structure that has blocked the synthesis.

For a molecule which replicates bidirectionally, such a block
does not exist or is not created. Whether a molecule replicates
unidirectionally or bidirectionally is determined probably by the
mechanisms that carry out synthesis of DNA fragments on both strands
in the region containing the origin of replication.

9. Transfer Replication

Col El is by itself non-transmissible but it can be transferred
to recipient bacteria with the help of a transmissible sex factor.
It has been proposed that one of the functions of the relaxation
complex is to initiate the transfer of the DNA (Helinski et al.,
1975). For some insertion derivatives of Col El, a correlation
was found between the loss of the relaxation property and the loss
of transmissibility (Inselburg, 1977).

It has been shown for F factor-mediated transfer of chromosomal
DNA that only the strand with the 5'-end is transferred irrespective
of the direction of transfer (Ohki and Tomizawa, 1968; Rupp and
Ihler, 1968; Vapnek and Rupp, 1970). If the 3'-end of an H-strand
of Col El DNA is exposed by dissociation of relaxation proteins,
the strand could be elongated from the end, displacing the H-strand
with the 5'-end to be transferred to the recipient bacterium. If,
in fact, the H-strand of Col El DNA is transferred from the 5'-end,
the overall direction of DNA replication should be opposite to that
observed for the normal replication (Fig. 9). This would provide
an example of another type of DNA synthesis experienced by the same
plasmid DNA. The synthesis may be triggered by exposure of the 3'-end
resulting from dissociation of the relaxation proteins. This could
occur without activation of the normal origin of replication (Fig.
9e) or could follow synthesis of a portion of the L-strand (Fig. 9f).
Synthesis of the L-strand may or may not continue. This model
might explain why the relaxation site is located very close to the
origin of replication in Col El DNA (Lovett et al., 1974; Sugino
et al. , 1975) and some R plasmids (Lovett et al., 1975).

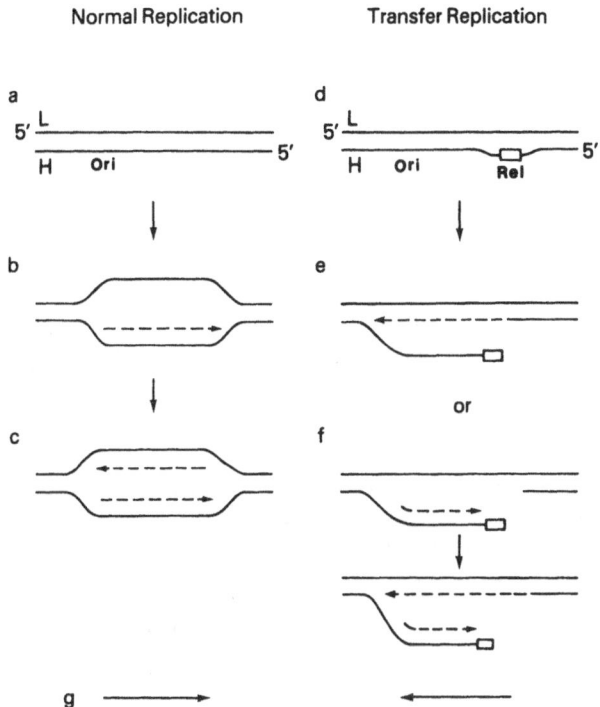

Normal Replication   Transfer Replication

Figure 9. *The early stages of normal and transfer replication. The*
early stage of normal replication (a to c) and transfer replication
(d to f) are schematically presented. The normal replication
initiates at the origin of replication (Ori) on the H-strand by
formation of a 6S L-fragment (b). The synthesis of 6S H-fragment
(c). In transfer replication, dissociation of the relaxation
proteins may exposre the 3'-end of the H-strand. The H-strand could
be elongated from the 3'-end (e) by displacing the 5'-end of the
strand. Alternatively the dissociation of the relaxation proteins
may be facilitated by the synthesis of a portion of the L-strand
initiated at the Ori site (f). Synthesis of the L-strand may or may
not be continued. The arrows in (g) indicate the overall direction
of replication.

10. Conclusion

There was no reason to presuppose that replication of Col E1
DNA in cell extracts would be a faithful representation of its
in vivo replication. It is rather surprising, therefore, that
the available evidence seems to show no major difference between
the *in vitro* and the *in vivo* replication of Col E1 DNA.

Replication of Col E1 DNA consists of various steps that occur sequentially. With this *in vitro* system, each of these steps could be studied independently from the others. Therefore, it is possible to obtain information that is specific to each step. Further biochemical analysis of each step, including the identification and purification of gene products and the reconstruction of a system with purified components, could in principle yield the complete biochemical mechanism of Col El DNA replication. It may be noted that mechanisms that may control or limit *in vivo* Col El DNA replication are not known. If there is a controlling function that directly affects replication, such a function could be studied *in vitro*.

Different steps of Col El DNA replication utilize different mechanisms for DNA synthesis.   The diversity of biochemical mechanisms of DNA synthesis by bacterial functions as observed with different species of single-stranded phage (Kornberg, 1975; Wickner and Hurwitz, 1975) is also evident for different steps of DNA synthesis of this single species of DNA. These results suggest that the synthesis of chromosomal DNA of *E.coli* also involves a variety of mechanisms of DNA synthesis. A possible advantage of having two types of mechanisms of DNA synthesis for replication of DNA has been discussed (see Section 4b). On the other hand, it would be disadvantageous for a cell to evolve too many different mechanisms of DNA synthesis. Therefore, it would be reasonable to suppose that a fair fraction of the basic mechanisms of DNA synthesis used by *E.coli* cells are also utilized for replication of Col El DNA.

ACKNOWLEDGEMENT

It is a great pleasure to express my sincere thanks to many colleagues and friends who provided me valuable comments on the manuscript.

REFERENCES

Bazaral, M. and D.R. Helinski (1968)   J. Mol. Biol.   36:185-194

Berkower, I., J. Leis and J. Hurwitz (1973)   J. Biol. Chem. 248: 5914-5921

Bird, R.E. and J. Tomizawa (1977)   J. Mol. Biol., in press

Blair, D.G., D.B. Clewell, D.J. Sherratt and D.R. Helinski (1971) Proc. Nat. Acad. Sci. 68:210-214

Blair, D.G., D.J. Sherratt, D.B. Clewell and D.R. Helinski (1972) Proc. Nat. Acad. Sci. 69:2518-2522

Cabello, F., K. Timmis and S.N. Cohen (1976) Nature 259:285-290

Clewell, D.B. and B.G. Evenchik (1973) J. Mol. Biol. 75:503-514

Clewell, D.B., B. Evenchik and J.W. Cranston (1972) Nature New Biol. 237:29-31

Clewell, D.B. and D.R. Helinski (1969)   Proc. Nat. Acad. Sci. 62:1159-1166

Clewell, D.B. and D.R. Helinski (1972) J. Bacteriol. 110:1135-1146

Collins, J., P. Williams and D.R. Helinski (1975) Molec. Gen. Genet. 136:273-289

Crosa, J.H., L.K. Luttropp and S. Falkow (1976) Nature 261:516-519

Darlix, J.-L. (1975) Eur. J. Biochem. 51:369-376

Fareed, G.C., C.F. Garon and N.P. Salzman (1972) J. Virol. 10: 484-491

Fareed, G. C., L. McKerlie and N.P. Salzman (1973) J. Mol. Biol. 74:95-111

Gefter, M.L. (1975) Ann. Rev. Biochem. 44:45-78

Geider, K. (1976) Current Topics Microbiol. Immunol. 74:55-112

Gellert, M., K. Mizuuchi, M.H. O'Dea, T. Itoh and J. Tomizawa (1977) Proc. Nat. Acad. Sci. 74: in press

Gellert, M., K. Mizuuchi, M.H. O'Dea and H. Nash (1976a) Proc. Nat. Acad. Sci. 73:3872-3876

Gellert, M., M.O'Dea, T. Itoh and J. Tomizawa (1976b) Proc. Nat. Acad. Sci. 73:4474-4478

Gilbert, W. (1976) In: RNA Polymerase. (R. Losick and M. Chamberlin, eds.) Cold Spring Harbor, New York.

Goebel, W. (1970) Biochim. Biophys. Acta 224:353-360

Goebel, W. (1970) Eur. J. Biochem. 15:311-320

Goebel, W. (1972) Nature New Biol. 237:67-70

Goebel, W. (1973) Biochem. Biophys. Res. Commun. 51:1000-1007

Goebel, W. (1974) Eur. J. Biochem. 41:51-62

Goebel, W. and H. Schrempf (1972) Biochim. Biophys. Acta 262:
    32-41

Grindley, N.D.F. and W.S. Kelley (1976) Molec. Gen. Genet. 143:
    311-318

Guiney, D.G. and D.R. Helinski (1975) J. Biol. Chem. 250:8796-
    8803

Helinski, D.R., M.A. Lovett, P.H. Williams, L. Katz, J. Collins,
    Y. Kupersztoch-Portonoy, S. Sato, R.W. Leavitt, R. Sparks,
    V. Hershfield, D.G. Guiney and D.G. Blair (1975) In: DNA
    Synthesis and Its Regulation (M. Goulian, P. Hanawalt and
    C.F. Fox, eds.) W.A. Benjamin, Menlo Park, Calif. pp 514-536

Henry, C.M., F.-J. Ferdinand and ·R. Knippers (1973) Biochem.
    Biophys. Res. Commun. 50:603-611

Hershfield, V., H.W. Boyer, L. Chow and D.R. Helinski (1976)
    J. Bacteriol. 126:447-453

Ikeda, J., A. Yudelevich and J. Hurwitz (1976) Proc. Nat. Acad.
    Sci. 73:2669-2673

Inselburg, J. (1974) Proc. Nat. Acad. Sci. 71:2256-2269

Inselburg, J. (1977) J. Bacteriol., in press

Inselburg, J. and M. Fuke (1971) Proc. Nat. Acad. Sci. 68:2839-
    2842

Inselburg, J. and P. Ware (1977) J. Bacteriol., in press

Jaenisch, R. and A. Levine (1973) J. Mol. Biol. 73:199

Kingsbury, D.T. and D.R. Helinski (1970) Biochem. Biophys. Res.
    Commun. 41:1538-1544

Kingsbury, D.T. and D.R. Helinski (1973a) Genetics 74:17-31

Kingsbury, D.T. and D.R. Helinski (1973b) J. Bacteriol. 114:
    1116-1124

Kingsbury, D.T., D.G. Sieckmann and D.R. Helinski (1973) Genetics
    74:1-16

Kornberg, A. (1969) Science 163:1410-1418

Kornberg, A. (1976) In: RNA Polymerase (R. Losick and
    M. Chamberlain, eds.) Cold Spring Harbor, New York, pp331-352

Kupersztoch, Y.M. and D.R. Helinski, (1973) Biochem. Biophys.
    Res. Commun. 54:1451-1459

Kurosawa, Y., T. Ogawa, S. Hirose, T. Okazaki and R. Okazaki (1975)
    J. Mol. Biol. 96:653-664

Lovett, M.A., D.G. Guiney and D.R. Helinski (1974) Proc. Nat.
    Acad. Sci. 71:3854-3857

Lovett, M.A. and D.R. Helinski (1975) J. Biol. Chem. 250:8790-
    8795

Lovett, M.A., L. Katz and D.R. Helinski (1974) Nature 251:337-340

Lovett, M.A., R.B. Sparks and D.R. Helinski (1975) Proc. Nat.
    Acad. Sci. 72:2905-2909

Miller, H.I., A.D. Riggs and G.N. Gill (1973) Identification and
    Characterization. J. Biol. Chem. 248:2621-2624

Novick, R.P., K. Smith, R.J. Sheehy and E. Murphy (1973) Biochem.
    Biophys. Res. Commun. 54:1460-1469

Ohki, M. and J. Tomizawa (1968) Cold Spring Harbor Symp. Quant.
    Biol. 33:651-658

Oka, A. and J. Inselburg (1975) Proc. Nat. Acad. Sci. 72:829-833

Oka, A. and M. Takanami (1976) Nature 264:193-196

Okazaki, R., T. Okazaki, K. Sakabe, K. Sugimoto, R. Kainuma, A.
    Sugino, and N. Iwatsuki (1968) Cold Spring Harbor Symp. Quant.
    Biol. 33:129-143

Roberts, J.W. (1976) In: RNA Polymerase. (R. Losick and
    M. Chamberlain, eds.) Cold Spring Harbor, New York. pp247-271

Razzell, W.E. and G. Khorana (1961) J. Biol. Chem. 236:1144-1149

Rupp, W.D. and G. Ihler (1968) Cold Spring Harbor Symp. Quant.
    Biol. 33:647-650

Sakakibara, Y. and J. Tomizawa (1974) Proc. Nat. Acad. Sci.
    71:802-806

Sakakibara, Y. and J. Tomizawa (1974) Proc. Nat. Acad. Sci. 71:
    1403-1407

Sakakibara, Y. and J. Tomizawa (1974) Proc. Nat. Acad. Sci.,
    71:4935-4939

Sakakibara, Y., K. Suzuki and J. Tomizawa (1976) J. Mol. Biol.
    108:569-582

Staudenbauer, W.L. (1974) Nucleic Acids Res. 1:1153-1164

Staudenbauer, W.L. (1976) Molec. Gen. Genet. 145:273-280

Staudenbauer, W.L. (1976b) Molec. Gen. Genet. 149:151-158

Sugino, Y., J. Tomizawa and T. Kakefuda (1975) Nature 253:652-654

Tacon, W. and D. Sherratt (1976) Molec. Gen. Genet. 147:331-335

Timmis, K., F. Cabello and S.N. Cohen (1974) Proc. Nat. Acad.
    Sci. 71:4556-4560

Timmis, K., F. Cabello and S.N. Cohen (1976) Nature 261:512-516

Tomizawa, J. (1975) Nature 257:253-254

Tomizawa, J., H. Ohmori and R.E. Bird (1977) Proc. Nat. Acad.
    Sci. 74:1865-1869

Tomizawa, J., Y. Sakakibara and T. Kakefuda (1974) Proc. Nat.
    Acad. Sci. 71:2260-2264

Tomizawa, J., Y. Sakakibara and T. Kakefuda (1975) Proc. Nat.
    Acad. Sci. 72:1050-1054

Tye, B.-K., P.-O. Nyman, I.R. Lehman, S. Hochhauser and B. Weiss
    (1977) Proc. Nat. Acad. Sci. 74:154-157

Vapnek, D. and D. Rupp (1970) J. Mol. Biol. 53:287-303

Wang, J.C. (1971) J. Mol. Biol. 55:523-533

Wickner, S. and J. Hurwitz (1975) In: DNA Synthesis and Its
    Regulation. (M. Goulian, P. Hanawalt and C.F. Fox, eds.)
    W.A. Benjamin, Menlo Park, Calif. pp227-240

Wickner, W., D. Brutlag, R. Schekman and A. Kornberg (1972)
    Proc. Nat. Acad. Sci. 69:965-969

A TWO-STAGE MECHANISM OF PLASMID DNA REPLICATION

Walter L. Staudenbauer

Max-Planck-Institut für Biochemie
Abteilung Hofschneider
D-8033 Martinsried bei München
Federal Republic of Germany

INTRODUCTION

A group of small plasmids from enterobacteria including the
ColE1 plasmid and the ampicillin resistance plasmid RSF1030 is
capable of extensive replication in the presence of chloramphenicol
(Clewell, 1972; Crosa et al., 1976). This replicative behaviour
has been termed "relaxed replication" and is defined operationally
by an increase in the ratio of plasmid to chromosomal DNA upon
inhibition of protein synthesis. Plasmid replication in the
presence of chloramphenicol is still sensitive to inhibition by
rifampicin indicating a direct role of E.coli RNA polymerase in
the replication process (Clewell et al., 1972).

As far as the three E.coli DNA polymerases (pol I, pol II and
pol III) are concerned, relaxed plasmid replication shows a strict
dependency on normal levels of pol I and these plasmids cannot be
maintained in the *polA1* amber mutant (Kingsbury and Helinski, 1973).
In contrast to chromosomal DNA replication ColE1 DNA synthesis
continues to a limited extent after thermal inactivation of *polC*
(dnaE) mutants thermosensitive in pol III (Goebel, 1972). However,
in view of the leakiness of *polC* mutants it has been suggested that
pol III might also be required for plasmid replication (Collins
et al., 1975).

Biochemical studies on relaxed plasmid replication have been
greatly facilitated by the development of cell-free *in vitro*
systems that can carry out the replication of exogenous plasmid
DNA (Sakakibara and Tomizawa, 1974; Staudenbauer, 1976a). For the
ColE1 plasmid it has been shown that replication starts at a fixed

827

origin and proceeds unidirectionally (Tomizawa et al., 1974). Plasmid
DNA added to extracts prepared from plasmid-free bacteria is
replicated in the absence of *de novo* protein synthesis (Tomizawa
et al., 1975). This indicates that relaxed plasmid replication does
not require any plasmid-specific protein and depends solely on
bacterial replication functions. Studies with extracts prepared
from E.coli mutants defective in bacterial DNA synthesis were
therefore performed to identify the replication proteins involved.
Evidence will be presented which indicates that plasmid replication
can be divided into two stages with different enzyme requirements.

EXPERIMENTAL PROCEDURES

Bacterial strains. E.coli BT1000 (*polA1*), BT1026 (*polA1
polC1026*), and BT10265 (*polA1 polB2 polC1026*) were kindly given to
us by Dr. F. Bonhoeffer, Tübingen, and Dr. V. Nüsslein, Köln. E.coli
AX727 (*dnaZts*) and AX727 (*dnaZ+*) were generously provided by
Dr. J.R. Walker, Austin. For genetic markers see Wechsler et al.
(1973) and Filip et al. (1974).

Preparation of extracts. Cells were grown in 4 l Difco
antibiotic medium M-3 supplemented with 1% nutrient broth under
aeration at 30°C. At an OD$_{600}$ = 1.0 chloramphenicol (150 µg/ml)
was added and aeration continued for 2 hours. Cultures were
harvested at room temperature, washed once with 25 mM Hepes pH 8.0 -
5 mM EGTA, resuspended in a final volume of 18.4 ml buffer, and
quick-frozen in liquid nitrogen. Frozen cells were thawed at 10°C
and 4.6 ml aliquots were transferred into Spinco Ti 50 poly-
carbonate tubes. 0.3 ml 2 M KCl and 0.1 ml of 15 mg/ml lysozyme
were successively added to the cells, which were then incubated in
an ice bath for 30 min. The suspension was quickly frozen in
liquid nitrogen, thawed by placing the tubes for 15 min in a 10°C
water bath, and centrifuged in a Spinco Ti 50 rotor for 30 min at
45,000 rpm (4°C). The supernatants were divided into small
aliquots and immediately frozen in liquid nitrogen.

Assay of plasmid DNA synthesis. Standard incubation mixtures
(25 µl) contained final concentrations of 40 mM Hepes pH 8.0, 100
mM KCl, 10 mM magnesium acetate, 1.5 mM ATP, 0.4 mM each of CTP,
GTP and UTP, 0.05 mM NAD, 0.05 mM cAMP, 0.025 mM each of dATP,
dCTP, dGTP and dTTP, 0.25 µg plasmid DNA, and 15 µl of extract
(about 250 µg protein). Either ^{3}H-dTTP (500 cpm x pmol^{-1}) or
^{32}P-dGTP (approximately 1000 cpm x pmol^{-1}) was employed as radioactive label. Incubations were carried out at 30°C or 37°C as
indicated. Incorporation was stopped by the addition of 0.2 ml 10%
trichloroacetic acid and the reaction mixtures were assayed for
acid-insoluble radioactivity.

Preparation of plasmid DNA and purification of E.coli DNA

polymerases have been described previously (Staudenbauer, 1976b).

RESULTS

As shown in Fig. 1, the replication of the ColE1 plasmid in cell-free extracts depends strictly on pol I. Extracts prepared from BT1000 (polA1) are almost totally deficient in plasmid DNA synthesis but can be complemented by addition of purified pol I. On the other hand, addition of comparable amounts of pol II does not restore ColE1 DNA synthesis. Identical results were obtained

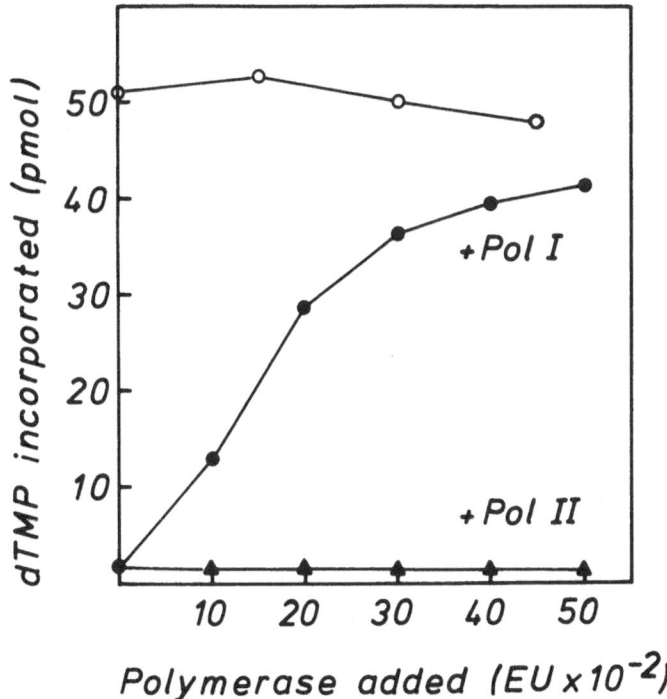

Figure 1. Complementation of polA1 extract by pol I. Standard incubation mixtures (25 µl) containing 0.25 µg ColE1 DNA were incubated for 40 min at 30°C with various amounts of pol I and pol II. One unit of DNA polymerase is defined as the amount catalyzing the incorporation of 10 nmoles of total nucleotides into activated salmon DNA in 30 min at 30°C under optimal assay conditions (Staudenbauer, 1976b). Symbols (O———O), BT1000 (polA+) extract plus pol I; (●———●), BT1000 (polA1) extract plus pol I; (▲———▲), BT1000 (polA1) extract plus pol II.

with an extract from an isogenic *polA1 polB2* double mutant
indicating that pol II is dispensible for plasmid DNA synthesis
in vitro.

To test whether pol III is involved in plasmid DNA replication
the temperature sensitivity of ColEl DNA synthesis in a BT1000
extract and a BT1026 (*polC1206*) extract was compared in the
presence of saturating amounts of pol I. As can be seen in Fig. 2A,
wild type extracts can carry out plasmid DNA synthesis at a normal
rate for 30 min at 37°C. By contrast, *polC1026* extracts show
already a slower rate of DNA synthesis at 30°C and a drastically
reduced amount of incorporation at 37°C (Fig. 2B). Moreover,
thermal inactivation of *polC1026* extracts prevents the synthesis of
closed-circular plasmid DNA and the incorporated radioactivity is
found in DNA pieces considerably shorter than one genome length
(Staudenbauer, 1976b). The residual DNA synthesis observed at 37°C
in a *polC1026* extract supplemented with pol I is increased almost
to wild type levels by addition of pol III holoenzyme isolated
according to the procedure of Wickner and Kornberg (1974). On the
other hand, addition of purified pol II has no stimulatory effect
in this system (Fig. 3). According to Wickner and Hurwitz (1976)
pol III holoenzyme preparations contain the *polC* and *dnaZ* gene

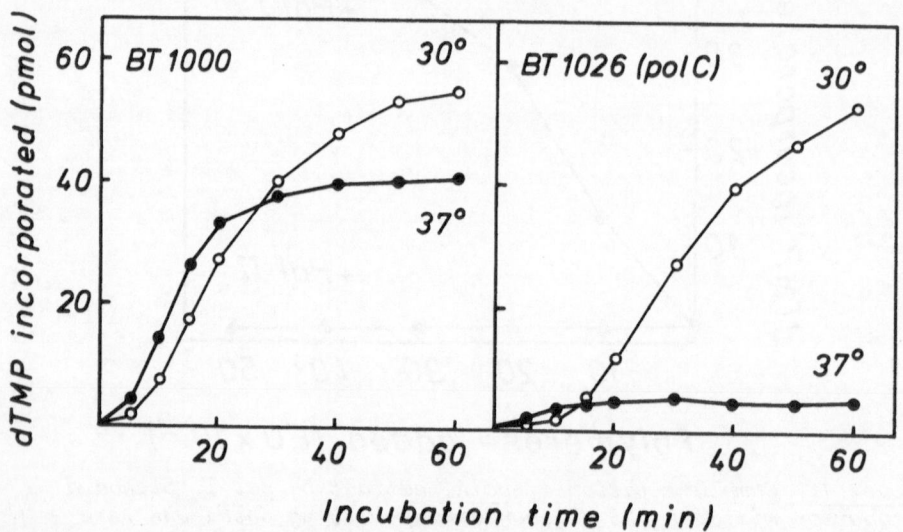

Figure 2. Temperature sensitivity of plasmid DNA synthesis. Kinetics of ³H-dTMP incorporation were determined by incubating standard reaction mixtures (250 μl) containing 2.5 μg ColE1 DNA with 150 μl extract plus pol I (5.0 units) at 30°C (O———O) or 37°C (●———●) respectively. At the times indicated 25 μl aliquots were removed and assayed for acid-insoluble radioactivity.

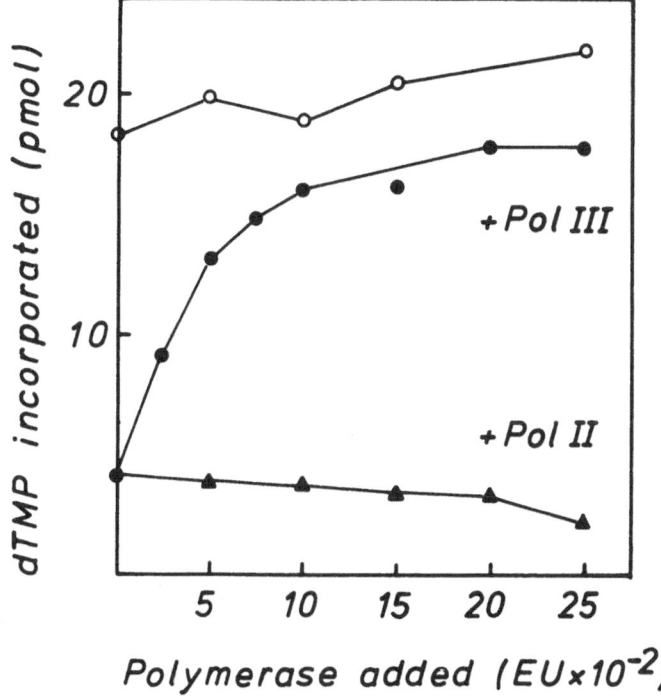

Figure 3. Complementation of polC1026 extract by pol III holoenzyme.
Standard incubation mixtures (25 μl) containing 0.25 μg RSF1030 DNA,
0.3 units pol I, and 15 μl extract were incubated for 40 min at 37°C
with various amounts of pol II or pol III holoenzyme. Symbols:
(O——O), BT1000 extract plus pol III;   (●——●), BT10265 extract
plus pol III; (▲——▲), BT10265 extract plus pol II.

products plus two other DNA elongation factors. It was therefore
of interest to investigate whether the *dnaZ* protein is also required
for plasmid DNA synthesis in vitro. Extracts prepared from E.coli
AX727 (*dnaZts*) were found to be totally defective in *dnaZ* function
(Staudenbauer, 1977). However, such extracts support a small but
significant amount of plasmid DNA synthesis. When the ColEl DNA
synthesized in a *dnaZ⁻* extract was heated to 90°C the radioactive
label was released as small fragments with a sedimentation coefficient of approximately 7S (Fig. 4A). This result indicates that the
label is incorporated into short DNA pieces which are non-covalently
associated with the template DNA. Most of the label incorporated in
the absence of a functional *dnaZ* protein can be chased into closed-
circular ColEl DNA upon addition of *dnaZ⁺* extract (Fig. 4B). It is

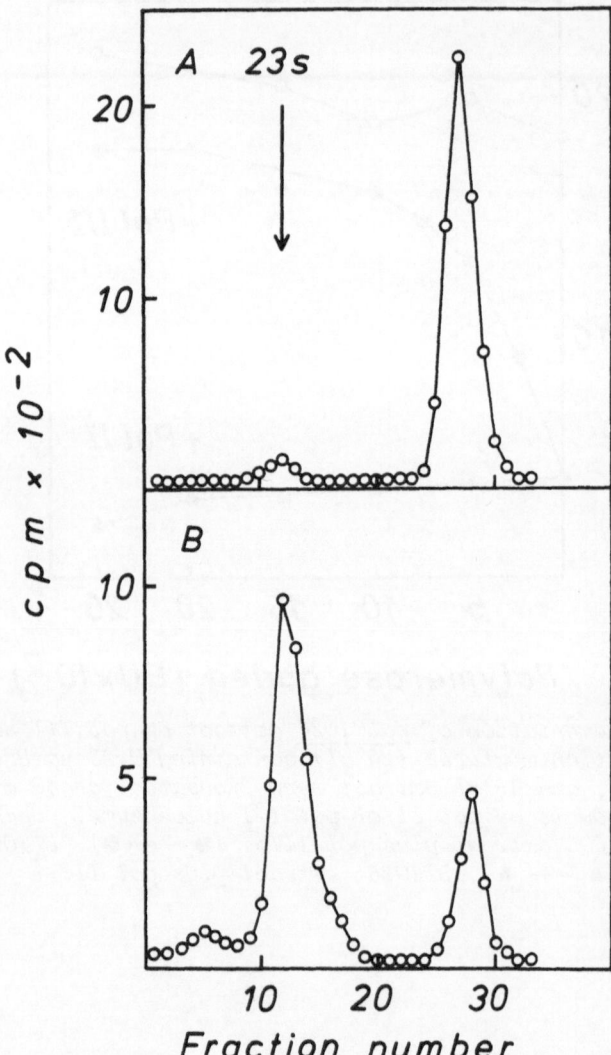

<u>Figure 4</u>. *Characterization of ColE1 DNA synthesized in a <u>dnaZ⁻</u>*
extract. A 2-ml standard incubation mixture containing 1.6 ml
AX727 (<u>dnaZts</u>) extract, 20 µg ColE1 DNA, and 10 µM ³²P-dGTP was
incubated for 30 min at 30°C. The incorporated label was then chased
by adding an equal volume of standard reaction mixture containing
AX727 (<u>dnaZ⁺</u>) extract, 10 µg/ml rifampicin, and 1 mM unlabeled
dGTP. Incubation was continued for 60 min at 30°C. DNA samples
extracted before (A) and after the chase (B) were heated for 5 min
at 90°C and analyzed by velocity sedimentation in neutral CsCl
gradients (density 1.2 - 1.4 g/ml) in a Spinco SW50 rotor for 150
min at 45,000 rpm (5°C). Sedimentation is from right to left. The
arrow indicates the position of ³H-labeled closed-circular ColE1 DNA
used as sedimentation marker.

therefore concluded that the 7S DNA pieces are derived from early
replicative intermediates which can be converted into fully
replicated molecules by addition of $dnaZ$ protein to the system.
Similar results were obtained with the RSF1030 plasmid (Staudenbauer,
1977).

Next we undertook to investigate the strand-specificity of the
label transferred from 7S DNA pieces into closed-circular ColEl DNA.
After linearization of the plasmid DNA by cleavage with the endonuclease EcoRl the denatured complementary strands were separated
by equilibrium centrifugation in a poly(U,G)-CsCl gradient. Under
these conditions the poly(U,G)-binding strand or "Crick" strand
bands at a higher density than the complementary "Watson" strand
(Hradecna and Szybalski, 1967). As can be seen from the density
profile in Fig. 5, most of the radioactivity incorporated into
early intermediate DNA in a $dnaZ^-$ extract can be chased into
"Watson" strand material upon addition of wild type extract. Only a
minor portion of the label is detected in the "Crick" strand. This
probably results from residual incorporation during the chase period.

To decide whether the 7S DNA pieces are derived from a unique
region of the ColEl DNA the early intermediates are cleaved with
endonuclease EcoRl and examined by electron microscopy (Fig. 6).
Numerous linear ColEl DNA molecules with a small internal loop were
observed. The majority of these loops appeared to have a D-loop
structure consisting of one single-stranded and one double-stranded
branch. Thirty molecules were randomly selected and the lengths of
the loops and the two linear unreplicated regions were measured.
Fig. 7 shows the line diagrams of these molecules aligned with the
short unreplicated region of DNA to the left. It was found that the
two branch points were located at 16.8 ± 1.0% and 23.9 ± 1.8% of the
total length of the molecules. The left branch point is thus
located in the same region of the molecule that has been previously
identified as the origin of ColEl replication in vivo (Inselburg,
1974). This conclusion was confirmed by physical mapping with
restriction enzymes (Staudenbauer and Beck, manuscript in
preparation).

DISCUSSION

The data presented in this paper taken together with earlier
results suggest that relaxed plasmid replication involves two
consecutive stages with different enzyme requirements. The features
of this model are summarized in Fig. 8.

Stage I: Upstream from the replication origin an RNA primer
is synthesized by RNA polymerase and then extended by pol I. Pol I
carries out the synthesis of a strand-specific "leader" DNA sequence
which is about 500 nucleotides long. The replication origin is

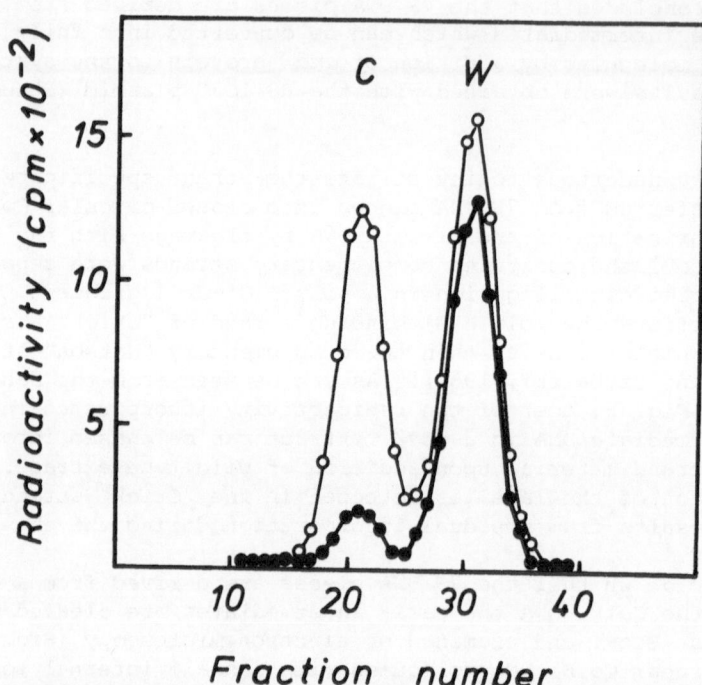

Figure 5. *Strand separation by poly(U,G)-CsCl equilibrium centri-*
fugation. Early intermediate DNA synthesized in an AX727 (dna2ts)
extract was labeled with ^{32}*P-dGTP. Incorporation was stopped by*
addition of 1 mM unlabeled dGTP and the incorporated label chased
into fully replicated molecules in the presence of wild type
extract as described in Fig. 4. Covalently closed plasmid DNA was
isolated and cleaved by treatment with endonuclease EcoR1 (Inselburg,
1974). The linearized molecules were denatured, annealed with
poly(U,G), and analyzed by CsCl density graident centrifugation
(Staudenbauer, 1976a). Density increases from right to left.
Symbols: (●——●), ^{32}*P-labeled DNA; (○——○),* ^{3}*H-labeled*
ColE1 reference DNA.

defined as the transition point from primer RNA to leader DNA.

Stage II: The D-loop structure of the early intermediate
formed during stage I is used as a substrate for the elongation
machinery of chromosomal DNA replication. Further plasmid DNA
synthesis on both template strands requires pol III holoenzyme
including the *dnaZ* protein. During this stage pol I has only a
supporting role and may be involved in gap-filling.

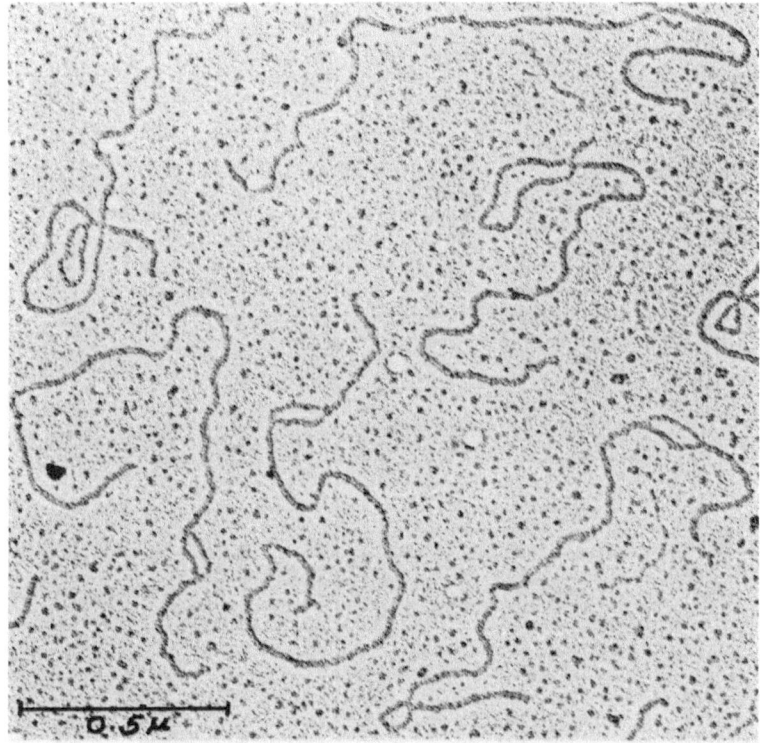

Figure 6. *Electron micrograph of early intermediates of ColE1 DNA treated with restriction endonuclease EcoR1. ColE1 DNA synthesized in a dnaZ extract was cleaved with endonuclease EcoR1, purified by CsCl density centrifugation, and dialyzed against 0.1 M Tris-HC1 pH 8.0 - 1 mM EDTA. Samples were spread for electron microscopy in the presence of 25% formamide and 1% cytochrome c as described by Davis et al. (1971). The grids were shadowed with platinum and examined in a Siemens Elmiskop 102.*

    The molecular signals which cause a switch from stage I to stage II of the plasmid replication cycle are not understood. One may speculate that the displaced single-stranded loop of the early intermediate DNA serves as attachment site for an ATP-dependent unwinding enzyme (Abdel-Monem et al., 1977). Introduction of single-stranded regions covered with DNA binding protein ahead of the leader DNA chain would then interfere with further DNA synthesis by pol I and facilitate pol III function (Molineux et al. 1974).

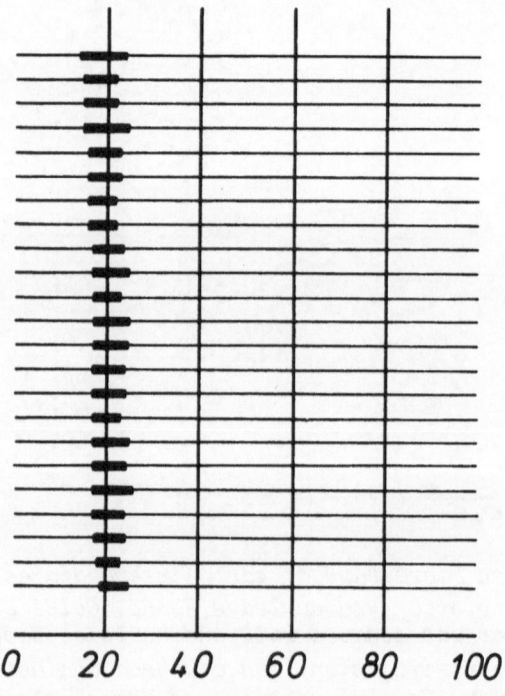

Percent of molecular length

Figure 7. Measurements of early intermediates of ColE1 DNA treated with restriction endonuclease EcoR1. Molecules were prepared for electron microscopy as described in Fig. 6. The linear structures generated by EcoR1 treatment of circular early intermediates are aligned with the short unreplicated region of DNA to the left. The location of the small loop is shown by thick lines. Measurements for each molecule are presented in terms of percent of total molecular length.

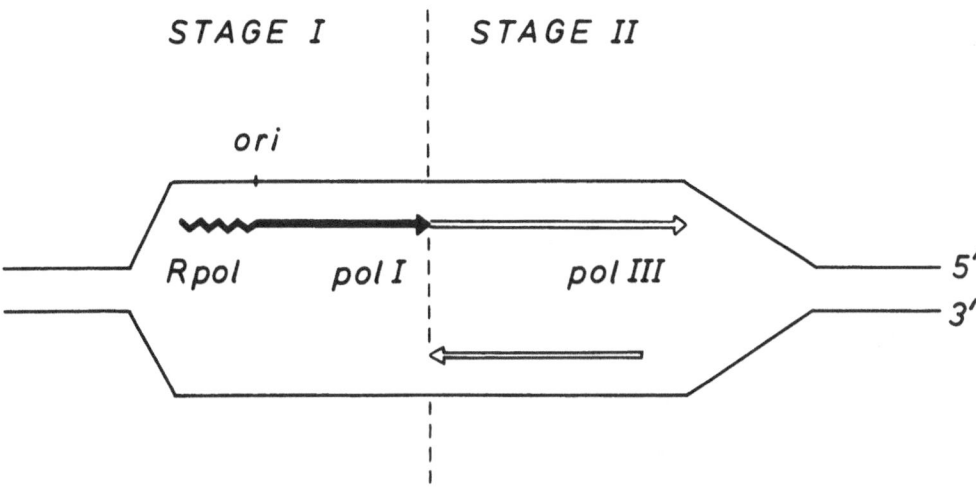

Figure 8. *Proposed model for relaxed plasmid replication. For convenience the plasmid DNA which replicates as a closed-circular structure (Oka and Inselburg, 1975) is shown in a linearized form. Symbols:* ∿∿∿∿ *, primer RNA synthesized by RNA polymerase (Rpol);* ➤ *, "leader" DNA synthesized by pol I;* ⟹ *, DNA synthesis carried out by pol III.*

ACKNOWLEDGEMENTS

I am grateful to Dr. P.H. Hofschneider for his interest and support of this work. I thank Miss E. Zehelein for skilful technical assistance. This work was supported by a grant from the Deutsche Forschungsgemeinschaft.

REFERENCES

Abdel-Monem, M., Lauppe, H., Kartenbeck, J., Dürwald, H. and Hoffman-Berling, H. (1977) J. Mol. Biol. 110:667-685

Clewell, D.B. (1972) J. Bact. 110:667-676

Clewell, D.B., Evenchik, B. and Cranston, J.W. (1972) Nature (Lond.) New Biol. 237:29-31

Collins, J., Williams, P. and Helinski, D.R. (1975) Molec. Gen. Genet. 136:273-289

Crosa, J.H., Luttropp, L.K. and Falkow, S. (1975) Proc. Natl. Acad. Sci. Wash. 72:654-658

Davis, R.W., Simon, M. and Davidson, N. (1971) Methods in
    Enzymology (eds. L. Grossman and K. Moldave) Vol. 21D,
    Academic Press, New York, pp413-428

Filip, C.C., Allen, J.S., Gustafson, R.A., Allen, R.G. and Walker
    J.R. (1974) J. Bact. 119:443-449

Goebel, W. (1972) Nature (Lond.) New Biol. 237:67-70

Hradecna, Z., and Szybalski, W. (1967) Virology 32:633-643

Inselburg, J. (1974) Proc. Natl. Acad. Sci. Wash. 71:2256-2259

Kingsbury, D.T. and Helinski, D.R. (1973) J. Bact. 114:1116-1124

Molineux, I.J., Friedman, S. and Gefter, M.L. (1974) J. Biol. Chem.
    249:6090-6098

Oka, A. and Inselburg, J. (1975) Proc. Natl. Acad. Sci. Wash.
    72:829-833

Sakakibara, Y. and Tomizawa, J. (1974) Proc. Natl. Acad. Sci.,
    Wash. 71:802-806

Staudenbauer, W.L. (1976a) Molec. Gen. Genet. 145:273-280

Staudenbauer, W.L. (1976b) Molec. Gen. Genet. 149:151-158

Staudenbauer, W.L. (1977) Molec. Gen. Genet., submitted

Tomizawa, J., Sakakibara, Y. and Kakefuda, T. (1974) Proc. Natl.
    Acad. Sci., Wash. 71:2260-2264

Tomizawa, J., Sakakibara, Y. and Kakefuda, T. (1975) Proc. Natl.
    Acad. Sci., Wash. 72:1050-1054

Tomizawa, J., Ohmori, H. and Bird, R.E. (1977) Proc. Natl. Acad.
    Sci.,Wash. 74:1865-1869

Wechsler, J.A., Nüsslein, V., Otto, B., Klein, A., Bonhoeffer, F.,
    Herrmann, R., Gloger, L. and Schaller, H. (1973) J. Bact.
    113:1381-1388

Wickner, S. and Hurwitz, J. (1976) Proc. Natl. Acad. Sci., Wash.
    73:1053-1057

Wickner, W. and Kornberg, A. (1974) J. Biol. Chem. 249:6244-6249

ADENOVIRUSES - MODEL FOR DNA REPLICATION IN MAMMALIAN CELLS - ROLE

OF VIRAL PROTEINS IN VIRAL DNA REPLICATION

M. Green[*] and W.S.M. Wold

Institute for Molecular Virology
St. Louis University School of Medicine
3681 Park Avenue
St. Louis, Missouri 63110, U.S.A.

I. INTRODUCTION

1. Human Adenoviruses-Model for DNA Replication in Mammalian Cells.

The synthesis of DNA in prokaryotic and eukaryotic cells is complex, involving numerous enzymes and other proteins that may function in multi-enzyme systems. A molecular description of DNA replication in *Escherichia coli* is only beginning to be realized, after fifteen years of intensive studies using small bacteriophage DNA molecules as probes. Full details, however, will no doubt require many years of additional study. Several of the approximately fifteen proteins that are involved in DNA replication in *E.coli* (excluding DNA precursor enzymes), as indicated by biochemical and/ or genetic studies, have not yet been identified. Moreover, functions have been established for only some of these proteins, and their complex interactions have only recently begun to be appreciated.

Because of their biochemical and genetic complexity, it has been suggested that attempts to understand mammalian cell DNA replication are premature, and should await elucidation of these processes in the simple, prokaryotic systems. However, in our view, animal DNA viruses (adenoviruses, papovaviruses, herpesviruses, poxviruses, parvoviruses) provide simple, powerful systems to study eukaryotic DNA replication. We have chosen adenoviruses (Ad), because cells replicating Ad genomes are particularly useful models to study DNA replication and its regulation for several reasons: (i) host cell DNA synthesis is blocked late after infection with human Ads, thus permitting the unambiguous analysis of viral

*A.S.I. Participant

DNA synthesis; (ii) isolated nuclei and subnuclear complexes have
been prepared from Ad2, 5, and Ad12-infected cells that synthesize
almost exclusively viral DNA sequences *in vitro*; (iii) DNA-negative
Ad2, 5, and Ad12 temperature sensitive (*ts*) mutants are available
that are very useful in understanding DNA replication; (iv) with
the exception of several viral-coded proteins, Ad DNA is replicated
mainly by cellular enzymes and other proteins and thus the study of
Ad DNA replication illuminates cellular mechanisms of DNA
replication; (v) of considerable practical importance, Ad infected
cultured human cells can be grown in very large quantities, and
thus purification and analysis of viral and cell-coded proteins
involved in DNA replication is feasible.

In this article, we review our current understanding of Ad
DNA replication, and summarize experiments from our laboratory
directed at understanding the properties of Ad-coded early proteins,
their possible role in viral DNA replication, and their association
with a sub-nuclear soluble "complex" that synthesizes Ad DNA *in
vitro*. For the sake of brevity, we cite mainly recent references.
An extensive review of adenovirus DNA replication can be found in
Wold et al. (1977).

2. Adenovirus 2 Productive Infection

Productive infection of permissive human KB cells (or HeLa
cells) in suspension culture by group C Ads (Ad1, 2, 5, 6) provides
one of the best understood eukaryotic virus-cell systems (Green
and Daesch, 1961; Wold et al., 1977). The Ad2 genome is a well
characterized linear duplex DNA molecule of 23×10^6 daltons
(see Fig. 1) which lacks duplex terminal repetitions, cohesive
ends, and circular permutations (Green et al., 1967; Green et al.,
1970). Two unusual features of Ad2 DNA, inverted terminal
repetitions and a protein covalently linked to the 5' termini, are
probably involved in mechanisms that initiate and terminate viral
DNA replication. The nucleotide sequence (102 bases) of the
inverted terminal repetition of Ad2 DNA has been established
recently. The Ad2 DNA molecule has been separated into r and l
strands (indicating rightward and leftward transcription,
respectively). Cleavage patterns with a variety of restriction
endonucleases have been established. Restriction fragments have
been used to map the region of the genome that code for viral mRNA
and proteins, and to map mutants, including three groups of DNA-
negative mutants. Most important, studies from several groups
have established the overall pattern of DNA replication and the
approximate location of origins and termini of viral DNA replication.

A model summarizing current concepts of the Ad productive
infection is illustrated in Fig. 2. Briefly, the virus replicates
with at least two stages of gene expression, "early", i.e., before

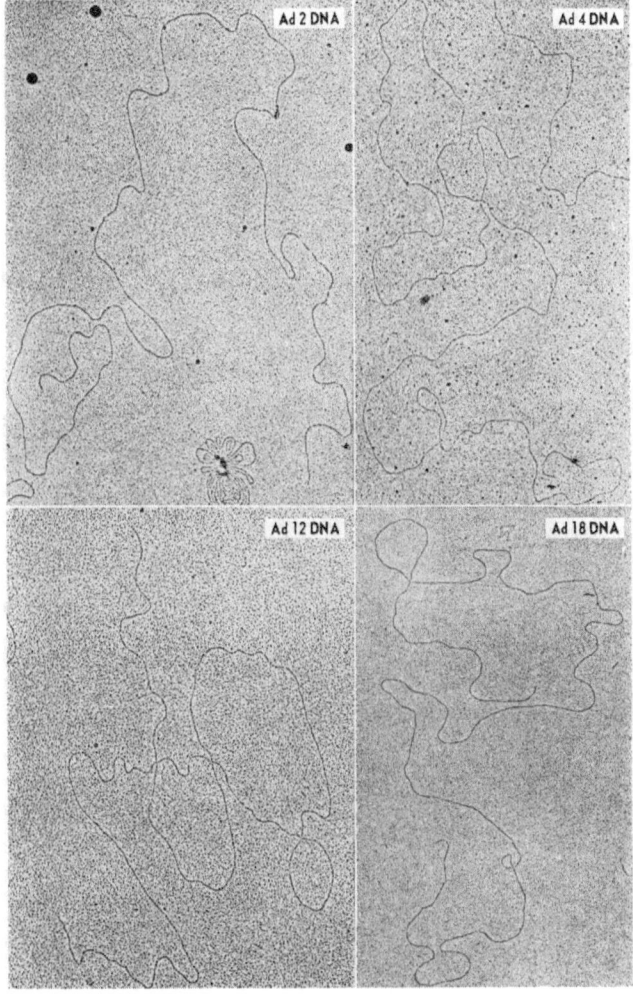

Figure 1. *Electron micrograph of linear molecules of Ad2, 4 and 12,*
and 18 DNAs. Magnification of photographic plate X 7000 (from Green
et al., 1967).

initiation of viral DNA synthesis, and "late". Early genes are
located in four noncontiguous gene blocks, two on each strand (Fig.
2). Early genes encode functions for viral DNA replication, cell
transformation, perhaps the switch from early to late stages of
gene expression, and possibly the block in synthesis of cell DNA
and other macromolecules. Late genes encode mainly virion structural
proteins, and probably several nonvirion regulatory proteins that
e.g., "shut-off" some early genes, or catalyze transport of virion
proteins from the cytoplasm into the nucleus where virions are
assembled. There is some evidence that an intermediate "early-late"

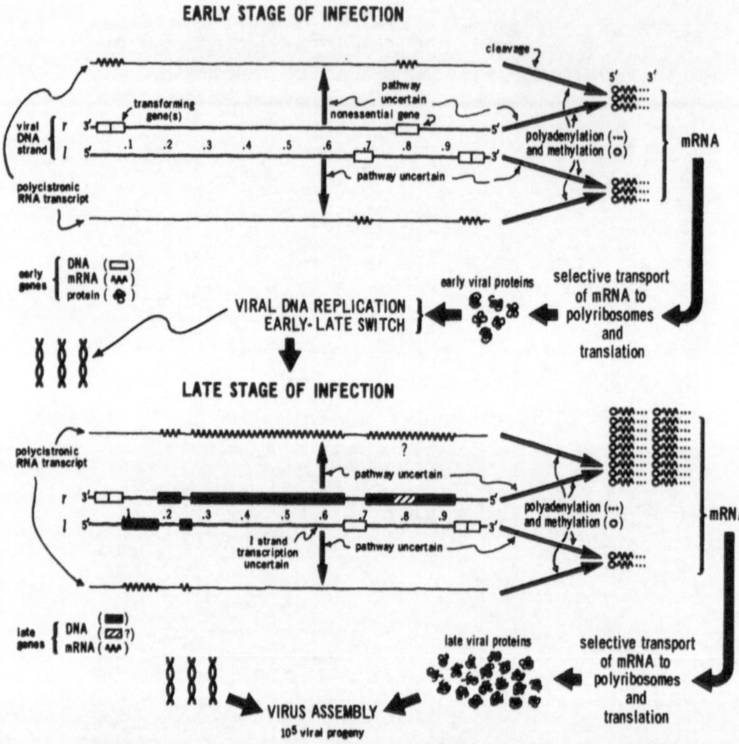

Figure 2. Schematic representation of current concepts regarding early and late stages of gene expression during productive infection of human KB cells by Ad2. Briefly, the following steps occur. (i) The virus penetrates the cell and the uncoated DNA is eventually localized in the nucleus. (ii) Early genes (light blocks) are transcribed into mRNAs (sawtooth lines), which are polyadenylated, methylated, and exported to the cytoplasmic polyribosomes where they are translated into early viral proteins. (iii) Some early proteins are transported back into the nucleus, where they initiate viral DNA replication, possibly turn on late genes and probably carry out other unknown functions that may include blocking host cell DNA synthesis. (iv) After viral DNA replication commences, late genes (black blocks) and class II early genes (hatched block) are expressed as late viral proteins, many of which are virion structural proteins. Class I early genes (light blocks shown in late stages of infection) are turned-off late. (vi) Virion proteins are transported back into the nucleus, where progeny virions are assembled. (From Wold, et al., 1977).

stage of gene expression may exist (e.g., 8-14 h postinfection), where genes directly related to DNA replication may be maximally expressed. The late stage of infection (e.g., 18 h postinfection) is particularly important for studies on viral DNA replication, because exclusively viral DNA is synthesized at this time (Pina and Green, 1969).

An infection of semipermissive or nonpermissive cells (e.g., rat cells) may result in stable cell transformation (10^{-6} frequency). Ad-transformed cells retain a portion of the Ad genome which includes "transformation gene(s)" in an integrated state. The transforming gene(s) (see Fig. 2) are expressed as mRNA and protein during early stages of productive infection. One or more viral-coded "transformation proteins" are believed to maintain cell transformation. An interesting possibility is that transformation proteins may function in viral DNA replication during the productive infection, and may maintain cell transformation by affecting cell DNA replication.

II.   CURRENT STATE OF KNOWLEDGE OF ADENOVIRUS DNA REPLICATION
      (see review by Wold et al., 1977 for detailed references)

1. Systems Employed to Study Adenovirus DNA Replication

Three systems have been utilized to study Ad DNA replication: whole cells, isolated nuclei, and sub-nuclear complexes. Whole cells have been useful for studies identifying origins and termini of replication, and studies with ts mutants. Isolated nuclei have the advantage that radionucleotide precursors equilibrate rapidly; however, initiation of DNA replication does not occur. Subnuclear complexes of two types have been studied that seem to carry out many steps in viral DNA synthesis. These complexes have the advantage that they are simple and can be easily manipulated *in* vitro; however, it is not known whether they are representative of the DNA replication machinery *in vivo*. No attempts have been made as yet to reconstruct DNA replication *in vitro* using purified proteins.

2. Viral-Coded Proteins Involved in Ad DNA Replication

In contrast to the replication of mammalian cell DNA which requires continuous synthesis of cell-coded protein(s), the replication of Ad DNA late after infection occurs in the presence of cycloheximide, an inhibitor of protein synthesis (Yamashita and Green, 1974). This is because viral proteins synthesized early after infection, i.e., in the absence of viral DNA replication, are required for viral DNA synthesis late after infection. At least 3 viral-coded proteins are required for replication of group C Ad DNA, since 3 DNA-negative complementation groups of Ad5 mutants (H5*ts*125, H5*ts*36, H5*hr*1) have been identified (Williams et al., 1974; Harrison and Williams, 1977). Three DNA-negative complementation groups, apparently also defective in the initiation of DNA synthesis, were reported for Ad12 *ts* mutants (Shimojo et al., 1974).

a. <u>Viral-coded DNA binding proteins</u>. Polypeptides of molecular
weights 70-75,000 (70-75K) and 40-50K that bind to single-stranded
DNA (and not RNA or double-stranded DNA) were isolated from cells
infected with Ad5 and Ad2; (van der Vliet and Levine, 1973;
Shanmugam et al., 1975); the 40-50,000 proteins appear to be sub-
species of the larger protein. The Ad2 DNA binding protein has been
purified to homogeneity and several of its properties studied
(see Section IV).

Temperature-shift experiments with H5*ts*125 suggest that this
protein functions in initiation of DNA replication (van der Vliet
and Sussenbach, 1975). Consistent with this, the 70-75K protein
single-stranded DNA binding protein (DBP) is found in an Ad2 nuclear
membrane (Yamashita and Green, 1974) and soluble (Arens et al.,
1977) DNA replication complexes. Immunofluoresence studies have
shown that DBP is localized in the nucleus during the active period
of Ad2 DNA replication (Sugawara et al., 1977b). DBP is produced
in large amounts, suggesting a stoichiometric rather than catalytic
role. At NPT, H5*ts*125 transforms cells more efficiently than wild
tyoe Ad5; thus DBP may play an indirect role in cell transformation.
Although some transformed cells produce the protein, most do not,
indicating that it is not a transformation-maintenance protein
(Levinson et al., 1976; Sugawara et al., 1977b).

A single-stranded DNA-binding protein of 60,000 has also been
isolated from Ad12-infected cells; this protein is defective in
the Ad12 early DNA-negative mutant H12*ts*275 (Rosenwirth et al.,
1975).

b. <u>Polypeptide linked to the 5'-termini of adenovirus DNA</u>.
Robinson et al. (1973) originally reported that Ad DNA readily
formed circles if the DNA was extracted in 4 M guanidinium hydro-
chloride, and without protease digestion. They suggested that a
protein must be firmly associated with the DNA termini, and that
this protein could play a role in DNA replication by circularizing
the Ad DNA. Although it now seems that circular forms probably
are not involved in Ad DNA replication, it has become clear that
the 5' termini of Ad DNA is linked, probably covalently (although
covalent linkage has not been proven chemically), to 5'-termini
via a dC moiety. Evidence supporting this conclusion is given
below (Keegstra et al., 1977; Rekosh et al., 1977; Sharp et al.,
1976). (i) DNA isolated without protease digestion readily forms
circles or oligomeric structures, and the DNA molecules are always
joined at the ends. Using procedures to visualize proteins in the
electron microscopy, a "knob" has been identified on circular
molecules, and two "knobs" on linear molecules, one at each end.
Proteolysis destroyed the ability of the DNA to circularize, and
removed the knobs. On the other hand, the ability of the DNA to
circularize is resistant to treatment with 4 M guanidinium hydro-
chloride, 4 M urea, 2 M perchlorate, 50% formamide, 10% pyridine,

1% SDS, 1% 2-mercaptoethanol, and chloroform-phenol, strongly
suggesting that the protein is covalently linked to the DNA. (ii)
Ad protein-DNA complexes will not enter 0.3-1.4% agarose gels when
subjected to electrophoresis, whereas protease-treated DNA readily
enters gels. If the protein-DNA complex is digested with
restriction endonucleases, and subjected to agarose gel electrophoresis, only end fragments remain at the top of the gels (Brown
et al., 1975; Keegstra et al., 1977; Sharp et al., 1976). This
indicates that a protein associated with Ad-terminal fragments
prevents fragments from entering agarose gels. (iii) Linear Ad DNA-
protein complexes are resistant to digestion by a progressive
nuclease (ATP-dependent DNase) that requires a free terminus for
activity, whereas protease-treated DNA is susceptible to digestion.
(iv) The majority of protease-treated DNAs cannot be labeled with
$(\gamma-^{32}P)$ATP after treatment with alkaline phosphatase and polynucleotide kinase (Carusi, 1977). These DNAs are resistant to
digestion with bacteriophage lambda exonuclease which hydrolyzes
DNA starting from 5' ends, whereas they can be digested with exonuclease III, which hydrolyzes DNA starting from 3' ends. This
indicates that the 5' but not 3' termini of Ad DNA are blocked. (v)
The 5' terminal residue to which the protein is linked is believed
to be dC (Steenbergh et al., 1975; Carusi, 1977; Rekosh et al., 1977).
(vi) The protein component of the protein-DNA complex can be
labeled with ^{125}I to yield a protein of about 55,000 daltons
(following DNA digestion).

The function of the 5' terminal protein is not known (and it
is not known whether it is an early or late viral protein). The
protein could have a structural role in circularizing the DNA in
the virion. It could also play a role in DNA replication, as
discussed below.

3. Replicative Forms and Viral DNA Intermediates

Ad DNA replication *in vivo* and *in vitro* is semiconservative
(Bellett and Younghusband, 1972; van der Vliet and Sussenbach, 1972;
Yamashita et al., 1975; Yamashita et al., 1977). "Okazaki"
fragments, 9-11S, have been identified in intact cells infected
with an avian adenovirus (CELO) (Bellett and Younghusband, 1972),
in Ad2 and Ad5 infected cells using inhibitors of DNA synthesis
(e.g., hydroxyurea), in isolated nuclei using low concentrations
of deoxyribonucleoside triphosphates, and *in vitro* with subnuclear
complexes (Winnacker, 1975; Vlak et al., 1975; Yamashita et al.,
1975). Since it is now clear that Ad DNA replication occurs by a
5' to 3' displacement mechanism from both DNA termini, there is no
theoretical requirement for Okazaki fragments in Ad DNA synthesis.
Therefore, further experiments are necessary to clarify what role,
if any, these fragments may play in DNA replication.

Replicating Ad DNA consists of linear genome-length molecules with extensive single-stranded regions; no circular or linear concatomers have been detected. The single-stranded DNA is derived from both strands of most of the viral genome (Lavelle et al., 1975; Tolun and Pettersson, 1975; Sussenbach et al., 1976; Flint et al., 1976). Temperature-shift studies with H5*ts*125 suggest that the single-stranded DNA is generated as the Ad genome is replicated (Flint et al., 1976). Electron microscopy studies of replicating viral DNA have detected both branched and linear forms with single-stranded regions (Sussenbach et al., 1974).

4. Origins and Termini of Ad DNA Replication

Within the past two years, the termini and origins of Ad DNA replication have been localized by several groups, using basically the method of Danna and Nathans (1972). In this experiment, viral DNA is pulse-labeled for a period (e.g., 10 min) less than required to synthesize full length strands. Mature native DNA molecules are then isolated, and the distribution of radioactivity in restriction fragments determined. With Ad DNA, much more label was found in terminal fragments than in middle fragments, indicating that replication termini are located at both ends of the genome (Schilling et al., 1975; Tolun and Pettersson, 1975; Horwitz, 1976; Sussenbach and Kuijk, 1977; Weingartner et al., 1976). Analysis of the strand specificity of the radioactivity in the terminal fragments (either by hybridization to separated fragment strands, or by direct separation of the labeled strands), showed that the label was in the l strand of right hand fragments, and the r strand of left hand fragments (Horwitz, 1976; Sussenbach and Kuijk, 1977; Weingartner et al., 1976). These results indicate that initiation of synthesis of r and l strands occurs at the right and left ends respectively of the parental strands. Consistent with this, Flint et al. (1976) have analyzed the single stranded DNA present in late infected cells, and found that r strand is at a progressively higher concentration than l strand in genome regions to the right of map position 28-41, whereas l strand predominates to the left of position 28-41. The exact site of initiation of DNA replication is not known, but it is believed to lie very close to or even at the termini.

5. Enzyme Involved in Adenovirus DNA Replication

Because multiple DNA polymerase activities are present in mammalian host cells and isolated nuclei, it is difficult to identify the specific enzymes involved in Ad DNA replication. Three major DNA polymerase enzymes have been identified in mammalian cells: DNA polymerase α, β and γ. The major DNA polymerase α (molecular

weight 1.2-2 x 10^5) as well as β (molecular weight 4.5 x 10^4)
have been purified to near homogeneity. The minor DNA polymerase
γ appears to represent several species with a molecular weight of
about 1.1 x 10^5; one of these species has been recently found to
be associated with mitochondria (Bolden et al., 1977).

The analysis of subnuclear replication complexes isolated from
Ad-infected cells may help clarify the role of various enzymes in
DNA replication. The major DNA polymerase in a nuclear membrane
complex capable of synthesizing Ad2 DNA Okazaki fragments *in vitro*
has been purified about 900-fold from Ad2-infected KB cells. The
enzyme was characterized as belonging to the class of mammalian DNA
polymerases (DNA polymerase γ) that can utilize poly(A)·**oligo(dT)**
as template primer (Ito et al., 1975). This suggests that γ DNA
polymerase may play a role in Ad DNA synthesis. A detailed
comparison of the complex from uninfected and Ad2-infected cells
revealed no differences in the DNA polymerase γ species with
regard to chromatographic properties and template specificity (Ito
et al., 1976). A minor DNA polymerase species has also been
identified).

A second subnuclear complex solubilized from nuclei of Ad2-
infected cells can complete the synthesis of viral DNA molecules *in
vitro*. The soluble complex contained DNA polymerase, RNA polymerase,
DNA ligase, and RNase activities (Arens et al., 1977). The role of
these various enzymes in the synthesis of Ad DNA is not yet
established.

6.   Current Model for Adenovirus DNA Replication

Recent studies have begun to clarify the mechanism of Ad DNA
replication. The Ad genome is replicated in the cell nucleus,
possibly in association with an organized replication complex, as
suggested by both biochemical and genetic studies. Several types
of subnuclear preparations that can synthesize Ad DNA sequences
in vitro have been described, although none have yet been shown
to both initiate and complete the synthesis of mature viral DNA.
Some of these preparations contain nuclear membrane material that
probably becomes associated during isolation, since recent studies
have suggested that Ad DNA is synthesized in the interior of the
nucleus and not on the nuclear membrane.

Evidence that both daughter strands of DNA are synthesized
discontinuously as relatively large (7-11S) Okazaki fragments has
been presented (see above). Evidence that the origin of DNA
replication is at the right and left hand ends is not only strong,
but attractive, since the inverted terminal repetition and
covalently bound protein at each terminus could serve as recognition
sites for a DNA polymerase or for specific initiation proteins.

A perplexing problem is the mechanism for completion of the 5' end of the daughter strands of linear Ad DNA. Since Ad DNA molecules contain neither duplex terminal repetitions nor cohesive ends, they cannot form the type of circular or concatemeric structures that are intermediates in the replication of linear bacteriophage DNA molecules (Watson, 1972). Two possible ways to overcome this dilemma have been suggested. Cavalier-Smith (1974) proposed a general model for completion of linear DNA molecules which requires palindromic sequences at DNA termini. According to the Cavalier-Smith model, after removal of the RNA primer at the 5' termini of daughter DNA molecules, the 3' terminus of the parental strand containing a putative palindrome could form a hairpin loop, the 3'-OH then serving as a primer to fill the gap left by removal of the RNA primer. After sealing the gap by polynucleotide ligase, the action of a specific endonuclease and DNA polymerase could then complete the maturation of the duplex Ad DNA. The sequences of the inverted terminal repetitions of Ad2 and Ad5 DNA have been established. The terminal repetitions consist of 102 (Ad2) or 103 (Ad5) bases. No palindromes were detected, nor any other structures comparable with the fold-back model of Cavalier-Smith. Most interesting, the inverted terminal repetitions are extremely A-T rich in the terminal half, and extremely G-C rich in the interior half. It has been suggested that this terminal A-T rich region, followed by the G-C rich region, could allow for facile "opening" of the termini to allow initiation of DNA replication to occur. Rekosh et al. (1977) proposed that the protein could function to prime DNA synthesis or to allow the completion of 5' termini. They suggest that the protein (possibly covalently attached to dC) could bind to the 3' end of the parental strand (perhaps at the inverted termini). The free 3'-OH on the dC attached to the protein could then prime 5' to 3' DNA synthesis utilizing the parental strand as template. This would yield a DNA covalently linked to the dC-protein as is found in virions. Their model, if correct, could provide a general solution for the perplexing problem of synthesis of 5' termini of linear duplex DNA molecules.

III.   IDENTIFICATION OF ADENOVIRUS 2-INDUCED EARLY PROTEINS

Genetic studies have established that at least three viral-coded early proteins are involved in viral DNA replication. Our long-range goal is to identify and purify Ad2-induced early proteins, and then to study their role in DNA replication. The first step in these studies is to identify which proteins are early viral proteins. Such identification studies are hampered by the large background of host cell protein synthesis during early stages of infection. We have therefore developed a procedure called "cycloheximide enhancement" to increase the synthesis of early viral proteins relative to the host cell proteins. Several years ago we showed that cycloheximide treatment increases the synthesis of early viral DNA

about 5-fold relative to cell mRNA (Parsons and Green, 1971). We
therefore reasoned that cycloheximide would also increase the
synthesis of viral proteins relative to host cell proteins because
viral mRNA was in greater abundance. Cells are incubated in the
presence of cycloheximide to accumulate viral mRNA, the cycloheximide
washed out, proteins labeled with ^{35}S-methionine, resolved by
sodium dodecyl sulfate (SDS)-polyacrylamide gel electrophoresis, and
visualized by autoradiography (Harter et al., 1976). By this
procedure, and by labeling in hypertonic medium (W.S.M. Wold,
unpublished data), we have identified 8 and possibly as many as 16
Ad2 induced (viral-coded or viral-induced cell-coded) proteins.
Fig. 3 illustrates a typical autoradiogram. We always observe in Ad2
infected cells, as prominent and reproducible bands (solid lines),
the single stranded DNA binding protein (DBP), 21K, 19K, 15K, 11.5K,
and 11K polypeptides (Fig. 3). Smaller quantities of other polypeptides of 55K, 52K, 42K, 20K, 18K, 17K, 13.5K, 12.5K, 8.8K, and 8.3K
(dahsed lines) are less prominent. Of interest, the DBP is a major
component of two DNA replication complexes (Yamashita and Green,
1974; Arens et al., 1977), and smaller quantities of 11K and possibly
21K, 15K, and 8.3K proteins are also found in association with these
complexes. The 21K protein is glycosylated, and appears to exist in
multiple forms. The DBP undergoes post-translational modification,
and is phosphorylated, as described below.

We have identified two proteins of great interest, 53,000 and
15,000 daltons, that are coded by the left end of the viral genome,
and are strong candidates as transformation proteins (Gilead et al.,
1976a). We are particularly interested to determine whether these
function in DNA replication, as this would provide unique insights
into both viral DNA replication and the mechanisms of viral cell
transformation. The 15,000 protein has been identified in a soluble
complex that synthesizes adenovirus DNA (see below). The role of
the 53,000 protein in DNA replication is unknown.

IV. PURIFICATION AND PROPERTIES OF THE ADENOVIRUS 2-CODED 70-75K DALTON SINGLE STRANDED DNA BINDING PROTEIN

As discussed above, DBP is synthesized in large amounts during
early stages of Ad2 DNA replication. the DBP gene is located at
positions 62-68 on the viral genome, and is transcribed from right
to left (see Fig. 2). This protein binds to single-stranded DNA,
but not double stranded DNA. Studies with H5ts125 mutants have
shown that this protein functions in initiation of DNA replication.
Below we described our studies with this interesting and important
protein.

1. Purification and Properties

The DBP has been purified by sequential DNA-cellulose,

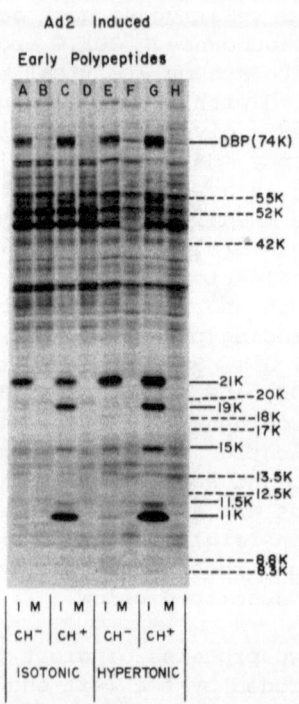

Figure 3. *Effect of hypertonic labeling and CH pretreatment on the relative synthesis of Ad2 early proteins and host polypeptides. Cells were labeled with 25 μCi/ml of (³⁵S) met from 7.5 to 8.5 h postinfection in normal medium (isotonic) and in medium with 250 mM NaCl (hypertonic). Total cell proteins were resolved by SDS poly-acrylamide gel electrophoresis (27 cm) and the dried gel auto-radiographed. Cells labeled in isotonic medium incorporated an average of 1.9 cpm/cell; infected and mock infected cells labeled in 250 mM NaCl incorporated an average of 1.6 and 0.3 cpm/cell respectively. Early proteins marked with a solid line are "major" ones, and those marked with dashed lines are "minor" ones.*

I: infected. M: mock infected.

Sephadex G-200, and DEAE-Sephadex chromatography, with a yield of 120 μg of DBP starting with 2 x 10⁹ infected cells (Sugawara et al., 1977a). Polyacrylamide gel electrophoresis of fractions obtained during purification indicate that after the DEAE-Sephadex step the protein is about 99% homogeneous (Fig. 4). By omitting the Sephadex G-200 step, 400-600 μg of DBP can be obtained, with about 95% homo-geneity. Two types of experiments suggest that the purified protein has retained its native conformation. (i) The purified

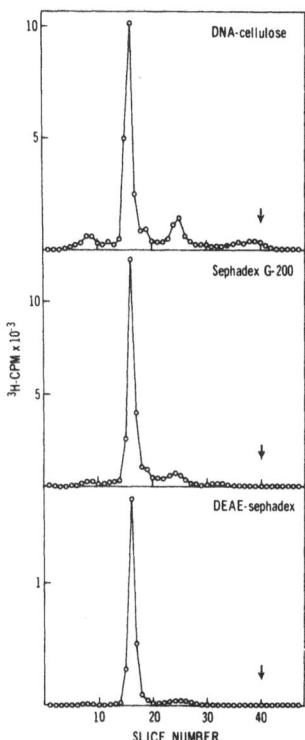

Figure 4. *Polyacrylamide gel electrophoresis of labeled DBP at various stages of purification. Samples of various preparations (2.9 x 10⁴ cpm of DNA-cellulose pool, 3.6 x 10⁴ cpm of G-200 pool, and 10⁴ cpm of DEAE-Sephadex pool) were subjected to electrophoresis on 6% SDS-polyacrylamide gels at 8 mA/gel for 6 h. The radioactivity present in 2-mm slices was determined. The arrow indicates the position of the tracking dye. (From Sugawara et al., 1977a).*

protein bound to and co-sedimented with single stranded Ad2 DNA (Fig. 5A), but not with double-stranded Ad2 DNA (Fig. 5B) during zonal centrifugation in sucrose gradients. (ii) Purified DBP was immunoprecipitated by a serum (from hamsters bearing tumors induced by inoculation of Ad1-SV40 hybrid virus) known to contain antibodies to the DBP (Fig. 6). These two properties, binding to single-stranded DNA and immunoprecipitation by specific sera, are typical of the protein before purification.

The molecular weight and hydrodynamic properties of DBP were studied. Zonal centrifugation in sucrose gradients and gel filtration indicated that the purified protein has a sedimentation coefficient of 3.4S and a Stokes radius of 5.2 nm (Fig. 7). Based on these two values, a molecular weight of 73,000 was calculated,

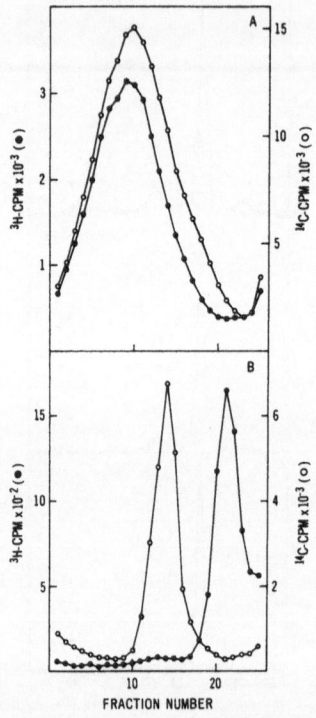

*Figure 5. Binding of purified DBP to single-stranded Ad2 DNA.
Purified ³H-labeled DBP was mixed with (A) denatured Ad2 ¹⁴C-labeled
DNA (7.2 µg of protein and 5.6 µg of DNA) and (B) native Ad2 ¹⁴C-
labeled DNA (3.0 µg of protein and 9.4 µg of DNA) in binding buffer.
After 1 h at 0°C, the reaction mixtures were layered on 5 to 30%
linear sucrose gradients made in binding buffer containing 0.02%
NP40 and centrifuged at 40,000 rpm for 3 h at 4°C in a Spinco SW50.1
rotor. Fractions were collected, and radioactivity was determined.
(From Sugawara et al., 1977a).*

in agreement with the estimate from sodium dodecyl sulfate-polyacrylamide gel electrophoresis. A frictional ratio of 1.88 was
calculated, suggesting that the Ad2 DNA binding protein does not
have a typical globulin protein structure. The bacteriophage T4 DNA
binding protein has similar properties.

                    2.   Immunofluorescence Studies

     High titer monospecific antiserum was prepared against the
purified 73K DBP, and used to study the synthesis of the protein
in productively infected cells (Sugawara et al., 1976b). The

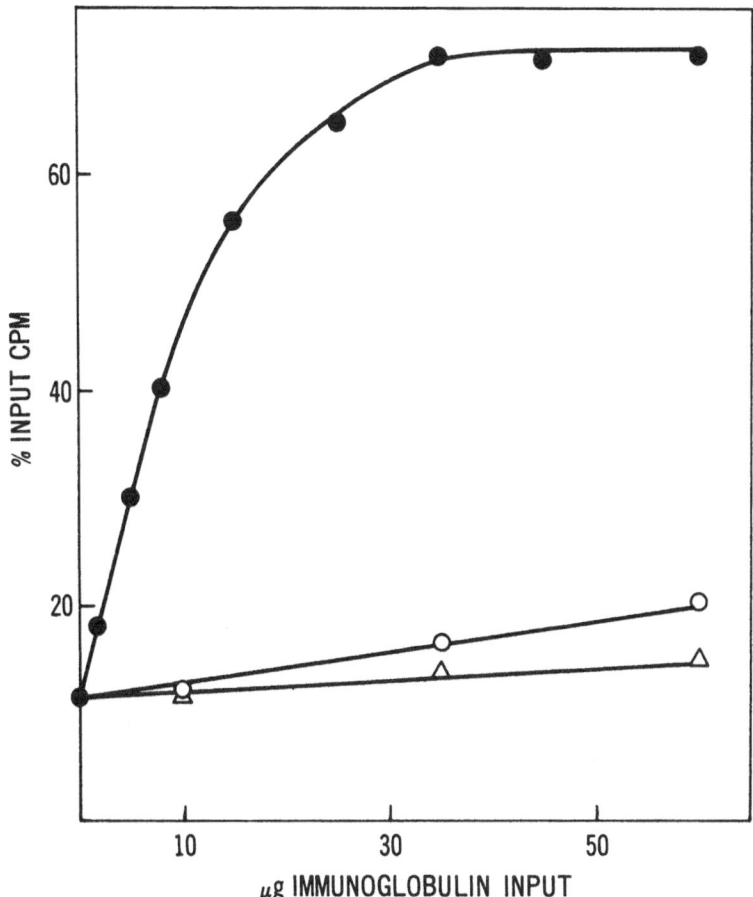

Figure 6.   *Radioimmune precipitation of purified DBP by Ad1-SV40 T*
serum.   Portions (5,400 cpm) of ^{35}*S-labeled DBP (DEAE-Sephadex pool)*
were mixed with varying amounts of immune IgG.   Normal IgG was added
to provide a final concentration of 80 µg of protein in each tube.
After 1 h of incubation at 37°C, goat anti-hamster IgG was added,
and the incubation was continued for 2 h at 37°C.   The immuno-
precipitates were sedimented and washed with PBS containing 1%
Triton X-100 and 0.5% sodium desoxycholate.   The precipitates
were dissolved in acetic acid (0.3 M), and the radioactivity was
determined.   Symbols: O, Ad1-SV40 T serum; O, 2 Ad12 T serum; Δ,
SV40 T serum. (From Sugawara et al., 1977a).

synthesis of this protein during productive infection is shown in
Fig. 8.   At 6 h postinfection, DBP is readily detected in the
cytoplasm of infected cells (Fig. 8b).   Following this period, the
nuclear fluorescence increased while the cytoplasmic fluorescence

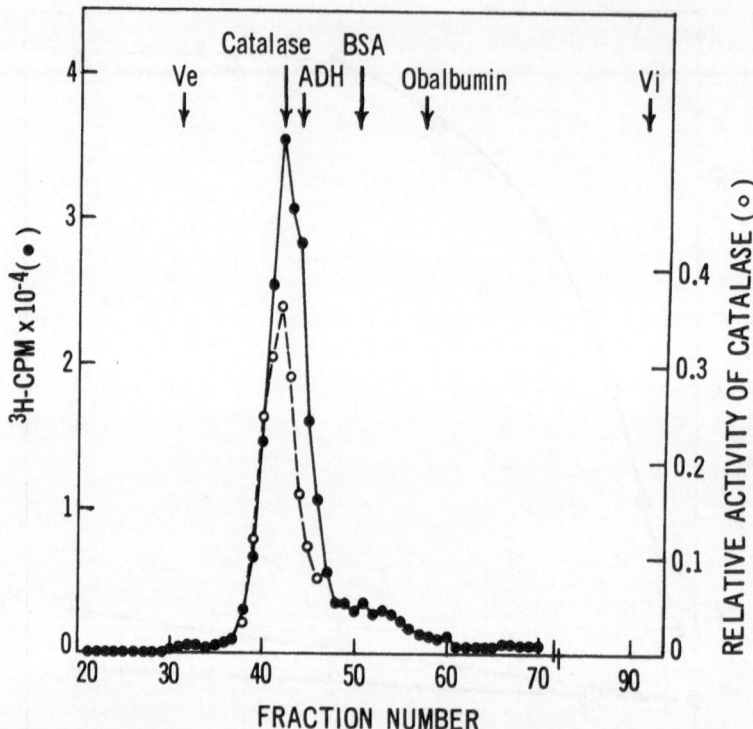

Figure 7. *Determination of the Stokes radius of DBP. Samples of 0.6 M eluate (0.4 ml; 5 x 10⁵ cpm) or marker proteins were applied to Sephadex G-200 columns (97 by 1.0 cm) equilibrated with 20 mM Tris-hydrochloride (pH 8.1) -0.6 M NaCl-1 mM 2-mercaptoethanol-1 mM EDTA-0.02% NP40 and eluted with the same buffer. Fractions (0.88 ml) were collected, and the elution position of DBP was determined by monitoring radioactivity. The excluded (Ve) and included (Vi) column volumes are indicated. Stokes radii of marker proteins were catalase (5.2 nm), alcohol dehydrogenase (4.5 nm), and bovine albumin (3.5 nm). (From Sugawara et al., 1977a).*

decreased, so that by about 14 h postinfection nuclear fluorescence was very prominent, as large characteristic globules, and cyto-plasmic fluorescence was much diminished (Fig. 8E and F). The appearance of nuclear fluorescence at 6-8 h postinfection coincided with the onset of DNA replication. These studies are consistent with the notion that DBP is synthesized in the cytoplasm, but functions in the nucleus in viral DNA replication.

Radioimmune inhibition studies have demonstrated that accumulation of DBP ceases at about 11-12 h postinfection (Gilead et al., 1976b). The immunofluorescence data shown in Fig. 8

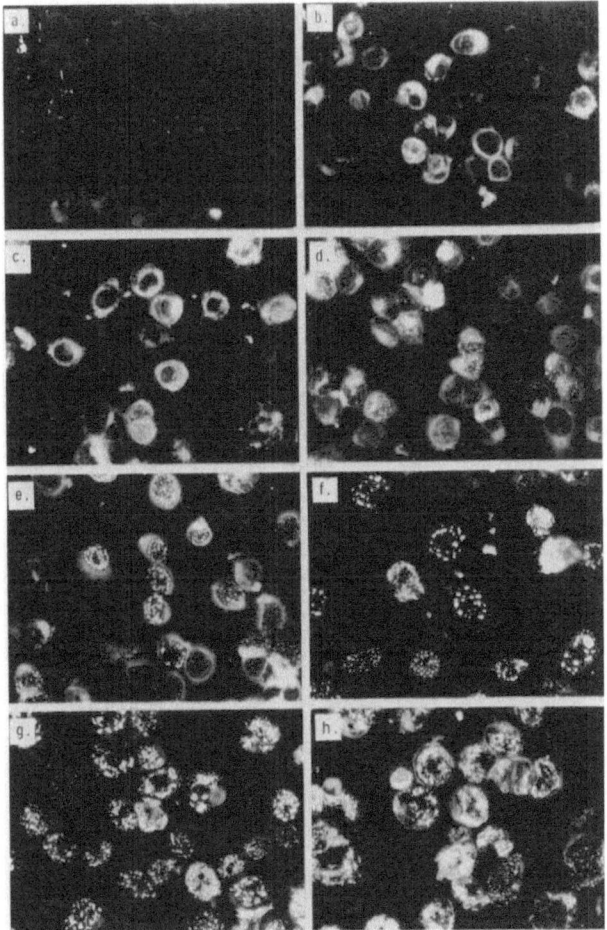

Figure 8. Immunofluorescence study of the time course of synthesis of DBP in Ad2 productively infected cells. KB cells were infected in suspension with 100 PFU of Ad2 per cell and seeded in tissue culture plates containing microscope slides. Slides were removed at the indicated times, fixed, incubated with anti-DBP guinea pig serum, and stained with fluorescein isothiocyanate-conjugated rabbit IgG anti-guinea pig IgG. (a) Mock-infected cells. (b) -(h) Infected cells: (b) 6 h p.i.; (c) 8 h p.i.; (d) 10 h p.i.; (e) 12 h p.i.; (f) 16 h p.i.; (g) 18 h p.i.; (h) 22 h p.i.; x 290. (From Sugawara et al., 1977b).

indicates that very little cytoplasmic DBP can be detected in later stages of infection, e.g. 16 h postinfection. These results suggested that at about 11-12 h postinfection the synthesis of the DNA binding protein may be "turned-off".

4.   Post-Translational Modification of the Single Stranded
DNA Binding Protein

Many cellular processes are controlled by post-translational
modifications, such as adenylation, methylation, acetylation,
glycosylation, and phosphorylation of key regulatory proteins.
Several groups have found the DBP is phosphorylated, and exists
in at least two forms that differ in mobility by sodium dodecyl
sulfate gel electrophoresis (Jeng et al., 1977; Levinson and Levine,
1977; Blair and Russell, 1977; Philipson et al., 1977). As shown
in Fig. 9, lanes A and M, after a 30 min pulse both ^{35}S-methionine-
labeled and ^{32}P-labeled DBP has an apparent molecular weight of
74,000 daltons, but during chase periods, the apparent molecular
weight of both ^{35}S- and ^{32}P-labeled protein increased to 77,000
daltons (lanes C and O). Both 74K and 77K dalton forms of ^{35}S-
labeled (lanes F and J) and ^{32}P-labeled (lanes R and V) protein
were immunoprecipitated by antibody against the DNA binding
protein. Both 74K and 77K forms of ^{35}S- and ^{32}P-labeled protein
bind to single-stranded DNA cellulose. The phosphate groups are
attached to serine residues (Fig. 10). Removal of the phosphate
groups, using alkaline phosphatase, does not grossly alter the
binding of the protein to single stranded DNA (Levinson and Levine,
1977). It should be of interest to establish the relationship
between phosphorylation of this protein, as well as other post-
translational modifications that affect its mobility in gels, and
its function in DNA replication.

V.   ASSOCIATION OF ADENOVIRUS 2-INDUCED EARLY PROTEINS WITH A
SOLUBLE COMPLEX THAT SYNTHESIZES VIRAL DNA *IN VITRO*

Yamashita et al. (1977) described a soluble complex from Ad2-
infected cells, that synthesizes exclusively Ad2 DNA by a semiconservative mechanism, and elongates nascent viral DNA chains into
full length (31S) DNA molecules (see Arens et al., this volume). The
partially purified complex contains cellular DNA polymerases α and
γ, plus other enzymes possibly involved in DNA replication (Arens
et al., 1977). Thus, this complex may represent, at least in part,
a multienzyme complex that replicates Ad2 DNA, and is a useful
system to study the mechanism of eukaryotic DNA replication. Below
we further characterize the soluble Ad2 DNA replication complex,
and show that several early proteins are present in the complex.

1.   Characterization of the DNA Synthesized *in vitro* by
the Soluble Adenovirus 2 DNA Replication Complex

Some of the properties of the complex have been described
(Yamashita et al., 1977; Arens et al., 1977; Arens et al., this

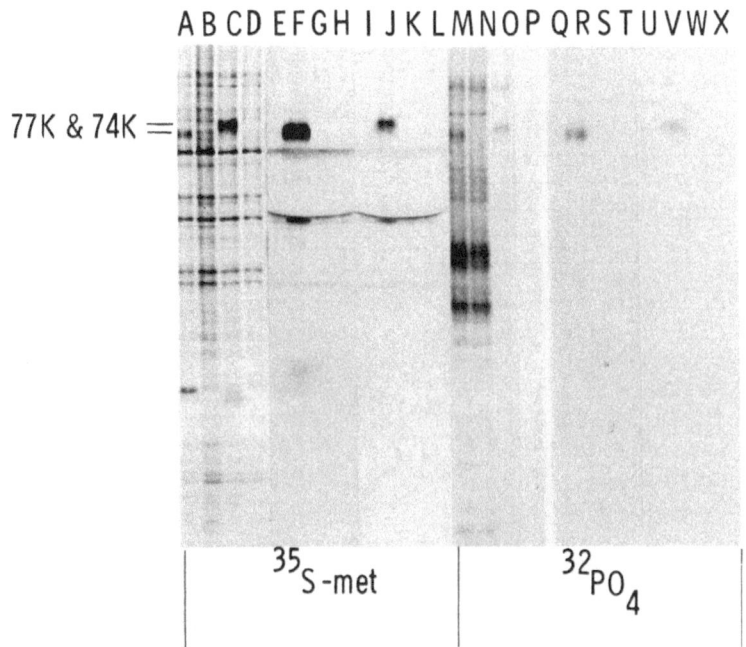

A B C D E F G H I J K L M N O P Q R S T U V W X

77K & 74K =

^{35}S-met ^{32}PO$_4$

Figure 9. *Autoradiogram of SDS-polyacrylamide gel electropherogram showing that Ad1-SV40 IgG immunoprecipitates both 74K
(seen after 30-min pulse) and 77K (seen after 30-min pulse
followed by 20 h chase) forms of* ^{35}*S- and* ^{32}*P-labeled DBP. Ad2-
infected and mock-infected cells were labeled with (*^{35}*S)methionine
or* ^{32}*PO$_4$ (from 7 to 7.5 h p.i., or 7 to 7.5 h followed by a 20 h
chase). Total cell extracts were prepared, and immunoprecipitation
experiments were performed with Ad2-SV40 and nonimmune hamster IgGs.
A through D and M through P, respectively, show* ^{35}*S- and* ^{32}*P-
labeled extracts electrophoresed before immunoprecipitation. Pulse-
labeled extracts are in A (Ad2 infected), B (mock infected), M (Ad2
infected), and N (mock infected). Pulse-chase labeled extracts
are in C (Ad2 infected), D (mock infected), O (Ad2 infected), and
P (mock infected). E through L (*^{35}*S-labeled extracts) and Q through
X (*^{32}*P-labeled extracts) show the results of the immunoprecipitation
experiments. (E and Q) infected pulse-labeled cells versus nonimmune hamster IgG; (F and R) infected pulse-labeled cells versus
Ad1-SV40 IgG; (G and S) mock-infected, pulse-labeled cells versus
nonimmune IgG; (H and T) mock-infected, pulse-labeled cells versus
Ad1-SV40 IgG; (I and U) infected chase-labeled cells versus non-
immune IgG; (J and V) infected chase-labeled cells versus Ad1-SV40
IgG; (K and W) mock-infected, chase-labeled cells versus nonimmune
IgG; and (L and X) mock-infected, chase-labeled cells versus Ad1-
SV40 IgG. (From Jeng et al., 1977).*

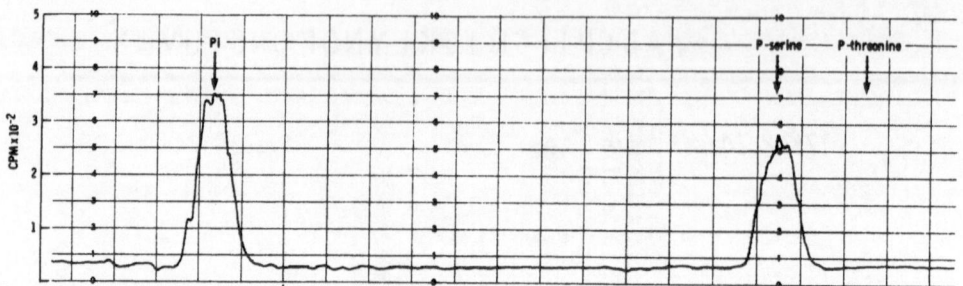

Figure 10. Identification of [32]*P-phosphoserine in pulse-chase labeled* [32]*P-DBP after immunoprecipitation by Ad1-SV40 IgG and preparative SDS-polyacrylamide gel electrophoresis. Ad2-infected cells were labeled with* [32]*PO$_4$, from 7 to 7.5 h p.i., then chased in normal MEM for 20 h.* [32]*P-labeled DBP was immunoprecipitated from cytoplasmic extracts by Ad1-SV40 IgG, then electrophoresed on SDS-polyacrylamide gels.* [32]*P-labeled DBP was eluted from the gel, acid hydrolyzed, mixed with carrier phosphoserine, and subjected to paper electrophoresis. The figure shows the correspondence between sample radioactivity and carrier phosphoserine present in the sample. "Marker" phosphothreonine and phosphoserine (i.e., not mixed with the sample) had the same electrophoretic mobilities as "carrier" phosphoserine and phosphothreonine in the sample. (From Jeng et al., 1977).*

volume). We wished to characterize further the DNA synthesized *in vitro* by the purified complex used for analysis of early proteins. DNA was labeled *in vitro* for 60 min with (^{3}H)TTP, extracted, mixed with ^{14}C-labeled Ad2 virion DNA, and cleaved with *Eco*Rl restriction endonuclease. The resulting *Eco*Rl fragments were resolved on the basis of size by agarose gel electrophoresis. As shown in Fig. 11, the ^{3}H-DNA synthesized by the complex was present in all 6 fragments, and the distribution of radioactivity corresponded very well to virion ^{14}C-DNA fragments. The fraction of radioactivity in each restriction fragment corresponded closely with the fraction of the virus genome represented by the fragment. These data have shown that all regions of the Ad2 genome are synthesized *in vitro* by the purified Ad2 DNA replication complex. Our results, together with those of Yamashita et al. (1977) and Arens et al. (1977) are consistent with the possibility that the soluble complex may represent a multienzyme complex that synthesizes Ad2 DNA *in vivo*.

2.   Identification of Adenovirus 2-Induced Early Proteins
          Associated with the Purified Complex

Early proteins were labeled with (^{35}S)methionine (in the presence of cytosine arabinoside), and then were allowed to

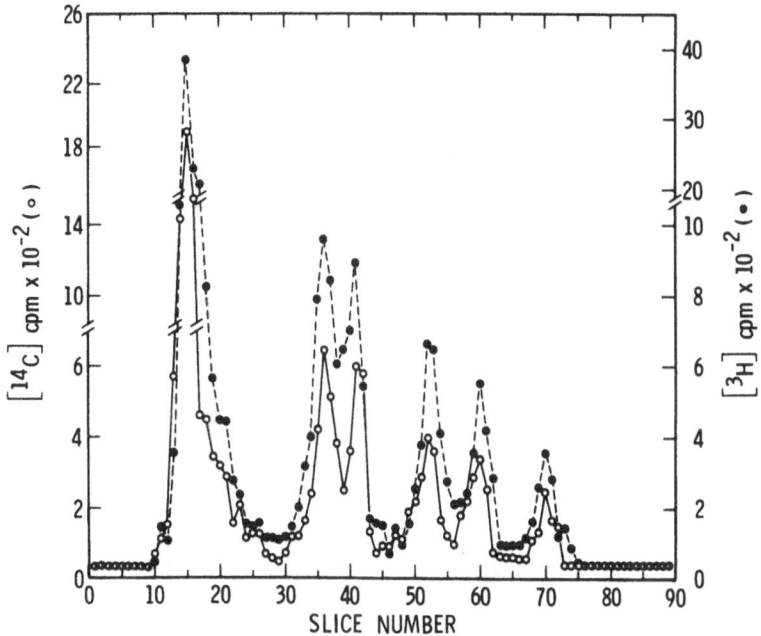

Figure 11. Co-electrophoresis of fragments generated by EcoR1
cleavage of ³H-labeled DNA synthesized *in vitro* by the purified
soluble Ad2 DNA replication complex, and of *in vivo* ¹⁴C-labeled
Ad2 virion DNA. DNA synthesized *in vitro* by the purified complex
was labeled with (³H)TTP for 60 min. The DNA was purified by
phenol-chloroform extraction followed by centrifugation through
neutral sucrose gradients. DNA sedimenting in a 31S peak (about
80% of the labeled DNA synthesized) was pooled. Virion DNA was
labeled *in vivo* with (¹⁴C)thymidine. *In vivo* and *in vitro* labeled
DNA were mixed, cleaved with EcoR1 restriction endonuclease, and
co-electrophoresed through 1.4% agarose gels (From Rho et al.,
in press).

become associated with the replication complex by incubating cells
for 8 h without cytosine arabinoside. As shown in Fig. 12C, the
crude complex prepared by this protocol had as much endogenous DNA
polymerase activity as a complex harvested 18 h p.i. from cells not
labeled or treated with drugs (Fig. 12A). The complex isolated
from cells at 10 h p.i., the end of the labeling period, had much
less polymerase activity (Fig. 12B). Very little polymerase was
detected in complexes from mock-infected cells. The crude ³⁵S-
labeled infected and mock-infected complexes were purified by
filtration through Bio-Gel A-50m. The infected cell complex had
the majority of the endogenous DNA polymerase activity in the
excluded fraction (Fig. 13A), whereas the mock-infected complex
had little polymerase activity in any fraction (Fig. 13B). These

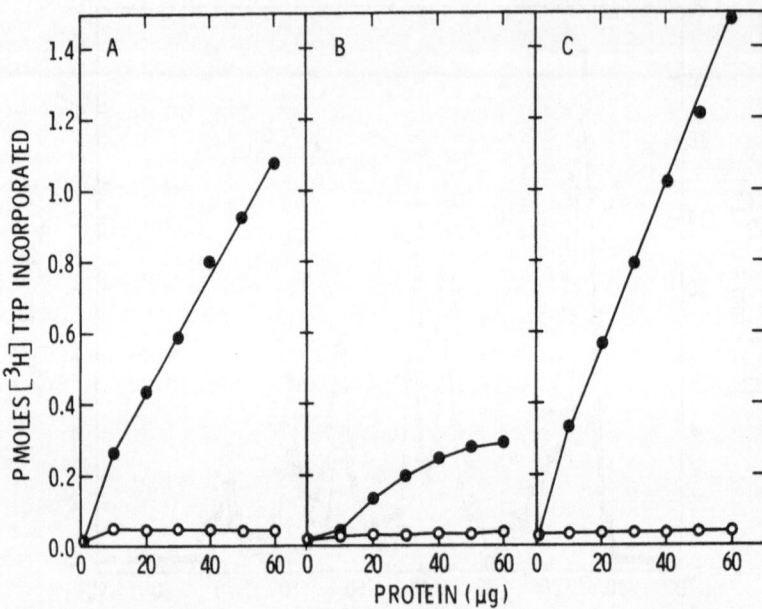

Figure 12. *Endogenous DNA polymerase activity in crude soluble Ad2 DNA replication complex from infected'(●) and mock-infected (○) cells. (A) Cells harvested 18 h postinfection, not treated with drugs and not labeled with (³⁵S)methionine. (B) Cells labeled at 7-10 h postinfection with (³⁵S)methionine and harvested at 10 h postinfection. (C) Cells labeled with (³⁵S)methionine at 7-10 h postinfection, then harvested at 18 h postinfection. Endogenous DNA polymerase activity was assayed as described by Yamashita et al. (1977).*

results suggest that our protocol for labeling early proteins does not adversely affect the formation of the Ad2 DNA synthesizing complex.

Fig. 14 illustrates the ³⁵S-labeled polypeptides present in a total cell extract of infected and mock-infected cells and in the purified complex. The following early polypeptides are seen in infected whole cell extracts (lane A): DBP, 21K, 19K, 15K, 11.5K, 11K and 8.3K. The purified soluble Ad2 DNA replication complex contained major amounts of DBP and 11K, and minor quantities of 21K, 15K, and possibly 8.3K (not shown). These early polypeptides remained associated with the complex after purification through a second Bio-Gel A-50m column (Fig. 14, lanes B), suggesting that they may be stably associated with a large complex. The infected cell-specific polypeptides of 40-50,000 daltons present in the purified complex (lane B) are the known sub-species (protease breakdown products) of DBP, as indicated by immunoprecipitation by mono-

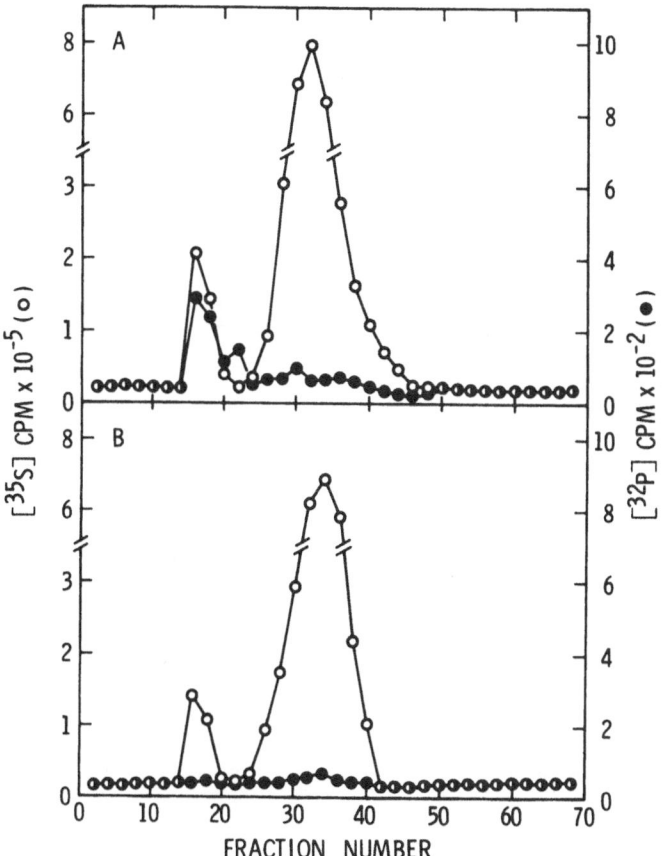

Figure 13. Endogenous DNA polymerase activity in Bio-Gel A-50m *excluded* (purified Ad2 DNA replication complex) and included (nucleoplasm) fractions. (A) Infected cells. (B) Mock-infected cells. (From Rho et al., in press).

specific antiserum against highly purified DBP (not shown).

Thus the major virus-induced early polypeptides in the purified complex were DBP and 11K, raising the possibility that these play a role in Ad2 DNA replication. DBP is a virus-coded early phosphoprotein that binds to single-stranded DNA (see above). It is certain that DBP functions in Ad2 DNA replication, as described in previous sections. The 11K is localized in the cell nucleus (Saborio and Oberg, 1976; Chin and Maizel, 1977; our unpublished data), consistent with a role in Ad2 DNA replication.

Small quantities of 21K, 15K, and possibly 8.3K were also present in the soluble complex, suggesting that these may also function, perhaps catalytically in virus DNA replication. We

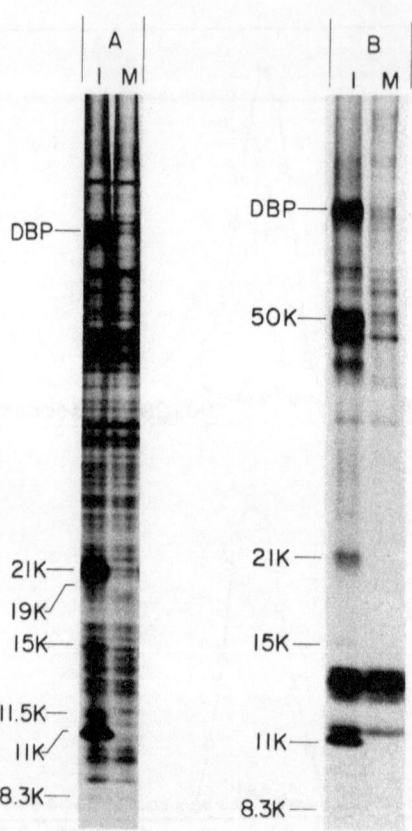

Figure 14.   Autoradiogram illustrating Ad2-induced early poly-peptides present in the purified soluble Ad2 DNA replication complex.
[35]*S-labeled infected (I) and mock-infected (M) cell complexes, and indicated cell fractions, were prepared and equal counts of each fraction were electrophoresed through 27 cm 5-20% gradient SDS-polyacrylamide slab gels.   Proteins were visualized by autoradio-graphy.   Lanes A: whole cell extracts.   Lane B: purified soluble Ad2 DNA replication complex passed through two Bio-Gel A-50m columns.   The radioactivity (cpm) in each of the fractions was as follows.   Cytoplasm: I, 272 x 10*[6]*; M, 285 x 10*[6]*.   Crude soluble complex: I, 35.5 x 10*[6]*; M, 46.5 x 10*[6]*.   Insoluble nuclear material: I, 51 x 10*[6]*; M, 52.2 x 10*[6]*.   Nucleoplasm: I, 24 x 10*[6]*; M, 21 x 10*[6]*. Purified soluble complex: I, 1.1 x 10*[6]*; M, 1.3 x 10*[6]*. (From Rho et al., in press).*

emphasize that because the complex contained only a small fraction of 21K, 15K, and 8.3K, we cannot exclude non-specific association of these polypeptides with the purified complex. The 21K, 15K and 11K are probably virus-coded, because they immunoprecipitated by antisera against Ad2-transformed rat cells (Wold and Green,

unpublished results), and because similar-sized polypeptides are
produced by cell-free translation of Ad2-specific early mRNA. The
21K protein is an early glycoprotein (Ishibashi and Maizel, 1974;
Jeng, Wold and Green, unpublished results).

The 15K protein may be the 15,000 candidate transformation
protein immunoprecipitated by antisera against Ad2-transformed cells.
It is not known whether the complex contains another candidate
transformation protein of 53,000 daltons.

Recent data (Yamashita, Arens and Green, unpublished data;
see Arens et al., this volume), suggest that DNA synthesized *in
vitro* is covalently attached to protein at both termini. Moreover,
preliminary evidence indicates that this is an early protein, i.e.,
it is synthesized in the absence of DNA replication. We did not
detect such a protein in our experiments, presumably because it has
an apparent molecular weight of 55,000 (Rekosh et al., 1977), and
there is a large "background" of host proteins in the 55,000 dalton
region of our gels (Fig. 14, lanes B).

VI. SUMMARY

Adenovirus 2 (Ad2) infection of permissive human KB cells in
culture provide an excellent model to study DNA replication in
eukaryotic cells. The well characterized Ad2 linear duplex DNA
of a molecular weight of 23×10^6 contains inverted terminal
repetitions and a covalently bound protein (55K) at each terminus;
thus both termini are identical. DNA replication is semiconservative
and occurs by a displacement mechanism (5' to 3' direction) in
which progeny strands initiate at or very near each parental 3' end
and terminates at their 5' ends. Ad2 DNA is synthesized maximally
during late stages of infection when host cell DNA synthesis is
blocked; thus analysis of viral DNA replication is uncomplicated by
cell DNA synthesis. Two sub-nuclear DNA replication complexes have
been isolated that perform various stages of viral DNA replication
in vitro. Virus producing cells can be obtained in large amounts
(50-100 liters) for the preparations of these complexes and for
the isolation of enzymes and other proteins that function in DNA
replication.

Our laboratory is focusing on Ad2 induced early proteins and
attempting to identify their roles in viral DNA replication and
cell transformation. A 73K single stranded DNA binding protein
(DBP), apparently necessary to initiate DNA replication, has been
identified as a major component of two DNA replication complexes
isolated from virus infected cells. We have purified DBP to near
homogeneity and have shown that it binds to single stranded but
not native Ad2 DNA. The purified protein has a high frictional
ratio, 1.88, indicating that it is not a typical globular protein.

Pulse chase experiments indicate that DBP exists in at least 2
apparent molecular weight forms, 74K and 77K, both of which are
phosphorylated (as phosphoserine), bind to single stranded DNA,
and are precipitated by antibody to purified DBP. By indirect
immunofluoresence with monospecific antiserum against highly
purified DBP, we observed that during the period of active viral
DNA replication, DBP is localized exclusively in the nucleus in
the form of discrete globules. DBP is not required for maintenance
of transformation, although it is synthesized in some, but not all
Ad2 transformed rodent cells. Synthesis of DBP late after infection
appears to be under negative regulation.

Three DNA negative Ad complementation groups have been
discovered (one group defective in DBP), indicating that in
addition to DBP, at least 2 other viral early proteins are involved
in DNA replication. DBP is the major component of a soluble DNA
replication complex isolated from nuclei of Ad2 infected KB cells.
This complex synthesizes all regions of the Ad2 DNA molecule *in
vitro*, as indicated by analysis of restriction fragments generated
after *Eco*Rl digestion of *in vitro* labeled DNA. In addition to DBP,
smaller quantities of other early Ad2 proteins including 11K
(exclusively localized in cell nucleus), 21K (glycoprotein), 15K
(candidate transformation protein), and possibly 8.3K are also
found associated with this complex. Studies are underway to purify
these proteins and to establish the specificity of their association
with this complex and whether they function in DNA replication.
Recently, we have identified 2 candidate transformation proteins,
53K and 15K. A matter of great interest is whether these proteins
are involved in DNA replication.

REFERENCES

Arens, M., Yamashita, T., Ito, K., Tsuruo,   T., Padmanabhan, R.
    and Green, M. In: DNA Synthesis; Present and Future (eds.
    I.J. Molineux and M. Kohiyama), Plenum Press, in press

Arens, M., Yamashita, T., Padmanabhan, R.,   Tsuruo, T., and Green,
    M.(1977) J. Biol. Chem., in press

Bellett, A.J.D. and Younghusband, H.B. (1972) J. Mol. Biol. 72:
    691

Bolden, A., Noy, G.P. and Weissbach, A. (1977) J. Biol. Chem.
    252:3351

Brown,   D.T., Westphal, M., Burlingham, B.T., Winterhoff, U. and
    Doerfler, W. (1975) J. Virol. 16:366

Carusi, E.A. (1977) Virology 76:380

Cavalier-Smith, T. (1974) Nature 250:467

Chin, W.W. and Maizel, J.V. Jr. (1977) Virology 76:79

Danna, K.J. and Nathans, D. (1972) Proc. Natl. Acad. Sci. U.S.A.
    69:309

Flint, S.J., Berget, S.M. and Sharp, P.A. (1976) Cell 9:559

Gilead, Z., Jeng, Y., Wold, W.S.M., Sugawara, K.. Rho, H.M., Harter,
    M.L. and Green, M. (1976a) Nature 264:263

Gilead, Z., Sugawara, K., Shanmugam, G. and Green, M. (1976b)
    J. Virol. 18:454

Green, M. and Daesch, G.E. (1961) Virology 13:169

Green, M.R., Mackey, J.K. and Green, M. (1977) J. Virol. 22:238

Green, M., Parsons, J.T., Pina, M., Fujinaga, K., Caffier, H. and
    Landgraf-Leurs, I. (1970) Cold Spring Harbor Symp. Quant. Biol.
    35:803

Green, M., Pina, M., Kimes, R., Wensink, P.C., MacHattie, L.A. and
    Thomas, C.A. (1967) Proc. Natl. Acad. Sci. USA 57:1302

Harrison, T., Graham, F. and Williams, J. (1977) Virology, 77:319

Harter, M., Shanmugam, G., Wold, W.S.M. and Green, M. (1976) J.
    Virol. 19:232

Horwitz, M.S. (1976) J. Virol. 18:307

Ishibashi, M. and Maizel, J.V. Jr. (1974) Virology 57:409

Ito, K., Arens, M. and Green, M. (1975) J. Virol. 15:1507

Jeng, Y.H., Wold, W.S.M., Sugawara, K., Gilead, Z. and Green, M.
    (1977) J. Virol. 22:402

Keegstra, W., van Wielink, P.S. and Sussenbach, J.S. (1977) Virology
    76:444

Lavelle, G., Patch, C., Khoury, G. and Rose, J. (1975) J. Virol.
    16:775

Levinson, A. and Levine, A.J. (1977) Virology 76:1

Levinson, A., Levine, A.J., Anderson, S., Osborn, M., Rosenwirth, B.
    and Weber, K. (1976) Cell 7:575

Linne, T., Jornvall, H. and Philipson, L.(1977) Eur. J. Biochem.
    76:481

Parsons, J.T. and Green, M. (1971) Virology 45:154

Pina, M. and Green, M. (1969) Virology 38:573

Rekosh, D.M.K., Russell, W.C., Bellett, A.J.D. and Robinson, A.J.
    (1977) Cell 11:283

Rho, H.M., Jeng, Y-H., Wold, W.S.M. and Green, M. Biochem. Biophys.
    Res. Comm., in press

Robinson, A.J., Younghusband, H.B. and Bellett, A.J.D. (1973)
    Virology 56:54

Rosenwirth, B., Shiroki, K., Levine, A.J. and Shimojo, H. (1975)
    Virology 67:14

Russell, W.C. and Blair, G.E. (1977) J. Gen. Virol. 34:19

Saborio, J.L. and Oberg, B. (1976) J. Virol. 17:865

Schilling, R., Weingartner, B. and Winnacker, E.L. (1975) J. Virol.
    16:767

Shanmugam, G., Bhaduri, S., Arens, M. and Green, M. (1975)
    Biochemistry 14:332

Sharp, P.A., Moore, C. and Haverty, J.L. (1976) Virology 75:442

Shimojo, H., Shiroki, K. and Yamaguchi, K. (1974) Cold Spring
    Harbor Symp. Quant. Biol. 39:533

Steenbergh, P.H., Sussenbach, J.S., Roberts, R.J. and Jansz, H.S.
    (1975) J. Virol. 15:268

Sugawara, K., Gilead, Z. and Green, M. (1977a) J. Virol. 21:338

Sugawara, K., Gilead, Z., Wold, W.S.M. and Green, M. ((1977b) J.
    Virol. 22:257

Sussenbach, J.S., Ellens, D.J., van der Vliet, P.Ch., Kuijk, M.G.,
    Steenbergh, P.H., Vlak, J.M., Ruzijn, H. and Jansz, H.S.
    (1974) Cold Spring Harbor Symp. Quant. Biol. 39:539

Sussenbach, J.S. and Kuijk, M.G. (1977) Virology 77:149

Sussenbach, J.S., Tolun, A., and Pettersson, U. (1976) J. Virol.
    20:532

Tolun, A. and Pettersson, U. (1975) J. Virol. 16:759

van der Vliet, P.C. and Levine, A.J. (1973) Nature New Biol.
    246:170

van der Vliet, P.C. and Sussenbach, J.S. (1976) Virology 67:415

van der Vliet, P.C. and Sussenbach, J.S. (1972) Eur. J. Biochem.
    30:548

Vlak, J.M., Rosijn, Th.H., and Sussenbach, J.S. (1975) Virology
    63:168

Watson, J.D. (1972) Nature New Biol. 239:197

Weingartner, B., Winnacker, E.L., Tolun, A. and Pettersson, U.
    (1976) Cell 9:259

Williams, J.F., Young, C.S.H. and Austin, P.E. (1974) Cold Spring
    Habor Symp. Quant. Biol. 39:427

Winnacker, E.L. (1975) J. Virol. 15:744

Wold, W.S.M., Green, M. and Büttner, W.(1977) Adenoviruses.
    In: The Molecular Biology of Animal Viruses (Ed. D.P. Nayak)
    Marcel Dekker, Publ., in press

Yamashita, T., Arens, M. and Green, M. (1975) J. Biol. Chem.
    250:3272

Yamashita, T., Arens, M. and Green, M. J. Biol. Chem., in press

Yamashita, T. and Green, M. (1974) J. Virol. 3:412

FUNCTION OF AN EARLY DNA BINDING PROTEIN IN THE REPLICATION OF ADENOVIRUS DNA

P.C. van der Vliet[*], W.M. Blanken, J. Zandberg and
H.S. Jansz

Laboratory for Physiological Chemistry
State University of Utrecht
Vondellaan 24 A, Utrecht, The Netherlands

INTRODUCTION

The human adenoviruses types 2 and 5 (Ad2 and Ad5) contain a linear, non-permuted double-stranded DNA of 23×10^5 dalton molecular weight (Green et al., 1967). The ends of the molecule have a 110 to 140 base pairs long inverted repetition (Wolfson and Dressler, 1972; Garon et al., 1972; Roberts et al., 1974), which allows formation of single-stranded circles when the DNA is denatured and renatured at low concentration.

Isolation of DNA from intact virions in the absence of proteolytic enzymes has revealed the presence of a protein which is bound strongly to both molecular ends, presumably covalently linked to the 5'-ends (Robinson et al., 1973; Keegstra et al., 1976; Carusi, 1977). Although the function of this terminal protein is not known, it has been speculated that it is involved in the initiation of DNA replication, either as a specific endonuclease (Wu et al., 1977) or as a priming protein (Rekosh et al., 1977).

Replication of adenovirus DNA begins 5-9 hr after infection of permissive host cells (e.g., human KB cells) and continues for at least 24 hr. The mechanism of replication shows a number of peculiar properties. Combined biochemical analysis of replicative intermediates (Sussenbach et al., 1972; van der Eb, 1973; Pettersson, 1973), electron microscopy (Ellens et al., 1974) and restriction enzyme analysis of termination sites (Weingärtner et al., 1976; Tolun and Pettersson, 1975; Sussenbach and Kuijk, 1977) has given rise to the following general picture (see Fig. 1).

[*] *A.S.I. Participant*

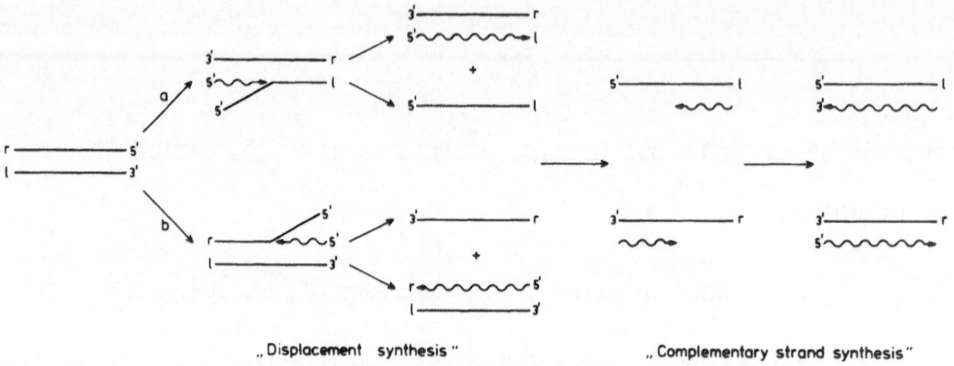

"Displacement synthesis"   "Complementary strand synthesis"

Figure 1.   *Replication Scheme for Adenovirus DNA*

*Replication starts either at the left end (a) or the right end (b)
of the molecule and proceeds by displacement of the complementary
strand (r-strand or l-strand, respectively). Since the ends
contain an inverted repetition, both termini may have identical
nucleotide sequences. It is therefore presumed that left and right
initiations occur according to the same mechanisms. Displacement
synthesis continues until a new daughter molecule is formed.
Subsequently the displaced single strand can be duplicated
("complementary strand synthesis"). A premature start of
complementary strand synthesis on internal origins, already during
displacement, has been observed in low frequency (not shown). As
yet there is no evidence that initiation can start right and left
at the same time in one molecule. This model, modified after
Sussenbach et al. (1972), does not account for the possible role of
a terminal protein during replication.*

―――――― *parental DNA*

⟿ *progeny DNA*

        DNA replication initiates at or very near to one of the
molecular ends, either the right or the left end, by displacement
of the 5'-terminated strand. The replication fork moves towards the
other end of the molecule, producing a displaced, single-stranded
DNA of genome length. This single-stranded DNA is derived from both
complementary strands in about equal amounts (Lavelle et al., 1975;
Sussenbach et al.,1976) which suggests an equal probability to
initiate either right or left, although other, electron microscopic
evidence indicates a preference for initiation at the right end
(Ellens et al., 1974). The displaced single strand is next converted
into a duplex molecule. According to electron microscopic data this

backward (complementary strand) synthesis can already start during
displacement, at internal initiation sites. This must be a minor
event, however, since restriction enzyme analysis of termination
sites shows a gradient towards the molecular ends with only a low
frequency of internal termination sites.

One of the interesting features of this replication scheme
is the pronounced asynchrony in the duplication of the two strands,
with the synthesis of one strand lagging way behind the other. One
might speculate that the long stretches of single-stranded DNA are
prone to nucleolytic attack and will be protected somehow in the
infected cells. Such a protection against nucleases might be
performed by single-strand specific DNA binding proteins (Alberts
and Frey, 1970; Molineux and Gefter, 1975).

Indeed, one of the early gene products of adenovirus is a
single-strand specific DNA binding protein. The genetic information
for this adenovirus-specific DNA binding protein is located between
0.58 and 0.70 on the physical map of adenovirus DNA (Lewis et al.,
1976). Its synthesis preceeds viral DNA synthesis and a constant
level of about 10^7 molecules per cell is found later in infection
(van der Vliet and Levine, 1973; Gilead et al., 1976). This high
amount is enough to cover all viral single-stranded DNA present in
infected cells.

Properties of the Adenovirus DNA Binding Protein (BP)

Selective extraction of infected cells followed by
chromatography on DNA cellulose leads to a 200 fold-purified,
90% pure preparation of Ad5 DNA binding protein. The apparent
molecular weight of the polypeptide in SDS-polyacrylamide gels is
72,000 daltons (72 K) which corresponds well to the native molecular
weight showing that the protein is a monomer. Its aberrant
behaviour during gel filtration (Sugawara et al., 1977) indicates
that it does not behave as a typical globular protein, a property
which is observed frequently for proteins that perform structural
rather than enzymatic functions. The 72 K DNA binding protein is
phosphorylated and is easily degraded to 42-48 K proteolytic
breakdown products which are not phosphorylated (Rosenwirth et al.,
1975); Linné et al., 1977). Both forms bind strongly to single
stranded DNA of various sources, without apparent sequence
specificity. No binding to double stranded DNA or RNA has been
observed.

The complex between single stranded DNA and binding protein
forms readily in the absence of metal ions or reducing agents. A
maximal protein to DNA ratio of $30(^W/_W)$ can be obtained which
corresponds to one 72 K molecule for 7 nucleotides. Such a
saturated complex has an extended configuration as shown by
sedimentation studies and electron microscopy (van der Vliet and

Keegstra, manuscript in preparation). This places the adenovirus
DNA binding protein in the class of T4 gene 32 and E.coli DNA binding
proteins which also keep single-stranded DNA extended, in contrast
to the gene 5 product of filamentous phages (Delius et al., 1972;
Alberts et al., 1972).

The large amount of this protein in infected cells, combined
with the presence of large regions of single-stranded DNA in
replicative intermediates makes the function of this protein an
intriguing question. This paper will summarize our results with
regard to its role in adenovirus DNA replication.

RESULTS AND DISCUSSION

From genetic analysis of adenovirus three DNA negative
complementation groups have been found so far (Williams et al.,
1974; Ginsberg and Young, 1976; Shiroki and Shimojo, 1974). One
of these has been allocated to the DNA binding protein gene and is
represented by the mutants H5ts125 (van der Vliet et al., 1975),
H12tsA275 (Rosenwirth et al., 1976) and $Ad_2^+ND_1ts23$. The latter
mutant was isolated by Williams and.co-workers after mutagenesis
of the adenovirus-SV40 hybrid $Ad_2^+Nd_1$ (Lewis et al., 1969) and does
not complement H5ts125 (Williams, personal communication).

DNA Binding Protein Synthesis in Mutant Infected Cells

We have analyzed two of the binding protein mutants, H5ts125
and ND_1ts23, in more detail. Cells were infected at the non-
permissive temperature (40°) and a cell free extract was
chromatographed on a DNA-cellulose column containing single-stranded
DNA. Mutant infected cells appear to contain at least 20-fold less
binding protein compared to wild type infected cells (Figure 2) or
cells grown at 32°.

Pulse-labelling of the binding protein in mutant infected
cells at 40°, however, shows a normal rate of synthesis. This
indicates that the mutant BP, while being synthesized normally,
is degraded much faster than wild type protein. Such an increased
breakdown was indeed observed in pulse-chase experiments (van der
Vliet et al., 1975). Thus the decreased pool of binding protein
in mutant infected cells is caused by an increased proteolytic
degradation of defective protein rather than by a lower rate of
synthesis. A fast breakdown of inactive proteins in ts-mutant
infected cells has also been observed for some late adenovirus
proteins (Russell et al., 1974).

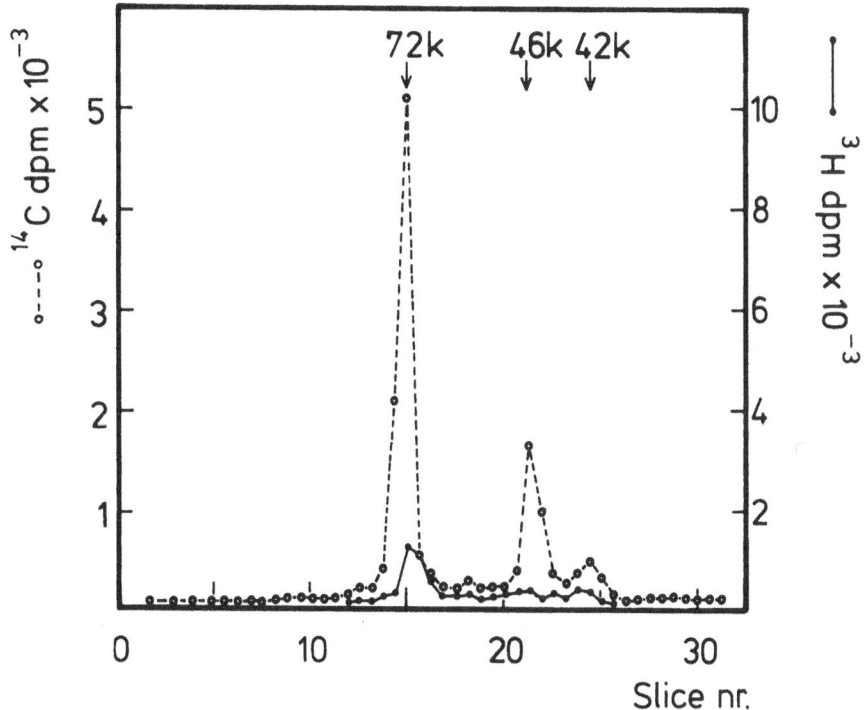

Figure 2.   *SDS-Polyacrylamide Gel Electrophoresis of Mutant*
              and Wild Type DNA Binding Proteins.

Cells were infected at 40° with either wild type Ad5 (●——●) or
ND_1ts23 (o---o),a DNA binding protein mutant.   Wild type infected
cells were labeled with ^{14}C-leucine and mutant infected cells with
^3H-leucine.   At 17 hr post infection the cells were mixed,
adenovirus DNA binding proteins were isolated by DNA cellulose
chromatography (van der Vliet and Levine, 1973) and electrophoresed.
In wild type infected cells three peaks are found, one at 72 K and
two at 46 K and 42 K, presumably proteolytic degradation products
(Rosenwirth et al., 1975).   From the $^{14}C/^3H$ ratio in the crude
protein extract (0.5) we calculate that mutant infected cells
contain 10-20 fold less DNA binding protein than wild type infected
cells.   A similar result was obtained with H5ts125 (van der Vliet
et al., 1975).

Viral DNA Synthesis in Cells Infected with Binding Protein
                            mutants

     H5ts125 infected KB cells grown at 40° do not replicate
adenovirus DNA (Ensinger and Ginsberg, 1972). When H5ts125 or

ND_1ts23 infected cells are grown at 32° and, during viral DNA
replication, are shifted to 40° a gradual inhibition in viral DNA
synthesis is observed, up to 90-95% after 45 minutes (Figure 3).

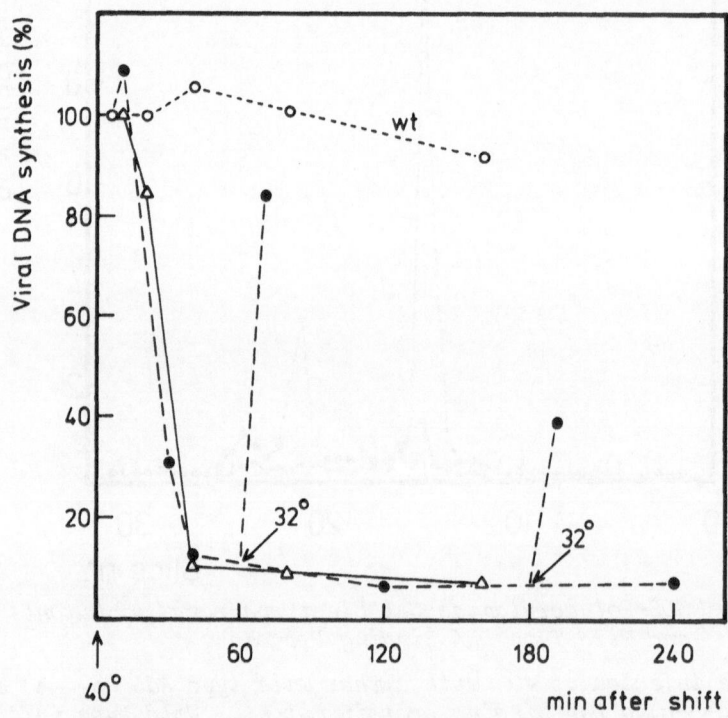

Figure 3.   *Reversible Block in Adenovirus DNA Synthesis in Cells
Infected with DNA Binding Protein Mutants*

*Cells infected with H5ts125 (o). ND_1ts23 (Δ) or wt (o) were grown
at 32° for 40 hr, when viral DNA synthesis is maximal. At t = 0 the
cultures were shifted to 40° and viral DNA synthesis was analyzed.
The results are expressed as % of synthesis at t = 0. To two of
the H5ts125 infected cultures cycloheximide was added at t = 0 and
at 60 min or 180 min they were shifted back to 32° (dashed lines)
which restores viral DNA synthesis to 83% and 39%, respectively.*

At this time most of the binding protein is still present in the
cells and is not degraded. DNA synthesis resumes to 83% of its
original level as soon as the temperature is shifted down to 32°
again, showing that the inactivation of the protein is reversible
(Figure 3). 3 hr after the shift a restoration from 6 to 39% is
still possible. The restoration of DNA synthesis occurs even in
the presence of cycloheximide which blocks protein synthesis. These
results indicate the presence of an excess pool of thermolabile
mutant protein that can be readily reactivated at 32° and is only

slowly diminished below the level required for viral DNA synthesis.

DNA Binding Protein Required for Initiation of DNA Synthesis

The shut off of viral DNA replication of H5ts125 and ND_1ts23 occurs within 40 min after the shift which is about the time required to complete one round of replication (Levine et al., 1976). This suggests that initiation, and not chain propagation, is blocked after a shift to 40°. Three independent observations confirm this hypothesis and show that H5ts125 as well as ND_1ts23 are indeed initiation mutants.

A. Analysis of pulse-labeled viral DNA at various times after the shift to 40° showed that both the amount of replicative intermediates (RI) and of mature DNA decreased equally. The ratio of labeled RI to labeled mature DNA did not change, even when the inhibition was 90% or more. A block in chain elongation would give rise to an increased amount of labeled RI compared to mature DNA. However, no such increase was observed (Ginsberg et al., 1974; van der Vliet and Sussenbach, 1975).

B. When pulse-labeled DNA was chased at 40°. the flow of replicative intermediates into mature DNA took place at equal rates either in wild type or mutant infected cells (Figure 4).

C. Isolated nuclei from adenovirus infected cells are capable of elongation and termination of replicative intermediates that existed in the infected cells prior to isolation of the nuclei (van der Vliet and Sussenbach, 1972; Winnacker, 1975), but are defective in initiation. Therefore isolated nuclei provide a system to study elongation in the absence of new initiations. H5ts125 or wild type infected cells were grown at 32°. Nuclei were isolated and incubated in the presence of ^{3}H-TTP to complete replicative intermediates. The rate of viral DNA synthesis appeared to be equal both in wild type and mutant infected nuclei (van der Vliet et al., 1977), which supports the notion that chain elongation is not blocked when the DNA binding protein is thermally inactivated.

Anti-DNA Binding Protein γ-Globulin Inhibits Chain Elongation

The experiments described above provide strong evidence for a function of the binding protein in the initiation process. They do not exclude, however, the existence of additional functions, e.g., stabilization of single stranded DNA or interaction with the DNA polymerizing machinery in the replication fork. Such hypothetical functions might remain undetected due to the limited number of

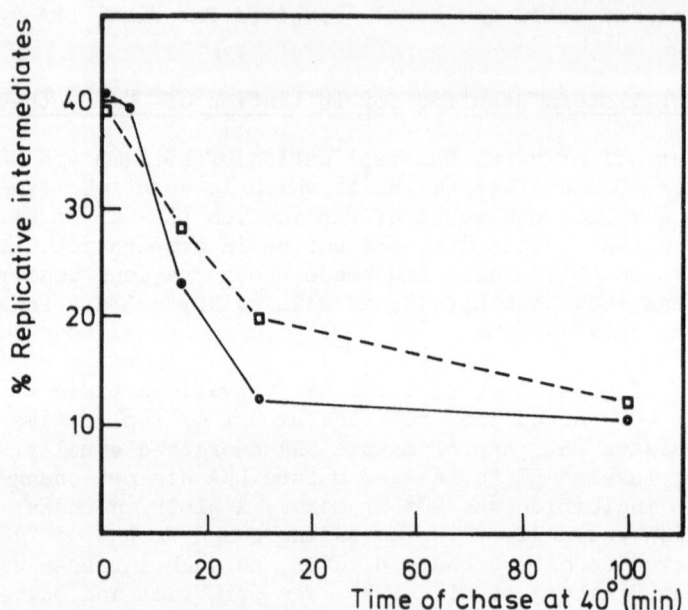

Figure 4. *Decrease in the Relative Amount of Replicating*
            Intermediates During a Chase at 40°.

Wild type (□) or ND1ts23 (●) infected cells were pulse-labeled
at 32° with ³H-thymidine, followed by a chase of the label at 40°.
Viral DNA was analyzed in neutral sucrose gradients and the % of
RI, defined as viral DNA that sediments faster than mature DNA, was
calculated. The flow of RI to mature DNA is a measure for the
rate of chain elongation. Similar results were found for H5ts125
(van der Vliet and Sussenbach, 1975).

mutants available in the binding protein gene. The available mutants
might block initiation but not the elongation function of the BP.
To obtain information about possible other functions, notably chain
elongation, we added rabbit anti-binding protein γ-globulin to
isolated nuclei from adenovirus infected cells. This resulted in
a 50-60% inhibition of viral DNA synthesis (Figure 5). A stronger
inhibiting effect was never observed, neither by addition of
detergents to permeabilize the nuclei nor by addition of the smaller
pepsin fragment of γ-globulin which retains its antigen binding
sites.

    The maximal inhibiting effect is observed at about 1 mg per ml
of γ-globulin. This concentration is high enough to precipitate
all the DNA binding protein present in the nuclei as revealed by an
indirect immunoprecipitation assay. Several lines of evidence
indicate that the decrease in viral DNA synthesis is indeed caused

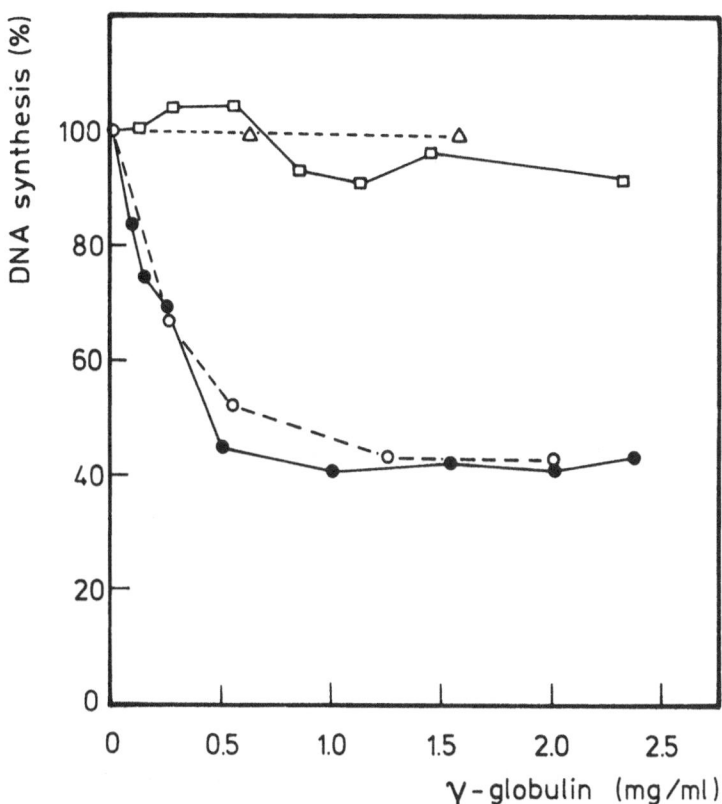

<u>Figure 5</u>. *Effect on DNA Chain Elongation by Addition of Anti-*
Binding Protein γ-Globulin to Nuclei from Adenovirus
Infected and Uninfected Cells.

Anti-BP γ-globulin (△,○,●) or control γ-globulin against DNA
binding proteins from uninfected cells (□) was added to isolated
nuclei and viral DNA synthesis was measured. The results are
expressed as % of the incorporation without γ-globulin.

●, □ : *wild type infected cells.   Nuclear synthesis at 32°*

○   : *H5ts125 infected cells.   Nuclear synthesis at 40°*

△   : *uninfected cells.   Nuclear synthesis at 32°*

by inactivation of the binding protein in the nuclei, and not by
other inhibiting factors that may be present in the γ-globulin
preparation.

  a. The inhibition is specific for adenovirus infected
      cells and does not occur in uninfected cells (Figure 5).

b.   Addition of purified Ad5 DNA binding protein
     destroyed the inhibiting effect of the γ-globulin
     (not shown).

c.   Antibody against DNA binding protein from
     uninfected cells, which might contaminate the
     γ-globulin preparation, did not decrease Ad5 DNA
     synthesis (Figure 5).

Since in isolated nuclei only replicative intermediates are
elongated we ascribe the inhibiti n of the DNA synthesis to a
decrease in chain elongation. Moreover, a similar inhibition is
observed in H5ts125 infected nuclei at 40° (Figure 5), a situation
where no initiations occur. This indicates an additional function
of the binding protein on chain elongation, besides the function
in initiation. Density labeling experiments employing BrdUTP have
shown that the decrease in chain elongation occurs in every
individual replication fork (van der Vliet et al., 1977) and is not
caused by a diminishing amount of replicating molecules. Both
strands replicate slower, indicating a general decrease in fork
movement in the presence of anti-BP γ-globulin.

Multifunctional Character of the Adenovirus DNA Binding Protein

Our results show that the early adenovirus DNA binding protein
performs at least two functions, and possibly three, in viral DNA
replication (Figure 6).

I.   Initiation of the first and subsequent rounds of
     replication requires an intact BP as shown by studies
     of the mutants H5ts125 and ND_1ts23. The role in
     initiation could involve interaction with other
     initiation proteins as well as with DNA. Host proteins
     can not take over this function, but in monkey cells
     infected with SV40 and SV40 A-gene product might
     substitute for the DNA binding protein (Levine et al.,
     1976).

II. The DNA binding protein performs some function during
     chain elongation, possibly by interaction with
     proteins (e.g., DNA polymerase) in the replication
     fork. This is indicated by the general decrease
     in fork movement observed in the presence of anti-
     BP γ-globulin. The antibody might work by
     interfering with the formation of an effective
     replication machinery or by disturbing the
     configuration of the DNA-BP complex. The reason
     for the limited (2-2.5 fold) inhibition by γ-globulin
     is not known. Either the BP functions only to

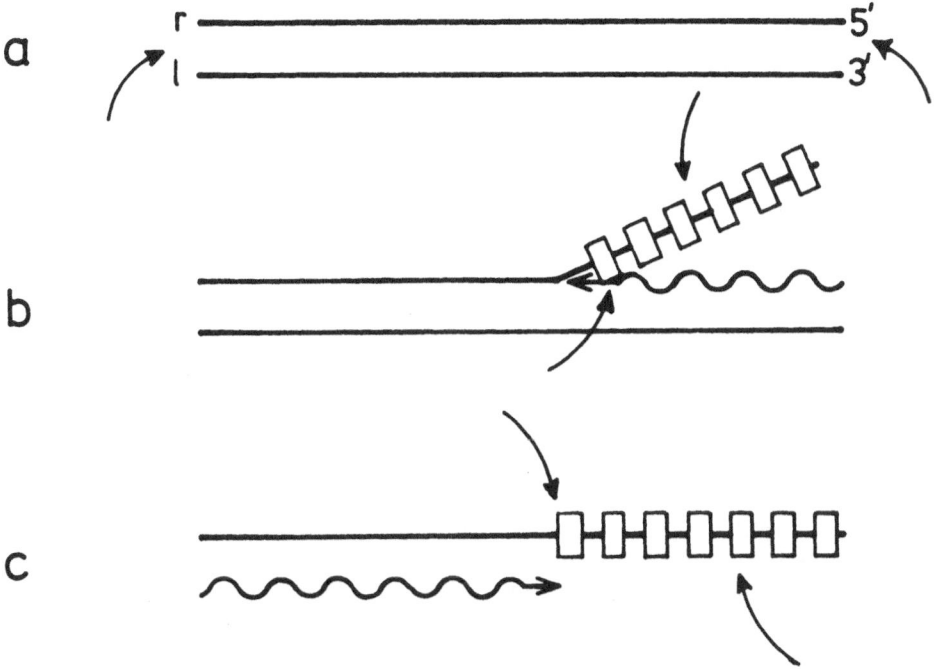

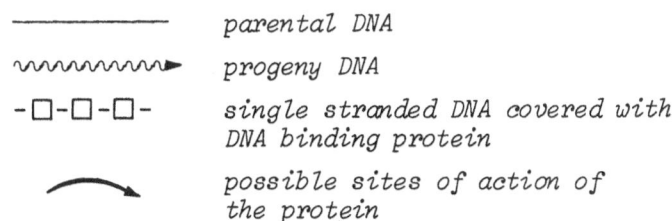

Figure 6. *Multifunctional Character of the Adenovirus DNA*
Binding Protein

Three types of adenovirus DNA molecules are shown, parental DNA
(a) or 2 types of replicative intermediates, one originating from
displacement synthesis (b) and one from complementary strand
synthesis (c). Three possible sites of action of the binding
protein are indicated by bended arrows. First, a role in
initiation at the molecular ends of double stranded DNA, a function
revealed by analysis of mutant phenotype. Second, interaction with
the replication fork in both types of replicative intermediates
(b and c). Third, stabilization of the single stranded DNA.

———————	*parental DNA*
∿∿∿∿∿∿▶	*progeny DNA*
-□-□-□-	*single stranded DNA covered with* *DNA binding protein*
⟶	*possible sites of action of* *the protein*

accelerate the elongation process or host proteins might substitute
for the adenovirus BP in this case.

III. A third function could be stabilization of the vast
      amount of single stranded DNA which is found in
      replicative intermediates. The DNA binding protein
      has been found as the major protein constituent in
      nucleo-protein complexes isolated from infected
      nuclei (Shanmugam et al., 1975; Kedinger, Willhelm
      and Briston, personal communication). Moreover,
      saturated complexes between ΦX174 ss-DNA and BP are
      slightly more resistant to the action of single-
      stranded specific nucleases than the uncovered DNA
      (unpublished observations). The protein could serve
      to maintain single stranded DNA in a favourable
      configuration, and thus permit effective copying of
      the displaced strand.

Multifunctional properties have also been observed for the T4
gene 32 DNA binding protein (Alberts and Frey, 1970; Mosig and
Breschkin, 1975) and for the SV40 A-protein, which protein is
presumably identical to the tumor antigen. In this respect it is
interesting to note that the adenovirus DNA-binding protein has also
been found in some adenovirus-transformed cell lines which contain
the DNA binding protein gene in an integrated form (Levinson et al.,
1976; Gilead et al., 1975; M. van Maarschalkerweerd, personal
communication). Whether the presence of the BP is purely
adventitious or whether it performs some function in the transformed
cells is unknown. The BP gene is not present in all transformed
cells and thus is not required for the maintenance of the
transformed state (Sambrook et al., 1974; Graham et al., 1974). It
is clear that isolation and analysis of more mutants in the BP gene
is required to establish firmly the multifunctional character of
this protein in adenovirus DNA replication.

ACKNOWLEDGEMENTS

We thank Dr. J.F. Williams for a gift of ND_1ts23 and
Dr. H.S. Ginsberg for H5ts125. Stimulating discussions with
Dr. J.S. Sussenbach are gratefully acknowledged. This study was
supported in part by the Netherlands Foundation for Chemical
Research (SON) with financial aid from the Netherlands Organization
for the Advancement of Pure Research.

REFERENCES

(1)   ALBERTS, B.M. and FREY, L. (1970) Nature, 227,1313

(2)   ALBERTS, B.M., FREY, L. and DELIUS, H. (1972) J. Mol. Biol.
      68,139

(3)   CARUSI, E.A. (1977) Virology 76,380

(4)   DELIUS, H., MANTELL, N.J. and ALBERTS, B.M. (1972) J. Mol.
       Biol. 67,341

(5)   ELLENS, D.J., SUSSENBACH, J.S. and JANSZ, H.S. (1974)
       Virology 61,427

(6)   ENSINGER, M.J. and GINSBERG, H.S. (1972) J. Virol. 10,328

(7)   GARON, C.F., BERRY, K. and ROSE, J. (1972) Proc. Nat. Acad.
       Sci., U.S.A. 69,2391

(8)   GILEAD, Z., SUGAWARA, K., SHANMUGAM, G. and GREEN, M. (1976)
       J. Virol. 18,454

(9)   GILEAD, Z., ARENS, M.Q., BHADURI, S., SHANMUGAM, G. and
       GREEN, M. (1975) Nature 254,533

(10)   GINSBERG, H.S. and YOUNG, C.H. (1976) Adv. Cancer Res. 23,91

(11)   GINSBERG, H.S., ENSINGER, M.J., KAUFMANN, R., MAYER, A.J. and
       LUNDHOLM, U. (1974) Cold Spring Harbor Symp. Quant. Biol.
       39,419

(12)   GRAHAM, F.L., van der EB, A.J. and HEYNEKER, H.L. (1974)
       Nature 251,687

(13)   KEEGSTRA, W., van WIELINK, P.S. and SUSSENBACH, J.S. (1977)
       Virology 76,444

(14)   LAVELLE, C., PATCH, C., KHOURY, G. and ROSE, J. (1975) J.
       Virol. 16,775

(15)   LEVINE, A.J., van der VLIET, P.C. and SUSSENBACH, J.S. (1976)
       Curr. Top. Microbiol. Immunol. 73,67

(16)   LEVINSON, A., LEVINE, A.J., ANDERSON, S., OSBORN, M.,
       ROSENWIRTH, B. and WEBER, K. (1976) Cell 7,575

(17)   LEWIS, A.M., LEVIN, M.J., WIESE, W.H., CRUMPACKER, C.S. and
       HENRY, P.H. (1969) Proc. Nat. Acad. Sci., U.S.A. 63,1128

(18)   LEWIS, J.B., ATKINS, J.F., BAUM, P.R., SOLEM, R., GESTELAND,
       R.F. and ANDERSON, C.W. (1976) Cell 7,141

(19)   LINNÉ, T., JÖRNVALL, H. and PHILIPSON, L. (1977) Eur. J.
       Biochem. in press

(20)   MOLINEUX, I.J. and GEFTER, M. (1975) J. Mol. Biol. 98,811

(21) MOSIG, G. and BRESCHKIN, A.M. (1975) Proc. Nat. Acad. Sci.,
      U.S.A. 72,1226

(22) PATTERSSON, U. (1973) J. Mol. Biol. 81,521

(23) REKOSH, D.M.K., RUSSELL, W.C., BELLETT, A.J.D. and ROBINSON,
      A.J. (1977) Cell, in press

(24) ROBERTS, R.J., ARRAND, J.R. and KELLER, W. (1974) Proc.
      Nat. Acad. Sci., U.S.A. 71,3829

(25) ROBINSON, A.J., YOUNGHUSBAND, M.B. and BELLETT, A.J.D. (1973)
      Virology 56,54

(26) ROSENWIRTH, B., ANDERSON, C. and LEVINE, A.J. (1976)
      Virology 69,617

(27) ROSENWIRTH, B., SHIROKI, B., LEVINE, A.J. and SHIMOJO, H.
      (1975) Virology 67,14

(28) RUSSELL, W.C., SKEHEL, J.J. and WILLIAMS, J.F. (1974) J. Gen.
      Virol. 24,247

(29) SAMBROOK, M., BOTCHAN, M., GALLIMORE, P., OZANNE, B.,
      PETTERSSON, U., WILLIAMS, J. and SHARP, P.A. (1974) Cold
      Spring Harbor Symp. Quant. Biol. 39,615

(30) SHANMUGAM, G., BHADURI, S., ARENS, M. and GREEN, M. (1975)
      Biochemistry 14,332

(31) SHIROKI, K. and SHIMOJO, H. (1974) Virology 61,474

(32) SUGAWARA, K., GILEAD, Z. and GREEN, M. (1977) J. Virol. 21,338

(33) SUSSENBACH, J.S., van der VLIET, P.C., ELLENS, D.J. and JANSZ,
      H.S. (1972) Nature New Biol. 239,47

(34) SUSSENBACH, J.S., TOLUN, A. and PETTERSSON, U. (1976) J.
      Virol. 20,532

(35) SUSSENBACH, J.S. and KUIJK, M.G. (1977) Virology   77,149

(36) TOLUN, A. and PETTERSSON, U. (1975) J. Virol. 16,759

(37) van der EB, A.J. (1973) Virology 51,11

(38) van der VLIET, P.C. and SUSSENBACH, J.S. (1972) Eur. J.
      Biochem. 30,584

(39)   van der VLIET, P.C. and LEVINE, A.J. (1973)   Nature New
       Biol. 246,170

(40)   van der VLIET, P.C., LEVINE, A.J., ENSINGER, M.J. and
       GINSBERG, H.S. (1975) J. Virol. 15,348

(41)   van der VLIET, P.C. and SUSSENBACH, J.S. (1975)   Virology
       67,415

(42)   van der VLIET, P.C., ZANDBERG, J. and JANSZ, H.S. (1977)
       Virology, in press

(43)   WEINGÄRTNER, B., WINNACKER, E.J., TOLUN, A. and PETTERSSON, U.
       (1976) Cell 9,259

(44)   WILLIAMS, J.F., YOUNG, H. and AUSTIN, P. (1974) Cold Spring
       Harbor Symp. Quant. Biol. 39,427

(45)   WINNACKER, E.J. (1975) J. Virol. 15,744

(46)   WOLFSON, J. and DRESSLER, D. (1972) Proc. Nat. Acad. Sci.,
       U.S.A. 69,3054

(47)   WU, M., ROBERTS, R.J. and DAVIDSON, N. (1977) J. Virol. 21,766

SEQUENCE SELECTIVITY IN THE INTEGRATION OF ADENOVIRUS TYPE 2 DNA

IN PRODUCTIVELY INFECTED CELLS

Ellen Fanning[*] and Walter Doerfler

Institute of Genetics
University of Cologne
Weyertal 121
Cologne, Germany

INTRODUCTION

Recombination between viral and cellular DNA leading to the covalent integration of adenovirus DNA sequences into the host genome appears to be a general phenomenon occurring in abortively infected (Doerfler, 1968; 1970), productively infected (Burger and Doerfler, 1974; Doerfler et al., 1974; Schick et al., 1976) and transformed cells (Bellett, 1975; Green et al., 1976; Groneberg et al., 1977; Fanning et al., 1977a).

The mechanism by which viral and cellular sequences are linked is not at all understood. Yet many features of the integration process are common to both abortively infected cells, from which transformed cells may arise, and productively infected cells. For example, in both systems viral DNA sequences are integrated in the form of fragments (Doerfler, 1970; Fanning and Doerfler, 1977; Baczko et al., 1977). There is evidence that viral DNA replication is not required for insertion (Doerfler, 1970; Fanning et al., 1977b), whereas the integrated form of adenovirus DNA cannot be detected when cellular DNA synthesis is completely blocked. These common features tempt speculation that the integration mechanism(s) may be similar in abortively and productively infected cells.

The study of integration in cells productively infected with adenovirus is facilitated by the large amounts of viral DNA linked to cellular DNA. When the total intracellular DNA from KB cells infected with adenovirus type 2 (Ad2) is separated into four size-

[*]A.S.I Participant. Present address: Fachbereich Biologie, University of Konstanz, Konstanz, Germany.

classes: >100S, 50-90S, 34S and <20S by zone velocity sedimentation
in alkaline sucrose density gradients, more than 1000 viral genome
equivalents per cell are found in DNA which sediments faster than
34S unit length Ad2 DNA (Fanning and Doerfler, 1977). A series of
experiments has demonstrated that the high molecular weight viral
DNA sequences represent viral DNA covalently joined to cellular DNA
sequences (Burger and Doerfler, 1974; Doerfler et al., 1974; Schick
et al., 1976; Fanning and Doerfler, 1977; Baczko et al., 1977).

This communciation deals with two aspects of the integration
mechanism: the role of cellular and viral functions in the
integration process and the question of specificity in the viral
sequences integrated.

MATERIALS AND METHODS

Cells and virus. Human KB cells are grown in monolayer or
suspension culture in Eagle's medium (Eagle, 1959) and 10% calf
serum (Flow Laboratories). Cell cultures were tested weekly
for mycoplasma contamination and only mycoplasma-free cultures
were used for the experiments reported here. Human adenovirus
type 2 (Ad2) was propagated in suspension cultures of KB cells
and purified by 3 cycles of equilibrium centrifugation in CsCl
density gradients as described previously (Doerfler, 1969).

Infection of KB cells with Ad2 and size-class analysis of
total intracellular DNA. KB cells were seeded in 60 mm diameter
dishes at 10^6 cells per dish. One day later, cells were infected
with CsCl-purified Ad2 at a multiplicity of 100 plaque-forming
units per cell. In some experiments, 6-^{3}H-thymidine was added
(40-60 μCi/ml) from 16-18 h post-infection. At 18 h post-infection,
the cells were washed 3-5 times with phosphate-buffered saline
(PBS), scraped off the plate in 1 ml of PBS, and 10^5 cells in
about 0.1 ml were lysed in a 1 ml layer of 0.5 M NaOH, 0.05 M EDTA
directly on top of a 5-20% alkaline sucrose gradient in an SW27
nitrocellulose tube. The alkaline sucrose solutions contained 5%
or 20% (w/v) sucrose (Schwarz/Mann) in 0.7 M NaCl, 0.3 M NaOH,
0.005 M EDTA. When unlabeled cells were used in the experiment,
one gradient was loaded with ^{3}H-labeled cells from a parallel
culture to allow determination of the location of the different
size classes of DNA in the gradients containing unlabeled DNA.
After complete lysis of the cells (18 h, 4°C), the samples were
centrifuged in the SW27 rotor at 4°C and 23,000 rpm for 7 h. DNA
from the four size-classes was pooled and processed for reassociation kinetics as described elsewhere (Fanning and Doerfler, 1977).

Preparation of restriction endonuclease fragments of ^{3}H-labeled viral DNA has been described previously (Fanning and Doerfler, 1976). Briefly, ^{3}H-labeled Ad2 DNA (2-3 x 10^7 cpm/OD$_{260}$) was cleaved with EcoRI or Bam H-I restriction endonuclease and the fragments were separated by electrophoresis in 1.5% polyacrylamide-0.8% agarose gels as reported earlier (Ortin et al. 1976).

DNA-DNA filter hybridization was carried out as described elsewhere (Fanning et al., 1976b).

Reassociation kinetics. The procedure used for reassociation kinetics (Britten and Kohne, 1968; Wetmur and Davidson, 1968) has been detailed elsewhere (Fanning and Doerfler, 1977). The single-strand specific nuclease S1, purified according to Vogt (1973), was used to assay the extent of renaturation as reported earlier (Fanning and Doerfler, 1977).

RESULTS

Role of the Cell in the Integration of Ad2 DNA
in Productively Infected Cells

As a first step in elucidating the mechanism by which viral DNA becomes covalently linked to the cell DNA, it is essential to determine whether viral or cellular functions, or both, are responsible for integration. Thus, several variables have been tested for their effect on the integration of Ad2 DNA into KB DNA, using zone velocity sedimentation in alkaline sucrose density gradients to separate the total intracellular DNA into 4 size-classes, followed by hybridization of each size-class to Ad2 DNA on filters. The extent of hybridization is a measure of the proportion of newly synthesized high molecular weight DNA composed of Ad2 sequences, and that, in turn, is a reliable estimate of integration, as established by numerous control experiments (cf. Doerfler et al., 1974; Burger and Doerfler, 1974; Schick et al., 1976; Fanning et al., 1977b).

Integration of Ad2 DNA depends on the growth density of the cell culture at the time of infection. When KB cells growing at different densities are infected with Ad2 (Fig. 1b,d,f) the proportion of the newly synthesized high molecular weight DNA containing Ad2 sequences is considerably reduced in confluent mono-layers. In both Ad2-infected and mock-infected KB cells (Fig. 1a, c, e), the amount of newly synthesized 50-90S DNA, as judged by the incorporation of ^{3}H-thymidine, decreases as the cell density increases. In Ad2-infected cells, the amount of 34S unit length Ad2 DNA is also reduced at high cell densities. The 50-90S size class in uninfected KB cells was previously shown to be an inter-mediate in cell DNA replication (Schick et al., 1976). The formation of high molecular weight viral DNA, then, appears to require active

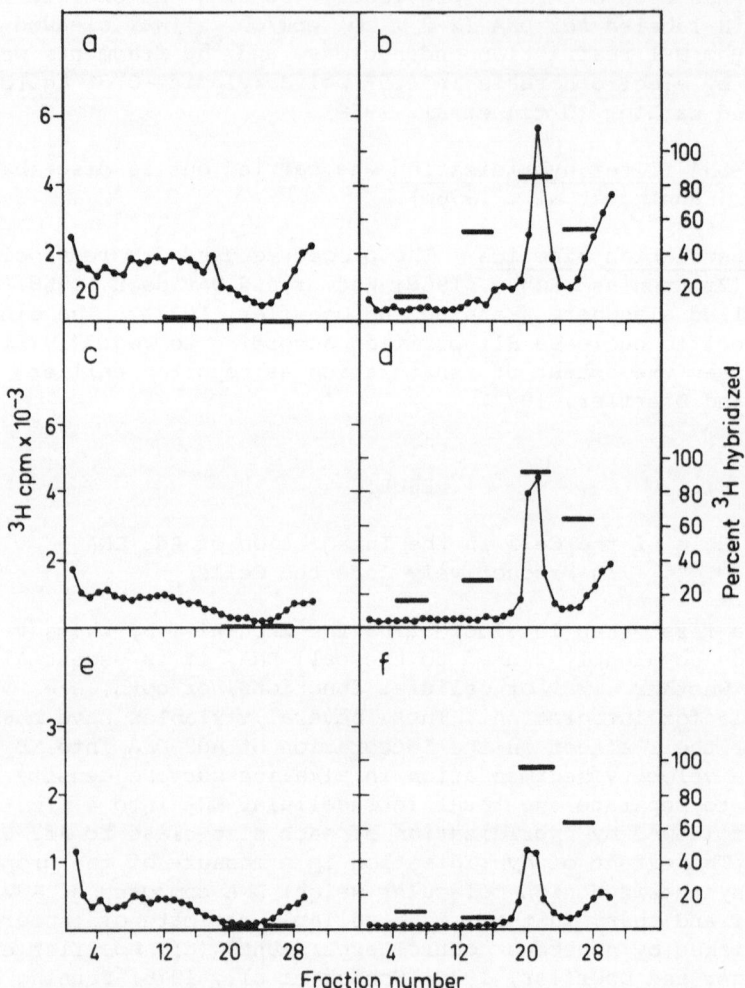

Figure 1. *Dependence of the amount of high molecular weight viral DNA and unit length Ad2 DNA on cell density.*

KB cells were seeded in 60 mm dishes at different densities: 1 x 10⁶ (a,b), 3 x 10⁶ (c,d), and 1 x 10⁷ (e,f) cells per dish. The cells were either mock-infected with PBS (a,c,e) or infected with Ad2 at multiplicities of 100 PFU/cell (b,d,f), labeled with ³H-thymidine (25 µCi/ml) from 16-18 h post-infection or post-mock-infection and analyzed by zone velocity sedimentation on alkaline sucrose density gradients (SW41, 35,000 rpm and 4°C for 3 h). The horizontal bars designate the pools of DNA hybridized to Ad2 DNA on filters, and the extent of hybridization is shown by the height of the bars. Direction of sedimentation is to the left.

cell growth, probably cell DNA replication.

The type of cell infected by Ad2 also influences the extent of integration. Ad2 DNA integrates into the genomes of HeLa, KB and human embryonic kidney cells at roughly the same frequency. In other human cells, however, including WI-38 and 2 lines of xeroderma pigmentosum fibroblasts XP4 and XP25, as well as in BHK21 and CHO cells, integration of viral DNA is greatly reduced (Fanning et al., 1976b).

Viral DNA replication does not appear to play any part in the integration process. Ad2 inactivated by exposure to ultraviolet light penetrates the host cell just as infectious virus does, but is no longer able to replicate its DNA. When inactivated Ad2 is used to infect KB cells, the formation of newly synthesized high molecular weight viral DNA is comparable to that in cells infected with unirradiated virus (Fanning et al., 1977b). These data support the notion that parental Ad2 DNA integrates into the cell genome, with little or no requirement for viral DNA replication. Other viral functions not inactivated by irradiation may, of course, be essential for integration to occur. Investigations using two Ad5 mutants which are temperature-sensitive in viral DNA replication have confirmed these results (Tyndall et al., 1977).

The Molecular Ends of the Ad2 Genome are Integrated Selectively

Additional information on the nature of the mechanism of integration may be gained by studying the viral sequences which are integrated (Fanning and Doerfler, 1977) and the cellular sequences to which they are joined (Schick et al., 1976; Baczko et al., 1977).

The viral DNA sequences in each of the four size-classes of DNA prepared from unlabeled Ad2-infected cells by zone velocity sedimentation in alkaline sucrose density gradients were quantitated using reassociation kinetics. The amount of viral DNA integrated into high molecular weight forms, 50-90S and >100S, remained relatively constant from 8 to 30 h post-infection, consistent with the idea that parental viral DNA is integrated (Fanning and Doerfler, 1977).

Each size-class of unlabeled DNA was also analyzed by re-association kinetics using ^{3}H-labeled EcoRI and Bam H-I restriction fragments of Ad2 DNA as probe to determine whether the entire genome was represented in each size-class. Each preparation of a particular size-class was also analyzed for viral sequences using ^{3}H-labeled Ad2 unit size DNA as the probe. The amount of each fragment of Ad2 DNA present was related to the amount of whole Ad2

DNA in the same preparation. In this way the relative abundance of each of the fragments in each of the size-classes has been determined. The EcoRI and Bam H-I restriction maps of the Ad2 genome are shown schematically in Fig. 2.

As expected, the EcoRI restriction fragments A-F are represented in about equimolar amounts in the 34S unit genome size Ad2 DNA peak (Fig. 3).

However, when the DNA from the 50-90S region of the alkaline sucrose density gradients was subjected to the same analysis, the result was very different. The right end of the Ad2 DNA molecule occurs at least ten to fifteen times more abundantly in the 50-90S DNA than the left terminus (Fig. 4). In fact, there is a striking gradient in the frequency of occurrence in 50-90S DNA of Ad2 DNA fragments from the left molecular terminus to the right molecular end. Similar results were obtained when the Bam H-I fragments were used in the analysis (Fanning and Doerfler, 1977).

In the <20S size-class of DNA the right half of the Ad2 DNA molecule is overrepresented by a factor of two to four, depending on the location on the viral DNA map.(Fanning and Doerfler, 1977).

Lastly, the >100S DNA prepared from Ad2-infected KB cells was also investigated by the technique described. In the >100S size-class the left half of the viral genome appears to be more abundant (Fig. 5), although the total amount of Ad2 sequences in the size-class is much smaller than in the 50-90S size-class (Fanning and Doerfler, 1977).

CONCLUSION

Any mechanism proposed to explain how adenovirus DNA becomes covalently linked to cellular DNA must consider that viral DNA replication is not essential, that parental viral DNA may become integrated, and that the machinery involved in cell DNA replication is probably involved. Moreover, the viral DNA appears to be inte-

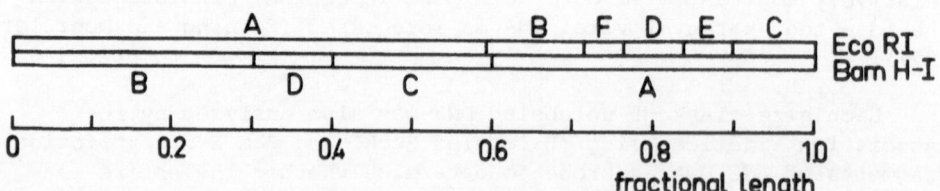

Figure 2. Map of the EcoRI and Bam H-I fragments of Ad2 DNA. The EcoRI fragments were mapped by Pettersson et al. (1973) and the Bam- H-I fragments by R. Roberts (personal communication).

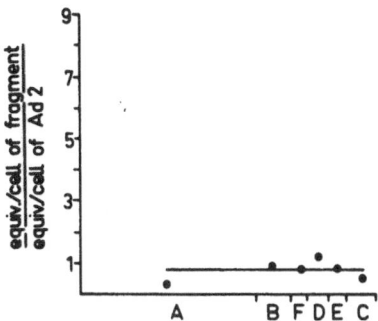

Figure 3. Relative abundance of each of the EcoRI fragments of
Ad2 DNA in the 34S DNA peak from Ad2-infected KB cells.
Unlabeled 34S (unit genome size) DNA was prepared 18 h post-
infection by zone velocity sedimentation on alkaline sucrose
density gradients.   This DNA was analyzed for viral DNA sequences
by reassociation kinetics using each of the EcoRI restriction
fragments of ³H-labeled Ad2 DNA as probe.   Each preparation of 34S
DNA was also analyzed using ³H-labeled Ad2 Virion DNA as probe.
The number of equivalents per cell of each fragment in 34S DNA was
divided by the number of equivalents per cell of whole genome in
the 34S size-class.   This relative abundance of the fragments is
shwon on the ordinate and the relative size and location of the
EcoRI fragments on the Ad2 genome is shown on the abscissa. Details
of the composition of the reassociation mixtures are given in
Fanning and Doerfler (1977), from which this figure was taken.

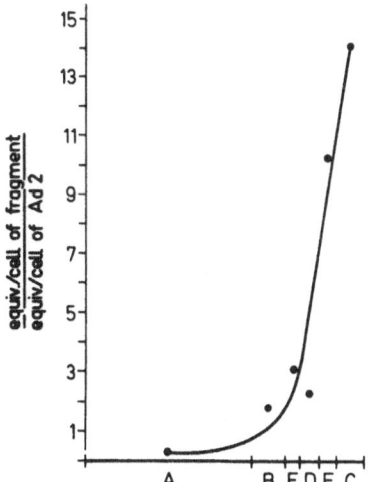

Figure 4. Relative abundance of each of the EcoRI fragments of Ad2
DNA in the 50-90S DNA from Ad2-infected KB cells.
Experimental conditions were as described in the legend to Fig. 3,
except that unlabeled 50-90S DNA was analysed.

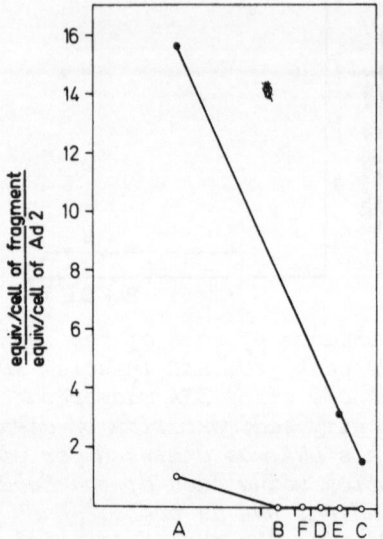

Figure 5. Relative abundance of each of the EcoRI fragments of the Ad2 DNA in the >100S DNA from Ad2-infected KB cells. Experimental details have been described in the legend to Fig. 3, except that unlabeled >100S DNA was analyzed.

-○- *Experiment 1*

-●- *Experiment 2*

grated in a fragmented form (cf. Baczko et al., 1977), with the molecular ends of the viral genome preferentially integrated. How this selectivity arises is not yet clear. Perhaps the protein covalently linked to the 5'-termini of the Ad2 DNA (Robinson et al., 1973) or undetected homologies between certain cell DNA sequences and the ends of the Ad2 genome could be responsible for the selective over-representation of the ends of Ad2 DNA in high molecular weight forms.

It is known that recombination frequencies in somatic cells and in cells in culture (e.g., Moore and Holliday, 1976) are generally rather low. The frequency of integration in adenovirus-infected cells, however, appears to be very high. Since cell DNA replication does not normally lead to extensive recombination in the uninfected cell, the virus infection must somehow induce the recombination machinery in the cell. Possibly the infection per se or some viral function expressed prior to viral DNA replication and not inactivated in irradiated virus could bring about this induction. Induction of several enzymes involved in DNA synthesis has, in fact, been demonstrated in cells abortively infected with adenovirus (Strohl, 1973). It would be interesting in this context to know whether the chromosome damage associated with adenovirus infection results from induced recombination either between viral and cellular DNA or between cellular DNA sequences.

ACKNOWLEDGEMENTS

We are indebted to Ute Winterhoff and Marianne Stupp for excellent technical assistance, Jochen Schick for performing DNA-DNA filter hybridizations, and I. Lühr for typing this manuscript.

We also wish to thank G. Mosig and H. Taylor for stimulating discussions.

This work was supported by the Deutsche Forschungsgemeinschaft (SFB 74) and the Federal Ministry of Research and Technology (01 VW 015 B13 BCT 112).

REFERENCES

Baczko, K., R. Neumann and W. Doerfler (1977) Virology, submitted

Bellett, A.J.D. (1975) Virology 65:427-435

Britten, R.J. and D.E. Kohne (1968) Science 161:529-540

Burger, H. and W. Doerfler (1974) J. Virol. 13:975-992

Doerfler, W. (1968) Proc. Natl. Acad. Sci. USA 60:636-643

Doerfler, W. (1969) Virology 38:587-606

Doerfler, W. (1970) J. Virol. 6:652-666

Doerfler, W., H. Burger, J. Ortin, E. Fanning, D.T. Brown, M. Westphal, U. Winterhoff, B. Weiser and J. Schick (1974) Cold Spring Harbor Symp. Quant. Biol. 39:505-521

Eagle, H. (1959) Science 130:432-437

Fanning, E. and W. Doerfler (1976) J. Virol. 20:373-383

Fanning, E. and W. Doerfler (1977) Virology 81:216-231

Fanning, E., K. Baczko, D. Kühn and W. Doerfler (1977a) 28. Coloquium-Mosbach, in press

Fanning, E., J. Schick and W. Doerfler (1977b) Virology, submitted.

Green, M.R., G. Chinna durai, J.K. Mackey and M. Green (1976)
    Cell 7:419-428

Groneberg, J., Y. Chardonnet and W. Doerfler (1977) Cell 10:101-
    111

Moore, P.D. and R. Holliday (1976) Cell 8:573-579

Ortin, J., K.H. Scheidtmann, R. Greenberg, M. Westphal and
    W. Doerfler (1976) J. Virol. 20:355-372

Pettersson, U., C. Mulder, H. Delius and P.A. Sharp (1973) Proc.
    Natl. Acad. Sci. USA 70:200-204

Robinson, A.J., H.B. Younghusband and A.J.D. Bellett (1973)
    Virology, 56:54-69

Schick, J., K. Baczko, E. Fanning, J. Groneberg, H. Burger and
    W. Doerfler (1976) Proc. Natl. Acad. Sci. USA 73:1043-1947

Strohl, W.A. (1973) Prog. Exp. Tumor Res. 18:199-239

Tyndall, C., H.B. Younghusband and A.J.D. Bellett (1977) J. Virol.
    24:000-000

Vogt, V.M. (1973) Eur. J. Biochem. 33:192-200

Wetmur, J.G. and N. Davidson (1968) J. Mol. Biol. 31:349-370

DNA REPAIR IN HUMAN CELLS: ENZYMES AND MUTANTS

Arturo Falaschi[*] and Antonia M. Pedrini[*]

Laboratorio di Genetica Biochemica ed Evoluzionistica
del Consiglio Nazionale delle Ricerche
Via S. Epifanio 14
27100 Pavia, Italy

A variety of chemical and physical agents may cause damage to
DNA, i.e., they may lead to the presence of chemical structures
different from those of the standard base pairs, or to chain breaks.
The cells of all organisms have a variety of repair mechanisms
available, still only partially understood even in prokaryote
systems [see the recent reviews of Hanawalt (1) and Grossman (2)].
Most of these mechanisms involve a certain amount of DNA synthesis,
which is operationally distinguishable from the synthesis occurring
at the growing point and is demonstrably due, in part at least, to
distinct molecules. Still, in bacteria, a certain overlapping can
be shown between the replication and repair systems, and also with
the related recombination process. Thus a given molecule mainly
involved in one process may, in particular situations, act in
another one. In mammalian systems the DNA repair mechanisms are

---

Abbreviations

AT = Ataxia Telangiectasia;   BS = Bloom's syndrome

FA = Fanconi's anemia;   HU = hydroxy urea;   MMC = mitomycin C;

MMS = methyl methanesulfonate;   NA-AAF = N-acetoxy-2-acetylamino-
fluorene;

4-NQO = 4-nitroquinoline-1-oxide;   SCE = sister chromatid exchange;

t' = 5,6 dihydroxy-dihydrothymine;   UDS = unscheduled DNA synthesis;

UV = ultraviolet light;   XP = Xeroderma pigmentosum.

---

[*]*A.S.I. Participants*

much less known than in microorganisms and many conflicting
observations limit the applicability of bacterial models; furthermore, important differences exist between the repair responses, e.g.,
of rodent and human cells, to UV, forbidding thus the simple
utilization of bacterial data for interpretations.

Among mammalian organisms, the only species in which mutants,
which are probably altered in repair processes, are available is
the human one: a number of inherited diseases seem in fact due to
genetic alterations of repair processes, which can be studied at
the cellular and molecular level. A study of the properties of these
mutants should throw light on the molecular basis of the repair
mechanisms and could allow a recognition of the function of some of
the enzymes of DNA mebatolism which are being described in human
cells.

We will briefly review here the properties of the enzymes or
other proteins of human cells whose functions may be involved in
repair processes, and the alterations observed, at the clinical,
cellular and molecular level, in the patients affected by some of
the genetic disorders possibly related to repair deficiency.

Enzymes of Human Cells Possibly Involved in DNA Repair

Table 1 reports a summary of the enzymic activities described
in human cells that could be acting at some step in repair; these
enzymes are at different stages of purification, ranging from a mere
description of a certain catalytical activity in crude extracts (such
as the activity removing γ-induced t' products (3) to complete
purification from cultured cells (such as the DNA polymerase α:
Korn et al., this volume). The human cells contain three DNA
polymerases, with similar catalytical properties, which are reviewed
by Falaschi and Spadari (this volume). The α-polymerase is probably
mainly involved in replicative synthesis, whereas a correlation of
the level of the β-polymerase with repair capacity has been observed
(4).

The properties of the polynucleotide ligase of human cells,
studied mainly by our group (5,6), were reviewed recently, together
with those of other mammalian enzymes, by Söderhäll et al. (7). The
function of such an enzyme in most repair models seems essential.

Among the nucleolytic enzymes the DNAse VI, an endonuclease
highly specific for single strand regions and producing 5'P ends,
could remove unpaired areas (8). The exonuclease described by
Grossman (9) could be adequate to enlarge a cut near a damaged
base, working exonucleolytically with both polarities on the incised
strand.   Enzymes which seem quite specific for damaged DNA are the

Table 1

ENZYME	REACTION
DNA POLYMERASE (α, β, γ)	
DNA LIGASE	
DNASE VI	
CORREXONUCLEASE	
DNA BINDING PROTEIN SPECIFIC FOR UV-IRRADIATED DNA (NON DIMERS)	
ENDONUCLEASE SPECIFIC FOR UV OR X-RAY BASE DAMAGE (NON DIMERS)	
ACTIVITY EXCISING DIMERS INCISED ON 5' SIDE	
N-GLYCOSIDASE SPECIFIC FOR MMS-TREATED DNA	
N-GLYCOSIDASE SPECIFIC FOR URACIL	
ENDONUCLEASE FOR APURINIC OR APYRIMIDINIC SITES	

endonuclease inicising UV irradiated DNA at sites other than dimers
(10,11) and the endonuclease specific for sites where the base has
been lost (12,13); to the latter enzyme the substrate could be
supplied by e.g., the N-glycosidase acting on MMS treated DNA (14)
or the one removing uracil from DNA (15). A dimer photoreactivating enzyme (16) has been described in some cell cultures.
A non enzymic protein binding specifically to UV-irradiated DNA at
sites other than dimers has been purified from the placenta (17).

An endonucleolytic activity specific for thymine dimers, like
the one described in prokaryotes, has *not* been found in human cells;
on the other hand, crude extracts of human cells are capable of
excising dimers after incision on the 5'side by the dimer specific
endonuclease from T4 phage (18); furthermore, a selective removal
of thymine dimers from UV irradiated chromatin by cell free extract
of HeLa cells has been observed; this implies the presence of a
specific endonuclease activity at present not yet directly assayable
(19). The thymine dimer excising system of human cell extracts
requires the presence of a protein factor which facilitates its
action (20).

Certainly many more enzymes and proteins await to be discovered
in human cells. The different enzymes could fit in a number of
models of DNA repair which we do not intend to review here, and for
which the evidence is still scanty. Is there a possibility to gain
information on the models by studying the cellular and molecular
properties of the mutants ?

                    Genetic Disorders Possibly Related to DNA
                              Repair Defects

The indication that a genetically determined disease may
involve DNA repair comes in the first place from clinical
observations: unusual sensitivity to sunlight or to ionizing
radiations and high proneness to develop malignant growth are
probably the most important items, and are present in all the
diseases object which are the object of this review. A further
important element is an increased tendency, either spontaneous or
induced, to chromosomal aberrations or SCE. The subsequent study
of isolated cells or cell cultures derived from such patients shows
enhanced sensitivity to the lethal effects of UV or ionizing
radiations or to a variety of DNA-damaging agents such as mutagens
and carcinogens.

The further study at the molecular level has given some
indication of the type of defect but, as of now, in no case has
it been possible to identify unambiguously the altered function.

Xeroderma pigmentosum (classical form). This is a rare
autosomal recessive disease characterized by sensitivity of the skin
to sunlight exposure (see reviews by Cleaver, 21,22). Clinical
manifestations include persistent erythema, abnormal pigmentation
and multiple skin malignancies in the sun-exposed areas and, in
some cases, neurological abnormalities. In the latter case, a
distinct clinical entity called De Sanctis-Cacchione syndrome is
identified as distinguished from a "classical" form. This
distinction does not seem to have significance at the genetic and
molecular level; we will thus call here "classical" form also the
De Sanctis-Cacchione syndrome, and distinguish both forms from the
"variant" (see below).

The cells from XP patients are sensitive to UV radiation, as
shown by a decreased colony-forming ability following UV treatment,
and to some carcinogens that link covalently to DNA (such as 4-NQO
and NA-AAF). They are normally resistant to X-rays and to monofunctional alkylating agents such as MMS.

The XP cells demonstrate typically their lowered repair
ability by a sharp reduction in repair replication, following the
action of the active agents just quoted. The repair replication
may be measured in different ways: UDS which shows by autoradiography
DNA synthesis of nuclei out of S phase; direct measure of UV-induced
thymidine incorporation in the presence of HU; measure of non semiconservative DNA synthesis (by BU incorporation and isopycnic
analysis or by BU-photolysis and size analysis). These measures are
instead normal after treatment with X-rays or with monofunctional
alkylating agents, such as MMS or N-methyl-nitrosourea.

The residual DNA repair replication ability shows different
inherited levels in different XP cell lines. The observation that
heterokaryons obtained by fusion between normal and XP fibroblasts
are now able to perform UDS thanks to the rapid diffusion of a
corrective substance from the normal to XP nuclei, leads to the use
of this technique for genetic analysis. On the basis of the
complementation studies it has been possible to demonstrate 5
different complementation groups denominated A to E.

In XP cells one can measure a reduced rate of excision of
pyrimidine dimers (the most important UV photo-product), of the
covalently bound acetylaminofluorene (23), and of O^6-alkylguanine
(24) (a minor alkylation product formed after treatment with the
monofunctional alkylating agents, such as N-methyl-N-nitrosourea
and N-ethyl-N-nitrosourea). The major alkylation product of the
same reagents, 7-alkylguanine, is lost at a similar rate in normal
and XP cells, as are the UV and γ-ray products of the t' type (25).

Single-strand breaks, which should be formed transiently during
dimer excision, appear rapidly after UV treatment as measured by DNA

alkaline elution and are resealed slowly in normal fibroblasts, whereas such breaks essentially do not appear in classical XP cells (26).

The defect in repair functions can be evidenced also by a pronounced reduction in host cell reactivation capacity after UV irradiation of the viruses which replicate in the cell nucleus (SV40, herpes virus, adenovirus 2).

The XP cells do not show any spontaneous increase in chromosomal aberration; these are increased (i.e., higher than after the same treatment on normal cells) by the exposure to those agents to which XP are abnormally sensitive (typically UV and 4-NQO). On the contrary the frequency of SCE, normal in untreated XP cells, increases abnormally not only after UV or 4-NQO, but also after treatment with monofunctional alkylating agents whose damage seems normally repaired in XP, as judged from repair replication measurements (27).

All the gathered evidence, summarized in Table II and III together with the other diseases, establishes clearly that some important derangement of repair processes is occurring in XP cells and points to the nucleotide excision process (typically pyrimidine dimer excision) as the site of the damage. In particular the slow rate of formation of breaks after UV irradiation as compared to normal controls indicates the incision step as the most likely site of the defect. The search for the molecule(s) responsible for this reaction and deficient in XP has not yet been successful. Positive evidence for a reduction in some enzymes (see Table IV and the last section for discussion) is still only partial and hard to reconcile with the *in vivo* data.

It is also worth pointing out a certain discrepancy between the lack of sensitivity of the XP cells to monofunctional alkylating agents (as measured by cell survival and repair synthesis) and the abnormal increase in SCE caused by these same agents. The recent observation (reported above) that a minor product of the mono-functional alkylating agents, O^6-alkylguanine, is removed at a reduced rate in XP cells, raises the possibility that this persisting damage may lead to increased SCE occurring in order to repair it. Thus, one can justify the normal repair replication rate of XP with monofunctional alkylating agents, since this is induced by the major alkylation products (which XP cells can remove normally); instead the lack of O^6-alkylguanine-induced repair synthesis may not be observable because of the low abundance of this structure, and the persistence of this altered structure in XP may lead to the use of a different repair mechanism. Such a process could be of the recombination type and lead to the observed increase in SCE. If such is the case, then perhaps the high level of SCE observed in pathological conditions would be evidence of the action of another repair mechanism.

Table II

Main Clinical Properties[a]

Clinical Alterations	Disease				
	XP (classical)	XP (variant)	BS	FA	AT
Skeleton anomalies	-	-	-	+	-
Defect in growth	-	-	+	+	+
Mental retardation	+	-		+	-
Bone marrow failure	-	-		+	
C.N.S. alterations or motor anomalies	+	-		+	+
Skin pigmentation anomalies	+	+	+	+	+
Immunoglobulin deficiency		-	+		+
Deficient response to mitogens		-			+
Tendency to cancer (generalized)	+	+	+	+	+
Tendency to leukosis	-	-	+	+	+
Sinlight sensitivity	+	+	+	-	-
Premature aging	-	-	-	-	+

[a]In this Table, and in the following ones, the + sign indicates the presence of a defect; the - sign stands for a normal situation.

Xeroderma pigmentosum (variant form). Clinically not distinguished from the previous, except for the mildness of the symptoms, which allows in some cases a relatively long life-span.

Table   III

Main Cellular and Molecular Properties

Cellular or molecular defect	XP (classical)	XP (variant)	BS	FA	AT
UV sensitivity	+	+	+	-	-
X or γ-rays sensitivity	-	-		+	+
MMC sensitivity	-			+	-
MMS sensitivity		-			-
Spontaneous increase in chromosome aberrations	-	-	+	+	+
Spontaneous increase in SCE	-	-	+	-	-
UV - excision repair defect	+	-		-	-
UV - post replication repair	-	+	-		
t' excision defect	-	-		+	-
Single strand break repair defect		+			-
Increased SV40 transformability	-	-	+		
Deficiency in dimer removal	+	-			
Deficiency in removal of γ damage					+
Deficiency in AAF removal	+	-		-	-

The distinctive feature from the classical form is a normal UV-
induced repair replication of the cultured cells; the dimer and AAF
excision process seem also unimpaired as well as the rate of
formation of breaks following UV (26).

Another type of repair process seems deficient in this case;

Table IV

Enzyme Levels Assayed

Enzyme deficiency		Disease		
	XP (classical)	XP (variant)	BS	FA
Photoreactivating enzyme deficiency	+			
Apurinic endonuclease deficiency	+	−		−
Correxonuclease deficiency	−	−		−
Defect in chromosomal protein favoring excision	+			
Single strand DNA DNAase defect	−		−	−
Ligase deficiency	−		−	
DNA polymerase deficiency α	−	−	−	−
β	−	−	−	−
γ	−	−	−	−
UV-inactivated DNA binding protein deficiency	−	−		
Deficiency in uracil-specific N-glycosidase	−			
Defect in DNAse specific for non dimer UV-products	−			

in fact XP variant cells are abnormally slow in converting into large size low MW DNA synthesized upon a template damaged by UV; this effect is enhanced by caffeine (28). In other words, XP variants are altered in the so-called post replication repair process, i.e., the process which allows an eventual bypass of the altered structure on the template strand by the newly synthesized chains during replicative synthesis.

The host cell reactivation of the viruses replicating in the nucleus is reduced here too, though less than in classical XP (29, 30). The frequency of mutation induction in cell cultures per lethal event is higher than in normal cells, whereas it is the same as normal in classical XP (31). This indicates that variant cells are probably more "error prone" in their residual repair process.

Bloom's syndrome. This is a rare autosomal recessive disorder characterized by pre-natal and post-natal growth retardation, sun sensitivity, facial telangiectasias, disturbed immune function and a predisposition to cancer (33). BS fibroblasts grow poorly, with a reduced plating efficiency with respect to controls. The average grain counts in BS nuclei are lower than in the controls suggesting that the rate of overall DNA synthesis is slowed down even in normal conditions. BS cells are more sensitive to UV radiation (measured as survival) but excision repair and post replication repair are not impaired (34). The newly synthesized DNA seems to have a lower molecular weight than that of control cells and the movement of the replicative fork (as measured by autoradiography) seems slower than the controls (35). These data suggest that the fundamental defect might involve a step in DNA replication. An increase of spontaneous chromosome breaks and rearrangements is observed (36,37).

The most striking feature of BS cells is an outstanding increase in the number of spontaneous SCE, which may be of the order of 90 per metaphase, vs a normal value of approximately 5 (38,39). This observation may be the consequence of the defect in replication that could lead to lesions or breaks in the newly synthesized chains, which could be subsequently repaired by recombinational exchanges between the sister chromatids. Thus, the SCE could in general be considered evidence of a "last resort" repair activity of the cells.

Fanconi's anemia. This is a rare hereditary disease transmitted as an autosomal recessive trait. It is characterized by hypo-plastic pancytopenia associated with short stature, dark pigment-ation, elevated levels of fetal hemoglobin, congenital malformations, predisposition to leukemia and other cancers.

Cells of Fanconi's anemia patients are moderately sensitive to γ rays and highly sensitive to the cross linking agent MMC (40) whereas they are normal in their sensitivity to UV (41). After UV treatment, the level of repair replication and the rate of dimer excision are normal (42); the repair replication is not enhanced also by MMC (43).

A characteristic feature of FA is the high tendency to spontaneous and induced chromosome aberrations. The most powerful inducing agents are X-rays (44) and the cross-linking drugs such as MMC (43,45,46). In contrast, the susceptibility to chromosome breakage by UV-light and by the monofunctional alkylating agents such as MMS, are within the normal level (45). MMC, so effective in inducing cell killing and chromosome aberrations, is less effective than in normal cells in the induction of SCE (47). This reduced increment in exchanges in the presence of MMC is accompanied by a reduced rate of progression through the cell cycle with a partial suppression of mitosis (45).

Sedimentation velocity of cellular DNA in alkaline sucrose gradients is not reduced during post incubation after treatment with MMC, contrary to what is observed in normal cells (40). In other words, no incision seems to occur near the cross-link in these cells, contrary to the normal ones.

A reduction in the ability to excise γ-ray products of the t' type is observed in extracts of FA cell (3).

Ataxia telangiectasia (Louis-Bar syndrome). This is an auto-somal recessive defect; affected individuals develop cerebellar ataxia and oculocutaneous telangiectasias in early childhood. Afflicted humans show an impaired cellular immunity and are pre-disposed to lymphorecticular malignancy. On receiving conventional radiotherapy for tumour treatment, they typically develop terminal complications.

The cells of AT patients show accordingly a greatly enhanced sensitivity to ionizing radiations. The rate of rejoining of γ-ray induced single and double strand breaks is normal (48,49). The ability to perform γ-stimulated repair replication is reduced in six strains which are genetically heterogeneous (as shown by cell fusion studies) and normal or elevated in four other strains (50). A defect in the repair of γ-modified base residues, as assayed by the disappearance *in vivo* of sites susceptible to endonucleolytic cleavage by a crude *M.luteus* extract, has been observed (51). The capacity to repair UV damage instead, namely excision repair of UV-photoproducts, is normal.

At the chromosomal level, increased numbers of spontaneous and induced chromosome breakage and rearrangements in cultured lympho-cytes with specific involvement of a D group chromosome are reported (52). No difference instead is observed in SCE between AT and normal individuals (53).

Concluding Remarks

From the data summarized above it seems plausible to consider these inherited syndromes as *bona fide* repair mutants: in all cases some important derangement of DNA metabolism is apparent; in Table V

a short-hand summary of the available evidence shows that we are
dealing with three UV-sensitive, and two ionizing radiation-sensi-
tive mutations.

The classical XP seems particularly deficient in incision of
DNA close to the dimer; the XP variant seems a good example of post-
replicative repair deficiency; in Bloom's syndrome the high
abundance in SCE orients the research towards the recombinational
repair pathways. Of the X-ray sensitive syndromes, Fanconi's
anemia is particularly interesting for the specific sensitivity
to cross-linking agents and Ataxia Telangiectasia stands out for a
great sensitivity to ionizing radiations.

The data obtained at the cellular level have oriented the
investigations towards certain enzyme activities. Thus, the
delayed removal of pyrimidine dimers, the reduced UV-repair
replication of classical XP (*vs* a normal repair replication after
X-rays) and the delayed formation of breaks pointed towards some
dimer-specific endonuclease, analogous to the one demonstrated in
bacteria and absent in some UV-sensitive bacterial mutants (2).
Such an enzyme has never been assayed in human cells, in spite of
considerable effort. An endonuclease specific for UV-damage other
than dimers is present in human cells, including the XP ones (54);
a number of other enzymes have been assayed in XP cells, with
altogether disappointing results (see Table V and references 9,15,
17,20, 55-61). In some cases a defect in some enzymes has been
reported (indicated with a + sign in the table); the reduction in
photoreactivating activity is hard to reconcile with the *in vivo*
data, both in normal cells (where photoreactivation could not be
shown) and in XP cells (where reduction in repair replication and in
incision rate would remain unexplained). The reports of defects in
apurinic endonuclease and in a protein favouring dimer excision
suffer from being limited only to some complementation groups,
whereas the basic cellular defect seems the same in all of them.

All other enzymes tested, including the three DNA polymerases
and ligase seem present at normal levels. It is worth stressing
that the classical XP defect must account also for the deficiency
in repair of several types of molecular damage other than the
pyrimidine dimers, namely the AAF-adducts and the O^6-alkylguanine
structures, so that the bacterial models (where a strictly dimer-
specific endonuclease is involved) cannot be applied as such.

In the XP variant, the deficiency in post-replication repair
evidenced by the delay in filling of gaps left in the newly
synthesized chains over irradiated template, points to a possible
role of some DNA polymerase activity in the defect. The three
known DNA polymerases are all present instead at normal levels and
also other tested enzymes are present.

Table V

Summary of Properties of the Most Studied Hereditary Syndromes
Involving DNA Repair

DISEASE	UV sensitive	X or γ sensitive	Cross-linking agents sensitive	Increase in spontaneous chromosome aberrations	Increase in spontaneous SCE	MOLECULAR DAMAGE
Xeroderma pigmentosum (classical)	+	–	–	–	–	Dimer and AAF excision - repair defect
Xeroderma pigmentosium (variant)	+	–	–	–	–	Defect in post replication repair (UV)
Bloom's syndrome	+	–	–	+	+	DNA synthesis defect
Fanconi's anemia	–	+	+	+	–	t' removal defect. Defect in cross-link repair
Ataxia telangiectasia	–	+	–	+	–	Defect in γ base damage removal

In Bloom's syndrome, the striking increase in SCE could indicate some major unbalance in the components of the recombinational apparatus; alternatively, considering also the evidence of a defect in DNA replication, one could assume that such a defect may leave damage in the newly synthesized strands, and the SCE would be evidence of recombinational events occurring to repair that damage. This would fit with the interpretation of the abnormal increase of UV-induced SCE in classical XP: here too, the damage persisting after replication could be removed by recombination: the recombination events evidenced by SCE would thus be a kind of delayed repair process. In any case, all the DNA polymerases and ligase (59 and our unpublished results) are present at normal levels in BS patients.

In Fanconi's anemia, the inability to repair cross-links seems particularly striking; the observed reduced rate of progression through the cell cycle is probably due to a block in growing chain advancement. The difficulty in completing chromosome synthesis could explain the lack of induction of SCE by MMC. The demonstrated inability of these cells to remove t' damage in the DNA may explain the γ-rays sensitivity; since the enzymic apparatus which accomplishes this removal is not described as yet, there is the possibility that the missing molecule may be responsible also for the cross-link removal defect: thus, e.g., either an N-glycosidase (which might offer a site for the apurinic endonuclease) or an endonuclease (acting on the product of the previous enzyme or directly on the damaged DNA) could be implied in both types of repair.

The level of an apurinic endonuclease has been found normal in FA cells (62) as well as that of two other DNAses (our unpublished results and L. Grossman, personal communication) and of the three DNA polymerases (59-61).

As for Ataxia Telangiectasia no data on enzyme levels are available; some unspecified product of γ-rays is not removed from DNA in AT cells; nothing is known of this structure, except that it is not of the t' type (63) and is not analogous to those produced by chemical carcinogens since no particular sensitivity to such agents is reported in AT (64).

Thus, at this point, no clear-cut enzyme defect has been recognized in any of the syndromes; this is not surprising if one considers the dearth of knowledge of the DNA enzymology in human cells, and the difficulty of working with this material, i.e., mainly fibroblast cultures; the fibroblasts in fact are available in low amounts, contain low levels of the enzymes of DNA metabolism (as compared to established cell lines) and cannot be directly compared to any "wild type" strandard. The scarcity of knowledge

on the distribution of levels of the enzymes of DNA metabolism
in the normal population makes the situation even more difficult.
As a consequence of this and of the relative scarcity of data at
the molecular level, it is not possible as yet to sketch plausible
pathways of repair in human cells. The low overlapping of
sensitivity to different agents in different diseases (64) points to
the existence of a variety of pathways relatively damage-specific.
In general, the bacterial models do not seem to be transposable to
the human sustem: we have already examined the non-correspondence
of the dimer excision mechanism of bacteria to the XP case; furthermore in bacteria a pronounced overlapping of sensitivity to different
agents is observed, contrary to the human mutants.

In spite of all these difficulties these syndromes remain a
most promising material in the field of mammalian DNA metabolism.
Several considerations invite a thorough effort at describing the
human DNA enzymology and genetics; in fact the common property of
high proneness to malignancy in all these diseases is a strong
indication in favour of the mutational origin of cancer: in all
the studied cases a higher mutation frequency is observed, with
respect to normal cells at a same dosage (even though only in the XP
variant a significant increase in mutation frequency per lethal
event is reported). This observation points to the extreme interest
in describing the mechanism by which the cell repairs most of the
potentially cancerogenic damage in normal conditions. Furthermore,
the overlapping of the repair pathways with recombination and
replication processes gives the hope that the future research in
this field may give insights also into these other essential
pathways. No other mammalian species offers at present this
opportunity to examine DNA metabolism from genetic and molecular
points of view.

Acknowledgement

We wish to thank Mrs. G. Santagostini for the excellent
secretarial assistance in the preparation of this manuscript.

References

1. Hanawalt, P. (1975) Molecular Mechanisms for Repair of DNA
    (eds. P. Hanawalt and R.B. Setlow) Plenum Press New York,
    p421

2. Grossman, L., A. Braun, R. Feldberg and I. Mahler (1975)
    Ann. Rev. Biochem. 44:19-43

3. Remsen, J.F. and P.A. Cerutti (1976) Proc. Natl. Acad. Sci.
    USA. 73:2419-2423

4.   Bertazzoni, U., M. Stefanini, G. Pedrali Noy, E. Giulotto,
     F. Nuzzo, A. Falaschi and S. Spadari (1976) Proc. Natl. Acad.
     Sci. USA   73:785-789

5.   Spadari, S., G. Ciarrocchi and A. Falaschi (1971) Eur. J.
     Biochem.   22:75-78

6.   Pedrali Noy, G., S. Spadari, G. Ciarrocchi, A.M. Pedrini and
     A. Falaschi (1973) Eur. J. Biochem.   39:343-351

7.   Söderhäll, S. and T. Lindahl (1976) FEBS Letters   67:1-8

8.   Pedrini, A.M., G. Ranzani, G.C.F. Pedrali Noy, S. Spadari
     and S. Falaschi (1976) Eur. J. Biochem.   70:275-283

9.   Doniger, J. and L. Grossman (1976) J. Biol. Chem.   251:4579-
     4587

10. Brent, T.P. (1975) Biochem. Biophys. Acta   407:191-199

11. Duker, N.J. and G.W. Teebor (1975) Nature   255:82-84

12. Brent, T.P. (1976) Biochem. Biophys. Acta   454:172-183

13. Linsley, W.S., E.E. Penhoet and S. Linn (1977) J. Biol. Chem.
     252:1235-1242

14. Brent, T.P. (1977) Nucleic Acid Res.   4:2445-2454

15. Sekiguchi, M., H. Hayakawa, F. Makino, K. Tanaka and Y. Okada.
     (1976) Biochim. Biophys. Res. Commun.   73:292-299

16. Sutherland, B.M. (1974)   248:109-112

17. Feldberg, R.S. and L. Grossman (1976) Biochemistry   15:2402-
     2408

18. Friedberg, E.C., J. Duncan, and J.E. Cleaver (1974) Radiation
     Res.   59:98

19. Duncan, J., H. Slor, K. Cook and E.C. Friedberg (1975)
     Molecular Mechanisms for Repair of DNA. (eds. P. Hanawalt and
     R.B. Setlow) Plenum Press, New York, p643-649

20. Mortelmans, K., E.C. Friedberg, H. Slor, G. Thomas and J.E.
     Cleaver (1976) Proc. Natl. Acad. Sci. USA   73:2737-2761

21. Cleaver, J.E., D. Bootsma, and E. Friedberg (1975) Genetics
     79:215-225

22. Cleaver, J.E., and D. Bootsma (1975) Ann. Rev. Genetics 9:
     19-38

23. Amacher, D.E. and M.W. Lieberman (1977) Biochem. Biophys.
     Res. Commun. 74:285-290

24. Goth-Goldstein, R. (1977) Nature 267:81-82

25. Hariharan, P.V. and P.A. Cerutti (1976) Biochim. Biophys.
     Acta 447:375-378

26. Fornance, A.J., Jr., K.W. Kohn and H.E. Kann, Jr. (1976)
     Proc. Natl. Acad. Sci. USA 73:39-43

27. Wolff, S., B. Rodin, J.E. Cleaver (1977) Nature 265:347-349

28. Lehman, A.R., S. Kirk-Bell, C.F. Arlett, M.C. Paterson,
     P.H.M. Lohman, E.A. de Weerd-Kastelein, and D. Bootsma (1975)
     Proc. Natl. Acad, Sci. USA 72:219-223

29. Day, R.S. III (1975) Nature 253:748-749

30. Abrahams, P.J. and A.J. van der Eb. (1976) Mutation Res.
     35:13-22

31. Maher, V.M., J.J. McCormick, P.L. Grover, and P. Sims (1977)
     Mutation Res. 43:117-138

32. de Weerd-Kastelein, E.A., W. Keijzer and P. Rainaldi (1977)
     Mutation Res. 46:163

33. German, J. (1969) J. Amer. J. Hum. Genet. 21:196-227

34. Giannelli, F., P.F. Benson, S.A. Pawsey and P.E. Polani (1977)
     Nature 265:466-469

35. Hand, R. and J. German (1975) Proc. Natl. Acad. Sci. USA
     72:758-762

36. German, J. and L. Pugliatti-Crippa (1966) Ann. de Génétique,
     9:143-154

37. Bertram, C.R., T. Koske-Westjhal, and E. Passarge (1976)
     Ann. Hum. Genet. 40:79-86

38. Chaganti, R.S.K., S. Schonberg and J. German (1974) Proc.
     Natl. Acad. Sci. USA 71:4508-4512

39. German, J., S. Schonberg, E. Louie and R.S.K. Chaganti (1977)
     Amer. J. Hum. Genet. 29:248-255

40. Fujiwara, Y. and M. Tatsumi (1975) Biochem. Biophys. Res.
     Commun. 31:592-598

41. Fujiwara, Y. and M.S. Sasaki (1977) Mutation Res. 46:120

42. Regan, J.D., R.S. Setlow, W.L. Carrier and W.H. Lee (1974)
     In: Proceedings of the IVth International Congress on Radiation
     Research and Symposium on Biological Medicine. North Holland,
     Amsterdam

43. Sasaki, M.S. (1975) Nature 257:501-503

44. Finkenberg, P., M.W. Thompson and L. Siminovitch (1974) Amer.
     J. Hum. Genet. 26:30A

45. Sasaki, M.S. and A. Tonomura (1973) Cancer Research 33:1829-
     1836

46. Auerbach, A.D. and S.R. Wolman (1976) Nature 261:494-496

47. Latt, S.A., G. Stetten, L.A. Juergens, G.R. Buchanan and
     P.S. Gerald (1975) Proc. Natl. Acad. Sci. USA 72:4066-4070

48. Taylor, A.M.R., D.G. Harnden, C.F. Arlett, S.A. Hartcourt,
     A.R. Lehmann, S. Stevens and B.A. Bridges (1975) Nature 258:
     427-429

49. Lehman, A.R. and S. Stevens (1977) Biochem. Biophys. Acta
     474:49-60

50. Paterson, M.C., B.P. Smith, P.A. Knight and A.K. Anderson
     Proc. VIIth International Congress on Photobiology, Rome
     (in press)

51. Paterson, M.C., B.P. Smith, P.H.M. Lohman, A.K. Anderson and
     L. Fishman (1976) Nature 260:444-446

52. Taylor, A.M.R., J.A. Metcalfe, J.M. Oxford, and D.G. Harnden
     (1976) Nature 260:441-443

53. Hatcher, N.H., P.S. Brinson and E.B. Hook (1976) Mutation Res.
     35:333-336

54. Bacchetti, A., A. van der Plas and G. Veldhuisen (1972)
     Biochem. Biophys Res. Commun. 48:662

55. Pedrini, A.M., L. Dalprà, G. Ciarrocchi, G.C.F. Pedrali Noy,
     S. Spadari, F. Nuzzo and A. Falaschi (1974) Nucleic Acids
     Res. 1:193-202

56. Kuhnlein, U., E.E. Penhoet and S. Linn (1976) Proc. Natl.
     Acad. Sci. USA 73:1169-1173

57. Bertazzoni, U., M. Stefanini, G.C.F. Pedrali Noy, F. Nuzzo
     and A. Falaschi (1977) Nucleic Acids Res. 4:141-148

58. Sutherland, B.M., R. Olivero, C.O. Fuselier and J.C. Sutherland
     (1976) Biochemistry 15:402-406

59. Parker, V.P. and M.W. Lieberman (1977) Nucleic Acids Res.
     4:2029-2037

60. Bertazzoni, U., M. Stefanini, G. Pedrali·Noy, A.I. Scovassi
     and A. Falaschi (1976) Proc. VIIth International Congress on
     Photobiology, Rome, p172

61. Scovassi, A.I., M. Stefanini, S. Spadari and U. Bertazzoni
     (1977) Atti. Ass. Genet. Ital. 22:239-240

62. Teebor, G. and N. Duker (1975) Nature 258:544-546

63. Remsen, J.F. and P.A. Cerutti (1977) Mutation Res. 43:139-146

64. Arlett, C.F. (1977) Mutation Res. 46:106

ENZYMES INVOLVED IN THE REPAIR OF DAMAGED DNA

Lawrence Grossman[*], Sheikh Riazuddin[**], Jay Doniger[1]
and Lester Hamilton

Department of Biochemistry, School of Hygiene and
Public Health
The Johns Hopkins University
615 North Wolfe Street, Baltimore, Maryland 21205, U.S.A.

I. INTRODUCTION

The progress that has been made in our understanding of the
enzymatic mechanisms by which modified nucleotides in DNA are
recognized and repaired is due, in part, to the developments that
have been made in photobiology and photochemistry. It was these
areas of activity that led to the identification of the primary
ultraviolet irradiation photoproducts formed in nucleic acids. The
initial discovery of pyrimidine cyclobutane dimers (1,2) led to
an understanding of their chemical properties which allowed
identification of such products in DNA irradiated either *in vivo* or
in vitro. As a consequence of this acquired technology, the fate
of these photochemical products during repair could be studied *in
vivo* and *in vitro*.

The mutagenic and lethal effects of UV irradiation can be
correlated with the formation of specific photoproducts, the
distribution and level of which can be controlled by monochromatic
UV light (3), either by directing photosteady states or by enzymatic
photoreversal (4). Such a relationship proved useful in considering

[*] *A.S.I. Participant*

[**] *The Nuclear Institute for Agriculture and Biology,
Lyallpur, Pakistan.*

[1] *Biology Division, Brookhaven National Laboratories, Upton,
Long Island, New York.*

the mutagenic and potentially carcinogenic effects of pyrimidine
dimers (5). Elucidation of these mechanisms was further substantiated by an understanding of the genetics which control
specific repair mechanisms and the gene products catalyzing them.
The molecular models created by the photobiologist are, therefore,
applicable to similar studies in carcinogenesis and mutagenesis. It
is for this reason that a discussion is presented of those nucleotide modifications which are sufficiently well understood and which
are amenable to analysis at the genetic, biological and biochemical
level.

II.   EXOGENOUSLY INDUCED DAMAGE

1. Alkylation Reactions   (Table 1)

The most common alkylating agents include alkyl sulfates, alkyl
sulphonates, alkyl halides, dialkyl nitrosamines, alkyl nitro-
soureas, acyl nitrosamides, mustards, diazo compounds, lactones,
epoxides, etc., many of which are mutagenic, carcinogenic or lethal
to cells (6).   In both RNA and DNA the predominant product of many
of the alkylating agents is the formation of 7-alkylguanine (7)
which because of the formation of the quarternary nitrogen leads to
labilization of the N-glycosidic bond eventuating in backbone
breakage. This is a *secondary* effect of many of the alkylating
agents. Some of the modifications are informationally *non-instruct-*

Table I

ALKYLATING AGENT	GUANINE			ADENINE		
	3.	O^6	7	I	3	7
Ⓡ·O-S-R (MMS, EtMs)	0.6	< 0.3	82	0.5	7	0.8
O=N-N-NH, O=C, Ⓡ (Me Nu, EtNu)	1.5	5	64	0,1	5	1.3

ional in which the depurinated site may lead to a variety of enzymatic effects, some of which will be discussed in later sections of this paper. Alkylation-induced *instructional* lesions may arise from the formation of O[6]-methyl guanine, leading to mis-pairing judging from incorporation studies during *in vitro* transcription and replication. Of the multitude of chemical changes affecting purines or pyrimidines some are directly repairable; or the secondary chemical events may be repaired if the primary structural changes are noninstructional.

2. Polycyclic Aromatic Hydrocarbons

Among the important classes of chemical carcinogens found covalently bound to tissue macromolecules are the polycyclic aromatic hydrocarbons, aromatic amines, nitrosamines, and afla-toxins. However, many of the individual compounds in these classes do not react directly with DNA, RNA or proteins. These chemically unreactive agents require prior metabolism to be chemically reactive with macromolecules.

From the work of Elizabeth and Hames Miller (8) it was established that many chemical carcinogens that are not themselves chemically reactive must be converted metabolically into electro-philic forms, which can then react with nucleophilic sites in cellular macromolecules, to initiate carcinogenesis (9).

Activation reactions may be initiated by an aryl hydrocarbon hydroxylase located in microsomal fractions as a group fo mixed-function oxidases which require NADPH, oxygen and the endogenous cytochromes P-450 and P-448. The overall result of this enzyme system's action is to convert these lipophilic compounds into hydrophilic ones (10). The irony of this system is that its role in the detoxification process renders the organism more vulnerable to the action of these agents. Furthermore, those enzymes are inducible (11) by many agents and appear to be under genetic control. For example, the aryl hydrocarbon hydroxylase in mice is expressed as a simple autosomal dominant trait (12).

The activation, therefore, of many of the aryl hydrocarbons results in a formation of powerful electrophilic species with extremely short half lives which react with a wide variety of cellular nucleophilic sites. These include amino and sulfhydryl groups associated with the nucleotides of nucleic acids and protein and which express their effects at a number of different loci - only some of which are at the DNA level.

Within the past few years, the more complicated chemistry of some of the powerful environmental carcinogens such as benzo(a) pyrene, dimethylbenzanthracene and acetylaminofluorene have been

studied in great detail. The reactive species and their adducts
have recently been sufficiently elucidated rendering these amenable for study at the enzymatic level (Fig. 1).

3. 2-Acetylaminofluorene (AAF) (Figure 1)

This potent carcinogen, activated by liver microsomal N-
hydroxylation to a proximal carcinogen, is converted by a sulfo-
transferase (13) to a highly reactive electrophilic sulfate ester,
which is considered to be the ultimate carcinogenic form (14).
This ester reacts covalently with DNA, RNA and proteins, thereby
initiating potential carcinogenic or mutagenic processes (15). The
ultimate carcinogen, the sulfate ester, cannot be prepared chemically
making it virtually impossible to examine. However, the synthetic
N-acetoxy-2-acetylaminofluorene derivative is more stable (16),
making it useful for studies relating its mode of action to acetylaminofluorene (17).

The major product of AAF with DNA is adduct formation with
the C-8 position of guanine forming 8-(N-2-fluorenylacetamido)
guanine (C^8-GAAF) (14,18). A minor guanine adduct has been
identified associated with the 2-amino group (19). This adduct
represents approximately 20% of the C^8-GAAF adduct in DNA. It
has also been suggested by Kriek that the 2-amino adduct, unlike
the C^8-derivative, is nonrepairable.

The unique chromatographic properties of the C^8-GAAF allows
for chromatographic identification of DNA digests. This problem,
therefore, is amenable to biochemical examination.

INTERACTION OF ACTIVATED N-AAF WITH GUANINE

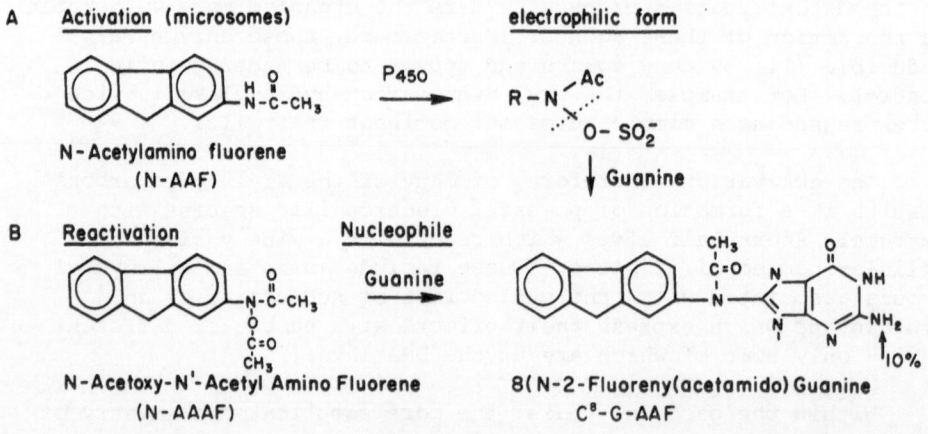

Figure 1

4. 7,12-Dimethylbenz(a)Anthracene (DMBA) (Figure 2)

The adducts that are formed between DMBA and its electro-
philic species with the guanine residues of poly rG have recently
been elucidated in which the putative products are resolvable in
HPLC (20) (Fig. 2).

5. Benzo(a)Pyrene (Figure 3)

This powerful carcinogen, a major component of cigarette
smoke forms, *in vivo*, an ultimate carcinogen which can exist as a
diolepoxide. The two isomeric forms (21) have been identified
as the (±)-7β, 8α-dihydroxy-9α,10α-epoxy-7,8,9,10,-tetrahydrobenzo
(a)pyrene and the (±)-7β,8α-dihydroxy-9β,10β-epoxy-7,8,9,10,-tetro-
hydrobenzo(a)pyrene isomer. These benzpyrene diolepoxides also
interact with the N^2 position of guanine in RNA. The absolute
stereochemistry of these conformations is not unequivocal and are
depicted in an arbitrary form. The crucial aspect, however, is
that the advances in the chemistry of benzopyrene interaction with
nucleic acid components lead to resolution by high pressure liquid
chromatography. The state of this chemistry, somewhat analogous
to the early days of photochemistry involving pyrimidine-pyrimidine
dimers is now at a point in which biochemical experimentation is
feasible.

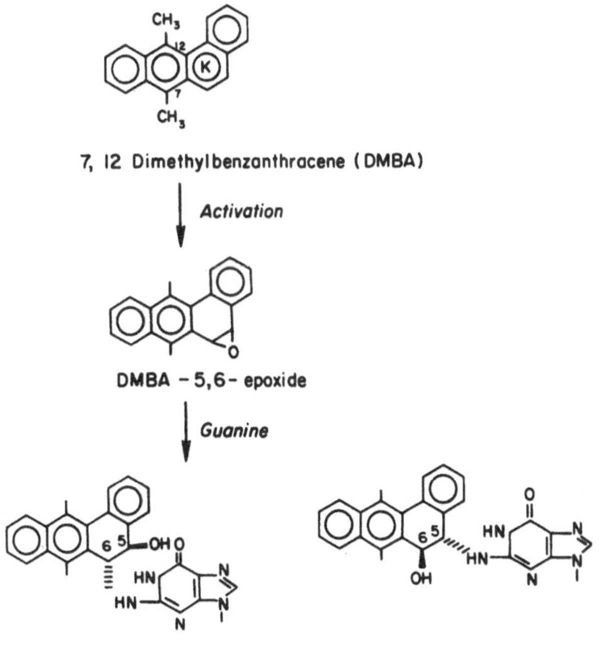

7, 12 Dimethylbenzanthracene (DMBA)

Activation

DMBA - 5,6- epoxide

Guanine

Figure 2. DMBA reactivity with DNA.

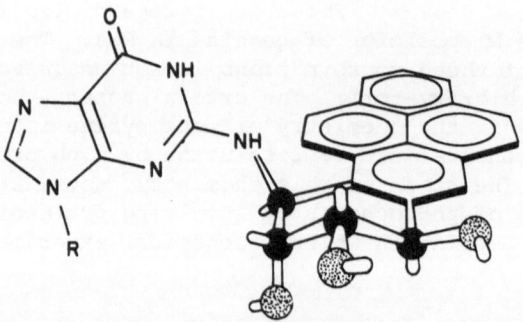

Figure 3.   Benz(a)pyrene activity with DNA.

III. ENDOGENOUS DAMAGE

The cellular environment of nucleic acids is such that they
do not exist in a static structural state, but rather that certain
regions (particularly in DNA) become sensitive to environmental
conditions of pH, ionic strength, temperature, etc. Furthermore,
the mechanical manipulations associated with purification and
characterization procedures may cause measureable adverse effects,
such as the stability of N-glycosidic bonds or loss of the amino
groups of cytosine through deamination. Realization of the
instability of purine N-glycosidic sites was first made by Greer
and Zamenhof (22) and extended by Lindahl (23). Assuming that DNA
is equally sensitive to pH changes *in vitro* and *in vivo*, it has
been calculated that depurination at 37°C and pH 7.4 proceeds at a
rate of $k = 3 \times 10^{-11} sec^{-1}$, and depyrimidination, under similar
environmental conditions, at $k = 1-5 \times 10^{-12} sec^{-1}$. From such
calculations it is apparent that the rates of replication of DNA must
influence the longevity of the purine N-glycosidic bond. These
calculations do not consider, for example, the possible stabilizing
affects of proteins nor of nearest neighbor effects.

IV. PHOTOCHEMICAL MODIFICATION (Figure 4)

The exposure of DNA or cellular materials to a variety of near and far ultraviolet irradiation results in a plethora of photochemical changes. The major photochemical change associated with ultraviolet irradiation of DNA in the wave length range from 200nm to 290nm is the formation of *intrastrand* crosslinks as a result of the formation of pyrimidine-pyrimidine cyclobutane dimers (24). Cytosine residues also form a water addition product, cytosine hydrates, which although relatively unstable do have a measureable half-life (25). Their presence induces transitional changes *in vitro* measured in transcription and replication systems (26). There are also undefined photochemical changes at very low doses which involve individual thymine residues (27).

The most abundant photoproducts are thymine-thymine dimers (T<>T) followed by the mixed thymine-cytosine dimers (T<>T) and then cytosine-cytosine pyrimidine dimers (C<>C) (28). Although *interstrand* crosslinks have been identified their quantum yield is approximately 1,000 times lower than that for the intrastrand pyrimidine dimers minimizing, therefore, their biological importance (29).

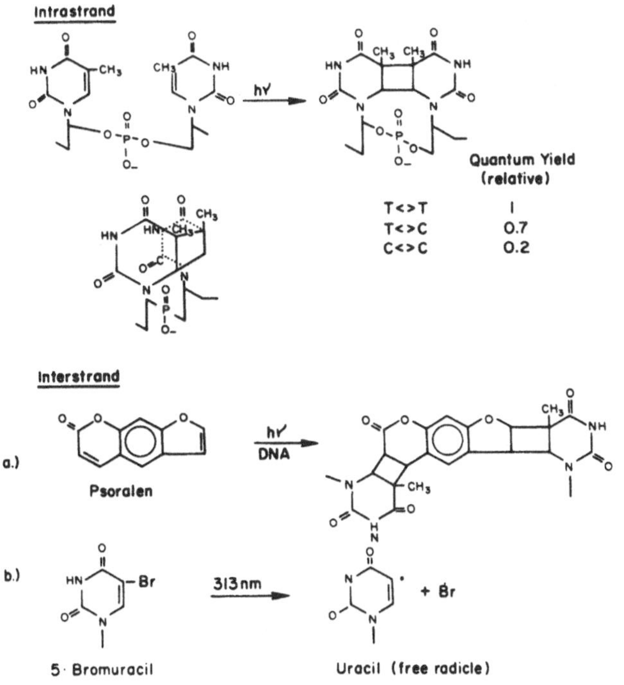

Figure 4. Induced DNA cross-links.

Interstrand crosslinks arise as a consequence of the action of psoralan and visible light (30) linking thymine residues on opposite strands. The use of psoralan in isolating and characterizing the various palindromes in DNA has recently shown to be of potential use (31).

The formation of interspecies crosslinks by 5-bromouracil substitution has proved to be useful as well because of the analogue's sensitivity to photolysis. Visible light irradiation of 5-bromouracil-containing DNA leads to its dehalogenation forming a radical which can abstract hydrogen from either the deoxypentose ring, leading to N-glycosidic bond hydrolysis; or from the amino groups of neighboring proteins. The resulting crosslinking between nucleic acids and proteins in the presence of photolyzing light has proved to be useful for stabilizing protein-nucleic acid inter- actions.

V. GENERAL REPAIR MECHANISMS (Figure 5)

The ability of a cell to survive the variety of non- instructional changes which lack hydrogen bonding properties must be repaired prior to replication. The model which can be invoked for studying repair of the aforementioned species of damage has in principle been derived from photobiology. The facile identification of pyrimidine-pyrimidine dimers, their identifiable chromatographic properties and ease of isolation has allowed for the temporal analysis of their effects during post-irradiation periods.

1. Photoreversal

Chronologically, the first repair mechanism was simultaneously discovered by Dulbecco (32) and by Kelner (33), who found that cells or virus-infected cells, when inactivated by ultraviolet light, were reactivated upon subsequent exposure to visible light reversing many of the damaging or mutagenic effects of this form of radiation. This phenomenon, referred to as *photoreversal* is due to a specific enzyme, *photolyase*, which in the presence of 340-370nm light, depending on the biological source, is able to monomerize the three species of pyrimidine dimers referred to (34). Although the precise mechanism of photoreactivation at the enzymatic level has not been fully characterized it is clear that those enzymes isolated in quantities sufficient for spectroscopic analysis do not possess an identifiable chromophore (35). The tight binding of photolyase to pyrimidine dimer regions in DNA results in a stable complex which in the presence of visible light results in mono- merization (36). The irradiated DNA-enzyme complex must possess its own unique chromophore in order for visible light to break the stable

I Photoreactivation (Kelner; Dulbecco)

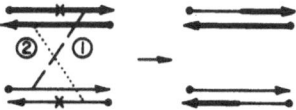

II Recombination "Sister Chromatid Exchanges" (Rupp and Howard
                                                        —Flanders)

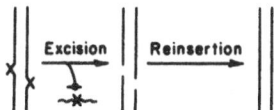

III Excision Repair (Setlow; Howard—Flanders)

Figure 5.   General UV repair mechanisms.

carbon-carbon bonds in pyrimidine dimers. Some insight into this
phenomenon is potentially available from recent work in which small
tryptophane containing peptides (37) or T_4 DNA-binding protein (38)
can affect a photoreactivation of dimers. The photoreactivities
observed either in crude extracts or perhaps in partially purified
preparations may, therefore, be adventitious and not associated with
a *bona fide* photolyase.

2.   Dilution Mechanisms

     The appearance of single strand gaps in DNA during post-
replication events in excision defective mutants is approximately
equivalent to the number of pyrimidine dimers present in DNA (39).
Following the appearance of such gaps there is evidence of sister
chromatid exchanges in which parental DNA is associated with newly
formed daughter strands, the extent of which is comparable to the
number of gaps observed during postreplication (40). The mechanisms

of recombination have yet to be elucidated to account for many of
these recA-dependent events which will be the subject of much
investigation for the next few years.

3.   Damaged Nucleotide Removal

Perhaps the best understood pathway; - one involving a
variety of nucleases has received the most attention and will be
discussed in greater detail. Excision either of nucleotides as
first described by Carrier and Setlow (41) and Boyce and Howard-
Flanders (42) or of bases as described by Lindahl (43) represent
the best understood pathways of repair. The initial recognition
by site-specific nucleases which act on the primary chemical or
photochemical damage is the rate-limiting step in the excision
process. The other pathways, involving N-glycosidic bond hydrolysis
by enzymes recognizing unique kinds of chemical changes, have
recently received a great deal of attention and will be discussed
at some length not only in this manuscript but also in that of
Dr. I.R. Lehman (this volume).

4.   General Endonucleolytic Mechanisms – Nucleotide Excision (Figure 6)

There have been a plethora of reports describing crude enzymes
or unfractionated extracts that act on damaged DNAs. There is a
need for some distinction at this juncture between enzymes that act
specifically at damaged sites and those that act as a consequence
of the distortion imposed upon the structure of DNA consequent to
some prior modification. There have been reports in the literature,
for example, that either the *Neurospora* endonuclease (44) or S_1
endonuclease act (45) on DNAs that have been heavily damaged with
ultraviolet light. The ultimate demonstration for the biological
role of such enzymes depends upon the identification of the
structural gene and its relationship to a specific gene product.
In lieu however of such a fine analysis, precautions should be taken
concerning the differences between a damage-specific and a damage-
dependent endonuclease. Fig. 6 depicts two kinds of endonucleases
that could act either in a specific or in a non-specific manner at
sites which are produced as a consequence of the secondary effects
of damage.

The requirement for an enzyme to participate in repair is that
the sites generated by that enzyme be amenable to excision by
bacterial polymerase-associated exonucleases, or by unique exonucleases which require proper termini for exonucleolytic hydrolysis. Furthermore, the sites which are generated should be on the
damaged strand only. An additional requirement is that a 5' phosphoryl site, suitable for ligation, be generated.

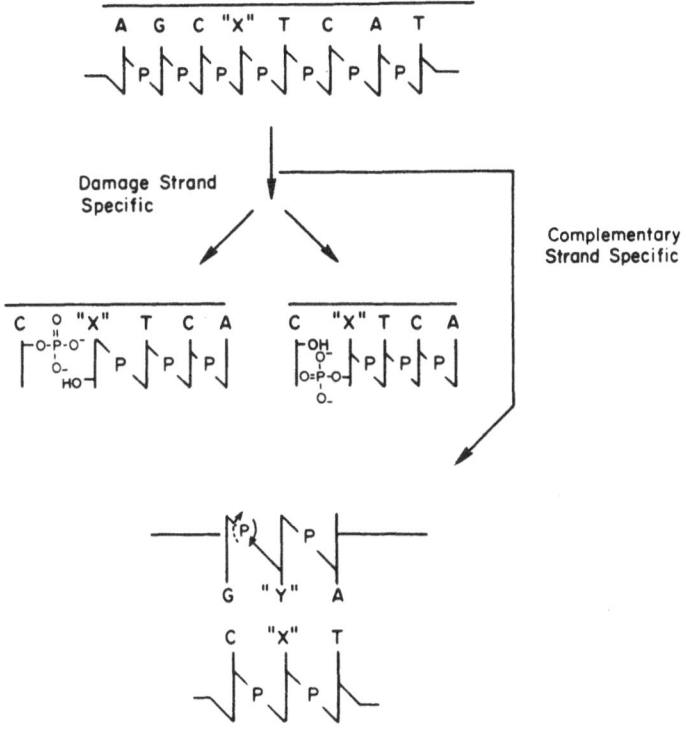

Figure 6. Damage dependent endonucleases.

It is necessary, furthermore, to perform complete dose-kinetics determining the stoichiometry of the numbers of endonucleolytic events associated with each pyrimidine dimer. Since many of the enzymes acting at initial stages in repair are not only rate limiting but possess relatively low turnover numbers it is anticipated that a damage-specific enzyme should have a stoichiometry approaching unity.

5. DNA N-Glycosidases (Figure 7)

Lindahl and coworkers (43,46) recently discovered a series of bacterial enzymes that recognize a modified nucleotide and remove the modified purine or pyrimidine base in the absence of phosphodiester bond hydrolysis. The general hydrolytic mechanisms catalyzed by these enzymes result in the formation of a deoxyribose moiety lacking an aglycone. The two enzymes that have been identified are from *Escherichia coli*, one of which recognizes the uracil which replaces thymine in the PBS II phage DNA, resulting in

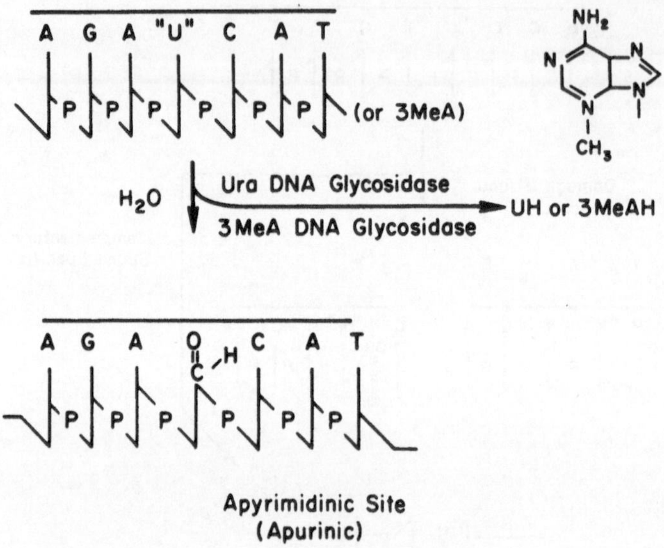

Apyrimidinic Site
(Apurinic)

Figure 7. N-Glycosidases.

an A:U pair, or in deaminated DNAs with resultant U:G nonhydrogen
bonded base pairs. This enzyme, the ura-N-glycosidase is, therefore, specific for the uracil residue generating an apyrmidinic
site.

A similar type enzyme specific for 3-methyladenine-containing
DNA and yielding an apurinic site has recently been highly
purified (46). These unique sites in DNA aRe sensitive to β-
elimination reactions and hence are alkali-labile. Unlike the
nucleoside phosphorylases and the NADases, the DNA N-glycosidase
reactions are neither phosphorolytic nor arsenolytic and are
irreversible. The role of these enzymes in *metabolic repair* has
been amply discussed by Dr. I.R. Lehman (this volume).

6. Nucleotide Excision -
Damage Specific Endonucleases (Figure 8)

There are three damage-specific endonucleases which act on
pyrimidine dimer-containing DNA. The T_4 bacteriophage V^+ gene
coded enzyme (47), the enzymes controlled by the *uvrA*, *uvrB*
structural genes (48) and the enzymes isolated from *Micrococcus
luteus* (49-51) behave in a similar manner. These enzymes are
specific for the damaged strand (52) resulting in phosphodiester
bond hydrolysis 5' to the damaged site generating 3'-hydroxyl and

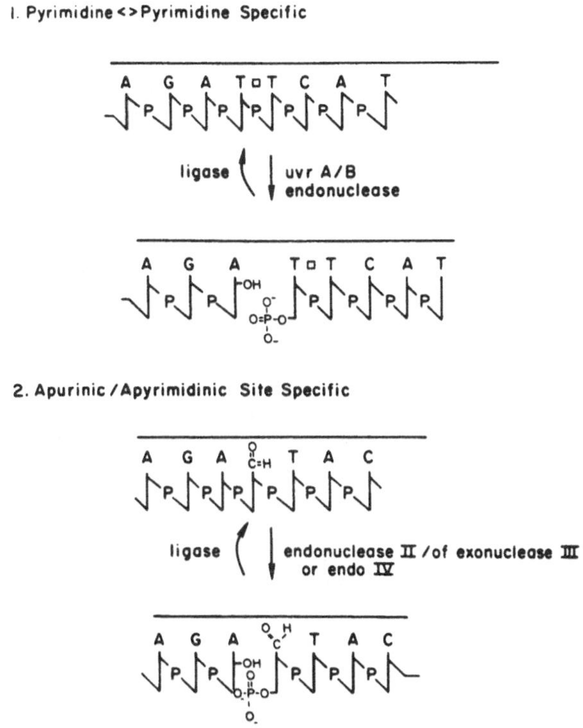

1. Pyrimidine <> Pyrimidine Specific

2. Apurinic / Apyrimidinic Site Specific

Figure 8. Damage specific endonucleases.

5'-phosphoryl groups (52). Such sites are removed with polynucleotide ligase by virtue of the juxtaposed 3' hydroxyl and 5'
phosphoryl group (50,53). Although polynucleotide ligase has been
used diagnostically in identifying these intermediate structures
it is possible this "reversibility" may play an important role *in*
vivo; this putative step being under the control of uvrC gene (54).

It is unclear why a controlling step is necessary at this
point in repair. However, definitive binding data at such sites
be competing enzymes, such as DNA polymerase I, and ligase is
needed. From preliminary information concerning the competition of
these two enzymes for nicked or incised sites, it can be suggested
that ligase can limit polymerization or strand displacement
reactions catalyzed by DNA polymerase I (55).

An endonucleolytic step is required to act specifically at
sites generated by the DNA N-glycosidases. These apyrimidinic or
apurinic sites act as substrates for the action of a number of
specific endonucleases. In *Escherichia coli* endonuclease II, which,
with its associated exonuclease III activity (56) is able to act at

such sites generating 5'-phosphoryl and 3'-hydroxyl termini, 5' to
the deoxyribose moiety (57). It has been suggested by Verly (58)
that the role of the associated exonuclease activity with endonuclease II in repair is for removal of a few juxtaposed nucleotides thereby limiting the action of ligase and thus *aborptive
repair*. Mutants of exo III/endonuclease II, (*xth⁻* mutants) possess
a unique residual endonuclease activity free of associated exonuclease which acts specifically at apurinic sites. Endonuclease
IV (59) has the capacity to act at such apurinic sites, presumably
acting 5' to the damage to generate suitable sites for excision
reactions.

A tabulation of the reported endonucleases from *E.coli* in
Table II attempts a unification of the terminology for a variety of
endonucleases that act at fairly specific sties (except endonuclease
I). Confusion has crept into the literature regarding the role of
endonuclease II and endonuclease IV in repair. However, distinctions can be made. Endonuclease II requires divalent cations and as
such is sensitive to chelating agents whereas the larger enzyme:
endonuclease IV, is insensitive to such agents. It is presumed
these are different proteins with similar endonucleolytic
specificity.

Two enzymes have been isolated which appear to be specific
for UV damage, one referred to as endonuclease III (60) and the
other is the gene product of uvrA and uvrB (48). The latter
enzyme(s) is absent in excision-defective mutants whereas endonuclease III, requiring more heavily irradiated DNA, is present in
excision-defective mutants. More will be said about such enzymes
which appear to be UV-damage-dependent rather than UV-damage-
specific.

Table II

Escherichia coli Endonucleases

Damage Specificity	ENDO I	(EXO III) ENDO II	ENDO III "UV"	ENDO IV	ENDO V	UV ENDO T<>T	
	None	Apu/Apy	Non-Dimer	Apu/Apy	Apu, OsO4 Uracil-DNA	C<>T	C<>C
Preferred Conformation	DS/SS	DS*	DS*	DS*	SS/DS*	SS*/DS* -	DS*
Molecular Wt. (M$_r$)		2800	2500	3200	1500	1500	1800
Sites Generated	5'-P	5'-P		5'-P	5'-P	5'-P	5'-P
Structural Gene(s)	endA endB	Xth	NA	NA	NA	uvrA	uvrB
Reference	Lehman	Weiss Verly Goldthwait	Radman	Lindahl	Linn	Grossman	

Endonuclease V recently reported by Gates and Linn (61) has a wide range of specificity; however its most interesting endo-nucleolytic property is its ability to act on DNAs containing uracil in place of thymine. The putative role of endonuclease V may be in metabolic repair under conditions in which uracil finds its way into DNA. The level of this particular enzyme in the recently isolated uracil N-glycosidase mutants (62) will be of interest.

VI. EXCISION OF DAMAGED NUCLEOTIDES (Figure 9)

1. DNA Polymerase-Associated Excision Enzymes

The bacterial polymerases with their associated exonucleo-lytic activities are involved in a number of important biological functions in addition to polymerization of nucleotides. Most notable is the 5' exonuclease activity specific for double stranded DNAs, which acts at nicks in an extremely efficient manner during concomitant polymerization. This *nick translational* activity of polymerase I functions by virtue of its nucleotide removal and reinsertion in a 5' → 3' direction. The ability of this associated exonuclease to function at internal rather than terminal phosphodiester bonds makes it suitable for the removal of 5'-terminal damaged nucleotides. This excision capability was first shown by Kelly et al. (63) to remove pyrimidine dimers from DNAs incised by nonspecific endonucleases. Similarly, damaged DNA treated initially by the correctional endonucleases from either *M. luteus* or from *E.coli*, followed by polymerase I loses the pyrimidine dimer-containing fragments as an initial step during the excision reaction (55). This is a very efficient process because of the coupling of polymerization and excision by this multifunctional enzyme. In the absence of concomitant polymeriza-tion, excision is fairly sluggish. This was best demonstrated with the proteolytic digestion products of polymerase I (64) in which the smaller molecular weight fragment, although able to excise pyrimidine dimers, did so at diminished rates. Addition of the larger 77,000 molecular weight fragment which contains the poly-merizing activity stimulated excision (65).

Polymerization III, according to Livingston and Richardson (66) possesses, in addition to its 3' to 5' *editing* function, a rather unique 5' exonuclease activity which requires that the 5' terminus exist in a noncomplementary state. Initiation of hydro-lysis occurs at such a structure and then proceeds into a duplex region in an uncontrolled way. From *in vivo* studies, the role of polymerase III in excision has not been convincingly implicated in repair.

I. Polymerase I Associated Exonuclease

- Acts at internal phosphodiester bonds
- Prefers duplex DNA
- Unaffected by esterified 5′ terminus

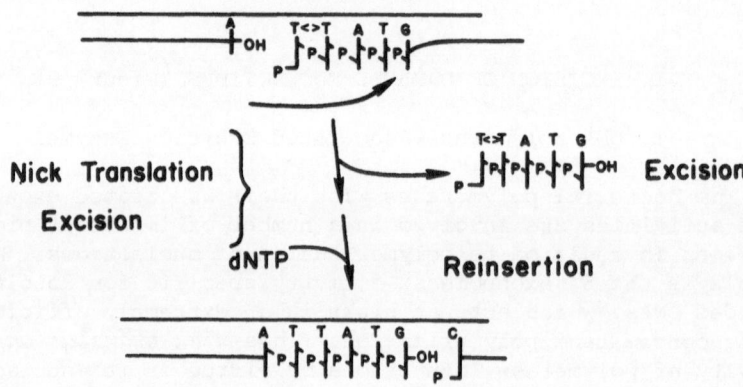

2. Polymerase III Associated Exonuclease

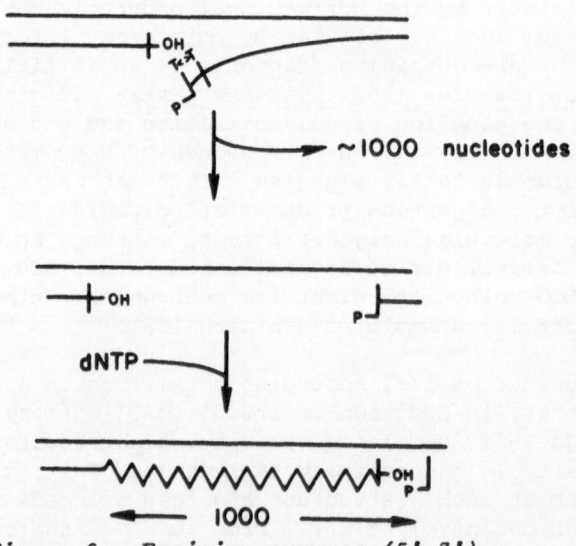

Figure 9.   Excision enzymes (5′→3′).

2. Unassociated Exonuclease (Figure 9)

A variety of unique exonucleases have been isolated from *M. luteus*, *E.coli* and from human placenta which act on single-stranded DNA but, interestingly, initiate hydrolysis from either the 3' or 5' terminus, generating small 5'-phosphorylated oligonucleotides (67,68,69). These single stranded-specific exonucleases are also able to act on irradiated-incised-duplex DNAs in an efficient manner. The placental correctional exonuclease can, in addition, digest at a nick in an undamaged duplex DNA in both a 3' and 5' direction generating a gap approximately 40 nucleotides in length (69). The bacterial enzymes, acting on incised-irradiated DNA generate a gap approximately 10 nucleotides in length (70), presumably because their activity is limited to single stranded-like regions of the incised DNAs.

3. Base – Nucleotide Incision Pathways (Figure 10)

The nucleotide incision and base removal mechanisms converge onto a common path. The initial removal of damage, catalyzed either by specific endonucleases acting at damaged sites or endonucleases acting at sites generated by DNA N-glycosidases lacking purines or pyrimidines are repairable by the efficient polymerase I excision-reinsertion pathway; finalized by strand restoration catalyzed by polynucleotide ligase. This most efficient pathway is mechanistically termed *short patch repair* (71). In polymerase I mutants the unassociated exonucleases are potentially capable of removing damaged oligonucleotides. By virtue of the rather short region available for either polymerase II or III in such mutants it can be suggested that a *gap expansion* step is necessary to allow for the processive binding and reinsertion by these two polymerases (72). This is the most likely pathway responsible for *long patch repair*. A discussion of the putative role of the rec BC exonucleases in this path has been made in a previous publication (73).

In either *short patch* or *long patch repair* accommodation must be made for preserving the integrity of the complementary strands during the transient stage in which vulnerable gaps exist following excision just prior to reinsertion. It has recently been shown that binding proteins inhibit the action of many nucleases which act at single stranded sites. It may be suggested that binding proteins can assume a protective role by inhibiting the action of nonspecific endonucleases at these sensitive sites (74,75).

RecA, *uvrA* double mutants are exquisitely sensitive to UV light, in which a single pyrimidine dimer is lethal (76). During postirradiation periods, the UV-sensitive single *recA* mutants result in an inordinant level of DNA degradation (*reckless*) which can be cured in recA and recB double mutants (77). These data suggest that

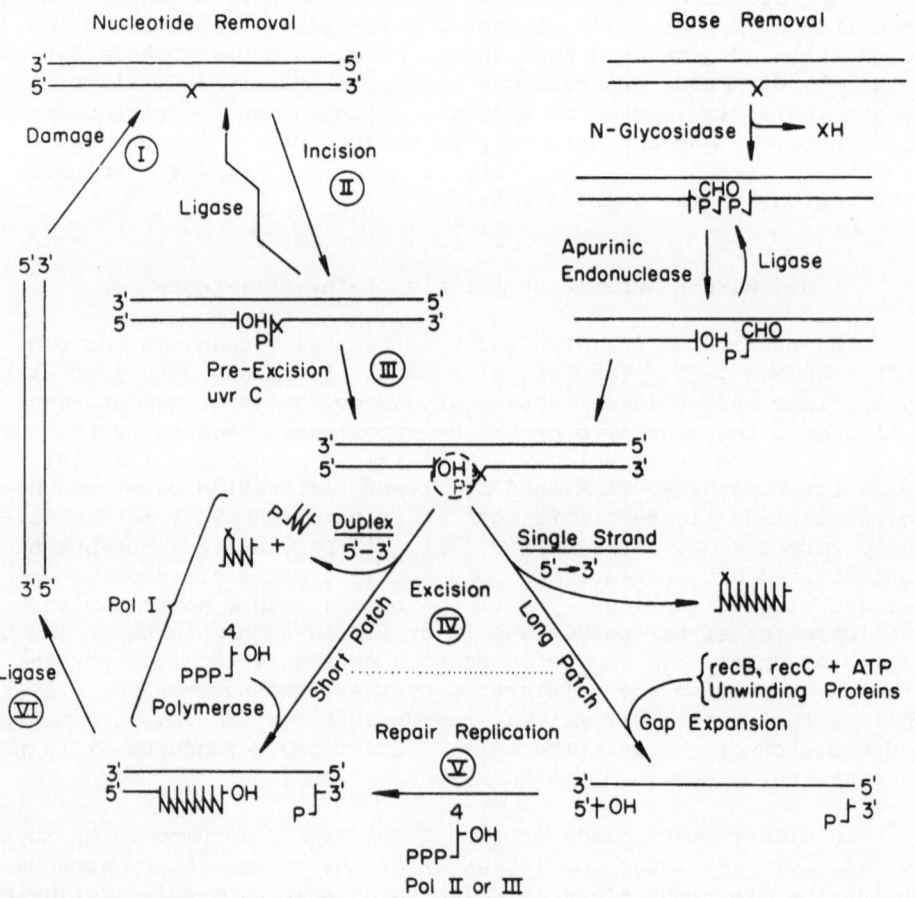

Figure 10.   DNA repair mechanisms.

the rec BC nuclease responsible for such degradation may be kept under control by the protective recA protein.

4. Phenotypes of Excision Defective Mutants (Table III)

In *E.coli* alterations in any of five genes result in mutants which share the phenotypic properties of being UV sensitive (UV⁻), mitomycin sensitive (M^S), and host cell reactivation negative (hcr-). These mutants, which map at discrete sites on the chromosome, are classified as *uvrA, uvrB, uvrC, uvrD and uvrE. UvrA, uvrB* and *uvrC* are unable to excise pyrimidine dimers from their DNA or from UV irradiated phage DNA. *uvrD* and *uvrE* have been included in this group of excision repair defective mutants chiefly on the basis of their reduced hcr activity.

uvrA, uvrB and *uvrC* appear to be very similar in their sensitivity to UV, their resistance to gamma (γ) irradiation and sensitivity to host cell reactivation, as well as induction of phage lambda. In addition, efforts to demonstrate sequential action of these gene products have yielded negative results (78). In spite of these phenotypic similarities recent experiments have shown that whereas uvrA and uvrB mutants lack an endonuclease specific for UV irradiated DNA, *uvrC* mutants contain wild type levels of this particular enzyme activity (48). Work reported with DNA extracted from *uvrC* mutants shows the presence of single-strand breaks after treatment with mitomycin (79) and UV irradiation (80). This evidence suggests that in mutnats of *uvrC* incision of damaged DNA does occur. The *uvrC* defect thus might possibly effect a step following incision.

The mutants classified as *uvrD* differ from *uvrA, uvrB* and *uvrC* according to their physiological properties. They are somewhat less sensitive to UV irradiation, sensitive to irradiation with gamma rays, show rapid and extensive degradation of DNA following exposure to UV irradiation and *uvrD⁻* is dominant over *uvrD⁺* (81).

Unlike *uvrA, uvrB* and *uvrC* mutants, which show no enhanced UV sensitivity as double mutants, the double mutant *uvrD, uvrB* is approximately three times as sensitive as *uvrB* alone. Since this double mutant shows a significant reduction (compared with the uvrD single mutant) in DNA degradation it has been suggested that *uvrD* function occurs after the *uvrB* step.

Table III

Phenotypic Expression of E. coli
Excision Defective Mutants

Characteristics	Wild Type	uvrA	uvrB	uvrC	pol A
Map Position	–	91'	17'	42'	85'
Survival { UV, mitomycin, psoralen }					
Host Cell Reactivation	–				
Mutation Frequency	–				
M_r DNA$_{ss}$		no change	no change	lig$^+$ no change / lig	normal
DNA Pyr<>Pyr Content		no removal	no removal	lig$^+$ no removal / lig	slow removal
Gene Product		Pyr<>Pyr endo	Pyr<>Pyr endo	?	DNA Polymerase I

5. Eukaryotic Excision Repair Mechanisms

Progress in elucidating prokaryotic repair mechanisms has been enhanced by the available mutants affecting various loci of the excision or reinsertion pathways of repair. In the absence of equivalent isogenic strains of cloned mammalian cell lines, those interested in eukaryotic repair pathways are exploiting related mutants - skin fibroblasts derived from patients with radiation sensitive diseases, such as xeroderma pigmentosum (XP) (Table IV).

Xeroderma pigmentosum is an infrequently occurring skin disorder which was shown more than 40 years ago to be genetically determined and to follow an autosomal recessive pattern of transmission (82,83). The biochemical defect in the enzymatic pathway of excision repair has been inferred from indirect experiments. The skin of homozygous affected individuals appear normal at birth but usually before age 3, severe changes subsequent to sun exposure appear, and then progress relentlessly. Freckles of varying sizes and degrees of brownness appear and are accompanied by increasing dryness, atrophy and number of keratoses. These changes in skin characteristically eventuate in some form, sometimes in multiple form, of malignant neoplasia of the skin and metastatic epithelioma, often causing death before the age of 30. Various other forms of benign and malignant tumors of ectodermal and mesodermal origin also occur with a much increased frequency in these infected individuals. All these abnormalities appear to be the consequence of exposure to sunlight.

Table IV
Phenotypic Expressions in
Xeroderma Pigmentosum

CELL CHARACTERISTICS	HOMOZYGOTE	HETEROZYGOTE
Growth (rate, extent)	Normal	Normal
Chromosome Structure	Normal	Normal
Sister Chromatid Exchange	Normal	Normal
Induced Chromosome		
Abnormalities	Increased	Normal
Killing $\begin{bmatrix} UV \\ 4NQO \\ BA \end{bmatrix}$	Increased	Slight Increase
HCR - Adenovirus	Decreased	Normal
Mutation Frequency	Increased	Normal
Biochemical Characteristics		
Pyrimidine Dimer Removal	Decreased	Normal
Unscheduled DNA Synthesis	Decreased	Normal
$\overline{M}_{ss}$ of DNA	None	Normal Decline
Defect	Endonucleolytic	None

Measurements of ultraviolet-induced pyrimidine dimers in
cellular DNA show that normal diploid human skin fibroblasts excise
up to 60% of the dimers in 24 hours, but that fibroblasts derived
from certain lines of xeroderma pigmentosum epithelial cells excise
less than 20% in 48 hours. Alkaline sucrose gradient sedimentation
experiments show that during the 24 hours after irradiation of
normal cells, large numbers of single strand breaks appear and then
as repair proceeds these breaks are mended. Such changes are not
seen in the same lines of xeroderma pigmentosum since such cells
apparently fail to start the incision process presumably because
they lack the required function of correctional endonucleases. These
inferential data are supported by the findings of Cleaver (84) in
which x-ray-induced phosphodiester bond breaks or those formed
after UV irradiation of 5-bromouracil containing DNA are repairable
in both normal and XP fibroblasts.

At least five complementation groups of XP and related cell
lines have been reported (85), indicating that the number of genes
controlling repair in mammalian cell systems must be at least as
numerous as the genes controlling UV sensitivity in *E.coli*. In
spite of the fact that cell lines from the various complementation
groups show some physiological defect in a UV repair system (Table
III & IV), the relationship between clinical systems and biochemical
abnormalities in repair has not been clearly established. This lack
of correlation is the result of a paucity of information dealing with
the enzymatic repair of DNA in normal mammalian cells. Perhaps most
significant is the fact that the first step in repair, an endonucleolytic activity specific for pyrimidine dimers, has not been
demonstrated to date in mammalian cells. The inherent difficulties
of this problem are aggravated by the fact that even in bacteria few
gene copies of a pyrimidine dimer specific correctional endonuclease
exists. Since the first step in repair appears to be the rate
limiting one, there are significant handicaps in isolating such a
limited number of gene products in bacterial cells. This problem
is further amplified in mammalian cells making the identification
of putative correctional endonucleases enormously difficult. Some of
the complications arising in assessing enzyme levels have recently
been alluded to concerning the nature of the DNA substrate. Chromatin appears to contribute, in some way, to controlling excision of
dimers by enzymatic activities present in crude extracts derived
either from XP cells or those from presumed heterozygotes (86).

It is clear, therefore, that work must be done with isogenic
cell lines developed from clones of normal human cells mutagenized
and screened for UV sensitivity with the ultimate hope of
classifying mutants according to there repair defects. The problems
of using complementation groups of XP lines is that noen of these
lines are isogenic and complementation is based exclusively on cell-
cell hybridization techniques with cell lines derived from many
different countries.

VII. POSTREPLICATION REPAIR (Figure 11)

A variety of modified nucleotides that cannot be enzymatically removed from DNA by base or nucleotide excision repair mechanisms found in the excision defective mutants can block the continuous progress of the DNA replication complex. This, however, does not prevent subsequent reinitiation of DNA synthesis at some point beyond a position of damaged nucleotides. When such modifications are present in a template strand, daughter strands are initially detected as segments of relatively low molecular weight (89). Their continuity is interrupted by gaps of about 1,000 nucleotides in length, the number of which corresponds approximately to the number and position of damage in the parental strand (39). These daughter strand gaps are secondary lesions caused by the replication of DNA containing primary damaged nucleotides. The type of enzymatic DNA repair shown in Figure 11 leads to sister chromatid exchanges. Although this phenomenon was first demonstrated in excision-defective

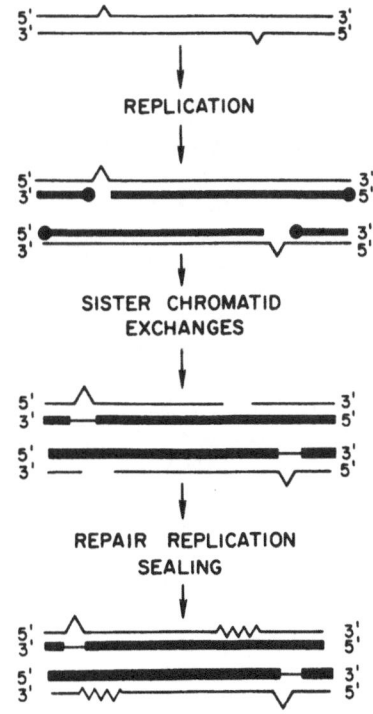

Figure 11. Postreplication repair.

strains, replication repair occurs in excision proficient strains capable of excising pyrimidine dimers (88). These data imply that the ability to perform excision repair does not prevent some non-instructive UV lesions from passing through a replication fork.

Postreplication repair is recombinational in principle, as specifically labeled parental DNA is found covalently inserted into the daughter strands. The number of such insertions corresponds to approximately the number of daughter strand gaps which are initially produced (40).

The postreplication repair mechanism in *E.coli* can occur perhaps by a number of distinct pathways all of which, however, require the recA$^+$ genotype (88).

1. UV Mutability [(Reviewed by Witkin) (89)]

The generalized mechanisms of pyrimidine dimer-removal by photoreversal or excision-repair are fairly accurate mechanisms since the numbers of UV-induced mutations is fairly low. However, excision-defective strains unable to excise pyrimidine photo-products, when exposed to low doses of UV irradiation show drastically reduced survival rates and the survivors produce a very high frequency of UV-induced mutations (90). This phenomenon shows that the UV-hypermutability of uvr$^-$ strains by unexcised pyrimidine dimers are much more mutagenic than those removed from the DNA by an excision pathway. The implication from these data is that postreplication repair in UV sensitive mutants are far more *error-prone* than the excision repair pathway. It may be implied from these data that the gap filling process is perhaps inaccurate. Such errors may be attributable to either the inability of certain polymerases to correctly interpret the reduced hydrogen bonding properties of pyrimidine dimers, or as suggested by Radman (91), the 3' → 5' exonucleolytic "editing" properties of polymerases are inhibited. The editing function of the 3'-exonuclease of DNA poly-merase I was originally described by Brutlag and Kornberg (92) and expanded by Bessman and his colleagues (93) to the T$_4$ DNA polymerase-associated 3' → 5' exonuclease. Removing non-complementary nucleo-tides incorporated into the growing end of a chain is a role assigned to this exonuclease. Supportive evidence for such a role has been demonstrated with the T$_4$ polymerase in which its exo-nucleolytic activity is inversely related to spontaneous mutation rates *in vivo* (93).

There is a great deal of evidence accumulating that the muta-bility by UV light is inducible and, like prophage induction, is repressed in normal cells but can be expressed in response to UV irradiation, or by mutants. UV mutagenesis depends upon *recA$^+$* and

$lexA^+$ gene products, which are also required for prophage induction by UV light and for the pleiotropic expression of the many UV-inducible functions. The inducibility by UV light of a number of phenomena was first observed by Lwoff et al. (94) who showed that UV radiation initiates mass induction of lambda prophage in iso-genic bacteria. Another early observation contributing to the inducibility of mutagenesis was made by Weigle (95) who showed that exposure of a host population of *E.coli* to low doses of UV radiation prior to infection substantially increased the survival of UV irradiated lambda phage and, further, resulted in a high level of phage UV-mutagenesis. It has been proposed that a single inducible error prone repair replication system, referred to as *SOS repair*, may be responsible for both the mutagenic reactivation of UV irradiated phage and for the error prone repair of bacterial DNA (96).

There is evidence that the Weigle effect can be induced indirectly (97). Indirect induction of λ prophage occurs when an unirradiated F⁻ lysogen is mated with the UV irradiated F⁺ or F' donor strain. The UV-damaged DNA of the F' episome, when transferred by conjugation into a recipient lysogen whose own DNA is undamaged, triggers prophage induction. Indirect induction of prophage is also promoted by transfer of either UV irradiated replicons such as colicin I factor or P1 bacteriophage. Indirect induction of the Weigle reactivation and mutagenesis was demonstrated by George et al. (97) who mated an unirradiated F⁻ host with UV irradiated donors before infection with UV irradiated lambda phage. Mutagenic reactivation of the phage occurred as efficiently as if the host itself had been irradiated. The wide ranging implications of such error prone repair, its inducibility by modified nucleotides and the relationship to recombination gene control is one of utmost importance and deserves consideration for future biochemical experimentation.

VIII.   RECOGNITION OF PYR<>PYR CORRENDONUCLEASE

1.   Structural Gene Control of Pyrimidine Dimer Specific Correctional Endonucleases

The initial step in the excision repair process in *E.coli* is controlled by an endonuclease activity which appears to be absent in both *uvrA* and *uvrB* mutants (48). The distal relationship of these two structural genes seems paradoxical in terms of the control of what appears to be a single enzyme. A similar situation exists in *M. luteus* although unmapped genetically. It is apparent that similar enzymatic activity is controlled by at least two genes in this organism (51). It seemed advantageous to investigate this

problem in M.luteus because of the greater concentration of such
endonucleolytic activity in this organism. An analysis of a
series of mutants and transformants of this organism indicates
(Table V) that the mutants which are obtained through nitrosoguanidine mutagenesis are UV sensitive and lack enzymatic activity
(DB7). Transformation of such mutant cells with wild-type DNA
produced a transformant which is UV sensitive but which has
demonstrable enzymatic activity when measured in crude extracts
(DB200). A further transformation of DB200 with wild type DNA
yields a revertant which is UV resistant and has fully restored
enzymatic activity (DB400). During the purification of the enzymes
on phosphocellulose chromatography (Fig. 12) wild type levels of
enzymes assayed with heavily irradiated DNAs were partially
resolved (peaks I and II). The mutant, DB7 possessed little or no
activity; DB200 the first transformant, although still UV sensitive,
possesses one peak restored to wild type levels. DB400, the
revertant possesses wild type levels of both peaks of activity.
The third, fourth and fifth peaks were used for reference for
further identification of enzyme activities which recognizes heavily
irradiated DNA (Fig. 13). Their persistence in all of the mutants
and transformants tested is diagnostic for their lack of involvement
in ultraviolet damage repair process in these organisms.

An analysis of the same enzymatic activities at doses which
induce no more than three to four pyrimidine dimers per genome is
seen in Fig. 13 in which the presence of peaks III, IV and V are
not evident until DNAs have been more heavily irradiated. The
important feature of this finding is that the first two peaks of
activity are both present when either heavily or lightly irradiated
DNAs are used as substrate.

The assay procedures employed for these experiments measure

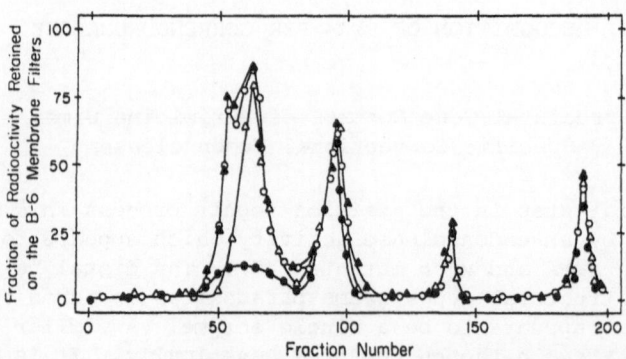

Figure 12. Chromatographic resolution of correndonuclease in M. luteus mutants.

Table V

Origin and Phenotypic Properties of Micrococcus Luteus Strains

Strain	Isolation	Resistance to Mitomycin (0.05µg/ml)	% Survival After UV Irradiation (50J/m^2)	% Recoveries of Py<>Py Correndonuclease I	II
Wild Type	ATCC 4698	+	100	100	100
DB-7	Wild Type Cells Mutagenized with NTG	-	0.01	2	5
DB-200	DB-7 Cells Transformed with Wild Type DNA	-	0.10	4	97
DB-400	DB-200 cells Transformed with Wild Type DNA	+	100	99	97

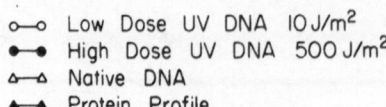

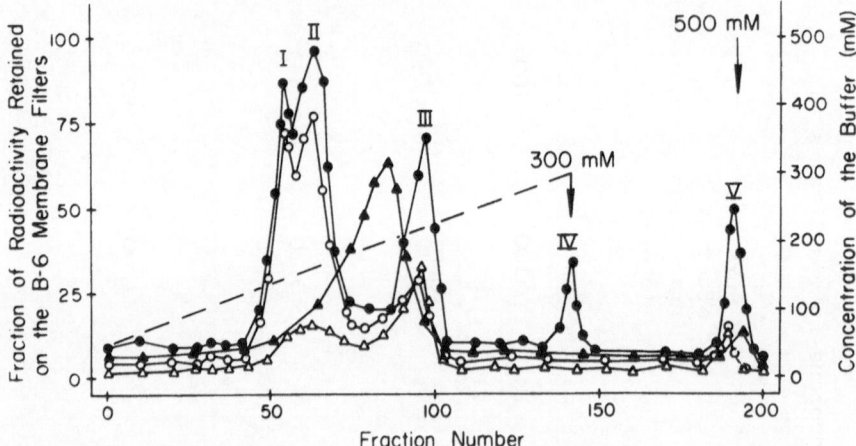

Figure 13. *Resolution of correndonucleases on P-11:-effects of low and high UV doses*.

the conversion of irradiated ØX174 RFI to RFII species (98). This has further virtue in permitting measurements of the binding properties of these enzymes to damaged DNA regions. Through an appropriate ionic environment the supercoiled DNA is not retained on nitrocellulose filters whereas protein is. It is useful to measure the enzyme-substrate complex in a manner analogous to that of repressor binding (99). The analysis of the phosphocellulose chromatograms for the binding properties of these enzymes (Fig. 14) indicates that the peaks I and II exhibit differential binding to irradiated RFI DNA. By taking advantage of this binding difference resolution of the enzymes is achieved by affinity chromatography (Fig. 15). Final purification of these two peaks on isoelectric focusing revealed that peak I differs in its isoelectric point of 8.7 from peak II which is more acidic with an isoelectric point of 4.7.

A comparison of the substrate saturation curves (Fig. 16) clearly shows that peaks I and II reach saturation with no more than 2 to 4 pyrimidine dimers per genome. Peak III, IV and V, however, required DNA substrates with higher levels of modification before saturation was obtained. These differences distinguish the *damage-specificity* of peaks I and II from the *damage-dependency* of the other three peaks of activity. Peak III has a wide range of specificity including λ irradiated DNA, alkylated-depurinated DNA,

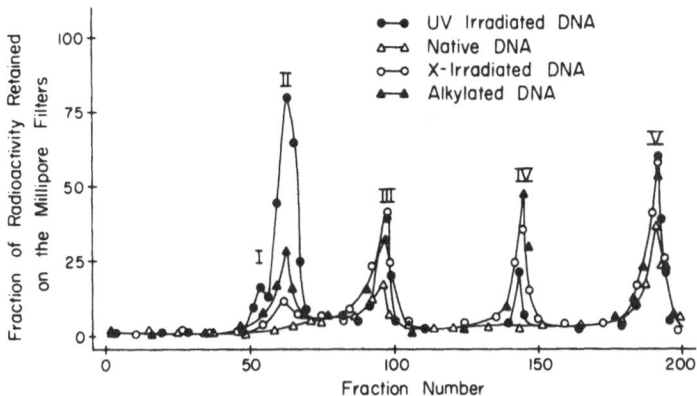

Figure 14. Binding properties of resolved correndonucleases

DNA treated with OsO_4 (which forms a thymine glycol derivative
analogous to a gamma irradiation product [100]) and undamaged
denatured DNA. This endonuclease probably recognizes the common
secondary effects arising from the diverse forms of damaging
agents.

That two endonucleases are specific for UV-irradiated DNA
demanded a resolution of their substrate specificities. The
ability of either peaks I or II to act specifically on pyrimidine-
pyrimidine dimers was examined. The action of photolyase on
irradiated DNAs (photolyase monomerizes T<>T, C<>T and C<>C
pyrimidine dimers), is able to reverse the action of the two enzymes
implicating pyrimidine dimers as the substrate on the DNA for these
enzymes. Incubation of irradiated DNA with this enzyme in the
absence of light, however, does not affect the substrate
capabilities of such DNA molecules (Fig. 17).

The question of whether the two enzymes are specific for a
unique species of pyrimidines was examined by irradiating DNA with
photosensitizers, in which tymine dimers are the major photo-
product and in C<>T and C<>C are absent (101). It was found
(Fig. 18) that both enzymes act on thymine-thymine dimers.
Pyrimidine dimer specificity was further examined more closely by
determining the effects of the separate endonucleases on excision
of the respective dimers from irradiated DNA. This was accomplished
by following the loss of pyrimidine dimers from DNA under conditions,
shown in Table VI, in which the endonuclease was subsequently treated
with either exonuclease VII or polymerase I for excision. The data
clearly suggest that neither enzyme I nor enzyme II show any unique
pyrimidine dimer specificity.

The question remained as to why there are two separate endo-

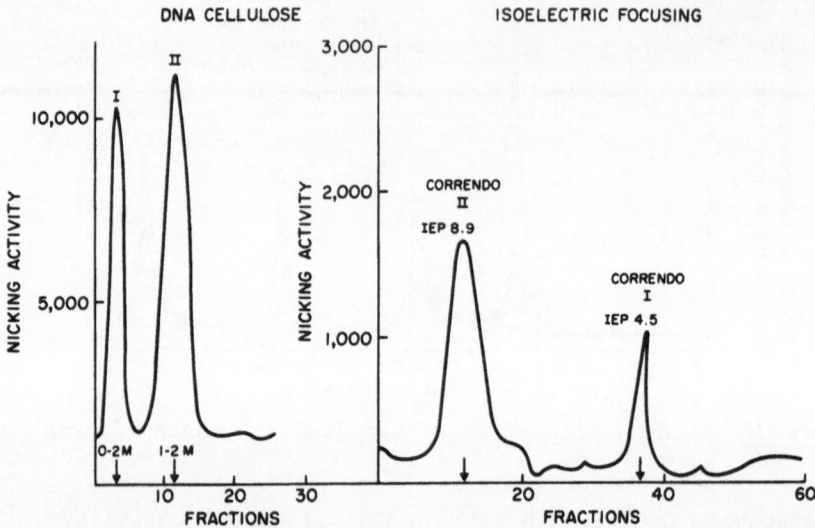

Figure 15. Resolution of Py<>Py correndonucleases by DNA-affinity and isoelectric focusing.

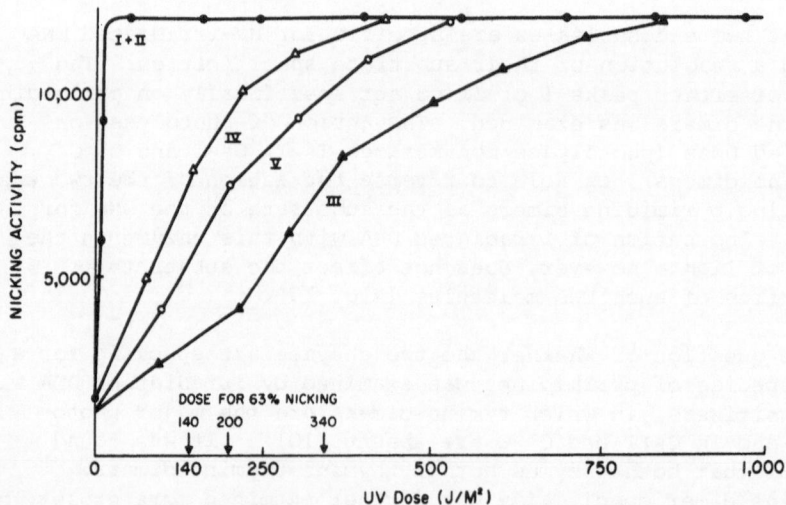

Figure 16. Ultraviolet dose saturation of correndonucleases for M. luteus.

nucleases seeming to act on the same substrate. Furthermore, genetic analyses indicated that both proteins are required to confer UV resistance on cells.

The possibility that these two enzymes may be acting in

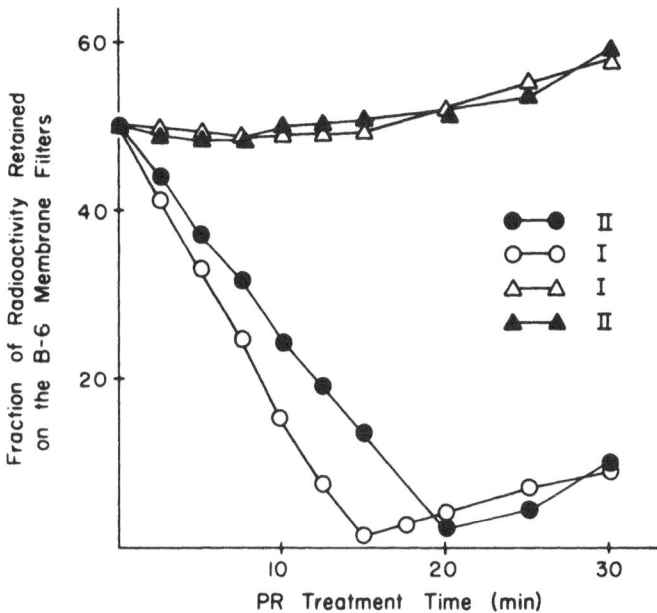

Figure 17. Effects of photoreactivation of UV irradiated RFI DNA as a substrate.

different regions of the DNA molecule was pursued. The irradiated DNAs were incused by each endonuclease and the structure of the sites generated in each case was studied. Three different experimental approaches were utilized, all of which are diagnostic for the appearance of single stranded-like regions at the site of incision. The first procedure utilized bacterial alkaline phosphatase (BAP) which is a heat stable and rather large protein. Weiss et al. (102) formerly showed that BAP cannot act directly at a nick and requires prior denaturation of DNA in order for the phosphomonoester group to become available to the action of BAP. The heat stability of the BAP permitted the study of the availability of such termini in an environment of changing temperature. The data in Fig. 19 shows the extent of phosphomonoester group availability to BAP as a function of temperature. What is apparent is that the phosphomonoester sites generated by correndonuclease I are significantly more accessible to the action of BAP than is that site generated by correndonuclease II. These data are consistent with two unique sites being generated.

It was assumed that a distinction between these termini could be made by examining the susceptibility of these nicks to polynucleotide ligase. Ligase can catalyze the formation of a phosphodiester bond even though the 3' juxtaposed nucleotide is non-

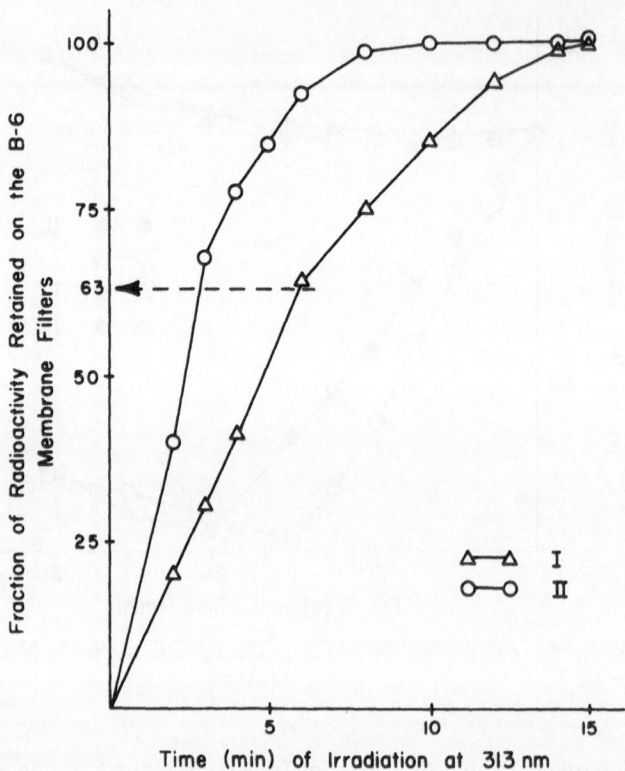

Figure 18. *The use of photosensitized RFI DNA as a substrate by Py<>Py correndonuclease I and II.*

Table VI
Excision of pyrimidine dimer species

TREATMENT	DIMER CONTENT			FRACTION EXCISED		
	T<>T	C<>T	C<>C	T<>T	C<>T	C<>C
UV IRRADIATION	8.5	8.05	3.15	-	-	-
Endo^{uv}I+polI	3.91	3.94	1.42	54	51	55
Endo^{uv}II "	3.49	3.22	1.10	59	60	65
Endo^{uv}I+Endo^{uv}II +polI	3.83	5.23	1.51	55	35	52
Endo^{uv}I+exoVII	4.59	4.67	1.61	46	42	49
Endo^{uv}II "	2.13	2.17	1.13	75	73	64
Endo^{uv}I+Endo^{uv}II +exoVII	2.38	5.6	1.57	72	30	50
Endo^{uv}I or Endo^{uv}II or pol I or exo VII	-	-	-	0.5 to 4.0		

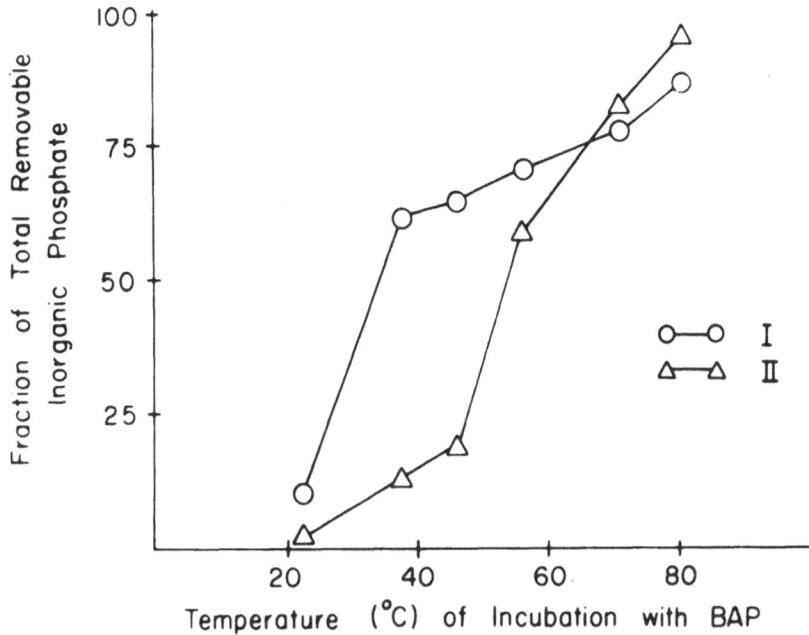

Figure 19. Dephosphorylation with BAP at various temperatures.

complementary (103). Rates of ligation were followed with DNAs
incised with either correndonuclease I, correndonuclease II, or
pancreatic DNase I. The data in Table VII clearly demonstrates
that distinctions exist between the two products of incision and
that DNA incised with enzyme I is converted less rapidly to RFI
in comparison to that generated by correndonuclease II. These
data are in keeping with the suggestion that the site generated by
correndonuclease I is more single stranded-like than that generated
by the second enzyme.

An additional approach to studying these structural differences,
employed the co-operative binding properties of the *E.coli* binding
protein to single stranded DNA. In order to assess the effects of
binding protein use was made of an observation in which single-
stranded DNA associated with binding protein is resistant to the
action of exonuclease VII (50). In these experiments the incised
irradiated DNA was preincubated with stoichemetric amounts of
binding protein and subjected to hydrolysis by exonuclease VII.
The data (Fig. 20) show that the site generated by enzyme I, when
25% hydrolyzed by exonuclease VII, causes complete shutoff of
further hydrolysis. The site generated by correndonuclease II is
quantitatively hydrolyzed although at a reduced initial rate. These
data support the suggestion that unique sites are generated by the
resolved correndonucleases.

Table VII
Rejoining of endonuclease incised DNA

Treatment	Radioactivity retained on the B-6 filters (cpm)		Fraction rejoined $\frac{A-B}{A}$ x 100
	-ligase (A)	+ligase (B)	
Pancreatic DNase I incised DNA	9,876	3,437	65
UV endonuclease I treated DNA	8,267	7,870	5
UV endonuclease II treated DNA	8,103	5,275	35

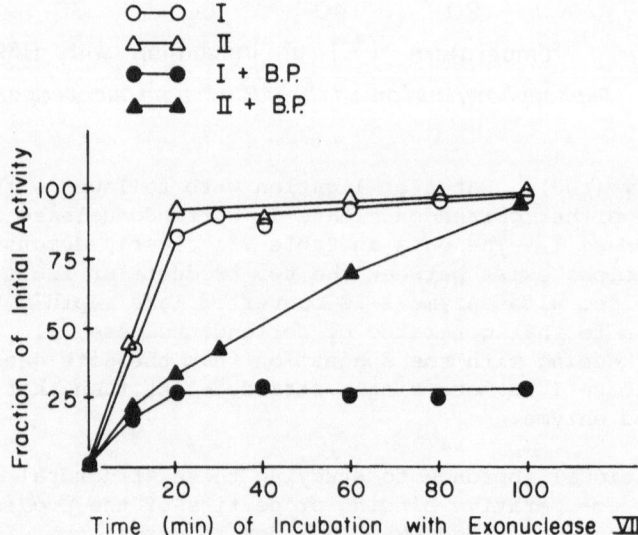

Figure 20. Hydrolysis of UV irradiated incised DNA by Exo VII.

Summary

Pyr<>Pyr correndonucleases I and II from *M.luteus* act
exclusively on thymine-thymine, cytosine-cytosine and thymine-
cytosine cyclobutane dimers in DNA, catalyzing incision 5' to the
damage, generating a 3' hydroxyl and 5' phosphoryl terminus. Both

enzymes initiate incision of pyrimidine dimers *in vitro* by correxo-
nucleases and DNA polymerase I. The respective incised DNAs,
however, differ in their ability to act as a substrate for polynucleotide ligase or bacterial alkaline phosphotase suggesting that
each endonuclease is specific for a conformationally unique site.
The possibility that their respective action generates termini
which represent different degrees of single strandedness is
suggested by the unequal protection by an *E.coli* binding protein
from the hydrolytic action of exonuclease VII.

IX. EXCISION IN MAMMALIAN CELLS

1. Introduction

Of many of the advantages of bacterial systems for the study
of the biochemistry of excision the availability of excision-
defective mutants is paramount. The isolation of enzymes whose
identification as specific gene products allows for identifying
the biological role of a specific enzyme in such a process. The
isolation and physiological characterization of the xeroderma
pigmentosum lines provides a potential for such parallel analysis
in mammalian cells. The recent acquisition of isogenic lines of
Chinese hamster ovary cells (CHO) also lends itself to further
biochemical analysis (104). The major difficulty with many of the
mammalian studies has been a sufficient source of putative wild
type biological material to initiate the isolation and characterization of enzymes that show potential for repair capabilities. For
this reason our laboratory has concentrated on using fresh placenta
obtained from normal patients. The human placenta which is divided
into a fetal and maternal side by a syncytium of trophoblasts found
on a barrier between the two sides of this organ. Such material can
yield up to 60 grams of relatively clean homogenous preparations
of trophoblasts for enzyme isolation. These cells are used as wild
type source material for comparison with the human cell lines
obtained from patients or from mutagenized human cells. The
isolation procedures for purification of repair enzymes involves
the separation of the trophoblasts into nuclear and cytoplasmic
fractions by standardized techniques. Fractionation of the extracts
from these various organelles is initiated by a batch-wise elution
from phosphocellulose ion exchange resin (P-11). This procedure
results in the resolution of two general peaks of enzyme activity.
The first peak contained a major portion of the exonuclease activity
and the second peak contains the β-DNA polymerase (Fig. 21). Endonuclease activities seem to be present in reconstitution experiments
(Fig. 22) when irradiated DNA was employed as a substrate. Neither
of the two separate peaks is able to affect incision of pyrimidine
dimers from irradiated DNA. However, a combination of the two peaks

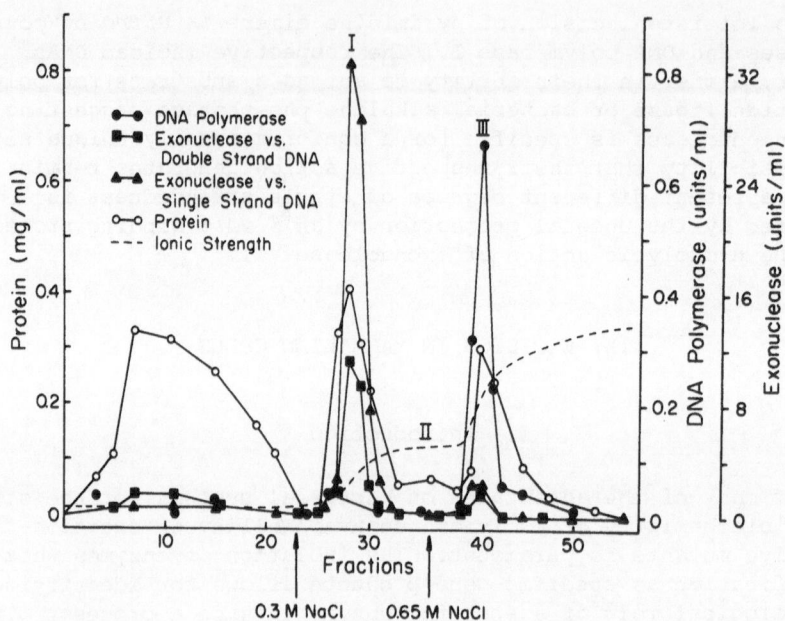

Figure 21. Nuclear enzymes N2 phosphocellulose I.

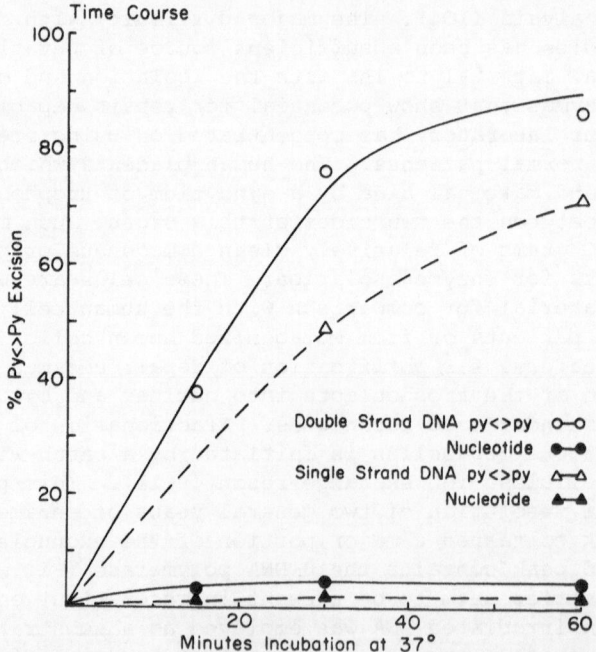

*Figure 22. Excixion of pyrimidine dimers from incised UV DNA by
fraction I exonuclease.*

does result in significant release of damaged nucleotides. It can be seen, furthermore, (Table VIII) that in addition to the reconstitution of excision, the two fractions are required for complete reinsertion of nucleotides following an incision-excision reaction. It is apparent therefore that the necessary ingredients for repair are present in these crude fractions.

The resolution of individual components is shown schematically in Fig. 23 in which the two separate fractions are resolved more

Table VIII
Stimulation of incorporation of nucleotides in UV damaged DNA by placental exonucleases

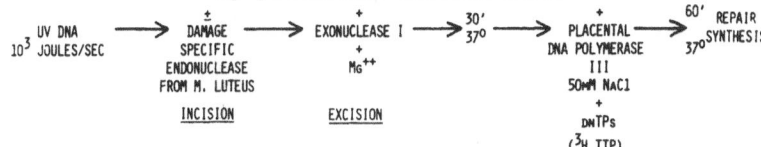

INCUBATION SEQUENCE		PMOLES NUCLEOTIDE INCORPORATED INTO DNA TEMPLATES				INCREASED SYNTHESIS BEFORE INCISION	INCREASED SYNTHESIS AFTER INCISION
EXCISION 30'AT 37°	INCORPORATION 60'AT 37°			INCISED			
		NATIVE	UV DNA	NATIVE	UV DNA		
I		0.7	0	0.4	0.9		
	III	1.5	0.9	0.6	2.2		
I	III	3.4	6.2	0.8	12.7	2.8	6.5
I & II		0.6	0	0.1	0.9		
I & II	III	4.1	6.1	3.3	13.2	2.0	7.1

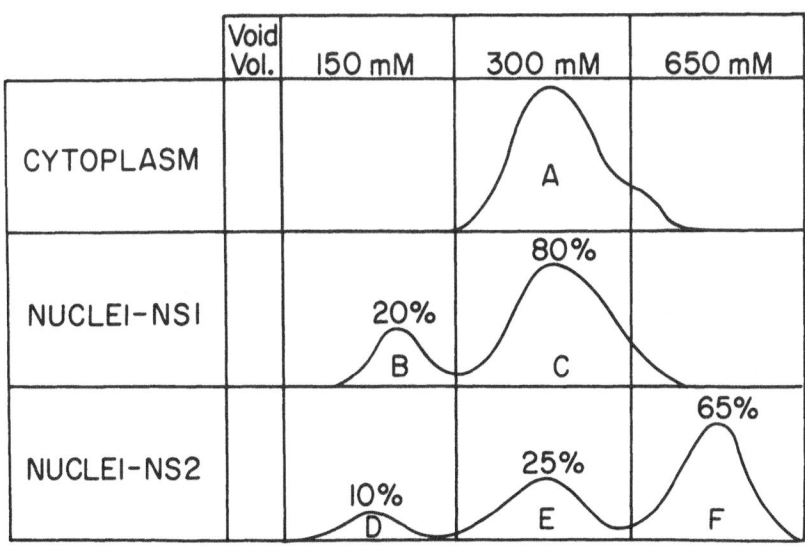

Figure 23. Distribution of placental exonucleases, potassium phosphate buffer, pH 7.5.

completely by gradient elution phosphocellulose chromatography.
The separated fractions are further purified by DNA cellulose
affinity chromatography and finally by gel exclusion chromatography.
Many of the exonucleases and β-DNA polymerases described in Table
VIII, have been purified at least 2,000 fold. The substrate
employed for following the directionality of the exonucleolytic
activities is shown in the upper portion of Table IX. It is clear
that there are a wide variety of unique exonucleases present in both
the nucleus and in the cytoplasm. The most interesting observation
is that none of the exonucleases present in the nucleus are capable
of excising pyrimidine dimers from UV-irradiated, incised DNA.
However, combining two of the fractions resulted in significant
release of these damaged nucleotides from the irradiated DNA. This
particular problem is now under intensive study in the laboratory.

The exonuclease purified from the cytoplasmic fraction,
referred to as correxonuclease, is extremely interesting and has
properties reminiscent of those associated with exonuclease VII of
E.coli (68) and the UV exonuclease from *M.luteus* (67). It is this
particular enzyme to which the rest of this discussion is addressed.

2.   Cytoplasmic Correxonuclease From Human Placenta(69)

This purified enzyme is specific for single stranded DNA
(Fig. 24). Furthermore, its ability to act on unirradiated and
irradiated single stranded DNA at the same rate and to the same
extent implies that the enzyme, like that isolated from *M.luteus*
(67) and *E.coli* (68) is probably acting at internal rather than
terminal phosphodiester bonds. A series of doubly labeled polymers
were prepared as substrates for this enzyme. (^{3}H) DNA partially
digested with exonuclease III was (^{32}P)-terminally labeled. Use of
such a substrate by the enzyme shows that it is able to act at the 5'
terminus immediately before release of internal tritiated nucleotides (Fig. 25). Using single stranded DNAs labelled at the 5'
terminus as substrate caused its release to the same extent and
rate as the internal tritiated nucleotides. Single stranded DNA
with approximately 7-(^{3}H) nucleotides at its 3' terminus shows
equal enzyme susceptibility at its 5' terminus. These data imply
that this enzyme like the bacterial enzymes is acting bi-directionally.

In order to test this in a more definitive manner a triply-
labeled polydeoxythymidylate was prepared with terminal nucleo-
tidyl transferase (105). This polymer was labeled with approximately
a 600 (^{14}C)-thymidylate primer to which approximately 2,000 unlabeled deoxythymidylates were added. This polymer was used as a
primer for the addition of 600 (^{3}H)-thymidylate residues at its 3'
terminus. The purified polymer was used as a substrate for the
cytoplasmic correxonuclease.

Table IX

Properties of placental exonucleases

Peaks	Organelle	Structure Single 3'	Single 5'	Duplex 3'	Duplex 5'	Excision Py <> Py %	Specificity
A	Cytoplasm	+	+	+	+	50.	Correxonuclease 5'D,S 3'D,S
B	Nucleus	−	+	−	+	0	5' D,S
C	"	−	+	−	−	0	5' S
D	"	+	−	+−	−	0	3' d,S
E	"	−	+	−	−	2.	5' S
F	"	−	−	−	+	4.	5' D
E + F	"		+		+	41.	Cooperative Excision

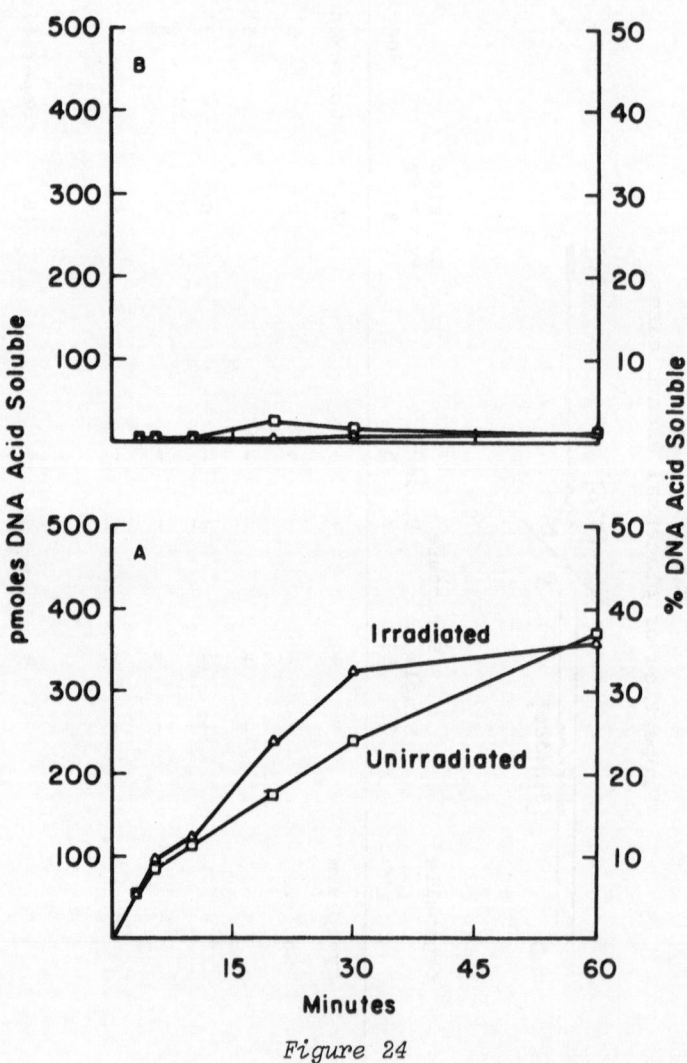

Figure 24

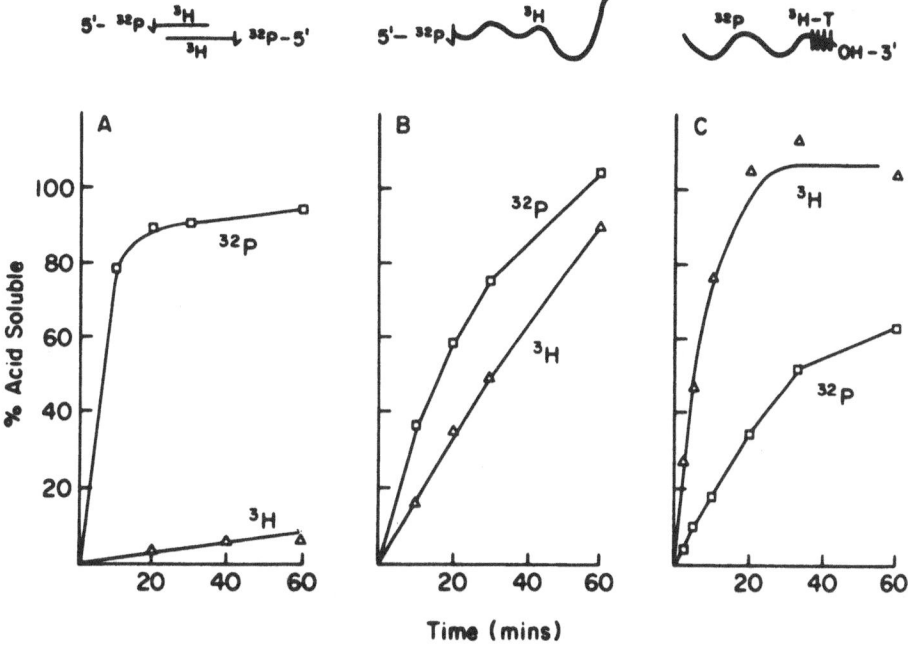

Figure 25. Terminal specificity of placental correxonuclease.

The initial rates of hydrolysis shown in the upper portion of
Fig. 26 demonstrates that the rate of hydrolysis from both termini
are idential. The removal of samples from such reaction mixtures
were subjected to gel electrophoresis. The results shown in the
lower portion of that figure indicate that a pentanucleotide is the
product of the reaction regardless of the terminus from which it
was hydrolyzed.

The question of whether this enzyme was capable of excising
pyrimidine dimers was examined using doubly-labeled DNAs. It was
possible, therefore, to measure the rate and extent of release of
pyrimidine dimers with associated nucleotides. From the data in
the upper portion of Fig. 27 it can be seen that the enzyme is able
to excise pyrimidine dimers at the same rate as nucleotide excision.
Since the enzyme can apparently hydrolyze bi-directionally on
single-stranded DNA it is possible that the enzyme was hydrolyzing
bi-directionally at such a nick. Support for this possibility is
seen in the lower portion of the figure in which irradiated DNA,
rather than being incised with the correctional endonuclease, was
nicked nonspecifically with pancreatic DNase. Since this enzyme
nicks randomly on the duplex it was expected that dimers would not
be released because this nonspecific endonuclease should hydrolyze
phosphoester bonds distal to the location of damaged nucleotides.
What was surprising, however, was that there was a significant

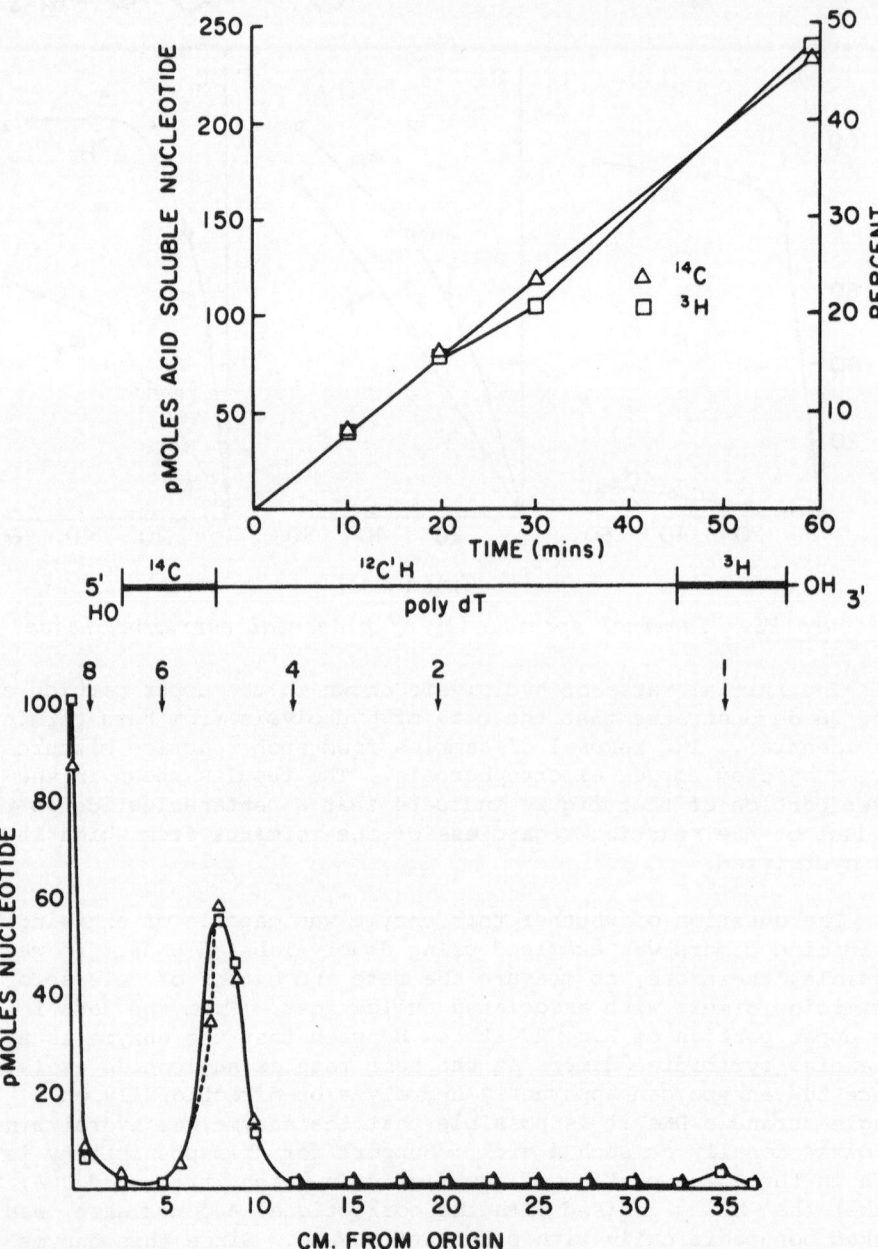

Figure 26. *Nature of excision products of (^{3}H) (^{14}C) poly cT by placental correxonuclease.*

release of nucleotides from such DNA under conditions in which
pyrimidine dimers were not released. These data strongly suggest
that even though this enzyme is specific for single stranded DNA it
can act at a nick on a duplex. Whether it is acting in a bi-directional manner was answered using a doubly labelled nicked DNA. The
data in Fig. 28 shows that the enzyme is acting bidirectionally at
the nick in this undamaged DNA.

The unique properties of this correctional exonuclease are
somewhat reminiscent of the genralized nucleolytic properties
associated with the $E.coli$ DNA polymerase I in that it possesses
the capabilities for both excision in a 5' to 3' direction as well
as editing in a 3' to 5' direction. Experiments are currently
under way in order to examine the nick translational capabilities
of such an enzyme in the presence of the β-DNA polymerase.

SUMMARY

An exonuclease which hydrolyzes single stranded DNA has been
purified from human placenta. It initiates hydrolysis at both the
3' and 5' termini of such DNA with equal facility yielding 5'
phosphorylated oligonucleotides averaging 5 nucleotides in length.
These oligonucleotides are released from both termini at equal
rates with the same size distribution. Although not detectably
active against intact native DNA this enzyme can initiate hydrolysis at single strand breaks creating a gap 30 to 40 nucleotides
long. If a pyrimidine dimer is adjacent to this break the enzyme
by virtue of its ability to act at internal phosphodiester bonds can
excise such dimers. We refer to this exonuclease as to human
correxonuclease because of its excision capability.

ACKNOWLEDGMENTS

Work reported in this contribution was supported by grants
to L.G. from the American Cancer Society (NP-8H), ERDA (EY-76-S-02-
2814), National Institues of Health (5 RO1 GM22846), and the
National Science Foundation (GB43550X).

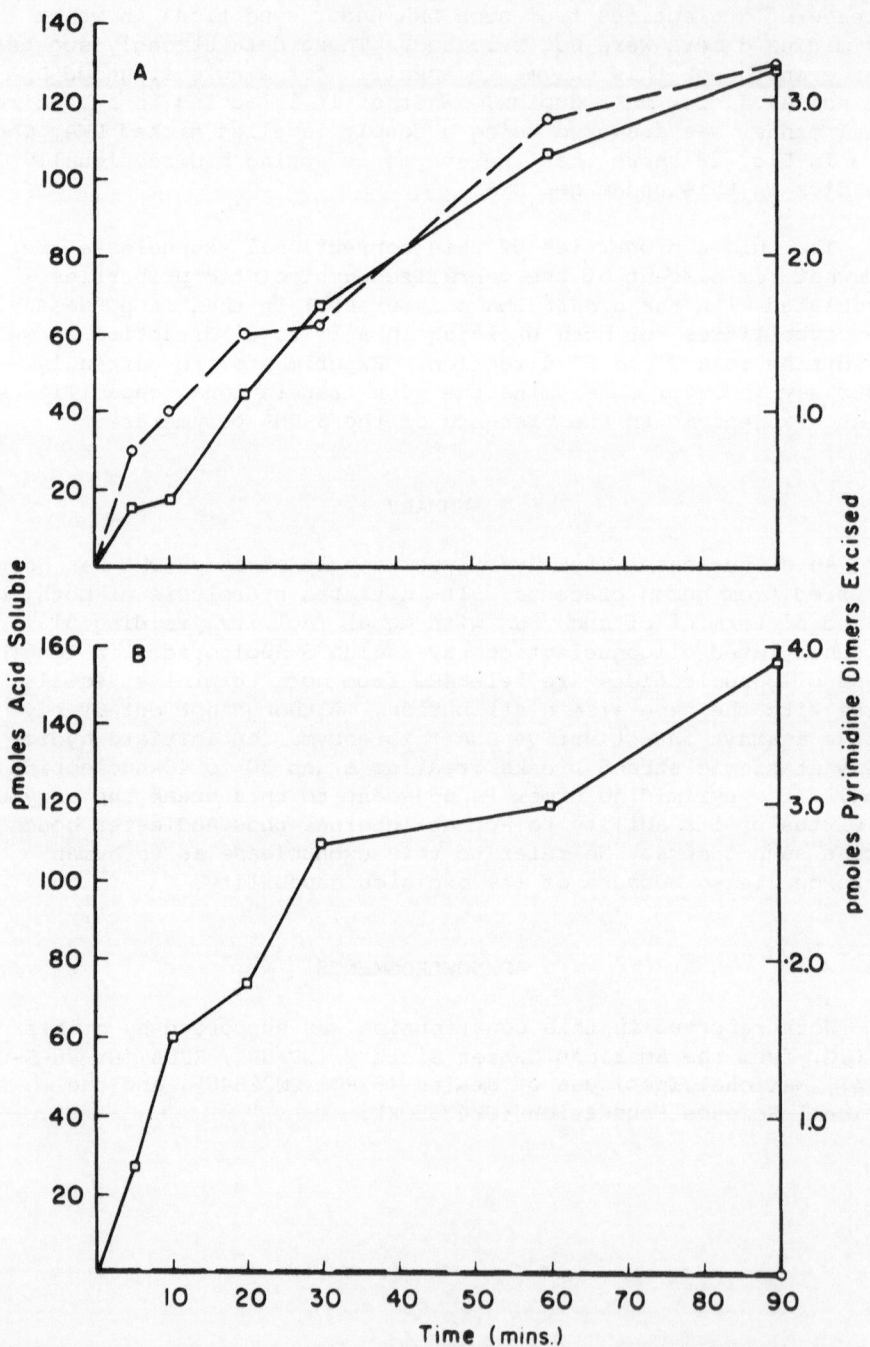

Figure 27.   *Py<>Py dimer and nucleotide excision by placental correxonuclease.*

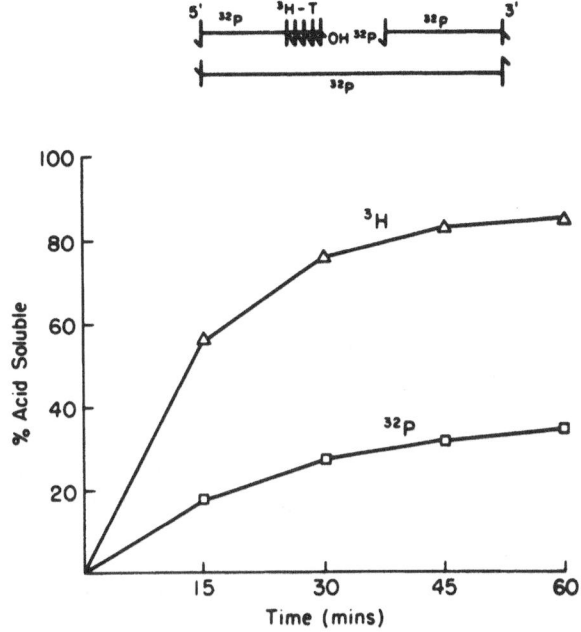

Figure 28. *Directionality of placental corExonuclease at nicks in DNA.*

REFERENCES

1. Beukers, B. and Berends, W. (1961) Biochim. Biophys. Acta 49: 181

2. Wang, S.Y. (1961) Nature 190:690

3. Setlow, R.B. and Setlow, J.K. (1962) Proc. Natl. Acad. Sci. 48:1250

4. Rupert, C.S. (1962) J. Gen. Physiol. 45:725

5. Hart, R.W. and Setlow, R.B. (1974) In: Molecular Mechanisms for Repair of DNA, Part B (eds. P.C. Hanawalt and R.B. Setlow) Plenum Press, p719

6.   Singer, B.(1975) In: Progress in Nucleic Acid Research and
      Molecular Biology, vol 15 (ed. W.E. Cohn) Academic Press, New
      York, p218

7.   Lawley, P.D. (1968) Nature 218:580

8.   Miller, E.C. and Miller, J.A. (1966) Pharmacol. Rev. 18:805

9.   Miller, J.A. (1970) Cancer Research 30:559

10.   Conney, A.H. (1972) Science 178:576

11.   Gillette, J.R. (1969) In: Microsomes and Drug Oxulations,
      Academic Press, New York, p547

12.   Nebert, D.W. and Gielen, J.E. (1972) Fed. Proc. 31:1315

13.   Miller, E.C. and Miller, J.A. (1974) In: Molecular Biology of
      Cancer, (ed. H. Busch), Academic Press, New York, p377

14.   Kriek, E., Miller, J.A., Juhl, U. and Miller, E.C. (1967)
      Biochemistry 6:177

15.   Fink, L.M., Nishimura, S. and Weinstein, I.B. (1970)
      Biochemistry 9:496

16.   Miller, E.C., Cooke, C.W., Lotlikar, P.D. and Miller, J.A.
      (1964) Proc. Am. Assoc. Cancer Res. 5:45

17.   Westra, J.G., Kriek, E. and Hittenhausen, H. (1976) Chem.
      Biol. Interactions 15:149

18.   Kriek, E. (1974) Biochim. Biophys. Acta 355:177

19.   Kriek, E. (1972) Cancer Res. 32:2042

20.   Jeffrey, A.M., Blobstein, S.H., Weinstein, I.B., Beland, F.A.,
      Harvey, R.G., Kasai, H. and Nakanishi, K. (1976) Proc. Natl.
      Acad. Sci. 73:2311

21.   Weinstein, I.B. (1976) In: Colloque International CNRS de
      Cancerogénèse Chimique, Menton, France.

22.   Greer, S. and Zamenhof, S. (1962) J. Mol. Biol. 4:123

23.   Lindahl, T., Nyberg, B. (1972) Biochemistry 11:3610

24.   Wacker, A., Dellweg, H. and Weinblum, D. (1960) Naturwissen-
      schaften 47:477

25. Grossman, L. and Rodgers, E. (1968) Biochem. Biophys. Res. Comm. 33:925

26. Ono, J., Wilson, R.G. and Grossman, L. (1965) J. Mol. Biol. 11:600

27. Feldberg, R.S. and Grossman, L. (1976) Biochem. 15:2402

28. Setlow, R.B. and Carrier, W.L. (1966) J. Biol. Chem. 17:237

29. Marmur, J. and Grossman, L. (1961) Proc. Natl. Acad. Sci. 47:778

30. Cole, R.S. (1971) Biochim. Biophys. Acta 254:30

31. Shen, J.C.-K. and Hearst, J.E. (1976) Proc. Natl. Acad. Sci. 73:2649

32. Dulbecco, R. (1950) J. Bacteriol. 54:329

33. Kelner, A. (1949) Proc. Natl. Acad. Sci. 35:73

34. Rupert, C.S. (1960) J. Gen. Physiol. 43:573

35. Muhammed, A. (1966) J. Biol. Chem. 241:516

36. Madden, J.J., Werbin, H. and Denson, J. (1973) J. Photochem. Photobiol. 18:441

37. Toulmé, J.J., Charlier, M. and Hélène, C. (1974) Proc. Natl. Acad. Sci. 71:3185

38. Hélène, C., Toulmé, F., Charlier, M. and Yaniv, M. (1976) Biochim. Biophys. Res. Commun. 71:91

39. Howard-Flanders, P., Rupp, W.D., Wilkins, B.M. and Cole, R.S. (1968) Cold Spring Harbor Symp. Quant. Biol. 33:195

40. Rupp, W.D., Wilde, C.E., Reno, D.L. and Howard-Flanders, P. (1971) J. Mol. Biol. 61:25

41. Carrier, W.L. and Setlow, R.B. (1964) Proc. Natl. Acad. Sci. 51:226

42. Boyce, R.P. and Howard-Flanders, P. (1964) Proc. Natl. Acad. Sci. 51:293

43. Lindahl, T. (1974) Proc. Natl. Acad. Sci. 71:3649

44. Weigand, R.C., Godson, G.N. and Radding, C.M. (1975) J. Biol.
     Chem. 230:8848

45. Shishido, K. and Ando, T. (1974) Biochem. Biophys. Res. Commun.
     39:1380

46. Riazuddin, S. and Lindahl, T. (1977) Institutional Congress on
     Environmental Mutagenesis, Edinburgh.

47. Friedberg, E.C. and King, J.J. (1969) Biochem. Biophys. Res.
     Commun. 37:646

48, Braun, A. and Grossman, L. (1974) Proc. Natl. Acad. Sci.
     71:1838

49. Riazuddin, S. and Grossman, L. (1977) J. Biol. Chem. 252:6280

50. Riazuddin, S. and Grossman, L. (1977) J. Biol. Chem. 252:6287

51. Riazuddin, S., Grossman, L. and Mahler, I. (1977) J. Biol.
     Chem. 252:6294

52. Braun, A., Radman, M. and Grossman, L. (1976) Biochem. 15:
     4116

53. Minton, K., Durphy, M., Taylor, R. and Friedberg, E.C. (1975)
     J. Biol. Chem. 250:2823

54. Seeberg, E. and Rupp, W.D. (   ) In: Molecular Mechanisms
     for the Repair of DNA (ed. P.C. Hanawalt) Plenum Press, New
     York, p439

55. Hamilton, L., Mahler, I. and Grossman, L. (1974) Biochem.
     13:1886

56. Weiss, B. and Milcarek, C. (1974) In: Molecular Enzymology
     (eds. L. Grossman and K. Moldave) Vol. 29 p180

57. Weiss, B. (1976) J. Biol. Chem. 251:1866

58. Verly, W., personal communication

59. Ljungquist, S. (1977) J. Biol. Chem. 252:2808

60. Radman, M. (1976) J. Biol. Chem. 251:1438

61. Gates, F.T. and Linn, S. (1977) J. Biol. Chem. 252:1647

62. Warner, H.R., Duncan, B.K., Mozer, T.J. and Thompson, R.B. (1977) Cold Spring Harbor Symp. Quant. Biol.   :95

63. Kelly, R.B., Cozarelli, N.R., Deutscher, M.P., Lehman, I.R. and Kornberg, A. (1970) J. Biol. Chem. 245:39

64. Setlow, P. and Kornberg, A. (1972) J. Biol. Chem. 247:232

65. Friedberg, E.C. and Lehman, I.R. (1974) Biochem. Biophys. Res. Commun. 58:132

66. Livingston, D.M. and Richardson, C.C. (1975) J. Biol. Chem. 250:470

67. Kaplan, J.C., Kushner, S.R. and Grossman, L. (1971) Biochemistry 10:3315

68. Chase, J. and Richardson, C.C. (1974) J. Biol. Chem. 249: 4553

69. Doniger, J. and Grossman, L. (1976) J. Biol. Chem. 251:4579

70. Kushner, S.R., Kaplan, J.C., Ono, H. and Grossman, L. (1971) Biochemistry 10:3325

71. Cooper, P.K. and Hanawalt, P.C. (1972) J. Mol. Biol. 67:1

72. Gefter, M.L. (1974) Prog. Nucl. Acid Res. 14:101

73. Grossman, L., Braun, A., Feldberg, R. and Mahler, I. (1975) Ann. Rev. Biochem. 44:19

74. Curtis, M.J. and Alberts, B. (1976) J. Mol. Biol. 102:793

75. Ruyechan, W.T. and Wetmur, J.G. (1976) Biochem. 15:5057

76. Howard-Flanders, P. and Boyce, R.P. (1966) Radiat. Res., 6: (Suppl.) 156

77. Clark, A.J., Chamberlin, M., Broyce, R.P. and Howard-Flanders, P. (1966) J. Mol. Biol. 19:442

78. Mattern, I.E., van Winden, M.P. and Rörsch, A. (1965) Mutat. Res. 2:111

79. Kato, T. (1972) Bacteriol. 112:1237

80. Seeberg, E. and Johansen, J. (1973) J. Mol. Gen. Genet. 123: 173

81. Ogawa, H., Shimada, K. and Tomizawa, J. (1968) J. Mol. Gen.
     Genet. 101:227

82. Seguira, J.H., Ingram, J.T. and Brain, R.T. (1973) In: Diseases
     of the Skin, Macmillan, New York, p1911

83. McKusick, V.A. (1966) Mendelian Inheritance in Man, Johns
     Hopkins Press, Baltimore, MD.

84. Cleaver, J.E. (1968) Nature 218:652

85. Kraemer, K.H., Coon, H.G., Petinga, R.A., Barnett, S.F., Rahe,
     A.E. and Robbins, J.H. (1975) Proc. Natl. Acad. Sci. 72:59

86. Mortelmans, K., Friedberg, E.C., Slor, H., Thomas, G. and
     Cleaver, J.E. (1976) Proc. Natl. Acad. Sci. 73:2757

87. Iyer, V.N. and Rupp, U.D. (1971) Biochim. Biophys. Acta 228:
     117

88. Smith, K.C. and Meun, D.H.L. (1970) J. Mol. Biol. 51:459

89. Witkin, E.M. (1976) Bacteriol. Revs. 40:869

90. Witkin, E.M. (1966) Science 152:1345

91. Radman, M. (1977) In: Réparation du DNA Mutagénèse Cancero-
     génèse Chimique, CNRS ps293-306

92. Brutlag, D. and Kornberg, A. (1972) J. Biol. Chem. 247:
     241

93. Muzyczka, N., Poland, R.L. and Bessman, M.J. (1972) J. Biol.
     Chem. 247:7116

94. Lwoff, A., Siminovitch, L. and Kjeldgaard, N. (1950) Ann.
     Inst. Pasteur 79:815

95. Weigle, J.J. (1953) Proc. Natl. Acad. Sci. 39:628

96. Borek, E. and Ryan, A. (1958) Proc. Natl. Acad. Sci. 44:374

97. George, J., Devoret, R. and Radman, M. (1974) Proc. Natl. Acad.
     Sci. 71:144

98. Center, M.S. and Richardson, C.C. (1970) J. Biol.Chem.
     245:6285

99. Riggs, A.D., Suzuki, J. and Bourgeois, S. (1970) J. Mol. Biol.
     48:67

100. Hariharan, P.V. and Cerutti, P.A. (1976) Biochem. Biophys. Acta   447:375

101. Lamola, A.A. (1969) Photochem. Photobiol.   9:291

102. Weiss, B., Live, T.R. and Richardson, C.C. (1968) J. Biol. Chem.   243:4530

103. Lehman, I.R. (1974)   Science, 186:790

104. Huang, P.C., personal communciation

105. Bollum, E.J. (1966) In: Procedures in Nucleic Acid Research (eds. G.L. Cantoni and D.R. Davies)   Vol. 1, Harper and Rowd, New York, pp577-583

IN VITRO CHARACTERIZATION OF THE *uvrA*[+], *uvrB*[+], *uvrC*[+]-CODED ATP-DEPENDENT UV-ENDONUCLEASE FROM *ESCHERICHIA COLI*

A. Seeberg[*][Δ], A.L. Steinum, E. Rivedal and J. Nissen-Meyer[†]

Norwegian Defence Research Establishment
P.O. Box 25
N-2007, Kjeller, Norway

INTRODUCTION

Excision repair of UV-induced pyrimidine dimers in the DNA proceeds by the sequential action of at least four different enzyme activities (for reviews see ref. 8,9). The repair is initiated by an endonuclease recognizing the damaged region and making a strand break (incision) adjacent to the dimer. The dimer is removed in a short olignucleotide by an exonuclease and the resulting gap is filled by a polymerase using the complementary strand as template. Finally, the strand is sealed by ligase.

Endonucleases responsible for the first step in this repair pathway have been extensively purified from bacteriophage T4 infected *Escherichia coli* (7,32) and from *Micrococcus luteus* (4,20), and have been characterized as small proteins (molecular weight ∿15.000 - 20.000) making 3'OH-5'P breaks at the 5' side of the dimers. In uninfected *E.coli*, however, where excision repair was

---

Abbreviations used: UV - ultraviolet; ATP - adenosine-5'-triphosphate; ADP - adenosine-5'-diphosphate; AMP - adenosine-5'-monophosphate; GTP - guanosine-5'-triphosphate; dNTP - deoxynucleoside-5'-triphosphate; NAD - nicotine amide adenine dinucleotide; UV-endonuclease - endonuclease specifically acting on ultraviolet irradiated DNA.

[*] *A.S.I. Participant*

[Δ] *To whom correspondence should be addressed*

[†] *Present address: Department of Biochemistry, University of Bergen, Norway*

first discovered (1,26), the incision enzyme has not been
satisfactorily characterized. The incision step of the repair in
this organism seems more complex than for T4 and *M.luteus*, because
genetic evidence indicates that correct incision requires at least
three different gene functions (23,27,29). Cells defective in
these functions, mutated in *uvrA*, *uvrB* or *uvrC*, were originally
characterized as being very UV sensitive and unable to excise dimers
from the DNA (10,19). Later studies of strand breakage in cellular
DNA following UV-irradiation of these mutants indicate that *uvrA*
and *uvrB* mutants have a major defect, whereas the *uvrC* mutant has
a partial defect in the incision step of the repair (11,23).

The incision enzymes so far characterized have been purified by
selecting proteins which have an endonucleolytic activity on UV-
irradiated DNA. This approach may not be feasible in the
purification of the $uvrA^+$, $uvrB^+$, $uvrC^+$-coded endonuclease, because
this activity results from the interaction between several gene
products (24) and will be host during purification if the proteins
separate. Essentially the same problem was encountered in the
purification of DNA replicating activity from *E.coli*, but the
difficulty was solved by using *in vitro* complementation assays to
purify separately each one of the participating proteins (22,31). We
have adopted this approach and established an *in vitro* complementation assay for the $uvrA^+$, $uvrB^+$ and $uvrC^+$ gene products. The
assay is based on the observation that the major UV-endonuclease
activity recovered in extracts from gently lysed cells depends on
all three gene products (24). In this communication we describe
further characterization of this UV-endonuclease activity as it
appears in cell extracts and the initial purification of the $uvrC^+$
product by means of the *in vitro* complementation assay system.

UV-endonucleases in crude extracts from *E.coli*

In a system with permeable cells, Waldstein et al. (30) first
showed that the uvr^+-gene dependent incision reaction in *E.coli*
requires ATP. This finding at least partially explained the
previous difficulty in detecting uvr^+ gene dependent UV-endonuclease activity in crude extracts (2,28). It was subsequently
shown that extracts from *E.coli* contain at least two different
endonuclease activities preferentially acting on UV-irradiated DNA
(24). One is ATP-independent and present in equal amounts in repair
proficient and excision deficient cells, while the other is ATP-
dependent and absent from *uvrA*, *uvrB* and *uvrC* mutated cells. The
latter activity seems 3-5 fold more efficient than the former (ref
24, Figs. 1 and 2). Fig. 1 shows dose response curves for endonuclease activity in crude extracts measured in the presence and
absence of ATP, using two different substrates purified in different
ways. Results obtained with ColEl and PM2 DNA are identical,
showing that the ATP-dependent UV-endonuclease activity is not

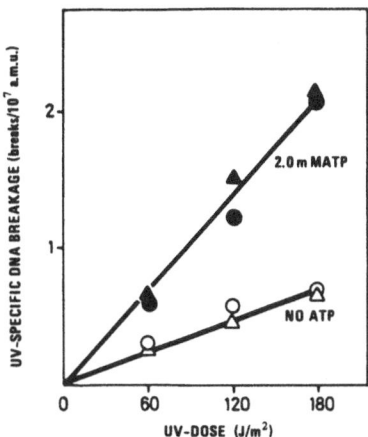

Figure 1. *UV-endonuclease activity in extracts from* Escherichia coli *K-12.*

Cell extract from strain AB1157 uvr[+] *prepared as previously described (24), was tested for UV-endonuclease activity in the absence (open symbols) and presence (closed symbols) of 2.0 mM ATP with covalently closed ColE1 DNA (circles) or bacteriophage PM2 DNA (triangles) as substrate.   Reaction mixtures contained 40 mM MOPS, pH = 7.5, 85 mM KC1, 15 mM MgSO$_4$, 1 mM EDTA, 1.mM dithiothreitol 0.05 μg DNA and extract from 10^8 cells.   Triplicate samples were incubated for 10 minutes at 37°C.   PM2 DNA was purified as described by Masamune et al. (13) and was the kind gift of Dr. Siv Ljungquist, Stockholm. ColE1 DNA was prepared as previously described (24).   The fraction of remaining covalently closed DNA was determined by filtration through nitrocellulose (21,2) and the number of breaks specifically induced in UV irradiated DNA was calculated as described by Kuhnlein et al. (12).*

specific for the ColE1 DNA originally used to demonstrate this activity (24).   Table 1 summarizes results obtained with different cofactors structurally related to ATP.   It is shown that neither AMP, ADP nor GTP can substitute for ATP, and there seems to be a specific requirement for ATP in the endonuclease reaction.

Extracts prepared from cells defective in polynucleotide ligase, show a higher activity than wild type extracts per dose unit to the DNA (Fig. 2).   This can probably be explained by repair of some breaks in wild type extracts due to repair completion despite the lack of exogenous DNA precursors and ligase cofactors in the reaction mixture.   When these compounds are added to the mixture together with wild type extracts, very few breaks are detected after incubation (Table 1).   This implies that ligase and polymerase are active in these extracts.

Table 1

Effect of different nucleotides on UV-endonuclease
activity in extracts from *E.coli*

Variations in reaction mixture	UV-endonuclease activity
Complete	(1)
+ ATP	3.8
+ GTP	0.8
+ ADP	0.9
+ AMP	1.5
+ ATP, 0.16 mM dNTP, 0.5 mM NAD	1.6

Cell extracts from strain AB1157 uvr^+ was tested
for UV-endonuclease activity as described in
legend to Fig. 1. Complete reaction mixture refers
to the response observed without ATP in the mixture,
using ColEl DNA irradiated with 180 J/m^2 as
substrate. This value is normalized to 1, and the
responses obtained with the addition of different
nucleotides are calculated relative to this value.
Concentration of nucleotides was 2.0 mM unless
otherwise indicated.

By means of purified UV-endonuclease from *Micrococcus luteus*
(generously provided by Dr. Tomas Lindahl, Stockholm) the number of
dimers produced per dose unit in the DNA was titrated. Assuming
that the *M.luteus* enzyme makes one incision for every dimer in the
DNA, and that breaks are distributed according to a Poisson
distribution in the circle population, one dimer is induced per 6
J/m^2 in ColEl DNA. This means that under optimal conditions, i.e.,
using extracts from lig^- cells (Fig. 2) only one break for every
20 dimers is induced by the ATP-dependent endonuclease activity.
Since fairly large doses therefore are needed to detect the endo[7]
nuclease, it might be questioned whether this activity is directed
against a secondary photoproduct less frequently produced. This
possibility seems unlikely, however, because the uvr^+-gene products

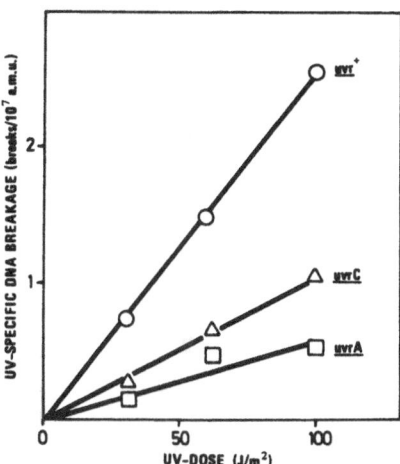

Figure 2. *UV-endonuclease activity in extracts from ligase deficient cells.*

Cell extracts from strains AB6006 uvr⁺ lig_ts7, AB6005 uvrA lig_ts7 and AB6003 uvrC lig_ts7 were assayed for UV-endonuclease activity in the presence of 2.0 mM ATP as described in legend to Fig. 1.

are directly responsible for the ATP-dependent endonuclease activity (ref 24, Fig. 2). Moreover, the characteristics of the ATP-dependent UV-endonuclease activity in extracts is in very good agreement with results obtained with permeable cells where the efficiency of nicking is considerably higher (25,30). The low efficiency of nicking in extracts can be explained if a small number of enzyme molecules are normally present in the cells and the activity becomes very diluted in the extracts.

In vitro complementation assay for the uvrA⁺, uvrB⁺ and uvrC⁺ gene products

Although extracts from the *uvrA*, *uvrB* and *uvrC* mutants are defective in the ATP-dependent UV-endonuclease (ref 24, Fig. 2), any combination of mutant extracts will complement to give normal activity similar to that observed in extracts from wild type cells (24). This shows that the ATP-dependent endonuclease results from the complementary action of the *uvrA⁺*, *uvrB⁺* and *uvrC⁺* gene products and that a mutation at any one of these loci will not prevent the formation of functional products from the other *uvr⁺* genes. This observation also implies that an *in vitro* complementation assay for the *uvr⁺*-gene products has been established, because any one of these products can be detected by its unique and selective ability to restore ATP-dependent UV-endonuclease activity in extracts

prepared from the corresponding mutants. Fig. 3 shows results
obtained by assaying for *uvrC*-complementing activity in a cell
extract from *uvrB* mutated cells. Addition of *uvrC*⁺ containing
fraction to cell extract from *uvrC* mutated cells results in a 3-
fold increase in the amount of UV-endonuclease activity observed.
The *uvrC*⁺ activity can be quantitated from the relative increase
observed with increasing amount of protein fraction (Fig. 3).
Similar results are obtained when assaying for *uvrA*-and-*uvrB*-
complementing activities, although the background values for UV-
endonuclease activity in these extracts are usually somewhat
lower than for *uvrC*.

DEAE-cellulose chromatography of *uvrC*-complementing activity

Braun and Grossman (2,3) have reported the purification of a
UV-endonuclease from *E.coli* that appears to be absent from *uvrA* and
uvrB, but present in *uvrC* mutated cells. This enzyme cannot be
detected in crude extracts, but appears after phosphocellulose
chromatography of the crude fraction. It has been assumed that this
enzyme, which is ATP-independent, represents the product(s) of the
uvrA and *uvrB* genes (8). It seemed therefore of most interest to
attempt the purification of the *uvrC*⁺ gene product by means of the
complementation assay.

Cell extract was applied to a DEAE cellulose column and

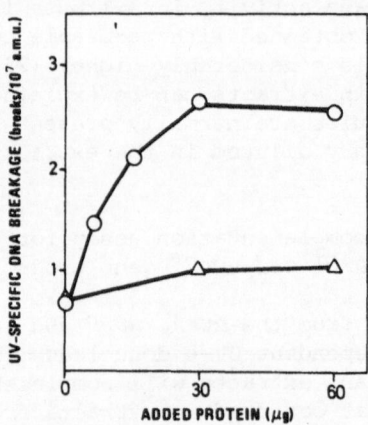

Figure 3. Assays for *uvrC*-complementing activity.

*Cell extract (∿40 µg) from strain AB1884 uvrC was supplemented with
various amounts of protein from a crude extract of AB1885 uvrB
(circles) or AB1884 as control (triangles). The mixtures were
tested for UV-specific endonuclease activity as described in legend
to Fig. 1. UV-dose to the DNA was 180 J/m². The amount of UV-
specific DNA breakage observed in the extract (30 µg protein) from
AB1885 was 0.60 breaks/10⁷ a.m.u.*

proteins were eluted with a salt gradient. Fractions were checked
for *uvrC*-complementing activity and a peak of *uvrC*[+] activity was
recovered eluting at about 0.3 M salt (Fig. 4). We conclude that
this activity reflects functional *uvrC*[+] gene product because (i)
the active fractions specifically complement *uvrC* mutant extract
(no complementation is observed with *uvrA* extract), and (ii) no
activity is observed when extract from *uvrC* cells is chromatographed in parallel experiments (data not shown). The *uvrC*[+]
product in this fraction does not exhibit UV-endonuclease activity
without complementation under standard conditions, and seems to be
completely separated from the *uvrA*[+] product since no *uvrA*[+] activity
is observed in any fraction from the DEAE cellulose column.

Ion exchange chromatography of *uvrA*- and *uvrB*-
complementing activities

To study the effect of the *uvrC*[+] on *uvrA*[+] and *uvrB*[+] products,
we have in a similar way attempted to isolate *uvrA*- and *uvrB*-
complementing activities (E.S. manuscript in preparation).
Surprisingly, it was discovered that *uvrA*[+] and *uvrB*[+] activities
are not locked in a complex, but can be separated completely by
ion exchange chromatography. The *uvrA*[+] activity elutes together
with the *uvrC*[+] activity upon DEAE-cellulose chromatography, whereas
the *uvrA*[+] activity can be recovered after phosphocellulose chromatography of the flow-through from the DEAE-cellulose. The DEAE-
cellulose and the phosphocellulose fractions thus isolated
complement to give ATP-dependent UV-endonuclease activity similar
to that observed in extracts from wild type cells, confirming that
functional *uvr*[+] gene products are recovered by the complementation
assay. Gel filtration of *uvrA*[+] and *uvrB*[+] activities in these
fractions indicate a molecular weight of about 100,000 for the *uvrA*[+]
product and 60,000 for the *uvrB*[+] product (E.S. manuscript in
preparation). It is difficult to reconcile these results with the
concept of a *uvrA*[+], *uvrB*[+]-coded endonuclease of molecular weight
12,000 (2,3).

DISCUSSION

In vitro complementation assays have previously been
successfully employed to purify proteins essential for DNA replication (15,22,23). The experiments described in this communication
show that this approach also can be useful for the purification of
DNA repair functions. So far, the complementation assay seems to
be the only method for assaying the *uvrC*[+] product. From the
preliminary characterization described above, conclusions concerning
the function of the *uvrC*[+] gene product cannot be made. However, the
results indicate that no UV-endonuclease activity is associated
with the *uvrC*[+] product.

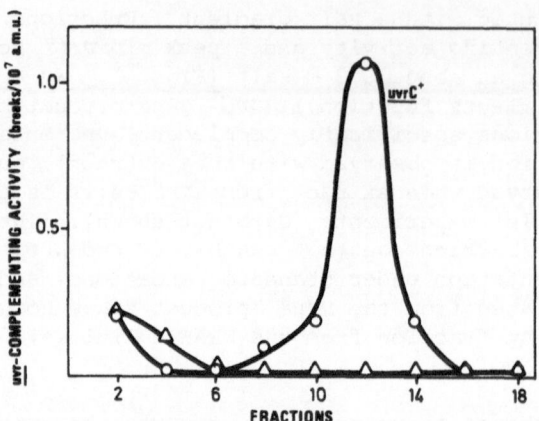

Figure 4. *DEAE-cellulose chromatography of uvrC-complementing activity*.

Cell extract (60 mg protein) from strain AB1157 uvr+ was applied to a DEAE-cellulose column (Whatman DE 52, 1 x 8 cm) previously equilibrated with 50 mM MOPS, pH = 7.5, 100 mM KCl, 1 mM EDTA, 1 mM dithiothreitol. Proteins were eluted with a linear salt gradient 100-400 mM KCl (40 ml) in the same buffer as above. Every second fraction was checked for uvrC- and uvrA-complementing activity as described in legend to Fig. 3. uvr-complementing activity is defined as the amount of UV-specific DNA breakage in complemented samples minus the background value observed in the mutant extract alone.

     Attempts to determine by gel filtration a molecular weight for the $uvrC^+$ product has been frustrated by great lability of this activity in the DEAE-cellulose fraction. So far, we have not found satisfactory means to stabilize this activity for further purification. However, to facilitate the purification of the $uvrC^+$ product, a transducing derivative of bacteriophage lambda carrying the *uvrC* gene has been isolated (Blingsmo, O.R. and Seeberg, E., unpublished), and we hope to amplify the amount of $uvrC^+$ product in the cell by means of this phage. Specialized transducing phages of lambda have previously been employed for the overproduction of polynucleotide ligase (17), ribonucleotide reductase (6), and products from the *trp* genes (14) in *E.coli*.

     T4 UV-endonuclease can replace the functions controlled by the *uvrA*, *uvrB* and *uvrC* genes in *E.coli* to reactivate irradiated viral infective DNA (29). This observation is compatible with the results indicating that the $uvrA^+$, $uvrB^+$ and $uvrC^+$ products determine the essential UV-endonuclease function in *E.coli*. However, *in vitro* treatment of irradiated viral DNA with T4 UV-endonuclease also

replaces the repair defect in *uvrD*-mutated cells (29). The *uvrD*
mutant of *E.coli* can excise dimers from the DNA (16) and show
normal levels of ATP-dependent UV-endonuclease activity in extracts
(P. Strike, personal communication). Therefore, the *uvrD* function
is probably acting subsequently to the endonuclease in the repair
(16). To reconcile these observations we suggest the schematic
model outlined in Fig. 5. The model suggests that irradiated
infective viral DNA pretreated with T4 endonuclease, is repaired
by the same pathway as used for the repair of irradiated bacteriophage T4 DNA. This repair pathway depends on the 5'-3' exonucleolytic activity of DNA polymerase I for excision of dimers (18),
whereas the normal host cell repair pathway does not (8). This
difference can be explained if the DNA regions incised by the T4
enzyme is structurally different from those created by the $uvrA^+$,
$uvrB^+$, $uvrC^+$-coded endonuclease. The DNA incised by the latter
enzyme could for instance contain single-strand gaps rather than
nicks. These structures may be substrates for the *uvrD* function
and other host cell enzymes, while repair of DNA incised by the
T4 endonuclease will depend on DNA polymerase I for excision and
resynthesis (18). This model may also explain why irradiated
infective viral DNA has a lower survival on *polA* hosts when
pretreated with T4 UV-endonuclease than without this treatment (29).
Preincised DNA can no longer be repaired by the host cell repair
pathway, and the need for DNA polymerase I is accentuated.

The $uvrA^+$, $uvrB^+$ and $uvrC^+$ products are all required for the
UV-endonuclease activity, but this does not exclude the
possibility that these factors have different functions which are

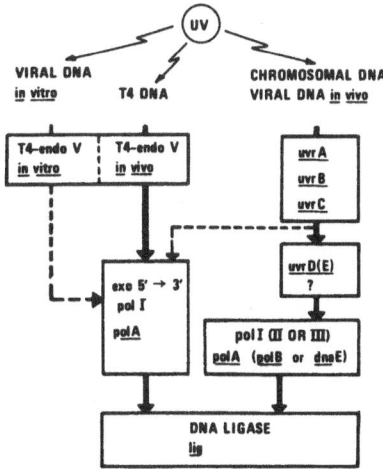

*Figure 5. Schematic model for host cell excision repair pathway
and T4 induced repair pathway in Escherichia coli.*

mutually dependent. If the number of molecules per cell is low,
then several turnovers may be needed before significant activity
can be detected. It could be speculated that three subsequent
steps, for instance recognition, incision and local unwinding, has
to be completed at one site before the repair is initiated at the
next site. However, evidence to support such speculations will
have to await further purification and characterization of the
separate $uvrA^+$, $uvrB^+$ and $uvrC^+$ gene products. The in $vitro$
complementation assay described above seems well suited for this
purpose.

ACKNOWLEDGEMENTS

This research was supported by travelling grants to E.S.
from the European Molecular Biology Organization (EMBO).

REFERENCES

1.   Boyce, R.P. and Howard-Flanders, P. (1964) Proc. Nat. Acad.
     Sci. USA   51:293-300

2.   Braun, A. and Grossman, L. (1974) Proc. Nat. Acad. Sci. USA
     71:1838-1842

3.   Braun, A., Radman, M. and Grossman, L. (1976) Biochem. 15:
     4116-4120

4.   Carrier, W.L. and Setlow, R.B. (1970) J. Bact. 102:178-186

5.   Cooper, P. (1977) Molec. Gen. Genet. 150:1-12

6.   Eriksson, S., Sjøberg, B.-M., Hahne, S. and Karlström, O.
     J. Biol. Chem. (in press)

7.   Friedberg, E.C. and King, J.J. (1971) J. Bact. 106:500-507

8.   Grossman, L., Braun, A., Feldberg, R. and Mahler, I. (1975)
     Ann. Rev. Biochem. 44:19-43

9.   Howard-Flanders, P. (1973) Brit. Med. Bull. 29:226-235

10. Howard-Flanders, P., Boyce, R.P. and Theriot, L. (1966)
     Genetics   53:1119-1136

11. Kato, T. (1972) J. Bact. 112:1237-1246

12. Kuhnlein, U., Penhoet, E.E. and Linn, S. (1976) Proc. Nat.
     Acad. Sci. USA   73:1169-1173

13. Masamune, Y., Fleischman, R.A. and Richardson, C.C. (1971)
    J. Biol. Chem. 246:2680-2691

14. Moir, A. and Brammar, W.J. (1976) Molec. Gen. Genet. 149:
    87-99

15. Morris, C.F., Sinha, N.K. and Alberts, B.M. (1975) Proc. Nat.
    Acad. Sci. USA 72:4800-4804

16. Ogawa, H., Shimada, K. and Tomizawa, J. (1968) Molec. Gen.
    Genet. 101:227-244

17. Panasenko, S.M., Cameron, J.R., Davis, R.W. and Lehman, I.R.
    (1977) Science 196:188-189

18. Pawl, G., Taylor, R., Minton, K. and Friedberg, E.C. (1976)
    J. Mol. Biol. 108:99-109

19. van de Putte, P., van Sluis, C.A., Van Dillwijn, J. and
    Rørsch, A. (1965) Mut. Res. 2:97-110

20. Riazuddin, S. and Grossman, L. J. Biol. Chem. (in press)

21. Riggs, A.D., Suzuki, H. and Bourgeois, S. (1970) J. Mol. Biol.
    48:67-83

22. Schekman, R., Weiner, J.H., Weiner, A. and Kornberg, A. (1975)
    J. Biol. Chem. 250:5859-5865

23. Seeberg, E. and Johansen, I. (1973) Molec. Gen. Genet. 123:
    173-185

24. Seeberg, E., Nissen-Meyer, J. and Strike, P. (1976) Nature
    263:524-526

25. Seeberg, E. and Strike, P. (1976) J. Bact. 125:787-795

26. Setlow, R.B. and Carrier, W.L. (1964) Proc. Nat. Acad. Sci.
    USA 51:226-231

27. Shimada, K., Ogawa, H. and Tomizawa, J.-I. (1968) Molec.
    Gen. Genet. 101:245-256

28. Takagi, Y., Sekiguchi, M., Okubo, S., Nakayama, H., Shimada,
    K., Uasuda, S., Nishimoto, T. and Yoshihara, H. (1968)
    Cold Spring Harbor Symp. Quant. Biol. 33:219-277

29. Taketo, A., Yasuda, S. and Sekiguchi, M. (1972) J. Mol. Biol.
    70:1-14

30. Waldstein, E.A., Sharon, R. and Ben-Ishai, R. (1974) Proc.
     Nat. Acad. Sci. USA 71:2651-2654

31. Wickner, R.B., Wright, M., Wickner, S. and Hurwitz, J. (1972)
     Proc. Nat. Acad. Sci. USA 69:3233-3237

32. Yasuda, S. and Sekiguchi, Y. (1970) Proc. Nat. Acad. Sci. USA
     67:1839-1845

THE *recA*[+] GENE PRODUCT OF *Escherichia coli*, AN INDUCIBLE PROTEIN

P.T. Emmerson[*], K.A. Powell, S.C. West and
P.K. Botcherby

Department of Biochemistry
The University
Newcastle-upon-Tyne, NE1 7RU, United Kingdom

The *recA*[+] gene is involved in genetic recombination (Clark and Margulies, 1965), DNA repair (Howard-Flanders and Boyce, 1966), prophage induction (Hertman and Luria, 1967), septum formation (Green, Greenberg and Donch, 1969), mutagenesis (Miura and Tomizawa, 1970), respiration (Swenson and Schenley, 1974), and ribosomal function (Powell and Emmerson, 1977).

The expression of the temperature sensitive mutation *tif-1* which maps at the *recA* locus (Castellazzi et al. 1977) causes prophage induction, lack of septum formation, repair of UV-irradiated phage and increased mutagenesis (Castellazzi et al. 1972a; Witkin, 1974; George et al. 1975). Expression of *tif* therefore mimics the effect of radiation although no evidence of inhibition of DNA synthesis has been found in *tif* strains. The *tif* mutation is suppressed by *recA* and *lexA* (Castellazzi et al. 1972b).

The *lexA* mutation causes a phenotype similar to *recA* except that *lexA* mutants are more or less recombination proficient (Moody et al. 1973) and do not degrade their DNA spontaneously (Mount et al. 1972). The *tsl* and *spr* mutations at the *lexA* locus are suppressors of the *lexA* phenotype (Mount et al. 1973; Mount, 1977). The phenotypes of *tsl* and *spr* mutants are similar to that of *tif-1*. Temperature sensitive *tsl* mutants are almost as UV-resistant as *lexA*[+] strains, although they continue to show abnormally low levels of mutagenesis following irradiation. The *spr* mutation results in constitutive expression of the phage induction and error-prone repair pathways without any inducing

A.S.I. Participant

treatment. Neither *spr* nor *tsl* are suppressed by *recA* except for the lambda induction pathway.

A protein of 40,000 molecular weight, protein X, is induced by DNA damage in a *recA⁺* cell and by temperature step up in a *tif* cell (Gudas and Pardee, 1975; West and Emmerson, 1977). Neither *recA* nor *lexA* strains induce protein X (Inouye, 1971; Gudas and Pardee, 1975). In *tsl lexA recA* strains there is low constitutive protein X synthesis at 30° and induction of protein X at 42°. Gudas and Pardee (1975) described a model for the control of protein X synthesis in which it was proposed that *tif* is the operator site for the gene coding for protein X and that the *lexA* and *recA* gene products respectively act as repressor and anti-repressor proteins on this operator site.

The two-dimensional gel electrophoretic technique of O'Farrell (1975) separates proteins by isoelectric point (pI) in the first dimension and by molecular weight in the second. This technique has been used to further study the induction of protein X (West and Emmerson, 1977; Emmerson and West, 1977). The results of this study are summarised below.

1.   Protein X produced in a wild-type *E. coli* after inducing treatment has a molecular weight of 40 kD and a pI of 6.0 (Fig. 1A).

2.   No visible protein X is produced in *recA* mutants.

3.   Protein X produced in the *tif-1* mutant at 42° has a molecular weight of 40 kD but a pI of 6.2.

4.   At 30° there is low constitutive synthesis of protein X of pI 6.2 in the *tif-1* mutant and synthesis is increased by inducing treatment even when the *tif* phenotype is suppressed (Fig. 1B). No protein X is produced in *tif recA* strains.

These results indicate that *tif-1* is a mis-sense mutation in the gene coding for protein X. Since genetic analysis of *tif-1* and a variety of *recA* mutations indicates that *tif-1* is a mutation of the *recA* gene (Castellazzi et al. 1977) it appears that protein X is a *recA⁺* gene product. The *recA⁺* gene product has recently been studied (McEntee et al. 1976) by labelling proteins synthesized in heavily UV-irradiated cells after infection with the transducing phage *λprecA⁺*. The *recA⁺* product was identified as a protein of molecular weight 43 kD which was synthesized in large amounts. We have repeated this experiment and examined the labelled proteins by the two-dimensional gel electrophoresis technique (Fig. 2). There is a prominent protein present in the *λprecA⁺*-infected cells which is absent in *λcI857*-infected cells.

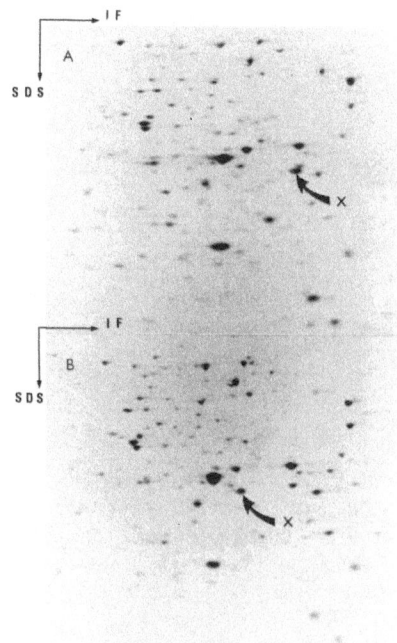

Figure 1. *Two dimensional gel electrophoresis of total cell proteins from (A) JM1 tif⁺ rec⁺ and (B) JM12 tif-1 rec⁺ following mitomycin C (MMC) treatment. Cells were grown to mid-log phase at 30° in low sulphate medium supplemented with 100 μg/ml guanosine and cytidine to prohibit tif expression. Mitomycin C (10 μg/ml) was added and the cells were then labelled with ³⁵S(100 μCi/ml) for 30 min. Labelled cells were collected by centrifugation, washed twice with 0.01 M tris-HC1, pH 6.8 and resuspended in 0.2 ml lysis buffer containing 9.5 M urea, 2% NP-40, 5% mercaptoethanol and 2% ampholines (1.6% pH range 5-8 and 0.4% pH range 3-10). Cells were lysed by three cycles of freezing and thawing and DNAse was added to a concentration of 50 μg/ml. Proteins were separated by the two-dimensional gel electrophoretic technique of O'Farrell (1975). This figure is a photograph of autoradiograms of two-dimensional 12% slab gels. Protein X is indicated on each autoradiogram.*

This protein has a molecular weight of 40,000 and a pI of 6.0. Superimposition of this autoradiogram with an autoradiogram obtained from an experiment in which wild-type cells were induced with mitomycin C confirms that this protein is protein X. McEntee et al. (personal communication) have also recently concluded that the recA⁺ product is protein X.

On the basis of these results a model has been formulated for the control of protein X synthesis (Emmerson and West, 1977). This model summarised in Fig. 3, accounts for the observed extent of protein X synthesis in *tif*, *recA*, *lexA*, *tsl* and *spr* mutants.

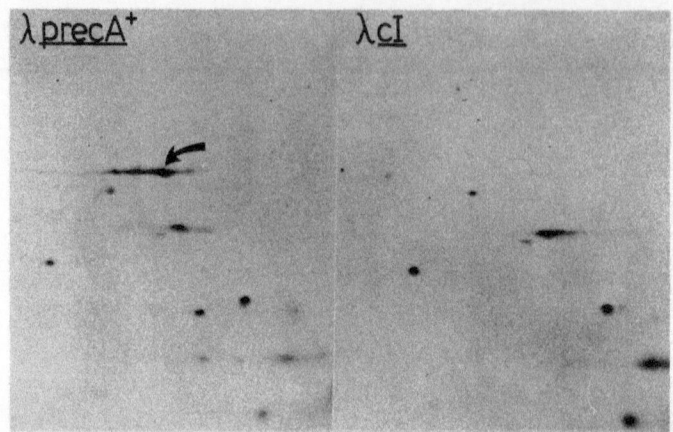

Figure 2. *Two-dimensional gel electrophoresis of* λprecA⁺- *and* λcI-*proteins*. *Strain 159* uvrA *was grown at 30⁰ in M9 medium to 2 X 10⁸ cells/ml, collected by centrifugation, and irradiated with a high dose of UV-light (2,400 J/m²). The cells were incubated at 30⁰ in M9 for 60 mins in the dark, concentrated and infected with the appropriate phage (m.o.i. of 10). Following absorption of the phage, the infected cells were incubated in M9 medium containing 5µCi/ml mixed (¹⁴C)-amino acids for 90 mins. Cell extracts were prepared and proteins separated as described in Fig. 1. This photograph shows the relevant portion of each autoradiogram. The* recA⁺ *gene product is indicated with an arrow. This protein appears in exactly the same position (molecular weight 40kD, pI 6.0) as does protein X induced in wild-type cells by mitomycin C treatment (Fig. 1A).*

In this model the *lexA⁺* gene codes for a repressor which binds to the operator region for the *recA⁺* gene. Protein X is not induced in *lexA* mutants because the defective *lexA* gene produces a super-repressor which either binds more tightly to the operator site or which cannot interact with the effector. The *tsl* and *spr* mutations, which map at the *lexA* locus, result in the production of a repressor which has a reduced affinity for the *recA* operator. In this model protein X, the *recA⁺* gene product, is involved in the control of its own synthesis. Protein X, together with an effector molecule is able to remove the *lexA⁺*-coded repressor from the operator site. The *tif* mutation results in the synthesis of a mutant form of protein X which does not require the presence of the effector to remove the repressor.

The model predicts that *tsl lexA recA* and *spr lexA recA* strains would constitutively produce a mutant protein X. We have examined one *spr lexA tif sfi recA* strain (Mount, 1977). Protein X is produced constitutively in large amounts in this strain at 30⁰ as

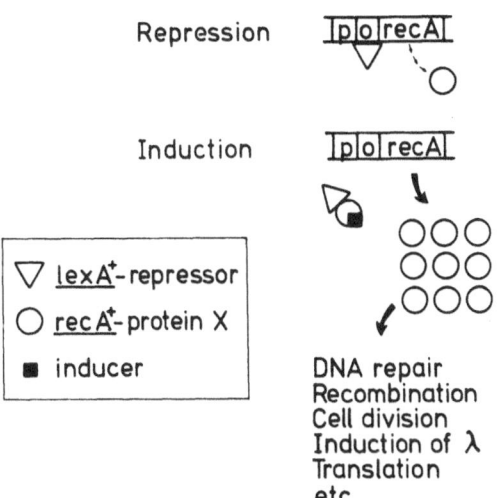

Figure 3. *Diagrammatic representation of a model to account for the control of synthesis of protein X, the recA⁺ gene product. The lexA⁺ gene codes for a repressor which binds to the operator site of the recA gene. On induction, the recA⁺-coded protein X and an unspecified effector molecule combine with the lexA⁺-coded repressor and remove it from the operator site. Thus, protein X controls its own synthesis. Removal of the repressor could occur in a number of ways (Emmerson and West, 1971). The tif-1 mutation is assumed to be in the recA gene and results in the synthesis of an altered form of protein X which at 42° is able to remove the repressor without the aid of the effector. Several important functions in which the recA⁺ product may be involved are indicated.*

predicted by the model, but it is at the pI (6.2) expected for protein X produced in *tif* mutants indicating that this particular *recA* mutation has not altered the pI of protein X (Fig. 4). However, only about one-third of mis-sense mutations result in a change in charge. Examination of large numbers of *lexA tsl recA* or *lexA spr recA* strains should reveal strains in which pI of protein X is altered by a *recA* mutation.

The model accounts for the roles of *recA* and *lexA* in protein X synthesis and might form the basis for an investigation of the many pleiotropic properties associated with *recA* and *lexA* mutants.

One explanation of the radiation sensitivity of *recA* mutants is that protein X controls DNA degradation during DNA repair (Gudas and Pardee, 1975; West and Emmerson, 1977). Protein X binds to DNA in the presence of Mg^{++} ions and appears to bind preferentially to single-stranded DNA (Gudas and Pardee, 1976). Other proteins are

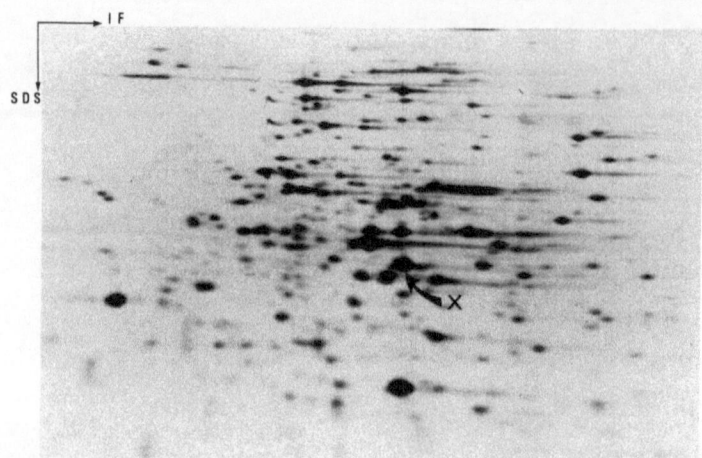

*Figure 4. Two-dimensional gel electrophoresis of total cell
proteins from DM1411 tif-1 sfiA 11 lexA3 spr-51 recA. Cells growing
exponentially at 30⁰ in low sulphate medium plus 100 µg/ml
guanosine and cytidine were labelled with ³⁵S(100 µCi/ml) for 30
min. Proteins were prepared and separated as described in Fig. 1.
This Figure is a photograph of an autoradiogram of a two-
dimensional 12% slab gel. Protein X (indicated) is produced
constitutively in large amounts at 30⁰ in this strain despite the
recA and lexA mutations.*

known which bind to single stranded DNA. Sigal et al. (1972) have
isolated a protein of molecular weight 22,000 from *E. coli* which
binds to single-stranded DNA preferentially. In addition, Taketo
and Kuno (1972) have described a protein present only in *E. coli*
strains which had been treated to inhibit DNA synthesis (e.g., with
nalidixic acid) which masked single-stranded DNA. This protein was
found in the supernatant fraction of *E. coli*. Protein X is present
to 90% in the supernatant fraction and 10% in the membrane fraction
of *E. coli*.

     Another protein which has analagous properties to protein X
is the gene 32 protein of the phage T4. Mutants defective in the
gene 32 protein are UV-sensitive and defective in DNA repair,
recombination and replication (Alberts and Frey, 1970). The gene
32 protein can catalyse the denaturation of DNA and stabilize
single-stranded regions of DNA by binding to single-stranded DNA.
Gene 32 protein synthesis is also stimulated by UV-irradiation
(Krisch and Van Houwe, 1976). However, both the DNA-binding
protein of Sigal *et al.* (1972) and gene 32 protein bind more
strongly to single-stranded DNA than does protein X.

     It is tempting to suggest that protein X is involved in
recombination and repair because of its ability to bind single-

stranded regions of DNA generated in these processes and perhaps
to protect them from nuclease attack.

Little is presently known about the initial molecular events
in genetic recombination. Preliminary experiments suggest that
protein X is not induced by the entry of Hfr DNA into *rec*+ cells
(Powell and Emmerson, unpublished results). Moreover, the basal
level of protein X in uninduced cells is very low and may be
insufficient to account for a role in recombination similar to that
ascribed to the phage T4 gene 32 product which is required
stoichiometrically rather than catalytically (Sinha and Snustad,
1971).

A possible role for protein X in DNA replication is suggested
by the inviability of *polA recA* double mutants (Gross et al. 1971).
Double mutants *polA12 recA*, which have a temperature sensitive
polA mutation, are viable at 30° but not at 42° (Monk and Kinross,
1972). At 42° these double mutants filament and show extensive
DNA breakdown (Monk et al. 1972). Recent experiments reveal
somewhat higher than normal levels of protein X in *polA1* mutants
in the absence of any inducing treatment (Botcherby and Emmerson,
unpublished results). Thus the viability of *polA1* cells may be
contingent upon their ability to synthesize protein X. In this
connection it is noteworthy that *tif-1* is one example of a *recA*
mutation (Castellazzi et al. 1977) which makes the cell non-viable
at 42°. It is possible that both overproduction and under-
production of protein X can be harmful to the cell under certain
circumstances. In wild-type cells a mechanism may exist which
limits the production of protein X. In our model production of X
would stop when the supply of the effector was reduced, for
example when DNA repair was complete. In *tif-1* and *tsl lexA*
mutants at 42°, where synthesis of X is independent of the
effector, uncontrolled X production could lead to filamentation
and cell death. For some, as yet unexplained reason, over-
production of X does not lead to filamentation and cell death if
the cell is also mutant at *sfiA* or *sfiB* (George et al. 1975; West
and Emmerson, 1977).

Protein X may have two different modes of action depending on
its concentration. At low concentrations it may have a role in
genetic recombination and removal of the *lexA*+ repressor and at
higher concentrations it may control DNA degradation, possibly by
binding to single-stranded DNA.

Of all *recA*+-dependent processes, perhaps the best understood
is the induction of prophage lambda. Induction of prophage lambda
by temperature step up in lysogens of *tif-1* mutants proceeds in
the presence of chloramphenicol (West, Powell and Emmerson, 1975)
and results in cleavage of the repressor (Roberts and Roberts, 1975;

Shinagawa et al. 1977). Thus the low constitutive level of protein X present in *tif* strains at 30° appears to be sufficient for repressor inactivation. The constitutive level of protein X therefore appears to be sufficient for the inactivation of both the *lexA* and lambda repressor proteins. One explanation of these results is that protein X is a protease which can cleave several different repressor molecules.

REFERENCES

Alberts, B. and Frey, L. (1970)   Nature 227:1313-1318

Castellazzi, M., George, J. and Buttin, G. (1972a) Molec. gen. Genet. 119:139-152

Castellazzi, M., George, J. and Buttin, G. (1972b) Molec. gen. Genet. 119:153-174

Castellazzi, M., Morand, P., George, J. and Buttin, G. (1977) Molec. gen. Genet.

Clark, A.J. and Margulies, A.D. (1965) Proc. Nat. Acad. Sci. (Wash.) 53:451-459

Emmerson, P.T. and West, S.C. (1977) Molec. gen. Genet.

George, J.M., Castellazzi, M. and Buttin, G. (1975) Molec. gen. Genet. 140:309-332

Green, M.H.L., Greenberg, J. and Donch, J. (1969) Genet. res. Camb. 14:159-162

Gross, J.D., Grunstein, J. and Witkin, E.M. (1971) J. Mol. Biol. 58:631

Gudas, L.J. and Pardee, A.B. (1975) Proc. Nat. Acad. Sci. (Wash.) 72:2330-2334

Gudas, L.J. and Pardee, A.B. (1976) J. Mol. Biol. 101:459-478

Hertman, I. and Luria, S.E. (1967) J. Mol. Biol. 23:117-123

Howard-Flanders, P. and Boyce, R.P. (1966) Radiat. Res. 6: (Supple.) 156-163

Inouye, M. (1971) Bacteriol. 106:539-542

Inouye, M. and Pardee, A.B. (1970) J. Biol. Chem. 245:5813-5819

Krisch, H.M. and Van Houwe, G. (1976) J. Mol. Biol. 108: 67-81

McEntee, K., Hesse, J.E. and Epstein, W. (1976) Proc. Nat. Acad. Sci., (Wash.) 74:300-304

Miura, A. and Tomizawa, J. (1969) Molec. gen. Genet. 103:1

Monk, M. and Kinross, J. (1972) J. Bacteriol. 109:971-978

Moody, E.E.M., Low, K.B. and Mount, D.W. (1973) Molec. gen. Genet. 121:197-205

Mount, D.W. (1977) Proc. Nat. Acad. Sci. (Wash.) 74:300-304

Mount, D.W., Kosel, C. and Walker, A. (1976) Molec. gen. Genet. 146:37-42

Mount, D.W., Low, K.B. and Edmiston, S.J. (1972) J. Bacteriol. 112:886-893

O'Farrell, P.H. (1975) J. Biol. Chem. 250:4007-4021

Powell, K.A. and Emmerson, P.T. (1977) Molec. gen. Genet. 154: 83-86

Roberts, J.W. and Roberts, C.W. (1975) Proc. Nat. Acad. Sci., (Wash.) 72:147-151

Shinagawa, H., Mizuuchi, K. and Emmerson, P.T. (1977) Molec. gen. Genet.

Sigal, N., Delius, H., Kornberg, T., Gefter, M.L. and Alberts, B. (1972) Proc. Nat. Acad. Sci. (Wash.) 69:3537-3541

Sinha, N.K. and Snustad, D.P. (1971) J. Mol. Biol. 62:267

Swenson, P.A. and Schenley, R.L. (1974) Int. J. Radiat. Biol. 25:51-60

Taketo, A. and Kuno, S. (1972) J. Biochem. 71:497-505

West, S.C. and Emmerson, P.T. (1977) Molec. gen. Genet. 151:57-67

West, S.C., Powell, K.A. and Emmerson, P.T. (1975) Molec. gen. Genet. 141:1-8

Witkin, E.M. (1974) Proc. Nat. Acad. Sci. (Wash.) 71:1930-1934

UV ENDONUCLEOLYTIC ACTIVITY IN EXTRACTS FROM MOUSE CELLS

Ingolf F. Nes[*] and Jon Nissen-Meyer

Department of Biochemistry
University of Bergen
Arstadveien 19
5000 Bergen, Norway

INTRODUCTION

The excision repair system removes lesions from the DNA molecule. The initial step in this process is the action of an endonucleolytic enzyme which recognizes the lesion and subsequently cuts the DNA at, or close to, the damaged site.

A number of endonucleases exist, both in prokaryotes and eukaryotes, which are more or less specific for different types of lesions. The most studied UV repair endonuclease is the pyrimidine dimer-specific enzyme of the bacterial and phage systems (1). The activity of an endonuclease with similar specificity has also been reported in cell free extracts of mammalian cells (2), but the enzyme has not been purified. Mammalian endonucleases that recognize UV induced lesions of unknown nature have, however, been purified and extensively characterized (3,4,5).

One way to elucidate the multiple facets of DNA repair is to study cells which show defects in the repair mechanism. Studies of human fibroblasts from patients with xeroderma pigmentosum, a genetic disease which affects the DNA repair, have exposed a surprising degree of complexity in the repair mechanism (6). So far, five complementing defects have been found (7).

Rodent cell lines permanently established in cell culture, have been shown to be defective in the "short patches" UV excision repair system, i.e., the cell does not excise pyrimidine dimers into acid soluble form (8). A defect in the incision or excision step

*A.S.I. Participant

of the repair process is the most obvious explanation for this
observation.

In this communication we report the presence of UV endonucleolytic activity in extracts from rodent cell lines permanently
established in cell culture, and partial characterization of this
activity in crude extracts.

RESULTS AND DISCUSSION

Detection of UV Endonucleolytic Activity in Mouse Cell Extracts

Mouse plasmacytoma cells (line MPC-11) and L cells (line 929)
were grown in suspension culture as previously described (10,11).
Extracts were prepared from these cells as schematically shown in
Fig. 1. UV endonucleolytic activity was assayed as described in
the legend to Fig. 2. Significant amounts of UV endonucleolytic
activity were found in nuclear extracts from both cell lines (Fig.
2). The number of UV specific nicks introduced per DNA molecule by
L 929 and MPC-11 cell extracts were 0.38 and 0.60 respectively.
The specificity of the endonuclease towards UV induced lesions is
further substantiated by the linear relationship between frequency
of DNA strand breaks and UV dose (Fig. 3). The data in this dose-
response curve also suggest a higher level of UV endonucleolytic
activity in MPC-11 than L cells.

The primary lesion induced by UV radiation in DNA at 254 nm,
is the pyrimidine dimer. If the distortion in the DNA is due to
the dimer and this is the recognition site for the enzyme, then it
only makes incisions adjacent to 0.5 - 1.0% of all dimers present.
This is a very low efficiency, and it may indicate that the UV
enzymic activity is induced by another lesion.

Levels of UV Endonucleolytic Activity in Mouse Cells in Relation
to that in HeLa and Vero Cells

It is known that mouse cells in culture remove very few, if
any, pyrimidine dimers by an excision repair mechanism which
releases the dimers into an acid soluble form. As indicated above,
this could be due to an enzymic defect in either the incision or
the excision step of the repair process.

A study was made to determine if this defect resulted in lower
UV endonucleolytic activity in extracts from mouse cells as compared
to those from cells proficient in excision repair. Moreover, a more
thorough investigation was performed in order to determine whether
or not there was a significant difference between UV endonucleolytic
activities in the nuclear extracts of L cells and MPC-11 cells. UV
endonucleolytic activities in extracts from L, MPC-11, HeLa and

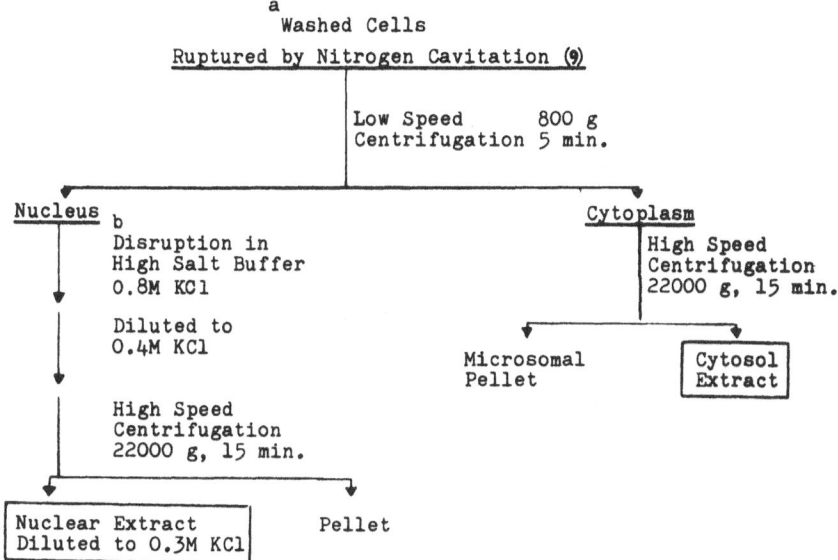

Figure 1.   Preparation scheme for Nuclear and cytosol extracts.
[a]*The cells were resuspended in 10mM Tris HCl pH = 7.3, 0.5 mM CaCl₂, 7.5 mM MgCl₂, 25 mM KCl, 10 mM mercaptoethanol and 7.5% sucrose.*
[b]*The nucleus fraction was sesuspended in a buffer containing 10 mM Tris HCl ph = 7.3, 1 mM EDTA, 10 mM mercaptoethanol, 7.5% sucrose and the given KCl concentration from the scheme.*

Vero cells were compared (HeLa and Vero cells were generously provided by Dr. Gunnar Haukenes, Bergen). HeLa cells were chosen as a control because they excise dimers into an acid soluble form (2), and extracts obtained from them have been reported to contain relatively large amounts of UV endonucleolytic activity (15).

The main conclusion from the data presented in Table 1 is that the mouse cell lines do not have less UV endonucleolytic activity than the other two cell lines. However, one should have in mind that this does not rule out the existence of other UV endonucleolytic activities which require different reaction conditions, or

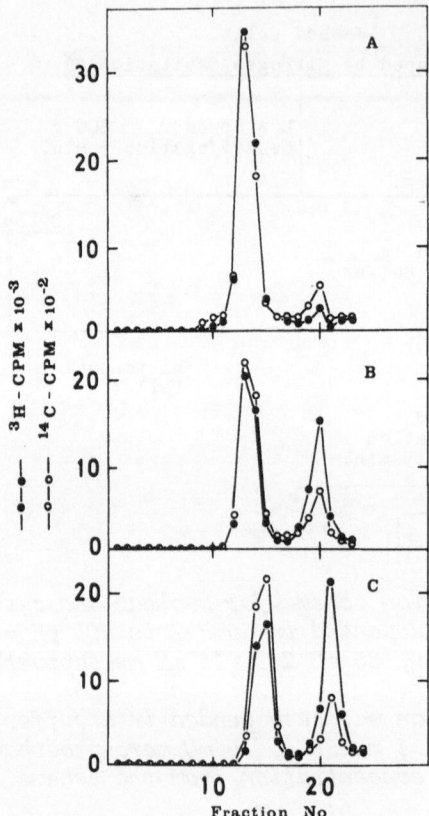

Figure 2. *Alkaline sucrose gradient centrifugation of UV irradiated*
^{3}H-DNA and nonirradiated ^{14}C-DNA, (RFI ØX$_{174}$ DNA) after incubation
with nuclear extracts for 15 min at 37°C.

A.   Control, no extract

B.   60 μg protein from L cells nuclear extract

C.   35 μg protein from MPC-11 cells nuclear extract

The assay contains 10 mM Tris HCl pH 8.0, 100 mM KCl, 3mM EDTA,
10 mM mercaptoethanol, 2.6 μg UV irradiated (10.000 erg/mm^2) ^{3}H-RFI
DNA (72.000 cp,), 5.2 μg ^{14}C-RFI DNA (6.200 cpm) and cell extract
in 100 μl reaction volume.

which exist in very low concentration in the cell. A second point
to be made is that the MPC-11 cells contain 2-3 times more of the
assayed activity than the other cells. The significance of this
observation is not known, but the MPC-11 cell line may be useful
for further studies.

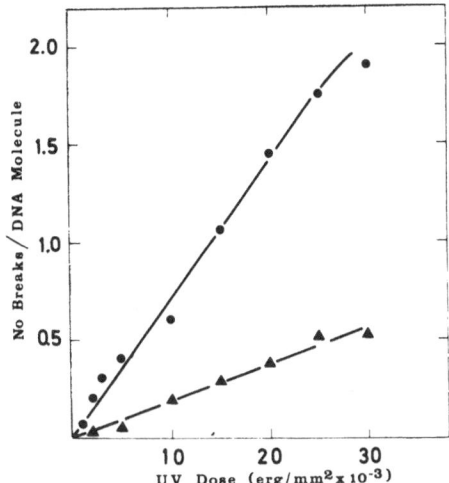

Figure 3. *Endonuclease susceptible sites as a function of UV dose.*

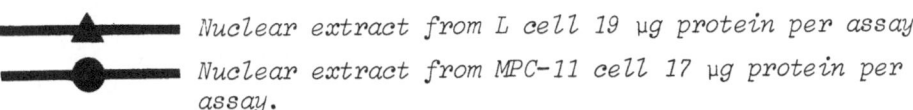

Nuclear extract from L cell 19 μg protein per assay

Nuclear extract from MPC-11 cell 17 μg protein per assay.

Assay conditions, 10 mM Tris HCl pH 8.0, 100 mM KCl, 3 mM EDTA, 10 mM mercaptoethanol, 0.25 μg UV irradiated RFI ØX174 ³H-DNA (6.800 cpm) and nuclear extract in 100 μl reaction volume. The reaction was incubated at 37°C for 15 min. The convertion of the DNA from RFI to RFII by nicking was detected by the nitrocellulose filter technique (14). Percentage of total radioactivity which bound to the filter was adjusted by a Poisson distribution to obtain the average number of nicks introduced per molecule (17). Non-enzymatic and unspecific endonuclease breaks have been subtracted.

The Main Portion of UV Endonuclease Activity in Mouse Cell Lines is Located in the Cell Nucleus

One would expect to find the DNA metabolizing enzymes to be localized in the cell nucleus. Despite this, different deoxy-ribonucleases have been found both in the cytoplasm and in the nucleus (12). Initially the α-DNA polymerase was found in the cytoplasm (13). Recently, however, it has been shown that this enzyme is present mainly in the nucleus (13) and that the extra-nuclear localization initially observed was probably due to leakage during the preparation. This prompted us to make an evaluation of the amount of extranuclear UV endonuclease activity found when isolating the nuclei.

Table 1

The total amount of UV endonucleolytic activity and protein in cytosol and nuclear extracts in four different mammalian cell lines*.

Cell Line	Enzymatic Activity (Units)**	Total Protein (mg)
HeLa	66	36
Vero	51	46
L	60	34
MPC-11	152	28

* The UV endonuclease assay is described in legend to Fig. 3.

** One UV endonuclease unit produces 1 pmole of nicks per 15 min when the DNA is irradiated with a UV dose of 10000 erg/mm^2.

Table II

The percentage distribution of UV endonuclease activity and protein between the nucleus and cytosol in MPC-11 and L Cells*.

	MPC-11 cells			L cells		
	Activity (%)	Protein (%)	(units/mg) Specific Activity	Activity (%)	Protein (%)	(units/mg) Specific Activity
Nucleus	60	36	9.2	70	40	3.2
Cytosol	40	64	3.1	30	60	0.9

* The UV endonuclease assay is described in legend to Fig. 3.

Cell free extracts were prepared according to the scheme in Fig. 1. The total enzymatic activity in the cytosol and nuclear extracts were quantitated and the results are presented in Table II. Data from both cell lines support the notion of an nuclear localization of the UV endonucleolytic activity. Nuclear extracts from MPC-11 cells and L cells contained 60% and 70% respectively of the total UV endonuclease activity observed. The specific activities of the enzyme in the two fractions (Table II) substantiate the view that the nucleus is the origin of the UV endonucleolytic activity. Moreover, from earlier observations we know that DNA inhibits the UV endonucleolytic activity. By removing cellular nucleic acid from the extracts by DEAE cellulose chromatography, the UV endonuclease activity in the nuclear extract increased 2-3 fold, while the activity in the cytosol increased only 1-2 fold. Consequently, the actual enzyme distribution is probably more in favour of a nuclear localization than that which is observed in the crude extracts according to Table II. There is a significant difference in MPC-11 and L cells in the distribution of activity between the cytosol and nuclear extract. This is most probably due to a more fragile nucleus in the MPC-11 cell (unpublished observation).

We suggest that the UV endonucleolytic activity in the cytosol extract is mainly due to leakage, and/or partial disruption of the nuclei during their isolation. This notion is supported by the fact that washing the nuclei in isotonic buffer reduced the enzymic level in the nuclei. Three washes lowered the UV endonucleolytic activity by about 50%, and the lost activity was recovered in the washing buffer.

Factors Effecting UV Endonuclease Activity
in Extracts from MPC-11 Cells

As was pointed out previously, crude extracts may contain repair endonucleases which are not detected because the proper assay conditions are not used. Moreover, the activity one does detect under a set of conditions may not necessarily be due to the endonuclease of major importance in the *in vivo* UV repair mechanism. Crude extracts most likely contain a number of endonucleases which recognize UV induced lesions and vary in their reaction requirements. Consequently, it is important to characterize how variations in reaction conditions may influence the UV endonucleolytic activity in crude extracts. The effect of KCl, NaCl and $MgCl_2$ on the UV endonucleolytic activity in cytosol and nuclear extracts from MPC-11 cells is shown in Table III. The data indicate that the same enzyme(s) is responsible for the activities in the nuclear and cytosol extracts, since the activities respond similarly to variations in the reaction conditions.

Table III

The effect of KCl, NaCl and MgCl$_2$ on the UV endonucleolytic Activity
in cytosol and nuclear extracts from MPC-11 cells

	% UV Endonuclease Activity			
	*Standard Reaction Mixture	-KCl	-KCl +0.1M NaCl	+ 10 mM MgCl$_2$
Nuclear Extract	100	25	100	195
Cytosol Extract	100	35	90	190

*Standard Reaction Mixture as described in the legend to Fig. 3.

NaCl and KCl show a similar stimulatory effect on the UV
endonucleolytic activity (Table III). Both salts also inhibit most
of the unspecific endonuclease activity present in the extract,
resulting in 90% of all endonuclease activity being specific for
UV irradiated DNA (data not shown). The reaction mixtures most
commonly used when assaying for UV endonucleolytic activity contain
little if any salt. In fact, the UV endonuclease purified from
calf thymus is inhibited by NaCl concentrations above 50 mM (4).

The presence of magnesium results in nearly a doubling of the
UV endonuclease activity (Table III). However, from the dose-
response curve (Fig. 4), it is seen that this stimulatory effect
is most pronounced at high UV doses. The effect of magnesium is
further complicated by the fact that it was found to be dependent
on the protein concentration. If seven times more protein from the
extract was used in the assay no stimulation of the UV endonucleolytic activity was observed.   The increase in UV endonucleolytic activity due to magnesium may result from the
stimulation of the same enzyme(s) responsible for the endonuclease
activity observed in the absence of magnesium, or it may be due to
a new UV endonuclease which requires magnesium. The purification
of these activities, which we are attempting at the present time,
should resolve this question.

In view of reports that the *uvr*-endonuclease in *E. coli* is ATP
dependent (18,19,20), we also studied the effect of ATP on the UV
endonucleolytic activity in the crude extracts from mouse cells.
Preliminary results seem to suggest that ATP may stimulate the UV
endonucleolytic activity in the presence of magnesium. However, the

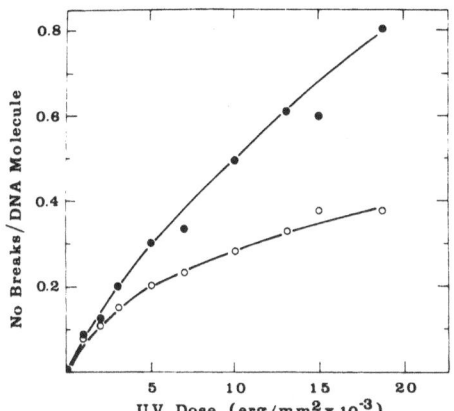

Figure 4. *Endonuclease susceptible sites as a function of UV dose.*
———O——— *Standard assay. See legend to Fig. 3.*
———●——— *Standard assay plus 10 mM MgCl$_2$.*
In both experiments 2.4 µg protein from MPC-11 nuclear extract,
was used.

data are uncertain because an ATP dependent "non-specific" endonuclease, which nicks both irradiated and unirradiated DNA,
interferes when assaying for the UV-specific activity. We are now
attempting to circumvent this problem by assaying for ATP dependent
UV endonucleolytic activity in partially purified extracts where
the unspecific ATP dependent endonuclease is not present.

ACKNOWLEDGEMENTS

        We are grateful to Professor K. Kleppe who initiated this work,
and to Dr. I.F. Pryme for help in preparing this manuscript. The
authors are Fellows of the Norwegian Cancer Society and this work
is supported by a grant from the same institution.

REFERENCES

(1) Grossman, L., Braun, A., Feldberg, R. and Mahler, I. (1975)
     Ann. Rev. Biochem. 44:19-43

(2) Duncan, J., Slor, H., Cook, K. and Friedberg, E.C. (1975)
     In: Molecular Mechanisms for Repair in DNA (eds. Hanawalt, P.
     C. and Setlow, R.B.) Basic Life Sciences Vol 5B, Plenum Press,
     New York, pp643-649

(3) van Lancker, J.L., Tomura, T. (1974) Biochim. Biophys. Acta
     353:99-114

(4) Bacchetti, S., Benni, R. (1975) Biochem. Biophys. Acta
     390:285-297

(5) Brent, T.P. (1976) Biochem. Biophys. Acta 454:172-183

(6) Robins, J.H. et al. (1974) Annals of Internal Medicine 80(2):
     221-245

(7) Kraemer, K.H. et al. (1975) Mut. Res. 33:327-340

(8) Ben-Ishai and Deleg, L. (1975) In: Molecular Mechanisms for
     Repair of DNA (eds. Hanawalt, P.C. and Setlow, R.B.) Plenum
     Press, New York, pp607-610

(9) Hunter, M.J. and Commerford, S.L. (1961) Biochem. Biophys.
     Acta 47:580-586

(10) Nissen-Meyer, J. and Eikhom, T.S. (1976) J. Mol. Biol. 101:
     211-221

(11) Nissen-Meyer, J. and Eikhom, T.S. I1976) Exptl. Cell Res.
     98:41-46

(12) Zøller, E.J. et al. (1975) Z. Naturforsch 30:781-784

(13) Foster, D.N. and Gurney, J.T. (1976) J. Biol. Chem. 251:7893-
     7898

(14) Braun, A. and Grossman, L. (1974) Proc. Nat. Acad. Sci. U.S.A.
     71:1838-1842

(15) Duker, N.J. and Teebor, G.W. (1975) Nature 255:82-84

(16) Lindahl, T. and Andersson, A. (1972) Biochem. 19:3618-3623

(17) Vinograd, J., Lebowitz, J., Radloff, R., Watson, R. and
     Laipis, P. (1965) Proc. Nat. Acad. Sci. U.S.A. 53:1104-1111

(18) Waldstein, E.A., Sharon, R. and Ben-Ishai, R. (1974) Proc. Nat.
     Acad. Sci. U.S.A.

(19) Seeberg, E. and Strike, P. (1975) J. Bact. 125:787-795

(20) Seeberg, E., Nissen-Meyer, J. and Strike, P. (1976) Nature,
     263:524-526

ON THE MECHANISM OF ULTRAVIOLET-INDUCED MUTAGENESIS

Giuseppe Villani[*], Serge Boiteux, Perrine Caillet-Fauquet,
Martine Defais[*], Giovanni Sponza[*], Silvio Spadari[**] and
Miroslav Radman

Département de Biologie Moléculaire
Université Libre de Bruxelles
B-1640, Rhode St. Genèse, Belgium

The purpose of this paper is to introduce a new approach to
the study of the mechanisms of induced mutagenesis. Following a
brief description of a series of experimental observations a
molecular mechanism is presented to account for the phenomenology
of UV-induced mutagenesis.

Genetics of Induced Mutagenesis

In relation to the mechanism of their mutagenic action, mutagens can be divided in two groups:

(a) Direct mutagens, which cause subtle modifications of DNA bases
in the template and/or in the precursors, do not inhibit DNA
synthesis and their mutagenic effect is probably caused by direct
mispairing (see mispairing schemes in the ref.1). Examples of such
mutagens are deaminating agents like hydroxylamine and bisulfite,
base analogs like 2-aminopurine and alkylating agents producing
O^6-alkyl-guanine such as ethylmethane sulfonate. No mutants are
known which eliminate mutagenesis induced by these agents, suggesting
little contribution of the cellular metabolism to this type of
mutagenesis.

(b) Indirect mutagens, which cause bulky DNA lesions and thus
provoke strong inhibition of DNA synthesis. Indirect mutagens

[*] *A.S.I. Participant*

[**] *Present address: Laboratorio di Genetica Biochimica ed
Evoluzionistica, C.N.R., via S. Epifanio 14, 27100, Pavia, Italy.*

require and influence the expression of specific cellular genes in
order to accomplish mutagenesis in *E.coli* (3,8,9). Examples of
such mutagens are ultraviolet and ionizing radiations and many
chemical mutagens and carcinogens such as mitomycin C, benzo(a)-
pyrene, aflatoxin B$_1$, etc (see 2). Considerable evidence (3) leads
to the conclusion that non coding-DNA lesions induce mutagenesis
in *E.coli* through the induction of new gene functions. This
requirement is clearly demonstrated by studies involving bacteriophage mutagenesis. UV-irradiation of λ and ØX174 results in killing
but no mutagenesis of phage following infection of unirradiated host
cells (4,5), whereas UV-irradiation of the host cells causes mutagenesis of both unirradiated and irradiated phage (6,7). This
induced mutagenic capacity, which also contributes to the survival
of damaged phage, is turned on following damage to DNA and then off
again once DNA synthesis is restored (6). Chloramphenicol inhibits
the expression of the phenomenon (6) thus indicating the necessity
for *de novo* protein synthesis.

A Biochemical Approach to UV-Induced Mutagenesis:
In Vivo Studies

In an attempt to study the biochemical mechanism of inducible
mutagenic DNA repair, ("SOS repair", 8,9,3), we took advantage of
the fact that the single stranded ØX174 DNA is subject to SOS
repair (10,5). Since neither excision nor recombination repair can
act upon single stranded DNA molecules, these two known repair
mechanisms are unlikely to be responsible for SOS repair. Furthermore, genetic evidence suggests that UV-induced mutations in ØX174
(which can be produced *only* if the host cell is irradiated as well)
are fixed in the first round of DNA synthesis, i.e., in the passage
from the single-stranded to the double-stranded form of ØX174 DNA
(ref.5). Synthesis of the full complement to the viral strand is
indispensable for the development of ØX174 phage, since this strand
is used as template for transcription and for rolling circle
replication. Therefore, the principal cause of killing ØX174 phage
by UV light may be the failure to synthesize a complete, uninterr-
upted strand complementary to the viral strand. The principal aim
of this study was to determine the extent of *in vivo* DNA synthesis
on viral DNA templates containing about five pyrimidine dimers, in
intact and in UV-irradiated host cells. Only parental phage DNA
was radioactively labeled, and in order to prevent the loss of
double-stranded ØX174 DNA due to rolling circle replication, ØX174
*A*am18 was used in HF4704, as suppressor-less host. Following
infection, intracellular ØX174 DNA was extracted at several time
intervals and subjected to the following three different analyses:
(a) isopycnic CsCl density gradient, (b) hydroxylapatite chromatography, (c) digestion by single stranded specific endonuclease S$_1$,
all aimed at providing a measure of the extent of DNA synthesis in
the first round of replication.

All three approaches used in this work clearly show that UV induced pyrimidine dimers block DNA synthesis on the viral strand. In addition, the size distribution of the double stranded fragments suggests that the chain elongation is arrested at the first pyrimidine dimer encountered. On the contrary, when host cells are irradiated and incubated to fully induce the error-prone repair system, a significant fraction of irradiated ØX174 molecules can be fully replicated (for results and experimental details see ref 11).

How could some irradiated ØX174 DNA molecules fully replicate in irradiated hosts ? It is hard to conceive of a mechanism for elimination of pyrimidine dimers (or other photolesions) from single stranded DNA, other than monomerization, as in photoreactivation. The latter process would be expected to lead to an increase in survival but not an increase in mutagenesis, which is a prominent feature of SOS repair.

In any case, our experiments were performed in the absence of exposure to light to avoid this possibility. Also, the host cells were deficient in excision repair.

A working hypothesis to account for our results and to propose a molecular mechanism for the induced mutagenesis of ØX174 phage is shown schematically in Fig. 1. DNA lesions, presumably pyrimidine dimers in this case, block the elongation of growing DNA chains (Fig. 1, molecule b). The presence of such replication-inhibiting DNA lesions in host DNA somehow turns-on the cellular SOS repair system (see schema in Fig. 5). This process requires *de novo* protein synthesis and one or more products allow the DNA replication machinery to proceed across pyrimidine dimers (Fig. 1, molecule c). We suggest that this process involves frequent incorporation of noncomplementary bases opposite pyrimidine dimers in the template strand (Fig. 1, molecule c). If both residues of the pyrmidine dimer have lost their capacity for Watson-Crick base pairing, as many as half of UV-induced mutations could arise as simultaneous changes of two neighbouring bases opposite a pyrimidine dimer (molecule c, double dots). In our hypothesis, erroneous DNA synthesis is supposed to make occasional errors also in undamaged regions of DNA (Fig. 1, molecule c, single dots).

A Biochemical Approach to UV-Induced Mutagenesis:
In Vitro Studies

(a) Inhibition of *in vitro* DNA synthesis by *E.coli* DNA polymerases I and III on UV-irradiated ØX174 DNA. Single stranded ØX174 phage DNA was primed by annealed fragments of ØX174 RF-I DNA. Up to 100 percent of the input non-irradiated ØX174 DNA template can be replicated in polymerization reactions by cell free extracts from *E.coli pol+* (Fig. 2a), *polA1-* (Fig. 2b) or by purified *E.coli* DNA

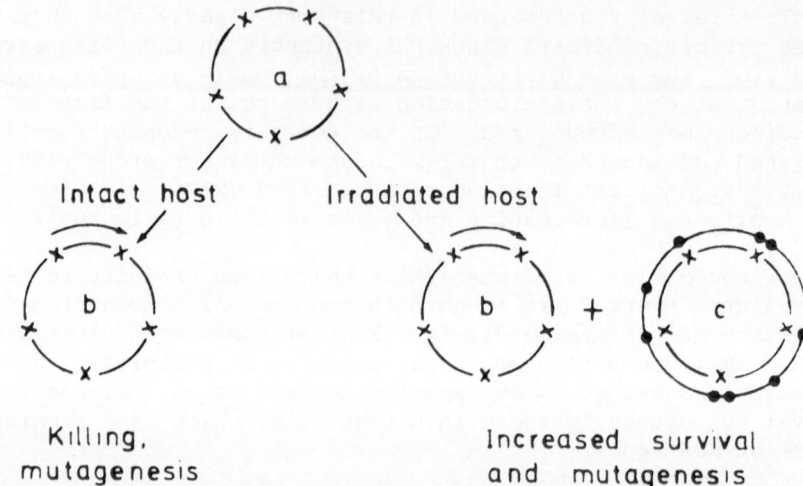

Intact host   Irradiated host

Killing,   Increased   survival
no mutagenesis   and   mutagenesis

*Figure 1. A scheme illustrating the mechanism of ultraviolet-
          induced mutagenesis and of SOS repair of ØX174 phage DNA.*

*Inducible mutagenic DNA repair (SOS repair) is interpreted here as
an error-prone DNA synthesis, which permits replication of UV
irradiated ØX174 DNA (i.e., survival) by inserting noncomplementary
nucleotides (—●—) opposite pyrimidine dimers (—✗—) and also
elsewhere in the DNA (mutagenesis). (For further explanations see
text).*

polymerase I (Fig. 2d) or its large fragment (Fig. 2c). In agreement
with previous studies (ref. 12,13,14), all four polymerization
reactions were similarly inhibited by UV-irradiation of ØX174 DNA
template (Fig. 2a,b,c,d). The dose response of the inhibition of
DNA synthesis suggests that each pyrimidine dimer is both an
absolute block to polynucleotide chain elongation *in vitro* and a
lethal hit *in vivo* as was previously shown by *in vivo* studies (11).
This conclusion is supported by the finding that *in vitro* monomerization of pyrimidine dimers by photoreactivation enzyme can
restore approximately 25% of the template activity (Fig. 2a). The
probable reason that the restoration of template activity was not
more efficient is that only one of the four possible isomers of
pyrimidine dimers induced by UV irradiation of the single stranded
DNA can be photoreactivated (16).

(b)   Fidelity of DNA synthesis on UV irradiated poly (dT) and poly
(dC) templates. Pyrimidine dimers are bulky structural deformations
in the DNA and are the principal UV-induced mutagenic lesions, hence
they must drastically affect hydrogen bonding properties of the
involved bases. If they were frequently copied, the overall effect

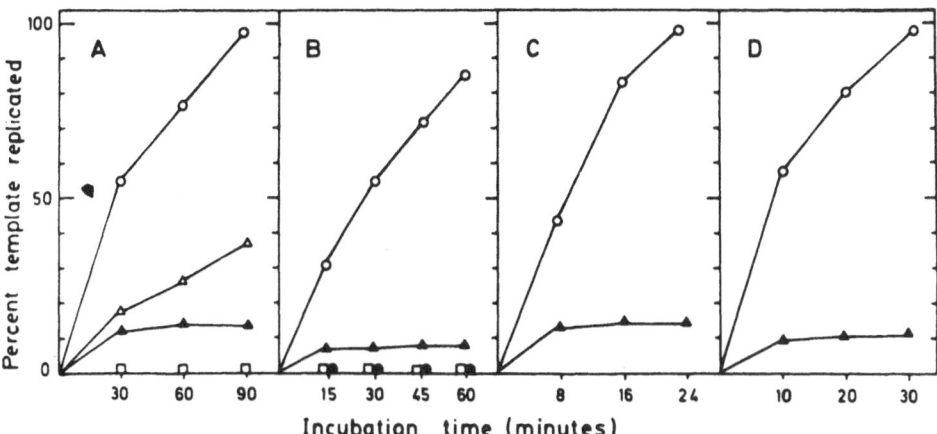

<u>Figure 2</u>.   Extent of DNA synthesis on intact and UV-irradiated
              primed ØX174 phage DNA.

Sources of DNA polymerase activity are: (A) cell-free extract of
E.coli pol⁺, (B) cell-free extract of E.coli pol⁻, (C) DNA polymerase I (large fragment) and (D) DNA polymerase I complete enzyme.
Polymerization reaction (A) is essentially carried out by DNA polymerase I and (B) by DNA polymerase III complex since reaction
conditions were optimal for this enzyme and the activity was
inhibited by 0.2 M KCl (22).   Knowing both the specific radioactivity of the ØX174 DNA template and the fraction covered by the
primer fragments, it was possible to express DNA synthesis as percent acid-insoluble template replicated.

Symbols: (O-O-O) intact template; (▲-▲-▲) UV-irradiated
template (100 J/m²); (△-△-△) UV-irradiated (100 J/m²) photoreactivated template; (□-□-□) reaction without template;
(●-●-●) reaction in the presence of 0.2 M KCl.

would be a higher rate of incorporation of incorrect bases than on
an unirradiated template. To study the fidelity of in vitro DNA
synthesis several fractions of different purity of DNA polymerases
I and III were tested with UV-irradiated and intact poly(dC) and
poly(dT) homopolymers as templates (for experimental details see
ref. 15). Very little, if any, increase in misincorporation was
observed on irradiated templates (Table I) inspite of the expected
large number of pyrimidine dimers. This result also suggests that
DNA polymerases I and III do not efficiently copy pyrimidine dimers.

Table I

Incorporation of complementary and non-complementary
nucleotides during DNA synthesis on untreated and
UV-irradiated poly(dC) and poly(dT) templates

Source of DNA Polymerase Activity (enzyme involved)	Template Primer	UV Fluence (J/m^2)	Correct Nucleo- tide	Incorrect Nucleotide	Error Frequency ($\frac{incorrect\ nucleotide}{correct\ nucleotide}$)
E.coli pol$^+$	poly(dC)oligo(dG)	0	dGTP	dATP	$\sim$ 1.4 10^{-3}
Cell Free		2500	dGTP	dATP	$\sim$ 1.0 10^{-3}
Extract	poly(dT)oligo(dA)	0	dATP	dGTP	$\sim$ 4.1 10^{-3}
(Polymerase I)		500	dATP	dGTP	$\sim$ 6.2 10^{-3}
E.coli pol$^+$	poly(dC)oligo(dG)	0	dGTP	dATP	4.5 10^{-4}
DNA cellulose		2500	dGTP	dATP	3.9 10^{-4}
Fraction	poly(dT)oligo(dA)	0	dATP	dGTP	2.0 10^{-3}
(Polymerase I)		500	dATP	dGTP	2.0 10^{-3}
Polymerase I	poly(dC)oligo(dG)	0	dGTP	dATP	4.3 10^{-4}
(Large Fragment)		2500	dGTP	dATP	6.1 10^{-4}
	poly(dT)oligo(dA)	0	dATP	dGTP	1.4 10^{-3}
		500	dATP	dGTP	2.0 10^{-3}
E.coli polA$^-$	poly(dC)oligo(dG)	0	dGTP	dATP	$\sim$ 2.1 10^{-3}
ammonium sulfate		1000	dGTP	dATP	$\sim$ 5.4 10^{-3}
fraction (Polymerase III)					

Deleting the correct nucleotide, no incorporation of
the incorrect one was detected, suggesting that terminal
labeling is not responsible for the observed misincorporation. Unlike the experiments with purified DNA
polymerases, when cell free extracts are used, high
blank values (no template added) for the incorrect
nucleotide are observed, decreasing the ratio signal/
blank to less than three. Although at least three
independent assays were performed with each cell free
extract, the indicated error frequencies are only
approximate ($\sim$). Blank values (no template added) were
subtracted from those of the complete reaction
mixtures.

(c) Possible involvement of the 3' to 5' exonuclease (proof-reading) in the blockage of DNA synthesis by UV photolesions. DNA polymerases are known to control the fidelity of DNA synthesis by selection of the correct nucleotide (17) and by possessing a "proof-reading" 3' → 5' exonuclease activity that can detect and excise 3' terminal mismatched bases (18). In the case of UV-irradiated DNA, one can assume that any base incorporated opposite pyrimidine dimers could be recognized and excised as a terminal mismatch by the 3' → 5' exonuclease activity associated with the *E.coli* DNA polymerases. This repeated excision could prevent further chain elongation, and hence account for the observed inhibition of DNA synthesis (Fig. 2). The above hypothesis results in two main predictions:

(1) The arrest of the chain elongation at a pyrimidine dimer should lead to increased production of free nucleoside monophosphates as a result of the 3' → 5' "proof-reading" exonucleolytic action. To test this prediction, both incorporation of nucleoside mono-phosphates from labeled triphosphates into primed ØX174 DNA and production of free nucleoside monophosphates from labeled tri-phosphates were followed in a single assay during synthesis of DNA polymerase I (large fragment) on unirradiated and irradiated DNA. This analysis was done with a purified DNA polymerase fraction that was free of DNA-independent triphosphatases. In addition, the large fragment of DNA polymerase I (21), missing the 5' → 3' exonuclease, but possessing the 3' → 5' exonuclease activity, was used to avoid possible production of free monophosphates by the 5' → 3' exo-nuclease degradation of newly incorporated nucleoside monophosphates even after a complete arrest of DNA synthesis on UV-irradiated ØX174 DNA. This repeated incorporation followed by excision will be referred to as "polymerase idling". On the contrary, DNA synthesis on nonirradiated DNA produced only small amounts of free mono-phosphates. To test whether DNA polymerase I turns over purine triphosphates more frequently than pyrimidine triphosphates, (^{3}H)dATP and (^{3}H)dTTP were used as precursors. No preference for dATP turn-over over dTTP turnover was observed with a UV-irradiated template (Fig. 3). The differences observed using unirradiated templates are not very accurate in view of the small amounts of free mono-phosphates produced (for experimental details, see ref 15).

(2) If the 3' → 5' exonuclease activity of *E.coli* DNA polymerases I and III removes nucleotides inserted opposite, or near, pyrimidine dimers, thereby effectively blocking replication past dimers, then a DNA polymerase lacking the 3' → 5' exonuclease may replicate a damaged template more efficiently than polymerases having this activity. Fig. 4 shows that this is true for homogeneous AMV reverse transcriptase (Fig. 4a) and for highly purified purified α DNA poly-merase from calf spleen (Fig. 4B),both of which are devoid of any detectable associated exonuclease activity (19-20) and which copied

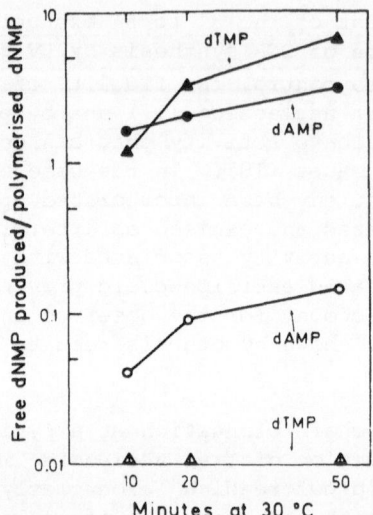

Figure 3. *Nucleotide turnover versus DNA synthesis by DNA*
polymerase I (large fragments) on intact and UV-
irradiated ØX174 DNA.

Open symbols are for intact template, closed symbols for UV-
irradiated template (for further explanations see text).

UV-irradiated ØX174 DNA far above the level copied by *E.coli* DNA
polymerase I.

A speculative Model for a Molecular Mechanism of
UV-Induced Mutagenesis

An attempt to summarize our present knowledge on the biochemical mechanism of UV-induced mutagenesis is given by the model
shown in Figure 5. Pyrimidine dimers in the DNA act as blocks for
DNA chain elongation (see Fig. 1 and 2). Repeated attempts of the
DNA polymerase to incorporate nucleotides opposite pyrimidine dimers
(Fig. 5 - 1), followed by exonucleolytic 3' → 5' excision, result
in an increase in the production of free nucleoside monophosphates
(Fig. 5 - 2) and hence prevent further DNA chain elongation. An
additional source of the nucleoside monophosphates would arise from
the breakdown of the DNA carrying UV-induced lesions (Fig. 5 - 2).
The nucleoside monophosphates produced by the DNA polymerase
"idling" may act as effectors for the "SOS induction" activating
protease(s) which cleave cellular repressors (3) and hence lead to
the induction of the operons governing indirect mutagenesis, cell
division, and oxygen consumption (see ref. 3). We propose that SOS
induction leads to an inhibition of the 3' → 5' exonuclease activity.

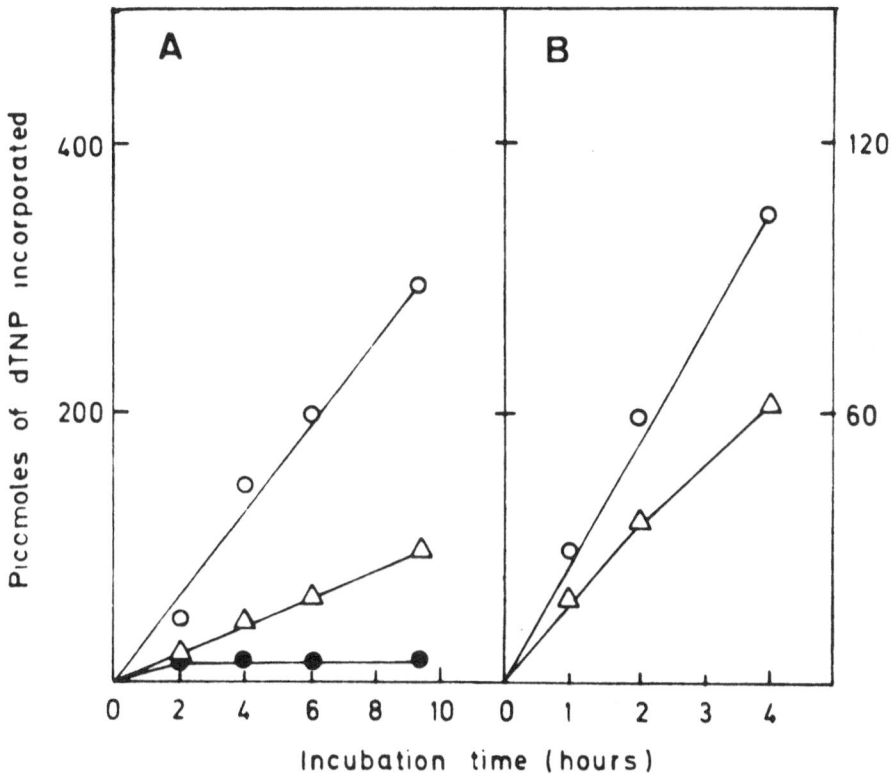

Figure 4.   *Extent of DNA synthesis by AMV DNA polymerase (reverse*
             transcriptase) and α-DNA polymerase on UV irradiated
             ØX174 DNA.

AMV DNA polymerase, purified to apparent homogeneity, was kindly
donated by Dr. L. Loeb; α-DNA polymerase from calf spleen was
purified up to a DNA cellulose step. Symbols: O-O-O incorporation
by AMV DNA polymerase (panel A) and calf spleen α-DNA polymerase
(panel B) on intact template; Δ-Δ-Δ incorporation by AMV DNA
polymerase (A) and α-DNA polymerase (B) on UV-irradiated template
(200 J/m²); ●-●-● incorporation by E.coli DNA polymerase I on
UV irradiated template (200 J/m²).

This would result in the decrease in production of free nucleoside
monophosphates (Fig. 5 - 4) permitting an error-prone DNA synthesis
to occur across the DNA lesions and turning off the inducing signal
(Fig. 5 - 5). Alternatively, free nucleoside monophosphates may act
directly as inhibitors of the 3' → 5' "proof-reading" activity
(Fig. 5 - 4[1]). Genetic evidence (see ref. 3) argues against the
simplicity of the second possibility. Furthermore the fidelity of
E.coli DNA polymerase I on either polyd(A-T) or intact and UV-

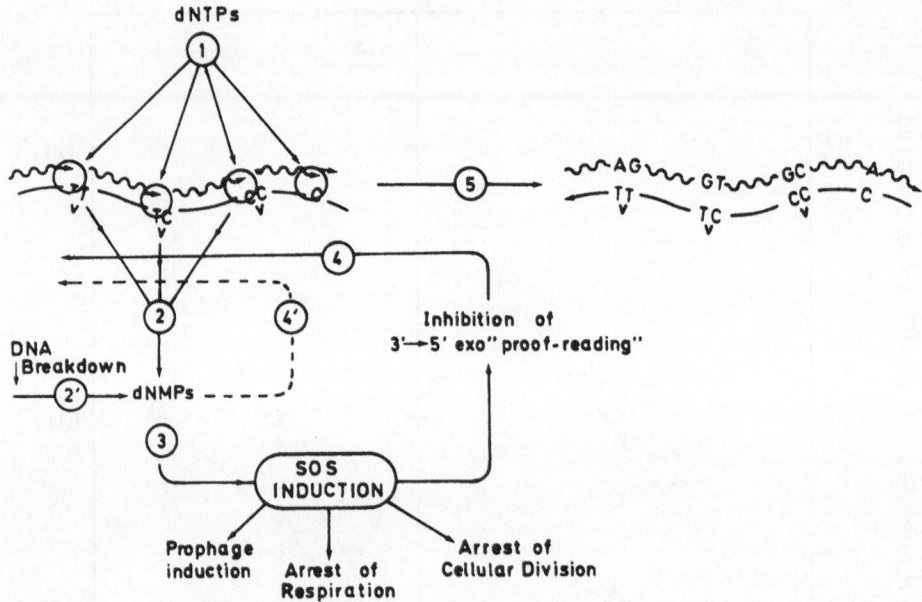

Figure 5. *A speculative model for a molecular mechanism of UV-induced mutagenesis.*

(See text for explanations).

irradiated polydC templates does not seem to be affected in the presence of a large excess of deoxynucleoside monophosphates (S. Boiteux, unpublished results, and L.Loeb, personal communication). The erroneous DNA synthesis due to the presumed inhibition of 3' → 5' exonuclease proof-reading is shown in our scheme (Fig. 5 - 5) to produce both single and double base changes opposite pyrimidine dimers and also few base changes on undamaged DNA segments. Recent sequencing of a series of UV-induced mutations in *E.coli lac* operon supports this possibility (J. Miller, personal communication).

REFERENCES

1. Drake, J.W. and Baltz, R.H. (1976) Ann. Rev. Biochem. 45: 11-37

2. Radman, M., Villani, G., Boiteux, S., Defais, M., Caillet-Fauquet, P. and Spadari, S. (1977) In: Origins of Human Cancer (eds., J.D. Watson and H. Hiatt) Cold Spring Harbor Laboratory (in press)

3. Witkin, E.M. (1976) Bacteriological Review 40:869-907

4. Weigle, J.J. (1953) Proc. Natl. Acad. Sci. USA 39:628-636

5. Bleichrodt, J.F. and Verheij, W.S.D. (1974) Molec. Gen. Genet. 135:19-27

6. Defais, M., Caillet-Fauquet, P., Fox, M.S. and Radman, M. (1976) Molec. Gen. Genet. 148:125-130

7. Ichikawa-Ryo, H. and Kondo, S. (1975) J. Mol. Biol. 97:77-92

8. Radman, M. (1974) In: Molecular and Environmental Aspects of Mutagenesis. (eds. L. Prakash, F. Sherman, M.W. Miller, C.W. Lawrence and H.W. Taber) C.C. Thomas, Springfield, Illinois, pp128-142

9. Radman, M. (1975) In: Molecular Mechanism for Repair of DNA. (eds. P.C. Hanawalt and R.B. Setlow) Part 1, Plenum Press, pp355-367

10. Tesmann, E. and Ozeki, T. (1960) Virology 12:431-439

11. Caillet-Fauquet, P., Defais, M. and Radman, M. (1977) J. Mol. Biol. 117:95-112

12. Masamune, Y. (1976) Molec. Gen. Genet. 149:335-345

13. Villani, G., Defais, M., Spadari, S., Caillet-Fauquet, P., Boiteux, S. and Radman, M. (1976) In: Research in Photo-Biology (ed. A. Castellani) Plenum Press, pp709-720

14. Poddar, K.R. and Sinsheimer, R. (1971) Biophys. J. 11:355-368

15. Villani, G., Boiteux, S. and Radman, M. Submitted to Proc. Natl. Acad. Sci. USA

16. Ben Hur, E. and Ben Ishai (1968) Biochim. Biophys. Acta 166:9-15

17. Gillin, F.D. and Nossal, N.G. (1976) J. Biol. Chem. 251: 5225-5232

18. Brutlag, D. and Kornberg, A. (1972) J. Biol. Chem. 247:241-248

19. Battula, N. and Loeb, L. (1976) J. Biol. Chem. 251:982-986

20. Weissbach, A. (1975) Cell 5:101-108

21. Setlow, P., Brutlag, D. and Kornberg, A. (1972) J. Biol. Chem. 247:224-231

22. Kornberg, A. (1974) In: DNA Synthesis. W.H. Freeman, San Francisco. p127

AN ADAPTIVE RESPONSE OF *E.COLI* TO LOW LEVELS OF ALKYLATING AGENT

Penelope Jeggo[*], Martine Defais[*], Leona Samson and
Paul Schendel

Imperial Cancer Research Fund
Mill Hill Laboratories
Burtonhole Lane, London, NW7 1AD, England

INTRODUCTION

N-methyl-N'-nitro-N-nitrosoguanidine, (MNNG), an alkylating
agent is a potent mutagen and carcinogen. Like N-methyl-N-
nitrosourea, (MNUA) and N-ethyl-N-nitrosourea, (ENUA), but in
contrast to most other common alkylating agents, it acts
selectively at the replication fork and produces a high proportion
of linked mutations (1,2). The mutagenicity of MNNG is probably
due to the methylation of DNA bases (possibly at specific sites
on bases) (3) but the mechanism by which the premutagenic lesions
become converted to stable mutations is not known, and the
characteristic action of MNNG at the replication fork is also not
understood.

Most of the above studies, and indeed the majority of studies
on mutagenesis, have used large doses of mutagen for short periods
of time. In order to consider mutagenesis in bacteria as a model
system for studying carcinogenesis we became interested in how cells
might respond to continuous exposure to low doses of mutagen; the
notion being that this might be more analogous to human exposure to
the carcinogens in our environment.

To investigate the long term exposure of cells to MNNG under
steady-state conditions we utilised a system developed by Helmstetter
and Cummings (4) termed the 'reverse filter technique'. Using this
technique we observed that cells growing in a low level of MNNG do
not accumulate mutations linearly with time, as might be expected,
but that mutations arise only in the first 30 to 60 mins of growth.
Later experiments have in fact shown that <u>Escherichia coli</u> cells

* *A.S.I. Participant*

growing in a low level of MNNG (1 µg/ml) develop a resistance both
to the killing and to the mutagenic effects of this mutagen. We have
termed this the MNNG-adaptive response. A preliminary account of
these results has appeared (5) in which it was shown that this effect
is inducible by the criteria that the response can be delayed by the
presence of chloramphenicol. We propose here to give a general
characterization of this adaptive response to MNNG.

MATERIALS AND METHODS

1. Detailed media plating conditions and MNNG treatment were
according to Jeggo, Defais, Samson and Schendel (in preparation).
A brief description is given in Samson and Cairns (5).

2. Experiment with ^{3}H-MNNG

   50 mls E.coli F26 (B/r his$^-$ thy$^-$) bacteria, adapted and non-
adapted, were grown to 1.6×10^8/ml, filtered rapidly through a
Millipore filter and resuspended in 1 ml medium. 25 µg ^{3}H-MNNG
(0.78 Ci/g) was added for 30 secs, followed by 25 µg cold MNNG.
After 5 mins MNNG treatment the cells were killed by the addition
of 5 mls ice-cold medium containing 10mM KCN, and 1 ml pyridine
added. The cells were precipitated by centrifugation, washed in
5 mls buffer (20mM Tris-1mM EDTA-10mM KCN) and finally resuspended
in 1 ml of the same buffer. The cells were lysed by treatment with
50 µl lysozyme (10 mg/ml) for 30 mins at 4°, followed by sonication
until complete lysis could be observed under the microscope. Enzyme
treatments were as follows:- pronase to a final concentration of
50 µg/ml for 60 mins at 37°, DNAse 1 to a final concentration of
50 µg/ml + 10mM Mg^{++} for 60 mins at 37°, pancreatic RNAse to a final
concentration of 10 µg/ml for 60 mins at 37°. CsCl gradient
centrifugation was followed according to Hanawalt and Cooper (6).

3. Weigle-reactivation

   Wild type λ and AB1157 bacteria were used in these experiments.
Media, plating conditions, and experimental procedure were those
described by Defais et al (7)

4. Measurement of DNA synthesis

   The thymine-requiring bacteria, F26 were grown in the presence
of 4 µg/ml thymine to a concentration of 4×10^7 cells/ml. The
culture was then divided into two, one half remained as a control
(non-adapted) and 1 µg/ml MNNG was added to the other half (adapted
culture). Both cultures were grown for a further 90 mins, when
^{3}H-thymine (24 Ci/mmole) was added to both a final concentration
of 0.2 µCi/ml. After 20 mins 50 µg/ml MNNG was added to half of
each culture, the other half being a control not receiving MNNG.

After a further 20 mins the MNNG was removed by rapid filtration through a Millipore filter (type HA 0.45 M) and resuspended in the same volume of prewarmed medium containing ^{3}H-thymine to the same specific activity. Control cultures were treated in the same way. TCA-precipitable counts were taken as a measure of DNA synthesis.

5. Cross-reactivity studies

Adaptation was performed by growing cultures for 90 mins in adaptive concentrations of each mutagen as follows:- 1 µg/ml MNNG; 5 µg/ml NQO; 0.01% MMS; 0.2% EMS; 10 µg/ml MNUA and 1 µg/ml ENNG. A range of challenge doses were analysed for each mutagen studied. The challenge doses given below were the highest analysed:- 100 µg/ml MNNG; 400 µg/ml NQO; 1% MMS; 2% EMS; 5µg/ml MNUA and 200 µg/ml ENNG.

RESULTS

The 'reverse filter technique' is a system particularly applicable for studying the long-term exposure of bacterial cells to mutagens. In this technique medium is pumped through an inverted Millipore filter, to which bacterial cells have been attached. The results of experiments using this technique showed that cells growing in a low level of MNNG (1 µg/ml) ceased to collect any further mutants after 1-2 generations. The filter experiments were designed to run for many days, but since this response was observed

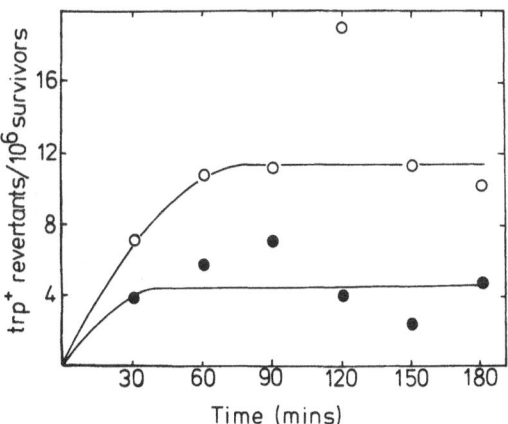

Figure 1. *Accumulation of trp$^+$ revertants by WP2 growing in low concentrations of MNNG. At time 0, a low concentration of MNNG was added to cultures of WP2 growing in minimal medium and at intervals thereafter 5mls samples were removed and the mutation frequency analysed. Samples were kept in log. phase by dilution with pre-warmed medium. (o——o) MNNG concentration 1 µg/ml. (●——●) MNNG concentration 0.5 µg/ml.*

relatively rapidly there seemed little advantage in using such a
complex technique for this study. Therefore the same experiment
was done using bacteria growing under standard conditions in liquid
medium and the results are shown in Fig. 1.   Preliminary
experiments had shown that cell growth rate and the rate of DNA
synthesis, as measured by ^{3}H-thymine incorporation, were unaffected
by 1 µg/ml MNNG. It can be seen from Fig. 1 that the E.coli B/r
strain WP2 growing in low concentrations of MNNG becomes resistant
to the mutagenicity of this low concentration of MNNG after 30 to
60 mins growth. We have termed this the MNNG-adaptive response.

        We next examined whether these cells, which were growing in a
low level of MNNG, were in fact more resistant to the mutagenicity
(and also to the killing effect) of a higher concentration. Thus
two parallel cultures, one grown in the absence of MNNG (a control
culture) and the second grown for 90 mins in the presence of
0.5 µg/ml MNNG (which we will call hereafter an MNNG-adaptive culture)
were given higher 'challenge' concentrations of MNNG for 5 mins, and
the survival and mutation frequency produced by these challenges were
estimated. The results for survival and mutation are shown in
Figs. 2 and 3 respectively. It can clearly be seen that MNNG-adapted

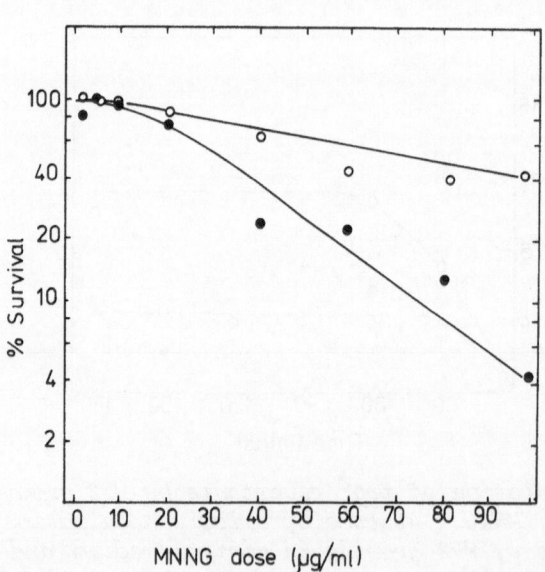

*Figure 2.   Survival of WP2 after varying challenge doses of MNNG
under non-adapted (●——●) and MNNG-adapted (○——○) conditions.
The adaptation procedure was 90 mins growth in 0.75 µg/ml MNNG.*

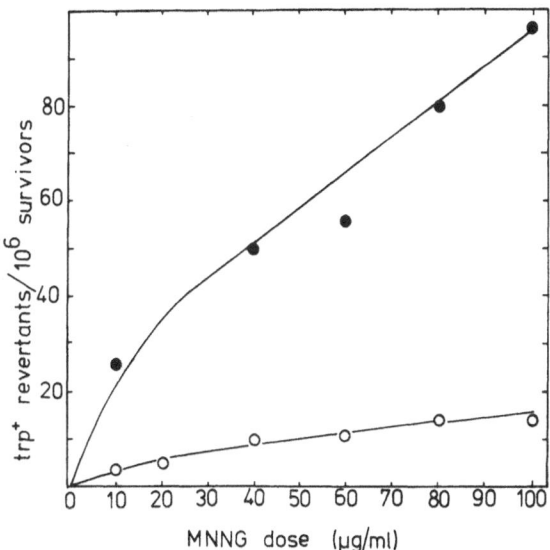

Figure 3. *Mutation responses of WP2 after varying challenge doses of MNNG under non-adapted (●—●) and MNNG-adapted (○—○) conditions. The adaptation procedure was 90 mins growth in 0.75 µg/ml.*

WP2 cells have acquired a resistance both to the killing and to the mutagenic effects of MNNG. Several E.coli K12 strains were also found to exhibit this response. Thus it is not a property peculiar to a specific B/r strain.

The resistance to MNNG might be achieved in one of two ways; the MNNG might be prevented from entering the cell, or more precisely, from reaching the DNA target site, or, lesions in the DNA may be produced to the same extent but processed in a more efficient and error-free manner. We have attempted to distinguish between these by measuring the degree of derivatisation of DNA by MNNG in adapted and non-adapted cultures. As described previously, MNNG is believed to act as a mutagenic agent by transferring its methyl group to DNA bases; the level of which can be measured by using MNNG bearing a ^{3}[H]-labelled methyl group. Thus the E.coli B/r strain F26, non-adapted or adapted by 90 mins growth in 1 µg/ml MNNG, was treated with ^{3}H-MNNG for 5 mins at 50 µg/ml and the degree of methylation of various cell components analysed as shown in Table 1. The samples that had been treated with pronase and RNase were also banded in CsCl gradients. Greater than 90% of the label was found at the buoyant density of DNA, and could be made TCA-soluble following DNAse digestion. The results given in this table demonstrate that the DNA in MNNG-adapted bacteria is methylated to approximately the same extent as in non-adapted bacteria. Thus the four fold difference in survival observed in these experiments is not due to a difference in the degree of DNA methylation.

TABLE 1. Analysis of adapted and non-adapted bacteria treated with ^{3}H-MNNG

treatment	non-adapted cpm/sample (% remaining)		adapted cpm/sample (%remaining)	
Pronase	931	(100)	939	(100)
RNase + Pronase	423	(45)	421	(45)
DNase + Pronase	540	(59)	608	(65)
RNase + DNase + Pronase	185	(20)	198	(21)
% survival	13%		64%	

Analysis of pooled fractions from CsCl gradient banding in region of DNA

Total cpm	2,220	1,750
Total μg DNA	141.4	103.5
cpm/μg DNA	15.7	16.9
% digestion by DNase	92%	98%

High doses of MNNG are known to block DNA replication, possibly because the cell cannot replicate past certain methylated bases (for example O6-methyl guanine). We have investigated whether this block also occurs in MNNG-adapted cultures by measuring the effect of MNNG upon the incorporation of ^{3}H-thymine and the results are shown in Fig. 4. Firstly, the results demonstrate that during adaptation DNA synthesis occurs at the full rate. Exposure to 50 µg/ml MNNG for 20 mins blocks DNA synthesis in non-adapted cultures within 1-2 mins of its addition and the synthesis does not resume until about 20 mins after its removal (Fig. 4). The rate of DNA synthesis in adapted cultures is reduced in the presence of this challenge dose of MNNG, but it resumes almost immediately after removal of the mutagen. From these results it is clear that adapted cultures overcome the MNNG-caused block in DNA replication very rapidly.

One explanation for these results could be the induction of a repair pathway able to process MNNG-derived lesions in a non-lethal error-free manner. In order to investigate further the nature of the adaptive response we have examined several strains mutant in genes known to be involved in DNA repair, for the presence of the adaptive response (Table 2). Since some of the strains tested were more

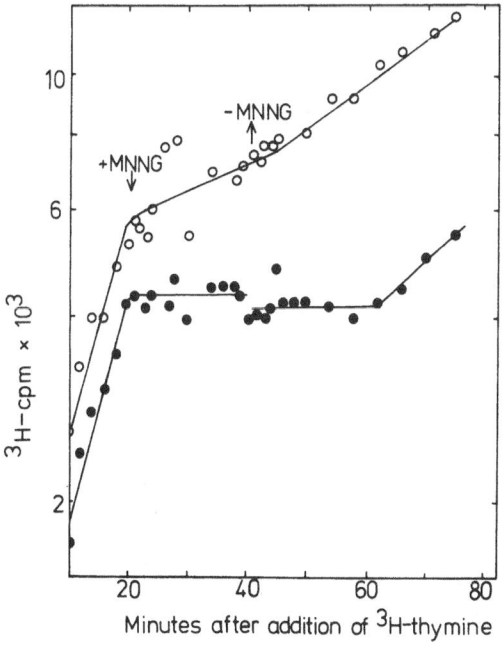

Figure 4. Rate of DNA synthesis in non-adapted (●——●) and MNNG-adapted (o——o) bacteria during and after a challenge of 50 µg/ml. See methods for experimental procedure.

TABLE 2.   Analysis of mutant strains for the MNNG-adaptive response

strain	defect	MNNG adaptive dose (μg/ml)	MNNG challenge dose (μg/ml)	genetic marker analysed	adaptive response killing	mutation
WP2	wild type	0.75	100	trp	+	+
WP2s	uvrA	0.75	100	trp	+	+
CM561	lexA	0.1	10	trp	+	+
CM611	uvrA lexA	0.1	10	trp	+	+
CM571	recA56	0.1	5	trp	+	+
WP67	uvrA-polA	0.075	5	trp	−	−
WP44s-NF	tif1 uvrA sfi					
	30°	0.75	100	trp	+	+
	42°	0.75	100	trp	+	−
AB1157	wild type	0.5	30	val^R arg his	+	+
AB3027/1	xthA14	0.5	30	val^R	+	+
BW9109	Δ(xthA-pncA)	0.5	30	val^R	+	+
AB2463	recA13	0.05	2	arg	+	+
BHL1	recA1	0.05	4	arg	+	+
DM455	recA99	0.02	1	val^R	+	+
KL108	recA12	0.1	5	his val^R	−	−

The adaptive response for all strains was measured over a range of challenge doses. The challenge dose given above is the MNNG concentration producing approximately 20% survival in a control (non-adapted) population. (+) for killing or mutation means that resistance to the killing or mutagenic effects of MNNG can be acquired during 90 min growth in the presence of the adaptive dose of MNNG. (−) means that resistance to the killing or mutagenic effects of MNNG cannot be acquired by growth in the presence of the adaptive dose.

sensitive to MNNG than the parent strain, it was necessary, in these cases to decrease the adaptive concentration. Generally, the adaptive dose was the highest concentration of MNNG that did not appreciably affect the growth rate.

Three strains showed aberrant responses in regard to adaptation. Firstly, a recA12 mutant was unable to show either function of the adaptive response. Secondly, the uvrApolA mutant examined could acquire resistance to the mutagenic effect of MNNG by growth in low concentrations but not to the killing effect of MNNG. Finally, the tif1 mutant at 42°, could adapt to the killing effect of MNNG but it could not acquire resistance to the mutagenic effect and indeed, whether adapted or not, was more readily mutated than an unadapted tif⁺ strain. All other strains examined were able to show the adaptive response.

In order to determine if the SOS response is also induced during MNNG adaptation, adapted cultures were analysed for their expression of the SOS functions (Table 3). UV-irradiated phage were found to plate with the same efficiency on MNNG-adapted and non-adapted bacteria, in contrast to the increased efficiency with which it plated on bacteria which have themselves been UV-irradiated prior to infection. Thus, adapted bacteria, unlike SOS-induced bacteria, do not show Weigle-reactivation. In addition, lambda

TABLE 3

	SOS REPAIR	MNNG ADAPTIVE RESPONSE
Inducible	+	+
Increased mutation	+	−
Weigle reactivation of λ	+	−
Increased λ mutation	+	−
λ induction	+	−
Filamentation	+	−

prophage are not induced during adaptation, as determined by the number of infective centres produced during adaptation of a λ lysogen. Finally adapted bacteria are not filamentous.

We have pursued some studies to ask whether this response can

be induced by, and produce resistance to other common mutagens. We
have investigated this cross-reactivity in two ways:- firstly, we
have determined whether MNNG adapted cultures are also rendered more
resistant to the killing and to the mutagenic effects of other
mutagens: and secondly, we have determined whether cultures grown on
low doses of other mutagens acquire resistance to those mutagens, or
to MNNG. The results are shown in Table 4. In summary, the MNNG
adaptive response was at least partially cross-reactive with all the
methylating and ethylating agents examined. However, MNNG
adaptation did not produce any resistance to UV, or to the UV-mimetic
mutagen, 4-nitroquinoline 1-oxide, and moreover neither of these
agents were able to induce resistance to MNNG.

DISCUSSION

In a previous publication (5) and here we have shown that E.coli
cells growing in a low level of MNNG (1 μg/ml) develop a resistance
both to the killing and mutagenic effects of this mutagen.

We have shown here, by the use of MNNG carrying a tritium
labelled methyl group, that the DNA is methylated to the same
degree in adapted and non-adapted cultures, which verifies that the
MNNG is not prevented from entering adapted cultures, or from
derivatising DNA. In the previous publication (5) we showed that the
survival of transforming plasmid DNA that had been alkylated 'in
vitro' was greater if it was assayed in cells that had been adapted
in MNNG. These two results strongly suggest that MNNG-derivatised
DNA is handled or actually processed in a less-lethal and less error-
prone manner in adapted bacteria.

An analysis of DNA synthesis during and after MNNG treatment
of adapted and non-adapted cultures shows that an exposure to MNNG
is far less inhibitory to the rate of DNA synthesis in adapted
cultures. This suggests that lesions in the fork region are
processed differently in adapted cultures as compared to non-adapted
cultures, although it should be mentioned that we have not shown that
the incorporation of ^{3}H-thymine observed in adapted cultures is ongoing DNA replication rather than some form of repair replication.

To investigate whether this response involves any of the
previously characterised pathways for DNA repair, we have examined
several strains mutant in gene products known to be involved in DNA
repair. Firstly, we have shown that the MNNG-adaptive response is
distinct from the SOS response. The SOS pathway represents a set of
diverse functions which are expressed in response to UV-irradiation
and certain forms of DNA damage, and appear to be under co-ordinate
control (8,9). These functions include Weigle-reactivation, error-
prone repair,(the sole mechanism for UV-induced mutagenesis) prophage
induction and filamentous growth. Two lexA strains showed the MNNG-

TABLE 4. Cross-reactivity of the adaptive response with other common mutagens

ADAPTIVE MUTAGEN	MNNG K	MNNG M	UV K	UV M	NQO K	NQO M	MMS K	MMS M	EMS K	EMS M	MNUA K	MNUA M	ENNG K	ENNG M
														CHALLENGE MUTAGENS
MNNG	+	+	−	−	−	−	+	+	not done	+	+	+	+	+
UV	−	−	not done											
NQO	−	−			−	−								
MMS	+	+					+	+						
EMS	+	+							not done	+				
MNUA	+	+									+	+		
ENNG	+	+											+	+

(+) means that resistance to the killing (K) or mutagenic (M) effects of the challenge mutagen can be acquired by 90 min growth in the low concentration of the adaptive mutagen. (−) means that resistance to the killing or the mutagenic effects of the challenge mutagen was not acquired by 90 min growth in low concentrations of the adaptove mutagen. Adaptive and challenge doses for each mutagen are given in Materials.

adaptive response although they are unable to induce any of the SOS
functions (10).   Four out of five recA strains examined were able
to acquire resistance to MNNG by adaptation yet SOS induction is
dependent on a recA$^+$ phenotype (10). Finally a tif strain was not
thermoinduced for the adaptive response, whereas the SOS response
is expressed constitutively in this strain at 42° (11,12).
Furthermore adapted cultures are not expressing any of the SOS
functions. These results show that MNNG adaptation can occur
independently of SOS induction.   Moreover, the ability of lexA
strains to adapt shows that the adaptive response does not merely
act by preventing the induction of error-prone SOS repair.

        An analysis of tif-1 cells, under conditions where the SOS
functions are constitutively expressed, showed that they could acquire
resistance to the killing effect of MNNG by adaptation but not to the
mutagenic effects.   Under these conditions mutagenesis is necessarily
complex, because SOS repair is error-prone and thus introduces
mutations, whilst adaptation is error-free and thus prevents
mutations.   The interaction of these two responses is at present
under investigation.

        Two pathways of excision repair have been reported in E.coli.
The first involves the uvr genes, which act to excise UV dimers. The
uvrA mutant examined was able to acquire the adaptive response,
showing that the response does not involve the uvr excision repair
pathway.   The second pathway for excising bases in E.coli involves
endonuclease II, which has been shown 'in vitro' to be active on
alkylated DNA and capable of releasing O^6-methyl guanine and N3-
methyl adenine from DNA that has been alkylated by MNUA (15). However
the two xthA mutants examined,both of which are deficient in
endonuclease II,could adapt showing that the response does not
involve this endonuclease, nor exonuclease III and the apurinic
acid endonuclease, the other two nucleases deficient in these strains
(16,17).
        The uvrApolA mutant examined was able to acquire resistance to
the mutagenic effect of MNNG by adaptation but not to the killing
effect.   The extreme sensitivity of polA mutants to the killing
effects of MNNG shows that DNA polymerase I is normally involved
in the repair of many potentially lethal lesions.   The inability of
polA strains to acquire resistance to the killing effects of MNNG
implies either that the adaptive response cannot handle those lesions
normally handled by DNA polymerase I (especially since the DNA
polymerase I dependent lesions outnumber the lesions handled by
adaptation), or that the adaptive response itself actually requires
DNA polymerase I.   Clearly, however, DNA polymerase I is not involved
in that part of the response producing resistance to the mutagenic
effect of MNNG.

        Finally, the recA12 mutant was unable to show either function
of the adaptive response.   However other recA mutants, recombination-

defective and defective in SOS induction, were able to adapt, suggesting that these recA functions are not involved in the adaptive response. We cannot at present eliminate that there may be an involvement of some, as yet uncharacterised, recA dependent function or that the recA12 mutant carries some additional mutation.

In conclusion, our results suggest that we are studying a previously uncharacterised response of cells which can be induced by growth on low levels of alkylating agents and renders them more resistant to the killing and to mutagenic effects of other alkylating agents but not to UV or to the UV mimetic mutagen NQO. We cannot eliminate a role of DNA polymerase I or the recA gene product in this response, but our results are consistent with the hypothesis that it is distinct from previously characterised pathways of DNA repair.

ACKNOWLEDGEMENTS

We wish to acknowledge Dr. J. Cairns and Dr. I. Molineux for considerable help and interest throughout the course of this work.

REFERENCES

(1) Guerola, N., Ingraham, J.L. and Cerda-Olmeda, E. (1971) Nature New Biol. 230:122

(2) Guerola, N. and Cerda-Olmeda, E. (1975) Mutat. Res. 29:145

(3) Lawley, P.D. (1974) Mutat. Res. 23:283

(4) Helmstetter, C.E. and Cummings, D. (1963) Proc. Nat. Acad. Sci. USA 50:767

(5) Samson, L. and Cairns, J. (1977) Nature 267:281

(6) Hanawalt, P.C. and Cooper, P.K. (1971) In: Methods in Enzymology (eds. Grossman, L. and Moldave, K.) Vol XXl. Part D pp221-229, Academic Press, New York.

(7) Defais, M., Caillet-Fauquet, P., Fox, M.S. and Radman, M. (1976) Molec. Gen. Genet. 148:125

(8) Radman, M. (1975) In: Molecular Mechanism for Repair of DNA (eds. Hanawalt, P. and Setlow, R.B.) Part A pp 355-367, Plenum Press, New York.

(9) Witkin, E.M. (1976) Bacteriol. Rev. 40:869

(10) Defais, M., Fauquet, P., Radman, M. and Errera, M. (1971) Virology 43:495

(11) Witkin, E.M. (1974) Proc. Nat. Acad. Sci. USA 71:1930

(12) Witkin, E.M. (1975) Molec. Gen. Genet. 142:87

(13) Boyce, R.P. and Howard-Flanders, P. (1964) Proc. Nat. Acad. Sci. USA 51:293

(14) Braun, A. and Grossman, L. (1974) Proc. Nat. Acad. Sci. USA 71:1838

(15) Kirtikar, D.M. and Goldthwait, A. (1974) Proc. Nat. Acad. Sci. USA 71:2022

(16) Kirtikar, D.M., Cathcart, G.R. and Goldthwait, D.A. (1976) Proc. Nat. Acad. Sci. USA 73:4324

(17) Yajko, D.M. and Weiss, B. (1975) Proc. Nat. Acad. Sci. USA 72:688

PLASMID-BORNE ERROR-PRONE DNA REPAIR

Carlo Monti-Bragadin[*], Nora Babudri and Sandra Venturini

Institute of Microbiology
University of Trieste
Trieste, Italy

R46 and some other plasmids of the N incompatibility group are described as coding for an error-prone DNA repair system, since they increase the number of spontaneous or induced mutants and protect bacteria from UV-killing (1-5). Similar effects have also been described for the colicinogenetic factor *col Ib* (6).

Plasmid-borne DNA repair is to be added to chromosomally determined repair and its characterization has obvious implications in understanding the DNA repair process as a whole. The first questions we would like to pose are:

(1)   how common are plasmids which code for a DNA repair system ?

(2)   do different plasmids code for a single DNA repair system or for many ?

(3)   how plasmid-borne DNA repair system(s) interact with a chromosomally determined system ?

Answers to these questions may suggest a strategy for future work on mutagenic DNA repair in *Escherichia coli.*

56 plasmids of independent origin have been tested for their ability to increase the number of trp^+ revertants in *Escherichia coli* strain WP2 (B/r trp$^-$) treated with various mutagens (Table 1) 33 plasmids belonging to many different incompatibility groups had little or no effect in this respect. However, 9 out of 10 N group plasmids, all the 7 members of the Iα group, R62 (a naturally

[*]*A.S.I. Participant*

Table   1

Plasmids tested for their effect on mutagenesis

Plasmids without effects	Plasmids with questionable effect	Plasmids which increase the number of revertants
R386 (FI)	R124 (FIV)	pKM101 (N)
pHH4 = R2 (FII)	R726 (H)	R46 (N)
R6 = R-Munich (FII)	R391 (J)	N3 (N)
pTM120 (FII)	R477 (S)	pTM558 (N)
pTM560 (FII)	øamp (Y)	pTM564 (N)
ColB-K98 (FIII)		pGM1 (N)
Hly-P212 (FVI)		pNR1013 (N)
R483 (Iβ)		pNR1014 (N)
R621a (Iγ)		S50 (N)
pTM556 (I not α)		R648 (Iα)
E47 (I not α)		R64drp (Iα)
RA1 (A)		R142 (Iα)
R57b (C)		R144 (Iα)
RP4 (P)		pTM551 (Iα)
pTM553 (P)		E8 (Iα)
S-a (W)		E43 (Iα)
R7K (W)		JR66a (Iω)
pTM561 (W)		R62 (I and N)
R387 (K)		R472 (L)
pTM557 (K)		R831 (L)
pTM559 (K)		R401 (T)
pTM562 (K)		R16 (O)
R811 (G)		R300 (unassigned)
R69-IP (M)		
R905 (V)		
R6K (X)		
P1CM (Y)		
E50 (N)		

(incompatibility group)

occurring hybrid between an I and an N group plasmid), JR66a (the only known member of the Iω group), R300 (a plasmid of unassigned incompatibility group) and the few members of the L, O and T groups, which we had available for testing (i.e., a total of 23 plasmids) greatly increased the number of revertants in *Escherichia coli* WP2.

The extent of this increase varied greatly for different plasmids: R46 gave the highest number of revertants; more than ten times the number given by *Escherichia coli* WP2 carrying no plasmid. Other members of the N group and some members of the I group increased the number of revertants to a lesser extent (7). The ratio between the activity of R46 and that of other plasmids remained approximately the s ame when methylmethanesulfonate (MMS), other methylating agents, nitroquinoline-N-oxide, or UV light were used as mutagenic agents.

When tested for their protective effect on UV-killing only slight differences between R46 and an I plasmid could be seen (Fig. 1). The increase in the number of revertants was similarly evident when plasmids were present in *uvrA*, *uvrD*⁻ and *polA*⁻ (3,7) strains of *Escherichia coli* WP2, and some UV protection from UV killing (3) was also evident in these strains (Fig. 2). Since the resistance of the wild type strain was not attained in any case we conclude that plasmids are not carrying the information missing in these repair deficient strains.

It was of special interest to test the effect of plasmids in *lexA*⁻ strains because of the *lexA*⁺-dependence of the mutagenic DNA repair known as SOS-repair (9).

Plasmids, which increase MMS-mutagenesis in *lexA*⁺ strains of *Escherichia coli* (Table 3) may be divided into two classes when tested in *Escherichia coli* WP2 lexA⁻ (Table 2). R46, pKM101, N3, R472 and R831 partially restore mutagenesis in this strain (class I), while the other plasmids leave *Escherichia coli* WP2 *lexA*⁻ non-mutable (class II) (7). However, plasmids of class I and II, as exemplified in Fig. 3 protect *Escherichia coli* WP2 *lexA*⁻ from UV-killing to a similar extent.

rnm mutants show a partially suppressed LexA phenotype, since they are only slightly more UV-sensitive than their *lexA*⁺ counterparts, but remain UV-non-mutable (10). Volkert et al. (10) suggest that the *lexA* gene product as altered by an *rnm* mutation, functions nearly as well as the *lexA*⁺ gene product in promoting UV-induction of the inhibitor of exonuclease V, but is essentially inactive in the induction of SOS-repair.

When present in an *rnm* strain (7), plasmids increase the UV-survival well over the level of the *lexA*⁺ strain (Fig. 4) and make

Table 2

Number of MMS-induced revertants in *Escherichia coli*
WP2 *lexA*⁻ carrying different plasmids

Plasmid	Inc Group	Micromoles of MMS per plate 5	1	0.2
None		1	1	0
pKM101	N	56	24	3
R46	N	66	39	8
N3	N	29	6	0
pTM558	N	5	0	0
pGM1	N	2	3	0
pNR1013	N	9	0	0
pNR1014	N	4	5	0
R62	I and N	11	4	0
R648	Iα	5	2	0
R64drp	Iα	5	4	0
R142	Iα	5	0	0
JR66a	Iω	4	2	1
R472	L	29	23	0
R831	L	109	91	15
R401	T	11	7	3
R16	O	13	11	2
R300	unassigned	10	7	2

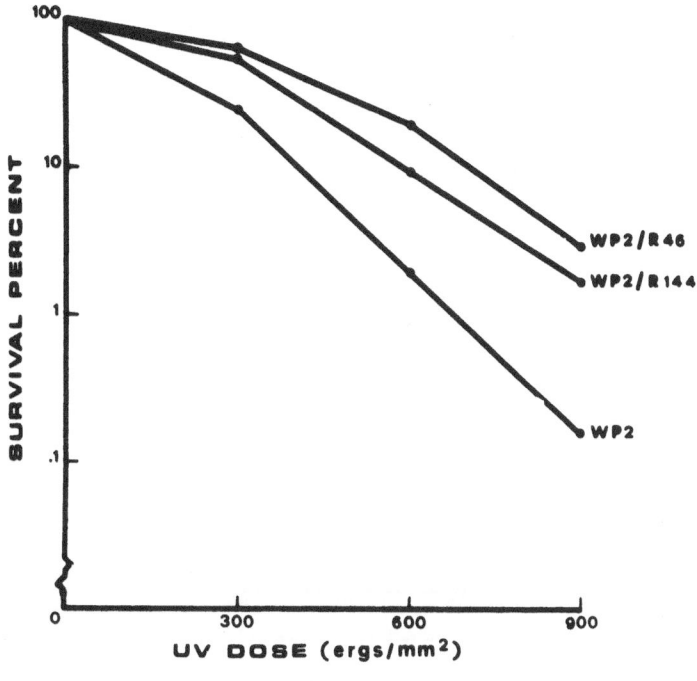

Figure 1

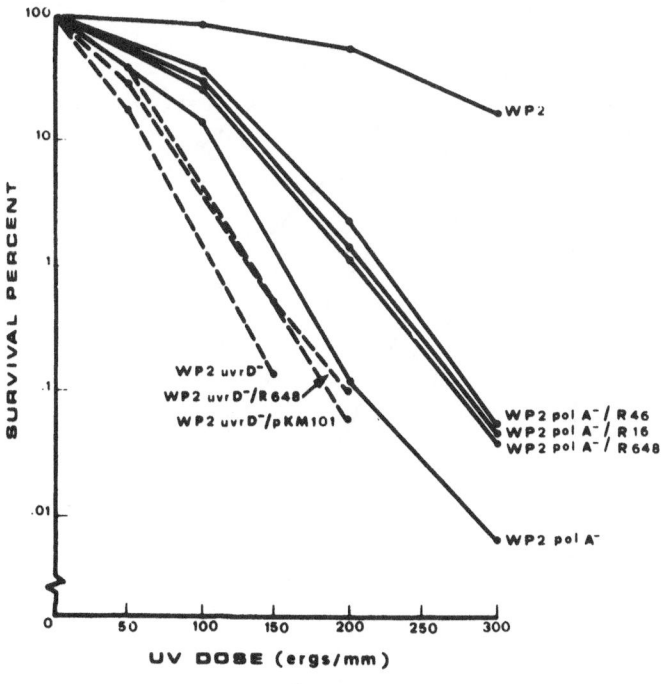

Figure 2

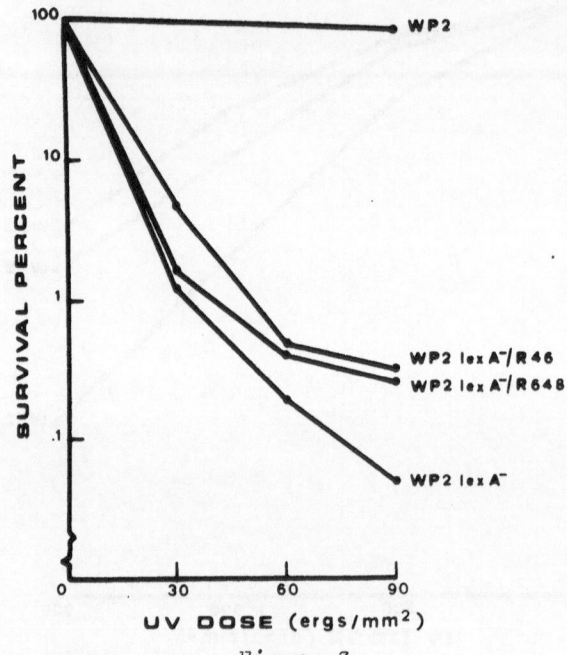

Figure 3

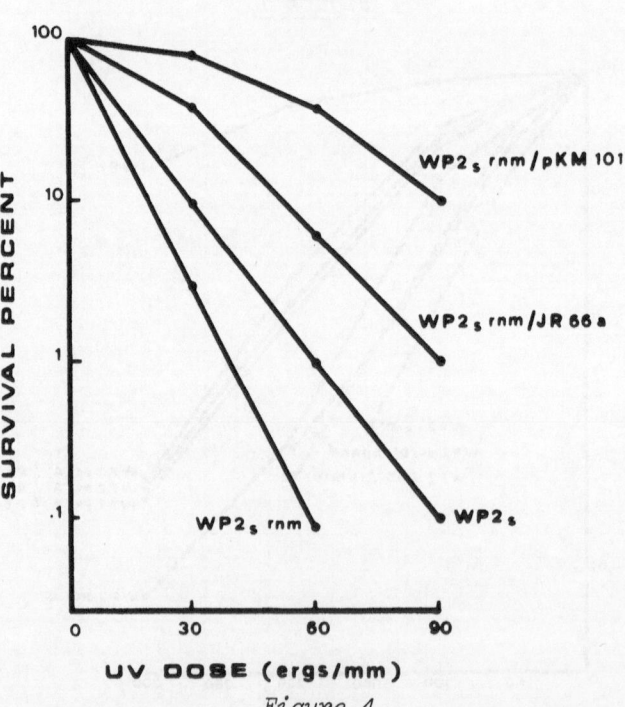

Figure 4

Table 3

Number of induced revertants in the *Escherichia coli*
carrying the LexA phenotype partially suppressed
by the *rnm* mutation

Bacterial strain	μmoles of MMS		UV dose (erg/mm^2)	
	1	5	0.5	1
WP2$_s$ (WP2uvrA)	22	57	2	55
WP2$_s$/R46	287	920	279	1200
WP2$_s$/JR66a	35	191	38	145
WP2$_s$ *rnm*	O	7	O	O
WP2$_s$ *rnm*/R46	162	579	53	103
WP2$_s$ *rnm*/JR66a	6	28	O	15

this strain MMS- and UV-mutable. This last effect is more pronounced
for class I plasmids, as defined above (Table 3).

It seems, therefore, that when the *rnm* mutation reduces the
uncontrolled exonuclease V activity of *lexA*$^-$ strains, the error-
prone, plasmid-borne repair system is well expressed in *lexA*$^-$
strains. Because of the dominance of the *lexA*$^-$ allele over *lexA*$^+$,
this explanation is more likely than to suppose that plasmids
restore SOS-inducibility.

In conclusion, our results show:

(1) that plasmids coding for an error-prone DNA repair system
     are widely represented in nature and belong to at least 5
     different incompatibility groups.

(2) that they probably code for different DNA repair systems
     of comparable repair efficiency, but error-prone to
     different degrees.

(3) that the plasmid-borne DNA repair system(s) are largely
     independent from chromosomally controlled repair functions.

REFERENCES

1.   McPhee, D.G.   (1973)   Applied Microbiol.   6:1004

2.   McCann, J., Spingarn, N.E., Kobori, J. and Ames, B.N. (1975)
     Proc. Nat. Acad. Sci. (Wash.)   72:979

3.   Monti-Bragadin, C., Babudri, N. and Samer, L. (1976) Molec.
     Gen. Genet.   145:303

4.   Mortelmans, K.E. and Stoker, B.A.D. (1976) J. Bacteriol.
     128:271

5.   Walker, G.C. (1977) Molec. Gen. Genet.   152:93

6.   Howarth, S. (1966) Mutation Res.   3:129

7.   Babudri, N. and Monti-Bragadin, C.   Molec. Gen. Genet., in
     press

8.   Venturini, S. and Monti-Bragadin, C. Mutation Res., submitted
     for publication

9.   Witkin, E.M. (1976) Bacteriol. Rev.   40:869

10. Volkert, M.R., George, D.L. and Witkin, E.M. (1976) Mutation
     Res.   36:17

DNA SYNTHESIS IN PERMEABLE CELL SYSTEMS FROM *SACCHAROMYCES*

CEREVISIAE

Wolfgang Oertel[*] and Mehran Goulian

Universität Würzburg
Institut für Genetik und Mikrobiologie
D-8700 Würzburg, West Germany
and
Department of Medicine
University of California, San Diego
La Jolla, California, U.S.A.

INTRODUCTION

Most of our knowledge concerning the mechanism and enzymology of eukaryotic DNA replication has been derived from animal cell systems (25,33,39,41,44,9,12,47) although simple eukaryotes, like yeast, offer some advantages. A haploid cell of *Saccharomyces cerevisiae* contains only three times as much DNA as the *E.coli* chromosome (19). This DNA is distributed to three different kinds of replication units: (1) Seventeen linear chromosomes, located in the nucleus and small enough to allow some kinds of physicochemical and biochemical analysis (7,37). (2) Large circular (25 μm) DNA molecules in the mitochondria, which for still unknown reasons, are mainly isolated as fragments (19,3,24) and (3): Small circular (2 μm), plasmid like "o"-DNA molecules, most likely associated with a cytoplasmic membrane fraction (4,5,17,18).

A distinct disadvantage of the animal cell system under study is the lack of mutants. *Saccharomyces cerevisiae*, however, has undergone extensive genetic analysis (31,32). A series of temperature sensitive DNA replication mutants has been isolated and partially characterized (19,20,21). Yeast cells, lacking one or all kinds of extrachromosomal DNA (15,18), are viable and provide a system to study nuclear DNA replication independently of the replication of the other DNA species. Technical

[*]*A.S.I. Participant*

difficulties, however, made research on DNA replication with yeast somewhat more difficult than with higher eukaryotes: The very tough cell wall prevents a fast and gentle lysis as required for the release of intact, active nuclei or unsheared DNA from the cell. In part this problem could be solved by the use of cell wall degrading enzymes (1,27). In addition it is not possible to label yeast DNA *in vivo* specifically with radioactive thymidine or bromodeoxyuridine, because yeast cells lack thymidine kinase (16). This can be overcome, however, by using dTMP uptake (45,14) or dTMP auxotrophic mutants (10). Unspecific long term labeling is possible in media containing ^{32}phosphate or ^{3}H uracil. Attempts to pulse label *in vivo*, however, are not satisfying with these methods because of the slow equilibration of the large nucleotide pools within the cell.

One way to solve this problem is to develop *in vitro* systems capable of utilizing deoxynucleoside triphosphates as precursors for DNA synthesis. So far three permeable cell systems from yeast have been described (2,22,49). Although DNA synthesis in one of these systems (22) is temperature sensitive *in vitro* if prepared from temperature sensitive DNA replication mutants (19,20,21), the labeled products may be mostly of mitochondrial origin (2). Only the system recently described by Yee et al. (49) was shown to synthesize both nuclear and mitochondrial DNA. In none of these systems were the products further characterized, most likely because of the very low *in vitro* DNA synthesis rate. Isolated yeast mitochondria were also found to use deoxynucleoside triphosphates for DNA synthesis (48); the product, however, is much smaller than *in vivo* labeled mitochondrial DNA and no evidence for a semiconservative mechanism was provided (48,50). We tried to develop improved *in vitro* DNA replication systems from *Saccharomyces cerevisiae* which are described and compared below.

MATERIALS AND METHODS

The haploid yeast strains used are *S. cerevisiae* A364A ρ^+, auxotrophic for adenine, uracil and various amino acids, and its temperature sensitive derivative ts13052ρ^+ *(CDC-8)*, defective in a gene involved in the propagation of DNA replication (20,21). A364Aρ^0-3 and ts 13052ρ^0-14 are derivatives of these mutants lacking mutochondrial DNA (34). DX2180-1aρ^+ (8) and its ρ^0- derivative DX2180-1aρ^0-2 (35) are prototrophic wild type strains.

The experimental procedures are either described in the legends to the figures and in (34) or will be detailed in a later publication (35).

RESULTS

The "Ether Cell" System

As *E.coli* cells (43), cells of *S. cerevisiae*, treated briefly,
at low temperature, with diethyl ether are permeable to nucleoside
triphosphates which are used for DNA-(dNTP) or RNA synthesis (rNTP);
(W. Oertel, M. Goulian, 1977) (35). Incorporation is dependent on
all four deoxynucleoside triphosphates, rATP, phosphoenol pyruvate,
Mg^{++}, a sulphydryl reagent and a buffer pH 7.8. The other three
ribonucleoside triphosphates are added routinely as RNA precursors.

Kinetics and Products of the DNA Synthesis in "Ether Cells"

The overall rate of DNA synthesis, the time period until
incorporation levels off, and the composition of the products are
very dependent on the yeast strain used. Generally "ether cells",
prepared from strains containing normal mitochondrial DNA ("ρ^+-
strains") incorporate deoxynucleotides with an at least 20 times
higher rate than its derivatives lacking mitochondrial DNA
("ρ^0-strains")(Fig. 1). An analysis of *in vitro* labeled DNA on
CsCl density gradients indicates that the product obtained with
ρ^+-strains is more than 95% mitochondrial DNA (Fig. II). Very
little nuclear DNA can be detected. With ρ^0-"ether cells" only
products with the density of nuclear DNA are synthesized (Fig. II).
The initial rate of *in vitro* mitochondrial DNA synthesis has been
calculated to be in the range of 10% (with strain A364Aρ^+) up to
nearly 100% (with DX2180-1aρ^+) of the rate of mitochondrial DNA
replication *in vivo*. In contrast, the initial rate of nuclear DNA
synthesis in the *in vitro* system is only 0.5-2% of the nuclear DNA
replication rate *in vivo*. An exception are strains A364Aρ^+ and
A364Aρ^0-3 in which the nuclear DNA replication is mostly inactivated
by the ether treatment. "Ether cells" from "petite" mutants,
which contain an altered mitochondrial DNA, incorporate deoxynucleotides with a much lower rate into mitochondrial DNA than
"ether cells" from normal ρ^+-strains (data not shown) (35).

DNA Replication Mutants

In temperature sensitive CDC-8 mutants (20,21) nuclear DNA
replication *in vivo* stops at the restrictive temperature immediately
whereas mitohcondrial DNA synthesis stops after a delay (7). With
the *CDC-8* mutants ts 13052ρ^0-14 and ts198ρ^0-4,lacking mitochondrial
DNA (35), temperature sensitivity can be demonstrated *in vitro* as
well (data not shown). A control experiment with ether treated
ρ^0-wild type cells confirmed that *in vitro* DNA synthesis is not
normally temperature sensitive. If ρ^+-strains of the two mutants
were tested after permeabilization with ether, no temperature

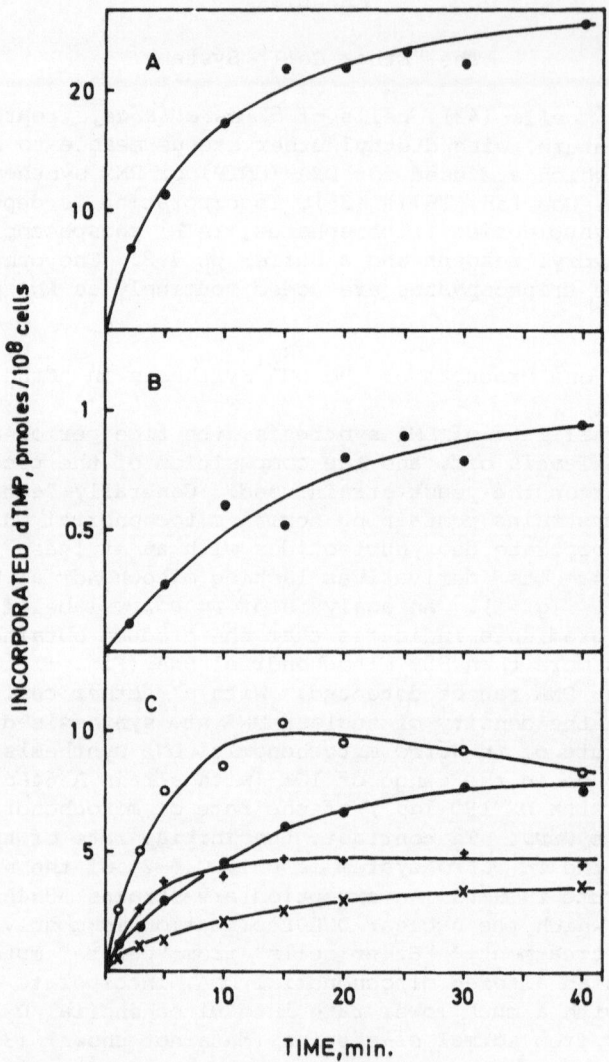

Figure 1. *Kinetics of DNA synthesis in "Ether Cells".*
Freshly harvested middle logarithmic phase cells were suspended in
one tenth of the original culture volume of buffer I (1M sorbitol,
0.1M KC1, 50 mM Tris-HC1 pH 7.5, 10 mM MgCl₂, 1 mM EGTA, 1 mM DTE)
and shaken at 0°C for 30 sec. with diethyl ether, (W. Oertel, M.
Goulian, 1977); (35) followed by removal from the ether-saturated
buffer by immediate centrifugation through a cushion of the same
buffer I. The reaction mixture, prepared from 250 μl "ether cells"
and 200 μl buffer I, was prewarmed (3 min.) to the temperature
indicated below and incorporation started by adding 50 μl of a ten
fold concentrated nucleotide-PEP-mixture. (The final concentrations

sensitivity of nucleotide incorporation could be observed (Fig. I;C). Since in ρ^+-"ether cells" almost only mitochondrial DNA and in the ρ^o-"ether cells" only nuclear DNA is synthesized (Fig. II;A,B), the *CDC-8* gene product is most likely not directly involved in mitochondrial DNA replication but only in the replication of nuclear DNA.

Mitochondrial DNA Replication in "Ether Cells"

The "ether cell" system of strain A364Aρ^+ was used to study some aspects of *mitochondrial* DNA replication, because nuclear DNA synthesis is almost completely inactivated: Incorporation of deoxynucleotides is optimal at a pH between 7 and 8 and is stimulated by KCl concentrations between 0.05 and 0.3mM Arabinocytosine triphosphate, an inhibitor of replicative but not repair DNA synthesis in other prokaryotic and eukaryotic systems (6,30), strongly inhibits incorporation of nucleotides into mitochondrial DNA. The same was found for ethidium bromide, a reagent which in yeast inhibits mitochondrial, but not nuclear DNA replication (15). Prelabeled or *in vitro* labeled DNA cannot be degraded in "ether cells" by externally added DNAses indicating an impermeability to proteins.

Semiconservative Synthesis of Mitochondrial DNA

To demonstrate that mitochondrial DNA synthesis in ρ^+-"ether cells" is semiconservative, purified DNA, density labeled with bromodeoxyuridine triphosphate, was sheared to a chain length of about 0.6-1 kilobase and banded in a CsCl density gradient (Fig. II;C). More than 50% of this material banded at "hybrid" density indicating that stretches of at least this length had been synthesized semiconservatively. In additional experiments of the same type but omitting shearing, some semiconservatively replicated DNA with a length up to 5 kilobase was detected.

0.1 mM of each dATP, dGTP, dCTP, rUTP, rGTP, rCTP, 0.01 mM ^3H labeled dTTP (4 mCi/μmole), 2 mM rATP and 10 mM). After the times indicated in the figure the reaction in 50 μl aliquots was terminated with EDTA. The incorporated dTMP/10^8 cells was calculated from the alkalistable, acid-insoluble radioactivity (34) in the cells, the specific activity, and the microscopically determined cell number.
A: Strain DX2180-1aρ^+ (incubated at 23°C).
B: DX2180-1aρ^o (23°C).
C: A364Aρ^+: (x) 23°C; (+)37°C; ts 13052ρ^+: (●)23°C; (o)37°C

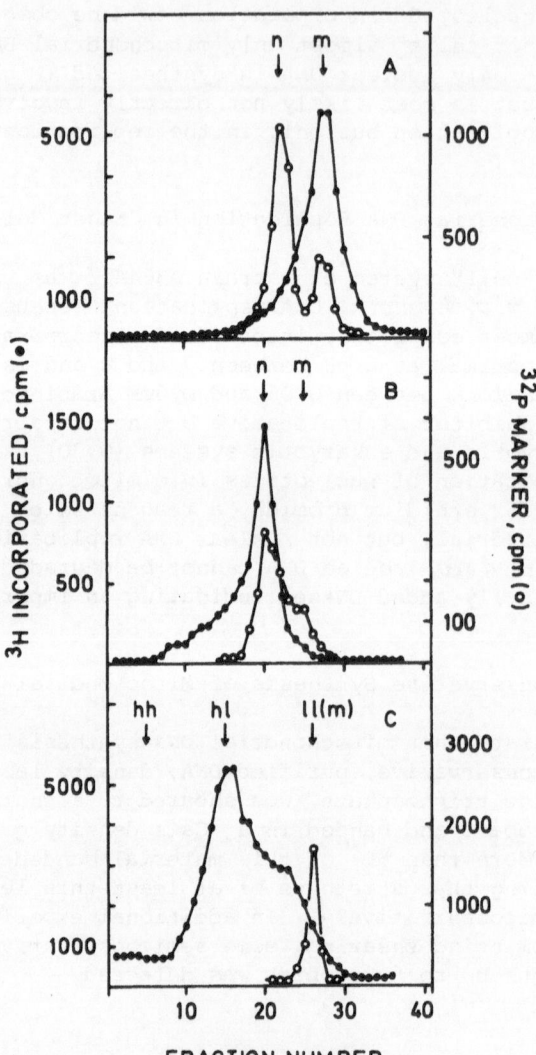

Figure II.   *Equilibrium Centrifugation Analysis of Normal and*
               Density Labeled DNA Synthesized in "Ether Cells".
The reaction mixtures (1 ml) were prepared with "ether cells" (2 x
10^9*/ml) of the strains specified below, as described in (35) and in*
the legend of Fig. I except that for experiments A and B the
specific activity of ^3H *dTTP was 18 mCi/μmole, in experiment C dTTP*
was replaced by dBuTP (0.1 mM) and the ^3H *label was in dCTP (0.01*
mM; 16 mCi/μmole). After 30 min. incubation at 23°C the reaction
was stopped with EDTA and the DNA isolated from the cells. The
isolation procedure described in (34,35) includes a brief treatment
with yeast cell wall degrading enzyme, lysis with "sarkosyl",
pronase digestion, phenol extraction and ethanol precipitation.

The Size of Mitochondrial DNA, Pulse Labeled in Vitro

To determine the size of the *in vitro* labeled mitochondrial DNA and to search for replication intermediates ("Okazaki-pieces") (36), DNA was pulse labeled with α^{32}PdTTP in ^{3}H uracil prelabeled A364Aρ^+- "ether cells" and then sedimented through neutral and laklaine sucrose gradients.

The *native, in vitro* pulse labeled (10 sec. to 15 min.) DNA is found in a broad peak at about 20S, corresponding to about 12% of the intact mitochondrial genome size.

Under denaturing conditions about 90% of the *prelabeled* DNA sediments faster than 30S; only about 10% is found in a very broad peak in the range of 8-30S with a maximum at 10-11S (Fig. III). The fast sedimenting prelabeled DNA bands in CsCl gradients like nuclear DNA whereas the slow material bands like mitochondrial DNA (data not shown).

In vitro products, pulse labeled for 10 seconds, were found exclusively in a peak coincident with the mitochondrial DNA but with the maximum at 8-9S. With longer pulse times material of this size accumulates without obvious further growth (Fig. III). A shoulder at about 5S was noticed in several experiments, and if present, it disappeared with a chase. This 5S material could be a replication intermediate but more convincing evidence is needed to confirm this.

Attachment to Preexisting DNA

To determine how many of the *in vitro* labeled mitochondrial DNA chains are covalently attached to preexisting DNA after a 30 min. incubation, and how many chains are free of preexisting material, DNA in ρ^+"ether cells" was density labeled with dBuTP, purified and denatured without shearing. About 30% of these DNA pieces

---

The DNA of sample C was sheared in addition by sonication to an average chain length of 0.6-1 kilobase. Before centrifugation ^{32}P labeled yeast mitochondrial and nuclear DNA were added as density markers. Density gradients were prepared from 8 g DNA solution and 10 g CsCl ($\underline{A;B}$) or 11 g CsCl ($\underline{B}$) and centrifuged in a Beckman type 65 rotor for 60 hours at 20°C and 35000 rpm ($\underline{A;B}$) or 4000 rpm (C). In each fraction alkali-stable, acid-insoluble radioactivity was determined by standard methods (34).
$\underline{A}$: strain A364Aρ^+
$\underline{B}$: DX2180-1aρ^o-2
$\underline{C}$: A364Aρ^+ (density labeled).

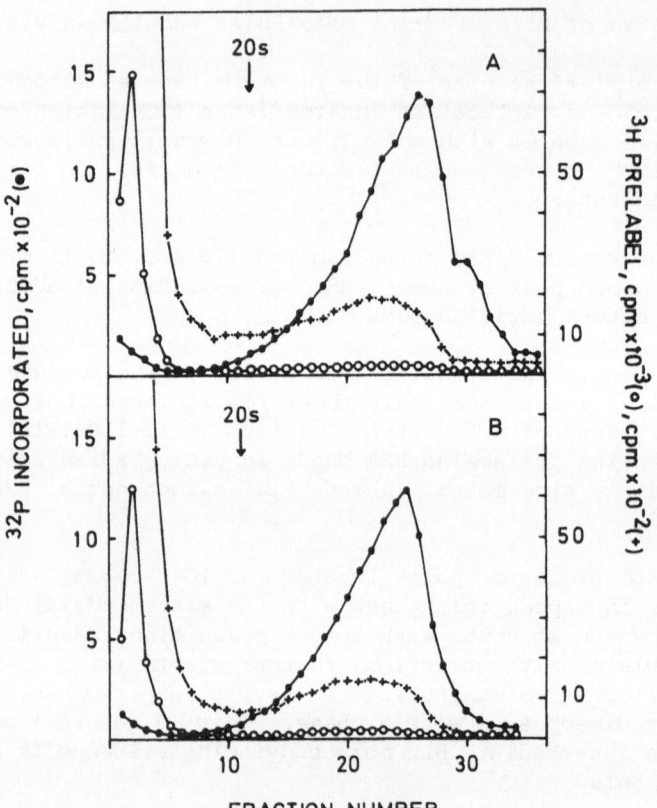

FRACTION NUMBER

Figure III. Alkaline Velocity Sedimentation Analysis of Mito-
            chondrial DNA Pulse Labeled in "Ether Cells".
A364Ap⁺ cells were prelabeled with ³H uracil (10μM; 0.5 mCi/μmole)
for six hours in YMM-medium (34) and permeabilized with ether.
The composition of the reaction mixture (1 ml; 4 x 10⁹ cells/ml)
is described in (35) and in the legend of Fig. I except that
α³²P dTTP (100 mCi/μmole) was used instead of ³H dTTP.
A: The reaction in a 400 μl aliquot was stopped after 10 sec. at
   23°C with EDTA.
B: After 10 sec. at 23°C a 250 fold excess of unlabeled dTTP over
   α³²P dTTP was added to a 400 μl aliquot. The reaction was
   continued for 3 min. and stopped.
The cells of both samples were briefly treated with cell wall
degrading enzyme (27) and lysed in 300 μl 0.3 M KOH. After a 15
hour incubation at 23°C (to hydrolyze ³H RNA) the lysates were
clarified by centrifugation and applied to alkaline sucrose
gradients (5-20%) (34). Centrifugation was performed in a Beckman
SW 40 rotor for 12½ hours at 20°C. The position of a ¹⁴C fd-phage
DNA marker (20S) was determined in a parallel tube. Alkali-stable,
acid-insoluble radioactivity was determined in each fraction
by standard methods (34).

obtained, band in CsCl density gradients like completely bromo-
uridine-substituted DNA and are thus free of preexisting material.
The rest is found between the normal and totally bromouridine-
substituted mitochondrial DNA. This material was obviously created
by addition of density labeled nucleotides, or perhaps whole density
labeled "Okazaki pieces", to preexisting DNA chains (35) (no figure
shown).

Nuclear DNA Synthesis in the "Ether Cell" System

To study nuclear DNA replication, Saccharomyces cerevisiae
DX2180-1αρ°-2 and ts13052ρ°-14 *(CDC-8)* were used. The *in vitro*
temperature sensitivity of temperature sensitive DNA replication
mutants is already a good indication that the observed DNS synthesis
is DNA replication and not repair. This was confirmed by further
pulse-chase and density labeling experiments which gave results
essentially similar to those obtained with the "permeable spheroplast" system (34) described next.

The "Permeable Spheroplast" System

Spheroplasts, prepared and cultured with osmotic support for
at least 2 hours as described by Hutchison et al. (26) incorporate
^{3}H uracil into DNA with about 10% of the rate measured in intact
cells growing without osmotic support (34). During attempts to
release intact nucleic from such spheroplasts by osmotic shock at
low temperature and in the presence of membrane stabilizing agents
(spermidine, Mg^{++}, Ca^{++}) it was noticed that the spheroplasts swell
and become less refractive under the phase microscope, but do not
burst. After stabilization by the readdition of sorbitol as an
osmotic support these spheroplasts proved to be permeable to
nucleotides. "Permeable sopheroplasts", as "ether cells", ·
incorporate nucleotides, if provided as triphosphates, into DNA
and RNA respectively. (W. Oertel, M. Goulian, 1977) (34). The
other r equirements are identical to the "ether cell system".

Kinetics and Products of the DNA Synthesis in
"Permeable Spheroplasts"

With the system prepared from cells containing normal mitochondrial DNA the incorporation of nucleotides is at least three
times higher than with the system from derivatives of the same
strains lacking mitochondrial DNA (Fig. IV;B,C) (23°C). The *in
vitro* labeled DNA from ρ°-cells has the density of double stranded
nuclear DNA (Fig. V;A). whereas the product of ρ^{+}-cells contains
additional mitochondrial DNA in the amounts expected from the

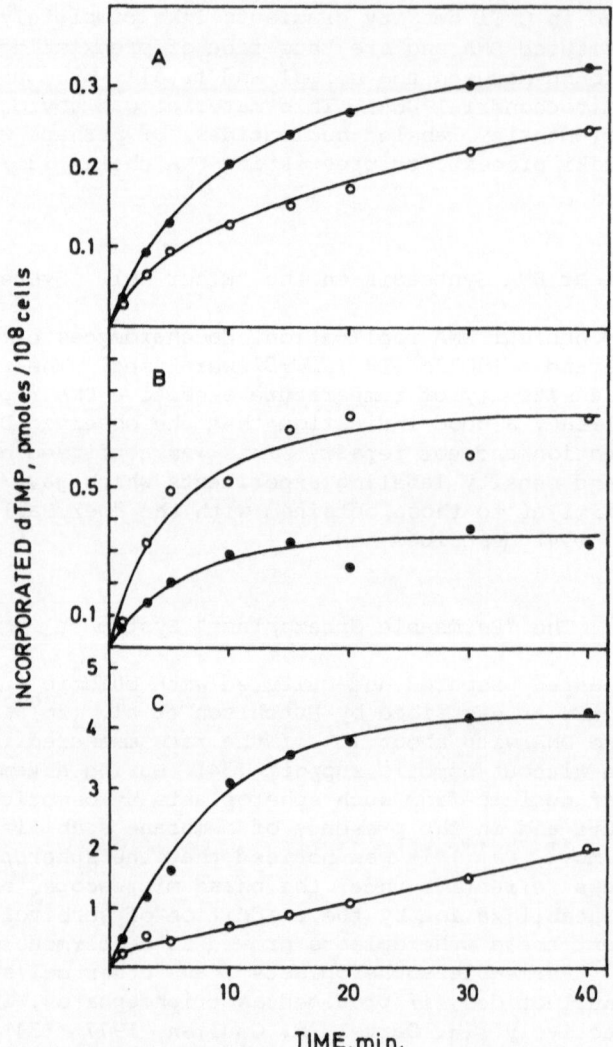

<u>*Figure IV.*</u> *Kinetics of Deoxynucleotide Incorporation into DNA of*
                 "Permeable Spheroplasts".
Mid-log cells of strains A364Ap^0-3 (A); ts13052p^0-14 (CDC-8) (B);
and ts13052p$^+$ (CDC-8) were converted to spheroplasts and cultured
in YMM-medium (34) with 1 M sorbitol as osmotic support for 2½
hours at 23°C. The spheroplasts were harvested by centrifugation
and osmotically shocked by suspension in a small volume of 75 mM
Tris-HCl pH 7.5, 15 mM MgCl$_2$, 1.5 mM CaCl$_2$, 7.5 mM spermidine-HCl,
1.5 mM DTE at 0°C. After stabilization by immediate readdition of
one half volume of 3 M sorbitol + 0.3 M KCl, the "permeable
spheroplasts" were separated from lysed cells by centrifugation
and resuspended at a final density of 2 x 10^9 cells/ml in the same

kinetic data (Fig. V;B).   The rate and extent of the nuclear DNA
synthesis is in the same order of magnitude as with "ether cells"
(0.5-1.5% of the in vivo rate) but the rate of mitochondrial DNA
synthesis is considerably lower (Fig. I; Fig. IV). If "permeable
spheroplasts" are prepared from strains in which nuclear DNA
replication is inactivated almost completely by ether, as A364A,
nuclear DNA synthesis stays active (34).

"Permeable Spheroplasts" from DNA Replication Mutants

In "permeable spheroplasts" derived from temperature sensitive
$CDC-8\rho^O$-mutants (20,21), the nuclear DNA replication is as
temperature sensitive in vitro as in vivo (34,35) (Fig. IV; B).
Again, as with the "ether cell" system, mitochondrial DNA synthesis
was found not to be temperature sensitive with $CDC-8\rho^+$-mutants
(Fig IV (C)) confirming that only nuclear DNA replication requires
the product of the defective gene.

Nuclear DNA Replication in "Permeable Spheroplasts"

The "permeable spheroplast" system prepared from strain
$A364A\rho^O$-3 was used to study nuclear DNA replication in yeast.

In vitro DNA synthesis is strongly inhibited by araCTP, ethidium bromide and SH-group blocking agents (34). The sensitivity
to ethidium bromide is in contrast to the insensitivity of nuclear
DNA replication in vivo. The system is permeable to smaller
proteins, at least to some degree, because in vitro labeled DNA
is degraded rapidly by externally added DNAses. The optimal
incorporation of deoxynucleotides into nuclear DNA depends on high
concentrations of rATP and phosphoenol pyruvate. The other ribonucleoside triphosphates are not required, but are added routinely.
We cannot exclude, however, that rATP is used only to regenerate
nucleoside triphosphates in the reaction mixture, and nucleotide
pools may have remained in the "permeable spheroplasts" sufficiently
to satisfy a possible ribonucleotide requirement.

Semiconservative Nuclear DNA Synthesis

To determine the length of the DNA stretches, being synthesized

---

*buffer with sorbitol (W. Oertel, M. Goulian, 1977) (34). The
reaction at 23°C (o) or 38°C (●) in 2 ml samples was started by
adding 0.2 ml of 10 x concentrated nucleotide-PEP-mixture (Legend
of Fig. I). At the times indicated 200 μl samples were withdrawn
and analyzed as described in (34) and the legend of Fig. I.*

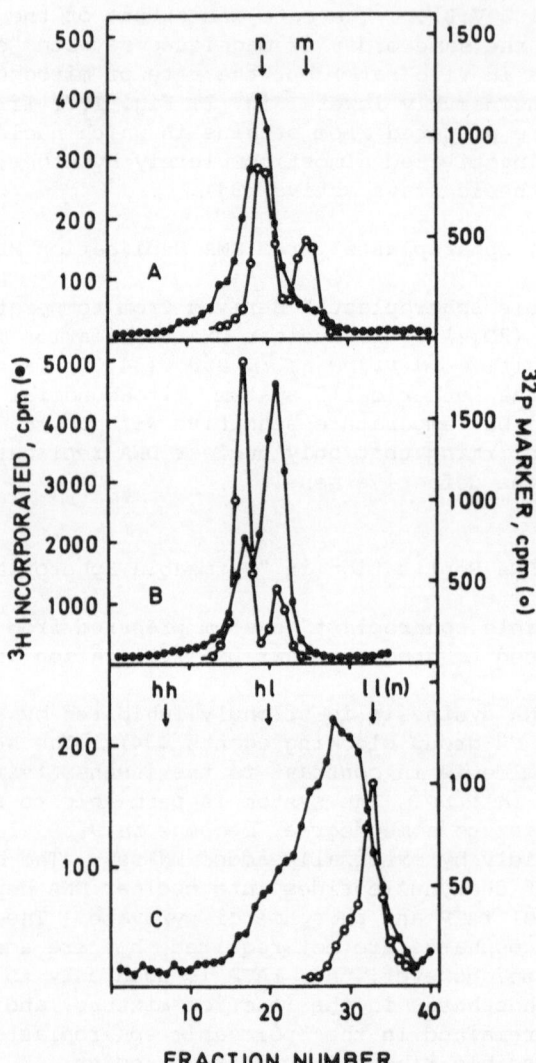

*Figure V. Equilibrium Centrifugation Analysis of Normal and
            Density Labeled DNA Synthesized in "Permeable Sphero-
            plasts".*

*The preparation of "permeable spheroplasts" and the reaction
conditions were as described in the legend of Fig. IV with the
exception that for experiments A and B the specific activity of
dTTP was 18 mCi/μmole, and that in experiment C dTTP replaced by
dBuTP (0.1 mM) and the radioactivity was in dCTP (0.01 mM; 1.6
mCi/μmole).   Samples A and B were incubated at 23°C for 30 min. and
sample C for 20 min. The purification of the DNA was as described
in the legend to Fig. II except that the treatment with cell wall
degrading enzyme was unnecessary. The density labeled DNA of sample*

by a semiconservative mechanism in the "permeable spheroplast"
system from strains lacking mitochondrial DNA, the DNA was density
labeled as described already for the "ether cells" and sheared to
an average chain length of about 1 kilobase. The result of an
analysis by density gradients (Fig. V;C) is that stretches up to
the whole fragment length are density labeled in one strand, but
the average length is only about 300 base pairs (34).

Discontinuous Nuclear DNA Synthesis

If the synthesis of nuclear DNA in yeast proceeds by a
discontinuous mechanism, as found in most other prokaryotic (36,40)
and eukaryotic organisms (25,33,39,41,44,9,12,47), the replication
intermediates labeled during a short pulse are expected to be small
DNA chains ("Okazaki pieces") (36), annealed to large DNA by
hydrogen bonds, which later are joined to the high molecular weight
DNA.

When DNA is analyzed, which was pulse labeled for 10-30 seconds
with $\alpha^{32}P$ dTTP in "permeable spheroplasts" prelabeled with ^3H
uracil, most of the ^{32}P labeled material sediments in alkali with
about 4S (Fig. VI; A). Some product is found sedimenting faster than
30S with the ^3H prelabeled bulk DNA. With longer pulse times or a
chase with unlabeled dTTP, the 4S pieces grow to a more heterogeneous size class sedimenting with about 6-8S, maximally 15S
and accumulate at this stage without being joined to the large
nuclear DNA (Fig. VI) (34).

By density labeling experiments, as described already for the
"ether cell" system, was determined that more than 75% of the
6-15S DNA chains which accumulate within a 20 min. incubation time,
contain preexisting DNA. Only at most 25% of the chains are free
of preexisting material as expected for *in vitro* initiated
replication intermediates (34).

Evidence for Primer DNA

Several authors reported recently that the initiation of

C was sheared by sonication to a chain length of about 1 kilobase.
Before centrifugation ^{32}P labeled nuclear and mitochondrial DNA were
added as density markers, as indicated in the figures. The
centrifugation conditions were as described in the legend to Fig.II.
A: Strain A364Aρ^O-3
B: A364Aρ$^+$
C: A364Aρ^O-3 (density labeled)

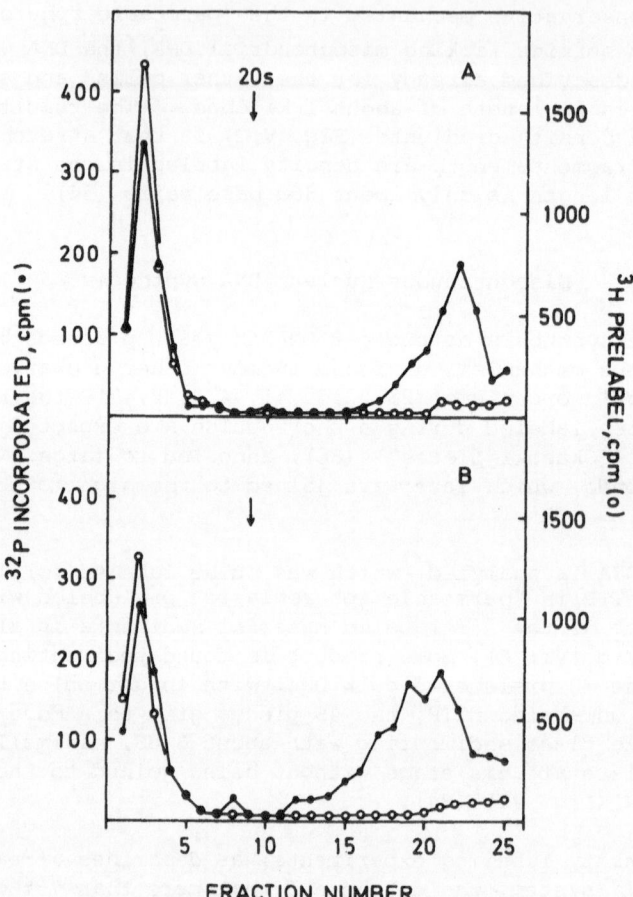

Figure VI. *Alkaline Velocity Sedimentation Analysis of Nuclear-DNA*
Pulse Labeled in "Permeable Spheroplasts".
A364Ap^O-3 cells were converted to spheroplasts as described (legend
of Fig. IV) and prelabeled for 2 hours at 23°C in YMMS-medium (34)
with ³H uracil (10 μM; 0.5 mCi/μmole). The reaction mixture (1.65
ml; 2 x 10⁹ cells/ml) was prepared and incubated at 23°C using
α³²P-dTTP (100 mCi/mole) instead of ³H dTTP as specified in the
legend of Fig. IV. After a 10 second pulse A: A 400 μl aliquot was
removed and the reaction terminated with EDTA; B: Unlabeled dTTP
was added in a 250 fold excess over α³²P dTTP to a 400 μl aliquot
and the reaction continued for another 15 minutes and then stopped.
Further processing and analysis in alkaline sucrose gradients was
as in the experiment described in Fig. III.

"Okazaki pieces" requires primer RNA (42,44,11,23,13,29,38,40). In
the "permeable spheroplast" system at least the fraction of the
in vitro labeled unjoined nuclear DNA pieces may have conserved a

piece of RNA covalently attached to their 5' end. To check this
possibility, heat denatured, *in vitro* pulse labeled DNA was
centrifuged in Cs_2SO_4 - or KI-gradients (no figure shown). More
than half of the pulse labeled DNA has a density higher than *in vivo*
long term labeled marker DNA. A density difference is not detected
when the DNA was treated with RNAse or alkali before centrifugation.
Although this indicates that indeed RNA is associated with DNA,
further analysis by gel electrophoresis under denaturing conditions
showed that most of the RNA is not covalently bound (data not
shown).

Better evidence for covalent association of RNA with DNA comes
from "nearest neighbour" transfer experiments (11,23,42,44,29,34).
After incorporation with one $\alpha^{32}P$ labeled deoxynucleoside triphosphate as the source of radioactivity, the products were alkali
treated and analyzed for the presence of ^{32}P labeled 3'(2') ribonucleoside monophosphates. Transfer of label occurs only when the
5' phosphate of a labeled deoxynucleotide is attached to the 3'
end of a ribonucleotide. The experiment was carried out separately
for each of the four deoxynucleotides. The results are listed in
Table I. All ribonucleotides are adjacent to all deoxynucleotides.
An unusually high transfer of label from dGTP to ribonucleotides and
a relatively high transfer from dCTP was found (34). Assuming that
after a 15 min. pulse at most 25% of the newly synthesized DNA
chains may have been initiated *in vitro* and grew to a chain length
of several hundred nucleotides, only the low transfer found with
$\alpha^{-32}P$-dTTP is in agreement with one transfer per chain. The very
high transfer with dGTP is very hard to explain only by priming
"Okazaki pieces".

Further fractionation of the $\alpha^{32}P$-dNTP labeled and denatured
RNA-DNA copolymers by isopycnic centrifugation in KI gradients
followed by analysis for ^{32}P transfer, yields in the case of $\alpha^{32}P$-
dGTP at least three different species of RNA-DNA copolymers with
different density, size and base composition at the RNA-DNA junction
(34). With $\alpha^{32}P$-dTTP only one of RNA-DNA copolymer was detected,
most likely the "Okazaki pieces" with primer RNA attached (34).
Further confirming evidence, however, is necessary because such
molecules can be created obviously by other mechanisms as well.

DISCUSSION

Advantages of the "ether cell" system are the speed and ease
of the permeabilization procedure and the high activity of the mitochondrial DNA synthesis with low, or without, simultaneous nuclear
DNA replication. To study nuclear DNA replication this system is
less suitable, however, with some strains lacking mitochondrial DNA
the incorporation of deoxynucleotides into DNA is reasonably high
and exhibits some characteristics of replicative DNA synthesis. In

Table I

^{32}P-labeled substrate	dATP	dGTP	dCTP	dTTP
Radioactivity, acid soluble after hydrolysis with KOH: %	13.2	42.7	20.0	1.7
Incorporated radioactivity transferred to ribonucleotides: %	1.2	16.4	5.4	0.54
Distribution of ^{32}P in (2')3'-rNMP's: % In:				
rAMP	73	32	32	34
rGMP	18	30	43	21
rCMP	3	18	9	14
rUMP	6	20	16	31

"Nearest Neighbour" Label Transfer from α^{32}P-dNTP's to 3'(2') rNMP's

"Permeable spheroplasts" were labeled for 15 min. at 23°C with each of the four α^{32}P dNTP's in separate incubations (4.5 x 10^9 cells/1.5 ml) using the strain and conditions described in the legend of Fig. IV except that the α^{32}P dNTP was at 10 μM instead of 100 μM. The spheroplasts were lysed with sarkosyl, and the total nucleic acids purified by pronase treatment, phenol-extraction, ethanol precipitation and chromatography on Sephadex G-100 (34). Special care was taken to avoid degradation of RNA. The nucleic acids were hydrolyzed with KOH, and the DNA (and K$^+$) precipitated with HClO$_4$ and removed by centrifugation. The neutralized, desalted supernatants of the four samples were analyzed for 3'(2') rNMP by paper chromatography (34). Acid soluble radioactivity after KOH hydrolysis was determined by measuring Cerenkov radiation in the pellet and the supernatant. Incorporated radioactivity transferred to ribonucleotides was calculated from the radioactivity in the DNA pellet and the deoxyoligonucleotide spot at the origin of the paper chromatograms and the radioactivity in the 3'(2') ribonucleotide spots.

some ρ^+ and ρ^0-strains nuclear DNA synthesis is inactivated almost
completely by ether. The impossibility to open the "ether cells"
fast and gently after incorporation without a short incubation
with cell wall degrading enzymes at elevated temperature, is a
disadvantage which still remains.

No such problems exist with the "permeable spheroplast" system.
With all yeast strains tested yet, nuclear DNA synthesis remains
active after permeabilization. Although the rate of the nuclear DNA
synthesis *in vitro* is not higher than with the "ether cells",
further processing is much easier because the cell wall is already
removed. Investigation of mitochondrial DNA replication should be
possible in "permeable spheroplasts" as well. The activity, however,
is lower than with "ether cells" and it was not possible yet to stop
nuclear DNA synthesis completely without affecting mitochondrial
DNA replication in this system. "permeable spheroplasts" are, in
contrast to "ether cells", at least partially permeable to smaller
proteins, as nucleases. This may provide the possibility to
complement an incomplete or defective system by externally added
protein factors.

The products of nuclear DNA synthesis in either system are
principally short pieces annealed by hydrogen bonds to one or both
strands of large preexisting DNA. The pieces are initially 100-200
nucleotides long and grow by continuous elongation, or joining to
each other, to an average chain length of about 600, maximally
1000 nucleotides, and accumulate (34). This resembles results with
in vitro systems from washed animal cell nuclei, in which the same
limited growth of 4S pieces without final joining to the large DNA
is observed (41). Reasons for the uneffective joining may be a
loss, inactivation, or a regulatory "switch off" of the DNA ligase,
DNA polymerase or hybridase activity (41). The defective ligation
step cannot be restored by addition of NAD^+, or by replacing
spermidine (an inhibitor of yeast DNA ligase) (W. Oertel; to be
published) by other membrane stabilizing agents. At least a
fraction of the pieces carries RNA covalently bound to the 5'end
(34); since, however, in the "permeable spheroplast" system
several species of RNA-DNA copolymers of unknown prigin are
synthesized, it is uncertain whether this RNA is really the primer.
The possibility that the high ^{32}P transfer observed with $\alpha^{32}P$-dGTP
is caused by incorporation of contaminating ribonucleotides into
RNA was largely excluded (34).

The size of the products of *mitochondrial* DNA synthesis in
the "ether cell" system appears not to increase with incubation
time although DNA synthesis within the pieces obviously proceeds.
The final length of the semiconservatively synthesized DNA
stretches within a particular piece reaches in most molecules in
average at least 10%, and in a significant portion (about 50%) up

to 100% of the full fragment length (35). These results are consistent with semiconservative synthesis on a template DNA which was initially much longer than the observed pieces, and subsequent fragmentation. There are indications that the synthesis in one or both strands proceeds by a discontinuous mechanism, but the evidence is not yet convincing. The consistently observed breakage of the mitochondrial template and probably of *in vitro* synthesized DNA, may not necessarily be an artefact of the system. *In vivo* mitochondrial DNA replication was found to appear dispersive rather than semiconservative, most likely because of continued breakage and rejoining after strand exchange by non classical recombination events (46). The "ether cell" system could be useful to study this unusual phenomenon. The nature of the defect in temperature sensitive CDC-8 DNA replication mutants (20,21) is not yet investigated in detail, but the results of density labeling experiments in "permeable spheroplasts" (34) are consistent with a defect in the elongation of the "Okazaki pieces" and not with a defect in the initiation or joining of these replication intermediates. The initial burst of nuclear DNA synthesis in the *in vitro* systems from *CDC-8* mutants (Fig. IV;B) at the restrictive temperature cannot be prevented by preincubation at the restrictive temperature (34), suggesting that something started *in vivo* is finished *in vitro*, or that the *CDC-8* gene product is slowly inactivated only if DNA replication proceeds. The slow stop of mitochondrial DNA replication in *CDC-8* mutants *in vivo* is most likely a secondary effect and not the result of the inactivation of the *CDC-8* gene product.

The possibility that the template of the *in vitro* products with the density of nuclear DNA is the small circular "o"-DNA rather than chromosomal DNA was largely excluded, because the native product sediments faster in neutral sucrose gradients than to expect for 2 μm circles and because it is mostly associated with the nucleus.

ACKNOWLEDGEMENTS

We gratefully acknowledge helpful discussions with Drs. Ben Tseng and Lynna Hereford. This work was supported by a NATO grant (430/402/802/3), a training fellowship and a research grant of the "Deutsche Forschungsgemeinschaft" (Oe 66/1-3), National Cancer Institute research grant CA 11705, and American Cancer Society grant NP-102.

REFERENCES

(1) Anderson, F.B., Millbank, J.W. (1966) Biochem. J. 99:682-687

(2)   Banks, G.R. (1973) Nature New Biol. 245:196-199

(3)   Blamire, J., Cryer, D.R., Finkelstein, D.B. and Marmur, J.
      (1972) J. Mol. Biol. 67:11-24

(4)   Clark-Walker, G.D. (1972) Proc. Nat. Acad. Sci. U.S.A. 69:
      388-392

(5)   Clark-Walker, G.D. and Miklos, G.C.G. (1974) Eur. J. Biochem,
      41:359-365

(6)   Cleaver, J.R. (1969) Rad. Research 37:334-343

(7)   Cryer, C.R., Goldthwaite, C.D., Zinker, S., Lam, K.B., Storm,E.,
      Hirschberg, R., Blamire, J., Finkelstein, D.B. and Marmur, J.
      (1973) Cold Spring Harbor Symp. Quant. Biol. 38:17-29

(8)   Duntze, W.,   Stötzler, D., Bücking-Throm, E., and Kalbitzer,
      S. (1973) Eur. J. Biochem. 35:357-365

(9)   Fareed, G.C., and Salzman, N.P. (1972) Nature New Biol. 238:
      274-277

(10) Fäth, W.W., Brendel, M., Laskowski, W. and Lehmann-Brauns, E.
      (1974) Molec. Gen. Genet. 132:335-345

(11) Flügel, R.M. and Wells, R.D. (1972) Virology, 48:394-407

(12) Francke, B., and Hunter, T. (1974) J. Mol. Biol. 83:99-121

(13) Francke, B. and Hunter, T. (1974) J. Mol. Biol. 83:123-130

(14) Game, J.C. (1976) Molec. Gen. Genet. 146:313-315

(15) Goldring, E.S., Grossman, L.I., Krupnick, D., Cryer, C.R.,
      and Marmur, J. (1970) J. Mol. Biol. 52:323-335

(16) Grivell, A.R., and Jackson, J.R. (1968) J. Gen. Microbiol.
      54:307-317

(17) Guerineau, M.C., Granchamp, C., Paoletti, C. and Slonimski, P.
      (1971) Biochem. Biophys. Res. Comm. 42:550-557

(18) Guerineau, M.C., Slonimski, P., and Avner, P.R. (1974) Biochem.
      Biophys. Res. Comm. 61:462-469

(19) Hartwell, L.H. (1974) Bacteriol. Rev. 38:164-198

(20) Hartwell, L.H. (1971) J. Mol. Biol. 59:183-194

(21) Hartwell, L.H. (1973) J. Bacteriol. 115:966-974

(22) Hereford, L.M. and Hartwell, L.H. (1971) Nature New Biol.
     234:171-172

(23) Hirose, S., Okazaki, R. and Tamanoi, F. (1973) J. Mol. Biol.
     77:501-517

(24) Hollenberg, C.P., Borst, P. and Can Bruggen, E.F.J. (1970)
     Biochem. Biophys. Acta 209:1-15

(25) Huberman, J.A. and Horwitz, H. (1973) Cold Spring Harbor
     Symp. Quant. Biol. 38:233-238

(26) Hutchison, H.T. and Hartwell, L.H. (1967) J. Bacteriol. 94:
     1697-1705

(27) Kitamura, K., Kaneko, T. and Yamamoto, Y. (1971) Arch. Biochem.
     Biophys. 145:402-404

(28) Lark, K.G., and Wechsler, J.D. (1975) J. Mol. Biol. 92:
     145-163

(29) Magnuson, G., Pigiet, V., Winnacker, E.L., Abrams, P. and
     Reichard, P. (1973) Proc. Nat. Acad. Sci. USA 70:412-415

(30) Masker, W.E., and Hanawalt, P.C. (1974) Biochem. Biophys.
     Acta 340:229-236

(31) Mortimer, R.K. and Hawthorne, D.C. (1966) Genetics 53:165-173

(32) Mortimer, R.K. and Hawthorne, D.C. (1973) Genetics 73:33-54

(33) Nuzzo, F., Brega, A. and Falaschi, A. (1970) Proc. Nat. Acad.
     Sci. U.S.A. 65:1017-1024

(34) Oertel, W. and Goulian, M. (1977) J. Bacteriol. (submitted
     for publication).

(35) Oertel, W. and Goulian, M. (1977) (manuscript in preparation)

(36) Okazaki, R., Okazaki, T., Sakabe, K. Sugimoto, K., Kainuma, R.
     Sugino, A., and Iwatsuki, N. (1968) Cold Spring Harbor Symp.
     Quant. Biol. 33:129-143

(37) Petes, D.T., Newlon, C.S., Byers, B. and Fangman, W.L. (1973)
     Cold Spring Harbor Symp. Quant. Biol. 38:9-16

(38) Ramareddy, G.V., Goulian, S.H. and Goulian, M. (1975) Biochem.
     Biophys. Acta 402:323-342

(39) Schandl, E.K. and Taylor, J.H. (1969) Biochem. Biophys. Res. Comm. 34:291-300

(40) Tomizawa, J., Sakakibara, Y. and Kakefuda, T. (1975) Proc. Nat. Acad. Sci. U.S.A. 72:1050-1054

(41) Tseng, B.Y. and Goulian, M. (1976) J. Mol. Biol. 99:317-337

(42) Tseng, B.Y. and Goulian, M. (1976) J. Mol. Biol. 99:339-348

(43) Vosberg, H.P. and Hoffman-Berling, H. (1977) J. Mol. Biol. 58:739-753

(44) Waquar, M.A., Minkoff, R., Tsai, A., Huberman, J.A. (1975) ICN-UCLA Symposia on Molecular and Cellular Biology, Vol. III (ed. M. Goulian, P. Hanawalt and C.F. Fox), W.A. Benjamin, Inc. Menlo Park, Calif., pp334-356

(45) Wickner, R. (1974) J. Bacteriol. 117:252-260

(46) Williamson, D.H. and Fennel, D.J. (1974) Molec. Gen. Genet. 131:193-207

(47) Winnaker, E.L., Magnusson, G. and Reichard, P. (1972) J. Mol. Biol. 72:523-537

(48) Wintersberger, E. (1966) Biochem. Biophys. Res. Comm. 25:1-7

(49) Yee, W.S., Hecker, R.W., Brunk, C.F. (1976) Biochim. Biophys Acta 447:385-390

(50) Zeman, L.J., Lusena, C.V. (1975) "Methods in Cell Biology" Vol. XII (ed. D.M. Prescott) Academic Press, New York, pp273-283

DNA REPLICATION IN SIMPLE EUKARYONS: IDENTIFICATION OF REPLISOMAL

COMPLEXES IN CELL EXTRACTS OF *Chlamydomonas reinhartii*

Stephen J. Keller[*], Thomas Yee and Ching Ho

Developmental Biology Program
Department of Biological Sciences
University of Cincinnati
Cincinnati, Ohio 45219, U.S.A.

The identification of replication proteins in eukaryons has generally been hampered by the complexity of eukaryon chromosomes, the unavailability of appropriate conditional lethal mutations, and an inability to define DNA replication *in vitro* as opposed to repair or recombination reactions. As a result, we have initiated a biochemical study on the unicellular green alga, *Chlamydomonas reinhartii*, since this organism is ideal for cell genetics, biochemical studies, and physiological manipulation in the laboratory (1). Density transfer experiments *in vivo* suggest that DNA replication is restricted to a limited portion of either the mitotic or meiotic cell cycles, that DNA repair occurs during normal replication, and that recombination events are limited to meiotic gametes (2-4). Vegetatively grown cells thus appear to contain primarily the enzymes necessary for replication. We have attempted to study the DNA replication proteins by preparing extracts from the cell-wall-less mutant, CW-15, of *Chlamydomonas* and by selecting experimental conditions which favor the formation of high molecular weight complexes which can interact with single stranded DNA. Our approach therefore relies upon observations which have been made on bacterial extracts which contain multienzymatic protein complexes that efficiently duplicate phage chromosomes (5,6).

MATERIALS AND METHODS

The cell-wall-less strain of *Chlamydomonas reinhardtii*, CW-15, was used throughout this study (7). Cells were routinely grown in a 30 liter fermentor in minimal sal media supplemented with 5% CO_2 in continuous light (2). Synchrony experiments were performed on cultures which were grown under the same conditions except that

[*]*A.S.I. Participant*

a 12-hour light - 12 hour dark alternating regime was used to
entrain the cells (8).   The cells were harvested when they
reached 1-4 x 10^6 cells per ml using a continuous flow centrifuge.
Cell pellets were washed once in a 0.01 M Tris-HCl, pH 8.0, 0.001M
EDTA, 0.002M β-mercaptoethanol (TME BUFFER) and lysed by hypertonic
shock by dialysis overnight against 3M NaCl in 3 x TME buffer. The
cell lysate was centrifuged at 105,000 x g for 2 hours to remove
insoluble debris and the supernatant was adjusted to 10% polyethylene glycol.   After standing for 30-60 minutes at 0°C, the
nucleic acid precipitate was removed by centrifuging at 15,000 x
g for 15 minutes. The supernatant was dialyzed against 3 changes
of 0.15M NaCl in TME buffer and applied to a ssDNA-Sepharose-6B
column which had been equilibrated in the same buffer (9).   The
column was washed with 5 column volumes of 0.15M NaCl in TME and
the DNA polymerase activity was eluted with 2 column volumes of
0.8M NaCl in TME.   The 0.8M eluate (fraction V) was either (a)
directly applied to a Sepharose-6B colum equilibrated in 0.8M NaCl
in TME buffer and the DNA polymerase activity found at a position
equivalent to a 3-5 x 10^5 daltons species. It is designated as
fraction VI; or (b) adjusted to 33% $(NH_4)_2SO_4$ and the resulting
protein precipitate collected by centrifugation at 20,000 x g for
30 minutes.   The precipitate was resuspended in 0.8M NaCl in TME
and chromatographed over a Sepharose-6B equilibrated in the same
buffer (fraction VII); or (c) adjusted to 33% $(NH_4)_2SO_4$, and the
protein precipitate was dialyzed against 0.15M NaCl in TME and then
applied to Sepharose-6B columns which had been equilibrated in the
same buffer.   DNA polymerase activity was determined by a filter
disc method of Meyer and Keller (10). DNAse activity was determined
by the same assay procedure but ^{3}H-DNA of *E.coli* at 4254 cpm/μg was
used as the substrate.   Single-stranded DNA binding was measured
on 2.5 cm nitrocellulose discs using the DMSO assay of Tsai and
Green with *E.coli* ^{3}H-DNA as the substrate (11).   Protein estimations
were made on each fraction by the Lowry procedure on acid
precipitates or directly by UV spectrophotometry (12,13). Other
methods used to analyze the protein fractions are described in the
text or figure legends.

RESULTS

Identification of high molecular weight complexes: Chromatography of the ssDNA bound proteins on Sepharose 6B resulted in the
comigration of DNA polymerase, DNAse, and ssDNA binding activities
(Fig. 1A). The molecular weight of 4 x 10^5 was extrapolated from a
standard protein curve.   Since chromatography was carried out in
0.8M NaCl in TME, non-specific protein aggregation or nucleic acid
association would seem unlikely.   Moreover the molecular weight
estimation of these activities is unaffected by pretreatment of the
sample with either RNAse A or DNAse I. The stability of protein
complexes in the ssDNA protein fraction has also been observed

under non-denaturing acrylamide electrophoresis. Over 70% of the
coomassie stained material is found in a single band at a relative
mobility which is coincident with the DNA polymerase activity (data
not shown). Sucrose gradient centrifugation in O.8M NaCl-TME -1%
PEG also indicated that DNA polymerase existed as a high molecular
weight complex. Two peaks of activity were noted at 12.6S and 6.3S
relative to an internal alkaline phosphatase standard. Assuming
that the conformation of the marker and the sample respond
similarly to the centrifugal conditions then the 12.6S peak would
correspond to a complex of 300,000 and the 6.3S to a molecular
weight of 100,000 (14). The difference between the sepharose 6B
size estimate and the centrifugal estimate could be explained by
either dissociation during centrifugation or by alterations in the
conformation.

     Since all the enzyme activities described above could reside
within a single prokaryon type DNA polymerase which self-associates
to form complexes, the polypeptide composition of the sepharose 6B
peak (Fig. 1A) was analyzed by SDS electrophoresis (16). After
denaturation, 8 coomassie-positive bands could be visualized on the
acrylamide gels (Fig. 2). The molecular weights were estimated from
a standard protein curve (Fig. 2B) and corresponded to 16,37,44
55,62,70,75 and 85 thousand daltons. Planimetry measurements of
each peak did not indicate that the proteins were present in any
simple ratio, although the sum of their molecular weights totalled
444,000. The composition would suggest heteropeptide complexes but
8 proteins can be organized into a large number of high molecular
weight aggregates.

     Dissociation of the HMW complexes: The associations that hold
the polypeptides together may be broken by concentrating the ssDNA
bound proteins by $(NH_4)_2SO_4$ and dialyzing the sample against O.15M
NaCl in TME buffer. Sepharose 6B chromatography in O.15M NaCl
resulted in the separation of the DNA polymerase, DNAse and ssDNA
binding activities (Fig. 1C). The DNA polymerase migrates as a
single peak at 45,000 daltons, whereas the DNAse activity migrates
at 60,000 daltons when ATP is included in the reaction assay and
at 45,000 daltons when ATP is absent. Single-stranded DNA binding
activity in O.15M NaCl could be observed at 32,000 and 16,000
daltons. The separation of the DNAse activity into two different
enzymes has been confirmed by preliminary characterization of the
sepharose 6B peaks. The 60,000 dalton fraction requires ssDNA, ATP
or dATP and is inhibited by methylene β γ ATP. In contrast the
45,000 dalton DNAse acts on either ssDNA or dsDNA and is inhibited
by O.5mM ATP. Each of the sepharose 6B peaks appears to be homogeneous for polypeptide size when analyzed by SDS acrylamide electrophoresis (Fig. 2). Although the 45,000 dalton fraction contains
both the DNA polymerase and the DNAse activities, the two may be
separated by electrophoresis, centrifugation and column chromatography (Yee, unpublished observations). The DNAse activity can also

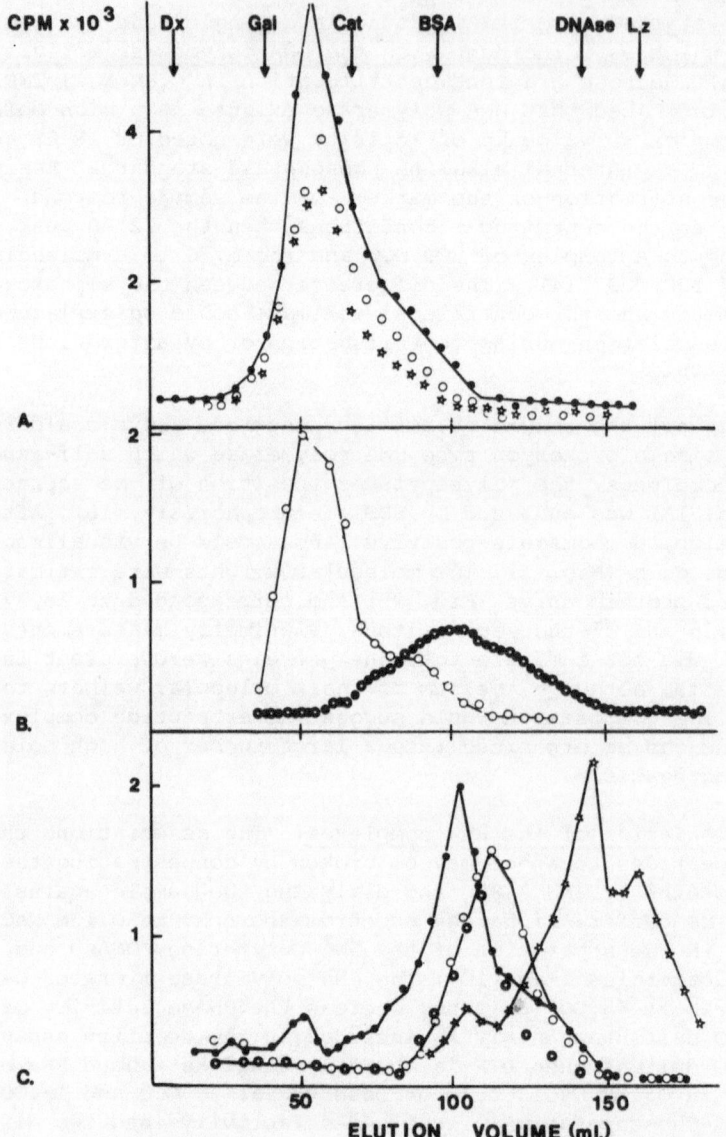

Figure 1. Sepharose-6B chromatograms of ssDNA-affinity-proteins
           (fr.V).

(A) 840 µg fr.V was chromatographed over 6B columns equilibrated
     in 0.8M NaC1-TME.

(B) 3 mg. fr.V was first precipitated with 33% $(NH_4)_2SO_4$ and
     chromatographed over 6B equilibrated in 0.8M NaC1-TME.

(C) 2 mg fr.V was precipitated with 33% $(NH_4)_2SO_4$ and chromato-

be removed from the DNA polymerase by maintaining the salt
concentration at O.8M NaCl in TME after the $(NH_4)_2SO_4$ precipitation
(Fig. 1B). The DNA polymerase still migrates at 4×10^5 daltons
on sepharose 6B, whereas the DNAse activity elutes in a broad peak
between 50,000 and 250,000 daltons. Since the molecular weight of
the DNA polymerase is unaffected by the removal of all the DNAse
activity, the two activities are probably part of different high
molecular weight complexes.

Functional characterization of the HMW DNA polymerases: The
DNA polymerase reactions associated with the sepharose 6B peaks in
Fig. 1A & B (fractions VI and VII) were characterized with respect
to ^{3}H-TTP incorporation and relaxation or breakage of supercoil
colicin E-1 DNA (Table 1 and Fig. 3). The two high molecular weight
polymerases had similar pH optima, divalent cation requirements,
sensitivity to sulfhydryl inhibitors, ethidium bromide inhibition
and affinity for TTP. The major differences between the two
fractions were noted in their template-primer requirements. Fraction
VI preferred ssDNA substrates 2-3 times more than activated DNA
substrates yet demonstrated similar binding affinities for either
DNA. In contrast, fraction VII preferred activated DNA substrates
5-6 times more than ssDNA substrates and demonstrated a slightly
lower affinity for DNA than fraction VI. The ability to
efficiently use ssDNA as a template primer by fraction VI therefore
must be the result of protein which has been removed or denatured
in the preparation of fraction VII. This result is substantiated indirectly by examining the heat stability at 46°C of fraction VI and
VII for ssDNA and activated DNA directed synthesis. Fraction VI
rapidly loses its ability to incorporate TTP using ssDNA in the
reaction mixture - 50% after 5 minutes at 46°C. When activated DNA
is used in the reaction mixture, both fraction VI and VII demonstrate
the same heat denaturation kinetics - losing 50% activity after 10
minutes at 46°C. The protein responsible for ssDNA utilization does
not appear to be the ssDNA binding protein since both preparations
contained the 16,000 peptide on SDS electrophoresis.

The two high molecular weight DNA polymerases also demonstrated
a number of differences when incubated with supercoiled colicin E-1
DNA (Fig. 3). Supercoiled colicin DNA migrates at O.51 in the O.8%
agarose gel columns whereas intact linear fragments produced by

---

C... *graphed on 6B equilibrated in 0.15M NaCl-TME. 3 ml fractions*
were collected for each column and aliquots were assayed for
DNA polymerase (open circles), ssDNA binding (stars), DNAse
(circled stars), and ATP-DNAse (closed circles). Protein
standards were run after identical experimental conditions. The
DNA polymerase peak in A is designated fr.VI, in B is designated
fr.VII, and in C is designated LMW in later studies.

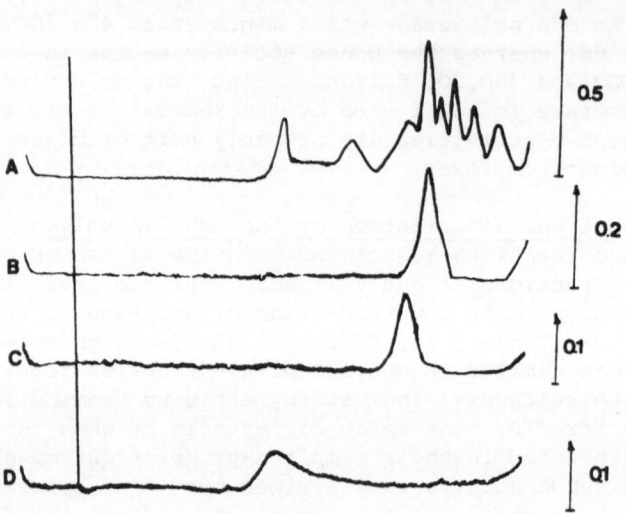

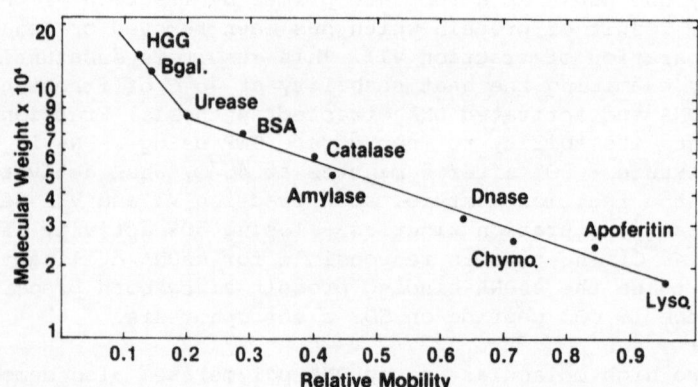

Figure 2. SDS gel electrophoresis of sepharose-6B column fractions.

Enzyme fractions from the sepharose columns in Fig. 1 were precipitated by 10% cold TCA; denatured, electrophoresed on 5 x 100 mM, 7% acrylamide gels and stained with Coomassie blue according to the procedure of Weber and Osborn (16). After destaining, the gels were scanned 600 nm with a linear transporter. Gel A. 50-55 ml of column A. Gel B. 98-103 ml of column C. Gel C. 120-125 ml of column C. Gel D. 145-155 ml of column C. Protein standards were run under identical conditions.

Table I

Physical and functional properties of *Chlamydomonas*
DNA polymerase fractions

	FR. VI	FR VII	LMW
1. Size, mw Sepharose-6B	$3-5 \times 10^5$	$3-5 \times 10^5$	0.45×10^5
2. Specific activity, nmoles dNMP/mg/hr			
aDNA	29.40	112.8	98.64
ssDNA	91.20	15.8	19.72
3. "Km" of DNA, µg/ml			
aDNA	8	17	30
ssDNA	5	n.d.	n.d.
dAT	16	17	30
4. Km of Mg^{++}, mM	15	8	5
5. Km of TTP, µM	0.63	0.63	2.85
6. pH optima, % maxima			
pH 7.0	(1.00)	0.63	0.75
pH 8.0	0.90	(1.00)	(1.00)
pH 9.0	0.65	0.71	0.75
7. (%) Inhibition by			
2 µM Ethidium Bromide	25	37	50
23 µM Rifamycins	50	n.d.	50
30 µM pCMB	20	51	60
2 mM NEM	58	52	60
8. Heat Inactivation - Time at $46^{\circ}C$ Required for 50% Activity:			
aDNA	10	10	5
ssDNA	5	n.d.	n.d.
9. Nucleotide Requirement	Stringent	Stringent	Relaxed

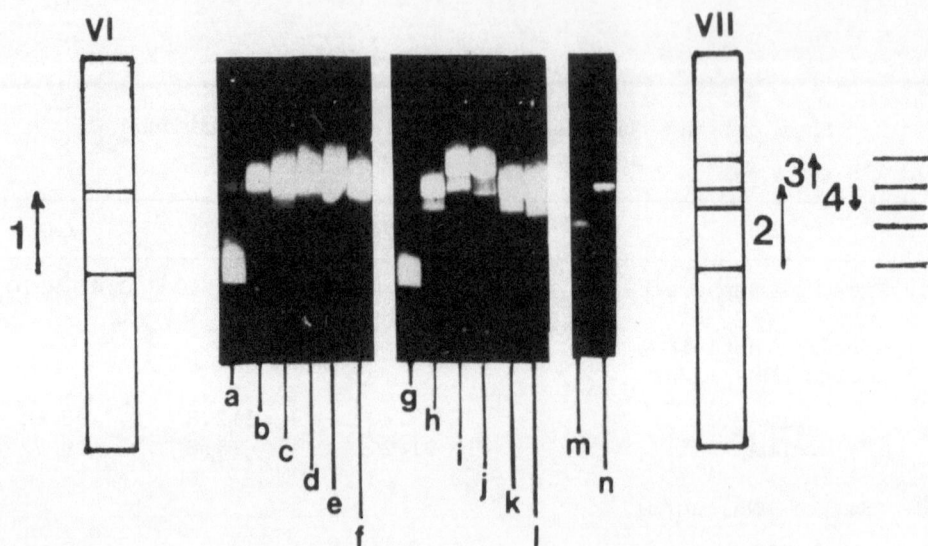

Figure 3. *Agarose electrophoresis of colicin E-1 assays.*

330ng of supercoil DNA was incubated at 35°C for 15 min. with either 4.6 μg fr. VI protein (gels b-f) or 3.9 μg of fr. VII protein (gels g-1) and: 0.1mM ATP (gels e,f,h,l); 0.1mM ATP and 0.01mM each dNTP (gels c,d,i,j); or 0.01mM each dNTP (gels b,h). Gel a is an enzymeless control. Gel g did not contain Mg⁺⁺. Gel m contained 110ng supercoil DNA treated with 1 μg E.cor-1 for 15 min. Gel n contained 110ng supercoil treated with 2μg RNAse H for 15 min. All reactions were terminated by the addition of SDS to 1%, and heating for 10 min. at 65°C. The samples were placed on 5 x 100 mm, 0.8% agarose columns, electrophoresed and ethidium bromide stained (19). The diagrams on each side summarize the effect of fractions VI or VII had on the colicin DNA. (1 and 2) supercoil → relaxed, (3) relaxed + ATP + dXTP → rep. relaxed (4) relaxed + dXTP → rep. linear.

EcoR-1 treatment migrate at 0.43 and relaxed circles produced by RNAse H treatment migrate at 0.32 (18,19). Supercoil colicin DNA can be completely relaxed with either fraction VI or VII in the presence of magnesium (Fig. 3). If ATP or dNTP is included in the reaction mixture, some of the relaxed circles are converted to partial supercoils or replication intermediates of a linear nature with fraction VII as evidenced by the appearance of a 0.37 DNA band (Gels h,k,l). In contrast, fraction VI colicin assays with ATP and dNTP displayed broad relaxed circle bands at O.32 (Gels b-f). Fraction VII assays containing dNTP and ATP demonstrated almost the complete conversion of relaxed circles to a new slow mobility of 0.22 (Gels i-j). The reduced mobility could be best explained by an opening of the relaxed circle in the form of a replication type intermediate. Incubation of fraction VII with only dNTP did not produce the O.22 band, indicating that ATP was necessary for this replicative structure (Gel h). Failure to resolve this structure in fraction VI may be the result of the ATP-dependent DNAse found in this fraction. If the replication intermediate can be verified as the O.22 band, then it would appear that both fractions VI and VII contain an ATP-dependent rep-type protein (20).

Cell cycle activities: The activities of DNA polymerase, ATP-dependent DNAse, and ssDNA binding proteins in fraction V were measured over the cell cycle by synchronizing cultures of Chlamydomonas by 12 hour light - 12 hour dark alternating regimes (8). Two one liter samples were removed every 3 hours from the fermentor, the cell count was determined by microscopic observation in an hemocytometer, and the cells were pelleted and fast-frozen in liquid nitrogen. After all the samples had been collected, the cell pellets were either processed for their total DNA content or for enzyme fraction V. The DNA polymerase was assayed in the presence of β, γ methylene ATP to minimize the possible stimulatory effects of accessory proteins. Each time-sample was subjected to an enzyme dilution so that under the assay conditions no substrates were limiting. As seen in Fig. 4, DNA polymerase increases from 3-9 hours in the light and 12-16 hours during the dark period. The double increase corresponds first to the period of chloroplast DNA synthesis and second to chromosomal DNA replication (21). This result might suggest that the same DNA polymerase is involved in both organelle and in nuclear replication or that the two potential polymerases co-purify throughout our preparation. The ATP-dependent DNAse rises only preceding chromosomal replication, thus implicating it as a chromosomal replicative protein. Surprisingly the ssDNA binding activity did not demonstrate any periodic increase over the cell cycle, which might suggest that there are several DNA binding proteins or that this protein has a broad cellular function which is not restricted to DNA replication (22). The failure to observe co-ordinate activities over the cell cycle certainly suggests that several types of complexes must exist within protein fraction V. Mixing experiments between hour 3 and hour 15 did not produce an unaccountable increase in polymerase, DNAse or binding activity so that activities probably are not the result of simple activation.

DISCUSSION

High molecular weight protein complexes containing the DNA polymerase, ATP-dependent DNAse, DNAse, ssDNA binding activity,

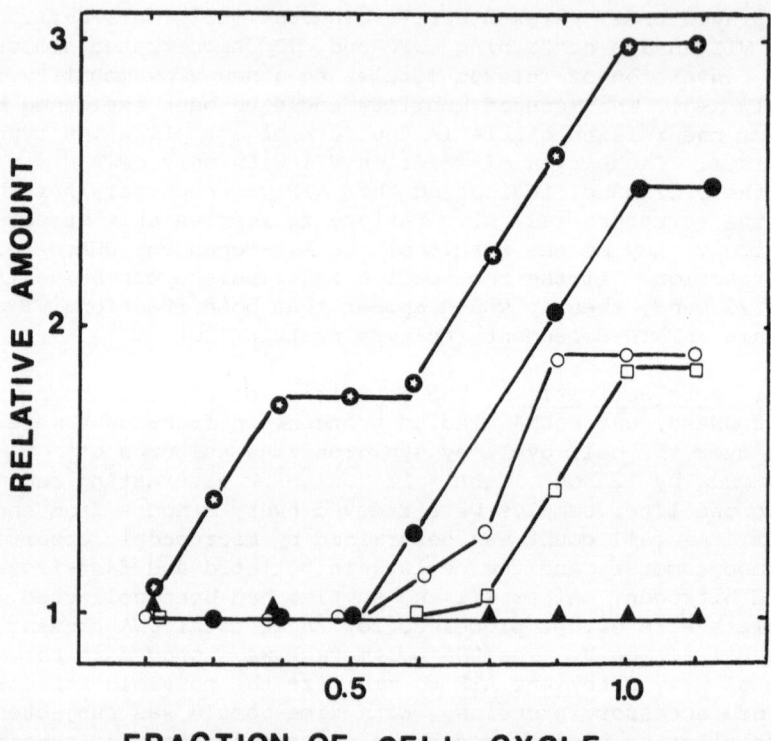

Figure 4. *Expression of replication activities in synchronized cultures*.

Fraction VI protein was prepared from frozen cell pellets that had been collected every 3 hours from a 12 hour light - 12 hour dark entrained culture of CW-15. Each sample was adjusted to a final volume of 5 ml and aliquots were removed to determine DNA polymerase activity (circled starts), ATP-DNAse (closed circles), and ssDNA binding (closed triangles). Each activity is expressed relative to the 0 fraction of the cell cycle. Cell counts (open squares) reflect observable daughter cells. Total DNA as measured by diphenylamine is represented by the open circles (2).

an ATP-dependent unwinding activity, and template-primer preference have been identified in extracts prepared by the gentle lysis of *Chlamydomonas*, CW-15. Despite comigration of these activities on sepharose-6B and native acrylamide gels in high salt and poly-ethylene glycol, macromolecular heterogeneity of the mixture is suggested by (1) the separation of ATP-dependent DNAse from the DNA polymerase without altering the molecular weight of the latter and (2) asynchrony of the expression of DNA polymerase, ATP-dependent DNAse and ssDNA binding activities over the cell cycle. The complexity of the high molecular weight fractions is supported by finding 8 different peptides under denaturing SDS electrophoresis. Dissociation of the high molecular complexes in low salt buffers and subsequent separation on sepharose-6B, indicated that all the ssDNA binding could be assigned to a 16,000 MW peptide. Since the macromolecular complexes were purified by their ability to bind ssDNA at 0.15M NaCl, the ssDNA binding protein must be an integral component of each complex.

T_4 gene 32 protein and *E.coli* DNA binding protein (DBP) have been reported to physically associate with homologous DNA polymerases and DNA ligases to form macromolecular complexes (23-25). In addition, genetic experiments with T_4 gene 32 indicate that this protein plays a role in recombination, repair and replication (26, 27). Gene 32 synthesis during T_4 infection appears to be auto-regulatory at translation and sensitive to titration by ssDNA *in vivo* (22). ssDNA binding proteins have been identified in two other eukaryons which are analogous to their bacterial predecessors in activity and affinity for replicative DNA polymerases (28,29).

The 60,000 mw, ATP-dependent DNAse has been purified to 90% homogeneity as estimated from SDS acrylamide gels. The enzyme appears to be a single stranded endonuclease, since it demonstrates lag type kinetics for the release of acid soluble cpm from ^{3}H-ssDNA substrates (S.J. Keller, unpublished results). Since colicin supercoils are not directly converted to linear DNA forms on long incubations with fraction VI, the ATP-DNAse must be coordinated with other proteins found in this fraction. Although the mechanism of the ATP-DNAse is not resolved, it is tempting to speculate that this protein may function as the knife at the replication fork (30). Alternatively the ATP-DNAse may function to prevent the accumulation of ssDNA chains from the replicating chromosome. Thus the enzyme would reduce the probability of recombination during normal duplication. A second DNAse activity has been associated with a 45,000 mw polypeptide, which is inhibited by 0.5mM ATP. Preliminary experiments suggest that the 45,000 DNAse is an exonuclease since it releases acid soluble cpm from ^{3}H-dsDNA without any delay in the reaction kinetics (Yee and Keller, unpublished results). ATP could therefore regulate the events of the replication fork by promoting strand separation and cleavage of the replicative branch. At low ATP concentrations, the DNA polymerase and the exonuclease could then be free to proof-read or excise mismatched bases. High ATP concentrations might stimulate replisome formation and strand separation. At present the role of ATP in eukaryon replication is unclear since (1) ATP stimulates the formation of replication complexes with ssDNA in rat liver nuclei and (2) ATP causes the dissociation of alpha and beta DNA polymerases from lymphocyte nuclei, thus inhibiting incorporation (32,33).

The low molecular DNA polymerase that has been purified from
the high molecular weight complexes demonstrates a number of
altered properties in the low molecular weight form. For example,
it has an increased Km for DNA and TTP, it is more heat labile than
the HMW forms, and it is more sensitive to inhibition by sulfhydryl
inhibitors and ethidium bromide. These differences can be
explained by assuming that other proteins in macromolecular complexes
are in association with DNA polymerase and therefore alter its
functionality. Alternatively, one could conclude that the other
proteins also have the capacity to interact with the inhibitors
or substrates thereby modifying the kinetic parameters. Metazoan
polymerases demonstrate similar problems in that small beta type
enzymes have been reported to be recovered from high molecular
weight complexes and have altered kinetic parameters (34-36).

In summary many of the events at the eukaryotic replication
fork appear to occur in an enzyme fraction that demonstrates macromolecular aggregates, affinity for single-stranded DNA, and contains
at least 8 different peptides. Until each peptide can be purified
to homogeneity, assigned a function and each of the complexes
reconstituted, the detailed mechanism of the replication fork will
be speculative.

ACKNOWLEDGEMENTS

This work was supported by Grants GB35440 and BMZ O2268AO1 from
the National Science Foundation and by the University of Cincinnati
Research Council.

REFERENCES

1.   Sager, R. (1972) Cytoplasmic Genes and Organelles. Academic
     Press, New York.

2.   Kates, J.R., Chiang, K.S. and Jones, R.F. (1968) Exp. Cell
     Res.   49:121-135

3.   Chiang, K.S. and Sueoka, N. (1967) J. Cell Physiol.   70
     (supple. 1):89-112

4.   Swinton, D. and Hanawalt, P.C. (1973) Biochem. Biophys. Acta
     294:385-395

5.   Morris, C.F., Sinha, N.K. and Alberts, B. (1975) Proc. Nat.
     Acad. Sci. USA   72:4800-4804

6.   Eisenberg, S., Scott, J.F. and Kornberg, A. (1976) Proc. Nat.
     Acad. Sci. USA   72:1171-1174

7.   Davies, D.R. and Plaskitt, A. (1971) Genet. Res. Camb. 17:
     33-43

8.   Lien, T. and Knutsen, G. (1976) Arch. Microbiol. 108:189-
     194

9.   Arndt-Jovin, D.J., Jovin, T.M., Bähr, W., Frischauf, A.-M.,
     and Marquardt, M. (1975) Eur. J. Biochem. 54:411-418

10. Meyer, R.R. and Keller, S.J. (1972) Anal. Biochem. 46:332-
     338

11. Tsai, R. and Green, H. (1973) J. Mol. Biol. 73:307-316

12. Lowry, O.H., Rosebrough, N.J., Farr, A.L. and Randall, R.J.
     (1951) J. Biol. Chem. 193:265-275

13. Layne, E. (1957) Methods in Enzymology 3:447-454

14. Keller, S.J., Ho, C., and Yuan, M.M. (1974) Fed. Proceedings
     33:1419

15. Davis, B.J. (1964) Ann. N.Y. Acad. Sci. 121:404-427

16. Weber, K. and Osborn, M. (1969) J. Biol. Chem. 244:4406-
     4412

17. Kornberg, A. (1974) DNA Synthesis. W.H. Freeman & Co., San
     Francisco.

18. Clewell, D.B. and Helinski, D.R. (1970) Biochemistry 9:
     4428-4440

19. Tanaka, T. and Weisblum, B. (1975) J. Bact. 121:354

20. Scott, J.R., Eisenberg, S., Bertsch, L.L. and Kornberg, A.
     (1977) Proc. Nat. Acad. Sci. USA 74:193-197

21. Oka, A. and Takanami, M. (1976) Nature 264:193-196

22. Krish, H.M., Bolle, A. and Epstein, R.H. (1974) J. Mol. Biol.
     88:89-104

23. Sigal, N., Delius, H., Kornberg, T., Gefter, M.L. and Alberts,
     B. (1972) Proc. Nat. Acad. Sci. USA 69:3537-3541

24. Molineux, I.J., Friedman, S. and Gefter, M.L. (1974) J. Biol.
     Chem. 249:6090-6098

25. Mosig, G. and Breschkin, A.M. (1975) Proc. Nat. Acad. Sci.
     USA 72:1226-1230

26. Tomizawa, J., Anraku, N. and Iwama, Y. (1966) J. Mol. Biol.
     21:247-253

27. Alberts, B.M., Amodio, F.J., Jenkins, M., Gutmann, E.D. and
     Ferris, F.L. (1968) Cold Spring Harbor Symp. Quant. Biol. 33:
     289-305

28. Herrick, G., Delius, H. and Alberts, B.M. (1976) J. Biol.
     Chem. 251:2124-2146

29. Banks, G.R. and Spanos, A. (1975) J. Mol. Biol. 93:63-77

30. Kornberg, A. (1969) Science 163:1410-1418

31. Banks, G.R., Holloman, W.K., Kairis, M.V., Spanos, V. and
     Yarranton, G.T. (1976) Eur. J. Biochem. 62:131-142

32. Genta, V.M., Kaufman, D.G. and Kaufmann, W.K. (1976) Nature
     259:502-503

33. Thompson, L.R. and Mueller, G.C. (1975) Biochem. Biophys.
     Acta 414:231-241

34. deRecondo, A.M. and Abadiendebat, J. (1976) Nuc. Acid Res.
     3:1823-1837

STUDIES WITH ENZYME COMPLEXES THAT SYNTHESIZE ADENOVIRUS 2 DNA

IN VITRO

Max Arens*, Tadashi Yamashita, Kazuhiko Ito, Takashi
Tsuruo, Raji Padmanabhan and Maurice Green*

Institute for Molecular Virology
St. Louis University Medical School
St. Louis, Missouri 63110, U.S.A.

INTRODUCTION

The genomes of the small (SV40 and polyoma) and medium-sized
(adenovirus) mammalian DNA viruses lend themselves readily to biochemical and structural analysis. Thus, these viruses have been
widely used as tools in model systems in much the same manner that
the small phages have been used for the understanding of
chromosomal replication in bacteria. To this end, nuclear and
subnuclear replication systems have been sought in order to
simplify and further investigate the process of eukaryotic DNA
replication. We have selected subnuclear complexes from adenovirus
2 (Ad2) infected KB cells on the basis of their ability to
synthesize Ad2 DNA *in vitro* in a DNA polymerase reaction mixture.
We have characterized the *in vitro* products and some of the proteins
and enzymes associated with these complexes.

MATERIALS AND METHODS

Cell Culture and Virus Infection. KB-III cells were cultured
and infected with 100 PFU/cell of adenovirus type 2 (Ad2) as
described (15).

Preparation of Ad2 DNA Replication Complexes.

I. Nuclear membrane replication complex (13). Infected cells were
harvested at 18 hpi, washed and resuspended in reticulocyte
standard buffer (RSB) at 10^6 to 10^7 cells/ml. Nuclei were
prepared by Dounce homogenization and centrifugation through 25%
sucrose in RSB. Nuclei were disrupted by brief sonication, sodium

A.S.I. Participants

citrate was added to 10% and the suspension was layered over a
discontinuous sucrose gradient consisting of (from top to bottom)
1.16, 1.18,1.20 and 1.22 g/ml of sucrose in TKM (50 mM Tris-HCl
(pH 7.5), 5 mM MgCl$_2$, 25 mM KCl and 3 mM dithiothreitol (DTT) plus
10% sodium citrate. After centrifugation the DNA replication
complex was located at the interface between sucrose densities
of 1.18 and 1.20 g/ml and was removed with a Pasteur pipette.

II. Soluble replication complex (14). Nuclei were prepared from
late infected cells as described above and suspended at 1-2 x 10^7
nuclei per ml in 50 mM Tris-HCl (pH 7.8) plus 25 mM KCl containing
2 mM Na$_2$HPO4 and 2 mM DTT. Sodium heparin was added with stirring
to 1 mg/ml final concentration and the suspension was placed in
ice for 30 min. Cell DNA was sheared by passage through a 17G
needle and the suspension was centrifuged at 45,000 x g for 1 h and
the supernatant was dialyzed. This crude complex was further
purified by exclusion chromatography on a 1.5 x 90 cm Bio Gel A
50m column as described (14). Eluted fractions with peak endogenous
DNA polymerase activity were stored at -70°C and were used without
further treatment.

In Vitro Synthesis of DNA. Endogenous DNA in either replication
complex was labeled *in vitro* in reaction mixtures containing 90 μM
each of dATP, dCTP, and dGTP, approximately 0.5 μM (^{3}H)dTTP, and
3 mM DTT in 50 mM Tris-HCl and MgCl$_2$ optimized for each complex.
In analytical assays, portions of the reaction mixture were spotted
on DE81 filters for counting and in preparative assays the DNA
products were purified by phenol extraction and ethanol
precipitation as described (13,14).

Dissociation and Fractionation of the Soluble Complex for
Enzyme Characterization. Fractions with peak endogenous DNA
polymerase activity from a Bio Gel A 50m column were pooled and BSA
was added to 50 μg/ml (1). After brief sonication, the solution was
dialyzed against saturated ammonium sulfate. The resulting
precipitate was collected and dissolved in buffer containing 0.5 M
NaCl. DEAE-Sephadex (equilibrated in the same buffer with 0.5 M
NaCl) was added and the slurry was placed in ice for 1 h with
occasional stirring. Sephadex was removed by centrifugation and the
supernatant was saved as the solubilized enzyme preparation (0.5 M
NaCl eluate) (1).

RESULTS

The characterization of the two Ad2 DNA replication complexes
has revealed many similarities but also many differences. The
results of the characterizations will be discussed in this section
and then will be compared and contrasted in the Discussion.

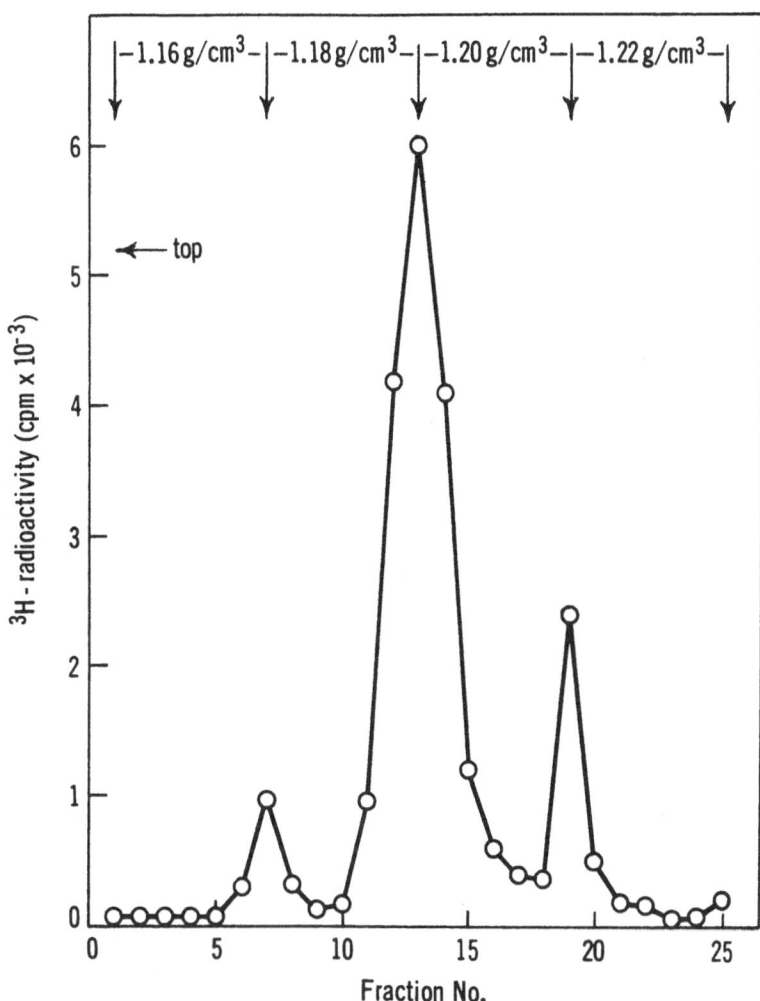

Figure 1. Association of newly synthesized DNA with nuclear membrane fractions isolated by the discontinuous sucrose gradient procedure. Nuclei prepared from KB cells pulse labeled with (^{3}H)thymidine for 5 min at 18 h postinfection were sonically disrupted and fractionated by the discontinuous sucrose gradient procedure. After centrifugation for 60 min at 25,000 rpm in a Spinco SW27 rotor, 1 ml fractions were collected and acid-precipitable radioactivity was determined.

Characterization of the Nuclear Membrane Replication Complex

General Characteristics. Ad2 infected KB cells were pulsed
labeled with (^{3}H)thymidine for 30 min at 18 hpi and then harvested,
washed and treated as described in Methods for preparation of the
membrane complex. A profile of the discontinuous sucrose gradient
used for preparation of the complex is shown in Fig. 1. About 80%
of the pulse labeled DNA sedimented to the 1.18 - 1.20 g/ml interface. When prepared from unlabeled cells and assayed in a standard
DNA polymerase reaction mixture, this fraction had endogenous DNA
polymerase activity that required all 4 dNTPs and Mg^{+2} for maximal
activity and was strongly inhibited by DNase, RNase, actinomycin D
and sodium pyrophosphate. The product of the in $vitro$ reaction was
determined by hybridization to be exclusively Ad2 viral DNA.

Analysis of in vitro DNA product. Ad2 DNA synthesized in $vitro$
in a scaled-up standard reaction mixture was purified and
analyzed in neutral and alkaline sucrose gradients. A neutral
gradient of the in $vitro$ labeled product revealed the sedimentation
pattern shown in Fig. 2. The majority of the product sedimented
at about 17-18S with a shoulder in the region of full length Ad2
DNA (31S). Although the in $vitro$ product is synthesized on a
large DNA template, the DNA that is made is itself small as shown
by analysis on alkaline sucrose gradients (Fig. 3). The DNA
labeled in $vitro$ is about 6-9S after a 5 min incubation and is not
further elongated even after 60 min.

Further analysis by CsCl equilibrium gradient centrifugation
of the in $vitro$ product after labeling with (^{3}H)dTTP confirmed its
density as being exactly that of Ad2 DNA. DNA labeled in $vitro$
with (^{3}H)BrdUTP and centrifuged to equilibrium in CsCl indicated
that the synthesis was semiconservative but that virtually no
reinitiation took place in $vitro$ since none of the product banded
at the density of fully substituted (HH) DNA.

Analysis of DNA Polymerase Activity in the Membrane Complex.
In order to study the properties of the DNA polymerase in the
replication complex it was necessary to remove the endogenous DNA
template. This was accomplished by treatment of the purified membrane
complex with 0.5% Triton X-100 and 0.5 M KCl (6) and then
centrifugation at 275,000 x g for 1 h to pellet the endogenous
DNA. The supernatant (containing 100% of the DNA polymerase activity)
was dialyzed to low salt, adsorbed and eluted batchwise with 0.6 M
KPO$_4$ buffer (pH 7.2) from phosphocellulose and again dialyzed to low
salt before adsorbing to DEAE-cellulose. Gradient elution from DEAE-
cellulose and assay of DNA polymerase activity yielded a peak of
activity at 0.1 M KCl with a small shoulder at 0.15 M KCl. The total
activity from an Ad2 membrane complex was pooled, concentrated and
analyzed on a glycerol gradient as shown in Fig. 4. The single
peak of activity with both activated calf thymus DNA and poly(A)·

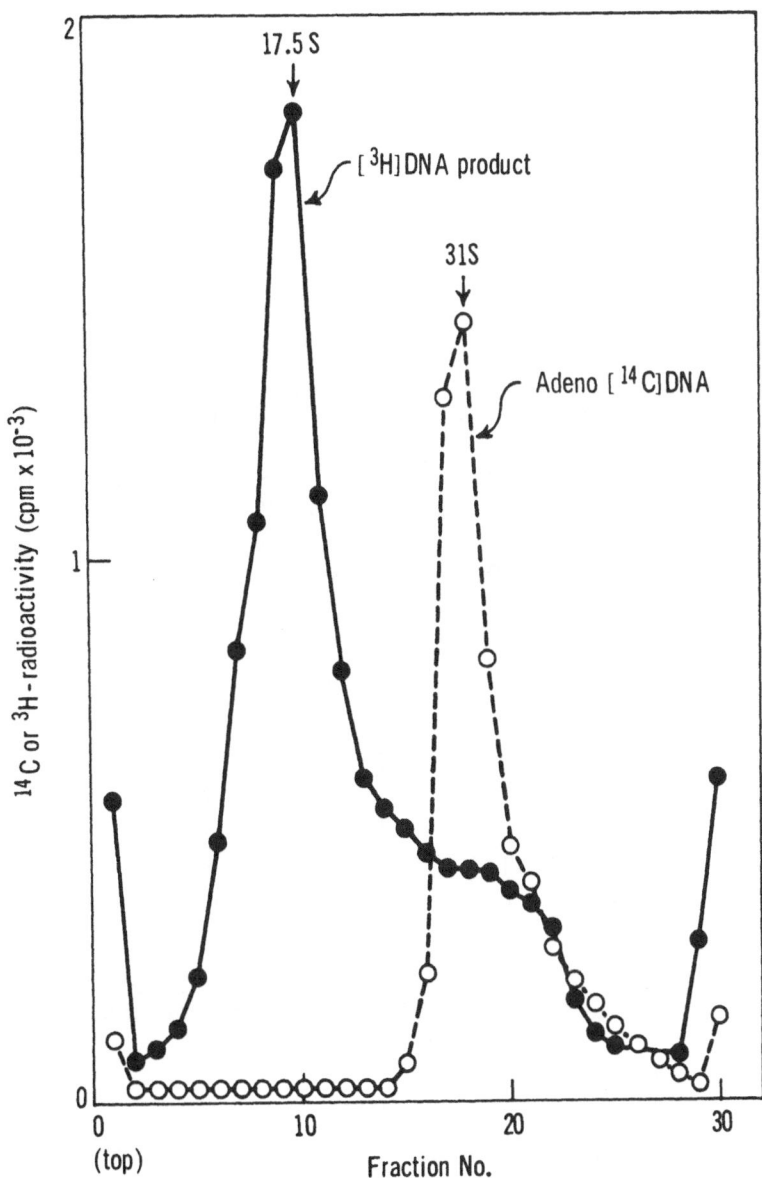

Figure 2. Rate zonal sedimentation in neutral sucrose gradients of (³H)DNA synthesized *in vitro* by the viral DNA membrane complex. DNA was synthesized *in vitro* for 60 min by the standard assay and purified by phenol extraction. The size of the DNA produced was analyzed by centrifugation in neutral sucrose gradients together with adenovirus 2 (¹⁴C)DNA as marker.

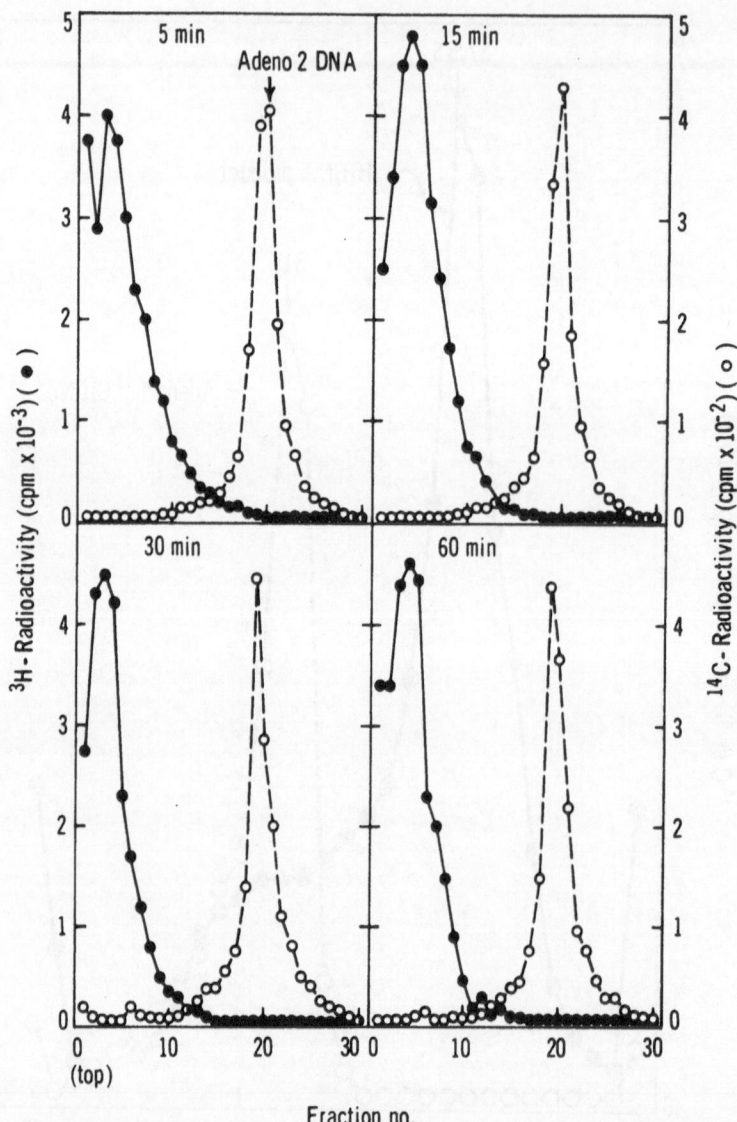

Fraction no.

Figure 3. Rate zonal sedimentation in alkaline sucrose gradients of
(³H)DNA synthesized in vitro by the adenovirus 2 DNA membrane
complex. The standard DNA polymerase assay was used with the
incubation times noted, and (³H)DNA was purified by phenol
extraction. The size of the DNA product was analyzed by centrifugation together with adenovirus 2 (¹⁴C)DNA as marker in alkaline
sucrose gradients. ●————● (³H)DNA product; ○------○ adenovirus 2 (¹⁴C)DNA.

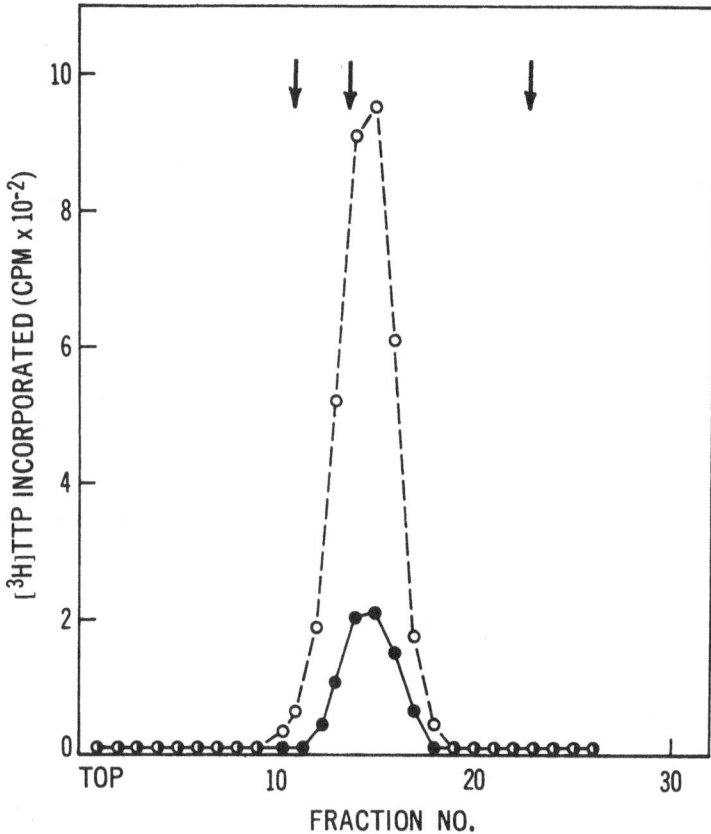

Figure 4. *Glycerol gradient sedimentation of DNA polymerase activity extracted from the membrane complex and purified over phosphocellulose and DEAE-cellulose column chromatography. Each fraction from the gradient was assayed with activated calf thymus DNA as template·primer (●) or with poly(A)·(dT)$_{12-18}$ as template-primer (○). Markers were BSA (4.4S), Escherichia coli alkaline phosphatase (6.3S) and beef liver catalase (11.2S) which were centrifuged in parallel gradients.*

oligo(dT) as template·primer and the measured sedimentation co-efficient of about 6.7S taken together virtually establish the identity of the enzyme as DNA polymerase γ (5,6). Further purifi-cation of DNA polymerase on DNA-cellulose showed a possible slight contamination with another DNA polymerase γ (6). We were also

unable to completely exclude the possibility of the presence of a
very small amount of DNA polymerase α in the purified preparations.

DNA polymerase activity from a membrane complex purified from
uninfected KB cells exactly as above yielded slightly different
results in that the glycerol gradient showed a split peak of
activity and DNA cellulose column analysis showed a larger
proportion of the contaminating DNA polymerase γ. Thus, we conclude
that the major activity in the uninfected complex is also DNA
polymerase γ.

Characterization of the Soluble Replication Complex

General Characteristics. The soluble replication complex
purified as described in Methods from Ad2 infected or uninfected
KB cells showed DNA polymerase activity as indicated in Fig. 5.
The complex from Ad2 infected cells had much higher endogenous
activity than that from uninfected cells, however, the uninfected
complex was slightly stimulated by the addition of exogenous DNA,
whereas the infected cell complex was slightly inhibited by added
activated calf thymus DNA.

Sodium citrate had a significant effect on both the activity
and conformation of the A d2 soluble replication complex.
Endogenous DNA polymerase activity was stimulated up to 7-fold by
the addition of sodium citrate to the reaction mixture with the
crude (not B10-Gel purified) complex. Citrate was optimal at
a concentration of 30 mM.

Sedimentation of the crude complex through sucrose gradients
revealed a correlation between conformation of the complex and the
citrate concentration. As shown in Fig. 6, sedimentation in the
absence of citrate was very rapid (Fig. 6A) but became slower
and heterogenous at low citrate (30 mM, Fig. 6B). A discrete
peak was observed at 135 mM (Fig. 6C) then the complex sedimented
homogenously at 340 mM citrate (Fig. 6D).

Analysis of the in vitro product. The Ad2 DNA labeled *in vitro*
by this replication complex was analyzed on neutral and alkaline
sucrose gradients. After a short (1 min) pulse label *in vitro*, the
neutral sucrose profile (Fig. 7A) very closely resembled those
observed after an *in vivo* pulse label of DNA. A large percentage of
the product sedimented considerably faster than the 31S marker but
there was no DNA that sedimented slower than 31S. After a pulse and
60 min chase *in vitro*, virtually all of the product cosedimented
with the viral DNA marker (Fig. 7B). Alkaline sucrose gradients of
products that were pulse-labeled (Fig. 8A) or pulse-labeled then
chased *in vitro* (Fig. 8B) were also found to be very revealing with
regards to the mechanism of *in vitro* synthesis. The pulse-labeled

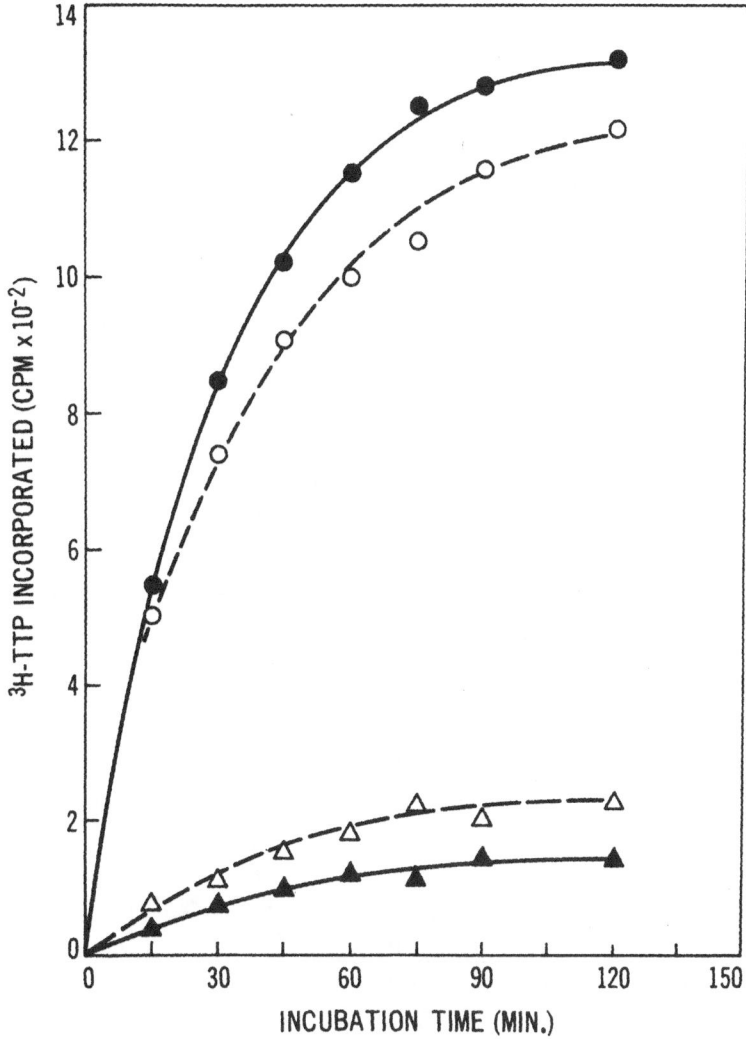

Figure 5. DNA polymerase activity of the crude soluble DNA
replication complex from uninfected (▲,△) and Ad2-infected
(●,○) cells. The complexes were prepared and DNA polymerase
activity was determined using the endogenous template (▲,●) or
adding activated calf thymus DNA (20 μg) (△,○).

DNA sedimented heterogenously between about 8 and 34S but most of
the radioactivity could be chased into full-length (34S) DNA after
60 min of incubation. Even after a short pulse *in vitro*, DNA of a
particular size class did not accumulate in the replication complex.

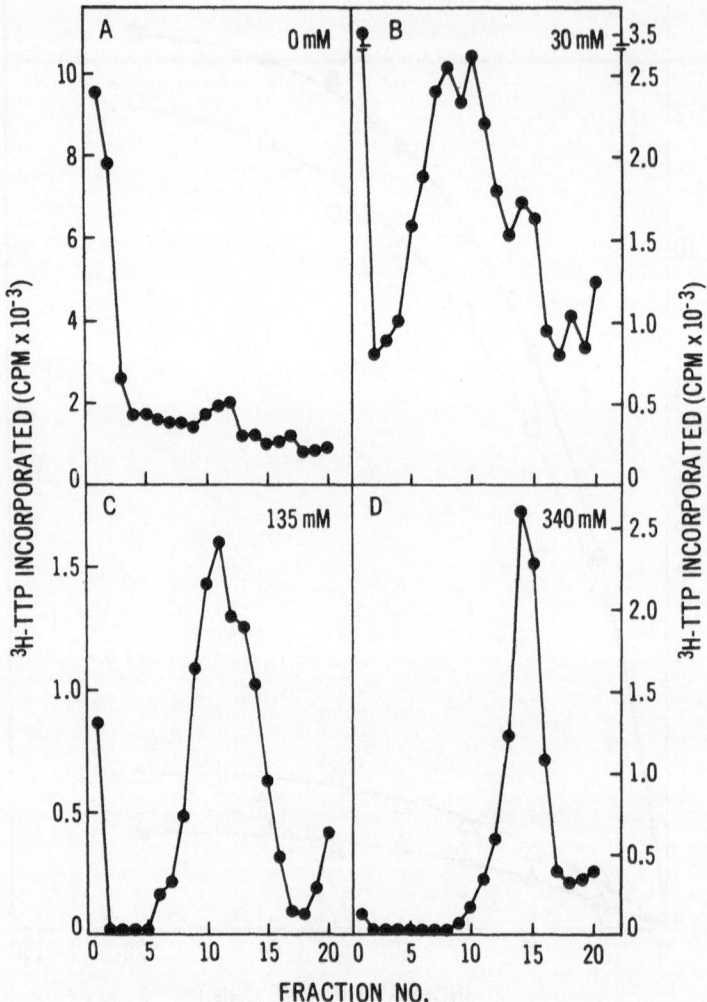

<u>Figure 6</u>. *Effect of sodium citrate on the sedimentation rate of the Ad2 DNA replication complex. Preparations of the crude soluble complex from Ad2-infected cells were treated with the indicated amounts of sodium citrate and layered onto a linear 15 to 30% (w/v) sucrose gradient containing the indicated amounts of sodium citrate. The gradients were centrifuged in a Spinco SW 41 rotor at 24,000 rpm for 16 h at 4°, fractionated, and aliquots were assayed for endogenous DNA polymerase activity. The final concentration of sodium citrate in the assay mixture was adjusted to 30 mM (gradients A, B and C) or 70 mM (gradient D). The Ad2 (³H)DNA marker run in separate tubes containing 135 mM and 340 mM sodium citrate sedimented as a symmetrical peak in Fraction 19.*

DNA was synthesized by the soluble replication complex during
a 60 min continuous labeling period with BrdUTP substituted for
dTTP and (^{3}H)dGTP for dGTP. The products were analyzed by centrifugation to equilibrium in CsCl as shown in Fig. 9. The shift of
over 90% of the BrdUTP-substituted DNA away from the LL region
confirms that the *in vitro* synthesis in the complex is semiconservative. The absence of radioactivity in the fully substituted
region (HH) indicates that reinitiation does not take place *in
vitro*. The density labeled product was also analyzed in alkaline
Cs$_2$SO$_4$ gradients. The results showed a peak shift of 50 mg/ml
from the L marker (Fig. 10). This would correspond to an overall
average substitution of about 75% for *in vitro* completed daughter
strands and thus confirms extensive elongation of strands *in vitro*
and also excludes the possibility of a repair-type of end-addition
mechanism,

In Vitro Termination by the Replication Complex. Although the
replication complex appears to carry out general synthesis of Ad2
DNA in a manner similar to that observed *in vivo*, it was of interest
to determine if the complex could perform *in vitro* a rather more
specific function of replication, namely, termination. To this end,
viral DNA was labeled for short times *in vitro* with (α-^{32}P)dTTP
then purified by phenol extraction. The completed (i.e., totally
double-stranded) molecules were separated from DNA containing single-
stranded regions by chromatography on BND-cellulose. The ds DNA
was digested with specific restriction endonucleases and the radioactivity measured in each of the DNA fragments was compared to the
expected radioactivity in the fragment assuming uniform labeling of
the DNA (i.e., equivalent to the actual physical percentage of
the fragment). As shown in Fig. 11, there is a gradient of radioactivity which increases dramatically toward the molecular ends of
the viral DNA. Analysis of the separated strands of fragments from
near the molecular ends has revealed a high degree of specificity
of strand labeling in that fragments from the right end are
labeled predominantly (80-90%) in the *l* strand and fragments from
the left end are labeled predominantly in the *r* strand (data not
shown). This pattern is the same as that observed for *in vivo*
termination (2,4,8,10) of adenovirus DNA synthesis and supports
bidirectional synthesis with initiation and termination sites at
the molecular ends. We thus conclude that *in vitro* synthesis is
also bidirectional and termination *in vitro* is indistinguishable
from the same process as it occurs *in vivo*.

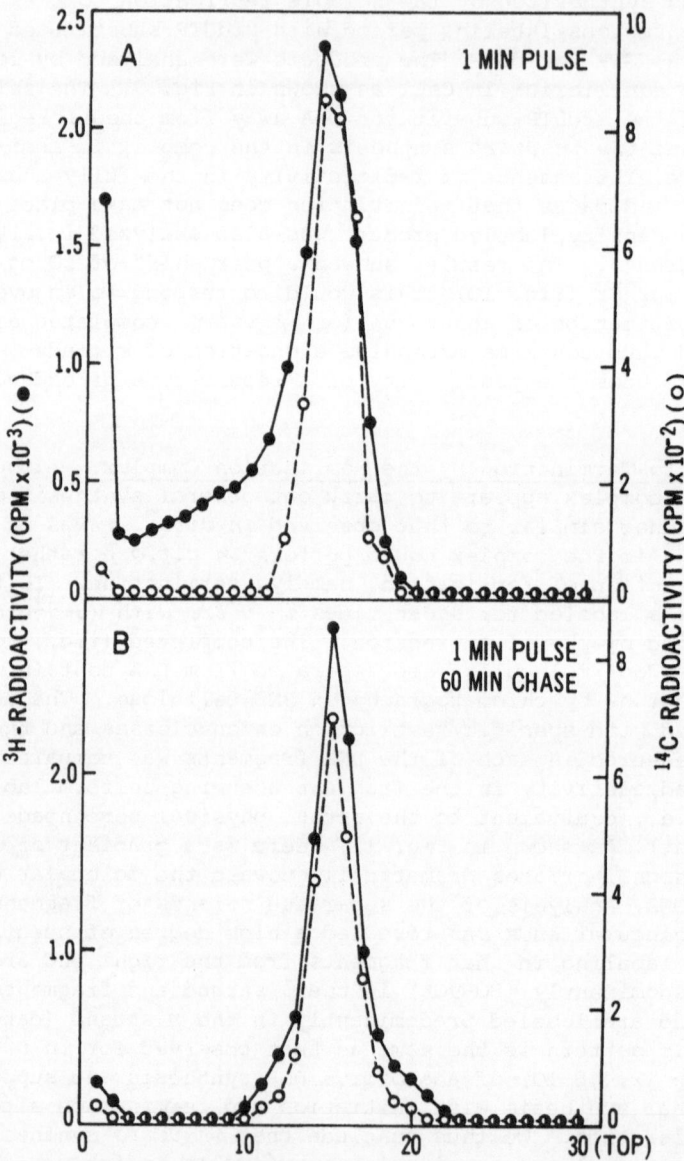

Figure 7. *Neutral sucrose gradient rate zonal sedimentation of Ad2 DNA synthesized in vitro*. *The DNA products were labeled (A) for 1 min with (^{3}H)dTTP, and (B) for 1 min followed by a 60-min chase with excess unlabeled dTTP (100 μM) using the endogenous reaction of the soluble complex. The (^{3}H)DNA was purified by phenol extraction. The size of the DNA products was analyzed by centrifugation in neutral sucrose gradients together with Ad2 (^{14}C)DNA as internal marker. (●———●, (^{3}H)DNA products; O ----- OAd2 (^{14}C)DNA).*

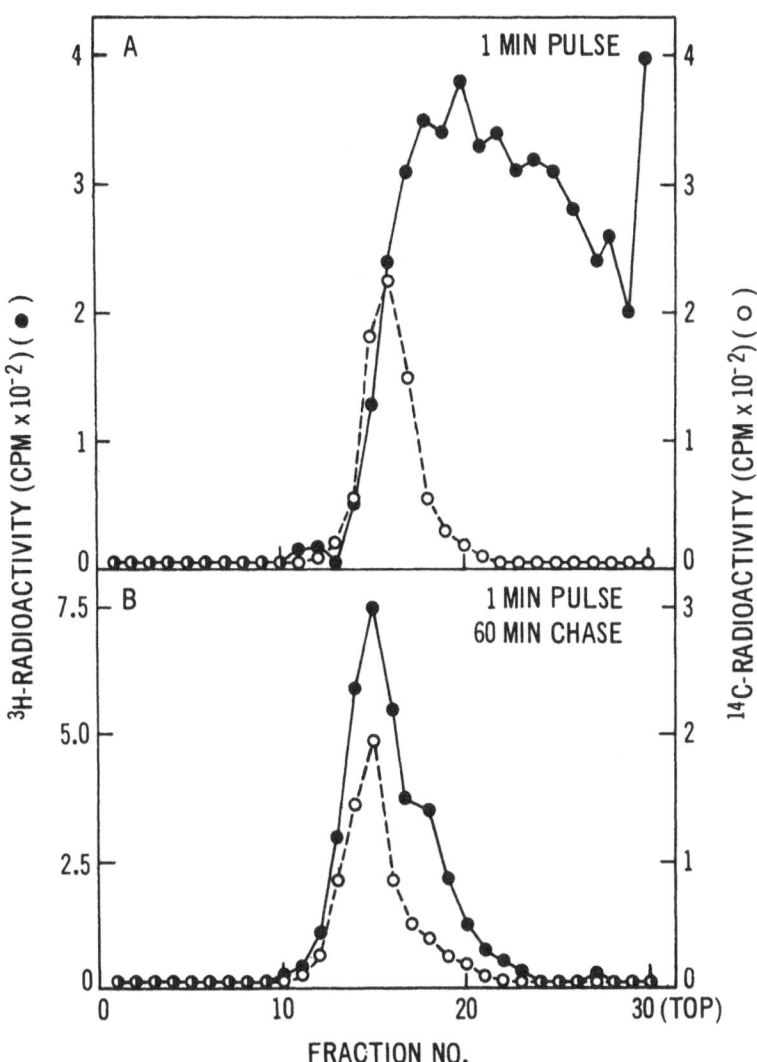

Figure 8. _Alkaline sucrose gradient rate zonal sedimentation of Ad2_
DNA synthesized in vitro. _The DNA preparations analyzed in Fig. 7_
were centrifuged in alkaline sucrose gradients with Ad2 (^{14}C)DNA
as an internal marker.

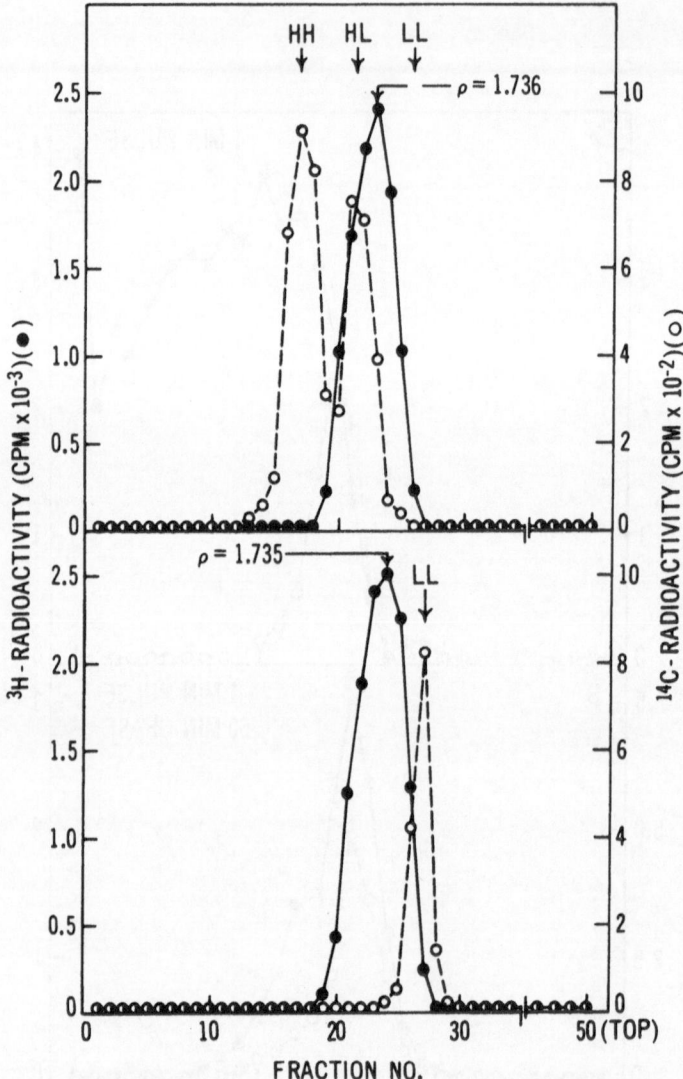

Figure 9. CsCl density gradient centrifugation of Ad2 DNA synthesized *in vitro* in the presence of BrdUTP. DNA was synthesized by the
soluble complex under standard conditions for 60 min except that
(^{3}H)dTTP and dGTP were replaced with BrdUTP and (^{3}H)dGTP,
respectively. DNA was purified and analyzed by equilibrium centrifugation in CsCl. Upper, the pattern for Ad2 DNA synthesized *in
vitro* in the presence of BrdUTP after centrifugation with Ad2 DNA
HH and HL used as internal density markers; lower, the same
preparation after centrifugation with Ad2 DNA LL as marker;
●———●, *in vitro* DNA products; ○-----○(^{14}C)Ad2 DNA density marker.

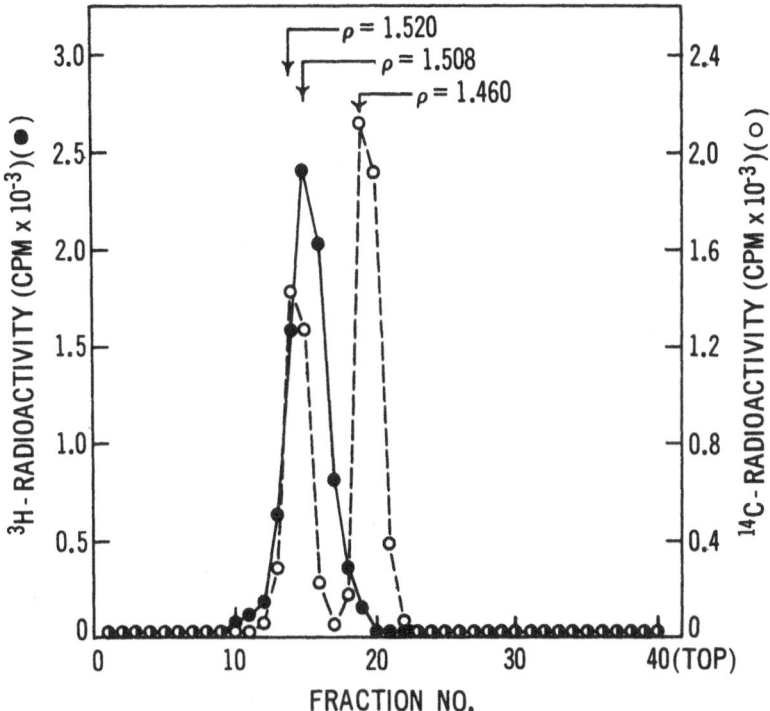

Figure 10. Semiconservative replication of DNA in vitro; alkaline Cs₂SO₄ equilibrium gradient. Alkaline Cs₂SO₄ gradient centrifugation of Ad2 DNA synthesized in vitro. The in vitro DNA products prepared with the soluble complex in Fig. 9 were analyzed by equilibrium density gradient centrifugation in alkaline Cs₂SO₄. ●━━━●*, in vitro DNA products;*○----○*Ad2 (¹⁴C) DNA density marker.*

Characterization of the Proteins in the Soluble Complex. Total proteins from the crude and Bio Gel purified complexes were separated on SDS-polyacrylamide gels. After staining with Coomassie blue, gels were scanned to reveal the profiles shown in Fig. 12. The crude complex contained at least 18 protein peaks (Fig. 12A) about 6 of which were relatively enriched in the complex after purification on a Bio Gel column (Fig. 12B). The most prominent polypeptides in the Bio Gel purified complex are the 75K protein and several low molecular weight proteins that may correspond to some of the early viral proteins (3). The 100K and 33K proteins which are present in the crude complex are selectively excluded during Bio Gel purification.

Proteins labeled from 4-18 h p.i. with (¹⁴C)leucine were electrophoresed in SDS-polyacrylamide gels as shown in Fig. 13.

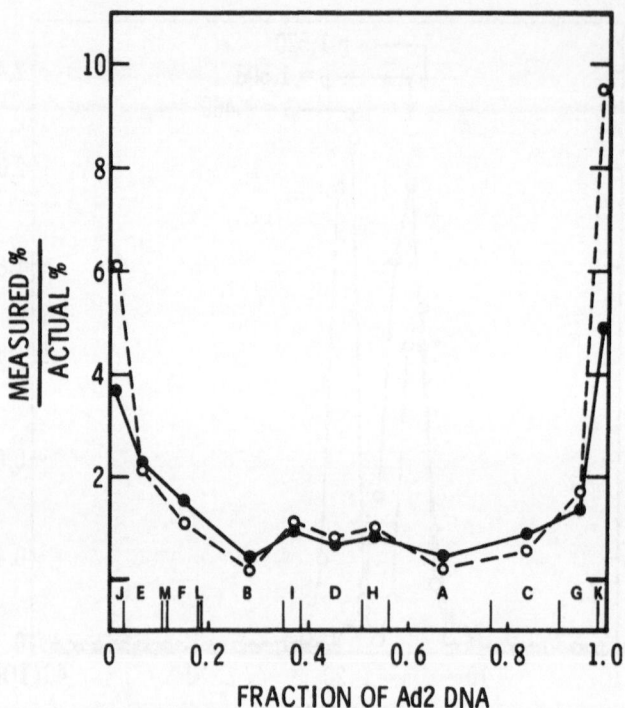

Figure 11.   *Relative incorporation in vitro of (α-³²P)dTTP into SmaI restriction fragments. DNA was labeled for 10 min (-O-) or 20 min (-●-) and processed by phenol extraction and then purification of dsDNA by BND-cellulose chromatography prior to digestion with the SmaI restriction endonuclease. Small letters on the abscissa represent SmaI restriction fragments of Ad2 DNA (C. Mulder, M. Green and H. Delius, personal communication). Fragments A and B were not separated in the gels and thus were treated collectively as one fragment. Fragments L and M together constitute about 1% of the genome and they were not considered in the calculations.*

Again, the crude complex had a very heterogeneous protein complement (Fig. 13A) but after passage over Bio Gel the 75K DNA binding protein was by far the major labeled protein in the replication complex (Fig. 13B). The labeling conditions used would probably not label very early proteins nor would they label small amounts of proteins to a high specific activity.

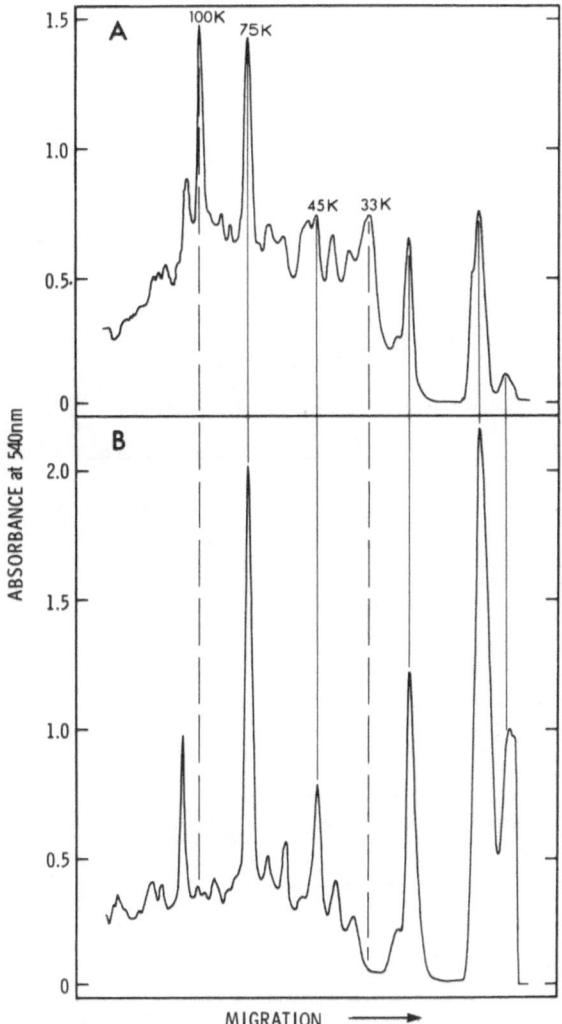

Figure 12. *Sodium dodecyl sulfate-polyacrylamide gel analysis of crude and Bio-Gel purified replication complex. The crude replication complex was prepared and purified on a Bio-Gel A 50m column. Aliquots of the crude and purified complexes, each containing 100 μg of protein, were precipitated with trichloroacetic acid, dissolved in electrophoresis buffer and electrophoresed in 6% polyacrylamide cylindrical gels. Stained gels were scanned at 540 nm using a Gilford spectrophotometer and linear transport mechanism. Molecular weights were determined from semilogarithmic plots using the following standard proteins: phosphorylase a, 94,000 daltons; bovine serum albumin, 67,000 daltons; human gamma globulin heavy chain, 50,000 daltons, ovalbumin, 43,500 daltons; human gamma globulin light chain, 25,000 daltons; cytochrome c, 12,400 daltons.*

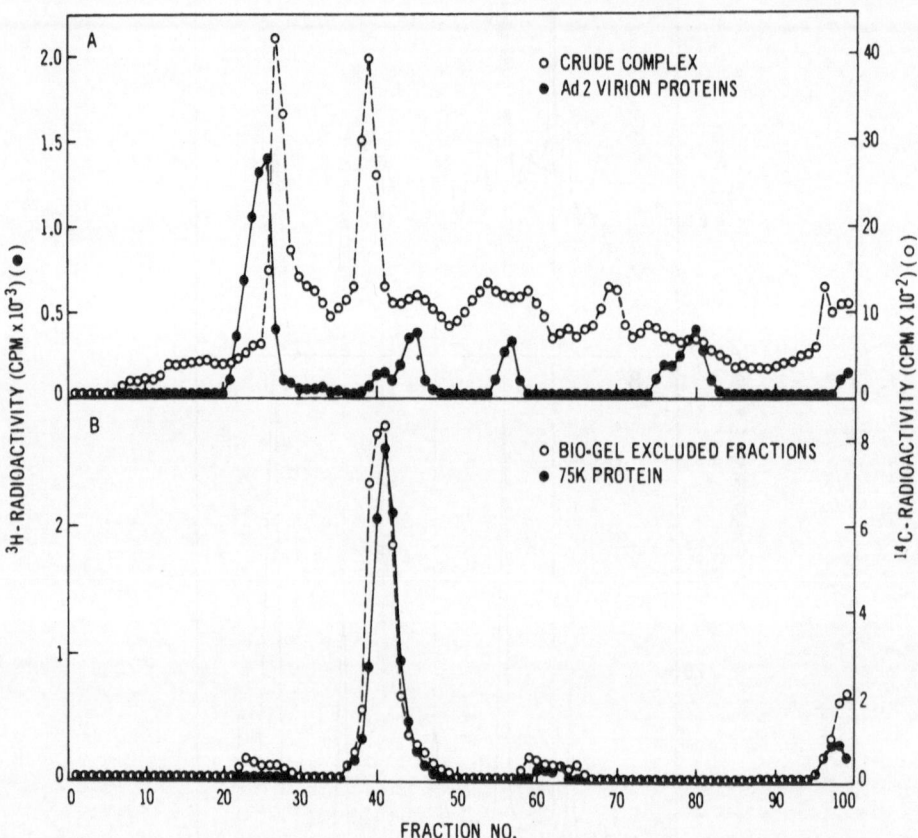

Figure 13. Sodium dodecyl sulfate-polyacrylamide gel analysis of crude and Bio-Gel purified (^{14}C)leucine labeled replication complex. The adenovirus 2 infected cells were labeled from 4-18 hours post infection with 2 µCi/ml of (^{14}C)leucine in leucine-free media. Aliquots of the crude and purified complexes were precipitated by trichloroacetic acid after addition of 50 µg of ovalbumin.

Characterization of DNA Polymerase Activity. Initial
characterization of the DNA polymerase activity solubilized from
the replication complex (see Materials and Methods) was performed
by the use of various template·primers, assays in high salt and
in the presence of specific inhibitors. Table I gives the results
of these assays and compares them with similar assays using KB
cell DNA polymerases α, β and γ. The activity from the soluble
complex does not directly correlate with any of the purified
enzymes with respect to all of the various criteria used for
comparison. The most notable exceptions are the activities with
poly(A)·(dT)$_{12-18}$ and the activities in the presence of 10 mM
N-ethylmaleimide (NEM). These two conditions taken together strongly
indicate the absence of DNA polymerase β (which is resistant to NEM)
and suggest the presence of DNA polymerases α and γ.

Fig. 14 shows the relative activities of different enzymes
in the standard reaction mixture and in the presence of various
concentrations of p-hydroxy-mercuribenzoate (p-HMB). Under these
conditions, DNA polymerase γ is virtually unaffected (some
preparations are slightly inhibited) whereas DNA polymerase α
is strongly inhibited (∿90%) at the higher concentrations of p-HMB.
The DNA polymerase activity solubilized from the replication complex
showed an intermediate inhibition pattern which paralleled that of
polymerase α but was inhibited about 30-40% less at all concentrations in the parallel portion of the curves. A curve
representing assays with the intact complex also fell between the
uninhibited γ and strongly inhibited α polymerases but, perhaps due
to conformational aspects of the complex, was inhibited less than
the solubilized enzymes. These observations again indicate the
presence of two enzymes in the solubilized preparation and, with
the exclusion of β as discussed for Table I, the data are
consistent with both α and γ polymerases being present in an
activity ratio of about 2 to 1. In fact, a mixture of purified
DNA polymerases α and γ (2 to 1) gave a curve very similar to that
for the solubilized enzymes shown in Fig. 13 (data not shown).

Further characterization of the two DNA polymerases present
in the soluble replication complex was accomplished by taking
advantage of the differences in K_m for triphosphates of the
different enzymes. Experimental determinations of the K_m for dTTP
of DNA polymerases α and γ yielded the values of 4 μM and 0.2 μM,
respectively. The double-reciprocal plot shown in Fig. 15, in
which the solubilized enzyme preparation was used, was biphasic
with one line indicating K_m for dTTP of 2 μM and the other line a
K_m of 0.6 μM. These data confirm the presence of DNA polymerases
α and γ in the Ad2 replication complex.

Demonstration of Other Enzyme Activities in the Replication
Complex. Although the 75K DNA binding protein and at least one and

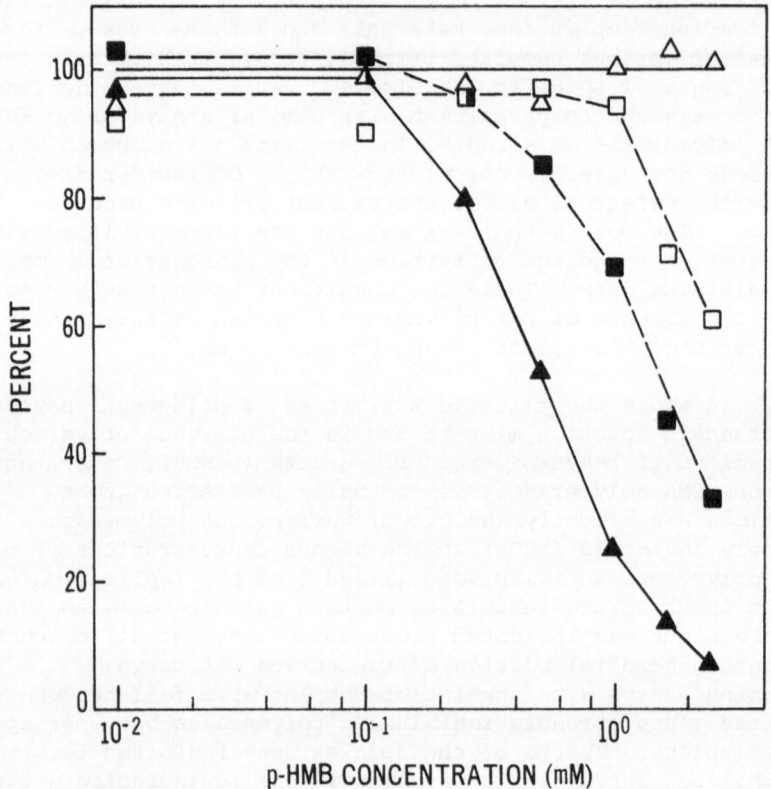

Figure 14. Effect of p-hydroxymercuribenzoate (p-HMB) on the
activity of various DNA polymerases. All DNA polymerase assays
were run exactly as described for the standard reaction under
Materials and Methods. Activated calf thymus DNA (10 μg) was used
in the reactions with solubilized or purified enzymes and the
reactions using the intact replication complex contained no
exogenous DNA. Activities in the absence of p-HMB (100% activities)
were: DNA polymerase α (▲), 3.4 pmoles of (^{3}H)dTTP incorporated
per reaction in 30 min at 37°; DNA polymerase γ (△), 3.6 pmoles;
solubilized enzyme preparation (■), 0.16 pmoles; intact complex
(□), 0.1 pmoles.

possibly both of the DNA polymerases are certainly involved in the
synthesis of Ad2 viral DNA, it was of further interest to determine
the presence of other enzymes in the replication complex that
might possibly play a role in DNA synthesis. The presence of other
ancillary enzymes would not necessarily implicate a function in
the synthesis of Ad2 DNA but their absence might at least indicate
they are not used in the *in vitro* synthesis of viral DNA. Enzymes

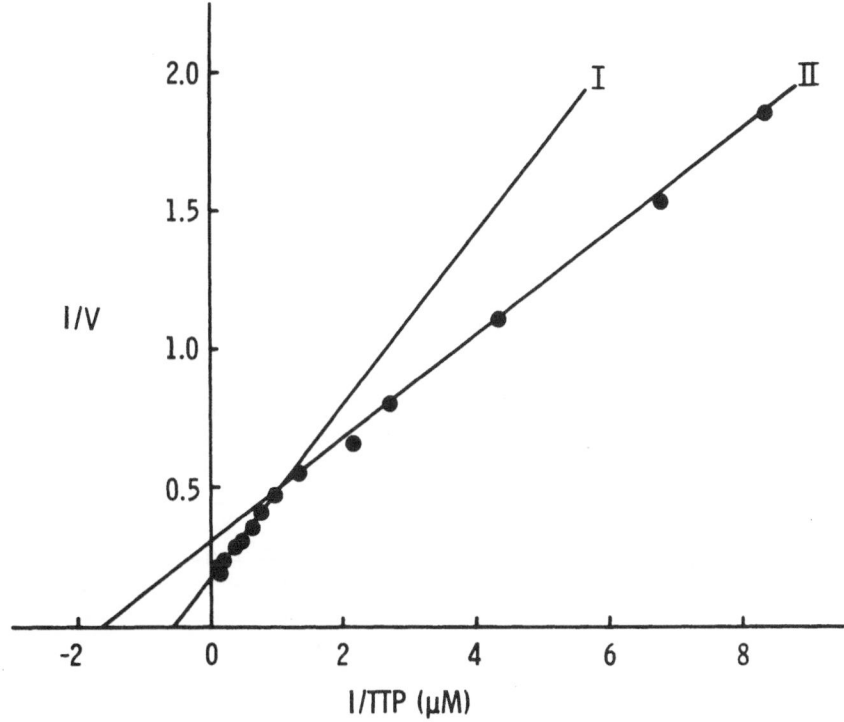

Figure 15. Double reciprocal plot of the DNA polymerase activities in the solubilized enzyme preparation (0.5 M NaCl eluate) from the replication complex. Assays were described under Materials and Methods in which the dTTP concentrations were varied as indicated. The apparent Km's determined from the graph were: I, 2 x 10⁶M and II, 0.6 x 10⁶M.

found to be present in the 0.5 M NaCl eluate from DEAE-Sephadex included RNA polymerases II and III, DNA ligase, an RNase activity and a 3'→ 5' exonuclease. On the other hand, DNA endonuclease and 5'→ 3' exonucleases were not found in detectable amounts. Quantitation of recovery of the total DNA polymerase activity, RNA polymerase activity and DNA ligase activity yielded the result that between about 0.5 and 2% of the total cellular activity of each of these enzymes was recovered in a solubilized protein fraction from the replication complex.

DISCUSSION

We have adopted a functional approach to investigate the

problem of the mechanism and enzymology of replication of Ad2 DNA. To this end, we have purified DNA-protein complexes that are active in the *in vitro* synthesis of Ad2 DNA.   We have investigated the general properties of these "replication complexes" with regards to their ability to mimic *in vivo* replication and have also attempted to dissect their protein and enzyme complements.   More recently we have begun to focus on very specific functions of the soluble Ad2 replication complex (e.g., termination of replication) and we are also developing an investigation into the interaction of proteins and DNA in the complex.

The Ad2 nuclear membrane replication complex, for reasons as yet undetermined, can only label short (6-9S) DNA pieces *in vitro*. The most likely explanation for this deficiency is either the lack of an enzyme or the inability of an enzyme to function properly in the isolated complex.   We have not extensively analyzed either the product or the enzymes from this particular complex but in view of the observation that the short DNA fragments are synthesized on a large (18-31S) template, it seems that the activity which is lacking is a DNA gap-filling function or a DNA ligating function. If the observation of short pieces being synthesized on a large template is not an artifact of the system, then it would be strong support for Okazaki type intermediates in the synthesis of full-length Ad2 DNA.   There has been other evidence for the existence of short intermediates in Ad2 DNA synthesis (11),12). Further analysis of the *in vitro* products of this complex by hybridization of short pieces (a) to themselves (b), to restriction fragments and (c) to separated DNA strands, would aid greatly in evaluating the significance of these data.   Another approach would be to supplement the membrane complex with various purified enzymes and cofactors and determine the effect on product size in alkaline gradients.   In view of the fact that DNA polymerase α is apparently not present in the complex, it is tempting to speculate that perhaps this enzyme is responsible for gap-filling between small nascent DNA pieces which are then sealed by ligase. These possibilities also remain for future investigations.

Other studies on proteins in the membrane complex have demonstrated the presence of the viral-coded 75K single-stranded DNA binding protein (9). This protein is present in very large amounts and presumably binds stoichiometrically with ssDNA to maintain it as such during its replication.

In contrast to the membrane complex, the soluble replication complex does elongate ssDNA during extensive incubation *in vitro*. For this reason we have recently turned our major attention toward this apparently more faithful and authentic complex. The synthesis of DNA *in vitro* by this soluble replication system is apparently the same as the *in vivo* process in all aspects so

far tested. Products sedimented in neutral gradients yielded some
large (>31S) replicative intermediates that could be chased into
31S. Single-stranded DNA labeled with a short pulse sedimented
heterogeneously in alkaline sucrose and the radioactivity could be
chased into full-length DNA. Synthesis was semiconservative and
elongation of strands was extensive. After short (10 to 20 min.)
pulses *in vitro*, restriction fragments from the molecular ends
were relatively more highly labeled than internal fragments and
also showed strand specificity in labeling (Arens,and Yamashita,
manuscript in preparation) which is consistent with bidirectional
synthesis of DNA with initiation and termination sites at each end
of the genom (2,4,8,10). Even the presence of specific proteins (7)
on the *in vitro* labeled DNA was similar to the *in vivo* situation
(Yamashita and Arens, unpublished data). Since the *in vitro* complex
appears to be representative of the *in vivo* situation, we hope to
be able to deal effectively with the enzymology of the system to
determine involvement of various activities.

    The analysis of enzymes in the soluble complex has so far been
revealing in that we feel we can essentially rule out the
possibility that DNA polymerase β is involved in Ad2 DNA synthesis.
The implication of α and γ (or both) is not yet clear but the
analysis of their involvement may possibly be more straightforward
using this complex than by using a nuclear or whole cell system.
The other enzymes shown to be present (1) may or may not be directly
responsible for a function related to Ad2 DNA synthesis. Since
viral mutants are few and cell mutants are not completely
characterized, probably only dissection and reconstitution of the
complex can answer these questions. It is possible that antibodies
against specific enzymes or early proteins would be very useful in
discerning functions of the antigen. However, the inaccessibility
of the antigen within a multienzyme complex might render results
from this approach difficult or impossible to interpret.

    The Ad2 replication complexes have been useful in the analysis
of enzymes and other proteins which are thought to be involved in
adenovirus replication. Extensive analysis of the products
synthesized *in vitro* by the soluble replication complex has
established the authenticity of the complex and we believe this
analysis has confirmed that it represents either all or part of a
true *in vivo* Ad2 replicating system. The products of the membrane
complex have not been investigated sufficiently to assess its
relationship to the authentic *in vivo* complex. Neither of these
complexes has realized its full potential as far as helping in an
understanding of the mechanism and enzymology of Ad2 DNA synthesis
The goal of future work is to realize this potential and begin to
dissect these complexes and reconstruct replication systems from
purified components.

REFERENCES

(1)   Arens, M., Yamashita, T., Padmanabhan, R., Tsuruo, T. and
      Green, M. (1977) J. Biol. Chem., in press

(2)   Ariga, H. and Shimojo, H. (1977) Virology, 78:415-424

(3)   Harter, M., Shanmugam, G., Wold, W. and Green, M. (1976)
      J. Virol. 19:232-242

(4)   Horwitz, M. (1976) J. Virol. 18:307-315

(5)   Ito, K., Arens, M. and Green, M. (1975) J. Virol. ·15:1507-
      1510

(6)   Ito, K., Arens, M. and Green, M. (1976) Biochem. Biophys.
      Acta 447:340-352

(7)   Rekosh, D., Russell, W., Bellet, A. and Robinson, A. (1977)
      Cell 11:283-295

(8)   Schilling, R., Weingartner, B. and Winnacker, E.L. (1975)
      J. Virol. 16:767-774

(9)   Shanmugam, S., Bhaduri, S., Arens, M. and Green, M. (1975)
      Biochemistry, 14:332-337

(10) Sussenbach, J.S. and Kuijk, M.G. (1977) Virology 77:149-157

(11) Vlak, J.M., Rozijn, Th. H., and Sussenbach, J.S. (1975)
      Virology 63:168-175

(12) Winnacker, E.L. (1975) J. Virol. 15:744-758

(13) Yamashita, T., Arens, M. and Green, M. (1975) J. Biol. Chem.
      250:3273-3279

(14) Yamashita, T., Arens, M. and Green, M. (1977) J. Biol. Chem.,
      in press

(15) Yamashita, T. and Green, M. (1974) J. Virol. 14:412-420

STUDIES ON REPLICATING SIMIAN VIRUS 40 CHROMOSOMES:

ASSOCIATION WITH THYMIDINE KINASE

M. Anwar Waqar[*] and J.A. Huberman

Department of Medical Viral Oncology,
Roswell Park Memorial Institute,
Buffalo, New York, 14263,   U.S.A.

INTRODUCTION

Although several enzymes are known which are likely to
participate in the process of DNA replication in mammalian cells,
so far not one such enzyme has been proved to be required for
replication (for review see reference 1). In the hope of
positively identifying some of the enzymes needed for mammalian
DNA replication, we have recently developed three subnuclear systems
for in vitro DNA synthesis from simian virus 40 (SV40)-infected
monkey cells (2,3). We chose to work with SV40 virus as a model
system for mammalian DNA replication because it is much easier to
characterize small homogeneous viral chromosomes than large,
heterogeneous cellular ones. We feel that SV40 DNA replication is
an excellent analog of cellular DNA replication because, with the
exception of the viral A gene product which is required for
initiation of viral DNA replication (4), all steps in the
replication of SV40 DNA are carried out by host cell enzymes. In
addition, SV40 DNA synthesis resembles cellular DNA synthesis in
many ways. In both cases DNA synthesis is bidirectional, Okazaki
fragments of 100-250 nucleotides are intermediates, and a large
fraction of the Okazaki fragments synthesized in vitro have
oligoribonucleotides which may serve as primers at their 5' ends
(reviewed in 1; see also 5).

One of our principal aims in developing subnuclear systems for
in vitro SV40 DNA replication has been the preparation of purified
replicating SV40 chromosomes, capable of replication in vitro in
the absence of nuclear structure or soluble factors. It is known
from previous studies in a number of other laboratories (6-12) that

[*]A.S.I. Participant

SV40 DNA replication takes place in a chromosome state, and that
SV40 chromosomes resemble cellular chromosomes in histone content
and structural features (6,13).

    We have in fact succeeded, by using a very low ionic strength
isolation procedure, in developing conditions where SV40 chromosomes
can be released from nuclei (2) and then partially purified (3) while
retaining considerable in vitro synthetic activity. The fact that
these chromosomes do retain synthetic activity means that some of
the enzymes required for in vitro DNA synthesis must remain bound
to the chromosomes.

    One of the goals of our ongoing research is the identification
and characterization of such chromosome-bound enzymes, and we are
collaborating with several other laboratories in this endeavor. At
this time three enzyme activities have been identified which
cosediment in sucrose gradients with replicating SV40 chromosomes.
These are DNA polymerase (Mary Jo Evans, unpublished; Howard
Edenberg, unpublished), DNA relaxing enzyme (David Kowalski,
unpublished) and thymidine (dThd) kinase (data presented below).
The potential role of the first two enzymes in DNA replication is easy
to imagine. The role of dThd kinase in DNA replication, if any, is
not clear, and we shall discuss these uncertainties in this paper.

    One of the major problems in assessing the significance of
the enzymes which we detect cosedimenting with replicating
chromosomes is that the replicating chromosomes are not yet
completely pure. We have been working on improved purification of
these chromosomes and we shall discuss our progress in this paper.

RESULTS

Comparison of Two Different Methods for Separating SV40 Chromosomes
from Cellular Chromatin

    In our published methods for preparation of SV40 chromosomes
(2,3) we used a modification of the low ionic strength chromatin
isolation procedure originally developed by Hancock (14). At 40
hours after infection of BSC-1 cells with SV40 virus, we lysed the
cells in a low ionic strength buffer (0.1 M sucrose, 0.2 mM NaKHPO,
pH 7.5), then washed their "chromatin bodies" by two cycles of
differential centrifugation. The "chromatin bodies" to which we
refer here have the appearance of very swollen nuclei in the phase
contrast microscope but can be seen by electron microscopy to lack a
nuclear membrane (2). We released SV40 chromosomes from the
chromatin bodies by mild Dounce homogenization (2,3). When the
released chromosomes were sedimented through a neutral sucrose
gradient (similar to that shown in Fig. 1a), the replicating
chromosomes (distinguished by brief pulse-labelling with ^{3}H-dThd

in vivo just prior to cell lysis) sedimented at about 61S (3), while the mature, non-replicating chromosomes (distinguished by overnight labeling with ^{14}C-dThd) sedimented at about 55S (3) (as in Fig. 1a). The OD_{260} and OD_{280} profiles of such gradients indicated that the majority of UV-absorbing material which sedimented in the chromosome region of the gradient cosedimented with mature SV40 chromosomes at 55S. However, a substantial amount of UV-absorbing material, as well as some ^{14}C and ^{3}H label, sedimented faster than the SV40 chromosomes (corresponding to fractions 1-12 in Fig. 1a). This region contains aggregated SV40 chromosomes and cellular chromatin fragments (3).

In the hope of eliminating the contaminating cellular chromatin fragments we tried a variation of our standard procedure. Instead of using Dounce homogenization to release SV40 chromosomes from the chromatin bodies, we tried extraction with the chromosome extraction buffer developed by Su and DePamphilis (11) (10 mM Hepes, pH 7.8, 5 mM KCl, 0.5 mM $MgCl_2$, and 0.5 mM dithiothreitol). In Fig. 1 are shown the results of an experiment in which a single preparation of chromatin bodies was split in half; one half was extracted for one hour with the Su and DePamphilis buffer, and the other half was subjected to mild Dounce homogenization in our standard procedure. Both extracts were then sedimented through neutral sucrose gradients. The sedimentation patterns obtained with both methods proved to be similar. The standard Dounce homogenization method (Fig. 1a) gave a higher yield but also produced more fast-sedimenting UV absorbing material. The extraction with buffer (Fig. 1b) produced a larger quantity of solubilized UV absorbing material which sedimented slowly (Fractions 22-26). The chromosomes recovered by both procedures proved to be equally active in DNA synthesis in vitro (data not shown). Because we obtained a higher yield with the standard Dounce homogenization procedure, we continued to use it for the other experiments described in this paper.

Attempts at Further Purification of Replicating SV40 Chromosomes

Because the OD_{260} and OD_{280} profiles in Fig. 1 show a peak at 55S (where mature SV40 chromosomes sediment) and not at 61S (where replicating chromosomes sediment) we conclude that there is a large excess of mature over-replicating chromosomes in those gradients. Thus the data of Fig. 1 suggest that the major

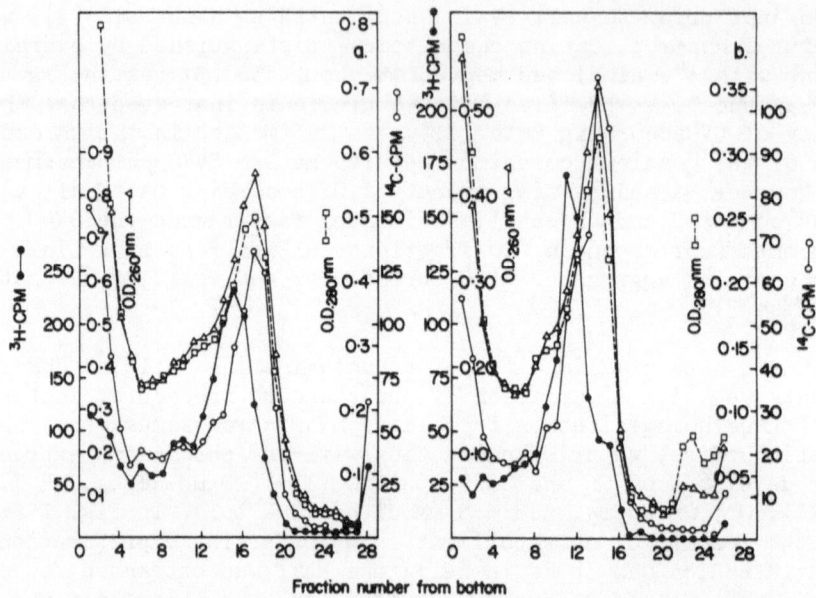

Fraction number from bottom

Figure 1. Comparison of Sedimentation Patterns of SV40 Chromosomes
in Neutral Sucrose Gradients After Isolation by Two
Different Procedures.

*Forty 100 mm Petri dishes of confluent BSC-1 cells were infected
with SV40 virus. At 24 hours after infection, [^{14}C]-dThd (0.08 µCi/
ml) was added to each dish. At 40 hours after infection, the dishes
were rinsed twice each with 5 ml of medium (37°C), and 5 ml of medium
containing [^{3}H]-thymidine (20 µCi/ml) was added to each. After 6 min,
the radioactive medium was removed, the dishes were rinsed with cold
PBS, and the chromatin bodies were prepared as described (3). The
"soluble system" was prepared from half of the chromatin bodies (3)
and SV40 chromosomes were extracted from the other half by resuspending the chromatin bodies in 1 ml of 10 mM Hepes, pH 7.8, 5 mM
KCl, 0.5 mM MgCl$_2$, and 0.5 mM dithiothreitol and incubating at 4° for
1 hour. Nuclei were removed by centrifugation at 10,000 rpm and 4°
for 10 min in an SW41 rotor. The supernatant, when viewed under
the phase contrast microscope, was free of nuclei. One ml of the
soluble system and one ml of the extracted nucleoprotein complexes
were layered over two separate gradients of 10-30% sucrose in 0.2
mM NaKHPO$_4$ (with a 1 ml shelf of 80% sucrose in the same buffer).
Centrifugation was at 40,000 rpm for 120 min at 4° in the SW41
rotor. Fractions were collected in the cold-room and 25λ aliquots
were spotted onto filter papers (Schleicher + Schuell #865E), which
were washed and counted as described (2,3). The OD$_{260}$ and OD$_{280}$
were determined on undiluted fractions of SV40 chromosomes using a
Beckman Model 25 Spectrophotometer.* ●——● ^{3}H cpm; ○——○ ^{14}C cpm;
△——△ OD$_{260}$; □——□ OD$_{280}$.

contaminant of the replicating chromosomes (61S region) is mature
chromosomes, and that cellular chromatin fragments are also a
significant contaminant. In our attempts to remove these
contaminants from the replicating chromosomes we have been
constrained by the continuous necessity for buffers of very low
ionic strength (so as to prevent loss of essential enzymes from
the replicating chromosomes). Gel filtration using agarose gels
is one method that seemed potentially helpful. Indeed, when pooled
replicating chromosomes from a neutral sucrose gradient like that
of Fig. 1a were subjected to gel filtration through Sepharose
4B (Fig. 2), complete recovery of ^{3}H cpm, ^{14}C cpm, and OD_{260} were
obtained, as well as complete recovery of _in vitro_ synthetic activity

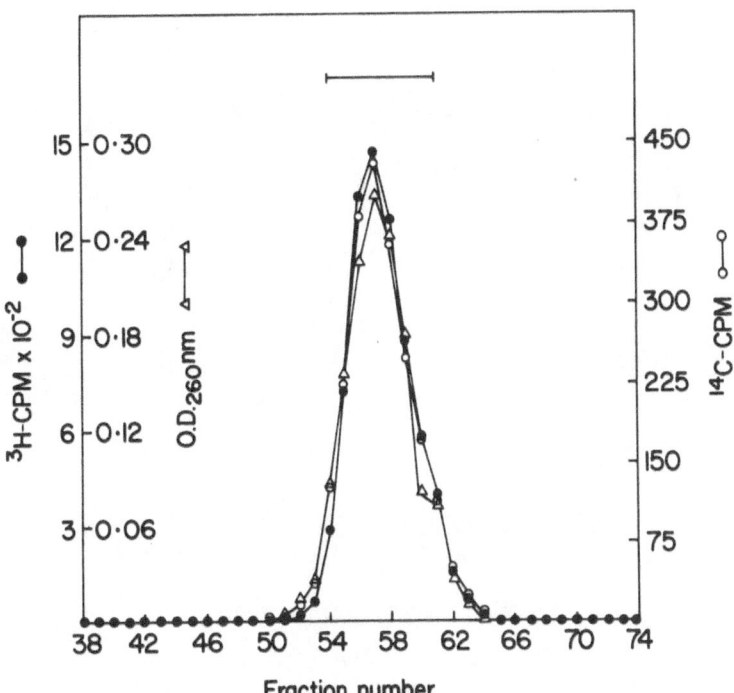

Figure 2. _Sepharose 4B Gel Filtration of SV40 Chromosomes._

_Partially-purified SV40 chromosomes were prepared from 40 Petri
plates (100 mm) of SV40-infected BSC-1 cells, as in Fig. 1. Sucrose
gradient fractions containing replicating and mature chromosomes
were pooled, dialyzed against buffer A' (0.1 M sucrose, 0.2 mM
NaKHPO$_4$, pH 7.5) to remove sucrose, and then passed through a column
of Sepharose 4B (2.5 cm diameter x 75 cm long). Aliquots from
each fraction were applied to filter papers which were then acid-
washed and counted in scintillation fluid as described (2,3). OD$_{260}$
was determined directly on each fraction. ●——● ^{3}H cpm;
○——○ ^{14}C cpm; △——△ OD$_{260}$_

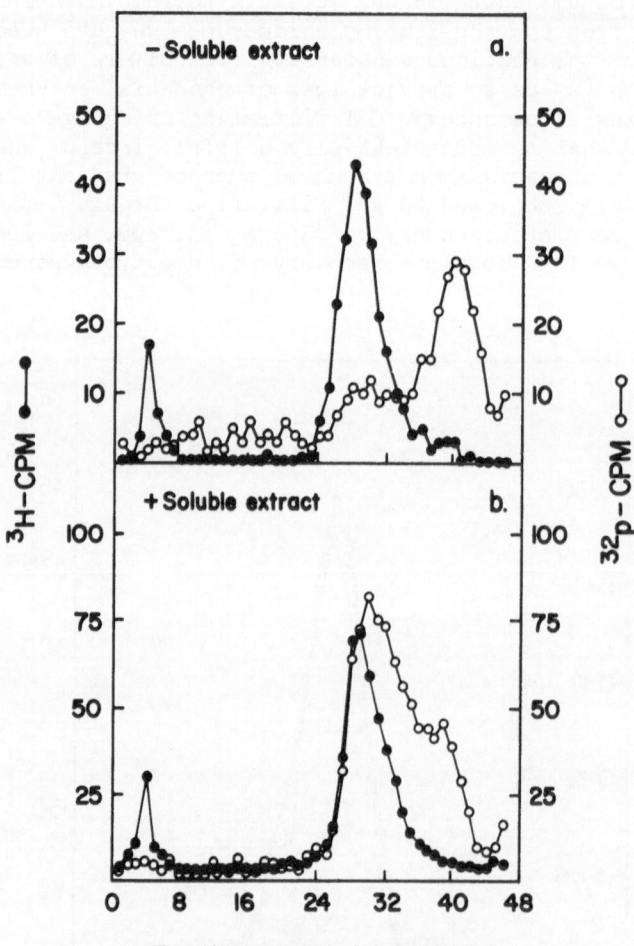

Two portions of the replicating chromosomes from the Sepharose 4B column (Fig. 2) were incubated in standard in vitro conditions (2,3) at 37°C in the standard reaction mixture, containing 5 µM α-^{32}P-dTTP (8.94 Ci/mmole), with or without soluble extract from HeLa cells (50% of the reaction mixture when present). Soluble extract was prepared according to the procedure reported previously (2). The reactions were carried out for 45 min. Reactions were stopped by addition of Sarkosyl (K and K Laboratories, Inc.) to 1%,

(see below). Unfortunately both replicating (marked by ^{3}H) and mature (marked by ^{14}C) chromosomes eluted in the void volume so no separation was obtained. We calculate that the SV40 chromosomes should elute in the fractionation range from Sepharose 2B, and experiments with Sepharose 2B are underway.

To test whether normal in vitro synthetic activity was recovered after passage of the chromosomes through Sepharose 4B, the chromosomes were incubated for 45 min under standard in vitro conditions with α-^{32}P-dTTP, both in the presence and in the absence of a soluble extract from HeLa cells which we have previously shown to stimulate both the extent of DNA synthesis and the joining of Okazaki fragments (2,3). The in vitro synthesized DNA was first purified on a neutral sucrose gradient (2,3), and then denatured and sedimented through an alkaline sucrose gradient (Fig. 3). In the absence of the soluble extract (Fig. 3a), the ^{32}P-labeled DNA synthesized in vitro sedimented in a peak at 4-6S and a faster-sedimenting shoulder up to about 16S (the leading edge of the large ^{3}H peak has been shown in other experiments to correspond to mature linear SV40 DNA which sediments at 16S). This distribution is identical to that obtained for SV40 chromosomes pooled from neutral sucrose gradients but not subjected to gel filtration (3). The 4-6S peak corresponds to Okazaki fragments while the more rapidly sedimenting material corresponds to continuously-synthesized chains and joined Okazaki fragments (2,3).

and the samples were layered onto 5-30% neutral sucrose gradients (in 1 M NaCl, 10 mM Tris-HCl, 1 mM EDTA, 0.1% Sarkosyl, pH 8.0). Centrifugation was carried out at 25,000 rpm and 10° for 18 hours in the SW 41 rotor. Fractions were collected and aliquots from each fraction were taken onto filter discs, washed three times with 1 M HCl, dried and counted. The 20-25S region of the gradients, containing SV40 DNA, was pooled and dialyzed against 10 mM EDTA, 10 mM NaCl, 10 mM Tris-HCl, pH 7.5, 0.1% Triton X100 to remove salts and sucrose, and finally concentrated by dialysis against 25% Carbowax (Union Carbide) in the same buffer. The volumes in each case were adjusted to 0.3 ml and 10N NaOH was added to a final concentration of 0.2 N and the samples were loaded over 10-30% alkaline sucrose gradients (0.7 M NaCl, 2.5 mM EDTA, 0.1% Sarkosyl, titrated to pH 12.1 with NaOH) with a 0.5 ml shelf of 80% alkaline sucrose. The samples were centrifuged at 30,000 rpm at 10° for 18 hours in the SW41 rotor. Fractions were collected and 50 µg of calf thymus DNA and 50 µg of bovine serum albumin (BSA) added to each before spotting onto filter paper discs. The entire volume of the fractions was spotted on the filter discs, dried, washed with 1 M HCl, and radioactivity was determined after addition of toluene-based scintillation fluid. Each sample was counted for at least 20 min.

When soluble extract was added to the incubation, much more DNA was synthesized, and the majority of the DNA now sedimented more rapidly than Okazaki fragments (Fig. 3b), consistent with the previously-demonstrated stimulation of joining of Okazaki fragments by soluble extract (2,3). Thus, in the properties of the DNA synthesized, as well as in the overall extent of in vitro DNA synthesis (data not shown), the chromosomes passed through Sepharose 4B behaved identically to chromosomes simply pooled from neutral sucrose gradients.

An alternative approach to further separation of mature and replicating chromosomes, while maintaining low ionic strength, is isopycnic centrifugation in nonionic density gradients, such as Metrizamide gradients. When SV40 chromosomes were pooled from a neutral sucrose gradient (like Fig. 1a), then centrifugated to equilibrium in a Metrizamide gradient (Fig. 4), the replicating chromosomes (marked by ^{3}H) banded at a slightly higher densith (1.235) than did the mature chromosomes (1.231). One possible interpretation of this density difference is that replicating chromosomes have a slightly higher protein to DNA ratio than mature chromosomes. Similar observations have previously been made for replicating and mature chromatin from eukaryotic cells (15-17), but not for SV40 or polyoma chromosomes (6,7,10). Unfortunately, in vitro DNA synthesis by chromosomes recovered from Metrizamide gradients is defective; no DNA is synthesized in the absence of soluble extract. For this reason we have not continued to use Metrizamide gradients for chromosome purification.

Additional techniques which have been tried in our laboratory for further purification of replicating chromosomes include agarose gel electrophoresis (Tsubota, unpublished), isoelectric focusing (Tsubota, unpublished), and hydrophobic chromatography (Waqar, unpublished). Agarose gel electrophoresis provided good separation between mature and replicating chromosomes, but in vitro synthetic activity was lost. The other two methods, so far, have failed to resolve mature and replicating chromosomes.

Thus our search for better methods of chromosome purification continues. Meanwhile, the results presented in the rest of this paper have been obtained with chromosomes pooled from neutral sucrose gradients; no further purification steps have been used.

Accumulation of Replicating Chromosomes by FdUrd Pretreatment

Because replicating chromosomes are much less abundant in SV40-infected cells than mature chromosomes, we have tried to increase the frequency of replicating chromosomes by treatment of SV40-infected cells with FdUrd. We expected that FdUrd would block dTTP biosynthesis, thus blocking replication fork movement, but the initiation of new rounds of replication might not be affected. Thus

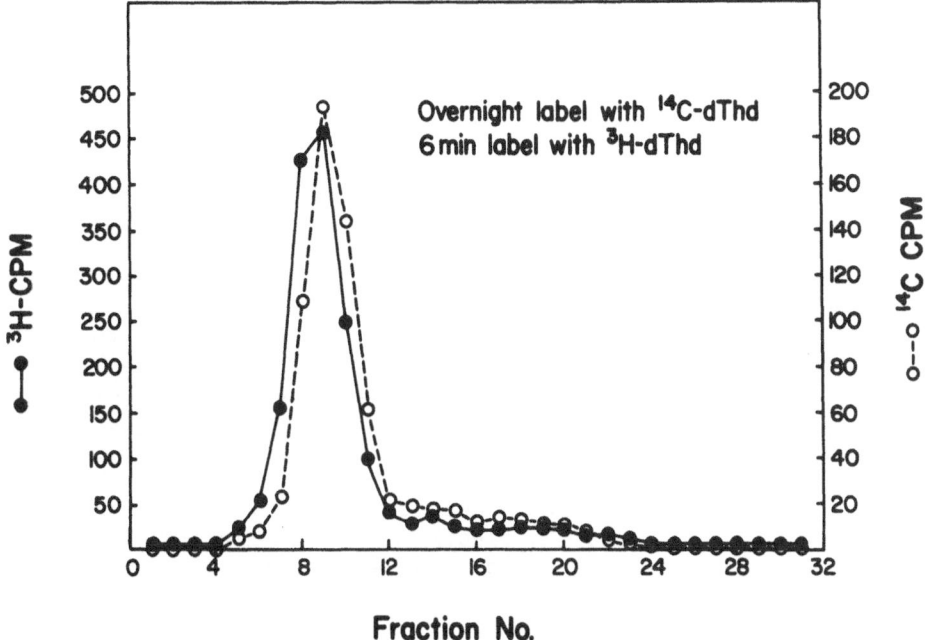

Figure 4. *Banding of SV40 Chromosomes in a Metrizamide Density*
          Gradient.

SV40 chromosomes which were labeled with ^{14}C and ^{3}H-dThd in vivo
as described in Fig.1 , and then partially purified on a neutral
sucrose gradient as in Fig.1 were pooled, dialyzed to remove
sucrose, and then adjusted to 35% Metrizamide in 0.2 mM $NaKHPO_4$,
pH 7.5 and to a volume of 1.7 ml. This solution was gently layered
over 1.7 ml of 45% Metrizamide in the same buffer in a polyallomer
tube for the SW65 rotor, and then 1.7 ml of 25% Metrizamide was
layered on top so as to complete a three-layer step gradient. This
step gradient was centrifuged at 50,000 rpm for 15 hours at 4°
in the SW65 rotor. Fractions were collected from the bottom of the
gradient, and aliquots of each fraction were spotted onto filter
papers which were acid-washed and counted as described (2,3).
Refractive index (not shown) was also determined on an aliquot
from each fraction.

when the FdUrd block was removed, we reasoned, we might find a
higher proportion of replicating chromosomes. Indeed, Edenberg,
Petes, Perlman and Huberman (unpublished) have shown that after a
22 min treatment with FdUrd, the proportion of SV40 DNA in
replicating form was increased 3-fold.

Salzman and Thoren (18) and Laipis and Levine (19) have shown
that when SV40-infected cells are pretreated with FdUrd, the DNA
synthesized just after removal of the FdUrd block sediments
abnormally slowly in alkaline sucrose gradients, suggesting that
FdUrd causes some abnormality in DNA synthesis in addition to lowering
the overall rate of DNA synthesis. Consequently we decided to check
to see whether the SV40 chromosomes prepared from FdUrd-pretreated
cells could carry out normal DNA synthesis in vitro. We prepared
SV40 chromosomes by our standard procedure (3) from infected cells
pretreated with FdUrd for 1 hour. Although these chromosomes
were capable of synthesizing DNA in vitro, and the synthesis was
stimulated about 6-fold by soluble extract from HeLa cells, the
synthesized DNA was abnormal in that it sedimented slowly in
alkaline sucrose gradients (Fig. 5), regardless of whether soluble
extract had been used during synthesis. These results suggest that
the abnormality detected by Salzman and Thoren (18) and Laipis and
Levine (19) is a defect in the proteins or DNA structure of the
replicating chromosome, and not a defect in the enzymes of the
soluble extract. The defect is clearly not a simple consequence of
dTTP pool size on polymerase action (18,19). Whatever the reason for
the defect, its presence was reason for us to abandon the use of
DfUrd for the other experiments described in this paper.

Detection of dThd Kinase Activity

Because recent experiments from Greenberg's laboratory (20)
suggested that dThd kinase might be part of a T4 phage DNA
replication complex, and because we could easily obtain expert help
from Ed Kowal (21) with dThd kinase assays, we decided to test
whether dThd kinase activity is associated with SV40 chromosomes.
We labeled SV40-infected cells with ^{14}C-dThd overnight to mark
mature SV40 chromosomes and then, at 40 hours after infection,
prepared SV40 chromosomes by our usual procedure (3). Replicating
and mature chromosomes were partially separated in a neutral
sucrose gradient (Fig. 6), and individual fractions were assayed
for dThd kinase activity. Such activity was detected sedimenting
slightly faster than mature chromosomes (Fig. 6), just where
replicating chromosomes are expected (Fig. 1). We could not label
the replicating chromosomes directly with ^{3}H-dThd in this experiment
because ^{3}H-dThd was used in the dThd kinase assay.

Preliminary data suggest that the dThd kinase activity is not
loosely bound to the replicating chromosomes. When the replicating
chromosomes were pooled from a neutral sucrose gradient like that
of Fig. 6 and resedimented in similar sucrose gradients containing

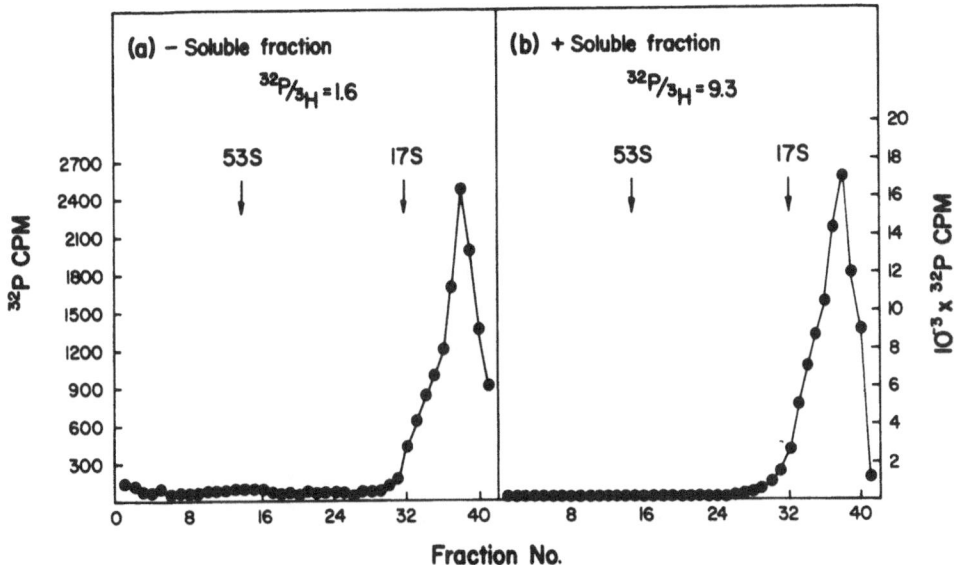

Figure 5.   *Size Distribution of DNA Synthesized in vitro by
             Replicating SV40 Chromosomes from Cells Treated with
             FdUrd.*

*^3H-dThd (5 μCi/ml) was added to infected cells for 16 h at 24 h
post-infection.   Then the medium was removed, the dishes were rinsed
twice in warm fresh medium containing dealyzed fetal calf serum, and
10 ml of warm medium with dialyzed serum and FdUrd (5 μM) were added.
After 60 min, the cells were harvested, and replicating SV40
chromosomes were prepared as described.   The replicating chromosomes
were incubated at 37° in the standard reaction mixture with 3 μM
α-^{32}P-dTTP (19 Ci/mmole) either with or without soluble extract for
30 min.   The reactions were stopped by adding Sarkosyl to 1%; DNA
was precipitated and rinsed twice with cold ethanol and dissolved
in 0.3 ml of 0.2N NaOH.   The solution was layered on 10–30% alkaline
sucrose gradients (0.3N NaOH, 0.7N NaCl, 0.1% Sarkosyl, 0.5 mM EDTA)
and centrifuged at 24,000 rpm, 10° for 18 hours in the SW41 rotor.
Fractions were processed as described.*

KC1 up to 1 M, the dThd kinase activity remained associated with
the replicating chromosomes.

Presence of TMP and TDP Kinase Activities

When the reaction products in the dThd kinase assay of
replicating SV40 chromosomes were characterized by paper chromato-

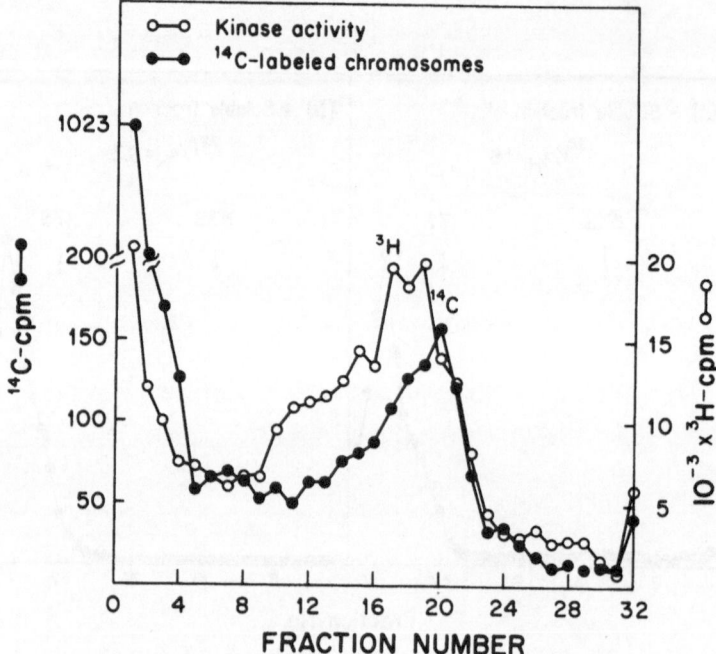

<u>Figure 6.</u>   *Sedimentation of SV40 Chromosomes Through a Neutral*
                 Sucrose Gradient and Assay for Thymidine Kinase Activity.

Twenty 100 mm Petri dishes of confluent BSC-1 cells were infected
with SV40 and 0.04 µCi/ml of [14C]-dThd was added to each plate at
24 hours after infection. The soluble system was prepared (3) and
1 ml of this soluble system was loaded over a gradient of 10-30%
sucrose in 0.2 mM NaKHPO₄, pH 7.5 (with a 1 ml shelf of 80% sucrose
in the same buffer). Centrifugation conditions, collection of the
gradient and counting of the aliquots were the same as described
earlier. For the assay of dThd kinase activity, 50 µl aliquots of
the gradient's fractions were incubated with 20 µM unlabeled dThd,
1 µM [3H]-dThd (51 Ci/mmole), 3 mM MgCl₂, 3 mM ATP, 4 mM phospho-
creatine, 0.5% BSA, 0.12 M Tris-HCl, pH 8.0, and creatine kinase (2
units/ml) in a final reaction mixture of 0.25 at 37° for 30 min.
Reactions were stopped by putting the reaction tubes in a boiling
water bath for 3 min and finally chilling the tubes in ice-cold
water. The tubes were centrifuged at 5,000 rpm and 4° for 10 min to
remove the precipitates. One hundred µl of the supernatant from each
reaction rube was pipetted onto a DEAE filter paper disc (Whatman,
DE81). The discs were washed in 1 liter of 1 mM ammonium formate for
5 min and then in 1 liter of distilled water for 5 min. The washings
with ammonium formate and distilled water were repeated once more.
Finally, the filters were rinsed in 100 ml of methanol, dried,
and counted with toluene-based scintillation fluid.

graphy (Fig. 7), small amounts of dTTP and dTDP were detected in addition to the large amount of product dTMP. Thus the 61S region of sucrose gradients like those in Figs. 1 and 6 contains small amounts of TMP kinase and TDP kinase activities.

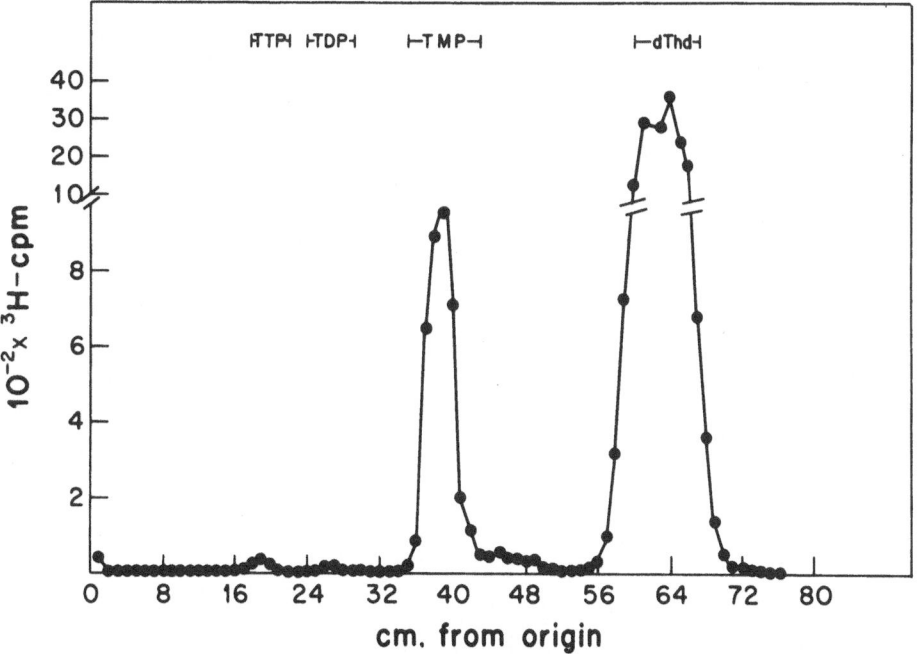

Figure 7. *Paper Chromatography of the Kinase Reaction Product.*

Peak fractions 17-19 from Fig. 6 were applied to Whatman 3 MM paper and chromatographed in isobutyric acid : 1 M NH₄OH : 0.1 M Na₂EDTA (100:60 : 1.6 V/V). Unlabeled dThd, dTMP, dTDP and dTTP were also added as optical density markers in the sample. The positions of deoxynucleoside and deoxynucleotides were located under ultraviolet light. Half cm strips were cut and counted with toluene-based scintillation fluid.

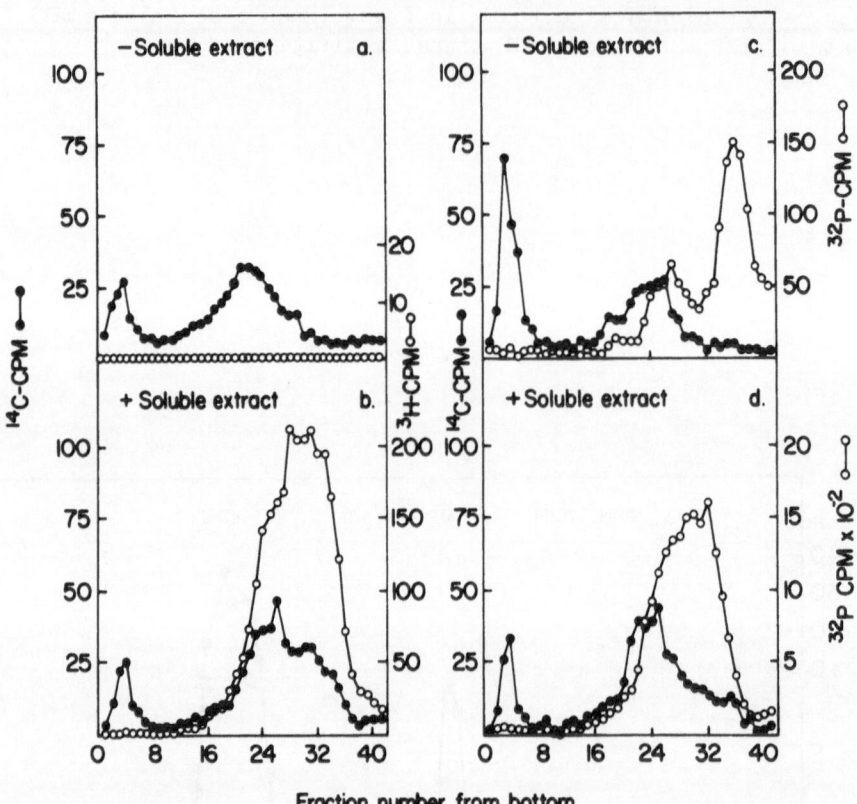

Figure 8.   *Comparison of Incorporation of ³H-dThd and α-³²P-dTTP*
                into Replicating SV40 Chromosomes.

Twenty 100 mm Petri dishes of confluent BSC-1 cells were infected
as described earlier and were labeled with [¹⁴C]-dThd (0.04 µCi/
ml for each plate) after 24 hours of infection. At 40 hours after
infection, the soluble system was prepared and centrifuged through
a gradient of 10-30% sucrose in 0.2 mM NaKHPO₄, pH 7.5.
centrifugation, gradient collection and gradient counting conditions
were the same as described earlier. The region of the replicating
chromosomes was pooled and divided into four lots. The first two
lots (a-b) were incubated at 37°C in the standard reaction mixture,
containing 10 µM [³H]dThd (47 Ci/mmole) with or without soluble
extract and the other two lots (c-d) in the standard reaction
mixture, containing 4 µM α-³²P-dTTP (62.3 Ci/mmole) with or without
soluble extract. Incubations were carried out for 45 min. Other
conditions were the same as described in Fig. 3.

Inability of Partially Purified Replicating Chromosomes to Incorporate ^{3}H-dThd

The significance of the TMP kinase and TDP kinase activities detected in Fig. 7 is not clear. But we thought it possible that these activities, when combined with the stronger dThd kinase activity of replicating chromosomes, might allow replicating chromosomes to convert sufficient ^{3}H-dThd to ^{3}H-dTTP so that detectable incorporation of ^{3}H-dThd into DNA could occur. To test this possibility we incubated partially purified (by our standard procedure, ref. 3) replicating chromosomes (prelabeled overnight with ^{14}C-dThd) with either ^{3}H-dThd or α-^{32}P-dTTP (but not both) as sole thymidine nucleotide source under our standard in vitro conditions, either in the presence or absence of soluble extract from HeLa cells. The size of the DNA product was analysed after denaturation by sedimentation through alkaline sucrose gradient (Fig. 8). The two peaks of ^{14}C-DNA in each gradient represent form I DNA (53S) and broken and intact strands of SV40 DNA (16S for intact linear strands, 18S for circular strands). In the absence of soluble extract, only α-^{32}P-dTTP was incorporated into DNA (Fig. 8c); ^{3}H-dThd was not incorporated (Fig. 8a). In the presence of soluble extract, both ^{3}H-dThd and α-^{32}P-dTTP were incorporated (Fig. 8b,d). Comparison of Fig. 8c with Fig. 8d demonstrates the usual stimulatory effect of soluble extract on the extent of synthesis and rate of joining of Okazaki fragments (compare with Fig. 3). We conclude from the data in Fig. 8 that partially purified replicating chromosomes do not have sufficient kinase activity to efficiently convert dThd to dTTP; however the soluble extract does have the necessary additional kinase activities and, in the presence of soluble extract, ^{3}H-dThd can be efficiently incorporated into DNA by replicating SV40 chromosomes.

DISCUSSION

The data we have presented (Figs. 6,7), and additional data not shown here, suggest that dThd kinase activity is associated with replicating SV40 chromosomes when they are prepared by our standard procedure. Before any firm conclusions can be drawn regarding the biological significance of this association, we must determine whether the association occurs in vivo, or whether it occurs artifactually only after the cells are lysed. This question is obviously an important one for every enzyme activity we may find associated with replicating chromosomes. Unfortunately it is a difficult question to answer. In the case of dThd kinase we plan to try to answer the question by carrying out the following experiment. We shall mix polyoma-virus-infected mouse cells which lack dThd kinase (TK$^-$) with uninfected wild-type mouse cells, then prepare polyoma replicating chromosomes from the mixed population.

If the replicating chromosomes do not have associated dThd kinase
activity, then we may conclude that such activity is not absorbed
to the chromosomes after cell lysis. This experiment is most
easily done with polyoma virus (which is related to SV40) because
polyoma infects mouse cells, and TK⁻ mutants of mouse cells are
available (26). The experiment has not yet been done. However, as
mentioned in Results, we have carried out preliminary experiments
on the effect of salt on the association between dThd kinase and
replicating chromosomes. These experiments suggest that dThd
kinase remains bound to replicating chromosomes even in 1 M KCl;
thus the association is a tight one. Because the dThd kinase
activities which have previously been measured in mammalian cells
are not chromatin-bound, the tight association which we see suggests
that we may have detected a different form of enzyme. Much work
remains to be done in characterizing the activity we have detected
and comparing it with known cellular activities.

Even though we have no information about the identity of the
dThd kinase activity or about the reality of its in vivo association
with replicating chromosomes, it is interesting to speculate briefly
about the role that a dThd kinase activity might have in DNA
replication. Although dThd is not on the normal pathway for in
vivo synthesis of dTTP, dThd kinase is one of many enzymes thought
to be involved in DNA synthesis which increase in activity when
cells enter S phase (22), or when cells are infected with SV40 or
polyoma virus (23-25). Such induction suggests that thymidine
kinase may be involved in DNA synthesis. Yet mouse cell lines
lacking dThd kinase grow at normal rates (26). Thus our
understanding of the role played by thymidine kinase during DNA
synthesis in mammalian cells is very incomplete.

Recently Greenberg's laboratory (20) has reported that, in
T4-infected E.coli, dThd kinase is one component of a multi-enzyme
complex which includes enzymes of dNTP synthesis as well as enzymes
of dNTP-polymerization and which participates in T4 phage DNA
replication. Could it be that such a multi-enzyme complex
participates in DNA replication in mammalian cells ? We plan to
test for the presence of other enzymes of dNTP synthesis associated
with replicating SV40 chromosomes.

In conclusion, the model system of replicating SV40 chromosomes
has already provided some preliminary information on the identity
of enzymes which may be involved in DNA replication. A great deal
of work remains to be done in characterizing the three enzyme
activities which have been detected, and still more work remains
to be done in testing for additional enzyme activities.

REFERENCES

(1) EDENBERG, H.J. and HUBERMAN, J.A. (1975) Ann. Rev. Genet.
     9,245

(2) EDENBERG, H.J., WAQAR, M.A. and HUBERMAN, J.A. (1976) Proc.
     Nat. Acad. Sci., U.S.A. 73,4392

(3) EDENBERG, H.J., WAQAR, M.A. and HUBERMAN, J.A. (1977)
     Nucleic Acids Res. (submitted for publication).

(4) TEGTMEYER, P. (1972) J. Virol. 10,591

(5) TSENG, B.Y. and GOULIAN, M. (1977) Fed. Proc. 36,654

(6) CREMISI, C., PIGNATTI, P.E., CROISSANT, O. and YANIV, M. (1976)
     J. Virol. 17,204

(7) GOLDSTEIN, D.A., HALL, M.R. and MEINKE, W. (1973) J. Virol.
     12,887

(8) GREEN, M.H. (1972) J. Virol. 10,32

(9) HALL, M.R., MEINKE, W. and GOLDSTEIN, D.A. (1973) J. Virol.
     12,901

(10) SEEBECK, T. and WEIL, R. (1974) J. Virol. 13,567

(11) SU, R.T. and DePAMPHILIS, M.L. (1976) Proc. Nat. Acad. Sci.,
     U.S.A., 73,3466

(12) WHITE, M. and EASON, R. (1971) J. Virol. 8,363

(13) GRIFFITH, J. (1975) Science 187,1202

(14) HANCOCK, R. (1974) J. Mol. Biol. 86,649

(15) LEVY, A., JAKOB, K.M. and MOAV, B. (1975) Nucleic Acids Res.
     2,2299

(16) FAKAN, S., TURNER, G.N., PAGANO, J.S. and HANCOCK, R. (1972)
     Proc. Nat. Acad. Sci., U.S.A. 69,2300

(17) WEINTRAUB, H. (1973) Cold Spring Harbor Symp. Quant. Biol.
     38,247

(18) SALZMAN, N.P. and THOREN, M.M. (1973) J. Virol. 11,721

(19) LAIPIS, P.J. and LEVINE, A.J. (1973) Virology, 56,580

(20) WOVCHA, M.G., CHIU, C.S., TOMICH, P.K. and GREENBERG, G.R. (1976) J. Virol. 20,142

(21) KOWAL, E.P. and MARKUS, G. (1976) Preparative Biochem. 6(5), 369

(22) STUBBLEFIELD, E. and MUELLER, G.C. (1965) Biochem. Biophys. Res. Commun. 20, 535

(23) DULBECCO, R., HARTWELL,L. and VOGT, M. (1965) Proc. Nat. Acad. Sci. U.S.A. 53,403

(24) HATANAKA, M. and DULBECCO, R. (1966) Proc. Nat. Acad. Sci., U.S.A. 56,736

(25) POSTEL, E.H. and LEVINE, A.J. (1975) Virology 63,404

(26) KIT, S., DUBBS, D. PIEKARSKI, L. and HSU, T. (1963) Exp. Cell Res. 31,297

SYNTHESIS OF DNA AND POLY(ADP RIBOSE) IN PERMEABILIZED EUKARYOTIC CELLS

Nathan A. Berger[*], Georgina Weber, Aaron S. Kaichi and
Shirley J. Petzold

The Hematology Oncology Division of the Department of
Medicine
The Jewish Hospital of St. Louis
Washington University School of Medicine
216 South Kingshighway, St. Louis, Mo. 63110, U.S.A.

Poly (Adenosine Diphospho Ribose) is a nucleic acid polymer
that has been found predominantly in eukaryotes and has been
implicated in the regulation of DNA synthesis (1,2). The repeating
subunit of this polymer is the adenosine diphospho ribose (ADPR)
moiety derived from nicotinamide adenine dinucleotide. The ADPR
moieties can also be covalently linked to acceptors such as proteins.
The synthesis of poly(ADPR) has usually been studied in isolated
nuclei, in chromatin preparations, or with enzyme extracts. These
systems have all provided a great deal of information on the
activity of poly(ADPR) polymerase; however, in order to understand
the physiological relation of poly(ADPR) to DNA synthesis, it is
desirable to work with a system which is as close to physiological
as possible.

We have been working with permeabilized, eukaryotic cells to
study some of the factors involved in the regulation of DNA
synthesis (3-6). When mouse L cells are subjected to a 15 minute
cold shock procedure in 0.01 M Tris/HCl pH 7.8, 0.25 M Sucrose,
1 mM EDTA, 4 mM $MgCl_2$ and 30 mM 2-mercaptoethanol, they become
permeable to exogenously supplied nucleotides (3). These cells can
be used for direct studies of the activities of their intrinsic
polymerases. For example, as shown in Fig. 1A, when permeable cells
are supplied with all four deoxynucleoside triphosphates, including
(^{3}H)dTTP in the presence of ATP, magnesium chloride and a monovalent cation such as sodium, the permeable cells use the exogenously supplied deoxynucleotides to synthesize DNA. Fig. 1B

[*]A.S.I. Participant

demonstrates that when the permeabilized cells are supplied with
(^{3}H)NAD and magnesium chloride, they synthesize the homopolymer
poly (ADP Ribose). In this system, the rate of DNA synthesis
plateaus after 20 to 30 minutes while poly (ADPR) synthesis
continues at a linear rate for at least 90 minutes.

   When DNA synthesis in permeable cells is compared to DNA
synthesis in isolated nuclei, it is apparent that the systems
differ in both rate and size of the newly synthesized material.
Fig. 2 demonstrates that DNA synthesis in the permeable cells is
active for 20 to 30 minutes and then levels off. In contrast,
isolated nuclei synthesize DNA at a slower initial rate and the
activity begins to level off after less than 10 minutes incubation.
Fig. 2 also demonstrates that the rate of DNA synthesis in
permeable cells is approximately 10 times greater than the rate in
isolated nuclei.

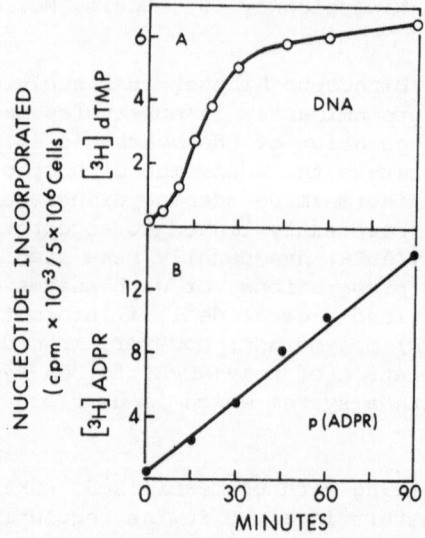

*Figure 1A. Radioactivity incorporated into DNA from (Me-^{3}H)dTTP. The
reactions were composed of 5 x 10^6 permeabilized cells in 100 μl of
.01 M Tris/HCl pH 7.8, 0.25 M Sucrose, 1 mM EDTA, 30 mM 2-mercapto-
ethanol and 4 mM MgCl$_2$ combined with 200 μl of DNA synthesis mix
containing 0.1 M HEPES pH 7.8, 0.2 M MgCl$_2$, 0.21 M NaCl, 15 mM ATP,
0.3 mM dATP, 0.3 mM dCTP, 0.3 mM dGTP and 0.3 mM (Me-^{3}H)dTTP (Sp Act
6 x 10^4 cpm/nmol). Reactions were incubated at 37°C.*

*Figure 1B. Radioactivity incorporated into poly (Adenosine Diphospho
Ribose) from [adenine-2, 8-(^{3}H)] NAD. The reactions were composed
of 5 x 10^6 permeabilized cells in the buffer described above
combined with 200 μl of poly (ADPR) mix containing 100 mM Tris/HCl
pH 7.8, 120 mM MgCl$_2$ and 1 mM (^{3}H)NAD (Sp Act 10 x 10^3 cpm/nmol).
Reactions were incubated at 30°C.*

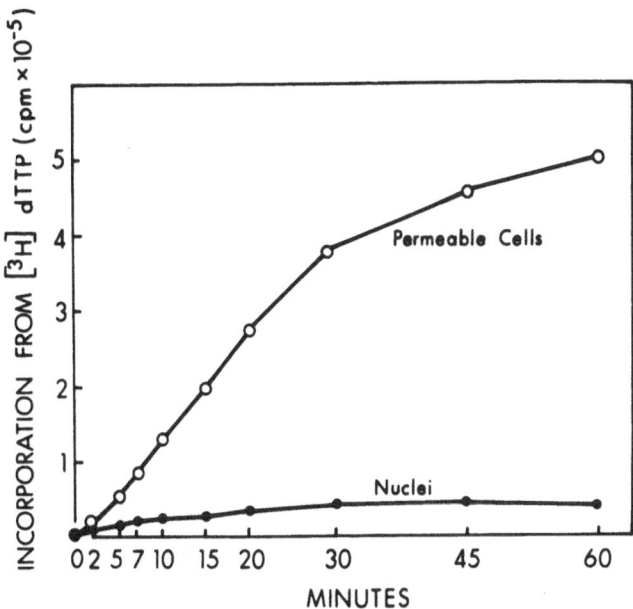

Figure 2. *Incorporation of (^3H) dTTP (Sp Act 43.7 x 10^6 cpm/nmol) into DNA by permeabilized L cells (O———O) or by L cell nuclei (●———●). Permeable cells and nuclei were prepared as previously described (6). The complete reaction system contains 1.25 x 10^6 permeabilized cells or 1.25 x 10^6 nuclei with DNA synthesis mix in a final volume of 75 μl.*

Fig. 3 presents the alkaline sucrose gradient pattern of the DNA synthesized by the permeable cells during a 30 minute incubation period. This newly synthesized DNA has a heterogenous distribution with two broad size classes centering around 26S and 71S. Pulse chase experiments were performed to determine if the 26S intermediates might be precursors for the 71S intermediates. For the first seven minutes, the permeable cells were incubated in DNA synthesis mix containing (^3H)dTTP. After seven minutes, (α-^{32}P)dTTP was added to the system and the incubation was continued for an additional 23 minutes. If the 26S intermediates were precursors of the 71S intermediates, then there should be an excess ratio of (^3H) to (^{32}P) labelled DNA in the 71S peak compared to the ratio of (^3H) to (^{32}P) in the 26S peak. In contrast, if both fragments are synthesized independently, then the ratio of (^3H) to (^{32}P) should be the same in both peaks. As demonstrated in Fig. 3B, the distribution of (^3H) and (^{32}P) is identical in both the 26S and 71S material thus eliminating the possibility of a precursor product relation. An alternative explanation for the synthesis of the different size classes of DNA is that the 71S material may represent continuous synthesis along one template strand, progressing in the direction of movement of the replication fork. The 26S pieces may represent the product of discontinuous synthesis along the

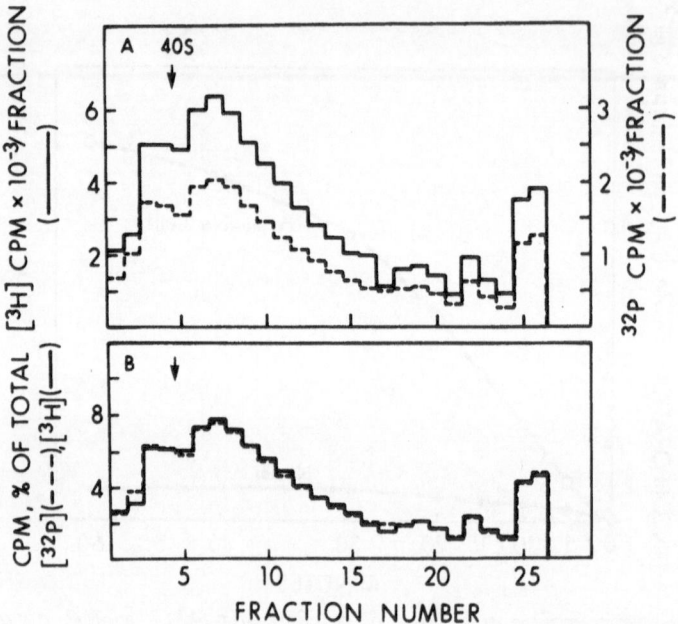

Figure 3. *5 to 20% alkaline sucrose gradient pattern of double
isotope, pulse chase labelled DNA synthesized by permeable cells.
Cells were permeabilized and incubated in DNA synthesis mix
containing 2.2 μm (^{3}H)dTTP (Sp Act 43.7 x 10^6 cpm/nmol). After 7
min, .038 nmol (α-^{32}P) dTTP (Sp Act 13.2 Ci/nmol) were added and
the incubation was continued for 23 min at 37°C. The cells were
lysed on top of the gradients for 12 to 18 hours at 4°C. They were
centrifuged in the SW 25.1 head of the Beckman L2 centrifuge at
22,500 rpm at 4°C for 6 hours. The arrow indicates the position of
a (^{32}P)labelled 40S λ phage marker included in another tube in the
same run. Panel A indicates the actual counts recovered from the
gradients. Panel B demonstrates the same counts expressed as per
cent of total for each isotope.*

opposite template strand. Thus, the 26S pieces may represent the
largest extension of Okazaki fragments before they are ligated into
a continuous DNA strand. Continuous synthesis along one template
strand with discontinuous synthesis along the opposite strand has
been well documented in bacterial systems (7,8).

     In contrast to the size of the DNA synthesized in the
permeable cells, Fig. 4 demonstrates that the DNA synthesized by
isolated nuclei is in the size range of 26S. Thus, permeabilized
L cells synthesize DNA at a faster initial rate and for a longer
period of time than do isolated L cell nuclei. The DNA
synthesized by the permeable cells reaches much higher molecular
weight than that synthesized by the isolated nuclei. The 71S
intermediates, synthesized by the permeable cells, are approximately
6.6 x 10^7 daltons which is in the size range of eukaryotic

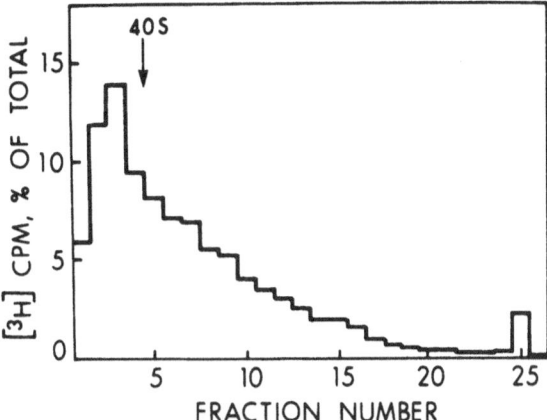

Figure 4.   *5 to 20% alkaline sucrose gradient of DNA synthesized*
by isolated L cell nuclei. Nuclei were isolated from logarith-
mically growing L cells and incubated for 30 min at 37°C with DNA
synthesis mix containing 2.2 μm (³H)dTTP (Sp Act 43.7 x 10⁶ cpm/
nmol). At the end of the reaction, nuclei were lysed directly on
top of preformed alkaline sucrose gradients and centrifuged as
described in the previous figure. The arrow indicates the location
of 40S (³²P) labelled λ phage marker.

replicons (9,10). We have demonstrated that permeable cells carry
on DNA synthesis from sites of DNA synthesis that were active in
vivo (6). We have also demonstrated that DNA synthesis in permeable
cells is semiconservative (3). Thus, it is probable that the
DNA synthesis in the permeable cells represents elongation of
replicons that were being synthesized in vivo. There is probably
no initiation of new replicons in the permeable cells and this
explains why the rate of DNA synthesis falls off after 20 to 30
minutes incubation at 37°C.

    Since DNA synthesis in the permeable cells has many
similarities to the process of DNA synthesis in intact cells, the
permeable cell system was used for an examination of poly (ADPR)
polymerase with the expectation that this system might maintain
some of the physiological relationships of the enzymes that were
present in the intact cells. Table I shows that the material
synthesized from the added (³H) NAD conforms to the nuclease
susceptibilities expected for poly (ADPR) (11). The polymer
synthesized during the 30 minute incubation with (³H) NAD was not
degraded by DNAse or RNAse. More than 95% of the radioactively
labelled material was degraded by venom phosphodiesterase. This
degradation pattern eliminates the possibility that the nucleotides
were incorporated into DNA, RNA or poly A. These results indicate
that virtually all of the material synthesized under these
conditions is poly (ADPR).

Table I

Nuclease treatment of product from poly(ADPR)
polymerase reaction in permeabilized L cells

Treatment	CPM (^{3}H) poly (ADPR) Remaining after Treatment
Boiled control	11,837
Incubation control	13,025
DNAse 100 µg/.3 ml	13,454
RNAse 500 µg/.3 ml	13,533
VPDase 600 µg/.3 ml	477

Log phase cells were permeabilized and incubated
with (^{3}H)NAD in poly (ADPR) synthesis mix for 30
mins at 30°C. The reaction was terminated by
boiling for 10 mins. Aliquots were transferred to
10% TCA, 2% Na$_4$P$_2$O$_7$ to measure incorporation of
(^{3}H)NAD into acid precipitable macromolecules. These
are labelled boiled control. Aliquots were also
incubated at 37°C with the indicated enzyme or with
buffer in the case of the incubation control. After
20 hours, the remaining macromolecular material was
precipitated with an excess of 10% TCA, 2% Na$_4$P$_2$O$_7$
and prepared for scintillation counting.

The activities of DNA polymerase and poly (ADPR) polymerase
can be separated by their different susceptibilities to a series
of inhibitors. Table 2 shows that analogs of nicotinamide as well
as thymidine, 5-bromodeoxyuridine and uridine all inhibit the poly
(ADPR) polymerase reaction in the permeable cells but have very
little effect on the DNA polymerase reaction. Caffeine and formycin,
which may compete with the adenine portion of NAD, were also able
to partially inhibit the poly (ADPR) polymerase reaction. They do
not inhibit the DNA polymerase reaction. In contrast, cytosine
arabinoside triphosphate, Cytembena, actinomycin and phosphonoacetic
acid all inhibit the DNA polymerase reaction without affecting the
poly (ADPR) polymerase reaction. Daunomycin complexes the DNA
template and causes complete inhibition of DNA synthesis and partial

Table 2

Comparison of agents affecting poly (ADPR)
polymerase and DNA polymerase in permeable L Cells

| Condition | Activity (% of control) | |
	p (ADPR) pol	DNA pol
Control	100	100
10 mM Nicotinamide	4.4	96.6
10 mM 5-CH$_3$-Nicotinamide	5.8	99.6
10 mM Thymidine	4.1	81.2
10 mM 5 BrdUrd	3.9	83.8
10 mM Uridine	27.9	87.4
1 mM Caffeine	50.6	110
1 mM Formycin	25.0	99.7
.1 mM Ara CTP	100.5	23.2
10 mM Cytembena	106	12.0
10 mM Phosphonoacetic Acid	94.1	11.6
.1 mM Daunomycin	61	9.5
50 μg/ml Actinomycin	86	12
.05% Triton X-100	319	161
.05% Triton X-100 + 100 μg DNAse I	1378	63

    Cells in mid log phase growth were used for all determinations
in this table. Agents to be tested were prepared in 50 mM Tris
HCl at a final pH of 7.8. They were added to the reaction
components just before addition of the cells. The concentration
of the inhibitor given in the table was the final concentration
in the complete reaction system. Poly (ADPR) synthesis reactions
were incubated at 30°C, DNA synthesis reactions at 37°C. All
incubations were for 30 mins. The amount of polymer synthesized in
the presence of each inhibitor is expressed as a percentage of the
amount synthesized in simultaneously performed, control, untreated
reactions.

inhibition of poly (ADPR) synthesis. This suggests that poly
(ADPR) synthesis is somewhat dependent on the conformation of the
DNA template.

It has previously been demonstrated that addition of DNAse to
HeLa cell nuclei causes a marked stimulation in the rate of poly
(ADPR) synthesis (2). We have previously demonstrated that when
permeable cells are incubated in the presence of .05% Triton X-100,
proteins such as DNAse have rapid access to the DNA in the cell
nucleus (4). We, therefore, added DNAse to the permeable cell
system in the presence of Triton X-100, and found that it produced
a drastic increase in the activity of poly (ADPR) polymerase. Fig.
5 presents the results of further studies on the DNAse stimulation
of poly (ADPR) polymerase activity. The endogenous activity of
poly (ADPR) polymerase measured in the absence of added DNAse was
higher in plateau phase cells than it was in logarithmically
growing cells. As the amount of added DNAse was increased, the
activity of poly (ADPR) polymerase increased until it reached a
maximum level. When the system was saturated with DNAse at
concentrations between 50 to 500 µg, both the log and plateau phase
cells demonstrated the same total activity for poly (ADPR)
polymerase. Fig. 5 also demonstrates that in the presence of 100

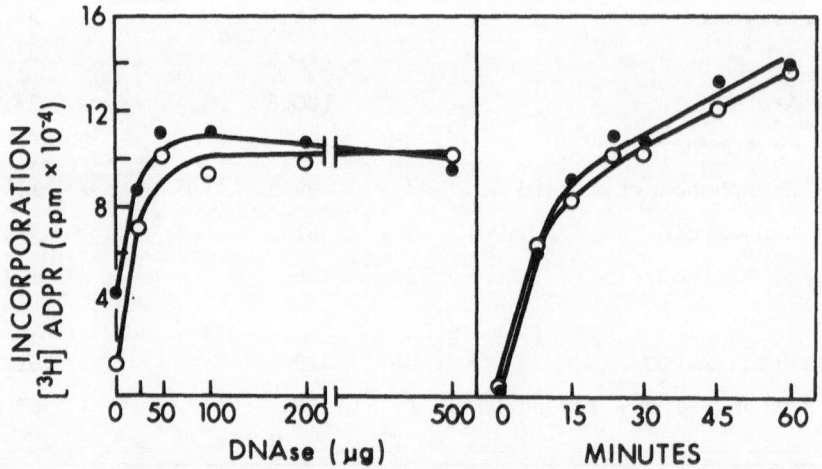

*Figure 5. Synthesis of poly (ADPR) in presence of DNAse I. Log
phase cells were at a density of 3 x 10⁵/ml. Plateau phase cells
were at 10 x 10⁵/ml. The reaction system contained 5 x 10⁶
permeabilized cells and (³H)NAD reaction mix. Triton X-100 and
DNAse were added to the mix just before the addition of the
permeabilized cells. The final concentration of Triton X-100 was
.05%. In the left panel, the total amount of DNAse in the reaction
system is indicated. All reactions were incubated for 30 min at
30° C. In the right panel all reactions contained 100 µg DNAse I and
were incubated for the indicated time period at 30° C. Cells from log
phase culture (O——O). Cells from plateau phase culture (●——●).*

µg DNAse, both the log and plateau phase cells synthesized poly
(ADPR) at the same rate. These studies indicate that permeable cells
can be used to measure two levels of activity for poly (ADPR) polymerase. The first or endogenous activity may be a measure of the
physiologic activity in the cells and may be dependent on the
availability of the endogenous acceptors. The second or total
activity can be measured by adding an excess of DNAse to uncover
the expression of the total available poly (ADPR) polymerase.

To determine whether the activity of poly (ADPR) polymerase
had any direct effect on DNA synthesis, mid log phase cells were
permeabilized and then incubated in one of three reaction systems.
The first contained only the components for DNA synthesis. The
second contained the components for poly (ADPR) synthesis, and the

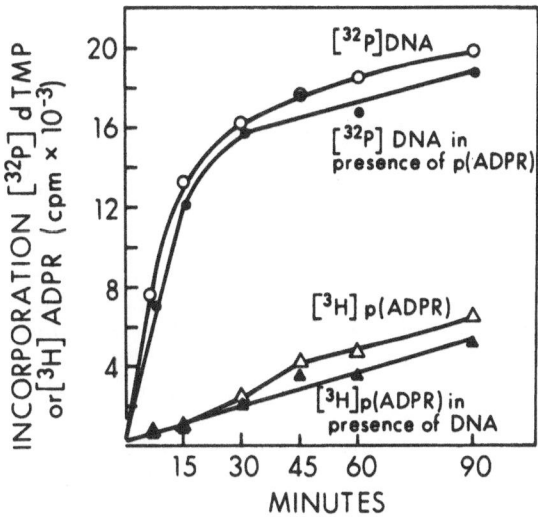

*Figure 6. Simultaneous synthesis of poly (ADPR) and DNA in
permeable cells. Cells from log phase culture at 3.1 x 10⁵/ml were
permeabilized and incubated in one of three reaction mixes. The
reaction system for DNA synthesis alone contained .02 µm (α-³²P)dTTP
(Sp Act 17.6 x 10³ cpm/pmol). The reaction system for poly (ADPR)
synthesis alone contained 160 µm (³H)NAD (Sp Act 10 x 10³ cpm/nmol).
The reaction system for the simultaneous synthesis of DNA and poly
(ADPR) contained 0.2 µm (α-³²P)dTTP (Sp Act 17.6 x 10³ cpm/pmol),
and 160 µm (³H)NAD (Sp Act 10 x 10³ cpm/nmol). The reactions also
contained all the other components required for DNA synthesis and
poly (ADP Ribose) synthesis. Incubations were for the indicated time
periods at 37°C. Synthesis of (³²P)DNA during independent DNA
synthesis reaction (O——O), synthesis of (³H) poly (ADPR) during
independent poly (ADPR) synthesis reaction (△——△). Synthesis of
(³²P)DNA (●——●) and (³H) poly (ADPR) (▲——▲) in presence of
simultaneous synthesis of DNA and poly (ADPR).*

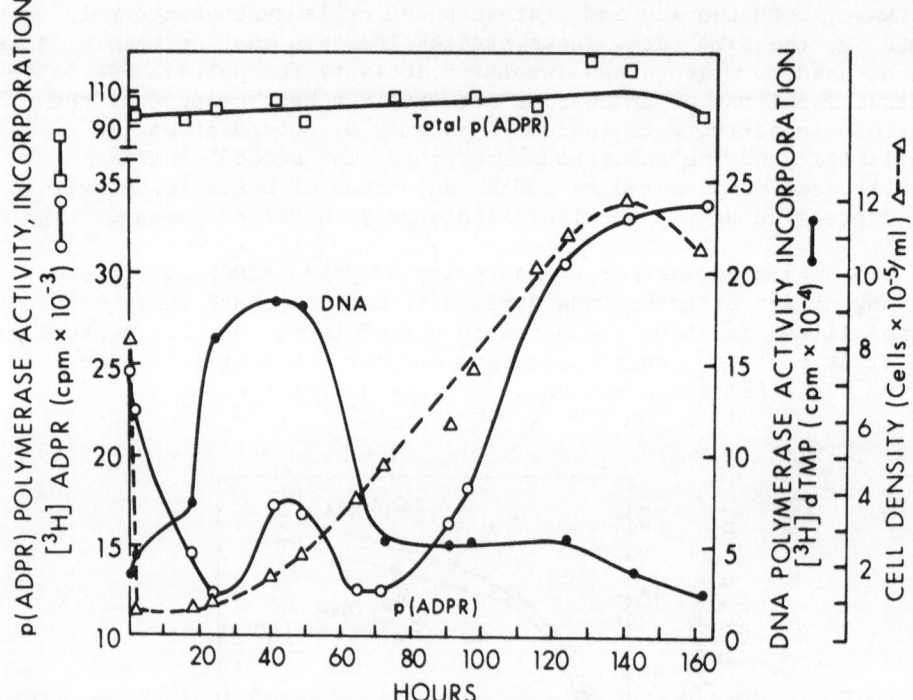

Figure 7. *Variation of DNA polymerase activity and poly (ADPR)*
polymerase activity with L cell growth. Plateau phase cells were
diluted to 10⁵/ml in a modified Eagles medium at time 0. Cell counts
were performed at subsequent time intervals (△——△). At each time
point aliquots of cells were removed, permeabilized, and incubated
with appropriate reaction mix for 30 min. DNA polymerase activity
(●——●), endogenous poly (ADPR) polymerase activity (○——○),
total poly (ADPR) polymerase activity measured in the presence of
added DNAse (□—— □).

third contained the components for both DNA synthesis and poly
(ADPR) synthesis. The results are presented in Fig. 6. They
demonstrate that the rate of poly (ADPR) synthesis is the same in
the presence or absence of DNA synthesis. Similarly, the rate of
DNA synthesis is the same in the presence or absence of poly (ADPR)
synthesis. Thus, when poly (ADPR) and DNA are synthesized
simultaneously, neither polymer has an effect on the rate of
synthesis of the other. Since DNA synthesis in the permeable cells
is apparently a measurement of elongation but not initiation, these
studies demonstrate that poly (ADPR) synthesis does not affect
elongation. It is still possible that poly (ADPR) synthesis could
effect the initiation of new origins of DNA synthesis.

In order to determine whether the activities of DNA polymerase
and poly (ADPR) polymerase change with different stages of cell

growth, cells were removed from culture at sequential time points, they were permeabilized and analyzed for activity of DNA polymerase, endogenous poly (ADPR) polymerase and total poly (ADPR) polymerase. The total poly (ADPR) polymerase was measured in the presence of added DNAse as described above. Fig. 7 demonstrates the fluctuations in enzyme activities as the cells pass from high density to logarithmic growth and back to plateau phase. Initially, when the cells were at high density, the activity of DNA polymerase was low while the endogenous activity of poly (ADPR) polymerase was high. Immediately after the cells were diluted, the activity of DNA polymerase began to increase and within 22 hours the activity of DNA polymerase reached four times the level of activity found in plateau phase cells. The activity of DNA polymerase reached its maximum in cells in mid log phase growth and then decreased again as the cells approached plateau phase. This is similar to the fluctuations noted previously for L cell DNA polymerase α (13).

The endogenous activity of poly (ADPR) polymerase also demonstrates marked variation with the cell growth status. The activity was highest in plateau phase cells and then fell to one third of that value during the active phase of DNA synthesis. As the cells progressed to high density, the endogenous activity of poly (ADPR) polymerase underwent a marked increase and continued to rise as long as the cells remained in plateau phase. While the endogenous poly (ADPR) polymerase activity demons-rated these marked oscillations with cell growth, the total poly (ADPR) polymerase activity remained relatively constant throughout the growth cycle. This suggests that the total amount of enzyme in the cell is relatively constant and is independent of growth status. It also suggests that the activity of this enzyme depends on fluctuations in the availability or amount of acceptors for poly (ADP Ribose).

In subsequent studies, we examined the effect of glucose deprivation, vaccinia virus infection and treatment with cytosine arabinoside on the activity of poly (ADPR) polymerase. In all these studies, the suppression of DNA synthesis was associated with a sharp increase in the endogenous activity of poly (ADPR) polymerase. These observations suggest that poly (ADPR) polymerase and its product poly (ADPR) may be important in the packaging, maintenance or protection of DNA during states of suppressed DNA synthesis.

ACKNOWLEDGEMENTS

This research was supported by grants from the National Leukemia Association, Inc., Garden City, New York, and the Jewish Hospital of St. Louis.

REFERENCES

1. Sugimura, T. (1973) Prog. Nuc. Acid Res. Mol. Biol. 13:127-151

2. Hayaishi, O. and Ueda, K. (1977) Ann. Rev. Biochem. 46:95-116

3. Berger, N.A. and Johnson, E.S. (1976) Biochim. Biophys. Acta 425:1-17

4. Berger, N.A., Erickson, W.P. and Weber, G. (1976) Biochim. Biophys. Acta 447:65-75

5. Berger, N.A. and Weber, G. (1977) J. Nat. Cancer Inst. 58:1167-1168

6. Berger, N.A., Petzold, S.J. and Johnson, E.S. (1977) Biochim. Biophys. Acta (in press)

7. Olivera, B.M. and Bonhoeffer, F. (1972) Nature New Biol. 240:233-235

8. Herrmann, R., Huf, J. and Bonhoeffer, F. (1972) Nature New Biol. 240:235-237

9. Huberman, J.A. and Riggs, A.D. (1968) J. Mol. Biol. 32:327-341

10. Gautschi, J.R., Kern, R.M. and Painter, R.B. (1973) J. Mol. Biol. 80:393-403

11. Nishizuka, Y., Ueda, K., Nadazawa, K. and Itayaishi, O. (1967) J. Biol.Chem. 242:3164-3171

12. Miller, E.G. (1975) Biochim. Biophys. Acta 395:191-200

13. Chang, L.M.S., Brown, M.K. and Bollum, F.J. (1973) J. Mol. Biol. 74:1-8

EFFECT OF *XENOPUS LAEVIS* OOCYTE AND EGG EXTRACTS ON SV40 DNA

G.P. Tocchini-Valentini[*], D. Gandini-Attardi,
E. Mattoccia, M.I. Baldi[*], and G. Carrara[*]

CNR - Laboratorio di Biologia Cellulare
Via Romagnosi 18A
Rome, Italy

Xenopus laevis oocytes and eggs provide an interesting system in which to study various aspects of DNA metabolism.

During oogenesis genetic recombination takes place (1). Young oocytes selectively replicate ribosomal DNA genes (2), whereas older oocytes are unable to carry out any kind of DNA replication (3). In fertilized eggs chromosomal DNA replication occurs at an exceptionally high rate; autoradiographic studies have shown that the exceptional rate of DNA synthesis is due to an unusually high number of origins of replication (4).

Microinjection techniques have been successfully employed to investigate the properties of this system. Graham et al.(5) have shown that brain nuclei, once injected into the cytoplasm of un-fertilized eggs are induced to synthesize DNA; this effect is not obtained when brain nuclei are injected into oocytes. In order to analyze the mechanism of induction of DNA synthesis by egg cyto-plasm, purified supercoiled polyoma DNA was injected and was found to stimulate the incorporation of thymidine (6). Moreover, Laskey and Gurdon (7) have shown using the same system that if heavy-light supercoiled polyoma DNA is injected, the density of part of the input DNA shifts to that of light double stranded DNA, suggesting synthesis of new DNA strands.

At this point it became clear that a next reasonable step would be to study the effect of soluble extracts on DNA synthesis (8,9).

We developed a soluble cell-free extract derived from stage 6

[*]*A.S.I. Participants*

oocytes of Xenopus laevis (8) and we chose to study its action on
SV40 DNA, a simple and well characterized molecule. The extract
when challenged with supercoiled SV40 DNA produced nicked circles
having a single strand scission, linear molecules of full unit size,
shorter length fragments and various forms of complex DNA (8).
Complex DNA consists of a repertoire of structures such as: figure
8 dimers, catenated dimers, circular dimers, catenated trimers,
late Cairns' structures, complex multimers and circular monomers
with tails. We reasoned that fractionation of the extract should
produce fractions lacking some of the activities necessary to produce
complex DNA and that therefore intermediate structures should
accumulate. In fact we observed that some fractions of the extracts
produced, from SV40 DNA, a high percentage of branched DNA molecules.
These branched molecules resemble the structures which were observed
by Broker and Lehman in E.coli cells infected at high multiplicity
by T4 polymerase and ligase deficient phages (10).

In this paper we describe the requirements and characteristics
of the reaction responsible for the formation of branched DNA
molecules and discuss the possibility that branched DNA is an intermediate in the formation of complex DNA.

We describe, moreover, an extract derived from Xenopus laevis
eggs, which complements chromatin bodies, derived from SV40 infected
cells (25).

Action of the Oocyte Extract on SV40 DNA

To produce the extract, stage 6 oocyte nuclei were manually
isolated and separate extracts from nuclei and from enucleated
cytoplasms were prepared. The nuclei-cytoplasms extract consists of
equal volumes of nuclear and cytoplasmic extracts (8).

The incubation mixture contained in a total of 100 µl: 50 mM
Tris HCl at pH 7.5, 100 µM dATP, dCTP, dGTP, 20 µM ATP, GTP, UTP,
CTP, 6 mM $MgCl_2$, 6 mM dithiothreitol, SV40 (^{3}H)DNA (10^4 cpm/µg) in
amounts specified under the figures, 25 µl of nuclear and 25 µl of
cytoplasmic extracts. After incubation for 30 min at 30° the
reaction was stopped by addition of sodium dodecyl sulfate and EDTA
at the final concentrations of 0.5% and 15 mM respectively, and the
mixture was applied to a gradient. Occasionally the incubation
mixture was extracted with phenol before being centrifuged.

Fig. 1 shows that the input supercoiled SV40 DNA, sedimenting
at 21 S, when incubated for 30 min at 30° with the nuclei-cytoplasms
extract in the presence of the standard mixture is converted to:
(a) Species sedimenting at the bottom of the gradient. The sample
was deproteinized before being applied to the sucrose gradient, and
therefore the sedimentation properties of this material suggest that

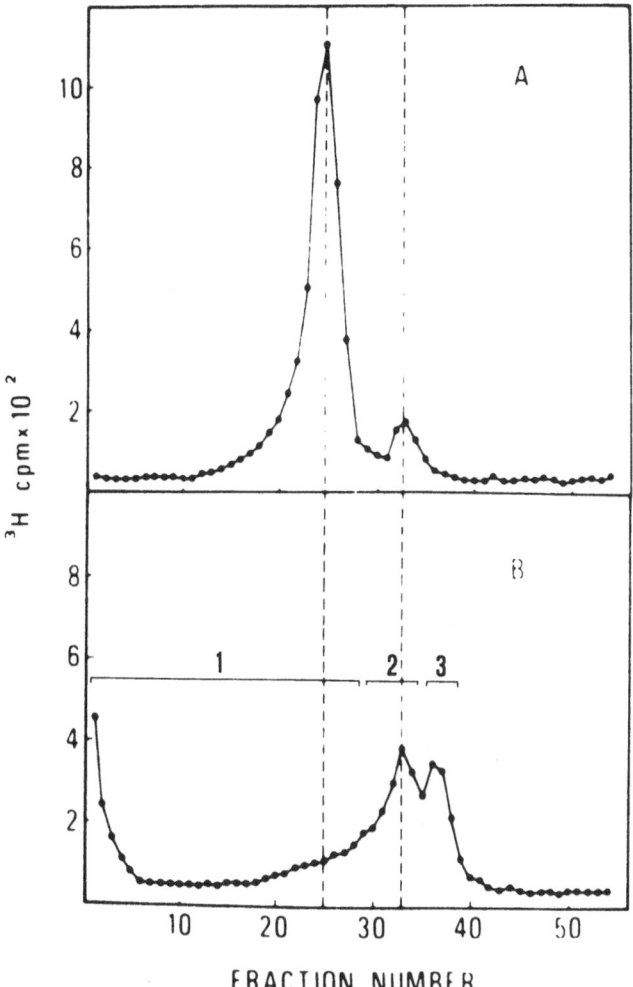

Figure 1. SV40 (^{3}H)DNA component I (supercoiled, 21 S) (3 μg, 28,000 cpm) was incubated under the conditions specified in the text without (A) and in the presence of (B) nuclei-cytoplasms extract. Sedimentation was on 15.30% neutral sucrose gradients in a Spinco SW41 rotor at 35,000 rpm at 4° for 16 hr. A 30 μl aliquot from each 0.120 ml fraction was precipitated with trichloroacetic acid and its radioactivity was determined. The remaining DNA was pooled into fractions 1,2 and 3 as shown, precipitated with ethanol, and examined with the E.M.

it contains oligomeric and/or catenated SV40 DNA (11,12). (b) Species sedimenting at 16 S, like open circular or relaxed circular structures which may be the product of nicking or of relaxation.

(c) Species sedimenting at approximately 15.4 S like full-length linear SV40 DNA.

We found that the presence of MG^{++} in the incubation mixture was required for the formation of all the species described above. The species sedimenting at the bottom of the gradient were not observed when either a nuclear or a cytoplasmic extract was used separately instead of the nuclei cytoplasms extract.

Analysis of the Products

To further investigate the nature of the various products the fractions of the gradient were combined in three different pools, marked pool 1, 2 and 3 respectively, as indicated in Fig. 1B.

Pool 1, when analyzed by E.M., showed an extraordinary repertoire of structures, such as: figure 8 dimers, catenated dimers, circular dimers, catenated trimers, Cairns' structures (15) (Fig. 2E) complex multimers, and circular monomers with tails. Table 1 shows the list of the structures found and their frequency of occurrence.

Table   1

Structure	No. of molecules observed
Figure 8	20
Concatenated dimers	100
Dimers	50
Concatenated trimers	50
Concatenated tetramers	5
Concatenated pentamers	5
Concatenated hexamers	5
Complex concatemers	200
Rolling circles	15
Cairns' structures	5

Complex structures were counted following the criteria established by Gordon et al. (13).

Total DNA from the incubation mixture observed at the E.M. before sucrose gradient fractionation contained 15% complex DNA computed as monomeric units, 35% open circles, 30% full-length linear molecules, and 5% fragments.

The material contained in pool 1, treated with the restriction enzyme EcoRI, which is known to cut SV40 DNA at a unique site (14), was converted to species sedimenting sharply at 14.5 S, like full-length linear SV40 DNA (data not shown). This finding indicates that the DNA contained in pool 1 consists of complex SV40 DNA and that there is no other kind of DNA associated with the viral molecules. Fig. 2 shows some typical examples of the complex forms of SV40 DNA present in pool 1.

As shown in Fig. 1A, the heavy sedimenting material constituting

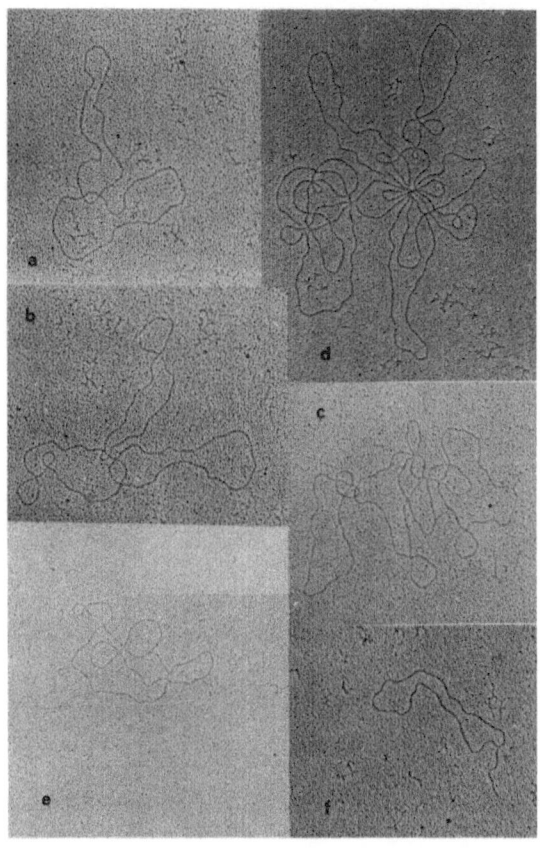

Figure 2. Electron micrographs of complex SV40 DNA: (a) catenated dimers; (b) catenated trimer, (c) catenated tetramer; (d) complex concatemer; (e) Cairns' structure, (f) rolling circle. x 61.000.

pool 1 is not present in the template DNA which has been incubated without the extract. The input DNA was also examined by E.M. after nicking by pancreatic DNase and only two out of 2000 molecules observed could possibly consist of complex DNA.

Pool 2 material when examined by E.M. was confirmed to consist of circular structures: their contour length measurements are reported in Fig. 3. Pool 2 material was analyzed in an alkaline sucrose gradient and Fig. 4 shows that two peaks, sedimenting at 18 S and 16S, were found in approximately equimolar amounts. The material sedimenting at 18S and 16 S corresponds to single-strand circular and single-strand linear SV40 DNA, respectively; therefore we can conclude that the material sedimenting at 16S in a neutral sucrose gradient (Fig. 1B) consists mainly of open circular molecules derived by the action of a nicking enzyme.

E.M. of pool 3 material confirmed that 95% of the molecules consist of full-length linear SV40 DNA; 5% of the molecules consist of fragments shorter than unit length. To establish that the linear molecules contained in pool 3 are really full length we have examined a mixture of pool 2 and pool 3 in the E.M. Comparative length measurements of the linear molecules with the circular ones on the same grid established that the linear products have the same contour length as the circular viral DNA (Fig. 3).

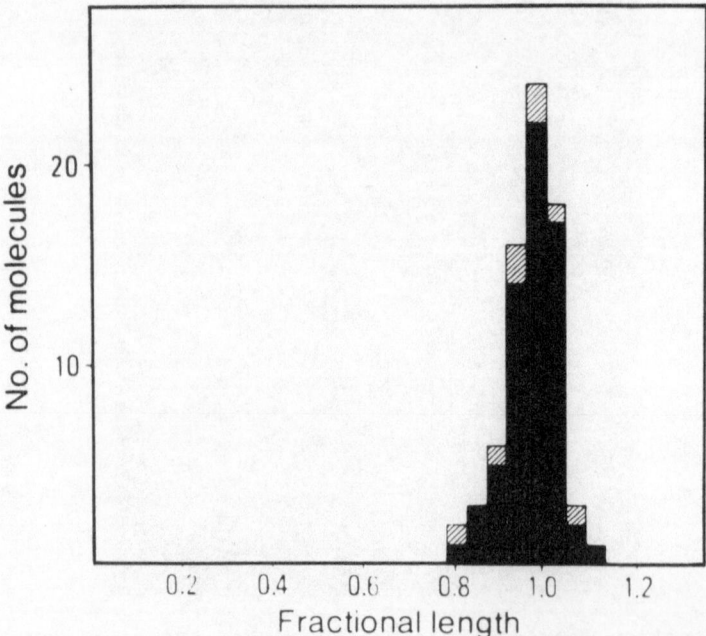

Figure 3. Histograms of length measurements of open circular SV40 DNA molecules contained in pool 2 (diagonal stripe) and linear molecules contained in pool 3 (black), mounted on the same grid.

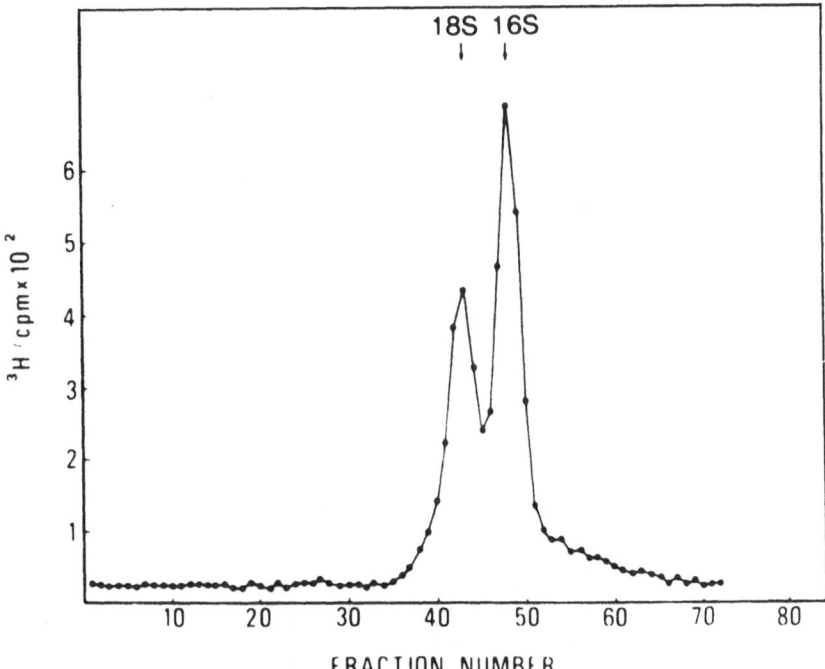

Figure 4. Alkaline sucrose gradient (5-20%) of pool 2 material (see Fig. 1B). Pool 2 material derived from a gradient like the one shown in Fig. 1B was ethanol precipitated, dissolved in 100 μl of 0.25 M NaOH, 0.1 M NaCl, 1 mM EDTA and centrifuged in a SW 41 rotor at 33,000 rpm at 4° for 15 hr. Fractions (80 μl) were collected, acid precipitated and counted. Arrows indicate the position of 18 S single-strand circular and 16 S single-strand linear molecules run separately as markers.

It is desirable to locate on the SV40 map (16) the site(s) at which the viral DNA is cut to generate the full-length linear molecules. Denaturation mapping (14) of full-length linear molecules produced by partially purified enzyme(s) indicates that the sites of cleavage are not randomly distributed.

Analysis by Agarose Gel Electrophoresis of the Products of the Cell-Free System

The nature of the products was further investigated by analysis by agarose gel electrophoresis. Fig. 5 shows that the cytoplasmic extract produces from supercoiled SV40 DNA material having the mobility corresponding to that of full-length linear SV40 DNA, material having the mobility corresponding to that of fully relaxed

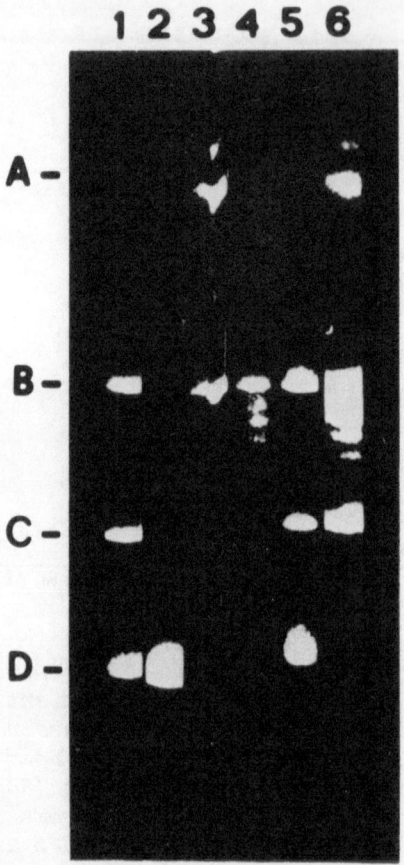

<u>Figure</u> 5. *Agarose gel electrophoresis of SV40 DNA treated with
nuclear and cytoplasmic extracts. Migration is toward the bottom.
A slowly migrating band corresponding to complex DNA, B, nicked
circular or fully relaxed DNA; C, linear full size DNA, D, supercoiled DNA. Lane 1, SV40 DNA partially digested with Hpa II
restriction enzyme. The reaction mixture contained, in a volume of
40 μl, 6.6 mM Tris HCl at pH 7.4, 6 mM MgCl₂, 6.6 mM mercaptoethanol, 10 mM NaCl, 0.5 μg of supercoiled SV40 DNA, and 5 μl of
enzyme. Incubation was carried out for 1 hr at 37°. Lane 3, complex DNA produced by combined nuclear and cytoplasmic extract.
Complex DNA was purified by sucrose gradient centrifugation as
described in ref.8. Lanes 2,4,5 and 6, reaction mixtures contained
in a volume of 100 μl, 50 mM Tris HCl at pH 7.5, 100 μM each dATP,
dCTP, dGTP and dTTP; 20 μM each ATP, CTP, GTP and UTP; 6 mM MgCl₂;
6 mM dithiothreitol; 0.5 μg of supercoiled DNA; and the following
extracts: 2 control; 4 nuclear extract (30 μl); 5, cytoplasmic
extract (30 μl); 6, nuclear extract (30 μl) plus cytoplasmic
extract (30 μl).*

or nicked circular DNA, and shorter length fragments which migrate
to the lower portion of the gel (not shown in the picture). When
the products of the nuclear extract are analyzed, the gel pattern
shows a series of bands migrating in the gel with mobilities intermediate between those of supercoiled DNA and fully relaxed circles.
Since this pattern is characteristic of a population of DNA molecules having a decreasing number of superhelical turns (17) this
finding suggests that an untwisting enzyme is present in our
extract and that this enzyme is located in the nucleus (18). When
the products of the combined nuclear and cytoplasmic extracts are
analyzed in gel electrophoresis, a pattern is observed that includes
all the same features observed in the products of the separate
nuclear and cytoplasmic extracts. Moreover, in the.upper portion
of the gel, a series of slowly migrating bands is observed. This
slowly migrating material, which requires for its formation the
combined action of the nuclear and cytoplasmic extracts,
corresponds to complex DNA. Slowly migrating bands are in fact
observed when complex DNA, purified by sucrose gradient sedimentation
(8), is examined by agarose gel.

Fractionation of the Nuclei-Cytoplasms Extract and Formation of Branched Molecules

The fractionation of the extract involved the preparation of
a high speed supernatant, precipitation with 60% saturated ammonium sulphate, and chromatography on DEAE-cellulose and phosphocellulose columns. The phosphocellulose column was eluted with
0-1.0 M KCl gradient.and the dialyzed fractions were incubated with
supercoiled SV40 DNA in the presence of Magnesium and ATP. The
phenol extracted DNA which was recovered from the incubation
mixtures was prepared for E.M. analysis.

The fractions eluting from the phosphocellulose column between
0.2-0.25 M KCl produced nicked circles, linear molecules and
branched DNA. The branched DNA was highly varied with respect to
form (circular and linear), length, number and arrangements of
branches. An analysis of the length of the molecules in the
branched structures, considering the length of the SV40 DNA,
indicated that more than one molecule is involved.

Fig. 6 shows some typical examples of the structures found.

Often several branches occurred in the same molecule; because
of the high dilution used for the E.M. preparation, these multiply
branched forms cannot be due to random association. The structures
are formed without added deoxytriphosphates and the possibility of
any remaining endogenous deoxytriphosphates is highly unlikely,
given the procedure used to prepare the fractionated extract.

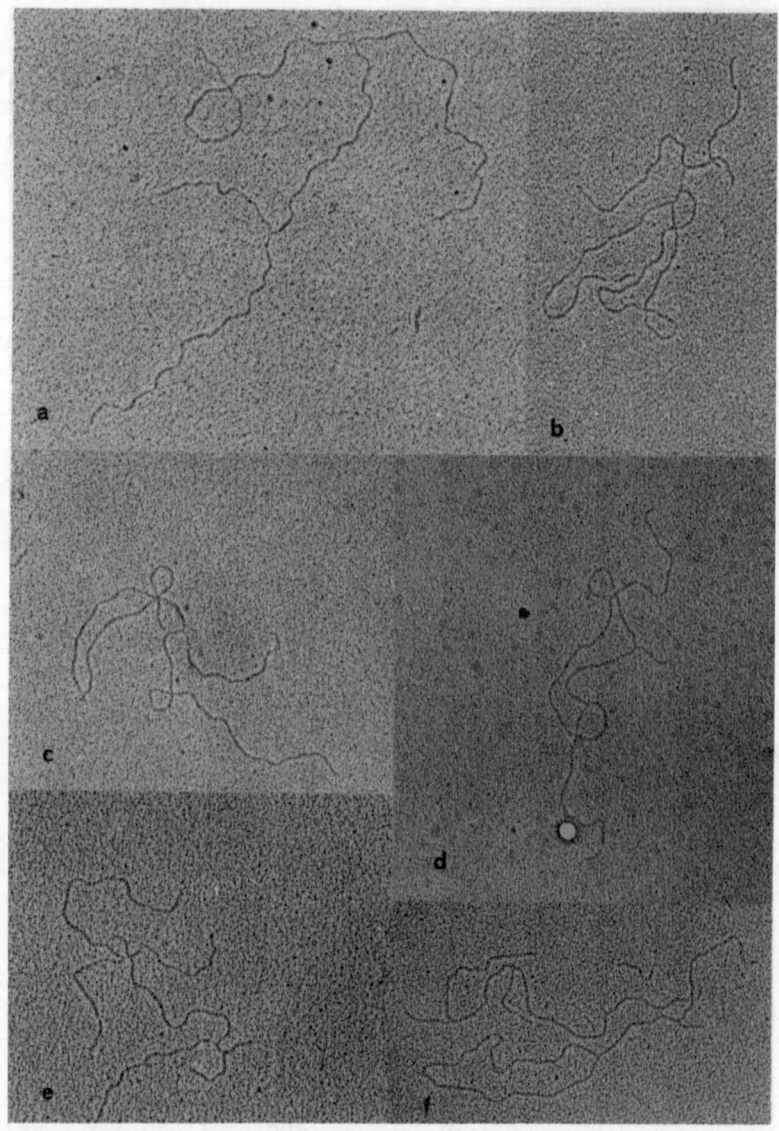

Figure 6. *Electron micrographs of branched molecules produced by incubating supercoiled SV40 DNA with a phosphocellulose fraction. Reaction mixtures contained in a volume of 50 μl: 50 mM Tris-HC1 pH 7.5, 6 mM MgCl$_2$, 6mM DTT, 0.1-0.5 μg of SV40 DNA and 25 μl of the phosphocellulose gradient fractions. After incubation for 60 min at 30°, the reaction was stopped by addition of SDS and EDTA to final concentrations of 1% and 15 mM respectively. 50 μl of distilled water were added to each reaction mixture and the DNA was extracted with phenol and ether and used directly for E.M. or after concentration by ethanol precipitation. Magnified 61.000 X.*

Branched DNA molecules are therefore formed in the absence of DNA synthesis.

Characteristics of the Reaction

The formation of branched molecules is absolutely dependent on the presence of magnesium. In the absence of ATP, we observed an extensive degradation of the DNA. We can therefore conclude that, in the presence of ATP, the degradative process is inhibited, but we could not demonstrate that the nucleotide triphosphate plays a role in the formation of branched molecules.

The addition of the four deoxytriphosphates does not increase the percentage of branched molecules, indicating once again that DNA synthesis is not required for the process.

Table 2 shows the effect of temperature and time of incubation on the formation of branched molecules. The percentage of branched molecules after 60 min of incubation at 37° is at least twice the percentage of branched molecules formed at 30°. The reaction is time dependent: the percentage of branched molecules formed after 60 min of incubation at 30° is increased four fold after an additional 60 min of incubation.

Since more than one molecule is involved in the branched structures it is reasonable to expect that the percentage of branched molecules should increase with increasing DNA concentration. This is indeed the case when we are dealing with concentrations up to 3 μg/ml. At higher DNA concentrations the percentage of branched molecules decreases, suggesting that, since the reaction is presumably catalyzed by many factors, some of these become limiting at increasing DNA concentrations.

Comparison of Different Forms of Substrate DNA

Supercoiled DNA is not required for the formation of branched molecules. Table 3 shows that four different forms of SV40 DNA have been compared as substrates. We have observed, as a matter of fact, that relaxed molecules prepared by the action of Xenopus laevis DNA-relaxing enzyme (18), are four times more efficient than the supercoiled form; nicked circles having a single strand scission and linear molecules produced by EcoRI react approximately with the same efficiency. We cannot offer an explanation for the relatively low efficiency. We cannot offer an explanation for the relatively low efficiency of supercoiled DNA. The data in Table 3 also indicate a correlation among multiple branches in a single molecule; when linear molecules were used as substrate, 20.5% of the 500 molecules observed were branched. An independent origin of each

Table 2

Effect of temperature and of time of incubation on the
formation of branched molecules

Temperature	Time (min)	Percentage Branched Molecules	No. of Molecules Scored
30	60	3.6	1110
	120	11.8	1180
37	60	8.9	1120

Supercoiled SV40 DNA at the concentration of 14 µg/ml was
incubated with the phosphocellulose fraction under the
conditions indicated in the text. Samples were prepared
for E.M. according to Davies et al. (19).

Table 3

Summary of electron microscopic analysis
of branched DNA molecules

DNA	Branched %	$\dfrac{\text{Multiply-Branched}}{\text{Branched}} \times 100$	No. of Molecules Scored
Supercoiled	5.6	22.2	1280
Open circles	23.8	47.6	500
Relaxed circles	20.6	46.1	500
Linear[+]	20.5	45.2	500

[+]Various size fragments were produced when linear molecules
were used as substrate. The percentage of branched molecules
was estimated taking into account only molecules equal to or
larger than full unit size.

The various forms of SV40 DNA, prepared as described (8,18),
were incubated with the phosphocellulose fractions under
the conditions indicated in the text at a concentration of
10 µg/ml for 2 hrs at 30°. Samples were prepared for E.M.
according to Davies et al. (19).

would have led to $(0.205)^2$ x 500 = 21 multiply-branched molecules;
instead, 46 multiply-branched molecules were observed. Similar
considerations can be applied to the cases in which other molecular
forms were used as substrates. The high value of the figure
corresponding to multiply-branched molecules suggests that two or
more branches may have been generated as part of a single exchange
event.Furthermore, multiple branches are often structurally very
close together.

     When linear SV40 DNA is used as substrate, a considerable
number of the branched molecules observed have an "H-like" configuration. Fig. 7a, b, c, d shows examples of "H-like" and "Y-like"
molecules. Unfortunately when linear DNA is used, a degradation
due to exonuclease(s) activity present in the fraction does not
always permit the recovery of intact molecules; therefore we have
been unable to map the cross bar of the "H-like" molecules relative
to the extremities.

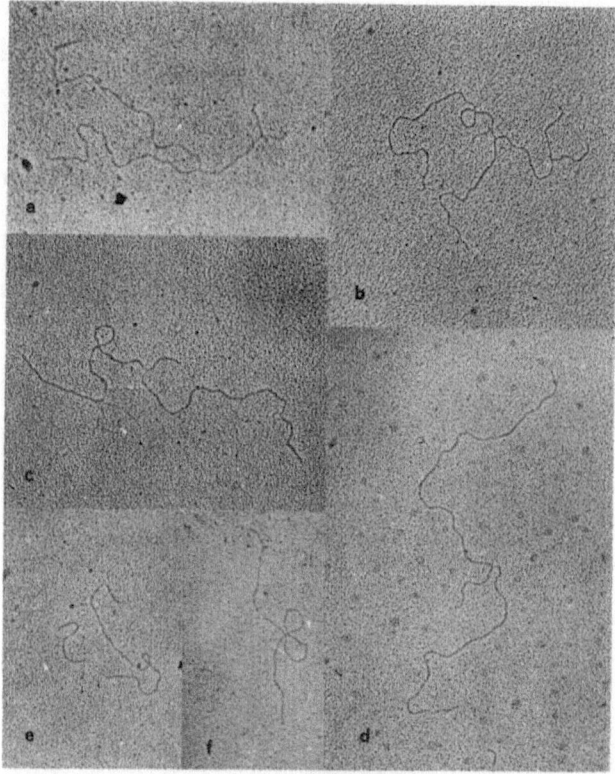

*Figure 7. Electron micrographs of branched molecules obtained by
incubating a phosphocellulose fraction under the conditions indicated
under Fig. 6 with linear SV40 DNA, produced by EcoRI, for 60 min at
30°. Magnified 61.000 X. a,b "H-type" recombinants. c,d "Y-type"
recombinants. e,f, gapped molecules.*

Size of the Paired Region

The regions between two branched points were measured and the
distribution of the lengths of these regions is shown in Fig. 8.
The lengths vary approximately from an apparent 0 to 0.5 fractional
length, with an average of 0.15 fractional length. Since SV40
chromosome contains 5.1 Kb pairs, the average length of the region
between two branched points corresponds to 765 nucleotides.

Interpretation of the Structures Formed

We would like to consider our results in the light of models
for the formation of branched DNA which are supported by evidence in
the phage T4 and phage T7 experimental systems (10,20). For
simplicity's sake, we will consider the "H-like" and "Y-like"
molecules shown in Fig. 7a,b,c,d. Structure of this kind could be
formed by a process involving an exonuclease which can initiate
hydrolysis from a nick produced by an endonuclease of the type
which we described in Xenopus laevis extract (18). An expanded
single strand region would allow for pairing with DNA homologous
in that region, thus yielding the "H-like" structures (Fig. 9).
For this to occur, strand breaks must be introduced on the opposite
strands of two molecules to be paired. We have frequently observed
gapped molecules in the electron micrographs of DNA incubated in
the presence of the fractionated extract. Fig. 7e,f shows examples
of such gapped molecules. "Y-like" molecules might originate from
the "H-like" structures through branch migration, a process in which

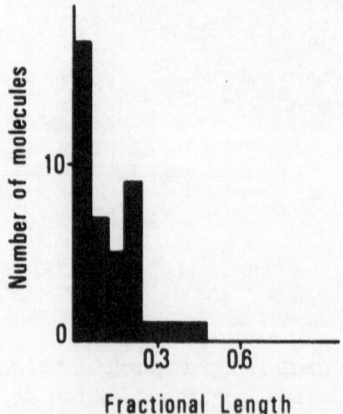

Figure 8. Histogram showing length distribution of the regions between two branched points of the doubly branched molecules.

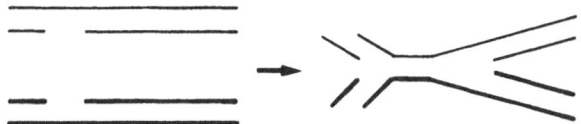

Figure 9. Schematic representation of DNA recombinant involving doubly branched molecules as intermediates.

there is a concerted dissociation of two sets of homologous parental DNA double helices and the reassociation of two sets of hybrid recombinant complements, with a resulting displacement of the branch point sequentially along the interacting DNA molecules. "Y-like" molecules might also result from the recombination of a terminal unpaired region with an internal site of a gapped molecule.

The soluble extract described in this paper appears to produce nicked circles, linear molecules of full unit size, shorter length fragments, and various forms of complex DNA from SV40 supercoiled DNA.

The fractionated extract is capable of producing recombinant molecules. The branched DNA molecules that we have observed are probably intermediates in the formation of complex DNA. Recombination between two gapped molecules (Fig. 10) produces a structure that by E.M. analysis is indistinguishable from a late Cairns' replicative intermediate. This type of structure could generate the various forms of complex DNA as is shown schematically in Fig. 10. However to conclusively prove that a precursor-product relationship exists between branched molecules and complex DNA, it would be necessary to convert the former in the latter by adding the factors, which have been removed by the purification, to the phosphocellulose fractions.

Our fractionated extract, in order to produce the structures we have observed should contain: (a) an endonuclease to produce nicks, (b) an exonuclease to convert nicks to gaps and termini to single strands, and most probably (c) an unwinding protein to protect single strands regions and to promote their complementary pairing. We are working at the purification and characterization of these activities. An unresolved question concerns the in vivo roles of these enzymes which in vitro produce the recombinant structures we have described.

Recombination which involves the interaction of two gapped molecules as indicated in Broker and Lehman's scheme (10) cannot give rise by maturation of the intermediate, to two reciprocal recombinant structures. It is clear therefore that our cell-free system is not faithfully mimicking the events that occur during

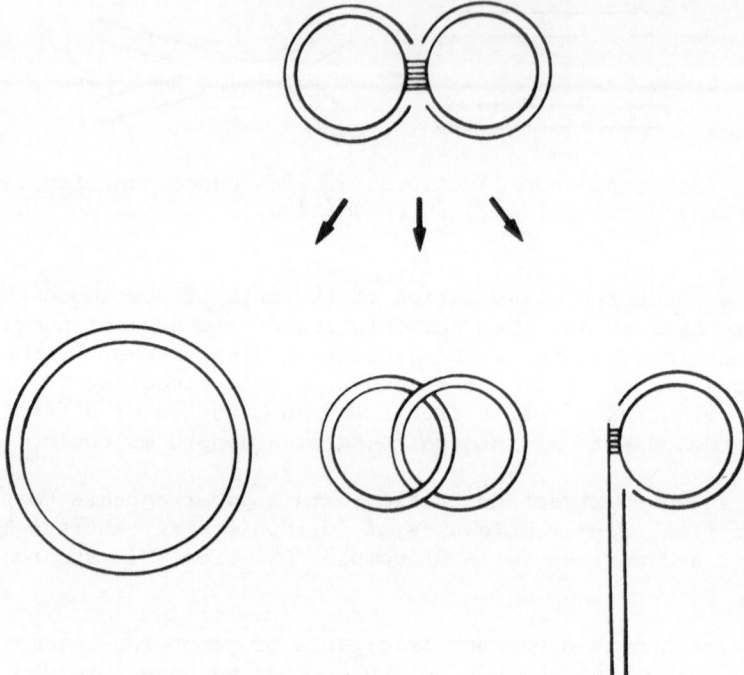

Figure 10. *Schematic representation of DNA recombinant involving two gapped circles as intermediates. Dimers, catenated dimers and circles with tail are shown as possible final products.*

meiotic recombination in vivo. Evidence exists indicating that in some systems recombination is carried out by a set of proteins which operate in a complex (21); if some of the proteins are inactive or absent from the complex, for example DNA polymerase and ligase, than it is possible that structures are formed which are not intermediates in the normal pathway. This should explain why Broker and Lehman found "H-like" intermediates and the same conclusion could be made about the structures we have found.

Alternatively, it could be that some special kind of recombination occurs in Xenopus laevis oocytes in which the inter-action of two input genomes leads to the construction of a single recombinant chromosome. For example, if a special mechanism exists for gene rectification (22), a circle of amplified ribosomal DNA could recombine with chromosomal DNA; by branch migration the amplified DNA could roll along the chromosomal sequences allowing extensive formation of heteroduplex regions. Structures of this kind could be involved in rectifying the sequence of multiple gene families by gene conversion (23). The process should end in such a way that rectified chromosomal DNA is recovered as a final product

and therefore the formation of only one of the possible linear
recombinants is necessary.

It is clear that our system, whatever is the in vivo role of
the enzymes we are considering, can be used to produce heteroduplexes
in vitro and to investigate, for example, the mechanism of gene
conversion in extract.

Action of Egg Extract on Chromatin Bodies

Egg extract was prepared from unfertilized eggs according to
Laskey et al. (24) except eggs were rinsed in TEMG buffer (50 mM
Tris HCl pH 7.5, 1 mM EDTA, 1.4 mM MSH, 20% (w/v) glyercol) before
homogenization. Chromatin bodies were prepared according to
Edenberg et al. (25).

The reaction mixture contained: 70 mM KCl, 7 mM $MgCl_2$ 1 mM
DTT 5 mM rATP, 0.40 mM each rGTP, rCTP and rUTP, 100 µM each dCTP,
dGTP and dATP; and (^{3}H) TTP 4 µM (46 Ci/mmole). Incubations were
performed at 30°C. Chase was performed by adding 100-fold excess
of cold TTP.

Characteristics of DNA Made "in vitro"

After 5 min incubation of the chromatin bodies, with or without
addition of egg extract some of the DNA synthesized sediments more
rapidly (between 22S and 28S) than form I DNA in neutral sucrose
gradients (Fig. 11), indicating that incorporation occurs into
replication intermediates. Therefore preexisting SV40 replicating
nucleoprotein complexes continue some aspects of replication "in
vitro".

After longer incubations (30 min) with chase, especially in
the presence of egg extract the peak of labeled DNA sediments more
slowly (Fig. 11). In order to determine the size of the DNA strands
made "in vitro", fractions 3 to 13, pooled from neutral sucrose
gradients, were sedimented through alkaline sucrose gradients. After
5 min of incubation and after 30 min plus chase, the DNA made in
the absence of egg extract contains labeled daughter strands ranging
from 4 S ("Okazaki pieces") to 16 S (genome length) in size (Fig.12).

In contrast, when the egg extract is added at 5 min, the
labeled DNA contains short strands and long strands, but with
longer incubations plus chase (30 min) the short strands chase into
longer strands. These results are similar to those obtained by
Edenberg et al. (25) and by Su and De Pamphilis (26) who supplemented
their subnuclear systems with a cytoplasmic preparation derived from
mammalian cells.

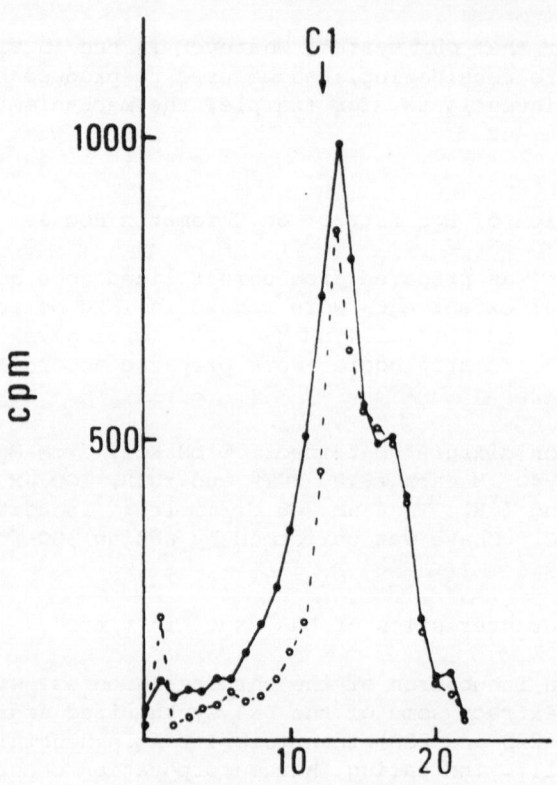

fraction number

Figure 11. *Sedimentation behavior of SV40 DNA synthesized "in vitro". Chromatin bodies, prepared from CV-1 infected cells 40 hours post infection with SV40 virus, were labeled "in vitro" for 5' at 30° C with (³H)TTP, and chased for 30' as specified in the text. After extraction, the DNA was layered on neutral sucrose gradients (10-45% sucrose in 0.1 M Tris-HCl pH 8.0 0.05 M EDTA 0.85 M NaCl) and centrifuged at 45,000 rpm for 225' at 22° C in the SV 50.1 rotor, an aliquot of each fraction was TCA precipitated and counted.*

●———● *5' incorporation "in vitro" in presence of the egg extract*

○------○ *5' incorporation "in vitro" in presence of the egg extract plus 30' chase.*

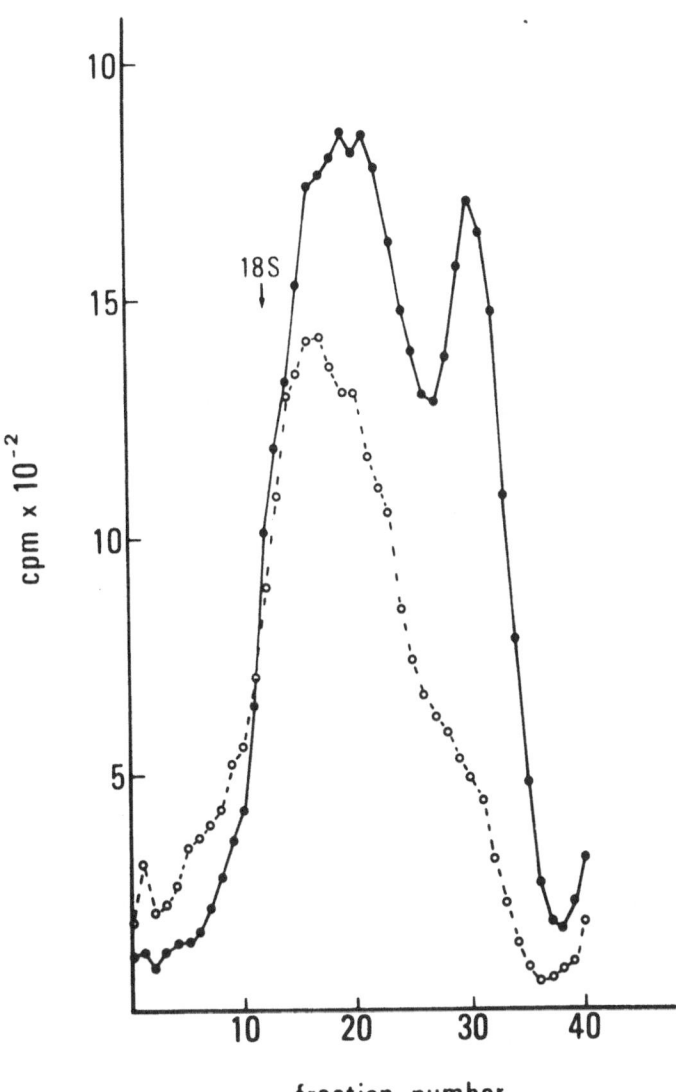

Figure 12. *Size analysis of SV40 DNA synthesized "in vitro".*
Fractions 3 to 13, pooled from neutral sucrose gradients, were
sedimented through alkaline sucrose gradients. The gradients (5-20%
sucrose in 0.25 M NaOH 0.75 M NaCl 0.01 M EDTA) were centrifuged at
49.000 rpm for 300' at 21ºC in the SW 50.1 rotor. Each fraction
was TCA precipitated and counted.

●————● *5' incorporation "in vitro" in presence of the egg extract*

○————○ *5' incorporation "in vitro" in presence of the egg extract*
          plus 30' chase

Therefore our Xenopus laevis egg extract can complement the chromatin bodies derived from SV40 infected cells. Xenopus laevis egg extract plus chromatin bodies mimic "in vitro" some aspects of DNA replication initiated "in vivo".

REFERENCES

1.   Balinsky, B.I. (1970) An Introduction to Embryology (W.B. Saunders Co.) chapter 4:59-98

2.   Brown, D.D. and Dawid, J.B. (1968) Science 160:272-279

3.   Gurdon, J.B. and Speight, V.A. (1969) Exp. Cell Res. 55:253-254

4.   Callen, H.G. (1973) Advances in Molecular Genetics. British Medical Bull. 29:192-195

5.   Graham, C.F., Arms, K. and Gurdon, J.B. (1969) Dev. Biol. 14:349-381

6.   Tocchini-Valentini, G.P., and Crippa, M. (1971) In: 2nd Lepetit Colloquium on Oncongenic Viruses (ed. L. Silvestri), North Holland Publishing Co., Amsterdam pp237-243

7.   Laskey, R.A. and Gurdon, J.B. (1973) Eur. J. Biochem. 37:467-471

8.   Gandini-Attardi, D., Martini, G., Mattoccia, E. and Tocchini-Valentini, G.P. (1976) Proc. Natl. Acad. Sci. USA 73:554-558

9.   Benbow, R.M. and Ford, C.C. (1975) Proc. Natl. Acad. Sci. USA 72:2437-2441

10. Broker, T.R. and Lehman, I.R. (1971) J. Mol. Biol. 60:131-149

11. Rush, M.G., Eason, R. and Vinograd, J. (1971) Biochim. Biophys. Acta 228:585-594

12. Jaenisch, R. and Levine, A. (1971) Virology 44:480-493

13. Gordon, N.C., Rush, M.G. and Warner, R.C. (1970) J. Mol. Biol. 47:495-503

14. Mulder, C. and Delius, H. (1972) Proc. Natl. Acad. Sci. USA 69:3215-3219

15. Cairns, J. (1963) Cold Spring Harbor Symp. Quant. Biol. 28:43-46

16. Tooze, J. (1973) The Molecular Biology of Tumor Viruses (Cold Spring Harbor Laboratory, Cold Spring Harbor, N.Y.).

17. Keller, W. (1975) Proc. Natl. Acad. Sci. USA 72:2550-2554

18. Mattoccia, E., Gandini Attardi, D. and Tocchini-Valentini, G.P. (1976) Proc. Natl. Acad. Sci. USA 73:4551-4554

19. Davis, R.W. and Davidson, N. (1971) In: Methods in Enzymology, (eds. L. Grossman and K. Moldave), Academic Press Inc., New York, Vol 21, pp413-428

20. Tsujimoto, Y. and Ogawa, H. (1977) J. Mol. Biol. 109:423-436

21. Mosig, G., Dannenberg, R. and Breschkin, A. (1975) Indian J. of Microb. 15:145-160

22. Buongiorno-Nardelli, M., Amaldi, F. and Lava-Sanchez, P. (1972) Nature Neur. Biol. 238:134-137

23. Tartof, K.D. (1975) Ann. Rev. Genetics 9:355-385

24. Laskey, R.A., Mills, A.D. and Morris, M.R. (1977) Cell 10: 237-243

25. Edenberg, H.J., Waqar, M.A. and Huberman, J.A. (1976) Proc. Nat. Acad. Sci. USA 73:4392-4396

26. Su, R.T. and De Pamphilis, M.L. (1976) Proc. Natl. Acad. Sci. USA 73:3466-3470

INDEX